Astrophysics and Space Science Library

For further volumes:
http://www.springer.com/series/5664

Astrophysics and Space Science Library

Lev Dorman

Solar Neutrons and Related Phenomena

Springer

Prof. Lev I. Dorman
Head of Israel Cosmic Rays & Space
Weather Center and Emilio Ségre Observatory
Affiliated to Tel Aviv University and Israel Space Agency, Israel
Chief Scientist of Cosmic Ray
Department of N.V. Pushkov IZMIRAN
Russian Academy of Sciences, Troitsk, Russia
P.O. Box 2217, Qazrin 12900, Israel
lid@physics.technion.ac.il, lid010529@gmail.com

ISSN 0067-0057
ISBN 978-90-481-3736-7 e-ISBN 978-90-481-3737-4
DOI 10.1007/978-90-481-3737-4
Springer Dordrecht Heidelberg London New York

Library of Congress Control Number: 2010930378

Cover figure: LASCO C2 image of solar coronal mass ejection January 8, 2002, 18:54 taken by SOHO (ESA & NASA). Copyright: SOHO (ESA & NASA)

Printed on acid-free paper

Springer is part of Springer Science+Business Media (www.springer.com)

The Sun during great solar flare with region where solar energetic particles are generated; their interaction with solar atmosphere produced solar neutrons and gamma rays (Photo from satellite. Adopted from Internet)

Vitaly Ginzburg (1916–2009)

Dedicated to the memory of my father-in-law Prof. Vitaly Lazarevich Ginzburg, Academician, Nobel Laureate

Preface

Short Historical Overview

In the 1940s, two phenomena in the field of cosmic rays (CR) forced scientists to think that the Sun is a powerful source of high-energy particles. One of these was discovered because of the daily solar variation of CR, which the maximum number of CR observed near noon (referring to the existence of continuous flux of CR from the direction of the Sun); this became the experimental basis of the theory that CR's originate from the Sun (or, for that matter, from within the solar system) (Alfvén 1954). The second phenomenon was discovered when large fluxes of high energy particles were detected from several solar flares, or solar CR. These are the so-called ground level events (GLE), and were first observed by ionization chambers shielded by 10 cm Pb (and detected mainly from the secondary muon-component CR that they caused) during the events of the 28th of February 1942, the 7th of March 1942, the 25th of July 1946, and the 19th of November 1949. The biggest such event was detected on the 23rd of February 1956 (see the detailed description in Chapters X and XI of Dorman, M1957).

The first phenomenon was investigated in detail in Dorman (M1957), by first correcting experimental data on muon temperature effects and then by using coupling functions to determine the change in particle energy caused by the solar-diurnal CR variation. After this, it became possible to estimate the influence of the geomagnetic field on the trajectories of CR particles, as well as to determine then the real direction of the daily solar CR anisotropy with regards to the Earth's magnetosphere in interplanetary space. It was shown that the generally accepted opinion of that time concerning the continuous flux of CR from the Sun was absolutely wrong, since the newly discovered direction turned out to be perpendicular to the Sun–Earth line. It furthermore became clear that the Sun is not a continuous source of CR; rather, CR particles must come from interstellar space, i.e. they are not solar, but galactic in origin.

The existence of the second phenomenon (the generation of high energy solar CR during chromospheric flares) lead to a very important consequential conclusion: in the solar atmosphere, many types of nuclear reactions must occur (distinct from those thermonuclear reactions that occur within the Sun), and must also generate secondary energetic particles. These include neutrons, first supposed by Biermann et al. (1951); gamma rays (as gamma ray lines from exited nuclei, from the decay of π^0 mesons, and from relativistic electrons bremsstrahlung). On the other hand, solar CRs must affect chemical and isotropic contents of the solar atmosphere, also as a result of the afore-mentioned nuclear reactions. It is well known that during thermonuclear reactions, deuterium will be fully destroyed at a temperature of $T = 1.2\times10^6$ K; likewise, lithium will be destroyed at $T = 3.2\times10^6$ K, and beryllium at $T = 3.6\times10^6$ K (this explains why these elements are so rare in the Universe). Nevertheless, these light elements were observed in the Sun's atmosphere, leading Shklovsky (1955) to suppose that these light elements are formed by nuclear reactions of CRs from solar flares with the matter in the solar atmosphere.

The problem of solar neutrons and related phenomena came to the forefront after the solar flares of August 1972 (when solar gamma rays were discovered), as well as after the flares of June 1980 and June 1982, when solar neutrons were discovered. However, many years before, forecasts and rough estimations were made of the expected nuclear reactions of solar energetic particles with the matter of the solar atmosphere, and of the generation of solar neutrons and gamma rays, in the frame of some simple models.

In the former USSR, I started, along with my colleagues in IZMIRAN (Moscow region) and in the Ionosphere Institute in Alma-Ata, to investigate statistical solar neutron effects in a high altitude neutron supermonitor. I continued to have an interest in the problem of solar neutrons in 1990–1991, when, together with Prof. D. Venkatesn, I worked at Calgary University (Canada) on the review on solar cosmic rays. This interest increased in 1996–1997, when I worked in Mexico (at the Geophysics Institute of UNAM) together with Prof. J.F. Valdes-Galicia on the problem of solar neutron propagation in the Earth's atmosphere, taking into account the so-called refraction effect.

Physical Motivation and Background

What is the physical sense, considered in this book, of the problem of solar neutrons and their related phenomena? The Sun is roughly 2×10^5 closer to Earth than the Earth is to the nearest other stars. This means that the fluxes of particles and γ-rays from the nuclear reactions of energetic particles in the Sun's atmosphere will have a magnitude about 4×10^{10} times bigger than those from the nearest stars. This also means that the investigation of nuclear reactions of energetic particles in stellar atmospheres must necessarily begin with the Sun. On the other hand, in laboratory conditions (using accelerators of energetic particles), a lot of nuclear reactions have been discovered and investigated in detail. For example, for the generation of neutrons, it is necessary to take into account (Lingenfelter et al. 1965a, b) the reactions initiated by energetic protons:

$He^4(p,pn)He^3$ (threshold kinetic energy $E_{tk} = 25.7$ MeV)
$He^4(p,ppn)H^2$ (32.6 MeV)
$He^4(p,ppnn)H^1$ (35.4 MeV)
$He^4(p,pn\pi)H^1$ (197.5 MeV)
$He^4(p,pnn\pi)H^1$ (207.0 MeV)
$H^1(p,n\pi^+)H^1$ (287.0 MeV)
$H^1(p,n\pi^+\pi)H^1$ (557.0 MeV)
$C^{12}(p,n)N^{12}$ (19.8 MeV)
$N^{14}(p,n)O^{14}$ (6.3 MeV)
$O^{16}(p,pn)O^{15}$ (16.5 MeV)
$Ne^{20}(p,pn)Ne^{19}$ (17.7 MeV)
as well as energetic α-particle initiated reactions
$H^1(\alpha,np)He^3$ ($E_{tk} = 102.8$ MeV)
$H^1(\alpha,ppn)H^2$ (130.3 MeV)
$H^1(\alpha,ppnn)H^1$ (141.5 MeV)
$He^4(\alpha,n)Be^7$ (38.8 MeV)
$He^4(\alpha,\alpha n)He^3$ (41.1 MeV)
$He^4(\alpha,np)Li^6$ (49.2 MeV)

Then, neutrons may be captured by H^1 with the formation of H^2 and the generation of a γ-quant of energy 2.223 MeV, or, they may be captured without the generation of a γ-quant, and may escape from the solar atmosphere. The escaped neutrons may decay into a proton, electron, and neutrino, or may reach the Earth's atmosphere, where they scatter and get partly absorbed, and where the so-called refraction effect (in which neutrons arrive at the detector not from the direction of the Sun, but in some direction between the Sun and the vertical, depending on neutron energy) is important. During nuclear reactions, a lot of excited and radioactive nuclei are formed that generate γ-ray lines, positrons, and other decay products (e.g., Ramaty and Lingenfelter 1973a, b). The decay of generated π^0 mesons gives energetic γ-rays (mostly above and more than 70 MeV), and the decay of $\pi^{\pm}$mesons produces energetic electrons and positrons. The bremsstrahlung of the generated relativistic electrons results in continuous γ-ray radiation.

It is important to note that the generation of neutrons, gamma-rays, positrons and other secondary particles is determined not only by the contents and energy spectrum of the accelerated charged particles during a solar flare, but also, it depends on chemical and isotopic contents, temperature, and vertical density distribution in the solar atmosphere, in the region where nuclear reactions occur along with the propagation of neutrons, positrons, and gamma-rays. This means that detailed experimental and theoretical investigations of solar neutron and/or gamma-ray events will give the unique possibility of obtaining direct information on the source function of solar CR, as well as on the properties of the solar atmosphere in the regions of generation, propagation, and interaction of neutrons and gamma-rays.

Plan and Structure of the Book

In the present book, we compiled and reviewed for the first time a huge experimental and theoretical body of material, constituting results published in the world's scientific literature for more than the past 50 years. This work concerns: solar neutrons and the products of their decay; solar gamma-rays generated together with neutrons in nuclear reactions of solar energetic particles in the solar atmosphere, propagation in the corona, as well as in interplanetary space and in the Earth's atmosphere. It is important to note that investigations of solar neutrons and related phenomena give not only unique information on accelerated solar particles directly at the source (including their chemical and isotopic composition), but also information on background plasmas, including their density and temperature distribution, and information on the mechanisms of energetic particle acceleration and propagation in the solar atmosphere. Let us note further that in Chapters 7–11, we consider in detail prominent solar neutron/gamma-ray events in chronological order, ending in the relatively recent events of 2005–2006. In our catalogue of the scientific literature, we were unable to find even a single weak solar neutron/gamma-ray event after this time. This dearth of events is probably related to the very low period of solar activity corresponding with the current, anomalously long solar minimum.

In Chapter 1, we consider the problem of solar neutrons and related phenomena as it was before the discovery (in 1972) of solar gamma-rays and (in 1980–1982) solar neutrons. The first supposition that high energy particles may be generated on the Sun (as a result of nuclear interactions of accelerated charged solar flare particles with solar atmospheric matter) was made in 1951, by L. Biermann, O. Haxel, and A. Schlüter. In the 1960s and at the beginning of the 1970s, many model calculations and flux estimations for solar neutrons and gamma-ray lines were made in key papers by E.L. Chupp, L.D. De Feiter, J.E. Dolan, G.G. Fazio, E.J. Flamm, W.N. Hess, K. Ito, R.E. Lingenfelter, H. Okazoe, R. Ramaty, H. Rasdan, Z. Svestka, and M. Yoshimori. They showed that detectable neutron and gamma-ray fluxes from major solar flares are expected in the Earth's vicinity. We also consider in Chapter 1 estimations of expected solar neutron and gamma-ray fluxes from some historically powerful flares that generated energetic charged particles. We examine the search for solar neutrons by balloon and space probe experiments as well as by ground measurements, and the search for solar gamma-rays. We underline that all attempts to search for solar neutrons and gamma-rays before the events of 1972, 1980 and 1982 gave only the upper limits for the fluxes from the Sun during quiet periods and chromospheric flare events.

Chapter 2 is devoted to the detailed description of the famous discovery by the 0.3–10 MeV gamma-ray detector on the OSO-8 satellite of solar gamma-rays from the flares of August 1972. The discovery was made by E.L. Chupp, P.P. Dunphy, D.J. Forrest, P.R. Higbie, C. Reppin, A.N. Suri, and C. Tsai, who, in the first 2 weeks of August 1972, first endeavored to use this instrument to look for gamma-quanta in the periods of intense X-ray emission of ≥ 4 erg.cm^{-2} s^{-1}

(class $\geq$M4) in the 1–8 Å band. Only upper limits of the 0.5, 2.2, 4.4 and 6.1 MeV gamma-quantum fluxes (mainly $\leq 5 \times 10^{-3}$ photon cm^{-2} s^{-1} for all gamma-ray lines) were obtained during this measurement period. However, during two very short periods (several minutes) during the flares of the 4th and 7th of August, 1972, real fluxes were measured for the positron annihilation line 0.51 MeV to an accuracy of 5σ, for the 2.2 MeV neutron capture line to an accuracy of 10σ, and for excited 4.4 and 6.1 MeV lines with an accuracy of about 3σ. This chapter is very short, but because of the great importance of these first positive results, we decided to leave them in their own unit, we devote the detailed description of solar gamma ray dicsovery to a separate chapter.

We describe in Chapter 3 the discovery of solar neutrons. This famous discovery, by E.L. Chupp, D.J. Forrest, J. Heslin, G. Kanbach, K. Pinkau, C. Reppin, E. Rieger, J.M. Ryan, and G.H. Share during the event of June 21, 1980 using the Gamma Ray Spectrometer aboard the SMM satellite, showed that for large energy loss events (10–140 MeV and 25–140 MeV), the measured pulses are not caused by gamma-rays, but rather, by neutrons. During the second event, on June 3, 1982, solar neutrons were simultaneously measured by both SMM and ground based neutron monitors for the first time (this seminal work appears in key papers by E.L. Chupp, H. Debrunner, E. Flückiger, D.J. Forrest, G. Kanbach, and G.H. Share).

Solar neutrons and/or solar gamma ray events observations on space probes we describe in Chapter 4 using results obtained on the satellites SMM and Hinotori, during the COMPTEL experiment on the Compton Gamma-Ray Observatory, obtained on satellite GAMMA-1. This chapter is based on the key papers of H. Aarts, V.V. Akimov, K. Bennett, R. Byrd, E.I. Chuikin, E.L. Chupp, H. Debrunner, P.P. Dunphy, S. Enome, G. Eymann, D. Forrest, C. Foster, M.I. Fradkin, G.M. Frye, Jr., A.M. Galper, J.E. Grove, L. Hanlon, T.L. Jenkins, C. Jensen, W.N. Johnson, G.V. Jung, R.L. Kinzer, J.D. Kurfess, L.V. Kurnosova, J. Lockwood, M. Loomis, M. McConnell, D. Morris, R.J. Murphy, H. Nakajima, S. Nakayama, V.E. Nesterov, H. Ogawa, R. Ramaty, G. Rank, J. Ryan, V. Schonfelder, G.H. Share, S. Stansfield, H. Steinle, M.S. Strickman, B.N. Swanenburg, R.A. Schwartz, K. Suga, K. Takahashi, S.A. Voronov, W. Webber, C. Winkler, M. Yoshimori. Let us note that on satellites, a significant π^{o}-decay peak at 70 MeV was observed for the first time during the solar neutron event of March 6, 1989 (key paper of P.P. Dunphy and E.L. Chupp).

In Chapter 5, we describe the problem of solar neutron propagation in the Earth's atmosphere, as well as the sensitivity of neutron monitors and other ground-based detectors of solar neutrons. Thanks to the charge invariance of neutrons and protons, it is important to note that for high-energy neutrons, we can use the coupling functions and integral multiplicities found for galactic and solar CR protons using theoretical calculations of cascades in the atmosphere (as well as from geomagnetic effects). In this way, the main results of the key papers of E.A. Brunberg, J.M. Clem, H. Debrunner, L.I. Dorman, E. Flückiger, N.I. Pakhomov, P. Stein were obtained. Important results were obtained not only for vertical particles but also for particles inclined at zenith angles of 15, 30, 45, 60, and 75° (L.I. Dorman and N.I. Pakhomov). The detailed Monte Carlo simulation of solar neutrons in the Earth's atmosphere and of the sensitivity of neutron monitors to them for vertically-arriving solar neutrons was made in a key paper by S. Shibata. Corresponding papers dealing with other, inclined zenith angles was published via L.I. Dorman, I.V. Dorman, and J.F. Valdes-Galicia.. Thus, the so-called refraction effect of solar neutrons, which depends on the arriving angle, the energy of the neutrons, and the atmospheric level at the place of observation, was determined with great accuracy. It was shown for the first time, in a key paper by D.F. Smart, M.A. Shea, and K. O'Bren, that this effect is very important for the interpretation of solar neutron observations made by neutron monitors and solar neutron telescopes.

Chapter 6 deals with statistical investigations of solar neutron events on the basis of ground observations. On the one hand, positive results were obtained from the Rome neutron monitor's (which sits at about sea-level) 5-min data in a paper by N. Iucci, M. Parisi, C. Signorini, M. Storini, and G. Villoresi. On the other hand, no positive visible effect was found on the basis of the high altitude Chacaltaya neutron monitor (discussed in a key paper by N.J. Martinic, A. Reguerin, E. Palenque, M.A. Taquichiri, M. Wada, A. Inoue, and K. Takahashi). We show that this negative result may have been caused mostly by choosing solar flares, which are characterized by great solar zenith angles. To check the statistical effect of solar neutrons, data from the high-altitude Tyan Shan neutron monitor are analyzed in detailed in key papers by V.M. Aushev, A.V. Belov, L.I. Dorman, V.N. Ishkov, O.N. Kryakunova, R.A. Saidaliev, Ya.E. Shvartsman, and A.G. Zusmanovich. It was shown that the statistical solar neutron effect exists if one chooses X-ray flares characterized by a small solar zenith angle with respect to the point of observations.

Chapter 7 is devoted to observations of solar neutron events by neutron monitors, solar neutron telescopes, and by other ground-based detectors, as well as to the interpretations of these results, all while taking into account observations of related phenomena. We start from the descriptions of the investigations of solar neutron events measured by the Tyan Shan high-altitude neutron supermonitor (as appears in the key paper of V. Antonova, V. Aushev, A. Belov, E. Eroshenko, O. Kryakunova, and A. Struminsky). In this chapter, we consider many solar neutron events, each of them having different peculiarities. A great volume of new information (including on the solar neutron refraction effect) was obtained during investigations of the largest event observed as of 2009 – the event of May 24, 1990 (as shown in key papers by T.P. Armstrong, E.I. Chuikin, A.T. Filippov, G.E. Kocharov, L.G. Kocharov, G.A. Kovaltsov, K. Murakami, Y. Muraki, A.N. Prikhod'ko, K.R. Pyle, M.A. Shea, S. Shibata, and D.F. Smart). Special interest is given to the solar neutron event of June 1, 1991, when surprisingly intense neutron emission was observed from a flare behind the limb of the Sun (as reported on in key papers by C. Barat, K.W. Delsignore, X.-M. Hua, B. Kozlovsky, N. Mandzhavidze, R.J. Murphy, R. Ramaty, G. Trottet, G.H. Share). Investigation of solar neutron events in association with the large solar flares of July 2000 and March–April 2001 (by E.O. Flűckiger, R. Bűtikofer, A. Chilingarian, G. Hovsepyan, Y. Muraki, Y. Matsubara, T. Sako, H. Tsuchiya, and T. Sakai) lead to the important conclusion that three categories of solar neutron events exist.

In Chapter 8, we consider the solar neutron decay phenomenon, discovered by P. Evenson, P. Meyer and K.R. Pyle by measuring the flux of 24–45 MeV protons observed on board the ISEE-3 spacecraft during the well-known event of June 3, 1982. This discovery highlighted the very important possibility of using measurements of neutron decay products to obtain additional information on solar neutron events. More detailed information on solar neutron decay protons (including on their generation and propagation into interplanetary space) was obtained during a much bigger solar neutron event on April 24, 1984. The first observation of electrons from solar neutron decay was made (also on the ISEE-3 spacecraft) during the event of June 21, 1980 (a key paper by W. Dröge, D. Ruffolo, and B. Klecker).

Chapter 9 is devoted to observations and interpretations of gamma-rays resulting from solar energetic particle interactions with the Sun's atmosphere. It was shown in the pioneering key papers of B.M. Kuzhevskij, E.I. Kogan-Laskina, and E.V. Troitskaia that one could determine the solar plasma density altitude profile in the region where solar neutrons are generated and propagated (up to the photosphere), using measurements of the time profile of the neutron capture gamma-ray line 2.223 MeV. The origin of long-duration solar gamma-ray flares (in which high-energy photon emission is present well beyond the impulsive phase, indicating the presence of either stored or continuously accelerated ions) was investigated in the key papers of J.M. Ryan. The present situation favors either the acceleration of protons and ions for long periods of time by second order Fermi acceleration in large coronal loops, or, alternatively, acceleration in large-scale, CME-associated reconnection sheets.

The possibilities of solar gamma-ray spectroscopy are demonstrated in key papers by M. Yoshimori, A. Shiozawa, and K. Suga through their investigations of the ^{3}He contents of the photosphere (^{3}He is thought to be produced primarily by the nuclear synthesis occurring in the early universe, and its abundance is used to place a constraint on cosmological models). Since the photospheric ^{3}He abundance cannot be determined by optical spectroscopy, observations of the neutron capture line at 2.223 MeV provide its only direct method of determination (It works as follows: neutrons, which are produced simultaneously with gamma-ray lines by the interactions of accelerated ions, diffuse into the photosphere, where the 2.223 MeV lines are emitted by neutron capture of hydrogen. Because of the time required for the neutrons to slow down and be captured, the 2.223 MeV line is produced about 100 s after their production, and the competing capture reaction ^{3}He(n,p)^{3}H affects this delay). The other example brought here is regarding the temporal variations of ambient plasma abundances in the acceleration region. This is done using measurements of low-FIP (First Ionization Potential) to high-FIP elements' gamma-ray line ratios (as seen in key papers by J.E. Grove, W.N. Johnson, G.V. Jung, R.L. Kinzer, J.D. Kurfess, R.J. Murphy, G.H. Share, A. Shiozawa, M.S. Strickman, K. Suga, and M. Yoshimori). This chapter describes quite a few solar gamma ray events, and each of these events is characterized by different important peculiarities (as detailed in the Contents).

In Chapter 10, important phenomena related to the problem of solar neutrons are considered, namely: positron generation during nuclear interactions of flare energetic particles with the solar atmosphere, and the generation of the 0.511 MeV annihilation line. Positrons are slowed down to ~10 eV where they either annihilate directly or form positronium atoms after thermalization. Direct annihilation and singlet state positronium emit two 511 keV photons, while triplet state positronium produces three gamma-rays (positronium continuum below 511 keV). Triplet positronium is broken up by collision if the ambient density is above 10^{14} cm^{-3}. Since a time profile of the 511 keV line depends on the density and lifetimes of β^+-decay nuclei, its temporal variation is complex, and depends on the peculiarities of solar flares. A ratio of 3γ to 2γ depends on the ambient density. The line width is a function of the temperature of the annihilation site. Therefore, detailed measurements and modeling of phenomena caused by solar positron generation and annihilation will give important information regarding not only solar energetic particles, but also the ambient plasma. This chapter is based mostly on the key papers associated with Yohkoh's observation of a gamma-ray flare on November 6, 1997 ((M. Yoshimori, S. Nakayama, H. Ogawa, N. Saita, A. Shiozawa, K. Suga, and H. Takeda), and on the RHESSI observation of the solar annihilation line from the July 23, 2002 solar flare (B.R. Dennis, H.S. Hudson, B. Kozlovsky, R.P. Lin, R.J. Murphy, R.A. Schwartz, G.H. Share, J.G. Skibo, A.Y. Shih, and D.M. Smith). B. Kozlovsky, R.J. Murphy, and G.H. Share, in a key paper, treat in detail positron production from the decay of radioactive nuclei produced in the nuclear reactions of accelerated ^{3}He (because of their large cross sections and low threshold energies, these reactions can significantly contribute to positron production in solar flares with accelerated particle compositions enriched in ^{3}He).

Chapter 11 describes the development of models and simulations for solar neutron and gamma-ray events. The detailed model of solar flare neutron production and the angular dependence of the 2.223 MeV capture gamma-ray line emission was developed in a key paper of X.-M. Hua and R.E. Lingenfelter. In key papers of X.-M. Hua and R.E. Lingenfelter, the special model for determining the ^{3}He/H ratio in the solar photosphere from flare gamma-ray line observations was also developed. Important models and simulations for the estimation of the intensity and directionality of flare-accelerated α-particles on the Sun using gamma-ray observations were developed in key papers by G.H. Share and R.J. Murphy. The method for estimating the spectral evolution of energetic protons in solar flares using gamma-ray observations and simulations was developed in a key paper by W.Q. Gan. Important methods and simulations of the estimation characteristics of energetic heavy ions on the Sun were developed in a key paper by G.H. Share and R.J. Murphy, using gamma-ray measurements. A model for the estimation by gamma-rays the ratio of interacted to interplanetary energetic protons in the case of diverging magnetic field lines with stochastic acceleration was developed in key papers by L. Kocharov, G. Kovaltsov, T. Laitinen, P. Mäkelä, and J. Torsti. The model for estimating the ratio of interacted to interplanetary energetic protons by gamma-ray measurements in the case of diverging magnetic field lines and parallel shock wave acceleration was developed in a key paper by R. Vainio, L. Kocharov, and T. Laitinen. The expected change with time of the angular distribution of gamma-ray fluxes from decay of π^0-mesons (generated by interactions of solar energetic particles with matter of solar corona and solar wind) was calculated in papers by L.I. Dorman. In order to estimate the ratio of interacted energetic particles to ejected into interplanetary space in high energy region during solar fare events, J.A. Lockwood, H. Debrunner, and J.M. Ryan developed a model using measurements of gamma-rays generated in π^0-decays. X.-M. Hua, B. Kozlovsky, R.E. Lingenfelter, R. Ramaty, and A. Stupp developed a both a model and a Monte Carlo simulation for estimating the angular and energy-dependence of neutron emissions from solar flare magnetic loops. In this Chapter, we also consider the expected production of light isotopes, which occurs because of nuclear interactions and acceleration in the flare region (as shown in a key paper by S.A. Balashev, M.F. Lytova, and V.M. Ostryakov). Important investigation of powerful solar flare characteristics by gamma rays from excited states of ^{12}C and various neutron capture lines was done in key paper of I.V. Arkhangelskaja, A.I. Arkhangelsky, L.I. Miroshnichenko, and E.V. Troitskaya.

The detailed **Contents** gives information on the problems discussed in the various parts of the book. Furthermore, there is a list of **Frequently Used Abbreviations and Notations**. After Chapter 11 there is an **Appendix**, which contains details of some complicated calculations, and then **Conclusions and Problems**, where we summarize the main results and propose some unresolved key problems that we feel are important for the development of this field of science. In the **References**, there are separate references for Monographs and Books (in the text they are marked by the letter M before the year of publication), as well as for each Chapter and Appendix. As an added convenience to the reader, there are also **Subject** and **Author indexes**.

I would be grateful for any comments and/or reprints that may be useful to our future research, and that can make the next edition of this book better and clearer. They may be sent by e-mail (lid@physics.technion.ac.il, lid010529@gmail.com) and by post to the address: Prof. Lev I. Dorman, Head of Israel Cosmic Ray and Space Weather Center and Emilio Ségre Observatory, P.O. Box 2217, Qazrin 12900, ISRAEL.

February 2010
Qazrin, Princeton, Moscow

Acknowledgements

It is my great pleasure to cordially thank:

My Teachers in science and in life – Prof. Evgeny Lvovich Feinberg (1912–2005) and Prof. Yuval Ne'eman (1925–2006);

Authors of papers and monographs reflected and discussed in this book;

My former students who became colleagues and friends – for many years of collaboration and interesting discussions – M.V. Alania, R.G. Aslamazashvili, V.Kh. Babayan, M. Bagdasariyan, L. Baisultanova, V. Bednaghevsky, A.V. Belov, A. Bishara, D. Blenaru, Ya.L. Blokh, A.M. Chkhetia, L. Churunova, T.V. Dzhapiashvili, E.A. Eroshenko, S. Fisher, E.T. Franzus, L. Granitskij, R.T. Gushchina, O.I. Inozemtseva, K. Iskra, N.S. Kaminer, V.L. Karpov, M.E. Katz, T.V. Kebuladse, Kh. Khamirzov, Z. Kobilinsky, V.K. Koiava, E.V. Kolomeets, V.G. Koridse, V. Korotkov, V.A. Kovalenko, Yu.Ya. Krestyannikov, T.M. Krupitskaja, A.E. Kuzmicheva, A.I. Kuzmin, I.Ya. Libin, A.A. Luzov, N.P. Milovidova, L.I. Miroshnichenko, Yu.I. Okulov, I.A. Pimenov, L.V. Raichenko, L.E. Rishe, A.B. Rodionov, O.G. Rogava, A. Samir Debish, V.S. Satsuk, A.V. Sergeev, A.A. Shadov, B. Shakhov, L.Kh. Shatashvili, G.Sh. Shkhalakhov, V.Kh. Shogenov, V.S. Smirnov, M.A. Soliman, F.A. Starkov, M.I. Tyasto, V.V. Viskov, V.G. Yanke, K.F. Yudakhin, A.G. Zusmanovich

For many years support of our research in the former USSR and in Russia– Alexander Evgenievich Chudakov, Georgy Borisovich Khristiansen, Vladimir Kuznetsov, Vladimir Vladimirovich Migulin, Michael Dmitrievich Millionshikov, Vladimir Nikolaevich Oraevsky, Nikolai Vasil'evich Pushkov, Irena Vjacheslavovna Rakobolskaya, Sergei Nikolaevich Vernov, Georgy Timofeevich Zatsepin

For interesting discussions and fruitful collaboration – H.S. Ahluwalia, T.M. Aleksanyan, V.V. Alexeenko, I.V. Alexeev, H. Alfvén, E.E. Antonova, W.I. Axford, J.H. Allen, G.A. Bazilevskaja, G. Bella, N.P. Ben'kova, M. Bercovitch, E.G. Berezhko, V.S. Berezinsky, J.W. Bieber, R.C. Binford, S.P. Burlatskaya, G. Cini Castagnoli, A.N. Charakhchyan, T.N. Charakhchyan, A. Chilingarian, J. Clem, E. Cliver, H. Coffey, J.W. Cronin, I. Daglis, E. Daibog, A. Dar, R. Davis, Jr., H. Debrunner, V.A. Dergachev, V.A. Dogiel, A.Z. Dolginov, S.S.

Dolginov, I.V. Dorman, L.O.C. Drury, M. Duldig, V.M. Dvornikov, D. Eichler, E. Etzion, Yu.I. Fedorov, P. Ferrando, E.O. Fluckiger, V. Fomichev, M. Galli, A.M. Galper, Yu.I. Galperin, V.L. Ginzburg, E.S. Glokova, N.L. Grigorov, O.N. Gulinsky, A.V. Gurevich, S.R. Habbal, J.E. Humble, N. Iucci, R. Kallenbach, G.S. Ivanov-Kholodny, G.E. Kocharov, I.D. Kozin, O.N. Kryakunova, G.F. Krymsky, K. Kudela, L.V. Kurnosova, V. Kuznetsov, A.A. Lagutin, A. Laor, A.K. Lavrukhina, Yu.I. Logachev, C. Lopate, H. Mavromichalaki, K.G. McCracken, B. Mendoza, M.A. Mogilevsky, I. Moskalenko, Y. Muraki, M. Murat, K. Mursula, V.S. Murzin, N. Nachkebia, K. Nagashima, G.M. Nikolsky, S.I. Nikolsky, V. Obridko, J. Pap, E. N. Parker, M. Parisi, S.B. Pikelner, L.P. Pitaevsky, M.K.W. Pohl, A. Polyakov, M.S. Potgieter, C. Price, N.G. Ptitsyna, V.S. Ptuskin, L.A. Pustil'nik, R. Pyle, A.I. Rez, S.I. Rogovaya, I.L. Rozental, S. Sakakibara, N. Sanchez, V. Sarabhai, I.A. Savenko, K. Scherer, V. Sdobnov, V.B. Semikoz, V.P. Shabansky, Yu.G. Shafer, G.V. Shafer, M. M. Shapiro, P.I. Shavrin, M.A. Shea, I.S. Shklovsky, Ya. Shwarzman, B.I. Silkin, J.A. Simpson, G.V. Skripin, D.F. Smart, A. Somogyi, T. Stanev, M. Stehlic, A. Sternlieb, P.H. Stoker, M. Storini, Yu.I. Stozhkov, A. Struminsky, A.K. Svirzhevskaya, S.I. Syrovatsky, P.J. Tanskanen, A.G. Tarkhov, I. Transky, E. Troitskaja, V.A. Troitskaya, B.A. Tverskoy, I.G. Usoskin, J.F. Valdes-Galicia, E.V. Vashenyuk, P. Velinov, D. Venkatesan, S.N. Vernov, E.S. Vernova, G. Villoresi, T. Watanabe, J.P. Wefel, G. Wibberenz, A.W. Wolfendale, V. Yakhot, G. Yom Din, A. K. Yukhimuk, N.L. Zangrilli, G.T. Zatsepin, G.B. Zhdanov, V.N. Zirakashvili, I.G. Zukerman

For constant support and the kind-hearted atmosphere during my education and long way in CR research – parents Isaac (1884–1954) and Eva (1894–1958), wife Irina, daughters Maria and Victoria, sisters Maria Tiraspolskaya and Mara Pustil'nik, brothers Abraham Argov (1914–2003) and Zuss (1916–1958), parents-in-law Olga Ivanovna Zamsha and Vitaly Lazarevich Ginzburg (1916–2009), son-in-law Michael Petrov, grand-children Elizabeth and Gregory; relatives in Israel – cousins Michal, David, Shlomo, Dickla, and nephews Raja, Ada, Lev, Dan, Dalia, Shlomo

For great help and collaboration in the period of my work in Israel which made possible to continue the research in CR – Yuval Ne'eman, Abraham Sternlieb, Aby Har-Even, Isaac Ben Israel, Zvi Kaplan, Lev Pustil'nik, Igor Zukerman, Michael Murat, Alexei Zusmanovich, Lev Pitaevsky, David Eichler, Matilda Elron, Ronit Nevo, Shushana Shalom, Sami Bar-Lev, Avi Gurevich, Nunzio Iucci, Giorgio Villoresi, Mario Parisi, Marisa Storini, John A. Simpson, W.I. Axford, Arnold W. Wolfendale, Victor Yakhot, Alexander Polyakov, Doraswamy Venkatesan, Harjit Ahluwalia, Jose F. Valdes-Galicia, Yasushi Muraki, Marc Duldig, Heleni Mavromichalaki, Anatoly Belov, Victor Yanke, Eugenia Eroshenko, Natalie Ptitsyna, Marta Tyasto, Olga Kryakunova

For great help in preparing many figures and full references – Igor Zukerman

For fruitful advice during preparing the book – Ramon Khanna, Tamara Biese, and Vaska Krabbe (Springer)

For great help in checking and improving English in this book – David Shai Applbaum and Ramon Khanna

The work of the Israeli-Italian Emilio Ségre Observatory is supported by the Collaboration of Tel Aviv University (ISRAEL) and "Uniroma Tre" University and IFSN/CNR (ITALY) – my great gratitude for foundation and supporting of this collaboration Nunzio Iucci, Isaac Ben Israel, Zvi Kaplan, Yuval Ne'eman, Mario Parisi, Lev Pustil'nik, Abraham Sternlieb, Marisa Storini, Giorgio Villoresi, and Igor Zukerman

Contents

Abbreviations

ACE	Advanced composition explorer satellite
CME	Coronal mass ejection
CR	Cosmic rays
Dst	Disturbance storm time index
E	Energy of CR particles
E_0	Energy of primary CR particle
ESOI	Israel-Italian Emilio Segre' Observatory
FEP	Flare energetic particles
FWHM	Full width at half maximum
GCR	Galactic cosmic rays
GR	Gamma rays
GLE	Ground level event of solar CR increasing
GRS	Gamma-ray spectrometer
H	Altitude
h	Atmospheric pressure
h_o	Pressure on the level of observations
IC	Ionization chamber, shielded by 10 cm Pb
ICME	Interplanetary coronal mass ejections
ICRC	International Cosmic Ray Conference
ICRC	Israel Cosmic Ray Center (1992–2002)
ICR&SWC	Israel Cosmic Ray and Space Weather Center (from 2003)
ICRS	International Cosmic Ray Service (proposed in 1991)
IEF	Interplanetary electric field
IGY	International Geophysical Year (July 1957 to December 1958)
IMF	Interplanetary magnetic field
IQSY	International Quiet Sun Year (1964–1965)
MC	Magnetic cloud

m w.e.	Meters of water equivalent
$m_i(R, h)$	Integral multiplicity: number of secondary CR particles of type i on level h from one primary CR particle with rigidity R on the top of atmosphere
m_o	Rest mass of particle
MT	Muon telescope
$N(R_c, h)$ or $I(R_c, h)$	CR intensity
NM	Neutron monitor or super-monitor
NM-64 or NM-IQSY	Neutron super-monitor of IQSY type
NM-IGY	Neutron monitor of IGY or Simpson's type
$R = pc/Ze$	Particle rigidity [GV]
$R(E_0, \theta_0, h) = m(E_0, \theta_0, h)/ m(E_0, \theta_0 = 0°, h/\cos\theta_0)$	Refraction effect of solar neutrons
R_C	Geomagnetic cut-off rigidity
r_E	Radius of the Earth
RHESSI	Reuven Ramaty High Energy Spectroscopic Imager Satellite
SA	Solar activity
SCR	Solar cosmic rays
SEP	Solar energetic particles
SGR	Solar gamma rays
SMM	Solar maximum mission
SNE	Solar neutron events
SNT	Solar neutron telescope
SW	Solar wind
SW	Space weather
$T(h)$	Vertical air temperature distribution
$W(R, h)$	Coupling function
Z or θ	Zenith angle
λ	Latitude
$\theta = \frac{\pi}{2} - \lambda$	The polar angle
φ	Longitude, azimuth

Chapter 1
Interactions of Charged, Accelerated Particles in the Solar Atmosphere, and the Generation of Secondary Energetic Particles and Radiation: Pioneer Results

In this chapter, we describe only the very first, pioneering results, those obtained before the famous discovery of solar gamma rays (which occurred during the events of August 1972) and the discovery of solar neutrons (which occurred on the 21 June 1980 and on the 3 June 1982). At those times, scientists concerned themselves mostly with problems involving the production of high energy neutrons and gamma ray lines, and did not much consider the problem of gamma ray generation as caused by the decay of neutral pions (generated by the interactions of accelerated protons of higher energy than 270 MeV with protons in the solar atmosphere). The first significant π^0-decay peak at 70 MeV was observed by the Gamma Ray Spectrometer (GRS) on the Solar Maximum Mission (SMM) satellite, during the solar neuron/gamma ray event of March 6 1989 (see Section 4.5; more detail about π^0-decay-driven gamma radiation is given in Sections 11.8 and 11.9). Let us note that the problem of energetic particle interactions in the solar atmosphere producing secondary radiation (both particulate and photon) is a specific case of the more general problem of cosmic ray interaction, propagation and acceleration in space plasmas, which is considered in detail in the Dorman (M2006) manuscript of similar title.

1.1 Interactions of Charged Accelerated Particles in the Solar Atmosphere and Expected Energetic Secondary Radiation from the Sun: The First Model Calculations

1.1.1 The Matter and Short History of the Problem

Biermann et al. (1951) discussed first a possible neutron flux producing by nuclear interactions of solar energetic protons with the photosphere. Subsequently, many model calculations and flux estimations for solar neutrons and gamma-ray lines have been made by Hess (1963a, b), Chupp and Razdan (1963), Chupp (1964),

L. Dorman, *Solar Neutrons and Related Phenomena*,
Astrophysics and Space Science Library 365, DOI 10.1007/978-90-481-3737-4_1,

Lingenfelter and Flamm (1964), Dolan and Fazio (1965), Lingenfelter et al. (1965a, b), Lingenfelter and Ramaty (1967), Ito et al. (1968), Lingenfelter (1969), Chupp (1971), Svestka (1971) and De Feiter (1971). They showed that in the Earth vicinity detectable neutron and gamma-ray fluxes from major solar flares are expected.

1.1.2 Expected Neutron Production in Solar Flare Acceleration Processes

In Hess (1963a, b) was supposed that accelerated charged particles escape isotropically from the flare region into the upward and downward hemispheres. This model with taking into account additional neutron production by charged trapped particles in the region of acceleration was used by Lingenfelter et al. (1965a) in Monte Carlo calculations to determine the intensity and spectrum of the solar neutron flux escaping into space. According to Lingenfelter et al. (1965a) the number of neutrons in the energy interval between E_n and $E_n + dE_n$ produced in the photosphere by flare-accelerated protons/α-particles will be

$$N_{p,\alpha}(E_n)dE_n = \int\limits_0^\infty I_{p,\alpha}\left(E_{p,\alpha}\right)\left\{\sum_k\sum_i M_k\left(E_{p,\alpha},E_{ni}\right)dE_{ni}\right\}dE_{p,\alpha}, \qquad (1.1)$$

where $I_{p,\alpha}\left(E_{p,\alpha}\right)$ is the number of protons/α-particles incident on the photosphere with kinetic energy $E_{p,\alpha}$. Summation over k in Eq. 1.1 includes all neutron producing reactions between flare-accelerated protons/α-particles and photospheric constituents (the relative solar abundances was used H–1.0, He–0.1, C–0.00053, N–0.0001, O–0.00092 and Ne–0.0005 in accordance with Goldberg et al. 1960). Summation over i in Eq. 1.1 includes the first ($i = 1$), the second ($i = 2$), and other order of protons/α-particles interactions in the solar photosphere. According to estimations of Lingenfelter et al. (1965a), for protons with energy below than 1 GeV essentially only the first interaction ($i = 1$); the remaining terms with $i \geq 2$ contribute only about 10% of the total neutron production by 1 GeV protons, about 40% of the production by 2 GeV proton and about 60% of the production by 4 GeV protons. The function $M_k\left(E_{p,\alpha},E_{ni}\right)$ is the number of neutrons generated in ith order interaction in the type k reaction with protons/α-particles with initial energy $E_{p,\alpha}$:

$$M_k\left(E_{p,\alpha},E_{ni}\right)dE_{ni} = \int\limits_{E_{tk}}^{E_{p,\alpha}} P_k\left(E'_{p,\alpha},E_{ni}\right)dE_{ni}m_k\sigma_{in}\left(E'_{p,\alpha}\right)_k \times \exp\left\{\left[Y\left(E_{p,\alpha}\right) - Y\left(E'_{p,\alpha}\right)\right]\Big/\overline{L}\left(E'_{p,\alpha}\right)\right\}\left(dr\Big/dE'_{p,\alpha}\right)dE'_{p,\alpha}, \qquad (1.2)$$

where E_{tk} is the threshold energy for the kth reaction,

$$E'_{p,\alpha} = E_{p,\alpha} - \int_0^r \left(dE_{p,\alpha}/dr\right)dr \tag{1.3}$$

is the energy of protons/α-particles on the depth r in the photosphere (in direction of particle propagation) and $dE_{p,\alpha}/dr$ is the differential energy loss of proton/α-particle in photosphere material, $Y(E_{p,\alpha})$ is the range in the photosphere of proton/α-particle with energy $E_{p,\alpha}$, $\overline{L}(E'_{p,\alpha})$ is the mean attenuation length over the energy interval from $E_{p,\alpha}$ to $E'_{p,\alpha}$ of a proton/α-particle in the photosphere, $m_k\sigma_{in}(E'_{p,\alpha})_k$ is the macroscopic cross section of a proton/α-particle with energy $E'_{p,\alpha}$ for the kth neutron-producing reaction in the photosphere, and $P_k(E'_{p,\alpha}, E_{ni})dE_{ni}$ is the probability that a neutron, produced in the kth reaction by proton/α-particle with energy $E'_{p,\alpha}$ will lie in the energy interval between E_{ni} and $E_{ni} + dE_{ni}$; in accordance with experimental data Lingenfelter et al. (1965a) approximated this function by the inverse of the excitation energy:

$$P_k\left(E'_{p,\alpha}, E_{ni}\right)dE_{ni} \approx dE_{ni}\Big/\left(E'_{p,\alpha} - E_{tk}\right). \tag{1.4}$$

In Lingenfelter et al. (1965a) were taken into account proton initiated reactions He^4(p,pn)He^3 (threshold energy $E_{tk} = 25.7$ MeV); He^4(p,ppn)H^2 (32.6 MeV); He^4(p,ppnn)H^1 (35.4 MeV); He^4(p,pnπ)H^1 (197.5 MeV); He^4(p,pnnπ)H^1 (207.0 MeV); H^1(p,nπ+)H^1 (287.0 MeV); H^1(p,nπ+π)H^1 (557.0 MeV); C^{12}(p,n)N^{12} (19.8 MeV); N^{14}(p,n)O^{14} (6.3 MeV); O^{16}(p,pn)O^{15} (16.5 MeV); Ne^{20}(p,pn)Ne^{19} (17.7 MeV); and α-particle-initiated reactions H^1(α,np)He^3 ($E_{tk} = 102.8$ MeV), H^1(α,ppn)H^2 (130.3 MeV); H^1(α,ppnn)H^1 (141.5 MeV); He^4(α,n)Be^7 (38.8 MeV); He^4(α,αn)He^3 (41.1 MeV); He^4(α,np)Li^6 (49.2 MeV).

Figure 1.1 shows cross sections for the production of 1 neutron in proton interactions with H, He, C, N, O and Ne versus the proton energy from 10 MeV up to 10 GeV.

Figure 1.2 shows the yield of neutrons per proton in proton reactions with H, He, in spallation and evaporation reactions of protons with C, N, O and Ne versus the proton energy. Calculations for α-particle interactions show that the yield from α-particle-induced reactions is less than 5% of the total neutron yield.

1.1.3 Expected Energy Spectra of Generated Neutrons

The energy spectra of generated neutrons $N_{p,\alpha}(E_n)$ according to Eq. 1.1 depends from the spectra of energetic particles accelerated in solar flare $I_{p,\alpha}\left(E_{p,\alpha}\right)$. In Lingenfelter et al. (1965a) this spectra was supposed in according with observations

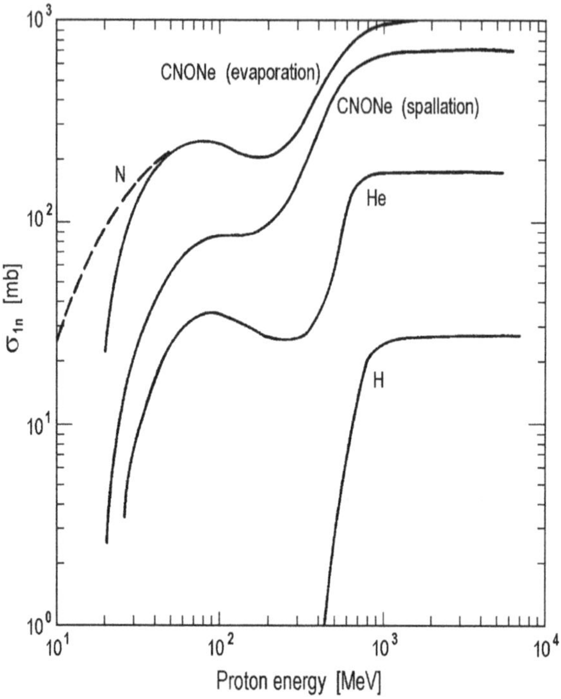

Fig. 1.1 Cross sections for the production of one neutron in proton interactions with H, He, C, N, O, and Ne (According to Lingenfelter et al. 1965a)

of solar energetic particles near the Earth's orbit (Webber 1962; Freier and Webber 1963) in the exponential form:

$$I_{p,\alpha}(E_{p,\alpha}) = I_{p,\alpha}(0)\exp(-R/R_0)dR/dE_{p,\alpha}, \tag{1.5}$$

where R is the rigidity of particles and R_0 is characteristic rigidity of spectra. The calculations of $N_{p,\alpha}(E_n)$ were made for $R_0 = 60$, 120, 200 and 300 MV. The results for $R_0 = 60$, 120 and 200 MV are shown in Fig. 1.3, and in Fig. 1.4 is shown the total neutron yield as a function of the characteristic rigidity R_0 of the incident protons (from 60 MV up to 300 MV).

1.1.4 On the Escaping of Solar Neutrons into Interplanetary Space

To calculate the escape probability of solar neutrons, Lingenfelter et al. (1965a) determined the neutron production as a function of depth within the photosphere (it was supposed that particles are accelerated in the flare region above the

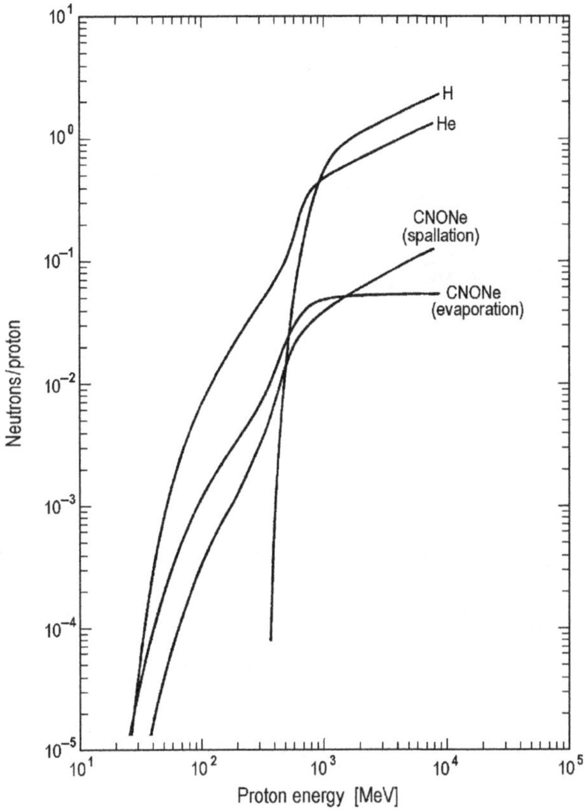

Fig. 1.2 Yield of neutrons per proton in solar material. The total yield of neutrons per proton is equal to the sum of the four curves (According to Lingenfelter et al. 1965a)

photosphere, from which they escape isotropically). According to Lingenfelter et al. (1965a), the neutron production depth distribution, i.e. the number of neutrons produced at a depth between x and $x + dx$ below the solar surface will be:

$$N_n(x)dx = \int_0^\infty I_{p,\alpha}\left(E_{p,\alpha}\right) dE_{p,\alpha} \int_{\mu_{\min}}^1 d\mu \exp\left(-x/\mu\bar{L}\right) \sum_k \sum_i m_k \sigma_{in}\left(E'_{p,\alpha}\right)_k dx, \quad (1.6)$$

where

$$\mu_{\min} = x/Y\left(E_{p,\alpha}\right);\; E'_{p,\alpha}\left(E_{p,\alpha,}, x, \mu\right) = E_{p,\alpha} - \int_0^{x/\mu} \left(dE_{p,\alpha}/dr\right)dr. \quad (1.7)$$

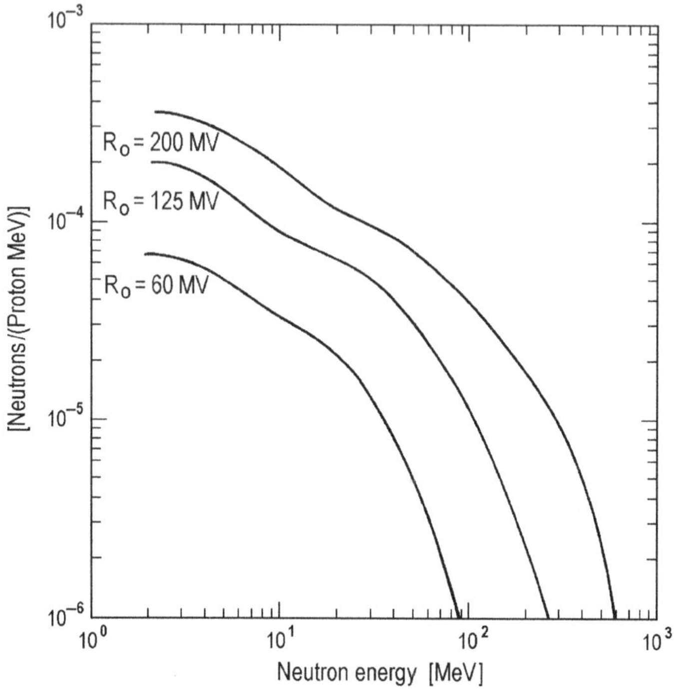

Fig. 1.3 Neutron production spectra calculated for incident protons with characteristic rigidities of $R_0 = 60$, 125 and 200 MV (According to Lingenfelter et al. 1965a)

The neutron escape probability was calculated from the depth distribution (Eq. 1.6) by the Monte Carlo method for 30,000 cases of neutrons with energy between 1 Mev and 1 GeV. It was found that neutron escape is essentially collisionless because more than half of the neutrons being produced within a one mean free path from the solar surface. Calculations of Lingenfelter et al. (1965a) show that for $R_0 = 125$ MV about 48% of neutrons produced with energies greater than 1 MeV were found to escape into interplanetary space with the spectrum about the same as shown in the Fig. 1.3 (curve for $R_0 = 125$ MV), but multiplying on factor 0.48 (the escape probability varies only a few percent over the range of R_0 from 60 to 200 MV).

1.1.5 The Solar Neutron Survival Probability of Reaching the Distance of 1 AU from the Sun Before Decaying

The expected spectrum of solar neutrons on the Earth's orbit will be deformed by neutron decay which sufficiently depends from the neutron energy E_n. This deformation determined by the solar neutron survival probability to reach the distance $r = 1$ AU from the Sun without decaying (see Fig. 1.5).

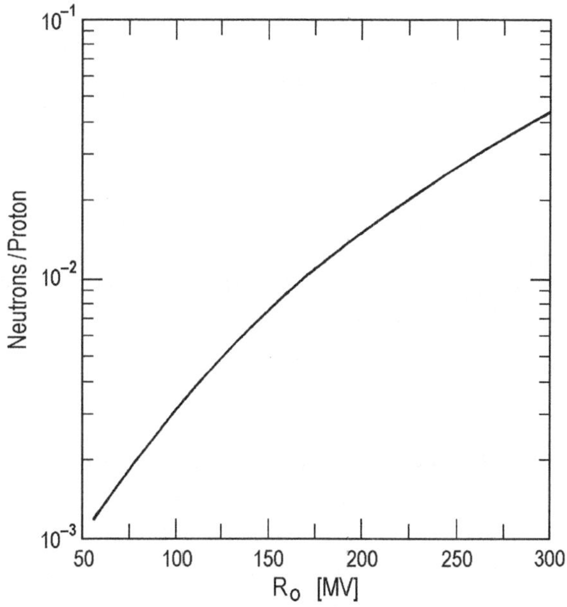

Fig. 1.4 Total neutron yield as a function of the characteristic rigidity R_0 of the incident protons. The yields are normalised to one incident proton with energy greater than 30 MeV (According to Lingenfelter et al. 1965a)

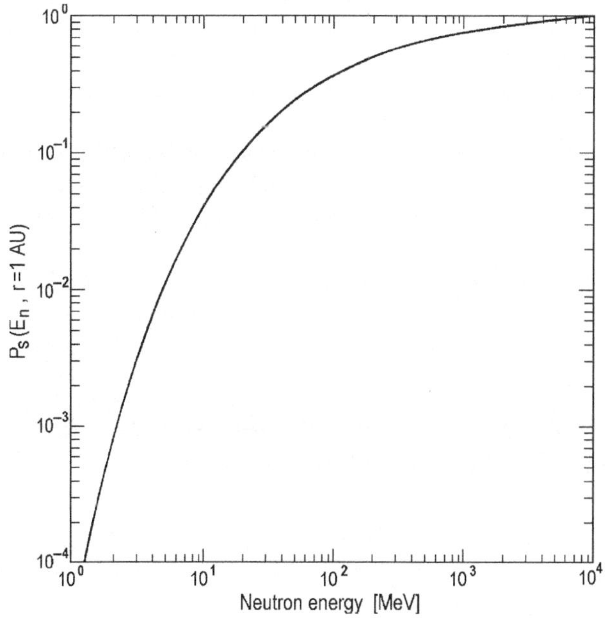

Fig. 1.5 The solar neutron survival probability $P_S(E_n, r)$ as a function of neutron energy E_n for $r = 1$ AU. This function given by Eq. 1.8, is the probability that a solar neutron will reach the distance $r = 1$ AU from the Sun before decaying (According to Lingenfelter et al. 1965b)

Figure 1.5 shows the solar neutron survival probability $P_s(E_n, r)$ to reach the distance $r = 1$ AU from the Sun before decaying in dependence from neutron energy. The curve in Fig. 1.5 is determined by equation:

$$\begin{aligned} P_s(E_n, r) &= \exp(-\Delta T(E_n)/\tau'(\tau, E_n)) \\ &= \exp\left(-\frac{r}{\tau c}\left(\left((E_n + m_n c^2)/m_n c^2\right)^2 - 1\right)^{-1/2}\right), \end{aligned} \tag{1.8}$$

where $\Delta T(E_n)$ is the traveling time of solar neutron with energy E_n on the distance $r = 1$ AU, $\tau'(\tau, E_n)$ is the neutron mean life time in the system of coordinates Sun-Earth, τ and m_n are the neutron mean life-time and mass in the rest system of coordinates.

1.1.6 On the Expected Spectrum of Solar Neutrons at the Distance of 1 AU from the Sun

According to Lingenfelter et al. (1965b), the expected spectrum of solar neutrons on the distance 1 AU from the Sun will be

$$N_n(E_n, r)dE_n = P_s(E_n, r)\frac{N_{esc}(E_n)}{4\pi r^2}dE_n, \tag{1.9}$$

where $P_s(E_n, r)$ is the solar neutron survival probability to reach the distance r from the Sun before decaying described by Eq. 1.8 and was shown in Fig. 1.5 and $N_{esc}(E_n)$ is the spectrum of solar neutrons escaping into interplanetary space (see Sections 1.1.4 and 1.1.5). Figure 1.6 shows the expected differential energy spectrum of solar neutrons calculated according to Eq. 1.10 and normalized to the flux 1 proton cm^{-2} of accelerated protons with energy $\geq$30 MeV. It was supposed that the spectrum of accelerated protons described by Eq. 1.5 with characteristic rigidity $R_0 = 60$, 125 and 200 MV. The dependence from R_0 of integral solar neutron flux at 1 AU is shown on Fig. 1.6.

In Fig. 1.7 is shown expected integral solar neutron flux at the distance $r = 1$ AU from the Sun versus characteristic rigidity R_0 and normalized as in Fig. 1.6.

1.1.7 Expected Time-Dependent Solar Neutron Flux at the Distance of 1 AU from the Sun

According to Lingenfelter et al. (1965b) the expected time-dependent solar neutron flux on the distance r from the Sun is related to the time integrated spectrum $N_n(E_n, r)$ by the expression

$$N_n(E_n, r, t) = N_n(E_n, r)dE_n/dt = N_n(E_n, r)\frac{\left(E_n^2 + 2m_n c^2 E_n\right)^{3/2}}{m_n^2 c^3 r}, \tag{1.10}$$

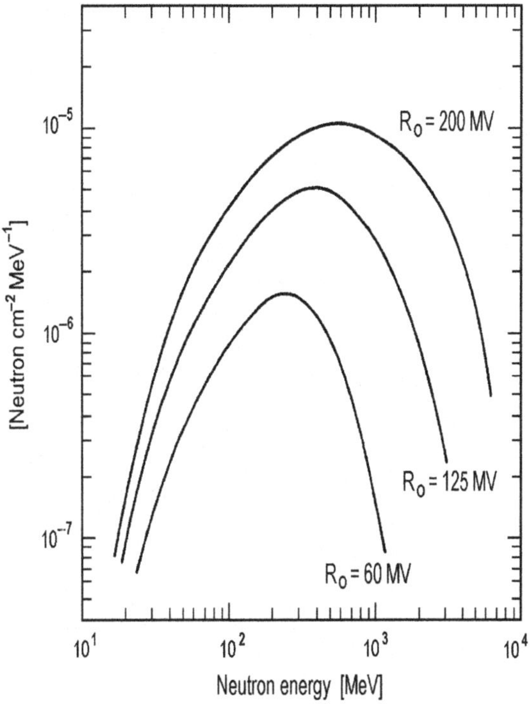

Fig. 1.6 The expected energy spectrum of solar neutrons at the distance $r = 1$ AU from the Sun, produced by an accelerated solar proton flux with characteristic rigidity $R_0 = 60$, 125 and 200 MV and normalized to the flux 1 proton cm^{-2} (From Lingenfelter et al. 1965b)

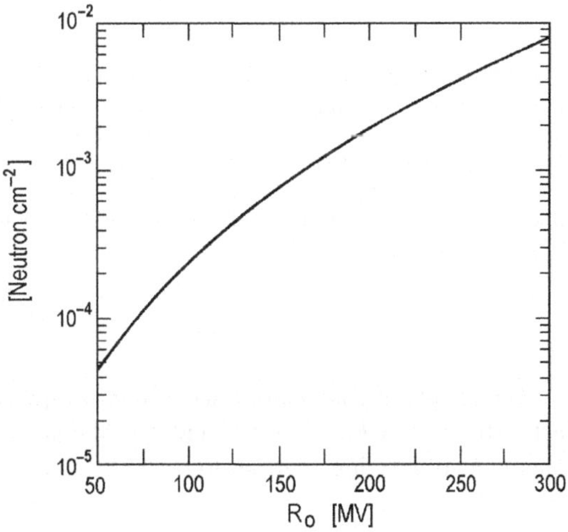

Fig. 1.7 Expected integral solar neutron flux at the distance $r = 1$ AU from the Sun versus characteristic rigidity R_0 and normalized as in Fig. 1.6 (From Lingenfelter et al. 1965b)

where the arriving time t for solar neutrons (starting from $t = 0$ at the time of neutron escaping from the Sun) is

$$\begin{aligned} t &= \frac{r}{c}\left(1 - \left(m_n c^2 / \left(E_n + m_n c^2\right)\right)^2\right)^{-1/2}, \\ dt/dE_n &= m_n^2 c^3 r \left(E_n^2 + 2 m_n c^2 E_n\right)^{-3/2}, \end{aligned} \tag{1.11}$$

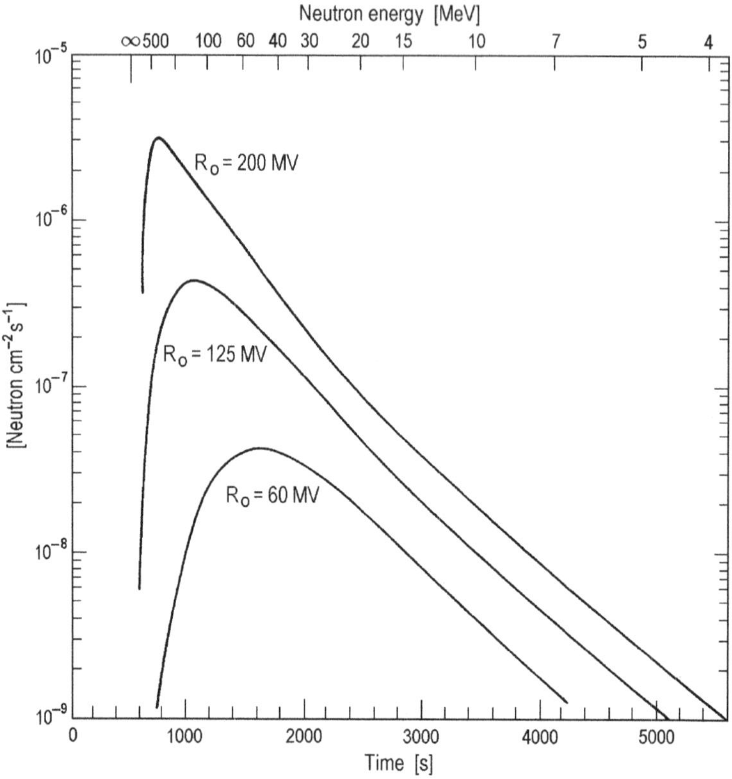

Fig. 1.8 The expected integral solar neutron flux at the distance $r = 1$ AU from the Sun as a function of time t (starting from $t = 0$ at the time of neutron escaping from the Sun). The upper scale gives the mean neutron energy arriving at times shown on the lower scale (According to Lingenfelter et al. 1965b)

and energy E_n in Eq. 1.10 of neutrons arriving the distance r at moment $\boldsymbol{t}$ can be determined from Eq. 1.11 as:

$$E_n = m_n c^2 \left(1 - (r/ct)^2\right)^{-1/2} - m_n c^2. \tag{1.12}$$

The results of calculations according to Eq. 1.10 are shown in Fig. 1.8 for characteristic rigidity $R_0 = 60$, 125 and 200 MV and $r = 1$ AU; there are shown also the energy E_n of arriving neutrons to $r = 1$ AU in dependence of t according to Eq. 1.12.

1.1.8 Expected Spectra of Decay Protons at the Distance of 1 AU from the Sun

According to Lingenfelter et al. (1965b), the spectra of protons resulting from decay of solar neutrons can be obtained by taking into account that the probability of

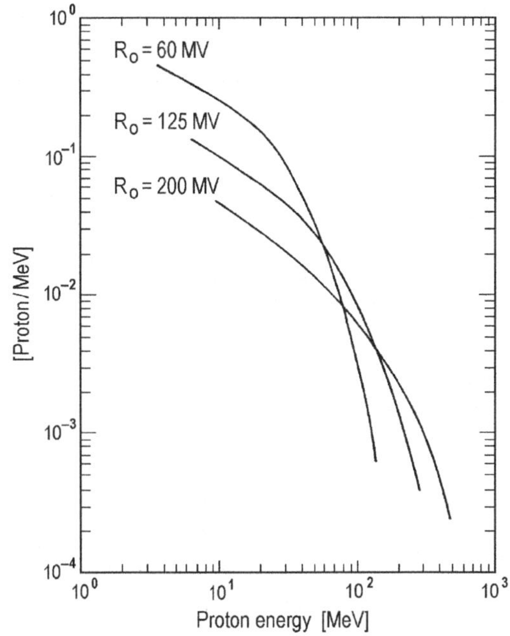

Fig. 1.9 The expected spectra of decay protons at the distance $r = 1$ AU from the Sun per one undecayed neutron (According to Lingenfelter et al. 1965b)

decay proton generation determined by the value $1 - P_s(E_n, r)$, where $P_s(E_n, r)$ is the solar neutron survival probability to reach the distance $r = 1$ AU from the Sun before decaying (see Eq. 1.8). Figure 1.9 shows the expected spectra of decay protons at 1 AU per 1 non-decayed neutron for characteristic rigidity of accelerated particle spectra $R_0 = 60$, 125 and 200 MV.

1.1.9 Probabilities That Solar Neutrons Will be Captured with the Generation of 2.2 MeV Gamma-Quanta, Will Decay or Will Escape from the Solar Atmosphere into Interplanetary Space in Dependence of Its Energy

Ramaty and Lingenfelter (1973a, b) calculated the probability of neutrons, generated in the solar atmosphere, to be captured with generation of 2.223 MeV gamma-quanta, to decay or to escape from the solar atmosphere into the interplanetary space. They suggested that neutrons are generated at the height of 10^9 cm above the photosphere and have isotropic distribution. Results for neutron energies at 1, 10 and 100 MeV are shown in Table 1.1.

On the basis of results listed in Table 1.1, Ramaty and Lingenfelter (1973a, b), calculated the expected 2.2 MeV gamma-ray flux at the Earth per 1 neutron, suggesting that all captured neutrons form 2.2 MeV gamma-quanta isotropically,

Table 1.1 Probabilities for neutron to capture with generation of 2.2 MeV gamma-quanta, to decay with generation energetic protons and electrons, or escape into interplanetary space (From Ramaty and Lingenfelter 1973a, b)

Process	$E_n = 1$ MeV	$E_n = 10$ MeV	$E_n = 100$ MeV
Capture	0.17	0.25	0.40
Decay	0.17	0.14	0.04
Escape	0.66	0.61	0.56

Table 1.2 Expected gamma-ray fluxes in the line 2.2 MeV at the Earth per one neutron in units photon × $(4\pi r^2)^{-1} = 3.6 \times 10^{-28}$ photon cm^{-2} (From Ramaty and Lingenfelter 1973a, b)

θ(°)	$E = 1$ MeV	$E = 10$ MeV	$E = 100$ MeV
0	0.16	0.21	0.16
60	0.15	0.17	0.09
75	0.13	0.14	0.05
85	0.08	0.06	0.01

also taking into account its Compton scattering by the overlying photosphere (each scattered photon is essentially lost from the 2.2 MeV line). The results are shown in Table 1.2, in dependence of angle θ between the vertical to the photosphere and the direction of observation.

Ramaty and Lingenfelter (1973a, b) also noted that the effect of the angle θ is largest for the highest neutron energies (because these particles are captured at the greatest depth in the photosphere), but the largest photon yield per neutron is not at the highest energy (although the capture probability increases monotonically with increasing neutron energy; see Table 1.1), – because of Compton scattering. The capture time for 100 MeV neutrons is $\leq$100 s but, for 1 MeV neutrons is about 200–300 s (it is a time lag between generation of neutrons and 2.2 MeV photons).

1.1.10 *Expected Continual Gamma-Ray Emission from Solar Flares*

The expected fluxes of gamma-rays in lines will be observed on the background of continual gamma-ray emission. Baisakalova and Kolomeets (1977) calculated the expected energy spectrum of continual gamma-rays in the 1–10 MeV energy range produced due to electrons bremsstrahlung, determined according to Cheng (1972) by the following equation:

$$I(E_\gamma) = \frac{n_i VZ(Z+1)r_0^2}{137\pi r^2 E_\gamma} \times \int_{E_{tr}(E_\gamma)}^{\infty} \left[\frac{4}{3}\left(1 - \frac{E_\gamma}{E}\right) + \frac{E_\gamma^2}{E^2}\right]\left[\ln\left(\frac{2E(E-E_\gamma)}{m_e c^2 E_\gamma}\right) - \frac{1}{2}\right] n(E)dE. \tag{1.13}$$

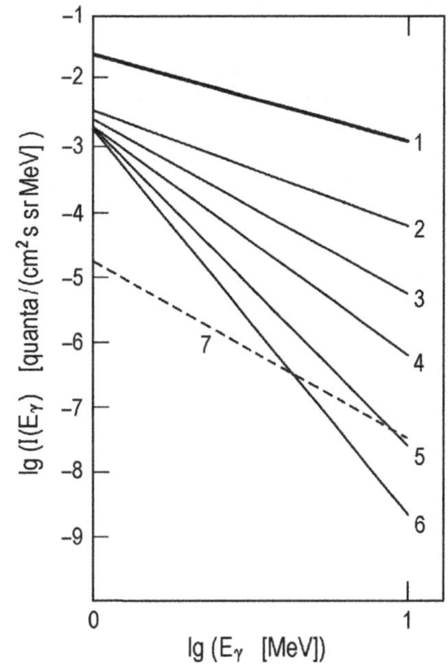

Fig. 1.10 Expected gamma-ray continual emission $I(E_\gamma)$ from the Sun. Curves 2–6 are calculated according to Eq. 1.13, bremsstrahlung spectra produced during flares by the power law electron spectra with exponents $\gamma = 2$, 3, 4, 5 and 6, respectively (at $n_i = 10^{10}\ \mathrm{cm}^{-3}$ and spectra from Eq. 1.15). Curve 7 is the expected gamma radiation from the quiet Sun. Curve 1 is the diffuse cosmic gamma-ray background, according to Trombka et al. (1973) (is shown for comparison) (From Baisakalova and Kolomeets 1977)

Here Z is the average charge of atoms in the solar atmosphere, $r_0 = e^2/m_e c^2$ is the classical electron radius, $r = 1$ AU, n_i is the ion density, V is the gamma-ray emitting volume, E_γ is the energy of gamma-quanta, E is the total energy of electrons and $n(E)$ is the differential energy spectrum of electrons. For the quiet period according to Lin et al. (1972) the spectrum of electrons is supposed in the form:

$$n(E) \propto \left(E - m_e c^2\right)^{-2.5\pm0.2} \tag{1.14}$$

for the energy interval $E \leq 20$ MeV. For the flare events it is supposed full number of energetic electrons $N_e = 10^{30}$, ion density $n_i = 3 \times 10^{13}\ \mathrm{cm}^{-3}$ and

$$n(E) \propto \left(E - m_e c^2\right)^{-\gamma}, \tag{1.15}$$

where $\gamma = 2, 3, 4, 5$ and 6. The results of expected continual bremsstrahlung solar radiation are shown in Fig. 1.10 in comparison with galactic diffuse gamma-ray background (according to Trombka et al. 1973).

1.1.11 Expected Gamma-Ray Fluxes in Lines from a Nucleus Excited by Accelerated Particles

The nuclear interactions in the solar atmosphere were taken into account by Lingenfelter and Ramaty (1967) to calculate the expected on the Earth gamma-ray fluxes in lines 4.43 MeV (from excitation of C^{12}) and in 2.31 MeV (from

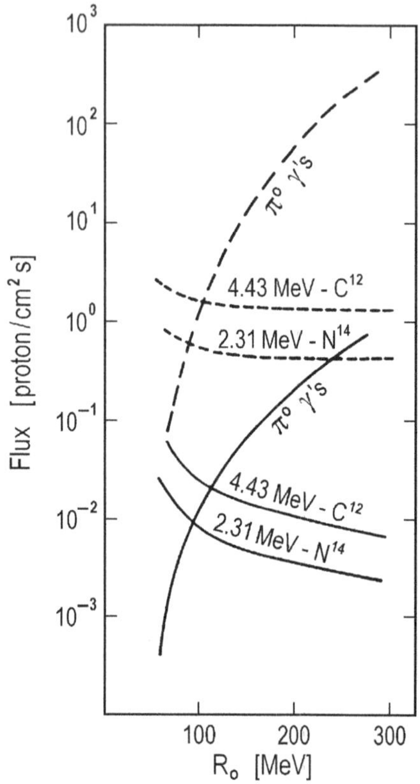

Fig. 1.11 Gamma ray fluxes at Earth for isotropic model: acceleration phase (*solid lines*) and slowing down phase (*dotted lines*) (According to Lingenfelter and Ramaty 1967) (From Chupp 1971)

excitation of N^{14}) in comparison with expected flux from pion decay for the same parameters of flare: the total number of accelerated particles with energy $\geq$30 MeV was 10^{34}, the density in region of acceleration 10^{10} cm^{-3}, time of acceleration 10^2 s and it was supposed that half of accelerated particles moved down in the photosphere. Results are shown in Fig. 1.11 (from Chupp 1971) in dependence of characteristic rigidity R_0 of accelerated particle spectra in the exponential form (Eq. 1.5).

Kuzhevskii (1968, 1969) calculated expected gamma-ray lines in suggestion that the energy spectra of accelerated particles has power character:

$$dN_p/dE_p = 5.6 \times 10^{32} E_p^{-2} \text{ protons.s}^{-1}.\text{MeV}^{-1}. \tag{1.16}$$

It was supposed also that the average acceleration or particle retention time is 200 s, the flare occupies an area of 10^{19} cm^2, and the interactions take place in a thickness 10^9 cm at a density 3×10^{13} cm^{-3}. Results of calculations are shown in Table 1.3.

As were pointed by Kuzhevskii (1968, 1969) and by Chupp (1971), if the accelerated particles are confined to a region of much lower density (for example, in the corona at density $\sim 10^9$ cm^{-3}) then the expected gamma-ray line fluxes will

Table 1.3 Expected on the Earth gamma-ray line fluxes from nuclides excited by accelerated protons in a class 3 flare (According to Kuzhevskii 1968, 1969)

Nuclide	Gamma-ray line (MeV)	Flux (photons cm^{-2} s^{-1})
O^{16}	6.14	6.2×10^{-2}
O^{16}	6.92 + 7.12	8.6×10^{-3}
O^{16}	8.87	2.0×10^{-5}
C^{12}	4.43	5.3×10^{-2}
C^{12}	12.7	1.1×10^{-3}
N^{14}	2.31	2.0×10^{-3}
N^{14}	3.95	5.6×10^{-5}
Si^{28}	1.37	2.6×10^{-3}
Si^{28}	1.78	2.7×10^{-3}
Si^{32}	2.24	2.0×10^{-3}
Ne^{20}	1.63	7.0×10^{-2}

Table 1.4 Expected fluxes of the line gamma-ray emission for the 'small' solar flares (photons cm^{-2} s^{-1}) (From Baisakalova and Kolomeets 1977)

γ	0.478 and 0.437 MeV	4.43 MeV	6.14 MeV
3	3.6×10^{-7}	8.4×10^{-7}	5.7×10^{-7}
4	7.6×10^{-7}	2.7×10^{-6}	1.9×10^{-7}
5	1.6×0^{-6}	1.3×10^{-5}	9.4×10^{-6}
6	3.9×10^{-5}	1.0×10^{-4}	7.1×10^{-5}

be several orders of magnitude lower than calculated in Table 1.3 and hence probably undetectable.

The nuclear interactions in the solar atmosphere was taken into account by Baisakalova and Kolomeets (1977) to calculate the gamma-ray fluxes in the 0.478, 0.431, 4.43 and 6.14 MeV lines due to excitation of the Li^7, Be^7, C^{12} and O^{16} nuclei respectively, according to the equation:

$$I\left(E_\gamma^*\right) = An_t \int D(E)dE \int d\cos\theta \int \sigma_t\left(E_\gamma^*, E'\right) \frac{dE'}{dE'/dx} \times \int \left(\frac{1}{L\cos\theta} + \frac{n\sigma\left(E_\gamma^*\right)}{\cos\alpha}\right) \exp(-Y(E) + Y(E'))d\cos\alpha. \tag{1.17}$$

Here A is the geometrical factor, n_t is the target nucleus density, $D(E)$ is the differential energy spectrum of incident particles, θ is the angle between moving particles and photons, α is the zenith angle of photons moving in the solar atmosphere, $\sigma_t(E_\gamma^*, E')$ is the excitation cross section for the target nucleus, $Y(E)$ is the ionisation range, L is the absorption length of incident particles, $\sigma(E_\gamma^*)$ is the cross section for absorption of the photon with energy E_γ^* in the solar atmosphere and n is the density of nuclei in the solar atmosphere. The results are shown in Table 1.4, for 'small' solar flares with full number of accelerated particles 10^{30} with energy spectra $\propto E^{-\gamma}$ with $\gamma = 3$, 4, 5 and 6.

The data on the cross sections of He^4 (α,p) $Li^{7*} \rightarrow \gamma + Li^7$, with energy of the emitted gamma-quantum $E_\gamma = 0.478$ MeV and He^4 (α,n) $Be^{7*} \rightarrow \gamma + Be^7$, with $E_\gamma = 0.431$ MeV obtained with accelerators at various energies have been used in Kozlovsky (1977) to estimate the gamma-ray intensity in the 0.478 and 0.431 MeV lines expected from solar flares. Comparisons are made with the expected flux in the line $E_\gamma = 4.4$ MeV from C^{12}. It has been found that the flux in the 0.478 MeV line from Li^7 should be approximately twice as small as that for 4.4 MeV from C^{12}. Since the observation data for the flare of August 4, 1972 shows that the gamma-ray flux in the 4.4 MeV line was 0.03 photon cm^{-2} s^{-1}, the expected flux in the 0.478 MeV line is 0.014 photon cm^{-2} s^{-1}. This is 20% in excess of the expected flux in the continuum in this energy range (the line width is estimated to be ~15 keV) for the spectrum $0.4 \times E_\gamma^{-3.42}$ photon cm^{-2} s^{-1}. It is noted in Kozlovsky (1977), that at a sufficiently high sensitivity of the instruments the solar gamma-quantum flux in the 0.478 and 0.431 MeV lines may be well detected.

1.1.12 Long-Lived Radioactive Nuclei Produced in the Solar Atmosphere During Chromospheres Flares and Possible Solar Gamma-Rays from the Quiet Sun

The long-lived radioactive nuclei produced in the solar atmosphere during chromospheres flares, according to Kuzhevskii (1977a, b), may be the cause of solar gamma-rays from the quiet Sun. The possible flux of gamma-quanta with $0.5\ \text{MeV} \leq E_\gamma \leq 3\ \text{MeV}$ near the Earth is estimated to be $10^{-7} \leq F(E_\gamma) \leq 10^{-5}$ quantum cm^{-2} s^{-1}.

1.1.13 Estimations of Expected Solar Neutron and Gamma-Ray Emissions in Suggestion That the Optical Emission of Solar Flares is Provided by Ionization Losses of Accelerated Charged Particles

Lingenfelter (1969) estimate the expected solar neutron and gamma-ray emissions from solar flares per unit of optical emission energy. His calculations are based on suggestion that the energy for the optical emission in solar flares is provided by ionisation losses of accelerated charged particles in the solar chromosphere (Gordon 1954; Parker 1963; Dubov 1963a, b). Gordon (1954) was the first suggested that the optical emission of solar flares could be provided by ionisation losses of accelerated particles and Parker (1963) estimated that at least for large flares the energy of accelerated particles could exceed that of the optical emission. According to detail study of energetic particle ionisation losses and optical emission processes in Dubov (1963a, b), the total energy and time dependence of the flare optical emission, principally in Hα and Lα, can be result of energy losses of accelerated charged particles in the flare region by ionisation and excitation.

The estimations of Lingenfelter (1969) based on two equations. The first one determined the instantaneous rate of production of neutrons and gamma-rays by nuclear interactions of accelerated particles in the flare region:

$$F_s(E_s) = \sum_i \int_0^\infty NI_i(E)V_i(E)\rho\eta\sigma_{is}(E)M_{is}(E,E_s)dE, \tag{1.18}$$

where index s denotes the type of expected emission (neutrons, gamma-ray nuclear lines or gamma-rays from pion decay), N is the total number of accelerated particles, $I_i(E)$ is the differential energy spectrum of accelerated particles of type i (normalized to 1), E is the kinetic energy of accelerated particles (in MeV), $V_i(E)$ is the velocity of accelerated particles of type i (in cm s^{-1}), ρ is the ambient gas density in the flare region (in g cm^{-3}), η is the atom density of the ambient gas (in atom g^{-1}), $\sigma_{is}(E)$ is the cross section for the nuclear interaction of an accelerated particle i with production of secondary neutrons or gamma-rays (in cm^2), and $M_{is}(E,E_s)$ is the probability that neutron or gamma-ray photon which results directly or indirectly from the interaction of accelerated particle i with energy E will have energy E_s.

The second equation determined the instantaneous energy loss rate (in erg s^{-1}) of accelerated particles in the same flare region:

$$W = 1.6\times 10^{-6}\sum_i \int_0^\infty NI_i(E)\,V_i(E)\rho\left[\eta\sigma_i(E)g_i(E) + (dE/dx)_i\right]dE, \tag{1.19}$$

where $\sigma_i(E)$ is the total nuclear interaction cross section (in cm^2), $g_i(E)$ is the mean energy loss per nuclear interaction (in MeV), and $(dE/dx)_i$ is the energy loss of accelerated particles by ionization and excitation in MeV/(g cm^{-2}).

According to Lingenfelter (1969), the ratio $F_s(E_s)/W$ determined the secondary neutron or gamma-ray production rate per unit solar flare power versus the energy E_s, or the yield per unit energy. This ratio is independent of the total number of accelerated particles N and of the ambient gas density ρ, but depends solely on the differential energy spectrum of accelerated particles $I_i(E)$ in the flare region (see Fig. 1.12).

Figure 1.12 shows the values of the ratio $F_s(E_s)/W$, expected on the distance 1 AU from the Sun (with taking into account decay of neutrons according to Section 1.1.4) for exponential spectrum of accelerated particles $I_i(E)$ in the form described by Eq. 1.5, as a function of the characteristic rigidity R_0 of accelerated particle spectrum (in interval from 60 up to 300 MV).

Lingenfelter (1969) notes that it is necessary to take into account that some part of accelerated particles escaping from the flare region and interacting in the photosphere may produce additional fluxes of neutrons and gamma-rays. Therefore, Fig. 1.12 shows only the lower limit of expected fluxes of neutrons and gamma-rays

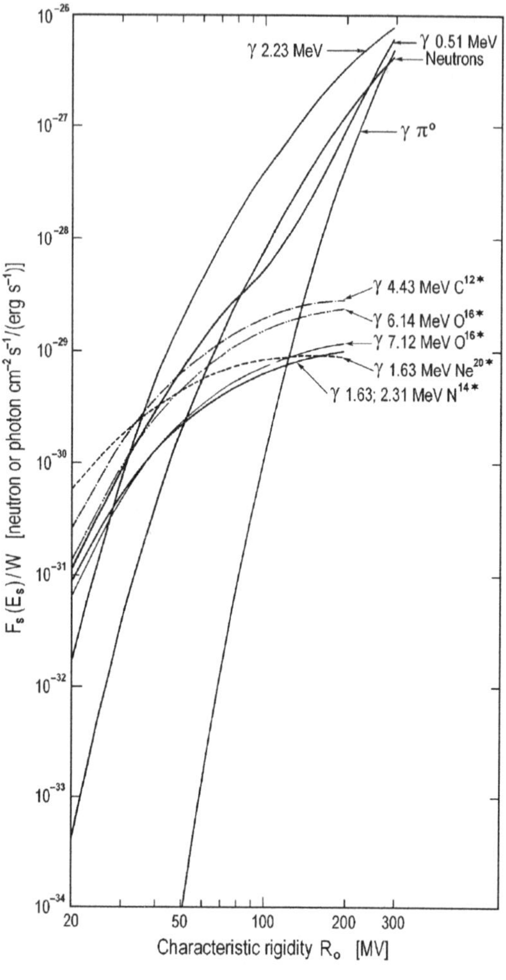

Fig. 1.12 The ratio $F_s(E_s)/W$ determined the secondary neutron or gamma-ray production rate per unit solar flare power versus the energy E_s, or the yield per unit energy (According to Lingenfelter 1969)

per unit energy of solar flare optical emission (if the suggestion on its generation mainly by accelerated particles in flare region is correct).

1.2 Estimations of Expected Solar Neutron and/or Gamma-Ray Fluxes from Historical Powerful Flare-Generated Energetic Charged Particles

Many calculations were made to estimate possible fluxes of solar neutrons expected in well known historical solar energetic events.

1.2.1 The Event of November 12, 1960

For the well known solar proton flare of November 12, 1960 (review of this event was given in books Dorman (M1963a, b) and Dorman and Miroshnichenko (M1968), the flux of solar neutrons in the period of acceleration phase is expected to be 2–20 neutron cm^{-2} s^{-1} and for the slowing down phase-10–33 neutron cm^{-2} s^{-1} (Lingenfelter and Ramaty 1967).

1.2.2 The Event of May 23, 1967

For the May 23, 1967 flare, the expected fluxes of neutrons and gamma-rays was calculated by De Feiter (1971). He based his calculations on Lingenfelter and Ramaty (1967) model, and supposed that the total energy of the particles with energies above 20 MeV, is

$$W_p = \int_{20\,\mathrm{MeV}}^{\infty} N_0 \exp(-R/R_0) E(R) \frac{dR}{dE} dE \geq 2.5 \times 10^{29} \text{ erg.} \qquad (1.20)$$

The calculations were made for $R_0 = 60$, 100 and 200 MV. Results of De Feiter (1971) of estimations of expected minimum numbers of gamma-quanta and neutrons at 1 AU per cm^2, integrated over time of the flare are shown in Table 1.5.

De Feiter (1971) noted that these estimates were based on the calculations of Lingenfelter and Ramaty (1967), where it was assumed that no decay of neutrons

Table 1.5 Expected minimum number of gamma-quanta or neutrons at 1 AU per cm^2 integrated over time for the flare of May 23, 1967 under the assumption that the continuum emission is produced by interaction of fast particles (the depth for interaction was supposed 1.6 g cm^{-2}) (According to De Feiter 1971)

E_γ or E_n (MeV)	Source	Acceleration phase			Slowing down phase		
		R_0 (MV)			R_0 (MV)		
		60	100	200	60	100	200
2.23	Neutron capture	107	286	574	147	760	3,720
0.51	Positron annihilation	34	48	170	82	268	3,190
10–100	π^0 meson decay	<1	8	106	<1	45	1,860
4.43	C^{12}	22	17	5	45	55	37
6.14	O^{16}	18	14	5	41	49	33
1.63	Ne^{20}	10	7	3	23	28	20
7.12	O^{16}	9	7	3	20	27	19
1.63 & 2.31	N^{14}	6	5	1	13	16	12
Neutrons	p(He), α(H)	22	55	298	65	343	2,130

took place before capture; therefore, the intensity of the neutron capture line at 2.23 MeV was overestimated. According to Ito et al. (1968), the neutron and gamma-ray emission from the slowing down phase of the solar flare event is strongly dependent on the emission direction of energetic protons.

1.2.3 The Events of August 4 and 7, 1972

Possible production of solar neutrons in flares of August 4 and 7, 1972 have been studied in Reppin et al. (1973). The intensity of the 2.2 MeV gamma-rays generated in the chromospheric flares at 06:22 GMT on August 4 and at 15:38 GMT on August 7, 1972 were measured from OSO-7. The 2.2 MeV gamma-ray flux was found to be $(2.2 \pm 0.2) \times 10^{-1}$ quant cm^{-2}, which corresponds to 6×10^{26} neutron–proton interaction events per 1 s on the assumption of isotropic ejection of gamma-quanta from the flare region. Considering the interaction cross section, the flare-generated neutron flux has been estimated to be $\sim 1.3 \times 10^{27}$ neutrons s^{-1}. The estimate shows that some 50% of the 10–100 MeV neutrons reach the Earth without decay. The flux of such neutrons near the Earth for the case of August 4, 1972 is expected to be $\sim 3 \times 10^{-3}$ neutron/(cm^2s). It is noted that such fluxes can well be detected using modern experimental techniques.

1.3 Search for Solar Neutrons Using Balloon and Space Probe Experiments

1.3.1 The Matter and Short History of the Problem

The nuclear reactions between accelerated particles and atoms in the solar atmosphere will produce neutrons. Theoretical investigations described in Section 1.1 stimulated numerous experimental attempts to detect solar neutrons and gamma-rays at the top of the Earth's atmosphere by balloons, in the interplanetary space on space-probes and by mountain based, low latitude, neutron monitors during quiet and active solar conditions. These calculations show that, although the neutrons are unstable, a marked number of neutrons will reach the region near the Earth's orbit.

In the period 1964–1972 many experimental attempts were undertaken to detect solar neutrons in the stratosphere by balloons and in the interplanetary space by space-probes during quiet and active solar conditions: Swetnick et al. (1952), Haymes (1959a, b, 1964a, b, 1965), Appa Rao et al. (1966), Bame and Asbridge (1966), Hess and Kaifer (1967), Webber and Ormes (1967), Kim (1968), Daniel et al. (1969), Forrest and Chupp (1969), Zych and Frye (1969), Cortellessa et al. (1970), Heidbreder et al. (1970), Kondo et al. (1970), Cortellessa et al. (1971), Daniel et al. (1971).

1.3.2 Search for Solar Neutrons at Quiet Conditions

Webber and Ormes (1967) have determined the upper limit values for the fluxes of the solar neutrons with $E_n > 60$ MeV, in a period corresponding to quiet geophysical conditions using the measurements from balloons launched on August 8, 1964 from the station Tukuman, Argentine. The two-element scintillating Cherenkov counter was used on the balloons as radiation detectors, measuring the proton energy spectrum in the range from 50 MeV to 1 GeV. The upper limit flux values of solar neutrons (by an order less than those presented in previous communications) were obtained by comparison of the secondary proton intensities at the atmospheric depth $\sim$12 g cm^{-2}, near the equator (which were measured during two flights). During the first flight the detector telescope was oriented vertically and during the second flight the zenith angle of the telescope (rotating over the azimuth with the period of 15 min) was $\sim$50°. As a result of this rotation, the telescope was directed in turn towards the Sun or normally to it. The absence of any flux increases in the secondary protons during time intervals, when the telescope was directed to the Sun, shows that the solar neutrons form the flux of the secondary protons less than 1 proton/(m^2 sr s), in the energy range of 60–320 MeV in the atmosphere above the detector's location. Including the efficiency of proton formation by solar neutrons in the air layer with the thickness of 12 g cm^{-2}, Webber and Ormes (1967) come to the upper limit flux value for the solar neutrons with $E_n \geq 100$ MeV equal to $\sim$24 neutron.m^{-2} s^{-1} (it was assumed that the neutron differential spectrum at the Earth was $\propto E_n^{-2}$). The limiting flux values of the neutrons in the other energy ranges were also roughly estimated; these values were calculated with the assumption that the observed protons were generated by neutrons in the same energy range. In particular, the upper limit neutron flux values in the ranges 60–100 MeV and 100–160 MeV were found 125 and 75 neutron.m^{-2} s^{-1}, respectively.

Heidbreder et al. (1970) noted that for the quiet Sun the upper limit of the solar neutron flux value was 12.5 neutron m^{-2} s^{-1}.

Measurements of neutrons were carried out by Eyles et al. (1971) over Kampala (Uganda, at 10°S) with the detectors, which were combined to a converter (where the neutrons knocked protons), a telescope (for detecting the knocked-out protons), an absorber and anti-coincidences counters to screen against charged particles. The measurements were taken on November 1, 1967 (at 7–10 g cm^{-2} altitude), March 21 (7–8 g cm^{-2}), May 29 (6.5 g cm^{-2}) and October 31, 1969 (6 g cm^{-2}). The upper limits were found to be $\sim$30 neutron/(m^2 s), for the mean flux of solar neutrons with $E_n \geq 50$ MeV in quiescent periods.

The energy spectrum and angular distribution of the 10–100 MeV neutron fluxes were measured by White et al. (1973) at 40°N, with airborne instruments at 5.4 g cm^{-2} of residual atmosphere. The equipment included a neutron telescope was in operation from 08:30 to 19:30 of solar local time on September 26, 1971. The telescope consisted of two tanks of a 0.5 m^2 effective area with 75 L of liquid scintillator in each and spaced 1 m apart. Each tank was divided into eight compartments with individual photomultiplier tubes so that the scattering of neutrons by protons in the upper and lower tanks measured on the basis of proton recoil moment made it possible to determine not only energy but also the direction of neutrons (the method of dual scattering). A special detector was used simultaneously to fix the Sun's position with respect to the telescope. The analysis of 10% of the data has shown that the upper limit of the 20–60 MeV solar neutron flux during quiet Sun on September 26, 1971, was ~10 neutron.m^{-2} s^{-1}. A more comprehensive analysis of above described experimental data was made in Moon et al. (1973, 1976). The upper limits of the neutron fluxes generated by quiet Sun have been found to be 1.1×10^{-4}, 2.6×10^{-4} and 5.9×10^{-4} neutron/(cm^2 s) in the 10–30, 30–50 and 50–100 MeV energy ranges, respectively.

The results of OGO-6 measurements of the 1–20 MeV solar neutron flux during the quiescent periods in June 7–December 23, 1969 are presented in Lockwood et al. (1974). The differences between the intensities near the equator at night and day hours have been found to be within statistical errors; the upper limit of the permanent flux of solar neutrons were found equal to 18 neutron/(m^2 s).

1.3.3 Search for Solar Neutrons in Periods of Flare Conditions

Bame and Asbridge (1966) have performed neutron measurements on the board of Vela satellites and found the upper limit flux 0.4 neutron cm^{-2} s^{-1} in the energy interval 1–100 MeV during solar flares of importance 3. Three measurements during solar flares of importance 2 have been made by Daniel et al. (1969) (upper limit flux 1.2×10^{-2} neutron cm^{-2} s^{-1} in the energy interval 15–150 MeV) and Kondo et al. (1970) (upper limit flux 1.0×10^{-2} neutron cm^{-2} s^{-1} and upper limit for integral flux 40 neutron cm^{-2} for the energy interval 6–100 MeV). Positive experimental evidence for a solar neutron flux 1.5×10^{-2} neutron cm^{-2} s^{-1} in the energy interval 50–500 MeV during a pre-flare phase reported by Daniel et al. (1971).

Eyles et al. (1971, 1972) search solar neutrons on balloons at heights of 6–10 g cm^{-2} over Kampala (Uganda, 10°S) at November 1, 1967, March 21, May 29 and October 31, 1969. It was possible to detect neutrons with $E_n \geq 50$ MeV and gamma-quanta with $E_\gamma \geq 20$ MeV. The upper limit of integral flux of neutrons with 50 MeV $\leq E_n \leq 350$ MeV for all times of solar flares was found to be 23 neutron cm^{-2} (for 1B flares) and 2.1 neutron cm^{-2} (for SN flares).

The search for solar neutrons in the energy interval 10–100 MeV was made by White et al. (1973) in balloon experiment at September 26, 1971 (5.4 g cm^{-2}, 40°S). The upper limit for neutron flux in the energy interval 10–60 MeV was found to be 10^{-3} neutron/(cm^2 s).

The solar neutrons were sought by Ifedili (1974) on the basis of OGO-6 measurements during some solar flares including the class 3B flares. The measurements were taken with a He^3 proportional counter, with a moderator surrounded with 22 counters connected to an anti-coincidence circuit to protect against the charged particles and neutrons produced by CR in the spacecraft body. The experiment has given only the upper limit of the possible flare-generated 1–20 MeV neutron flux (5×10^{-2} neutron.cm^{-2} s^{-1}) within a 95% reliability interval.

The obtained values proves to be very close to the expected theoretical flux, so it can be hope that the solar neutron flux could have been detected with a somewhat higher sensitivity of the instrument. On the other hand, if the obtained upper estimates are used to calculate the gamma-quantum flux produced in the Sun's atmosphere as a result of the reaction of neutron capture by hydrogen, such flux will prove to be closer to that observed actually from OSO-6 and OSO-7 during flares. This circumstance also indicates that the real neutron flux is close to the obtained upper estimates.

1.3.4 Comparison of the First Attempts to Search for Solar Neutrons Using Balloons with the Expected for August 4, 1972

Table 1.6 from Moon et al. (1973) shows the summary of the first measurements of solar neutrons in quite conditions and in periods of solar flares. For comparison in the last row are shown the expected flux in the August 4, 1972 event according to Reppin et al. (1973) (see Section 1.3.3).

1.4 Search for Solar Neutrons Using Ground Measurements

Sufficiently more exact estimations of solar neutrons from large flares were made by Kirsch (1973). He used the data of continues CR observations by high-mountain neutron monitors and supermonitors, which have very low statistical errors (see Table 1.7)

Since the solar neutrons are not deflected by interplanetary field and by the Earth's magnetic field, the analysis was carried out by the method Chree of epoch superposition with selection of Class 3 flares observed near local noon (in the interval ± 3.5 h). For recalculation from observed neutron monitor data to the top of the atmosphere and to the source of solar neutrons on the Sun, Kirsch (1973) used the results of Lingenfelter et al. (1965a) and Lingenfelter and Ramaty (1967) to calculate transport of solar neutrons from the Sun to the top of the Earth's atmosphere. It was suggest that the proton spectrum on the Sun have a form

$$D(R) \propto \exp(-R/R_0), \quad R_0 \geq 50 \text{ MV}, \tag{1.21}$$

and was taking into account the absorption of neutrons in the atmosphere as well as the results of Alsmiller and Boughner (1968) on generation and propagation in the

Table 1.6 The first attempts to measure solar neutron fluxes on balloons (in units neutron $cm^{-2}\,s^{-1}$) and comparison with expected in the August 4, 1972 event (According to Moon et al. 1973)

Method	Height (g cm^{-2})	Energy (MeV)	Quiet Sun, flux	Solar flare, imp.	Solar flare, flux	Reference
Plastic scintillator	4	25–120	$<2 \times 10^{-2}$	2B	$<4 \times 10^{-2}$	Forrest and Chupp (1969)
Spark chamber	7	12–100		1	$<1.7 \times 10^{-2}$	Zych and Frye (1969)
Proton recoil liquid scintillator	13	6–100		2B	$<3 \times 10^{-2}$	Kondo et al. (1970)
Proton recoil nuclear emulsion	6.6	20–160	$(2.2 \pm 2.5) \times 10^{-2}$			Kim (1968)
Plastic scintillator	25	25–120		2B	$<1.2 \times 10^{-2}$	Daniel et al. (1969)
Double scattering in spark chamber	7	100–400	$<1.25 \times 10^{-2}$			Heidbreder et al. (1970)
Proton recoil scintillator	3	50–350	$<5 \times 10^{-3}$			Eyles et al. (1971)
Proton recoil plastic scintillator	4.5	10–200	$<5.5 \times 10^{-3}$			Cortellessa et al. (1971)
Double scattering using pulse height	5	20–60	$<1.0 \times 10^{-3}$			White et al. (1973)
and time of flight		10–100	$<5 \times 10^{-4}$			Moon et al. (1973)
From 2.2 MeV gamma-ray line for August 4, 1972	5	10–100		3B	$\sim 3 \times 10^{-3}$	Reppin et al. (1973)

Table 1.7 Neutron monitor stations which were used for solar neutron search by Kirsch (1973)

Station and equipment	Geographic coordinates	Cutoff rigidity (GV)	Height of the station (m)	Mean air pressure (g cm^{-2})	Mean counting rate (h^{-1})	Statistical error per 1 h (%)
Chacaltaya, Bolivia (12 NM-64, S = 14×10^4 cm^2)	16.3°S, 68.2°W	13.3	5,200	550	5.6×10^6	0.042
Chacaltaya, Bolivia, (IGY monitor, S = 4×10^4 cm^2)	16.3°S, 68.2°W	13.3	5,200	550	$>0.24 \times 10^6$	<0.202
Mina Aguilar, Argentine, (IGY monitor, S = 4×10^4 cm^2)	23.2°S, 65.7°W	12.6	4,000	640	0.24×10^6	0.202
Pic-du-Midi, France, (9 NM-64, S = 10×10^4 cm^2)	42.9°N, 0.25°E	5.6	2,860	730	2.3×10^6	0.065
Jungfraujoch, Switzerland, (IGY monitor, S = 6×10^4 cm^2)	46.5°N, 8°E	4.8	3,550	660	0.54×10^6	0.136

Table 1.8 Expected solar neutron fluxes (neutron cm^{-2} s^{-1}) and parameters F_{550} and F_{730} for R_0 =125 MV (According to Kirsch 1973)

Neutron energy (MeV)	Flux at 0 g cm^{-2}	Flux at 550 g cm^{-2}	F_{550}	Flux at 730 g cm^{-2}	F_{730}
50 ± 5	23	0.12	5.2×10^{-3}	8.3×10^{-3}	3.6×10^{-4}
100 ± 5	42	2.00	4.7×10^{-2}	0.29	6.9×10^{-3}
200 ± 5	34	11.0	0.324	3.2	0.094
300 ± 5	25.4	14.5	0.570	4.6	0.181
400 ± 5	15.9	13.0	0.816	4.3	0.270
500 ± 5	7.5	6.50	0.866	2.2	0.293
$\sum$	147.8	47.12		14.59	

atmosphere of secondary neutrons with $E_n < 20$ MeV. These results give possibility to determine the parameter F of solar neutron transport attenuation coefficient in the atmosphere relative to the primary flux. The results for 550 g cm^{-2} (F_{550}) and for 730 g cm^{-2} (F_{730}) for different energy intervals of the primary solar neutrons as well as the expected solar neutron fluxes are shown in Table 1.8 (Kirsch 1973).

For the energy interval 50–500 MeV, in the case of R_0 = 125 MV from Table 1.8 it was found

$$F_{730}(R_0 = 125 \text{ MV}) = \sum \left(730 \text{ g.cm}^{-2}\right) \Big/ \sum \left(0 \text{ g.cm}^{-2}\right) = 0.098, \qquad (1.22)$$

$$F_{550}(R_0 = 125 \text{ MV}) = \sum \left(550 \text{ g.cm}^{-2}\right) / \sum \left(0 \text{ g.cm}^{-2}\right) = 0.32. \qquad (1.23)$$

For $R_0 = 60$ MV these parameters were found

$$\overline{F_{550}}(R_0 = 60MV) = 0.0625, \ \overline{F_{730}}(R_0 = 60) = 0.0165. \qquad (1.24)$$

Figure 1.13 shows the results of Chree analysis of the data of the supermonitors on stations Chacaltaya and Pic-du-Midi during solar flares of the importance 3, which are not classified as proton flares by Croom (1971). For Chacaltaya the 16 flares are: November 14, 1966; February 13, March 6, 26, and 31, May 8, July 28 and September 28, 1967; July 12, August 8 and October 27, 1968; June 5, 11, and 13, July 5 and November 7, 1969. For Pic-du-Midi the 24 flares are: April 15, September 20, October 9, November 14, 1966; February 24, March 30 and 31, May 5, 8, 15 and 19, June 2 and 23, September 1, 17 and 28, December 27, 1967; July 12, September 29, 1968; January 24, April 21, May 24, November 26, 1969; March 1, 1970.

I prepared Fig. 1.14 where are shown the same data, but without trend, i.e.:

$$\langle N_i \rangle = N_i - (N_{i+1} + N_i + N_{i-1})/3. \qquad (1.25)$$

From the analysis of Fig. 1.13, Kirsch (1973) came to the conclusion that Chacaltaya and Pic-du.Midi neutron supermonitors, with an accuracy of 3σ did

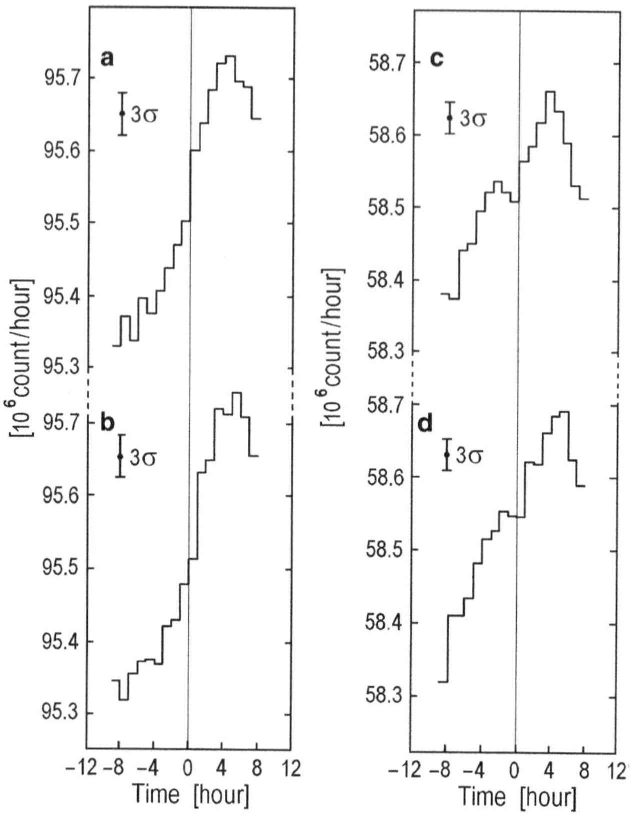

Fig. 1.13 Superposition of neutron supermonitor measurements during flares of importance 3, which are not classified as proton flares by Croom (1971). (**a**) 16 flare days from Chacaltaya with the 0 h according to the middle of the optical flare; (**b**) 16 flare days from Chacaltaya with the 0 h according to the beginning of the optical flare; (**c**) 24 flare days from Pic-du-Midi with the 0 h according to the middle of the optical flare; (**d**) 24 flare days from Pic-du-Midi with the 0 h according to the beginning of the optical flare (From Kirsch 1973)

not show any evidence of solar neutron flux. It gives for the upper limit of solar neutron flux with 99% confidence level the value

$$I_{sn} \leq \frac{3}{qG\overline{F}} (N/n\Delta T)^{1/2}, \tag{1.26}$$

where N is the background counting rate (s^{-1}), $\Delta T = 3600$ s is the observation time, n is the number of superimposed events (16 for Chacaltaya and 24 for Pic-du-Midi), q is the detection efficiency (according to Hatton and Carmichael 1964, for 5 and 10 MeV neutrons, $q = 0.05$ and 0.08, accordingly), G is the geometric factor of the neutron supermonitor, parameter $\overline{F}$ for different levels of observation was determined by Eqs. 1.22–1.24.

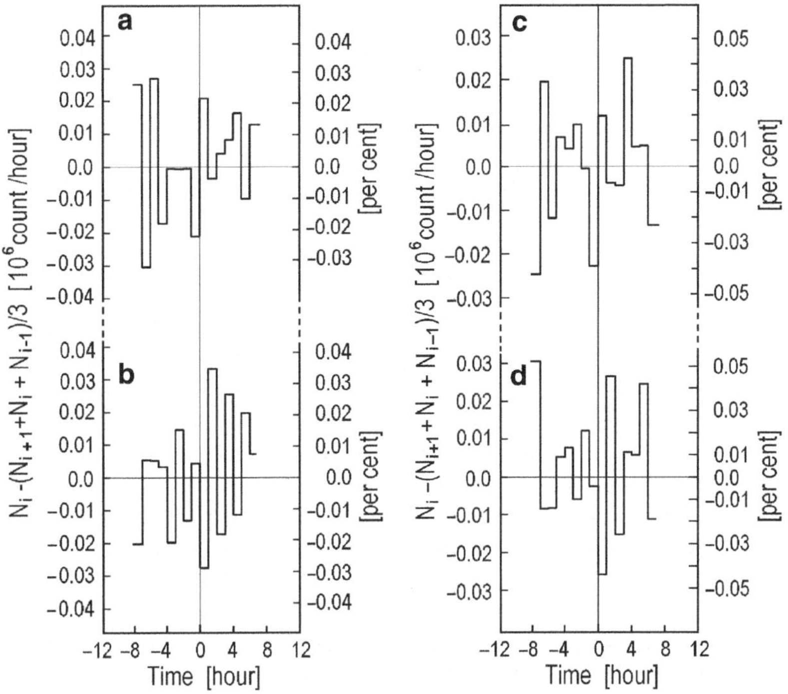

Fig. 1.14 The same as in Fig. 1.13, but without trend: data shown in Fig. 1.13 are recalculated in accordance with Eq. 1.25

For Chacaltaya measurements $3\sigma = 3 \times 10^4$ neutron h^{-1}, so

$$I_{sn} \leqslant 2.2 \times 10^{-3} \text{ neutron cm}^{-2}\text{s}^{-1}, \quad \text{for } R_0 = 125 \text{ MV}, \tag{1.27}$$

$$I_{sn} \leqslant 1.1 \times 10^{-2} \text{ neutron cm}^{-2}\text{s}^{-1} \quad \text{for } R_0 = 60 \text{ MV} \tag{1.28}$$

and for Pic-du-Midi ($3\boldsymbol{\sigma} = 2.3 \times 10^4$ neutron h^{-1})

$$I_{sn} \leqslant 5 \times 10^{-3} \text{ neutron cm}^{-2}\text{s}^{-1} \quad \text{for } R_0 = 125 \text{ MV}, \tag{1.29}$$

$$I_{sn} \leqslant 3 \times 10^{-2} \text{ neutron cm}^{-2}\text{s}^{-1} \quad \text{for } R_0 = 60 \text{ MV}. \tag{1.30}$$

I have four comments to this important result of Kirsch (1973). The first is connected with Fig. 1.13. If we consider this figure more carefully and compare with Fig. 1.14 we can see that the increase in Chacaltaya monitor at 0 h (Panel (a) in Figs. 1.13 and 1.14) or at +1 h (Panel (b) in Figs. 1.13 and 1.14) is unusually large (by about three to four times larger than at other time moments). This peculiarity can be caused by solar neutron flux, which gives an increase of about 5×10^4

neutron h^{-1} (about 0.05% on average for 16 flares). This corresponds to the solar neutron flux for $E_n \geq 50$ MeV on the top of the atmosphere (with accuracy $\pm 1\boldsymbol{\sigma}$):

$$I_{sn} = (3.7 \pm 0.7) \times 10^{-3} \text{ neutron.cm}^{-2}.\text{s}^{-1} \text{ for } R_0 = 125 \text{ MV} \tag{1.31}$$

or in percents of averaging counting rates of galactic CR at Chacaltaya,

$$I_{sn} \approx (0.05 \pm 0.01)\% \ . \tag{1.32}$$

The second comment is as following: for a more exact result it is necessary to extract solar-daily variations from observed superimposed time variation curves on Fig. 1.13; it is possible to do by using original data (we tried to do this in Fig. 1.14).

The third comment concerns the spectrum of solar protons and generated neutrons in the exponential form (Eq. 1.21). For detection by neutron supermonitors the important components are the high energy neutrons which are generated by high energy protons, but according to numerous experimental data (see review in Dorman and Miroshnichenko M1968; Dorman M1978; Dorman and Venkatesan 1993; Miroshnichenko M2001) accelerated protons have the power form $\propto E_n^{-\gamma}$ which is much more hard spectra than exponential spectra described by Eq. 1.21.

The fourth comment concerns to the time period of solar neutron flux increase. Kirsch (1973) used 1 h time intervals for this period; but, as we know now, the real average duration of solar neutron flux is only 5–10 min. This was shown for events in June 1980 and June 1982 (see Chapter 4) and by statistical analysis by Dorman et al. (1993) on the basis of 5-min data of Tyan-Shan mountain neutron supermonitor and by other authors by statistical investigations and for individual events (see Chapters 6 and 7). If the real duration of solar neutron flux is for example, 10 min, we expect the average solar neutron differential flux on the top of the atmosphere for $E_n \geq 50$ MeV from flares of importance 3 to be much larger:

$$I_{sn} \approx (2.2 \pm 0.4) \times 10^{-2} \text{ neutron.cm}^{-2}.\text{s}^{-1} \text{ for } R_0 = 125 \text{ MV}, \tag{1.33}$$

which corresponds to the integral flux

$$\int I_{sn}\, dt \approx (13.3 \pm 2.5) \text{ neutron.cm}^{-2} \text{ for } R_0 = 125 \text{ MV}. \tag{1.34}$$

In this case the percent average counting rate increase relative to the normal level of galactic cosmic rays will be

$$I_{sn} \approx (0.30 \pm 0.05)\%, \tag{1.35}$$

which is in a good agreement with the results of statistical analysis of Belov et al. (1989), Aushev et al. (1993) (see Chapter 6).

1.5 The First Attempts at Solar Gamma-Ray Search

1.5.1 Solar Gamma Ray Flux in Quiet Geophysical Conditions

As was mentioned (Section 1.3.2), in the measurements of solar neutrons on balloons by Eyles et al. (1971) over Kampala (10° S), the detectors also can detect gamma-quanta with energy $E_\gamma \geq 20$ MeV. The upper limit for the flux of gamma-quanta in this energy interval in quiescent periods was found to be about 4×10^{-4} photon cm^{-2} s^{-1}. During the balloon flight over Churchill (Canada), in quiet time July 12, 1971, the measurements of the angle distribution of gamma-rays with energy $E_\gamma \geq 10$ MeV was carried out by Kim (1975). Throughout this day there were no intense solar flares and the sunspot number was 55. The gamma-ray intensity was determined by the number of electron–positron pairs in photo-emulsions exposed on the balloon. The background intensity of gamma-rays were found to be $(1.6 \pm 0.21) \times 10^{-2}$ photon cm^{-2} s^{-1}. The upper limit for the gamma-ray flux with $E_\gamma \geq 10$ MeV from the quiet Sun (after extracting background gamma radiation) was found to be 3.1×10^{-4} photon cm^{-2} s^{-1}.

1.5.2 On the Rigid X-Radiation from Solar Flares

The short information on solar rigid x-radiation is important to understand what type of gamma-ray fluxes (continued x-ray spectrum) can be expected. For example, we consider the results of the OGO-5, Explorer-35 and Explorer-41 measurements of x-rays with energies $E_X \geqslant 10$ keV during the explosive phases of small solar flares (Kane 1973). The characteristic times of increases and decreases of x-rays with $E_X \approx$ 40 keV were 3 and 6 s (the characteristic times decreased with increasing E_X). The energy spectrum in the interval 10–100 keV has a form $\propto E_X^{-\gamma}$ and the power exponent γ decreased with increasing x-ray flux and increased during the x-ray intensity decay, so that the minimum value of $\gamma_{\min}$ was observed at the moment of x-ray maximum. In all cases $2.5 \leq \gamma_{\min} \leq 4.5$. The 10–100 keV electron spectrum has been found to be $\propto E_e^{-\phi}$, where $\varphi = \gamma - 0.75$. This fact confirms the assumption that the small flares can directly generate the electrons whose bremsstrahlung is just the observed rigid x-rays. The electrons are accelerated up to the maximum energies within ≤ 1 s and the duration of the entire process of acceleration in the flares is about 100 s; the probable location of the acceleration region is the lower corona, where the concentration is $\leq 10^9$ cm^{-3}.

1.5.3 The First Searches for Gamma-Ray Fluxes from Solar Flares

In the mentioned above (see Section 1.4.2) balloon experiments of Eyles et al. (1971) were found also the upper limits of gamma-ray fluxes with $E_\gamma \geq 20$ MeV from chromospheric flares of class 1B and SN: 6 and 0.3 photon cm^{-2} respectively (Eyles et al. 1972).

1.6 The First Attempts to Search for Solar Positrons

Lingenfelter and Ramaty (1967), Ramaty and Lingenfelter (1973a, b), and Ramaty (1986) have considered in detail positron production and annihilation during solar flares. For solar energetic particle spectra typical of those observed at Earth, they found that the dominant positron source is β^+ decay of nuclear interaction products such as ^{11}C, ^{13}N and ^{15}O. According to these investigations Hurford et al. (1973) examined the production mechanism of positrons with kinetic energy 0.2 MeV $\leq E_{e^+} \leq$ 2 MeV in the periods of chromospheric flares. Consideration was given to the following nuclear reactions between the accelerated protons and the C, N and O nucleus in the solar atmosphere:

$$\begin{aligned}&^{12}\mathrm{C}\ (\mathrm{p,pn})^{11}\mathrm{C}\left(\mathrm{E_p}\geqslant 20.2\ \mathrm{MeV}\right), ^{14}\mathrm{N}(\mathrm{p,2p2n})^{11}\mathrm{C}\left(\mathrm{E_p}\geqslant 13.1\ \mathrm{MeV}\right),\\&^{14}\mathrm{N}\ (\mathrm{p,a})^{11}\mathrm{C}\left(\mathrm{E_p}\geqslant 2.9\ \mathrm{MeV}\right), ^{16}\mathrm{O}(\mathrm{p,3p3n})^{11}\mathrm{C}\left(\mathrm{E_p}\geqslant 28.6\ \mathrm{MeV}\right),\\&^{11}\mathrm{C}\ \rightarrow\ ^{11}\mathrm{B}+\mathrm{e}^+ + \nu_\mathrm{e}\left(\mathrm{E_{e^+}}\leqslant 0.97\ \mathrm{MeV},\ \mathrm{T}_{1/2}=20.5\ \mathrm{min}\right).\end{aligned} \tag{1.36}$$

$$\begin{aligned}&^{14}\mathrm{N}(\mathrm{p,pn})^{13}\mathrm{N}\left(\mathrm{E_p}\geqslant 11.3\ \mathrm{MeV}\right), ^{16}\mathrm{O}(\mathrm{p,2p2n})^{13}\mathrm{N}\left(\mathrm{E_p}\geqslant 5.54\ \mathrm{MeV}\right),\\&^{13}\mathrm{N}\ \rightarrow\ ^{13}\mathrm{C}+\mathrm{e}^+ + \nu_\mathrm{e}\left(\mathrm{E_{e^+}}\leqslant 1.19\ \mathrm{MeV},\ \mathrm{T}_{1/2}=9.96\ \mathrm{min}\right)\end{aligned} \tag{1.37}$$

$$\begin{aligned}&^{14}\mathrm{N}\ (\mathrm{p,n})^{14}\mathrm{O}\left(\mathrm{E_p}\geqslant 6.4\ \mathrm{MeV}\right),\\&^{14}\mathrm{O}\ \rightarrow\ ^{14}\mathrm{N}+\mathrm{e}^+ + \nu_\mathrm{e}\left(\mathrm{E_{e^+}}\leqslant 1.86\ \mathrm{MeV},\ \mathrm{T}_{1/2}=1.18\ \mathrm{min}\right),\end{aligned} \tag{1.38}$$

$$\begin{aligned}&^{16}\mathrm{O}\ (\mathrm{p,pn})^{15}\mathrm{O}\left(\mathrm{E_p}\geqslant 16.54\ \mathrm{MeV}\right),\\&^{15}\mathrm{O}\ \rightarrow\ ^{15}\mathrm{N}+\mathrm{e}^+ + \nu_\mathrm{e}\left(\mathrm{E_{e^+}}\leqslant 1.73\ \mathrm{MeV},\ \mathrm{T}_{1/2}=2.07\ \mathrm{min}\right).\end{aligned} \tag{1.39}$$

In Eqs. 1.36 through 1.39 protons considered as energetic particles and nucleus C, N and O as targets. Significant part of positrons will be generated also by the energetic C, N and O nucleus, in interaction with protons as targets (in this case the threshold energies will be the same, but in units MeV/nucleon).

The source of positrons described in Eqs. 1.36 through 1.39 is the main for generation gamma-line 0.51 MeV. The second source of positrons is the $\pi^+ \rightarrow \mu^+ \rightarrow e^+$ decay which is important for higher energies (10–100 MeV) and is not significant for the formation of 0.51 MeV gamma-ray line. According to Hurford et al. (1973), in the frame of the thick target model, when positrons are the result from the interactions of energetic particles moving downwards into the Sun, the energy loss and magnetic field considerations suggest that only a small part of the 1 MeV positrons would escape from the Sun. But, if the positrons are produced by the interactions of accelerated particles as they leave the Sun, and if the amount of matter transferred is small (thin target model), it is possible that a significant fraction of the 1 MeV produced positrons may escape. The results of these calculations of the expected low-energy positron flux and fluency (during all time in the event) are compared with IMP-7 measurement data obtained during four solar flares: October 29, November 25 and 28, and December 26, 1972 (see Table 1.9).

In the frame of thin target model and assuming that all positrons escape and propagate to the Earth, in a manner similar to protons, Hurford et al. (1973) calculated an expected positron fluency of $\approx 6 \times 10^2$ positron cm^{-2} sr^{-1} for the biggest event, October 29, 1972, which is about two orders smaller than the upper limit of 8×10^4 positron cm^{-2} sr^{-1} for this event (see Table 1.9).

The attempts to measure solar positrons was made also by Mewaldt et al. (1975), during five rich by ^{3}He events of solar cosmic rays on IMP-7, in 1973–1974 (two of them events were also detected on IMP-8 in 1974). The efficiency for identifying positrons by Caltech Electron/Isotope Spectrometer was determined by calculation and calibration: $\approx 3 \times 10^{-3}$ for IMP-7 and $\approx 9 \times 10^{-3}$ for IMP-8. The results of these observations are summarised in Table 1.10.

Comparison of positron and ^{3}He fluxes expected according to Ramaty and Lingenfelter (1973a, b), taking into account the particularities of propagation of ^{3}He and positrons, leads Mewaldt et al.(1975) to the conclusion that we can expect,

Table 1.9 Measured fluency of protons and electrons and upper limits for positrons and for ratio $e^+/(e^+ + e^-)$ in events October–December of 1972 (According to Hurford et al. 1973)

Events	Duration of measurements (h)	Fluency (particle cm^{-2} sr^{-1})			Ratio
		Protons 1.2–27.5 MeV	Electrons 0.2–2.0 MeV	Positrons 0.2–2.0 MeV	$e^+/(e^+ + e^-)$ 0.2–2.0 MeV
October 29, 1972	48	$\approx 3 \times 10^8$	$\approx 8 \times 10^6$	$<8 \times 10^4$	$<10^{-2}$
November 25, 1972	36	$\approx 2 \times 10^5$	$\approx 2 \times 10^6$	$<2 \times 10^3$	$<10^{-3}$
November 28, 1972	48	$\approx 10^6$	$\approx 4 \times 10^5$	$<5 \times 10^3$	$<1.3 \times 10^{-2}$
December 16, 1972	48	$\approx 10^5$	$\approx 9 \times 10^5$	$<7 \times 10^4$	$<2 \times 10^{-3}$

Table 1.10 Upper limits for fluency of positrons (0.16–1.6 MeV), electrons (0.16–16 MeV), and ^{3}He (2.9–15 MeV/nucleon) in events February 1973–May 1974 from measurements on IMP-7 and IMP-8 by the Caltech Electron/Isotope Spectrometer (From Mewaldt et al. 1975)

Events	Satellite	Fluency (particle cm^{-2} sr^{-1})		
		Positrons	Electrons	^{3}He
20–24 h, February 14, 1973	IMP-7	$<2 \times 10^3$	$\approx 10^5$	$\approx 3 \times 10^2$
0–15 h, June 29, 1973	IMP 7	$<10^3$	$\approx 10^5$	$\approx 3 \times 10^2$
3–24 h, September 5, 1973	IMP-7	$<2 \times 10^3$	$\leq 10^4$	$\approx 3 \times 10^2$
12 h, February 20–24 h, February 21, 1974	IMP-7, IMP-8	$<1.2 \times 10^3$	$\approx 3 \times 10^5$	$\approx 6 \times 10^2$
8–24 h, May 9, 1974	IMP-7, IMP-8	$<4 \times 10^2$	$\leq 2 \times 10^4$	$\approx 6 \times 10^2$
Five event sum	IMP-7	$<6.6 \times 10^3$	$\approx 5 \times 10^5$	$\approx 2 \times 10^3$
Two event sum	IMP-7, IMP-8	$<1.6 \times 10^3$	$\approx 3 \times 10^5$	$\approx 1.2 \times 10^3$

near the Earth's orbit, comparable fluxes for positrons and ^3He. Table 1.10 (see columns for positrons and ^3He) shows that this suggestion is not in contradiction with experimental data of five events listed in this table. Mewaldt et al. (1975) note that for expected positron flux and ratio $e^+/^3He$ on the Earth's orbit can be important also possible preferential trapping by the solar magnetic fields.

References for Preface and Chapter 1

Alfvén H (1954) On the origin of cosmic radiation. Tellus 6(3):232–253

Alsmiller RG, Jr, Boughner RT (1968) Solar Neutron Transport in the Earth's Atmosphere. J Geophys Res 73(15):4935–4942

Appa Rao MVK, Daniel RR, Vijayalakshmi B, Bhatt VL (1966) Evidence for the possible emission of high-energy neutrons from the Sun. J Geophys Res 71(7):1781–1785

Aushev VM, Dorman LI, Kryakunova ON, Shvartsman YaE, Zusmanovich AG (1993) Possible registration of solar neutrons by high-mountain neutron monitor. In: Proceedings of the 23rd international cosmic ray conference, vol 3, Calgary, pp 155–158

Baisakalova AB, Kolomeets EV (1977) The gamma ray line and continuum emission from the Sun during the flares and the quiet periods. In: Proceedings of the 15th international cosmic ray conference, vol 5, Plovdiv, pp 76–79

Bame SJ, Asbridge JR (1966) A search for solar neutrons near solar minimum. J Geophys Res 71(19):4605–4616

Belov AV, Dorman LI, Zusmanovich AG, Ishkov VN, Saidaliev RA (1989) Small increases of cosmic ray intensity according to neutron monitor data. In: Somov BV (ed), Investigations of the Solar Plasma, Ashkhabad, Russion, pp 188–199

Biermann L, Haxel O, Schlüter A (1951) Neutral cosmic rays produced in the Sun. Z Naturforsch A 6(1):47–48

Cheng CC (1972) Theoretical studies of the flux and energy spectrum of gamma radiation from the Sun. Space Sci Rev 13(1):3–123

Chupp EL (1964) Production of gamma rays and neutrons in solar flare regions. In: Wilmot N Hess (ed) The physics of solar flares. Proceedings of the AAS-NASA symposium, 1963. Washington, DC, pp 445–450

Chupp EL (1971) Gamma ray and neutron emissions from the Sun. Space Sci Rev 12(4):486–525

Chupp EL, Razdan H (1963) Detection of solar surface nuclear reactions during active periods. Trans Am Geophys Union 44, No. 1, P27, AGU, Washington, D.C., 75–75

Cortellessa P, Di Benedetto P, Paizis C (1970) A search for solar neutrons near solar maximum. Sol Phys (Netherlands) 14(2):427–439

Cortellessa P, Di Benedetto P, Paizis C (1971) A search for solar neutrons near solar maximum, II. Sol Phys (Netherlands) 20(2):474–490

Croom DL (1971) Solar microwave bursts as indicators of the occurrence of solar proton emission. Sol Phys (Netherlands) 19(1):152–170

Daniel RR, Gokhale GS, Joseph G, Lavakare PJ, Sekhon BS (1969) A search for energetic neutrons emitted during solar flares. Sol Phys (Netherlands) 10(2):465–471

Daniel RR, Gokhale GS, Joseph G, Lavakare PJ (1971) Improved estimate of solar neutron flux measured in an earlier experiment. J Geophys Res 76(13):3152–3153

De Feiter LD (1971) On the flux of gamma rays from solar flares. Sol Phys (Netherlands) 19(1):207–208

Dolan JE, Fazio GG (1965) Gamma-ray spectrum of Sun. Rev Geophys 3(2):319–343

Dorman LI, Venkatesan D (1993) Solar cosmic rays. Space Sci Rev 64(3–4):183–362

Dubov EE (1963a) On the possible mechanism of chromospheric flare emission. Izvestia Krym Astrofyz Observ 29:86–96 (in Russian)

Dubov EE (1963b) On a possible mechanism of chromospheric flare luminescence. Dokl Akad Nauk SSSR, Moscow 150(6):1246–1248

Eyles CJ, Linney AD, Rochester GK (1971) A search for neutrons of solar origin using balloon borne detectors 1967–70. In: Proceedings of the 12th international conference on cosmic rays, vol 2, Hobart, p 462–469

Eyles CJ, Linney AD, Rochester GK (1972) A search for neutrons of solar origin using balloon borne detectors 1967–69. Sol Phys (Netherlands) 24(2):483–497

Forrest DJ, Chupp EL (1969) Upper limit for the solar neutron flux in the energy interval 20-120 MeV. Sol Phys (Netherlands) 6(3):339–350

Freier PS, Webber WR (1963) Exponential rigidity spectrums for solar-flare cosmic rays. J Geophys Res 68(6):1605–1629

Goldberg L, Muller EA, Aller LH (1960) The Abundances of the Elements in the Solar Atmosphere. Astrophys J Suppl 5:1–137

Gordon IM (1954) To the problem on the physical nature of chromospheric flares. Dokl Acad Nauk SSSR 94(5):813–816

Hatton CJ, Carmichael H (1964) Experimental investigation of the NM-64 neutron monitor. Canad J Phys 42(12): 2443–2472

Haymes RC (1959a) High altitude neutron intensity diurnal variation. Phys Rev 116(5): 1231–1237

Haymes RC (1959b) High altitude neutron intensity diurnal variation. Ph.D. thesis, New York University

Haymes RC (1964a) Fast neutrons in the Earth's atmosphere 1. Variation with depth. J Geophys Res 69(5):841–852

Haymes RC (1964b) Fast neutrons in the Earth's atmosphere 2. Time variations at high altitudes. J Geophys Res 69(5):853–859

Haymes RC (1965) Terrestrial and solar neutrons. Rev Geophys Space Phys 3(3):345–364

Heidbreder E, Pinkau K, Schonfelder V (1970) Measurement of the distribution in energy and angle of high-energy albedo neutrons and determination of an upper limit for the solar neutron flux. J Geophys Res 75(31):6347–6349

Hess WN (1963a) Neutrons in space. In: Proceedings of the 5th inter-American seminar on cosmic rays, La Paz, Bolivia, p 17–24

Hess WN (1963b) Neutrons in Space, Technical Report NASA-TN-D-1696, NASA, Washington DC, pp 1–8

Hess WN, Kaifer RC (1967) The OSO-1 solar neutron experiment. Solar Phys 2(2):202–210

Hurford GJ, Mewaldt RA, Stone EC, Vogt RE (1973) Measurements of the flux of low-energy solar-flare positrons. In: Proceedings of the 13th international cosmic rays conference, vol 2, Denver, CO, pp 1613–1617

Ifedili SO (1974) A search for solar neutrons during solar flares. Sol Phys (Netherlands) 39(1):233–241

Ito K, Okazoe H, Yoshimori M (1968) Production of neutrons and gamma rays by nuclear interactions in solar flares. Canad J Phys 46(10, Part 4):S780–S783

Kane SR (1973) Acceleration of electrons during the flash phase of solar flares. In: Proceedings of the 13th international cosmic rays conference, vol 2, Denver, pp 1607–1612

Kim CY (1968) A search for high-energy solar neutrons. Canad J Phys 46(10, Part 4):S753–S756

Kim CY (1975) A search for high energy gamma rays from a quiet Sun. In: Proceedings of the 14th international cosmic rays conference, vol 5, Munich, pp 1650–1654

Kirsch E (1973) Estimation of an upper limit for the solar neutron emission during large flares. Sol Phys (Netherlands) 28(1):233–246

Kondo I, Nagase F, Yasue H (1970) Upper limit of solar neutron flux associated with solar flare on September 28, 1968, Report of ionosphere and space research in Japan, Science Council Jap 24(2):147–161

Kozlovsky B (1977) Narrow lines from $\alpha\alpha$ reactions. In: Proceedings of the 15th international cosmic ray conference, vol 5, Plovdiv, pp 88–93

Kuzhevskii BM (1968) Nuclear reactions and the line spectrum of solar-flare gamma rays. Astronom Z 45(4):747–751 (in Russian)

Kuzhevskii BM (1969) Nuclear reactions and the line spectrum of solar-flare gamma rays. Sov Astron 12(4):595–598

Kuzhevskii BM (1977a) Radioactive nuclei and the gamma-ray background of the quiet sun. Pis'ma v Astronom Z 3(1):26–27 (in Russian)

Kuzhevskii BM (1977b) Radioactive nuclei and the gamma-ray background of the quiet sun. Soviet Astron Lett 3(1):13–14

Lin RP, Anderson KA, Cline TL (1972) Detection of interplanetary electrons from 18 keV to 1.8 MeV during solar quiet times. Phys Rev Lett 29(15):1035–1038

Lingenfelter RE (1969) Solar flare optical, neutron and gamma-ray emission. Sol Phys (Netherlands) 8(2):341–347

Lingenfelter RE, Flamm EJ (1964) Solar neutrons and the Earth's radiation belts. Science 144 (3616):292–294

Lingenfelter RE, Ramaty R (1967) High energy nuclear reactions in solar flares. In: Shen BSP (ed) High energy nuclear reactions in astrophysics. W.A. Benjamin, New York/Amsterdam, pp 99–158

Lingenfelter RE, Flamm EJ, Canfield EH, Kellman S (1965a) High-energy solar neutrons 1. Production in flares. J Geophys Res 70(17):4077–4086

Lingenfelter RE, Flamm EJ, Canfield EH, Kellman S (1965b) High-energy solar neutrons 2. Flux at the Earth. J Geophys Res 70(17):4087–4095

Lockwood JA, Webber WR, Hsieh L (1974) Solar flare proton rigidity spectra deduced from cosmic ray neutron monitor observations. J Geophys Res 79(28):4149–4155

Mewaldt RA, Stone EC, Vogt RE (1975) A search for solar flare positrons. In: Proceedings of the 14th international cosmic ray conference, vol 5, Munich, pp 1668–1673

Moon S, Simnett GM, White RS (1973) A search for solar neutrons from 10-100 MeV. In: Proceedings of the 13th international cosmic ray conference, vol 2, Denver, CO, pp 1583–1588

Moon S, Simnett GM, White RS (1976) Upper limits to the quiet-time solar neutron flux from 10 to 100 MeV. Astrophys J 207(2, Part 1):630–638

Parker EN (1963) The solar-flare phenomenon and the theory of reconnection and annihilation of magnetic fields. Astrophys J Suppl Ser 8(77):177–211

Ramaty R (1986) Nuclear processes in solar flares. In: Sturrock PA, Holzer TE, Mihalas D, Ulrich RK (eds) The physics of the Sun, vol 2. D. Reidel, Dordrecht, pp 291–323

Ramaty R, Lingenfelter RE (1973a) Nuclear gamma rays from solar flares. In: Proceedings of the 13th international cosmic ray conference, vol 2, Denver, CO, pp 1590–1594

Ramaty R, Lingenfelter RE (1973b) Nuclear gamma rays from solar flares. In: Ramaty R, Stone RG (eds) High energy phenomena on the Sun, Proceedings of a Symposium, Greenbelt, 1972, NASA SP-342. Washington, DC, pp 301–314

Reppin C, Chupp EL, Forrest DJ, Suri AN (1973) Solar neutron production during the events on 04 and 07 August 1972. In: Proceedings of the 13th international cosmic ray conference, vol 2, Denver, CO, pp 1577–1582

Shklovsky IS (1955) To the problem on contents in the solar atmosphere lithium, beryllium, and deuterium. DAN SSSR, 105(5): 931–934

Svestka Z (1971) On the flux of neutrons from flares. Sol Phys (Netherlands) 19(1):202–206

Swetnick MJ, Neuburg HAC, Korff SA (1952) Diurnal effect on cosmic-ray neutrons at high altitudes. Phys Rev 86(4, minor contributions):589

Trombka JI, Metzger AE, Arnold JR, Matteson JL, Reedy RC, Peterson LE (1973) The cosmic γ-ray spectrum between 0.3 and 27 MeV measured on Apollo 15. Astrophys J 181(3, Part 1):737–746

Webber WR (1962) Some features of the response of neutron monitors to low-energy particles incident on the top of the atmosphere. Canad J Phys 40(7):906–923

Webber WR, Ormes JE (1967) An upper limit on the quiet time solar flux at energies >60 Mev. J Geophys Res 72(13):3387–3394

White RS, Moon S, Preszler AM, Simnett GM (1973) Earth albedo and solar neutrons. Space Res 13, Berlin 2:683–687

Zych AD, Frye GM Jr (1969) Upper limits for high-energy neutrons from two importance 1 solar flares. J Geophys Res 74(14):3726–3729

Chapter 2
The Events of August 1972 and the Discovery of Solar Gamma-Radiation

As mentioned in Section 1.2, before the events of August 1972, all attempts to search for solar gamma-rays during quiet periods and during chromospheric flare events gave only upper limits for the fluxes from the Sun.

2.1 Main Peculiarities of the August 1972 Events

The data observations from high energy particles, radio, X-ray, gamma-ray emissions and interplanetary shock waves generated by the flares of August 4 and 7, 1972 are used by Hudson et al. (1975) to study the non-thermal processes in great solar flares. The data observed from the X-rays of the August 4 flare have shown that the $\geq$25 keV electrons with a total energy of $\geq 2 \times 10^{32}$ erg were generated during the explosive phase. The transfer of the $\approx$500 Gs magnetic field energy in a $\approx 10^{29}$ cm^3 volume to the fast electron energy is realised in the flare with high efficiency. The transferred energy is sufficient for the lower corona and the upper chromosphere region to be suddenly heated, for plasma to be ejected and for powerful interplanetary shock waves to be generated (it is also possible that in such a way the fast electrons are also responsible for Hα emissions and for chromospheric emissions of white light).

The results of observations of the relativistic electrons with an energy $\geq$20 MeV generated in the class 3B chromospheric flare on August 7, 1972 have been analyzed by Maccagni et al. (1973). These electrons were detected at a distance of $\approx 33\, r_E$ from the Earth. The initial electrons were detected at 15:35–15:37 UT on the 7th of August, within 5–10 min after the arrival of $\leq 4 \times 10^9$ eV protons. The maximum bursts of the 9.4, 19.0 and 37.0 MHz radio emissions from the Sun were observed at 15:14 UT, the X-ray maximum at 15:13 UT and the Hα intensity maximum was observed as late as 15:26 UT. The acceleration of relativistic particles (electrons and protons) lasted for 8–10 min up to 15:15 UT. The analysis of the radio emissions and the X-ray bursts have shown that the electrons were accelerated in a pulsed mode, that the duration of each pulse was less than a minute

L. Dorman, *Solar Neutrons and Related Phenomena*,
Astrophysics and Space Science Library 365, DOI 10.1007/978-90-481-3737-4_2,

and that four pulses were produced within 10 min. The delay of the electron arrival relative to the moment of arrival of the relativistic protons indicates that the electrons resided in the solar corona during a period of 5–10 min after acceleration.

2.2 The Discovery of Solar Gamma-Radiation

The 0.3–10 MeV gamma-ray detector on OSO-7 was used by Chupp et al. (1974a, b, 1974) to look for gamma-quanta in the periods of intense X-ray emission of ≥ 4 erg cm^{-2} s^{-1} (class $\geq$ M4) in the 1–8 Å band during the first 2 weeks of August 1972. Excluding two short periods of several minutes each, only the upper limits of the 0.5, 2.2, 4.4 and 6.1 MeV gamma-quantum fluxes (mainly $\leq 5 \times 10^{-3}$ photon cm s^{-1} for all gamma-ray lines) were obtained during the measurement period (see Table 2.1).

Figure 2.1 shows the gamma-ray spectrum for 6:23–6:32 UT of August 4, 1972. From Table 2.1 and Fig. 2.1 it is possible to see that at 6:24–6:33 UT on August 4, 1972 the reliable fluxes in the lines 0.5, 2.2, 4.4 and 6.1 MeV were in the first time detected:

$$I_\gamma(0.5\ \text{MeV}) = (7.0 \pm 1.5) \times 10^{-2}\ \text{photons} \times \text{cm}^{-2} \times \text{s}^{-1} \quad (2.1)$$

$$I_\gamma(2.2\ \text{MeV}) = (2.2 \pm 0.2) \times 10^{-1}\ \text{photons} \times \text{cm}^{-2} \times \text{s}^{-1} \quad (2.2)$$

$$I_\gamma(4.4\ \text{MeV}) = (3 \pm 1) \times 10^{-2}\ \text{photons} \times \text{cm}^{-2} \times \text{s}^{-1} \quad (2.3)$$

$$I_\gamma(6.1\ \text{MeV}) = (3 \pm 1) \times 10^{-2}\ \text{photons} \times \text{cm}^{-2} \times \text{s}^{-1} \quad (2.4)$$

The corresponding values at 15:38–15:47 UT on August 7, were 1.5–4 times smaller: $(3.7 \pm 0.9) \times 10^{-2}$, $(4.8 \pm 1.0) \times 10^{-2}$, 2×10^{-2}, and 2×10^{-2} photon cm^{-2} s^{-1} in 0.5, 2.2, 4.4 and 6.1 MeV lines. The two periods (06:24–06:33 UT on August 4 and 15:38–15:47 UT on August 7) correspond to the development maximum of class 3B optical flares and to the $>5 \times 10^{-1}$ erg cm^{-2} s^{-1} X-ray flux in the 1–8 Å band (class >X5). During these periods the 60–120 keV X-ray flux also reached its maximum ($\approx$600 photon cm^{-2} s^{-1}).

According to Hudson et al. (1975), the data on gamma-ray emission in the August 1972 events, have shown that the fast protons and heavier nuclei appeared within several minutes after the electrons. Some of the 5% of the generated electrons and $\geq$99% of the fast protons and nuclei moved in the direction from the photosphere where they were ejected into the interplanetary space.

Numerous measurements of solar gamma-rays were carried out during the powerful chromospheric flares in August 1972. For example, the gamma-ray bursts associated with the flares of August 4 and 7, 1972, were detected on Prognoz-2 (Vedren et al. 1975). The gamma-ray bursts associated with the flares of

Table 2.1 Gamma-ray line emissions from McMath region 11976 during August 1972 (According to Chupp et al. 1974a, b)

Events				Flux (photon $cm^{-2}\ s^{-1}$)			
Optical class	Date Aug. 1972	Observation intervals (UT)	$I_{max}(c/s)$60–120 keV	0.5 MeV	2.2 MeV	4.4 MeV	6.1 MeV
1 B	2	18:06–18:30	81 at	$<1.4 \times 10^{-2}$	$<1.3 \times 10^{-2}$	$<7.4 \times 10^{-3}$	$<8.2 \times 10^{-3}$
		18:40–18:55	18:40 UT	$<1.8 \times 10^{-2}$	$<1.7 \times 10^{-2}$	$<1.5 \times 10^{-2}$	$<1.1 \times 10^{-2}$
2 B	2	19:38–20:30	310 at	$<1.1 \times 10^{-2}$	$<1.0 \times 10^{-2}$	$<8.3 \times 10^{-3}$	$<6.3 \times 10^{-3}$
		21:34–22:02	21:44 UT	$<1.8 \times 10^{-2}$	$<1.8 \times 10^{-2}$	$<1.0 \times 10^{-2}$	$<8.4 \times 10^{-3}$
3 B	4	05:38–06:21	7,520 at	$<1.1 \times 10^{-2}$	$<1.0 \times 10^{-2}$	$<7.7 \times 10^{-3}$	$<6.2 \times 10^{-3}$
		06:24–06:33	06:26 UT	$(7 \pm 1.5) \times 10^{-2}$	$(2.2 \pm 0.2) \times 10^{-1}$	$(3 \pm 1) \times 10^{-2}$	$(3 \pm 1) \times 10^{-2}$
1 B	7	03:55–04:07		$<1.8 \times 10^{-2}$	$<1.6 \times 10^{-2}$	$<9 \times 10^{-3}$	$<1.1 \times 10^{-2}$
3 B	7	14:03–14:58	4,855 at	$<9.7 \times 10^{-3}$	$<8.9 \times 10^{-3}$	$<7.6 \times 10^{-3}$	$<6.1 \times 10^{-3}$
		15:38–15:47	15:39 UT	$(3.7 \pm 0.9) \times 10^{-2}$	$(4.8 \pm 1) \times 10^{-2}$	$<2 \times 10^{-2}$	$<2 \times 10^{-2}$

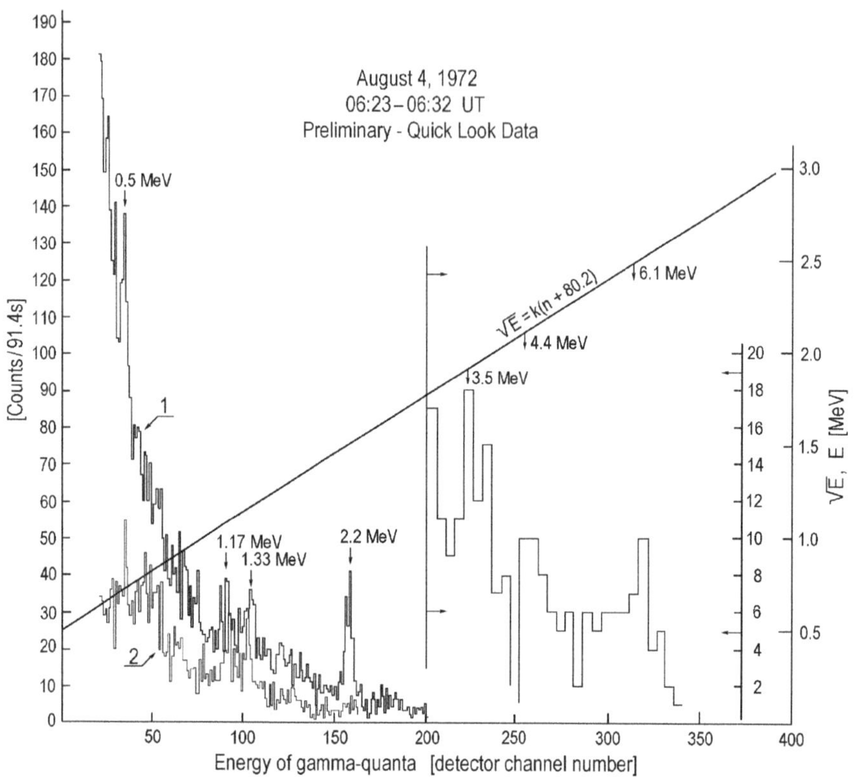

Fig. 2.1 The event of August 4, 1972 at 06:23–06:32 UT. The solar (*curve 1*) and background (*curve 2*) gamma-ray spectra during the rise of the 3B flare (According to Chupp et al. 1974a, b)

August 4 and 7, 1972, have been studied in detail in Albern et al. (1974, 1975). The study of the time dependence and spectral characteristics of the data obtained with the observations of the optical and radio emissions have revealed the important characteristic phases of flare development. The gamma-ray increases have been found to be closely associated with the appearance of high energy protons in the solar corona. A possible connection between shock waves to the mechanism of high energy particle acceleration has been noted.

References for Chapter 2

Albern F, Vedren Zh, F. Kambu F, Kudriavtsev MI, Likin OB, Melioranskii AS, Nazarova NI, Pankov VM, Pisarenko NF, Savenko IA (1974) Flare observed on August 7, 1972. Kosmicheskie Issledovaniia 12(6):930–935 (in Russian)

Albern F, Vedren Zh, Kambu F, Kudriavtsev MI, Likin OB, Melioranskii AS, Nazarova NI, Pankov VM, Pisarenko NF, Savenko IA (1975) Flare observed on August 7, 1972. Cosmic Res 12(6):842–846

Chupp EL, Forrest DJ, Higbie PR, Suri AN, Tsai C, Dunphy PP (1973a) Solar gamma ray lines observed during the solar activity of August 2 to August 11, 1972. Nature 241(5388):333–335

Chupp EL, Forrest DJ, Suri AN, Reppin C (1973b) Gamma ray emissions from McMath region 11976 during August 1972. In: Proceedings of the 13th international cosmic rays conference, vol 2, Denver, CO, pp 1595–1601

Chupp EL, Forrest DJ, Suri AN (1974) Gamma ray observations during the August 1972 solar activity. In: Correlated interplanetary and magnetospheric observations. Proceedings of the seventh ESLAB symposium, Saulgau, 1973. D. Reidel, Dordrecht, pp 519–531

Hudson HS, Jones TW, Lin RP (1975) Nonthermal processes in large solar flares. In: Kane SR (ed) Solar gamma-, X-, and EUV radiation. Proceedings of the symposium, Buenos Aires, 1974. D. Reidel, Dordrecht, pp 425–426

Maccagni D, Perotti F, Villa G, Cherki G, Raviart A, Treguer L (1973) Acceleration and injection of highly relativistic electrons during the August 7, 1972, solar flare. Nature 246(5431):300–301

Vedren Z, Likin OB, Melioranskii AS, Pisarenko NF, Savenko IA, Talon R, Shamolin VM (1975) Study of gamma-radiation splashes, observed on August 4 and 7, 1972 from Prognoz-2 satellite in time of flashes. Izv Akad Nauk SSSR, Ser. Fiz. 39(2):272–280 (in Russian)

Chapter 3
The Events of June 1980 and June 1982, and the Discovery of Solar Neutrons

As shown in Section 1.2, before the well-known events of June 7 and 21, 1980 and June 3, 1982, the numerous attempts to search for solar neutrons during quiet periods and chromospheric flare events yielded only upper limits for solar neutron fluxes.

3.1 Main Peculiarities of the June 1980 and June 1982 Solar Events

3.1.1 Energetic Particle Observations on Helios-1

Important investigations of the differential energetic spectra of protons for solar gamma-ray/neutron events of June 7 and 21, 1980 and June 3, 1982 were made by McDonald et al. (1985) on the basis of Helios-1 observation data.

The event of June 7, 1980 (solar co-ordinates of chromospheric flare N12, W74; index 1B) was observed when Helios-1 was at a distance of 0.37 AU from the Sun. The difference in helio-longitudes was 14° and the maximum pulse phase was at 3:12 UT. The differential energetic spectrum for protons in the interval 3–30 MeV obtained in the form $\propto E^{-\gamma}$, where $\gamma = 2.3$; for electrons in the interval 1–2 MeV was found $\gamma = 0.7$. For the same event the ratios H/He ≈ 4, ^{3}He/^{4}He ≤ 0.02.

The event of June 21, 1980 (chromospheric flare N19, W88; 1B) was observed by Helios-1 when it was at a distance of 0.54 AU from the Sun and the difference in helio-longitudes was 33°. The maximum pulse phase was at 1:18 UT. The spectrum of protons in the interval 8–200 MeV is characterized by $\gamma = 2.6$ and for electron in the interval 3–6 MeV by $\gamma = 0.25$; the ratios Fe/O $= 0.9 \pm 0.2$, H/He ≈ 29 and ^{3}He/^{4}He $= 0.030 \pm 0.013$.

The event of June 3, 1982 (chromospheric flare S09, E72; 2B) was observed by Helios-1 at 0.57 AU from the Sun, the difference in the helio-longitudes was only 3° and the maximum of pulse phase was at 11:43 UT. The differential energy spectrum of protons in the interval 3–200 MeV is characterized by $\gamma = 1.2$ and for electrons in

L. Dorman, *Solar Neutrons and Related Phenomena*,
Astrophysics and Space Science Library 365, DOI 10.1007/978-90-481-3737-4_3,

the interval 3–6 MeV by $\gamma = 1.0$; the ratios Fe/O = 2.5 ± 0.5, H/He ≈ 132, $^3He/^4He = 0.020 \pm 0.014$.

All the above mentioned events are characterised by the following peculiarities: (1) great electron content, (2) relative hard energy spectra (small γ), (3) several hours before each event a small pre-increase of solar energetic particles was observed.

3.1.2 Peculiarities of Particle Injection in the Three Events of June 7, 1980

Neustock et al. (1985) investigated the characteristics of energetic particle injection in acceleration processes for three solar flare events at 1:17 UT, 3:12 UT and 7:25 UT on June 7, 1980 on the basis of observations of electrons with energy ≥0.3 MeV, protons and nucleons with energy ≥1.3 MeV/nucleon (it was used the equipment from Kile University on board Helios-1). During this time Helios-1 was at a distance of 0.37 AU from the Sun, inside a 'hole' in the solar wind (with very low density and very regular magnetic field), near the region directly connected by interplanetary magnetic field force lines with the active solar region No. 16886; where mentioned above three chromospheric flares occurred.

Because the magnetic field was very regular, it became possible to re-calculate the data to the Sun, by taking into account the transport time of particles in dependence with the energy and velocity. The obtained results can be formulated by the following:

1. With an error of ≤1 min, the start of the injection of electrons with energy ≈ 0.5 MeV in all three events were simultaneous with hard X-ray bursts.
2. The duration of electron injections was 15–20 min, which is much longer than the hard X-ray bursts.
3. There are no significant correlation between solar electron flux in the interplanetary space and emissivity of X-rays in the band 1–8 Å and gamma rays with an energy ≥0.3 MeV.
4. Protons and nucleons were detected with sufficient lag time caused by big lag time of ejection into interplanetary space. Table 3.1 shows the comparison of time ejection into interplanetary space electrons in the energy interval 0.3–4.0 MeV and protons in energy interval 4–37 MeV as well as spectral index (in the energy spectrum $\propto E^{-\gamma}$).
5. Observed in the event at 3:12 UT only one electron ejection and three consequent injections of protons and nucleons (with more and more softer spectrum, see Table 3.1).
6. The particles with higher energy/nucleon ejected from the Sun later, what means that the particles are accelerated longer than particles with smaller energy/nucleon.
7. It is no enough accuracy of experimental data and in recalculations from Helios-1 to the Sun to made definite choose between two models discussed in Neustock et al. (1985): (a) simultaneous acceleration of electrons and protons with consequent lagging in the flare magnetic trap, and (b) the acceleration of protons and

Table 3.1 Time ejection intervals and spectral index γ for electrons and protons in the events of June 7, 1980 (According to Neustock et al. 1985)

Events	Electrons		Protons	
	Time ejection (UT)	Spectral index	Time ejection (UT)	Spectral index
1:17 UT	1:13–1:54	3.50 ± 0.11	1:38–3:17	2.29 ± 0.25
3:12 UT	3:07–3:30	3.48 ± 0.07	4:16–4:49	2.67 ± 0.13
			5:13–6:46	3.04 ± 0.15
			7:12–7:45	3.66 ± 0.24
7:25 UT	7:20–8:04	3.93 ± 0.05	7:45–8:58	3.18 ± 0.13

nucleons in much higher layers of the solar atmosphere than electrons (may be by coronal shock waves).

3.1.3 *Isotopic and Chemical Composition of Energetic Particles in the Events on June 21, 1980 and June 3, 1982*

Van Hollebeke et al. (1985) investigated the isotopic and chemical composition of energetic particles in the gamma ray/neutron events of June 21, 1980 and June 3, 1982 on the basis of observations on Helios-1. During the first event Helios-1 was at a distance of 0.54 AU from the Sun and $\approx 33°$ from the magnetic force line connected to the flare, and during the second event these values were 0.57 AU and $\approx 3°$. The ratios of nucleus fluxes in the energy interval 10–47 MeV/nucleon for June 21, 1980 and June 3, 1982 events are shown in Table 3.2 (for comparison purposes also the average ratios of many other events are given).

Table 3.2 shows that the considered events characterized by enhancement of Fe compare to O (Fe/O about 14 times larger for June 21, 1980 and about 40 times for June 3, 1982 events than the average ratio for many events). Contents of isotopes for these two events for the energy interval 3.3–39 MeV/nucleon were obtained: ^{2}H/H $< 5.10^{-4}$ and $<4.10^{-4}$; ^{3}H/H$<4.5.10^{-4}$ and $<4.10^{-4}$. For the energy range 30–50 MeV/nucleon it was obtained: ^{3}He/^{4}H $= 0.030 \pm 0.013$ and 0.020 ± 0.014 (measurements were made at 01:30–12:00 UT June 21, 1980 and at 11:30–18:00 UT June 3, 1982). Van Hollebeke et al. (1985) came to conclusion that for these events not only the nuclear interactions of protons but also the nuclear interactions of He and Fe must be important for the generation of neutrons.

3.1.4 *Peculiarities in the Radio Emission During the June 3, 1982 Event*

Spectral and radiometric observations in diapason 160–1,000 MHz during the June 3, 1982 event were analysed by Karlicky et al. (1986). In the radio-emission the flare started by a group of U-bursts in the 160–380 MHz, after which consequent wide stripe continuum with a sharp front. Important particularity of this event is the

Table 3.2 Ratios of nucleus fluxes for the events of June 21, 1980 and June 3, 1982 as well as an average ratios for many events (According to Van Hollebeke et al. 1985)

Ratios	June 21, 1980	June 3, 1982	Average ratios
H/He	29	132	60
He/O	79 ± 10	102 ± 14	53 ± 5
C/O	0.53 ± 1.0	0.38 ± 1.0	0.45 ± 0.02
Ne/O	0.46 ± 0.13	0.87 ± 0.20	0.13 ± 0.01
Mg/O	0.19 ± 0.06	0.62 ± 0.18	0.18 ± 0.01
Si/O	0.4 ± 0.1	0.20 ± 0.08	0.15 ± 0.01
Fe/O	0.91 ± 0.02	2.5 ± 0.5	0.066 ± 0.006

fast formation of II-type bursts in the 160–480 MHz (the time interval between the pulse phase and the II-type bursts was 35 s). On the initial stages of flare radio emissions in the high frequency region (640–1,000 MHz) was characterized by a positive frequency drift. After this on the background of continuum was detected as different structures: wide-stripe pulsations with time-scale of $\approx$0.3 s in 480–800 MHz, a narrow-stripe with dm-spikes in 640–720 MHz and bursts with an intermediate drift in 800–1,000 MHz. As an explanation of these results it was supposed that flare energy-release take place in the magnetic loop with average electron density $n_e \approx 4 \times 10^9\ \mathrm{cm}^{-3}$ (the top of the loop must be at a height of $\approx 9 \times 10^9$ cm, where $n_e \approx 1.1 \times 10^9\ \mathrm{cm}^{-3}$). The positive drift of radio emission on these high frequencies shows that some part of the flare energy propagates downward, to the photosphere (important for the generation of neutrons and nuclear gamma-ray lines). The data on the bursts with intermediate drifts show that the magnetic field strength, in the bottom of the loop was H $\approx$12 Gs. The shock responsible for the burst of type II, must have had an initial velocity of $\approx$1,740 km/s and characterized by magnetic Mach number $\geq$2.9. The pulsations on the final stage of the type IV burst was considered as a manifestation in radio-diapason of the processes of magnetic reconnection, which leads to restoration of the closed magnetic field configuration and to formation of the magnetic loops in the post-flare system.

3.2 Observations of Gamma Radiation During the Events of June 1980 and June 1982 as Evidence of Neutron Generation

3.2.1 The Event of June 7, 1980: Time Behaviour of Acceleration Processes

According to Forrest et al. (1981) during chromospheric flare that started at 03:12 UT on June 7, 1980 was detected by a gamma ray spectrometer on SMM, a significant flux in gamma rays were observed in the line 2.223 MeV from the reaction $^1\mathrm{H}(n,\gamma)^2\mathrm{H}$. This gamma radiation and its time development show that the high energy particles (kinetic energy $E \geq 30$ MeV) interact with the background atmosphere and generate fast neutrons during $\approx$45 s. This time interval was

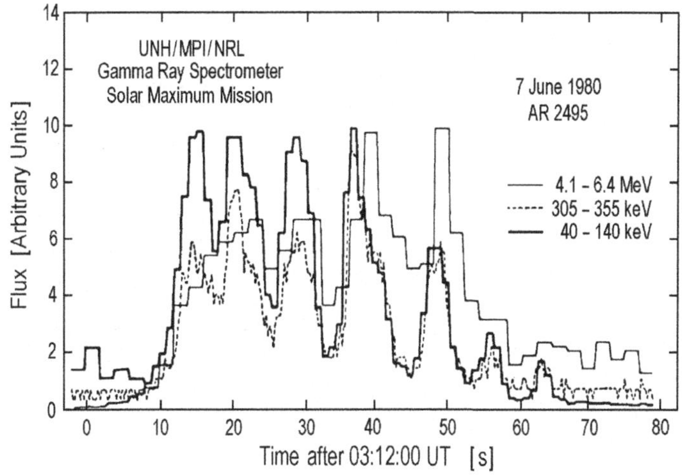

Fig. 3.1 The event of June 7, 1980. Counting rate versus time for various channels of the SMM gamma-ray spectrometer during the gamma-ray line flare at 3.12 UT on 7 June 1980 (According to Forrest et al. 1981)

characterized by impulse hard X-ray emissions. It was also observed gamma-ray lines 4.43 MeV (from the excitation of $^{12}C^*$) and 6.13 MeV (from $^{16}O^*$). The time behaviour of gamma-rays shows that during the process of proton and ion acceleration (≈45 s) there were four quasi-periodical bursts with duration of several seconds each. A comparison of time profiles of X-ray and gamma-ray emissions (Fig. 3.1) shows that electrons and ions accelerated approximately in the same short time intervals, about several seconds.

3.2.2 The Event of June 21, 1980: Gamma-Radiation and the First Satellite Observation of High Energy Solar Neutrons

According to Chupp et al. (1982), during the event of June 21, 1980 measured by instruments on SMM, data of gamma-rays and neutrons of several 100 MeV were obtained as evidence for acceleration of ions with GeV energies. Figure 3.2 shows the time history of this event in the channels of the gamma ray spectrometer on SMM for large energy loss events of 10–140 and of 25–140 MeV.

During the impulsive phase, lasting ≈1 min, it was detected high energy photons probably resulted from neutral pion decay and electron bremsstrahlung. After this, in a period of about 17 min a flux of solar neutrons was detected. It was supposed that solar neutrons were produced in a δ-function pulse at the time of the first impulse burst (in this case neutrons arrive at the Earth's orbit in accordance with their energy dependent time-of-flight, see the scale side of Fig. 3.2). Figure 3.2 shows that solar neutrons of at least 500 MeV kinetic energy were present with a spectrum extending down to ≈50 MeV.

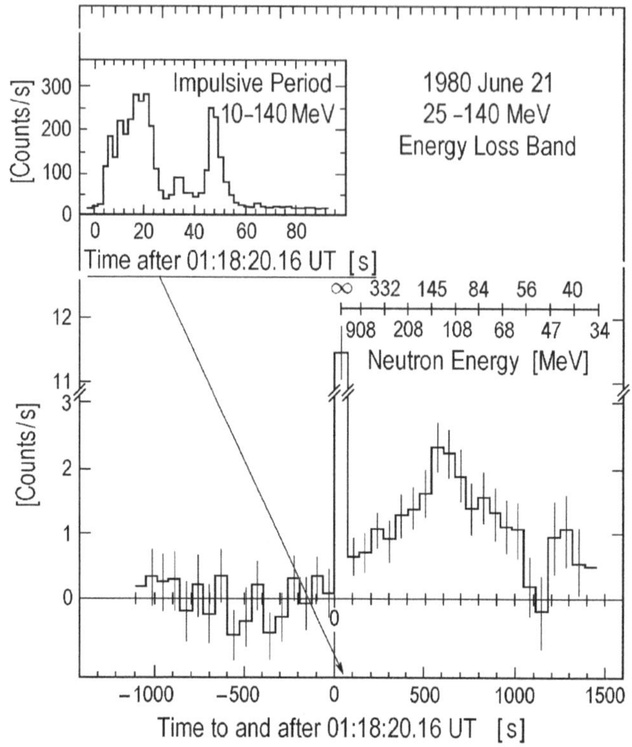

Fig. 3.2 The event of June 21, 1980. *Lower panel*: excess counting rate versus time in the SMM GRS is shown for electron equivalent energy loss events 25–140 MeV before and after start of the impulsive flare at 01:18:20 UT; the neutron energy scale (center) assumes the neutrons left the Sun before 01:18:20 UT (with taking into account the light travel time 507 s). *Inset*: the total counting rate versus time after 01:18:20 UT for electron equivalent energy loss events 10–140 MeV during impulsive phase only (According to Chupp et al. 1982)

3.2.3 The Event of June 3, 1982: Two-Pulses of High Energy Gamma-Radiation and Evidence of Solar Neutron Generation

Figure 3.3 (Chupp 1983) shows the time profile for large energy loss events (> 40 MeV) on SMM during the flare at 11:43 UT on June 3, 1982 (chromospheric flare coordinates 9°S, 72°E).

In this case the impulsive phase of gamma-rays arrived during the two major peaks in their duration $\leq$2 min. More detailed analysis of these observations shows that the first neutrons arrived during the second major peak and later, up to $\approx$16 min. This indicates that neutrons of $\approx$ 1 GeV energy were produced at the Sun in the first major impulse burst. Some of the neutrons are in the solar atmosphere in the post-impulsive phase; evidence of this according to Chupp (1983) gives the gamma-ray line spectrum near 2.223 MeV during 11:46:33–11:52:01 UT June 3,

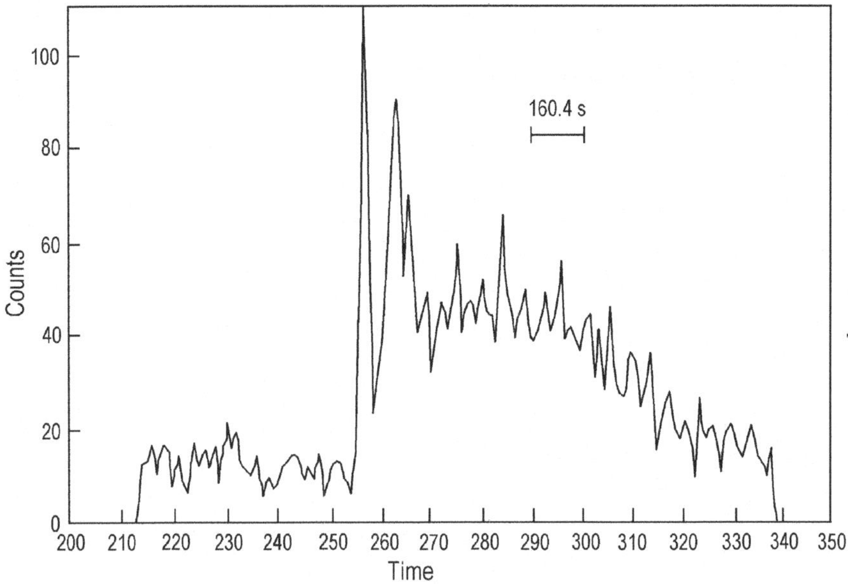

Fig. 3.3 The event of June 3, 1982. The raw count (uncorrected) for large energy loss events (> 40 MeV) is shown during the intense solar flare at 11:43 UT on June 3, 1982. A characteristic high energy neutron signal is seen after the impulsive spikes (According to Chupp 1983)

1982, several minutes after the impulsive phase (see Fig. 3.4). Chupp (1983) came to the conclusion that ions with energies ≥1 GeV were produced within a time scale of seconds, early in the impulsive phase of gamma radiation.

3.3 The Discovery of Solar Neutrons by the Gamma Ray Spectrometer on the Solar Maximum Mission

3.3.1 The 1:18 UT June 21, 1980 Event: the First Reliable Solar Neutron Observation

Chupp et al. (1982) reported the first reliable observation of solar neutrons by gamma ray spectrometer (GRS) of the solar maximum mission (SMM) near the Earth, with a greater than 13σ statistical significance. The GRS consists of seven 7.6×7.6 cm NaI (Tl) detectors, recording gamma-rays from 0.3 to 9.0 MeV. These detectors were used in conjunction with a 25 cm diameter and 7.5 cm thick CsI (Na) crystal which acted as a high energy detector of neutral events with a time resolution ≈2.05 s and protected by a 4π anticoincidence shield from charged particles (Forrest et al. 1981). The combined NaI and CsI detectors provide an effective thickness of 50 g/cm^2 or ≈0.3 interaction length for neutrons with energy ≥200

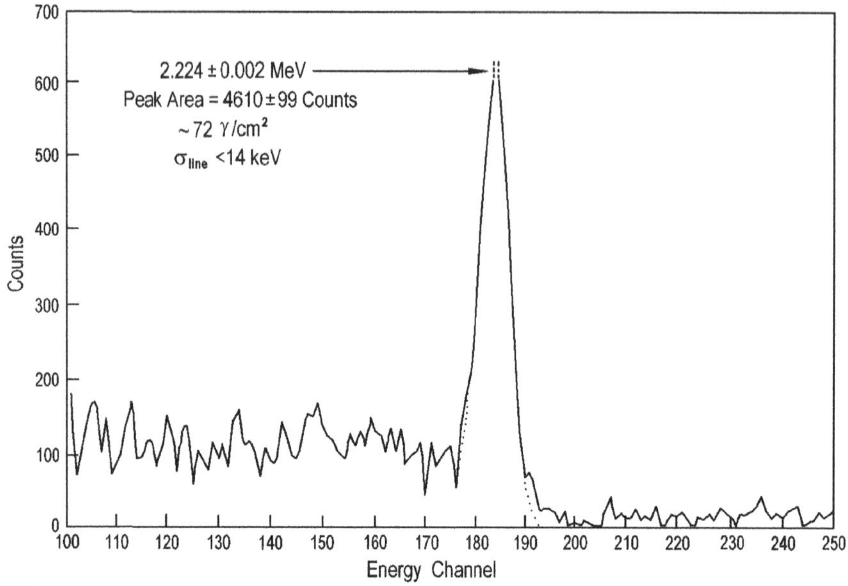

Fig. 3.4 The event of June 3, 1982. SMM gamma ray spectrometer data for the gamma-ray line spectrum near 2.223 MeV in the post impulsive phase (11:46:33–11:52:01 UT) (According to Chupp 1983)

MeV and ≈0.4 interaction length for gamma-rays with energies >10 MeV (in-flight calibration was carried out by placing independent front and back plastic shield elements in coincidence with charged particle detector).

In order to determine the time-dependent high energy background (25–140 MeV) during and after the impulsive phase of the flare, Chupp et al. (1982) studied the effect of changes in geomagnetic cut-off rigidity for the same part of the satellite orbits on days before and after the flare (see Fig. 3.5).

In Fig. 3.5 the solid curve shows the best fit exponential to the non-flare orbit count rates and was used for the flare orbit background. The two high points at ≈13.6 GV occurred during the impulsive phase of the flare, which was due to the high energy gamma-ray and X-ray emissions from the meson decay and relativistic electron bremsstrahlung.

What is important is that the anomalous extended emission observed after the impulsive phase (Fig. 3.5) was detected only in the energy loss band greater than 25 MeV and continued for ≈17 min. Chupp et al. (1982) described the evidence which leads to the conclusion that this delayed excess radiation was due to solar neutrons (see Fig. 3.6): the impulsive phase excess spectrum (curve a) and the background spectrum (curve b) are similar and appear to decrease monotonically with increasing energy (as expected for energetic gamma-rays) but, the post-impulsive phase excess spectrum (curve c) has a different shape, consistent with that expected from energetic neutrons interacting in alkali halide scintillators.

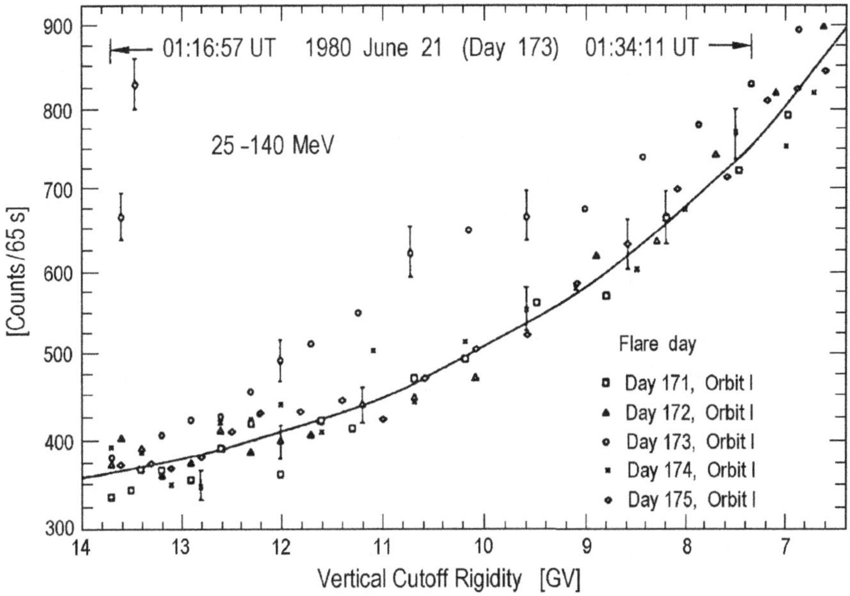

Fig. 3.5 The event of June 21, 1980. The count rate in the high-energy detector is shown for electron equivalent energy loss events between 25 and 140 MeV versus the cut-off rigidity at the satellite for the initial SMM orbit on 5 successive days. The flare occurred in orbit 1 on June 21, 1980 at 01:18:20 UT when the satellite was at a cut-off rigidity for vertical direction of 13.6 GV (an orbital period was 96 min) (According to Chupp et al. 1982)

This important circumstance supports the results in Fig. 3.6 on properties of solar gamma-radiation and solar neutrons from the flare on June 21, 1980 (see Section 3.2.2). The detection efficiency by CsI detectors were determined by cross-sections for neutrons which increase with energy from 13 $\pm$ 6 mb per atom for 14 MeV to 660 $\pm$ 150 mb per atom for 150 MeV neutrons; the total effective areas for neutron detection at 50, 100, 200, 400 and 600 MeV are 11, 40, 103, 150 and 153 cm^2, respectively. On the basis of these data Chupp et al. (1982) determined that the peak counting rate corresponds to an average flux at the Earth as being $(3.8 \pm 0.6) \times 10^{-2}$ at $\approx$130 MeV. The total number of neutrons emitted from the Sun is $\approx 3 \times 10^{28}$ neutrons/sr^1 above 50 MeV.

3.3.2 The 11:43 UT June 3, 1982 Event: Simultaneous Observations of Solar Neutrons on SMM and by a High Mountain Neutron Supermonitor

The second solar neutron event observed by GRS on SMM and by neutron supermonitor on Jungfraujoch (Chupp et al. 1983) was caused by a chromospheric flare,

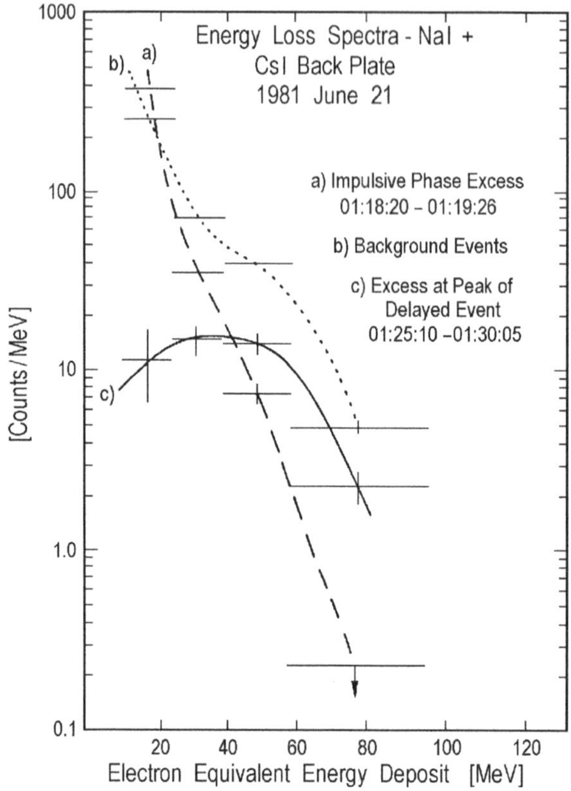

Fig. 3.6 The event of June 21, 1980. The differential number of events in the combined NaI(Tl) and CsI(Na) high-energy detector elements is shown for each electron equivalent energy loss window. Curve a – the impulsive phase excess spectrum; curve b – the background spectrum; curve c – the post-impulsive phase excess spectrum (According to Chupp et al. 1982)

started in Hα-radiation on June 3, 1982 at 11:38 UT. For comparison data on X-ray fluxes with energies of 80–140 keV and gamma-ray fluxes with energies 4.1–6.4 MeV and >25 MeV was used. The start of detection of high energy gamma-ray fluxes was at 11:43:29 UT. The first high energy solar neutrons was detected by the neutron supermonitor at 11:44 UT and the average flux of neutrons with energy from ≈0.5 GeV up to several gigaelectron volt in the period 11:44–11:47 UT was 0.08–0.10 neutrons/cm^2/s. The first neutrons on SMM with energy of 300–400 MeV were detected at 11:48 UT with the flux ≈0.2 neutrons cm^{-2} s^{-1}. The maximum of solar neutron flux on SMM was detected at 11:58 UT (the effective energy ≈75 MeV, flux ≈0.7 neutrons/cm^2/s). The energy of solar neutrons was determined by the same time-of-flight method as for the event 21 June 1980, with the suggestion that the time-dependence of neutron generation and ejection into interplanetary space, described by δ-function and the moment of generation coincide with the time of high energy gamma-ray generation (impulsive phase of gamma-radiation time-profile, see Section 3.2.3).

The recalculation to the source was made by taking into account the neutron decay in dependence of its energy. It was determined that for neutron kinetic energy $E_K = 10^3$ MeV, the differential flux from the Sun was $\approx 10^{24}$ neutrons/MeV/sr, for $E_K = 10^2$ MeV the flux was $\approx 10^{27}$ neutronsMeV/sr. The expected flux of solar neutrons at $E_K = 10$–20 MeV was found to be $\approx 3.9 \times 10^{28}$ neutrons/MeV/sr, which was determined on the basis of measurements of gamma-ray line 2.223 MeV formed by the interaction ${}^1\mathrm{H}(n,\gamma){}^2\mathrm{H}$. According to Chupp et al. (1983), the spectrum of ejected solar neutrons in the June 3, 1982 event can be approximated by the power law $\propto E^{-\gamma}$ with $\gamma \approx 5.5$ (the same γ that was obtained for the June 21, 1980 event); for the generation of measured number of neutrons with energy $E_K >$ 50 MeV is necessary to generate $\approx 4 \times 10^{32}$ protons with energies $E_K > 100$ MeV.

3.4 June 3, 1982: the First Solar Neutron Event Observed by Neutron Supermonitors

The mountain Jungrfranjoch neutron supermonitor was the first which data was used for the June 3, 1982 event investigation (Debrunner et al. 1983). Figure 3.7 shows the 1- and 5-min data of this neutron supermonitor in comparison with the neutron counting rate by gamma-ray spectrometer on SMM. In the 5-min records the two values at 11:45–11:50 UT and at 11:50–11:55 UT show an enhancement of 6.7 $\boldsymbol{\sigma}$ and 4.0 $\boldsymbol{\sigma}$, respectively, with respect to the average counting rate. The 1-min

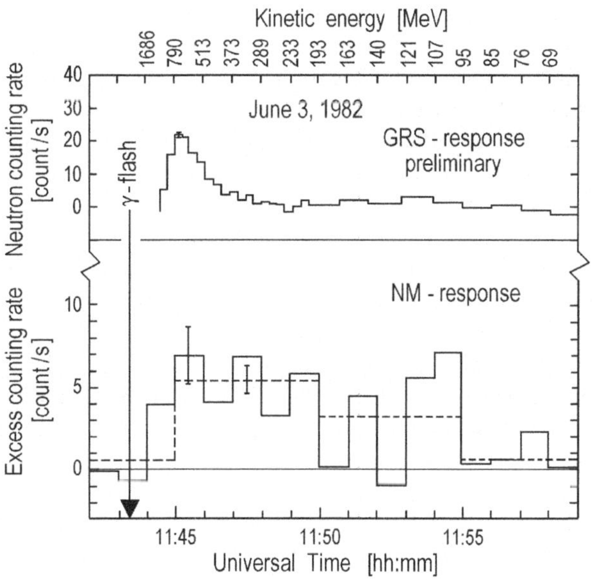

Fig. 3.7 The event of June 3, 1982. SMM gamma ray spectrometer counting rate and neutron monitor excess counting rate at Jungfraujoch (According to Debrunner et al. 1983)

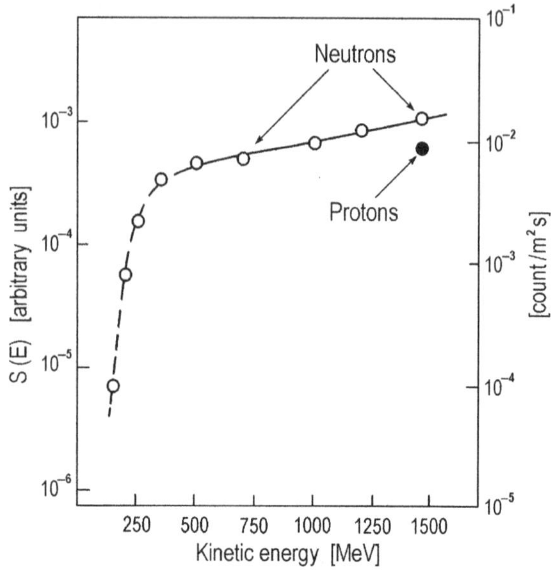

Fig. 3.8 Yield of an IGY-NM at $h = 498$ mm Hg for primary nucleons with zenith angle $\theta_s = 25°$ (According to Debrunner et al. 1983)

records locate the onset of the solar neutron event in the interval 11:44–11:45 UT (the following six records were increased by 1.8–3.8 σ). After 11:50 UT, only the counting rate between 11:53 UT and 11:55 UT differs significantly from the average (5.0 σ). The upper part of Fig. 3.7 shows the expected neutron energy with the suggestion that all neutrons are ejected from the Sun at the moment of gamma-flash (11:43:25 UT) as δ-function.

The yield function for detecting neutrons by supermonitor (Fig. 3.8) was calculated with a Monte Carlo program, simulating nucleonic cascades in the atmosphere (Debrunner and Brunberg 1968), for pressure of 498.0 mm Hg and zenith angle $\theta = 25°$, which corresponds to conditions for Jungfraujoch station on June 3, 1982.

In Fig. 3.8 the point S for protons ($E = 1454$ MeV) was used for calibration of yield function for neutrons. By these data and by the time-flight-energy conversion Debrunner et al. (1983) found the integral and differential neutron intensity at the top of the atmosphere in the time interval 11:44:22–11:50:00 UT at June 3, 1982 (see Table 3.3).

The Jungfraujoch station had the best conditions in the world, to detect solar neutrons from the event of June 3, 1982. The second station with little poorer conditions was the mountain station Lomicky Stit (Efimov et al. 1983). The 5-min data from this station shows an increase in the period 11:45–11:50 UT at June 3, 1982 on 2.9% (Fig. 3.9).

The result shown in Fig. 3.9 is in a good agreement with the measurements from the gamma-ray-spectrometer on SMM and by the neutron supermonitor at Jungfraujoch. According to Efimov et al. (1983), the detected flux of neutrons I_n is connected with the flux I_γ of 2.2 MeV gamma-quanta (generated in the Sun's atmosphere by the reaction $n + p \rightarrow d + \gamma$) in dependence of the characteristic

Table 3.3 The event June 3, 1982: solar neutron intensity at the top of the atmosphere (According to Debrunner et al. 1983)

Time interval, UT	Energy interval (MeV)	Intergral neutron Intensity [neutrons/ (m^2/s)]	Differential neutron intensity [neutrons/ $(m^2/s/MeV)$]
11:44:21.8–11:45:00	1187–790	7.2×10^2	1.8
11:45:00–11:46:00	790–513	1.1×10^3	3.9
11:46:00–11:47:00	513–373	8.1×10^2	5.8
11:47:00–11:48:00	373–289	2.1×10^3	24.9
11:48:00–11:49:00	289–233	1.6×10^3	29.5
11:49:00–11:50:00	233–193	7.3×10^3	181.5

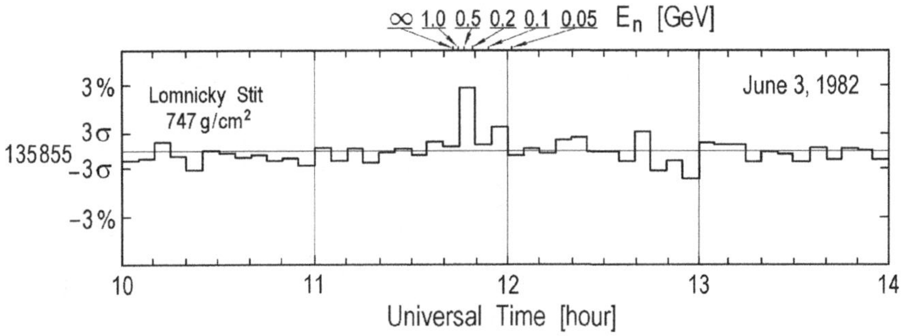

Fig. 3.9 The event of June 3, 1982. The counting rate versus time for the Lomnicky Stit neutron monitor (According to Efimov et al. 1983)

rigidity of accelerated particles $R_0 D((R) \propto \exp(-R/R_0)) : I_n/I_\gamma \approx 1$ corresponds to $R_0 = 100$ MV, $I_n/I_\gamma \approx 2$ corresponds to $R_0 = 200$ MV. It was found that for neutron supermonitor Lomnicky Stit (atmospheric pressure 747 g/cm^2) one detected solar neutron at $R_0 = 200$ MV correspond on the top of the atmosphere to the fluency 5×10^{-2} neutrons/cm^2. In the period 11:45–11:50 UT at June 3, 1982 there was detected a 4,000 impulses of solar origin (2.9% amplitude increase) which corresponds to the fluency $I_n \approx 200$ neutrons/cm^2. On SMM was detected for this event in the line 2.2 MeV, $I_\gamma = 72$ quanta/cm^2, that $I_n/I_\gamma \approx 2.7$ which corresponds to $R_0 \approx 250$ MV (or for power spectrum $D(R) \propto R^{-\gamma}$ with $\gamma \approx 2$–3).

The effect of June 3, 1982 was detected also by the neutron supermonitors at Rome and Tsumeb (see Figs. 3.10 and 3.11), according to Iucci et al. (1985) and Stoker (1987), correspondingly.

Table 3.4 is a summary of the observations of this event by neutron supermonitors.

Figure 3.12 (Iucci et al., 1985) shows the dependence of amplitude increase at 11:45–11:50 UT, from the depth of the atmosphere h_s in the direction of the Sun at 11:45 UT at June 3, 1982, normalized to one level of intensity. It was in the first time determined important parameter of solar neutron propagation in the Earth's atmosphere: effective attenuation length $\Lambda \approx 110$ g/cm^2. This value of Λ is in good

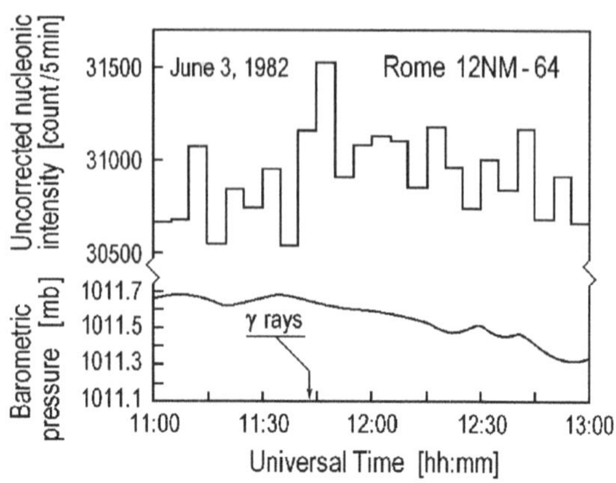

Fig. 3.10 The event of June 3, 1982. The 5-min data of Rome neutron supermonitor (According to Iucci et al. 1985)

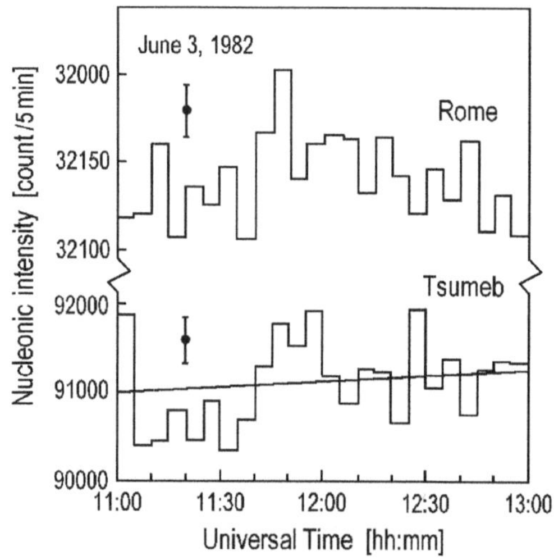

Fig. 3.11 The event of June 3, 1982. The 5-min data of Rome and Tsumeb neutron supermonitors (According to Stoker 1987)

accordance with calculations of Shibata (1994) of solar neutron interaction and propagation processes in the Earth's atmosphere (see Chapter 6).

3.5 Some Constraints for Neutron Generation Processes on the Sun Follow from June 21, 1980 and June 3, 1982 Event Data

Chupp et al. (1985) on the basis of more exact data on the efficiencies of gamma-quanta and neutron registration by gamma ray spectrometer on SMM, reconsidered the observation results of events June 21, 1980 and June 3, 1982 (the efficiencies were calculated by the Monte Carlo method). For the event June 21, 1980 the results

Table 3.4 Ground level recordings of solar neutrons, June 3, 1982 (From Stoker 1987)

Station	R_C (GV)	h at 11:45 UT (mm Hg)	θ_s at 11:45 UT (°)	h_s at 11:45 UT (g/cm^2)	Increase at 11:45–11:50 UT (%)
Jungfraujoch	4.63	498.0	24.50	745	3.88 ± 0.59
Lomnicky Stit	4.00	563.0	29.90	883	2.71 ± 0.33
Rome	6.30	760.7	20.95	1,108	2.46 ± 0.69
Kiel	2.30	761.4	23.30	1,225	1.07 ± 0.69
Tsumeb	9.20	664.0	43.60	1,247	0.75 ± 0.41

R_C – cutoff rigidity in the vertical direction
h – atmospheric pressure
θ_s – zenith angle of the Sun
h_s – atmospheric depth in direction to the Sun

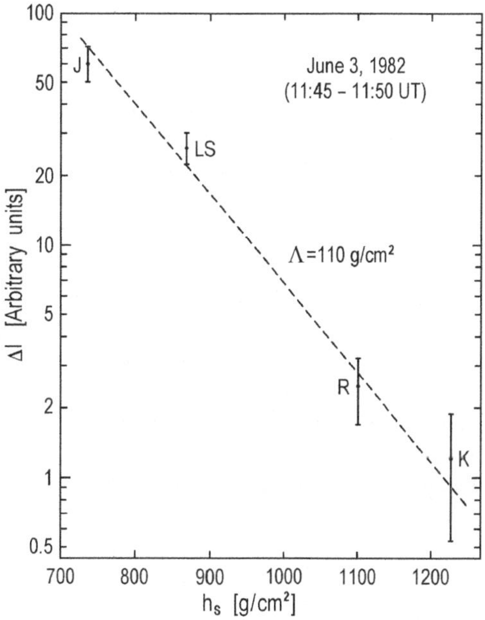

Fig. 3.12 The event of June 3, 1982. The absolute amplitude ΔI for neutron monitors on stations Jungfraujoch (J), Lomnicky Stit (LS), Rome (R) and Kiel (K) versus atmospheric depth in direction to the Sun h_s (in g/cm^2). The ΔI are computed from observed ΔI/I by applying the latitude and altitude cosmic ray intensity I variations (normalization to one level of intensity) (According to Iucci et al. 1985)

of new analysis confirmed the suggestion that neutrons had energies 50–250 MeV and was generated momentary as δ-function (but it was not excluded that neutron ejection function can be the same as the time-profile of gamma-burst); the energy spectrum of protons which generate solar neutrons can be described by power law $\propto E_k^{-3.6}$ (acceleration by shock waves) or by Bessel function with the parameter of acceleration $\alpha T \approx 0.02$ (Fermi stochastic acceleration). For the event of June 3, 1982 data on SMM and Jungfraujoch neutron supermonitor can be explained by the neutron generation function described the ejection of neutrons not a momentary but, a long-time ejection (≥20 min). In regards to the spectrum of neutrons, Chupp et al. (1985) came to the conclusion that the better approximation could be obtained for the spectrum more hard than Bessel function at $\alpha T \approx 0.05$. These results for the event of June 3, 1982 were developed in Chupp et al. (1987) on the basis of solar gamma-ray and neutron observations on SMM and by ground-based neutron super-monitors. In this paper Chupp et al. (1987) came to the conclusion that neutron generation in the solar atmosphere by protons with energy several gigaelectron volt lasted a very short time (not more than 16 s) in the region with matter density $n_i \geq 10^{14}\ \mathrm{cm}^{-3}$. The spectrum of ejected neutrons from the Sun in the energy region $\geq$0.1 GeV can be described by the power law

$$D(E_k) \propto E_k^{-2.4} \tag{3.1}$$

or in the form

$$D(E_k) \propto E_k^{3/2} \exp\left[-(E_k/0.016)^{1/4}\right], \tag{3.2}$$

where E_k is the kinetic energy in GeV, the power law lasted up to energy E_{kc} where 2 GeV $\leq E_{kc} \leq$ 4 GeV. Integral emissions of solar neutrons for $E_k \geq 0.1$ GeV was equal to $\approx 8 \times 10^{28}$ neutron/sr.

Murphy and Ramaty (1984) calculated the expected spectrum of generated solar neutrons in the frame of thick target model and at the suggestion on isotropic distribution of accelerated charged particles, with energy spectrum described by Bessel function (stochastic acceleration) and by power function (shock wave acceleration). Comparison with experimental data of events June 21, 1980 and June 3, 1982 shows that proton fluxes inside the solar atmosphere in the region of neutron generation was much bigger than neutron fluxes ejected into the interplanetary space, but, the energy spectra are very similar. Murphy and Ramaty (1984) came to the conclusion that for the event June 21, 1980 was predominate stochastic mechanism of acceleration, but for the June 3, 1982 event both mechanisms (stochastic and shock wave acceleration) were important.

References for Chapter 3

Chupp EL (1983) High energy particle acceleration in solar flares – observational evidence. Sol Phys (Netherlands) 86(1–2):383–393

Chupp EL, Forrest DJ, Ryan JM, Heslin J, Reppin C, Pinkau K, Kanbach G, Rieger E, Share GH (1982) A direct observation of solar neutrons following the 0118 UT flare on 1980 June 21. Astrophys J 263, Part 2(2):L95–L99

Chupp EL, Forrest DJ, Share GH, Kanbach G, Debrunner H, Flueckiger E (1983) Solar neutrons from the impulsive flare on 1982 June 3 at 1143 UT. In: Proceedings of the 18th international cosmic ray conference, vol 10, Bangalore, pp 334–337

Chupp EL, Forrest DJ, Vestrand WT, Dubrunner H, Flueckiger EO, Cooper JF, Kanbach G, Reppin C, Share GH (1985) Time extended production of neutrons during a solar flare. In: Proceedings of the 19th international cosmic ray conference, vol 4, La Jolla, CA, pp 126–129

Chupp EL, Debrunner H, Fluckiger E, Forrest DJ, Golliez F, Kanbach G, Vestrand WT, Cooper J, Share G (1987) Solar neutron emissivity during the large flare on 1982 June 3. Astrophys J 318, Part 1(2):913–925

Debrunner H, Brunberg EA (1968) Monte Carlo calculation of nucleonic cascade in atmosphere. Canad J Phys 46(10), Part 4:S1069–S1072

Debrunner H, Flückiger EO, Chupp EL, Forrest DJ (1983) The solar cosmic ray neutron event on June 3, 1982. In: Proceedings of the 18th international cosmic ray conference, vol 4, Bangalore, pp 75–78

Efimov YuE, Kocharov GE, Kudela K (1983) On the solar neutrons observation on high mountain neutron monitor. In: Proceedings of the 19th international cosmic ray conference, vol 10, Bangalore, pp 276–278

Forrest DJ, Chupp EL, Ryan MM, Reppin C, Rieger E, Kanbach G, Pinkau K, Share G, Kinzer G (1981) Evidence for impulsive ion acceleration during the 0312 UT flare of 1980 June 7. In: Proceedings of the 17th international cosmic ray conference, vol 10, Paris, pp 5–8

Iucci N, Parisi M, Signorini C, Storini M, Villoresi G (1985) Suggestions for improving the efficiency of ground-based neutron monitors for detecting solar neutrons. In: Proceedings of the 19th international cosmic ray conference, vol 4, La Jolla, CA, pp 134–137

Karlicky M, Aurass H, Mann G (1986) Radio observations of the solar neutron flare of 3 June, 1982. Sol Phys (Netherlands) 104(1):165–168

McDonald FB, Van Hollebeke MAI, Trainor JH (1985) HELIOS 1 energetic particle observations of the solar gamma ray flare events of 7, 21 June 1980 and 3 June 1982. In: Proceedings of the 19th international cosmic ray conference, vol 4, La Jolla, CA, pp 98–101

Murphy RJ, Ramaty R (1984) Solar-flare neutrons and gamma-rays. Adv Space Res 4(7):127–136

Neustock HH, Wibberenz G, Iwers B (1985) Injection of energetic particles following the gamma-ray flares on June 7, 1980, as observed on HELIOS 1. In: Proceedings of the 19th international cosmic ray conference, vol 4, La Jolla, CA, pp 102–105

Shibata S (1994) Propagation of solar neutrons through the atmosphere of the Earth. J Geophys Res 99(A4):6651–6665

Stoker PH (1987) Search for solar neutron recordings by the 18NM64 at Tsumeb during the 21st solar cycle. In: Proceedings of the 20th international cosmic ray conference, vol 3, Moscow, pp 98–100

Van Hollebeke MAI, McDonald FB, Trainor JH (1985) Energetic particle observations of the solar-gamma ray/neutron flare events of 3 Jun 1982 and 21 June 1980 isotopic and chemical composition. In: Proceedings of the 19th international cosmic ray conference, vol 4, La Jolla, CA, pp 209–212

Chapter 4
Space Probe Observations of Solar Neutron and Gamma Ray Events

4.1 The First Solar Neutron and Gamma Ray Event Observed on SMM

We described above some of the very important solar neutron observations made during the events of June 21, 1980 and June 3, 1982 by the gamma ray spectrometer on the SMM satellite. SMM observed many – in fact, more than 100 – gamma-ray events, and only a few of these were gamma-ray/neutron events (neutron events constitute only a few percent of all gamma-ray events). Chupp (1988) mentioned that solar neutrons give very important information regarding the spectrum of accelerated particles at the source as well as on the ratio of ^{3}He/H in the photosphere of the Sun.

4.2 Solar Neutron and Gamma Ray Events Observed on the Satellites SMM and Hinotori

By the satellites SMM and Hinotori were observed many gamma-ray events. Yoshimuri (1989) by comparison with data of Tokyo neutron supermonitor showed that some of these events can generate neutrons also. It was investigated five events: June 7, 1980 (maximum energy of gamma-quanta on Hinotori was $E_{\gamma max} = 8\ MeV$); June 21, 1980 ($E_{\gamma max} = 50 MeV$); November 6, 1981 ($E_{\gamma max} = 10 MeV$); November 26, 1982 ($E_{\gamma max} = 15 MeV$); and April 25, 1984 ($E_{\gamma max} = 70 MeV$). By the method of superposition Chree, it was shown that in average for these five events was observed an increase in Tokyo neutron supermonitor of 7.3σ, where σ is the standard deviation (see Fig. 4.1). In more details these events were compared with ground-based observations and are discussed in Chapter 3 (June 21, 1980) and Chapter 7 (other events).

L. Dorman, *Solar Neutrons and Related Phenomena*,
Astrophysics and Space Science Library 365, DOI 10.1007/978-90-481-3737-4_4,

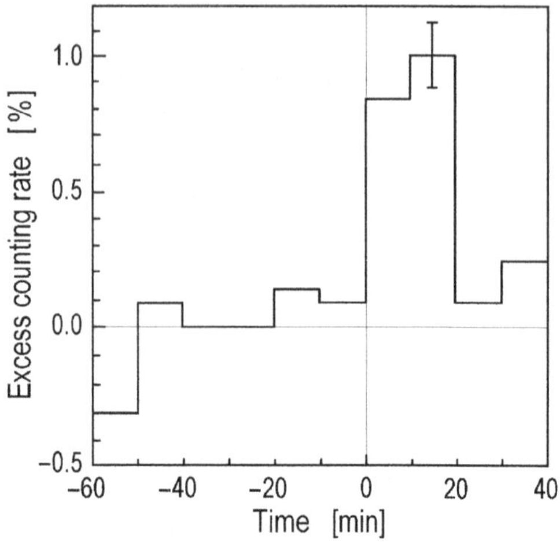

Fig. 4.1 Epoch analysis of Tokyo neutron monitor 10-min data for five solar neutron events observed on SMM and/or Hinotori as high energy gamma-ray events (According to Takahashi 1989)

4.3 Solar Neutron and Gamma Ray Events Observed By the COMPTEL Experiment at the Compton Gamma-Ray Observatory

The high energy neutron fluxes continue registration was made on the COMPTEL experiment by the Compton Gamma-Ray Observatory (Morris et al. 1993). In the first, this continues registration measurements were important for determining the background for main measurements of gamma-rays by GRO. In the second, this continues registration was important as the satellite patrol of solar neutron events. For this continue registration they used one of seven liquid-scintillator modules as uncollimated neutron detector, with an energy threshold of 12.8 MeV (the cylinder with radius 13.8 cm and depth 8 cm, are viewed by eight photomultipliers). The neutron measurements of Morris et al. (1993) show a good agreement with balloon measurements of neutron albedo in dependence if cutoff rigidity.

According to Jenkins et al. (1991) for neutron measurements on the gamma-ray observatory (GRO) can be used also each of four detectors of the oriented scintillation spectrometer experiment (OSSE). These detectors could be used for gamma-ray and neutron registrations. Each detector consists of a 13 in. diameter phoswich with central part surrounded by a 3.3 in. thick annular NaI (Tl) anticoincidence counter; the phoswich is composed of a 4 in. thick NaI (Tl) crystal above a 3 in. thick CsI (Na) crystal. An engineering model of the detectors was exposed by neutrons at Indiana University Cyclotron Facility. The neutron beam for calibration was produced by the $^7\text{Li}(p,n)^7\text{Be}$ reaction using protons with energies of 30, 55, 90, 140 and 200 MeV.

Several neutron detectors on space probes are a good guarantee that many gamma-ray/neutron solar events will be detected and a more exact and full information on solar acceleration and nuclear interactions processes will be accessible.

4.4 The Solar Neutron and Gamma Ray Event of April 25, 1984

This event was the third for which on a space probe was directly observed solar neutrons (Chupp 1990). Figure 4.2 shows a time history of the relative gamma ray spectrometer rates on SMM, in several energy bands.

The intense impulsive structure occurs within less than 2 min in the energy range from 80 keV to over 40 MeV. The total prompt nuclear line fluency in the 4–8 MeV energy band was $\approx$200 photons cm^{-2}. For the delayed lines the fluency values were $\approx$400 photons cm^{-2} at 0.511 MeV and $\approx$700 photons cm^{-2} at 2.223 MeV (the fluencies of these delayed lines were over a factor of 2 larger than in the event of June 3, 1982). Chupp (1990) mentioned that since the GRS has a poor energy discrimination for neutron events and the time-of-flight method can not be used to determine the neutron spectrum (because of the time extended neutron production

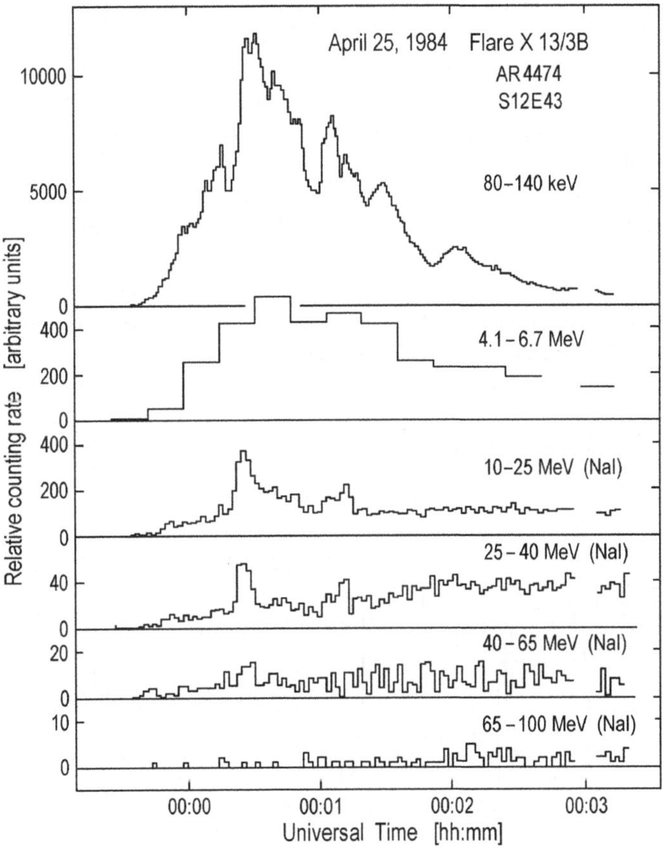

Fig. 4.2 The solar neutron event of April 25, 1984. The relative count rate time history is shown for several energy bands during solar flare of importance 3, as observed by the gamma ray spectrometer on SMM (According to Chupp 1990)

in April 25, 1984 event), the interpretation of the GRS neutron counts was not direct.

The first evidence that high energy neutrons were produced in the event on April 25, 1984 was obtained by observation of neutron decay protons on the ICE spacecraft by Evenson et al. (1985b, 1990) (see Chapter 8). This event was also detected by ground based neutron monitors (see Section 7.14).

4.5 The Solar Neutron and Gamma Ray Event of March 6, 1989

This event was caused by X15/3B solar flare at 13:50 UT on March 6, 1989, located at 35°N, 69°E. According to Dunphy and Chupp (1991) the high energy data for this event was obtained from the SMM GRS high energy matrix (this mode of the GRS treats seven 7.6 × 7.6 cm NaI (Tl) main channel scintillators as one layer of a two-layer detector; the second layer is a 7.5 cm thick × 25 cm diameter CsI (Na) back shield). The high energy matrix records the energy loss events in the range 10–100 MeV with time resolution 2.048 s. Figure 4.3 shows the time history of the ratio of the gamma-ray flux in the 60–110 MeV energy interval to the flux in the 10–60 MeV interval.

The sudden increase of the ratio of the gamma-ray flux in the 60–110 MeV energy interval to the flux in the 10–60 MeV interval at about 14:07 UT gives evidence of π^0-decay photons which produce a much harder spectrum than electron bremsstrahlung. Figure 4.4 shows the spectrum observed before 14:06 UT when the continuum spectrum, presumably due to electron bremsstrahlung is dominated, and after 14:06 UT when there is clear evidence of a π^0-decay peak.

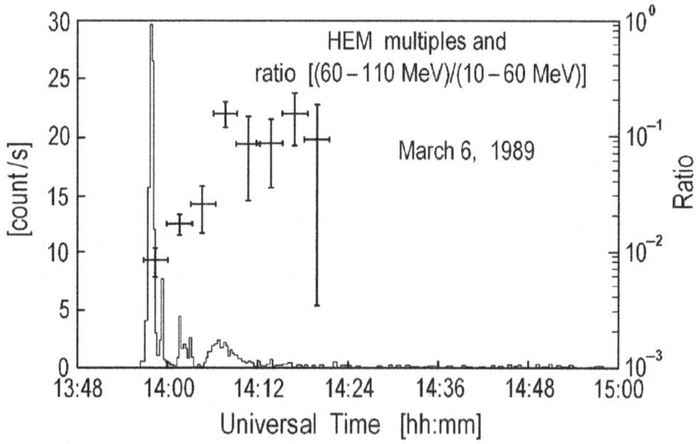

Fig. 4.3 The solar neutron event of March 6, 1989. Time history of the GRS HEM 'multiple' events (events in both layers >25 MeV). Also shown is the ratio of gamma-ray fluency above 60 MeV to fluency below 60 MeV (*heavy line* and *right axis*). The ratio increases significantly at 14:06 UT (According to Dunphy and Chupp 1991)

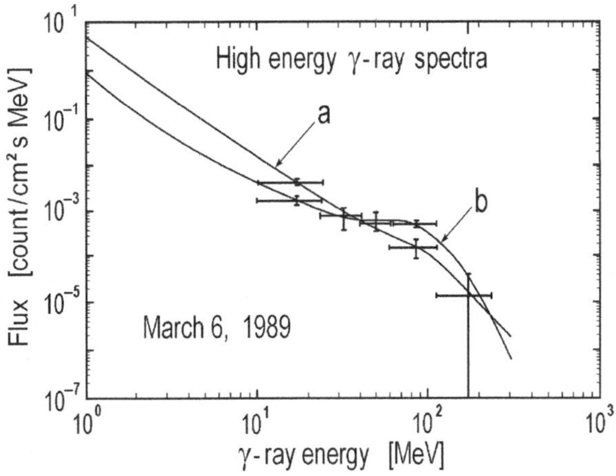

Fig. 4.4 The solar neutron event of March 6, 1989. Gamma-ray spectra for two time intervals during flare: (**a**) 14:03–14:06 UT shows only weak evidence for π^0- decay peak at 70 MeV; (**b**) 14:06–14.09 UT shows significant π^0-decay peak at 70 MeV. Data are fit with exponential continuum and peak at 70 MeV (According to Dunphy and Chupp 1991)

Dunphy and Chupp (1991) noted that the time history of the π^0-decay radiation indicates that ~60% of the pion production and therefore, the bulk of the high energy neutron production, occurred between 14:06–14:12 UT. The other important discovery on the neutron time production can be obtained from the High Energy Matrix of GRS, which has enough energy resolution to constrain time-of-flight dependence from energy production time of neutrons over a broad time period. These data limits the likely period of neutron production to 14:06–14:15 UT.

From the observations by the main channel detector of GRS, the 2.223 MeV gamma-ray line (from the capture of neutrons by hydrogen), it was shown that most of the neutrons were produced between 14:00 and 14:09 UT. The resulting neutron spectrum at the Earth, for the time of neutron maximum production (14:07 UT), is shown in Fig. 4.5.

Figure 4.5 shows an envelope around the most likely spectrum for two limiting-case production times 14:04 and 14:13 UT. The total neutron fluency for energies >50 MeV at the Earth's orbit was ~50 neutrons cm^{-2}; this implies a neutron emissivity for energy >50 MeV at the Sun would be $\sim 4 \times 10^{28}$ neutrons sr^{-1} (if the neutrons were emitted isotropically). For the period of significant pion production (14:06–14:08 UT) the total fluency in the π^0-decay peak was 12.1 $\pm$ 2.6 photons cm^{-2}. In this time period the fluency in the 2.223 MeV gamma-ray line was 43.8 $\pm$ 2.2 photons.cm^{-2} and the fluency in the 4–7 MeV energy band, from nuclear deexcitation of gamma-rays was 42.7 $\pm$ 4 photons cm^{-2}. According to Dunphy and Chupp (1991) from these data it can be found the ratios of the π^0-peak fluency and 2.223 MeV peak fluency (after correcting for limb-darkening effects) to the 4–7 MeV fluency, which can be used to find the proton spectrum at the Sun which produced them. The relationship between these ratios and solar

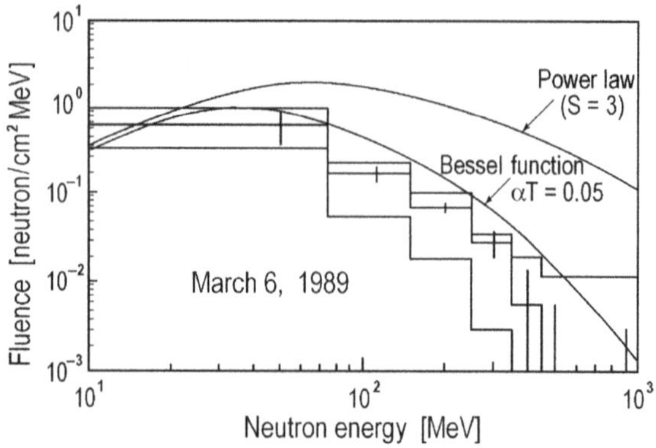

Fig. 4.5 The solar neutron event of March 6, 1989. Time integrated neutron spectrum at the Earth from the flare of March 6, 1989. Spectrum and 'envelope' depend on production time (see Section 4.5). Curves labeled 'Power law' and 'Bessel function' are spectra predicted from gamma-ray data using those proton spectral shapes in an isotropic, thick-target model (According to Dunphy and Chupp 1991)

proton spectra with power-law and Bessel function shapes was found by Murphy and Ramaty (1984). A comparison with this model gives the single Bessel-function proton spectrum with intensity

$$N_p(>30\ MeV) = (4.5 \pm 1.2) \times 10^{32} \text{ protons} \tag{4.1}$$

and a shape parameter of acceleration

$$\alpha T = 0.062 \pm 0.007.Z \tag{4.2}$$

The closest to the experimental data is a single power law proton spectrum

$$D(E_\text{k}) \propto E_\text{k}^{-\gamma} \tag{4.3}$$

which corresponds to the intensity

$$N_p(>30MeV) = (3 \pm 2) \times 10^{32} \text{ protons} \tag{4.4}$$

with power-law index

$$\gamma = 3.2 \pm 0.4. \tag{4.5}$$

Table 4.1 shows the comparison of parameters for three solar neutron events observed on SMM.

Table 4.1 Comparison of three solar neutron events detected on SMM (According to Dunphy and Chupp 1991)

Event	Heliocentr. angle (°)	Fluency (photon cm^{-2}) 2.2 MeV	4–8 MeV	Neutron emissivity (>50 MeV), (neutron sr^{-1})	Observed π^0-decay gamma-rays	Proton spectral shape, αT
June 21, 1980	89	3.1 ± 0.2	98 ± 2	3×10^{28}	No	0.020
June 3, 1982	72	314	305 ± 30	2×10^{29}	Yes	0.035
March 6, 1989	77	43.8 ± 2.2	45.4 ± 6.6	1×10^{29}	Yes	0.062

4.6 The Solar Neutron and Gamma Ray Event of June 4, 1991

An X12.0 class solar flare occurred at 3:37 UT in NOAA region 6659 (30°N, 70°E) on June 4, 1991. In this case, intense emission of gamma rays was observed by the instruments BATSE and OSSE onboard the CGRO satellite (Ramaty et al. 1994; Murphy et al. 1997, 2007). The gamma ray lines at 2.2 and 4.4 MeV were clearly observed by CGRO/OSSE, and solar neutrons were also observed. Murphy et al. (2007) analyzed these OSSE data in detail, and obtained many parameters of Hua et al. (2002) model as shown in Table 4.2, which explain observed 2.2 MeV gamma ray line data.

Murphy et al. (2007) used the 4.4 MeV line time history as the ion acceleration release time history. For the accelerated ion composition, ambient composition, atmospheric model, and photospheric ^{3}He/H ratio, Murphy et al. (2007) used typical values estimated from observations of previous flares. Although they could not obtain flare loop length from observed data since there is no imager of the X-ray or gamma ray at that time, they obtained loop length by fitting observed 2.2 MeV time history, combination with values of the level of pitch angle scattering within the loop (λ), magnetic convergence ratio (δ), and spectral power-law index (s). By using Hua et al. (2002) model with these parameters, they calculated neutron time history at the OSSE, and by comparing observed neutron data. They determine the upper cutoff energy of accelerated ions to fit the observed neutron data. That gave 125MeV, but such a value is too low energy to observe solar neutrons on the Earth (see Section 7.21.9, where it will be shown that from comparison with neutron monitor observation data, the upper cutoff for solar neutrons must be about two times bigger.

4.7 The Solar Neutron and Gamma Ray Event of June 9, 1991

According to Ryan et al. (1993), this solar gamma-ray/neutron event was detected by COMPTEL experiment on Compton Gamma Ray Observatory (see Section 4.3). The event of June 9, 1991 was caused by X10/3B class solar flare with the coordinates of 34°N, 04°E, that was observed by all instruments on the Compton

Table 4.2 Parameters of the Hua et al. (2002) model program for 1991 June 4 event (According to Murphy et al. 2007)

Accelerated ion composition (impulsive)	$\alpha/p = 0.5$ $^3He/^4He = 1$
Ambient composition (coronal)	He/H = 0.1 Ne/O = 0.25
Atmospheric model	Avrett (1981)
Photospheric ^{3}He/H	3.7×10^{-5}
Acceleration release time history	4.4 MeV γ-ray line profile
Loop length	11,500 km
Flare heliocentric angle	74.5°
Pitch angle scattering (λ)	300
Magnetic convergence (δ)	0.20
Power index (s)	4.0
Cutoff energy (E_c)	125 MeV

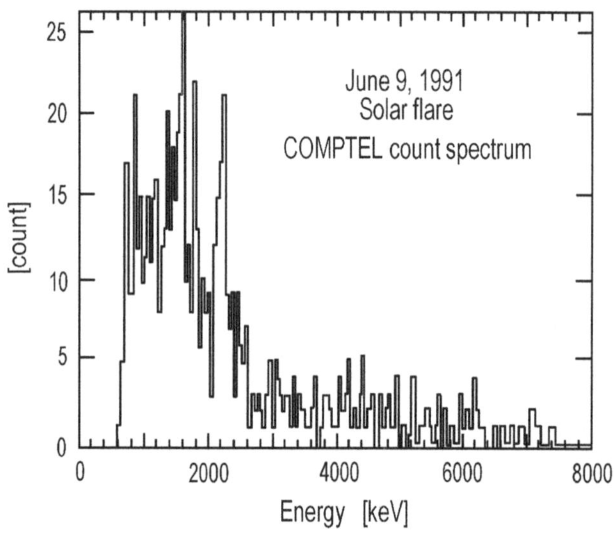

Fig. 4.6 The solar neutron event of June 9, 1991. The raw telescope mode count spectrum from the June 9, 1991 solar flare (According to Ryan et al. 1993)

Gamma Ray Observatory. According to GOES observations the X-ray flux started at 01:34 UT with maximum at 01:43 UT on June 9, 1991. The gamma-ray onset according to COMPTEL observations occurred at 01:36 UT and the impulsive phase lasted until 01:42 UT. Figure 4.6 shows the gamma-ray emission spectra during the impulsive phase; the strong neutron capture line of 2.2 MeV can be seen (other lines are also present with less statistical significance).

According to Ryan et al. (1993), solar neutron events from 01:55 to 02:22 UT were selected (this time period avoids troublesome period around the impulsive phase with large dead times and other instrumental effects). The time 01:55 UT

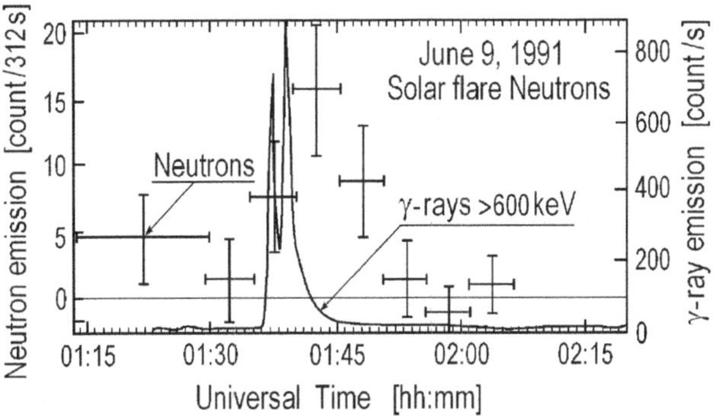

Fig. 4.7 The solar neutron event of June 9, 1991. Neutron and gamma-ray emission time-profiles, plotted at the time corresponding to a photon arrival time (From Ryan et al. 1993)

corresponds to a 50 MeV neutron produced at the flare start 01:36 UT or a 60 MeV neutron from the flare maximum 01:39 UT. The energy of each neutron was used to compute its production time at the Sun. The observed gamma-ray emission time profile and expected neutron emission time profile for the event of June 9, 1991 are shown in Fig. 4.7.

4.8 The Solar Neutron and Gamma Ray Event of June 15, 1991

The event of June 15, 1991 (Debrunner et al. 1993) was caused by X12/3B class solar flare, with heliographic coordinates 33°N, 69°W. The gamma-rays and neutrons could be observed only during the decay phase beginning at ~09:00 UT, approximately 50 min after the flare onset (before this CGRO was on the night side of the Earth). The 0.8–10 MeV gamma-ray observations on COMPTEL during the time interval 09:01.67–09:36.67 UT shows an exponential decay with a decay time of 13.9 ± 1.8 min in good agreement with the 50 MeV–2GeV gamma-ray observations on the space probe GAMMA-1 (Akimov et al. 1991), indicated that gamma-ray emission was decaying exponentially from 08:37:37 to 08:50:00 UT with an e-folding time of 9.8 ± 1.0 min.

The COMPTEL gamma-ray measurements at 0.8–10 MeV during the time interval of 09:01:67–09:36:67 UT also reveals an exponential decay with a decay time of 13.9 ± 1.8 min. Comparison with background periods 08:09.2–08:44.2 and 09:42.3–10:17.3 UT on June 14, 1991, with near the same geophysical and orbital parameters, leads Debrunner et al. (1993) to the conclusion that 43.5 ± 11.8 solar neutrons were detected by COMPTEL originating from the solar flare event on June 15, 1991.

Taking into account the observed exponential decay of gamma-ray emission and suppose that the same was for neutron emission, Debrunner et al. (1993) determined the time integrated neutron emissivity of the solar flare event of June 15, 1991 (see Fig. 4.8).

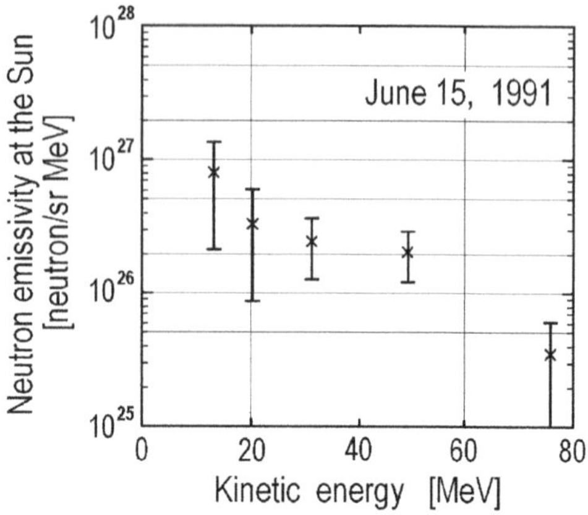

Fig. 4.8 The solar neutron event of June 15, 1991. The time integrated directional neutron emissivity (According to Debrunner et al. 1993)

If the solar neutron emissivity at $E_n \leq 100$ MeV is described by a power law in energy $Q(E_n) \propto E_n^{-\gamma}$, a spectral index $\gamma \approx 1.5$ is found by Debrunner et al. (1993). In the higher energy region the increasing γ up to $\sim$2.5 is expected (from data of GAMMA-1) and it can be a cause that this event was not observed by ground neutron monitors.

4.9 The Solar Neutron and Gamma Ray Event of November 6, 1997

Yoshimori et al. (2001) investigate high-energy particle production in the 1997 November 6 flare as viewed from gamma rays and neutrons on the basis of satellite Yohkoh data. According to Yoshimori et al. (2001), Yohkoh observed hard X- and gamma-rays from a X9.4 flare on November 6, 1997. Strong gamma-rays were emitted in 11:52–11:56 UT (peak phase). After that, weak and extended gamma-ray production lasted for 600 s (extended phase). The OSSE aboard CGRO detected neutrons associated with this flare between 12:08 and 12:28 UT. The neutron count-rate time profile exhibit a gradually decrease with time. Yoshimori et al. (2001) derive the proton spectra and the timing of particle acceleration to explain the observed neutron time profile. The proton spectra of $E^{-3.5}$ in the peak phase and of $E^{-3.0}$ in the extended phase give a good fit to the observed neutron time profile. Different aspects of this event will be considered in more details in Chapters 7 and 9.

References for Chapter 4

Akimov VV, Afanassyev VG, Belaousov AS, Blokhintsev ID, Kalinkin LF, Leikov NG, Nesterov VE, Volsenskaya VA, Galper AM, Chesnokov VJ, Kirillov-Ugryumov VG, Lutchkov BI, Ozerov YV, Popov AV, Rudko VA, Runtso MF, Voronov SA, Zemskov VM, Fradkin MI, Kurnosova LV, Rusakovitch MA, Topchiev NP, Chuikin EI, Tugaenko VY, Tian TN, Ishkov VN, Gros M,

Grenier I, Barouch E, Wallin P, Bazer-Bachi AR, Lavigne J-M, Olive J-F, Juchniewicz J (1991) Observation of high energy gamma-rays from the Sun with the GAMMA-1 telescope (E > 30 MeV). In: Proceedings of the 22nd international cosmic ray conference, vol 3, Dublin, pp 73–76

Avrett EH (1981) Reference model atmosphere calculation – the Sunspot sunspot model. In: The physics of sunspots; Proceedings of the conference, Sacramento Peak Observatory, NM, July 14–17, 1981. Sacramento Peak Observatory, Sunspot, NM, (A83-18101 06-92), pp 235–255, 257

Chupp EL (1988) Solar neutron observations and their relation to solar flare acceleration problems. Solar Phys (Netherlands) 118(1–2):137–154

Chupp EL (1990) Emission characteristics of three intense solar flares observed in cycle 21. Astrophys J Suppl Ser (USA) 73(2):213–226

Debrunner H, Lockwood JA, Ryan JM, McConnell M, Schonfelder V, Aarts H, Bennett K, Winkler C (1993) Neutrons from the 15 June 1991 solar flare. In: Proceedings of the 23rd international cosmic ray conference, vol 3, Calgary, pp 115–118

Dunphy PP, Chupp EL (1991) High-energy gamma rays and neutrons from the solar flare of 1989 March 6. In: Proceedings of the 22nd international cosmic ray conference, vol 3, Dublin, pp 65–68

Evenson PA, Hovestadt D, Meyer P, Moses D (1985a) The energy spectra of solar flare electrons. In: Proceedings of the 19th international cosmic ray conference, vol 4, La Jolla, CA, pp 74–77

Evenson PA, Kroeger R, Meyer P (1985b) Further observations of protons resulting from the decay of neutrons ejected by solar flares. In: Proceedings of the 19th international cosmic ray conference, vol 4, La Jolla, CA, pp 130–133

Evenson PA, Kroeger R, Meyer P, Reames D (1990) Solar neutron decay proton observations in cycle 21. Astrophys J Suppl Ser (USA) 73(2):273–277

Hua X-M, Kozlovsky B, Lingenfelter RE, Ramaty R, Stupp A (2002) Angular and Energy dependent Neutron Emission from Solar Flare Magnetic Loops. Astrophys J Suppl Ser 140:563–579

Jenkins TL, Frye GM Jr, Stansfield S, Kinzer RL, Jensen C, Kurfess JD, Johnson WN, Strickman M, Byrd R, Foster C (1991) Response of the OSSE GRO detector to 30-200 MeV neutrons. In: Proceedings of the 22nd international cosmic ray conference, vol 3, Dublin, pp 760–763

Morris DJ, Aarts H, Bennett K, Byrd R, Eymann G, Lockwood JA, McConnell ML, Ryan JM, Steinle H (1993) Neutron flux measurements in near-Earth orbit with COMPTEL. In: Proceedings of the 23rd international cosmic ray conference, vol 3, Calgary, pp 809–812

Murphy RJ, Ramaty R (1984) Solar-flare neutrons and gamma-rays. Adv Space Res 4(7):127–136

Murphy RJ, Share GH, Grove JE, Johnson WN, Kinzer RL, Kurfess JD, Strickman MS, Jung GV (1997) Accelerated particle composition and energetics and ambient abundances from gamma-ray spectroscopy of the 1991 June 4 solar flare. Astrophys J 490 (Part 1, 2):883–900

Murphy RJ, Kozlovsky B, Share GH, Hua X-M, Lingenfelter RE (2007) Using gamma-ray and neutron emission to determine solar flare accelerated particle spectra and composition and the conditions within the flare magnetic loop. Astrophys J Suppl Ser 168:167–194

Ramaty R, Schwartz RA, Enome S, Nakajima H (1994) Gamma-ray and millimeter-wave emissions from the 1991 June X-class solar flares. Astrophys J 436 (Part 1, 2):941–949

Ryan J, Bennett K, Debrunner H, Forrest D, Hanlon L, Lockwood J, Loomis M, McConnell M, Morris D, Rank G, Schonfelder V, Swanenburg BN, Webber W, Winkler C (1993) Neutron and gamma ray measurements of the solar flare of 1991 June 9. In: Proceedings of the 23rd international cosmic ray conference, vol 3, Calgary, pp 103–106

Takahashi K (1989) Detection of solar neutrons by ground-based neutron monitor in 1989. Space Sci Rev 51(1–2):123–146

Yoshimuri M (1989) Observational studies of gamma-rays and neutrons from solar flares 1989. Space Sci Rev 51(1–2):85–115

Yoshimori M, Suga K, Nakayama S, Ogawa H, Share GH, Murphy RJ (2001) High-energy particle production in the 1997 November 6 flare as viewed from gamma rays and neutrons. In: Proceedings of the 27th international cosmic ray conference, vol 8, Hamburg, pp 3029

Chapter 5
Solar Neutron Propagation in the Earth's Atmosphere, and the Sensitivity of Neutron Monitors and Other Ground Based Detectors to Solar Neutrons

The main parameters determining the nature of solar neutron propagation in the Earth's atmosphere are the neutron energy, the zenith angle of neutron arrival (i.e. the zenith angle of the Sun), and the atmospheric depth in the direction of the Sun. The integral multiplicity of ground-based detectors to solar neutrons depends upon these parameters. The integral multiplicity and sensitivity of ground-based detectors to solar neutrons are determined by the interactions of neutrons with air nuclei, which include: generation of secondary particles; energy change; scattering; absorption. The radioactive decay of neutrons is negligible, because the neutron's propagation time through the atmosphere is very small compared to its lifetime.

5.1 The Atmospheric Depth for Ground Solar Neutron Observations in the Direction of the Sun

This depth h_s towards the Sun is very important because, the normalized amplitude of increase is proportional to $exp\left(-h_s/\Lambda\right)$, where Λ is determined by the process of neutron propagation in the atmosphere. According to Shibata (1994), for neutrons with energy $\geq$200 MeV, the value $\Lambda \approx 110$ g/cm^2 taking into account the scattering of neutrons in the atmosphere; up to $\sim$1,000 MeV this value does not depend on the neutron energy (see in more details in Section 5.5).

The position of the Sun, where the Sun's zenith angle is 0° (subsonic point) is determined by the latitude λ_s and longitude f_s, where:

$$\lambda_s = 23.5^\circ \times \cos(2\pi(t - \text{June}, 21)/365), \tag{5.1}$$

$$f_s = 15^\circ(12 - t_s). \tag{5.2}$$

L. Dorman, *Solar Neutrons and Related Phenomena*,
Astrophysics and Space Science Library 365, DOI 10.1007/978-90-481-3737-4_5,

where t is the day of the solar neutron event, t_s is the time (UT) of event. The zenith angle $\theta_s(t, t_s, \lambda, f)$ of the Sun for the solar neutron event in time t, t_s in the point of observation with geographic latitude λ and longitude f will be

$$\theta_s(t, t_s, \lambda, f) = \arccos(\cos(\lambda - \lambda_s)\cos(f - f_s)), \tag{5.3}$$

where λ_s and f_s are determined by Eqs. 5.1 and 5.2.

In the approximation of plane-parallel layer of the Earth's atmosphere, the depth in the Sun's direction $h_s(h, t, t_s, \lambda, f)$ for observation on the atmospheric depth **h** will be

$$h_s(h, t, t_s, \lambda, f) = h/cos\,\theta_s = \frac{h}{cos(\lambda - \lambda_s)cos(f - f_s)}, \tag{5.4}$$

where λ_s and f_s are determined by Eqs. 5.1 and 5.2. If we take into account the spherical character of this layer with effective depth $H \approx h/\rho(h) \approx 10^6$cm (approximately does not depend from h), we obtain sufficient difference with Eq. 5.4 only for great $\theta_s(t, t_s, \lambda, f)$:

$$h_s(h, \theta_s, H) = \frac{h(r_e + H)}{H}\left[\cos\left(\arcsin\left(\frac{r_e}{r_e + H}\sin\theta_s\right)\right) - \frac{r_e}{r_e + H}\cos\theta_s\right], \tag{5.5}$$

where r_e is the radius of the Earth and $\theta_s(t, t_s, \lambda, f)$ is determined by Eq. 5.3. Comparison between h_s determined by Eqs. 5.4 and 5.5 is shown in Table 5.1 for $h = 1{,}000$ g/cm^2. It is shown also the ratio 'spherical/plane' of h_s determined by Eq. 5.5 to h_s determined by Eq. 5.4. From Table 5.1 we can see that the relative error in the determining of h_s by (5.4) is smaller than 0.1% for $\theta_s \leq 50°$, between 0.1% and 0.21% for $50° \leq \theta_s \leq 60°$, between 0.21% and 0.57% for $60° \leq \theta_s \leq 70°$,

Table 5.1 Comparison between h_s as determined by Eqs. 5.4 and 5.5 for h = 1,000 g/cm^2 and the ratio 'spherical/plane' (in the first approximation up to mountain altitudes not depends from h)

θ_s	h_s (g/cm^2) Eq. 5.4	h_s (g/cm^2) Eq. 5.5	$\frac{\text{Spherical}}{\text{Plane}}$
50°	1,555.9	1,554.4	0.9990
60°	2,000.0	1,995.7	0.9979
70°	2,924.0	2,907.3	0.9943
80°	5,760.0	5,621.0	0.9759
82.5°	7,662.8	7,345.0	0.9585
85°	11,481.0	10,496.0	0.9142
87.5°	22,936.0	17,476.0	0.7619
90°	∞	35,990	0

between 0.57% and 2.4% for $70^\circ \le \theta_s \le 80^\circ$, between 2.4% and 4.2% for $80^\circ \le \theta_s \le 82.5^\circ$, between 4.2% and 6.6% for $82.5^\circ \le \theta_s \le 85^\circ$, between 6.6% and 23.8% for $85^\circ \le \theta_s \le 87.5^\circ$. These estimations show at what values of θ_s is necessary to use Eq. 5.5.

5.2 Integral Multiplicity and the Sensitivity of Neutron Monitors, Muon and Electron-Photon Detectors to High Energy Solar Neutrons (Obtained by Using Experimental Data of the Geomagnetic Effects of Charged Particles)

It is necessary to take into account that primary high energy neutrons in the nuclear interactions in the Earth's atmosphere generate not only secondary neutrons but also protons and pions; the decay of pions generates muon and electron-photon components. It means that high energy solar neutrons can be detected by not only neutron monitors, but also by muon detectors (by ground and underground muon telescopes, by ionization chambers shielded by 10 cm Pb) and by detectors of electron-photon component.

5.2.1 On the Possibility for Use, for High-Energy Solar Neutrons Data, of Ground Detectors Coupling Functions, Integral Multiplicity and Sensitivity Obtained for Charged Particles of Galactic and Solar Cosmic Rays

For high energy solar neutrons (more than few GeV) it is necessary to take into account all four processes because in this case the propagation determined by nuclear-meson cascade and calculations became very complicated. From other side, if we take into account the charge invariance of neutron and proton relative to nuclear interactions, we can expect that in the first approximation the propagation of high energy neutrons in the Earth's atmosphere will be the same as protons with the same energy (when we can negligible by ionization energy losses on the average path for nuclear interaction, i.e. about

$$\Delta E_p = \left(2\text{MeV}/\left(\text{g/cm}^2\right)\right) \times 60\text{g/cm}^2 = 0.12\ \text{GeV} \tag{5.6}$$

in comparison with kinetic energy of proton). It means that in the first approximation for research of high energy solar neutron propagation in the atmosphere we can use results of experimental determinations of these parameters (as well as results of theoretical calculations of nuclear-meson cascade in the Earth's atmosphere, see Section 5.3) made for charged particles of galactic and solar cosmic rays.

5.2.2 Possible Use, for High-Energy Solar Neutrons Data, of Experimental Determinations of Galactic and Solar Cosmic Ray Coupling Functions and Integral Multiplicity

As was shown in Dorman (M1957, M1963a, b, M1974) by measurements of latitude effect of neutron, muon, electron-photon or some other cosmic ray components it is possible to determine experimentally polar coupling functions $W_o^i(R_k,h)$ of these components for cosmic rays arrived mostly from vertical direction:

$$W_o^i(R_k,h) = -\frac{\partial N^i(R_k,h)}{N_o^i(h)\partial R_k}, \tag{5.7}$$

where index i determined the type of component (neutron, muon, electron-photon or some other cosmic ray component), k determined the point with vertical cut-off rigidity R_k and intensity $N^i(R_k,h)$ on the level with atmospheric pressure h and $N_o^i(h)$ is the cosmic ray intensity of i-th component at $R_k = 0$. In Eq. 5.7 R_k changed from 0 to the maximum of vertical cut-off rigidity on the Earth $R_k \approx 15$ GV. From other side according to Dorman (M1957, M1963a, b)

$$W_o^i(R,h) = \frac{D(R)m^i(R,h)}{N_o^i(h)}, \tag{5.8}$$

where $D(R)$ is the differential rigidity spectrum of cosmic ray particles and $m^i(R,h)$ is the integral multiplicity of i-th component on the level h versus of primary particle rigidity R. In Eq. 5.8 R changed from 0 to possible maximum rigidity in cosmic rays (it means about ∞). Coupling functions $W_o^i(R,h)$ are normalized according to condition

$$\int_0^\infty W_o^i(R,h)dR = 1 \tag{5.9}$$

and have following main properties:

$$1)\quad W_o^i(R,h) \to 0 \quad at \quad R \to 0 \tag{5.10}$$

because at $R \to 0$ in Eq. 5.8 $m^i(R,h) \to 0$ and $D(R) \to 0$;

$$2)\quad W_o^i(R,h) \propto R^{-(\gamma-\beta_i(h))} \quad \text{at} \quad R >> R_{k\max} \tag{5.11}$$

because at $R >> R_{k\max}$ in Eq. 5.8 rigidity spectra of primary particles $D(R) \propto R^{-\gamma}$ and $m^i(R,h) \propto R^{\beta_i(h)}$, where $\beta_i(h)$ is very near to 1 according to direct measurements of $m^i(R,h)$ for electron-photon, muon and neutron components in EAS (external atmospheric showers) and according to theoretical calculations of nuclear-meson cascade in the Earth's atmosphere from high energy primary particles (see review in Khristiansen M1975; Dorman M1975a, b).

From comparison of Eqs. 5.7 and 5.8 we obtain:

$$m^i(R_k, h) = -\frac{\partial N^i(R_k, h)/\partial R_k}{D(R_k)}. \tag{5.12}$$

Equation 5.12 shows that by measurements of cosmic ray intensity of i-th component $N^i(R_k, h)$ on the level with atmospheric pressure h versus the vertical cut-off rigidity R_k and by measurements in the same time of the differential rigidity spectrum of primary cosmic ray particles $D(R)$ we can experimentally determine $m^i(R, h)$ – the integral multiplicity of i-th component on the level h versus primary particle rigidity R at least in the interval from 0 to about 15 GV (by using East–West geomagnetic effect for inclined cosmic ray measurements it is possible to extend this interval in about two to three times). We must take into account that for solar neutrons according to Eq. 5.6 it can be used only the rigidity interval greater than about 2–3 GV. It is necessary to take into account also that in Eq. 5.12 $N^i(R_k, h)$, R_k and $D(R)$ must be measured in accordance units. For example, if the intensity$N^i(R_k, h)$ was measured in time intervals Δt by i-th type detector with effective area S in units i-th particles/(Δt $\times$ S) and R_k is in GV, then measured on satellites or on balloons $D(R)$ must be recalculate to the units particles $\times$ $S^{-1}.\Delta t^{-1}.GV^{-1}$. The integral multiplicity $m^i(R, h)$ according to Eq. 5.12 determines the sensitivity of i-th type detector on the level with atmospheric pressure h to primary charged particles versus the rigidity from 0 to about 15 GV, and determines also the sensitivity of the same i-th type detector to primary solar neutrons versus neutron energy which corresponds proton rigidity greater than 2–3 GV.

On the basis of many latitudinal measurements of $N^i(R_k, h)$ and measurements of time variations of galactic and solar cosmic rays in dependence of cut-off rigidity R_k and for several levels of observation h were determined coupling functions $W^i_o(R, h)$ for many types of detectors (see review in Dorman M1957, M1963a, b, M1974, M1975a, b, M2004). From these data it is also possible to determine according to Eq. 2.3 the integral multiplicity $m^i(R, h)$:

$$m^i(R, h) = \frac{N^i_o(h) W^i_o(R, h)}{D(R)}. \tag{5.13}$$

5.2.3 *Cosmic Ray Geomagnetic Effects and the Determination of Coupling Functions*

In Fig. 5.1 are shown coupling functions on the geomagnetic latitude 50° for neutron, total ionization and muon components for different levels of observation

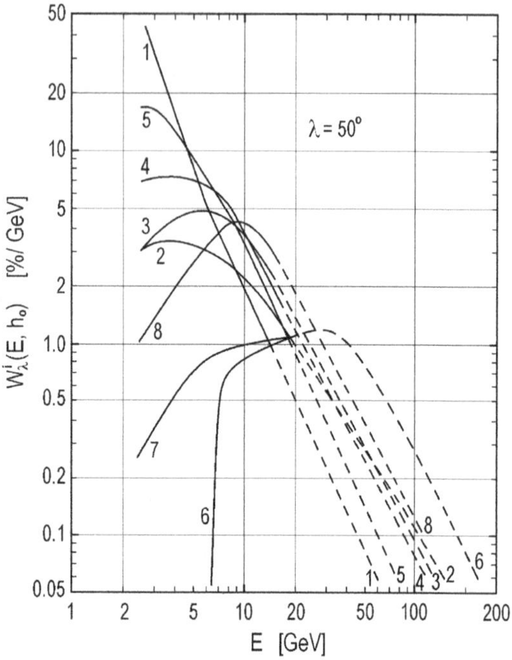

Fig. 5.1 Coupling functions for geomagnetic latitude 50° versus total energy of primary particles: **1**, **2** and **8** for ionizing component at top of the atmosphere, at altitude 4.3 and 10 km; **3**, **4** and **5** for neutron component at sea level, at level with **h** ≈ 700 g/cm^2 and at altitude 10 km; **6** and **7** for muon component at sea level corrected and not corrected for the influence of latitude meteorological effects (According to Dorman M1957)

obtained from data on geomagnetic effects and in Fig. 5.2 for ionization chamber shielded by 10 cm Pb and for unshielded vertical telescope on sea level obtained by taking into account observation data of the great solar cosmic ray event of February 23, 1956 in Dorman (M1957).

Webber and Quenby (1959) from geomagnetic effects with taking into account results of Quenby and Webber (1959) of trajectory calculations of vertical cut-off rigidity (with including effects of penumbra) found the differential sensitivities dN/dR (the same as not normalized according to Eq. 5.9 coupling functions) for neutron component on the levels 312, 680 and 1,000 g/cm^2, and for ionizing component on the levels 312 and 1,000 g/cm^2 (see Fig. 5.3).

For determining coupling functions can be used results of latitude measurements of different cosmic ray components obtained by Simpson (1948), Simpson and Uretz (1949, 1953), Rose and Katzman (1956), Rose et al. (1956), Kodama and Miyazaki (1957), Skorke (1958), Carmichael et al. (1965), Dorman et al. (1967d, e, 1968a, b), Blokh et al. (1974), Raubenheimer and Stoker (1974), Aleksanyan et al. (1982), Potgieter et al. (1980), Stoker (1993), Clem and Dorman (2000). Coupling functions for vertical direction of primary particle arriving was found on the basis of latitude measurements of neutron, ionizing and muon components by Dorman et al. (1966, 1967a, 1970a, b), Granitsky et al. (1966, 1967, 1968b), Avdeev et al. (1973), Villoresi et al. (2000), Iucci et al. (2000), Dorman et al. (2000a), and review in Chapter 3 in Dorman (M2004).

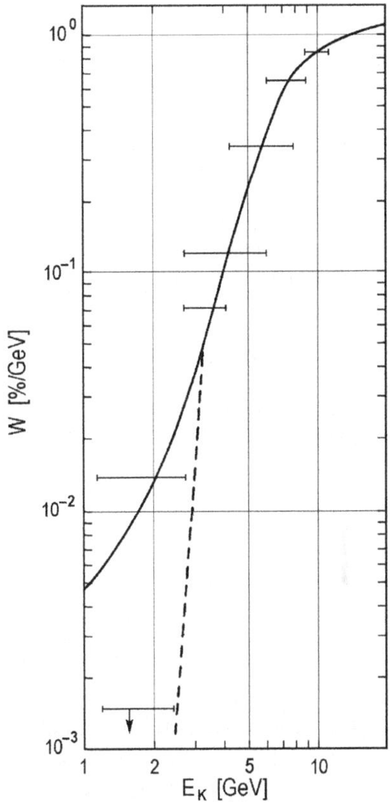

Fig. 5.2 Coupling functions in the small energy region for shielded by 10 cm Pb ionization chamber (*full curve*) and unshielded telescope measuring the total ionizing component (*dashed curve*). Determined on the basis of observation data in the event February 23, 1956 in Dorman (M1957)

5.2.4 Analytical Approximation of Coupling Functions and Integral Multiplicities

As it was shown in Dorman (1969, 1972, 1974), experimental data on coupling functions can be approximated with good accuracy analytically by the function:

$$W_o^i(R,h) = a_i(h)k_i(h)R^{-(k_i(h)+1)}\exp\left(-a_i(h)R^{-k_i(h)}\right), \tag{5.14}$$

where parameters $a_i(h)$ and $k_i(h)$ are determined by some statistical method from comparison of geomagnetic effects experimental data with the approximation function derived from Eq. 5.14 by Eq. 5.7 and characterized the dependence of relative cosmic ray intensity $N^i(R,h)/N_o^i(h)$ from rigidity R:

$$N^i(R,h)/N_o^i(h) = 1 - \exp\left(-a_i(h)R^{-k_i(h)}\right). \tag{5.15}$$

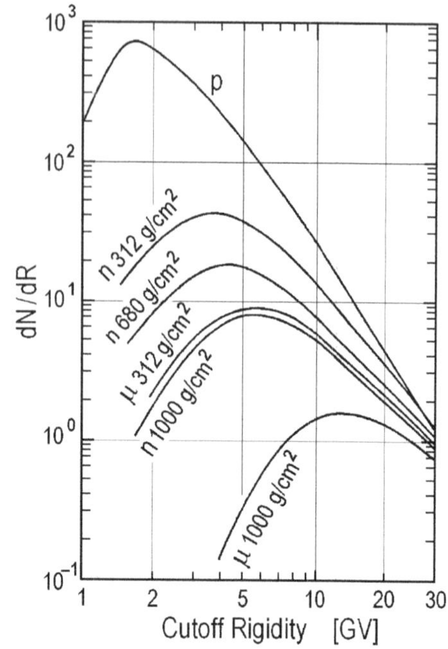

Fig. 5.3 The differential sensitivities *dN/dR* obtained by Webber and Quenby (1959) from geomagnetic effects with taking into account results of Quenby and Webber (1959) of trajectory calculations of vertical cut-off rigidity (with including effects of penumbra) for ionizing component (**μ**) at levels 312 and 1,000 g/cm^2, for neutron component (**n**) at levels 312, 680 and 1,000 g/cm^2 and for the primary radiation (**p**) (According to Webber and Quenby 1959)

It is easy to see that Eq. 5.14 at any positive values of parameters $a_i(h)$ and $k_i(h)$ satisfied all discussed above conditions (see Eqs. 5.9, 5.10, and 5.11) for coupling functions $W_o^i(R,h)$. The physical meaning of the parameter $k_i(h)$ can be seen from the comparison of Eq. 5.14 with Eq. 5.11:

$$k_i(h) = \gamma - \beta_i(h) - 1, \tag{5.16}$$

where $\beta_i(h)$ characterized the asymptotic dependence of integral multiplicity $m^i(R,h)$ from R in the high energy region and can be determined independently from latitude measurements:

$$\beta_i(h) = \gamma - k_i(h) - 1. \tag{5.17}$$

The comparison of $\beta_i(h)$ determined from latitude measurements according to Eq. 5.17 and determined from EAS measurements shows a good agreement in the frame of statistical errors (Dorman M1975a, M2004); this shows that in the first approximation the analytical description (5.14) of $W_o^i(R,h)$ is correct also for high energy region. If $W_o^i(R,h)$ is measured in units GV^{-1} and R in GV, then the parameter $a_i(h)$ have dimensions GV^{k} and parameter $k_i(h)$ is free dimensions. Parameters $a_i(h)$ and $k_i(h)$ determined the position of maximum value of $W_o^i(R,h)$:

$$R^i_{\max}(h) = [a_i(h)k_i(h)/(k_i(h)+1)]^{1/k_i(h)} \tag{5.18}$$

$$W^{i}_{o\max}(h) = (k_i(h)+1)\left(\frac{k_i(h)+1}{a_i(h)k_i(h)}\right)^{1/k_i(h)} \exp\left(-\frac{k_i(h)+1}{k_i(h)}\right). \tag{5.19}$$

Let us take into account that cosmic ray primary differential rigidity spectrum on the top of the atmosphere can be described by the powerful function multiplied on the function approximately characterized solar cycle modulation (Dorman M1975b; Zusmanovich M1986):

$$D(R) = B \times (b+R)^{-\gamma} \exp(-M/R), \tag{5.20}$$

where B is a constant with dimensions in accordance with $N^{i}_{o}(h)$, constant $b \approx 1$ GV characterized the change of spectrum in small energy region and constant M is the modulation parameter. According to Dorman and Dorman (1967a, b), Zusmanovich (M1986) and Belov et al. (1990), the parameter $M \approx 0.3\text{–}0.4$ GV in the minimum of solar activity. In the maximum of solar activity the modulation for positive and negative particles of cosmic rays will be different and will depend on the direction of the general magnetic field of the Sun (caused by drift effects): $M \approx 1.6$ GV for protons, nuclei and positrons in cycles 1955–1965, 1975–1985, 1995–2005 and for electrons and antiprotons in cycles 1965–1975, 1985–1995; $M \approx 1.2$ GV for protons, nuclei and positrons in solar cycles 1965–1975, 1985–1995 and for electrons and antiprotons in cycles 1955–1965, 1975–1985, 1995–2005. Therefore Eq. 5.20 approximately reflects the 11-year and 22-year modulation in cosmic ray flux in dependence of energy. On the basis of Eqs. 5.13, 5.14, and 5.20 we obtain the analytical approximation for the integral multiplicity:

$$\begin{aligned} m^{i}(R,h) = {} & \frac{a_i(h)k_i(h)N^{i}_{o}(h)}{B} R^{-(k_i(h)+1)}(b+R) \\ & \times \exp\left(-a_i(h)R^{-k_i(h)} + MR^{-1}\right) \end{aligned} \tag{5.21}$$

It is necessary to take into account that in Eq. 5.21 parameters $a_i(h), k_i(h), N^{i}_{o}(h)$ and modulation parameter M must be determined in the same period of solar activity cycle as considered solar neutron event.

5.2.5 Experimental Data on Coupling Functions for the Neutron Component Represented in Analytical Form

In the first the analytical approximation described by Eq. 5.15 to the cosmic ray latitude dependence was used in Dorman et al. (1970a) for data obtained by aircraft neutron monitor on the pressure level about 300 mb in January–February, 1966 (i.e. near the minimum of solar activity). It was found $a_i(h) = 6.49$ and $k_i(h) = 1.41$. The example of determining parameters $a_i(h)$ and $k_i(h)$ from latitudinal measurements is given by Fig. 5.4. In Fig. 5.4 is shown the rigidity dependence of neutron component intensity $N^i(R_k, h)$ of the 1976 sea level latitude survey of Potgieter et al. (1980), normalized to $N^i_o(h) = 100$, with a least square regression fitting to the measurement data by so called Dorman function described by Eq. 5.15 in Moraal et al. (1989).

For the 1976 sea level latitude survey it was found $a_i(h) = 8.427$, $k_i(h) = 0.8935$. In Stoker (1993, 1994) and Stoker et al. (2000) were recalculated data of many latitude expeditions on sea level and on aircraft altitudes to the conditions of minimum of solar activity and determined the Dorman function's parameters $a_i(h)$, $k_i(h)$ and $N^i_o(h)$ (normalized to $N^i_o(h) = 100$ for sea level) for pressure levels from 760 up to 226 mm Hg (see Table 5.2).

In Table 5.3 are represented results of determination parameters $a_i(h)$ and $k_i(h)$ for the 3NM-64 neutron supermonitor, found by Aleksanyan et al. (1979a) on the basis of 1967–1977 latitude expeditions on the research vessel 'Academician Kurchatov' with a least square regression fitting to experimental data by function described by Eq. 5.15. Then in accordance with Eqs. 5.18 and 5.19 were found values $R^i_{\max}(h)$ and $W^i_{o\max}(h)$.

By data listed in Tables 5.2 and 5.3 can be found integral multiplicity for neutron component according to Eq. 5.21.

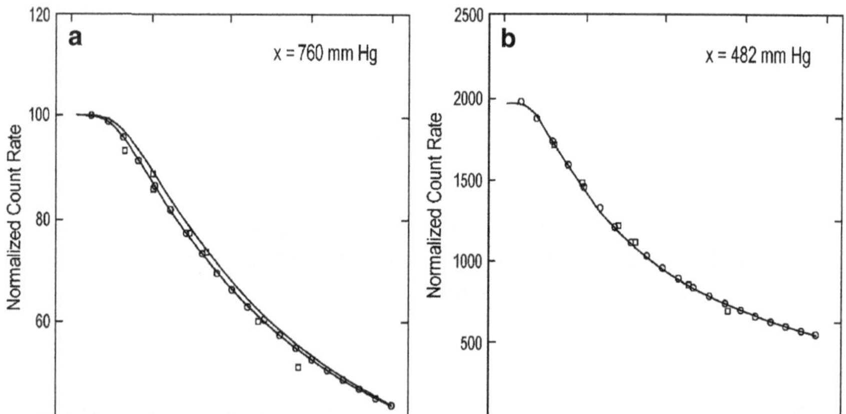

Fig. 5.4 (**a**) The rigidity dependence of neutron component intensity $N^i(R_k, h)$ of the sea level latitude survey approximated by Dorman function described by Eq. 5.15 for 1976 (*lower curve*) and for 1986–1987 (*upper curve*); (**b**) The same for the pressure level 482 mm Hg (survey for 1976) (According to Stoker 1993)

Table 5.2 Parameters $a_i(h)$, $k_i(h)$ and $N_o^i(h)$ of the Dorman function for different pressure levels (According to Stoker 1994)

Pressure (mm Hg)	Notes	$a_i(h)$	$k_i(h)$	$N_o^i(h)$
760	1	8.4275	0.8935	100.00 (fixed)
730	2	8.179 ± 0.02	0.9063 ± 0.0003	136.7 ± 0.2
696	3	8.095 ± 0.08	0.9363 ± 0.0035	195.0 ± 0.3
660	4	8.050 ± 0.12	0.9475 ± 0.0045	284.4 ± 0.6
605	5	7.884 ± 0.14	0.9737 ± 0.006	510.0 ± 1.7
575	6	7.824 ± 0.16	0.9882 ± 0.006	701.5 ± 2.5
550	7	7.778 ± 0.20	1.0002 ± 0.007	916 ± 4
518	8	7.739 ± 0.22	1.0162 ± 0.008	1,286 ± 6
504	9	7.723 ± 0.25	1.0232 ± 0.009	1,492 ± 10
482	10	7.707 ± 0.27	1.0347 ± 0.010	1,883 ± 14
450	11	7.693 ± 0.29	1.0520 ± 0.011	2,636 ± 18
406	12	7.713 ± 0.32	1.0779 ± 0.012	4,169 ± 28
350	13	7.775 ± 0.37	1.1144 ± 0.014	7,402 ± 74
295	14	7.914 ± 0.46	1.1572 ± 0.017	12,813 ± 155
258	15	8.104 ± 0.55	1.1898 ± 0.020	18,358 ± 300
226	16	8.286 ± 0.65	1.2209 ± 0.023	24,874 ± 450

Notes: 1 and 2. For sea level and near sea level stations (760–730 mm Hg); 3. For Mt. Wellington (725 m); 4. For Calgary (1,128 m), Potchefstrum (1,351 m), Tsumeb (1,240 m); 5. For Mt. Washington (1,909 m), Hafelekar (2,290 m), 6. For Mexico City (2,274 m), Sulphur Mt. (2,283 m); 7. For Lomnicky Stit (2,635 m); 8. For Huancayo (3,400 m), Haleakala (3,052 m), Mt. Norikura (2,770 m); 9. For Tyan Shan (3,340 m), Climax (3,400 m), South Pole (2,820 m); 10. For Jungfraujoch (3,570 m); 11. for Vostok (3,488 m); 12. For Chacaltaya (5,200 m); 13. For 20,000 ft altitude; 14. For 24,000 ft altitude; 15. For 27,000 ft altitude; 16. For 30,000 ft altitude.

Table 5.3 Parameters $a_i(h)$ and $k_i(h)$ of Dorman function, and values $R_{\max}^i(h)$ and $W_{o\max}^i(h)$ for the sea level NM-64 neutron supermonitor (According to Aleksanyan et al. 1979a)

Year	$a_i(h)$	$k_i(h)$	$R_{\max}^i(h)$ (GV)	$W_{o\max}^i(h)$ (%/GV)
1967	7.45	0.86	3.6	4.6
1968	7.57	0.85	4.3	4.4
1969	7.79	0.83	4.5	4.1
1971	7.73	0.81	5.0	4.2
1974	7.28	0.83	4.2	4.8
1975	7.48	0.85	4.3	4.91
1976	7.92	0.87	4.5	4.86
1977	8.10	0.88	4.5	4.88

5.2.6 On the Coupling Functions for the Lead-Free Neutron Monitor

For solar neutron measurements can be useful neutron monitor without lead which is more sensitive to small energy neutrons. The latitude measurements and determination of coupling functions for this type of neutron monitor were made in many expeditions by Mischke et al. (1973), Aleksanyan et al. (1977, 1979b, 1985), Dorman et al. (1979, 1984), Stoker and Raubenheimer (1985). In Table 5.4 are

Table 5.4 Parameters $a_i(h)$ and $k_i(h)$ of Dorman function and values $R^i_{\max}(h)$ and $W^i_{o\max}(h)$ for the sea level 3NM-64 neutron supermonitor and 4NM-D lead-free monitor (According to Aleksanyan et al. 1979b)

Type of monitor	$a_i(h)$	$k_i(h)$	$R^i_{\max}(h)$ (GV)	$W^i_{o\max}(h)$ (%/GV)
3NM-64	8.10 ± 0.05	0.88 ± 0.02	4.5	4.85
Lead-free	7.13 ± 0.04	0.85 ± 0.02	4.0	5.19

listed results of determination parameters $a_i(h)$ and $k_i(h)$, obtained by Aleksanyan et al. (1979b) with a least square regression fitting to experimental data of sea level survey on the research vessel 'Academician Kurchatov' in 1977 by function described by Eq. 5.15. Then in accordance with Eqs. 5.18 and 5.19 were found values $R^i_{\max}(h)$ and $W^i_{o\max}(h)$. Measurements were made by 3NM-64 neutron supermonitor and by lead-free monitor (four large neutron proportional counters SNM-15 with $^{10}BF_3$ gas in 18 mm thick polyethylene tubes, mounted above neutron supermonitor).

The parameters $a_i(h)$ and $k_i(h)$ found by Stoker (1993, 1994) for 1NM-64 neutron supermonitor and for 1NM lead-free monitor on the basis of experimental data of 1976 sea level survey in the frame of statistical errors are in agreement with data in Table 5.4.

5.2.7 On the Coupling Functions for Multiple Neutrons in a Neutron Supermonitor

For solar neutron research can be useful also measurements of multiple neutrons because its sensitivity dependence from primary particle energy is different and with increasing of measured multiplicity monitor became more sensitive to higher energy. The latitude measurements of multiple neutrons were made by Dyring and Sporre (1965, 1966), Kodama and Ohuchi (1968), Granitsky et al. (1968a), Aleksanyan et al. (1979a). Figure 5.5 shows the latitude effects of different multiplicities from $m = 1$ to $m \geq 6$ obtained by Kodama and Ohuchi (1968) with a 3NM-64 neutron supermonitor in the period December 1966 to April 1967 on the journey Japan–Australia–Antarctica–South Africa–Ceylon–Japan (the corrections for barometric effect were made by barometric coefficients determined during stops of the ship at Syowa, Freemantle and Tokyo). Table 5.5 listed the results of determination parameters $a_i(h)$ and $k_i(h)$ for the different neutron multiplicities $m = 1, 2, 3, 4, 5$, and ≥ 6 in the 3NM-64 neutron supermonitor, found by Aleksanyan et al. (1979a) with a least square regression fitting to experimental data by function described by Eq. 5.15. There are used data obtained in latitude expeditions on the research vessel 'Academician Kurchatov' in 1975, 1976 and 1977 (near minimum of solar activity). Then in accordance with Eqs. 5.18 and 5.19 were found values $R^i_{\max}(h)$ and $W^i_{o\max}(h)$. For comparison in Table 5.5 were shown also values of mentioned

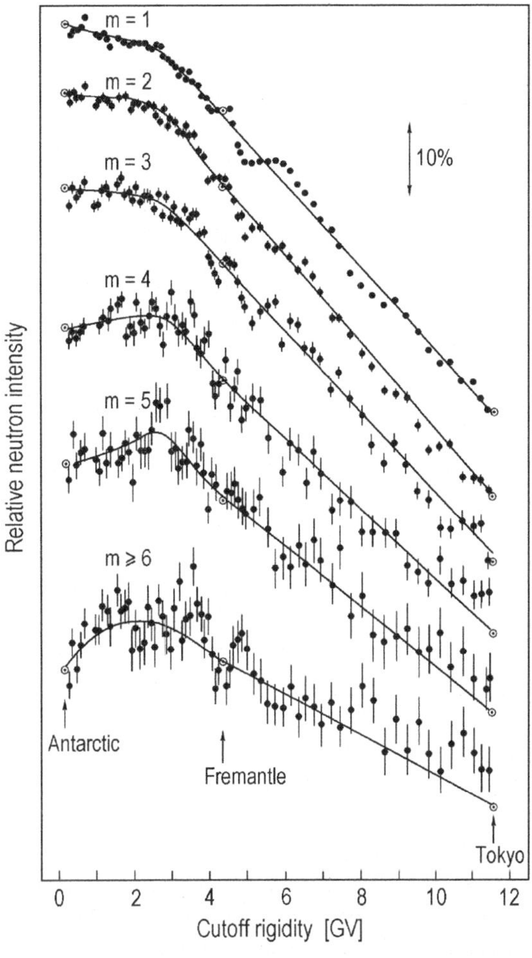

Fig. 5.5 The latitude effects of different multiplicities from $m = 1$ to $m \geq 6$ obtained with a 3NM-64 neutron supermonitor in the period December 1966 to April 1967 on the journey Japan–Australia–Antarctica–South Africa–Ceylon–Japan (According to Kodama and Ohuchi 1968)

above parameters for total neutron component. The statistical errors of data increase very much with increasing of multiplicity m, but nevertheless it is possible to see from Table 5.5 the tendency of increasing $R^i_{\max}(h)$ and decreasing $W^i_{o\max}(h)$ with increasing m. This means that with increasing of multiplicity m the neutron supermonitor became more sensitive to primary particles with higher rigidity (or in case of solar neutrons became more sensitive to neutrons with higher energy).

In described above results of determination parameters $a_i(h)$ and $k_i(h)$ in the analytical approximation (see Eq. 5.15) it was supposed that $N^i_o(h)$ can be measured experimentally and we have only two parameters to determine, for example, by the method of least squares:

Table 5.5 Parameters $a_i(h)$ and $k_i(h)$ in Dorman function and values $R^i_{\max}(h)$ and $W^i_{o\max}(h)$ for the different neutron multiplicities and for total neutron component of the sea level 3NM-64 supermonitor (According to Aleksanyan et al. 1979b)

Year	Multiplicity m	$a_i(h)$	$k_i(h)$	$R^i_{\max}(h)$ (GV)	$W^i_{o\max}(h)$ (%/GV)
1975	1	6.94 ± 0.05	0.84	3.9	5.21
	2	7.74	0.89	4.3	5.28
	3	7.89	0.83	4.6	4.34
	4	6.28	0.66	4.0	3.3
	5	7.82	0.66	5.5	2.4
	≥6	10.0	0.59	9.5	1.1
	Total component	7.48	0.85	4.3	4.91
1976	1	7.49 ± 0.04	0.84	4.3	4.76
	2	9.7	1.00	4.8	5.58
	3	10.61	0.99	5.4	4.96
	4	11.31	0.95	6.0	4.1
	5	10.32	0.85	6.0	3.3
	≥6	8.96	0.69	6.5	2.2
	Total component	7.92	0.87	4.5	4.86
1977	1	7.67 ± 0.05	0.87	4.3	5.05
	2	8.13	0.89	4.6	5.0
	3	9.36	0.89	5.3	4.27
	4	7.68	0.73	5.0	3.2
	5	6.06	0.55	4.0	2.3
	≥6	6.08	0.36	4.0	0.8
	Total component	8.10	0.88	4.5	4.88

$$\ln\left[\ln\left(N^i_o(h)/\left(N^i_o(h) - N^i(R,h)\right)\right)\right] = \ln a_i(h) - k_i(h)\ln R, \tag{5.22}$$

which follows from Eq. 5.15.

The problem is that in many cases the data of measurements in polar zones are absent and we must consider $N^i_o(h)$ as third unknown parameter. As it was mentioned in Aleksanyan et al. (1985), it appears impossible to determine simultaneously three parameters $N^i_o(h)$, $a_i(h)$ and $k_i(h)$ in Eq. 5.15 by the method of least squares because of divergence of the iteration process.

More successive method was used by Aleksanyan et al. (1985) for analysis experimental data obtained in 1982 on the research vessel 'Academician Kurchatov' by 2NM-64 neutron supermonitor (total component and m = 1, 2, 3, 4; data for $m = 5$ and $m \geq 6$ were also obtained but they had too great statistical errors) and by 4NM-D (leed-free monitor). In accordance with Demidovich et al. (M1962) it was used the expansion of Eq. 5.15 in a Taylor series with an accuracy to first-order terms at the point of the initial approximation of parameters:

$$\begin{aligned} N = {} & \left[N_0\left(1 - \exp\left(-aR^{-k}\right)\right)\right]_0 + \left[1 - \exp\left(-aR^{-k}\right)\right]_0 \Delta N_0 \\ & + \left[N_0 R^{-k}\exp\left(-aR^{-k}\right)\right]_0 \Delta a + \left[N_0 a R^{-k}\exp\left(-aR^{-k}\right)\ln R\right]_0 \Delta k, \end{aligned} \tag{5.23}$$

where the expressions in the square brackets are taken at the point of the initial approximation $N_0 = N_0^{(0)}, a = a^{(0)}, k = k^{(0)}$. Then one minimizes the sum of the squares of the differences of the calculated values of N according to Eq. 5.23 and experimental values of N:

$$S = \sum (q\Delta N_o + N_o \Delta a - N_o ar \ln \rho \Delta k - N_o q - N) = \min, \tag{5.24}$$

where

$$q = 1 - \exp\left(-aR^{-k}\right), \rho = R^{-k} \exp\left(-aR^{-k}\right). \tag{5.25}$$

The condition $\partial S/\partial N_0 = 0$, $\partial S/\partial a = 0$, $\partial S/\partial k = 0$ of the minimum in Eq. 5.24 gives the system of equations:

$$\begin{cases} \Delta N_0 \sum q^2 + N_0 \Delta a \sum qr - N_0 a \Delta k \sum qr \ln R = \sum q(N - N_0 q), \\ \Delta N_0 \sum qr + N_0 \Delta a \sum r^2 - N_0 a \Delta k \sum r^2 \ln R = \sum r(N - N_0 q), \\ \Delta N_0 \sum qr \ln R + N_0 \Delta a \sum r^2 \ln R - N_0 a \Delta k \sum (r \ln R)^2 = \sum r(N - N_0 q) \ln R, \end{cases} \tag{5.26}$$

the solution of which $\Delta N_0, \Delta a, \Delta k$ determines the next approximation of the parameters:

$$N_0^{(1)} = N_0^{(0)} + \Delta N_0, a^{(1)} = a^{(0)} + \Delta a, k^{(1)} = k^{(0)} + \Delta k. \tag{5.27}$$

This cycle of operations was repeated for subsequent approximations until stable values of the parameters N_o, a, k are obtained. Results are shown in Fig. 5.6 (analytical approximation to experimental data on latitude effect) and in Table 5.6 (parameters of analytical approximation Eq. 5.15 and values determined the position of the maximum of coupling functions according to Eqs. 5.18 and 5.19).

Table 5.6 shows a sufficient difference in the coupling functions for total component of 2NM-64 supermonitor and 4NM-D lead-free monitor (see Section 5.2.6), and for 2NM-64 the moving of the position of maximum of coupling function to higher rigidity with increasing neutron multiplicity m.

5.2.8 On the Coupling Functions for Cosmic Ray Inclined Telescopes

In the relation with the problem of integral multiplicity for solar neutrons arriving to the Earth atmosphere under different zenith angles, it will be possible to use coupling functions obtained from geomagnetic effects by Rao and Sarabhai (1961) for inclined muon telescopes from West and East under zenith angle 45° (on the basis of unidirectional measurements of the cosmic ray latitude effect by Johnson and Read (1937)). Coupling functions for inclined muon component on the basis of expeditions measurements of latitude and East–West geomagnetic effects was determined also in Dorman and Masaryuk (1967) for zenith angle 33° and in Dorman et al. (1967b, c) for zenith angles 0°, 33° and 53°.

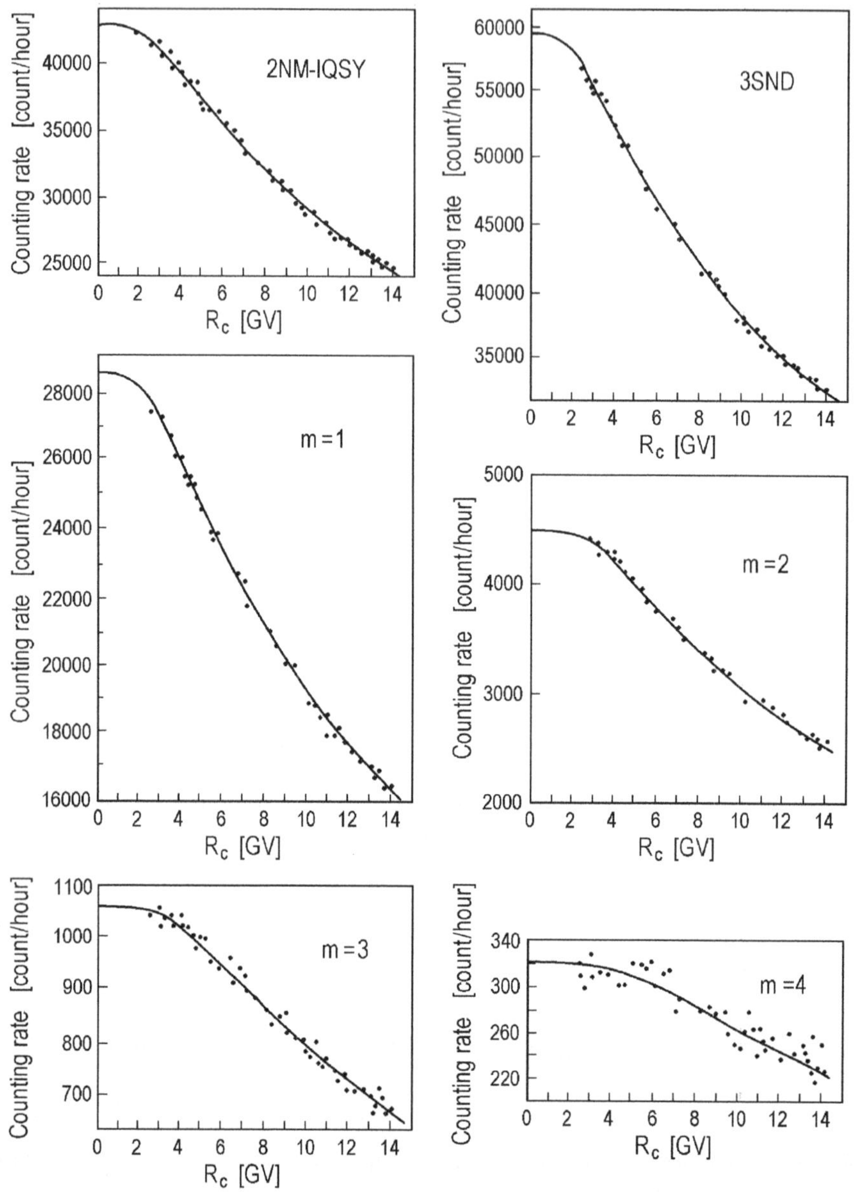

Fig. 5.6 Experimental data (*points*) obtained in 1982 on the research vessel 'Academician Kurchatov' by 2NM-64 neutron supermonitor (total component and $m = 1, 2, 3, 4$) and by 4NM-D (lead-free monitor). *Curves* represent analytical approximation by Dorman function described by Eq. 5.15 (According to Aleksanyan et al. 1985)

Table 5.6 Parameters $N_o^i(h)$, $a_i(h)$ and $k_i(h)$ in Dorman function and values $R_{\max}^i(h)$ and $W_{o\max}^i(h)$ for 2NM-64 supermonitor (total neutron component and m = 1, 2, 3, 4) and for 4NM-D lead-free monitor (According to Aleksanyan et al. 1985)

Detector	Component	$N_o^i(h)$	$a_i(h)$	$k_i(h)$	$R_{\max}^i(h)$ (GV)	$W_{o\max}^i(h)$ (%/GV)
2NM-64	Total	42,930	8.32	0.866	4.76	4.55
	$m = 1$	28,592	7.16	0.808	4.21	4.58
	$m = 2$	4,481.8	10.26	0.951	5.43	5.99
	$m = 3$	1,055.5	13.48	0.978	6.96	3.76
	$m = 4$	325.6	20.30	1.060	9.15	3.23
4NM-D	Total	59,076	6.55	0.800	3.80	4.99

5.2.9 Determination of Integral Multiplicity and the Sensitivity of Ground Based Detectors to Geomagnetic Effects Measurements

After founding coupling functions on the basis of geomagnetic effects measurements (see Section 5.2.3) it is possible by help of Eq. 5.12 or Eq. 5.13 to determine the integral multiplicity $m^i(R,h)$. In case when it was used a least square regression fitting to experimental data by Dorman function (Eq. 5.15) and it was found parameters $a_i(h)$ and $k_i(h)$ for the analytical approximation (Eq. 5.14) of the coupling function (see Sections 5.2.5, 5.2.6, 5.2.7, and 5.2.8), it is possible to determine $m^i(R,h)$ by using Eq. 5.21.

Figure 5.7 shows the results of Webber and Quenby (1959) of determination of specific yield function $S^i(R,h)$ (the same as integral multiplicity $m^i(R,h)$) for neutron component on levels 312, 680 and 1,000 g/cm^2 and for ionizing component on levels 312 and 1,000 g/cm^2.

More detail data on $m^i(R,h)$ for sea level neutron component with using data on event February 23, 1956 are shown in Fig. 5.8.

Figure 5.9 shows the integral multiplicity $m^i(R,h)$) for neutron component on sea level according to Lockwood et al. (1974) found on the basis of many sea level latitude cosmic ray surveys and on the data of cut-off rigidity distribution according to trajectory calculations of Shea et al. (1965). For comparison there are shown also $m^i(R,h)$ obtained by Lockwood and Webber (1967), Kodama and Inoue (1970) from cosmic ray latitude surveys, and by Debrunner and Flückiger (1971)) from calculations of cosmic ray propagation in the Earth's atmosphere.

In Fig. 5.10 is shown the comparison between $m^i(R,h)$ for neutron supermonitor NM-64 and for lead-free monitor according to Stoker (1994).

A big analysis of 21 data sets of sea level surveys by neutron monitors, 6 data sets at high altitude surveys on atmospheric depth 307, 602, 680 and 843 g/cm^2 in the period from 1954 to 1986 (when neutron monitor counting rate $J_{MtW}(t)$ changes from 1990 to 2465 in units 64 counts/h) and one set of primary spectrum observed in the minimum of solar activity in 1965 was made by Nagashima et al. (1990).

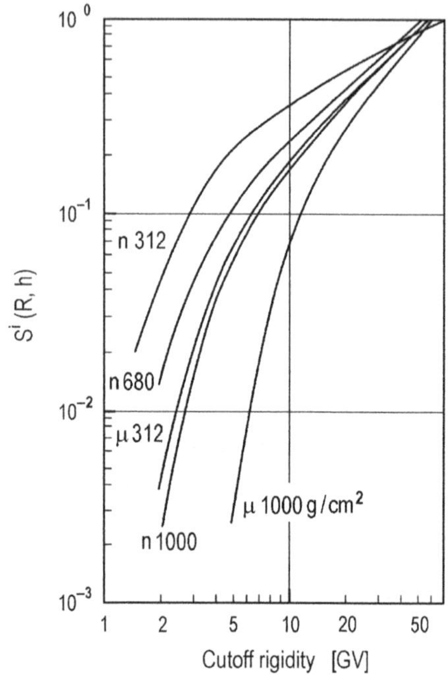

Fig. 5.7 The specific yield function $S^i(R,h)$ (the same as integral multiplicity $m^i(R,h)$) for neutron component (**n**) on levels 312, 680 and 1,000 g/cm^2 and for ionizing component (**μ**) on levels 312 and 1,000 g/cm^2 (According to Webber and Quenby 1959)

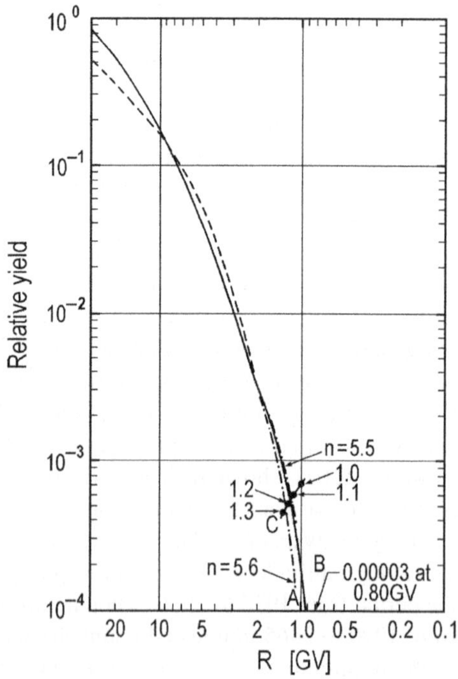

Fig. 5.8 Data on $m^i(R,h)$ for sea level neutron component with using data on event February 23, 1956: *dashed curve* from latitude effect analyses by Webber and Quenby (1959), *full curve* with taking into account penumbra corrections, *point-dash curve* extrapolation using data on February 23, 1956 event (According to Webber 1962)

Fig. 5.9 Specific yield function (the integral multiplicity $m^i(R,h)$) for neutron component on sea level (*full curve*) found in Lockwood et al. (1974) on the basis of many sea level latitude cosmic ray surveys and on the data of cut-off rigidity distribution according to trajectory calculations of Shea et al. (1965). For comparison there are shown also $m^i(R,h)$ obtained by Lockwood and Webber (1967), Kodama and Inoue (1970) from cosmic ray latitude surveys, and by Debrunner and Flückiger (1971) from calculations of cosmic ray propagation in the Earth's atmosphere (According to Lockwood et al. 1974)

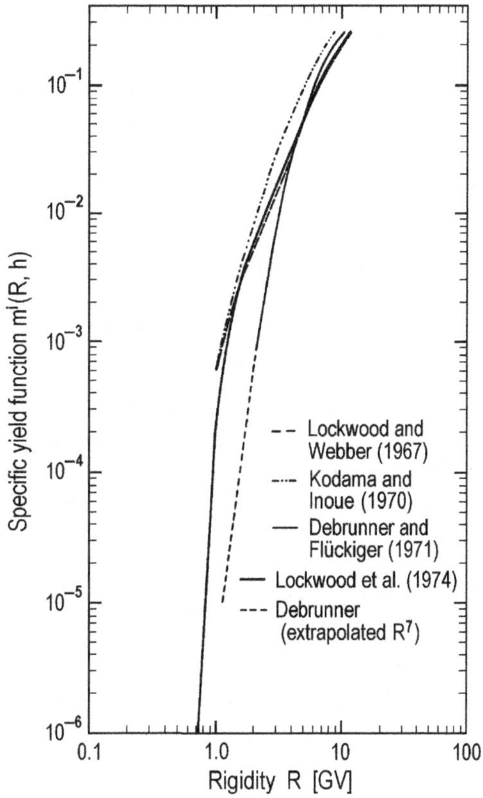

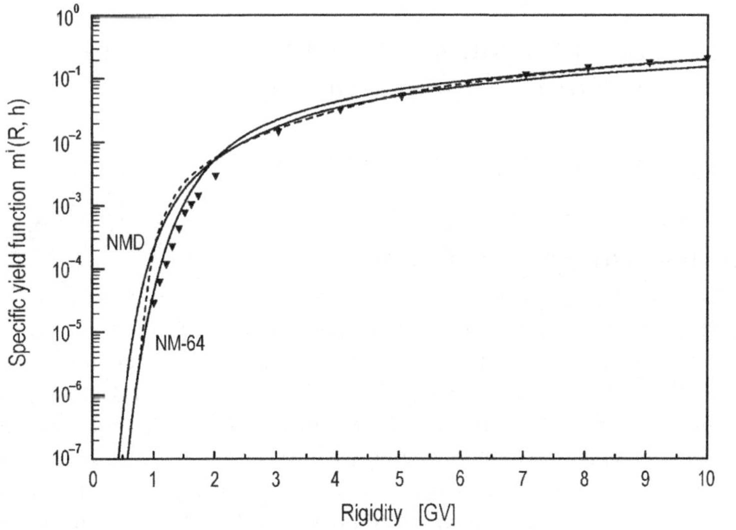

Fig. 5.10 The comparison between $m^i(R,h)$ for neutron supermonitor NM-64 and for lead-free monitor (According to Stoker 1994)

In this analysis by the nonlinear least-squares method were found 14 parameters $A_1 - A_3, B_1 - B_5, C_1 - C_6$ determined simultaneously:

1. The unmodulated rigidity spectrum $D_0(R)$ expressed in terms of the total energy $E = E_k + mc^2$ according to formula

$$D_0(E)dE = A_1\left(A_2 + \left(E/mc^2\right)\right)^{-A_3} d\left(E/mc^2\right), \tag{5.28}$$

2. The modulation function $F(R,t)$, assumed to be of exponential type as

$$F(R,t) = \exp\left(-\frac{B_1 + B_2[1 - (J_{MtW}(t)/J_{MtW}(1965))]^{B_3}}{B_4 + (R/GV)^{B_5}}\right), \tag{5.29}$$

   where $J_{MtW}(1965) = 2{,}465$ in units 64 counts/h, and
3. The yield function (integral multiplicity) $Y(E,H)$ expressed in terms of the total energy $E = E_k + mc^2$ and relative pressure $H = h/(1033\text{g.cm}^{-2})$ as

$$Y(E,H) = \left(E/mc^2\right)^{C_1} \exp\left(-C_2 H^{C_3} - C_4 H^{C_5}\left(\ln\left(E/mc^2\right)\right)^{-C_6}\right). \tag{5.30}$$

The values of parameters $A_1 - A_3, B_1 - B_5, C_1 - C_6$ are found in Nagashima et al. (1990) as following:

$$A_1 = (1.21 \pm 0.02) \times 10^4, A_2 = 0, A_3 = 2.585 \pm 0.004 \tag{5.31}$$

for the unmodulated rigidity spectrum according to Eq. 5.28,

$$\begin{aligned} &B_1 = 1.15 \pm 0.02, B_2 = 14.9 \pm 0.8, B_3 = 1.12 \pm 0.03, \\ &B_4 = 0.097 \pm 0.009, B_5 = 1.02 \pm 0.02 \end{aligned} \tag{5.32}$$

for the modulation function according to Eq. 5.29, and

$$\begin{aligned} &C_1 = 0.0 \pm 0.1,\ C_2 = 2.2 \pm 2.5,\ C_3 = 1.62 \pm 0.96,\ C_4 = 12.7 \pm 5.6, \\ &C_5 = 0.50 \pm 0.09,\ C_6 = 0.42 \pm 0.21 \end{aligned} \tag{5.33}$$

for yield function (integral multiplicity) according to Eq. 5.30.

From Eqs. 5.31, 5.32, and 5.33) one can see that the parameters for the unmodulated rigidity spectrum according to Eq. 5.28 and for the modulation function according to Eq. 5.29 determined with very high accuracy, the relative errors not exceed few percents (the unmodulated rigidity spectrum and spectrum for 1965 as well as the modulation function for 1965 are shown in Fig. 5.11). If we take into account the values of parameters determined by Eq. 5.32 for the modulation

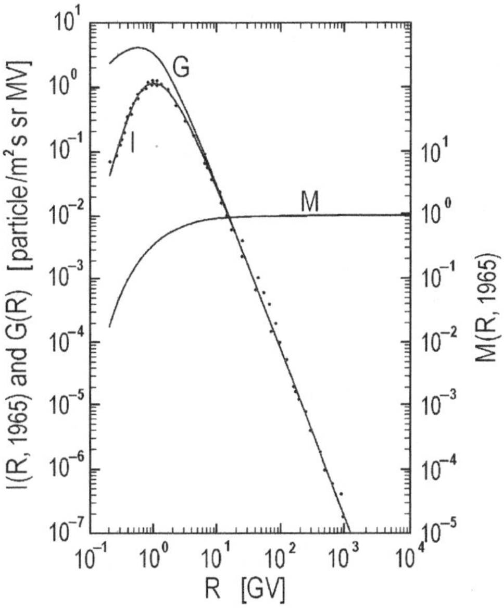

Fig. 5.11 The unmodulated rigidity spectrum and spectrum for 1965 as well as the modulation function for 1965 (According to Nagashima et al. 1990)

function in the approximation described by Eq. 5.29, we can see that it is about the same what we used in Section 5.2.4.

From other side, the parameters for the yield function (integral multiplacity) are determined with relative errors from 20% to more than 100%. We don't know exactly the cause of these great errors and it was not discussed in Nagashima et al. (1990). From our opinion it can be caused by the possible great difference of the character of dependence yield function (integral multiplicity) from primary particle rigidity and from observation pressure level choused in Eq. 5.30 from real one. In any case it will be very important to compare found in Nagashima et al. (1990) yield function (integral multiplicity) with calculated theoretically and determined experimentally by other authors. The obtained results on the yield function and comparison with theoretical calculations made by Dorman and Yanke (1981), and by Murakami (1988) are shown in Fig. 5.12 (from Nagashima et al. 1990).

In Fig. 5.12 the theoretical and experimental curves are normalized at $h = 1{,}033$ g/cm^2 and at $E = 10$ GeV. It is necessary to note that experimental curves in Fig. 5.12 are on several orders lower than obtained theoretically in Dorman and Yanke (1981) and than obtained experimentally by other authors (see Figs. 5.7, 5.8, 5.9, 5.10, 5.11, 5.12, 5.13, 5.14, and 5.15). The made normalization of experimental and theoretical curves in Nagashima et al. (1990) is not correct because the yield function (integral multiplicity) have an absolute sense: number of second particles on observation level per one primary particle. This normalization can only help to compare the character of yield function dependence from primary particle energy.

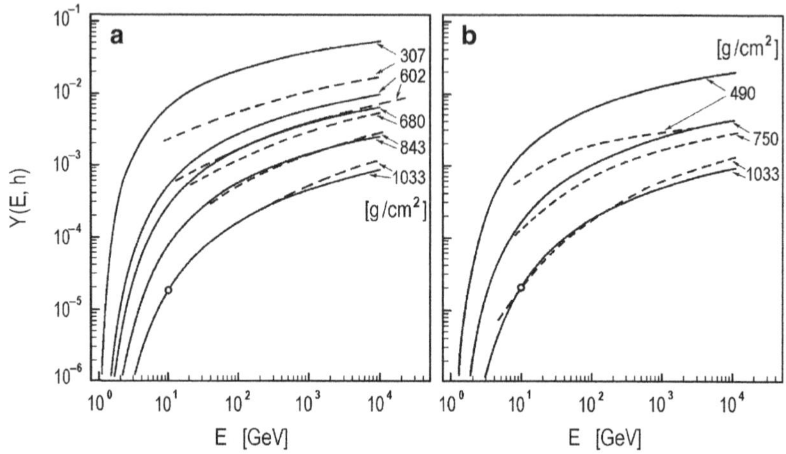

Fig. 5.12 The yield function obtained from latitude surveys (*full curves*), in comparison with theoretical calculations (*dashed curves*) made by Murakami (1988) – panel (**a**), and by Dorman and Yanke (1981) – panel (**b**) (From Nagashima et al. 1990)

5.3 The Integral Multiplicity and Sensitivity of Neutron Monitors and Detectors of Charged Particles to High Energy Solar Neutrons, Obtained by Using Results of Theoretical Calculations of Meson-Nucleon Cascades of Galactic Cosmic Rays

As it was discussed in Section 5.2.1, because of the charge invariance of neutron and proton relative to nuclear interactions, we can expect that in the first approximation the propagation of high energy neutrons in the Earth's atmosphere will be the same as protons with the equal energy. It means that for research of high energy solar neutron propagation in the atmosphere we can use also the results of theoretical calculations of nuclear-meson cascade in the Earth's atmosphere made for charged particles of galactic and solar cosmic rays.

5.3.1 Calculations of Integral Multiplicity for the Muon Component

The review of these calculations was done in Dorman (M1963b, M1974, M1975a, M2004). In Dorman (1961) were made calculations of meson-nucleon cascade in the Earth's atmosphere with taking into account ionization losses of charged particles, decay and nuclear interactions of charged pions and decay and ionization losses of muons. It was supposed that all secondary particles moved in the same direction as primary particle (one-dimensional model, see results in Fig. 5.13).

Figure 5.13 shows the expected integral multiplicity for muon component with energy threshold $\Delta E\mu = 0.1$ GeV as a function of energy of primary particle E_0 for four models of energy distribution in the elementary act (see Table 5.7).

Figure 5.13 shows that the integral multiplicity sufficiently depend from the choose of elementary act model; it moved to higher energy with decreasing of fraction of energy given to pions. The cause of this is easy to understand if we take into account that muon component on sea level is caused by pions generated in the first or in few first interactions of primary particle with air atoms.

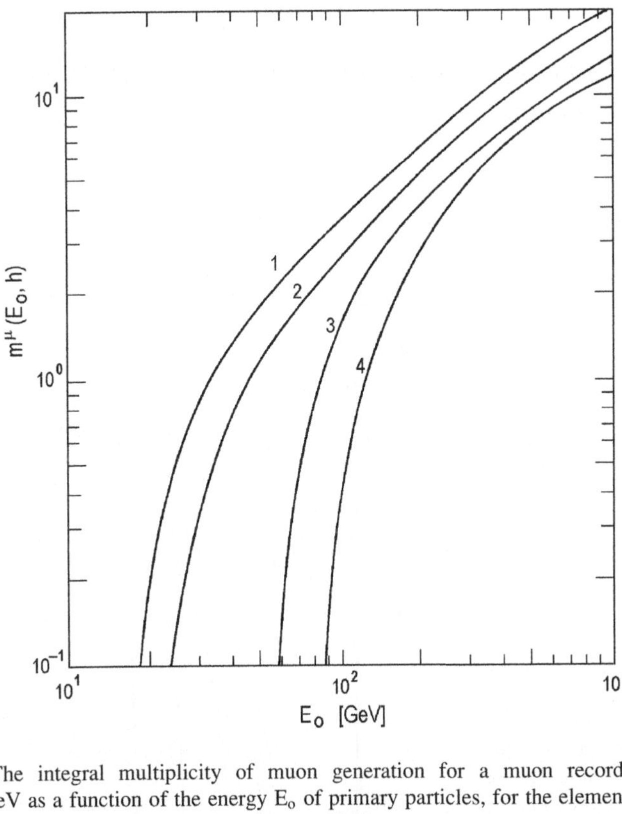

Fig. 5.13 The integral multiplicity of muon generation for a muon recording threshold $\Delta E = 0.1$ GeV as a function of the energy E_o of primary particles, for the elementary processes assumed in the Table 5.7 (According to Dorman 1961)

Table 5.7 Assumptions on the energy distribution in the elementary act, made in Dorman (1961) for meson-nucleon cascade in the atmosphere calculations

Fraction of energy given to:	(1)	(2)	(3)	(4)
Leading nucleon	0	0.2	0.4	0.7
Secondary nucleons	0.5	0.4	0.3	0.15
Pions	0.5	0.4	0.3	0.15

In Fig. 5.14 is shown the dependence of integral multiplicity for muon component at sea level in dependence of the low threshold of muon recording ΔE_{μ} (from 0.1 up to 10 GeV) in assumption that the leading nucleon gets 40%, the **δ**-nucleon gets 10% and the pions 50% in the elementary act.

In Fig. 5.15 are shown results of calculations Krymsky et al. (1965)) of integral multiplicity for primary particle arriving in vertical direction and at zenith angle 60° for muon component observation on sea level and underground at 7, 20 and 60 m water equivalent. These estimations were made in the frame of model Peters (1963) under assumption that at nucleon-nucleon interaction half of kinetic energy go for excitation and that the number of neutral pions is half of charged pions.

5.3.2 *Calculations of Integral Multiplicity for the Neutron Component*

Calculations of the integral multiplicity for the neutron component observations at various altitudes using the method of discontinues Markov processes was made in

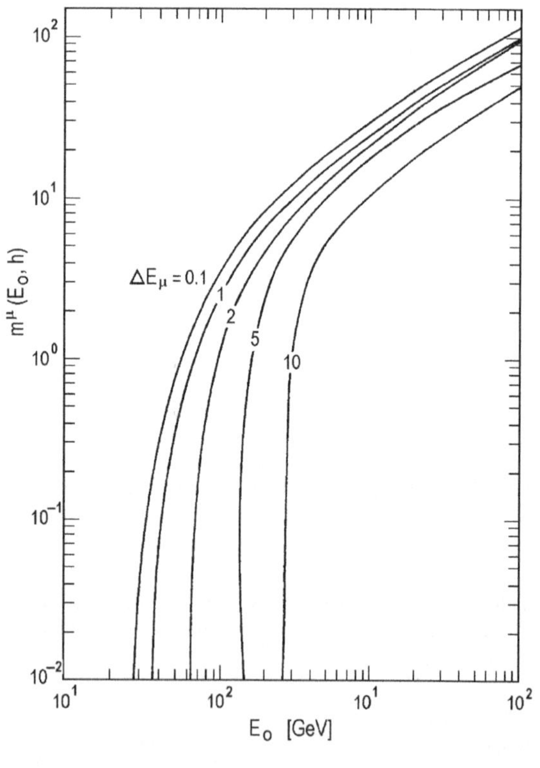

Fig. 5.14 The integral multiplicity of muon generation as a function of the primary particle energy E_0 for muon recording threshold $\Delta E_{\mu} = 0.1$; 1, 2, 5, 10 GeV at an assumption that in the elementary act the leading nucleon gets 40%, the **δ**-nucleon 10% and pions 50% of the energy of the incoming particle (According to Dorman 1961)

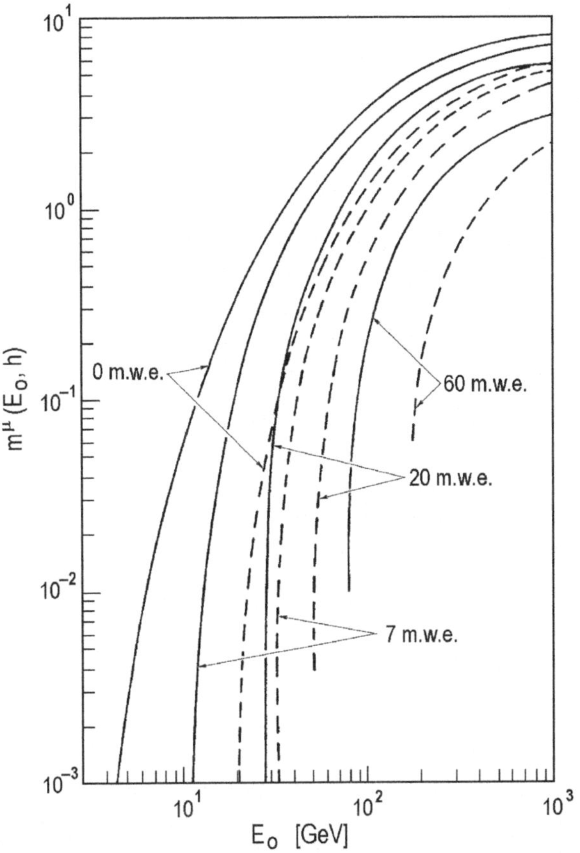

Fig. 5.15 The integral multiplicity for primary particle arriving in vertical direction (*full curves*) and at zenith angle 60° (*broken lines*) for muon observations on sea level (0 m w.e.) and on the depth underground 7, 20 and 60 m w.e (According to Krimskij et al. 1965)

Dorman and Rishe (1973), and with including data on three-dimensional elementary act in Viskov et al. (1974). According to Dorman and Yanke (1981), the differential spectrum of neutron component $N(E_0, E, h, \theta_0)$ on the level h from one primary particle with energy E_0 arrived on the atmosphere boundary at zenith angle θ_0 will be determined by 'high-energy nucleons' $N^n(E_0, E, h, \theta_0)$ and by δ-nucleons $N^\delta(E_0, E, h, \theta_0)$ in elementary act:

$$N(E_o, E, h, \theta_o) = N^n(E_o, E, h, \theta_o) + N^\delta(E_o, E, h, \theta_o), \tag{5.34}$$

where $N^n(E_0, E, h, \theta_0)$ and $N^\delta(E_0, E, h, \theta_0)$ can be determined by the system of equations:

$$\frac{\partial N^n(E_0,E,h,\theta_0)}{\partial h} = -\frac{N^n(E_0,E,h,\theta_0)}{\Lambda_n(E)\cos\theta_0} + \frac{\partial}{\partial E}\left(\frac{\beta(E)}{\cos\theta_0}N^n(E_0,E,h,\theta_0)\right) + \int_E^{E_0}\frac{N^n(E_0,E',h,\theta_0)}{\Lambda_n(E')\cos\theta_0}f_{nn}(E',E)dE' \tag{5.35}$$

$$\frac{\partial N^\delta(E_0,E,h,\theta_0)}{\partial h} = -\frac{N^\delta(E_0,E,h,\theta_0)}{\Lambda_\delta(E)\cos\theta_0} + \frac{\partial}{\partial E}\left(\frac{\beta(E)}{\cos\theta_0}N^\delta(E_0,E,h,\theta_0)\right) + \int_E^{E_o}\frac{N^n(E_0,E',h,\theta_0)}{\Lambda_n(E')\cos\theta_0}f_{n\delta}(E',E)dE', \tag{5.36}$$

where $f_{nn}(\mathrm{E}',E)$ is the secondary nucleon spectrum resultant from nuclear-atom interaction of nucleon with energy E'(in accordance with Gaisser (1976), the inelasticy coefficient is about 0.5), $f_{n\delta}(\mathrm{E}',E)$ is the same for δ-nucleons, and $\beta(E)$ is the energy loss. After determining $N(E_0,E,h,\theta_0)$ according to Eq. 5.34 it is easy to calculate integral multiplicity for neutron monitor according to expression:

$$m^n(E_0,\theta_0,h) = \int N(E_0,E,\theta_0,h)P(E)dE, \tag{5.37}$$

where $P(E)$ is the differential sensitivity of detector (the probability to detect neutron with energy E). The results of calculations of $m^n(E_0,\theta_0,h)$ with $P(E)$ according to Dorman et al. 1981 (see below, Section 5.3.3), are shown in Fig. 5.16 for levels of observation 1,030, 760, 490 and 40 g/cm^2 for zenith angle $\theta_0 = 0°$ in dependence of energy of primary particles E_0 from 1 up to 10^4 GeV. For sea level it is shown also what gave separately 'high-energy nucleons' $N^n(E_0,E,h,\theta_0)$ and **δ**-nucleons $N^\delta(E_0,E,h,\theta_0)$.

Figure 5.17 shows the dependence of $m^n(E_0,\theta_0,h)$ from h in the interval from 0 (boundary of the atmosphere) up to sea level 1,030 g/cm^2 for different primary energy E_0 from $E_0 = 1$ GeV up to $E_0 = 10^4$ GeV; for $E_0 = 10$ GeV it is shown also what gave separately 'high-energy nucleons' $N^n(E_0,E,h,\theta_0)$ and **δ**-nucleons $N^\delta(E_0,E,h,\theta_0)$.

Figure 5.18 shows the dependence of the partial barometric coefficient $\beta_p(E_o,\theta_o,h) = \partial \ln m^n(E_o,\theta_o,h)/\partial h$ of the integral multiplicity $m^n(E_0,\theta_0,h)$ for $\theta_0 = 0°$ and for h = 1,030, 760, 490 and 40 g/cm^2 in dependence of energy of primary particles E_0.

On Fig. 5.19 are shown values $m^n(E_0,\theta_0,h)$ for sea level in dependence of zenith angle θ_0 for different primary energy E_0 from E_0 =1 GeV up to $E_0 = 10^4$ GeV.

Dorman et al. (1981) estimate the accuracy of obtained results by the calculation accuracy (about 1%) and by the errors due to uncertainties in some parameters (6–8%); the errors may also be due to the choose of the energy sensitivity function

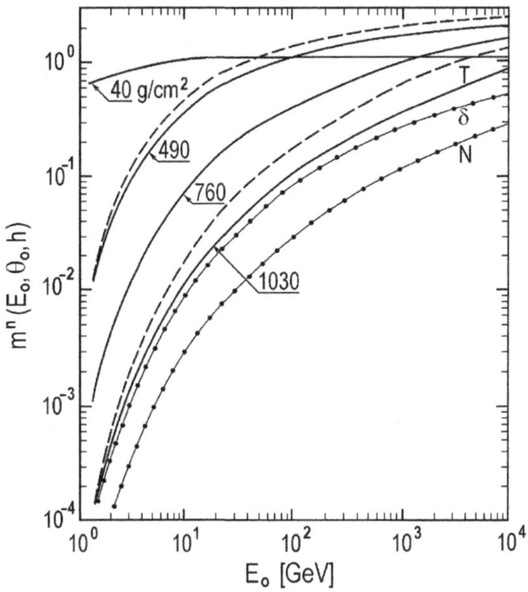

Fig. 5.16 The dependence of the integral generation multiplicity of neutrons $m^n(E_0, \theta_0, h)$ from energy of primary particles E_0 from 1 up to 10^4 GeV for levels of observation $h = 1{,}030$, 760, 490 and 40 g/cm^2 and for zenith angle $\theta_0 = 0°$. For the sea level are shown separately the integral multiplicity caused by 'high-energy nucleons' (curve **N**) and by δ-nucleons (**δ**) as well as total integral multiplicity (**T**). The dashed lines show the integral multiplicity for inelastic coefficient 0.45 (According to Dorman and Yanke 1981)

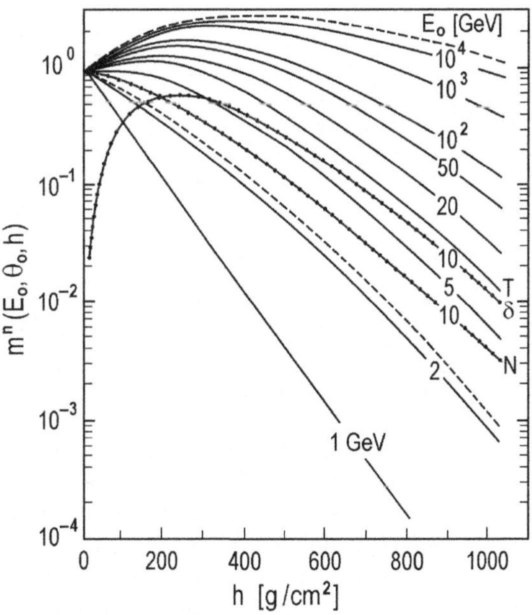

Fig. 5.17 The same as in Fig. 5.16 but for the dependence of the integral generation multiplicity of neutrons $m^n(E_0, \theta_0, h)$ for zenith angle $\theta_0 = 0°$ from h (in the interval from 0 up to 1,030 g/cm^2) for primary particle energy $E_0 = 10^4$, 10^3, 10^2, 50, 20, 10, 5, 2, and 1 GeV. For $E_0 = 10$ GeV are shown also what gave separately 'high-energy nucleons' $N^n(E_0, E, h, \theta_0)$ (curve **N**) and **δ**-nucleons $N^\delta(E_0, E, h, \theta_0)$ (curve **δ**) (According to Dorman and Yanke 1981)

Fig. 5.18 The barometric coefficient $\beta(E_0, \theta_0, h)$ of the integral multiplicity (partial barometric coefficient) for $\theta_0 = 0°$ and for $h = 1{,}030$, 760, 490 and 40 g/cm^2 in dependence of energy of primary particles E_0; other nominations as for Fig. 5.16 (According to Dorman and Yanke 1981)

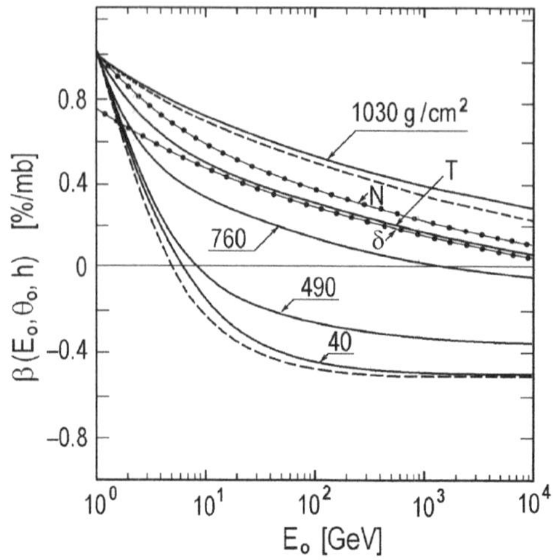

Fig. 5.19 The dependence of the integral generation multiplicity of neutrons $m^n(E_0, \theta_0, h)$ on the sea level ($h = 1{,}030$ g/cm^2) from the zenith angle θ_0 for $E_0 = 10^4$, 10^3, 10^2, 10, 5, and 2 GeV. For $E_0 = 100$ GeV are shown also what gave separately high-energy nucleons' $N^n(E_0, E, h, \theta_0)$ (curve **N**) and **δ**-nucleons $N^\delta(E_0, E, h, \theta_0)$ (curve **δ**) (According to Dorman and Yanke 1981)

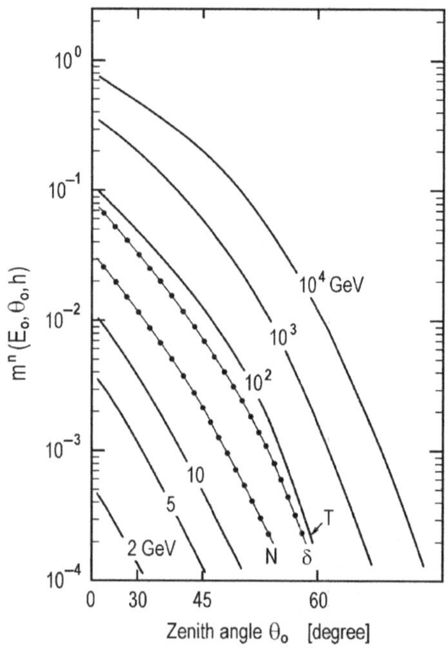

$P(E)$ and to the selection of particular model of cosmic ray interactions and propagation in the atmosphere. Therefore for controlling of obtained results in Dorman et al. (1981) were calculated also expected coupling functions and barometric coefficients for minimum and maximum of solar activity in dependence of cut-off rigidity (by help of the partial barometric coefficient $\beta(E_0, \theta_0, h)$ according to Dorman (M1972)) and it was shown that they are in good agreement with obtained from observations.

5.3.3 Calculations of Integral Multiplicity for Multiple Neutrons in the NM-64 Supermonitor

According to Dorman et al. (1981) the integral multiplicity $m_k^n(E_0, \theta_0, h)$ for k-multiple neutrons in the NM-64 supermonitor will be determined by the same equation as Eq. 5.37, but instead of the function $P(E)$ will be $P_k(E)$ which denote the probability that the incidence neutron with energy E will be detected as k-multiple neutron:

$$M_k^n(E_0, \theta_0, h) = \int N(E_0, E, \theta_0, h) P_k(E) dE, \tag{5.38}$$

where $N(E_0, E, \theta_0, h)$ was determined in Dorman and Yanke (1981) and is the same as in Section 5.3.2, expression Eq. 5.34 and Eqs. 5.35 and 5.36. The function $P_k(E)$ according to Dorman et al. (1981) is determined by expression:

$$P_k(E) = \sum_{m=k}^{\infty} \int_0^E dE' \int_0^{E'} d\text{э}\, G(E, E') H_m(E') D(E', \text{э}) B_m^k(\text{э}). \tag{5.39}$$

Here E' is the excitation energy of the residual nucleus in the neutron monitor from interaction with incident neutron with energy E; the distribution of E' in dependence of E is described according to Metropolis et al. (1958) and Hayakawa (M1973) by the function

$$G(E, E') = (2\pi)^{-1/2} \overline{E'}(E) \exp\left(-\frac{1}{2} \cdot \left(\frac{E'}{\overline{E'}(E)} - 1\right)^2\right), \tag{5.40}$$

where $\overline{E'}(E) = a \times \lg E - b$ (here $\overline{E'}(E)$ and E are in MeV, parameters $a = 69$ MeV and $b = 61$ MeV for $E \leq 350$ MeV; $a = 350$ MeV and $b = 770$ MeV for $350 \leq E \leq$ 1,000 MeV; $a = 580$ MeV and $b = 1{,}460$ MeV for $E \geq 1{,}000$ MeV).

The function $H_m(E')$ determined the probability that from the residual nucleus excited to the energy E' will be evaporate m neutrons. According to Dostrovsky et al. (1958) and Hayakawa (1973) $H_m(E')$ can be approximated by the expression (E' in MeV):

$$H_m(E') = \frac{10}{\overline{m}\sqrt{2\pi}} \exp\left(-50\left(\frac{m}{\overline{m}} - 1\right)^2\right); \; \overline{m} = 0.6E' \text{ for Pb.} \tag{5.41}$$

The function $D(E', \text{э})$ determined the differential energy spectrum of evaporated neutrons with energy э in dependence of nucleus excitation energy E'. According to Dostrovsky et al. (1958) and Hayakawa (1973) the spectrum of evaporated neutrons is sufficiently close to the Maxwellian form:

$$D(E', э) = 23(э/E')\exp\left(-5э\Big/\sqrt{E'}\right). \tag{5.42}$$

The last function in Eq. 5.39 $B_m^k(э)$ determined the probability that from m evaporated neutrons with energy э will be detect k neutrons. If the neutron deceleration in neutron monitor between evaporation and detection is disregarded, then according to Dorman et al. (1981) this function will be:

$$B_m^k(э) = C_m^k(\Phi(э))^k(1-\Phi(э))^{m-k}, \tag{5.43}$$

where $\Phi(э)$ determined the effectiveness of neutron supermonitor to detect evaporated neutron with energy э (in MeV) and according to Pearce and Fowler (1964) can be approximated by expression:

$$\Phi(э) \approx 0.079 - 0.031 \times \lg э - 0.01 \times (\lg э)^2. \tag{5.44}$$

Equation 5.38 with taking into account Eqs. 5.39, 5.40, 5.41, 5.42, 5.43, 5.44 and $N(E_0, E, \theta_0, h)$ from paper Dorman and Yanke (1981) determined the integral multiplicity $M_k^n(E_0, \theta_0, h)$ for k-multiple neutrons in the NM-64 supermonitor. On the basis of founded $P_k(E)$ for multiple neutrons it is possible to determine $P(E)$ for total neutron component:

$$P(E) = \sum_{k=1}^{\infty} A_k P_k(E), \tag{5.45}$$

where A_k depend from the relation between dead time T_d of monitor electronic scheme and life time T_l of neutrons in the monitor: $A_k = k$ for very small dead time T_d and $A_k = 1$ for great T_d. According to Carmichael (M1964), for the standard NM-64 neutron supermonitor $T_d = 10$ mks and $T_l = 329$ mks. For these parameters according to Dorman et al. (1981) $A_k = 1.0, 1.97, 2.91$ and 3.83 for $k = 1, 2, 3$ and 4. If, for example, $T_d = 100$ mks, then $A_k = 1.0, 1.74, 2.35$ and 2.84 for $k = 1, 2, 3$ and 4. Results of determination of $P_k(E)$ and $P(E)$ are shown in the Fig. 5.20 in comparison with the results of Pakhomov and Sdobnov (1977), where the process of neutron deceleration in the monitor was taken into account.

Figure 5.20 shows that the neutron deceleration in the monitor is especially important at calculations of differential sensitivity $P(E)$ for small energy neutrons arrived to monitor.

Figures 5.21 and 5.22 shows the results of calculations $m_k^n(E_0, \theta_0, h)$ as well as $m^n(E_0, \theta_0, h)$ for sea level and level 541 g/cm^2. For sea level is shown also the comparison with $m_k^n(E_0, \theta_0, h)$ determined by Debrunner and Flückiger (1971)) for multiple neutrons and with $m^n(E_0, \theta_0, h)$ for total neutron component determined in Pakhomov and Sdobnov (1977).

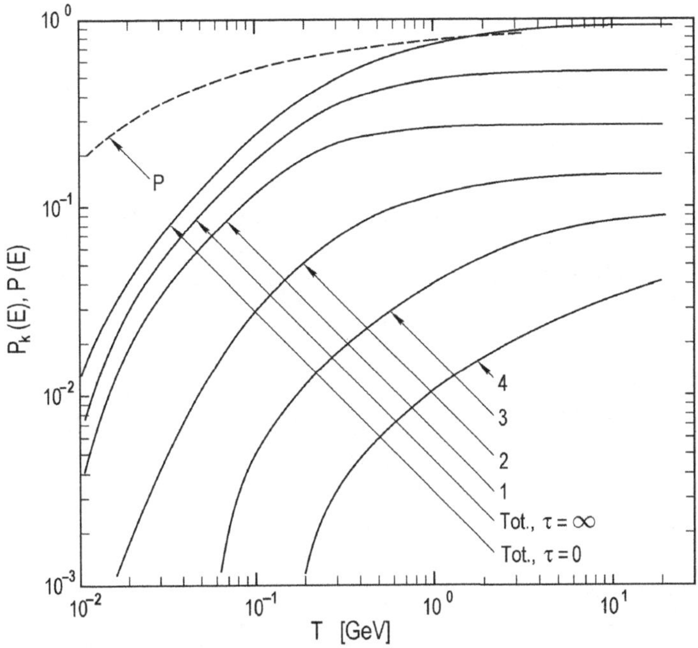

Fig. 5.20 Determination of $P_k(E)$ for neutron multiplicity $k = 1, 2, 3, 4$ and $P(E)$ for the total neutron component $\sum_\infty$ (when $T_d >> T_l$) and $\sum_0$ (when $T_d << T_l$) in dependence of incident neutron energy E in the interval from 10 MeV up to 10 GeV. For comparison is shown $P(E)$ for total neutron component obtained in Pakhomov and Sdobnov (1977), where the process of neutron deceleration in the monitor was taken into account (dashed curve **PS**) (According to Dorman et al. 1981)

From Figs. 5.21 and 5.22 can be seen that there are good agreement between different calculations as well as good agreement between calculated in Dorman et al. (1981) coupling functions with determined one on the basis of cosmic ray latitude observations according to Aleksanyan et al. (1979c).

5.4 Calculations of Integral Multiplicity and Sensitivity of the Neutron Monitor at Various Depths Depending on the Zenith Angle of Primary Particle Incidence

In connection with the problem of solar neutron propagation in the atmosphere there are important calculations of integral multiplicity for protons arrived to the Earth's atmosphere at different zenith angles.

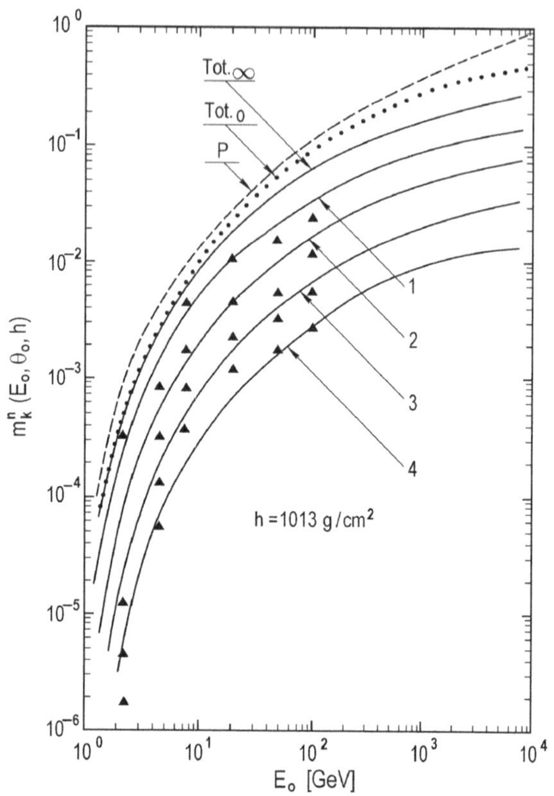

Fig. 5.21 Results of calculations of neutron integral multiplicities $m_k^n(E_0, \theta_0, h)$ for sea level ($h = 1{,}013$ g/cm^2) and zenith angle $\theta_0 = 0°$ for $k = 1, 2, 3$ and 4 as well as $m^n(E_0, \theta_0, h)$ for the total neutron component $\sum_\infty$ (when $T_d >> T_l$) and $\sum_o$ (when $T_d << T_l$) in dependence of primary particle energy E_0 in the interval from 1 up to 10^4 GeV. Is shown also the comparison with $m_k^n(E_0, \theta_0, h)$ determined by Debrunner and Fluckiger (1971) (*triangles*) for multiple neutrons and with $m^n(E_0, \theta_0, h)$ for total neutron component determined in Pakhomov and Sdobnov (1977) (dashed curve **PS**) (According to Dorman et al. 1981)

5.4.1 Calculations of Integral Multiplicity for Primary Protons with Energies 3 and 10 GeV

The integral generation multiplicities of neutron component at atmosphere depths 710, 806, 965 and 1,033 g/cm^2 from 3 to 10 GeV protons incident on the atmospheric boundary at zenith angles 0°, 15°, 30°, 45°, 60° and 75° have been calculated by Dorman and Pakhomov (1979). To obtain the integral generation multiplicities, the Monte Carlo method was used in simulating a pion-nucleon cascade in the atmosphere (the cascade evaporation model was used as the basis of simulation of inelastic pion-nucleous interactions according to Barashenkov and Tokayev M1972). Small energy neutron propagation was described in terms of transport approximation according to Abagyan et al. (M1964). The ionization energy loss in propagation of charged particles was calculated by using the phenomenological expressions from Vzorov (1969) and the standard atmospheric model was used to simulate the conditions in the Earth's atmosphere. As result of these calculations was found the differential multiplicity $N(E_0, E, h, \theta_0)$ described the differential energy spectrum of secondary neutrons on the pressure level h, generated by one primary particle with energy E_0 and incident on the atmospheric

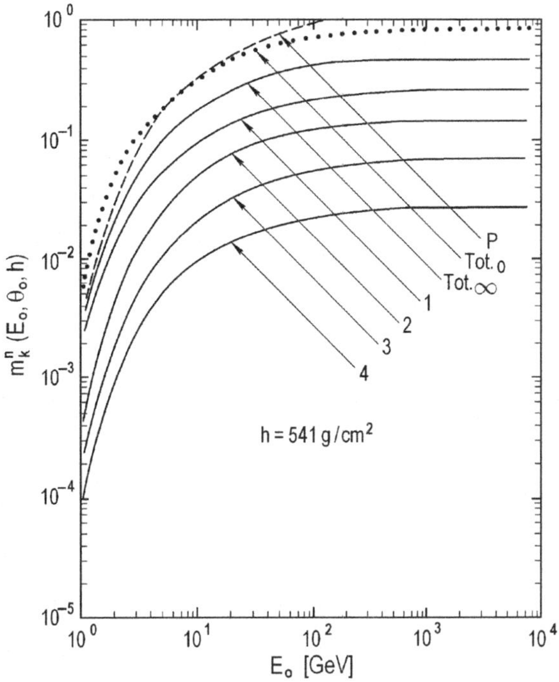

Fig. 5.22 The same as in Fig. 5.21 but for $h = 541$ g/cm^2 (From Dorman et al. 1981)

Table 5.8 Integral multiplicity for NM-64 neutron supermonitor for $E_0 = 3$ GeV in dependence of h and θ_0 (According to Dorman and Pakhomov 1979)

h (g/cm^2)	Zenith angle θ_0						γ
	0°	15°	30°	45°	60°	75°	
710	9.5×10^{-2}	8.2×10^{-2}	5.6×10^{-2}	2.6×10^{-2}	1.0×10^{-2}	2.6×10^{-3}	2.8
806	5.6×10^{-2}	4.4×10^{-2}	2.4×10^{-2}	1.1×10^{-2}	3.5×10^{-3}	8.2×10^{-4}	3.4
965	2.3×10^{-2}	1.9×10^{-2}	7.9×10^{-3}	1.9×10^{-3}	1.1×10^{-3}	5.4×10^{-5}	4.6
1,033	1.3×10^{-2}	1.2×10^{-2}	5.2×10^{-3}	1.5×10^{-3}	5.2×10^{-4}	1.6×10^{-6}	5.0

boundary at zenith angle θ_0. The integral multiplicity for NM-64 neutron supermonitor $m^n(E_0, \theta_0, h)$ will be determined then by Eq. 5.37 in the Section 5.3.2, but with $P(E)$ according to Luzov et al. (1971) and Pakhomov and Sdobnov (1977). The results of calculation $m^n(E_0, \theta_0, h)$ according to Eq. 5.37 are listed in Table 5.8 for $E_0 = 3$ GeV and in Table 5.9 for $E_0 = 10$ GeV.

5.4.2 *On the Dependence of Integral Multiplicities $m^n(E_0, \theta_0, h)$ with Atmospheric Depth*

The dependencies of integral multiplicity from atmospheric depth h for primary protons with energy $E_0 = 3$ and 10 GeV and $\theta_0 = 0°$, 15°, 30°, 45°, 60° and 75° are shown in Fig. 5.23.

Table 5.9 Integral multiplicity for NM-64 neutron supermonitor for $E_0 = 10$ GeV in dependence of h and θ_0 (According to Dorman and Pakhomov 1979)

h (g/cm^2)	Zenith angle θ_0						γ
	0°	15°	30°	45°	60°	75°	
710	2.7×10^{-1}	2.5×10^{-1}	1.7×10^{-1}	6.1×10^{-2}	1.7×10^{-2}	5.1×10^{-2}	3.2
806	1.6×10^{-1}	1.4×10^{-1}	8.4×10^{-2}	2.6×10^{-2}	8.0×10^{-3}	1.7×10^{-3}	3.7
965	6.8×10^{-2}	5.8×10^{-2}	2.5×10^{-2}	6.1×10^{-3}	2.4×10^{-3}	1.1×10^{-4}	4.9
1,033	4.9×10^{-2}	3.9×10^{-2}	1.5×10^{-2}	3.1×10^{-3}	1.4×10^{-3}	3.9×10^{-5}	5.4

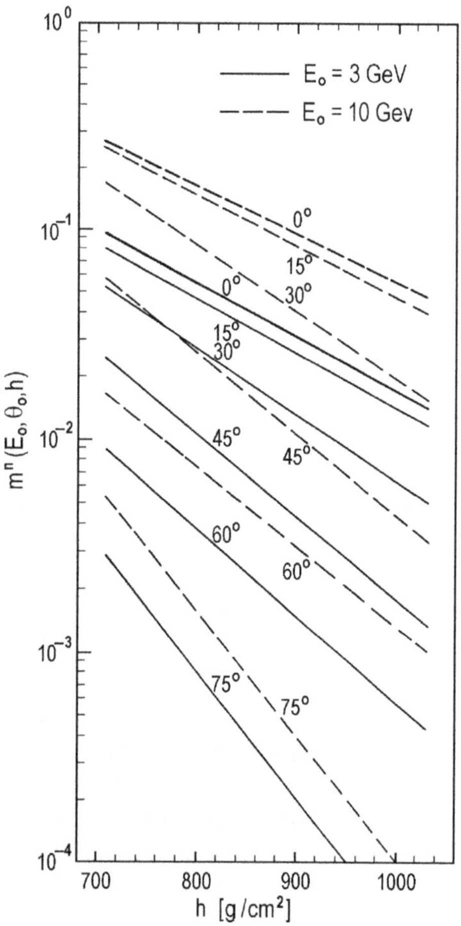

Fig. 5.23 Integral multiplicities $m^n(E_0, \theta_0, h)$ for the total neutron component in dependence of h for primary particle energy $E_0 = 3$ and 10 GeV and $\theta_0 = 0°, 15°, 30°, 45°, 60°$ and 75° (According to Dorman and Pakhomov 1979)

It can be seen from Fig. 5.23 that for each θ_0 and each E_0 the dependence of $m^n(E_0, \theta_0, h)$ from h in the interval from 710 to 1,033 g/cm^2 is approximately exponential:

$$m^n(E_0, \theta_0, h) \propto \exp(-h/\Lambda(E_0, \theta_0)), \quad (5.46)$$

where the values of attenuation coefficient $\Lambda(E_0, \theta_0)$ for the integral multiplicities are listen in the Table 5.10.

In Table 5.10 there are four types of determining $\Lambda(E_0, \theta_0)$:

1. Obtained from Eq. 5.46 by comparison values of $m^n(E_0, \theta_0, h)$ at levels 710 and 1,033 g/cm^2 in Tables 5.8 and 5.9
2. Obtained by multiplying $\Lambda(E_0, \theta_0 = 0)$, determined in **(1)**, on $\cos\theta_0$;
3. Obtained by averaging $\Lambda(E_0, \theta_0)$ determined from Eq. 5.46 by comparison values of $m^n(E_0, \theta_0, h)$ at each level of observation (710, 806, 965 and 1,033 g/cm^2) with all other (on the basis of data listed in Tables 5.8 and 5.9)
4. Obtained by multiplying $\Lambda(E_0, \theta_0 = 0)$, determined in **(3)**, on $\cos\theta_0$

5.4.3 *On the Dependence of Integral Multiplicities $m^n(E_0, \theta_0, h)$ with Zenith Angle θ_0*

As it was shown in Dorman and Pakhomov (1979) the dependence of the neutron monitor integral multiplicities $m^n(E_0, \theta_0, h)$ from the zenith angle θ_0 for each level of observation h and each primary energy E_0 can be approximated by function

$$m^n(E_0, \theta_0, h) = m^n(E_o, \theta_0 = 0^\circ, h) \times (\cos\theta_0)^{\gamma(E_0,h)}. \qquad (5.47)$$

The values of $\gamma(E_o, h)$ are listen in the last columns in Tables 5.8 and 5.9 for $E_0 = 3$ and 10 GeV, correspondingly. The dependence of $\gamma(E_o, h)$ from h is shown in Fig. 5.24.

The dependencies of $\gamma(E_o, h)$ from h shown in Fig. 5.24 can be approximated by linear function:

$$\gamma(E_o, h) = A(E_o) + B(E_o)h, \qquad (5.48)$$

where for $E_0 = 3$ GeV:

Table 5.10 The values of attenuation coefficient $\Lambda(E_0, \theta_0)$ in g/cm^2 for $E_0 = 3$ and 10 GeV in dependence of θ_0

E_0 (GeV)	Type	Zenith angle θ_0					
		0°	15°	30°	45°	60°	75°
3	(1)	162.4	168.1	135.9	113.2	109.2	43.7
	(2)	162.4	156.9	140.6	114.8	81.2	42.0
	(3)	162.8	168.1	138.9	135.7	110.6	51.1
	(4)	162.8	157.2	141.0	115.1	81.4	42.1
10	(1)	189.3	173.9	133.0	108.4	67.3	66.3
	(2)	189.3	182.8	163.9	133.8	94.6	49.0
	(3)	190.5	173.9	133.1	108.1	89.5	67.3
	(4)	190.5	184.0	165.0	134.7	95.2	49.3

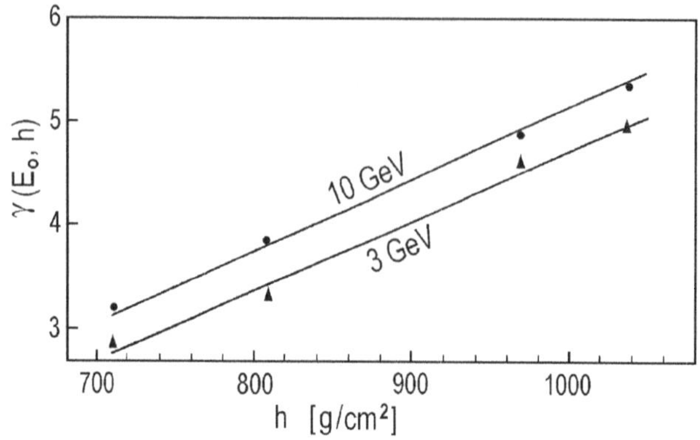

Fig. 5.24 The exponent $\gamma(E_o, h)$ in Eq. 5.47in dependence of h for primary particle energy $E_0 = 3$ and 10 GeV (According to Dorman and Pakhomov 1979)

$$A(E_o) = -2.15; B(E_o) = 6.90 \times 10^{-3}(\mathrm{g/cm^2})^{-1} \tag{5.49}$$

and for $E_0 = 10$ GeV:

$$A(E_o) = -1.93; B(E_o) = 7.17 \times 10^{-3}(\mathrm{g/cm^2})^{-1}. \tag{5.50}$$

5.4.4 *On the Dependence of Integral Multiplicities from Atmospheric Depth and Zenith Angle*

We can combine Eqs. 5.46 and 5.47 and determine approximately the dependence of integral multiplicities $m^n(E_0, \theta_0, h)$ from atmospheric depth h and from the zenith angle θ_0 simultaneously:

$$\begin{aligned} m^n(E_0, \theta_0, h) = m^n(E_0, \theta_0 = 0°, h = h_0) \times (\cos\theta_0)^{\gamma(E_0,h)} \\ \times \exp(-(h - h_0)/\Lambda(E_0, \theta_0)), \end{aligned} \tag{5.51}$$

where $m^n(E_0, \theta_0 = 0°, h = h_0)$ is the integral multiplicity for neutron component observation on the level $h = h_0$ from primary particles arrived in vertical direction. In Eq. 5.51 power index $\gamma(E_o, h)$ is determined by Eqs. 5.48, 5.49, and 5.50 and values of attenuation coefficient $\Lambda(E_0, \theta_0)$ were given in Table 5.10.

5.4.5 On the Testing of Dorman and Pakhomov (1979) Calculations of the Integral Multiplicities Using Solar Neutron Observation Data

Efimov et al. (1993) tested the calculations of Dorman and Pakhomov (1979) of the integral multiplicities $m^n(E_0, \theta_0, h)$ for primary protons with energies 3 and 10 GeV by solar neutron data obtained for the events of June 3, 1982 and May 24, 1990. These events were observed by many ground based neutron monitors, characterized with different levels of observation h and different zenith angles θ_0 at the time of event. From Tables 5.8 and 5.9 as well as from Fig. 5.24 is possible to seen that the average $\gamma(E_o, h)$ for levels from 700 to 1,030 g/cm^2 and for particle energy from 3 to 10 GeV have value about 4. Figure 5.25 (from Efimov et al. 1993) shows the dependences of observed solar neutrino flux N divided on $(\cos\theta_0)^{\gamma(E_0,h)}$ (where $\gamma(E_0, h) = 4$) in dependence of observation level depth h (in g/cm^2). From Fig. 5.25 it can be seen that all experimental data for both events are in good agreement with Eq. 5.51 and shows that for the event of June 3, 1982 the average $\Lambda \approx 96$ g/cm^2 and for the event of May 24, 1990 it was $\Lambda \approx 111$ g/cm^2. The testing made by Efimov et al. (1993) of calculations Dorman and Pakhomov (1979) shows that in the first approximation we can use the results on the $m^n(E_0, \theta_0, h)$ obtained for protons also for research of high energy solar neutrons detected by neutron monitors and other ground based detectors.

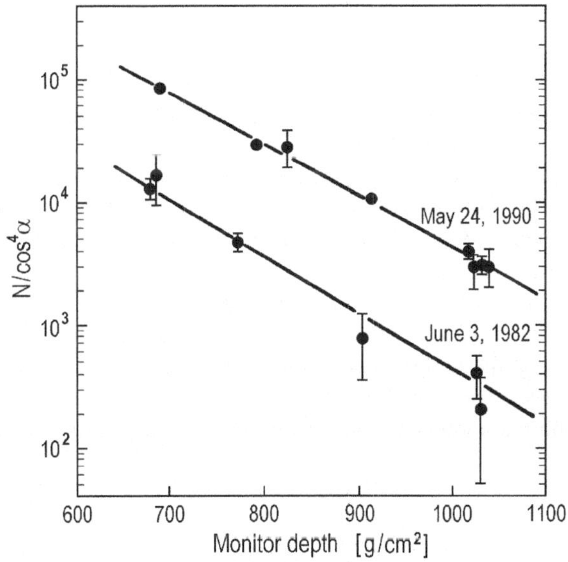

Fig. 5.25 The testing of Dorman and Pakhomov (1979) calculations of the integral multiplicities $m^n(E_0, \theta_0, h)$ using solar neutron observation data: neutron monitor responses to solar neutron events of June 3, 1982 (at 11:45–11:50 UT) and of May 24, 1990 (at 20:50–20:55 UT). Lines are $\exp(-h/\Lambda)$, where $\Lambda \approx 96$ g/cm^2 for event June 3, 1982 and $\Lambda \approx 111$ g/cm^2 for event May 24, 1990 (According to Efimov et al. 1993)

5.5 On the Small Energy Solar Neutron Propagation in the Earth's Atmosphere and the Sensitivity of Neutron Monitors and Other Ground Based Detectors to Solar Neutrons

5.5.1 Calculations of Solar Small Energy Neutron Propagation in the Atmosphere to an Depth 300 g/cm²

These calculations specially for the case of small energy solar neutrons was made by Alsmiller and Boughner (1968). They considered solar neutron propagation down to an atmospheric depth of 300 g/cm^2, but with ignoring of very important elastic scattering of nucleons on nuclei at energies above 25 MeV. As was shown by Shibata (1994), this ignoring of elastic scattering lead to the great underestimated flux of solar neutrons (especially in the low energy region, see below, Section 5.5.3).

5.5.2 The First Calculations of Neutron Monitor 'Sensitivity' to Small Energy Solar Neutrons

Debrunner et al. (1983, 1989) based their calculations of neutron monitor 'sensitivity' to solar neutrons on the Debrunner and Brunberg (1968) Monte Carlo simulation of atmospheric nucleonic cascade in the atmosphere with taking into account inter-nuclear cascades. In general the Monte Carlo method can be used to simulate the development of the nucleonic-meson cascade in the atmosphere to determine $N_{kij}(h, E_0, \theta_0)$ – the number of nucleons of type k (neutrons and protons) in the kinetic energy interval $E_i \pm \Delta E_i$ and in the zenith angle element $\theta_j \pm \Delta\theta_j$, which is produced on the average at the atmospheric depth h by a primary neutron penetrating into the Earth atmosphere with kinetic energy E_0 and zenith angle θ_0. On the basis of $N_{kij}(h, E_0, \theta_0)$ it can be found the sensitivity $S_n(h, E_0, \theta_0)$ of the neutron monitor on the vertical depth h to primary neutron with kinetic energy E_0 and zenith angle θ_0. The probability that a primary neutron will produce a secondary nucleon with parameters k, i, j striking the neutron monitor in the atmospheric depth h then is

$$P_{kij}(h, E_o, \theta_o) = \frac{N_{kij}(h, E_o, \theta_o)}{\sum\limits_k \sum\limits_i \sum\limits_j N_{kij}(h, E_o, \theta_o)} \times \left[1 - \exp\left(-\sum_k \sum_i \sum_j N_{kij}(h, E_o, \theta_o)\right)\right]. \quad (5.52)$$

The response of the neutron monitor to the secondary nucleons in the atmosphere is described by the parameters:

1. The probability

$$I_k(v, E_i, T) = (v_k(E_i, T))^v (1 + v_k(E_i, T))^{-(v+1)} \tag{5.53}$$

that a neutron or proton of the energy E_i in lead of thickness T (measured in units of the mean free path) will produce v low energy neutrons (here $v_k(E_i, T)$ is the average number of neutrons produced in lead, according to Shen 1968);

2. The probability

$$W_m^v = \binom{v}{m} \text{э}^m (1-\text{э})^{v-m} \tag{5.54}$$

of detecting m out of v low energy neutrons, where $\text{э} = 0.056$ is the probability to detect one low energy neutron in monitor;

3. The thickness of lead in neutron monitor T_{jqr} in dependence of direction of particle arriving, weighted according to their relative areas by factor α_{jqr}.

With S being the total area of neutron monitor its sensitivity to primary neutrons can then be expressed by

$$\begin{aligned} S_n(h, E_0, \theta_0) = & \sum_{m=1}^{\infty} m \sum_k \sum_i \sum_j \sum_q \sum_r S \cos \theta_0 P_{kij}(h, E_0, \theta_0) \alpha_{jqr} \\ & \times \left[1 - \exp(-T_{jqr})\right] \sum_{v=m}^{\infty} I_k(v, E_i, T_{jqr}) W_m^v. \end{aligned} \tag{5.55}$$

For a beam of solar neutrons with intensity $\Phi_n(E_0, t)$ (expressed in units of neutrons/m^2/s) and with impact zenith angle θ_0 the response of the neutron monitor (in units cps) will be

$$N_n(h, \theta_0, t) = \int \Phi_n(E_0, t) S_n(h, E_0, \theta_0) dE_0. \tag{5.56}$$

Results of calculation of $S_n(h, E_0, \theta_0)$ for a standard 6NM-64 neutron monitor ($S = 6.21$ m^2) for $h = 650$, 850 and 1,033 g/cm^2 for different neutron energy from 0.1 to 10 GeV and for zenith angle $\theta_0 = 0°$ are shown in the Table 5.11.

Figure 5.26 shows the sensitivity of a 6NM-64 neutron monitor to solar neutrons penetrating vertically into the Earth's atmosphere in dependence of neutron energy up to 3,000 MeV for observation levels 650, 850 and 1,033 g/cm^2.

These important calculations according to Eqs. 5.52, 5.53, 5.54, 5.55, and 5.56 were continued in Debrunner et al. (1990). Figure 5.27 shows the expected sensitivity of a 6 NM-64 neutron monitor at sea level to solar neutrons with $E_0 = 750$ and 2,000 MeV in dependence of θ_0.

Table 5.11 The sensitivity $S_n(h, E_0, \theta_0 = 0°)$ of a 6 NM-64 neutron supermonitor on the depth h to solar neutrons penetrating into the Earth atmosphere with kinetic energy E_0 and zenith angle $\theta_0 = 0°$ (According to Debrunner et al. 1983, 1989)

E_0 (GeV)	$h = 650$ (g/cm^2)	$h = 850$ (g/cm^2)	$h = 1{,}033$ (g/cm^2)
0.1	4.0×10^{-6}		
0.2	1.4×10^{-3}	1.1×10^{-4}	3.0×10^{-6}
0.3	4.1×10^{-3}	3.9×10^{-4}	3.2×10^{-5}
0.5	7.2×10^{-3}	6.3×10^{-4}	7.8×10^{-5}
0.75	8.0×10^{-3}	8.2×10^{-4}	1.0×10^{-4}
1.0	9.8×10^{-3}	9.8×10^{-4}	1.2×10^{-4}
2.0	4.3×10^{-2}	5.8×10^{-3}	5.3×10^{-4}
3.0	1.15×10^{-1}	1.8×10^{-2}	2.6×10^{-3}
5.0	3.0×10^{-1}	5.3×10^{-2}	1.05×10^{-2}
7.5	4.9×10^{-1}	1.05×10^{-1}	2.2×10^{-2}
10.0	6.2×10^{-1}	1.45×10^{-1}	2.8×10^{-2}

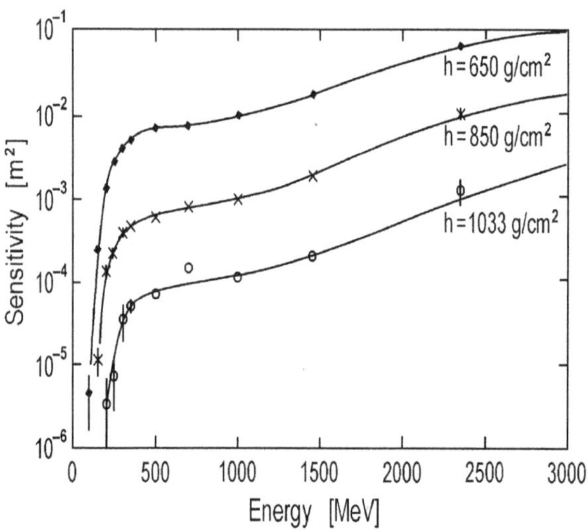

Fig. 5.26 Sensitivity of 6NM-64 neutron monitor to solar neutrons penetrating vertically into the Earth's atmosphere in dependence of neutron energy up to 3,000 MeV for observation levels 650, 850 and 1,033 g/cm^2. *Full curves* are smooth fits to calculated data points. *Error bars* represent uncertainties due to statistical fluctuations in the Monte Carlo simulations (According to Debrunner et al. 1989)

From Fig. 5.27 can be seen that the dependence from θ_0 for $E_0 = 2{,}000$ MeV can be approximated by the function:

$$S_n(h = 1033\,g/cm^2, E_0, \theta_0) = S_n(h = 1033\,g/cm^2, E_0, \theta_0 = 0°) \times \exp(10.33(1 - 1/\cos\theta_0)). \quad (5.57)$$

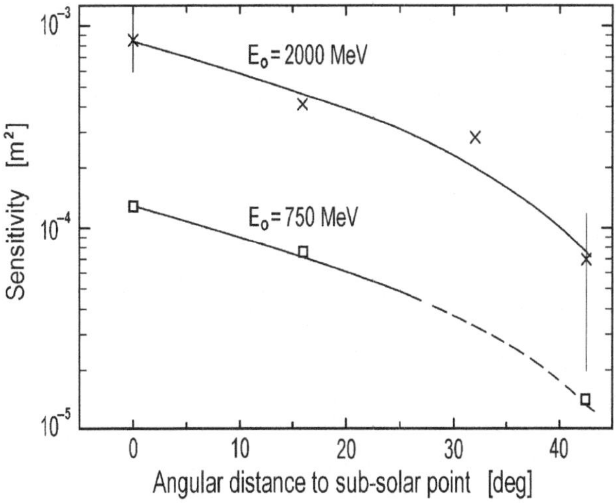

Fig. 5.27 Sensitivity of a 6NM-64 neutron monitor at sea level to solar neutrons with kinetic energies $E_0 = 2{,}000$ MeV (*upper curve*) and $E_0 = 750$ MeV (*lower curve*), in dependence of the angular distance of the station from the sub-solar point (According to Debrunner et al. 1990)

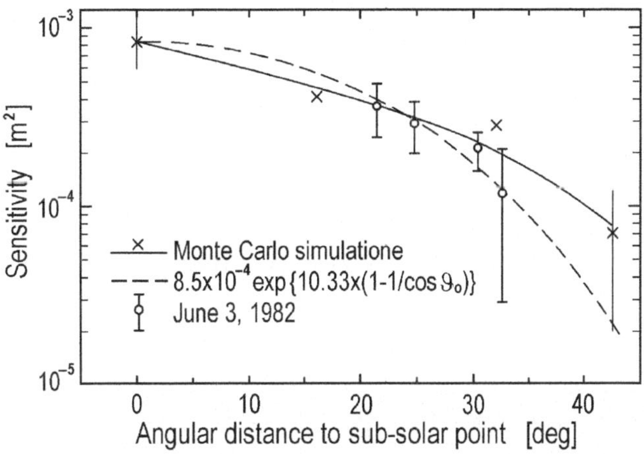

Fig. 5.28 Comparison of $S_n(h = 1033 \text{ g/cm}^2, E_0 = 2000 \text{ MeV}, \theta_0)$ as obtained from the Monte Carlo simulations with the approximation according to Eq. (5.51), and with experimental points deduced from the neutron monitor observations of the 3 June 1982 solar neutron event (According to Debrunner et al. 1990)

Figure 5.28 shows the comparison of $S_n(h, E_0, \theta_0)$ determined by Monte Carlo simulation for sea level and $E_0 = 2{,}000$ MeV in dependence from θ_0 with experimental data of the event 3 June 1982 as well as with expected from Eq. 5.57.

5.5.3 A One-Dimensional Model of Solar Neutron Propagation in the Earth's Atmosphere and the Relative Role of Elastic Scattering

To show the relative role of elastic scattering in the solar neutron propagation through the atmosphere, Shibata (1994) considered a simple model of one-dimensional propagation:

$$dN(h)/dh = -N(h)/\Lambda + N(h)(\sigma_{el}/\Lambda\sigma_{tot}), \tag{5.58}$$

where h is the depth of atmosphere along the direction to the Sun (in g/cm^2), Λ is the mean free path for collisions. According to Barashenkov et al. (1969) and Del Guerra (1976) for kinetic energy of neutrons $E \geq 150$ MeV $\sigma_{tot} \approx 3 \times 10^{-25}$ cm^2 and $\sigma_{el} \approx 10^{-25}$cm^2 for carbon with A = 12 (the dependencies of σ_{el}, σ_{inel} and σ_{tot} from neutron energy in the interval 10–1,000 MeV are shown in Fig. 5.29).

On the basis of these results Shibata (1994) came to conclusion that for $E \geq 150$ MeV and for mean air nucleus (<A> = 14.4) the expected value of

$$\Lambda \approx (12/14.4)^{3/2}\left[14.4/\left(6.02 \times 10^{23} \times 3 \times 10^{-25}\right)\right] \approx 70.6\text{g/cm}^2 \tag{5.59}$$

and the solution of Eq. 5.58 will be

$$N(h) = N_0 \exp[-(1 - \sigma_{el}/\sigma_{tot})h/\Lambda] = N_0 \exp(-h/\Lambda_{eff}), \tag{5.60}$$

where

$$\Lambda_{eff} = \Lambda/(1 - \sigma_{el}/\sigma_{tot}) \approx 106\text{g/cm}^2. \tag{5.61}$$

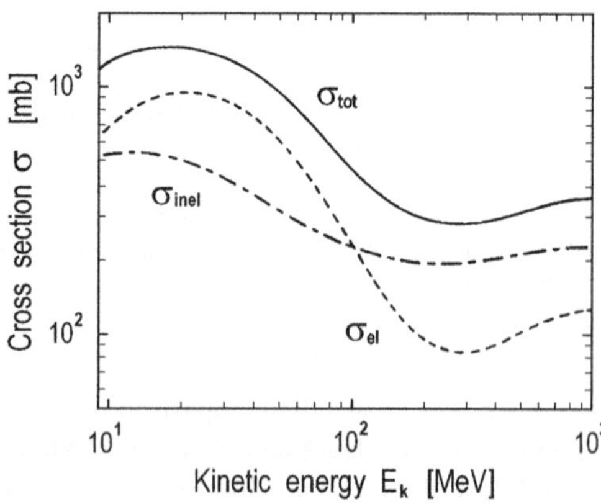

Fig. 5.29 The energy dependence of neutron-carbon scattering cross sections according to Barashenkov et al. (1969) and Del Guerra (1976). The *solid curve* indicates the total cross section σ_{tot}, and the *dashed curve* shows the cross section of elastic scattering σ_{el}, and *dot-dashed curve* indicates the cross section of inelastic scattering σ_{inel} (From Shibata 1994)

This value is in accordance with experimental data on solar neutron event June 3, 1982 and others (see Chapter 7). According to Eqs. 5.60 and 5.61 on the depth 700 and 1,000 g/cm^2 in the direction of propagation will reach a parts of solar neutrons 1.36×10^{-3} and 7.99×10^{-5}, accordingly. If we do not take into account elastic scattering, i.e. $\sigma_{el} = 0$, then $\Lambda_{eff} = \Lambda = 70.6$ g/cm^2 and these depths of the atmosphere will reach much smaller part of solar neutrons: 4.94×10^{-5} for $h = 700$ g/cm^2 (27.4 times smaller) and 7.06×10^{-7} for $h = 1{,}000$ g/cm^2 (113.3 times smaller). Table 5.12 shows the comparison of the attenuation of solar neutrons reaching the depth 300, 500, 700, 800, 900, 1,000, 1,100, 1,200 and 1,300 g/cm^2 for $\Lambda_{eff} = 106$ and 70.6 g/cm^2, and expected ratio $I_n(106)/I_n(70.6)$.

Table 5.12 shows that the taking into account elastic scattering is very important and this importance increase very much with increasing of the atmosphere's depth in direction of solar neutron propagation (from 4.13 to 468.5 for increasing this depth from 300 to 1,300 g/cm^2).

According to Shibata (1994), the results obtained by Debrunner et al. (1989, 1990) on the neutron monitor sensitivity to solar neutrons (see Section 5.5.2) must be corrected taking into account elastic scattering between nucleons and air nuclei; these results are under-estimated, because the contribution of elastic scattering was ignored (as in Alsmiller and Boughner, 1968). This contribution increases with decreasing of neutron energy (because of increasing of average angle scattering) and became very important at energies below 1 GeV (Section 5.5.5).

5.5.4 *Deflection by Scattering and Expected Angular Distribution of Secondary Nuclear Active Particles in the Atmosphere Depending on E_0 and Depth h*

In Section 5.4.3 was shown that elastic scattering sufficiently reduced the attenuation of the flux of solar neutrons and secondary produced nuclear-active particles. According to Shibata (1993), the deflection by scattering and expected angular distribution $n(h, \theta)$ of solar neutrons and secondary nuclear-active particles in the atmosphere (where angle θ is measured from the incident direction of solar neutrons

Table 5.12 Expected attenuation of solar neutrons reaching different depths for Λ_{eff} with and without taking into account elastic scattering

Depth (g/cm^2)	$\Lambda_{eff} = 106 g/cm^2$	$\Lambda_{eff} = 70.6$ g/cm^2	$I_n(106)/I_n(70.6)$
300	5.90×10^{-2}	1.43×10^{-2}	4.13
500	8.94×10^{-3}	8.40×10^{-4}	10.6
700	1.36×10^{-3}	4.94×10^{-5}	27.4
800	5.28×10^{-4}	1.99×10^{-5}	44.0
900	2.05×10^{-4}	2.91×10^{-6}	70.6
1,000	7.99×10^{-5}	7.06×10^{-7}	113.3
1,100	3.11×10^{-5}	1.71×10^{-7}	181.9
1,200	1.21×10^{-5}	4.15×10^{-8}	291.9
1,300	4.72×10^{-6}	1.01×10^{-8}	468.5

at the top of the atmosphere) can be determined as following. The relation between $n(h,\theta)$ and number $N(h)$ of neutrons on the level h (which used in one-dimensional model, Section 5.5.3) will be

$$N(h) = \int_0^\pi n(h,\theta)d\theta, \quad n(h,\theta) = \sum_{k=0}^{\infty} n_k(h,\theta), \tag{5.62}$$

where $n_k(h,\theta)$ denotes the number of neutrons after k collisions and can be determined by equation

$$\partial n_k(h,\theta)/\partial\theta = -\lambda^{-1} n_k(h,\theta) + \lambda^{-1}(\sigma_{\mathrm{el}}/\sigma_{\mathrm{tot}}) \int_0^\pi P(\theta,\theta')\, n_{k-1}(h,\theta')d\theta, \tag{5.63}$$

where $P(\theta,\theta')$ is the probability for a neutron incident at angle θ' to be scattered on angle θ. Since the angular distribution is determined only by the number of collisions k and function $P(\theta,\theta')$ and does not depend on the depth h explicitly, the number of neutrons $n_k(h,\theta)$ can be simply expressed by a product of two functions with separate variables h and θ: $n_k(h,\theta) = N_k(h)F_k(\theta)$. Then instead of Eq. 5.63 it will be coupled equations:

$$\partial N_k(h)/\partial h = -\lambda^{-1}N_k(h) + \lambda^{-1}(\sigma_{\mathrm{el}}/\sigma_{\mathrm{tot}})N_{k-1}(h); N_0(h) = \exp(-h/\lambda), \tag{5.64}$$

$$F_k(\theta) = \int_0^\pi P(\theta,\theta')F_{k-1}(\theta')d\theta'; \quad F_0(\theta) = \delta(\theta). \tag{5.65}$$

Equation 5.64 described the attenuation of neutrons in the atmosphere and Eq. 5.65 expresses the broadening of the angular distribution of the neutrons caused by a collisions. The solutions of Eqs. 5.64 and 5.65 according to Shibata (1994) will be

$$N_k(h) = (\sigma_{\mathrm{el}}/\sigma_{\mathrm{tot}})^k (k!)^{-1} (h/\lambda)^k \exp(-h/\lambda), \tag{5.66}$$

$$F_k(\theta) = \int_0^\pi d\theta^{(k-1)} \int_0^\pi d\theta^{(k-2)} \ldots \int_0^\pi d\theta^{(1)} P\left(\theta,\theta^{(k-1)}\right) P\left(\theta^{(k-1)},\theta^{(k-2)}\right) \ldots P\left(\theta^{(1)},0\right). \tag{5.67}$$

For comparison with a result of one-dimensional model (Section 5.5.3), Eq. 5.66 can be rewritten in the form:

$$N_k(h) = (k!)^{-1} \left(\frac{h\sigma_{\mathrm{el}}}{\lambda\sigma_{\mathrm{tot}}}\right)^k \exp\left(-\frac{h\sigma_{\mathrm{el}}}{\lambda\sigma_{\mathrm{tot}}}\right) \exp\left(-\left(1-\frac{\sigma_{\mathrm{el}}}{\sigma_{\mathrm{tot}}}\right)\frac{h}{\lambda}\right), \tag{5.68}$$

Table 5.13 The dependence of the standard deviation $\Delta\theta$ from the energy of neutrons E (According to Shibata 1994)

E (MeV)	20	50	100	200	300	400	500	600	800
$\Delta\theta$ (grad)	24.5	16.2	12.4	10.1	8.3	7.1	6.2	5.5	4.6
E (GeV)	1.0	1.2	1.5	2	3	4	6	8	10
$\Delta\theta$ (grad)	4.0	3.5	3.0	2.4	1.8	1.4	1.0	0.8	0.6

where the last term is the same as in Eq. 5.60 and the remaining part reflects additional solar neutron attenuation caused by the broadening of the angular distribution.

According to review of measurements of neutron scattering compiled in Shibata (1993), the measured angular distribution can be approximated by the Gaussian distribution with the some standard deviation $\Delta\theta$. The dependence of $\Delta\theta$ from neutron energy can be calculated by semi-empirical approximated formulas in Shibata (1993); results are shown in Table 5.13.

Table 5.13 shows that with increasing of neutron energy the standard deviation $\Delta\theta$ characterized the neutron scattering in the atmosphere, decrease sufficiently. The expected angular distribution of solar neutrons arrived in vertical direction ($\theta_0 = 0°$) on the level of observation h will be also Gaussian type with the standard deviation

$$<\Delta\theta> = \Delta\theta\left(\frac{h\sigma_{\mathrm{el}}}{\lambda\sigma_{\mathrm{tot}}}\right)^{1/2}. \qquad (5.69)$$

For example, for observation at mountain altitude ($h = 776$ g/cm^2) and at sea level ($h = 1{,}000$ g/cm^2), the parameter $\frac{h\sigma_{\mathrm{el}}}{\lambda\sigma_{\mathrm{tot}}} = 3.66$ and 4.72 and the angular spread of solar neutrons with energy 500 MeV ($\Delta\theta = 6.2°$) will be expected with the standard deviation $<\Delta\theta> = 11.9°$ and 13.5° for mountain and sea level observations, correspondingly. If the arriving zenith angle of solar neutrons is θ_0, that in Eq. 5.69 instead of h will be the air mass depth in direction of Sun: $h_S = h/\cos\theta_0$. Table 5.14 shows the dependence of $<\Delta\theta>$ from solar neutron energy E_0 and from the air mass in direction to the Sun h_S.

5.5.5 *Model for Monte Carlo Simulation of Small Energy Neutron Propagation Through the Earth's Atmosphere*

In Shibata (1994) the following processes between a nucleon n (neutron or proton) and an air nucleus A were considered in the Monte Carlo simulation of small energy neutron propagation through the Earth's atmosphere: 1) elastic scattering $n + A \rightarrow n + A$ (see Section 5.5.3), 2) charge exchange process $n + A \rightarrow n' + A$, 3) inelastic scattering $n + A \rightarrow n' + A' + X$, and 4) energy loss of protons by ionization in the air. All particles are considered as stable (the lifetime of neutrons 889 s is greater than the neutron's propagation time in the atmosphere at least six to seven orders),

Table 5.14 The expected standard deviation <$\Delta\theta$> (in degrees) for the angular spread of solar neutrons for the different air mass depth in direction of Sun h_S in dependence from the energy of solar neutrons E_0

h_S (g/cm^2)	E_0 (MeV)							
	50	10	20	30	50	80	1,000	1,500
60	27.2	20.9	17.0	14.0	10.4	7.7	6.7	5.0
65	28.4	21.7	17.7	14.5	10.9	8.1	7.0	5.3
70	29.4	22.5	18.4	15.1	11.3	8.4	7.3	5.5
75	30.5	23.3	19,0	15.6	11.7	8.7	7.5	5.6
80	31.5	24.1	19.6	16.1	12.0	8.9	7.8	5.8
85	32.4	24.8	20.2	16.6	12.4	9.2	8.0	6.0
90	33.4	25.5	20.8	17.1	12.8	9.5	8.2	6.2
95	34.3	26.2	21.4	17.6	13.1	9.7	8.5	6.4
1,000	35.2	26.9	21.9	18.0	13.5	10.0	8.7	6.5
1,200	35.5	29.5	24.0	19.7	14.8	10.9	9.5	7.1
1,500	43.1	33.0	26.9	22.1	16.5	12.2	10.6	8.0
2,000	49.8	38.1	31.0	25.5	19.0	14.1	12.3	9.2

that the density effects will be negligible and the atmosphere can be considered as absorber characterized only by one parameter: air mass depth in direction of neutron propagation.

According to Shibata (1994), the calculation is started by injecting one neutron with energy E_0 (in MeV) at an angle $\theta_0 = 0°$ at the top of the atmosphere ($h = 0$). Then all collision processes of neutrons and protons with air nuclei are simulated by the Monte Carlo method with including all mentioned above processes. The incident neutron and all produced nucleons are followed until their energies drop below a given threshold level $E_{th} = 20$ MeV or until they arrive at a given limit of atmospheric depth. At every 100 g/cm^2 atmospheric level the kind of particles, their energy and angle of propagation (relative to the direction of primary neutron arriving) are recorded. The propagation of neutrons in the atmosphere according to Shibata (1994) is shown schematically in Fig. 5.30.

In Shibata (1994) for each incident energy one million neutrons are injected, and the energy and angular distribution of particles are obtained at each atmospheric depth. Results of calculations of Shibata (1994) are shown in Figs. 5.31, 5.32, 5.33, 5.34, 5.35, and 5.36. Figure 5.31 shows the expected attenuation of solar neutrons with primary energies from 50 up to 900 MeV as a function of atmospheric depth up to 1,100 g/cm^2.

Let us note that in Fig. 5.31 the attenuation of neutrons is the same as integral multiplicity $m^n(E_0, \theta_0, h)$ as we used in previous sections (number of detected particles from one primary particle, see Sections 5.2 and 5.3).

The local mean attenuation length

$$\lambda_{att} \approx -(h_2 - h_1)/[\ln(m^n(E_0, \theta_0, h_2)) - \ln(m^n(E_0, \theta_0, h_1))] \qquad (5.70)$$

between levels h_1 and h_2 was determined for atmospheric depth near 700 g/cm^2; it is shown in Fig. 5.32.

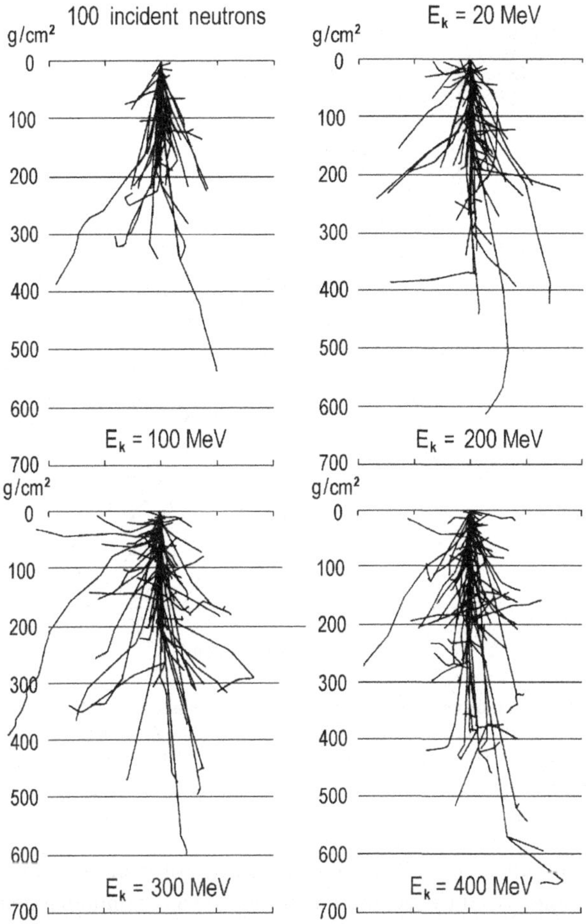

Fig. 5.30 The simulation of the propagation of solar neutrons in the atmosphere for incident neutrons with $E_0 = 100$, 200, 300 and 400 MeV. The incident neutron and all produced nucleons (neutrons and protons) are followed until their energies drop below $E_{th} = 20$ MeV. The unit of both the ordinate and the abscissa corresponds to 100 g/cm^2 (According to Shibata 1994)

From Fig. 5.32 it can be seen that λ_{att} increases from ~50 g/cm^2 for $E_0 = 50$ MeV to $\lambda_{att} \approx 110$ g/cm^2 for $E_0 \geq 250$ MeV and up to 1,000 MeV does not depend from E_0.

The attenuation of neutrons (integral multiplicity) on levels from 100 up to 1,100 g/cm^2 in dependence of E_0 is shown in Fig. 5.33.

The example of expected energy spectra of neutrons from one incident neutron with primary energy $E_0 = 500$ MeV on the levels from 100 to 1,100 g/cm^2 is shown in Fig. 5.34 (the same as differential multiplicity $N(E_0, E, h, \theta_0)$, see Section 5.3.3).

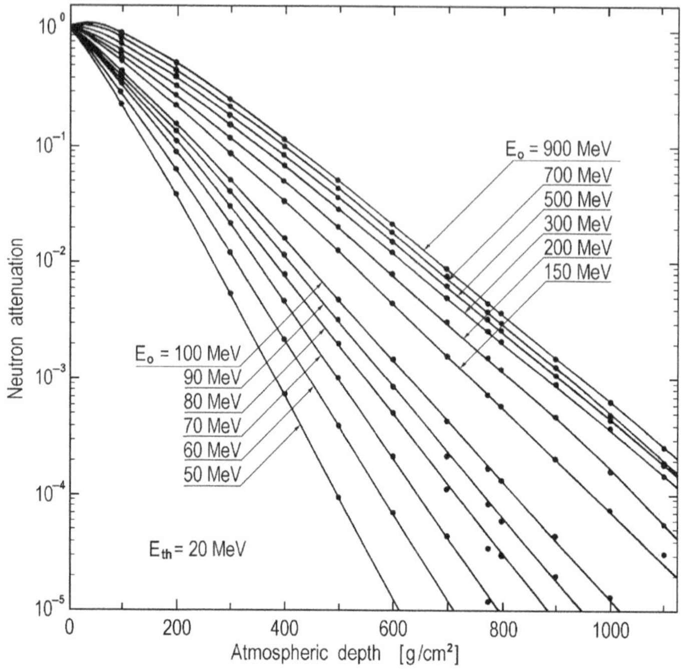

Fig. 5.31 The attenuation of neutrons as a function of atmospheric depth. The *curves* correspond to the incident energies of neutrons at the top of the atmosphere: $E_0 = 900, 700, 500, 300, 200, 150, 100, 90, 80, 70, 60$, and 50 MeV from the *topmost curve*. The threshold level $E_{th} = 20$ MeV (According to Shibata 1994)

According to Shibata (1994), the energy spectra of neutrons at an observational altitude are characterized by two main features: (1) a continuum extending over the whole energy range which can be approximated by an exponentially decreasing function

$$N(E_0, E, h, \theta_0) \propto \exp(-E/165), \tag{5.71}$$

where E is the neutron energy in MeV; (2) a narrow peak close to the incident energy which corresponds to the part of primary neutrons penetrating through the atmosphere without any collision, or which undergo only elastic scattering processes (punch through). In Fig. 5.35 are shown expected spectra of neutrons in dependence of E_0 for the level $h = 776$ g/cm^2 (these data can be used for calculations of neutron monitor sensitivity to solar neutrons).

Figure 5.36 shows the results of Shibata (1994) calculations of the sensitivity $S_n(h, E_0, \theta_0)$ of neutron monitors of IGY type and of NM64 type in comparison with results of Chupp et al. (1987) and Debrunner et al. (1989).

In connection with Fig. 5.36 Shibata (1994) notes that the results of Chupp et al. (1987) for IGY type neutron monitor and of Debrunner et al. (1989) for NM64

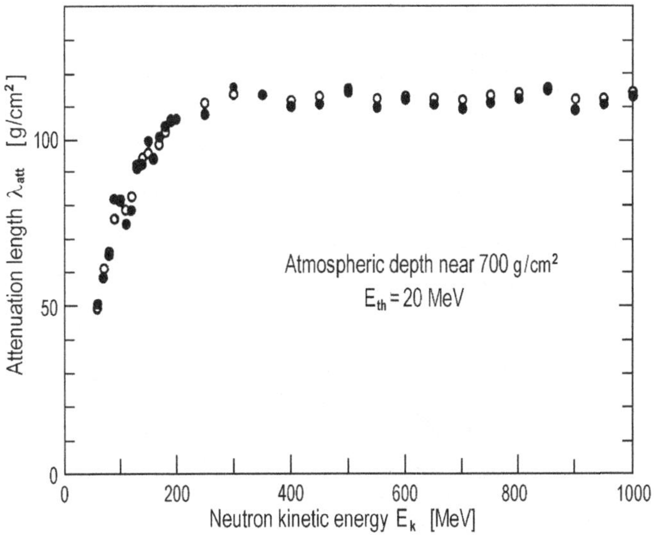

Fig. 5.32 The local mean attenuation length λ_{att} near the atmospheric depth of 700 g/cm^2. The values of λ_{att} are calculated using Eq. 5.70 with *open circles* representing $h_1 = 600$ g/cm^2, $h_2 = 776$ g/cm^2 and *solid circles* representing $h_1 = 700$ g/cm^2, $h_2 = 776$ g/cm^2 (According to Shibata 1994)

neutron monitor are in contradiction with conclusion of Hatton (1971) that the detection efficiency of the NM64 neutron monitor is higher than that of IGY type neutron monitor. From other hand, the calculations of Shibata (1994) of the sensitivity $S_n(h, E_0, \theta_0)$ of these two types of neutron monitors (see Fig. 5.36) are in good agreement with results of Hatton (1971). The energy dependence of the neutron monitors sensitivity above 300 MeV of primary energy is about the same in Chupp et al. (1987), Debrunner et al. (1989) and Shibata (1994), but below 300 MeV even normalized Debrunner et al. (1989) sensitivity is much lower than Shibata (1994) calculations by Monte Carlo method. This discrepancy appears for both NM64 and IGY neutron monitors and caused by the fact that the contribution of elastic scattering in Chupp et al. (1987) and Debrunner et al. (1989) has not been taken into account. It is especially important for small energy neutrons, as it was mentioned above (Section 5.4.3).

5.6 On the 'Refraction' Effect in Solar Neutron Propagation Through the Earth's Atmosphere

Smart et al. (1995) note that the assumption of straight-ahead transport of solar neutrons at an oblique angle through the Earth's atmosphere with exponential density gradient appears to be flawed for case of mono-directional flux of incident neutrons.

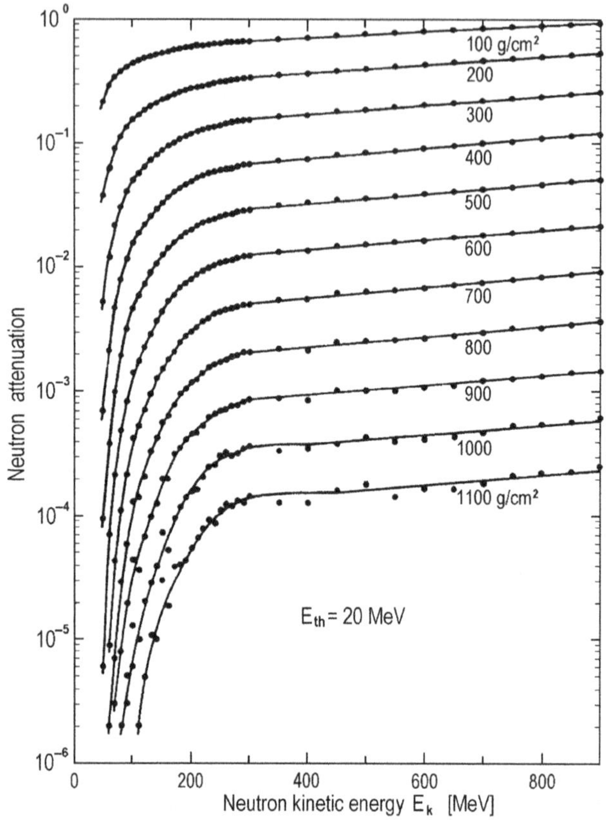

Fig. 5.33 The attenuation of neutrons (integral multiplicity $m^n(E_0, \theta_0, h)$) in the atmosphere as a function of the incident energies of neutrons at the top of the atmosphere. The threshold level $E_{th} = 20$ MeV. The label under each curve corresponds to the atmospheric depth (According to Shibata 1994)

5.6.1 Apparent Air Mass Along a Line of Sight Through the Atmosphere to the Sun and Contradiction with the Theory of Solar Neutron Propagation in the Atmosphere for the Event of 24 May, 1990

This event is the largest ground-level increase attributable to solar neutrons yet reported (Pyle et al. 1991; Shea et al. 1991; see also below, Section 7.17) and was detected also by neutron monitors for which the apparent air mass along a line of sight through the atmosphere to the Sun where much more than 1,000 g/cm^2. In Smart et al. (1995) all data were recalculated to the counting rate of an equivalent 12-NM-64 neutron monitor taking into account the latitude of observation and type of detector. Figure 5.37 shows a 12-NM-64 equivalent counting rate increase for the

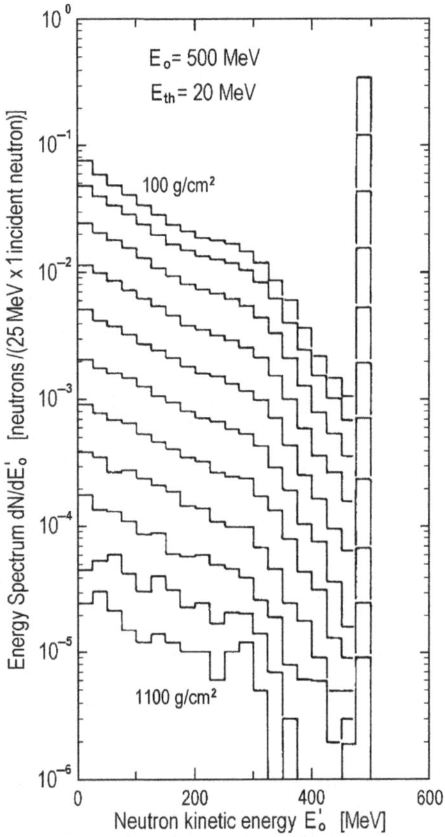

Fig. 5.34 The energy spectra of neutrons at various atmospheric depth (the same as differential multiplicity $N(E_0, E, h, \theta_0)$, see Section 5.3.3). The incident energy of primary neutrons is $E_0 =$ 500 MeV, the threshold level $E_{th} = 20$ MeV (According to Shibata 1994)

24 May 1990 event in dependence of line-of sight air mass. From Fig. 5.37 is possible to obtain an apparent attenuation length. The results of this procedure gave a totally unreasonable apparent attenuation length of 208 g/cm^2 which is about 2 times bigger than expected from the theory of solar neutron propagation through the atmosphere (see Section 5.5.5), and even bigger than attenuation path for galactic cosmic rays (about 145 g/cm^2).

5.6.2 The Illustrative Model of the Solar Neutron 'Refraction' Effect in the Earth's Atmosphere

To explain this discrepancy Smart et al. (1995) supposed that in the atmosphere with down directed air mass density gradient the zenith angle of the main direction of oblique solar neutron flux will be little decrease after each scattering and in this case the total air mass passed by solar neutrons can be much smaller and the attenuation length became also smaller (see Fig. 5.38).

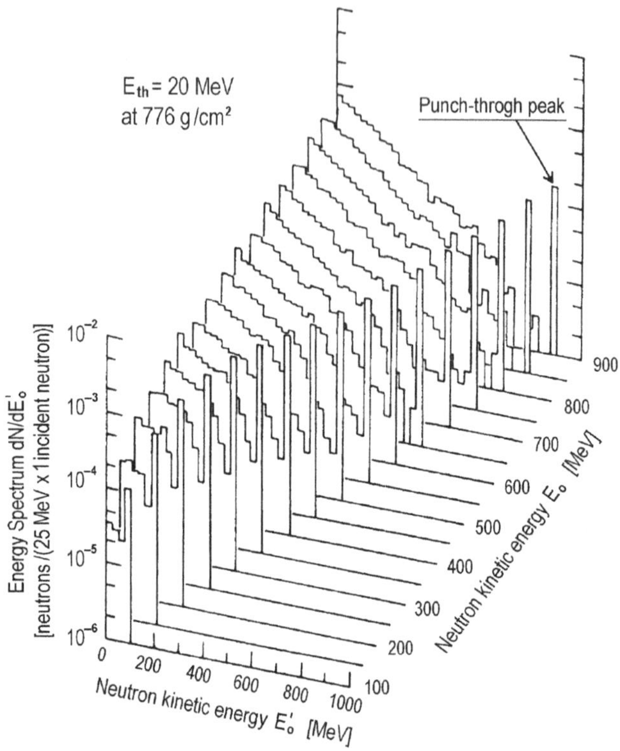

Fig. 5.35 The energy spectra of neutrons (differential multiplicity $N(E_0, E, h, \theta_0)$) at an atmospheric depth $h = 776$ g/cm^2, at the various incident neutron energies E_0 labeled on the right; E is the energy of the secondary neutrons (According to Shibata 1994)

Table 5.15 shows how will be reduce the air mass passing by solar neutrons taking into account 'refraction' effect (h_r) in comparison with air mass in direction of Sun (h_S) in dependence of the zenith angle of the Sun θ_S.

5.7 Simulation and Analytical Description of the Refraction Effect

5.7.1 The Matter and Short History of the Problem

As we note above, neutrons are one of the many products of solar flare processes, they are of particular importance since they are produced directly by nuclear interactions of the flare-accelerated ions with the ambient gas of the solar atmosphere. Since the first confirmed observations of solar neutrons at the Earth's surface (Chupp et al. 1987; see also above, Chapter 3) there has been growing interest in

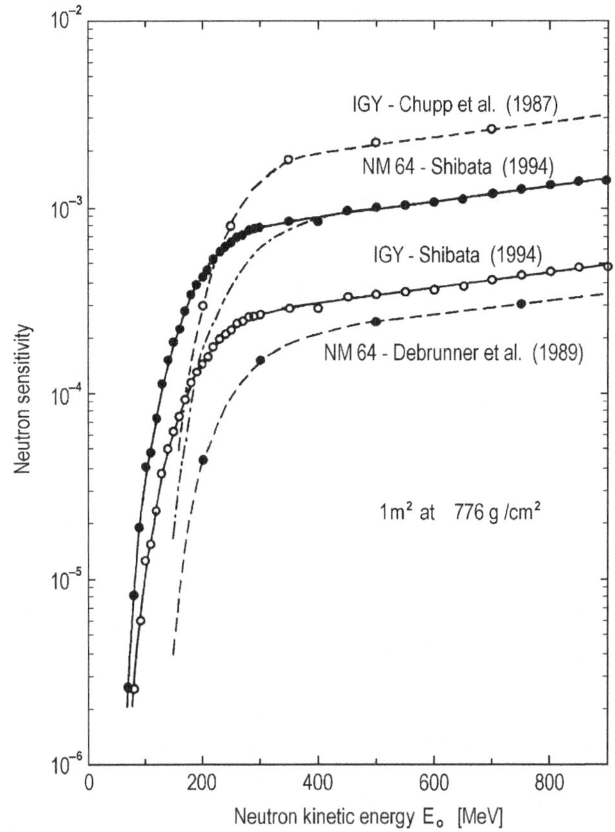

Fig. 5.36 Sensitivity of neutron monitors $S_n(h, E_0, 0_0)$ of IGY and NM64 types to solar neutrons normalized to 1 m^2 area (as in Section 5.4.2). The *abscissa* indicates the kinetic energy E_0 of incident neutrons at the top of the atmosphere. *Open circles* indicate the IGY neutron monitor, and *solid circles* indicate the NM64 neutron monitor. *Solid curves* depict results of Shibata (1994), and *dashed curves* depict the results of Chupp et al. (1987) and Debrunner et al. (1989). The *dot-dashed curve* indicates the result of the NM64 neutron monitor of Debrunner et al. (1989) but increased by a factor of 4 to normalize with the curve calculated by Shibata (1994) for the NM64 neutron monitor at higher energies (According to Shibata 1994)

their detection by the world network of CR observatories. Several research groups investigated the processes affecting neutron propagation in the Earth's atmosphere and the way in which the different kinds of neutron detectors react to this solar flux (see, e.g., Efimov and Terekhov 1988; Debrunner et al. 1990; Shibata 1993, 1994). To date these efforts have shown some success in understanding atmospheric neutron transport and detection at the Earth's surface. The event of May 24, 1990, the largest solar neutron event ever recorded by neutron monitors, is an enormous challenge to the existing explanations of the phenomena (see e.g. Debrunner et al. 1997 and references therein). The apparent attenuation length λ obtained under the assumption of straight-ahead transportation of solar neutrons

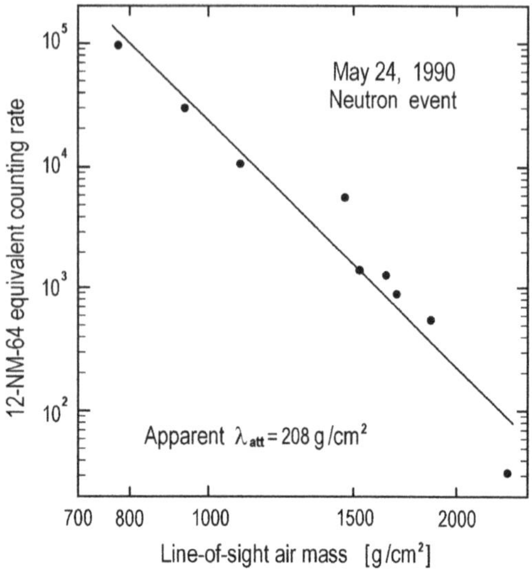

Fig. 5.37 The 12-NM-64 equivalent counting rate increase in dependence of line-of-sight air mass for the solar neutron event of May 24, 1990. The determined apparent attenuation length $\lambda_{att} = 208$ g/cm^2 (According to Smart et al. 1995)

through the Earth's atmosphere is 208 g/cm^2 (Debrunner et al. 1993; Smart et al. 1995). This is in contradiction with the calculated value of 110 g/cm^2, based on measurements of neutron cross-section scattering in carbon (Shibata 1993, 1994; see above, Section 5.4). If we calculate λ following the qualitative approach for the neutron refraction suggested by Smart et al. (1995), the solar neutron attenuation length λ may be reduced to values around 110 g/cm^2.

Several authors have performed careful numerical simulations. Debrunner et al. (1989, 1990), in an attempt to estimate the sensitivity of neutron detectors, did Monte Carlo calculations from which several characteristics of the nucleon cascade may be extracted. Shibata (1993, 1994) used the Monte Carlo method with a nuclear interaction model based on accelerator experiments and was able to reproduce observations of two solar neutron events. Because of an asymmetry in the distribution of neutrons caused by the refraction effect (as is discussed below), the transport of solar neutrons through the Earth's atmosphere for arrival at inclined zenith angles is essentially different from that for vertical arrival. This is in contradiction with the assumption of Shibata (1993, 1994) in this respect. This effect was noticed by Debrunner et al. (1993) and referred to as 'lateral spreading'. However, in order to fit observations they had to take the upper limit of the Monte Carlo simulations of Debrunner et al. (1990). The authors gave no further explanation. In an additional attempt to clarify the problems associated with solar neutron propagation through the Earth's atmosphere, Dorman et al. (1999a) simulate this process, taking into account multiple scattering and absorption; an analytical solution is also presented.

Dorman et al. (1999a) present results of a numerical simulation of small-angle neutron multi-scattering and attenuation in the Earth's atmosphere. A range of initial zenith angles, θ_0 from 0° up to 90°, and different atmospheric depths h are

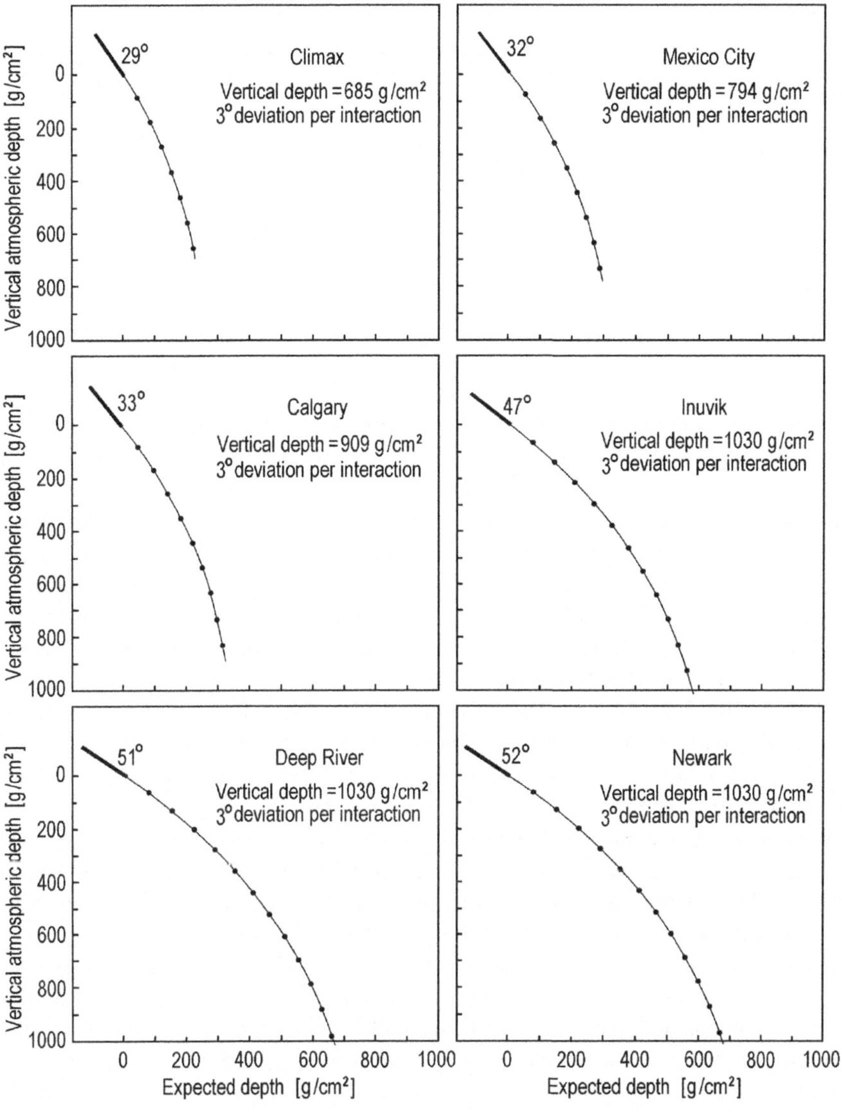

Fig. 5.38 Conceptual illustration of the bending away from the straight-ahead direction for a mono-direction beam for some of the neutron monitors observing the 24 May 1990 solar neutron event. In each panel the heavy line indicates the initial line-of-sight direction from the Sun to the neutron monitor. The dotted line represents the possible path if at each interaction there is a 3° scattering in the direction of the density gradient (From Smart et al. 1995)

considered. It was shown that the angular distribution of neutrons remains symmetrical only for vertical arrival, $\theta_0 = 0°$. For inclined arrival, the distribution becomes asymmetric; this asymmetry grows with increasing initial zenith angle θ_0. The asymmetry of the angular distribution enhances the solar neutron refraction effect.

Table 5.15 Expected h_r in comparison with h_S for the event May 24, 1990 (From Smart et al. 1995)

Station name	h (g/cm^2)	θ_S (deg)	h_S (g/cm^2)	h_r (g/cm^2)
Climax	685	28	777	721
Mexico City	793	31	929	844
Calgary	909	33	1,090	960
Inuvik	1,030	47	1,525	1,182
Deep River	1,015	51	1,618	1,226
Newark	1,033	51	1,664	1,237
Mt. Washington	821	55	1,441	1,062
Durham	1,033	55	1,826	1,228
Goose Bay	1,030	62	2,258	1,390
Mt. Norikura	741	77	3,508	1,356

It follows from Dorman's et al. (1999a) simulation that this effect is caused by the stronger attenuation of neutrons scattered to zenith angles larger than the arrival angle. This effect makes the effective zenith angle become smaller as the solar neutrons propagate downwards. The refraction angle of our simulation is not constant, but it is a growing function of θ_0 and h. An analytical description of the problem is also performed starting from simple assumptions of the parameters involved in the solution. We modify this simple model in successive steps to take into account additional details of the solar neutron propagation, until it reaches reasonable agreement with our numerical simulation estimates.

Our results show that: (1) The result of Shibata (1993, 1994) with respect to the expected solar neutron flux for inclined arrival of neutrons must be corrected, and (2) The 'refraction' effect of solar neutrons suggested by Smart et al. (1995) and referred to by Debrunner et al. (1993) as 'lateral spreading' exists but is not constant; it depends on the variables θ_0 and h. The causes of this effect are also different from those assumed by Smart et al. (1995). To reach these conclusions, we present the general approach to the problem in Section 5.7.2. Section 5.7.3 describes the method to simulate neutron propagation in the Earth's atmosphere. Results of the simulation for the expected fluxes for vertical and inclined arrival are discussed in Section 5.7.4. In Section 5.7.5 we present a detailed analytical solution to the problem and make comparisons with the numerical simulation results. Finally, we make our summary and conclusions in Section 5.7.6.

5.7.2 *Expected Zenith Angle Distributions of Scattered Neutrons*

From the point of view of neutron propagation, the atmosphere may be considered homogeneous since the neutron decay time is much longer than the time required to cross the atmosphere. Additionally, as the only relevant parameter for neutron transport in the atmosphere is the path in g/cm^2, we use a planar and uniform atmosphere model. With this in mind, consider a solar neutron arriving at the top of the atmosphere with an initial zenith angle θ_0. At some point P the neutron interacts with an atom of air and is scattered by an angle Δ, where $0 \leq \Delta \leq \pi$. After the interaction the scattered angle Δ and the azimuth angle φ ($0 \leq \varphi \leq 2\pi$) describe the motion of the neutron. Under the assumptions that the probability distribution will be, (1) homogeneous and equal to $1/2\pi$ in φ space and, (2) Gaussian in Δ space, we have

$$f(\Delta,\phi)d\phi d\Delta = \left[(2\pi)^{3/2}\sigma\right]^{-1}\exp(-\Delta^2/2\sigma^2)d\phi d\Delta, \tag{5.72}$$

where σ is the angular standard deviation of the neutron distribution that depends on the neutron energy, and:

$$\int_0^{2\pi} d\phi \int_0^{\pi} f(\Delta,\phi)d\Delta \to \int_0^{2\pi} d\phi \int_0^{\infty} f(\Delta,\phi)d\Delta = 1. \tag{5.73}$$

The upper limit π in the integral over Δ may be replaced by ∞ in Eq. 5.73 if $\sigma << 1$ (where σ is in radians). We determined the values of σ (see Table 5.16) for different neutron energies from the semi-empirical formulas of Shibata (1993).

Because the attenuation of neutrons is determined by the zenith angle θ, we may estimate the expected zenith angle distribution on the basis of Eq. 5.72. To make an explicit use of θ, in the calculations, we do a change of variables in Eq. 5.72 to replace Δ and φ, by $\theta - \theta_0$ and ϕ as

$$\theta - \theta_0 = \Delta\cos\phi,\quad \varphi = \Delta\sin\phi,\quad \Delta^2 = (\theta - \theta_0)^2 + \varphi^2. \tag{5.74}$$

According to Table 5.16, $\sigma << 1$ (in radians) for neutron energies $E \geq 100$ MeV. Using this result, we introduce Eq. 5.74 into Eq. 5.72 and integrate over ϕ to arrive at

$$\begin{aligned} f(\theta - \theta_0) &= \frac{1}{\pi\sigma\sqrt{2\pi}} \int_{-\infty}^{\infty} \exp\left(-\frac{(\theta-\theta_0)^2 + \varphi^2}{2\sigma^2}\right) d\varphi \\ &= \frac{1}{\pi}\exp\left(-\frac{(\theta-\theta_0)^2}{2\sigma^2}\right). \end{aligned} \tag{5.75}$$

Table 5.16 The standard deviation σ for the angular distribution of scattered neutrons as a function of neutron energy E. Calculated according to the semi-empirical formulas of Shibata (1993)

E (MeV)	σ (deg)	E (GeV)	σ (deg)
20	24.5	1.0	4.0
30	16.2	1.2	3.5
100	12.4	1.5	3.0
200	10.1	2	2.4
300	8.3	3	1.8
400	7.1	4	1.4
500	6.2	5.6	1.0
600	5.5	8	0.8
800	4.6	10	0.6

Normalizing $f(\theta - \theta_0)$ to 1, we get

$$f_n(\theta - \theta_0) = \frac{1}{\sigma\sqrt{2\pi}} \exp\left(-\frac{(\theta - \theta_0)^2}{2\sigma^2}\right). \tag{5.76}$$

Here the subscript n indicates that the function has been normalized, i.e. $\int_{-\infty}^{\infty} f_n(\theta - \theta_0)d\theta = 1$. Therefore, if $\sigma << 1$, the zenith angle distribution of scattered neutrons will also be a Gaussian with the same standard deviation σ. Consequently, the average zenith angle deviation may be estimated from the first moment of $f_n(\theta - \theta_0)$. Any elementary interaction is completely equivalent; hence Eq. 5.76 can be generalized to represent the angular distribution of any successive neutron interaction. In this case, $\theta - \theta_0$ changes to $\theta - \theta_k$, and it represents the deviation from the direction of a secondary neutron propagating at an angle θ_k before the k-th interaction.

Before going into the details of the simulations, let us divide the whole $\theta - \theta_k$ space into three regions: (1) region ***a*** is from $-\infty$ to $-\chi$, (2) region ***b*** is from $-\chi$ to $+\chi$, and (3) region ***c*** is from $+\chi$ to $+\infty$. If $\int f_n(\theta - \theta_0)d\theta = 1/3$ in each region, then $\chi = 0.430685\sigma$, and we obtain for each of the three regions the following:

$$\begin{aligned} \langle\theta - \theta_0\rangle_a &= -\delta = -1.09082\sigma, \quad \langle\theta - \theta_0\rangle_b = 0, \quad \langle\theta - \theta_0\rangle_c = +\,\delta \\ &= 1.09082\sigma. \end{aligned} \tag{5.77}$$

5.7.3 Description of the Numerical Simulation

Dorman et al. (1999a) have produced a model of the interaction of an individual neutron with an atmospheric atom assuming that all interactions are equivalent; we now assume that neutrons interact only at specific depths $h_{i,j(i)}$ for which the total mass depth is $L\left(h_{i,j(i)}\right) = i\lambda$. Here λ, is the transport path; the number of interactions at $i = 1, 2, 3, \ldots$ is $j(i) = 1,\ 2,\ 3,\ \ldots\ 3^i$ (see below). For simplicity we take the same value of λ for both attenuation and scattering processes ($\lambda = 110$ g/cm^2, as obtained by Shibata, 1993, 1994). According to Eq. 5.77, after each interaction the neutron will either: (1) decrease its previous zenith angle by an angle δ, (2) remain at the same zenith angle, or (3) increase its zenith angle by an angle δ. For example, according to Eq. 5.77, if we choose $\delta = 5°$, then $\sigma = 4.584°$. From Table 5.16 we see that this value of σ corresponds to a neutron energy $E \approx 800$ MeV.

The Dorman's et al. (1999) numerical simulation starts with one neutron at the top of the atmosphere and stops when either the neutron has crossed the selected level of observation, h, or the zenith angle reaches $\pm 90°$. We determine the zenith angle distributions of fluxes in the vertical plane that contains the line joining the neutron arrival point with the Sun. The arrival angle at the top of the atmosphere is θ_0. The first interaction will be at a depth

$$h_1 = \lambda \cos \theta_0. \tag{5.78}$$

After the first interaction the original flux is divided into three and attenuated; therefore we obtain three fluxes, each of intensity $3^{-1} \times \exp(^{-}\lambda)$. Since the fluxes are normalized, the intensity is taken to be dimensionless. The corresponding zenith angles of each of these are:

$$\theta_{1,1} = \theta_0 - \delta, \quad \theta_{1,2} = \theta_0, \quad \theta_{1,3} = \theta_0 + \delta. \tag{5.79}$$

The second interactions will be at the depths

$$h_{2,1} = h_1 + \lambda \cos \theta_{1,1}, \quad h_{2,2} = h_1 + \lambda \cos \theta_{1,2}, \quad h_{2,3} = h_1 + \lambda \cos \theta_{1,3}, \tag{5.80}$$

where h_1 is determined by Eq. 5.78 and $\theta_{1,1}, \quad \theta_{1,2}, \quad \theta_{1,3}$ are determined by Eq. 5.79. After i interactions we obtain 3^i fluxes with the same intensity $3^{-i} \times \exp(-i\lambda)$ in directions characterized with zenith angles $\theta_{i,j(i)}$, where $j(i) = 1, 2, 3, \ldots 3^i$. These fluxes will be generated at depths $h_{i,j(i-1)}$ and the levels of the following interactions will be

$$h_{i+1,j(i)} = h_{i,j(i-1)} + \lambda \cos \theta_{i,j(i)}. \tag{5.81}$$

The recurrent formula to calculate $\theta_{i,j(i)}$ is

$$\theta_{i,j(i)} = \theta_{i-1,j(i-1)} + (l-2)\delta, \tag{5.82}$$

where $l = 1$, 2 or 3 and can be estimated from the relation

$$j(i) = 3 \times [j(i-1) - 1] + l \tag{5.83}$$

(see Fig. 5.39 for rough illustration of neutron ‘shower’).

Parameters i and $j(i) \leq 3^i$ determine the paths of the solar neutrons from the top of the atmosphere, the heights of the past $h_{i,j(i-1)}$ and following $h_{i+1,j(i)}$ interaction, the intensity at each height $I_{i,j(i)}$, and the zenith angles $\theta_{i,j(i)}$. Now we can calculate the intensity at any height h that corresponds to the flux $I_{i,j(i)}$ crossing this level. This means that h is between the simulation levels $h_{i,j(i-1)}$ and $h_{i+1,j(i)}$. Therefore the flux of solar neutrons that crossed the level h with zenith angle $\theta_{i,j(i)}$ will be

$$I_{i,j(i)} = 3^{-i} \times \exp\left\{-\lambda\left[i + \left(h - h_{i,j(i-1)}\right) / \left(h_{i+1,j(i)} - h_{i,j(i-1)}\right)\right]\right\}. \tag{5.84}$$

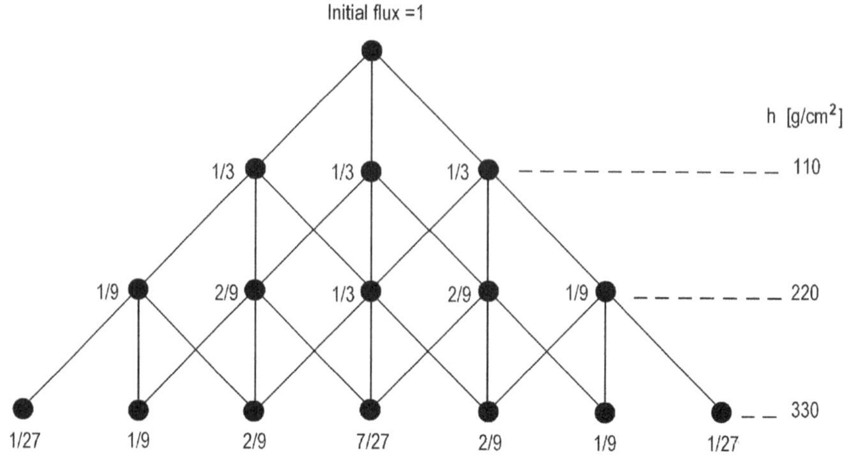

Fig. 5.39 Schematic representation of the process simulating the solar neutron 'shower' in the atmosphere. Every fraction must be multiplied by the corresponding attenuation factor to get the actual neutron flux at each point (see text) (From Dorman et al. 1999a)

After adding all the fluxes for the same θ at depth h we obtain the normalized angular distribution $f_n(h, \theta_0, \theta)$ of the solar neutron flux with arrival zenith angle θ_0. The expected total flux of solar neutrons, $F(h, \theta_0)$, may be estimated adding all these contributions. Therefore the effective zenith angle $\theta_e(h, \theta_0)$ of the solar neutron distribution may be approximated as

$$\theta_e(h, \theta_0) = F^{-1}(h, \theta_0) \int_{-\pi/2}^{\pi/2} f_n(h, \theta_0, \theta)\theta d\theta. \tag{5.85}$$

It is important to note that Dorman's et al. (1999) approximation to the problem of atmospheric neutron transport does not explicitly consider charge exchange processes, inelastic scattering or proton ionization energy losses. Shibata (1994) did Monte Carlo calculations of the neutron cascade in the atmosphere where all the aforementioned processes were included. He considered atmospheric density effects to be negligible and represented the atmosphere as an absorber characterized by one parameter only, namely the air mass path in the direction of propagation, as it is also done in Dorman et al. (1999a). From an extensive calculations, Shibata (1994) found that the attenuation length λ equals 110 g/cm^2, independent of energy in the range from 250 to 1,000 MeV; this value includes all the processes considered in his work. We use in our calculations the attenuation length obtained by Shibata (1994), and in this manner we make an implicit consideration of all the nonelastic processes considered by him.

5.7.4 Expected Fluxes and Zenith Angle Distributions for Vertical Arrival of Solar Neutrons

To illustrate the process, we consider the case of vertical arrival ($\theta_0 = 0°$) of solar neutrons with $E \sim 800$ MeV. Since the attenuation length is constant in a wide energy range (see Section 5.7.3), for smaller energies the only difference will be a bigger angle scattering, implying a wider distribution at each atmospheric depth level. Otherwise, the results will be essentially the same. According to the procedure described in Section 5.7.3, the first interaction occurs at a depth $h_1 = \lambda = 110$ g/cm^2 and produces three fluxes at $\theta_{11} = -5°$, $\theta_{12} = 0°$, and $\theta_{13} = +5°$. The second interactions will occur for θ_{11} and θ_{12} at $h_{21} = h_{23} = (110 + 110 \times \cos 5°)$ g/cm^2 and for θ_{12} at $h_{22} = 220$ g/cm^2. Continuing this process according to Section 5.7.3, we may estimate the expected angular distribution of solar neutrons at different atmospheric levels. In Fig. 5.39 was shown a diagram that represents the numerical simulation process to develop the solar neutron 'shower'. The small differences in the interaction levels for vertical and inclined flux propagation are ignored in Fig. 5.39. Every fraction must be multiplied by $\exp(-i)$, where i =1, 2, 3 corresponds to each of the three levels of the interaction shown to get the actual flux at every point. Results of the expected flux and zenith distribution at the depths 400, 500, 600, and 700 g/cm^2 are shown in Table 5.17 and illustrated in Fig. 5.40. Because of the symmetry of the angular distribution, Table 5.17 shows only positive angles, but the fluxes are always the total fluxes (positive and negative angles).

The predicted fluxes for vertical arrival are the same as those obtained by Shibata (1993, 1994) as expected, since we have used the same λ. The zenith angle distributions of solar neutrons are explicitly obtained in Dorman et al. (1999a) for the first time and are a tool that may be used for comparison with actual observations.

5.7.5 Expected Fluxes and Zenith Angle Distributions for Inclined Arrival of Solar Neutrons

Dorman et al. (1999a) simulated monoenergetic (E = 800 MeV) solar neutron scattering and attenuation for inclined arrival by the method described in Section 5.7.3

Table 5.17 The expected fluxes and zenith angle distributions[a] of solar neutrons for $\theta_0 = 0°$ and observation levels h = 400, 500, 600, and 700 g/cm^2 (From Dorman et al. 1999a)

h (g/cm^2)	+30°	+25°	+20°	+15°	+10°	+5°	0°	Flux
400				9.36E-4	2.87E-3	5.80E-3	6.80E-3	0.02602
500			1.20E-4	4.97E-4	1.27E-3	2.07E-3	2.47E-3	0.010396
600		1.48E-5	7.89E-5	2.47E-4	5.10E-4	7.78E-4	8.87E-4	0.004145
700	1.76E-6	9.68E-6	4.16E-5	1.07E-4	2.01E-4	2.91E-4	3.28E-4	0.001643

[a]Only positive angles are shown but the flux is total. Read 9.36E-4 as 9.36×10^{-4}.

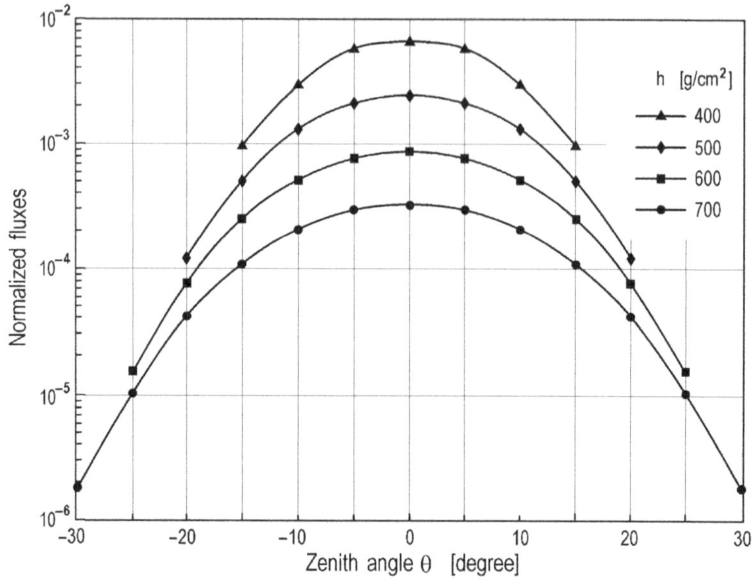

Fig. 5.40 Numerical simulation expected zenith angle distribution of solar neutrons at the atmospheric depths of 400, 500, 600 and 700 g/cm^2 for vertical arrival at the top of the atmosphere (From Dorman et al. 1999a)

and computed the zenith angle distributions of fluxes in the vertical plane that contains the line joining the arrival point with the Sun. The results are shown in Figs. 5.41, 5.42, 5.43, 5.44, and 5.45 for $\theta_0 = 15°$, 30°, 45° 60°, and 75°.

The distributions shown in Figs. 5.41, 5.42, 5.43, 5.44, and 5.45 are not symmetrical relative to the arrival directions of the solar neutrons. This asymmetry is due to the greater attenuation of neutrons scattered to larger zenith angles. The asymmetry shown by Figs. 5.41, 5.42, 5.43, 5.44, and 5.45 increases with the arrival zenith angle, and for a given θ_0 the asymmetry grows as the neutrons reach deeper layers of the atmosphere. As we will see below, a consequence of this asymmetry is that the effective zenith angle of the solar neutron distribution (Eq. 5.85) will always be smaller than the arrival zenith angle. For small θ_0 the effect is also small; for $\theta_0 = 15°$ it takes 800 g/cm^2 to have a flux which has a factor two difference at $\pm$ 25° of the maximum. However, as θ_0 becomes larger, the effect becomes dramatic: At $\theta_0 \geq 60°$, even the maximum of the distribution is shifted toward smaller angles in the first 100 g/cm^2. After only 200 g/cm^2 the distribution is so asymmetric that it has a difference of an order of magnitude at $\pm 15°$ away from the arrival angle (see Figs. 5.41 and 5.44). In Table 5.18 we present the results of calculations of the total expected flux $F(h, \theta_0)$ and the effective zenith angle $\theta_e(h, \theta_0)$ for solar neutrons arriving at $\theta_0 = 15°$ and 60°.

Table 5.19 shows the same results for solar neutrons arriving at $\theta_0 = 30°$, 45°, and 75°.

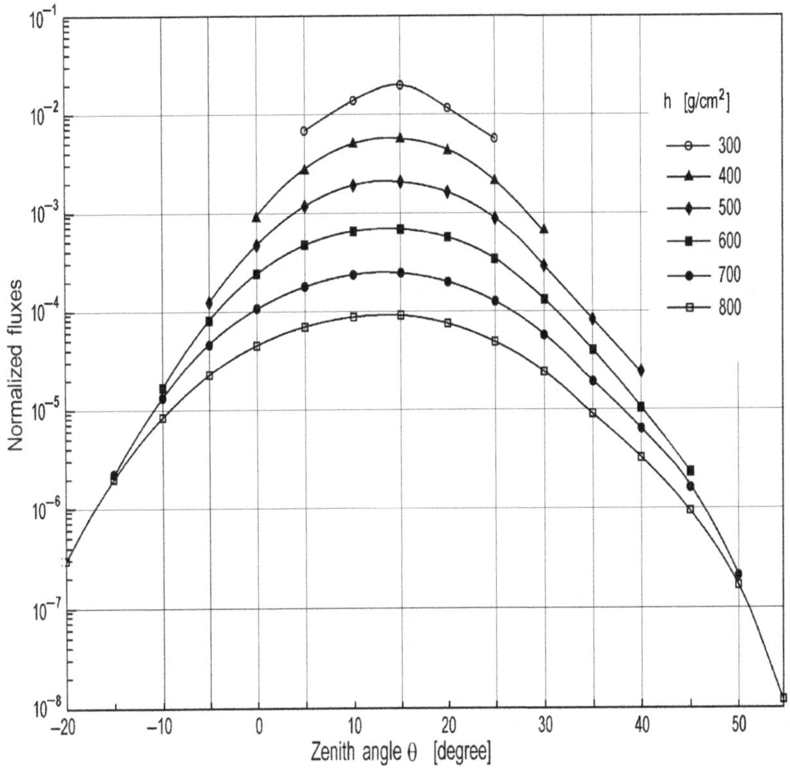

Fig. 5.41 Numerical simulation of expected zenith angle distribution for solar neutron arrival at $\theta_0 = 15°$ in the vertical plane that contains the Sun–Earth line at the depths $h = 300, 400, 500, 600, 700$ and 800 g/cm^2 (From Dorman et al. 1999a)

From Tables 5.18 and 5.19 one can easily appreciate the decrease in $\theta_e(h, \theta_0)$ with atmospheric depth for all arrival angles. This decrease becomes more important as the arrival angle is larger; for an incidence at 15°, θ_e decreases by 1.8° at $h = 800$ g/cm^2. In contrast, θ_e decreases by 8° in only 400 g/cm^2 when the incidence is at 60°. Significant differences are also appreciable in the attenuation of the flux as the neutrons arrive at larger zenith angles: It takes 300 g/cm^2 for the flux arriving at 15° to decrease by a factor of 100, while at 75° incidence this is achieved in only 100 g/cm^2. These results demonstrate that the neutron propagation in the Earth's atmosphere is highly dependent on the zenith angle with which the flux arrives. Therefore, for each observation site the zenith angle of the Sun at the time of the event and the atmospheric depth of the detector location must be considered individually. Extrapolations of a particular situation to a more general case are, in general, not valid.

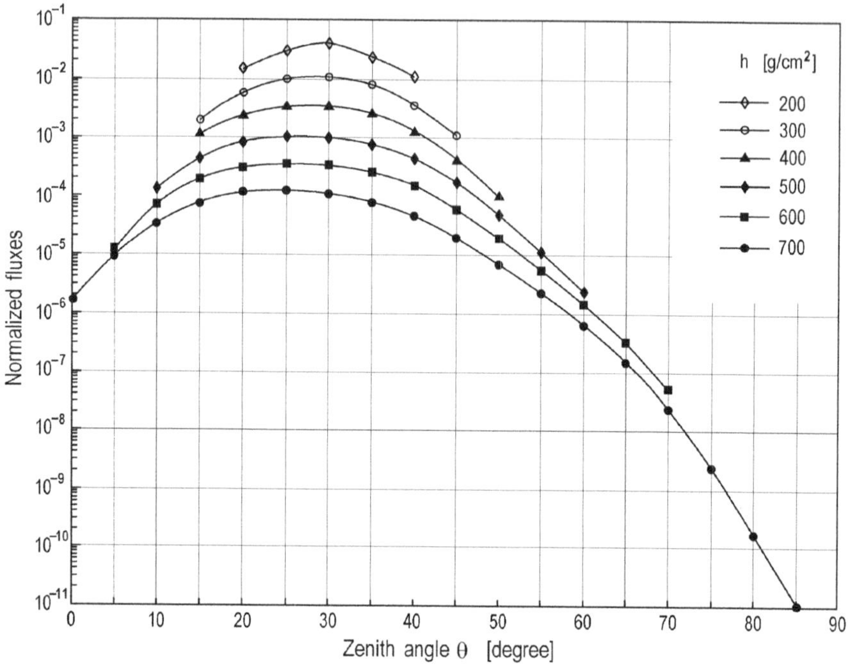

Fig. 5.42 The same as in Fig. 5.41, but for $\theta_0 = 30°$ and $h = 200, 300, 400, 500, 600$, and 700 g/cm^2 (From Dorman et al. 1999a)

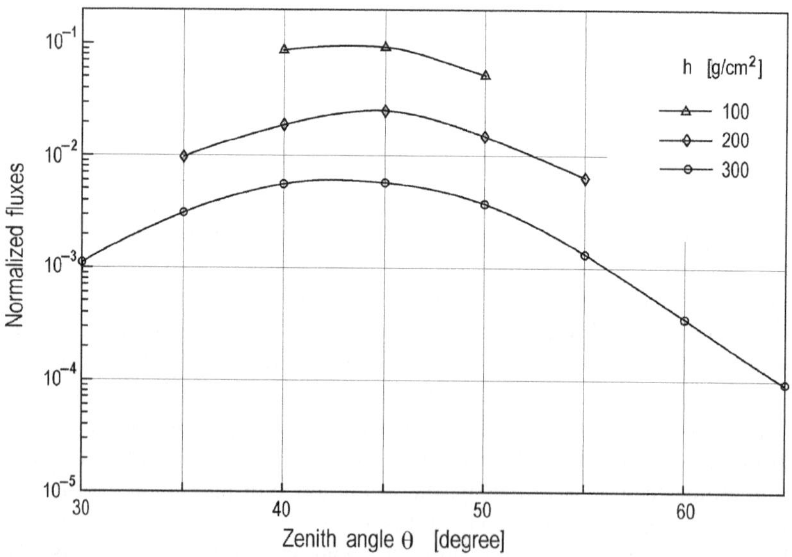

Fig. 5.43 The same as in Fig. 5.41 but for $\theta_0 = 45°$ and $h = 100, 200$ and 300 g/cm^2 (From Dorman et al. 1999a)

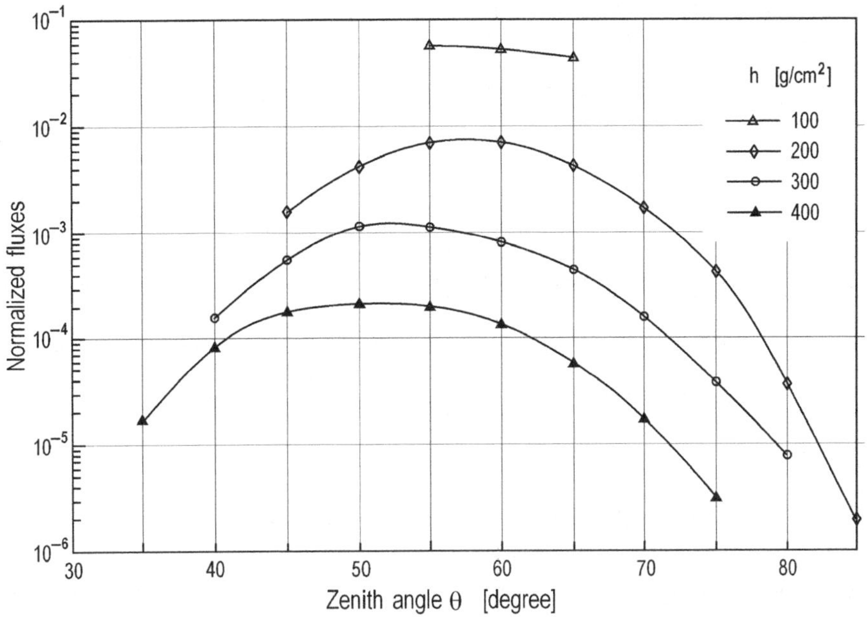

Fig. 5.44 The same as in Fig. 5.41 but for $\theta_0 = 60°$ and $h = 100$, 200, 300 and 400 g/cm^2 (From Dorman et al. 1999a)

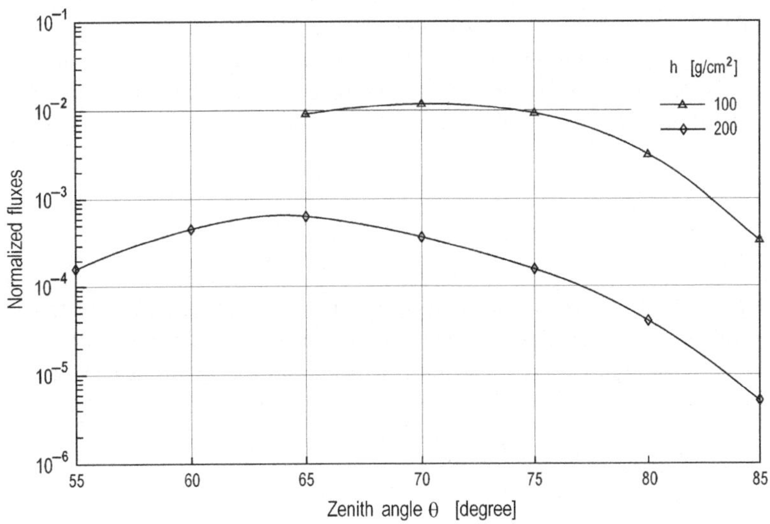

Fig. 5.45 The same as in Fig. 5.41 but for $\theta_0 = 75°$ and $h = 100$ and 200 g/cm^2 (From Dorman et al. 1999a)

Table 5.18 The expected flux $F(h, \theta_0)$ and effective zenith angle $\theta_e(h, \theta_0)$ of solar neutrons for $\theta_0 = 15°$ and $60°$ (From Dorman et al. 1999a)

θ_0(deg) =	15						60			
h (g/cm^2) =	200	300	400	500	700	800	100	200	300	400
$F(h, \theta_0)$=	1.5E-1	5.9E-2	2.3E-2	8.8E-3	1.3E-3	5.1E-4	1.6E-1	2.6E-2	4.6E-3	9.3E-4
$\theta_e(h, \theta_0)$=	14.98	14.97	14.58	14.31	13.55	13.15	59.58	57.52	54.72	52.0

Table 5.19 The expected flux $F(h, \theta_0)$ and effective zenith angle $\theta_e(h, \theta_0)$ of solar neutrons for $\theta_0 = 30°$, $45°$ and $75°$ (From Dorman et al. 1999a)

θ_0(deg) =	30						45		75
h (g/cm^2) =	200	300	400	500	600	700	200	300	100
$F(h, \theta_0)$=	1.2E-1	4.2E-2	1.5E-2	5.2E-3	1.8E-3	6.3E-4	7.5E-2	2.1E-2	3.1E-2
$\theta_e(h, \theta_0)$=	29.79	29.41	28.84	28.08	27.30	26.30	44.36	43.36	70.87

5.7.6 Analytical Approach: General Solution

As an alternative approach, Dorman et al. (1999a) solved analytically the problem of solar neutron multi-scattering and attenuation in a homogeneous atmosphere. Several attempts to have proper estimates of the parameters of the solution are considered in order to reproduce the results of our numerical simulations. This led us to a better understanding of how the various parameters behave and their importance in the atmospheric transport of solar neutrons. It also resulted in a simple and useful approach to show the importance of the refraction effect. In Section 5.7.2 we showed that for angular scattering with standard deviation $\sigma << 1$ (in radians), the zenith angle distribution of solar neutrons in the atmosphere is described by a Gaussian distribution with the same standard deviation σ. If we modify Eq. 5.72 to consider also neutron attenuation, we obtain the expected zenith angle distribution $f(\theta, \theta_0, h)$ of solar neutrons at depth h arriving with initial zenith angle θ_0 as

$$f(\theta, \theta_0, h) = \left(\sigma_e \sqrt{2\pi}\right)^{-1} \exp\left[-(\theta - \theta_0)^2 \big/ 2\sigma_e^2\right] \exp(-L_e/\lambda), \tag{5.86}$$

where L_e, is the effective average mass crossed by neutrons and we have replaced σ by σ_e to indicate the effective standard deviation for neutron multiscattering. Generally speaking, σ_e and L_e increase with h, and L_e should also be a function of θ_0 and $\theta(h)$. The expected flux $F(\theta_0, h)$ and the effective angle $\theta_e(\theta_0, h)$ will be

$$F(\theta_0, h) = \int f(\theta, \theta_0, h) d\theta, \quad \theta_e(\theta_0, h) = \int f(\theta, \theta_0, h) \theta d\theta / F(\theta_0, h). \tag{5.87}$$

Equations 5.86 and 5.87 are a full solution of the problem of solar neutron propagation in the atmosphere. However the determination of σ_e and L_e is not straightforward. Here we will consider several approximations to determine σ_e and L_e; in each case we will compare the resulting zenith angle distributions with the results of our numerical simulation.

5.7.7 Analytical Solution: The First Approximation

As a first approximation we assume, according to Shibata (1994), that for inclined arrival of solar neutrons at the depth h, the propagation will be the same as that for vertical arrival but to an equivalent depth $h/\cos\theta_0$. Therefore the effective standard deviation of the angular distribution and the effective average mass crossed by neutrons are $\sigma_e = \sigma\sqrt{h/\lambda\cos\theta_0}$ and $L_e = h/\cos\theta(h)$.

We introduced the values of Og and L, into Eq. 5.86 to obtain the zenith angle distribution of solar neutrons that we compare with the results of the numerical simulation in Fig. 5.46.

The comparison in Fig. 5.46 is done for an initial zenith angle $\theta_0 = 30°$ to an atmospheric level $h = 600$ g/cm^2. A good agreement with the simulation is found for $20° \leq \theta \leq 30°$. For other zenith angle regions we observe a great difference with respect to the results of the simulation. This difference is not symmetrical with respect to $\theta_0 = 30°$, that is, for $\theta \leq 15°$ the analytical approximation gives a larger flux than the simulation does (about 10 times larger at $\theta = 5°$), and for $\theta \geq 35°$ the

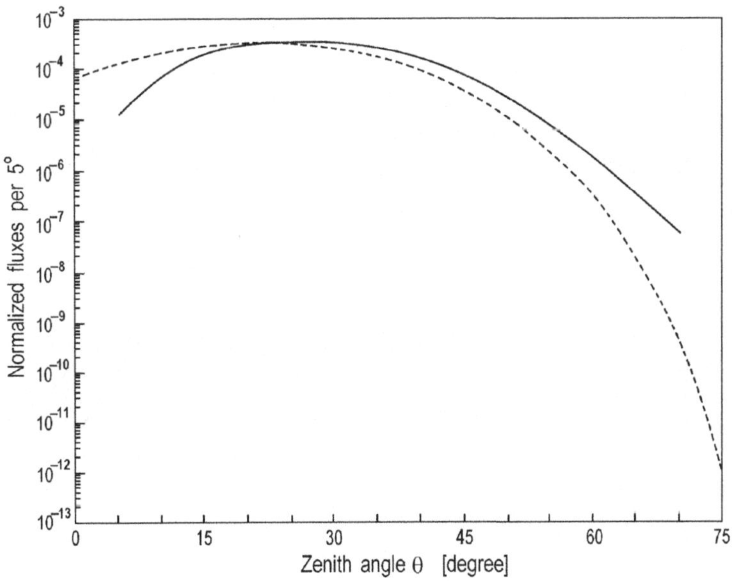

Fig. 5.46 Comparison of the results of the numerical simulation (*continues line*) with calculations based on Eq. 5.86 for the first approach (*dashed line*) for $\theta_0 = 30°$ and $h = 600$ g/cm^2 (From Dorman et al. 1999a)

analytical approximation gives a smaller flux than the simulation does (about 150 times less at $\theta \geq 70°$). From these results it is apparent that for $\theta < \theta_0$ the average path of neutrons is underestimated by our assumption, but for $\theta > \theta_0$, L_e is overestimated. The reason for this is that in this approximation the path of neutrons between interactions is a straight line.

5.7.8 Second Approximation: $\sigma_e(h, \theta_0) = \sigma_0/(h/\lambda \cos\theta_0)$ and L_e as the Arc of a Circle Connected by Arrival and Detection Points

As a next step, we assume that the average neutron path is an arc of a circle connected of arrival and detection points rather than a straight line, and L_e is modified accordingly. With this new assumption for L_e, a comparison with the results of the numerical simulation for $\theta_0 = 30°$ and $h = 600$ g/cm^2 is shown in Fig. 5.47.

As can be seen from Fig. 5.47, the asymmetry in the discrepancy disappears, and the differences became smaller than those with the first approximation, i.e., from 150 to 10 for $\theta = 70°$. Nevertheless, the analytical calculations still predict fluxes that are several times in magnitude larger than those of the numerical simulation at $\theta = (30–25)°$ and $\theta = (30 + 25)°$. These differences arise mainly due to the underestimate of the mass crossed by neutrons L_e with θ far from θ_0.

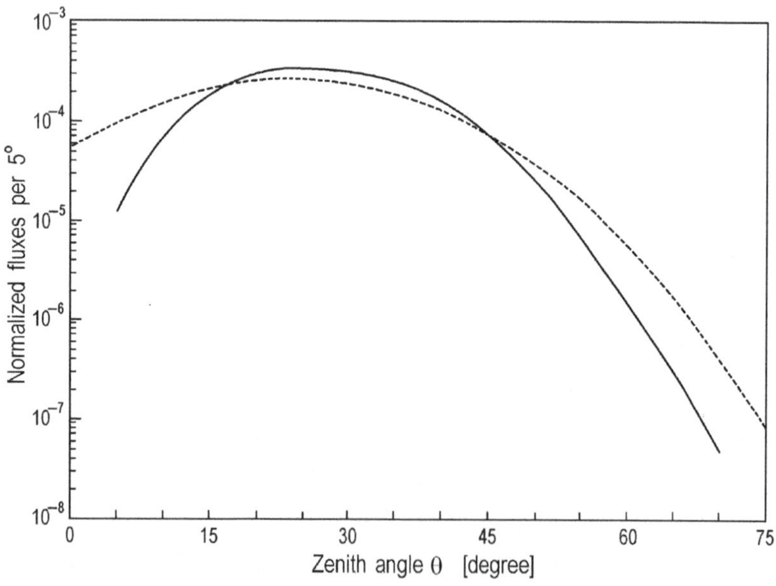

Fig. 5.47 The same as in Fig. 5.46, but for the Second approach (From Dorman et al. 1999a)

5.7.9 Third and Fourth Approaches: σ_e and L_e as for Multiple Scattering

To have a better estimate of the average mass crossed by neutrons, in a **third approach** we sought an improved representation of parameters σ_e and L_e in a diffusive approximation. To have a better understanding of the corresponding mathematical expression for L_e, consider the solar neutron one-dimensional diffusion in zenith angle space. The displacement for one interaction will be σ, and the average time between two interactions $\Delta t = \lambda/\rho v$, where λ is the transport path for scattering (in g/cm^2), ρ is the density of air, and v is the velocity of neutrons. The diffusion coefficient in zenith angle space will be

$$D_\theta = \sigma^2/\Delta t = \sigma^2 \rho v/\lambda \tag{5.88}$$

and the angular displacement after time $t = L'/\rho v$ will be

$$|\theta' - \theta_0| \approx \sqrt{D_\theta t} = \sigma\sqrt{L'/\lambda}. \tag{5.89}$$

Let us suppose that L'/L_e is proportional to h'/h; here L' is the mass path crossed by neutrons at depth h'. Therefore the neutron trajectory zenith angle θ' at depth h' for neutrons with arrival angle θ_0 and final angle $\theta(h)$ will be

$$\theta'(h') = \theta_0 + (\theta(h) - \theta_0)\sqrt{h'/h}. \tag{5.90}$$

From Eq. 5.90 can be seen that at the boundary of atmosphere, at $h' \to 0$ we obtain that $\theta'(h')_{h' \to 0} = \theta_0$. At the level of observations, at $h' \to h$ we obtain that $\theta'(h')_{h' \to h} = \theta(h)$. In some intermediate level h' we obtain that $\theta'(h') \subset [\theta_0, \theta(h)]$. The final average length

$$L_e(\theta, \theta_0, h) = \int_0^h \frac{dh'}{\cos(\theta'(h'))}. \tag{5.91}$$

From Eq. 5.91 follows natural solution that at $L_e(\theta, \theta_0, h)_{\theta \to \theta_0} = h/\cos\theta_0$. The resulting zenith angle distributions were still unsatisfactory when compared with those predicted by our numerical simulations. The main problem of the third approach is that for particles scattered at large zenith angles it either overestimates the amount of matter crossed or, more likely, we have to consider that σ_e is not only a function of h and θ_0 but is also dependent on θ.

In the **fourth approach** we assume that σ_e is determined not only by h and θ_0 but also by the final zenith angle θ. To do this, we use the effective length $L_e(\theta, \theta_0, h)$, as in Eq. 5.91, and

$$\sigma_e(\theta, \theta_0, h) = \sigma\sqrt{L_e(\theta, \theta_0, h)/\lambda}. \tag{5.92}$$

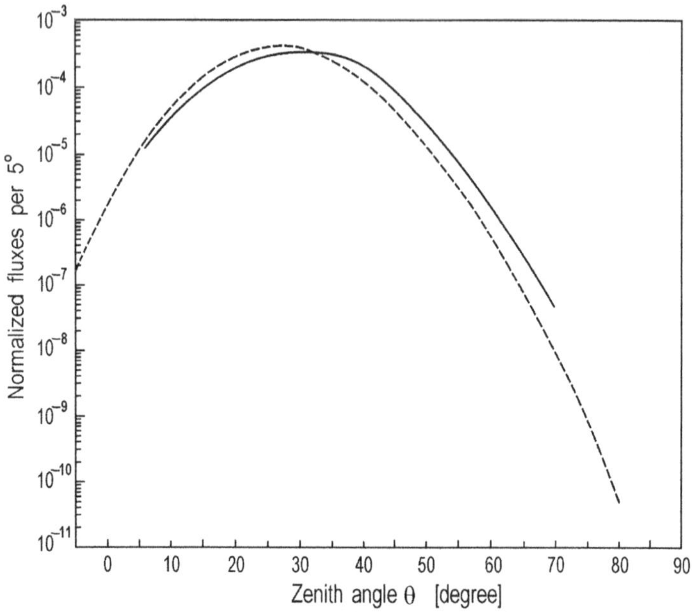

Fig. 5.48 Same as Fig. 5.46 but for the fourth approach (From Dorman et al. 1999a)

In Fig. 5.48 we compare the calculations of this approach with the results of the numerical simulation for a neutron flux arrival at 30° zenith angle that has reached an atmospheric depth of 600 g/cm^2.

We can see from Fig. 5.48 that for $\theta \leq 40°$ there is a very good agreement with the results of the simulation (differences are less than a factor 2). For great θ the analytical calculations still give smaller fluxes than the numerical simulation does but the discrepancy is now much smaller: for example, for $\theta = 70°$ it decreased from a factor 150 in the first approach to ~5 in the fourth approach. A comparison of the results of the fourth approach with those of the numerical simulation for $\theta_0 = 15°$, 45°, and 60° is shown in Fig. 5.49.

From Fig. 5.49 can be seen that for$\theta_0 = 15°$ the simulation and analytical results are almost identical (Panel **a**). For $\theta_0 = 45°$ some differences become noticeable at $\theta > 60°$ (Panel **b**), but their effect in the total flux is less than 1%. For larger arrival zenith angles as $\theta_0 = 60°$ (Panel **c**), great differences in the angular distributions are appreciable for the largest zenith angles. As the neutron fluxes at great $|\theta|$ fall very quickly, their role in determining the total flux $F(\theta_0, h)$ and the effective zenith angle $\theta_e(\theta_0, h)$ is not important when we perform the integrals of Eq. 5.87: for example, F(30°,600) = 0.0018007 according to numerical simulation compared with 0.0018010 according to the fourth approach. We also compared the results of the fourth approach with those of our numerical simulations for $\theta_0 = 0°$, 15°, 45°, 60°, and 75°, and we obtained always differences of less than 1% in the expected fluxes. Therefore our fourth approach serves us well to calculate $F(\theta_0, h)$ and

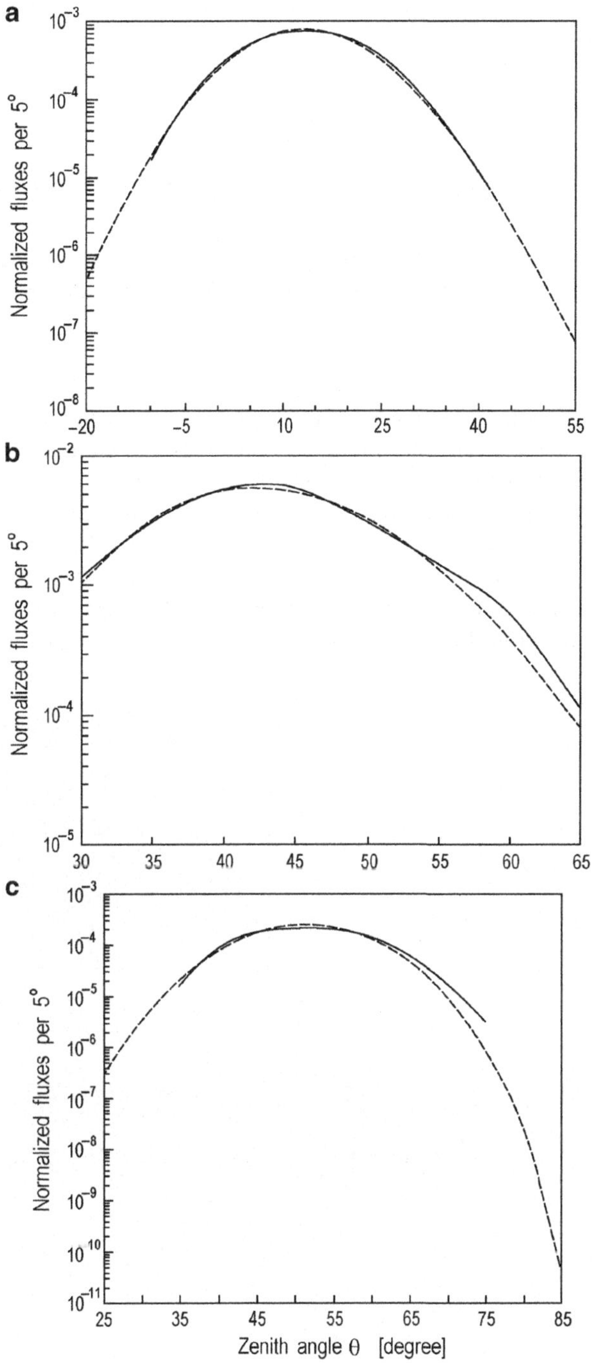

Fig. 5.49 Comparison of the zenith angle distribution described by the Fourth Approach (*dashed lines*) with the results of the numerical simulation (*continues lines*): (**a**) for $\theta_0 = 15°$ and $h = 600$ g/cm^2, (**b**) for $\theta_0 = 45°$ and $h = 300$ g/cm^2, (**c**) for $\theta_0 = 60°$ and $h = 400$ g/cm^2 (From Dorman et al. 1999a)

$\theta_e(\theta_0, h)$ for θ_0 in the range from 0° to 75° and h up to at least 600 g/cm^2 for the largest arrival angles.

5.7.10 Refraction Effect in the Expected Solar Neutron Flux

To illustrate the importance of the refraction effect for determining the flux of solar neutrons, in Fig. 5.50 we present the distribution of the ratio-factor

$$R(\theta_0, h) = F(\theta_0, h)/F(0°, h/\cos\theta_0). \tag{5.93}$$

Equation 5.93 represents the ratio of the solar neutron flux according to our calculations to that expected according to Shibata (1994), see Fig. 5.50.

As can be seen from Fig. 5.50, parameter $R(\theta_0, h)$ is about unity for small zenith angles ($\theta_0 \leq 25°$) at all atmospheric depths. This is the range where our calculations agree with those of Shibata (1994); in this region the refraction effect is not important for solar neutron propagation. In contrast, for larger θ_0, $R(\theta_0, h)$ greatly increases in a wide range of atmospheric depths: e.g. $R(\theta_0, h) \sim 10^8$ for $\theta_0 = 75°$ at h = 1,000 g/cm^2. From Fig. 5.50 we obtain a rough estimate of the importance of

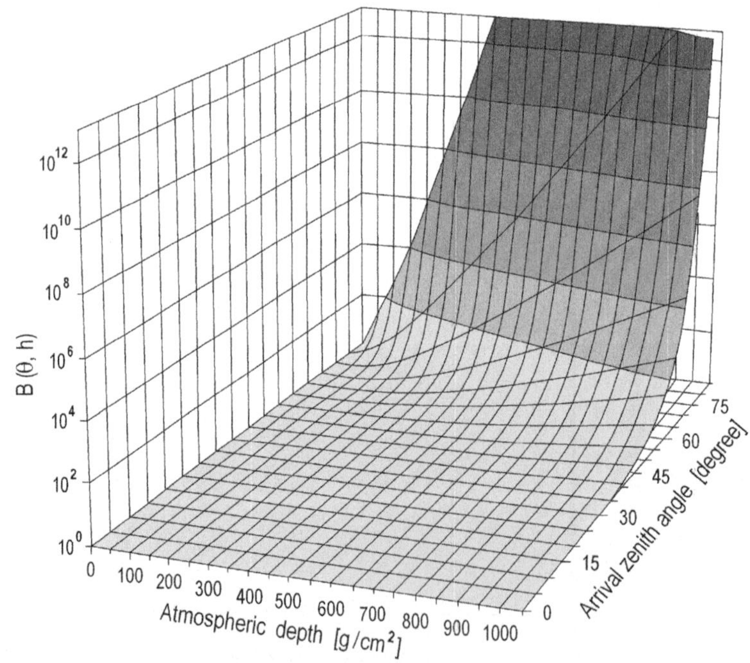

Fig. 5.50 The ratio of solar neutron fluxes according to Dorman et al. (1999a) to that expected according to Shibata (1994) (From Dorman et al. 1999a)

the refraction effect in the measured flux of solar neutrons: At mountain altitudes (~600 g/cm^2) the effect starts to be appreciable for $\theta_0 \geq 55°$, but at sea level for the same θ_0, $R(\theta_0, h)$ is already around 10. Therefore, the sea level cosmic ray stations observing a solar neutron event in the early morning or late afternoon hours may see a neutron flux much larger than that predicted by Shibata (1994).

5.7.11 Refraction Effect and the Effective Zenith Angle of Solar Neutron Arrival Deep in the Earth's Atmosphere

The other important parameter in the refraction effect is the refraction angle:

$$\Delta(\theta_0, h) = \theta_0 - \theta_e(\theta_0, h). \tag{5.94}$$

In Fig. 5.51 we show results of our estimates of refraction angle $\Delta(\theta_0, h)$.

For constructing Fig. 5.51 we use the values of $\theta_e(\theta_0, h)$ calculated according to the fourth approach. The refraction angle $\Delta(\theta_0, h)$ is very small for small θ_0 but

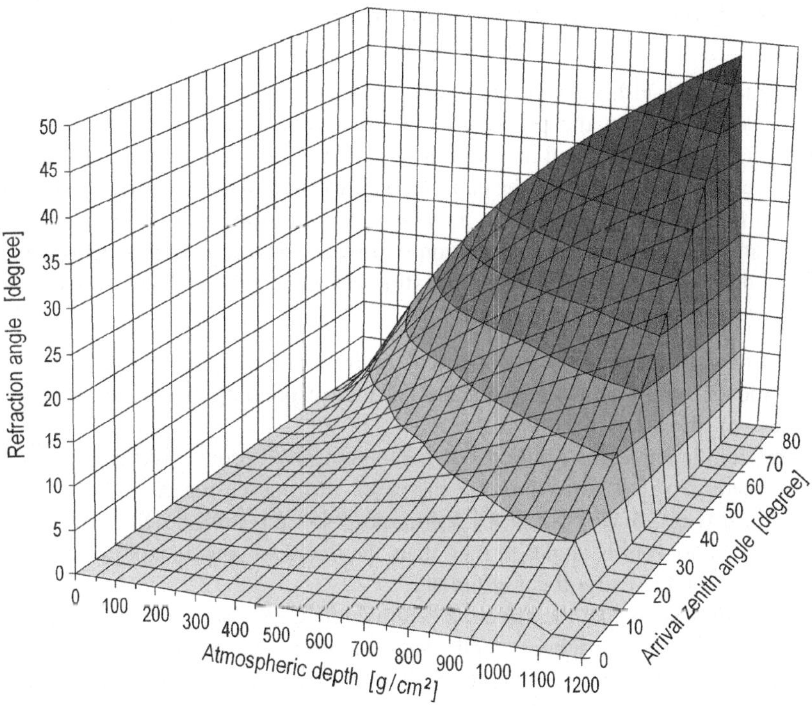

Fig. 5.51 The refraction angle for solar neutrons $\Delta(\theta_0, h)$ according to Eq. 5.94 (From Dorman et al. 1999a)

increases considerably with increasing θ_0 at practically all atmospheric depths; that is, for $\theta_0 = 75°$, $\Delta(\theta_0,h) = 32°$ at 750 g/cm^2 and $\Delta(\theta_0,h) = 48°$ at 1,000 g/cm^2. These are extreme examples, but also a moderate θ_0 ($\sim$30°) can produce a $\Delta(\theta_0, h)$ that is more than 10° for a sea level location.

The refraction effect demonstrated here has important consequences for the appropriate use of a solar neutron telescope. The telescope must be aligned with the effective zenith angle $\theta_e(\theta_0, h)$. This direction is, in general, different from θ_0, the direction of the Sun. This refraction effect is caused by the greater amount of matter crossed by neutrons scattered to larger zenith angles. If a neutron telescope has an aperture of $\sim 20°$, the ratio of the expected fluxes, $R(\theta_0, h)$, will be ≈ 1 for small θ_0 but will increase considerably with increasing θ_0 and h, and could have values of several orders of magnitude for large arrival angles, even at mountain altitudes (see Fig. 5.50). For such large ratios the effect would also be noted in neutron monitors where the aperture angle is some larger. This result implies that stations located at early morning hours at the time of a solar neutron emission are more likely to observe it than was previously thought, as no quantitative estimate of the refraction effect was available, in accordance with suggestion of Smart et al. (1995).

5.7.12 Summary of Main Results and Conclusions

In Dorman et al. (1999a) were presented results of a numerical simulation and an analytical solution to the problem of solar neutron propagation in the Earth's atmosphere. We have taken into account neutron multi-scattering and attenuation. We were able to calculate (1) the expected neutron zenith angle distribution, (2) the neutron flux relative to the flux at the top of the atmosphere, and (3) the effective zenith angle of the neutron distribution. Estimates were done for several neutron arrival directions θ_0 and observation atmospheric levels h. The main obtained results can be summarized as follows:

1. For a vertical arrival the symmetrical zenith angle distributions relative to the incidence zenith angle are predicted. The corresponding fluxes are the same as those of Shibata (1993, 1994).
2. The zenith angle distributions of solar neutrons become asymmetric for an inclined arrival. This asymmetry increases with increasing θ_0 and observation depth h (Figs. 5.41, 5.42, 5.43, 5.44, and 5.45 and Table 5.18). The neutrons scattered to zenith angles larger than the incident angle experience greater attenuation and cause this asymmetry.
3. Our results show that the effective zenith angle $\theta_e(\theta_0, h)$ (Tables 5.18 and 5.19) decreases with atmospheric depth. Solar neutrons are spread or refracted when they travel down through the atmosphere as was indicated by Debrunner et al. (1993) and suggested by Smart et al. (1995). We demonstrate that the cause of this phenomenon is the different amount of matter crossed by particles scattered to angles lower or greater than the arrival angle and not the atmospheric density

gradient as was suggested by Smart et al. (1995). This produces a refraction effect that is not constant but that makes the value of $\theta_e(\theta_0, h)$ increase with increasing θ_0 and h.

4. A multiple scattering representation of the parameters σ_e and L_e was found to give an analytical solution that agrees reasonably well with the results of the numerical simulations. The analytical solution provides 99.9% of the total solar neutron flux predicted by the simulations.
5. On the basis of our analytical solution we estimate the ratio $R(\theta_0, h)$ of our predicted fluxes to those of Shibata (1994) and the refraction angle $\Delta(\theta_0, h)$. These results show the drastic differences that our solution has compared to the predictions of Shibata (1994) at large arrival zenith angles. We showed that the refraction effect, as characterized by $R(\theta_0, h)$ and $\Delta(\theta_0, h)$ is especially important for great θ_0 and h.
6. The refraction effect calculated here has important consequences for the correct pointing of a solar neutron telescope where a careful differentiation must be done between the direction to the Sun (zenith angle θ_0) and the corresponding effective zenith angle $\theta_e(\theta_0, h)$.
7. Neutron monitors located at early morning hours at the time of a solar neutron emission are more likely to observe the event than was previously thought.

The reviewed and discussed in Section 5.7 results should be applicable to any solar neutron event observed by detectors on the Earth's surface. A limitation of the approach used here is that we have not taken into account the decrease of the average energy of the incident solar neutrons during their propagation through the Earth's atmosphere. This will lead to an increase of the scattering angle σ as the neutron cascade penetrates to deeper layers of the atmosphere while it degrades in energy. The treatment of this problem is considerably more complicated, and we intend to pursue it in Section 5.8. Nevertheless, as the energy loss of neutrons will enhance the scattering angle, the results obtained here represent a lower limit to the refraction effect. We expect the correction to be important only for the lowest energy solar neutrons.

5.8 Analytical Description of Solar Neutron Propagation in the Earth's Atmosphere by Considering Scattering, Attenuation and Energy Change

5.8.1 The Matter of the Problem

Dorman et al. (1999b) developed analytical calculation on solar neutron propagation in the atmosphere for different initial zenith angles, by taking into account not only scattering and attenuation, as it was done in Dorman et al. (1999a) and described in the previous Section 5.7, but also neutron energy change (leading to the increase of scattering angles, which is especially important for small energy solar neutrons). We test the commonly used conception that solar neutron propagation through the atmosphere for some initial zenith angle θ_0 of the Sun is the same as for vertical direction; only a different atmospheric depth $h/\cos\theta_0$ should be considered. Our calculations of small angles multi-scattering of neutrons, by considering attenuation and energy change for different initial zenith angles, show that this suggestion is not correct. By taking into account the neutron energy change, we show that the asymmetry in solar neutron propagation and refraction effect increases with decreasing the solar neutron energy.

In Dorman et al (1997), Dorman and Valdes-Galicies (1997), Dorman et al. (1999a) we extended the investigation of solar neutron propagation done by Shibata (1994) for vertical initial incidence, in the case of solar neutrons arriving at different initial zenith angles $\theta_0 = 0°$, $15°$, $30°$, $45°$, $60°$, and $75°$. We did calculations of the angular distribution of arriving neutrons at different atmospheric levels by taking into account neutron scattering and attenuation. In a first approximation we supposed that the scattering angle in single interactions does not change during solar neutron propagation through the atmosphere. It means to neglect the energy change of neutrons and to obtain a lower limit of refraction effect. In this Section 5.8 we will take into account the energy decrease of neutrons in scattering processes and the fact that each single effective scattering angle increases during solar neutron propagation through the atmosphere. We will show that the energy decrease of solar neutrons leads to increase the refraction effect.

5.8.2 *Decrease of Solar Neutron Energy During Scattering in the Atmosphere*

In each elastic scattering with atoms of air (oxygen or nitrogen) the energy of neutrons E decreases, as an average, proportionally to the coefficient 0.8793. If a solar neutron arrives to the boundary of the atmosphere with initial energy E_0 and initial zenith angle is θ_0, the energy of neutron at a level h and at zenith angle θ will be

$$\ln(E(E_0, \theta_0, \theta, h)) = \ln(E_0) + (L_e(\theta_0, \theta, h)/\lambda) \times \ln(0.8793), \tag{5.95}$$

or

$$E(E_0, \theta_0, \theta, h) = E_0 \times (0.8793)^{L_e(\theta_0, \theta, h)/\lambda}, \tag{5.96}$$

where $L_e(\theta_0, \theta, h)$ (in g/cm^2) is the effective average path of neutrons propagating from the boundary of atmosphere ($h = 0$) to the level h at zenith angle θ (determined in Dorman and Valdes-Galicia 1997; Dorman et al. 1999a), and λ is the average path for neutron scattering and attenuation. According to Shibata (1994) λ is about 110 g/cm^2 and is practically independent on energy of neutrons in a very broad interval.

5.8.3 Dependence of Single Effective Scattering Angle on Neutron Energy

In Fig. 5.52 is shown the dependence of single effective scattering angle of neutrons in the Earth's atmosphere $\sigma(E)$ (in radians) on $\ln(E)$ (where neutron energy E is in GeV), based on data compiled in Shibata (1994). This dependence can be approximated by a parabolic function:

$$\sigma(E) = 0.010337(\ln(E))^2 - 0.047038\ln(E) + 0.068585 \tag{5.97}$$

with correlation coefficient 0.99753.

5.8.4 Expected Solar Neutron Angular Distribution in the Atmosphere

Taking into account Eqs. 5.95, 5.96, and 5.97 we obtain, according to Dorman and Valdes-Galicia (1997), that the expected solar neutron angular distribution in the atmosphere at the depth h will be

$$f(E_0, \theta_0, \theta, h) = \left(\sqrt{2\pi}\sigma(E)\sqrt{L_e/\lambda}\right)^{-1} \exp\left(-\frac{(\theta - \theta_0)}{2\sigma(E)(L_e/\lambda)}\right) \times \exp(-L_e/\lambda), \tag{5.98}$$

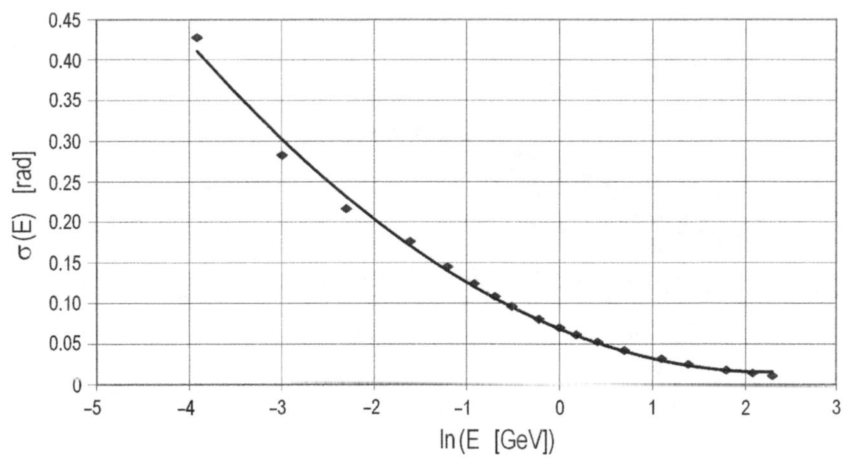

Fig. 5.52 Single effective scattering angle of neutrons in the Earth's atmosphere $\sigma(E)$ (in radians) on $\ln(E)$ (where neutron energy E is in GeV), based on data compiled in Shibata (1994) (From Dorman et al. 1999b)

where $E(E_0, \theta_0, \theta, h)$ is determined by Eq. 5.95 or 5.96 and $\sigma(E)$ by Eq. 5.97, and $L_e(\theta_0, \theta, h)$ was determined by Dorman and Valdes-Galicia (1997) and described above, in Section 5.8. Now we need to take into account that really $L_e(\theta_0, \theta, h)$ depends also from neutron energy E_0, i.e. instead of $L_e(\theta_0, \theta, h)$ we need to use $L_e(E_0, \theta_0, \theta, h)$. By numerical simulation of solar neutron propagation in the atmosphere for different initial zenith angles θ_0, Dorman et al. (1997, 1999a) determined the solar neutron angular distributions at different levels h and compared these distributions with results obtained by the analytical approximation of different four approaches (see Sections 5.7.7-5.7.9). It was found a good agreement in the case of constant neutron energy for 4-th approach, reflected in Eq. 5.98. Therefore, we think that Eq. 5.98. could describe well the solar neutron angular distribution also when the change of neutron energy is included. We determined solar neutron angle distributions for $E_0 = 0.1, 0.2, 0.3, 0.4, 0.5, 0.65, 0.8, 1.0, 1.2, 1.5, 2, 3, 5, 7, 10, 12$, and 15 GeV, and for $\theta_0 = 0°, 15°, 30°, 45°, 60°, 75°$ at the levels h from 50 up to 1,050 g/cm^2 in steps of 50 g/cm^2.

5.8.5 *Solar Neutron Angular Distributions in the Atmosphere for Vertical Initial Arrival*

As an example, we show in Fig. 5.53 the angular distributions for $E_0 = 0.1$ GeV and $\theta_0 = 0°$. The values on the ordinate axis show the expected flux per one initial neutron per unit of surface and per unit of time, and per 5° of zenith angle.

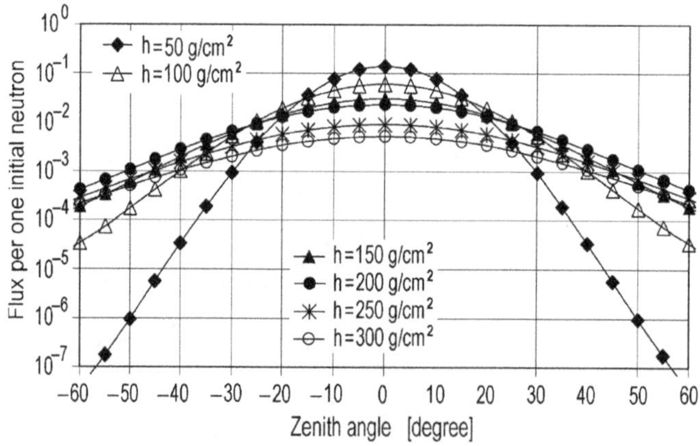

Fig. 5.53 Expected angular distributions for $E_0 = 0.1$ GeV and $\theta_0 = 0°$ and different levels h with taking into account scattering, attenuation and energy change of solar neutrons during propagation through the Earth's atmosphere (From Dorman et al. 1999b)

5.8.6 Solar Neutron Angular Distributions in the Atmosphere for Arrival at the Initial Zenith Angle 30°

The angular distributions are shown for this case in Fig. 5.54 for $E_0 = 1.0$ GeV and $\theta_0 = 30°$.

5.8.7 Solar Neutron Angular Distributions in the Atmosphere for Arrival at the Initial Zenith Angle 75°

The angular distributions are shown for this case in Figs. 5.55 and 5.56 for initial energies 1 and 10 GeV, respectively.

5.8.8 Effective Zenith Angle of Arriving Neutrons

On the basis of determined angular distributions $f(E_0, \theta_0, \theta, h)$ (see examples in Figs. 5.53, 5.54, 5.55, and 5.56) Dorman et al. (2000) calculate expected effective zenith angle $\theta_e(E_0, \theta_0, h)$ as:

$$\theta_e(E_0, \theta_0, h) = \int f(E_0, \theta_0, \theta, h)\theta d\theta / F(E_0, \theta_0, h), \tag{5.99}$$

where

$$F(E_0, \theta_0, h) = \int f(E_0, \theta_0, \theta, h) d\theta. \tag{5.100}$$

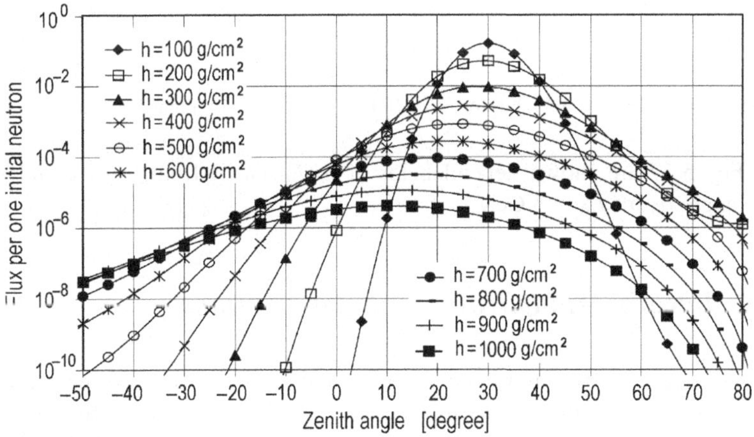

Fig. 5.54 Same as in Fig. 5.53, but for $E_0 = 1.0$ GeV and $\theta_0 = 30°$ (From Dorman et al. 1999b)

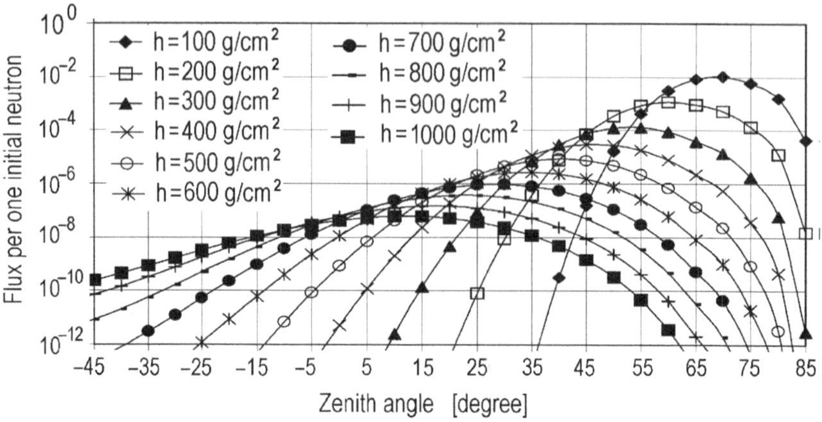

Fig. 5.55 Same as in Fig. 5.53, but for $E_0 = 1.0$ GeV and $\theta_0 = 75°$ (From Dorman et al. 1999b)

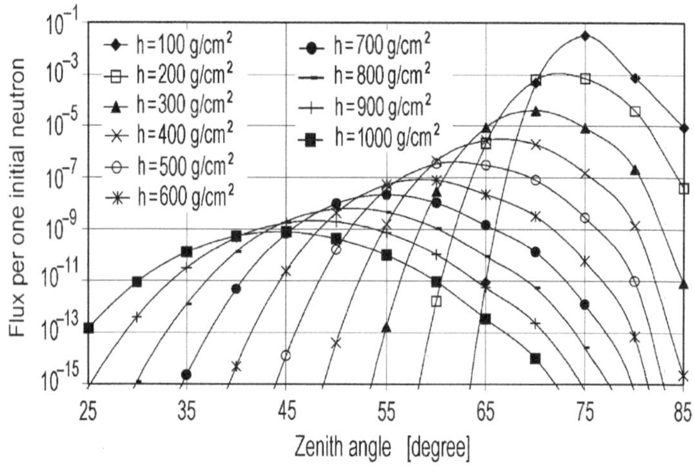

Fig. 5.56 Same as in Fig. 5.53, but for $E_0 = 10$ GeV and $\theta_0 = 75°$ (From Dorman et al. 1999b)

In Fig. 5.57 are shown dependencies of $\theta_e(E_0, \theta_0, h)$ from level h and initial zenith angle θ_0 for $E_0 = 3$ GeV.

5.8.9 Integral Multiplicities for Solar Neutrons

Dorman et al. (2000) determined the expected total neutron flux per one solar neutron, i.e. integral multiplicity $m(E_0, \theta_0, h) \equiv F(E_0, \theta_0, h)$ for E_0 from 0.1 to 15 GeV at different depth h in dependence of the initial zenith angle θ_0 according to Eq. 5.100 (where $f(E_0, \theta_0, \theta, h)$ is determined by Eq. 5.98), taking into account the change of neutron energy according to Eqs. 5.95 or 5.96 and change of effective scattering angle according to Eq. 5.97. As an example, results for the initial neutron energy $E_0 = 3$ GeV are shown in Fig. 5.58.

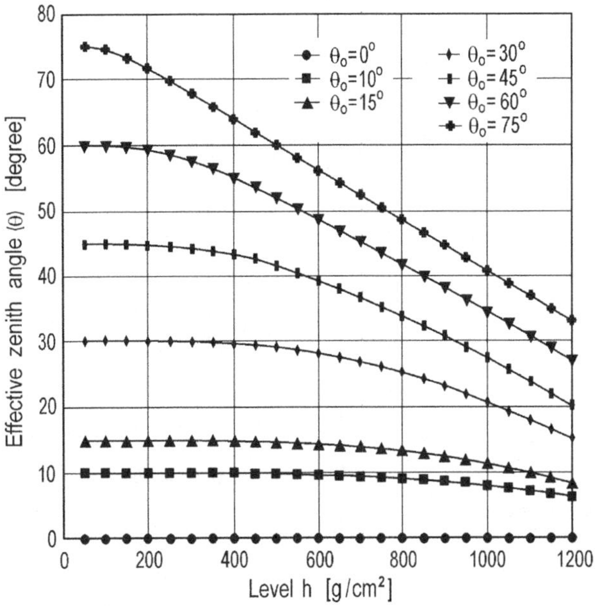

Fig. 5.57 Effective zenith angle $\theta_e(E_0, \theta_0, h)$ versus level h for different initial zenith angle θ_0 and $E_0 = 3$ GeV (From Dorman et al. 2000)

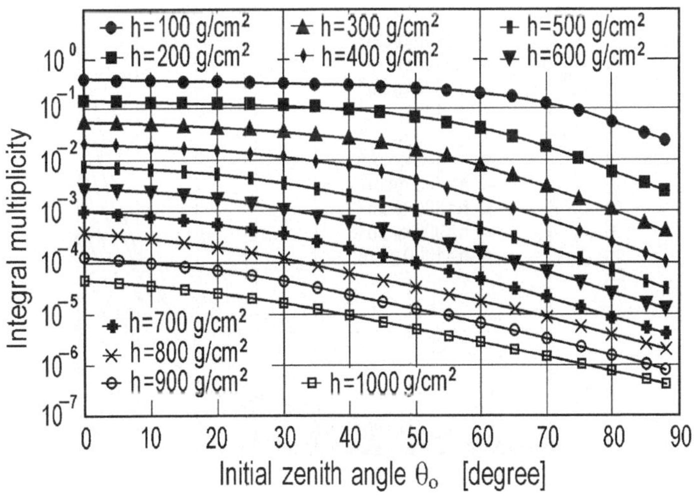

Fig. 5.58 Expected integral milltiplicily $m(E_0, \theta_0, h)$ for solar neutrons in dependence of initial zcnilli angle θ_0 at different h from 100 to 1,000 g/cm^2 for $E_0 = 3$ GeV (From Dorman et al. 2000)

5.8.10 Refraction Effect in Dependence from θ_0 and h

To illustrate the importance of the refraction effect for determining the flux of solar neutrons, in Figs. 5.59 and 5.60 we present the ratio of the expected solar neutron integral multiplicity, taking into account attenuation, scattering and neutron energy

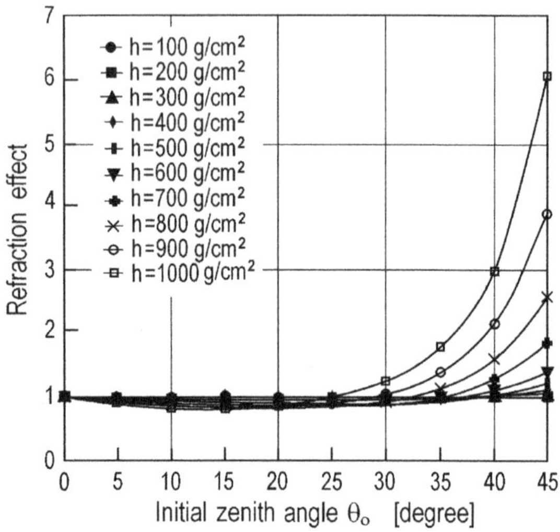

Fig. 5.59 Refraction effect $R(E_0, \theta_0, h)$ *vs* θ_0 from 0° to 45° (From Dorman et al. 2000)

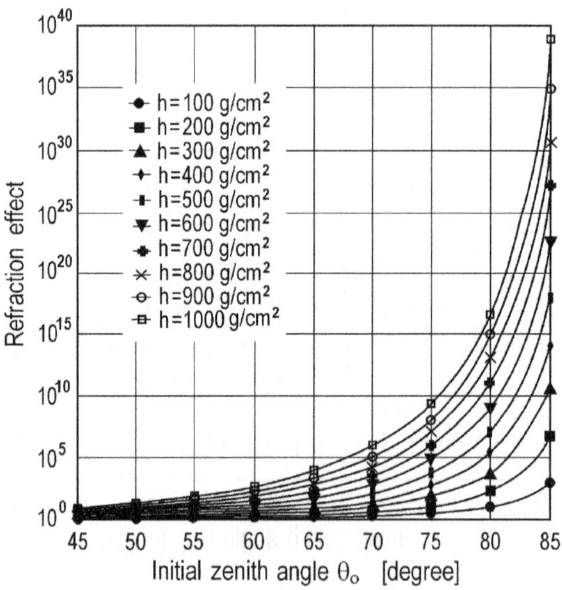

Fig. 5.60 Refraction effect $R(E_0, \theta_0, h)$ *vs* θ_0 from 45° to 85° (From Dorman et al. 2000)

change, to the usually used as integral multiplicity for vertical initial direction but on the depth $h/\cos\theta_0$

$$R(E_0, \theta_0, h) = m(E_0, \theta_0, h)/m(E_0,\ 0°, h/\cos\theta_0) \tag{5.101}$$

in dependence from initial zenith angle θ_0 at different h from 100 to 1,000 g/cm^2 for $E_0 = 3$ GeV. Equation 5.101 represents the ratio of the solar neutron flux according to our calculations to that expected according to Shibata (1994); compare with Eq. 5.93 and Fig. 5.50 in the previous Section 5.7 without taking into account the change of neutron energy during their propagation through the Earth's atmosphere.

5.8.11 *Comparison of Results With and Without the Inclusion of the Neutron Energy Change*

To show the importance of the including of the neutron energy change in calculations of solar neutron propagation through the atmosphere, we compare results for integral multiplicity for both cases. In Fig. 5.61 we show this comparison for the initial neutron energy $E_0 = 3$ GeV.

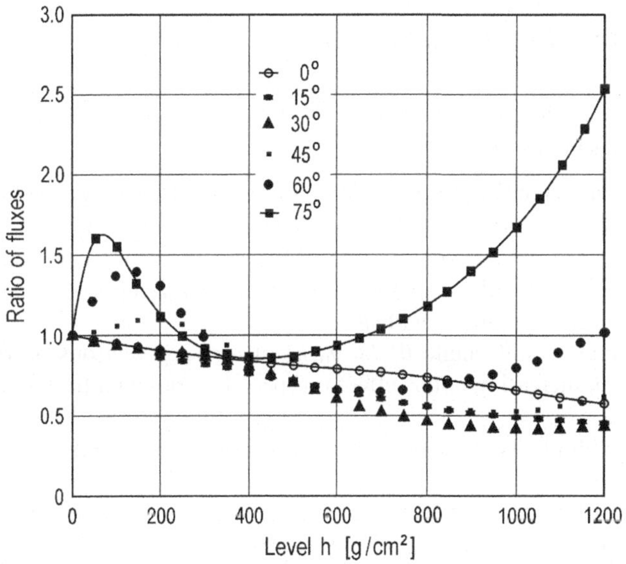

Fig. 5.61 Ratios between fluxes of solar neutrons computed by taking into account neutron energy change and not taking into account the change of neutron energy. These ratios are given for initial energy $E_0 = 3$ GeV and for different initial zenith angles θ_0 from 0° to 75° (From Dorman et al. 2000)

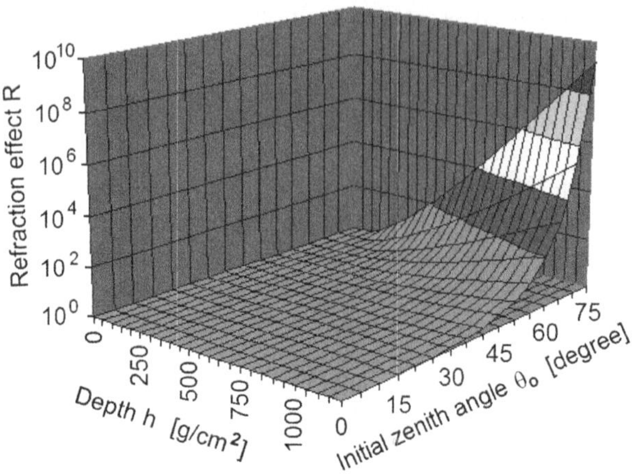

Fig. 5.62 Refraction effect of solar neutrons in the Earth's atmosphere for $E_0 = 3$ GeV

From Fig. 5.61 can be seen that the importance of energy changing including in calculations increased with increasing both of atmospheric deptli and initial zenith angle. Calculations of tlie same ratios for other initial neution energies show that this importance increases with decreasing of neutron initial energy.

5.8.12 Main Results and Conclusion

The main results described in Section 5.8 (based on papers Dorman et al. 1999b, 2000) can be summarized as following:

1. The comparison of Section 5.8 results with those of Dorman et al. (1997, 1999a), described in previous Section 5.7, shows that, by including the change of neutron energy in our previous model, the displacement of angular distributions to vertical direction with increasing h and with increasing θ_0 (refraction effect) becomes bigger and more evident.
2. Tlie effective zenith angle $\theta_e(E_0, \theta_0, h)$ of arriving neutrons decreases sufficiently with increasing h; this effect is especially important for big initial zenith angles θ_0.
3. Integral multiplicity of solar neutrons decreases sufficiently with increasing h and θ_0.
4. The value of refraction effect $R(E_0, \theta_0, h)$ increases very much with increasing h and θ_0: for sea level for E_0 =3 GeV it reaches about 6 at $\theta_0 = 45°$ and about 10^{10} at $\theta_0 = 75°$ (see Figs. 5.61 and 5.62).
5. The importance of energy changing included in calculations increases with increasing atmospheric depth h and initial zenith angle θ_0. This importance increases also with decreasing neutron initial energy.

References for Chapter 5

Aleksanyan TM, Babayan VK, Bednazhevsky VM, Blokh YL, Dorman LI, Mashkov YE (1977) The study of the cosmic ray neutron component using a leadless monitor. In: Proceedings of the 15th International Cosmic Ray Conference, Plovdiv, vol 9, pp 113–113

Aleksanyan TM, Bednazhevsky VM, Blokh YL, Dorman LI, Starkov FA (1979a) Coupling coefficients of the neutron component for the stars of various multiplicities inferred from the measurements on board research vessel "Academician Kurchatov". In: Proceedings of the 16th International Cosmic Ray Conference, Kyoto, vol 4, pp 315–320

Aleksanyan TM, Blukh YL, Dorman LI, Starkov FA (1979b) Examination of the differences in the coupling coefficients and barometric coefficients of 3 NM-64 neutron supermonitor and lead-free monitor. In: Proceedings of the 16th International Cosmic Ray Conference, Kyoto, vol 4, pp 321–324

Aleksanyan TM, Bednazhevsky VM, Blokh YL, Dorman LI, Libin YI, Khodzhaev AN, Yanke VG, Sitnov AM, Lemberger AM, Starkov FA, Chechenov AA (1979c) Some problems of the methods of experimental study of cosmic ray variations. Izv Akad Nauk SSSR, Ser fiz 43 (12):2624–2627, in Russian

Aleksanyan TM, Dorman IV, Dorman LI, Babayan VK, Belov AV, Blokh YL, Kaminer NS, Korotkov VK, Libin IY, Manshilina AA, Mashkov YE, Mymrina IV, Rogovaya SI, Sitnov AM, Yudakhin KF, Yanke VG (1982) Experimental studies of the geomagnetic effects in cosmic-rays and spectral effect of growth before magnetic storms. Izv Akad Nauk SSSR, Ser fiz 46(9):1689–1691 (in Russian)

Aleksanyan TM, Dorman LI, Yanke VG, Korotkov VK (1985) Coupling functions for lead and lead-free neutron monitors from the latitudinal measurements performed in 1982 in the survey by research vessel "Academician Kurchatov". In: Proceedings of the 19th International Cosmic Ray Conference, La Jolla, vol 5, pp 300–303

Alsmiller RG Jr, Boughner RT (1968) Solar neutron transport in the Earth's atmosphere. J Geophys Res 73(15):4935–4942

Avdeev EA, Bednazhevsky VM, Blokh YA, Viskov VV, Dorman LI, Manshilina AA (1973) Studies of the coupling coefficients of the cosmic ray neutron component for the period of 1967–1971. In: Proceedings of the 13th International Cosmic Ray Conference, Denver, vol 2, pp 843–850

Barashenkov VS, Gudima KK, Toneev VD (1969) Cross sections for fast particles and atomic nuclei. Fortschr Phys 17(10):683–725

Belov AV, Gushchina RT, Dorman LI, Sirotina IV (1990) Rigidity spectrum of cosmic ray modulation. In: Proceedings of the 21th International Cosmic Ray Conference, Adelaide, vol 6, pp 52–55

Blokh YL, Dorman LI, Dorman IV, Bednaghevsky VM, Viskov VV, Manshilina AA (1974) Research of cosmic ray latitude effects. Izvestia Acad Sci USSR, Ser Phys 38(9):1970–1973

Carmichael H, Bercovitch M, Steljes JF, Magidin M (1965) Latitude survey in North America. In: Proceedings of the 9th International Cosmic Ray Conference, London, vol 1, pp 553–556

Chupp EL, Debrunner H, Fluckiger E, Forrest DJ, Golliez F, Kanbach G, Vestrand WT, Cooper J, Share G (1987) Solar neutron emissivity during the large flare on 1982 June 3. Astrophys J 318 ((Part 1, 2)):913–925

Clem JM, Dorman LI (2000) Neutron monitor response functions. Space Sci Rev 93 (1–2):335–359

Debrunner H, Brunberg EA (1968) Monte Carlo calculation of nucleonic cascade in atmosphere. Canad J Phys 46(Part 4, 10):S1069–S1072

Debrunner H, Flückiger E (1971) Calculation of the multiplicity yield function of the NM-64 neutron monitor. In: Proceedings of the 12th International Cosmic Ray Conference, Hobart, vol 3, pp 911–916

Debrunner H, Flückiger EO, Chupp EL, Forrest DJ (1983) The solar cosmic ray neutron event on June 3, 1982. In: Proceedings of the 18th International Cosmic Ray Conference, Bangalore, vol 4, pp 75–78

Debrunner H, Flückiger EO, Stein P (1989) On the sensitivity of neutron monitors to solar neutrons. Nucl Instr Meth Phys Res A 278(2):573–575

Debrunner H, Fluckiger E, Stein H (1990) On the sensitivity of a NM-64 standard neutron monitor at sea level to solar neutrons in dependence of the angular distance of the station from the sub-solar point. In: Proceedings of the 21st International Cosmic Ray Conference, Adelaida, vol 5, pp 129–132

Debrunner H, Lockwood JA, Ryan JM (1993) Solar neutron and proton production during the 1990 May 24 cosmic-ray flare increases. Astrophys J 409(Part 1, 2):822–829

Debrunner H, H Debrunner, JA Lockwood, C Barat, R Bütikofer, JP Dezalay, E Flückiger, A Kuznetsov, JM Ryan, R Sunyaev, OV Terekhov, G Trottet, N Vilmer (1997) Energetic neutrons, protons and gamma rays during the 1990 May 24 solar cosmic ray event. Astrophys J 479, Part 1(2): 997–1011

Del Guerra A (1976) A compilation of n-p and n-C cross sections and their use in a Monte Carlo program to calculate the neutron detection efficiency in plastic scintillator in the energy range 1–300 MeV. Nucl Instrum Methods 135(2):337–352

Dorman LI (1961) Calculations of integral generation multiplicity of muons in the atmosphere for observations under various absorber thicknesses. Cosmic Rays, Vol. 3, Acad Sci USSR, Moscow 3:174–198 (in Russian)

Dorman LI (1969) Geophysical effects and properties of the various components of the cosmic radiation in the atmosphere. In: Proceedings of the 11th International Cosmic Ray Conference, Budapest, Volume of Invited Papers and Rapporteur Talks, 381–442

Dorman IV, Dorman LI (1967a) Solar wind properties obtained from the study of the 11-year cosmic ray cycle. J Geophys Res 72(5):1513–1520

Dorman IV, Dorman LI (1967b) Propagation of energetic particles through interplanetary space according to the data of 11-year cosmic ray variations. J Atmosph Terr Phys 29(4): 429–449

Dorman LI, Masaryuk EA (1967) The measurements on the latitude effect and and the coupling coefficients of the inclined cosmic ray component. Cosmic Rays 8:153–158 (in Russian)

Dorman LI, Pakhomov NI (1979) The dependence of the integral generation multiplicity of neutron component at various depths in the atmosphere on zenith angle on primary particle incidence. In: Proceedings of the 16th International Cosmic Ray Conference, Kyoto, vol 4, pp 416– 420

Dorman LI, Rishe LE (1973) Calculations of the integral generation multiplicity, coupling coefficients and partial barometric coefficients for the cosmic ray neutron component observations at various altitudes using the method of discontinues Markov processes including non-ordinary of nuclear-meson cascade. In: Proceedings of the 13th International Cosmic Ray Conference, Denver, vol 2, pp 835–842

Dorman LI, Valdes-Galicia JF (1997) Analytical approximation to solar neutron scattering and attenuation in the Earth's atmosphere. In: Proceedings of the 25th International Cosmic Ray Conference, Durban, vol 1, pp 29–32

Dorman LI, Yanke VG (1981) The coupling functions of NM-64 neutron supermonitor. In: Proceedings of the 17th International Cosmic Ray Conference, Paris, vol 4, pp 326–329

Dorman LI, Okulov YI, Kaminer NS, Manshilina AA (1966) Determination of coupling coefficients of the cosmic ray neutron component on the base of ship "Zarya" data. Izv Akad Nauk SSSR, ser fiz 30(11):1873–1874

Dorman LI, Kovalenko VA, Milovidova NP (1967a) Latitude distribution and coupling coefficients for cosmic radiation neutron and general ionizing components. Izv Akad Nauk SSSR Ser fiz 31(8):1387–1389 (in Russian)

Dorman LI, Kovalenko VA, Milovidova NP (1967b) The latitude distribution, integral multiplicities and coupling coefficients for the neutron, total ionizing and hard components of cosmic rays. Il Nuovo Cimento B 50(1):27–39

Dorman LI, Kovalenko VA, Milovidova NP (1967c) The effect of east-west asymmetry in cosmic rays inferred from crossed telescope measurements in 1965. Izv Akad Nauk SSSR, Ser fiz 31(8):1390–1392 (in Russian)

Dorman LI, Granitsky LV, Gushchina RT, Davidov AE, Dorman IV, Kovalenko VA, Madyarov RZ, Milovidova NP, Khachikyan VS (1967d) Preliminary results of cosmic ray distribution study in the East part of USSR by expedition of Siberia IZMIRAN. Izv Akad Nauk 31 (8):1393–1394

Dorman LI, Granitsky LV, Bortnik SG, Novikova GA, Raikhbaum IM (1967e) Investigations of cosmic ray neutron component on atmospheric depth 260–315 mb. Izv Akad Nauk 31 (8):1322–1324

Dorman LI, VA Kovalenko, NP Milovidova, SB Chernov (1968a) Preliminary results of cosmic ray hard component latitude measurements. In: Proceedings of the All-Union Cosmic Ray Phys Conference, Tashkent, Part II(2), FIAN USSR, Moscow, pp 125–129 (in Russian)

Dorman LI, VA Kovalenko, NP Milovidova (1968b) Cosmic ray intensity distribution on the territory of the USSR. In: Proceedings of the All-Union Cosmic Ray Physics Conference, Tashkent, Part II(3), FIAN USSR, Moscow, 23–27 (in Russian)

Dorman LI, Fedchenko SG, Granitsky LV, Rishe GA (1970a) Coupling and barometric coefficients for measurements of cosmic ray variations at altitudes of 260–400 mb. Acta Phys Acad Sci, Hungaricae 29(suppl 2):233–236

Dorman LI, Kovalenko VA, Milovidova NP, Chernov SB (1970b) The cosmic ray intensity distribution on the territory of USSR and the coupling coefficients. Acta Phys Acad Sci, Hungaricae 29(suppl 2):359–363

Dorman LI, Dorman IV, Libin IY, Khodzhaev AN (1979) Coupling coefficients and latitude effect of the cosmic ray neutron component inferred from the expedition measurements with a lead-free monitor. In: Proceedings of the 16th International Cosmic Ray Conference, Kyoto, vol 4, pp 325–328

Dorman LI, Yanke VG, Korotkov VK (1981) Energy sensitivity of supermonitors to the multiple neutrons and the coupling functions. In: Proceedings of the International Cosmic Ray Conference, vol 4, pp 330–333

Dorman IV, Dorman LI, Korotkov VK (1984) Libin IY (1984) Expeditionary investigations by neutron monitor without lead. Izv Akad Nauk SSSR Ser fiz 48(11):2123–2125

Dorman LI, Dorman IV, Valdes-Galicia JF (1997) Simulation of solar neutron scattering and attenuation in the Earth's atmosphere for different initial zenith angles. In: Proceedings of the 25th International Cosmic Ray Conference, Durban, vol 1, pp 25–28

Dorman LI, Valdes-Galicia JF, Dorman IV (1999a) Numerical simulation and analytical description of solar neutron transport in the Earth's atmosphere. J Geophys Res 104 (A10):22417–22426

Dorman LI, Dorman IV, Valdes-Galicia JF (1999b) Analytical description of solar neutron propagation in the atmosphere for different initial zenith angles by considering scattering, attenuation and energy change. In: Proceedings of the 26th International Cosmic Ray Conference, Salt Lake City, vol 7, pp 367–370

Dorman LI, Villoresi G, Lucci N, Parisi M, Tyasto MI, Danilova OA, Ptitsyna NG (2000) Cosmic ray survey to Antarctica and coupling functions for neutron component near solar minimum (1996–1997), 3. Geomagnetic effects and coupling functions. J Geophys Res 105 (9):21047–21056

Dorman Lev I, Irina V Dorman, Valdes-Galicia JF (2000b) Propagation of inclined solar neutrons: scattering, energy decrease, attenuation, and refraction effect. In: Mewaldt RA, Jokipii JR, Lee MA, Möbius E, Zurbuchen TH (ed) Acceleration and transport of energetic particles observed in the heliosphere: ACE 2000 Symposium, AIP Conference Proceedings, vol 528, Melville, New York, pp 173–176

Dostrovsky I, Rabinowitz P, Bivins R (1958) Monte Carlo calculations of high-energy nuclear interactions. I. Systematics of nuclear evaporation. Phys Rev 111(6):1659–1676

Dyring E, Sporre B (1965) On the multiplicity effect in an IGY neutron monitor. In: Proceedings of the 9th International Cosmic Ray Conference, London, vol 1, pp 478–480

Dyring E, Sporre B (1966) The latitude effect of the neutron multiplicity as detected by a shipborne neutron monitor. Arkiv for Geofysik 5(1):67–77

Efimov YE, Terekhov MM (1988) Specific yield functions of neutron supermonitors for solar neutrons. Geomagn Aeron 28(5): 832–835 (in Russian). In English: Geomagn Aeron (USA) 28(5): 713–715

Efimov YE, Kocharov LG, Kovaltsov GA, Usoskin IG (1993) Detection of solar neutrons by the neutron monitor network. In: Proceedings of the 23rd International Cosmic Ray Conference, Calgary, vol 3, pp 159–162

Gaisser TK (1976) Proceedings of Second Symposium on High Energy Cosmic Ray Modulation, Tokyo, p 33

Granitsky LV, Dorman LI, Dorman IV, Kovalenko VA, Kovalenko MA, Karpov VP, Neermolov AP, Novikova GA, Madyarov RZ, Okulov YI, Syrovatsky SI (1966) Preliminary results of investigation of space-rigidity cosmic ray intensity distribution by Siberia Academy of Sciences of USSR in 1965. Izv Akad Nauk SSSR Ser fiz 30(11):1870–1872 (in Russian)

Granitsky LV, Gushchina RT, Davydov AE, Dorman IV, Dorman LI, Kovalenko VA, Madyarov RZ, Milovidova NP, Khachikyan VS (1967) Preliminary results of a study of distribution of cosmic rays in eastern part of USSR. Izv Akad Nauk SSSR Ser fiz 31(8):1393–1394

Granitsky LV, Dorman LI, Rishe GA (1968a) Latitude measurements of neutron multiplicity. In: Proceedings of All-Union Cosmic Ray Physics Conference, Tashkent, Part II, vol 3, FIAN USSR, Moscow, pp 10–12

Granitsky LV, Dorman LI, Kovalenko VA (1968b) Latitude effects and coupling coefficients of nucleonic component at sea level according to measurements in 1965–1968. In: Proceedings of the All-Union Cosmic Ray Physics Conference, Tashkent, Part II, vol 3, FIAN USSR, Moscow, pp 121–125

Hatton CJ (1971) The neutron monitor, In Progress in Physics of Cosmic Ray and Elementary Particles, ed. by J.G. Wilson and S.A. Wouthuysen, 10, North-Holland Publ. Co., Amsterdam, 3–100

Iucci N, Villoresi G, Dorman LI, Parisi M (2000) Cosmic ray survey to Antarctica and coupling functions for neutron component near solar minimum (1996–1997), 2. Determination of meteorological effects. J Geophys Res 105(A9):21035–21045

Johnson TH, DV Read (1937) Undirectional measurements of the cosmic ray latitude effects, Phys. Rev., 51, No. 7, 557–564

Kodama M, Inoue A (1970) Differential rigidity response of different multiplicities in the NM-64 neutron monitor. Acta Phys Acad Scient, Hungaricae 29(suppl 2):577–581

Kodama M, Miyazaki Y (1957) Geomagnetic latitude effect of the cosmic-ray nucleon and meson components at sea level from Japan to the Antarctic. Rept Ionosphere Res Japan 11:99–115

Kodama M, Ohuchi T (1968) Latitude survey of neutron multiplicity using a shipborne NM-64 neutron monitor. Canad J Phys 46(10):1090–1093

Krymsky GF, Krivoshaplin PA, Skcripin GV (1965) Angular dependence of coupling coefficients of the muon component of cosmic rays. In: Proceedings of the 9th International Cosmic Ray Conference, London, vol 1, pp 503–505

Lockwood JA, Webber WR (1967) Differential response and specific yield functions of cosmic-ray neutron monitors. J Geophys Res 72(13):3395–3402

Lockwood JA, Webber WR, Hsieh L (1974) Solar flare proton rigidity spectra deduced from cosmic ray neutron monitor observations. J Geophys Res 79(28):4149–4155

Luzov AA, Mathjukhin UG, Sdobnov VE (1971) Calculation of the energy sensivity of the neutron monitor NM64. In: Studies in Geomagnetism, Aeronomy, and Solar Physics, Nauka, Moscow, vol 20, pp 356–362

Metropolis N, Bivins R, Storm M, Miller JM, Friedlander G, Turkevich A (1958) Monte Carlo calculations on intranuclear cascades. II. High-energy studies and pion processes. Phys Rev 110(1):204–219

Mischke CFW, Stoker PH, Duvenage J (1973) The neutron moderated detector and the determination of rigidity dependence of protons from the 1/2 September 1971 solar flare. In: Proceedings of the 13th International Cosmic Ray Conference, Denver, vol 2, pp 1570–1575

Moraal H, Potgieter MS, Stoker PH, Van der Walt AJ (1989) Neutron monitor latitude survey of cosmic ray intensity during the 1986/1987 solar minimum. J Geophys Res 94(A2):1459–1464

Murakami K (1988) Private communication to K. Nagashima (see in Nagashima et al., 1990)

Nagashima K, Sakakibara S, Murakami K, and Morishita I (1990) Response and yield functions of neutron monitor, galactic cosmic-ray spectrum and its solar modulation derived from all the available world-wide survey. In: Proceedings of the 21st International Cosmic Ray Conference, Adelaida, vol 7, pp 96–99

Pakhomov NI, Sdobnov VE (1977) Calculations of the neutron supermonitor parameters. In: Studies in geomagnetic aeronomy and solar physics, Nauka, Moscow, vol 42, pp 185–189

Pearce RM, Fowler AG (1964) Calculated spectral sensitivity of IGY and IQSY neutron monitors. J Geophys Res 69(21):4451–4455

Peters B (1963) Meson production at high energy and secondary cosmic radiation. Proc. 8th Inter. Cosmic Ray Conf., Jaipur, India, 1963, 5, 423–442

Potgieter MS, Raubenheimer BC, Stoker PH, Van der Walt AJ (1980) Modulation of cosmic rays during solar minimum. 2. Cosmic ray latitude distribution at sea-level during 1976. S Africa J Phys 3:77–89

Pyle KR, Shea MA, Smart DF (1991) Solar flare generated neutrons observed by neutron monitors. In: Proceedings of the 22nd International Cosmic Ray Conference, Dublin, vol 3, pp 57–60

Rao UR, Sarabhai V (1961) Time variations and directional cosmic ray intensity at low altitudes, I. Comparison of daily variation of the intensity of cosmic rays incident from east and west. Proc Phys Soc A Math Phys Sci 263(1312):101–117

Raubenheimer BC, Stoker PH (1974) Various aspects of the attenuation coefficient of a neutron monitor. J Geophys Res 79(34):5069–5076

Rose DC, Katzman J (1956) The geomagnetic latitude effect on the nucleon and meson component of cosmic rays at sea level. Canad J Phys 34(1):1–19

Rose DC, Fenton KB, Katzman J, Simpson JA (1956) Latitude effect of the cosmic ray nucleon and meson components at sea level from the Arctic to the Antarctic. Canad J Phys 34 (9):968–984

Shea MA, Smart DF, McCracken KG (1965) A study of vertical cutoff rigidities using sixth degree simulations of the geomagnetic field. J Geophys Res 70(17):4117–4130

Shea MA, Smart DF, Pyle KR (1991) Direct solar neutrons detected by neutron monitors on 24 May 1990. Geophys Res Lett 18(9):1655–1658

Shen M (1968) Neutron Production in Lead and Energy Response of Neutron Monitor. Suppl. Nuovo Cimento 6, 1177–

Shibata S (1993) Propagation of solar neutrons through the atmosphere of the Earth. Ph.D thesis, Faculty of Science, Nagoya University, Japan

Shibata S (1994) Propagation of solar neutrons through the atmosphere of the Earth. J Geophys Res 99(A4):6651–6665

Simpson JA (1948) The latitude dependence of neutron densities in the atmosphere as a function of altitude. Phys Rev 73(11):1389–1391

Simpson JA, Uretz RB (1949) On the latitude dependence of nuclear disintegrations and neutrons at 30, 000 feet. Phys Rev 76(4):569–570

Simpson JA, Uretz RB (1953) Cosmic-ray neutron production in elements as a function of latitude and altitude. Phys Rev 90(1):44–50

Skorke S (1958) Breiteneffekt der Nukleonen – und Mesonenkomponente der Ultrastrahlung in Meereshohe im Indischen und Atlantischen Ozean. Z f Physik 151(5):630–645

Smart DF (1996) Solar cosmic rays. Il Nuovo Cimento C 19(5):765–775

Smart DF, Shea MA, O'Bren K (1995) Comment on the minimum mass path of a uni-directional solar neutron particle beam through the Earth atmosphere. In: Proceedings of the 24th International Cosmic Ray Conference, Rome, vol 4, pp 171–174

Stoker PH (1993) Cosmic ray latitude distribution at neutron monitor and aircraft altitudes. In: Proceedings of the 23th International Cosmic Ray Conference, Calgary, vol 3, pp 785–788

Stoker PH (1994) Relativistic solar proton events. Space Sci Rev 73(1–4):327–385

Stoker PH, Raubenheimer BC (1985) The neutron moderated detector and groundbased cosmic ray modulation studies. In: Proceedings of the 19th International Cosmic Ray Conference, La Jolla, vol 5, pp 502–505

Stoker PH, Dorman LI, Clem JM (2000) Neutron monitor design improvements. Space Sci Rev 91 (1–2):361–380

Villoresi G, Dorman LI, Iucci N, Ptitsyna NG (2000) Cosmic ray survey to Antarctica and coupling functions for neutron component near solar minimum (1996–1997), 1. Methodology and data quality assurance. J Geophys Res 105(A9):21025–21034

Viskov VV, Dorman LI, Yanke VG (1974) Calculations of integral multiplicities and coupling coefficients. Izv Akad Nauk SSSR, Ser fiz 38(9):1978–1981

Vzorov IK (1969) Empirical dependence range energy for 100–100000 MeV protons. In: Technical Report JINR-P1-4442, Joint Institute for Nuclear Research, Dubna, USSR, pp 1–7 (in Russian)

Webber WR (1962) Some features of the response of neutron monitors to low-energy particles incident on the top of the atmosphere. Can J Phys 40(1):906–923

Webber WR, Quenby JJ (1959) On the deviation of cosmic ray specific yield functions. Philos Mag 4(41):654–664

Chapter 6
Statistical Investigations of Solar Neutron Events on the Basis of Ground Observations

For many weak solar neutron events, the statistical errors inherent in neutron monitors and supermonitors are so big relative to the expected amplitudes as to make it impossible to study individual cases. Data averaging for many events, using some statistical methods (the superposed epoch method, for example), gives the possibility of a decrease in this error sufficient enough to determine some average effects for many solar neutron events. The first attempts at this method were made many years ago – before the discovery of solar neutrons, in fact – on the basis of 1-h data (see Section 1.4). Positive results, however, were only obtained later by using 5-min data.

6.1 Statistical Investigations of Solar Neutron Events Using the Rome Neutron Supermonitor 5-min Data for 1978–1982

The first statistical investigation of solar neutron events with positive results were made by Iucci et al. (1984) on the basis of 5-min data for the period 1978–1982 from the Rome neutron supermonitor (41.9°N, 12.5°E, altitude 80 m above sea-level, cut-off rigidity for the vertical direction 6.32 GV). It was investigated eight gamma-ray line flares occurred when the zenith angle of the Sun at Rome was within the interval 20–40° (see Table 6.1).

The results of the superposed epoch analysis shows an appreciable increase of about 4σ in the 5-min interval immediately following the gamma-line time or hard X-ray maximum (see Fig. 6.1).

For controlling this result Iucci et al. (1984) investigated the average effect of 14 solar gamma-ray line flares occurred when zenith angle of the Sun at Rome was $\theta_s > 50°$ or behind the horizon (Table 6.2).

The result of this controlled analysis (Fig. 6.2) does not show any appreciable increase in the counting rate after the gamma-line time and confirms that the increase shown in Fig. 6.1 was indeed produced by neutral particles.

L. Dorman, *Solar Neutrons and Related Phenomena*,
Astrophysics and Space Science Library 365, DOI 10.1007/978-90-481-3737-4_6,

Table 6.1 Gamma-ray line flares used for superposed epoch analysis by Iucci et al. (1984), for flares when for Rome $20° \le \theta_s \le 40°$

Date	Flare location	Gamma-ray line or hard X-ray maximum (UT)	Cosmic ray epoch time (UT)
July 11, 1978	18°N, 43°E	10:51–10:52	10:50
August 8, 1979	22°S, 73°E	12:48	12:45
June 4, 1980	14°S, 59°E	6:55	6:55
July 1, 1980	12°S, 38°W	16:27	16:30
April 24, 1981	18°N, 50°W	14:.00	14:00
June 3, 1982	9°S, 72°E	11:43	11:45
June 6, 1982	9°S, 25°E	16:33	16:35
June 15, 1982	22°S, 66°E	15:11	15:10

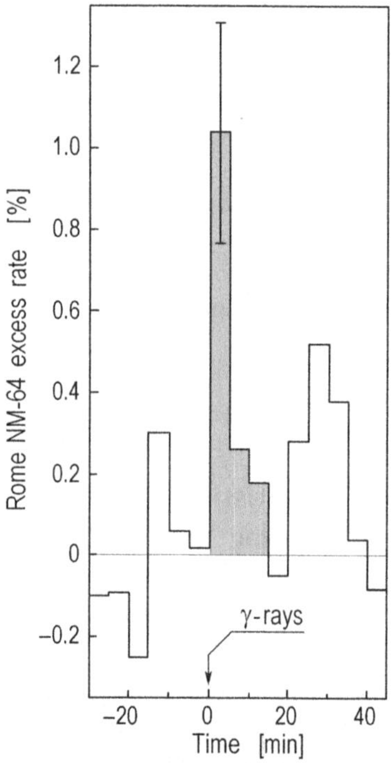

Fig. 6.1 Superposed epoch analysis of the Rome NM-64 5-min data for eight gamma-ray line flares (see Table 6.1) that occurred when the Sun's zenith angle was $20° \le \theta_s \le 40°$ (According to Iucci et al. 1984)

Iucci et al. (1984) extended the statistical search for other possible solar neutron events to the largest hard X-ray events which are not accompanied by observed gamma-lines above 2 MeV: owing to the fact that for some events the limb darkening effect (Wang and Ramaty 1974; Ramaty et al. 1983; Yoshimori et al. 1983) could be responsible for the lack in the gamma-ray line observation. These solar flares are listed in Table 6.3.

The result of superposed epoch analysis for the hard X-ray flares is shown in Fig. 6.3, the 15 min average neutron monitor intensity after the maximum of the hard X-ray flux appears to be well above 3σ, relative to the 0-level.

Table 6.2 Solar gamma-ray line flares that occurred when Rome had $\theta_s > 50°$ (From Iucci et al. 1984)

Date	Flare location	Gamma-ray line or hard X-ray maximum (UT)	Cosmic ray epoch time (UT)
Nov. 5, 1979	15°S, 43°E	23:47	23:45
Nov. 9, 1979	12°S, 02°W	03:05	03:05
Nov. 10, 1979	15°S, 13°W	06:42	06:40
Dec. 19, 1980	16°S, 53°E	22:14	22:15
June 7, 1980	14°N, 70°W	03:11–03:12	03:10
June 21, 1980	17°N, 91°W	01:18–01:19	01:20
Nov. 6, 1980	12°S, 72°E	03:44–03:45	03:45
Nov. 7, 1980	07°N, 11°W	02:04	02:05
April 1, 1981	43°S, 52°W	01:47	01:45
April 4, 1981	44°S, 85°W	05:03	05:05
May 13, 1981	11°N, 58°E	04:18	04:20
Oct. 7, 1981	19°S, 88°E	23:01	23:00
Oct. 14, 1981	06°S, 86°E	17:06	17:05
Oct. 15, 1981	21°S, 09°W	04:44	04:45

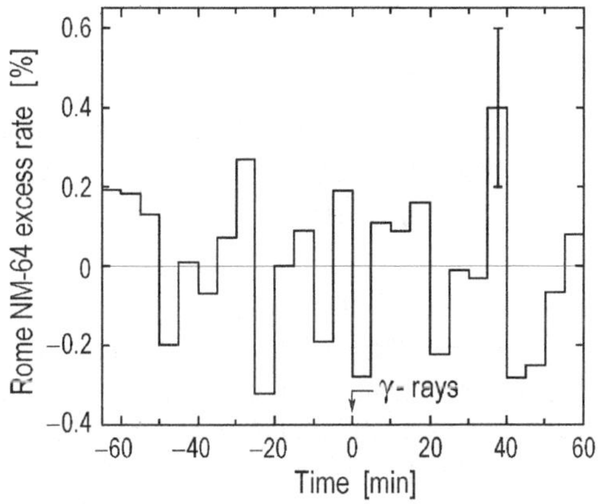

Fig. 6.2 Superposed epoch analysis of the Rome NM-64 5-min data for 14 gamma-line flares (see Table 6.2) that occurred when the Sun was inclined >50° from the zenith or behind the horizon (According to Iucci et al. 1984)

Table 6.3 Hard X-ray solar flares not accompanied by observed gamma-ray lines above 2 MeV and occurred when for Rome was $20° \leq \theta_s \leq 40°$

Date	Flare location	Maximum of hard X-ray (UT)	Cosmic ray epoch time (UT)
Aug. 18, 1979	10°N, 90°E	14:12	14:15
Aug. 20, 1979	05°N, 76°E	09:14	09:15
Sept. 15, 1979	15°S, 57°E	10:27	10:25
April 10, 1980	12°N, 42°W	09:19	09:20
April 2, 1981	43°S, 68°W	11:06	11:05
April 10, 1981	11°N, 53°E	11:09	11:10
April 18, 1981		10:51	10:50
April 26, 1981	12°N, 74°W	11:48	11:45
May 4, 1981	14°N, 16°E	08:38	08:40
May 5, 1981	17°N, 01°W	14:10	14:10

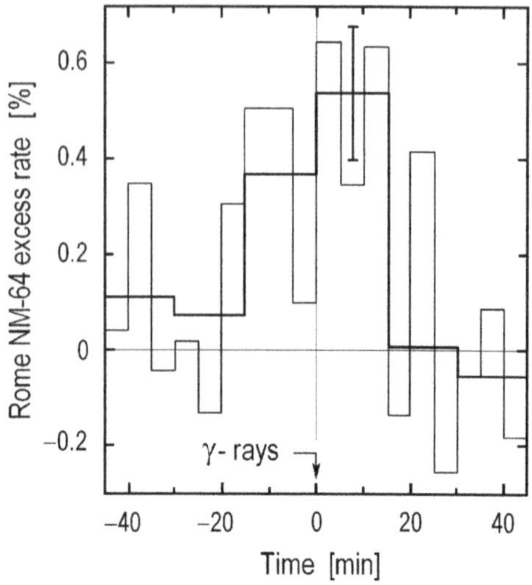

Fig. 6.3 Superposed epoch analysis of the Rome NM-64 5-min data for ten hard x-ray flares without gamma-lines above 2 MeV (see Table 6.3) that occurred when the Sun's zenith angle was $20^\circ \le \theta_s \le 40^\circ$ (According to Iucci et al. 1984)

6.2 Statistical Investigation of Solar Neutron Events Using the Chacaltaya Neutron Supermonitor 5-min Data for 1980–1982

Martinic et al. (1985) reported the results of superposed epoch analysis to search solar neutron effects in NM-64 equipment, 5-min data for 1980–1982, from Chacaltaya (16.32°N, 68.15°W, cut-off rigidity for the vertical direction 13.1 GV, 5,300 m above sea-level, $h = 552$ g/cm^2). Investigations were carried out for three separate groups of solar flares, accompanied by hard X-ray and/or gamma-ray emissions from observations made by satellite SMM, HONOTORI and others. In the first group with expected largest solar neutron effect was selected visible from Chacaltaya solar flares with helio-longitude near the limb (>70°); this group is shown in Table 6.4.

Figure 6.4 shows the result of superposed epoch analysis for these flares. It can be seen that the observed increase of about 0.15%, is inside statistical errors and that such "increases" in a 3 h period are several. A negative result was obtained by Martinic et al. (1985) also for 21 visible gamma-ray and/or X-ray flares far from the limb, with longitudes <70° (see Table 6.5 and Fig. 6.5).

The negative result obtained on the basis of high-altitude Chacaltaya, NM-64 data surprised me very much (in comparison with the positive result obtained on the basis of the sea-level NM-64 data, see Section 6.1). To understand the cause of this discrepancy between Martinic et al. (1985) and Iucci et al. (1984), we determined

Table 6.4 List of 11 visible solar flares from Chacaltaya, near the limb (helio-longitude >70°) (From Martinic et al. 1985)

Date	Flare location	Onset time (UT)	λ_s	f_s	θ_s	h_s (g/cm^2)
April 28, 1980	14°S, 89°W	20:30	14.1°N	127.5°W	59.4°	1,084
Dec. 23, 1980	13°S, 90°E	21:15	23.5°S	138.8°W	75.2°	2,161
April 26, 1981	12°N, 74°W	11:44	13.4°N	4.0°E	72.2°	1,806
April 26, 1981	13°N, 79°W	17:39	13.4°N	84.8°W	16.9°	577
Sept. 9, 1981	13°S, 83°W	20:05	4.5°N	121.2°W	54.0°	939
Sept. 15, 1981	18°N, 78°W	21:13	2.1°N	138.2°W	70.7°	1,670
Oct. 14, 1981	06°S, 86°E	17:05	9.3°S	76.2°W	26.8°	618
Dec. 7, 1981	06°S, 90°E	14:0	22.9°S	42.5°W	45.7°	790
Feb. 8, 1982	13°S, 88°W	12:49	15.5°S	12.2°W	61.6°	1,161
July 21, 1982	23°N, 88°W	18:22	20.4°N	95.5°W	27.6°	623
Dec. 8, 1982	07°N, 8°E	14:37	22.9°S	39.2°W	45.2°	783

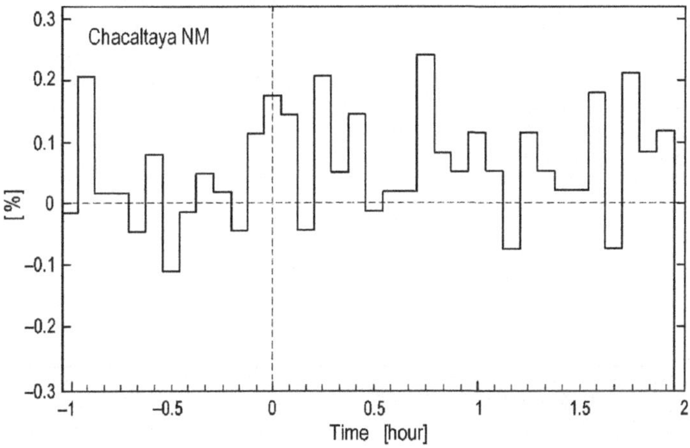

Fig. 6.4 Superposed epoch analysis of the Chacaltaya NM-64 5-min data for 11 visible gamma-ray or/and X-ray solar flares with helio-longitudes >70° (see Table 6.4). The onset of the flares is shown with a *dash vertical line* (According to Martinic et al. 1985)

for each solar flare in Tables 6.4 and 6.5 the position of the sub-solar point (latitude λ_s and longitude f_s, see Section 6.1) and then we also found for Chacaltaya NM-64 the zenith angle of the Sun, θ_s and the atmospheric depth in the Sun's direction h_s. From Table 6.4 we can see that from 11 visible flares, with helio-longitude >70° only five events had $\theta_s \leq 50°$ ($h_s \leq 860$ g/cm^2) and five events had $h_s \geq 1{,}000$ g/cm^2 (up to more than 2,000 g/cm^2). From the 21 visible flares at Chacaltaya, with helio-longitudes <70° (Table 6.5) we have only ten events with $\theta_s \leq 50°$and for six events $h_s \geq 1{,}000$ g/cm^2 (up to 3,000 g/cm^2). This may be the main cause of the negative result obtained by Martinic et al. (1985).

Table 6.5 List of 21 visible solar flares from Chacaltaya far from the limb (helio-longitude <70°) (From Martinic et al. 1985)

Date	Flare location	Onset time (UT)	λ_s	f_s	θ_s	h_s (g/cm^2)
April 26, 1980	17°S, 61°E	20:31	13.5°N	127.7°W	59.6°	1,091
May 21, 1980	14°S, 15°W	20:54	20.2°N	133.5°W	65.4°	1,326
July 1, 1980	12°S, 37°W	16:26	23.2°N	66.5°W	7.1°	556
Oct. 9, 1980	17°S, 50°E	11:23	7.5°S	9.2°E	79.2°	2,946
Nov 2, 1980	28°S, 58°W	14:18	15.8°S	34.5°W	45.2°	783
Nov 8, 1980	8°S, 57°E	14:52	17.5°S	43.0°W	41.2°	734
Nov 11, 1980	11°S, 71°W	17:43	18.3°S	85.7°W	38.3°	703
Dec. 18, 1980	7°N, 11°W	19:21	23.5°S	110.2°W	55.2°	967
Feb. 17, 1981	20°N, 20°W	21:46	12.6°S	146.5°W	79.8°	3,117
Feb. 26, 1981	12°S, 53°E	14:24	9.3°S	36.0°W	40.2°	723
March 23,1981		16:13	0.9°N	63.2°W	16.2°	575
April 10, 1981	11°N, 53°E	11:09	7.6° N	12.8°E	81.1°	3,569
April 24, 1981	18°N, 50°W	14:00	12.7°N	30.0°W	38.3°	703
May 5, 1981	17°N, 1°W	14:09	16.2°N	32.2°W	35.9°	681
July 26, 1981	15°S, 27°E	13:53	19.4°N	28.2°W	40.0°	721
July 28, 1981	10°S, 18°W	20:05	18.9°N	121.2°W	53.1°	919
Oct 16, 1981	19°S, 36°W	12:48	10.1°S	12.0°W	60.1°	1,107
Feb 15, 1982	0°N, 27°E	13:21	13.2°S	20.2°W	54.4°	948
June 15, 1982	22°S, 66°E	15:11	23.4°N	47.7°W	43.9°	766
Dec 15, 1982	10°S, 15°E	16:30	23.4°S	67.5°W	39.7°	717
Dec 17, 1982	8°S, 21°W	18:56	23.4°S	104.0°W	51.4°	885

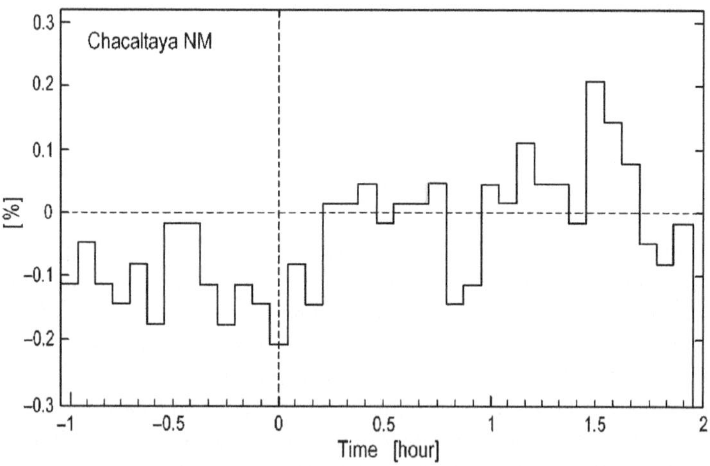

Fig. 6.5 Superposed epoch analysis of the Chacaltaya NM-64 5-min data for 21 visible gamma-ray or/and X-ray solar flares with helio-longitudes <70° (see Table 6.5). The onset of the flares is shown with a *dash vertical line* (According to Martinic et al. 1985)

6.3 Statistical Investigations of Solar Neutron Events Using the Tyan Shan Neutron Supermonitor 5-min Data

6.3.1 *Solar Neutron Events and the Behavior of the Cosmic Ray Variations* δ *and Its Augmentations* Δ

The Tyan Shan 18 NM-64 supermonitor (43.2°N, 76.9°E, cut-off rigidity for the vertical direction 6.7 GV, altitude 3,340 m above sea level, mean atmospheric depth in vertical direction $h = 680\ \text{g/cm}^2$) shows a high counting rate ($\approx 4 \times 10^5$ in 5-min, in correspondence with $\sigma = 0.15\%$ statistical accuracy for 5-min data) and is convenient when seeking for the minor cosmic ray variations associated with solar neutron events. One hundred and thirty-six solar flares with rigid (>0.3 MeV) electromagnetic radiation were analysed by Belov et al. (1987a, b; 1989), with the aim of examining only powerful flares with a strictly determined periods of intensive energy release (on the basis of Catalogue of Rieger (1982) according to observations on SMM). The events with ordinary solar cosmic ray increases on the Earth's surface (caused by charged particles) were excluded. The 5-min intervals within which a gamma-burst onset was observed were taken to be 0th. Consideration was given to the behaviour of the variation δ (and its augmentation Δ) of the counting rate N in the r.m.s. deviation units ($\sigma = \overline{N}^{-1/2}$):

$$\delta_k = \left(N_k - \overline{N}\right)/\sqrt{N}, \quad \Delta_k = \delta_k - \delta_{k-1}. \tag{6.1}$$

The mean counting rate from -7th to -2nd 5-min intervals were taken to be the basic value. Figure 6.6 shows the average behaviour of cosmic ray variations near a gamma-burst. The $|\Delta|$ value may be taken to be the measure of the cosmic ray fluctuations with ~5-min characteristic time. The highest mean $\overline{|\Delta|}$values for the 136 events are observed in the 0th–1st and 5th–6th 5-min intervals, i.e. at the moment when the solar neutron increases and the anomalous proton increases may be expected. However, these $|\Delta|$ variations are due not only to the solar cosmic ray increases, as indicated by the behaviour of the mean variation of δ.

The cosmic ray fluctuations during the 0th–1st 5-min intervals are mainly due to the flares of a great power (≥M4) in X-ray range. The high $|\Delta|$ values in the 5th–6th 5-min intervals coincide most often with flares of a relatively low power. It is of interest that the helio-latitude distribution of the flares coinciding with substantial cosmic ray fluctuations in the 0th–1st 5-min intervals ($|\Delta| \geq 4$ and especially $|\Delta| \geq 5$) is very different from the helio-longitude distribution of the flares during which the effects in cosmic rays were absent (see Fig. 6.7).

A separate neutron group was composed of 13 powerful (≥M4) flares during which the solar neutron observation conditions at Tyan Shan station were most favourable ($h_s < 1{,}000\ \text{g/cm}^2$). Figure 6.8 shows the changes in the distribution of the values of the variations of δ.

From Fig. 6.8 can be seen that the initial distribution (the -6th to -2nd 5-min intervals) is close to normal with variance $a \approx 2\sigma^2$, in correspondence with the

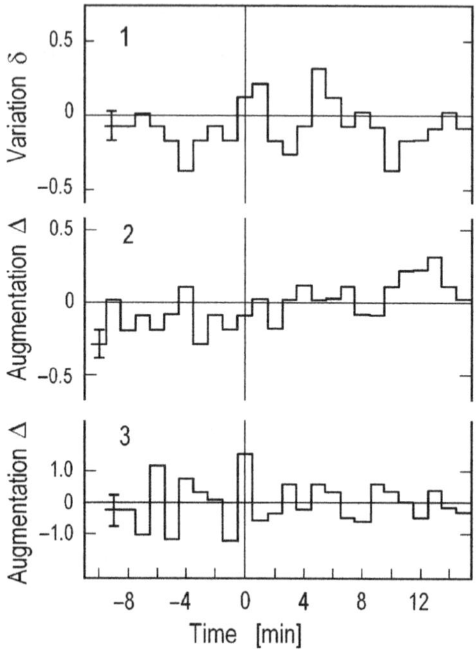

Fig. 6.6 Averaged behavior of cosmic ray variation δ and of its augmentation Δ during solar flares: 1 – δ for all 136 events, 2 – Δ for all 136 events, 3 – Δ for 13 flares of class ≥ M4 when $h_s < 1{,}000$ g/cm^2 (According to Belov et al. 1987a, b)

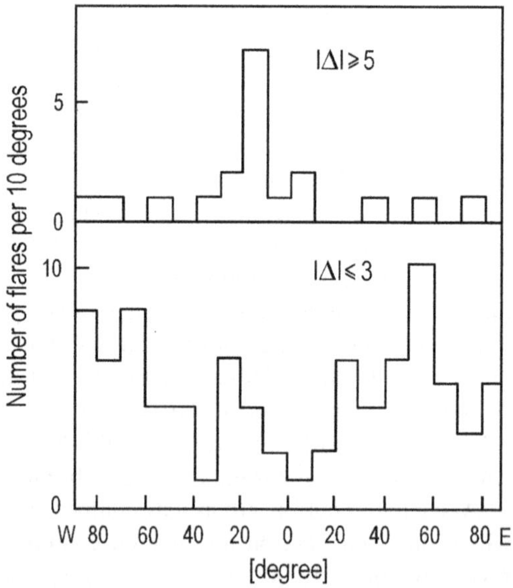

Fig. 6.7 Helio-longitude distribution of the flares coinciding with cosmic ray variations during the 0th–1st 5-min intervals (According to Belov et al. 1987a, b)

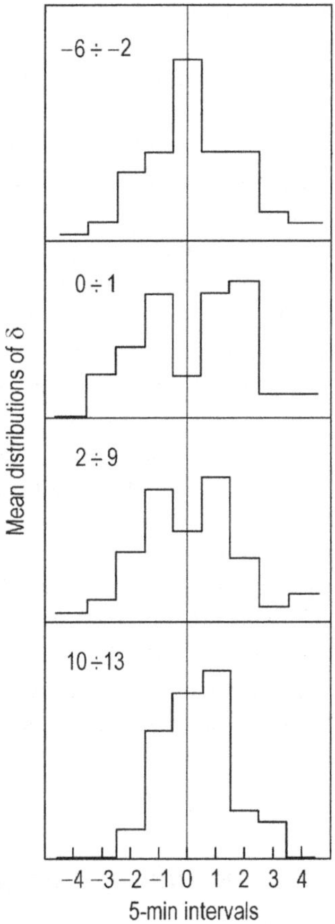

Fig. 6.8 Mean distributions of the value of variation δ for 13 evens from the 'neutron' group with $h_s < 1{,}000$ g/cm^2 (According to Belov et al. 1987a, b)

variance expected considering the multiple neutrons. During the 0th–1st 5-min intervals the distribution becomes bimodal and its difference from the initial distribution is statistically significant. Later (the 2nd–9th and 10th–13th 5-min intervals), the distribution approaches the initial form. The maximum mean augmentation $\Delta = 1.3$ in the neutron group was observed during the 0th 5-min interval (see Fig. 6.8). During the 0th–1st intervals the increase of $a > 2\sigma^2$ was observed in ten events out of 13. To these events, the April 27, 1981 event should probably be added during which the augmentation $\Delta = 3.6$ occurred in the 1st interval.

Belov et al. (1987a, b, 1989) came to conclusion that none of the substantial (>1%) cosmic ray variations attributable directly to a solar flare was observed in any of the events. At the same time, the minor (<1%), but not random cosmic ray variations are often observed during powerful solar flares. It may be thought that a fraction of the increases observed at Tyan Shan station near the gamma-burst moments are due to the arrival of flare generated neutrons. An example of such an event is August 10, 1981 when, during a 1B/M5 flare, the neutron monitor detected a 0.9% (5.2σ) increase which occurred at local noon ($h_s \approx 760$ g/cm^2) and coincided with a gamma-burst. However, the arrival of neutrons can not explain all cosmic ray variations near gamma-bursts. Such an explanation is inapplicable to the negative fluctuations and to several substantial increases at $h_s > 1{,}000$ g/cm^2.

One can hardly speak also of any statistically significant evidence for the existence of anomalous proton increases, but may well note several large (3–5 σ) increases occurring in the 15–20 min after a gamma-burst onset, i.e. when such increase should be expected. On June 5, 1982, for example, in ~25 min after a burst (06:15:30–06:16:42 UT) a cosmic ray increase (>5σ) commenced and lasted for 5–10 min. Thus, the 0.1–1.0% variations can be discriminated in the high-mountain neutron monitor data obtained during powerful solar flares. The relevance of the variations to the flare power and longitude indicated that they are due to the events occurring on the Sun.

Suffice it to say that in 13 out of 16 events with powers >M4 in the X-ray range and helio-latitude ranging in the interval 0–30°W, we have $|\Delta| \geq 5$. There is some evidence indicative of a higher probability for the effects from solar flare-generated neutrons to be detected at ground-based high-mountain stations during the periods with gamma-bursts produced by solar activity. However, effects of different nature are likely to take into account in this case. To explain them, it is necessary to consider not only the solar cosmic ray increases of various types, but also the impact of flares on the primary and secondary cosmic rays generated and propagated to the Earth's surface through the solar flare phenomena in the upper atmosphere, ionosphere, and magnetosphere.

6.3.2 Results of Superposed Epoch Analysis

The statistical investigation of solar neutron events on the basis of Tyan Shan neutron supermonitor 5-min data was continued by Aushev et al. (1993). The results are shown in Figs. 6.9 through 6.12 (a consecutive 5-min intervals from 0th point are given on abscissa axis; on ordinate axis there are deviations from average in percent). Cosmic ray data averages over 34 events when splash solar radiation was in interval ±3 h from local noon at Tyan Shan station are shown in Fig. 6.9.

From Fig. 6.9 can be seen that in considered case there is not any increase during splash. Figure 6.10 shows the average data for nine events from solar proton fluxes observed in space; there is no increase in this case also.

When flares are selected at ±1 h near local noon (Fig. 6.11, 11 events), the increase at the moment of splash, with an amplitude of 0.13 ± 0.06% observed. There is an increase with amplitude 0.23 ± 0.05% for flares of X-class (eight events) as it is clear seen from Fig. 6.12.

When the Prognoz-9 (Abrosimov et al. M1987) registered hard X-ray emission from the Sun at relatively convenient periods for the registration of solar neutrons at Tyan Shan station Aushev et al. (1993) have used these data in the analysis. It was

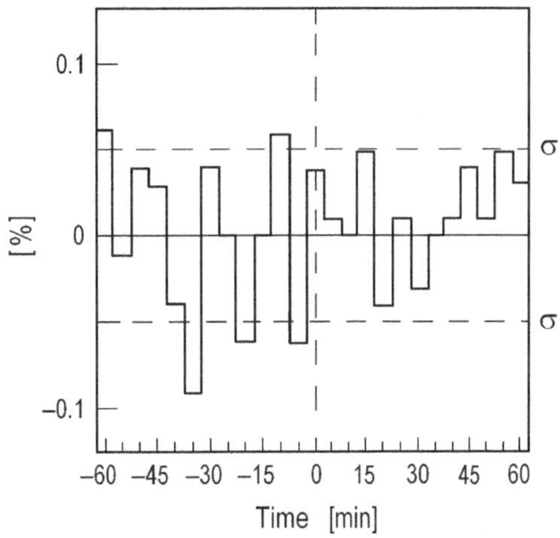

Fig. 6.9 Superposed epoch analysis of the Tyan Shan 18-NM-64 supermonitor 5-min data over 34 gamma-ray events detected on SMM when splash solar radiation was in the interval ±3 h near local noon of Tyan Shan station (According to Aushev et al. 1993)

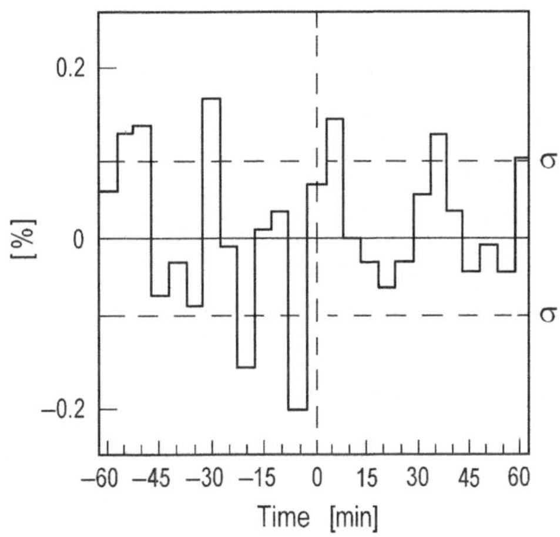

Fig. 6.10 Superposed epoch analysis of the Tyan Shan 18-NM-64 supermonitor 5-min data for nine events from 34 events considered in Fig. 6.9 and accompanied by energetic protons in space (According to Aushev et al. 1993)

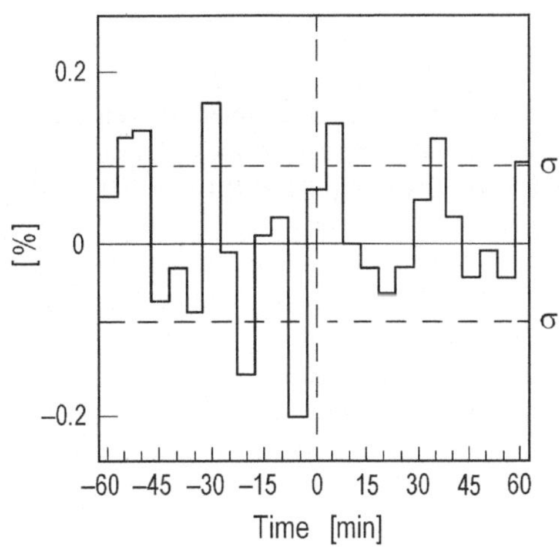

Fig. 6.11 Superposed epoch analysis of the Tyan Shan 18-NM-64 supermonitor 5-min data for 11 events from previous 34 events considered in Fig. 6.9, but occurring at ±1 h near local noon (According to Aushev et al. 1993)

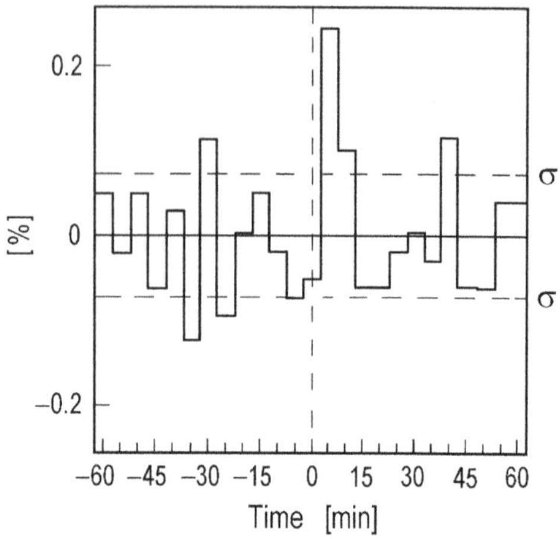

Fig. 6.12 Superposed epoch analysis of the Tyan Shan 18-NM-64 supermonitor 5-min data for eight events from 11 events in Fig. 6.11, but only for flares of X-class (According to Aushev et al. 1993)

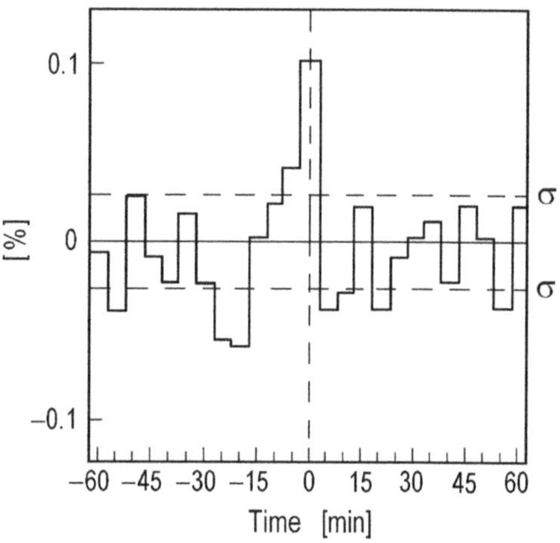

Fig. 6.13 Superposed epoch analysis of the Tyan Shan 18-NM-64 supermonitor 5-min data for 20 hard X-ray events detected on Prognoz-9 and occurred within the interval ±3 h to local noon of Tyan Shan station (According to Aushev et al. 1993)

observed that in average over 40 events there were no increases in cosmic rays. But for 20 events, when flares took place in the interval ±3 h to local noon, the increases observed had amplitude 0.10 ± 0.03% (Fig. 6.13). From these 20 events were selected three events (25 July 1983, 21 September 1983, 3 February 1984), when flares had helio-longitude above 60° and observed during ±1 h to local noon (Fig. 6.14). In this case the average amplitude of increase in the 5-min interval was 0.28 ± 0.06%.

Aushev et al. (1993) came to conclusion from adduced results, that the neutron fluxes with great probability are generated in the solar flares with large emissions of

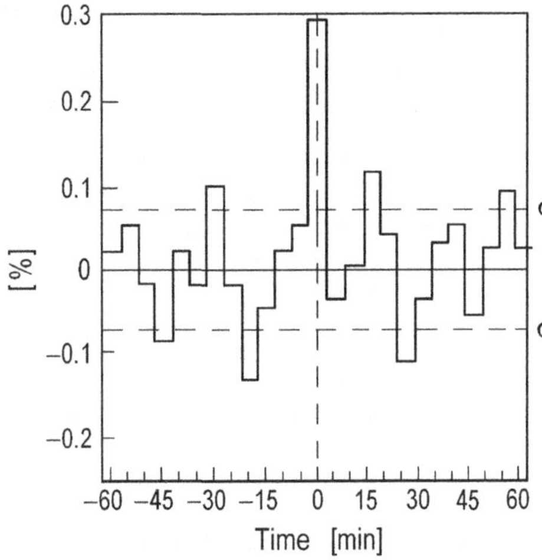

Fig. 6.14 Superposed epoch analysis of the Tyan Shan 18-NM-64 supermonitor 5-min data for three events from 20 events analyzed in Fig. 6.13, but occurring at ±1 h near local noon and when flares had helio-longitudes above 60° (July 25, 1983; September 21, 1983 and February 3, 1984) (According to Aushev et al. 1993)

hard electromagnetic radiation and the greatest amplitudes of cosmic ray increases in he first 5-min interval were observed in the case of large flux of hard X-ray emissions and X-ray flares with large helio-longitudes.

6.3.3 Extended Superposed Epoch Analysis Using GOES and CGRO Satellite Data on X-Ray Events

Aushev et al. (1999) created two catalogues of events when the Tyan Shan NM might observe solar neutrons. The first is based on the GOES satellite data and accounts 776 events over the period of 1974–1999 when all flares had >M1 importance and all proton flares followed by optical flare >2 class and/or with >1 h duration. The second uses data of hard X-ray and gamma-ray solar emission obtained by the BATSE detector aboard the CGRO satellite (1991–1999) with magnitude >1,000 imp/s and accounts 341 events might be effective for detecting by Tyan Shan NM. Aushev et al. (1999) calculates possible response of the Tyan Shan NM (the parameter Np) in percent above background to the neutron flux of the prominent neutron event of 3 June 1982, i.e. what should be observed if a flux of solar neutrons near the Earth would be like on June 3, 1982. Choosing, as a threshold, the response Np equal to 0.03, more than 750 events in CR intensity mostly favourable for observation in Tyan Shan were selected.

Aushev et al. (1999) did the preliminary statistical analysis by the epoch method. They applied this method for two data sets: one contains all events from the list (more than 750 events), another group with expected effect more than 1% contains

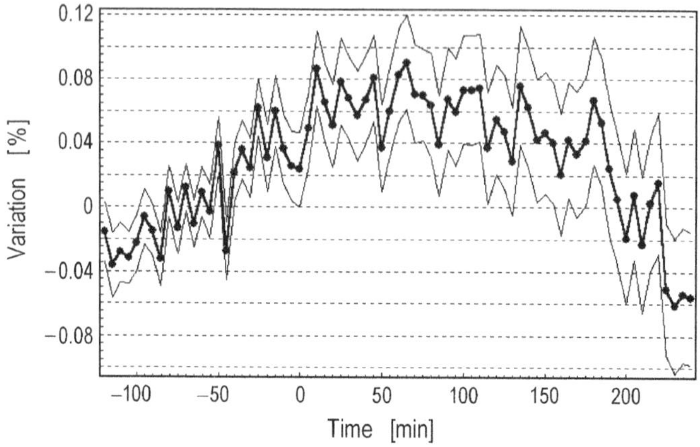

Fig. 6.15 Mean variation of the Tyan Shan NM count rate for the second set of possible neutron events (contains 286 events). Zero time on the bottom corresponds to onset of X-ray events (From Aushev et al. 1999)

286 events. Data of all events within of each data set were centered on the X-ray onset of parent solar event and then were averaged. Results obtained for both groups are similar, but the effect is better pronounced for the second group (see Fig. 6.15).

In Fig. 6.15 corridor around the main curve corresponds to two standard statistical errors. One can see the clear enhancement close to the X-ray event onset, although its magnitude is not large, about 0.07%. It is interesting and gives some reasons for speculations that maximums of averaged variations are observed at the X-ray onset and about 30 min later. However, Aushev et al. (1999) mentioned that GLEs and small Forbush effects were not removed from the data used to plot this picture and 5-min data of the 89–91 years were not corrected for pressure variations, although it can not change the result significantly.

6.4 Statistical Investigation of Solar Neutron Events Using the 5-min Data of the Neutron Supermonitors of the Sayan Spectrograph

6.4.1 Short Description of the Sayan Spectrograph

The Sayan spectrograph includes three neutron supermonitors at cut-off rigidity in the vertical direction 3.6 GV: Hulugaisha (6NM-64, $h = 712$ g/cm^2), Chasovie Sopki (12NM-64, $h = 806$ g/cm^2) and Irkutsk (18NM-64, $h = 965$ g/cm^2).

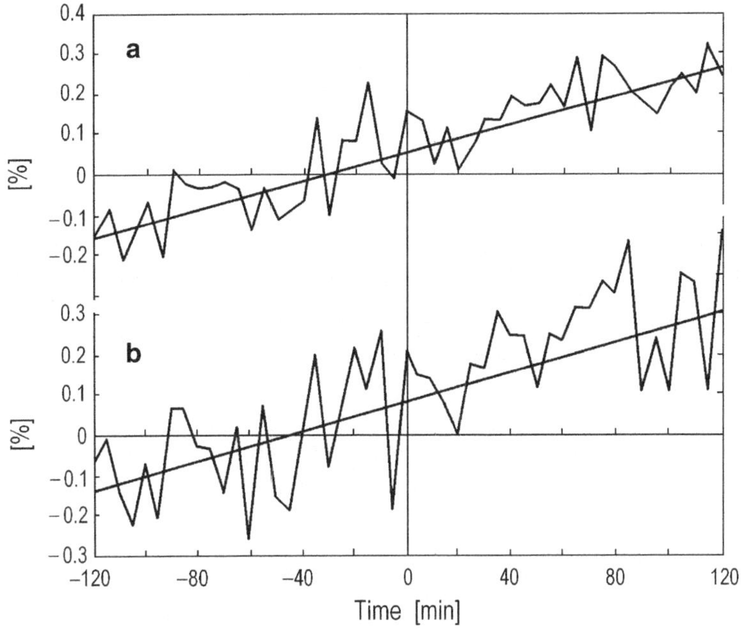

Fig. 6.16 Superposed epoch analysis of 5-min data of Hulugaisha NM not corrected on atmospheric pressure: for 49 Hα flares and sub-flares (*top panel* **a**) and for 18 Hα flares of power ≥1 (*bottom panel* **b**). All selected flares are occurred on helio-longitudes ≥70° when zenith angle of the Sun $\theta_s \leq 45°$. *Straight lines* – trends caused by daily variation of atmospheric pressure (From Pakhomov et al. 1987)

6.4.2 Superposed Epoch Analysis for Hα Flares

By the method of superposed epoch Pakhomov et al. (1987) analyzed 5-min data of Sayan spectrograph's NM. In Fig. 6.16 are shown original 5-min data of Hulugaisha NM not corrected on atmospheric pressure for 49 Hα flares and sub-flares (top panel a) and for 18 Hα flares of power ≥1 (bottom panel b). All flares were occurred on solar longitudes ± (70–90°) and the zenith angle of the Sun in the point of registration did not precede 45°.

From Fig. 6.16 can be seen that obtained results are negative, CR variations near 0th 5-min interval in both cases do not precede the statistical fluctuations.

6.4.3 Solar Neutrons from Flares Accompanied by X-Rays of Class ≥M2.0

According to Pakhomov et al. (1987), for 11 Hα flares on helio-longitudes ± (55–90°), occurred in 1980–1984, within ±3.5 h near noon and accompanied by hard X-rays of class ≥M2.0 it was found by superposed epoch analysis that the average effect is 0.60 ± 0.15% (see Fig. 6.17).

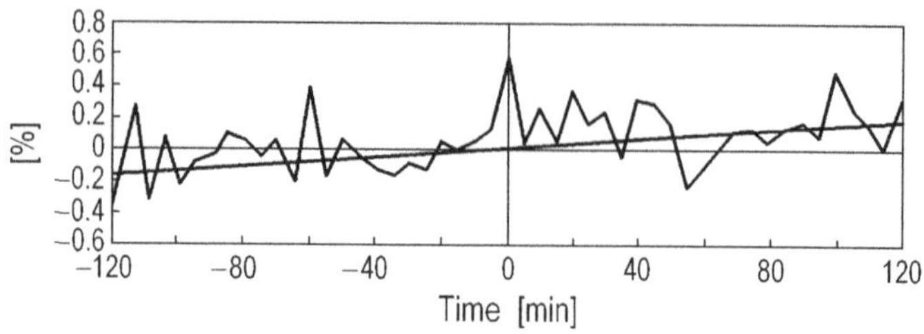

Fig. 6.17 The same as in Fig. 6.16, but for 11 Hα-flares on solar longitudes ± (55–90°), occurred in 1980–1984, within ±3.5 h near noon and accompanied by hard X-rays of class ≥M2.0. *Straight line* – trend caused by daily variation of atmospheric pressure (From Pakhomov et al. 1987)

6.4.4 Solar Neutrons from Impulsive Bright Flares Accompanied by Hard X-Rays of ≥M2.0 Class

As noted Pakhomov et al. (1987), among 11 events analyzed in Fig. 6.17 there are several individual events which will be consider in Chapter 7:

1. June 7, 1980 (Hulugaisha, 6NM-64, $\theta_s = 32°$, $h_s = 850$ g/cm^2, amplitude of increase A = 0.6 ± 0.5%; Chasovie Sopki, 12NM-64, $h_s = 979$ g/cm^2 and A = 0.2 ± 0.5%)
2. April 4, 1981 (for Hulugaisha, $\theta_s = 49°$, $h_s = 1{,}099$ g/cm^2, A = 1.3 ± 0.5% for 5-min interval 4:57–5:02 UT, and A = 0.9 ± 0.5% for 5:02–5:07 UT)
3. June 12, 1982 (Hulugaisha, $\theta_s = 32°$, $h_s = 850$ g/cm^2, A ~ 0.5%, Chasovie Sopki, $h_s = 979$ g/cm^2, A = 0.65 ± 0.5%)
4. August 14, 1982 (Hulugaisha, $\theta_s = 38°$, $h_s = 913$ g/cm^2, A = 0.7 ± 0.5%)
5. May 20, 1984 (Hulugaisha, $\theta_s = 44°$, $h_s = 1{,}004$ g/cm^2, A = 1.0 ± 0.5%, Chasovie Sopki, $h_s = 1{,}153$ g/cm^2, A = 0.4 ± 0.5%)

These five events were caused by impulsive bright flares accompanied by hard X-rays of ≥M2.0 class. The superposed epoch analysis for these five impulsive events gives an average effect on Hulugaisha 6NM-64 $\overline{\mathrm{A}} = 0.9 \pm 0.2\%$.

6.4.5 Solar Neutrons from Gradual Bright Flares Accompanied by Hard X-Rays of ≥M2.0 Class

According to Pakhomov et al. (1987), in the considered group of 11 Hα-flares there were five flares with prolonged phase of increase for more than 5 min (gradual flares): July 10, 1982; September 1, 1982; May 8, 1983 and April 24 and 27, 1984. The average superposed effect on Hulugaisha 6NM-64 for these gradual events gives $\overline{\mathrm{A}} = 0.5 \pm 0.2\%$, about two times smaller than for the impulsive flares.

6.5 Upper Limits on High-Energy Solar Neutron Fluxes from Gamma Ray Flares According to Yangbajing NM Observations

6.5.1 The Matter of the Problem

Tsuchiya et al. (2007, 2008) noted that solar neutrons are one of the best tools to deeply understand acceleration mechanism of ions in solar flares, because they are produced via nuclear interactions between the accelerated ions and the solar ambient plasma. High-altitude detectors, including neutron monitors (Stoker et al. 2000) and neutron telescopes (Muraki et al. 2007), provide a good opportunity to detect ≥50 MeV solar neutrons (see also Chapter 4 in Dorman M2004). Actually, such high-altitude detectors have accomplished positive detections of the high-energy solar neutrons associated with X-class solar flares. On the other hand, none, including space satellites, has succeeded in detecting solar neutrons from C- and M-class flares. Near the Earth satellites actually detected bremsstrahlung photons from many C- and M-class flares, revealing that such flares can accelerate electrons to high energies. This fact allows to consider that solar neutrons may be produced in such less intense flares, because usually it is very natural that ions are simultaneously accelerated at the same time as electrons. Therefore, a systematic search for solar neutrons from the less intense flares, as well as very strong ones, are important to strictly constrain some key parameters of the ion acceleration process, including numbers, pitch angle distributions, and energy spectra of accelerated ions. Tsuchiya et al. (2007, 2008) hence perform a systematic search for solar neutrons from 164 flares of M- and X-classes. This search was made by the Yangbajing NM.

6.5.2 Main Properties of the Yangbajing NM

The Yangbajing NM has been in operation at Yangbajing (90.522°E, 30.102°N in Tibet, China, since October 1998 (Kohno et al. 1999; Miyasaka et al. 2001; Tsuchiya et al. (2007)). Installed at an altitude of 4,300 m above sea level, it has an advantage of a much reduced air mass, 606 g/cm^2. Furthermore consisting of 28 NM64 type detectors, it has a total area of 32 m^2 which is currently the largest one in the world-wide NMs. In addition to these advantages, the Yangbajing NM has the highest geomagnetic cutoff rigidity, 14 GV, among the world NMs. These conditions allow the Yangbajing NM to be one of the most sensitive detectors for solar neutrons.

6.5.3 Flare Distribution

The BATSE aboard CGRO observed more than 7,000 solar flares in the hard X-ray range above ~25 keV until its re-entry to the atmosphere on 4th June 2000. Among them 164 events satisfy conditions with the GOES peak flux higher than 10^{-5} W/m^2 which corresponds to the GOES class of M1, detected over a period from October 1998 up to June 2000. The 164 events are composed of 157 M- and seven X-class flares, constituting the 'preliminary sample'. By the extracting a set of flares with the zenith angles of the Sun at the flare onset time smaller than 60° from the 'preliminary sample', was prepared the 'final sample' consisting of two X- and 16 M-class flares as shown in Fig. 6.18.

6.5.4 *Emission Profiles of Solar Neutrons*

Tsuchiya et al. (2007) search the Yangbajing NM data for possible enhancements associated with solar flares. Thus they prepared 5-min count histories from the Yangbajing NM, ranging over ±1.5 h from the onset time determined by the BATSE data and corrected them for atmospheric pressure changes. In order to quantitatively constrain neutron signals from individual flares, we need to define for each flare an 'On time window', i.e., a time interval when solar neutrons might arrive at the Yangbajing NM, and use the remaining two time intervals (before and after the flare) to calculate the background. For this aim, a time profile of the solar

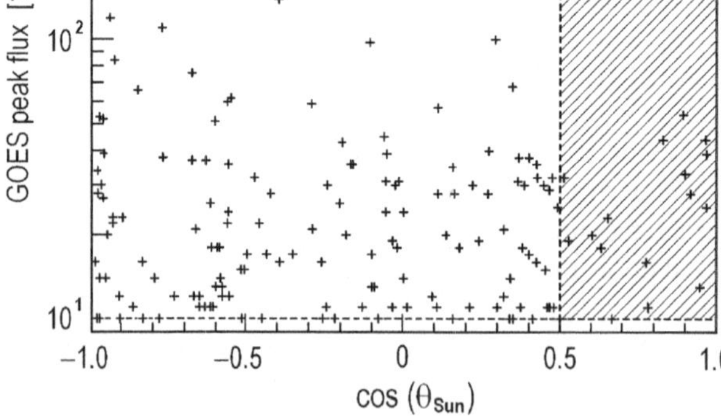

Fig. 6.18 The cosine of the zenith angle of the Sun at Yangbajing θ_{Sun} at onset time of individual flares in the 'preliminary sample', plotted against their GOES peak flux in units of 10^{-6} W/m^2. The *vertical* and *horizontal dashed lines* correspond to $\theta_{Sun} = 60°$ and the GOES flux of 1×10^{-5} W/m^2, respectively. The 18 events in the hatched area constitute the 'final sample' (From Tsuchiya et al. 2007)

neutron production at the Sun in each flare as well as the minimum and maximum kinetic energies of solar neutrons are assumed. Then the 'On time window' opens at the arrival of the most energetic neutrons ejected at the beginning of the neutron emission at the Sun, and closes at that of the least energetic ones ejected at the end of the production interval. In the work of Tsuchiya et al. (2007), the most energetic neutron assumes the kinetic energy of 10 GeV, while the least energetic one 50 MeV. The following two neutron-emission time profiles are employed: δ-emission and continuous emission. The δ-emission simply means that solar neutrons are radiated from the Sun instantaneously at the BATSE hard X-ray emission peak, while the continuous-emission profile assumes that neutrons are continuously and constantly emitted from the Sun throughout the BATSE hard X-ray emission. Using these injection profiles, Tsuchiya et al. (2007) can define the 'On time window' for each injection profile. Examples of the 'On time window' specified by the two profiles are shown in Fig. 6.19.

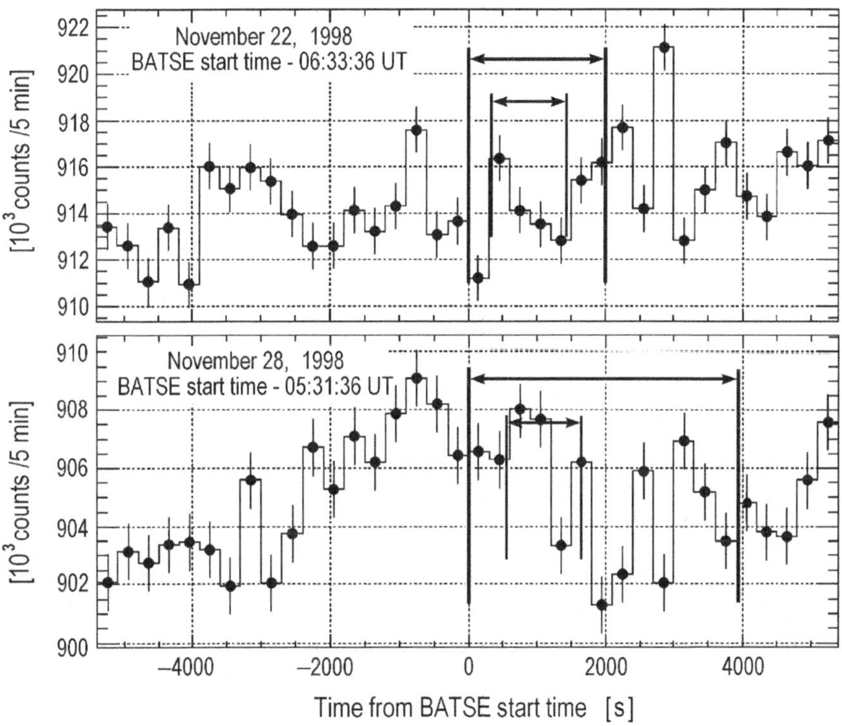

Fig. 6.19 The 5-min counting-rate histories of the Yangbajing NM for two X-class flares. *Left* and *right panels* show the counting-rate histories for an X3.7 (981122) and X3.3 (981128) flares, respectively. Abscissa in all panels shows the time measured from the BATSE start time, ranging over ± 5,400 s, where zero corresponds to the BATSE start time. The intervals of individual 'On time windows' specified by the δ- and continuous-emission profiles are shown by *horizontal thin* and *thick arrows*, respectively (From Tsuchiya et al. 2007)

6.5.5 Results of Statistical Significances of Neutron Signals

Tsuchiya et al. (2007) evaluated the statistical significances of the 164 preliminary sample flares. The results are shown in Fig. 6.20, implying all null detections.

By fitting a Gaussian curve with each histogram in Fig. 6.20, Tsuchiya et al. (2007) obtained two important consequences. One is that the obtained Gaussian centroids are consistent with 0 within the fitting errors. Hence, there is no evidence for apparent neutron signals from the 164 flares. The other point is that the obtained standard deviations are consistent with 1.0, suggesting that the significance scatter among the 164 flares can be fully described by statistical fluctuations.

6.5.6 The Flux Upper Limits of the 'Final-Sample Flares'

Because of null detections, Tsuchiya et al. (2007) calculated 99% confidence level flux upper limits for individual final sample flares assuming that an energy spectrum of solar neutrons at the Sun is a power-law form with the spectrum index of 3, 4, and 5.

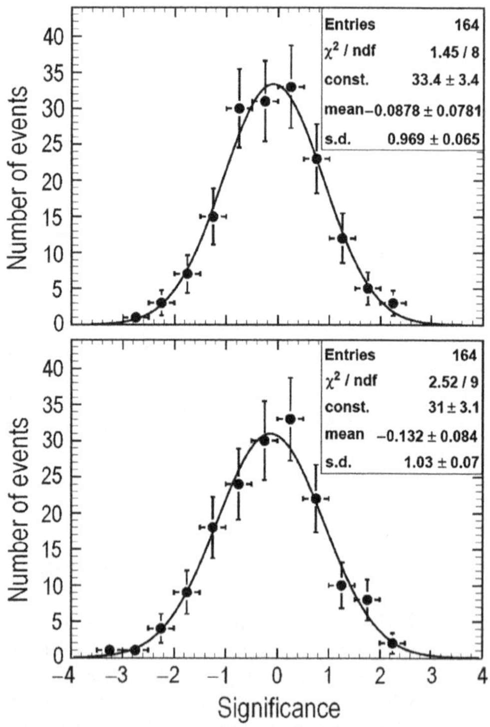

Fig. 6.20 Significance distributions of neutron signals from the 'preliminary-sample flares', calculated assuming the two neutron-emission profiles. *Top panel* corresponds to the δ-emission model while bottom one corresponds to the continuous-emission model. *Vertical error bars* are Poissonian. The best-fit gaussian curves to individual histograms are drawn by *solid lines* (From Tsuchiya et al. 2007)

Figure 6.21 shows the upper limits on the >50 MeV solar-neutron flux for the two (X3.7 and X3.3) of the final sample flares. For comparison, the previous positive detections are also plotted.

As can be seen in Fig. 6.21, a strong positive correlation between the GOES peak flux and the absolute neutron flux appears. Figure 6.22 shows all upper limits

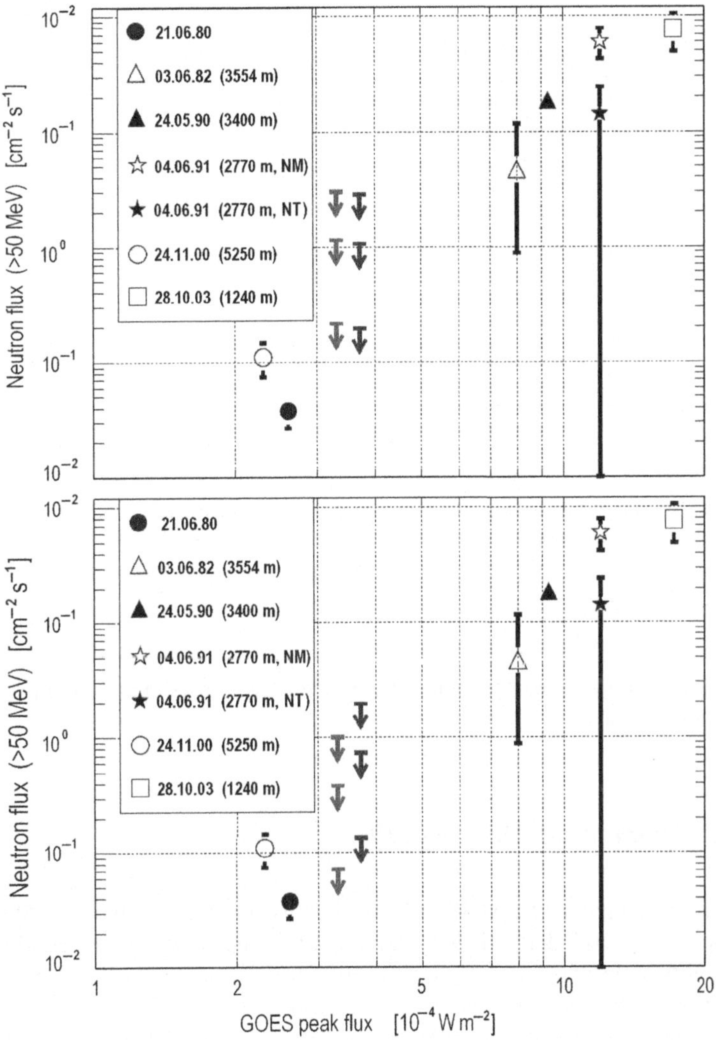

Fig. 6.21 The 99% confidence level upper limits on the >50 MeV solar-neutron flux at the top of the Earth atmosphere, from 981122 (*red arrows*) and 981128 (*blue arrows*) flares, are plotted as a function of F_{GOES}. Past positive detections are also plotted. The horizontal axis gives the GOES peak flux in units of 10^{-4} W/m^2. *Top* and *bottom panels* correspond to δ- and continuous-emission, respectively. Upper limits from low to high correspond to assumed power-law index of solar neutrons at the Sun of 3, 4, and 5, respectively (From Tsuchiya et al. 2007)

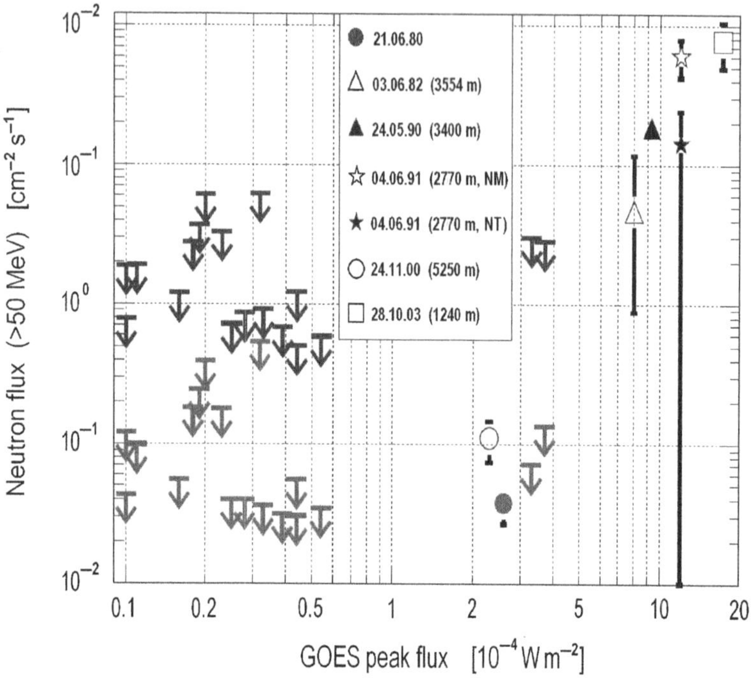

Fig. 6.22 The same as Fig. 6.21, but including all flares in the final sample. Individual *red arrows* show the most stringent upper limit for individual final sample flares, while the least stringent ones are shown by *blue arrows*. The detail information on each data point is shown in Fig. 6.21 (From Tsuchiya et al. 2007)

on >50 MeV neutron flux for each final sample flare, consisting of the most and the least stringent ones.

6.5.7 Summary of Main Results

Utilizing the Yangbajing NM data obtained between 1998 October and 2000 June, Tsuchiya et al. (2007) derived 99% confidence level upper limits on $\geq$50 MeV solar neutron flux from two flares of X-class and 16 flares of M-class with reasonable assumptions. The most stringent upper limits are obtained assuming that the spectrum index is 3 and the emission model is the continuous-emission, while the least stringent ones are calculated assuming that the spectrum index is 5 and the emission model is the δ-emission. The most stringent upper limits seem to indicate that the $\geq$50 MeV neutron flux associated with M-class flares may be below $\sim$0.03 cm^2/s at one AU, with a 99% confidence level.

References for Chapter 6

Aushev V, Antonova V, Belov A, Eroshenko E, Kryakunova O, Struminsky A (1999) Search for Solar Neutron Events in Alma-Ata NM Data, Proc. 26th Intern. Cosmic Ray Conf., Salt Lake City, 1999, 6, 50–53

Aushev VM, Dorman LI, Kryakunova ON, Shvartsman YaE, Zusmanovich AG (1993) Possible registration of solar neutrons by high-mountain neutron monitor. In: Proceedings of the 23rd international cosmic ray conference, vol 3, Calgary, Contributed papers, pp 155–158

Belov AV, Ishkov VN, Zusmanovich AG, Kudela K (1987a) Observation of solar neutrons by high-altitude neutron monitors. Geomagn Aeronomiya 27(3):375–382 (in Russian)

Belov AV, Dorman LI, Ishkov VN, Saidaliev RA, Zusmanovich AG (1987b) Variations of secondary neutrons during powerful solar flares. In: Proceedings of the 20th international cosmic ray conference, vol 3, Moscow, pp 90–93

Belov AV, Dorman LI, Zusmanovich AG, Ishkov VN, Saidaliev RA (1989) Small increases of cosmic rays intensity according to neutron monitors data. In: Somov BV (ed) Investigations of the solar plasma. Ylym, Ashkhabad, pp 188–199 (in Russian)

Iucci N, Parisi M, Signorini C, Storini M, Villoresi G (1984) Detection of solar neutrons by ground-based neutron monitors. Il Nuovo Cimento C 7(6):732–740

Kohno T, Miyasaka H, Matsuoka M, Yamada Y, Shimoda S, Takahashi K, Sakamoto E, Munakata K, Yuda T, Ding LK, Tan YH, Zhang CS, Shi ZZ, Wang H, Guo HW, Meng XR, Jia HY, Mu J, Zhang NJ (1999) Program of solar neutron observation with NM-64 counters at Yangbajing in Tibet. In: Proceedings of the 26th international cosmic ray conference, vol 6, Salt Lake City, CA, pp 62–65

Martinic NJ, Reguerin A, Palenque E, Taquichiri MA, Wada M, Inoue A, Takahashi K (1985) Search for solar neutrons using NM-64 equipment. In: Proceedings of the 19th international cosmic ray conference, vol 4, La Jolla, CA, pp 138–141

Miyasaka H, Shimoda S, Yamada Y et al. (2001) Recent status of YBJ neutron monitor observation. In: Proceedings of the 27th international cosmic ray conference, vol 8, Hamburg, pp 3050–3052

Muraki Y, Tsuchiya H, Fujiki K, Masuda S, Matsubara Y, Menjyo H, Sako T, Watanabe K, Ohnishi M, Shiomi A, Takita M, Yuda T, Katayose Y, Hotta N, Ozawa S, Sakurai T, Tan YH, Zhang JL (2007) A solar neutron telescope in Tibet and its capability examined by the 1998 November 28th event. Astropart Phys 28(1):119–131

Pakhomov NI, Shapovalova LA, Yanchukovsky AL (1987) Recording of solar flare-produced neutrons at the Sayan spectrographic facility. In: Kocharov G (ed) Processes on the surface and in the interior of the Sun. Nauka, Leningrad, pp 25–41

Rieger E (1982) Gamma-ray measurements during solar flares with the gamma-ray detector on solar-maximum – an overview. In: HINOTORI symposium on solar flares, Tokyo, 1982, Institute of Space and Astronautical Science, pp 246–262

Ramaty R, Murphy RJ, Kozlovsky B, Lingenfelter RE (1983) Gamma-ray lines and neutrons from solar flares, Sol. Phys. (Netherlands), 86, No. 1-2, 395–408

Stoker PH, Dorman LI, Clem JM (2000) Neutron monitor design improvements. Space Sci Rev 91 (1–2):361–380

Tsuchiya H, Miyasaka H, Takahashi E, Shimoda S, Yamada Y, Kondo I, Makishima K, Zhu F, Tan Y, Hu H, Tang Y, Zhang J, Lu H, Meng X (2007) Upper limits on the solar-neutron flux at the Yangbajing neutron monitor from BATSE-detected solar flares. Astron Astrophys 468 (3):1089–1097

Tsuchiya H, Miyasaka H, Takahashi E, Shimoda S, Yamada Y, Kondo I, Makishima K, Zhu F, Tan Y, Hu H, Tang Y, Zhang J, Lu H, Meng X (2008) Upper limits on high-energy solar neutrons from satellite-detected flares with the Yangbajing neutron monitor. In: Proceedings of the 30th international cosmic ray conference, Merida, Mexico, 2007, vol 1, pp 49–52

Wang HT, Ramaty R (1974) Solar Neutron propagation and 2.2 MeV gamma-ray line production in the solar atmosphere, Solar Phys. 36, No. 1, 129–137

Yoshimori M, Okudaira K, Hirasima Y, Kondo I (1983) Gamma-ray observations from HINOTORI, Sol. Phys., 86, No. 1-2, 375–382

Chapter 7
Observations of Solar Neutron Events by Ground Based Detectors, and Their Interpretation

7.1 Investigations of Solar Neutron Events by the Tyan Shan High Altitude Neutron Supermonitor

7.1.1 The Matter and Short History of the Problem

According to Aushev et al. (1999), the Tyan Shan high altitude neutron monitor has a favourable location and good data statistics for solar neutron observations; however, only one enhancement of June 15, 1991 has been attributed to primary solar neutrons during 1976–1998 years. The work of Aushev et al. (1999) presents a search for other possible solar neutron events in the Tyan Shan NM data. The catalogue of major x-ray events registered by the GOES satellites and the available information on the hard x-ray and gamma solar emission obtained aboard the SMM and CGRO satellites were used in this search. For each event from the catalogue, Aushev et al. (1999) calculated an expected response of the Alma-Ata neutron monitor to the flux of solar neutrons like that observed on June 3, 1982. Possible candidates were chosen by comparing expectations with the real NM count rate. Some particular candidates and the statistics of the NM count rate close to onset of energetic solar events are discussed.

As noted Aushev et al. (1999), the observation at the Earth of solar protons and neutrons, generated during powerful solar flares (in combination with x-ray and gamma-ray data) allows us to obtain unique information on the Sun's flare process and particle acceleration mechanisms. The first solar neutrons were observed near the Earth by the Gamma Ray Spectrometer on the Solar Maximum Mission (SMM) on June 21, 1980 (Chupp et al. 1982). The first simultaneous measurements of solar neutrons by space and ground based detectors were made during the prominent event on June 3, 1982 (Chupp et al. 1987; Debrunner et al. 1983). Solar neutron measurements provide more direct information about the source of acceleration, than proton and electron observations do, and may provide the crucial information to test the models of particle acceleration. For instance, solar neutron observations gave the decisive argument for multi-step acceleration mechanisms during the solar

L. Dorman, *Solar Neutrons and Related Phenomena*,
Astrophysics and Space Science Library 365, DOI 10.1007/978-90-481-3737-4_7,

flare event on June 4, 1991 (Struminsky et al. 1994). The identification of the solar neutron enhancement in NM data among other CR variations is a complex problem. Solar neutrons propagate straight-line in the interplanetary space and are closely associated with solar gamma and x-ray radiation, so using some additional information we may try to identify the onset of ground-level neutron event. The detection of solar neutrons is mostly probable near local noon at mid- and low-latitude mountain neutron monitors. There are only a few such CR stations in the world network (about 10), and high mountain (3,340 m) neutron monitor at Tyan Shan (Institute of Ionosphere, MN-ANRK) is among them. It is the only station with such capability to detect solar neutrons in this longitudinal region. The combination of its geomagnetic cutoff rigidity (6.7 GV), altitude (3,340 m) and high statistical accuracy (18NM-64) makes this station enable to record solar neutrons of the energy order of 300 MeV. However, only one enhancement registered on June 15, 1991 has been attributed to primary solar neutrons up to now (Usoskin et al. 1995). Belov et al. (1987) studied variations of high-altitude Tyan Shan and Lomnicky Stit NMs around onsets of x-ray events in order to find their response to solar neutrons. The work of Aushev et al. (1999) presents the search for other possible solar neutron events in the Tyan Shan NM data obtained during the period of 1976–1998 years.

7.1.2 Data and Method of Analysis

Aushev et al. (1999) underlined that NM registers solar neutrons, if their intensity is sufficient at the Earth orbit and the NM looks in the right direction. Aushev et al. (1999) created two catalogues of events when the Tyan Shan NM might observe solar neutrons. The first is based on the GOES satellite data and accounts 776 events over the period of 1974–1999 when all flares >M1 importance and all proton flares followed by optical flare >2 class and/or with >1 h duration. The second uses data of hard x-ray and gamma-ray solar emission obtained by the BATSE detector aboard the CGRO satellite (1991–1999) with magnitude >1,000 imp/s and accounts 341 events might be effective for detecting at Alma-Ata. A complex database was created on the basis of these catalogues and corresponding data from the Tyan Shan neutron monitor. A special computer program was elaborated to work with this database. It allows sorting data by date, magnitude of expected or observed effect, amplitude of solar x-ray or gamma-ray event and so on. The program calculates possible response of the Tyan Shan neutron monitor (the parameter Np) in percents above background to the neutron flux of the prominent neutron event of 3rd June, 1982, i.e. what should be observed if a flux of solar neutrons near the Earth would be like on June 3, 1982. Choosing, as a threshold, the response Np equal to 0.03, more than 750 events in CR intensity mostly favourable for observation in Tyan Shan were selected. Therefore, it was assumed that the flux of solar neutrons can be about 100 times greater than that observed on June 3, 1982.

Analyzing variations of real NM count rate and comparing them with expectations Aushev et al. (1999) may select possible neutron enhancements among other CR effects of magnetosphere, interplanetary, and solar origin. As additional information we used in our analysis data of other NM's, observations of hard x-ray and gamma-ray solar emission aboard the SMM and CGRO satellites.

7.1.3 The List of Possible Neutron Events During 1979–1998

According to Aushev et al. (1999), the considered effects, in general, are not clear and large; however, it looks very interesting and promising in some cases. To Aushev et al. (1999) opinion the most probable candidates for registration of solar neutrons are: 1979.07.24 (Year.Month.Day), 07:59 (UT – onset of the x-ray event); 1980.04.07, 05:18; 1982.08.09, 06:37; 1989.09.04, 04:37, 06:25; 1989.09.14, 06:59; 1989.10.02, 08:41; 1990.09.17, 07:50; 1991.06.01, 04:00; 1991.06.06, 07:05; 1991.06.12, 07:00; 1991.06.15, 06:33; 1991.06.17, 08:09; 1991.07.11, 08:35; 1991.07.14, 08:27; 1991.07.22, 09:36; 1991.08.05, 05:24; 1991.10.27, 05:36; and 1998.05.08, 05:53. Aushev et al. (1999) discussed on some particular events from this list in Sections 7.1.5–7.1.7, and on the mean statistical effect in Section 6.1 (see above, Chapter 6). For comparison in Section 7.1.4 are shown event caused by solar protons but not neutrons.

7.1.4 The Event of 7 May, 1978, Caused by Solar Protons But Not Neutrons

In Fig. 7.1 is shown for comparison an example of the event caused by solar protons but not neutrons: the relatively large enhancement observed on May 7, 1978. The Np parameter for this event was small and GLE 31, occurred at that time, was recorded by the NM network. Although it was not big (about 2%), but in 5-min data both isotropic and anisotropic stages are pronounced very clearly.

7.1.5 The Event of 24 July, 1979 (Onset of the x-Ray Event at 07:59 UT) as an Example of Possible Neutron Enhancement

In Fig. 7.2 is shown example of possible neutron enhancement, which amplitude is much more than statistical variations, is the event on July 7, 1979 with the x-ray event at 07:59 UT.

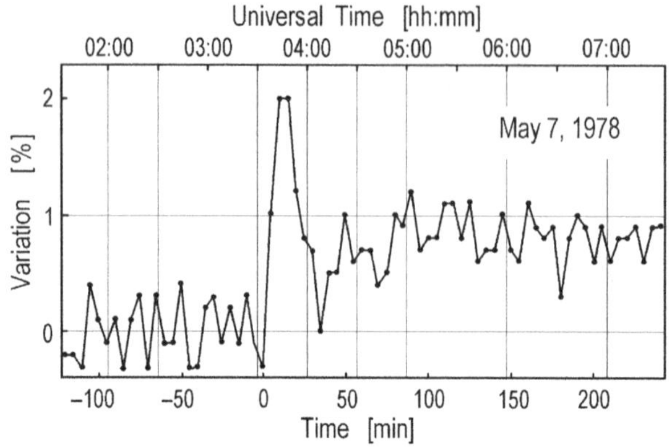

Fig. 7.1 Variations of the Tyan Shan NM 5-min count rate at 7 May, 1978. Zero time on the bottom corresponds to the onset of solar x-ray event. From Aushev et al. (1999)

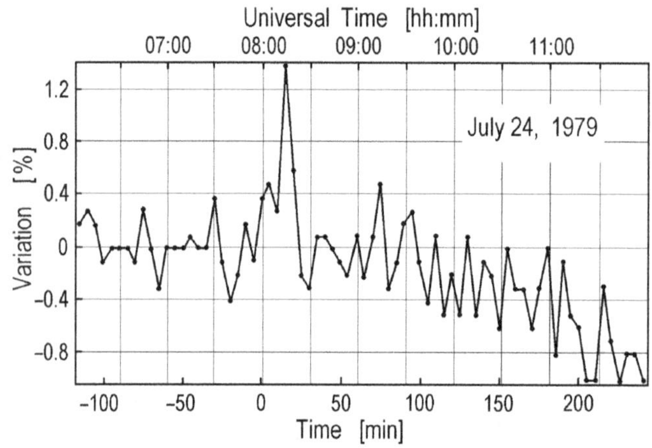

Fig. 7.2 Variations of the Tyan Shan NM 5-min count rate at 24 July, 1979. Zero time on the bottom corresponds to the onset of solar x-ray event. From Aushev et al. (1999)

7.1.6 The Event of 17 June, 1991 (Onset of the x-Ray Event at 08:09 UT) as an Example of Possible Neutron Enhancement and the Shock Wave Effect

This event is shown in Fig. 7.3. Here we see two effects. First increase in the time of solar flare onset (08:27 UT) is possible solar neutron effect and second increase followed by decrease is shock wave effect (SSC was in 10:10 UT) and beginning of Forbush-decrease.

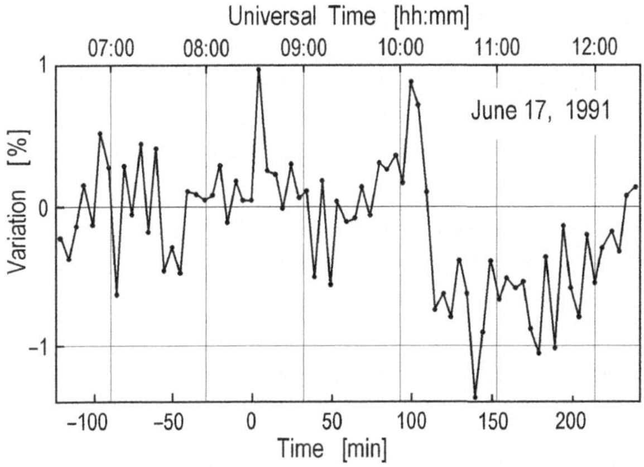

Fig. 7.3 The same as in Fig. 7.1, but for event June 17, 1991. From Aushev et al. (1999)

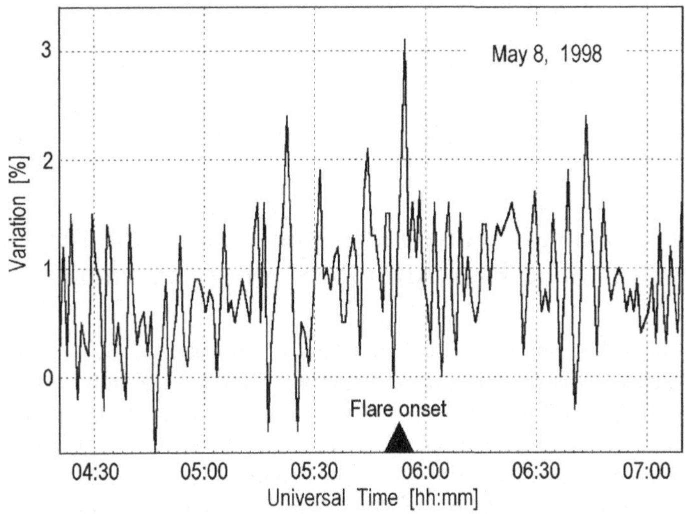

Fig. 7.4 Variations of the Tyan Shan NM 1-min count rate at May 8, 1998. From Aushev et al. (1999)

7.1.7 The Event of 8 May, 1998 (Onset of the x-Ray Event at 05:53 UT)

Aushev et al. (1999) noted that the effect observed on May 8, 1998 is very interesting from many points of view (see Fig. 7.4). Counting rate increase occurred on Tyan Shan NM at the moment coinciding with the x-ray flare onset (05:53 UT). It had not great amplitude (1.5%) and was detected only due to the high statistics

(1,200 imp/s) and good time resolution (1 min data) of the Tyan Shan neutron monitor. This event appears to be originated from the same active region that had produced on May, 2 the small and very anisotropic GLE. On May, 8 this region was on the western limb. At the moment of x-ray flare a significant brightening in UV and radio emission (the Nobeyama observatory) was observed on the western limb. At the same time BATSE recorded a long-duration gamma-event with complex time profile. This set of measurements shows that the great solar energetic event has occurred on the Sun, but only its top has been observed in x-rays.

7.1.8 Main Results and Discussion

Aushev et al. (1999) analyzed data of the Tyan Shan NM obtained during 1976–1998. This NM is one of the best instrument for ground based observations of solar neutron events. Unfortunately no one powerful solar event with large x-ray and gamma-emission was favorable for solar neutron detection by the Tyan Shan NM during the considered period. Aushev et al. (1999) did not find any new large enhancement might be attributed to solar neutrons. However, they found several small enhancements with amplitude greater than statistical variations just in time for neutron favorable observations in Tyan Shan NM, so solar neutrons might cause them. The expected effect for other high altitude NMs is much less in these cases.

7.2 Possible Solar Neutron Event on May 9, 1980 Detected by Tyan Shan High Altitude Neutron Supermonitor

This possible event was caused by solar flare of importance 1B, with coordinates S21, W32, accompanied according to SMM observations by gamma-ray burst M7.2 starting at 7:12:17 UT on May 9, 1980. For this event according to Section 5.1, Eqs. 5.1 and 5.2 $\lambda_S = 17.4°$N and $f_S = 71.9°$E. For Tyan Shan station (43.3°N, 76.9°E, $h = 680$ g/cm^2) for this event according to Eqs. 5.3 and 5.4 $\theta_S = 26.2°$ and $\cos\theta_S = 0.897$ that $h_S = 758$ g/cm^2. The increase was observed only in the first 5 min, with amplitude $0.45 \pm 0.15\%$. The consideration of distribution of the nearest neutron monitors for the event of May 9, 1980 shows that this effect at other stations is expected much smaller than that at Tyan Shan high altitude station (3,440 m) and could not be detected.

7.3 The Event of June 7, 1980

This event was mentioned as possible solar neutron event on the basis of 2.22 MeV gamma-ray line by the Gamma Ray Spectrometer on SMM (Chupp et al. 1981; Prince et al. 1983). From time-varying 2.22 MeV line intensity Chupp et al. (1983,

1984, 1985) obtained the total neutron yield of >1 MeV solar neutrons as 3 10^{29}. But, in this event the solar neutrons have not been recorded by the SMM GRS. Detailed analysis of this event, including data of ground based neutron supermonitor at Tokyo was made in Takahashi et al. (1987, 1990) and Takahashi (1989), and by using data of Tixie Bay neutron supermonitor in Filippov (1987, 1990).

This event was caused by solar flare of importance SN and in gamma-rays as M7.3 started at 3:11:24 UT, June 7, 1980 (location 14°N, 70°W). Figure 7.5 (from Takahashi 1989) shows the 4.1–6.4 MeV gamma-ray flux starting at the first impulse at 3:11:57 UT and reaches a maximum after 30 s (see the upper part of Fig. 7.5). As can be seen from Fig. 7.5 (panel a), the 2.22 MeV gamma-ray flux starting at 30 s after the arrival of the 4.1–6.4 MeV flux, reaches a maximum after

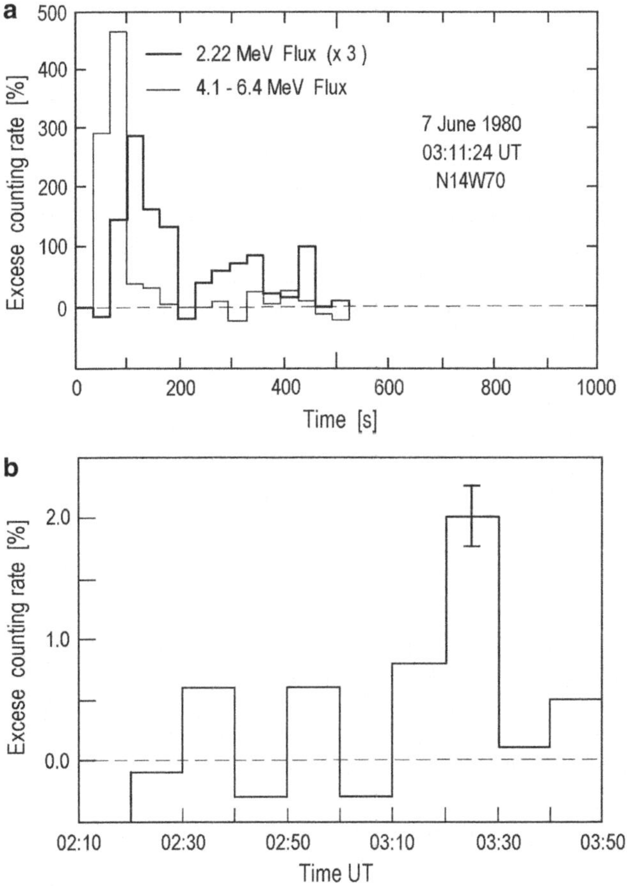

Fig. 7.5 The solar neutron event of June 7, 1980. (**a**) Time profiles for 2.22 MeV and 4.1–6.4 MeV gamma-ray emission (Prince et al. 1983); (**b**) Excess counting rates of the Tokyo 36 NM-64 10-min data. According to Takahashi (1989)

30 s after the maximum of the 4.1–6.4 MeV flux and the emission in this line continues for 6 min. These facts indicate that the solar neutrons are emitted for 7 min from 3:11:57 UT. Figure 7.3. shows also the bottom part of the time profile in the 10-min data of the Tokyo neutron supermonitor (36 NM counters, average counting rate 9.07 10^5 h^{-1}; 35.8°N, 139.8°E, $R_C = 11.5$ GV, altitude 20 m above sea level). It can be seen that two values, at 3.10–3.20 and 3.20–3.30 UT have an enhancement of 3.1$\boldsymbol{\sigma}$ and 7.8$\boldsymbol{\sigma}$, respectively, where for the 10 min interval $\sigma =$ 0.264%. Assuming the impulsive emission of solar neutrons with a power-law spectrum form on index 3 at the gamma-ray maximum, the ratio of the value 3.10–3.20 UT to that of 3.20–3.30 UT requires (according to Takahashi 1989) an upper cutoff energy of 190 MeV in these neutrons (if the solar neutron emissions were continuum this ratio requires the upper cutoff energy of 240 MeV).

According to Filippov (1987, 1990), this event was detected even by Tixie Bay neutron supermonitor (18 NM counters at sea-level). Figure 7.3 shows the 5 min data of Tixie Bay neutron supermonitor in comparison with the 10 min data of Tokyo neutron supermonitor for the event of June 7, 1980.

As we can see from Fig. 7.6 that at 3.10–3.15 UT it was increasing by 1.8 ± 0.6% and at 3.15–3.20 UT the amplitude of increase was 1.0 ± 0.6%. For the interval 3.10–3.20 UT the average was 1.4 ± 0.4% (in Tokyo for this period the amplitude was 0.8 ± 0.25%). For the next 10 min interval (3.20–3.30 UT) in Tixie Bay no increase was observed (0.2 ± 0.4%), although in Tokyo the observed increase was 2.0 ± 0.3%. According to Eqs. 5.1 and 5.2 for this event $\lambda_S =$ 22.9°N, f_S = 130°E. Figure 7.3 shows the distribution of neutron monitors for the solar event of June 7, 1980 relative to the position of the Sun.

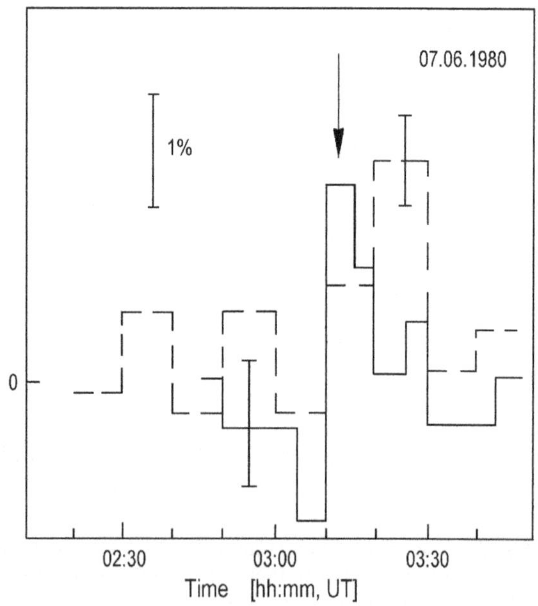

Fig. 7.6 The solar neutron event of June 7, 1980. Excess counting rates of the Tixie Bay 18 NM-64 5- min data according to Filippov 1990 (*solid line*) and of the Tokyo 36 NM-64 10-min data according to Takahashi 1989 (*dashed line*)

Let us compare the results obtained at Tokyo and Tixie Bay neutron supermonitors, with taking into account that at the moment of the event of June 7, 1980 the zenith angle of the Sun for Tokyo was $\theta_S = 16.2°$, $h_S = 1{,}073$ g/cm^2 and for Tixie Bay was $\theta_S = 48.8°$, $h_S = 1{,}564$ g/cm^2. The amplitude of increase at Tixie Bay must be smaller than at Tokyo according to the difference h_s (Trixie Bay)–h_s (Tokyo) $= 491$ g/ cm^2 , i.e. in exp(491/Λ) times, where Λ is the attenuation length of solar neutrons in the Earth's atmosphere. According to Shibata (1994) for solar neutrons with kinetic energy >200 MeV the value $\Lambda \approx 110$ g/cm^2. This value does not strongly depend from E_n up to 1,000 MeV. In our case exp(491/Λ) $\approx$ 86.8 times. We must also take into account the small difference in the background intensity of neutrons at Tokyo and Tixie Bay from the latitudinal effect of galactic cosmic rays (10%). It means the observation result on increasing intensity at Tixie Bay (for the interval 3:10–3:20 UT), was not real although the observed amplitude of increase was more than 3σ ($1.4 \pm 0.4\%$). Thus we come to the conclusion that direct neutrons in the event of June 7, 1980 can not be detected by Tixie Bay station. The possibility exists that the observed effect could be caused by protons from neutron decay in the interplanetary space (Filippov 1990), needs in careful additional consideration. For the next 10-min interval (3:20–3:30 UT) the Tixie Bay results are not in contradiction with Tokyo. It would be very important to control the situation in the event of June 7, 1980 from data of stations shown in Fig. 7.7. Mount Norikura had the best position ($\theta_S = 15.3°$, $\cos\theta_S = 0.964$ and $h_S =$ 761 g/cm^2) but the 5 and 10 min data were not available. Here we expect

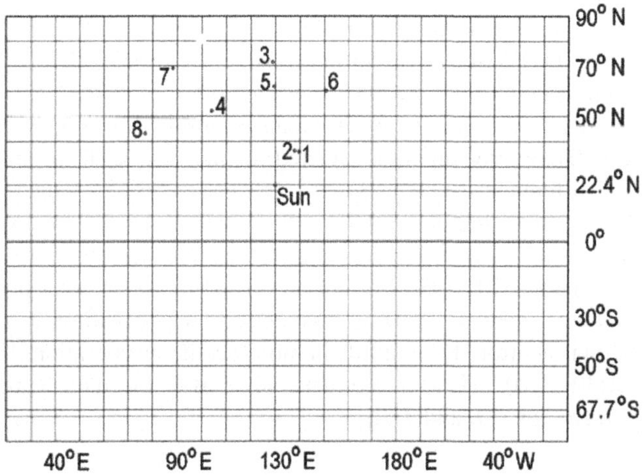

Fig. 7.7 Solar neutron event of June 7, 1980. Distribution of neutron monitors at the moment 03:20 UT of June 7, 1980: (**1**) Tokyo (35.8°N, 139.8°E; $\theta_S = 16.2°$, $h_S = 1{,}073$ g/cm^2), (**2**) Mt. Norikura (36.1°N, 137.9°E; $\theta_S = 15.3°$, $h_S = 761$ g/cm^2), (**3**) Tixie Bay (71.7°N, 128.9°E; $\theta_S =$ 48.8°, $h_S = 1{,}564$ g/cm^2), (**4**) Irkutsk (52.2°N, 104.2°E; $\theta_S = 34.0°$, $h_S = 1{,}206$ g/cm^2), (**5**) Yakutsk (62.0°N, 129.7°E; $\theta_S = 39.1°$, $h_S = 1{,}289$ g/cm^2), (**6**) Magadan (60.1°N, 151.0°E; $\theta_S = 42.0°$, $h_S =$ 1,384 g/cm^2), (**7**) Norilsk (69.3°N, 88.1°E; $\theta_S = 59.1°$, $h_S = 2{,}008$ g/cm^2), (**8**) Tyan Shan (43.2°N, 76.9°E; $\theta_S = 55.7°$, $h_S = 1{,}208$ g/cm^2)

h_s (Mt. Norikura) – h_s (Tokyo) $= -312$ g/cm^2 ; it means that at Mt. Norikura the amplitude of increase must be more than at Tokyo in exp(312/Λ) $\approx$ 17.1 times. Since the background intensity from galactic cosmic rays at Mt. Norikura will be in exp(398.4/145) $\approx$ 7.8 times bigger than that at Tokyo, so the real increase we would expect is 2.2 times. At Irkutsk ($\cos\theta_S = 0.829$), Yakutsk ($\cos\theta_S = 0.776$), Magadan ($\cos\theta_S = 0.744$), Tyan Shan ($\cos\theta_S = 0.563$), and Norilsk ($\cos\theta_S = 0.513$) all reported that there was a cosmic ray increase in the event of June 7, 1980.

7.4 The Event of June 21, 1980

This event was the first that solar neutrons were discovered from observations on SMM (Chupp et al. 1982; see Chapter 3). This event was caused by a solar flare of importance 1B with the co-ordinates 20°N, 90°W accompanied by gamma-ray burst of class X2.6, with energy flux 1,152.8 cm^{-2} MeV^{-1} and spectrum component -2.5 (according to SMM observations). The gamma-ray burst started at 01:18:20 UT on June 21, 1980. For this event $\lambda_s = 23.5$°N and $\lambda_s = 23.5$°N and $f_s = 160.4$°E. The Tokyo neutron supermonitor for this event was characterised by $\theta_s = 23.9°$, cos $\theta_s = 0.915$ and $\theta_s = 23.9°$, cos $\theta_s = 0.915$ and $h_s = 1,129$ g/cm^2. The results of observation of this event by Tokyo neutron supermonitor in comparison with the observations on SMM are shown in Fig. 7.8 (Takahashi 1989).

According to Filippov (1990), this event was also observed at Tixie Bay neutron supermonitor ($\theta_s = 55.4°$, cos $\theta_s = 0.568$ and $h_s = 1,813$ g/cm^2), see Fig. 7.9.

Let us now compare the results obtained from Tokyo and Tixie Bay. In the interval 1:20–1:30 UT Tokyo monitor observed an increase 0.39 $\pm$ 0.26%, but we have to take into account that solar neutrons can arrive only after 1:25 UT, so we expect this increase was caused by the solar neutrons arriving in the interval 1:25–1:30 UT. Tixie Bay monitor detected an increase with an amplitude 1.5 $\pm$ 0.75%. The difference h_S(Tixie Bay) – h_S (Tokyo) = 684 g/cm^2, what gives the expected decrease in amplitude at Tixie Bay compared to Tokyo, exp(684/Λ) $\approx$ 500 times. We came to the conclusion that the intensity increase observed at 1:25–1:30 UT on June 21, 1980 can not be caused by direct solar neutrons. The hypothesis that this increase at Tixie Bay neutron supermonitor was caused by protons from solar neutron decay needs in more detail consideration. In regards to the interval 1:30–1:40 UT at Tokyo and Tixie Bay the data again is not in contradiction (see Figs. 7.8 and 7.9).

7.5 The Event of November 6, 1980

On the cosmic ray intensity increase in this event was reported by Takahashi et al. (1987, 1990) and Takahashi (1989). Figure 7.10 shows (according to Prince et al. 1983) the time profile of 4.1–6.4 MeV gamma-ray flux.

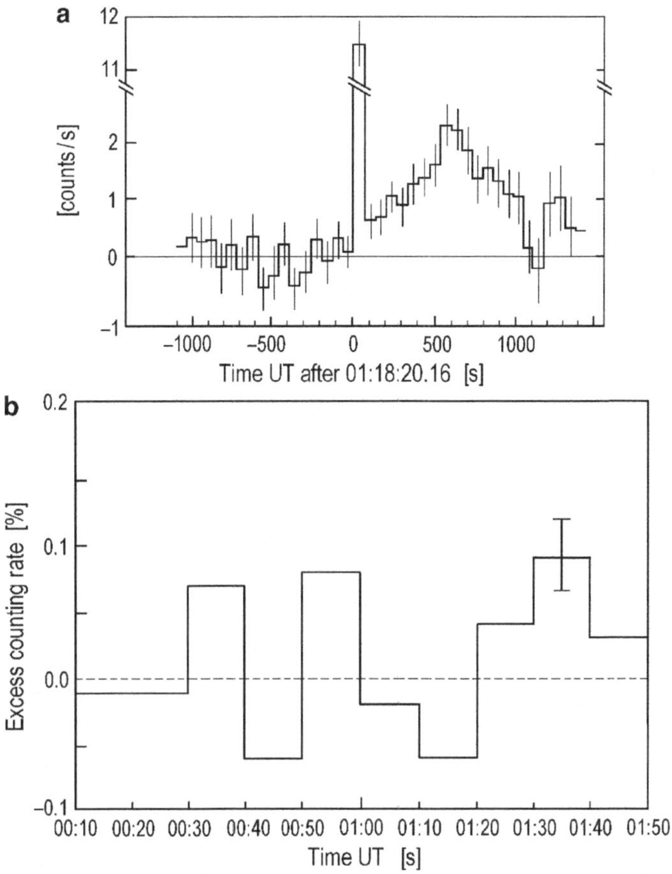

Fig. 7.8 The solar neutron event of June 21, 1980. (**a**) Time profiles for gamma-ray flux with energies greater than 25 MeV and solar neutrons from SMM (according to Chupp et al. 1982), (**b**) excess counting rates of the Tokyo 36 NM-64 10-min data (according to Takahashi 1989)

From Fig. 7.10 can be seen that the first impulse in 4.1−6.4 MeV flux started at 03:44:40 UT and reached maximum after 30 s, the decreased gradually during the following 5 min. Figure 7.10 shows also the time-profile of the 2.22 MeV gamma-ray line, starting 30 s after 4.1−6.4 MeV flux and reaches a maximum after 1 min after the maximum of the 4.1−6.4 MeV flux. The 2.22 MeV flux was followed by high level emissions up to 03:50:20 UT, for about 5 min and 40 s (03:44:40–03:50:20 UT). It means that during this time neutrons were generated in the solar atmosphere. According to Takahashi (1989), these neutrons were detected by the Tokyo neutron supermonitor (Fig. 7.11). For the moment 03:50 UT of November 6, 1980 according to Eqs. 5.1 and 5.2 $\lambda_S = 17.0°$S, $f_S = 122.5°$E. For Tokyo according to Eq. 5.3 $\theta_S = 54.7°$, $\cos\theta_S = 0.577$ and $h_S = 1{,}789$ g/cm^2.

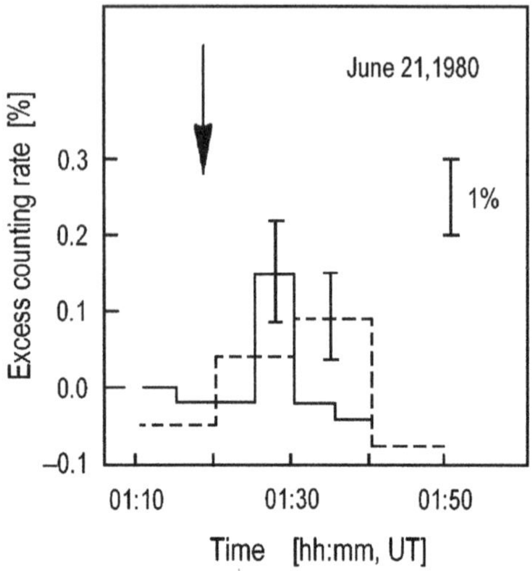

Fig. 7.9 The solar neutron event of June 21, 1980. Excess counting rates of the Tixie Bay 18 NM-64 5- min data according to Filippov 1990 (*solid line*) and of the Tokyo 36 NM-64 10-min data according to Takahashi 1989 (*dashed line*)

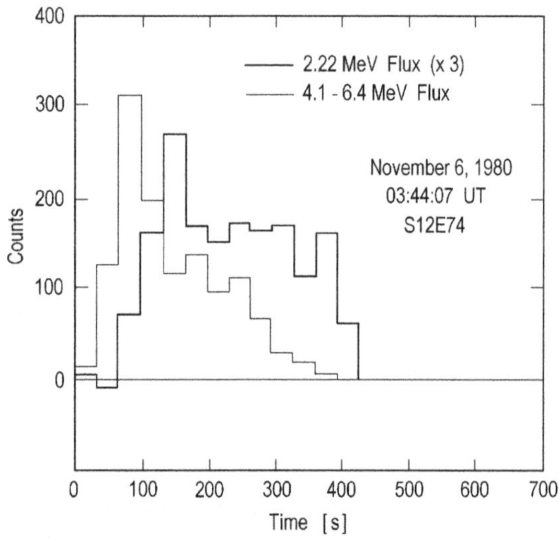

Fig. 7.10 The solar neutron event of November 6, 1980. Time profiles for 2.22 MeV and 4.1–6.4 MeV gamma-ray emission (according to Prince et al. 1983)

It can be seen from Fig. 7.11 that at 03:40–03:50 and 03:50–04:00 UT there were enhancements of 5.4σ and 2.7σ, respectively. According to Takahashi et al. (1990), in this event solar neutrons are emitted with spectrum index 3 during the interval 03:44:40–03:50:20 UT in the energy range 100–850 MeV; in this case the observed ratio of the enhancements in the time intervals 03:40–03:50 and 03:50–04:00 UT could be explained very well.

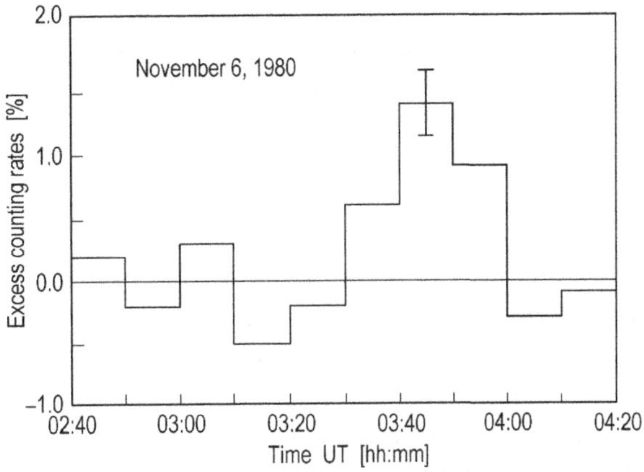

Fig. 7.11 The solar neutron event of November 6, 1980. Excess counting rates of the Tokyo 36 NM-64 10-min data (according to Takahashi 1989)

7.6 Possible Solar Neutron Events on the 1st and 2nd of April 1981, as Detected by Tyan Shan High Altitude Neutron Supermonitor

The possible event on 1st April, 1981 was caused by solar flare of importance 2B, with co-ordinates S09, E68, and accompanied according to SMM observations by gamma-ray burst X2.7, with energy flux 9.07 cm^{-2} MeV^{-1} and spectrum exponent $\gamma = 2.2$. The gamma-ray burst started at 05:33:20 UT on April 1, 1981. For this event $\lambda_{\theta s} = 4.1°N$ and $F_s = 96.6°E$. The Tyan Shan neutron supermonitor for this event was characterized by $\theta_s = 43.1°, \cos\theta_s = 0.731, h_s - 930$ g/cm^2 and shows an increase only in the first 5 min interval with amplitude $0.30 \pm 0.15\%$.

The possible solar neutron event at 2nd April, 1981 was caused by solar flare of importance 1N, with co-ordinates S43, W63, accompanied, according to SMM observations, by gamma-ray burst X2.2, which started at 11:04:53 UT on April 2, 1981. For this event $\lambda_{\theta s} = 4.5°N$ and $f_s = 13.8°E$. The Tyan Shan neutron supermonitor for this event was characterised by $\theta_s = 69.4°\cos\theta_s = 0.353, h_s = 1{,}926$ g/cm^2 and shows an increase only in the first 5-min interval, with an amplitude $0.30 \pm 0.15\%$.

7.7 The Solar Neutron Event on April 4, 1981

7.7.1 Solar Ground Observations

As noted Pakhomov et al. (1987), solar Hα flare on 4 April, 1981 is an impulsive very short (only about 3 min) flare started at 05:02 UT, reached the maximum intensity at 05:03 UT, and finished at 05:05 UT. Solar coordinates of solar Hα flare were S44°, W85°, it had the surface 245 millionths parts of solar disc and was classified as 2B. It was accompanied by intensive radio-emission in continuum in frequency range from 100 MHz up to 36 GHz with energy flux $\left(10^4 - 10^5\right) \times 10^{-22}$ W/(m^2·Hz).

7.7.2 Observations on the Satellite Hinotori of Hard x- and γ-Radiations

Pakhomov et al. (1987) noted that according to observations on Hinotori satellite (Yoshimori et al. 1983), the powerful impulsive emission of hard x-rays were observed in different energetic intervals with the same start time 05:02 UT at 4 April, 1981 (see Fig. 7.12). From Fig. 7.12 can be seen that maximums of x-ray pulses for higher energies x-rays occurred little later and they are shorter than for smaller energies (e.g., for highest energy interval 152–359 keV only about 1 min, from 05:02 up to 05:03 UT).

According to observations on Hinotori satellite (Yoshimori et al. 1982), emission of hard x-rays in impulsive solar flare of 4 April, 1981, caused by bremsstrahlung of energetic electrons, was accompanied by about simultaneously emission of γ-ray lines 0.845 and 4.44 MeV from nuclear reactions of excitation of Fe and C, correspondingly (see Fig. 7.13).

On the basis of Hinotori satellite data on simultaneously observations of hard x-rays and γ-ray lines, Yoshimori et al. (1982) came to conclusion that in this impulsive solar flare electrons, protons, and ions are accelerated practically simultaneously and main part of them moved down to chromosphere with production of hard x-rays and γ-ray lines.

7.7.3 Satellite Observations of Energetic Protons and Estimation of Relative Part of Particles Escaping into Interplanetary Space

As noted Pakhomov et al. (1987), sufficient increasing of solar proton flux from flare 4 April, 1981 in energy interval 16–500 MeV was detected on meteorological satellite Himawari. From other hand, Yoshimori et al. (1982) estimated total number of protons with energy more than 10 MeV generated in this flare from observed intensity of γ-ray lines in suggestion that protons are distributed in the source isotropic. Comparison of obtained value with observed number of protons on satellite Himawari led Yoshimori et al. (1982) to conclusion that only about 10% of solar energetic protons are escaping into interplanetary space. From this follows that solar neutrons also must be generated in the flare 4 April, 1981.

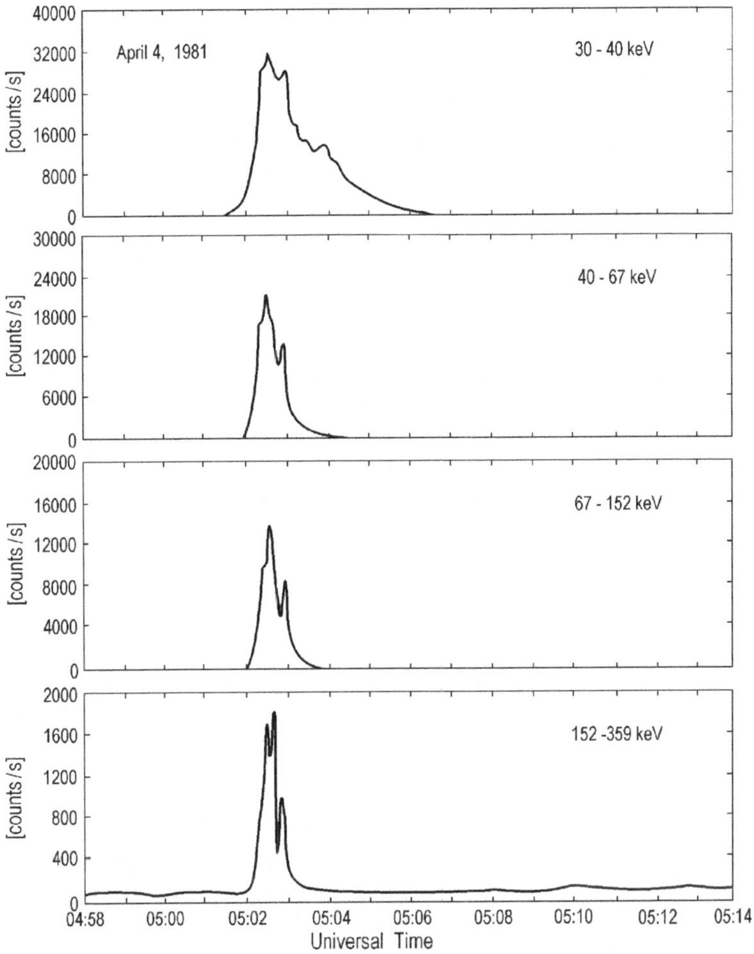

Fig. 7.12 Time profiles of hard x-rays during the solar flare 4 April, 1981 according to Hinotori satellite observations. From Pakhomov et al. (1987)

7.7.4 6NM-64 Observations at Station Hulugaisha

Figure 7.14 shows 5-min data of registration neutron component by 6NM-64 neutron supermonitor on high-altitude mountain station Hulugaisha. During the event at 05:02 UT of 4 April, 1981 on the station Hulugaisha zenith angle of the Sun was $\theta_S = 49°$.

According to estimation of Pakhomov et al. (1987), the intensity of 6NM-64 on Hulugaisha (altitude 3 km, the air depth in direction to the Sun was 1,099 g/cm^2) increased on 1.3 ± 0.5% in time interval 04:57–05:02 UT (solar neutrons with energy more than 1 GeV) and on 0.9 ± 0.5% in time interval 05:02–05:07 UT (solar neutrons with energy 250 MeV–1 GeV). At the same time intervals 12NM-64

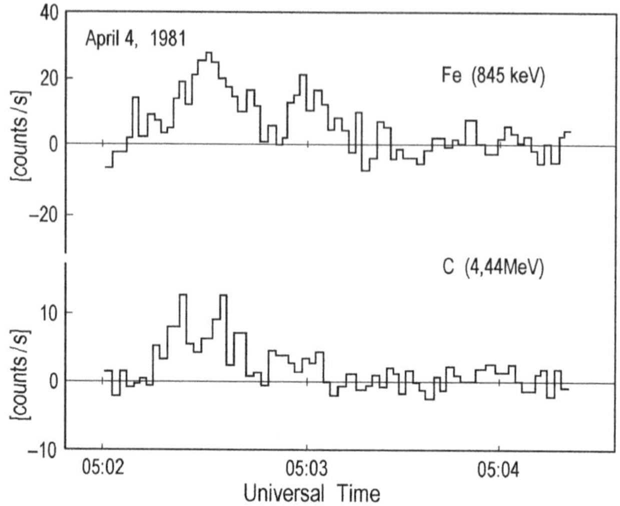

Fig. 7.13 Time profiles of γ-ray lines 0.845 MeV and 4.44 MeV from the impulsive solar flare of 4 April, 1981 according to observations on Hinotori satellite. From Pakhomov et al. (1987)

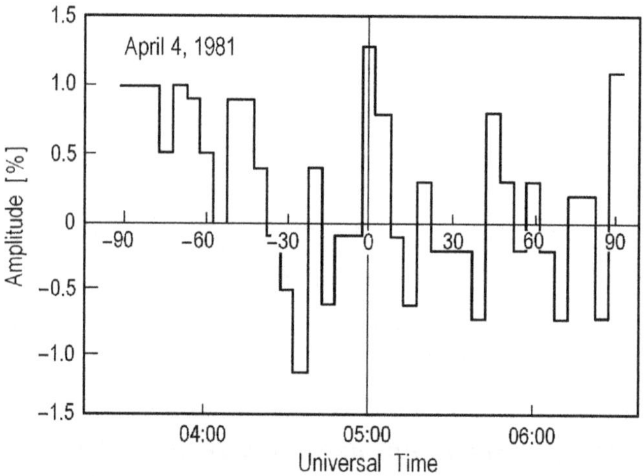

Fig. 7.14 Registered effect of neutron intensity increasing on station Hulugaisha during solar flare 4 April, 1981 according to 5-min data. From Pakhomov et al. (1987)

at station Chasovie Sopki (altitude 2 km, the air depth in direction to the Sun was 1,265 g/cm^2) did not show any CR intensity increase. Pakhomov et al. (1987) came to conclusion that solar neutrons are effectively absorbed in the air layer with depth 166 g/cm^2 (between 1,099 and 1,265 g/cm^2).

7.8 The Possible Event on April 27, 1981

This event was caused by solar flare of importance 1N, with co-ordinates N16, W90, accompanied, according to SMM observations, by gamma-ray burst X5.5, with energy flux 281.2 cm^{-2} MeV^{-1} and spectrum exponent −2.7. The gamma-ray burst started at 08:03:28 UT on April 27, 1981. For this event $\lambda_s = 13.7°N$ and $f_s = 59.1°E$. The Tyan Shan neutron supermonitor for this event was characterised by $\theta_s = 69.4°$, 34.0°, cos $\theta_s =$ 0.829 and $h_s =$ 820 g/cm^2 and shows an increase only in the first 5-min interval with amplitude 3σ, where $\sigma = 0.15\%$.

7.9 Possible Solar Neutron Event on July 19, 1981

This event was caused by solar flare of importance 2B, with co-ordinates S28, W56, accompanied, according to SMM observations, by gamma-ray burst X2.7, with energy flux 35.0 cm^{-2} MeV^{-1} and spectrum exponent −3.3. The gamma ray burst started at 05:58:44 UT on July 19, 1981. For this event $\lambda_s = 20.8°N$ and $f_s = 90.45°$. The Tyan Shan neutron supermonitor for this event was characterised by $\theta_s = 26.0°$, $\cos\theta_s = 0.899$, $h_s = 756$ g/cm^2 and shows an increase only in the first 5-min interval, with amplitude 4σ, where standard deviation $\sigma = 0.15\%$.

7.10 Possible Solar Neutron Event on August 10, 1981

This event was caused by solar flare of importance 1B, with co-ordinates S13, W15, accompanied, according to SMM observations, by gamma-ray burst M4.8, starting at 06:58:58 UT on August 10, 1981. For this event $\lambda_s = 15.3°N$ and $f_s = 75.3°E$. The Tyan Shan neutron supermonitor for this event was characterises by $\theta_s = 27.9°$, cos θ_s = 0.883, $h_S =$ 770 g/cm^2 and shows an increase only in the first 5-min interval, with amplitude $0.9 \pm 0.15\%$.

7.11 Full Differential Flux of Solar Neutrons at the Top of the Earth's Atmosphere During the Event of 3 June 1982

This was the first solar neutron event detected simultaneously on the space probe SMM and by many ground based neutron supermonitors (see Chapter 3). This event was caused by solar flare of importance 2B, with co-ordinates S9, E72, accompanied by gamma-ray burst X8.0, with energy flux 2,748.8 cm^{-2} MeV^{-1} and spectrum exponent −2.0. The gamma-ray burst started at 11:42:44 UT on June 3, 1982. For this event λ_s = 22.4°N and f_S = 4.3°E. The Tyan Shan neutron supermonitor for this event was characterised by $\theta_s = 73.8°$, $\cos\theta_s = 0.280$, $h_s = 2{,}429$ g/cm^2

and although on this big value of h_s it shows an increase in the first 5 min interval (after about 3 min of gamma-ray burst start), with amplitude 3 $\boldsymbol{\sigma}$, where $\boldsymbol{\sigma} = 0.15\%$. In Chapter 3 it was described the results of this event with many European neutron monitors with much smaller h_s.

On the basis of all these data Shibata et al. (1993) determined that in the event of June 3, 1982 the full differential flux of solar neutrons at the top of the Earth's atmosphere was

$$F(E_n) = \left(4.6^{+0.7}_{-0.6}\right) \times 10^5 (E_n/100\ \text{MeV})^{-3.7\pm0.2}\ \text{neutrons} \cdot \text{MeV}^{-1} \cdot \text{m}^{-2}, \quad (7.1)$$

where E_n is the kinetic energy of solar neutrons.

7.12 Possible Solar Neutron Events on June 5 and 12, 1982

The possible event of 5 June, 1982 was caused by solar flare of importance SB, with co-ordinates S10, E40, accompanied, according to SMM observations, by gamma-ray burst X1.1, with energy flux 2.75 $\text{cm}^{-2}\ \text{MeV}^{-1}$ and spectrum exponent −2.5. The gamma-ray burst started at 06:15:30 on June 5, 1982. For this event $\lambda_s = 22.6°$N and $fs = 86.1°$E. The Tyan Shan neutron supermonitor for this event was characterised by $\theta_s = 22.5°$, $\cos\theta_s = 0.924$, $h_s = 736$ g/cm^2 and shows an increase in the second 5 min interval, with amplitude 3σ, where $\sigma = 0.15\%$. In this event it was also the second increase with amplitude 1.05% after about 1 h caused by high energy protons.

According to Pakhomov et al. (1987), solar flare of 12 June, 1982 was an impulsive Hα-flare of power 1B and solar coordinates N10°, E85°. It started at 05:12 UT, with maximum at 05:14 UT and the end at 06:00 UT. This flare has accompanied by hard x-ray emission of class X3.6 (the flux has maximal amplitude 3.6 10^{-4} W/m^2 for 1−8 Å and 10^{-4} W/m^2 for 0.5−4 Å). This flare was accompanied also by radio emission of III-type in m-diapason (the energy flux in cm-diapason was 2 10^{-20} W/m^{2}Hz). Zenith angle of the Sun was during the flare $\theta_s = 22.5°$, $\cos\theta_s = 0.924$, and air depth for solar neutrons in direction to the Sun were 850, 979, and 1,161 g/cm^2 for Hulugaisha, Chasovie Sopki, and Zui, correspondingly. Experimental data of 6NM-64 monitor on station Hulugaisha (3 km altitude) and 12-NM-64 monitor on station Chasovie Sopki (2 km altitude) show increasing in counting rates on 0.8 ± 0.5% and 0.65 ± 0.5%, correspondingly. At the same time 18-NM-64 monitor on station Zui did not show any increasing in counting rate.

7.13 Possible Solar Neutron Events on July 6 and August 14, 1982

The event of 6 July, 1982 was caused by solar flare of importance 3B with co-ordinates N17, E73, accompanied, according to SMM observations, by gamma-ray burst X9.8, with energy flux 250.37 $\text{cm}^{-2}\ \text{MeV}^{-1}$ and spectrum exponent −2.8.

The gamma-ray burst started at 07:35:26 UT on July 6, 1982. For this event $\lambda_s = 22.4°N$ and $f_s = 66.1°E$. The Tyan Shan neutron supermonitor for this event was characterised by $\theta_s = 23.3°$, $\cos\theta_s = 0.918$, $h_s = 741$ g/cm^2 and shows an increase in the first 5 min interval, with amplitude about 4σ, and an increase in the second 5 min interval with amplitude about 3σ, where $\sigma = 0.15\%$, is a standard deviation for 5 min interval registration.

According to Pakhomov et al. (1987), solar flare of 14 August, 1982 was an impulsive Hα-flare of power 1B and solar coordinates N08°, W55°. It started at 03:57 UT, with maximum at 04:02 UT and the end at 06:17 UT. The energy flux in cm-diapason was 10^{-22} W/m^{2}Hz. Zenith angle of the Sun was during the flare $\theta_s = 22.5^0, cos\,\theta_s = 0.924$, and air depth for solar neutrons in direction to the Sun were 913 and 1,053 g/cm^2 for Hulugaisha and Chasovie Sopki, correspondingly. Experimental data of 6NM-64 monitor on station Hulugaisha (3 km altitude) shows increasing in counting rates on $0.7 \pm 0.5\%$. At the same time 12-NM-64 monitor on station Chasovie Sopki (2 km altitude) did not show any increasing in counting rate.

7.14 The Solar Neutron Event on November 26, 1982

This event was caused by a solar flare of importance 2B, with co-ordinates S10, W87, accompanied, according to SMM observations, by gamma-ray burst X4.5, with energy flux 122.83 cm^{-2} MeV^{-1} and spectrum exponent −2.9. The solar flare in Hα″ started at 02:29:04 UT; gamma-ray emission 4–8 MeV started at 02:32:40 UT, formed a gentle peak during the interval 02:33–02:36 UT and terminated at 02:37:30 UT (see Fig. 7.15).

For this event $\lambda_s = 21.4°S$ and $\lambda_s = 21.4°S\, and\, f_s = 141.2°E$. The Tokyo neutron supermonitor for this event was characterised by $\theta_s = 54.2°$, $\cos\theta_s = 0.542$, $h_s = 1,585$ g/cm^2, $h_s = 1,585$ g/cm^2 and according to Takahashi (1989) and Takahashi et al. (1990) shows an increase in the 10 min interval 02:30–02:40 UT with an amplitude of 3.1σ, which corresponds to $1.4 \pm 0.45\%$ (see Fig. 7.16).

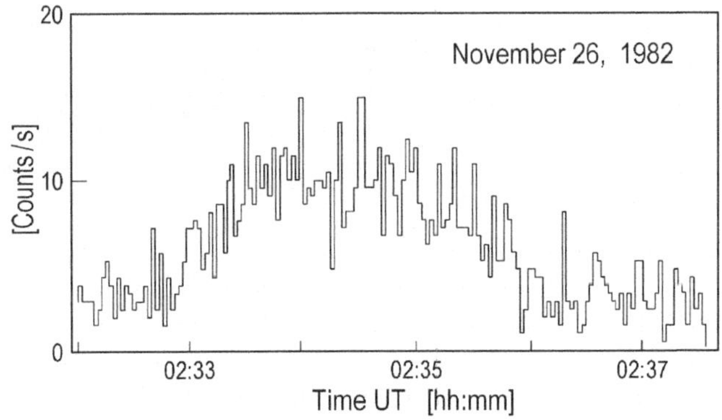

Fig. 7.15 The solar neutron event of November 26, 1982. Time profile for 4–8 MeV gamma-ray emission (from Takahashi 1989)

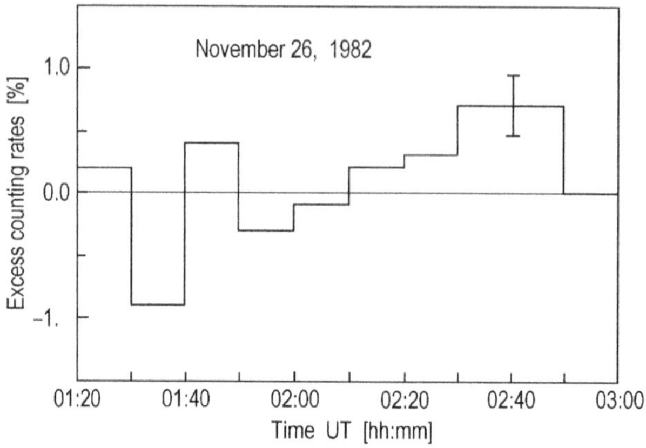

Fig. 7.16 The solar neutron event of November 26, 1982. Excess counting rates of the Tokyo 36 NM-64 10-min data (according to Takahashi 1989)

Fig. 7.17 The solar neutron event of November 26, 1982. Excess counting rates of the Tixie Bay 18 NM-64 5-min data according to Filippov 1990 (*solid line*) and of the Tokyo 36 NM-64 10-min data according to Takahashi 1989 (*dashed line*)

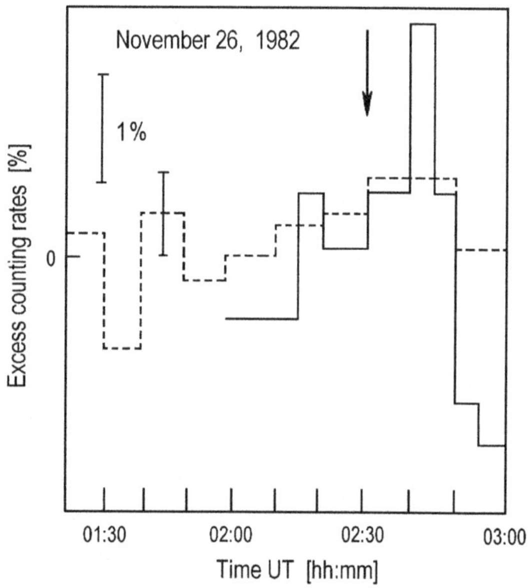

If the increase in Fig. 7.16 was caused by solar neutrons then it must start, not earlier than at 02:35 UT, when a peak in the 4–8 MeV gamma-ray emission was observed on SMM. If it is true, the real increase was in the interval 02:35–02:40 UT with amplitude 2.8 ± 0.45%. In the next 10 min interval at 02:40–02:50 UT there was an increase observed also with an amplitude 1.4 ± 0.45%.

Filippov (1990) reported on solar neutron detection in this event by Tixie Bay neutron supermonitor (see Fig. 7.17).

For this event Tixie Bay station was characterised by $\theta_s = 93.0°$, as solar neutrons can not come to Tixie Bay directly. The observed small increase at Tixie Bay neutron supermonitor can be caused by protons from neutron decay. On Tyan Shan neutron supermonitor ($\theta_s = 79.3°$, $\cos\theta$subs $= 0.186$, $h_s = 3{,}656$ g/cm^2) there was no any increase in the first and second 5 min intervals, but after about 20 min an increase was observed with an amplitude 0.75 ± 0.15%, with duration of 30 min. This increase probably was caused by high energy protons accelerated in the same solar flare.

7.15 The Solar Neutron Event on 25 April, 1984, and the Possible Neutron Event on May 20, 1984

7.15.1 Solar Ground and SMM Observations of Flare 25 April 1984

This event was caused by solar flare of importance 3B, with co-ordinates S11, E45, accompanied (according to SMM observations by GRS) by gamma-ray flux with energy up to 70 MeV (Boulder x-ray flux classification X13). The high-energy gamma-ray emission started at 00:00:00UT on April 25, 1984 and reached the first impulsive peak within 30 s, the second gradual peak around 00:03 UT according to Yoshimori et al. (1987) could have been caused by solar neutrons (see Fig. 7.18). For this event $\lambda_s = 13.1°N$ and $f_s = 180°$E (see Fig. 7.19).

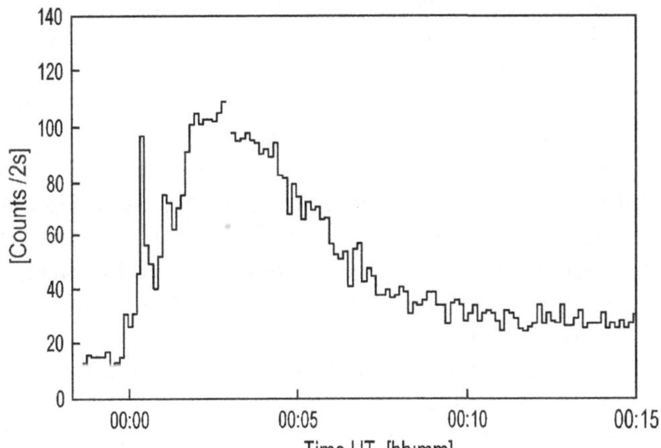

Fig. 7.18 The solar neutron event of April 25, 1984. Time profile for gamma-ray flux with energies greater than 25 MeV, detected by GRS (from Takahashi 1989)

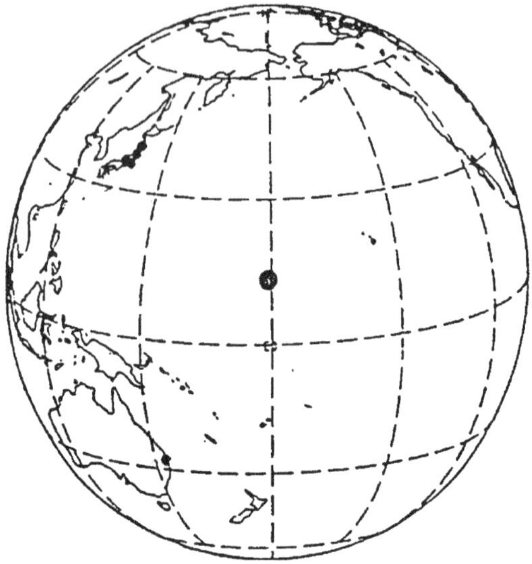

Fig. 7.19 The solar neutron event of April 25, 1984. Map of possible neutron monitor detection sites. The map is oriented on the subsolar point (*large solid circle*). According to Smart et al. (1990)

7.15.2 Solar Neutron Observations by NM During the Event 25 April, 1984

The Tokyo neutron supermonitor for this event was characterised by $\theta_s = 45.2°$, $\cos\theta_s = 0.705$, $h_s = 1,466\ \text{g/cm}^2$ and (according to Takahashi 1989; Takahashi et al. 1990) shows an increase of about 3.1σ, (where $\sigma = 0.26\%$ as standard deviation) with respects to the background, in the interval 00:00–00:10 UT, coinciding with the second gradual peak of gamma-ray enhancement (see Fig. 7.20). Takahashi (1989) came to conclusion that the observed intensity increase $0.8 \pm 0.26\%$, in the Tokyo neutron supermonitor in the time interval 00:00–00:10 UT was caused by solar neutrons with energies of 120–650 MeV.

According to Filippov (1990), the April 25, 1984 event was detected also by Tixie Bay station neutron supermonitor (see Fig. 7.21).

For this event Tixie Bay station was characterised by $\theta_s = 70.9°, \cos\theta_S = 0.327, h_s = 3,149\ \text{g/cm}^2$. The value of h_s for Tixie Bay is so big that it is very difficult to consider the intensity increase with amplitude $3.3 \pm 0.5\%$, in the interval 00:00–00:05 UT on the Tixie Bay supermonitor as caused by direct flux of solar neutrons. May be, this increase was caused by protons from solar neutron decay.

Shea et al. (1987), Smart et al. (1990) investigated this event on the basis of Japanese (Tokyo-Itabashi, Fukushima, Morioka, Mt. Norikura) and Australian (Brisbane) neutron monitors data with $\theta_s < 50°$ (see Fig. 7.22).

Figure 7.15 shows the observed averaged effect on neutron monitors Fukushima, Tokyo-Itabashi and Morioka by 10-min data (unfortunately the data from Mt. Norkura and Brisbane was only 1 h data, but they do not contradict the results of Fig. 7.22).

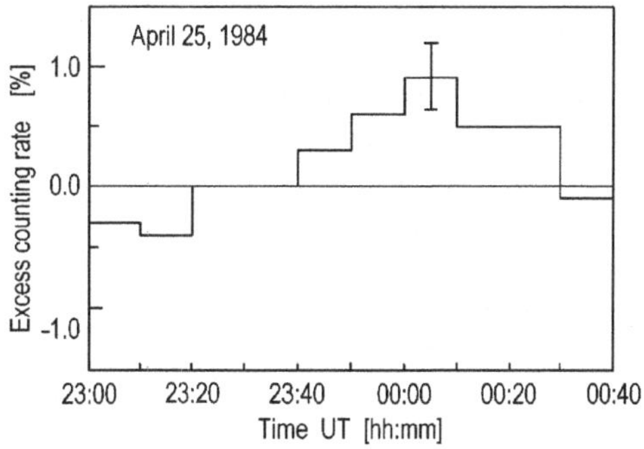

Fig. 7.20 The solar neutron event of April 25, 1984. Excess counting rates of the Tokyo 36 NM-64 10-min data (according to Takahashi 1989)

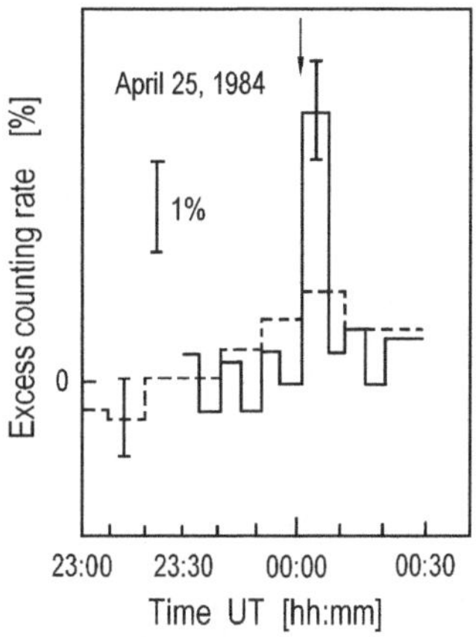

Fig. 7.21 The solar neutron event of April 25, 1984. Excess counting rates of the Tixie Bay 18 NM-64 5-min data according to Filippov 1990 (*solid line*) and of the Tokyo 36 NM-64 10-min data according to Takahashi 1989 (*dashed line*)

7.15.3 Comparison of the Event of 25 April, 1984 with the Event of June 3, 1982

Figure 7.23 shows the comparison with the event of June 3, 1982. From this figure it can be seen that the event of April 25, 1984 was much more powerful than the event of June 3, 1982. Unfortunately, the event of April 25, 1984 occurred over the Pacific Ocean and hence the characteristics λ_s, f_s were not recorded by neutron monitors located at $\theta_s < 40°$ and $h_s < 1{,}000$ g/cm^2.

7.15.4 Possible Solar Neutron Event of 20 May, 1984

According to Pakhomov et al. (1987), at 20 May, 1984 was observed solar impulsive Hα flare of class 1B, beginning time 01:24 UT, maximum at 01:27 UT and the

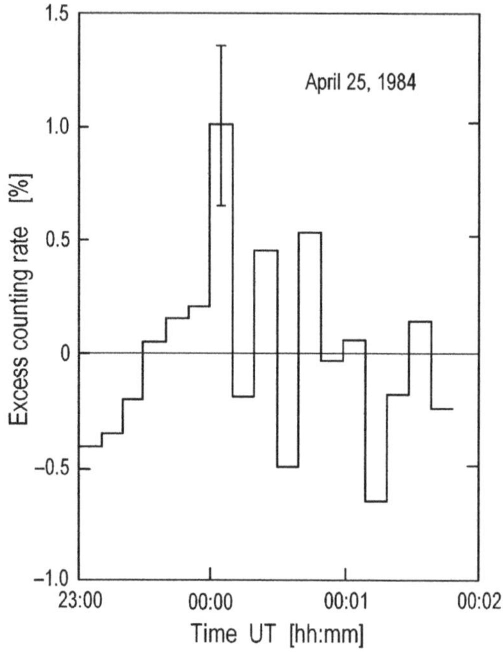

Fig. 7.22 The solar neutron event of April 25, 1984. Normalized composite increase showing the probable solar neutron 'signal' in the 10-min data from the Tokio-Itabashi, Morioka, and Fukushima neutron monitors (according to Smart et al. 1990)

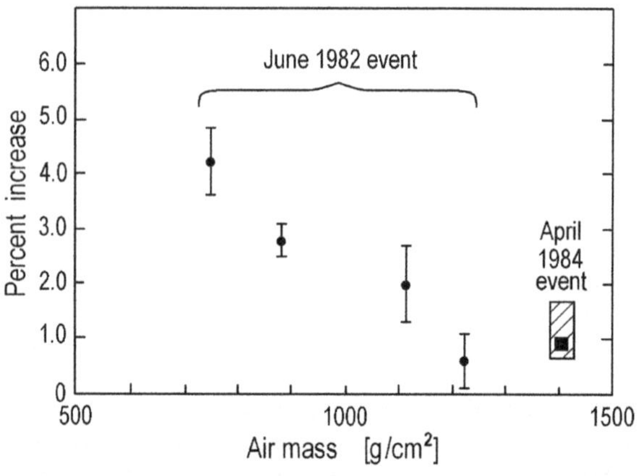

Fig. 7.23 The solar neutron event of April 25, 1984. Comparison of solar neutron events at June 21, 1982 and April 25, 1984 (according to Smart et al. 1990)

end at about 01:34 UT. Solar coordinates of flare are S10°, E65°. The solar flare in Hα was accompanied by hard x-ray emission of class M2.9. Zenith angle of the Sun $\theta_S = 44°$ and air depth in direction to the Sun was 1,004 and 1,153 g/cm^2 for stations Hulugaisha and Chasovie Sopki, correspondingly. The observed increasing in counting rate of 6NM-64 monitor on station Hulugaisha was 1.0 $\pm$ 0.5% and of 12NM-64 monitor on station Chasovie Sopki – in the frame of statistical errors (only 0.4 $\pm$ 0.5%)

7.16 On the Possible Solar Neutron Event on December 16, 1988

This possible event was caused by solar flare of importance 1B/X4.7, starting at 08:38 UT on December 16, 1988, with evidence of emission of 511 keV and 2.2 MeV gamma-ray lines (see Fig. 7.24). This event was characterised by $\lambda_s = 23.4°S$, $f_S = 50.5°E$ (for the moment 08:38 UT).

Stoker and Krüger (1990) tried to find some effects of this event (see Fig. 7.24) in South Africa neutron supermonitors 6NMD (without lead) and 18 NM64 at Tsumeb $\theta_s = 24.8°$ and by neutron supermonitor at Potchefstrum ($\theta_s = 15.6°$). From Fig. 7.24 it can be seen that there is no evidence of any effects in the interval 08:40–09:00 UT in all three detectors (may be only in Tsumeb the neutron supermonitor shows a slight increase at 08:40–08:45 UT with an amplitude 1–2σ). It is necessary to control this event by other observational data.

7.17 The Largest Solar Neutron Event on May 24, 1990

7.17.1 Properties of Flare and x-Ray and γ-Ray Data

According to Pyle et al. (1991), Shea et al. (1991), Kocharov et al. (1993), Filippov and Prikhod'ko (1993), Shibata et al. (1993), this event was the largest solar neutron

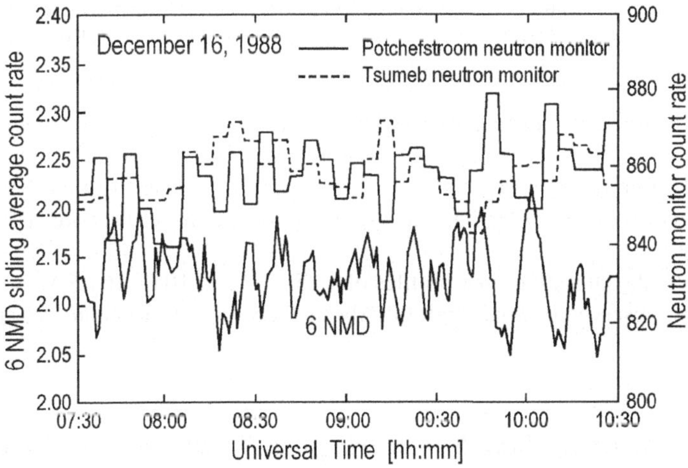

Fig. 7.24 Solar neutron event of December 16, 1988. The 5-min counting rates of the Potchefstroom and Tsumeb neutron supermonitors 18 NM-64 and the 5-min sliding average of the 1-min counting rate of the 6NMD neutron monitor at Tsumeb from 07:00 UT to 10;00 UT on December 16, 1988 (according to Stoker and Krüger 1990)

Table 7.1 Characteristics of the flare May 24, 1990

Radiation	Onset (UT)	Max (UT)	End (UT)	Importance
Hα	20:46	20:49	21:45	1B
1–8 Å x-rays	20:45.9	20:49.5	21:45	X9.3
2,695 MHz	20:46.2	20:48.6	>20:57	13,290 SFU[a]
2,800 MHz	20:45.7	20:48.6	>21:45	18,150 SFU
8,800 MHz	20:45.3	20:45.3	>21:07	45,250 SFU
15,400 MHz	20:45.5	20:48.2	>21:11	41,990 SFU
Type IV	20:48		>21:20	
Type II	21:00		21:07	

[a]SFU = 10^{-22} W/m^{2}Hz

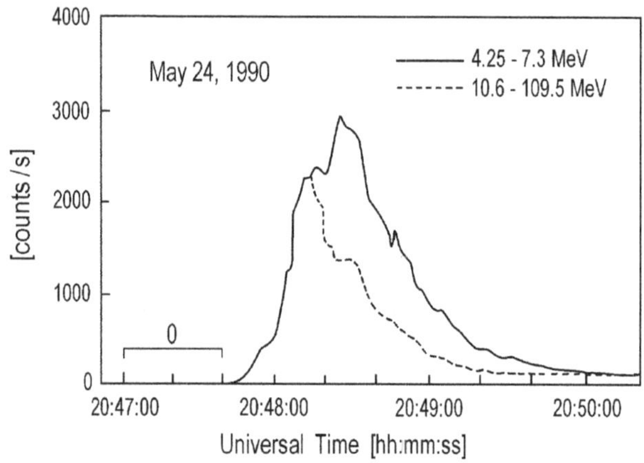

Fig. 7.25 The solar neutron event of May 24, 1990. The 4.25–7.3 MeV and 10.6–109.5 MeV gamma-ray flux observed by the GRANAT spacecraft (from Kocharov et al. 1993)

event compared to the events observed before. This event was caused by the flare of optical class 1B with heliographic co-ordinates N36, W76, accompanied by extremely impulsive x-ray emissions (Boulder x-ray classification X9.3). The characteristics of this solar flare according to Shea et al. (1991) are listed in the Table 7.1.

Figure 7.25 shows the time profiles of 4.25−7.3 and 10.6−109.5 MeV gamma-ray fluxes, observed on spacecraft GRANAT (according to Kocharov et al. 1993).

7.17.2 Position of the Sun During the Event, and NM Data

The sub-solar point at 20:50 UT on May 24, 1990 was located at the geographic coordinates $\lambda_s = 20.84°N$, $f_s = 133.47°E$. Table 7.2 lists the North American stations which had smallest θ_s and detected largest amplitude increase.

Table 7.2 North American stations observing the 24 May, 1990 solar neutron event (from Shea et al. 1991)

Station	Type of NM	Cut-off (GV)	Hei-ght (m)	θ_s (°)	h_S (g/cm^2)	Obser 5-min (%)	Norm 5-min (%)
Climax	IGY	2.99	3,400	28.8	782	25.3	25.3
Mexico city	9NM64	8.60	2,274	32.8	938	24.9	19.3
Calgary	18NM64	1.08	1,128	33.8	1,095	14.1	14.6
Mt. Washington	IGY	1.46	1,909	55.9	1,466	3.9	4.0
Inuvik	18NM64	0.17	21	47.5	1,526	4.5	4.7
Deep river	48NM64	1.14	145	51.8	1,640	3.5	3.6
Newark	9NM64	2.09	50	52.3	1,690	2.8	2.9
Durham	8NM64	1.58	3	56.2	1,857	2.0	2.1
Goose Bay	18NM64	0.64	46	63.4	2,300	0.1	0.1

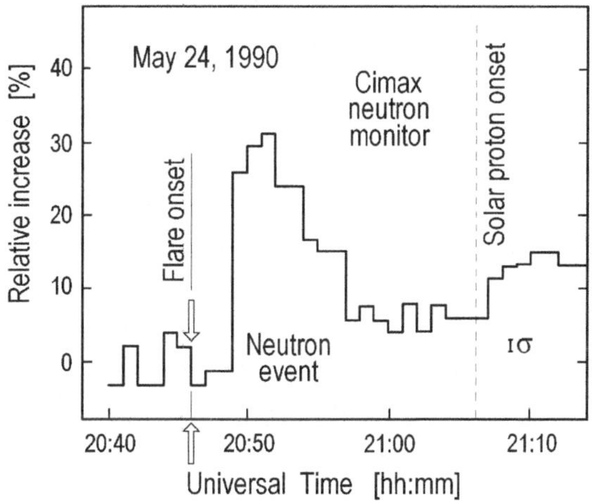

Fig. 7.26 Solar neutron event of May 24, 1990. One-min averages of the cosmic ray time-intensity profile for the Climax neutron monitor; the intensity is a percentage above the hourly average counting rate recorded between 19:00 UT and 20:00 UT on 24 May, 1990 (according to Shea et al. 1991)

In the Table 7.2 the observed amplitudes of increase as well as the amplitudes of normalised increase (to the Climax background intensity) are shown. This normalisation was made only partly with taking into account the background neutron intensity dependence from cut-off rigidity, but not from altitude. The largest amplitude of increase (31.4% in 1-min data at 20:51–20:52 UT) was observed by Climax neutron monitor (see Fig. 7.26 for 1-min data).

Figure 7.27 shows the time profiles of the 5-min data for the North American stations Climax (h_S = 782 g/cm^2), Mexico City (938 g/cm^2), Calgary (1,095 g/cm^2), Mt. Washington (1,466 g/cm^2), Inuvik (1,526 g/cm^2), Deep River (1,640 g/cm^2), Newark (1,690 g/cm^2), Durham (1,875 g/cm^2) and Goose Bay (2,300 g/cm^2).

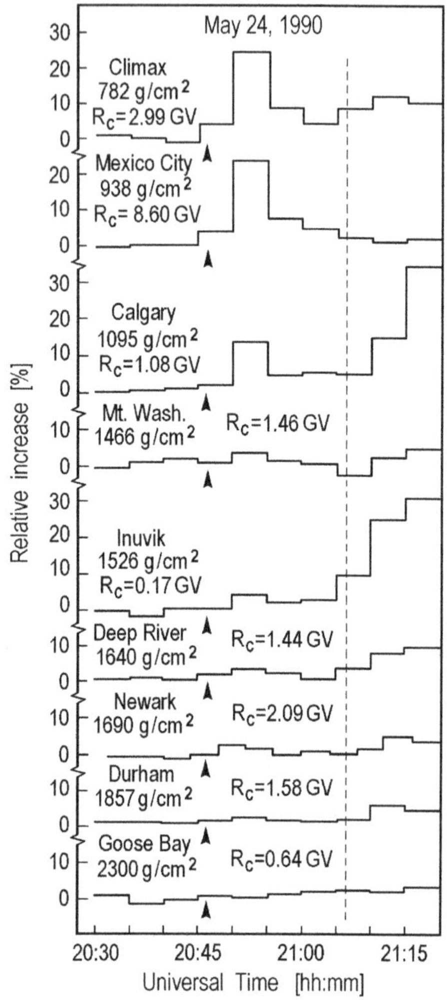

Fig. 7.27 Solar neutron event of May 24, 1990. Cosmic ray intensity recorded by North American neutron monitors during the onset of the solar particle event on 24 May, 1990. The data are as a percentage above the hourly average background recorded in the period 19:00–20:00 UT of 24 May. The graphs are ordered by increasing air mass along the line of sight to the Sun. The dark arrowhead indicates the onset of the 1-8Å x-ray emission; the *dashed line* indicates the onset of the anisotropic high energy proton event. According to Shea et al. 1991

7.17.3 Comparison with the Event of June 3, 1982

In Fig. 7.28 is shown the comparison of dependence of increase and amplitude from h_S for the events of June 3, 1982 and May 24, 1990. According to Shea et al. (1991), the arrival of solar neutrons at the Earth on May 24, 1990 3 min after the onset of soft x-rays and at the same time as the recorded maximum in both soft x-ray emission and Hα indicate that the first solar particles were produced within the first 2 min of the impulsive x-ray event and prior to the maximum in solar x-ray emission from the Sun (very rapid acceleration of ions to GeV energies).

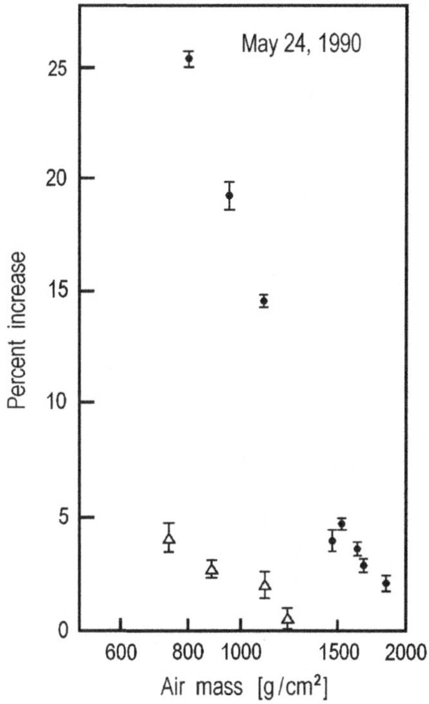

Fig. 7.28 Solar neutron event of May 24, 1990. Normalized amplitude of the increase (*dots*) in the time interval 20:50–20:55 UT 24 May, 1990, taken from Table 7.1, plotted as a function of the air mass along the line of sight from the Sun to the observing location. The amplitude of the increase observed by neutron monitors for the 3 June, 1982 solar neutron event is illustrated by *triangles*. According to Shea et al. (1991)

7.17.4 Peculiarities of Effects Observed by NM and Muon Telescopes in Yakutsk

Very strange effects in this event was observed in Yakutsk, where direct solar neutrons can not arrive (at 20:50 UT on May 24, 1990 for Yakutsk $\theta_s > 90°$). According to Filippov and Prikhod'ko (1993) the neutron supermonitor in Yakutsk observed two increasing events, first at 20:45 UT, by the neutron supermonitor (about 1.5%) and by muon telescopes on the ground and underground (7 m w.e) and the second after 15 min by the neutron supermonitor (see Fig. 7.29).

Because solar neutrons can not arrive directly to Yakutsk, the observed effect must be caused by so called 'refraction effect' (see above, Chapter 6) or by high energy charged particles from neutron decay (what is much more probably). The second increase can be caused by the same population of charged particles which generated from the solar neutrons for the first increase, but these charged energetic particles are arrive to the Earth after one-two scattering the interplanetary space.

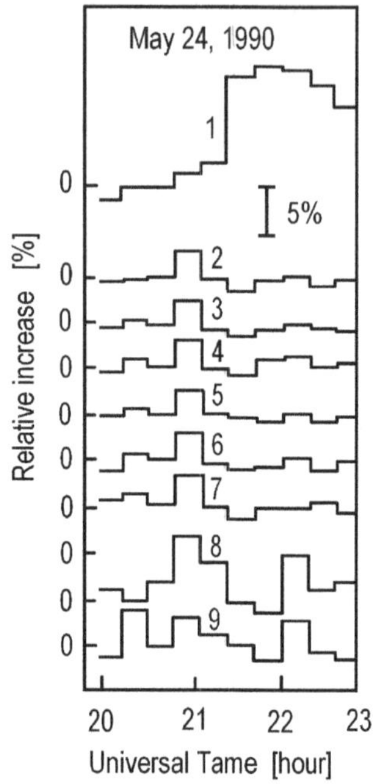

Fig. 7.29 The solar neutron event of May 24, 1990. 15-min data in Yakutsk: (**1**) neutron supermonitor, (**2**) vertical muon telescope on the ground (0 m w.e.), (**3**) vertical muon telescope underground (7 m w.e.), (**4**) muon telescope at 0 m w.e. in direction 30°N, (**5**) muon telescope at 0 m w.e. in direction 30°S, (**6**) muon telescope at 7 m w.e. in direction 30°N, (**7**) muon telescope at 7 m w.e. in direction 30°S, (**8**) muon telescope at 0 m w.e. in direction 60°N, (**9**) muon telescope at 0 m w.e. in direction 60°S. According to Filippov and Prikhod'ko (1993)

7.17.5 Determining of the Full Differential Flux of Solar Neutrons at the Top of the Earth's Atmosphere

On the basis of the NM data of the solar neutron event on May 24, 1990, Shibata et al. (1993) determined the full differential flux of solar neutrons at the top of the Earth's atmosphere during this event:

$$F(E_n) = (7.6 \pm 0.4) \times 10^5 \times (E_n/100\ \text{MeV})^{-2.6\pm0.1}\ \text{neutrons} \cdot \text{MeV}^{-1} \cdot \text{m}^{-2}, \tag{7.2}$$

where E_n is the kinetic energy of solar neutrons.

7.17.6 Comparison of Full Differential Fluxes of Solar Neutrons at the Top of the Earth's Atmosphere During the Events of 24 May, 1990 and 3 June, 1982

Comparison of Eq. 7.2 with Eq. 7.1 shows that the May 24, 1990 event was much greater and sufficiently harder than the June 3, 1982 event: at 24 May, 1990 the coefficient in the full differential flux of solar neutrons at the top of the Earth's atmosphere was $(7.6 \pm 0.4) \times 10^5$ and power index was -2.6 ± 0.1 instead of $\left(4.6^{+0.7}_{-0.6}\right) \times 10^5$ and -3.7 ± 0.2 at June 3, 1982 event.

7.18 Solar Neutron Event on February 22, 1991 and Possible New Special Class of Solar Flares, Very Deep in the Chromosphere (Near Photosphere)

This possible event was reported by Muraki et al. (1991a). They used the data of modern solar neutron telescope, operated at Mt. Norikura (2,770 m above sea level) since October 17, 1990 (the detail description of neutron telescope see in Chapter 4 in Dorman, M2004). The neutron telescope looks at particles coming from the direction of the Sun, within either $\pm 11°$ or $\pm 22°$. This telescope measures the energies of solar neutrons in three deposit energy bands: low (50–360 MeV), medium (280–500 MeV) and high ($\geq$ 390 MeV). The detector was inclined to 54° on February 22, 1991 and according to Muraki et al. (1991a) a very clear event was observed by the solar neutron telescope at 03:59 UT ($\lambda_s = 10.8°S$, $f_S = 120.2°E$, $\theta_s = 49.4°\cos\theta_s = 0.651, h_s = 1,128\ \mathrm{g/cm^2}$). The experimental data are given in the Table 7.3.

Figure 7.30 shows the time profiles of this event in different channels of solar neutron telescope. The problem for this event is that there was no significant solar flare with great x-ray or gamma-ray emissions as in other solar events.

Muraki et al. (1991a) suggests that it can be a special class of solar flares, very deep in the chromosphere (near photosphere) with well generation of solar neutrons which can transport to the Earth, but due to strong attenuation in the solar atmosphere in radio waves, optical and x-ray wave branches these large solar flares (which occurred deep in the solar atmosphere) might not be observed at these wave lengths (if this type of solar flare and solar neutron event is real, this discovery must open a new field of solar physics).

7.19 Solar Neutron Events on March 22 and May 3, 1991

7.19.1 Main Properties of Solar Flare at March 22, 1991

According to Pyle and Simpson (1991) the solar neutron event on March 22, 1991 was detected by Haleakala neutron supermonitor (18 NM-64, 20.72°N, 156.28°W;

Table 7.3 Mt. Norikura neutron telescope data for the event at 03:59 UT on February 22, 1991 (According to Muraki et al. 1991a)

Channel	0	1	2	3	4	5	6	7
E_n (MeV)	50–360	280–500	>390	50–360	280–500	>390	280–500	>390
Effect (%)	13 ± 2.2	78 ± 14.2	14 ± 10	13 ± 5.4	29 ± 19.3	−2 ± 7	37 ± 6.4	10 ± 4

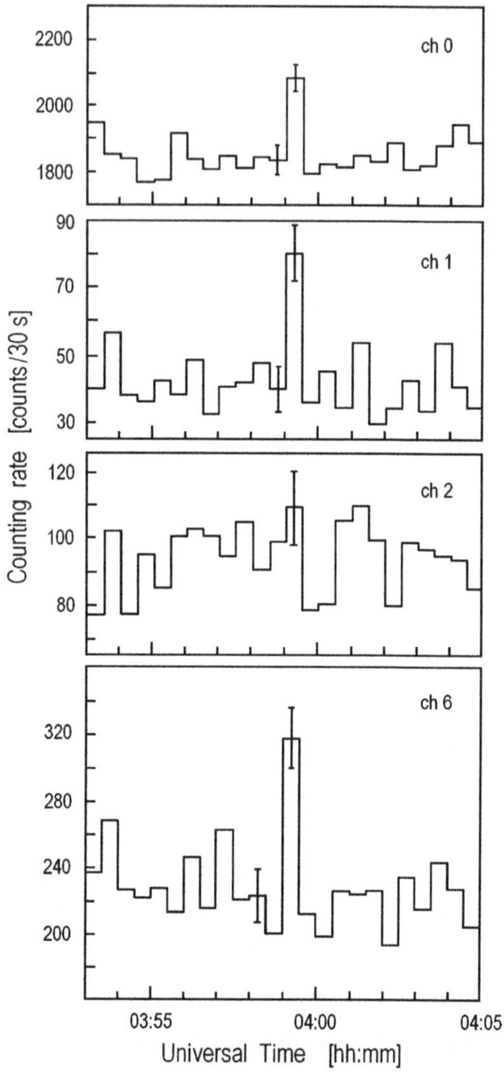

Fig. 7.30 Solar neutron event of February 22, 1991. Time profiles of counting rates detected by Mt. Norikura neutron telescope (from top to bottom, channel numbers 0, 1, 2, and 6). According to Muraki et al. (1991a)

Table 7.4 Properties of solar flare caused the solar neutron event on March 22, 1991 (According to Pyle and Simpson 1991)

Radiation	Onset (UT)	Max (UT)	Importance
Hα	22:43	22:45	3B
1–8 Å x-rays	22:42		X9.4
2,695 MHz	22:43.0	22:44.0	36,000 SFU
8,800 MHz	22:43.0	22:44.0	37,000 SFU

3,030 m above sea-level; $R_C = 12.91$ GV). This event was caused by a 3B solar optical flare at S26, E28 (accompanied with X9.4 solar x-ray bursts) which started in Hα at 22:43 UT on March 22, 1991 (see Table 7.4).

7.19.2 Solar Neutron Observation from 22 March, 1991 by Neutron Monitor on Mt. Haleakala

For this event $\lambda_s = 0°$, $f_S = 142.2°$W. Figure 7.31 shows the time profiles of neutron component detected by Haleakala neutron monitor (2-min data) in comparison with 0.5–4.0 Å x-rays (1-min data) and with fluxes of >850 MeV protons

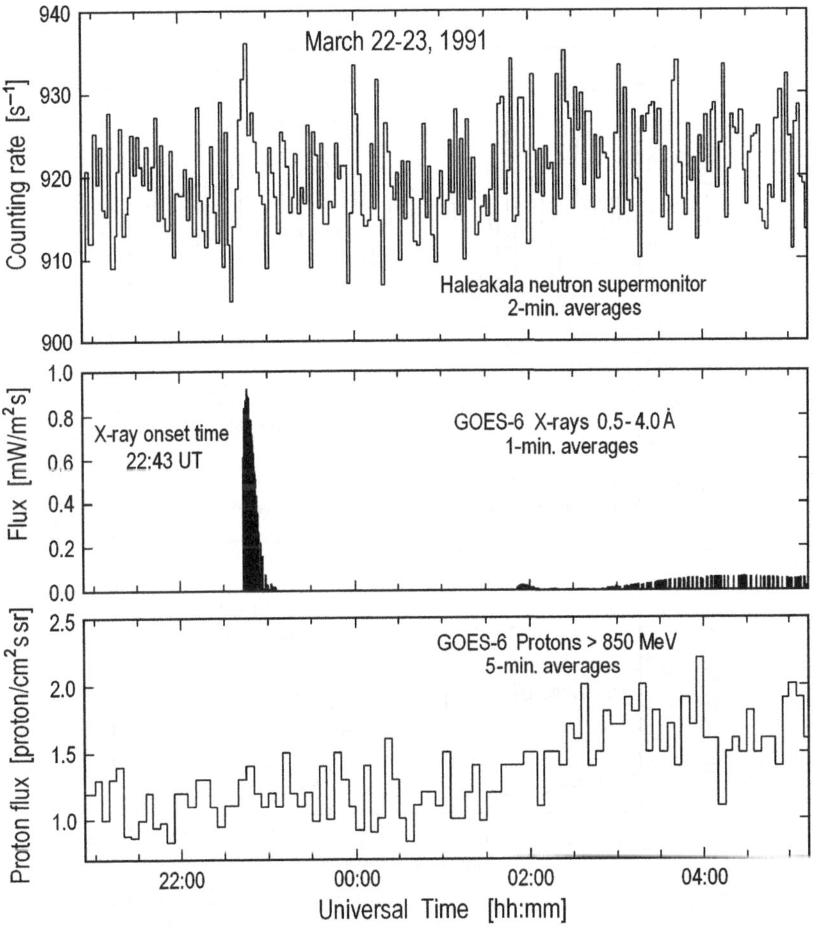

Fig. 7.31 The solar neutron event of March 22, 1991. Time profiles of neutron intensity at Haleakala, x-rays and protons with energies >850 MeV at GOES-6 from 21:30 UT of March 22 to 04:30 UT of March 23, 1991. According to Pyle and Simpson (1991)

(5-min data) from GOES-6, in the period from 21:30 UT March 22 to 04:30 UT March 23, 1991.

7.19.3 Comparison of the Solar Neutron Event of March 22, 1991 with the Gamma-Ray Event Observed by IMP-8 and x-Ray Event Observed by GOES-6

Figure 7.32 gives the same information, with better time resolutions on solar neutron event detected by Haleakala neutron monitor accompanied by gamma-ray event observed by IMP-8 and x-ray event observed by GOES-6.

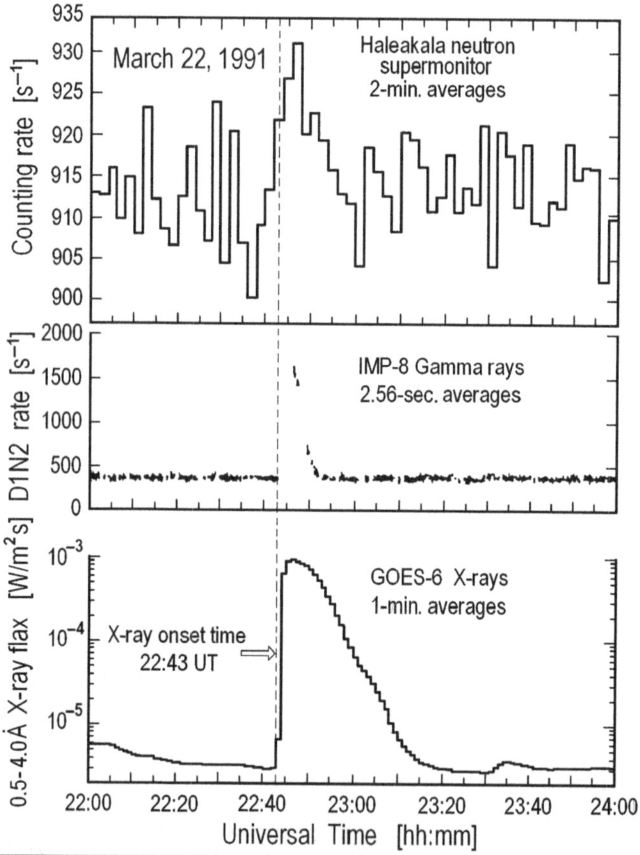

Time interval (UT)	22:(42-44)	22:(44-46)	22:(46-48)	22:(48-50)	22:(50-52)
Neutron energy [MeV]	>1100	440-1100	250-440	200-250	170-200

Fig. 7.32 The solar neutron event of March 22, 1991. Time profiles of neutron intensity at Haleakala, gamma-rays at IMP-8 and x-rays at GOES-6. In the bottom is shown the energies of neutrons in dependence of time detection. According to Pyle and Simpson (1991)

7.19.4 Full Differential Flux of Solar Neutrons at the Top of the Earth's Atmosphere During the Solar Neutron Event of March 22, 1991

For this event Shibata et al. (1993) estimated the expected full differential flux of solar neutrons at the top of the Earth's atmosphere by the following equation:

$$F(E_n) = (0.11 \pm 0.01) \times 10^5 (E_n/(100\ \text{MeV}))^{-2.4\pm0.1}\ \text{neutrons} \cdot \text{MeV}^{-1} \cdot \text{m}^{-2} \tag{7.3}$$

Shibata et al. (1993) noted that this event was characterised with very small neutron fluxes at E_n100 MeV (about 40 times smaller than that of June 3, 1982 and about 70 times smaller than that of May 24, 1990). The cause of this can be connected with position of solar flare on the solar disk: the event of March 22, 1991 was caused by near central flare (26°S, 28°E), while the other two events were caused by near limb flares (9°S, 72°E for June 3, 1982 and 33°N, 78°W for May 24, 1990).

7.19.5 Possible Solar Neutron Event on May 3, 1991

According to Muraki et al. (1991a) the Mt. Norikura neutron telescope detected a sudden sufficient increase at 04:15 UT, May 3, 1991 (see Fig. 7.33 and Table 7.5).

For this event

$$\lambda_s = 15.6°N,\ f_s = 116.2°E,\ \theta_s = 29.5°,\ h_s = 844\ \text{g/cm}^2. \tag{7.4}$$

The neutron telescope for this event was inclined to 27°. Since the main excess was observed in the interval of 0.5 min, Muraki et al. (1991a) examined the possibility that such fluctuations might happen through instability of the triggering electronics. The checking of the triggering electronics shows that the increase at 04:15 UT, May 3, 1991 was made by real particles coming from the outside and not by electronic noise. In the event May 3, 1991 (as in the event of February 22, 1991, see above, Section 7.18) there were only insufficient solar flares and Muraki et al. (1991a) supposed that this event was caused by solar flare which occurred deep within the solar atmosphere, with neutron generation but without a strong burst of electromagnetic radiation.

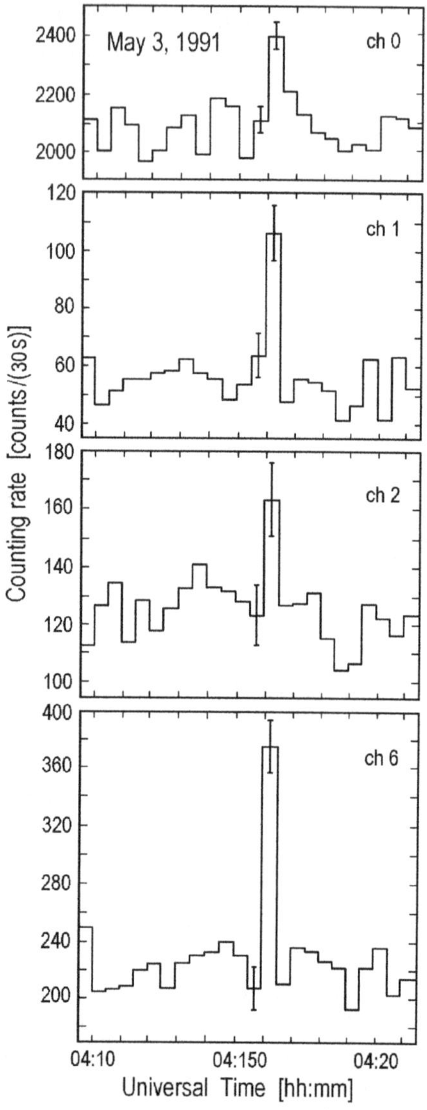

Fig. 7.33 The solar neutron event of May 3, 1991. Time profiles of counting rates detected by Mt. Norikura neutron telescope (from *top* to *bottom*, channel number 0, 1, 2, and 6). According to Muraki et al. (1991a)

Table 7.5 The Mt. Norikura neutron telescope data for the event at 04:15 UT on May 3, 1991 (According to Muraki et al. 1991a)

Channel	0	1	2	3	4	5	6	7
E_n (MeV)	50–360	280–500	<390	50–360	280–500	<390	280–500	<390
Effect (%)	16 ± 2.2	104 ± 13.9	31 ± 9.4	18 ± 5.6	33 ± 17.4	0	72 ± 6.8	8 ± 3.5

7.20 Solar Neutron Event at 1st June 1991: Surprisingly Intense Neutron Emission from a Flare Behind the Limb of the Sun

7.20.1 The Matter and Short History of the Problem

According to Murphy et al. (1999), the Oriented Scintillation Spectrometer Experiment on board the Compton Gamma Ray Observatory observed a strong flux of neutrons from the behind-the-limb flare that occurred on 1991 June 1. This is surprising if the neutrons were produced by thin-target interactions in the Sun's corona, as is suggested by γ-ray observations of this flare made by Granat/PHEBUS. Murphy et al. (1999) compare neutron and γ-ray observations of the June 1 flare with thick-target emissions observed from a flare 3 days later, where the interactions took place in the chromosphere and photosphere. A very hard spectrum for the accelerated particles is required to account for the number of neutrons observed on June 1 if they were produced by thin-target interactions in the corona.

As noted Murphy et al. (1999), most flares that emit detectable γ-ray and neutron emissions occur on the visible disk of the Sun. While it is generally believed that particle acceleration in flares takes place in the corona, interactions producing γ-rays and neutrons are thought to occur predominantly in compact regions at the foot-points of magnetic loops. These thick-target interactions are believed to occur in the lower chromosphere or upper photosphere at densities greater than $10^{12}\ \mathrm{cm}^{-3}$ (Chupp 1984). Because thick-target γ-ray line and neutron production is relatively efficient, such emission from flares occurring on the solar disk is expected to overwhelm any thin-target emission from particles traversing the low density coronal portions of magnetic loops. If the flare is located at or behind the limb, however, part or all of the thick-target emission could be occulted, which would lead to a higher coronal contribution. Observations of γ-rays from over-the-limb flares are rare.

Murphy et al. (1999) noted that an example is the 1981, April 27 flare where Hα observations suggested that it occurred between 0° and 5° beyond the limb, which corresponds to an occulting height of less than 3,000 km above the photosphere (Takakura et al. 1983). In their x-ray$-\gamma$-ray timing analysis of this flare, Hulot et al. (1992) found that greater than 30% of the strong γ-ray line flux observed with the Gamma Ray Spectrometer (GRS) aboard the Solar Maximum Mission (SMM) satellite could have been produced in the corona. Another example of a behind-the-limb γ-ray producing flare was observed with SMM/GRS on September 29, 1989 (Vestrand and Forrest 1993). Vestrand and Forrest (1993) placed the location of the flare between 5° and 15° beyond the west limb of the Sun, which corresponds to an occultation height between 3,000 and 20,000 km. What was extraordinary for this flare was the detection of a strong 2.223 MeV neutron capture line. This line is formed when flare-produced neutrons are captured on ambient H. Because the density must be high enough for the neutrons to slow down and be captured before decay, the line is only produced deep in the photosphere (Hua and Lingenfelter

1987a). For flares close to the limb of the Sun, the 2.223 MeV line is therefore strongly attenuated relative to the nuclear deexcitation lines produced at higher altitudes. This led Vestrand and Forrest (1993) to propose that energetic particles propagated from beyond the limb to impact over a spatially extended area, including part of the visible disk, thus producing thick-target emission. Vestrand and Forrest (1993) suggested that this might occur in the presence of a giant magnetic loop that reached around the limb to place a footpoint in the visible hemisphere. They also suggested this could result from a large, diffuse interaction region extending over the limb, as might be produced by 'back-diffusion' of particles accelerated over a large range of helio-longitudes by a coronal mass ejection. For this flare, then, the thick-target γ-ray line emitting region again dominated any coronal radiation from the loops of the flaring region behind the limb.

As underlined Murphy et al. (1999), one of the most intense flares ever recorded emanated from Active Region 6659 on 1991 June 1 when it was located $6^\circ - 9^\circ$ beyond the east limb of the Sun (Barat et al. 1994), which corresponded to an occultation height between 3,000 and 7,000 km above the photosphere. The flare was well observed by the Solar x-Ray/Cosmic Gamma-Ray Burst Experiment aboard the interplanetary spacecraft Ulysses (Kane et al. 1995), located 22° beyond the east limb of the Sun. Kane et al. (1995) estimated that the total energy released in the flare by greater than 20 keV electrons was 10^{34} erg, with an energy release rate of 2% of the total solar luminosity. Thus the June 1 flare was the largest flare of Cycle 22, and it prompted Kane et al. (1995) to question whether the material and energy resources of the active region were adequate.

As noted Murphy et al. (1999), the June 1, 1991 flare was also observed by the PHEBUS instrument on Granat satellite (Barat et al. 1994). In spite of the flare location, hard x-rays and γ-rays were observed through the peak of emission. The observed spectrum showed excesses above a bremsstrahlung power law in the 1.1–1.8 and the 4.1–7.6 MeV regions consistent with nuclear line emission. No significant 2.223 MeV neutron-capture line was detected. The June 1 flare could have had geometry similar to the September 29, 1989 flare mentioned above, with a separate or extended region of thick-target emission on the visible disk producing the γ-rays. The lack of a detectable 2.223 MeV line could be understood if the region were confined near the limb, as in other γ-ray-producing limb flares where the line is strongly attenuated, such as the June 21, 1980 (Chupp et al. 1982) and April 27, 1981 (Murphy et al. 1990) flares observed with SMM (the June 21, 1980 flare also produced neutrons similar to the June 1, 1991 flare.) If so, the June 1, 1991 γ-ray emission would be expected to have the spectral characteristics of thick-target emission, as observed in such limb flares; however, the June 1 γ-ray spectrum measured by PHEBUS was very different in that the (1.1–1.8 MeV)/(4.1–7.6 MeV) ratio was exceptionally high (Ramaty et al. 1997). This led Ramaty et al. (1997) to argue that such a high ratio is only possible from thin-target interactions occurring in the corona with accelerated particles having heavy elements (greater than ^{16}O) abundance enhancements typical of energetic particles in impulsive events observed in interplanetary space (see more details in Section 9.10). This supported the conclusion of Barat et al. (1994) of a coronal source for the γ-rays based on the lack of evidence of a large-scale magnetic loop system connecting the active region behind the limb to the visible disk. Such thin-target interactions, however, are considerably less efficient in producing γ-rays than are thick-target interactions. If the bulk of the emission from the June 1, 1991 flare was from the corona as suggested, the strong γ-ray signal detected apparently resulted from its phenomenal size.

According to opinion of Murphy et al. (1999), an unexpected feature of the June 1, 1991 flare is that energetic neutrons were detected in low Earth orbit by the Oriented Scintillation Spectrometer Experiment (OSSE) on board the Compton Gamma Ray Observatory (CGRO). Flare neutrons are produced by interactions of particles with energies typically greater than about 50 MeV/nucleon. Neutrons that are initially directed upward have a good chance of escaping from the solar atmosphere without scattering, and those that are initially directed downward lose most of their energy in the atmosphere with little chance of escape (Hua and Lingenfelter 1987b). Such downward-directed neutrons either decay at the Sun or are captured by nuclei at the Sun (see the 2.223 MeV line discussion above). The escaping neutron flux at Earth is extended in time because of velocity dispersion and is attenuated owing to neutron decay. Because the probability of scattering in the corona is small, and because any scattered neutrons would be of such low energy (Hua and Lingenfelter 1987b) that they would not survive the 1 AU distance, we do not expect the observed neutron flux to contain a significant contribution from neutrons produced by occulted thick-target interactions.

As underlined Murphy et al. (1999), detection of neutrons from flares is rare, even from thick-target disk flares, and thin-target interactions are even less efficient in producing neutrons than the thick-target interactions of disk flares. Thus the strong neutron flux observed from the 1991 June 1 flare, if also produced by the thin-target interactions believed responsible for the γ-ray emission, is surprising. In paper of Murphy et al. (1999) are describe the OSSE detection of neutrons from this flare and determine whether thin-target interactions can account for the high flux. It was done this by comparing the neutron counts and γ-ray fluence observed from the June 1, 1991 flare with the same quantities observed by OSSE from the thick-target disk flare that occurred 3 days later on June 4 while OSSE was pointed at the Sun. The June 4 flare was well observed by OSSE, and γ-rays over a broad spectral range (from 0.1 to 100 MeV) and neutrons were detected (Murphy et al. 1997; DelSignore 1995; see also below, Section 7.21). The greater than 15 MeV γ-ray emission observed from June 4 flare was much more impulsive than the long lasting 4–7 MeV nuclear emission, falling below detectability within 250 s after the γ-ray peak. The spectrum of the high-energy γ-ray emission showed no evidence for a pion-decay feature, which implies that the emission was predominantly electron bremsstrahlung. A strong neutron signal was detected following the high-energy γ-ray emission and continued until the Sun was occulted by the Earth.

Murphy et al. (1999) underlined that the 1991 June 1 and June 4 flares comparison will establish a neutron-production efficiency for June 1 relative to a known thick-target disk flare. This relative efficiency will then be compared with predicted thin-target neutron efficiencies calculated at a variety of accelerated-particle spectral indexes and the predicted thick-target efficiency calculated at the known spectral index determined (Murphy et al. 1997) for the June 4 flare. The comparison will show that thin-target interactions of accelerated particles are not adequate to account for the number of neutrons observed from the June 1 flare unless the spectrum is extremely hard (a power-law index less than 2).

7.20.2 The OSSE Response to Solar Neutrons

According to Murphy et al. (1999), the four 10 750 cm^2 thick NaI detectors provide OSSE with significant stopping power for high-energy neutrons. The effective area of each of these detectors is about 120 cm^2 (DelSignore 1995). Both high-energy γ-rays and neutrons can be detected and are distinguished by their different pulse shapes in NaI (Share et al. 1978); the rise time to maximum amplitude for energy losses by the low-energy protons, alpha particles, etc., produced by neutron interactions is significantly shorter than that for energy losses by the electrons produced by γ-ray interactions. Pulse-shape discrimination is therefore used to distinguish neutron and γ-ray events, and their energy-loss spectra are separately accumulated in 16 channels, up to about 220 MeV for neutrons and about 150 MeV for γ-rays (Johnson et al. 1993; DelSignore 1995). Murphy et al. (1999) note that because the scintillation light output of a neutron interaction is less than that of a γ-ray of the same energy, a given neutron spectral channel corresponds to a higher energy loss than that of the equivalent γ-ray channel. The dotted histogram of Fig. 7.34 shows the distribution of pulse durations for channel 3 of the 16- channel spectra obtained at the peak of the June 4 flare when the emission was due to γ-rays only.

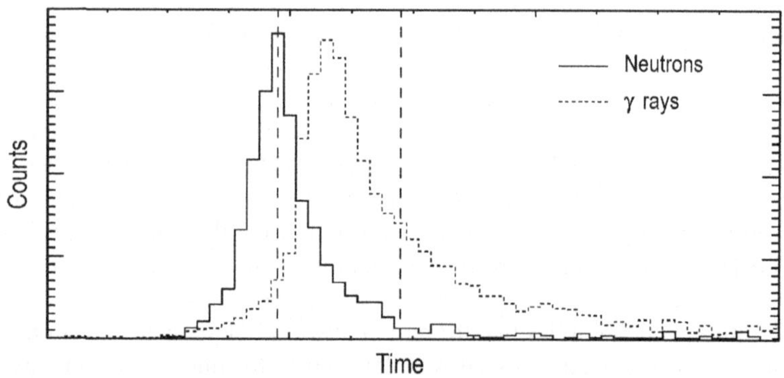

Fig. 7.34 Pulse duration distributions of neutron (*solid line*) and γ-ray (*dotted line*) events for channel 3 of the 16-channel energy-loss spectra. Channel 3 corresponds to energy losses of 36–49 MeV for neutrons and 26–36 MeV for γ-rays. The *dashed lines* indicate the duration acceptance window for γ-rays; all events with duration times falling to the left of the window are categorized as neutrons. From Murphy et al. (1999)

The solid histogram in Fig. 7.34 shows the distribution of durations for channel 3 obtained late in the flare after the high-energy γ-ray emission had dropped below detectability and only neutrons were present (channel 3 corresponds to 26–36 MeV energy losses for γ-rays and 36–49 MeV energy losses for neutrons). The dashed lines indicate the pulse-duration acceptance window for γ-rays; all events with durations falling to the left of the window are categorized as neutrons. The ability to separate neutrons and γ-rays is demonstrated, although there is some overlap that increases with energy. Ground-based and in-orbit calibrations are used to correct for this overlap. A prototype OSSE detector was exposed to neutrons produced at the Indiana Cyclotron (Jenkins et al. 1991).

7.20.3 Observations During the June 1, 1991 Solar Flare

An overview of the June 1, 1991 flare according to Murphy et al. (1999) is shown in Fig. 7.35. The top curve shows the GOES 1–8 Å soft x-ray emission that was saturated because of the intense flux. The bottom curve shows the count rate of hard x-rays (greater than 27 keV) observed by Ulysses (Kane et al. 1995).

The data points in Fig. 7.35 show the 1.1–1.8 and 4.1–7.6 MeV nuclear γ-ray fluxes (excess above a bremsstrahlung power-law continuum) observed by Granat/

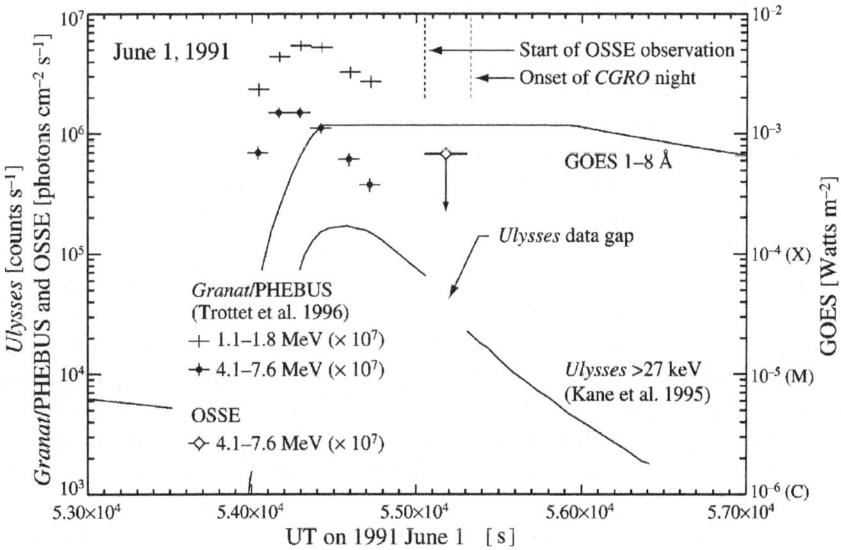

Fig. 7.35 Overview of the 1991 June 1 solar flare event. The *upper line* is the GOES 1–8 Å soft x-ray emission, the *bottom line* is the Ulysses >27 keV hard x-ray emission, and the data points are the Granat/PHEBUS 1.1–1.8 MeV and 4.1–7.6 MeV nuclear γ-ray excesses. The open diamond is the upper limit of the 4.1–7.6 MeV nuclear excess observed by OSSE. Also shown are the start time of the CGRO/OSSE observation and the onset of CGRO night. From Murphy et al. (1999)

PHEBUS (Trottet et al. 1996). The γ-ray data are binned into 128 s intervals from 14:59:38 to 15:13:09 (53,978–54,789 s) UT. The line emission appears to peak during the third interval starting at 15:03:54 (54,234 s) UT, about 250 s before the peak of the hard x-rays observed by Ulysses. OSSE data are not available during this peak portion of the flare because CGRO was in the high radiation zone of the South Atlantic Anomaly (SAA) at that time. When observations resumed at 750 s after the peak of the γ-ray emission (55,049 s UT), OSSE was pointed 100° away from the Sun and observing the black hole candidate Cygnus X-1. Earth occultation of the Sun occurred about 260 s later (55,312 s UT). The OSSE observation interval is indicated in Fig. 7.35 by dotted lines. OSSE has some response to γ-rays above 1 MeV even at this large offset angle (20% of the on-axis response). Because of SAA contamination, however, a reliable background-subtracted spectrum could not be obtained, so a 4–7 MeV nuclear line flux could not be directly derived. The 4–7 MeV count rate did show a sudden decrease at the time of satellite night, which suggests that γ-ray emission from the Sun was detected. The magnitude of the change in this count rate was converted to a photon flux using an estimate for the 100° off-pointed effective area at 4–7 MeV. The resulting flux is plotted in Fig. 7.35. It is shown as an upper limit to the 4–7 MeV nuclear γ-ray flux because (1) it includes an unknown contribution from electron bremsstrahlung, and (2) there may be some contamination of the count rate at these energies due to energy losses of higher energy solar neutrons. A fit to the OSSE data at this time for the narrow 2.223 MeV neutron-capture line indicated no significant flux, with a 2 σ upper limit of 1 10^{-2} photons·$cm^{-2}s^{-1}$. Surprisingly, a decrease at satellite night in the rate of greater than 15 MeV events attributed to neutrons was also observed. Plotted in Fig. 7.36 are the background-subtracted rates in the neutron channel.

According to Murphy et al. (1999), background was obtained from observations approximately 15 orbits after the flare observation when orbital conditions (location and aspect to the Earth) were similar. Murphy et al. (1999) were not able to use the average of rates 15 orbits both before and after the flare as is usual because parameters affecting the neutron–γ-ray separation were set differently on the previous day. This one-sided background prediction was tested during flare quiet times and successfully produced null difference spectra. The uncertainties shown in Fig. 7.36 are statistical. After satellite night, the background-subtracted rate fell to zero, as is expected for a solar source. A key question is whether this high-energy emission is indeed due to neutrons. Because the OSSE neutron–γ-ray distinction via pulse-shape discrimination is not perfect, a fraction of neutron events are counted as neutrons and the complement as γ-rays. The analogous situation holds true for γ-ray events. The true neutron and γ-ray energy-loss spectra in channels 3–7 (36–100 MeV neutron and 15–65 MeV γ-ray energy losses) can be recovered by using the separation constants developed by DelSignore (1995). (Above channel 7, the neutron–γ-ray distinction is too poor to allow reliable separation.) The separated channels 3–7 neutron and γ-ray energy-loss spectra for Detector 1 are presented in Fig. 7.37 and clearly show that the high energy emission detected by OSSE from the June 1 flare is indeed predominantly due to neutrons.

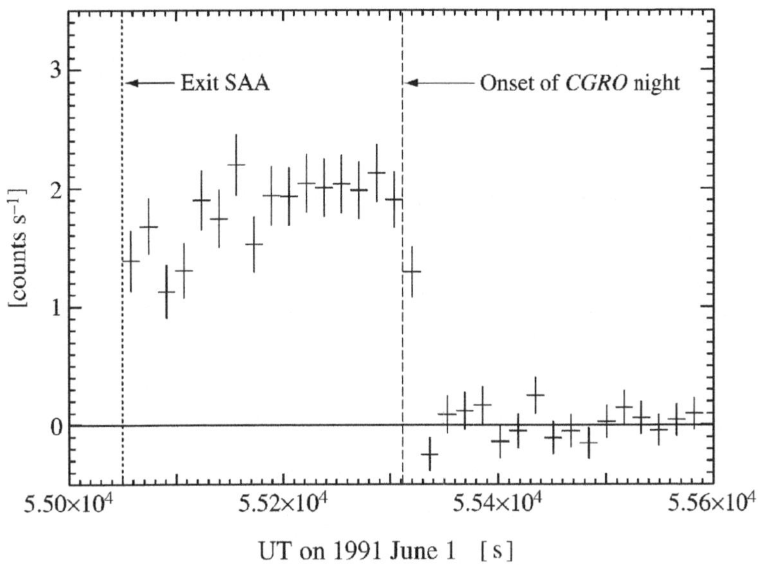

Fig. 7.36 OSSE June 1, 1991 background-subtracted mean count rate per detector for events attributed to neutrons. The *dotted line* indicates when OSSE resumed observations of Cygnus X-1 after SAA exit, and the *dashed line* shows the onset of CGRO night. From Murphy et al. (1999)

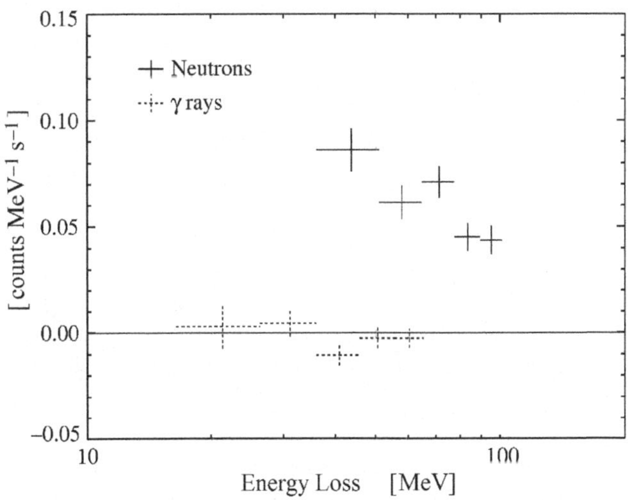

Fig. 7.37 Separated channels 3–7 neutron and γ-ray energy-loss spectra derived from the OSSE June 1 high-energy data. Channels 3–7 correspond to 36–100 MeV neutron and 15–65 MeV γ-ray energy losses. From Murphy et al. (1999)

As noted Murphy et al. (1999), no high-energy γ-ray data are available during the peak of emission, but the OSSE data indicate that any greater than 15 MeV γ-ray emission late in the June 1 flare was below detectability by the OSSE detectors.

7.20.4 Comparison Observations of June 1, 1991 and June 4 Solar Flares

Time profiles of nuclear γ-rays and count rates attributed to neutrons from both the June 1 and the June 4 flares are shown in Fig. 7.38.

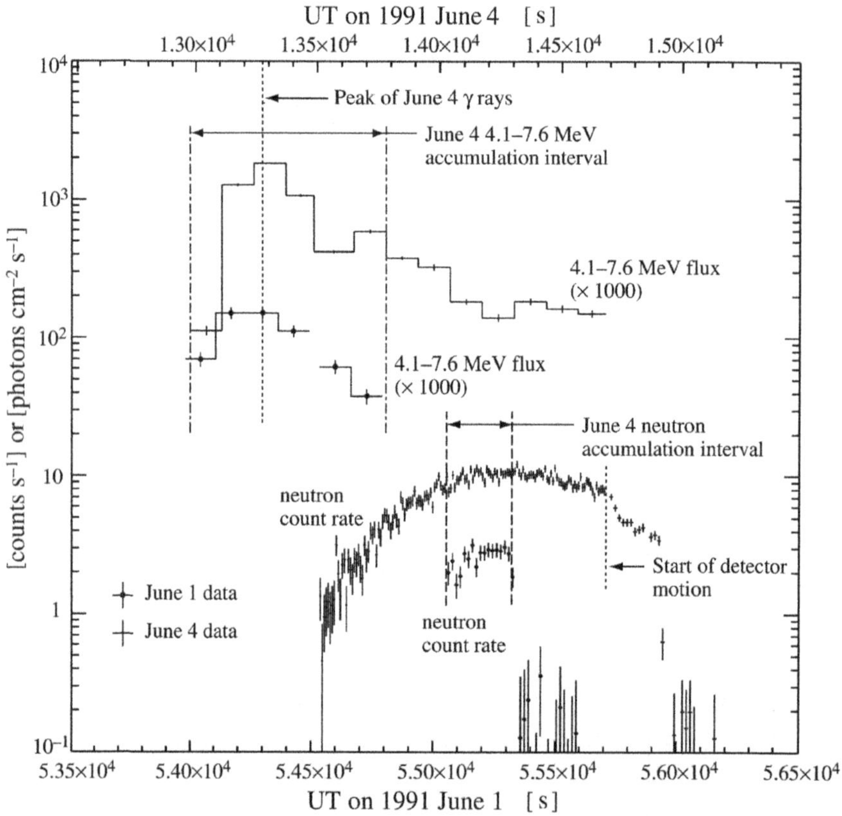

Fig. 7.38 Nuclear γ-ray and neutron observations of the June 1, 1991 and June 4, 1991 flares. The bottom time axis is for the June 1 flare and the top time axis is for the June 4 flare. The June 1 neutron data have been corrected for the OSSE off-axis response. The peak of the June 4 γ-ray emission is shown by the *dotted line*, and the June 4 time axis has been shifted so that this emission peak coincides with the midpoint of the time bin containing the peak of the June 1 γ-ray emission. Also shown are the accumulation intervals of the June 4 OSSE γ-ray and neutron data used in the analysis and the start time of detector motion away from the Sun during the June 4 flare. From Murphy et al. (1999)

In Fig. 7.38, the bottom time axis is for the June 1 flare and the top time axis is for the June 4 flare. The 4.1–7.6 MeV nuclear γ-ray fluxes observed by Granat/PHEBUS (Trottet et al. 1996) for the June 1 flare and by OSSE for the June 4 flare are binned into 128 and 131 s intervals, respectively. The background-subtracted neutron count rates observed by OSSE for the June 1 flare are binned into 16 s intervals. For the June 4 flare they are binned into 8 s intervals when pointed at the Sun and 16 s intervals after OSSE rotated away from the Sun to point at Cygnus X-1. Earth occultation of the Sun on June 4 occurred 230 s after repointing. Background estimation for the June 4 neutron data was derived using data obtained 15 orbits before and 15 orbits after the observation. The June 4 neutron count rate fell to zero at satellite night, as is expected for a solar source. The time axis for the June 4 flare data has been shifted so that the peak of the June 4 γ-ray emission (dotted line at 13,270 s UT on June 4) coincides with the midpoint of the time bin containing the peak of the June 1 γ-ray emission.

As noted Murphy et al. (1999), because the OSSE detectors were pointed 100° away from the Sun during the June 1 observation but were pointed at the Sun during the June 4 observation, the June 1 neutron count rates must be corrected relative to those of June 4 to account for the reduced detector efficiency. The laboratory calibration of Jenkins et al. (1991) indicated that the response at 100° is 70% of the on-axis response. Murphy et al. (1999) confirm this using the June 4 neutron data. The dotted line in Fig. 7.38 associated with the June 4 neutron data indicates the end of solar pointing and the onset of motion as the OSSE detectors rotated 100° away from the Sun to acquire Cygnus X-1. Rotation was complete in about 50 s. Linear fits to the data obtained both before and after detector movement were performed. Extrapolations of these fits to the midpoint of detector motion imply that the off-pointed effective area is 70% ± 13% of the on-pointed area, which confirms the calibration. The June 1 neutron count rates shown in Fig. 7.38 have been corrected for this reduced OSSE off-axis response. The uncertainties plotted are statistical; the systematic uncertainty due to this off-axis correction will be added in quadrature with the statistical uncertainty in the analysis that follows.

As underlined Murphy et al. (1999), since the comparing neutron count rates with 4–7 MeV fluences have been obtained with different instruments, the relative γ-ray calibration of the two detectors must also be considered. Coincident data are available for the June 11, 1991 flare (Barat and Trottet 1997) when both instruments observed the peak of emission and OSSE was in the same observing configuration as for the June 4 flare. The mean 4–7 MeV nuclear γ-ray fluences derived with the two instruments were consistent to within 12% (less than 1.3 σ difference). It will be also include this additional uncertainty in the analysis that follows.

7.20.5 Main Results and Discussion of the June 1, 1991 Event, and Comparison with June 4, 1991 Event

Murphy et al. (1999) compare the high-energy neutron count rates and the 4–7 MeV nuclear γ-ray fluences for the June 1 and June 4, 1991 flares. For solar neutrons, Murphy et al. (1999) integrated the June 4 rates over a 280 s interval at approximately the same time following the June 4 γ-ray peak, as was available for the June 1 flare (see Fig. 7.38). Because the neutron flux is changing slowly at the time of interest, a small shift in this accumulation interval will not significantly affect the results. The total integrated neutron counts are 1,430 $\pm$ 270 for the June 1 flare (corrected for off-axis angular response), and 4,720 $\pm$ 45 neutrons for the June 4 flare. The 4–7 MeV γ-ray time profiles of the two flares are similar (see Fig. 7.38), although the fluxes are significantly different. During the 800 s interval defined by the dash-dotted lines in Fig. 7.38, the measured 4.1–7.6 MeV nuclear fluences are 89 $\pm$ 7 photons cm^{-2} for the June 1 flare (Trottet et al. 1996), and 649 $\pm$ 4 photons cm^{-2} for the June 4 flare. The uncertainties are statistical.

Murphy et al. (1999) define a neutron efficiency, η, as the ratio of observed neutron counts to observed 4.1–7.6 MeV nuclear γ-ray fluence. For the June 1 flare, $\eta_{\mathrm{June}\ 1} = 16.2 \pm 5.1$, and for the June 4 flare, $\eta_{\mathrm{June}\ 4} = 7.3 \pm 0.1$. In estimating the June 1 efficiency, it was included the systematic uncertainty associated with the 4–7 MeV inter-calibration of the OSSE and Granat/PHEBUS detectors, and the off-axis neutron response correction as discussed above. The ratio of these two efficiencies, $\eta_{\mathrm{June}\ 1}/\eta_{\mathrm{June}\ 4}$. shows that the *June 1 thin-target flare* was more efficient in producing neutrons than the *June 4 thick-target flare* by a factor of 2.2 $\pm$ 0.5. Murphy et al. (1999) note that if the June 1 OSSE neutron observation did not occur near the peak of the neutron count rate as did the corresponding June 4 neutron observation (see Fig. 7.38), this factor would be higher, which implies that the value of 2.2 is actually a lower limit.

Then Murphy et al. (1999) compare this neutron efficiency ratio of the two flares, $\eta_{\mathrm{June}\ 1}/\eta_{\mathrm{June}\ 4}$. = 2.2 $\pm$ 0.5 with predicted ratios of thin- and thick-target neutron efficiencies, $\eta_{\mathrm{thin}}/\eta_{\mathrm{thick}}$. for power-law accelerated particle spectra. For the ambient material Murphy et al. (1999) assume a coronal composition (Reames 1995). Gamma-ray measurements of other flares (Murphy et al. 1991, 1997; Ramaty et al. 1995, 1996) have shown that the ambient flare material typically has the low first ionization potential elemental abundance enhancements found in the corona and solar wind. For the accelerated particles, Murphy et al. (1999) assume the mean composition (Ramaty et al. 1996) of impulsive flares observed in interplanetary space (with accelerated α/p ratios of either 0.1 or 0.5). This impulsive-flare composition is considerably different from photospheric or coronal abundances, having significant enhancements of Ne, Mg, Si, and Fe, in addition to ^{3}He/α = 1. In the analysis of the June 4 flare (Murphy et al. 1997) was found that such enhanced high-Z accelerated particle composition is necessary to achieve agreement between the two techniques of determining accelerated particle spectral indexes: the neutron capture ^{12}C 4.44 MeV line fluence ratio and the ^{16}O 6.13 MeV–^{20}Ne 1.63 MeV line fluence ratio. This was consistent with the April 27, 1981 flare analysis of Murphy et al. (1991) and the 19 SMM/GRS flares analysis of Ramaty et al. (1996). In their analysis of the Granat/PHEBUS γ-ray data from the June 1 flare, Ramaty et al. (1997) also found that the accelerated particles showed

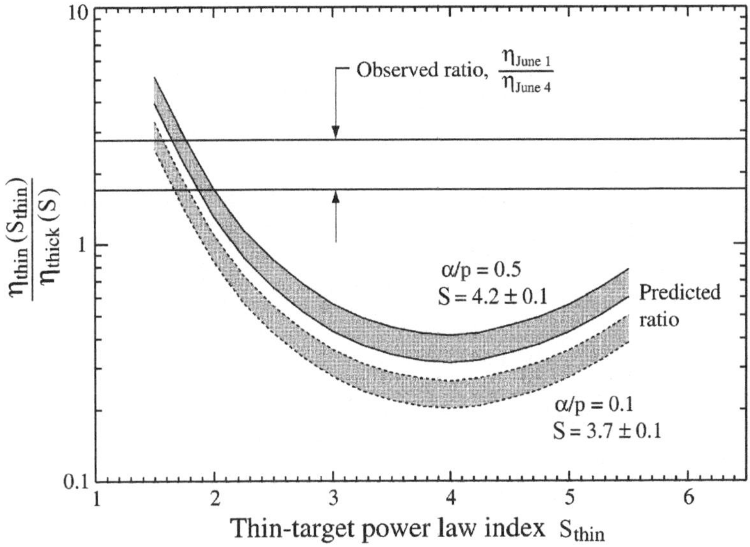

Fig. 7.39 Calculated ratios of thin-target to thick-target neutron efficiencies as a function of accelerated particle spectral index S_{thin} calculated at either of two thick-target spectral indexes S derived by Murphy et al. (1997) for the June 4 flare: $S = 4.2 \pm 0.1$ (assuming $\alpha/p = 0.5$; *full lines*) and $S = 3.7 \pm 0.1$ (assuming $\alpha/p = 0.1$; *dotted lines*). Also shown is the observed June 1 to 4 June efficiency ratio, $\eta_{June\ 1}/\eta_{June\ 4}$. From Murphy et al. (1999)

heavy element enhancements. Murphy et al. (1999) use the calculations of Ramaty et al. (1996) for the thick-target neutron yield. For the thin-target neutron yields and the thin- and thick-target 4–7 MeV nuclear γ-ray yields, Murphy et al. (1999) use results of new calculations. Shown in Fig. 7.39 is the predicted ratio $\eta_{thin}(S_{thin})/\eta_{thick}(S)$, where $\eta_{thin}(S_{thin})$ is the thin-target neutron efficiency calculated as a function of accelerated particle spectral index S_{thin}, and $\eta_{thick}(S)$ is the thick-target neutron efficiency calculated at either of two spectral indexes derived by Murphy et al. (1997) for the June 4 flare: $S = 3.7 \pm 0.1$ and $S = 4.2 \pm 0.1$. These June 4 indexes were derived using the measured fluence ratio of the 2.223 MeV neutron capture and 4.44 MeV ^{12}C lines, and the thick-target calculations of Ramaty et al. (1996). The harder index ($S = 3.7$) was derived assuming $\alpha/p = 0.1$, and the softer index ($S = 4.2$) was derived assuming $\alpha/p = 0.5$. The lower value of α/p results in a lower neutron efficiency that must be balanced by a harder energetic particle spectrum.

In Fig. 7.39 the observed June 1 to June 4 neutron efficiency ratio $\eta_{June\ 1}/\eta_{June\ 4}$ is compared with the predicted efficiency ratios. This comparison shows that thin-target interactions of particles can produce sufficient neutrons only if the accelerated-particle spectrum is extremely hard, with an index less than 1.8 (for $\alpha/p = 0.1$, $S = 3.7$) and less than 2.0 (for $\alpha/p = 0.5$, $S = 4.2$).

7.20.6 Summary and Conclusions

Murphy et al. (1999) summarized obtained results on the flare June 1, 1991 event and results on comparing with the flare June 4, 1991 event as following:

1. Because the magnetic loop foot-points associated with the June 1 behind-the-limb flare were hidden from view, the nuclear γ-ray emission observed (Barat et al. 1994) with Granat/PHEBUS has been interpreted as originating from the corona over the active region.
2. Furthermore, because of the exceptionally high ratio of the 1.1–1.8 MeV to 4.1–7.6 MeV excesses observed from this flare, the emission is thought (Ramaty et al. 1997) to have resulted from thin-target interactions of accelerated particles with impulsive-flare abundances, which lends support to the coronal hypothesis.
3. CGRO/OSSE also observed neutrons from this flare. Neutron production in flares is rare, even from the thick-target interactions of disk flares. Since thin-target interactions are even less efficient than thick-target interactions in producing neutrons, the strong neutron signal observed from this flare is surprising.
4. It was compared the neutron counts observed by OSSE and the nuclear γ-ray fluence observed by Granat/PHEBUS from the June 1 flare with the same quantities observed with OSSE alone from the June 4 flare that occurred on the disk.
5. The comparison establishes a neutron-production efficiency for the June 1 thin-target flare relative to a known thick-target disk flare.
6. This relative efficiency was compared with predicted ratios of thin-target neutron efficiencies (calculated at a variety of spectral indexes) to thick-target efficiencies (calculated at the spectral indexes determined previously for the June 4 flare; Murphy et al. 1997).
7. The calculations were performed assuming average impulsive-flare abundances (Ramaty et al. 1996) for the accelerated particles, and coronal abundances (Reames 1995) for the ambient medium.
8. The comparison in Fig. 7.39 shows that thin-target interactions of accelerated particles are not adequate to account for the number of neutrons observed from the June 1 flare unless the accelerated-particle spectrum is extremely hard, with a power-law index $S_{thin} < 2.0$.
9. In the thin-target analysis of the Granat/PHEBUS 1.1–1.8 and 4.1–7.6 MeV γ-ray observations of the June 1 flare, Ramaty et al. (1997) obtained a spectral index as hard as $S_{thin} \approx 2.5$ for the 10–30 MeV accelerated particles (assuming average impulsive-flare abundances), and possibly harder than 2 (assuming the highest observed heavy-element enhancements).
10. The described analysis of the OSSE neutron data has shown that such a hard spectrum continues to energies greater than 50 MeV.
11. For the first three of the intense flares of 1991 June (June 1, 4, and 6), the Sun was outside the field of view of the EGRET spark chamber. After the target of

opportunity was declared, CGRO was reoriented, which allowed the spark chamber to observe the June 9 and 11 flares.

12. The June 11, 1991 flare in particular was remarkable in that the high-energy γ-ray emission continued for 8 h (Kanbach et al. 1993).
13. For both the June 1, 1991 and June 4 flares, the OSSE > 15 MeV data indicate that by 750 s after the peak of the nuclear γ-ray emission, the high-energy γ-ray emission had fallen to below detectability.
14. If these flares had produced high-energy γ-ray emission as intense and temporally extended as that observed by EGRET from the June 11 flare, OSSE would have detected it at a significance of 8 σ from the June 4 flare, and less than 2 σ from the June 1 flare (due to the reduced 100° off-pointed sensitivity).
15. Also, the spectrum of the June 4 emission at the peak of the flare was dominated by electron bremsstrahlung (DelSignore 1995; Murphy et al. 1997) rather than pion-decay radiation as in the June 11 flare.
16. Thus the high-energy emission of at least the June 4 flare does not appear to be similar to that of the June 11 flare.
17. Because no high-energy γ-ray data are available during the peak of the June 1 flare, it is not known whether such emission, if present, was due to electron bremsstrahlung or pion decay.
18. The very hard accelerated particle spectrum derived for this flare, however, would suggest a strong pion-decay component.

7.21 Solar Neutron Event on June 4, 1991

7.21.1 Main Properties of the Flare and Satellite Observations of 2.223 MeV Gamma-Ray Line and 17 GHz Microwave Intensity

As was mentioned by Muraki et al. (1991b), on June 1, 1991, a very active region appeared on the solar surface (NOAA 6659) and it moved on to June 17, 1991 to the opposite side of the Sun. Within this duration six large solar flares with x-ray class >10 were observed by the GOES x-ray detector, on June 1st (at 15:09 UT start and at 15:29 UT maximum), June 4th (03:37 and 03:42 UT), June 6th (00:54 and 01:12 UT), June 9th (01:37 and 01:40 UT), June 11th (02:09 and 02:29 UT) and June 15th (06:33 and 08:31 UT). The four flares (June 4, 6, 9, 11) were observed near noon on Japanese local time.

The solar neutron event of June 4, 1991 ($\lambda_s = 22.5°N$, $fsubs = 125°E$) was detected at Mt. Norikura ($(\theta_s = 18.6°,\ h_s = 775\,g/\mathrm{cm}^2)$) by the solar neutron telescope (effective area 1 m^2) and other neutron and muon detectors in Japan. This event was caused by a large solar flare 3B/X12, which occurred at 30°N, 70°E on the Sun. The x-ray burst monitor on Ginga satellite detected the start of x-ray flux at 03:37:32 UT (the GRO satellite observed the gamma-ray burst at

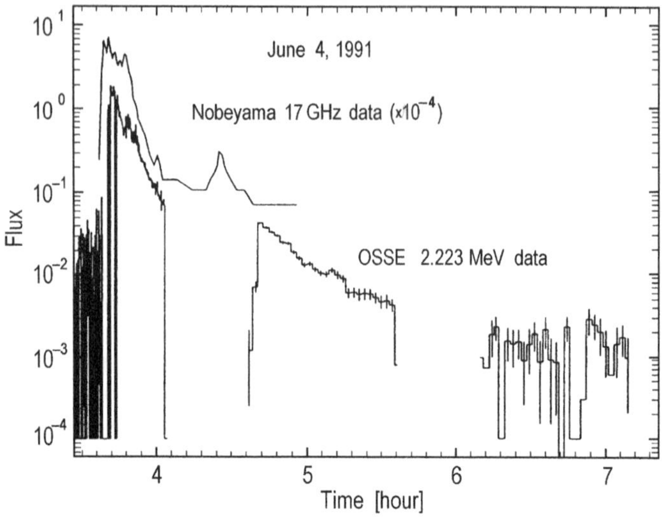

Fig. 7.40 Solar neutron event of June 4, 1991. Combination of 2.223 MeV gamma-ray line intensity as observed by OSSE on CGRO (according to Murphy et al. 1993) and as observed by Nobeyama 17 GHz microwave intensity. From Struminsky et al. (1994)

03:41:00 UT), also a radio-burst began at 03:37:30 UT and reached maximum at 03:41:10 UT

This event is shown in Fig. 7.40 as observed by the satellite GRO (time profile of 2.223 MeV line intensity) and Nobeyama (17 GHz microwave intensity).

7.21.2 Observations by Solar Neutron Telescope at Mt. Norikura and Estimation of Solar Neutron Fluency at the Top of the Atmosphere

Figure 7.21 shows the data of the solar neutron telescope (for the channels 0 and 3; 50–360 MeV). According to Fig. 7.41, the highest peak (about 5σ) was observed between 03:46 and 03:49 UT (these small energy neutrons must be produced near the start time of solar flare, 03:37 UT). A 3σ excess was also detected in an intermediate energy channel (280–500 MeV), but solar neutrons with energy 330 MeV arrive at Earth only 4 min later then light. Therefore Muraki et al. (1991b) conclude that these neutrons were not produced at the start time (03:37 UT) but at the solar flare maximum time (03:42 UT).

On the basis of the solar neutron telescope data at Mt. Norikura, Shibata et al. (1993) determined the integrated for all time of event June 4, 1991 solar neutron fluency at the top of the atmosphere by equation:

$$F(E_n) = (4.6^{+8.7}_{-3.0}) \times 10^5 \, (E_n / 100\,\text{MeV})^{-4.9\pm1.7} \qquad (7.5)$$

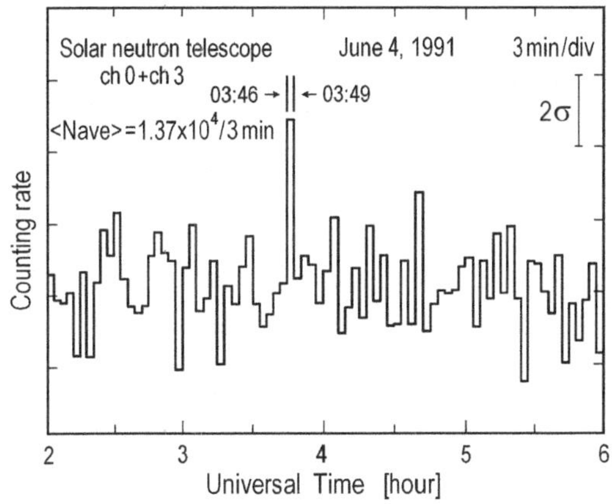

Fig. 7.41 The solar neutron event of June 4, 1991. Time profiles of counting rates detected by Mt. Norikura solar neutron telescope. According to Muraki et al. (1991b)

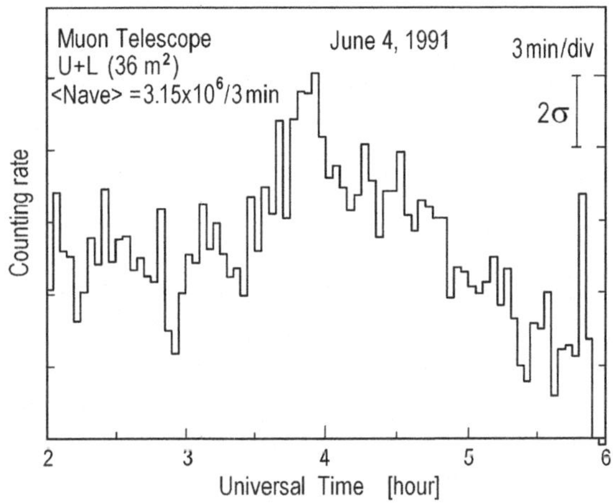

Fig. 7.42 The solar neutron event of June 4, 1991. Time profile of counting rate of muon telescope at Nagoya. According to Muraki et al. (1991b)

7.21.3 Observations of Muons Generated in the Earth's Atmosphere by Solar Neutrons

Figure 7.42 shows the time-profile of the counting rate of the muon telescope at Nagoya (effective area 36 m^2). These data shows that in this event high energy neutron can generate muons in high altitude layers of the Earth's atmosphere, which can reach ground level and can be detected by muon telescopes.

According to Shibata et al. (1993), from muon telescope data follows that;

$$F(E_n) = \left(4.6^{+0.9}_{-0.8}\right) \times 10^5 (E_n/100 \text{ MeV})^{-6.9\pm0.5} \text{neutrons} \times \text{MeV}^{-1} \text{ m}^{-2}. \quad (7.6)$$

Important evidence of muon generation by solar neutrons associated with the solar flare on June 4, 1991 was obtained by observations on Akeno Giant Air

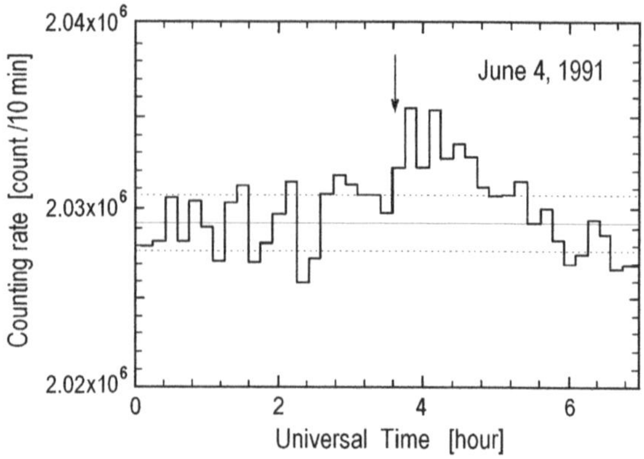

Fig. 7.43 The solar neutron event of June 4, 1991. Variations of counting rates summed over 83 scintillators (182.6 m^2). According to Chiba et al. (1992a, b)

Shower Array (about 900 m above sea-level). The total area of scintillators used by Chiba et al. (1992a, b) for this analysis was 182.6 m^2 and the excess count rate was found as $0.35\,^{+0.05}_{-0.10}\,per\;m^2\,s$ (see Fig. 7.43).

Chiba et al. (1992a, b) came to conclusion that the excess signals in the scintillators can be interpreted as muon production by solar neutrons of energies 5–10 GeV and the average flux in this energy interval is expected 0.2–1.0 per m^2 s for more than 20 min after the start of the solar flare.

7.21.4 Observations by Neutron Monitors in Japan

According to Takahashi et al. (1991), the Mt. Norikura neutron monitor was enlarged by three times and by this the solar neutron signal to cosmic ray noise ratio have been improved by a factor of 2. A new recording system can record events with neutron multiplicities of 1, 2, 3, 4, 5 and $\geq$6, connected with neutron energy. Figure 7.44 shows the time-profile for the Mt. Norikura neutron 5-min counting rate on June 4, 1991.

From Fig. 7.44 can be seen that there are two large peaks: at 03:40–04:10 and 04:15–05:05 UT. For the first peak the maximum of the enhancement 03:50–03:55 is 9.6σ (from 10-s data Takahashi et al. 1991 estimated that the onset time of the neutron burst was at 03:43 UT). Comparison between total counting rates at Mt. Norikura and Tokyo, with different h_s is shown in Fig. 7.45 (correlation).

7.21.5 Determination of the Solar Neutron Attenuation Mean Free Path in the Earth's Atmosphere

Results of determination of attenuation mean free path of solar neutrons for the first and second peak (see above, Fig. 7.44) are shown in Fig. 7.46.

It was found that for the first peak the attenuation mean free path of solar neutrons are 120 g/cm^2 and for the second peak 100 g/cm^2.

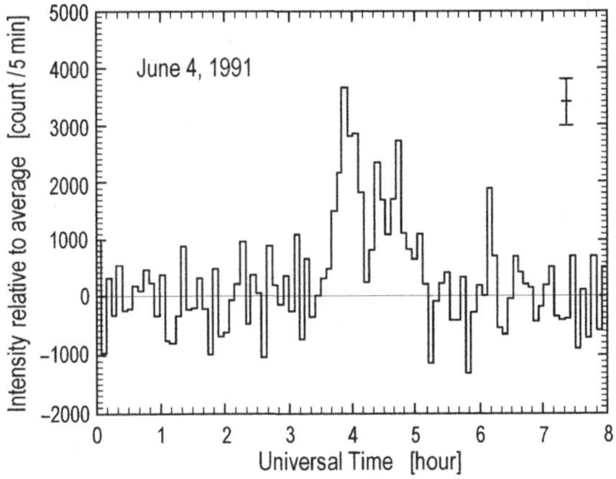

Fig. 7.44 The solar neutron event of June 4, 1991. Time profile for the Mt. Norikura neutron monitor 5-min counting rate relative to average. According to Takahashi et al. (1991)

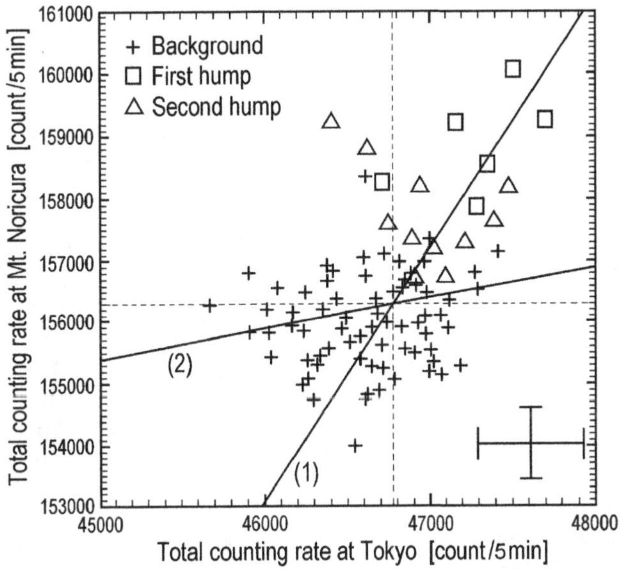

Fig. 7.45 The solar neutron event of June 4, 1991. Correlation between total neutron counting rates at Mt. Norikura and at Tokyo. **1** and **2** – the best fit lines for the enhancement and for background. According to Takahashi et al. (1991)

7.21.6 Energy Spectrum of Solar Neutrons on the Top of the Atmosphere

By using the neutron monitor sensitivity to solar neutrons obtained by Debrunner et al. (1989) it is possible to determine the energy spectrum of solar neutrons, expected on the top of the atmosphere (Fig. 7.47).

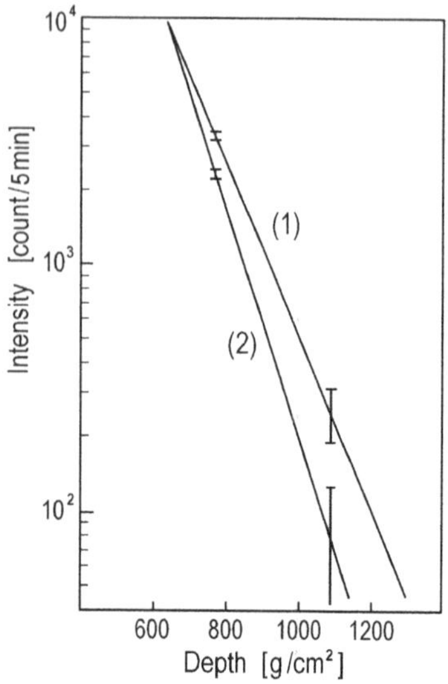

Fig. 7.46 The solar neutron event of June 4, 1991. Determination of the attenuation mean free pah for the first hump (1) and for the second hump (2) on the basis of neutron monitor data at Mt. Norikura and Tokyo. According to Takahashi et al. (1991)

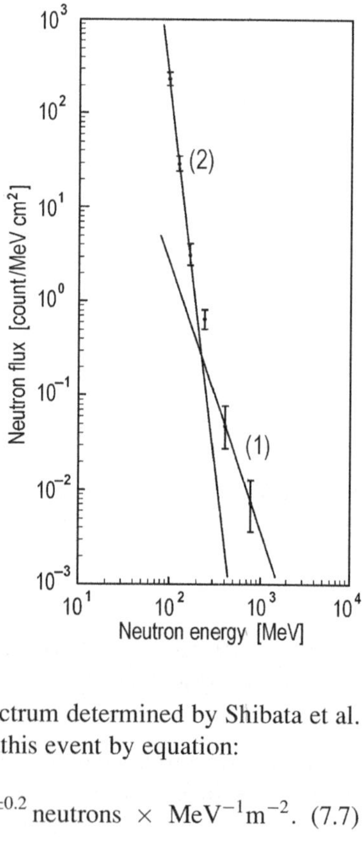

Fig. 7.47 Solar neutron event of June 4, 1991. The spectrum of solar neutrons on the top of the atmosphere determined on the basis of neutron monitor data. The two lines, 1 and 2, have power-law indices of 2.7 and 7.9, respectively. According to Takahashi et al. (1991)

This spectrum can be compared with the spectrum determined by Shibata et al. (1993) on the basis of neutron monitor data for this event by equation:

$$F(E_n) = (3.2 \pm 0.2) \times 10^5\,(E_n/100\text{ MeV})^{-7.0\pm0.2}\,\text{neutrons} \times \text{MeV}^{-1}\text{m}^{-2}. \quad (7.7)$$

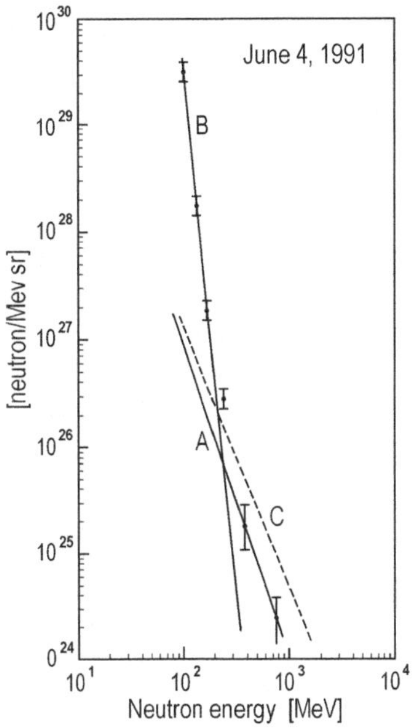

Fig. 7.48 The solar neutron event of June 4, 1991. The time integrated directional solar neutron emissivity spectrum at the Sun. The lines A and B have power-law indices of 3.0 and 7.1, respectively. The line C is the result for the solar neutron event of June 3, 1982. According to Takahashi et al. (1991)

7.21.7 *Emissivity Spectrum of Solar Neutrons on the Sun*

The obtained spectrum on the top of the atmosphere (Fig. 7.47) can be recalculated to the Sun, with taking into account the neutron decay (see Fig. 7.48), where the obtained spectrum is shown in comparison with the spectrum for the solar neutron event on June 3, 1982 obtained by Chupp et al. (1987).

7.21.8 *Possible Time-Profile of Solar Neutron Generation on the Sun and Energy Spectrum at the Source*

A detailed investigation of possible time-profile of solar neutron generation on the Sun and energy spectrum at the source was made by Struminsky et al. (1994), on the basis of assumption that the total neutron production on the Sun in the energy range 0.1−3.0 GeV, has the same time-profile as for the gamma-ray 2.223 MeV line intensity (keep in mind that this line is delayed by about 100 s according to Ramaty and Kozlovsky 1975). The production time-profile of neutrons is supposed in the form:

$$N(E_n, t) = N_0(\alpha) E_n^{-\alpha} \Gamma(t), \tag{7.8}$$

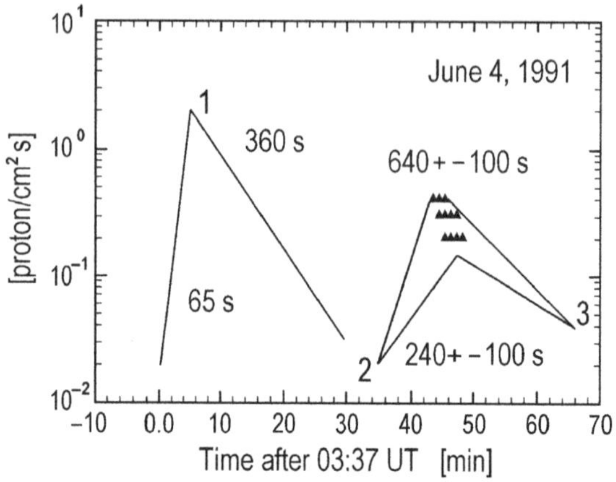

Fig. 7.49 The solar neutron event of June 4, 1991. The approximation of OSSE CGRO gamma-ray 2.223 MeV intensity at Earth orbit during the satellite day time and expected intensity during night time: (1) the first maximum 2 photons·cm^{-2} s^{-1} at 03:42 UT; (2) the start of the second enhancement 0.02 photons·cm^{-2} s^{-1} at 04:12 UT; (3) the beginning of second orbit 0.04 photons·cm^{-2} s^{-1} at 04:42 UT; *Triangles* mark the possible position of the second maximum. According to Struminsky et al. (1994)

where t is sunlight delayed time at Earth's orbit, $\Gamma(t)$ is the time-profile of the 2.223 MeV gamma-ray line intensity at the Earth's orbit (approximation of $\Gamma(t)$ from Fig. 7.40 is shown in Fig. 7.49).

The parameters α and $N_0(\alpha)$ can be different for the first and second period of generation and are determined by comparison of expected neutron monitor enhancement $\Delta I(t_1, t_2)$ during the time interval (t_1, t_2) caused by neutron production function (Section 7.21.4) using the expression

$$\Delta I(t_1,t_2) = r^{-2}\int_{t_1}^{t_2} dt \int_{0.1\,\mathrm{GeV}}^{3\,\mathrm{GeV}} dE_n N\left(E_n, t - \frac{r}{v_n}\right) P_s(E_n, r) S(E_n), \qquad (7.9)$$

where r is the Sun–Earth distance,

$$v_n = c\left(1 - \left(m_n c^2/(E_n + m_n c^2)\right)^2\right)^{1/2}, \qquad (7.10)$$

is the velocity of neutrons,

$$P_s(E_n, r) = \exp\left(-\frac{r/v_n}{\tau_n (E_n + m_n c^2)/m_n c^2}\right) \qquad (7.11)$$

is the neutron survival probability (see Section 1.1.4), τ_n is the time decay of the rest neutrons, $S(E_n)$ is the sensitivity of the neutron monitor (used were two types of $S(E_n)$ determined by Debrunner et al. 1989 and by Shibata 1994; see details on

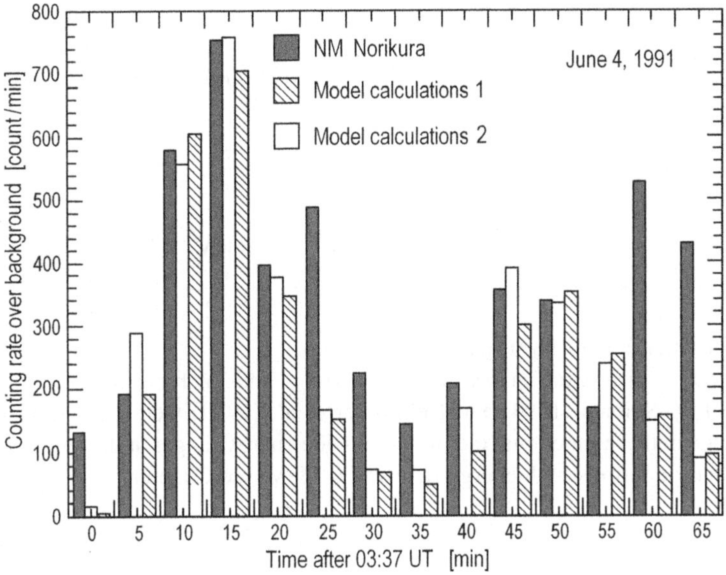

Fig. 7.50 The solar neutron event of June 4, 1991. Comparison of Mt. Norikura neutron monitor data and model calculations using sensitivity of neutron monitor to solar neutrons from: (1) Shibata (1994), (2) Debrunner et al. (1989). According to Struminsky et al. (1994)

calculations of $S(E_n)$ in Chapter 5). The best fit analysis of Mt. Norikura neutron monitor count rates, using a χ^2-test and linear correlation should give the possibility to determine the unknown parameters α and $N_0(\alpha)$ for each neutron production period. The comparison is shown in Fig. 7.50. For Shibata (1994) the sensitivity function best fit with correlation coefficient 0.97 and $\chi^2 = 0.76$ gives $\alpha_1 = 3.5$ and $N_0(\alpha_1) = 7.0\,10^{23}$ neutrons (s·sr·GeV)$^{-1}$ for the first period of generation and $\alpha_2 = 2.25 \pm 0.25$, $N_0(\alpha_2) = 6.7\,10^{24}$ neutrons (s·sr·GeV)$^{-1}$ for the second generation. For Debrunner et al. (1989) sensitivity function were obtained correlation coefficient 0.98, $\chi^2 = 0.62$, $\alpha_1 = 5.25$, $N_0(\alpha_1) = 5.5\,10^{23}$ neutrons (s·sr·GeV)$^{-1}$ and $\alpha_2 = 3.75 \pm 0.25$, $N_0(\alpha_2) = 11.3\,10^{24}$ neutrons (s·sr·GeV)$^{-1}$.

The time integrated directional emissivity $J(E_n)$ at the Sun and the total energy W, of neutrons in the 0.1–3.0 GeV energy band for the first period of generation are found:

$$J_{1S}(E_n) = .1 \times 10^{27} E_{\mathrm{n}}^{-3.5}\ \text{neutrons} \cdot \text{sr}^{-1} \cdot \text{GeV}^{-1};\quad \text{W}_{1S} = 9.1 \times 10^{27}\ \text{ergs} \tag{7.12}$$

for Shibata (1994) sensitivity function, and

$$J_{1D}(E_n) = 8.3.10^{26} E_{\mathrm{n}}^{-5.25}\ \text{neutrons} \cdot \text{sr}^{-1} \cdot \text{GeV}^{-1};\quad \text{W}_{1D} = 4.6 \times 10^{26}\ \text{ergs} \tag{7.13}$$

for Debrunner et al. (1989) sensitivity function.

For the second period of generation Struminsky et al. (1994) obtained

$$J_{2S}(E_n) = \left(2.7^{+0.6}_{-0.5}\right) \times 10^{27} E_{\mathrm{n}}^{-2.25\pm0.25}\,\mathrm{neutrons}\cdot\mathrm{sr}^{-1}\cdot\mathrm{GeV}^{-1};\ \mathrm{W}_{2\mathrm{S}} = 3.0\times10^{27}\,\mathrm{ergs}, \quad (7.14)$$

for Shibata (1994) sensitivity function, and

$$J_{2D}(E_n) = \left(4.7^{+0.4}_{-0.8}\right) \times 10^{27} E_{\mathrm{n}}^{-3.75\pm0.25}\,\mathrm{neutrons}\cdot\mathrm{sr}^{-1}\cdot\mathrm{GeV}^{-1};\ \mathrm{W}_{2\mathrm{D}} = 0.22\times10^{27}\,\mathrm{ergs} \quad (7.15)$$

for Debrunner et al. (1989) sensitivity function.

The great difference in the results of $J(E_n)$ and W for sensitivity functions obtained by Shibata (1994) and Debrunner et al. (1989) were caused by the big difference in these functions (factor of 4 in wide energy ranges and even much more for primary neutrons with energy below 200 MeV; see Chapter 5).

7.21.9 CGRO Satellite and Neutron Monitor Observations of Solar Neutrons

According to Watanabe et al. (2007), in association with the X12.0 flare on June 4, 1991 solar neutrons were observed in space by OSSE onboard the CGRO satellite (see in detail above, Chapter 4, Section 4.6) and by ground-based detectors, such as the 12 m^2 NM at Mt. Norikura (Japan). The γ-ray lines were also observed by CGRO/OSSE, and it can be use the 4.4 MeV line time histories as the ion acceleration release time history. Using these γ-ray line emissions, Murphy et al. (1997, 2007) calculated predicted time-dependent neutron spectra arriving at Earth using the solar-flare are magnetic-loop transport and interaction model of Hua et al. (2002). Using the OSSE neutron response function, they compared predicted count rates with the observed OSSE count rates and found that the upper limit for solar neutron energy was 125 MeV (see above, Section 4.6).

Watanabe et al. (2007) compare solar neutron signals observed by the Norikura neutron monitor with fluxes predicted from the gamma ray observation. The parameter that can be changed from Muraki et al. (2007) is the upper cutoff energy of the accelerated ions. When to use E_C = 125 MeV, then predicted profile significantly underestimates the observed neutron-monitor count rate. Thus, at first, Watanabe et al. (2007) recalculate the cutoff energy for the OSSE neutron data. When E_C = 210 MeV, predicted neutron profile is well fitted to the observed data of OSSE as shown in Fig. 7.51, and the reduced χ^2 is 1.18.

Next, Watanabe et al. (2007) calculate the resulting neutron-monitor count rates due to these arriving neutron spectra by using the solar neutron atmospheric attenuation ratio from the Shibata (1994) program and the neutron monitor efficiency calculated by Clem and Dorman (2000). Watanabe et al. (2007) compare the

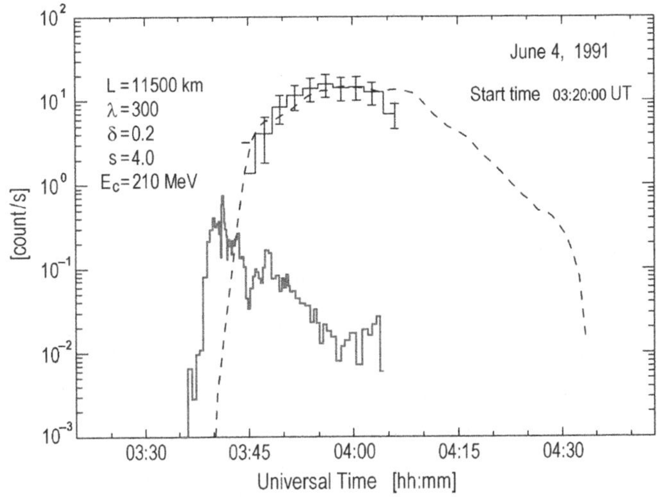

Fig. 7.51 The observed and predicted neutron time histories by the OSSE on June 4, 1991. The *black solid line* is the observed counting rates from OSSE. The *dashed line* represent the predicted result of Hua et al. (2002) program with cutoff energy 210 MeV. The *red line* represents the 4.4 MeV line time history as the ion acceleration release time history. From Watanabe et al. (2007a)

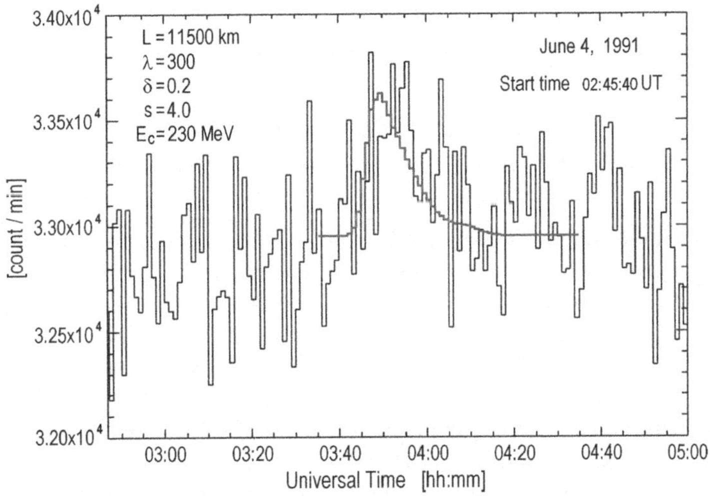

Fig. 7.52 The observed and predicted neutron time histories by the Norikura neutron monitor on June 4, 1991. The *black line* is the observed 1-min counting rates from the Norikura neutron monitor. The *red line* represents the predicted result of Hua et al. (2002) program with cutoff energy 230 MeV. From Watanabe et al. (2007)

predicted count rate with the observed count rate in Fig. 7.52. When $E_C = 230$ MeV, the predicted neutron profile is well fitted to the observed neutron monitor data as shown in Fig. 7.52, and the reduced χ^2 is 0.78.

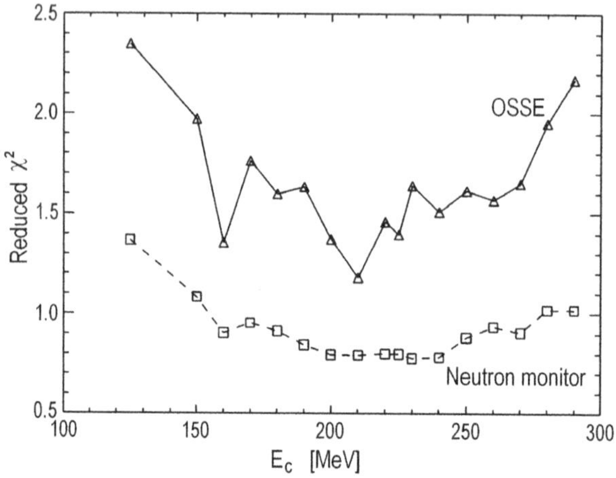

Fig. 7.53 The reduced χ^2 distribution for the fit to the OSSE and NM solar neutron data for each cutoff energy E_C. From Watanabe et al. (2007)

From the reduced χ^2 distribution of this fit as shown in Fig. 7.53, Watanabe et al. (2007) obtain errors of these cutoff energy as $E_C = 210 \pm 10$ MeV for fitting to the OSSE data, and as $E_C = 230^{+20}_{-50}$ MeV for the neutron monitor data.

7.22 Solar Neutron Event on June 6, 1991

7.22.1 Observations by the Solar Neutron Telescope at Mt. Norikura

As was mentioned above (Section 7.21), this event was caused by solar flare from the same active region, NOAA 6659 which generate discussed above the event of June 4, 1991. According to GOES x-ray observation this solar flare started at 00:54 UT and reached maximum in x-ray intensity at 01:12 UT on June 6, 1991 (for this event $\lambda_s = 21.7°N$, $f\text{subs} = 162°E$).). According to Muraki et al. (1991b), the solar neutron telescope at Mt. Norikura ($\theta_s = 27.9°, h_s = 830.7\,\text{g/cm}^2$) detect the increase in the counting rate with amplitude 3.3σ (33 ± 10%) in the highest energy channel ($E_n \geq$ 390 MeV) at the time 01:12–01:15 UT (see Fig. 7.54).

Muraki et al. (1991b) mentioned that a significant excess was also observed at Mt. Norikura by the muon telescope at the time 01:15–01:21 UT, June 6th, 1991.

7.22.2 Comparison of Satellite Gamma Ray and NM on Mt. Norikura and Mt. Haleakala Observations

According to Watanabe et al. (2003a), during the event of June 6, 1991 solar neutrons have been detected simultaneously by the neutron monitor located at

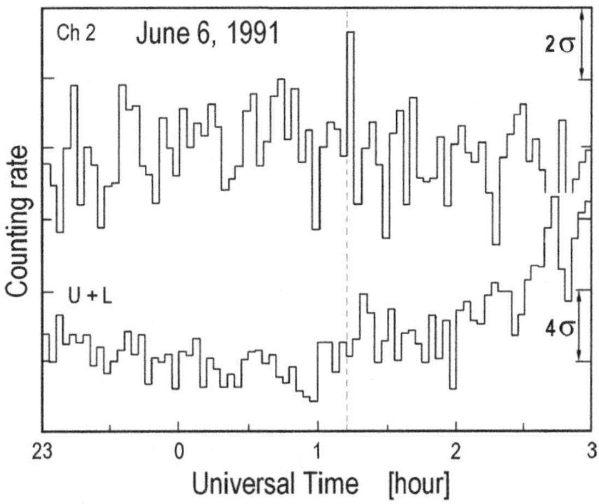

Fig. 7.54 The solar neutron event of June 6, 1991. Time profiles of counting rates for channels of solar neutron telescope at Mt. Norikura. According to Muraki et al. (1991b)

Mt. Norikura, in Japan, and Mt. Haleakala, in Hawaii. The Mt. Norikura station is located at 36.1°N, 137.5°E, and 2,770 m above sea level, where the vertical air mass is 730 g/cm^2. At the flare start time, the zenith angle of the Sun was 26.0° and the air mass along the line of sight to the Sun was 812 g/cm^2. The neutron monitor installed at Mt. Norikura is 12 NM-64. The counting rate is recorded every 10 s. The Mt. Haleakala station is located at 20.7°N, 203.7°E, and 3,030 m above sea level, and the vertical air mass is 707 g/cm^2. At the flare start time, the zenith angle of the Sun was 44.5° and the air mass along the line of sight to the Sun was 991 g/cm^2. The neutron monitor installed at Mt. Haleakala is 18 NM-64. The counting rate is recorded every 1 min.

The time profiles of neutrons observed by the two neutron monitors are shown in Fig. 7.55. A clear excess was found between 1:12 and 1:42 UT at Mt. Norikura, and between 1:12 and 1:27 UT at Mt. Haleakara. At Mt. Norikura, the statistical significance of the strongest excess is 3.96σ during 1:12–1:17 UT, and the total significance for 30 min between 1:12 and 1:42 UT, is 5.16σ. At Mt. Haleakara, the statistical significance of the strongest excess is 3.03σ during 1:22–1:27 UT, and the total significance for 15 min between 1:12 and 1:27 UT, is 4.28σ.

Watanabe et al. (2003a) noted that, in principle, there is a possibility that these excesses came from energetic ions because the NM can also detect energetic ions. But, there is no evidence that the enhancement was produced by energetic ions since the measurements by the other stations in the worldwide network of neutron monitors and the proton channel of GOES satellite showed no enhancement. Therefore, these signals shown in Fig. 7.55 must have come from solar neutrons.

As underlined Watanabe et al. (2003a), an X12.0 class solar flare occurred at 0:54 UT in NOAA region 6659 on June 6, 1991. The location of the active region was N33°, E44°. In this flare intense emission of 1–10 MeV gamma rays and 80 GHz mm-waves was observed by the Burst and Transient Source Experiment (BATSE) onboard the Compton Gamma Ray Observatory (CGRO) and the

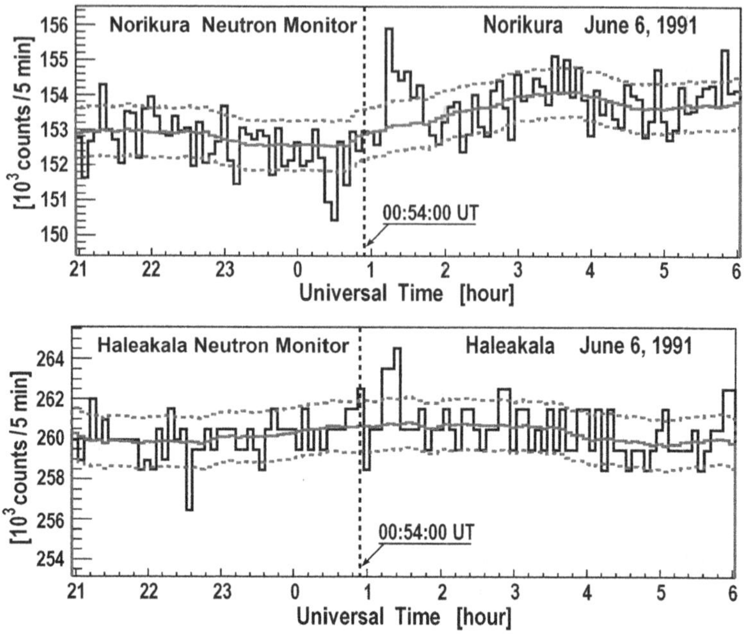

Fig. 7.55 The time profile of neutrons detected by two neutron monitors on June 6, 1991. The *vertical axis* is the counting rate per 5 min. The *solid smooth line* is the averaged background, and *dashed lines* are $\pm 1\sigma$ from the background. The upper figure is the data from Norikura neutron monitor, and the lower one is the data from Haleakala. According to Watanabe et al. (2003a)

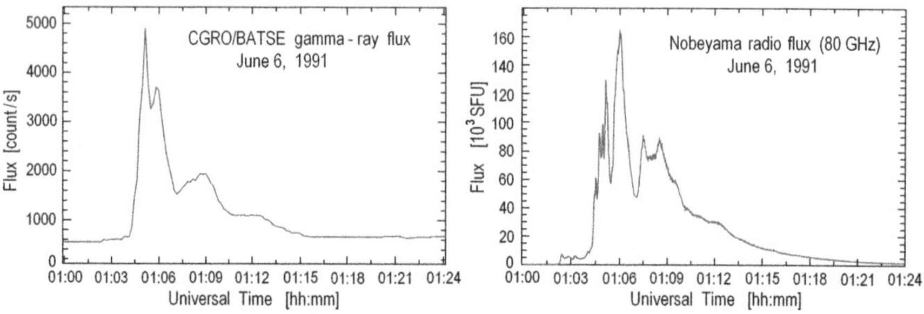

Fig. 7.56 The time profile of 1 – 10 MeV gamma ray flux observed by the CGRO/BATSE (left), and the time profile of 80 GHz millimeter-waves intensity observed by the NoRP (right) on June 6, 1991. According to Watanabe et al. (2003a)

Nobeyama Radio Polarimeters (NoRP) at the Nobeyama Radio Observatory, in Japan, respectively (see Fig. 7.56). In the γ-ray data, two large peaks are seen at 1:05 and 1:06 UT, and one small peak is seen around 1:09 UT. In the 80 GHz data, several intense emissions are seen, the largest peak is 1:06 UT, which is the same time as of gamma ray peak.

According to Watanabe et al. (2003a), if solar neutrons were produced at 1:05 or 1:06 UT instantaneously, then the energy of the neutrons detected by the Norikura neutron monitor between 1:12 and 1:42 UT is calculated to be 208–16 MeV. The energy of the neutrons detected by the Haleakala neutron monitor between 1:12 and 1:27 UT is calculated to be 215–40 MeV. Since neutrons suffer violent attenuation in the Earth's atmosphere, low energy neutrons, less than 50 MeV, cannot arrive at the ground (Shibata 1994; see also Chapter 5). Hence, the excesses detected by the Norikura neutron monitor between 1:27 and 1:42 UT cannot be explained by this assumption. Therefore, for this event, it is necessary to consider the possibility of the extended production of neutrons. Watanabe et al. (2003a) conclude that since in Fig. 7.55, the excess flux at the Norikura neutron monitor continued for 30 min, it is fairly probable that solar neutrons were produced continuously during the production of gamma rays and radio emissions.

7.23 Solar Neutron Events on November 6, 1997

7.23.1 Observations at Chacaltaya by the Solar Neutron Telescope from the Flare X9.4/2B at 11:49 UT

According to Matsubara et al. (1993), a new type of solar neutron detector has been in operation since September 1992 at Mt. Chacaltaya, 5,250 m above sea level and near the equator (16°21′S and 68°08′W). The height and the location of Chacaltaya provide excellent conditions for detecting solar neutrons. The main detector in this experiment consists of four 1 m^2 plastic scintillators with a thick- ness of 40 cm. Neutrons are detected when protons are produced inside plastic scintillators by (n, p) reactions. The pulse height obtained by each photomultiplier is discriminated with four threshold levels corresponding to the energy of a recoil proton of 40, 80, 120, and 160 MeV. Single counts of these four channels in each scintillator are recorded every 10 s. Typical counts in units $(10\ \mathrm{s}\ 4\ \mathrm{m}^2)^{-1}$ are 40,000 (>40 MeV), 22,000 (>80 MeV), 12,000 (>120 MeV), and 5,100 (>160 MeV) respectively. The anticounter system to veto charged particles consists of 17 scintillators, each of which has an area of 220 46 cm and a thickness of 1 cm. The top of the main detector is covered by five antiscintillators and each side by three antiscintillators. The backgrounds due to charged particles are reduced to one half of the above values at lower energies and one third at higher energies. A schematic view of the detector is shown in Fig. 7.57.

As noted Matsubara et al. (1999), the Sun has become active since the end of 1997, and three X-class flares were detected in November, 1997. The solar flare which occurred on November 6, 1997, was the largest among all those that occurred during solar cycle 23. The maximum flux of x-rays was observed at 11:55 UT, which corresponds to 7:55 am in Bolivia. Neutrons with an energy of >100 MeV are expected to be attenuated by a factor of 10^{-4} when they are observed by the detector

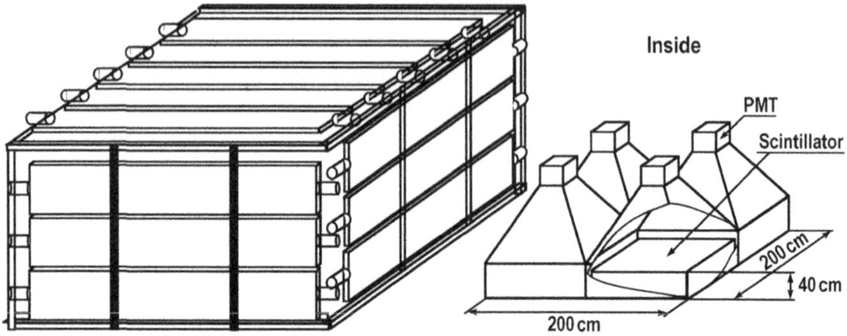

Fig. 7.57 A schematic view of the solar neutron telescope at Chacaltaya. From Matsubara et al. (1999)

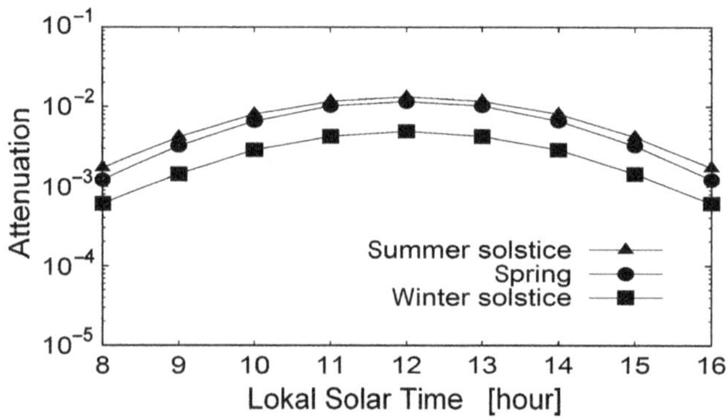

Fig. 7.58 Attenuation of neutrons with energy >500 MeV at Chacaltaya according to calculations by Tsuchiya et al. (1999). From Matsubara et al. (1999)

at Chacaltaya, if neutrons propagate in the atmosphere undeflected. However, if a systematic scattering effect into the direction of the atmospheric density gradient is taken into account, neutrons and protons will traverse a significantly smaller mass path than the line-of-sight path (Smart et al. 1995; see in details above, Chapter 5). This was case for the solar neutron event on May 24, 1990 (Debrunner et al. 1993, 1997; see in details above, Section 7.17). Therefore it is worthwhile to search for neutron signals in the detector of Chacaltaya on November 6, 1997. According to calculations by Tsuchiya et al. (1999), the attenuation is expected to be 10^{-3} for neutrons with an energy of >500 MeV at Chacaltaya as shown in Fig. 7.58.

As noted Matsubara et al. (1999), the GOES satellite recorded a large solar flare (X9.4) on November 6, 1997, successively after the X2.1 class flare on November 4. This flare occurred at 11:49 UT and the x-ray flux achieved its peak at 11:55 UT. The position of the flare was on the solar surface S16°, W43°. Enhancement of the proton flux was seen by the GOES satellite after 12:00 UT. Ground level enhancements were also seen by several neutron monitors. This flare showed the largest photon counts in four energy bands of x-rays (14–23, 23–33, 33–53, and 53–94 keV) among all the flares ever detected by the Yohkoh satellite (Sato et al. 1999, 2000). The flux of hard x-rays peaked between 11:53 and 11:54 UT, slightly earlier than the peak of soft x-rays measured by GOES, the energy of which corresponds to 1.6–12 keV. Gamma rays were also detected up to 20 MeV by the Yohkoh Gamma Ray Spectrometer. In the gamma ray spectrum, the neutron capture line of 2.2 MeV and the excited line of carbon 4.4 MeV were detected (Yoshimori et al. 1999).

In the analysis of Matsubara et al. (1999), the signal counts corresponding to neutrons, which were vetoed by charged particles, were summed every 3 min. The statistical excess of every 3 min counting rate was compared with the 1 h average taken during ±30 min from the particular time. The excess of each counting rate was calculated by subtracting the 1 h average and dividing it by the statistical fluctuation from the average. Excesses thus obtained are shown in Fig. 7.59 as a function of local time.

Four panels in Fig. 7.59 correspond to four different energy thresholds, ch1: >40 MeV, ch2: >80 MeV, ch3: >120 MeV, and ch4: >160 MeV. Dotted vertical lines are the onset of the flare (11:49 UT). Horizontal lines show interval $\pm\sigma$. It is shown that $\geq 3\ \sigma$ excesses were obtained for four different energy thresholds at Bolivian local time 7:51–7:54 am (11:51–11:54 UT). This time interval contains the onset of gamma ray lines (11:53.5 UT).

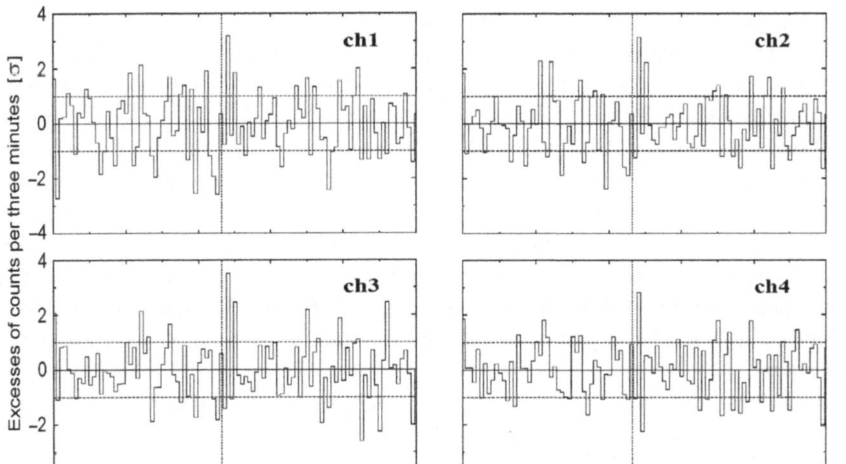

Fig. 7.59 Excesses of counts per 3 min in Chacaltaya neutron telescope are shown in units of σ. Dotted *vertical lines* are the onset of the flare (11:49 UT). *Horizontal lines* show interval $\pm 3\ \sigma$. From Matsubara et al. (1999)

As noted Matsubara et al. (1999), these excesses are not statistical confirmation of a solar neutrons detection. However, neutrons with energy $\geq$1 GeV should have been detected, if these excesses were real and neutrons were produced at the same time as the onset of gamma ray lines. In the case of the X9.3 event on May 24, 1990, the apparent attenuation length of neutrons was 208 g/cm^2 because of the atmospheric density gradient effect (Smart et al. 1995; see in details Section 7.17 and Chapter 5). The detection efficiency of the Bolivian detector to neutrons is 20% for neutrons, energies of which are above 300 MeV (Tsuchiya et al. 1999). By adopting these values, the integrated flux of neutrons with energy $\geq$1 GeV at the top of the atmosphere, is calculated to be 2.2 10^6 m^{-2}. This value is the same as that calculated for the event on May 24, 1990 (Shibata et al. 1993; see in details Section 7.17).

7.23.2 Observations of Neutrons in Association with C4.7 Solar Flare Which Started at 11:31 UT and Continued Until 11:44 UT of 6 November, 1997

As noted Tsuchiya et al. (2001a), at 11:49 UT on 1997 November 6th, a X9.4/2B solar flare was observed at S18W63 on the solar surface. In space, several satellites detected phenomena associated with the solar flare. Satellite Yohkoh detected strong impulsive hard x-ray and gamma-ray (up to 20 MeV) emissions. The neutron capture line (energy of 2.223 MeV) was also detected by Yohkoh as confirmation of the production of solar neutrons at the solar surface (Yoshimori et al. 1999). Furthermore, other gamma-ray line emissions were also detected, which are due to the deexcitation processes of C (4.443 MeV) and O (6.129 MeV) nucleus. The detection of solar neutrons from this flare was considered in the previous Section 7.23.1.

Tsuchiya et al. (2001a) underlined that the Compton Gamma Ray Observatory (CGRO) was in South Atlantic Anomaly (SAA) during the peak phase of X9.4 solar flare, therefore no data for the peak phase were available, but BATSE on board CGRO detected hard X ray emissions 10 min before X9.4 solar flare. This emission was thought to be due to the C4.7 solar flare which started at 11:31 UT and continued until 11:44 UT. As shown in Fig. 7.60, the Bolivian solar neutron telescope at Mt. Chacaltaya detected a clear signal in association with this C4.7 solar flare.

From Fig. 7.60 can be seen that after the BATSE flare onset time (11:34:02 UT), a clear signal is seen in all channels with energy thresholds (>40, >80, >120 and >160 MeV) between 11:41 and 11:43 UT. The statistical significances for this solar neutron signal for each channel were 5.7 σ, 6.8 σ, 5.0 σ, and 3.2 σ respectively. In order to obtain the neutron spectrum at the top of the atmosphere, 30 s counting rates for >40, >80 and >120 MeV were used. In Fig. 7.61, the spectrum given has been calculated from the data for >80 MeV. The attenuation of solar neutrons in the atmosphere and the detection efficiency of the Bolivian solar neutron detector were

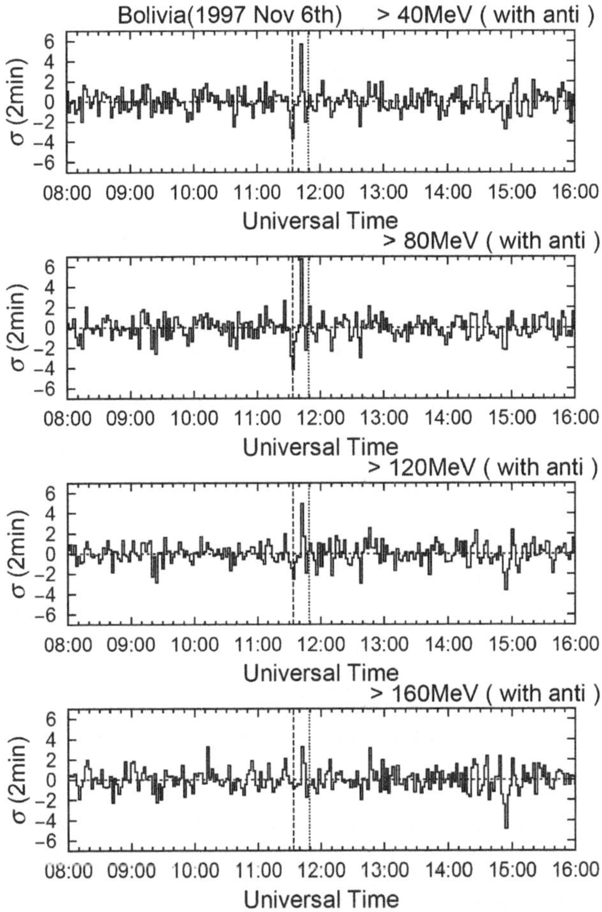

Fig. 7.60 The statistical significance of the 2 min average counting rate between 8 and 16 UT. The *vertical dashed line* in all graphs indicates the flare onset time determined by BATSE/CGRO, which corresponds to 11:34:02 UT. The second *vertical dotted line* shows the X9.4 solar flare onset time (11:49 UT). From Tsuchiya et al. (2001a)

calculated using a Monte Carlo simulation based on the Shibata (1994) model (see in details above, Chapter 5). The spectrum has an index of -3.3 ± 1.6 if the data is fitted with a power law.

As noted Tsuchiya et al. (2001a), this event displays a very new feature in comparison with previous solar neutron events. In the past results, solar neutrons have been detected on the ground only in association with >X8 solar flares. Neutrons produced at the limb of the Sun have higher probability to arrive at the Earth than those produced at the center of the Sun (Hua and Lingenfelter 1987a, b). On the other hand, it seems that 2.223 MeV photons (neutron capture line gamma rays) emitted at the limb have a lower probability of arriving at the Earth than those

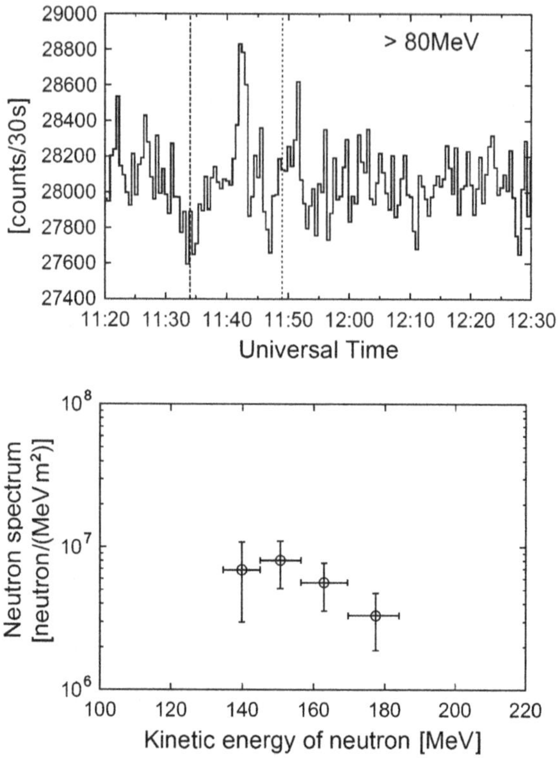

Fig. 7.61 Thirty second counting rate for the >80 MeV threshold energy channel is shown in the top panel. The bottom panel shows the spectrum derived for the top of the atmosphere. In the top panel, the dashed and dotted *vertical lines* indicate the BATSE flares start times related to C4.7 solar flare and the X9.4 flare respectively. Two spikes can be clearly seen after the occurrence of C4.7 and X9.4 flares. From Tsuchiya et al. (2001a)

produced at the center of the Sun. This phenomenon is the result of attenuation of 2.223 MeV photons by Compton scattering in the solar atmosphere. This phenomenon is called "limb darkening" (Wang and Ramaty 2000). Also, 2.223 MeV photons are produced in deeper regions of the solar surface in comparison with the place where the original low energy neutrons are produced, because these must be decelerated before being captured by ambient protons. The flare which was observed on November 6, 1997 was not located at the limb, but Tsuchiya et al. (2001a) have assumed that in considered event 2.223 MeV photons are produced deeper in the solar atmosphere and are strongly attenuated by Compton scattering. So, a calculation for escape probability of neutrons and 2.223 MeV photons from the solar atmosphere has been made, the result of which is shown in Fig. 7.62. In this calculation, the standard composition of the solar atmosphere was used (Reames 1999).

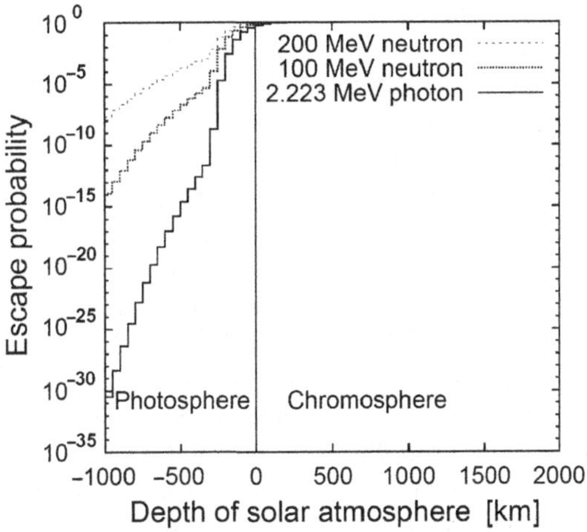

Fig. 7.62 Calculated escape probability of neutrons and 2.223 MeV photons from the solar atmosphere. The *horizontal axis* represents the depth from the boundary between the photosphere and the chromosphere. From Tsuchiya et al. (2001a)

As underlined Tsuchiya et al. (2001a), from Fig. 7.62, it can be understand that there is no big difference between neutrons and 2.223 MeV photons in the case that the production regions of neutrons and 2.223 MeV photons are in the chromosphere. However, once the production of neutrons and 2.223 MeV photons takes place in the photosphere, the escape probability for neutrons and 2.223 MeV photons apparently changes. The probability for 2.223 MeV photon in the solar atmosphere decreases rapidly as the production level becomes deeper. According to calculations of Hua and Lingenfelter (1987a, b), the production of 2.223 MeV photons must occur approximately 100–200 km deeper in comparison with the neutron production region. Taking account of this effect, the escape probability of the 2.223 MeV photons should be much smaller. For example if the production of neutrons occurs at $z = -300$ km and the 2.223 MeV photons at $z =$ 400 km, the ratio of the escape probability of neutrons (100 MeV) and photons (2.223 MeV) is estimated as $4.0\ 10^8$. As noted Tsuchiya et al. (2001a), in the event the solar flare occurred at the position S16°, W63° on the solar surface. Moreover, no neutron capture line was detected during C4.7 solar flare (11:31 – 11:44 UT) although Yohkoh detected the neutron capture line in the X9.4 solar flare. Therefore, there is a possibility that neutrons were produced in association with C4.7 solar flare and neutron capture line also was emitted, but they might be masked by the thick solar atmosphere because of deep production region of neutrons and 2.223 MeV photons.

7.24 Possible Solar Neutron Event on November 23, 1998

According to Hoshida et al. (1999), three large solar flares were consecutively observed in November 22nd, 23rd and 28th of 1998. At that time, the Sun was observed at the zenith angle of 50°, and respective local time at Yangbajing (Tibet) was near noon. There are analyzed data and examined whether the neutron telescope at Yangbajing detected solar neutrons or not. No signal was found for the November 22, 1998, but positive excesses were observed for 23rd and 28th flares. For the event of November 23rd, the statistical significance of the excess was at the level of 3.4σ. The statistical significances of Ch. 1 (>40 MeV), Ch. 2 (>80 MeV), Ch. 3 (>120 MeV), and Ch. 4 (>160 MeV) are 3.4 σ, 2.9 σ, 2.7 σ, and 1.7 σ respectively (see Fig. 7.63). According to the Nobeyama radioheliograph, the increase of the radio intensity was seen from 06:32 UT. A rapid increase of the radio intensity was observed at 06:34 UT. The peak intensity was reported at 06:35:30 UT. The GOES satellite shows that the peak time was 06:53 UT. Hoshida et al. (1999) have assumed the acceleration time of ions at 06:33:30 UT, which corresponds to the most rapid increase time of the radio intensity. The vertical solid line in Fig. 7.63 surely corresponds to this time. Unfortunately, the Yohkoh satellite and CGRO satellite were just in the shadow of the Earth. So it is not possible to compare details of the time profile with theirs. Since these events were real, the solar neutron telescopes newly installed in the world give rise to very fruitful results in forthcoming observations. The Riken neutron monitor was recently transferred to Tibet. It has an area of 28 m^2 and also observed an excess at about 3.5 σ level (Kohno and Miyasaka 1999). Hoshida et al. (1999) noted that the 28 m^2 Riken neutron monitor shows about the same level positive excess as the 9 m^2 neutron telescope.

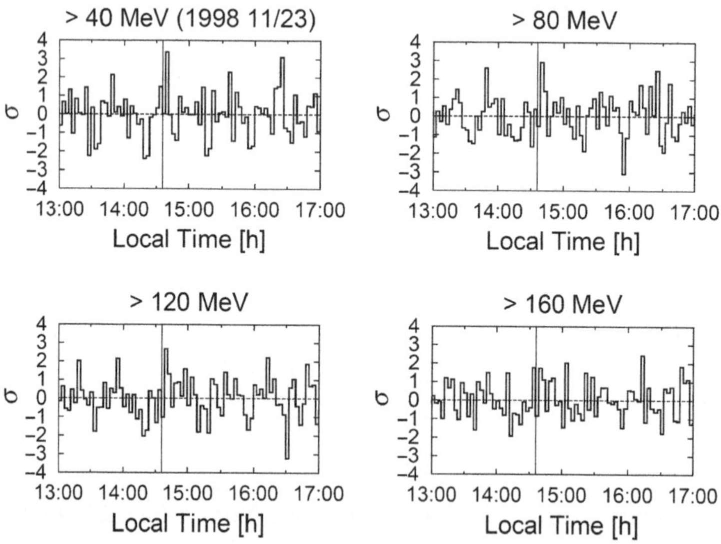

Fig. 7.63 The statistical significance of the counting rate excess of Tibet solar neutron telescope at different channels during possible neutron event at November 23, 1998. Abscissa axis and ordinate axis represent the local time and the significance of time variation, respectively. The *vertical solid line* in the graphs shows the assumed acceleration time of ions (06:33:30 UT or 14:33:30 LT). From Hoshida et al. (1999)

Hoshida et al. (1999) came to conclusion that the Tibet solar neutron telescope possibly observed solar neutrons in association with solar flare of November 23, 1998 with low X-class flare such as X = 2.2.

7.25 Solar Neutron Event on November 28, 1998

7.25.1 Observations by the Tibet Solar Neutron Telescope

According to Hoshida et al. (1999) and Muraki et al. (2007) a few days later the possible neutron event at November 23, 1998 (discussed in previous Section 7.24), again a large flare was observed at the Sun. The meridian passage of the Sun was again above Tibet. The GOES satellite shows the peak time of the flare was 06:09 UT about. In this event, the Nobeyama radio heliograph, Yohkoh satellite and CGRO satellite were all successfully observing. No strong line gamma-rays were detected by the Gamma-Ray Spectrometer on board Yokoh (Yoshimori 1999). The Yohkoh Soft x-ray Telescope shows quite complex loop structures of the solar surface. A clear magnetic field polarization was seen by the Nobeyama radioheliograph. A clear magnetic sharing was also observed by Mitaka magnetograph (Sakurai 1999). However the Riken neutron monitor detected a minor excess (1.7 σ) at the corresponding time (Kohno and Miyasaka 1999). At Tibet solar neutron telescope, a channel (>120 MeV) indicated 1.8 σ level enhancement. The statistical significance is not so clear. Therefore Hoshida et al. (1999) compared the signals coming from the south direction with the north direction. The data coming from the acceptance $\pm 23.5^{\circ}$ in the east-west direction were used for the data analysis. Towards the north-south direction, Hoshida et al. (1999) compared the data between 20.5° and 28.6°. As shown in panel b of Fig. 7.64, a sharp enhancement is seen in the south direction, on the contrary, no enhancement in the north direction (see panel a of Fig. 7.64).

The observation of enhancement of the south direction suggests for Hoshida et al. (1999) that the excess was produced by neutrons from the Sun. The statistical significance was 4.5σ, but the bottom counter counts twice sometimes in the method, so that the actual significance of the excess should be reduced slightly (then it turns out to be 3.5σ). Hoshida et al. (1999) came to conclusion that newly installed Tibet solar neutron telescope demonstrates very high sensitivity in comparison with traditional neutron monitors and to confirm obtained result shown in Fig. 7.64, it is necessary to finish Monte Carlo simulation, taking into account the side neutron showers; this Monte Carlo simulation is necessary to explain why solar neutrons are weakly detected by the lowest channel of solar neutron telescope and also by the Riken neutron monitor.

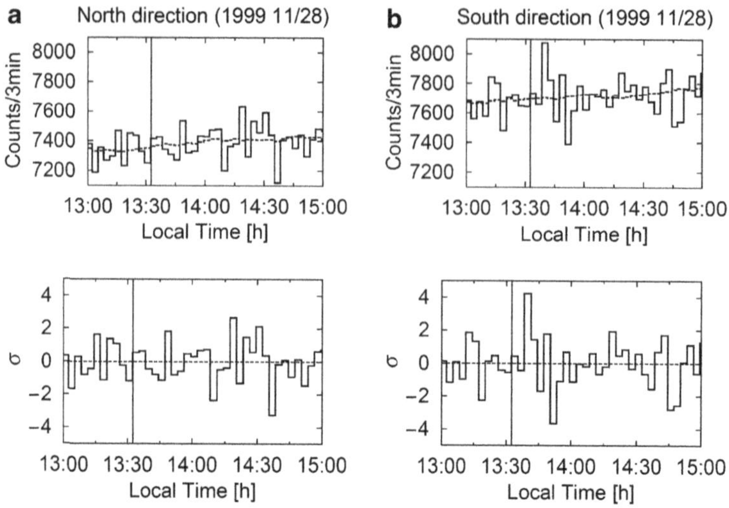

Fig. 7.64 The raw data (top panels) and their statistical significances (bottom panels) for November 28, 1998. The statistical significance from the south direction shows a 4 σ level enhancement. The *vertical solid line* in the graphs represents the acceleration time of ions which was assumed at 05:32:30 UT or 13:32:30 of the local time. From Hoshida et al. (1999)

7.25.2 Arrival of Solar Neutrons at Large Zenith Angle and the Refraction Effect

For explaining observation results discussed in the previous Section 7.25.1, Tsuchiya et al. (2001b) and Muraki et al. (2007) accounted the refraction effect which is especially sufficient for propagation of solar neutrons arriving at big zenith angle (Smart et al. 1995; Dorman and Valdes-Galicia 1997; Dorman et al. 1997a, b, 2000; Valdes-Galicia et al. 2000; see also in details Chapter 5). Tsuchiya et al. (2001b) noted that Valdes-Galicia et al. (2000) applied the refraction effect to 1990 May 24th solar flare, the same event as Smart et al. (1995) investigate, and found that the increases of the neutron monitors were consistent with their calculations taking into account the refraction effect. The refraction effect makes the path length of solar neutrons shorter than otherwise expected. For example, in this event, the air mass along the line of sight from the Sun to the detector was 997 g/cm^2, if we use a simple formula 600/cos (53°). However, if after each scattering in average at 6°, the total path length turns out to be 747 g/cm^2. Therefore, even on the basis of a simple model, it is found that the path length is shorter than the result obtained without considering the refraction effect. Indeed, Monte Carlo simulations taking account of the refraction effect show an interesting result concerning the attenuation of solar neutrons at Yangbajing. In Fig. 7.65 Tsuchiya et al. (2001b) present the results of a Monte Carlo simulation.

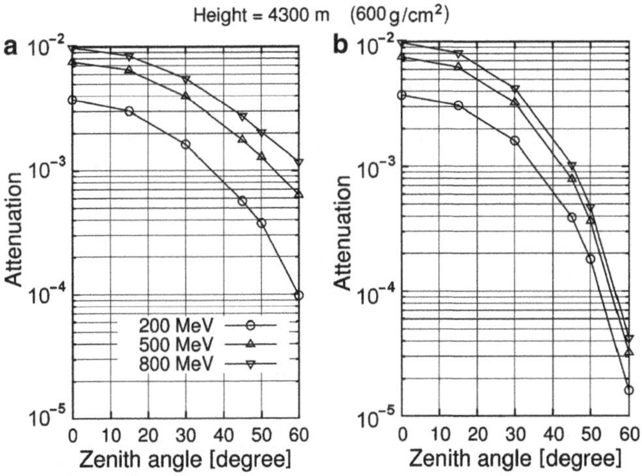

Fig. 7.65 Attenuation curves expected for solar neutrons at Yangbajing. Panel (**a**) represents the result of Monte Carlo simulation taking into account of the refraction effect, while panel (**b**) corresponds to the result without taking account of the refraction effect. From Tsuchiya et al. (2001b)

As shown in Fig. 7.65, the attenuation of solar neutrons with an incident angle of less than 30° is almost the same in both calculations. However, for solar neutrons with the incident angle over 30°, the results of the calculations deviate strongly. In this event, the incident angle of solar neutrons was 53°. It was found from the present calculation that solar neutrons with the energies of 200, 500 and 800 MeV had a higher survival probability than the previous estimations. For neutrons with an angle of incidence 53°, the probabilities are three, seven, ten times higher than they would be without including the refraction process. Thus, it appears likely that solar neutrons arrived at Yangbajing with a flux detectable by the neutron telescope even with an incident angle 53°.

7.25.3 Determining of the Solar Neutron Spectrum at the Top of the Earth's Atmosphere

The energy spectrum of solar neutrons at the top of the Earth's atmosphere was derived using 1 min counting rate data from the southern direction. The results are given in Fig. 7.66.

When the spectrum of neutrons was derived, Tsuchiya et al. (2001b) assumed that protons and electrons were accelerated at the same time and the production of neutrons took place at the rise time of X ray emission (30–60 keV) impulsively. Thus, the neutron production time was assumed to have occurred at 05:37:50 UT.

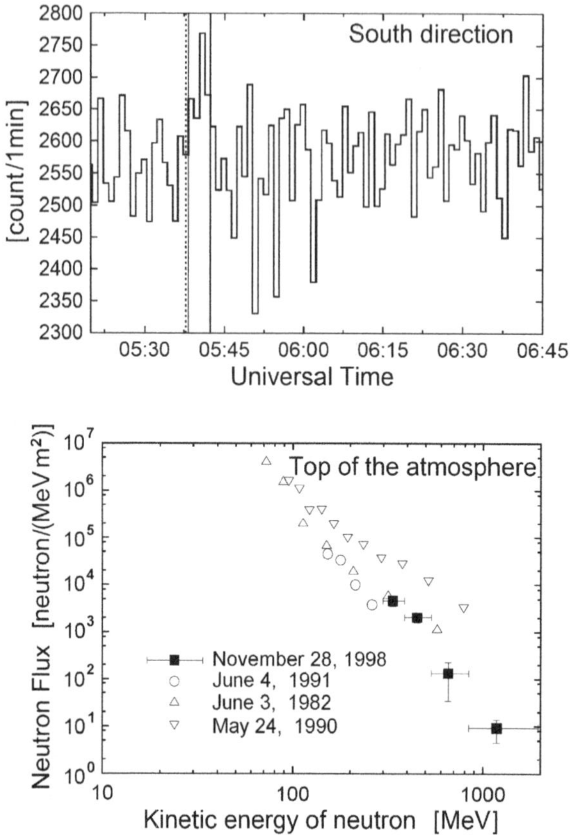

Fig. 7.66 Top panel represents 1 min counting rate for the event November 28, 1998. In the bottom panel, the neutron spectrum obtained by these data was shown in comparison with past events. It can be see that present spectrum was hard but the intensity was low. In the top panel, the *vertical dashed line* indicates the production time of neutrons (assumed to be 05:37:50 UT). The energy spectrum of neutrons was derived using the events detected in the period defined by the two *solid lines*. From Tsuchiya et al. (2001b)

The time is shown in the top panel of Fig. 7.66 by the dashed lines. The counts within the two vertical solid lines were used for deriving the energy spectrum of the neutrons. The interval corresponds to the time between 05:38:19 and 05:41:19 UT. This implies that solar neutrons with the energy between 400 MeV and 2 GeV were observed. The threshold energy of the higher part of the detector was set at 230 MeV, so the observed ranges of neutron energies are consistent with each other.

In Tsuchiya et al. (2001b) it has been demonstrated that solar neutrons can penetrate the thick atmosphere and arrive at the detector if account is taken of a 'refraction' effect. They came to conclusion that the refraction effect is significant not only for this event, but could be important for other solar neutron observations which have been made in the early morning, late evening and also in the winter season. In the time interval between 05:38 and 05:41 UT when a clear signal was obtained, radio and hard x-ray emissions were also detected. This correlation between x-ray and radio data suggests that the commencement of the proton and electron acceleration happens at the same time. If the simultaneous acceleration and impulsive production of neutrons took place at the solar surface, solar neutrons detected by the Tibet solar neutron telescope must have energy between 400 MeV and 2 GeV. The detection of solar neutrons in this event was made possible by the measurement of the arrival direction of solar neutrons. Therefore, it was demonstrated that the ability to measure the arrival direction of solar neutrons is very useful for solar neutron observations. Finally, Tsuchiya et al. (2001b) came to conclusion that solar neutrons were produced in such a moderate level solar flare (X3.3) and arrived at Yangbajing due to the refraction effect in the atmosphere. This was determined by using the directional information from the Tibet solar neutron telescope.

7.26 Search for Solar Neutrons in Association with Large Solar Flares in July 2000 and in March–April 2001: Three Categories of Solar Neutron Events

7.26.1 Data Analysis

According to Flűckiger et al. (2001), large solar flares were observed on July 12, 2000, at 10:18 UT, and on July 14, 2000, at 10:03 UT, at heliographic locations N17°, E27° and N22°, W07°, respectively. The x-ray intensities measured by GOES for these events were reported as X1.9 and X5.7, respectively. The best sites for solar neutron detection of these flares by the international solar neutron network were Gornergrat and Mt. Aragats. From March 29 to April 15, 2001, nine large solar flares were observed by the GOES satellites with magnitude $X > 1.0$. For six of them, the most suitable positions to detect solar neutrons by detectors of the global solar neutron network were again either Gornergrat or Mt. Aragats. Therefore Flűckiger et al. (2001) have concentrated their efforts on analyzing the data from the July 2000 and the March and April 2001 events, including two more $M > 5$ solar flares in the 2001 time period.

Unfortunately BATSE data from the CGRO satellite are no longer available. Flűckiger et al. (2001) therefore assumed that the ion acceleration occurred at the beginning of the flare as indicated by the GOES x-ray data. This naturally assumes that the ions were accelerated at the same time as the electrons, and of course this is not necessarily true. It should be noted that the GOES start time is defined when 4 min of continuous enhanced intensity were observed in the x-ray channel. So the beginning of ion acceleration should be considered as being at or before the GOES start time. Flűckiger et al. (2001) have searched for an excess in the Gornergrat and

Mt. Aragats solar neutron detectors data within ±8 min of the GOES start time. The scoring is summarized in Table 7.6.

In Table 7.6 Flűckiger et al. (2001) present the GOES start time, the statistical significance of the neutron channel count rate, the atmospheric depth taking into account the refraction effect (Smart et al. 1995), and finally the situation of the Yohkoh satellite (whether it could observe the flare or if it was in the shadow of the Earth or in the South Atlantic Anomaly). As a result, Flűckiger et al. (2001) find an excess at (more or less) the 3σ at least at one of the two stations in almost every event. No count rate enhancement due to solar neutrons could be identified for the flare on July 14, 2000. As can be seen from Table 7.6 it is quite difficult to get a solar neutron event with a high statistical significance at two stations. This can be understood by considering the magnitude of the events and the difference in the zenith angle of the Sun at the different locations. Observations at different longitudes provide with another strong confirmation of the arrival of solar neutrons, even when the signal is weak. Flűckiger et al. (2001) have presented the data of the April 12, 2001 event in Fig. 7.67.

Flűckiger et al. (2001) noted that the time of both solar neutron detectors is based on GPS time signals. They believe that these clocks are exact within ±30 s to GPS time at any given moment. The data at the two solar neutron telescopes are sampled every 10 s.

7.26.2 Simultaneous Observations

As noted Flűckiger et al. (2001), simultaneous observations of solar neutrons were achieved previously at Mt. Norikura (4.6σ) and Haleakala stations in Hawaii (5.7σ) for the June 6th, 1991 flare (see above Section 7.22). The events considered here include probably the second (July 12, 2000) and third (April 12, 2001) successful two-station observations, although the statistical significance was only at the 3σ level. Probably this low statistical significance arises because of the relative weakness of the corresponding flares (X1.9 and X2.0) in comparison with the June 6, 1991 flare (X12.0). It is not certain whether the observed x-ray intensity reflects the intensity of the accelerated protons. But in comparison with the enhancements observed on June 6, 1991 at Mt. Norikura and Haleakala, and taking into account the difference in the attenuation length and the difference in the x-ray intensities, the enhancements in the neutron detector count rates recorded on April 12, 2001, are consistent with the observation of solar neutrons.

7.26.3 The Estimation of the Start Time of Ion Acceleration

Flűckiger et al. (2001) discuss the start time of ion acceleration based on the data in Table 7.6. When there are compared the onset times of the excesses in the neutron

Table 7.6 The scoring of main data (From Flűckiger et al. 2001)

Date	X-ray class	Flare location	Start time (UT)	Score			Statistical significance		Atmospheric depth	
				Gornergrat	Mt. Aragats	Yohkoh	Gornergrat	Mt. Aragats	Gornergrat	Mt. Aragats
2000.7.12	X1.9	N17E27	10:18 (10:15)	□*1 (10.12)	□*2 (10.15)	□	3.3σ (4 min)	3.5σ (1 min)	716	704
2000.7.14	X5.7	N22WO7	10:03 (10:03)	×	×	Δ*3	–	–	724	705
2001.3.29	X1.7	N20W19	09.57 (09:55)	⨀ (10:01)	Δ (09:54)	□	4.2σ (1 min)	2.2σ (3 min)	850	803
2001.4.1	M5.5	?	10.55	Δ (10:42)	□ (10:43)	Δ	2.8σ (2 min)	3.2σ (1 min)	808	826
2001.4.2	X1.4	N17W60	10.04 (10.05)	Δ (09:57)	⨀ (09:57)	□	2.7σ (2 min)	4.7σ (25 min)	831	794
2001.4.2	X1.1	N17W61	10.58	□ (11:03)	Δ (11:12)	×	3.1σ (1 min)	2.4σ (5 min)	804	826
2001.4.9	M7.9	S21W04	15:20 (15:24)	⨀ (15:09)	evening	□	4.9σ (6 min)	–	Large angle	–
2001.4.10	X2.3	S23W09	05:06 (04:59)	Morning	⨀ (05:06)	×	–	4.5σ (6 min)	–	Large angle
2001.4.12	X2.0	S21W04	09:39 (B11:12)	□ (09:38)	□ (09:39)	Δ	3.0σ (1 min)	3.4σ (1.5 min)	821	767
2001.4.15	X14.4	S20W85	13:19 (13:36) *4	Δ (13:19) *5	Evening	□	2.8σ (6 min) *6	–	808 (g/cm^2)	–

⨀ Observed at 4.5 σ or higher significance
□ Observed at 3.0–4.5 σ significance
Δ Claimed signals with less than 3 σ
× not seen
*1 A peak was observed 6 min before GOES start time
*2 A peak was observed 3 min before GOES start time
*3 The triangle mark of Yohkoh represents partially observed events
*4 The number within the brackets implies the start time determined from the optical flare
*5 The number within the brackets represents the neutron start time
*6 The number within the brackets shows the duration of the observed excess

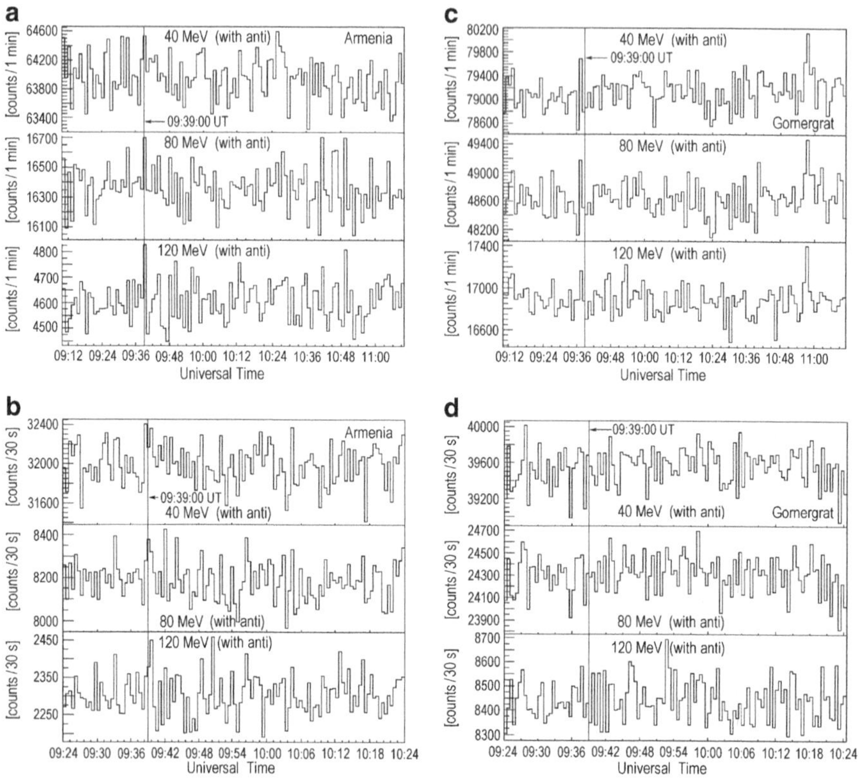

Fig. 7.67 Time profiles of the neutron channel counting rates of the April 12th, 2001 event at Mt. Aragats, Armenia (Panel **a**: 1 min values; Panel **b**: 30 s values) and Gornergrat, Switzerland (Panel **c**: 1 min values; Panel **d**: 30 s values). The *vertical line* in each panel represents the flare start time 09:39:00 UT obtained from the GOES x-ray data. From Flűckiger et al. (2001)

channels with the start times of the corresponding flares in x-rays or Hα, the differences in the start times can be classified into three groups. In the events of 29 March and 2 and 10 April, 2001 the neutron excess starts about 5–7 min after the start time of the flare in x-rays or Hα. Flűckiger et al. (2001) call these events as *'Category I'*. Since the excesses were observed in the channels >80 and >120 MeV, the initial neutrons that arrived at the detector must have had an energy >250 MeV. Neutrons with energy of 250 MeV take 5 min more than light when they travel from the Sun to the Earth. So this group could be explained by the hypothesis that neutrons were produced together with electrons in the flare. However for the event on April 10, 2001, if it is assumed the acceleration start time of electrons to be 05:06 UT (x-rays), then this event should belong to the 'Category II'.

According to Flűckiger et al. (2001), *'Category II'* events have the typical feature that the neutron excess starts at the same time as the increase in X- and γ-ray emissions. Flűckiger et al. (2001) provide two possible explanations for this group, to which the events of 1st, 2nd (earlier one), 12th, and 15th April, 2001 are belong. One hypothesis is that the neutrons causing the increase in the detector count rate have an energy around 250 MeV and that the acceleration of ions started earlier than that of electrons. Another possibility is that the neutrons might have a higher energy (1 GeV) and that they reach the Earth at almost the speed of light. Since there are not have an absorber under the telescopes, it is unable to distinguish between the two cases. As noted Flűckiger et al. (2001), a precise analysis of the 15 April, 2001 event using the Yohkoh soft x-ray images will clarify this question.

'Category III' events, according to Flűckiger et al. (2001), are those in which the neutron excess begins about 6–10 min before the x-ray or optical flare start time. The events of 12 July, 2000 and 9 April, 2001 belong to this category. However, in the July 12, 2000 event, if it will be take the flare's start time as 10:15 instead of 10:18 UT, the difference is not so big, and the event might be reclassified as a 'Category II' event. The GOES start time may not always reflect the correct time of commencement of particle acceleration. Further careful study of the 'Category III' events using the Yohkoh x-ray image data must be very interesting.

7.26.4 Summary of Main Obtained Results: Three Categories of Solar Neutron Events

In conclusion Flűckiger et al. (2001) summarized obtained main results as following:

(i) The international solar neutron network has probably succeeded in detecting several solar neutron events in association with large solar flares in July 2000 and March–April 2001.
(ii) These solar neutron events can be classified into three categories, one of which is consistent with a picture that ions are accelerated at the same time as electrons ('Category I').
(iii) However, there are seen several candidates for solar neutron events where the observation would require that the ion acceleration must have started before the electron acceleration ('Category II' and 'Category III').
(iv) But no event was observed which suggests that ion acceleration started at the time of the flare maximum.

7.27 Solar Neutron Event on November 24, 2000

7.27.1 Observations of x- and γ-Ray Fluxes

According to Watanabe et al. (2003b), three X-class solar flares occurred successively in NOAA active region 9236 on November 24, 2000. At 14:51 UT, an X2.3 class flare was observed which was the largest among these three. The location of

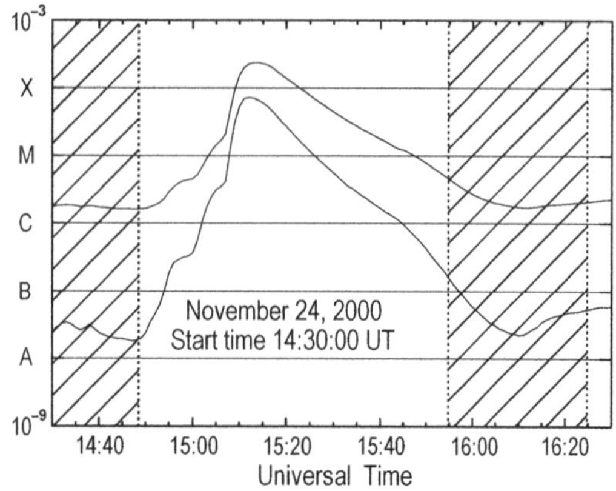

Fig. 7.68 Time profile of the soft x-ray flux observed by the GOES satellite at the solar flare that occurred on November 24, 2000 (14:30–16:30 UT). The start time of this X2.3 flare was 14:51 UT. *Oblique lines* mean night for the Yohkoh satellite. The upper profile expresses the x-ray flux in the wavelength range 1.0–8.0 Å, and the lower one is for 0.5–4.0 Å. From Watanabe et al. (2003b)

the active region was N22°, W7° at the time of the flare, and the flare was a disk flare. The time profile of the soft x-ray flux in this flare observed by the GOES satellite is shown in Fig. 7.68. Although the start time of this flare defined by the GOES satellite was 14:51 UT, the time when the soft x-ray flux became a maximum was 15:13 UT, which was more than 20 min after the start time. In Fig. 7.68 it can be distinguish three bumps starting at 14:51, 15:00, and 15:07 UT, respectively. Each of them arrives at C, M, and X class at its maximum. The time profile of the hard x-rays observed by Yohkoh/HXT is shown in Fig. 7.69.

Watanabe et al. (2003b) noted that the hard x-ray flux suddenly increased from 15:07:30 UT, when the new bright spot appeared in the soft x-ray image. The hard x-ray image observed by Yohkoh/HXT is shown in Fig. 7.70 on the soft x-ray image observed by Yohkoh/SXT.

Watanabe et al. (2003b) underlined that at the time when the new soft x-ray source appeared, the hard x-ray source turned up at the footpoint of the soft x-ray flare loop. Therefore, hard x-rays were produced in parallel with soft x-rays around 15:08 UT. Moreover, the energy spectrum of the hard x-rays was very hard between 15:07:30 and 15:09:30 UT. High-energy electrons can emit hard x-rays by bremsstrahlung. Therefore, the onset time of hard x-rays can indicate the time when the particle acceleration took place. But hard x-rays are produced by high-energy electrons, not by ions. Therefore, it cannot be asserted that at the time when a large amount of hard x-rays are produced, ions are accelerated and solar neutrons are produced. It is more useful to examine information on γ-rays to estimate the acceleration time of ions. Around 15:08 UT, a large amount of γ-rays was observed by Yohkoh/GRS. Figure 7.71 shows the energy spectrum of γ-rays at that time.

As underlined Watanabe et al. (2003b), in Fig. 7.71 a clear signal of 2.223 MeV neutron capture γ-ray line is seen on the background bremsstrahlung component. Therefore, it is evident that neutrons were produced around 15:08 UT on November 24, 2000. Furthermore, between 4 and 7 MeV, weak signals of γ-ray lines produced by deexcited ions C (4.443 MeV) and O (6.129 MeV) are seen too. Consequently, it is certain that ions were accelerated at that time. In principle, from these γ-ray lines

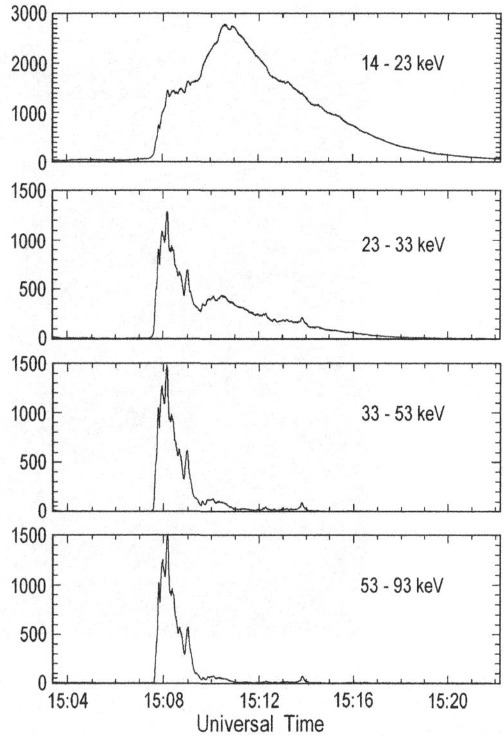

Fig. 7.69 Time profile of hard x-rays observed by Yohkoh/HXT on November 24, 2000. The *vertical axis* is the count rate per sub-collimator. Hard x-rays were suddenly emitted from 15:07:30 UT. From Watanabe et al. (2003b)

data, it can be calculate the amount of produced neutrons and the energy spectra of the accelerated ions independently of the neutron measurements (Ramaty et al. 1996). However, at this event the fluxes of γ ray lines produced by deexcited ions were too weak. It is difficult to estimate the amount of γ-rays produced by each deexcitation process because of a contamination of electron bremsstrahlung. For this reason, the neutron spectrum was not derived from the γ-ray lines data to compare it with that determined by the neutron monitor data. The time profiles of 2.223 MeV neutron capture γ-ray line and 4–7 MeV γ-rays observed by the Yohkoh/GRS around 15:08 UT on November 24, 2000 are shown in Fig. 7.72. The bremsstrahlung component is not subtracted. The 4–7 MeV γ-ray time profile included line γ-rays produced by deexcited ions C (4.443 MeV) and O (6.129 MeV). Therefore, the 4–7 MeV γ-ray time profile is approximately equal to the C + O line and bremsstrahlung γ-ray time profile. As can be seen in Fig. 7.72, the duration of 2.223 MeV γ-ray emission is longer than that of 4–7 MeV γ-rays. The 2.223 MeV γ-rays are produced when thermal neutrons are captured by hydrogen. High-energy neutrons are produced simultaneously with γ-ray lines of deexcited ions by the interaction of accelerated ions and the solar atmosphere. On the other hand, the 2.223 MeV γ-rays are produced about 100 s after the production of the high-energy neutrons, because of the time required for neutrons to slow down and be captured. Therefore, the time profile of 2.223 MeV γ-rays is dilated compared

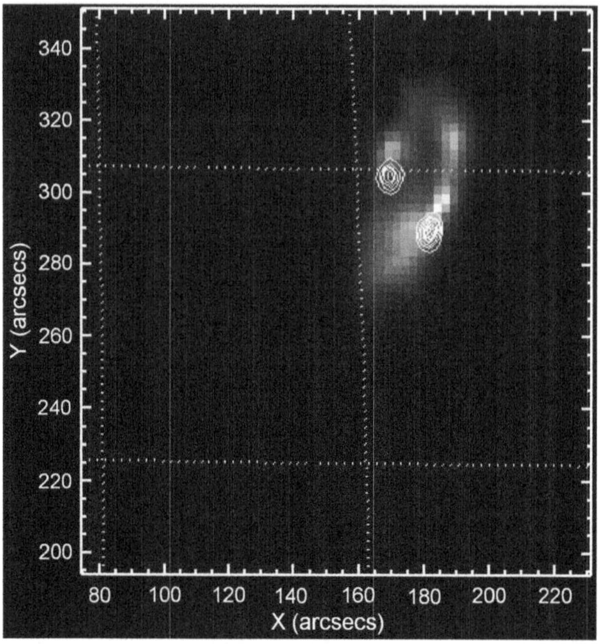

Fig. 7.70 Soft x-ray and hard x-ray images observed by Yohkoh/SXT and Yohkoh/HXT at 15:07:53 UT on November 24, 2000. The hard x-ray image is drawn by contours on the soft x-ray image. The hard x-ray source is at the footpoint of the soft x-ray flare loop. From Watanabe et al. (2003b)

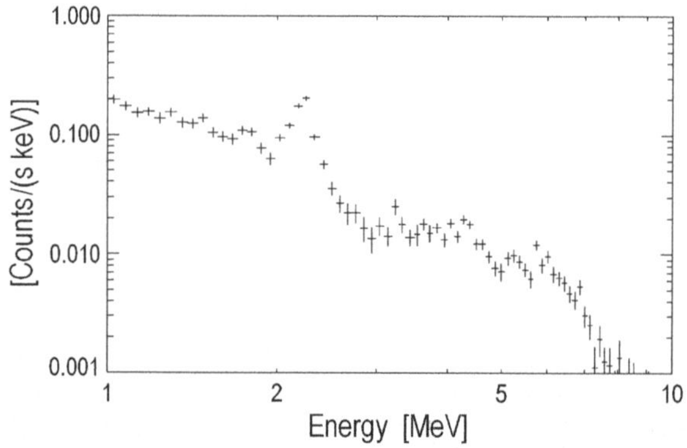

Fig. 7.71 Background-subtracted γ-ray spectrum observed by the Yohkoh/GRS on November 24, 2000. There is a clear signal of 2.223 MeV γ-ray line and weak signals of de-excited nuclei γ-ray line between 4 and 7 MeV on the background bremsstrahlung component. From Watanabe et al. (2003b)

with that of 4–7 MeV γ-rays. Hence, it can be consider that the high-energy neutrons were produced at the same time as deexcited nuclei γ-rays rather than 2.223 MeV γ-rays.

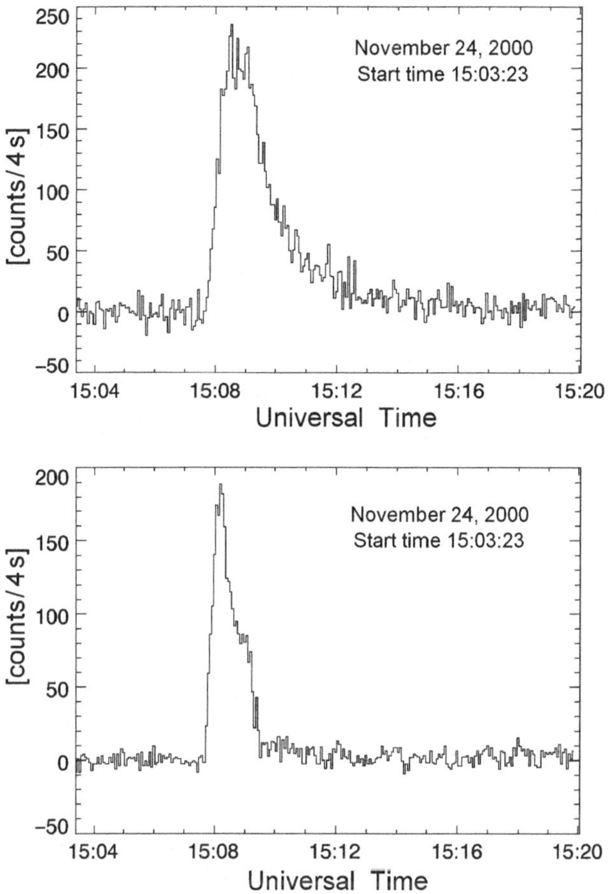

Fig. 7.72 Time profiles of 2.223 MeV neutron *capture line* γ-ray and 4–7 MeV γ-rays on November 24, 2000. The upper figure is the 2.223 MeV time profile, and the lower one is the 4–7 MeV time profile. The bremsstrahlung component is not subtracted. The 4–7 MeV γ-ray time profile includes line γ-rays produced by deexcited ions, C (4.443 MeV) and O (6.129 MeV). In both of them, the *vertical axis* is the count rate per 4 s. The time profile of neutron capture γ-rays is more expanded than that of 4–7 MeV γ-rays. From Watanabe et al. (2003b)

7.27.2 Observations of Solar Neutrons at Mt. Chacaltaya

According to Watanabe et al. (2003b), at 15:08 UT on November 24, 2000 the Sun was over Bolivia and the neutron monitor installed at Mt. Chacaltaya, Bolivia, was at the most suitable place for observing solar neutrons. This station is located at E292.0°, S16.2°, 5,250 m above sea level, and the vertical air mass is 540 g·cm^{-2}. At this time, the zenith angle of the Sun was 17.47° and the air mass for the line of sight to the Sun was 566 g·cm^{-2}. The neutron monitor installed at Mount

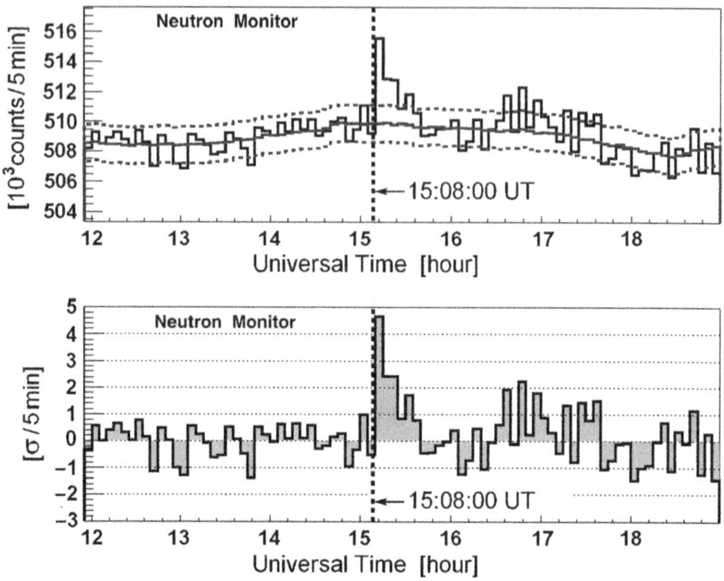

Fig. 7.73 Five-minute pressure-corrected count rate observed by the Bolivia neutron monitor on November 24, 2000. The top panel is the time profile observed by the neutron monitor and background. The *solid smooth line* is the averaged background, and *dashed lines* are $\pm 1\sigma$ from the background. The bottom panel is the statistical significance. From 15:10 to 15:25 UT, a clear excess made by solar neutrons was observed. The total statistical significance for 15 min is 5.5σ. From Watanabe et al. (2003b)

Chacaltaya is 13.1 m^2 in area and of the NM-64 type. The count rate is recorded every 10 s. The time profile of neutrons observed by the neutron monitor is shown in Fig. 7.73.

In Fig. 7.73 a clear excess can be seen between 15:10 and 15:25 UT. The statistical significance of each bin is 4.7σ at 15:10–15:15 UT, 2.4σ at 15:15–15:20 UT, and 2.4σ at 15:20–15:25 UT. The total significance for 15 min, between 15:10 and 15:25 UT, turned out to be 5.5σ. Watanabe et al. (2003b) noted that there is a possibility that these excesses came from energetic ions because the neutron monitor can observe energetic ions. But there is no evidence that the enhancement was produced by energetic ions since the measurements by the other stations in the worldwide network of neutron monitors showed no enhancement. In addition, the cutoff rigidity at Mt. Chacaltaya is too high (12.53 GV), so if the observed enhancement caused by charged energetic particles, the expected enhancements at middle and high latitudes are expected very big what strong contradict to real observation data on the worldwide network of neutron monitors. Therefore, the signals in Fig. 7.73 must have come only from solar neutrons.

As noted Watanabe et al. (2003b), neutron monitors cannot measure the energy of neutrons; therefore, it is not possible directly derive the energy spectrum of solar neutrons. However, using the time-of-flight (TOF) method by assuming the emission time of solar neutrons, this can be derived. To this end, we must postulate the production time of solar neutrons. As shown above, the production time of deexcited nuclei γ-ray lines are taken to be that of solar neutrons. Unfortunately, the excess obtained in this event is not strong enough to examine the time profile of the signals in detail. Therefore Watanabe et al. (2003b) do not touch on the possibility of an extended production of neutrons and simply assume that neutrons were produced instantaneously in this event. If solar neutrons were produced at 14:51 UT, which was the start time of soft x-rays, then the energy of the neutrons detected between 15:10 and 15:25 UT is calculated to be 47–19 MeV. Since neutrons suffer violent attenuation in the Earth's atmosphere, such low-energy neutrons cannot arrive at the ground. On the other hand, the assumption of the production time as 15:08 UT, which is the time of hard x-ray and γ-ray emissions were seen, gives 772–57 MeV, which is much more reasonable. Consequently, the production time of solar neutrons is set at 15:08 UT hereafter.

7.27.3 Checking of the Detection Efficiency of the Neutron Monitor

According to Watanabe et al. (2003b), from the time profile obtained by the neutron monitor, the flux $F(E_n)$ of solar neutrons at the top of the Earth's atmosphere can be calculated by the formula

$$F(E_n) = \Delta N / (\in \times S \times \Delta E_n), \tag{7.16}$$

where ΔN is the excess count contributed by solar neutrons and $\in$ is the detection efficiency of the neutron monitor. Here $\in$ includes the attenuation of solar neutrons through the Earth's atmosphere, S is the area of the neutron monitor, and ΔE_n is the energy range corresponding to one time bin. The detection efficiency of the neutron monitor and the attenuation of solar neutrons depend on the energy of the neutrons. The detection efficiency of the neutron monitor was calculated by Hatton (1971) and by Clem and Dorman (2000). Experimentally, that has been measured directly by the accelerator experiment (Shibata et al. 2001). The results are shown in Fig. 7.74, together with calculations by Hatton (1971) and by Clem and Dorman (2000).

From Fig. 7.74 can be seen that although the two calculations are consistent with the experimental result in the energy range 100–400 MeV, there is a big discrepancy between the calculations of Hatton (1971) and Clem and Dorman (2000) outside this experimental range. As noted Watanabe et al. (2003b), the deviation of the calculation by Clem and Dorman (2000) from the experimental results is $\pm 5\%$, whereas that by Hatton (1971) is $\pm 10\%$. Consequently, in Eq. 7.16 Watanabe et al. (2003b) adopted the efficiency calculated by Clem and Dorman (2000) because it was closer to the results of Shibata et al. (2001) than calculations by Hatton (1971).

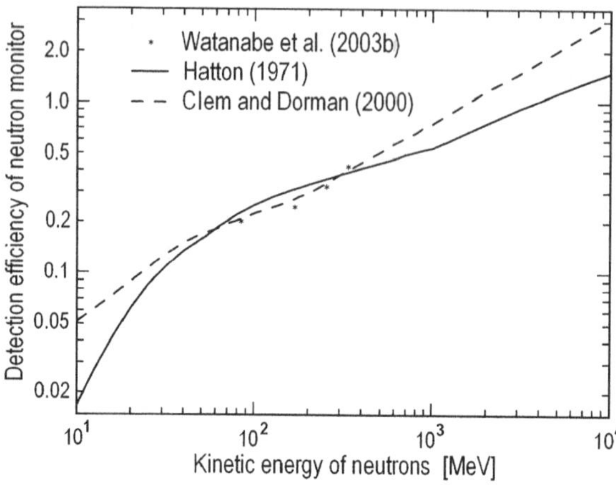

Fig. 7.74 Detection efficiency of the neutron supermonitor NM-64. The results of the accelerator experiments by Shibata et al. (2001) are compared with theoretical predictions by Hatton (1971) and by Clem and Dorman (2000). From Watanabe et al. (2003b)

7.27.4 Attenuation of Solar Neutrons in the Atmosphere During the 24th of November 2000 Event

As underlined Watanabe et al. (2003b), the attenuation of solar neutrons in the Earth's atmosphere was calculated by Debrunner et al. (1989) and Shibata (1994) by Monte Carlo simulations. Hereafter they are called the Debrunner model and the Shibata model, respectively. There is a big discrepancy between the two models. In order to examine which model is correct, an accelerator experiment was conducted at the Research Center for Nuclear Physics, Osaka University (RCNP). It was found that the Shibata model could explain the experimental result (Koi et al. 2001). Consequently, Watanabe et al. (2003b) adopted the Shibata model in calculating the propagation of neutrons in the air. The attenuation of solar neutrons in the Earth's atmosphere at the level of Mt. Chacaltaya at 15:08 UT on 24 November, 2000 is shown in Fig. 7.75.

From Fig. 7.75 can be seen that solar neutrons whose kinetic energy are below 100 MeV are strongly attenuated by the Earth's atmosphere.

7.27.5 Energy Spectrum at the Solar Surface and Total Energy of Emitted Solar Neutrons

As underlined Watanabe et al. (2003b), to derive the energy spectrum of neutrons at the solar surface from the flux at the top of the Earth's atmosphere, the survival probability of neutrons between the Sun and the Earth must be taken into account. The result is shown in Fig. 7.76.

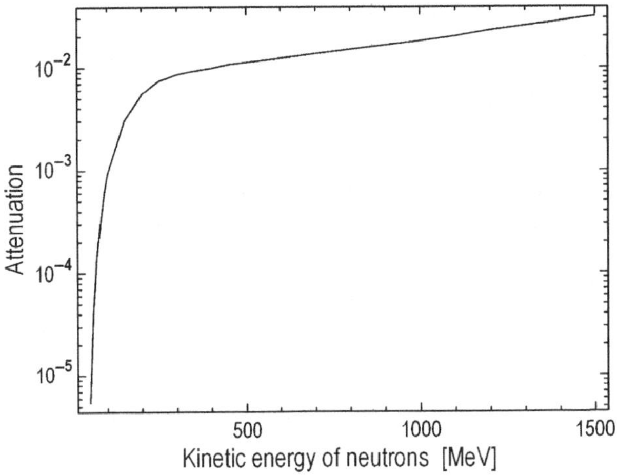

Fig. 7.75 Attenuation of solar neutrons through the Earth's atmosphere at Mt. Chacaltaya at 15:08 UT on November 24, 2000 according to Shibata (1994) model. The *horizontal axis* is the kinetic energy of the neutrons at the top of the Earth's atmosphere. From Watanabe et al. (2003b)

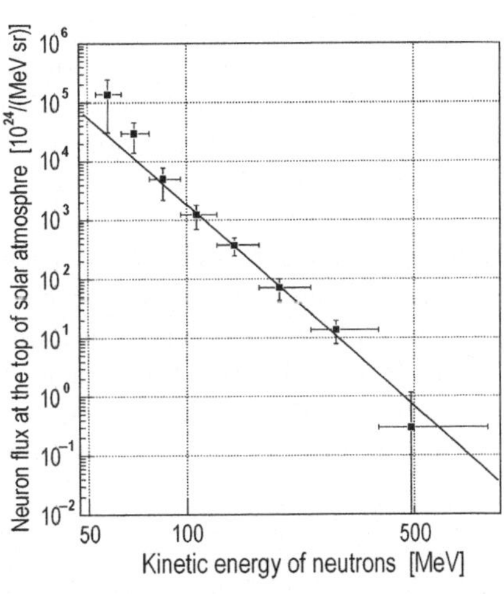

Fig. 7.76 Energy spectrum of solar neutrons on the solar surface for the flare that occurred on November 24, 2000. From Watanabe et al. (2003b)

In Fig. 7.76, the vertical errors come only from statistical ones. This spectrum was derived from 2 min count rate. By fitting these data points with a power law of the form

$$F_S(E_n) = C \times (E_n/100\,[\text{MeV}])^{\alpha}, \tag{7.17}$$

The energy spectrum of solar neutrons was obtained, where C is the flux of neutrons at $E_n = 100$ MeV, and α means the power-law index. The fitting region is chosen above 100 MeV, because in this region the errors from neutron attenuation in the Earth's atmosphere are small. The obtained result is as follows:

$$F_S(E_n) = (1.8 \pm 0.8) \times 10^{27} \left[\text{MeV}^{-1}\ \text{sr}^{-1}\right] (E_n/100\,[\text{MeV}])^{-4.9 \pm 0.7}. \tag{7.18}$$

The total energy flux of solar neutrons that were emitted from the Sun in the energy range between 50 and 800 MeV was calculated by simply integrating Eq. 7.18:

$$W_S = 7.4 \times 10^{25}\ \text{erg sr}^{-1}. \tag{7.19}$$

7.27.6 Comparison with Other Solar Neutron Events

Watanabe et al. (2003b) by using results of Shibata et al. (1993), compared obtained results for the 24 November, 2000 event with several other solar neutron events (see Table 7.7). Using the Shibata et al. (1993) values, neutron energy spectra at the solar surface of these solar neutron events can be calculated. Calculated values are shown in Table 7.7 together with the results of the 24 November, 2000 event.

As noted Watanabe et al. (2003b), the solar neutron event on 4 June, 1991 was observed by three different detectors, but only the neutron monitor value is shown in Table 7.7. The solar neutron event on 22nd March, 1991 is the weakest event in Table 7.7. Because the Haleakala neutron monitor has a large area, this event was detected. Comparing the solar neutron events observed in former cycles with that observed on 24 November, 2000 event, the latter event is fainter than previous events. Because of the thin air mass at Mount Chacaltaya (540 g·cm^{-2} for vertical air mass), even faint signals can be detected. Table 7.7 shows most solar neutron events observed in former solar cycles come from limb flares. But, the 24 November, 2000 event came from disk flare. Solar neutrons are released tangentially to the solar surface (Hua and Lingenfelter 1987a, b; Hua et al. 2002). Therefore, solar neutrons are thought to be detected easily from limb flares.

Table 7.7 Comparison of the solar neutron event of 24 November, 2000 with other events (From Watanabe et al. 2003b)

Date	June 3, 1982	May 24, 1990	March 22, 1991	June 4, 1991	November 24, 2000
Time (UT)	11:43	20:44	22:44	3:37	14:51
Observatory	Jungfraujoch (Switzerland)	Climax (USA)	Haleakala (Hawaii)	Mount Norikura (Japan)	Mount Chacalyaya (Bolivia)
Height (m)	3,475	3,400	3,030	2,770	5,250
X-ray class	X8.0	X9.3	X9.4	X12.0	X2.3
Sunspot location	S09°E72°	N36°W76°	S26°E28°	N30°E70°	N22°W07°
Detector	12IGY	12IGY	17NM64	12NM64	12NM64
Fiux at 100 MeV ($\times 10^{28}$ neutrons MeV^{-1} sr^{-1})	2.6 ± 0.7	4.3 ± 0.4	0.06 ± 0.01	1.8 ± 0.2	0.18 ± 0.08
Power index	−4.0 ± 0.2	−2.9 ± 0.1	−2.7 ± 0.1	−7.3 ± 0.2	−4.9 ± 0.7

The result in the solar neutron event in solar cycle 23, however, is not explained by this argument. The production direction of solar neutrons in the solar atmosphere has to be treated in more detail, taking into account the position of the solar flare at the solar surface and the loop structure of each event.

Watanabe et al. (2003b) underlined that the spectrum of accelerated ions can be calculated from the neutron spectrum using the spectrum of escaping neutrons produced by the accelerated ions (Hua and Lingenfelter 1987a, b). From the neutron spectra shown in Table 7.7, the number of protons above 30 MeV would be larger than 10^{33} in all events, under the assumption that there is no turnover of the spectrum. This is larger than the number of protons accelerated in the very large flare on June 4, 1991 (Murphy et al. 1997). Estimating the number of solar neutrons below 100 MeV accurately, however, is impossible because these low-energy neutrons suffer from high attenuation through the Earth's atmosphere. Possibly that the turnover of the spectrum in the low-energy range results in a smaller estimate of the number of accelerated protons. The observation of solar neutrons below 100 MeV by satellite experiments is indispensable in order to determine the total number of protons accelerated in a solar flare.

7.28 The Solar Neutron and GLE Event of April 15, 2001

7.28.1 The Event of the 15 April, 2001 and the Importance of Solar Neutron and GLE Event Investigation

The event of 15 April, 2001 was caused by a strong solar flare at the west limb of the solar surface (S20°, W85°) with intensity of x-rays measured by the GOES satellite as X14. It was investigated in details by Muraki et al. (2006, 2007, 2008a, b). As underlined Muraki et al. (2008a, b), more than 25 years have passed since the discovery of solar neutrons by the detector onboard the SMM satellite (Chupp et al. 1982) and detectors located on high mountains (Chupp et al. 1983, 1987; Efimov et al. 1983). Initially detection of solar neutrons was a very rare event. At that time, we did not know whether or not solar neutrons were produced impulsively or produced by gradually accelerated ions. Even today, we cannot say that we understand well the particle acceleration mechanism at the Sun, for example whether they are accelerated by a shock mechanism (Ramaty and Murphy 1987; Ohsawa and Sakai 1987; Hoshino et al. 1992; Tsuneta and Naito 1998; Ohyama and Shibata 2000; see also in details Chapter 4 in Dorman M2006) or by a DC mechanism (Litvinenko 1996).

An effort of Muraki et al. (2006, 2007, 2008a, b) to collect more solar neutron and GLE events has provided partial answers to these questions. The ions are observed to be accelerated simultaneously up to high energies with electrons, because when solar neutron events are observed, hard x-rays and γ-rays are detected (Watanabe et al. 2003b, 2006a, b).

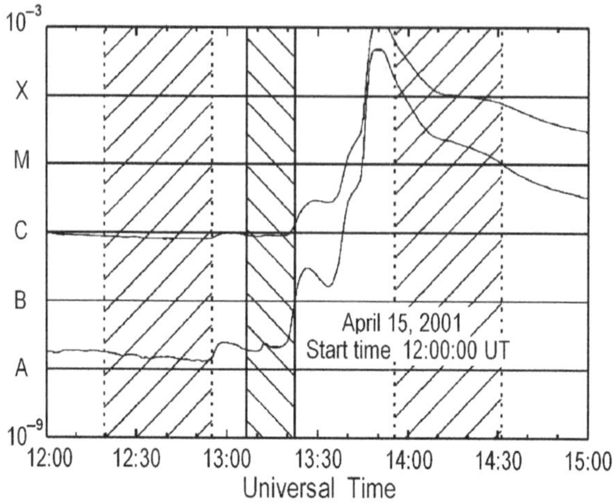

Fig. 7.77 The x-ray intensity time profile observed by the GOES satellite. The upper curve shows the wavelength band, 0.1–0.8 nm, while the lower curve shows the band, 0.05–0.4 nm. The x-ray intensity increased in three steps; at 13:28 UT (C4), 13:40 UT (M1), and 13:48 UT (X10). The shadows represent the South Atlantic Anomaly (negative slope) for the Yohkoh satellite and shadow of the Earth (positive slope) for the Yohkoh orbit, respectively. The *vertical axis* units are W/m^2. The levels A, B, C, M, and X refer to x-ray event levels for the 0.1–0.8 nm band. From Muraki et al. (2008b)

7.28.2 Satellite x-Ray and γ-Ray Observations

According to the observation of the GOES satellite, the solar flare started at 13:19 UT and reached maximum at 13:50 UT. However, as shown in Fig. 7.77, the flare increased in intensity by three steps from a level of C4 at 13:28 UT, to M1 at 13:40 UT, and to X10 at 13:48 UT. The x-rays increased abruptly from M4 to X10 within 3 min between 13:45 and 13:48 UT. The Yohkoh satellite could observe the flare from the initial stage at 13:22 UT through the maximum until 13:56 UT.

The shadows in Fig. 7.77 shown represent the South Atlantic Anomaly of the Yohkoh satellite and the shadow of the Earth for the Yohkoh orbit. Results of γ-ray measurements on Yohkoh satellite are shown in Fig. 7.78.

7.28.3 NM Observations at Mt. Chacaltaya

According to Muraki et al. (2008a), the 10 min time profiles observed by the Chacaltaya neutron monitor and the plastic neutron counter with >40 MeV show a typical daily variation and the minimum value was observed around 11 a.m. local time, while high-energy parts of the plastic counter show only pressure variations. As shown in Fig. 7.79, in the 5 min time profile of the neutron monitor, it can be seen a clear peak starting from 13:51 UT. From Fig. 7.79 can be seen that in the data of

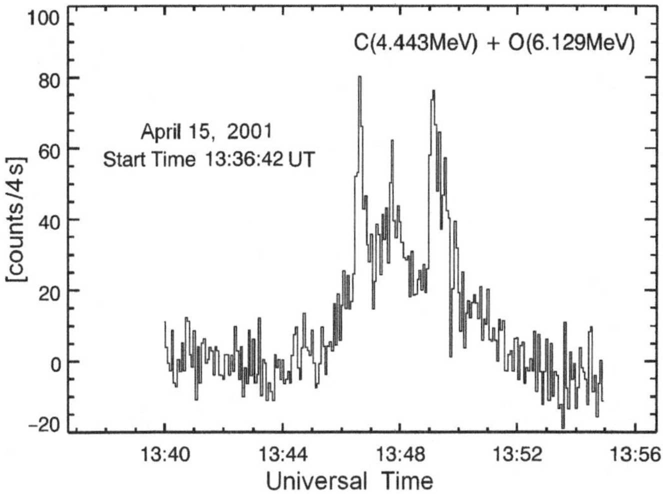

Fig. 7.78 The gamma-ray time profile observed by Yohkoh/GRS. The counting rate observed in the energy range between 4 and 7 MeV for carbon and oxygen gamma-ray lines. From Muraki et al. (2008b)

the plastic neutron counter, only marginal (2σ level) enhancements were observed. The 5 min data of the neutron monitor tell us that the excess continued for more than 30 min, while the data of the plastic scintillator show an excess in the highest-energy channel (E_n > 160 MeV) in a 1 min interval only. The background may mask the high-energy low flux part of the signal. The statistical significance of the excess observed by the neutron monitor in the 24 min from 13:51 and 14:15 UT is 3.6σ.

7.28.4 NM Observations at Mt. Jungfraujoch and Mt. Gornergrat

In Fig. 7.80 are plotted the time profiles of the counting rates of two channels recorded by the Mt. Gornergrat neutron telescope and in Fig. 7.81 are displayed the 5 and 1 min values for the 3-NM-64 and the 18-IGY neutron monitors located at Mt. Jungfraujoch. Their altitudes are 3,475 and 3,570 m, respectively. Panel b in Fig. 7.81 indicates that the GLE started from 14:02 UT.

7.28.5 NM Observations at the South Pole

Figure 7.82 shows the neutron monitor data observed at the South Pole in the Antarctic, located at altitude 2,820 m. As can be seen from Fig. 7.82, the increase of the counting rate started at 14:00 UT, nearly the same time observed by the neutron monitors at Mt. Jungfraujoch. The vertical geomagnetic cutoff rigidity for protons at Mt. Jungfraujoch and the South Pole are 4.44 and 0.09 GV, respectively.

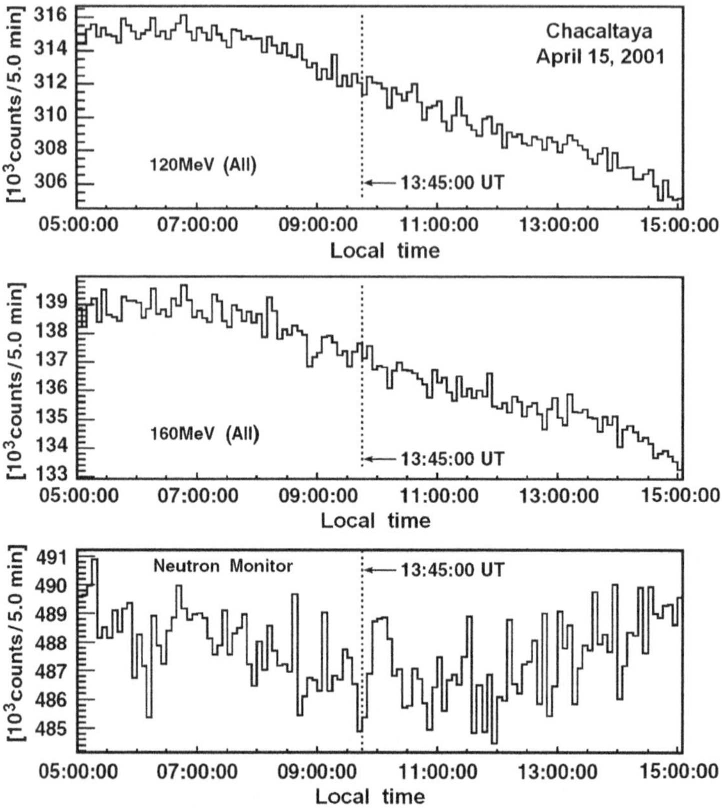

Fig. 7.79 Five minutes values observed by the Chacaltaya plastic neutron detector (*top* and *middle panel*) and by the neutron monitor (*bottom panel*) in April 15, 2001. A clear enhancement was seen in the data of the neutron monitor starting around 13:48 UT. The *horizontal axis* represents the local time in Bolivia. From Muraki et al. (2008a)

7.28.6 NM Observations at Mt. Aragats

Figure 7.83 represents the time profile of the neutron counter located at Mt. Aragats. No remarkable enhancement due to the GLE was seen at Mt. Aragats. Muraki et al. (2008) have calculated the asymptotic directions of CR arriving at selected neutron monitor stations. The results are shown in Fig. 7.84. The atmospheric depth to the Sun was 719 and 808 g/cm^2 at Chacaltaya and Gornergrat, respectively, but at Mt. Aragats it was 1,006 g/cm^2.

Muraki et al. (2008b) noted that according to their empirical knowledge, signals of solar neutrons in current detectors were not observed when the atmospheric depth is deeper than 800 g/cm^2 (Flückiger et al. 2005). For energetic solar charged particle effect is important value of cutoff rigidity: for the Chacaltaya NM, the cutoff rigidity is estimated as 12.1 GV. The Chacaltaya neutron detectors did not

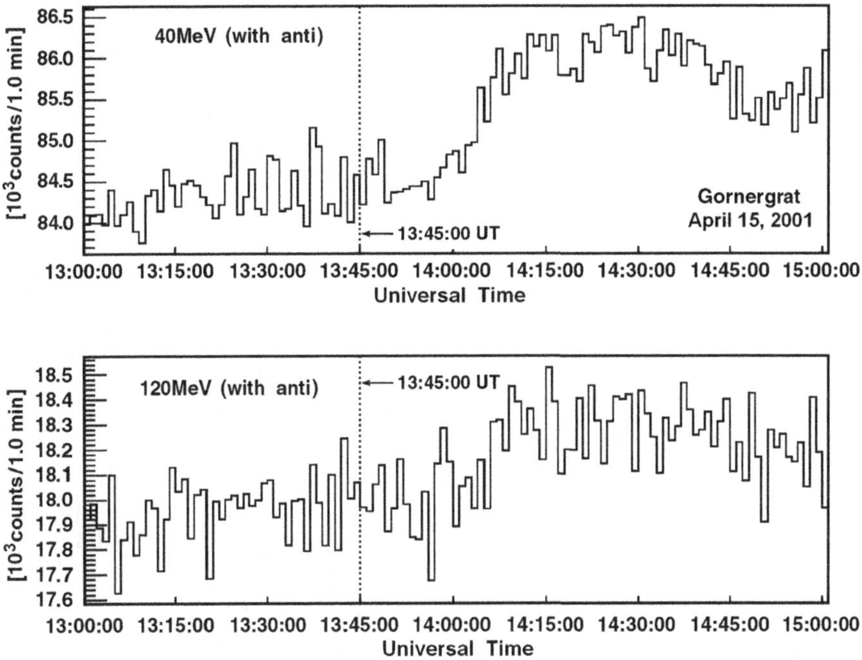

Fig. 7.80 The time profiles of the counting rates of two channels observed by the Mt. Gornergrat neutron telescope for the event at April 15, 2001. *Top panel* represents the intensity of neutrons above 40 MeV and *bottom panel* represents the intensity above 120 MeV. From Muraki et al. (2008b)

see any remarkable enhancement due to 'traditional' GLE that enhances rapidly and decays slowly for more than 2 h (Dorman and Miroshnichenko M1968; Dorman M1978; Dorman and Venkatesan 1993; Miroshnichenko M2001). This suggests that the low energy solar protons were rejected by the geomagnetic field and could not reach the Chacaltaya observatory at 5,250 m. Muraki et al. (2008a) conclude therefore that the enhancement observed by the Chacaltaya NM between 13:51 and 14:02 UT must be mainly due to solar neutrons.

7.28.7 The Sequence of Observations and Results of Data Analysis for Solar Neutrons

The sequence of observations is shown in Fig. 7.85.

Figure 7.78 was shown the time profile of the gamma-rays in the energy range of 4–7 MeV observed by the Yohkoh/GRS during 13:40–13:55 UT. Muraki et al. (2008b) assume that ions were accelerated with the same time profile as gamma-rays, i.e. the ions were accelerated from 13:45 UT, as shown in Fig. 7.78 and

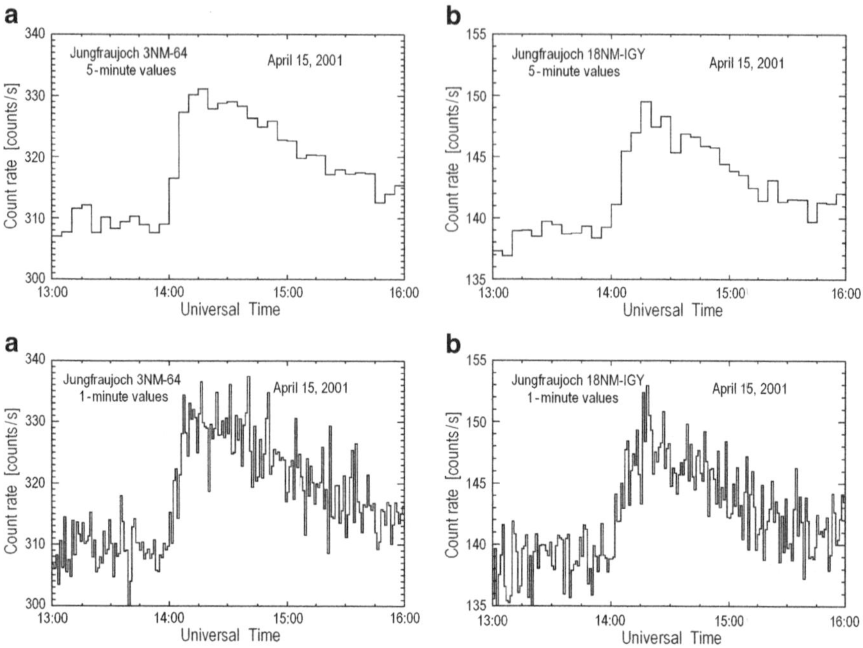

Fig. 7.81 The 5 and 1 min values of the neutron monitors located at Mt. Jungfraujoch: (**a**) data were obtained by 3-NM-64 and (**b**) data are from 18-NM-IGY neutron monitor. The altitudes of the locations are 3,470 and 3,570 m, respectively. From Muraki et al. (2008a)

Fig. 7.82 The 1 min data of the neutron monitor located at the South Pole. The altitude of the South Pole is 2,820 m. The data in the plot represents the data after correction for atmospheric pressure. From Muraki et al. (2008b)

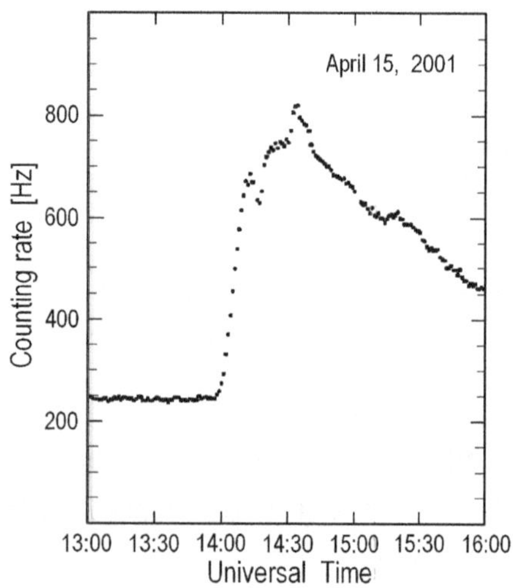

particle acceleration continued until 13:51 UT, the end of strong emission of gamma-rays. Also Muraki et al. (2008b) assume that the energy spectrum of solar neutrons can be expressed by a power law.

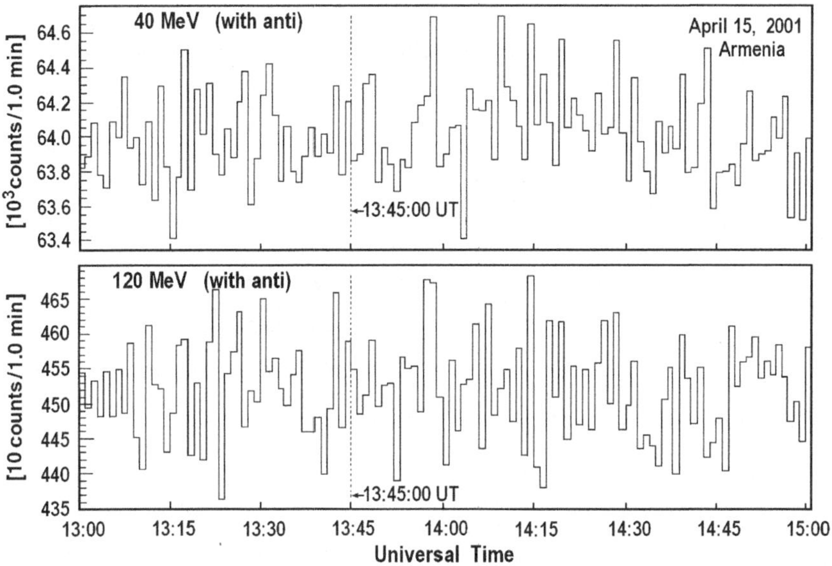

Fig. 7.83 The time profile of the counting rate of the neutral channel observed by the Mt. Aragats neutron counter. *Top panel* represents the intensity of neutral particles above 40 MeV and *bottom panel* represents the intensity above 120 MeV. The Mt. Aragats detector did not see the increase due to the GLE. From Muraki et al. (2008a)

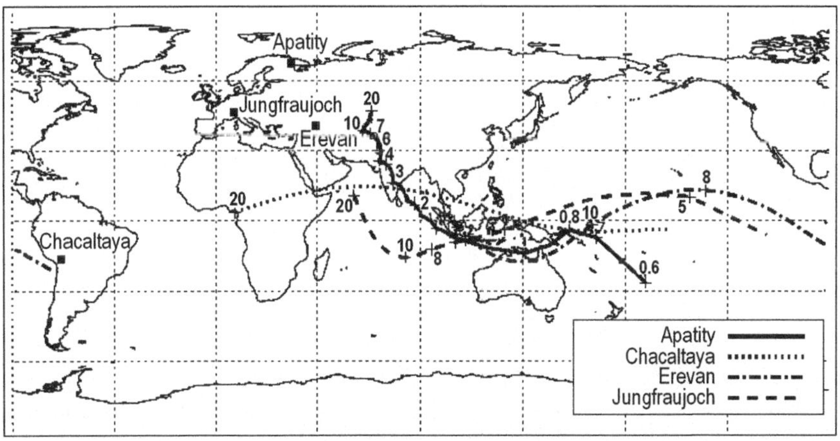

Fig. 7.84 Asymptotic directions of vertically incoming cosmic ray particles at the NM stations Apatity (*solid line*), Mt. Chacaltaya (*dotted line*), Mt. Aragats (*dashed-dotted line*), and Jungfraujoch (*dashed line*) at 13:45 UT on April 15, 2001. The numbers close to the lines give the rigidity of the particles in GV. From Muraki et al. (2008b)

Figure 7.86 shows the reduced χ^2-values of fitting the measurements of the Chacaltaya NM by assuming a power law spectrum with indices γ in the range from -3.0 to -9.0. The dependence of the reduced χ^2-value from the power index, γ,

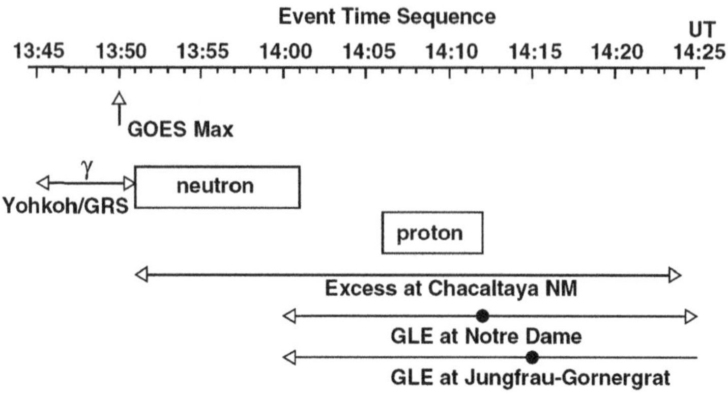

Fig. 7.85 The event time profile observed by each detector as a function of universal time. Following the detection of gamma-rays with energy 4–7 MeV by the Yohkoh GRS detector, high-energy neutrons and protons were detected by the Chacaltaya NM. The excess as a function of time used for determining the statistical significance is indicated by the box. The GLE commenced at 14:01 UT and continued for 2 h, however, the high-energy portions with energy >30 GeV continued for only 25 min. The *small filled circles* indicate the peak time of the GLE recorded at each station. Each time corresponds to the time at the Earth. From Muraki et al. (2008a)

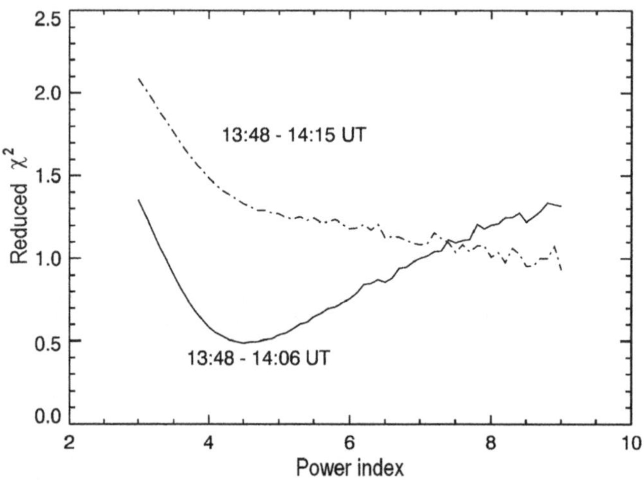

Fig. 7.86 The reduced χ^2-value of the fitting. When Muraki et al. (2008b) attempt to fit the impulsive production model to the data between 13:48 and 14:15 UT, χ^2-values does not show a minimum (*dashed-dotted line*). However, when they fit the data between 13:48 and 14:06 UT, the χ^2-value has a minimum value at $\gamma \approx -4.0$ (*solid line*)

was made by fitting the measurements in the time intervals 13:45–14:15 and 13:48–14:06 UT.

As can be seen from Fig. 7.86, the reduced χ^2-value for the fit in the time range 13:48–14:06 UT has a minimum at $\gamma \approx -4.0$. The result of fitting the data between 13:51 and 14:06 UT with $\gamma \approx -4.0$ is presented in Fig. 7.87 by filled circles.

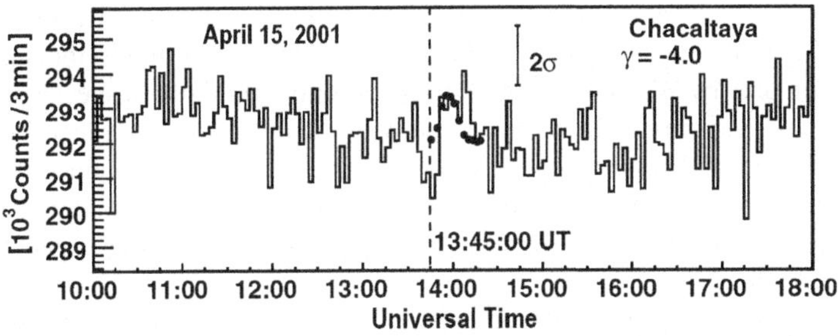

Fig. 7.87 The results of fitting the 3 min values of the Chacaltaya neutron monitor at the time interval between 13:51 UT and 14:15 UT. It was assumed that neutrons were produced with the same production time profile of gamma-rays as shown in Fig. 7.78. In the time 14:06–14:12 UT, another peak can be recognized in the data of the neutron monitor. This second peak may be induced by the high-energy solar protons with $E > 12$ GeV, since during this time the GLE was observed. From Muraki et al. (2008a)

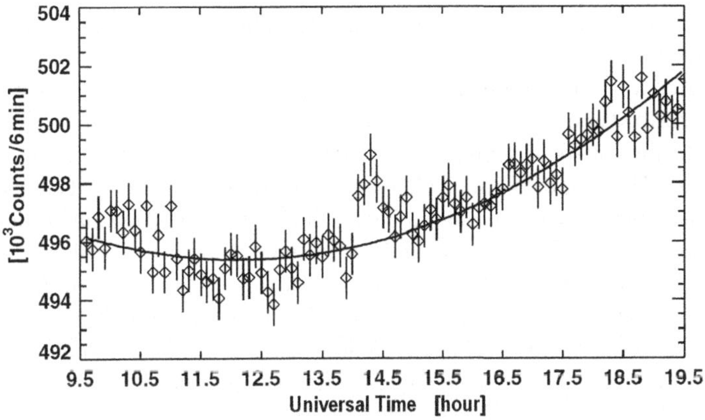

Fig. 7.88 The 6 min counting rates above background observed by GRAND (*boxes with error bars*). A 6.1σ enhancement was seen during 14:00–14:30 UT. The line represents the background. Muons produced by protons with energies above ≈20 GeV can be detected by GRAND. The median-energy for primary protons is estimated as 30 GeV. From Muraki et al. (2008b)

From Fig. 7.87, a deviation from the fit can be seen around the peak at 14:06 UT. Therefore it should be used a modified model to fit the observed data. In the solar flare of May 24, 1990, the arrival time of neutrons is clearly separated from the arrival time of protons (Debrunner et al. 1993; Muraki and Shibata 1996). However, in the flare of April 15, 2001, the time difference between arrival times of the neutrons and the protons was very short. Since the flare occurred at the west limb (W85), protons arrived at the Earth rapidly. This complicates the situation. Muraki et al. (2008) assume that from 13:51 to 14:02 UT only neutrons are involved in the enhancement observed at Chacaltaya as shown in the box of Fig. 7.85. However, after 14:02 UT, high-energy protons may be included in the data. The hypothesis is strongly supported by the data obtained by GRAND as shown in Fig. 7.88.

Muraki et al. (2008) attempted to separate the proton and neutron components in this time interval, since the GLE started at 14:02 UT. According to the analysis of past neutron data by Watanabe (2005), the differential production spectrum of solar neutrons can be expressed by a power law with the index γ in the range from -3.0 to -4.0. The present result is consistent with this analysis. We obtained the proton flux after 14:06 UT by subtracting the neutron flux from the raw data. The procedure was as follows. The event time profile was fit to the expected curve with a power law.

The data points of Fig. 7.88 represents the expected points for neutrons. The residuals from these plots were regarded as a proton component. The number of protons is estimated to be 3,500 events for 6 min during 14:06–14:12 UT. The statistical significance is 3.0σ while the statistical significance of the neutron component is also 3.0σ. A multiplication factor of 1.52 has been applied to the neutrons inside the neutron monitor.

Statistical significances in Muraki et al. (2008) were determined as follows. Figure 7.89 (panel a) shows the 5 min values recorded by the Chacaltaya neutron monitor together with the fitted background (solid line) and its one sigma deviations (dashed lines).

The latter were determined by allowing for the multi-counting correction factor of 1.52 described above. The result is shown in panel c in Fig. 7.89. The shaded data points indicate statistical significances from 13:51 to 14:03 UT. Their significance is nearly 2σ. The four data points yield a combined excess of $\approx 4\sigma$. Those statistical significances were obtained using a background of 292,000 counts during 3 min. This background level was estimated by a running average for the 5 min value (Fig. 7.89) excluding possible peak regions. However, Muraki et al. (2008) could draw a background line at the level of 291,600 counts for 3 min. Then the statistical significance would be 4.0σ for neutrons and 3.7σ for protons. Muraki et al. (2008) have used the more conservative value for the background.

7.28.8 Comparison with Data on Solar Protons

Figure 7.90 displays the integral flux of solar protons at the top of the atmosphere.

In Fig. 7.90 the low-energy part of the flux was obtained by the GOES satellite. Muraki et al. (2008b) assumed that the peak value of the GOES data reflects the energy spectrum of protons at the acceleration site. The observed peak intensities for ground level detectors are plotted at their corresponding cutoff rigidities. Muraki et al. (2008b) have used the neutron monitors located at different rigidities as a spectrometer. For the neutron monitor located at Apatity and at Jungfraujoch and Chacaltaya, they have used the multiplication factor of neutrons 1.4 and 1.52, respectively (Hatton 1971). In Fig. 7.90, the error bars represent the statistical error only. The data points for Apatity (cutoff rigidity 0.65 GV), Jungfraujoch (4.44 GV) and Chacaltaya (12.1 GV) shown in Fig. 7.90 are obtained after the correction on the absorption effect of protons and neutrons in the atmosphere. The ratios of

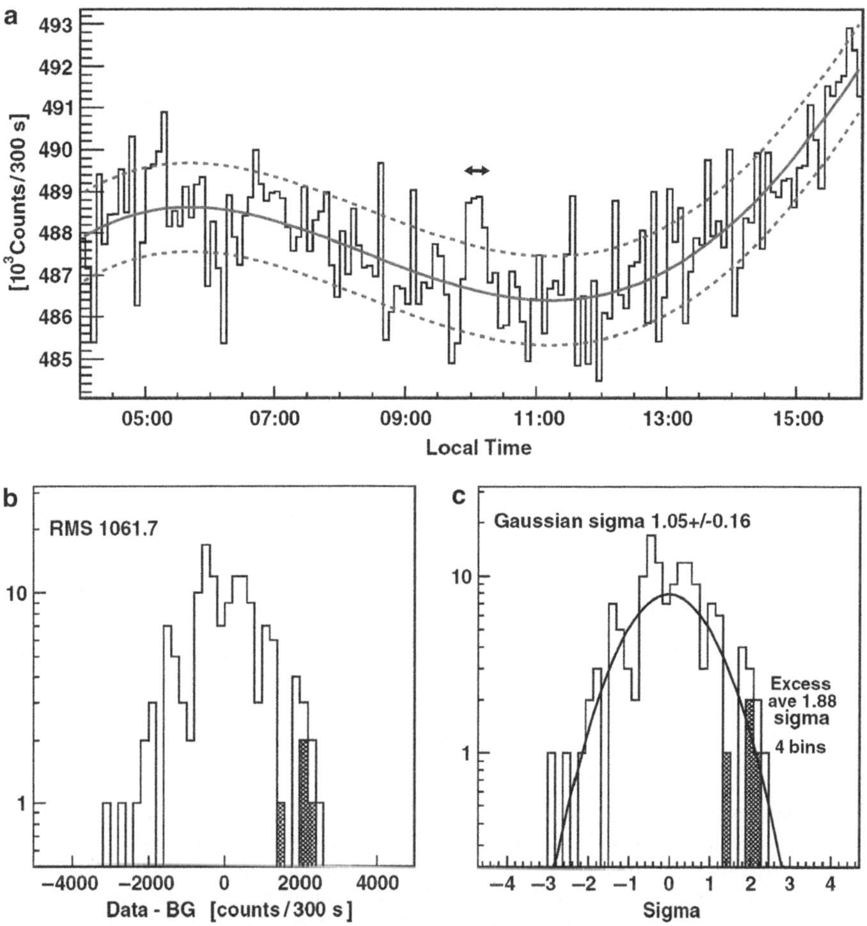

Fig. 7.89 Panel **a**: 5 min values recorded by the Chacaltaya neutron monitor. The *solid line* shows the mean counting rate fitted by a 5th order polynomial outside the signal region. The *dashed lines* are one sigma limits. Panel **b** shows the deviation of each point from the fitted background plotted on a semi-logarithmic scale. Panel **c** shows deviations from the mean on a semi-logarithmic scale together with the Gaussian distribution (*solid line*). When the deviations in Panel **b** are reduced by a factor 1.52, the deviations follow the Gaussian distribution. The four data points near 1.88σ yield a combined excess of $\approx 3.7\sigma$. From Muraki et al. (2008a)

secondary to primary nucleons are $3\ 10^{-4}$, 0.1 and 0.33, respectively (Shibata 1994; Capdevielle and Muraki 1999; see also above, Chapter 5). Muraki et al. (2008) have corrected for the detection efficiency of the NMs at Chacaltaya and Jungfraujoch estimated to be 30%. However, for the NM at Apatity, they have estimated it to be 20%, since the detection efficiency of the neutron monitor is lower at lower-energy, 100 MeV (Hatton 1971; Clem and Dorman 2000; Shibata et al. 2001; see in details above, Chapter 5).

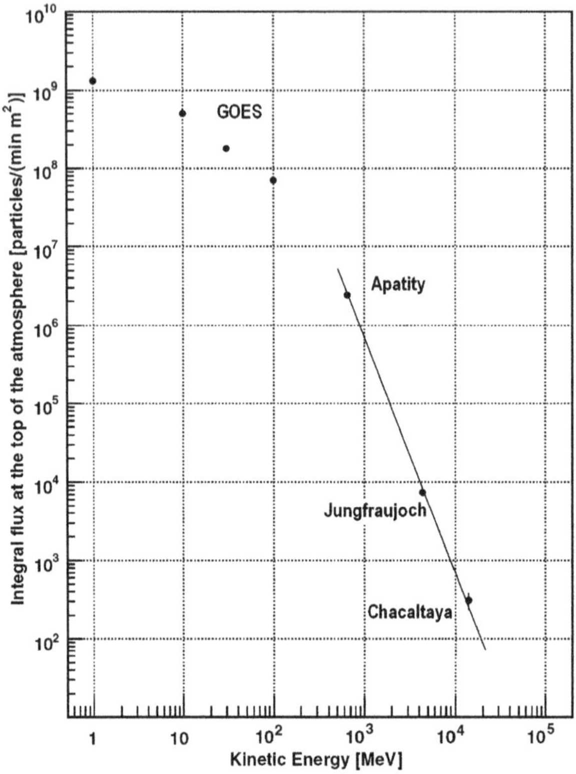

Fig. 7.90 The integral proton flux produced by the flare of April 15, 2001. The *vertical axis* represents the integral proton flux per unit m^2 and per minute. The *horizontal axis* represents the energy of protons. The data points correspond to the maximum flux measured by the GOES detector and the NM station, respectively. The maximum time of the GOES detector is observed on April 16 at about 3:00 UT for >1 MeV protons; and on April 15 at 19:23 UT, 16:15 UT, and 15:27 UT for protons, >10, >30 and >100 MeV, respectively; while for the neutron monitor data for Apatity 14:05 UT, for Jungfraujoch 14:18 UT and for Chacaltaya 14:06 UT, respectively. To the data points observed by the ground level detector, the rigidity at the observation point was plotted. The line corresponds to expected values from the power index $\gamma = -3.0$. From Muraki et al. (2008b)

Figure 7.91 represents the integral flux of solar protons for various GLEs. In Fig. 7.91, data of the Easter event are presented by black circles. It is impressive that protons were accelerated beyond 12 GeV and the flux can be expressed by a simple power law with an index of $\gamma = -2.75 \pm 0.15$ in the energy range between 650 MeV and 12 GeV. The flux of the highest-energy point observed at Chacaltaya is already one order less than the flux of the galactic CR. Since the altitude of the observatory is very high, the attenuation of solar protons was small, so it could be possible to detect the signals from this small flux.

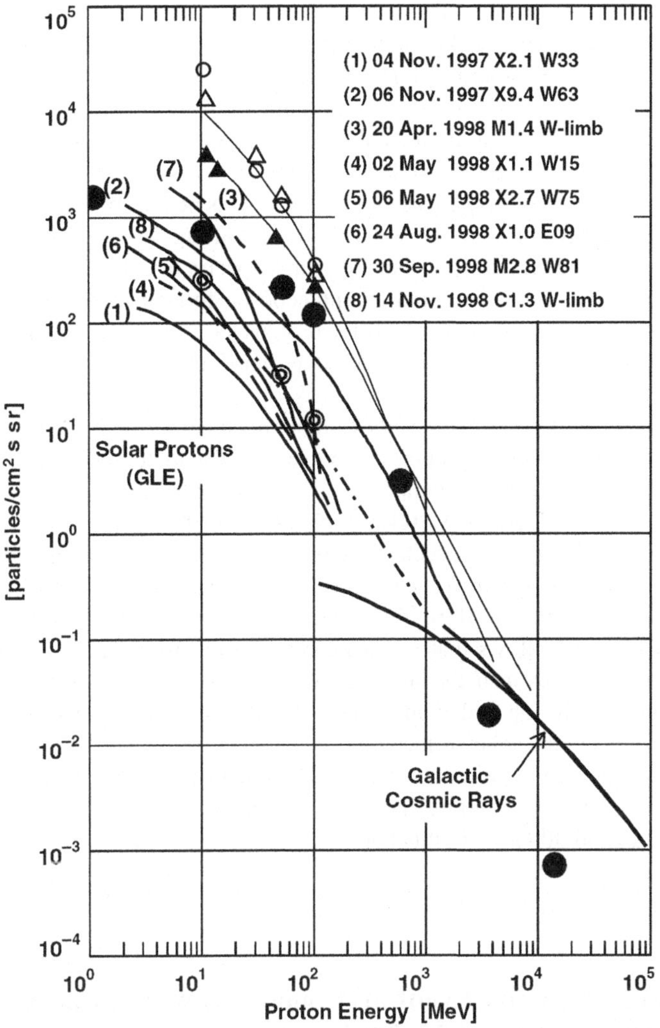

Fig. 7.91 The integral flux of solar protons for different GLEs according to Miroshnichenko (M2001). The *vertical axis* is in units of particles/(cm^2·s·str). The Easter event is presented by the black circles as •. The *black triangles* and *thin line* show the data observed in 1989, September 29th flare (X9.8). The *open round mark* and *thin line* corresponds to the data of the Bastille day event of 2000 (X5.7). The *double circle* and the *open triangle* represent the data on April 18, 2001 flare (C2.2) and November 8, 2000 flare (M7.4), respectively. From Muraki et al. (2008b)

As noted Muraki et al. (2008), the detection of muon signals by GRAND implies that protons were accelerated beyond 30 GeV (possibly over the energy 56 GeV) in this flare (Allkofer 1983; Alania et al. 2003; Cattani et al. 2003; Karpov et al. 2003, Grieder M2001). The GRAND, located adjacent to the University of Notre Dame, has recorded a 6.1σ excess in the time interval 14:00–14:30 UT shown in Fig. 7.88. The counting rate above background at the peak was 2,900 $\pm$ 730 muons for this 6 min bin or 24 $\pm$ 6 protons/min/m^2 for the threshold energy of 30 GeV where GRAND response function is 0.5 muon/proton (D'Andrea and Poirier 2003). For this energy of the incoming protons, the flux is estimated (the data was divided by 0.5). The shape of the time profile is similar to that of the Chacaltaya neutron monitor. The observed flux of particles at GRAND is near a straight line projection of the lower-energy points on this logarithmic plot of Fig. 7.91. Muraki et al. (2008) noted that it may be the first time that the energy spectra of both high-energy protons and neutrons have been measured simultaneously, though there are a number of reports on the energy spectra of GLEs protons (see review in Miroshnichenko M2001). The excess shown for GRAND is very small in comparison with the other GLEs.).

In comparison of described results with obtained earlier by Bieber et al. (2004), Muraki et al. (2008) came to important conclusion. The matter is that in Bieber et al. (2004) were analyzed data obtained by the Spaceship Earth network (see description in Chapter 4 in Dorman M2004). Bieber et al. (2004) come to suggestion that protons most probably were accelerated only up to 1 GeV by a CME-driven shock wave. But according to Muraki et al. (2008b), solar protons are found in data of the Chacaltaya NM with cutoff rigidity about 12 GV. According to Muraki et al. (2008b), from this follows that at least some part of protons must be accelerated in the magnetic loops near the solar surface and those protons must hit the solar atmosphere from the top as they moved in a downward direction. The injection time of those protons to the solar surface must be between 13:45 and 13:51 UT when the gamma-ray lines were observed by the Yohkoh/GRS. From this, according to Muraki et al. (2008b), follows that solar protons were accelerated over 12 GeV (probably beyond 30 GeV) in this flare.

7.29 Solar Neutron Event in Association with a Large Solar Flare on August 25, 2001

7.29.1 The Matter of Problem

As noted Watanabe et al. (2003c), more than 60 X-class flares occurred in solar cycle 23. Among them, solar neutrons were detected by the NM installed at Mt. Chacaltaya, Bolivia, only in association with an X2.3 flare on November 24, 2000 (Watanabe et al. 2003b; see above, Section 7.27), and with an X5.3 flare on August 25, 2001. In Watanabe et al. (2003c) are described the solar neutron event of August 25, 2001. The NM data for this solar flare have been analyzed and compared with x-ray and γ-ray data obtained by the Yohkoh satellite.

7.29.2 NM Observations at Mt. Chacaltaya in Comparison with x-Ray and γ-Ray Data Obtained by the Yohkoh Satellite

According to Watanabe et al. (2003c), an X5.3 class solar flare occurred at 16:23UT in NOAA region 9591 on August 25, 2001. The location of the active region was S17°, E34° and this flare was a disk flare. The soft x-ray flux observed by the GOES satellite was at a maximum at 16:45 UT, which was 22 min after the flare onset time. Since it took a considerable time for the soft x-ray flux to reach the level of X5.3, the neutron production time cannot be determined from the soft x-ray data alone. Large fluxes of hard x-rays and γ-rays were also observed in this event. Figure 7.92 shows the time profile of hard x-rays observed by the Yohkoh/HXT, the γ-ray spectrum and the time profile of γ-rays around 2.223 MeV observed by the Yohkoh/GRS on August 25, 2001.

At the start time of an X5.3 flare on August 25, 2001, the Sun was directly over Bolivia and the NM installed at Mt. Chacaltaya, Bolivia, was the most suitable place for observing solar neutrons. This station is located at 292.0°E, 16.2°S, 5,250 m above sea level, and the vertical air mass is 540 g/cm^2. At this time, the zenith angle of the Sun was 26.5° and the air mass for the line of sight to the Sun was 603 g/cm^2.

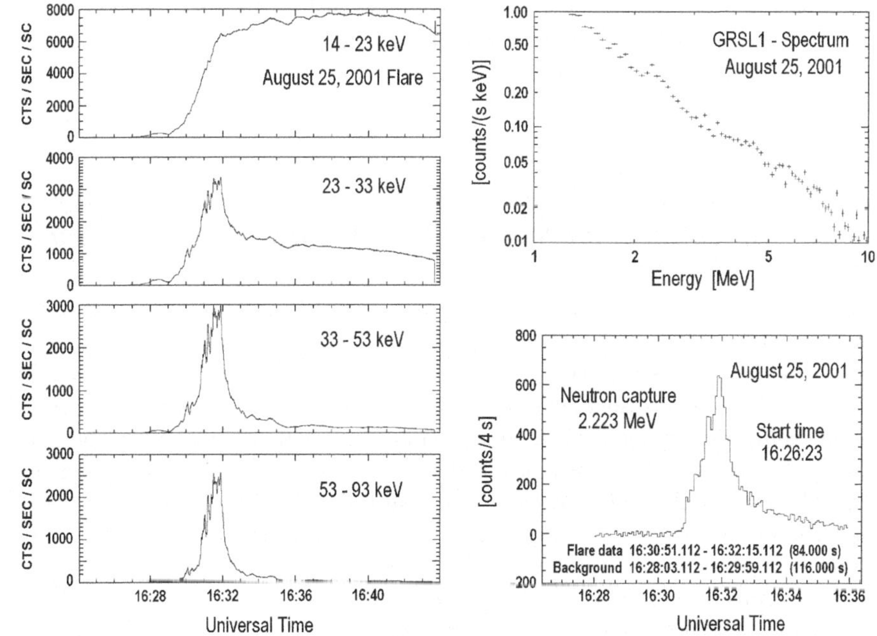

Fig. 7.92 The hard x-ray and γ-ray data observed by Yohkoh on August 25, 2001. The left side shows the time profiles of hard x-rays. The upper figure on the right shows the γ-ray spectrum, and the lower one the time profile of γ-rays around 2.223MeV. From Watanabe et al. (2003c)

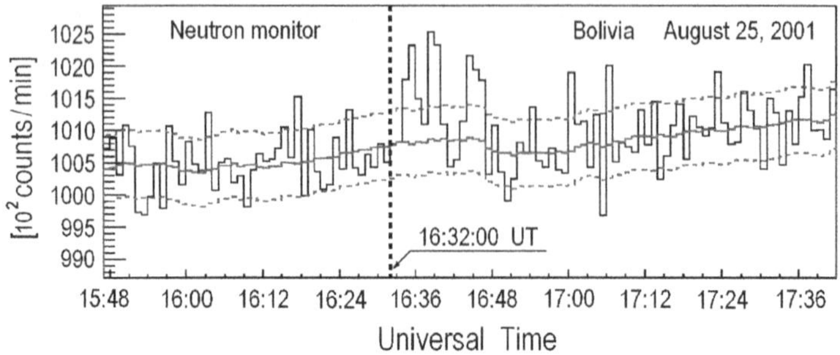

Fig. 7.93 The time profile of neutrons detected by the Bolivia neutron monitor on August 25, 2001. The *solid smooth line* is the averaged background, and *dashed lines* are $\pm 1\sigma$ from the background. From Watanabe et al. (2003c)

The NM installed at Mt. Chacaltaya is 13.1 m^2 in area and of the NM64 type. The counting rate is recorded every 1 min. The time profile of neutrons observed by the neutron monitor is shown in Fig. 7.93. A clear excess was found between 16:34 and 16:49 UT. The statistical significance of these excesses is 4.7σ at 16:34–16:39, 1.4σ at 16:39–16:44, and 2.5σ at 16:44–16:49 UT. The total significance for the period of 15 min, between 16:34 and 16:49 UT, is 4.7σ. There is a possibility that these excesses came from energetic ions which can also be detected by the neutron monitor. However, there is no evidence that this was the case since measurements by other stations in the worldwide neutron monitor network showed no enhancement. Accordingly, these signals must have come from solar neutrons.

7.29.3 Estimation of the Differential Neutron Ejection Energy Spectrum at the Sun and Total Energy in Solar Neutrons

As noted Watanabe et al. (2003c), in the γ-ray spectrum in Fig. 7.92, a weak signal of 2.223 MeV neutron capture line γ-rays is seen in the bremsstrahlung component. However, the γ-ray lines produced by de-excited ions are not seen because this solar flare was electron rich. Solar neutrons are produced simultaneously with line γ-rays of de-excited ions by interactions of accelerated ions with the solar atmosphere. In this event, the γ-ray lines of de-excited ions were not observed, however, large fluxes of hard x-rays and γ-rays are evident in Fig. 7.92. The time profiles of the high energy channel of hard x-rays and γ-rays are similar, and the fluxes of the hard x-rays and γ- rays were largest at 16:32 UT. From these data, it can be deduce that solar neutrons were produced at 16:32 UT. On this basis, the energies of neutrons which cause the increases recorded by the NM are estimated to be 612.2 MeV at 16:34 UT, and 54.6 MeV at 16:49 UT. From the time profile of the neutrons, Watanabe et al. (2003c) calculated the energy spectrum of solar neutrons at the solar surface. They use the detection efficiency of the neutron monitor calculated by Clem and Dorman (2000), and the attenuation of solar neutrons in the Earth's

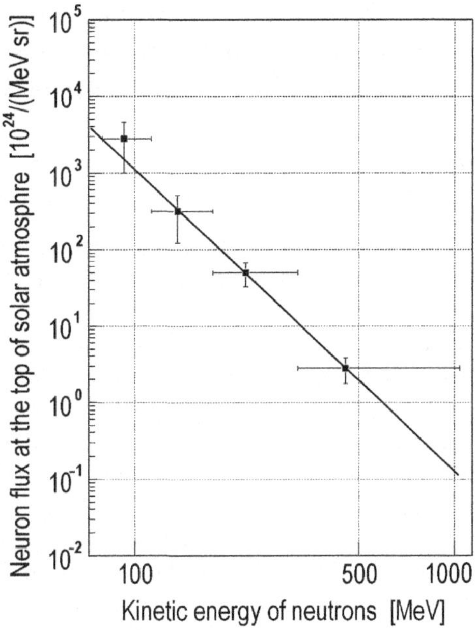

Fig. 7.94 The spectrum of solar neutrons at the solar surface produced in association with the solar flare that occurred on August 25, 2001. The *vertical error bars* represent only statistical errors. From Watanabe et al. (2003c)

atmosphere as calculated by Shibata (1994) by Monte Carlo simulations. To derive the energy spectrum of neutrons at the solar surface from the flux at the top of the Earth's atmosphere, the survival probability of neutrons between the Sun and the Earth must also be taken into account. The result is shown in Fig. 7.94. This spectrum was derived from the 3 min counting rate. Data points in Fig. 7.94 were fitted with a power law of the form

$$D_S(E_n) = C \times (E_n/100[\mathrm{MeV}])^{\alpha}, \tag{7.20}$$

To obtain the energy spectrum of the solar neutrons. Here C is the flux of neutrons at 100 MeV, and α is the power law index. The fitting region is chosen to be above 100 MeV, where the errors resulting from the uncertainty of neutron attenuation in the Earth's atmosphere are small. The values obtained were as follows:

$$D_S(E_n) = (1.1 \pm 0.9) \times 10^{27} \times (E_n/100[\mathrm{MeV}])^{-3.9\pm0.7} \left[\mathrm{MeV}^{-1}\ \mathrm{sr}^{-1}\right]. \tag{7.21}$$

For this fit, the value of $\chi^2 = 0.007$.

The total energy flux of solar neutrons which were emitted from the Sun in the energy range between 50–600 MeV was calculated by Watanabe et al. (2003c) as

$$W_S(50 - 600\ \mathrm{MeV}) = 3.4 \times 10^{25}\ \mathrm{erg/sr}, \tag{7.22}$$

what was obtained by simply integrating of Eq. 7.21.

7.30 Detection of Solar Neutrons During the Event in Association with the 24 September, 2001 Flare, Using Tibet Solar Neutron Telescope

7.30.1 The Matter and Short History of the Problem

As noted Sako et al. (2003), solar neutrons are among the best probes for investigating ion acceleration during a solar flare. Solar neutron telescopes are designed to observe specifically solar neutrons (Tsuchiya et al. 1999). They have anti-coincidence counters, a direction measurement capability, and also a calorimeter to derive information on the energy spectrum. Sako et al. (2003) present observations of a solar neutron event, using the Tibet solar neutron telescope, associated with a solar flare that occurred on 24 September, 2001. The obtained results are compared with Monte Carlo calculations. Solar neutrons have been detected to be associated with an X2.6 class solar flare that occurred at 9:32 UT, 24 September, 2001. Neutrons with kinetic energies more than 300 MeV were detected with a statistical significance of 4.6σ just 2 min after the flare onset time. The arrival directions of neutrons were well collimated with the direction of the Sun after taking atmospheric scattering into account. An important peculiarity of this event is that the solar flare seems not to be accompanied by hard x-rays and gamma-rays around 9:32 UT.

7.30.2 Properties of a Flare and Conditions for Solar Neutron Observations

According to Sako et al. (2003), the solar flare began at 9:32UT on 24 September, 2001 in NOAA region 9632 (S16°, E23°) and reached the x-ray class of X2.6. No significant increase of hard x-rays and gamma-rays was found by the Yohkoh satellite around the start time. A proton event was observed a few hours after the start time. At the flare start time, solar neutron telescopes in Armenia, Switzerland and Tibet (China) were in the day time zone. Although the zenith angle to the Sun was smallest in Armenia, the different properties of the telescopes, i.e., altitude, effective area, direction measurement capability, made the Tibet neutron telescope as the most suitable for the observation of neutrons from this flare. The position of the Sun was 30° in elevation and 69° west from south at the site, which is at Yangbajing, Tibet, China (4,300 m above sea level, 30.1°N, 90.5°E, 603 g/cm^2 vertical air mass).

7.30.3 The Tibet Solar Neutron Telescope and Observation Data

According to Sako et al. (2003), the Tibet solar neutron telescope is composed of 3 3 m 40 cm plastic scintillator and four layers of orthogonally-aligned proportional counters. Each layer contains 30 proportional counters. Incoming neutrons are converted into charged particles in the scintillator. The direction of a recoiled particle is measured using proportional counters. Nine directions both in EW ($\pm 50^\circ$ from the zenith) and NS ($\pm 41^\circ$), i.e., a total of 81 directions, are monitored. The counting rate of each channel is recorded every 10 s. Because a total of 20 cm thickness of wood absorber is inserted, the estimated energy threshold for vertically incident events is 230 MeV. Charged particles are rejected using veto of proportional counters surrounding the scintillators (details of the detector may be found in Katayose et al. 1999) Because high energy neutrons arrive at the Earth within a few minutes if their emission is instantaneous, the data is binned every 2 min. Standard deviations from the running average are defined to 1σ and a significance map of the 81 directions at 9:34 UT is derived, as shown in the top-left of Fig. 7.95, where an excess concentration towards the west is seen.

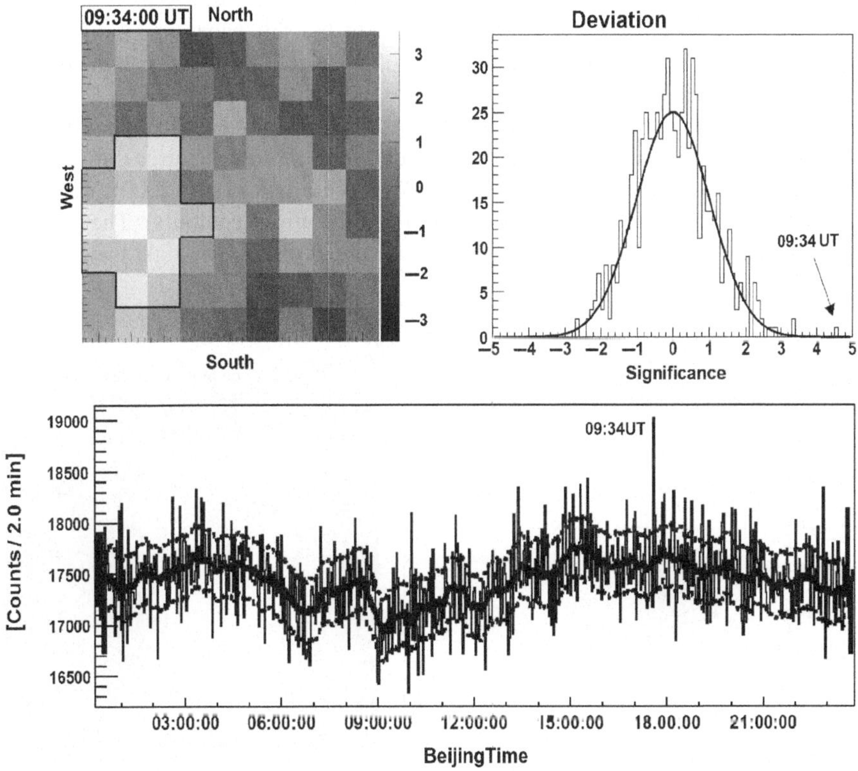

Fig. 7.95 The event observed on 24 September, 2001. *Top-left*: A significance map of the 81 directions. *Top-right*: A significance distribution of the day for the combined counting rate of the 14 selected directions. *Bottom*: 2 min counting rate of the combined 14 directions. From Sako et al. (2003)

As noted Sako et al. (2003), such an excess is not found for other times and directions. 14 channels (enclosed by dark lines in Fig. 7.95) were selected and the combined counting rate for the day is shown at the bottom with the running average and 1σ deviation. A distribution of the significance of the day is shown in the top-right, confirming that the estimation of the significance is correct. The most significant excess (4.58σ) is marked as '9:34 UT', which is also the time of the significance map.

7.30.4 Comparison with Simulation Results

According to Sako et al. (2003), the response of the Tibet solar neutron telescope was simulated for a possible solar neutron spectrum. In the simulation, Sako et al. (2003) assume neutrons are released with a power law spectrum (index –4.5) in kinetic energy at the Sun, in-flight decay was included. The attenuation in the Earth's atmosphere was calculated using Shibata (1994) model and the response of the detector was obtained using Geant 3.21. In the left panel of Fig. 7.96, the angular distribution of neutrons at the top of the detector is shown with the field of view of the Tibet solar neutron telescope.

The filled star in Fig. 7.96 indicates the position of the Sun at the flare start time. The open star shows the expected arrival direction of solar neutrons after being scattered in the atmosphere according to Valdes-Galicia et al. (2000): it is well overlapped on the results of the Monte Carlo calculation (contours). The position and spread are also very similar to the observational result. The right panel shows

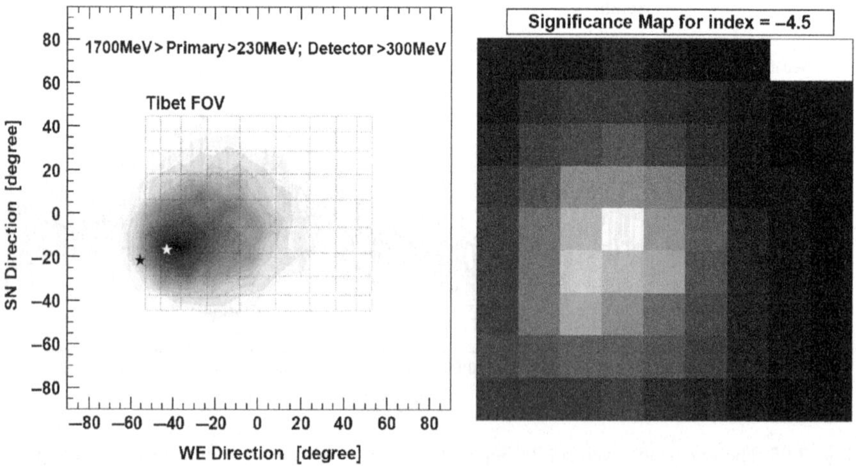

Fig. 7.96 Simulated arrival directions of neutrons at the top of the detector are shown in the left panel; in the right – after including the detector response. From Sako et al. (2003)

an expected significance map (graduation linearly and arbitrary scaled) of the event. After including the detector response, the arrival direction moves to the zenith and some difference with respect to the observations is found. Because Sako et al. (2003) compare the map in terms of significance, it is affected by the background counting rate. In the simulation, the background rate has also been simulated. In the next step, we must verify the consistency of the background estimation.

7.30.5 Peculiarities of the Solar Neutron Event in Association with the 24 September, 2001 Flare

From comparison of observations with simulation results, Sako et al. (2003) came to conclusion that really solar neutrons have been observed in association with the flare on 24 September, 2001 at a significance level of 4.6σ, using the Tibet solar neutron telescope. The arrival direction is well collimated with the expected solar position after taking atmospheric scattering into account. Some difference is found when considering the response of the detector. Sako et al. (2003) noted that it will be necessary to investigate the detector response to background particles. If the neutrons were emitted at the flare start time (9:32 UT), the observed excess found in 9:34–9:36 UT is caused by neutrons with kinetic energies 335–650 MeV. This is a quite reasonable value compared with past observations and the detector threshold. Although Watanabe et al. (2003b) indicated that solar neutrons seem to be accompanied by hard x-rays and gamma-rays, no such emissions were observed around 9:32 UT in this event. This suggests an existence of a variety of acceleration/emission mechanisms among the solar flares.

7.31 Solar Neutron Event of October 28, 2003

7.31.1 The Matter and Short History of the Event Observation and Research

According to Watanabe et al. (2006a, b, 2007) and Le et al. (2007), during the period when the Sun was intensively active at the end of October and beginning of November 2003 a remarkable solar neutron event at October 28, 2003 was observed by the ground-based neutron monitors in association with an X17.2 large flare. This solar flare was a remarkable event in 23rd solar cycle not only it was a large event, but many phenomena were observed in association with this flare. It is particularly worth noting the large flux of relativistic particles at the Earth (Veselovsky et al. 2004; Panasyuk et al. 2004). Among these particles were solar neutrons that were observed by the ground-based neutron monitor before the main ground-level enhancement (GLE). This solar neutron event has already been reported and discussed by Bieber et al. (2005). Watanabe et al. (2006b) compare neutron data with γ-ray data observed by the International Gamma-Ray Astrophysics Laboratory (INTEGRAL) satellite and derive the energy spectrum of neutrons using these γ-ray data. Solar neutrons were detected with high statistical significance (6.4σ) by the neutron monitor at Tsumeb, Namibia. The detector on board the INTEGRAL satellite observed a high flux of hard x-rays and γ-rays at the same time in this event. By using the time profiles of the γ-ray lines, Watanabe et al. (2006a) can explain the time profile of the NM counting rate. It appears that neutrons were produced at the same time as the γ-ray emission. In Watanabe et al. (2007) are considered physics of ion acceleration in the solar flare on 2003 October 28 which determine gamma-ray and neutron production. The propagation and decay of relativistic solar neutrons in the event of October 28, 2003 is considered in details by Le et al. (2007).

7.31.2 Satellite Observations of γ-Rays and Determining the Time of Solar Neutron Production

According to Watanabe et al. (2006a), an X17.2-class solar flare occurred at 9:51 UT (time observed at Earth; same definition is used hereafter) on 2003 October 28 located in NOAA active region 10486 at S16°, E08°. From 10:36 to 11:06 UT, an interval that includes the start time of intense emission of soft x-rays from the X17.2 flare, the RHESSI (Reuven Ramaty High Energy Solar Spectroscopic Imager) satellite was, unfortunately, in the South Atlantic Anomaly (SAA). However, intense emission of high-energy γ-rays was seen in the data after 11:06 UT, indicating that strong particle acceleration occurred during this flare. On the other hand, large fluxes of hard x-rays and γ-rays were observed by the INTEGRAL satellite shortly after 11:00 UT. Figure 7.97 shows the bremsstrahlung and line γ-ray time profiles from INTEGRAL. In the top panel of Fig. 7.97, two peaks of intense emission of bremsstrahlung γ-rays are seen at around 11:03 UT and 11:05 UT. However, there is only one peak (around 11:05 UT) in line γ-ray time profiles as shown in the second to fourth panels in Fig. 7.97. This more or less coincides with the second peak in the bremsstrahlung γ-rays.

Figure 7.98 shows γ-ray spectra between 11:02 and 11:03 UT and between 11:03 and 11:15 UT. From 11:02 to 11:03 UT, when the first peak of bremsstrahlung γ-rays was seen, there is no line γ-ray component. The γ-ray lines were clearly seen in the γ-ray spectrum after 11:03 UT, consistent with the line γ-ray time profiles shown in Fig. 7.97. Thus, it appears that the time profiles of ion and electron acceleration were quite different at this event, and ion acceleration either did not occur or was quite weak during the first peak of the bremsstrahlung γ-rays. Watanabe et al. (2006b) assume that the ion acceleration started only after 11:03 UT. In Fig. 7.97, note that the 2.2 MeV neutron capture γ-ray line peaks around 11:06 UT and has a long decay time. The 4.4 and 6.1 MeV γ-ray lines of de-excited C and O ions peak around 11:05 UT, giving about a 1 min gap between the two peak

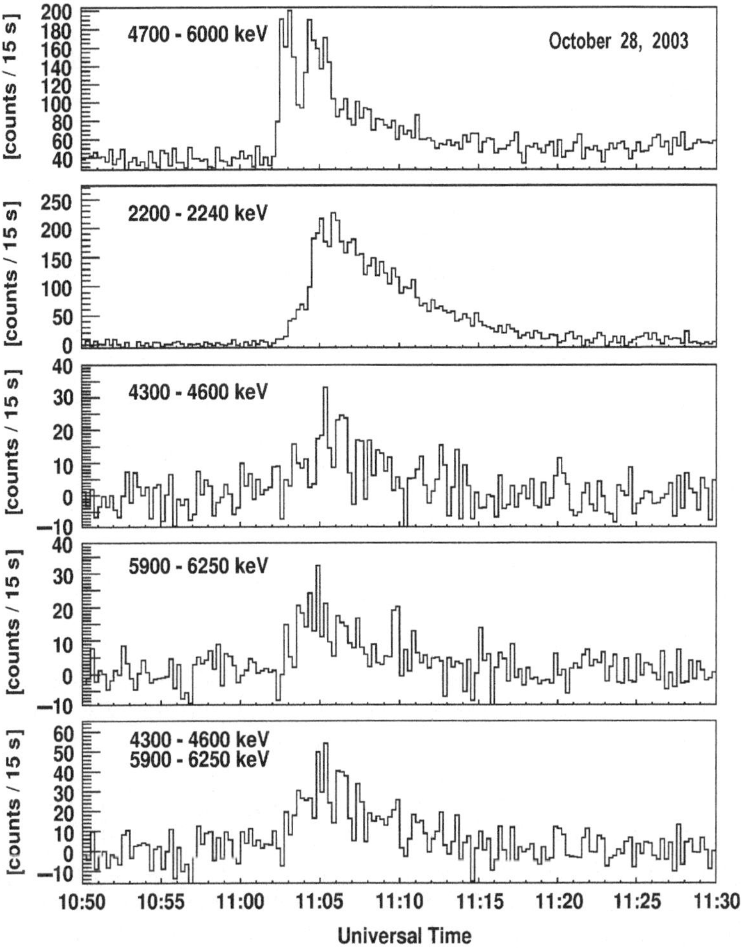

Fig. 7.97 *Top panel*: Time profile of bremsstrahlung γ-rays observed by the INTEGRAL satellite on October 28, 2003. *Second to fourth panels*: Line γ-ray time profiles observed by the INTEGRAL satellite on October 28, 2003. The bremsstrahlung component has been subtracted. *Second panel*: Time profile of the 2.2 MeV neutron capture γ-rays. *Third panel*: Profile of 4.4 MeV γ-rays of C nuclei. *Fourth panel*: Profile of 6.1 MeV γ-rays of O nuclei. *Bottom panel*: Sum of the data in the third and fourth panels. From Watanabe et al. (2006a)

times. In general, neutron capture γ-rays are delayed from γ-ray lines of de-excited ions, since it takes time for high-energy neutrons to slow down and be captured by protons (Wang and Ramaty 1974). Thus, it is evident that solar neutrons were produced at this flare, and they were probably produced at the same time that 4.4 and 6.1 MeV γ-ray lines were emitted. Hereafter, Watanabe et al. (2006a) assume that solar neutrons were produced around 11:05 UT.

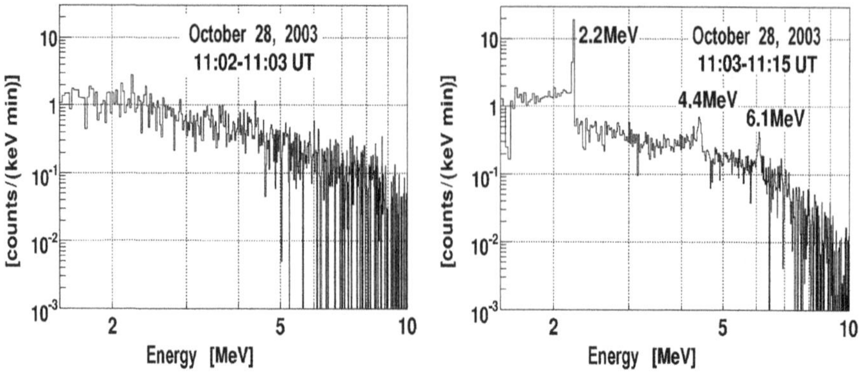

Fig. 7.98 Spectra of γ-rays between 1.5 and 10 MeV observed by INTEGRAL at 11:02–11:03 UT (*left*) and 11:03–11:15 UT (*right*) on 2003 October 28, with background subtracted. Right: Clear signals of 2.2, 4.4, and 6.1 MeV γ-rays appear superimposed on the bremsstrahlung component. From Watanabe et al. (2006b)

7.31.3 Solar Neutron Observations

According to Watanabe et al. (2006b), at 11:05 UT on 2003 October 28 (when solar neutron were generated, see previous Section 7.31.2), the Sun was located over Africa. Among the international network of solar neutron telescopes, Gornergrat in Switzerland and Aragats in Armenia had a possibility of observing solar neutrons from this flare. On the other hand, Tsumeb observatory (17.6°E, 19.1°S; 1,240 m above sea level) was located just under the Sun at this time. The altitude of Tsumeb Observatory is a little bit low; however, the air mass for the line of sight to the Sun was thinner than that of any other observatory because the zenith angle of the Sun was only 9.5°. Solar neutrons were clearly observed by the Tsumeb neutron monitor (Bieber et al. 2005). The 5 min counting rate of the Tsumeb NM is shown in Fig. 7.99 (top). Clear excesses are seen between 11:05 and 11:15 UT and between 11:20 and 11:25 UT. The statistical significances of these excesses are 4.8 σ for 11:05–11:10 UT, 4.2 σ for 11:10–11:15 UT, and 3.4 σ for 11:20–11:25 UT. The total significance for the 10 min between 11:05 and 11:15 UT is 6.4 σ. At the same time, high-energy protons were produced in association with this flare, and large ground-level enhancements (GLEs) occurred around the world. In opinion of Watanabe et al. (2006a), it can be excluded the possibility that excesses observed at Tsumeb came from energetic ions by considering the time profile of the NM at Lomnicky Stit (20.2°E, 49.2°N; 2,634 m above sea level), together with that of Tsumeb's NM (Fig. 7.99, bottom). The start time of the first excess of the Tsumeb NM is about 10 min earlier than the event at Lomnicky Stit, while the second excess at Tsumeb is consistent with this event at Lomnicky Stit. Thus, it appears that the second excess at Tsumeb came from energetic ions and the first excess was from solar neutrons.

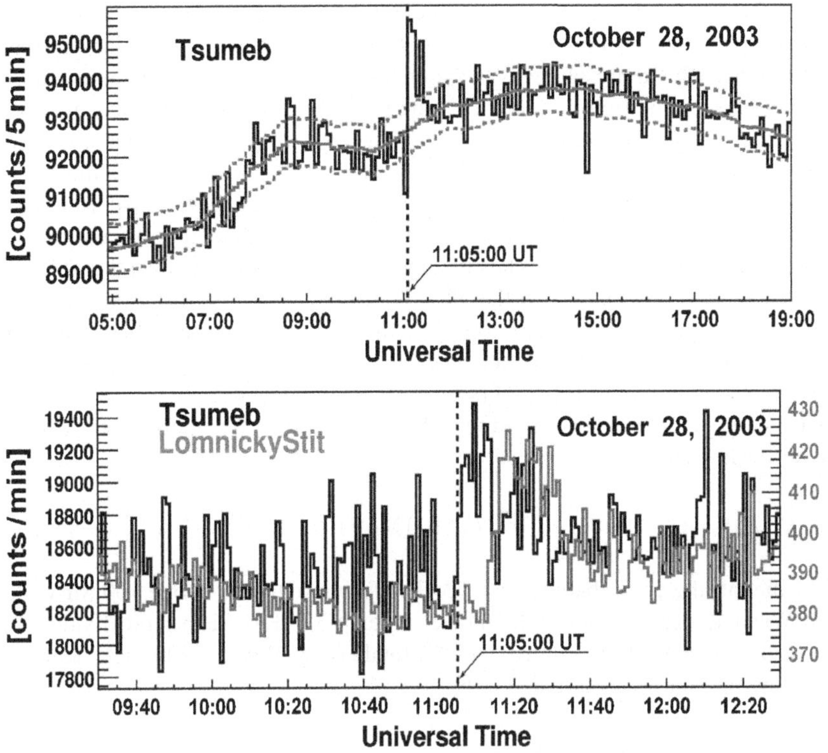

Fig. 7.99 *Top*: 5 min counting rate observed by the Tsumeb NM on October 28, 2003. The *smooth solid line* is the averaged background, and the *dashed lines* are $\pm 1\ \sigma$ from the background. Bottom: 1 min counting rate of the Tsumeb neutron monitor (*black line*) and time profile of the Lomnicky Stit NM counting rate (*gray line*). The solar neutron event in the Tsumeb data started well before the GLE event seen at Lomnicky Stit. From Watanabe et al. (2006a)

7.31.4 Analysis of Observational Data and the Determining of the Ejected Solar Neutron Energy Spectrum

Using observational data presented in Sections 7.31.2 and 7.31.3, Watanabe et al. (2006b) calculate the energy spectrum of the solar neutrons even though NMs cannot measure the energy of neutrons. For this they used the time of flight (TOF) method and assumed the emission time of solar neutrons from the Sun the same as of γ-rays, i.e. that the neutrons were produced at 11:05 UT, when line γ-ray emission peaked, and that the energy of the neutrons responsible for the excesses recorded by the neutron monitor is greater than 100 MeV. From the time profile of the neutrons, Watanabe et al. (2006a) calculate the energy spectrum of solar neutrons at the solar surface by the following formula:

$$D_S(E_n) = \Delta N / (\varepsilon S \Delta E_n P), \tag{7.23}$$

where ΔN is the number of excess counts contributed by solar neutrons and ε is the detection efficiency of the NM. Here ε includes the attenuation of solar neutrons through the Earth's atmosphere; S is the area of the NM, ΔE_n is the energy range corresponding to one time bin, and P is the survival probability of solar neutrons traveling from the Sun to the Earth. To obtain the ε in Eq. 7.23, Watanabe et al. (2006b) calculated the attenuation of solar neutrons by the Earth's atmosphere using the Shibata (1994) model. Solar neutrons with energy less than 100 MeV are strongly attenuated by the Earth's atmosphere, so the detection of neutrons by the Tsumeb monitor directly implies that the spectrum extended beyond 100 MeV. For the detection efficiency ε of the NM, Watanabe et al. (2006a) used the result calculated by Clem and Dorman (2000). Using these observational and simulation results, Watanabe et al. (2006a) calculated the energy spectrum of neutrons at the solar surface according to Eq. 7.23 (see Fig. 7.100).

The energy spectrum shown in Fig. 7.100 was fitted in Watanabe et al. (2006b) by a power law as:

$$D_S(E_n) = (3.1 \pm 1.0) \times 10^{27} (E_n/100\,\mathrm{MeV})^{-3.6 \pm 0.3}\,\mathrm{MeV}^{-1}\,\mathrm{sr}^{-1}. \qquad (7.24)$$

For this fit, the χ^2 probability is 0.53. The fitting region is above 100 MeV. As noted Watanabe et al. (2006a), the power index -3.6 ± 0.3 is a typical value for solar neutron events observed thus far. The total energy flux of >100 MeV neutrons emitted by the Sun was estimated to be

$$W_S(E_n > 100\ \mathrm{MeV}) = 3.1 \times 10^{25}\,\mathrm{ergs} \cdot \mathrm{sr}^{-1}. \qquad (7.25)$$

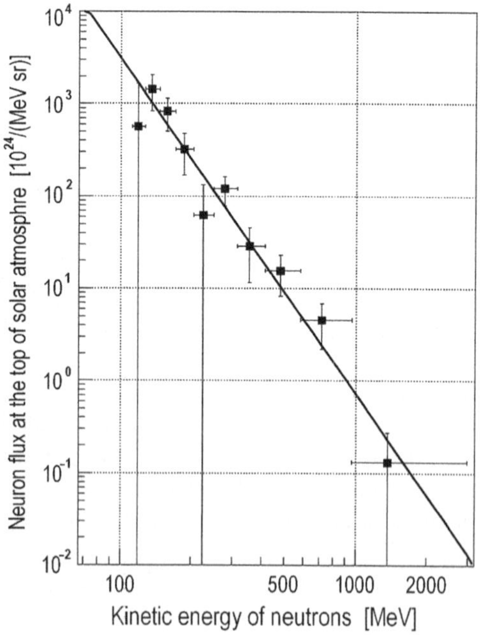

Fig. 7.100 Energy spectrum of neutrons at the solar surface on October 28, 2003 event. From Watanabe et al. (2006b)

7.31.5 Simulation in the Frame of Impulsive Models

As underlined Watanabe et al. (2006b), in the analysis method described above in Section 7.31.4, the energy spectrum is calculated by dividing the response into several bins, each characterized by a mean energy. For the survival probability of solar neutrons, as well as the attenuation of neutrons and detection efficiency of the NM, the values at these discrete energies are used. In order to calculate the energy spectrum of the solar neutrons in detail, Watanabe et al. (2006a) include an assumption about the time profiles of solar neutrons but still assume a power-law spectral index at the solar surface. Using this method, Watanabe et al. (2006b) investigate whether the neutrons are produced continuously. To clarify the consistency with the conventional method, they begin by assuming that the neutrons are produced impulsively. In this simulation, the power index of the neutron spectrum at the Sun is changed from -1.5 to -7.0 with a step of 0.1, while the energy range of the incident neutrons is confined to 50–1,500 MeV. The time profile of neutrons detected by the NM is calculated using the neutron attenuation in the Earth's atmosphere given by the Shibata (1994) model and the detection efficiency of the neutron monitor as calculated by Clem and Dorman (2000). The decay of neutrons between the Sun and the Earth is also taken into account. The result of this simulation can then be compared with the observational data, normalizing the simulated counting rate (N) to the observed excess counting rate (N_0).

Figure 7.101 (top) shows the reduced χ^2-distribution of the fit of the simulated counting rate to the observed excess of the Tsumeb NM obtained from the following formula:

$$\chi^2 = \sum_{i=1}^{n} \frac{(N_i - N_{0i})}{N_{0i}}. \tag{7.26}$$

In this fitting, data obtained from 11:05 to 11:15 UT are used. The χ^2 has its smallest value when the spectral index is around -3.5.

From Fig. 7.101 (bottom) can be seen that when the spectral index is -3.5, the simulated result reproduces well the observed results. From this fitting, the energy spectrum is determined by Watanabe et al. (2006a) as follows:

$$D_S(E_n) = (3.1 \pm 1.0) \times 10^{27} (E_n/100\ \mathrm{MeV})^{-3.5+0.4/-0.2}\ \mathrm{MeV}^{-1}\ \mathrm{sr}^{-1}. \tag{7.27}$$

This is comparable to the result obtained using the simpler method shown in Eq. 7.24, but the total energy flux of solar neutrons with energy range between 50 and 1,500 MeV is

$$W_S(E_n > 50\ \mathrm{MeV}) = \left(9.8\,^{+1.2}_{-0.9}\right) \times 10^{25}\ \mathrm{ergs} \cdot \mathrm{sr}^{-1}. \tag{7.28}$$

about a factor of 3 higher than the other estimate. This is because of the lower cutoff energy of the neutron spectra.

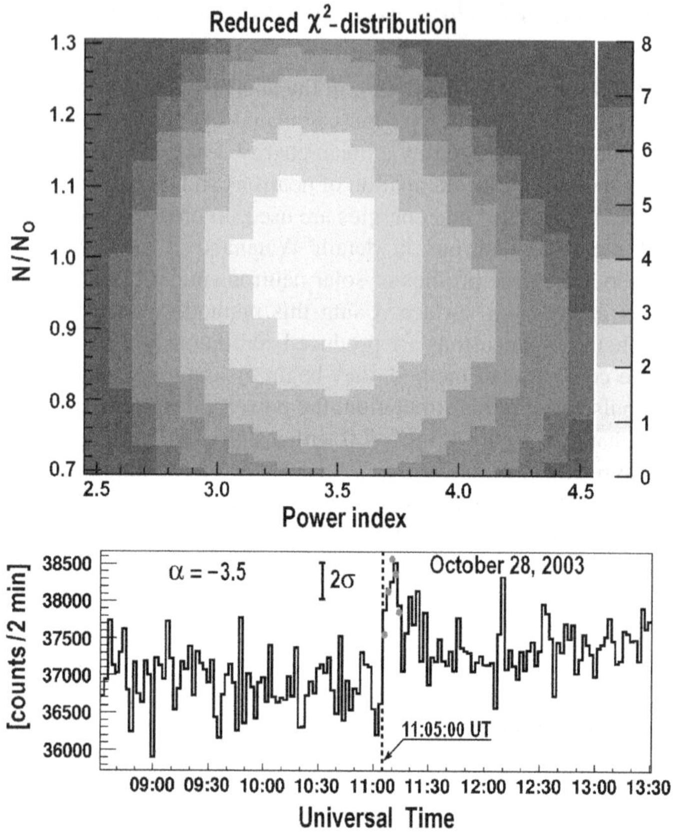

Fig. 7.101 *Top*: Reduced χ^2-distribution of the fit of the simulated counting rate to the observed excess of the Tsumeb NM. A 2 min counting rate is used in this calculation. The X-axis represents the power index of the simulated time profiles, and the Y-axis corresponds to the ratio of the simulated counting rate (N) to the observed one (N_0). In this fitting, data obtained during 11:05–11:15 UT are used. *Bottom*: 2-min counting rate (*solid line*) observed by the Tsumeb NM on 2003 October 28, together with simulated time profile (points) for which solar neutrons are assumed to be produced instantaneously at 11:05 UT, when power index is −3.5. From Watanabe et al. (2006a)

7.31.6 Simulation by Neutron Production with γ-Ray Time Profile

Watanabe et al. (2006b) next simulated the neutron time profiles detected at Tsumeb NM by assuming that neutrons were produced with a time spread, since extended production of line γ-rays was observed by INTEGRAL satellite. For this calculation, they used the summed time profile of 4.4 and 6.1 MeV γ-rays as shown in the bottom panel of Fig. 7.97 as the production model of solar neutrons. These are the γ-ray lines of carbon and oxygen, which indicate the time profile of ion

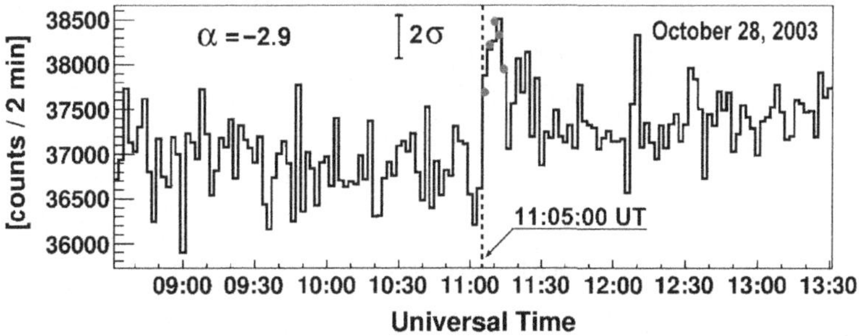

Fig. 7.102 Observed and simulated time profiles of the Tsumeb NM counting rate. The *solid line* is the observed 2-min counting rate, and points indicate the best-fit simulated time profile for solar neutrons assumed to be produced with the same time profile as γ-ray lines. From Watanabe et al. (2006b)

acceleration. Watanabe et al. (2006a) used the data observed from 11:02:45 to 11:10:00 UT. The spectral index of neutrons at the Sun is varied from -1.1 to -7.0 in steps of 0.1. The energy range of the neutrons is again taken to be 50–1,500 MeV. The χ^2 for the fit were calculated by using data obtained from 11:05 to 11:15 UT. The χ^2 has its smallest value for the spectral index $\alpha = -2.9$. The simulated result reproduces the observed result most closely when the spectral index $\alpha = -2.9$ as shown in Fig. 7.102.

When the spectral index is $\alpha = -2.9$, $\chi^2 = 0.69$, which provides the minimum χ^2 for the simulated time profiles. From this fit, the spectral index is found to be $\alpha = -2.9 \pm 0.3$. The best fit spectral index is harder than the index derived on the assumption that the neutrons were produced impulsively, but the total energy flux of the neutrons is now estimated to be

$$W_S(E_n > 50\ \text{MeV}) = \left(6.2\,^{+0.5}_{-0.6}\right) \times 10^{25}\ \text{ergs} \cdot \text{sr}^{-1}, \tag{7.29}$$

not very much different from the result for impulsive production.

7.31.7 *Simulation on the Basis of the Time History of the 2.223 MeV Neutron Capture Line and the Spectra of Neutrons Escaping from the Sun*

For simulation on the basis of time history of the 2.2 MeV neutron capture line Watanabe et al. (2007) used the transport model of Hua et al. (2002). By using this program, the neutron spectrum at the Earth's atmosphere can be estimated directly from γ-ray line data. Hua et al. (2002) model is a simulation program that calculates the time history of the 2.2MeV neutron capture line and the spectra of neutrons

Table 7.8 Parameters of the Hua et al. (2002) program for October 28, 2003 event (From Watanabe et al. 2007)

Accelerated ion composition (impulsive)	$\alpha/p = 0.5$, $^3He/^4He = 1$
Ambient composition (coronal)	He/H = 0.1, Ne/O = 0.25
Atmospheric model	Avrett (1981)
Photospheric $^3He/H$	3.7×10^{-5}
Acceleration release time history	4.4 & 6.1 MeV γ-ray line profile
Loop length	85,000 km
Flare heliocentric angle	23.0°

escaping from the Sun. A number of parameters must be set to use Hua et al. (2002) program. Many of these parameters can be derived from the observed data. For the acceleration release time history, Watanabe et al. (2007) used the 4.4 and 6.1 MeV γ-ray line time profile observed by the INTEGRAL and RHESSI satellites. The flare heliocentric angle was determined from the 2.2 MeV line γ-ray image obtained from RHESSI data (23.0 degrees). The loop length was also obtained with the 2.2 MeV image: Watanabe et al. (2007) measured the separation of the two foot-point sources and calculated the loop length assuming it is semicircular, giving 85,000 km. For the accelerated ion composition, ambient composition, atmospheric model and photospheric $^3He/H$ ratio, Watanabe et al. (2007) used typical values estimated from observations of previous flares. These parameters are summarized in Table 7.8.

Watanabe et al. (2007) derived the remaining parameters (λ, δ, and s) by comparing 2.2 MeV line time histories calculated with Hua et al. (2002) code with the observed time history. λ is the level of pitch-angle scattering within the loop, δ is the magnetic field convergence index, and s is the index of the power law energy spectrum of the accelerated ions at the solar surface. Watanabe et al. (2007) varied λ from 20 to 40,000, and δ from 0.00 to 0.45. If δ is 0.0, there is no convergence, and if δ is 0.45, there is strong convergence. Two-parametric confidence contours for combinations of λ and δ are shown in Fig. 7.103.

When λ is 100, δ is 0.10, and s is -3.86, χ^2 is minimal and the predicted time profile of the 2.2 MeV line fits the observed data quite well as shown in Fig. 7.104. By using these parameters, Watanabe et al. (2007) calculate predicted time-dependent neutron spectra arriving at the top of the Earth's atmosphere as shown in Fig. 7.105. At first, high energy neutrons arrive, followed by lower energy neutrons. They calculate resulting NM count rates due to these arriving neutron spectra by using the solar neutron atmospheric attenuation ratio calculated by the Shibata (1994) program and the neutron monitor efficiency calculated by Clem and Dorman (2000). Watanabe et al. (2007) compare the predicted count rate with the observed count rate in Fig. 7.106.

From Fig. 7.106 can be seen that the calculation count rate significantly underestimates the observed NM count rate, with a reduced χ^2 of 7.2. For the λ and δ derived from the 2.2 MeV time history, too many neutrons are going downwards toward the solar atmosphere, and to few are going upwards toward the Earth.

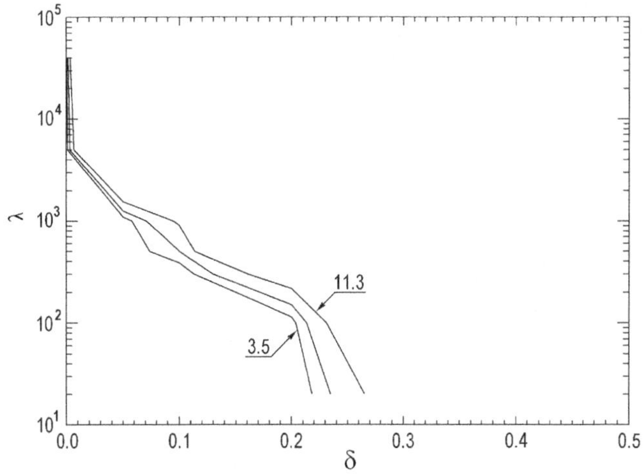

Fig. 7.103 Two-parametric confidence contours for combinations of λ and δ resulting from the time history analysis of the October 28, 2003 flare. The spectral index at minimum χ^2 was −3.86. From Watanabe et al. (2007)

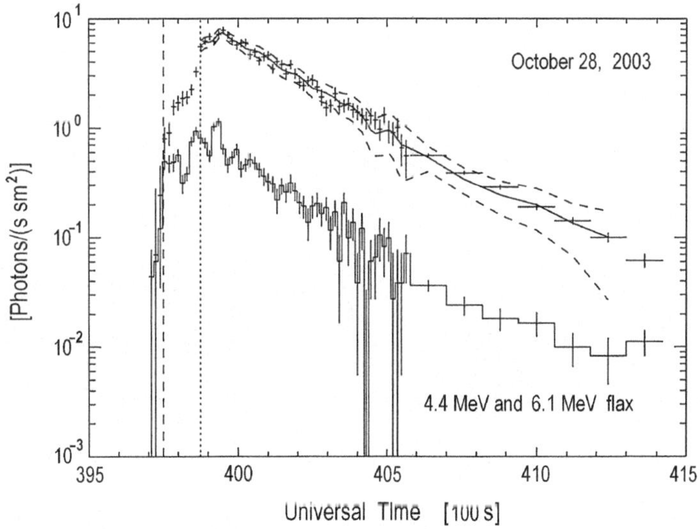

Fig. 7.104 Comparison of the 2.223 MeV neutron capture line time history measured with INTEGRAL and RHESSI satellites from the October 28, 2003 flare with a time history calculated by using Hua et al. (2002) program (*solid curve*). Also shown is the time history of the 4.4 and 6.1 MeV line flux assumed to represent the accelerated ion release time history. The dashed curves represent the $\pm 1\sigma$ uncertainties for the calculated curves. From Watanabe et al. (2007)

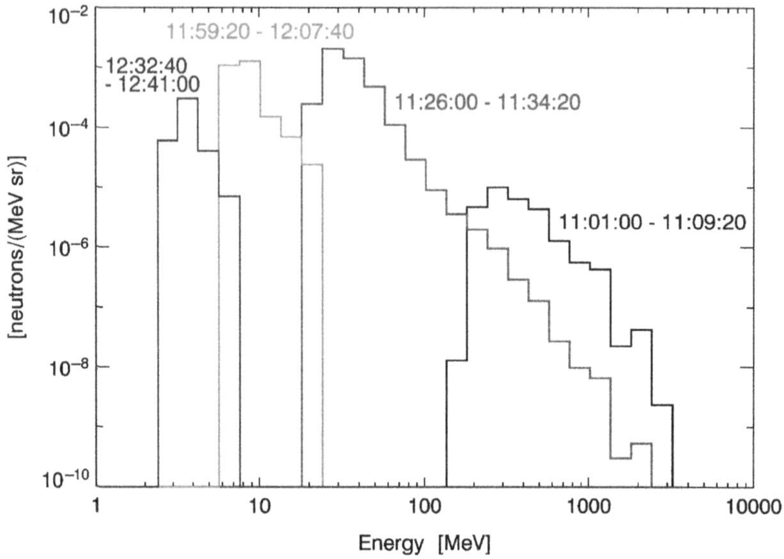

Fig. 7.105 The time dependent neutron spectra at the top of the Earth's atmosphere. From Watanabe et al. (2007)

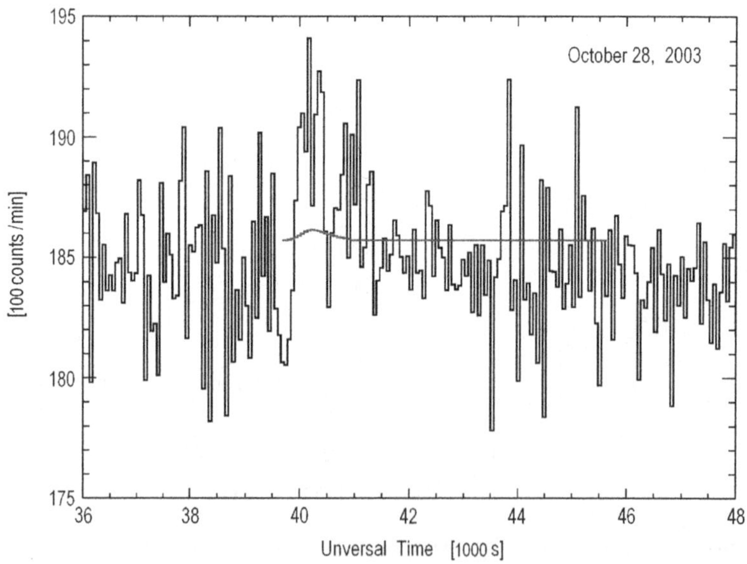

Fig. 7.106 Observed and predicted NM count rates on October 28, 2003. The *black line* is the observed 3-min count rate from the Tsumeb NM. The *red line* represent the predicted count rate. From Watanabe et al. (2007)

7.31.8 *Summarizing of the Main Results on the Event of 28 October, 2003*

Watanabe et al. (2007) summarized main obtained results as following:

(i) There are compared observed neutron count rates for the 2003 October 28 solar flare obtained with the Tsumeb NM with calculated count rates based on γ-ray data obtained with INTEGRAL and RHESSI satellites.
(ii) It was found that the predicted rate is significantly less than the observed rate when the values of λ, δ, and s derived based on the 2.2 MeV neutron capture line are used.
(iii) However, these values were derived assuming a loop length of 85,000km based on the RHESSI image of γ-ray foot-points obtained late in the flare; if the foot-point separation increased with time during this flare, as x-ray foot-points are often seen to do, this loop length would represent only an upper limit.
(iv) If the separation were less (and the loop length thus smaller) during the peak of the flare when most of the neutrons were produced, the derived values of λ and δ would change; the neutron angular distribution would then be less downward directed which would increase the predicted neutron flux at Earth.

7.32 Solar Neutron Event on November 2, 2003

7.32.1 *Intensive Activity of the Sun from Late October to the Beginning of November 2003*

According to Watanabe et al. (2006a), the Sun was intensely active from late October to the beginning of November 2003. During this period, three extensive active regions (NOAA Nos. 484, 486 and 488) appeared on the Sun and produced 11 X class solar flares (see Table 7.9).

Among 11 X class solar flares, two notable solar neutron events on November 2 and 4 occurred in association with two solar flares. Neutrons were detected by the neutron monitors located at Mt. Chacaltaya, Bolivia and Haleakala, Hawaii, respectively (see below, Section 7.32.3 for event of 2nd November and Section 7.33 for event of 4th November).

7.32.2 *Gamma-Ray Observations During the November 2, 2003 Event*

As noted Watanabe et al. (2005b, 2006a), at 17:03 UT on November 2, 2003, an X8.3 class solar flare occurred in NOAA active region 486. The start time was

Table 7.9 List of X class solar flares observed by the GOES satellite from late October to the beginning of November 2003 (from Watanabe et al. 2006a)

Date	GOES start time (UT)	GOES soft x-ray class	Active region NOAA No.	Active region location
October 19, 2003	16:29	X1.1	484	N08 E58
October 23, 2003	08:19	X5.4	486	S21 E88
October 23, 2003	19:50	X1.1	486	S17 E84
October 26, 2003	05:57	X1.2	486	S15 E44
October 26, 2003	17:21	X1.2	484	N02 W38
October 28, 2003	09:51	X17.2	486	S16 E08
October 29, 2003	20:37	X10.0	486	S15 W02
November 2, 2003	17:03	X8.3	486	S14 W56
November 3, 2003	01:09	X2.7	488	N10 W83
November 3, 2003	09:43	X3.9	488	N08 W77
November 4, 2003	19:29	X28	486	S19 W83

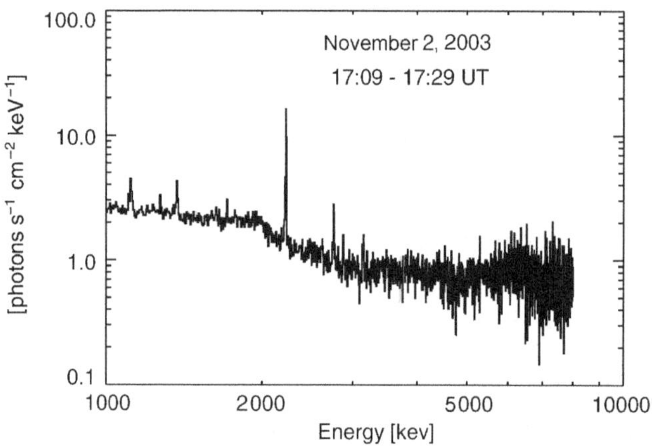

Fig. 7.107 Gamma-ray spectrum observed by the RHESSI satellite on November 2, 2003. The 2.223 MeV neutron capture γ-ray line is clearly seen. From Watanabe et al. (2006a)

17:03 UT and the flux reached its maximum at 17:25 UT, 22 min after the onset of the solar flare. During this time, large amounts of hard x-ray and γ-ray emissions were observed by the RHESSI satellite. Fig. 7.107 shows the energy spectrum of γ-rays observed by the RHESSI satellite on November 2, 2003. In this figure, an intense emission of 2.223 MeV neutron capture γ-ray line is clearly seen. Thus, it is evident that solar neutrons were produced at this flare. However, there is no such clarity in line γ-rays of the de-excited ions between 4 and 7 MeV. Figure 7.108 shows time profiles of the 2.223 MeV and the 4–7 MeV γ-rays. The 4–7 MeV γ-ray reached its peak at 17:17 UT, and the neutron capture γ-ray peaked at 17:18:40 UT. There is a 100-s gap between the peak times. It takes about 100 s for high energy neutrons to slow down and be captured by protons (Wang and Ramaty 1974). From this data, it can be conclude that solar neutrons were produced at 17:17 UT.

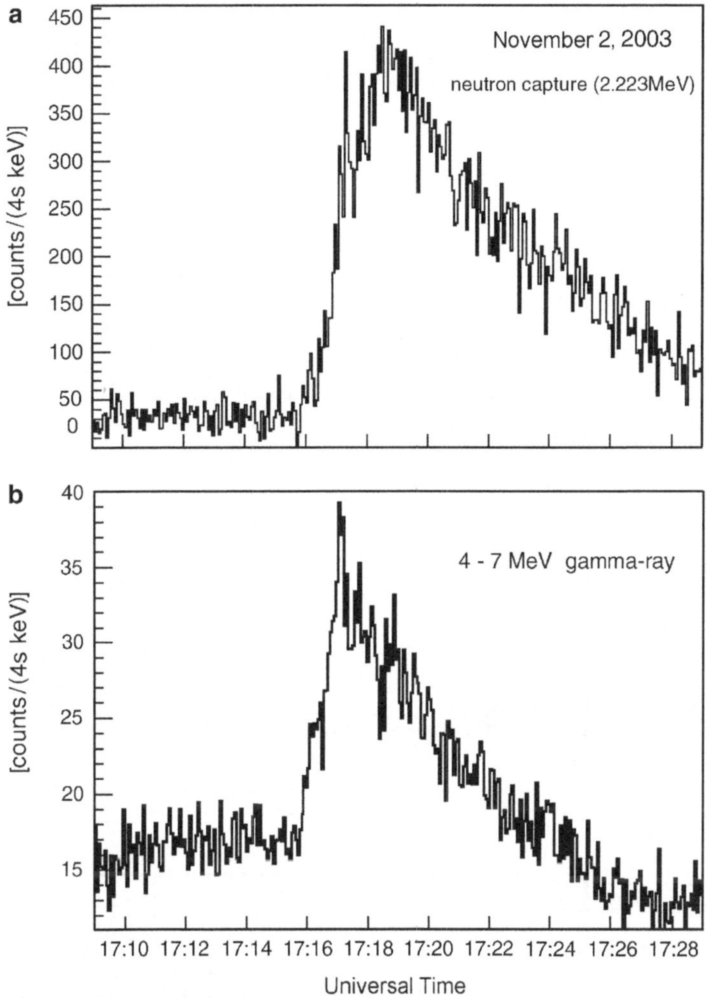

Fig. 7.108 The 2.223 MeV and 4–7 MeV γ-ray time profiles observed by the RHESSI satellite on November 2, 2003. The peak time of the 2.223 MeV γ-ray delay about 100 s from that of the 4–7 MeV γ-ray. From Watanabe et al. (2006a)

7.32.3 Solar Neutron Observations During the November 2, 2003 Event

As noted Watanabe et al. (2005b, 2006a), from upon the point of occurrence of this flare, the Sun was positioned over Bolivia. Solar neutrons were detected by the neutron monitor at Mt. Chacaltaya, Bolivia (292.0°E, 16.2°S, 5,250 m a.s.l.). At 17:17 UT, the zenith angle of the Sun was 11.5° at Mt. Chacaltaya, and the air mass for the line of sight to the Sun was 551 g/cm^2. Unfortunately, during this time, blackouts often occurred at Chacaltaya Observatory. Although the data of the neutron monitor after 18:00 UT were not recorded due to a power outage, clear excesses were detected by the neutron monitor between 17:24 and 17:50 UT (Panel **a** of Fig. 7.109). The statistical significance of this event was 4.7 σ (Panel **b** of Fig. 7.109). Following the production of high energy protons in association with

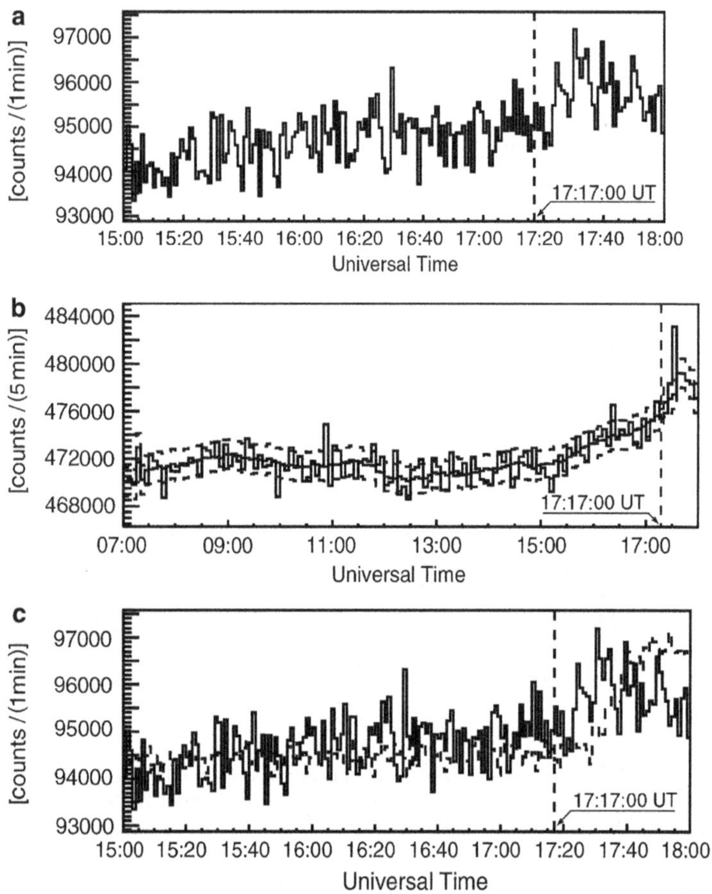

Fig. 7.109 Time profiles of the neutron monitor located at Mt. Chacaltaya, Bolivia on November 2, 2003. (**a**) 1-min count rate observed by the Bolivia neutron monitor on November 2, 2003. The data after 18:00 UT were not recorded due to the blackout. Clear excesses are seen between 17:24 and 17:50 UT. (**b**) 5-min count rate. The *solid smooth line* is the averaged background and *dashed lines* are ±1σ from the background. The statistical significance at 17:30–17:35 UT bin was 4.7σ. (**c**) 1-min count rate of the Bolivia neutron monitor (*solid line*) and the time profile of the McMurdo neutron monitor (*dashed line*). Solar neutron event of the Bolivia neutron monitor started before the GLE event of the McMurdo neutron monitor. From Watanabe et al. (2006a)

this flare, large ground level enhancement (GLE) occurred around the world. Thus, there is a possibility that excesses came from energetic ions. However, since the cutoff rigidity of 12.53 GV at Mt. Chacaltaya is very high, it is difficult for ions to reach this point. In addition, Watanabe et al. (2006a) plotted the time profile of the McMurdo neutron monitor together with that of the Bolivia neutron monitor (Panel **c** of Fig. 7.109). The time profile of the GLE event of the McMurdo neutron monitor has a shape typical of the GLE event on November 2, 2003. The GLE event started after 17:30 UT, about 5 min after the start time at the neutron monitor of Mt. Chacaltaya. Thus, it can be infer that these excesses of the neutron monitor at Mt. Chacaltaya were due to the solar neutrons.

7.32.4 Solar Neutron Energy Spectrum at the Sun

Assuming that solar neutrons were produced at 17:17 UT when large fluxes of hard x-rays and γ-rays were observed by the RHESSI satellite, the energy of the solar neutrons detected by the Bolivia neutron monitor is estimated by Watanabe et al. (2006a) to be between 50 and 180 MeV. Solar neutrons in this energy range can penetrate to the ground through 551 g/cm^2 of atmosphere. From the time profile of the neutrons, Watanabe et al. (2006a) have calculated the energy spectrum of solar neutrons at the solar surface using the Shibata program for atmospheric attenuation (Shibata 1994) and efficiency of neutron monitor calculated by Clem and Dorman (2000). Figure 7.110 shows the energy spectrum of solar neutrons produced at the solar surface assuming that the solar neutrons were produced at 17:17 UT. Energy spectrum is well fitted by a power law with an index of -7.0. The total energy flux of solar neutrons detected between 50 and 180 MeV is 2.7 10^{25} erg/sr.

In the next phase, Watanabe et al. (2006a) simulated the time profile of solar neutrons assuming the power law index of the solar neutrons at the solar surface. In Fig. 7.111, dots show simulation results obtained by varying power law indexes together with the data observed by the Bolivia neutron monitor. The peak time of the simulated time profile when the power law index equals -3.0 is earlier than the observed data (Panel **a** in Fig. 7.111). The softer a power law index is, the closer the simulated time profile is to the observed one (Panel **a**–**c** in Fig. 7.111). The nearest data obtained are when the power law index is -7.0 (Panel **c** in Fig. 7.111), and much softer indexes do not give a better fit (Panel **d** in Fig. 7.111). This is comparable to the observation result.

As concluded Watanabe et al. (2006a), the energy spectrum of solar neutrons at the solar surface can be described by a power law using the energy of solar neutrons E_n as follows:

$$D_S(E_n) = (2.6 \pm 1.4) \times 10^{26} \times \left(\frac{E_n}{100\ \text{MeV}}\right)^{-7.0\pm1.3} \left[\text{MeV}^{-1}\ \text{sr}^{-1}\right]. \quad (7.30)$$

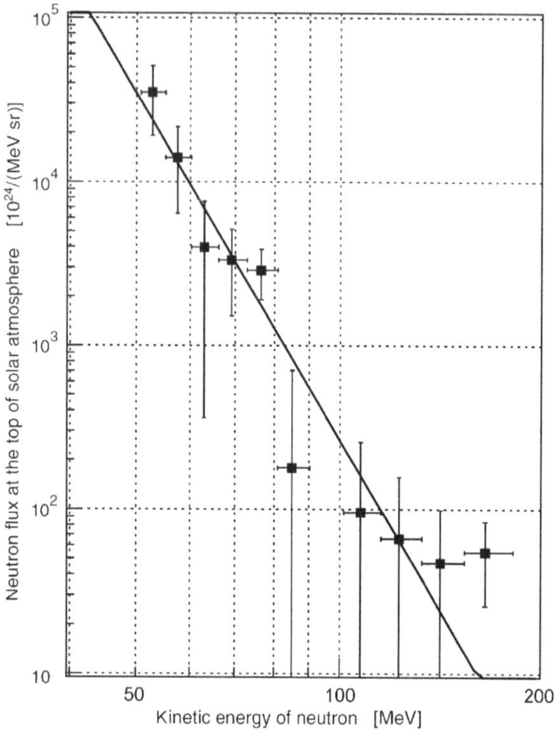

Fig. 7.110 Energy spectrum of solar neutrons on the solar surface for the flare which occurred on November 2, 2003. From Watanabe et al. (2006a)

7.33 Solar Neutron Event on 4 November, 2003

7.33.1 Satellite Observations of x- and γ-Rays

According to Watanabe et al. (2005c, 2006b), on November 4, 2003 an X28-class solar flare occurred at 19:29 UT, located in NOAA active region 10486 at S19°, W83°. This is the largest solar flare on record. At around 19:42 UT, intense emission of soft x-rays was detected by GOES (Geostationary Operational Environmental Satellite) such that the detection was saturated. After 19:42 UT, intense emission of hard x-rays and γ-rays was observed by the INTEGRAL spacecraft. Unfortunately, at this time, the RHESSI spacecraft was on the night side of the Earth. Figure 7.112 shows the energy spectrum of γ-rays observed by INTEGRAL.

As noted Watanabe et al. (2005c, 2006b), in this event, although the components of the line emission produced by de-excited ions, C (4.4 MeV) and O (6.1 MeV), were not prominent, the 2.2 MeV neutron capture line can be clearly seen. Intense bremsstrahlung x-rays and γ-rays were also observed. Figure 7.113 shows the time profiles of γ-rays for different energy bins that contain line γ-ray components produced as a result of the ion acceleration.

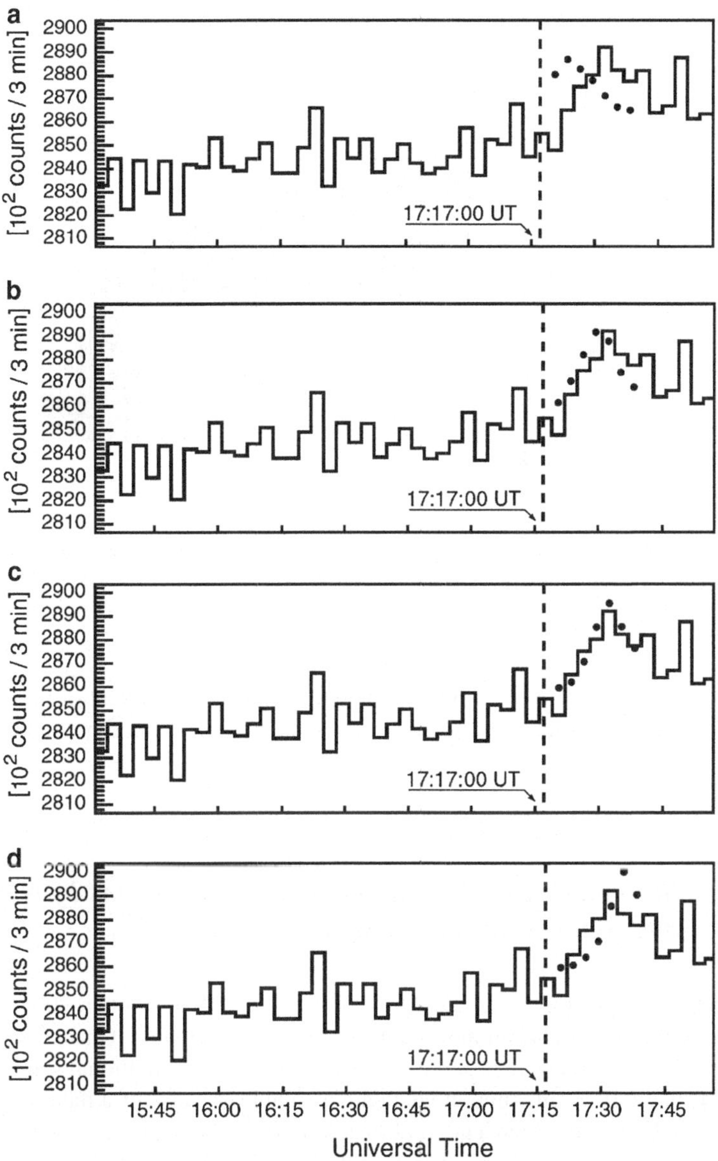

Fig. 7.111 Three-min count rate observed by the Bolivia neutron monitor on November 2, 2003 together with simulated time profiles when power law indexes are: (**a**) −3.0; (**b**) −5.0; (**c**) −7.0; (**d**) −9.0. When the power law index is −7.0 the simulated time profile is well-fitted observed data. From Watanabe et al. (2006a)

From Fig. 7.113 can be seen that there is a delay of the 2.2 MeV neutron capture γ-ray emission from that of the line γ-ray components produced by excited ions of C and O. Watanabe et al. (2006b) assume that ion acceleration occurred at the same

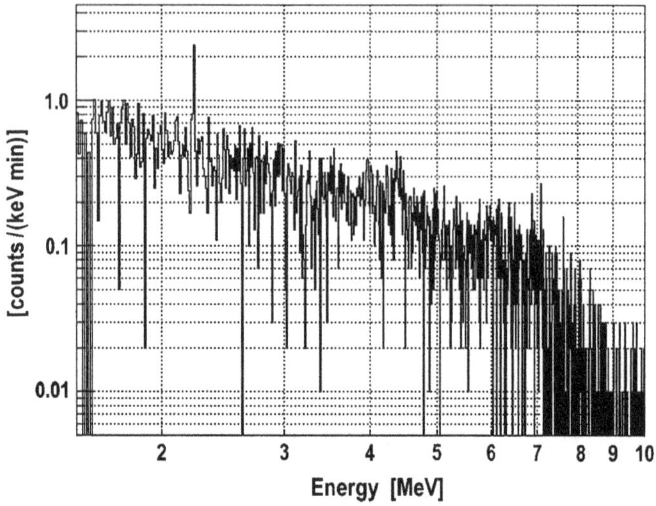

Fig. 7.112 Spectrum of γ-rays between 1.5 and 10 MeV observed by INTEGRAL satellite from 19:40 to 19:50 UT on November 4, 2003, with background subtracted. A signal produced by 2.2 MeV γ-rays appears superimposed on the bremsstrahlung component. There is weak evidence for 4–7 MeV γ-ray lines. From Watanabe et al. (2006b)

time as the γ-ray lines were emitted, although the main component of these γ-rays is bremsstrahlung. And they assume that solar neutrons were produced at the same time.

7.33.2 Simultaneous Observations of Solar Neutrons by NM and Solar Neutron Telescope

As underlined Watanabe et al. (2006b), at 19:45 UT of 4 November, 2003, the Sun was located between Hawaii and South America. Although the Chacaltaya observatory was the best place to observe solar neutrons in the international solar neutron telescope network at this time, no data are available because of a data gap. Sierra Negra would also have been a good place to observe solar neutrons, but the Mexico solar neutron telescope had not started continuous observation at that time. Thus, it was necessary to examine data from the Hawaii observatory, which was the third closest of the observatories able to observe solar neutrons. At 19:45 UT, the zenith angle of the Sun was 49.9° at Mauna Kea and 50.5° at Haleakala. The air mass along the line of sight to the Sun was 947 and 1,112 $g{\cdot}cm^{-2}$, respectively. The other suitable location was Mexico City, where the zenith angle of the Sun was 40.52° and the air mass along the line of sight to the Sun was 1,026 $g{\cdot}cm^{-2}$. Attenuation of solar neutrons by the Earth's atmosphere above these observatories is calculated using the Shibata (1994) model. The Haleakala and Mexico City observatories have

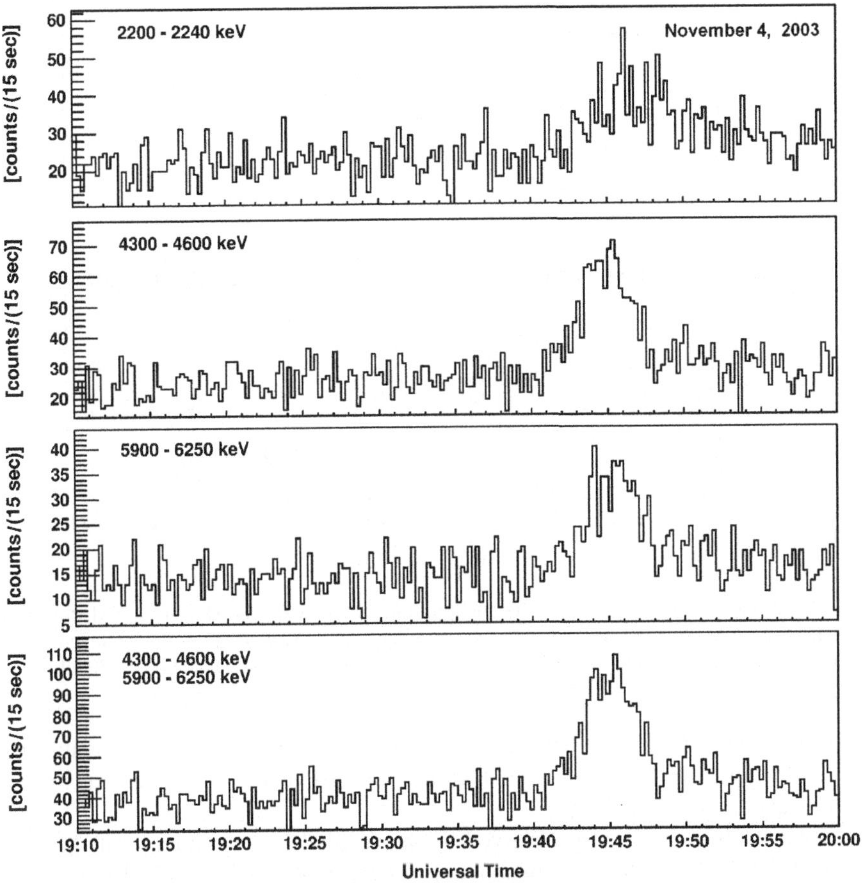

Fig. 7.113 Time profiles of γ-ray lines observed by the INTEGRAL satellite on November 4, 2003. The bremsstrahlung component has not been subtracted. *Top panel*: Time profile of the 2.2 MeV neutron capture γ-rays. *Second panel*: 4.4 MeV γ-rays of C nuclei. *Third panel*: 6.1 MeV γ-rays of O nuclei. *Bottom panel*: Sum of the data in the second and third panels. Although these time profiles contain line γ-ray components, which indicate the time profile of ion acceleration, the dominant component is bremsstrahlung. From Watanabe et al. (2006b)

nearly the same attenuation, while Mauna Kea is located at the best place to observe solar neutrons. However, simultaneous signals were found in both the Haleakala and the Mexico City neutron monitors. Solar neutrons were observed by the 18NM-64 neutron monitor at Haleakala, Hawaii (203.7°E, 20.7°N; 3,030 m above sea level). Figure 7.114 (top) shows the 5 min averages of the counting rate observed on November 4, 2003. At this time, the sampling interval of the Haleakala neutron monitor was 10 s.

From Fig. 7.114 clear excesses were seen after 19:45 UT, continuing for 15 min. The statistical significances of these excesses are 4.5 σ for 19:46:20–19:51:20 UT, 5.3 σ for 19:51:20–19:56:20 UT, and 3.1 σ for 19:56:20–20:01:20 UT. The total

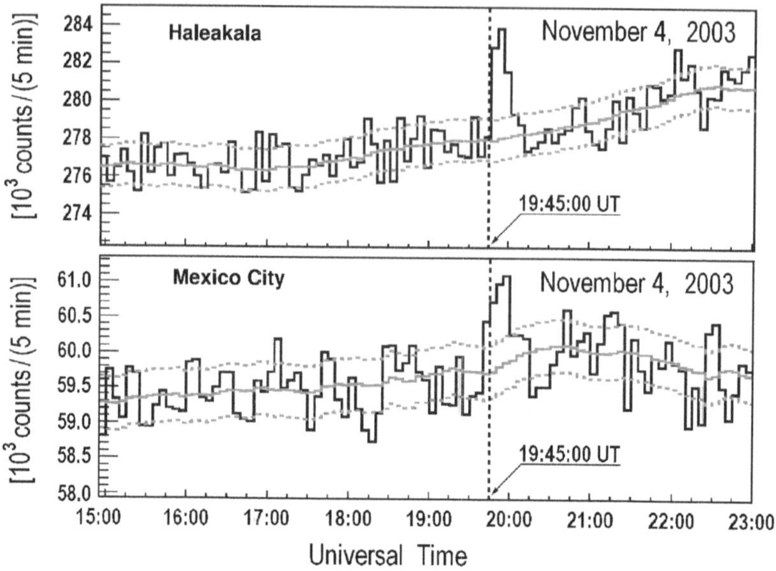

Fig. 7.114 Five minute averages of the counting rate observed by the Haleakala NM (*top*) and Mexico City NM (*bottom*) on November 4, 2003. The *smooth solid line* is the averaged background, and the *dashed lines* are $\pm 1\ \sigma$ from the background. From Watanabe et al. (2006b)

significance for the 15 min between 19:46:20 and 20:01:20 UT is 7.5 σ (Watanabe et al. 2006b note that this time interval was just taken to get the maximum significance).

Solar neutrons were also observed by the 6NM64 neutron monitor at Mexico City (260.8°E, 19.33°N; 2,274 m above sea level), as shown in Fig. 7.114 (bottom). At this time, the sampling interval of the Mexico City NM was 5 min. Clear excesses were seen after 19:45 UT, which was the same time as the excesses observed by the Haleakala NM. The statistical significances of these excesses are 2.6 σ for 19:45–19:50 UT, 3.1 σ for 19:50–19:55 UT, and 3.3 σ for 19:55–20:00 UT. The total significance for the 15 min between 19:45 and 20:00 UT is 5.2 σ.

One would expect that Mauna Kea (203.7°E, 19.8°N; 4,200 m above sea level) should be a better place to observe neutrons in this event than Haleakala and Mexico City. This is the location of the Hawaii solar neutron telescope with an area of 8 m^2, constructed from proportional counters and plastic scintillators, but only a minimal excess was seen after 19:45 UT in the PMT_L, PMT_H, and layer 1 with anti-coincidence channels in this telescope as shown in Fig. 7.115.

In Fig. 7.115 are shown data from the PMT_L and PMT_H are channels of scintillation counter that detect neutrons (recoil protons), the energy thresholds of which are 12 and 20 MeV, respectively. The layer 1 with anti-coincidence is a proportional counter channel, which is located under the scintillation counters. This apparent discrepancy between NM and the solar neutron telescope is discussed in Watanabe et al. (2006b) considering the surrounding environment of the detector.

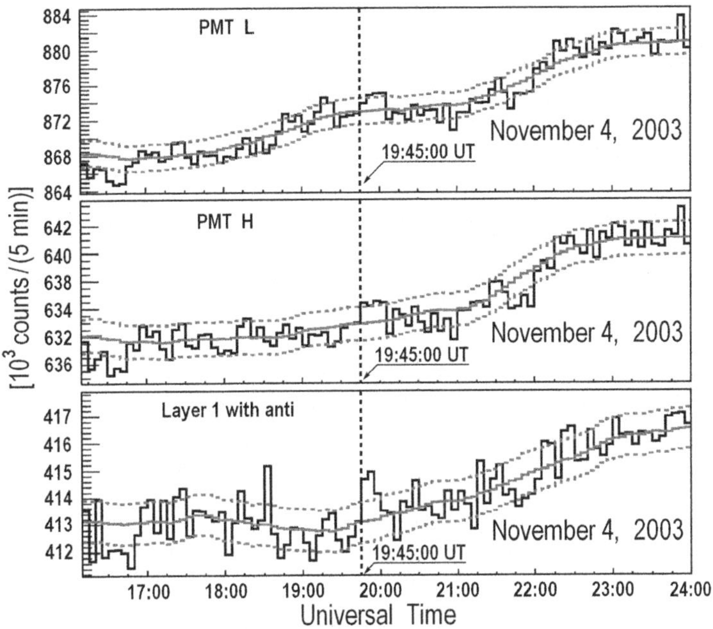

Fig. 7.115 Five minute averages of the counting rate of PMT_L, PMT_H, and layer 1_with_anti channels of the Hawaii solar neutron telescope on November 4, 2003. The *solid smooth line* is the averaged background, and the *dashed lines* are $\pm 1\sigma$ from the background. From Watanabe et al. (2006b)

7.33.3 Analysis of Obtained Experimental Results and Ejected Spectra

In order to understand the Mauna Kea result, Watanabe et al. (2006b) first use the other observational data to estimate the energy spectrum of the solar neutrons. They begin with the data from the Haleakala NM because it recorded the largest excess with the best time resolution. Watanabe et al. (2006b) determine the neutron energy by using the TOF method, assuming that all solar neutrons were produced at 19:45 UT (the peak of the intense emission of high-energy γ-rays observed by INTEGRAL as shown in Fig. 7.113). Under this assumption, the energy of neutrons observed by the Haleakala NM between 19:51:20 and 19:56:20 UT ranged from 59 to 913 MeV. To derive the energy spectrum of neutrons at the solar surface from the observed time profile by the NM, the survival probability of neutrons between the Sun and the Earth, the attenuation of solar neutrons passing through the Earth's atmosphere, and the detection efficiency of the NM must be taken into account. Attenuation is calculated using the Shibata (1994) model, and the detection efficiency is using calculated by Clem and Dorman (2000). Using these results, Watanabe et al. (2006b) calculated the energy spectrum of neutrons at the solar

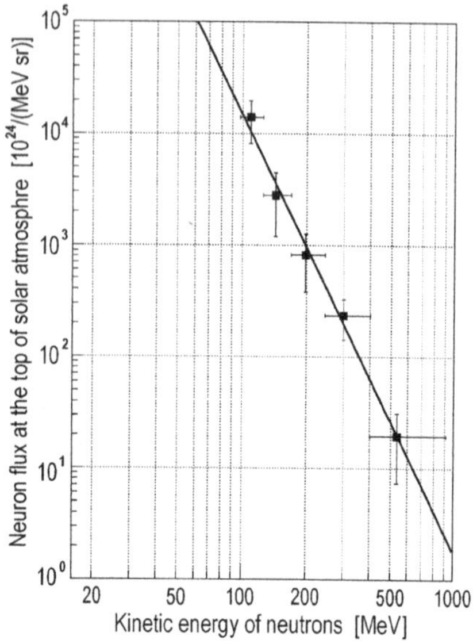

Fig. 7.116 Energy spectrum of neutrons at the solar surface on November 4, 2003 calculated from the data of the Haleakala NM. From Watanabe et al. (2006b)

surface using the same method as was described above for the event 28 October 2003 in Section 7.31.4. The result is shown in Fig. 7.116.

According to Watanabe et al. (2006b), the spectrum shown in Fig. 7.116 was derived from 2 min averages of the counting rate, where the vertical errors that are shown are only statistical errors. The energy spectrum is well fitted by a power law as

$$D_S(E_n) = (1.5 \pm 0.6) \times 10^{28} (E_n/100 \text{ MeV})^{-3.9\pm0.5} \text{ MeV}^{-1} \text{ sr}^{-1}. \qquad (7.31)$$

The fitting region is chosen as 100 MeV and above because there errors from neutron attenuation in the Earth's atmosphere are small. For this fit, the χ^2 probability is 82%. The obtained spectral index about −4 is typical for solar neutron events observed thus far. The total energy flux of neutrons emitted from the Sun in the energy range 59–913 MeV is estimated to be

$$W_S(913 \text{ MeV} > E_n > 59 \text{ MeV}) = 3.4 \times 10^{26} \text{ ergs} \cdot \text{sr}^{-1}, \qquad (7.32)$$

7.33.4 Simulation by Using the Impulsive Model

Watanabe et al. (2006b), by using the same method as was described above, in Section 7.31.5, and assuming their spectral index at the solar surface according to Eq. 7.31, the time profiles of solar neutrons were simulated on the assumption that solar neutrons were produced impulsively. Watanabe et al. (2006b) examine the

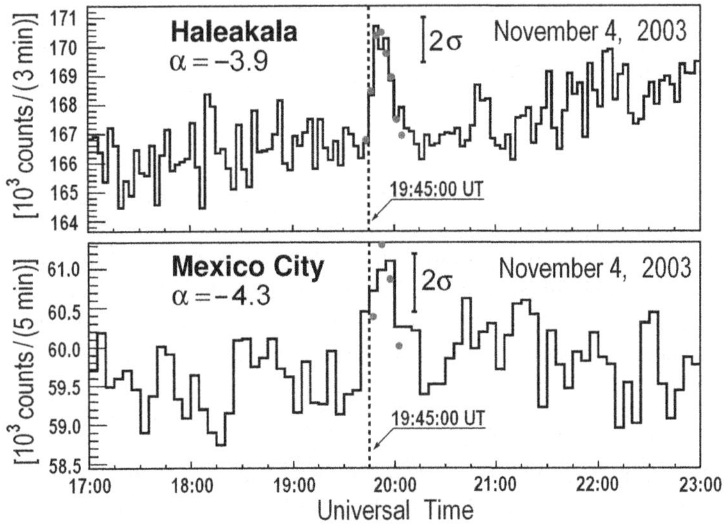

Fig. 7.117 Best-fit simulated time profiles (points) when the spectral index is −3.9 for Haleakala (top) and −4.3 for Mexico City (bottom), superposed on the observed counting rate. The start time of the simulated time profile is 19:45 UT, corresponding to the peak time of γ-ray emission. From Watanabe et al. (2006b)

χ^2 of the fit of the simulated counting rate to the observed excess of the Haleakala NM obtained from Eq. 7.26. In this fitting, data obtained from 19:45 to 20:06 UT are used. The χ^2 has its smallest value when the spectral index is around −3.9 (see Fig. 7.117, top).

From this fitting, the energy spectrum is determined as follows:

$$D_S(E_n) = \left(2.1^{+0.2}_{-0.1}\right) \times 10^{28} \left(\frac{E_n}{100\ \mathrm{MeV}}\right)^{-3.9^{+0.1}_{-0.2}}\ \mathrm{MeV}^{-1}\ \mathrm{sr}^{-1}. \tag{7.33}$$

This is consistent with the result obtained using the simpler method shown in Eq. 7.31. The total energy flux of solar neutrons within the energy range 50–1,500 MeV is

$$W_S(1,500\ \mathrm{MeV} > E_n > 50\ \mathrm{MeV}) = \left(6.7^{+0.5}_{-0.4}\right) \times 10^{26}\,\mathrm{ergs} \cdot \mathrm{sr}^{-1}, \tag{7.34}$$

about the same order as the calculated value from Eq. 7.31.

Watanabe et al. (2006b) have done the same analysis for the Mexico City NM. In this fitting, data obtained during 19:45–20:05 UT are used. The χ^2 has its smallest value when the spectral index is around −4.3 (see Fig. 7.117, bottom). From this fitting, the energy spectrum is determined as

$$D_S(E_n) = (1.6 \pm 0.2) \times 10^{28} (E_n/100\ \mathrm{MeV})^{-4.3\pm0.4}\ \mathrm{MeV}^{-1}\ \mathrm{sr}^{-1}. \tag{7.35}$$

Although the spectral index is softer than the result of the Haleakala NM, it is consistent with Eq. 7.31. The total energy flux of solar neutrons with energies between 50 and 1,500 MeV is calculated to be

$$W_S(1,500 \text{ MeV} > E_n > 50 \text{ MeV}) = (5.4 \pm 0.7) \times 10^{26}\,\text{ergs} \cdot \text{sr}^{-1}, \tag{7.36}$$

about the same order as the result obtained from the analysis of Haleakala NM data.

Watanabe et al. (2006b) then simulated the time profile of neutrons that should be observed from the Hawaii solar neutron telescope using the energy spectrum of incident neutrons obtained from the data of the Haleakala NM and described by Eq. 7.31, namely (they did not attempt to derive a spectrum from the data of telescope because excesses of the solar neutron telescope are small):

$$D_S(E_n) = 1.5 \times 10^{28}(E_n/100 \text{ MeV})^{-3.9} \text{ MeV}^{-1} \text{ sr}^{-1}. \tag{7.37}$$

The detection efficiency of the Hawaii solar neutron telescope was calculated using Geant3, FLUKA-COLOR model. In this calculation, the Hawaii solar neutron telescope is surrounded by 20 cm concrete walls, since it is situated within the building housing the SUBARU telescope. Figure 7.118 shows the detection efficiencies of the Hawaii solar neutron telescope for neutrons and γ-rays.

The simulated result for the layer 1_with_anti channel of the Hawaii solar neutron telescope, which recorded the largest excess, is shown in Fig. 7.119.

From Fig. 7.119 can be seen that the simulation result is consistent with the observed excess. Because of the high counting rate from the non-hadronic CR

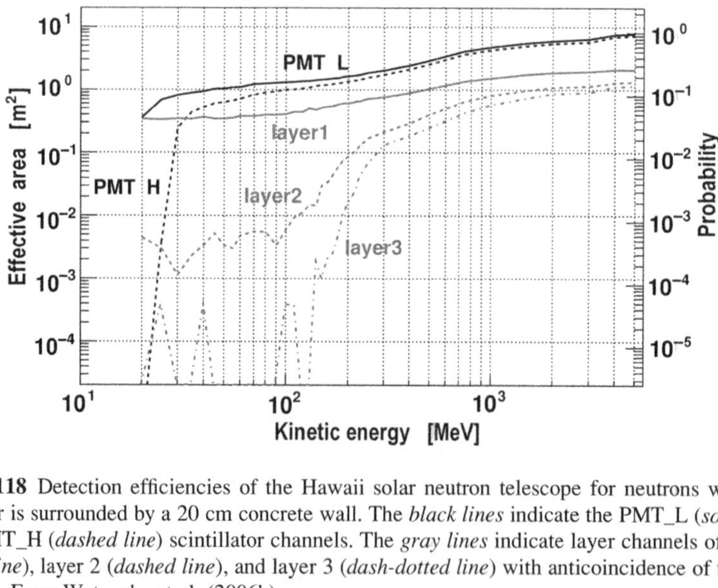

Fig. 7.118 Detection efficiencies of the Hawaii solar neutron telescope for neutrons when the detector is surrounded by a 20 cm concrete wall. The *black lines* indicate the PMT_L (*solid line*) and PMT_H (*dashed line*) scintillator channels. The *gray lines* indicate layer channels of layer 1 (*solid line*), layer 2 (*dashed line*), and layer 3 (*dash-dotted line*) with anticoincidence of the anti-counter. From Watanabe et al. (2006b)

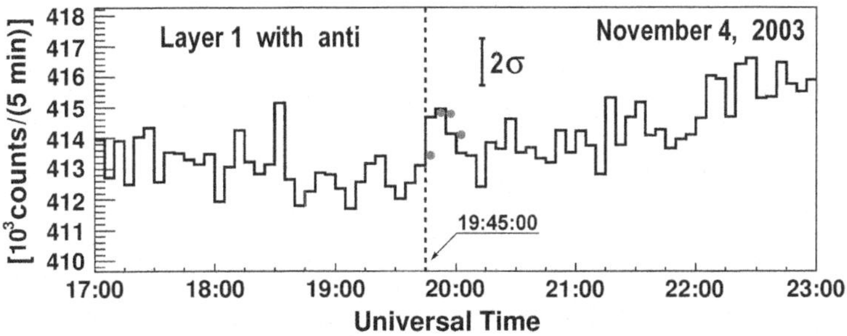

Fig. 7.119 Simulated time profile (*points*) of 5 min counting rate of layer 1_with_anti channel of the Hawaii solar neutron telescope on November 4, 2003, superposed on the observational data. The energy spectrum of the incident neutrons is according Eq. 7.37, which was obtained from the data of the Haleakala NM. The start time of this time profile is 19:45 UT, corresponding to the peak time of γ-ray emission. From Watanabe et al. (2006b)

component (γ-rays, muons, and so on) in the Hawaii solar neutron telescope, these excesses are not statistically significant, although they correspond to the same total flux of solar neutrons observed by the Haleakala NM. It is however possible that solar neutrons actually produced the little hump in the data.

7.33.5 Simulation by Neutron Production Using the γ-Ray Profile

Next Watanabe et al. (2006b) simulated the neutron time profiles detected at Haleakala and Mexico City by assuming that neutrons were produced with a time spread. The calculation method is the same as in Section 7.31.6. For this calculation, Watanabe et al. (2006b) used the γ-ray time profile observed by the INTEGRAL satellite during 19:42–19:48:00 UT as the production time profile of solar neutrons, as shown in the bottom panel in Fig. 7.113. The χ^2 of the fit between observed and simulated time profiles of the Haleakala and Mexico City neutron monitors were calculated. In this fitting, data obtained from 19:42 to 20:06 UT are used for the Haleakala NM and from 19:45 to 20:05 UT for the Mexico City NM. For the Haleakala data, when the power index is -3.9 (Fig. 7.120, top), χ^2 giving the minimum value among the simulated time profiles. The spectral index is determined to be -3.9 ± 0.2.

For the Mexico City data, when the power index is -4.3 (Fig. 7.120, bottom), χ^2 giving the minimum value among the simulated time profiles and a spectral index that is determined to be -4.3 ± 0.4. The best-fit spectral indices are the same as those derived by assuming that the neutrons were produced impulsively. The total energy fluxes of neutrons are estimated to be

$$W_S(1,500\ \text{MeV} > E_n > 50\ \text{MeV}) = (7.0 \pm 0.5) \times 10^{26}\ \text{ergs} \cdot \text{sr}^{-1}, \qquad (7.38)$$

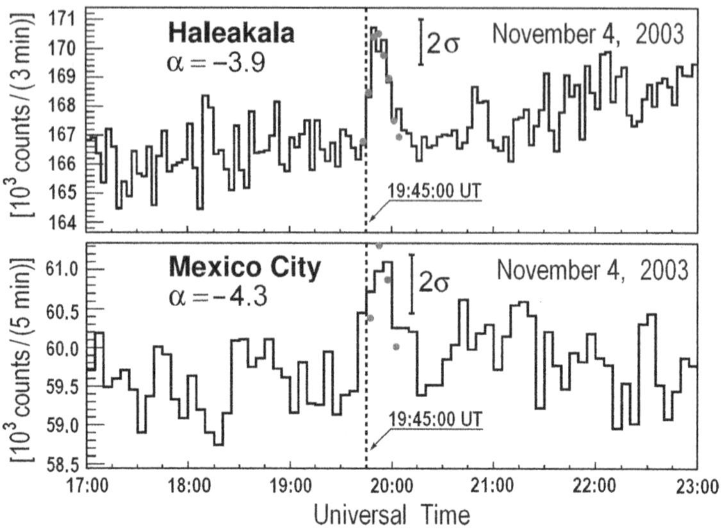

Fig. 7.120 Simulated time profiles (*points*) when the spectral index is −3.9 for Haleakala (*top*) and −4.3 for Mexico City (*bottom*), superposed on the observed counting rate of the Haleakala NM. Points are the simulated time profile for solar neutrons, assuming that they were produced with the same time profile as the high-energy γ-rays shown in Fig. 7.33 From Watanabe et al. (2006b)

from the Haleakala NM data and

$$W_S(1,500 \text{ MeV} > E_n > 50 \text{ MeV}) = \left(5.7^{+0.7}_{-0.8}\right) \times 10^{26} \text{ ergs} \cdot \text{sr}^{-1}, \tag{7.39}$$

from the Mexico City NM data.

7.33.6 Discussion and Main Results for the Event of the 4 November, 2003

Watanabe et al. (2006b) discussed and summarized the main results for the solar neutron event of 4 November, 2003 as following.

1. Solar relativistic neutrons were detected in association with the X28 solar flare simultaneously by the NMs at Haleakala and Mexico City and also by the solar neutron telescope at Mauna Kea.
2. During this event, intense emissions of high-energy γ-rays were observed by the INTEGRAL satellite. In order to investigate the production time of solar neutrons, there are compared the solar neutron data with the γ-ray data obtained from INTEGRAL.
3. In the November 4 event, the time profiles of γ-ray lines, which are assumed represent the time profile of solar neutron production, cannot be obtained

independently, since the bremsstrahlung component was strong and the line γ-ray components were buried in bremsstrahlung.

4. However, from the time profile of the 2.2 MeV neutron capture γ-rays, it appears that the time profile of ion acceleration was approximately the same as that of bremsstrahlung emissions. Assuming that solar neutrons were produced at the time when these γ-rays were emitted, it could possible to explain the observed excesses.
5. If assume that solar neutrons were produced impulsively at 19:45 UT on November 4, when the γ-ray lines peak, it can be derive the energy spectrum of solar neutrons at the solar surface from the NM data (Eq. 7.31).
6. For the 4 November, 2003 event, in order to examine whether all excesses observed by the Haleakala and Mexico City NMs and the solar neutron telescope at Mauna Kea can be expressed by one energy spectrum consistently, and for more detailed analysis, it was simulated time profiles of solar neutrons for these detectors and compared with observed time profiles. All of the simulation results are consistent with Eq. 7.31. Thus, it could explain all observations with a consistent spectrum.
7. Although it can fit the data by assuming that solar neutrons are produced impulsively, it is more natural to assume that solar neutrons are produced continuously over a finite time. It was therefore modeled the time profiles of solar neutrons by assuming that the neutrons were produced with the same time profile as γ-ray lines from excited ions. The spectral indices derived by assuming that neutrons are produced continuously tend to be harder than those derived by assuming that neutrons are produced impulsively. For the November 4 event, the result was that the index obtained using the line γ-ray time profile is clearly harder (−2.9) than that obtained using the impulsive model (−3.5). Therefore, for this event, the observations were explained by assuming that solar neutrons were produced with the same time profile as γ-ray lines observed by the RHESSI satellite (Share et al. 2004) and INTEGRAL satellite (Tatischeff et al. 2005).
8. The spectrum of accelerated ions can be calculated from the neutron spectrum using the spectrum of escaping neutrons produced by the accelerated ions (Hua and Lingenfelter 1987a, b; Hua et al. 2002). From the neutron spectra shown in Eq. 7.31, the number of protons above 30 MeV would be about 10^{32} sr^{-1} under the assumption that there is no turnover of the spectrum. This is a typical value for solar neutron events observed thus far.

7.34 Solar Neutron Event of the 7th of September 2005

7.34.1 The Matter of Problem and Short History of Observations and Research

As noted Sako et al. (2006, 2008), strong signals of energetic neutrons associated with the solar flare of 7 September 2005 were detected by using solar neutron telescopes located at Mt. Chacaltaya in Bolivia and Mt. Sierra Negra in Mexico, as well as by NM located at Mt. Chacaltaya and Mexico City. In Sako et al. (2006) was shown that the observed time profiles indicate a continuous emission of solar neutrons. In Sako et al. (2008), Watanabe et al. (2007) and González et al. (2008) were presented results of a unique analysis of solar neutron telescopes data to obtain energy spectrum of solar neutrons. The solar neutron telescopes channels with different response functions in energy enabled to obtain spectrum without any assumption on the emission time profile. The obtained result is consistent with the independent of Sako et al. (2006) studies, which indicate a differential power law index of 3. This supports the conclusion of Sako et al. (2006) that the emission of neutrons and γ-rays started at the same time and neutrons were continuously emitted. A marginal inconsistency about the existence of a spectral cutoff at around 500 MeV may indicate some more complicated scenario. In Watanabe et al. (2007) was considered possible ion acceleration and neutral emission mechanisms for 7 September, 2005 flare.

7.34.2 *Observation Data*

According to Sako et al. (2006), strong neutron signals were detected in association with the solar flare occurred on 7 September, 2005. The x-ray flux peaked at 17:40 UT and was classified as X17.0. The flare occurred in AR10808, which was located at S06°, E89°. Ground-based observations were made by solar neutron telescopes in Mexico (Sierra Negra) and Bolivia (Chacaltaya), as well as by NM in Mexico (Mexico city) and Bolivia (Chacaltaya). Parts of the observed data by the Mexico solar neutron telescope (Valdes-Galicia et al. 2004) are shown in Fig. 7.121.

According to Valdes-Galicia et al. (2004) and Sako et al. (2008), the Mexico solar neutron telescope has plastic scintillators of a 4 m^2 area and 30 cm thickness covered by proportional counters in anticoincidence (so called anti-counters). Scintillator signals are discriminated by four different energy thresholds (30, 60, 90 and 120 MeV) and they are designated to be S1, S2, S3, S4, respectively. Scintillator signals without associated anti-counter signal are regarded as generated by neutral particles and designated like S1-anti. Four layers of proportional counters are installed below the scintillators and coincident signals with S1-anti are named as L1-anti, L2-anti, L3-anti and L4-anti. In Fig. 7.121, 2-min counting rates of S1-anti and L1-anti are presented. It can be seen a strong excess in the S1-anti (top panel) but only a marginal increase in L1-anti (bottom panel).

7.34.3 *Response Functions for Different Channels*

To estimate the primary neutron energy spectrum, Sako et al. (2008) integrated the counts of each channel recorded between 17:30 and 18:30 UT as indicated by arrows in Fig. 7.121). Background counts are estimated from 3rd order polynomial

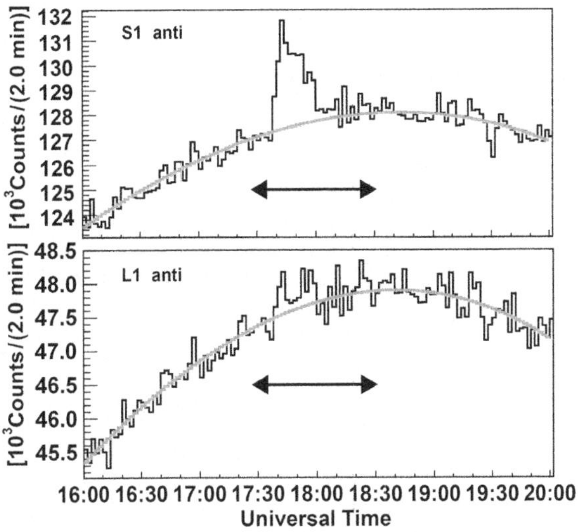

Fig. 7.121 Two-minutes counting rates of the Mexico solar neutron telescope. *Top panel*: Neutral particles measured at the lowest threshold scintillator channel. *Bottom panel*: Neutral particles penetrated into the first layer of the proportional counters below the scintillator. Histograms show the observed count and the smoothed curves indicate the estimated background. *Arrows* indicate the time window to integrate the total excess count. From Sako et al. (2007a)

fit, as plotted in the Fig. 7.121, excluding period 17:30 – 18:30 UT. The statistical significances (excess/σ) of each channel are 16.6, 12.0, 9.9, 6.2, 2.7, 1.7, 1.6 and 1.8 for S1-anti, S2-anti, S3-anti, S4-anti, L1-anti, L2-anti, L3-anti, L4-anti, respectively. So the detections in the scintillator channels are significant but marginal in the lower proportional counter channels. Because the responses of each channel have different energy dependence, the relative count among channels must be sensitive to the energy spectrum of primary neutrons. To have a general idea of this energy dependence, the detection efficiencies of each channel are shown in Fig. 7.122 as functions of neutron kinetic energy at the top of the detector.

7.34.4 Determining the Solar Neutron Differential Energy Spectrum

In this calculation, Sako et al. (2008) used GEANT3 code and supposed distributed neutrons uniformly over the detector in the vertical direction. To simulate the case of the 7th September, 2005 event, Sako et al. (2007a) assumed that the neutrons had a power law energy spectrum at the Sun and taken into account β-decay of neutrons in flight and attenuation in the earth's atmosphere according to Shibata (1994) including the actual atmospheric depth and the solar zenith angle at Sierra Negra at

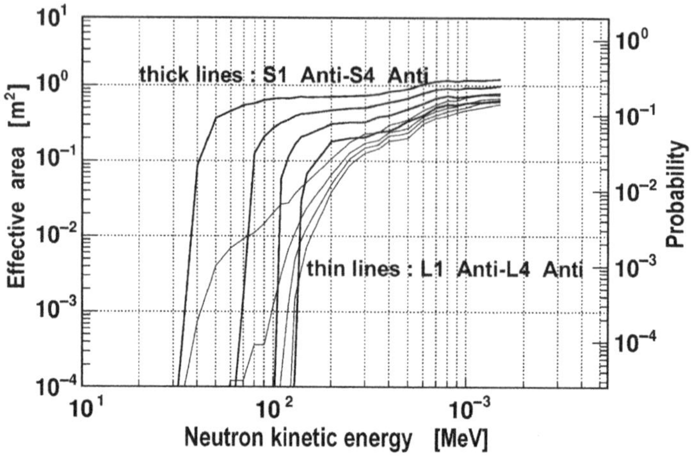

Fig. 7.122 Response function of various channels of the Mexico SNT. Detection efficiencies (*right axis*) of each channel are plotted as functions of the neutron kinetic energy at the top of the detector. In the calculation, neutrons were uniformly distributed over the detector with a vertical injection angle. From Sako et al. (2008)

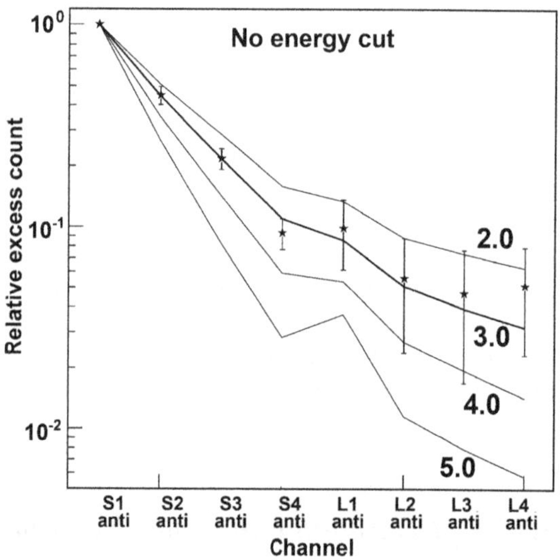

Fig. 7.123 Relative excess counts of various channels. Plots are results of the 7 September, 2005 event with statistical errors. *Solid lines* show expected counts for different neutron spectra. Neutron spectra are assumed to be power laws at the Sun with indexes indicated at each line. From Sako et al. (2008)

the flare time. The relative counts normalized to S1-anti are plotted in Fig. 7.123 for some differential power law indexes, 2–5, assumed in the simulation. The plots with error bars are the results of observation. Error bars are calculated based only on the statistical errors.

From Fig. 7.123, can conclude that the most likely power law index is 3. This is consistent with the results of Sako et al. (2006) and Watanabe et al. (2007) where indexes were estimated to be 3.2 and 3.1, respectively, based on the analyses of the Bolivia NM data and γ-ray profile. This is very important that Sako et al. (2008) have obtained the index without any assumption of the emission time and duration that was necessary in studies of Sako et al (2006). The consistent results of two independent analyses with different data sets support the assumptions and the conclusions in the Sako et al (2006) analyses. Those are (1) emission of the neutrons started at a same time with the γ-ray radiations, (2) the neutrons were emitted continuously.

Sako et al. (2008) note that in the previous work Sako et al. (2006), a single power law model could not fit the observed counting profile very well. It was concluded that a truncated spectrum at 400–500 MeV can fit the data very well. Sako et al. (2008) tested spectra with a cutoff at 500 MeV and results are shown in Fig. 7.124.

As noted Sako et al. (2008), the result shown in Fig. 7.124 indicates a slightly harder index than 3.0, but the difference is within the statistical errors and not significant.

7.34.5 Discussion on Possible Truncated Solar Neutron Spectra at 500 MeV

Sako et al. (2008) have analyzed the solar neutron data observed by the Mexico solar neutron telescope on 7 September, 2005. The data showed excesses in several channels those have different response functions in energy. They demonstrated the relative excess counts of these channels are sensitive to the spectrum of the primary neutrons. The comparison between the observed data and results of simulation

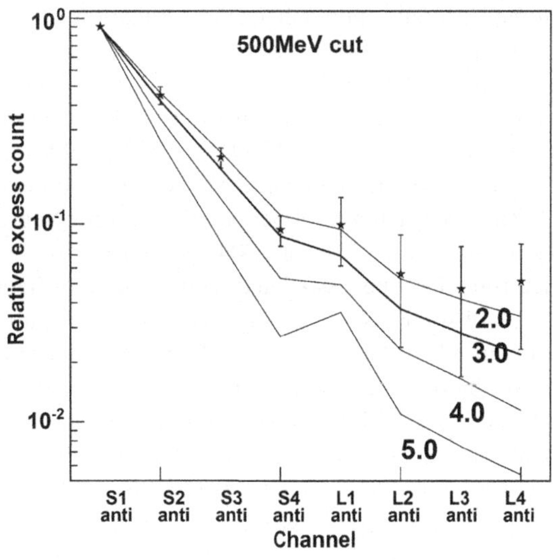

Fig. 7.124 Same as Fig. 7.123, but spectra truncated at 500 MeV. From Sako et al. (2008)

indicated that the differential power law index of the energy neutron spectrum was likely to be 3, that is consistent with the completely independent analysis from the Bolivia NM data and γ-ray profile. This supports the assumptions and conclusions made in Sako et al. (2006). However, Sako et al. (2008) analysis preferred a single power law spectrum though the previous work indicated a cutoff at around 500 MeV. Because the result is marginal and sensitive to the detector calibrations, it is necessary first carefully treat the systematics. If the discrepancy becomes true, a possible solution is a time evolution of the neutron spectrum that is ignored in the analyses of the Sako et al. (2006, 2008) works. High energy cutoff was introduced to explain the slow rise in the NM counting profile. However, if there are high energy component as indicated in Sako et al. (2008) study, they must be emitted not in the fast phase but some later time. If such delayed high energy neutrons exist, it can be found with analyses like Sako et al. (2008) paper but for time interval slices. As noted Sako et al. (2008), it is possible to test this and other kinds of models in further detail with fruitful data of the Mexico and Bolivia solar neutron telescopes as well as by data of Mexico and Bolivia NM. Time dependent analyses of all the available data sets will lead to the final goal to decide the energy spectrum and emission profile of neutrons at the Sun only from the neutron observations. Consequently, it will be possible to understand a behavior of high energy ions near the solar surface.

7.34.6 Energy Spectrum for the Solar Neutron Event of the 7 September, 2005, Derived from the Solar Neutron Telescope at Sierra Negra

Some different approach in the estimation of the energy spectrum for the solar neutron event of the 7 September, 2005, derived from the same solar neutron telescope data considered in Sections 7.34.4 and 7.34.5, were made by González et al. (2008) using the attenuation model of Dorman and Valdes-Galicia (1997), Dorman et al. (1997a, b, 1999a, b, 2000), Valdes-Galicia et al. (2000) as well as the detector efficiency calculation by Valdes-Galicia et al. (2004).

The solar neutron telescope at Mt. Sierra Negra in Mexico (19.0°N, 97.3°W) is taking data since June of 2004. A solar neutron event was registered by this solar neutron telescope, associated with the flare of September 7, 2005, at the minimum phase of solar cycle 23. The solar neutron event of 7 September, 2005, associated with an X17 solar flare was the first event registered by the solar neutron telescope at Mt. Sierra Negra. Furthermore, it was registered by the NM in Mexico City, the Chacaltaya's NM and solar neutron telescope. By using the method described in Watanabe et al. (2003b), Sako et al. (2006) calculated the energy spectrum of neutrons at the Sun, which was fitted by a power law as

$$D_S(E_n) = 6.1 \times 10^{27} (E_n/100\ \mathrm{MeV})^{-3.8}\ \mathrm{MeV}^{-1}\ \mathrm{sr}^{-1}. \tag{7.40}$$

For these calculations Sako et al. (2006) used data from NM on Chacaltaya, the Shibata (1994) program for atmospheric attenuation and the efficiency of the NM calculated by Clem and Dorman (2000). In paper González et al. (2008) are presented an alternative spectrum for this solar neutron event, using different models of attenuation and detector efficiency.

An X17 solar flare occurred on 7 September, 2005 was registered by GOES satellite. The soft-x-ray emission started at 17:17 UT, reached its maximum at 17:40 UT, and decayed to half-maximum at 18:03 UT. This flare was classified as an East limb flare (S06°, E89°), and it occurred in AR10808. The GOES satellite did not detect a significant increase in charged particles (<0.1 particles s^{-1} cm^{-2} sr^{-1} above 100 MeV). Terasawa et al. (2005) obtained hard x-ray data of the Geotail satellite, showing that the >50 keV x-ray emission peaked at 17:36:40 UT (Fig. 7.125).

At the time of maximum x-ray emission detected by GOES satellite (17:40 UT), the solar neutron telescope at Mt. Sierra Negra was suitable place to observe solar neutrons. The solar zenith angle was 17.5°, and the air mass in the line of sight to the Sun was 603 g cm^{-2}. It was assumed that the ion acceleration occurred at the same time as the line γ-rays were emitted, although the main component of these γ-rays is bremsstrahlung, and that solar neutrons were also produced at the same time (Watanabe et al. 2005a). The Sierra Negra solar neutron telescope can discriminate four different energy thresholds, which correspond to energy deposit of >30, >60, >90, and >120 MeV (Valdes-Galicia et al. 2004). Figure 7.126 show the 2 min counting rates observed by the Sierra Negra solar neutron telescope on September 7, 2005.

In order to determine the energy spectrum of solar neutrons during this event, González et al. (2008) use 3 min average data. They determine the neutron energy

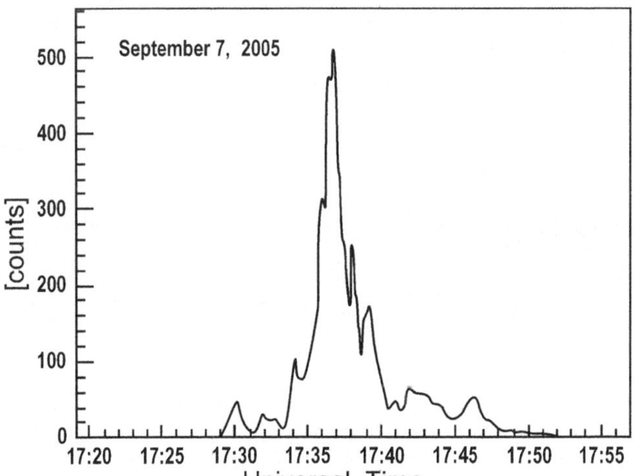

Fig. 7.125 Hard x-ray time profile observed by Geotail satellite on September 7, 2005. The maximum emission of x-rays was at 17:36:40 UT. From González et al. (2008)

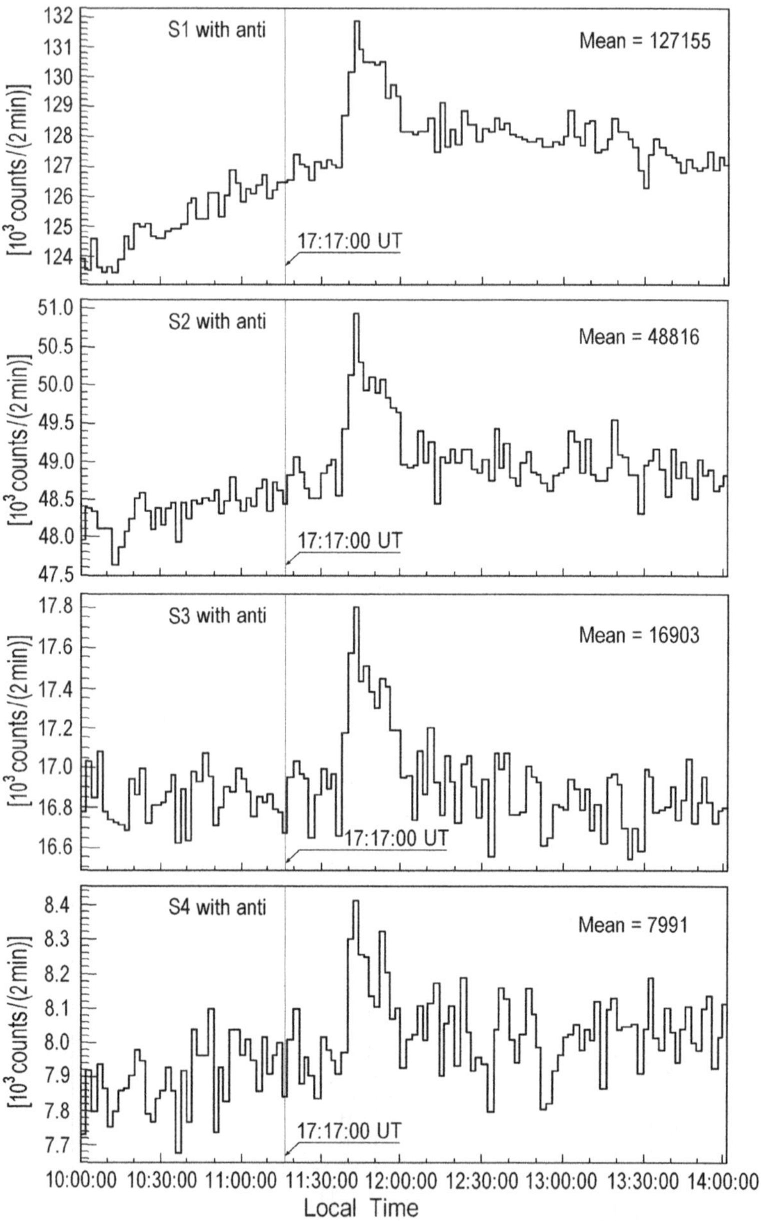

Fig. 7.126 Two minute averages of the counting rate observed by solar neutron telescope at Sierra Negra. The time is in local time, which corresponds to 6 h less than universal time. S1_with_anti, S2_with_anti, S3_with_anti, and S4_with_anti, correspond to deposit energy >30, >60, >90, and >120 MeV, respectively. 17:17:00 UT is the time of beginning of solar flare. From González et al. (2008)

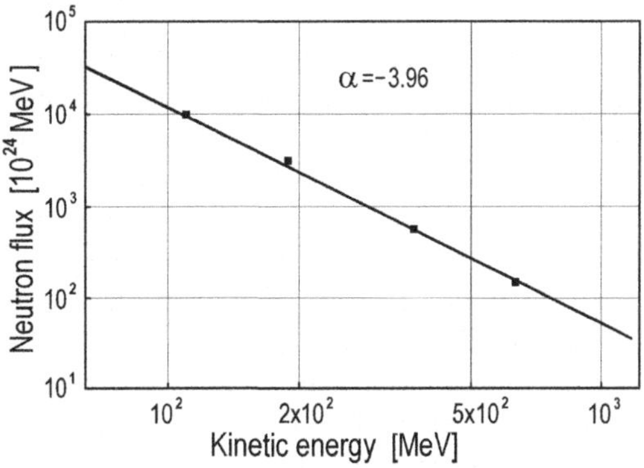

Fig. 7.127 The energy spectrum of solar neutrons for the flare of the 7th September, 2005. From González et al. (2008)

by using the time of flight method, assuming that all the solar neutrons were produced at 17:36 UT, the peak time of the intense emission of high energy γ-rays observed by Geotail satellite as shown in Fig. 7.125. To derive the energy spectrum of neutrons at the solar surface, González et al. (2008) calculated the survival probability of neutrons between the Sun and the Earth, the attenuation of solar neutrons passing through the Earth's atmosphere using the model by Dorman and Valdes-Galicia (1997), and the detection efficiency of the solar neutron telescope by Valdes-Galicia et al. (2004). Using these results, González et al. (2008) calculated the energy spectrum of solar neutrons for the flare of 7 September, 2005. The spectrum is shown in Fig. 7.127.

By fitting data points in Fig. 7.127 with a power law, the energy spectrum of solar neutrons was obtained. The fitting region is chosen as 100 MeV and above, because there errors from neutron attenuation in the Earth's atmosphere are small (Watanabe et al. 2005a). The energy spectrum is well fitted by a power law as:

$$D_S(E_n) = 5.3 \times 10^{27}(E_n/100\ \text{MeV})^{-3.96}\ \text{MeV}^{-1}\ \text{sr}^{-1}. \tag{7.41}$$

what is in good accordance with Eq. 7.40 obtained in Sako et al. (2006) from NM data in the frame of the Shibata (1994) atmospheric attenuation model and the detector efficiency by Clem and Dorman (2000). González et al. (2008) came to conclusion that two different independent approaches based on different experimental data give practically the same final results on the energy spectrum of solar neutrons in the event of 7 September, 2005.

7.34.7 Ion Acceleration and Neutral Emission Mechanisms for the 7 September, 2005 Flare

As underlined Watanabe et al. (2007), in association with an X17.0 flare on September 7, 2005 intense emissions of hard x-rays and γ-rays were registered by satellites, and relativistic neutrons were observed by ground-based detectors. The neutron signal continued for more than 20 min with high statistical significance. The long decay of the signals suggests that ions were continuously accelerated or trapped in the emission site. Watanabe et al. (2007) also find that γ-rays were emitted over a corresponding extended period. Only when Watanabe et al. (2007) incorporate the high-energy γ-ray emission time history became possible to explain the long-lasting neutron emission. As noted Watanabe et al. (2007) it is not possible explain neutron emissions in the frame of Hua et al. (2002) model.

According to Watanabe et al. (2007), at 17:17 UT on 2005 September 7, when solar active region 10808 (S06°, E89°) appeared at the East limb, the first and most energetic X-class flare occurred with a soft x-ray magnitude of X17.0. During this flare, intense emissions of hard x-rays and γ-rays were observed by INTEGRAL and RHESSI satellites (Fig. 7.128).

As noted Watanabe et al. (2007), RHESSI satellite was both in the South Atlantic Anomaly and on the night side of the Earth during most of the event. However, some emission was observed around 17:45 UT between South Atlantic

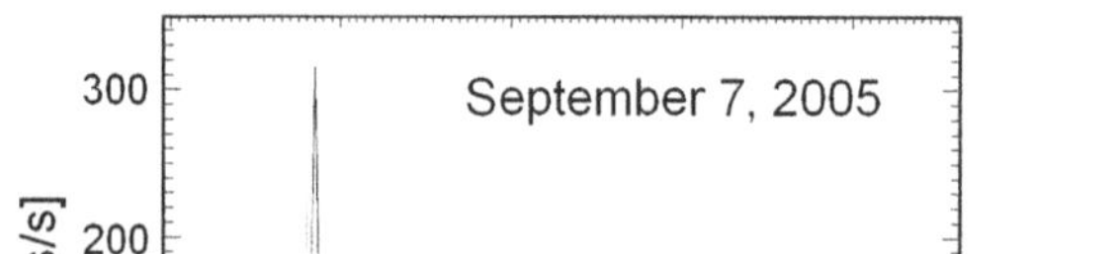

Fig. 7.128 The x-ray data observed by the INTEGRAL and RHESSI satellites on September 7, 2005. *Top panel* shows hard x-ray light curve observed by the INTEGRAL satellite between 200 and 300 keV. *Bottom panel* shows spectrogram observed by RHESSI. The RHESSI was in South Atlantic Anomaly until 17:43 UT and at the night side of the Earth from 17:47 UT. From Watanabe et al. (2007)

Anomaly and night: γ-ray lines at 0.511, 2.2, 4.4, and 6.1 MeV were observed and there was evidence for π^0-decay radiation. On the other hand, strong emission of hard x-rays and γ-rays during the whole flare time was observed by the INTEGRAL satellite. As shown the top panel of Fig. 7.129, the 4.4 MeV Carbon γ-ray line was clearly observed by the INTEGRAL satellite. It does not clearly show the 2.2MeV neutron-capture line, but this is consistent with the limb location of this flare. By fitting spectra, Watanabe et al. (2007) obtained the time history of 4.4MeV line γ-rays as shown bottom panel of Fig. 7.129.

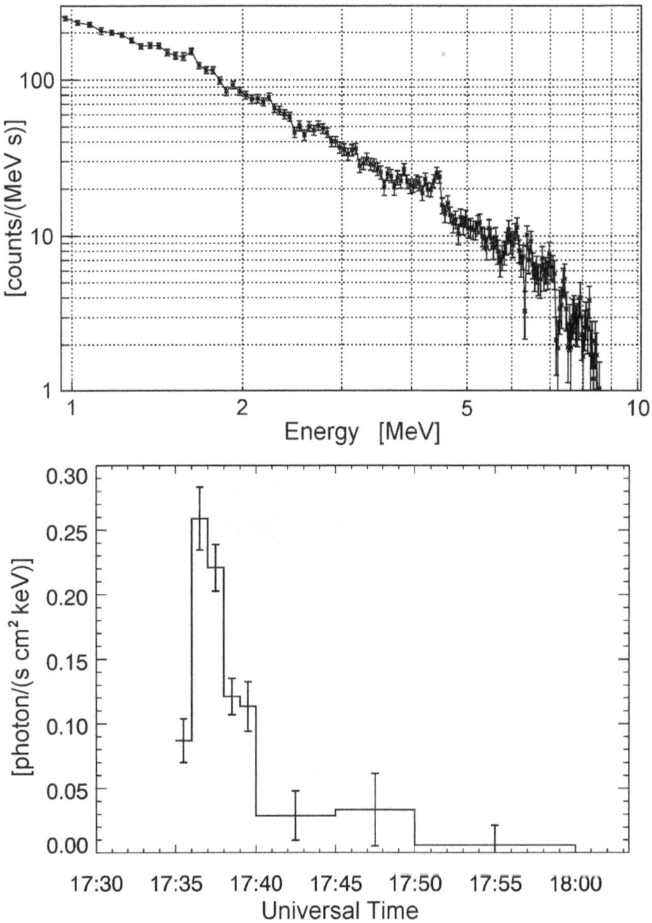

Fig. 7.129 The γ-ray data observed by the INTEGRAL satellite on September 7, 2005. *Top panel* shows the γ-ray spectrum observed from 17:35–18:00 UT, and 4.4 MeV line appear superimposed on the bremsstrahlung component. *Bottom panel* shows the time profile of line γ-ray components of Carbon nuclei. From Watanabe et al. (2007)

7.34.8 Soft and Hard x-Ray Images of the 7th September, 2005 Flare Loops

As underlined Watanabe et al. (2007), since such deexcitation γ-rays are prompt, this time history is a good approximation of the ion acceleration time history. Ions were accelerated not only during the impulsive phase, but also during the decay phase. The 4.4 MeV γ-ray emission does not return to the background level until at least 18:00 UT. During this decay phase, hard x-ray and γ-ray emissions were also observed by RHESSI satellite. The x-ray image obtain with RHESSI is shown in Fig. 7.130.

A soft x-ray loop is seen above the limb with only one hard x-ray foot-point visible. Although only one image was obtained from RHESSI satellite data, Watanabe et al. (2007) obtained many images from the GOES/SXI data and the soft x-ray loop could be seen rising up from the limb. Figure 7.131 shows the altitude of the soft x-ray emission center from the limb recorded by the GOES/SXI

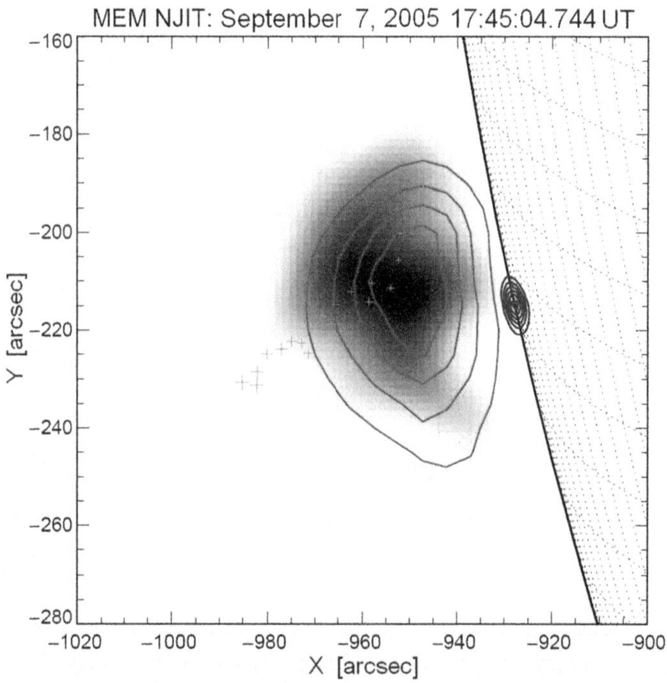

Fig. 7.130 The x-ray image observed by the RHESSI and GOES/SXI satellites on September 7, 2005. Red image shows 5–25 keV soft x-ray image obtained by RHESSI satellite, red contour shows soft x-ray image obtained by GOES/SXI satellite, blue contour shows 100–150 keV hard x-ray image obtained by RHESSI satellite, and green + marks show the center locations of the soft x-ray emission obtained by the GOES/SXI satellite. From Watanabe et al. (2007)

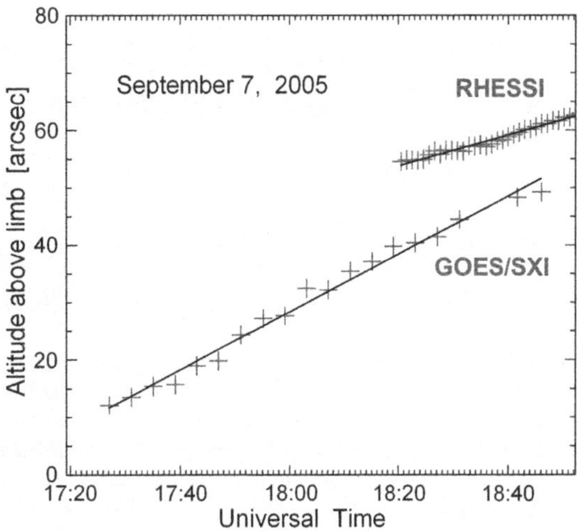

Fig. 7.131 The altitude of the soft x-ray emission center from limb observed by the GOES/SXI and RHESSI satellites on September 7, 2005. From Watanabe et al. (2007)

and RHESSI satellites (the difference of the two position is due to the different energies detected by RHESSI and GOES/SXI).

7.34.9 Checking the Solar Neutron Spectrum Using NM Data

As it was described above, on the ground, relativistic neutrons were observed by the NM at Mt. Chacaltaya and Mexico City and by the solar neutron telescopes at Chacaltaya and Mt. Sierra Negra (Sako et al. 2006). The statistical significance of all neutron signals was more than 10σ and the detection lasted for more than 20 min. This was the largest solar neutron event observed in solar cycle 23. Watanabe et al. (2007) analyzed the neutron data obtained with the Bolivia NM since this detector recorded the largest signal. To estimate the energy spectrum of the emitted neutrons, it was used the same method as in Watanabe et al. (2006b). This method calculates time dependent arriving neutron spectra from an emitted neutron spectrum assumed to be a power law. Watanabe et al. (2007) used the atmospheric attenuation ratio of solar neutrons calculated by Shibata (1994) and efficiency of the neutron monitor calculated by Clem and Dorman (2000) to convert these spectra into count rates and compare with the observed neutron data. Watanabe et al. (2007) used the time profile of the 4.4 MeV line γ-rays (Fig. 7.129) as a hypothetical production time profile for the solar neutrons. Good agreement was obtained (reduced $\chi^2 = 1.8$) for a neutron spectral index of -3.1, as shown in Fig. 7.132. The extended presence of high-energy ions implied by the line emission is required to explain the extended neutron emission. Such ions could have been continuously accelerated or trapped in the emission site.

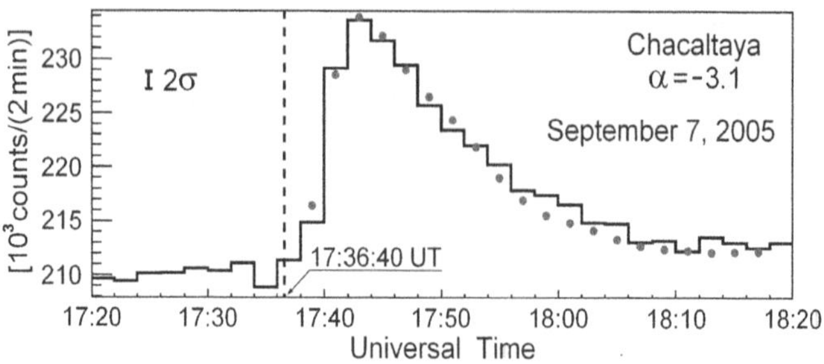

Fig. 7.132 The observed and simulated time profiles of neutrons on September 7, 2005. The *black solid line* is the observed 2-min counting rates from the Bolivia NM. The red points represent the simulated time profiles for solar neutrons assumed to have been produced with the same time profile as the 4.4 MeV γ-rays with spectral index -3.1. The high-energy cutoff of the solar neutron energy is assumed to be 500MeV. From Watanabe et al. (2007)

7.34.10 Using the Simulation Program of Hua et al. (2002) to Estimate the Spectrum of Accelerated Ions

Watanabe et al. (2007) next used the simulation program of Hua et al. (2002) to estimate the spectrum of accelerated ions. By using this program, neutron spectra arriving at the Earth's atmosphere can be estimated from line γ-ray data. These arriving spectra are converted into NM count rates using the atmospheric attenuation according to Shibata (1994) and neutron monitor efficiency according to Clem and Dorman (2000), as above. By comparing these rates with the observed rate, Watanabe et al. (2007) can determine the accelerated ion spectrum. A number of parameters must be set to use Hua et al. (2002) program. For the accelerate ion composition ($\alpha/p = 0.5$, $^3He/^4He = 1$), ambient composition ($He/H = 0.1$, $Ne/O = 0.25$) and atmospheric model (Avrett 1981), Watanabe et al. (2007) used typical values estimated from observations of previous solar neutron events. For the acceleration release time history, Watanabe et al. (2007) used the 4.4MeV line γ-ray time history observed by INTEGRAL and RHESSI satellites (Fig. 7.129). The loop length was calculated from Figs. 7.130 and 7.131. At 17:37:30 UT (the peak time of γ-ray emission), the altitude of the soft x-ray emission center was about 16.7 arcsec (11,700 km) from the limb. Using this altitude as the radius of a semicircular flare loop, the loop length is 36,800 km. The flare heliocentric angle is also determined from the images as 89.0 degree. Watanabe et al. (2007) derive the remaining parameters (λ, δ, and s) by systematically varying them and comparing the predicted count rates with the observed count rate. Figure 7.133 shows the best fit to the observed data obtained with λ (level of pitch-angle scattering within the loop) of 1,000, δ (magnetic field convergence index) of 0.35, and an accelerated ion power-law index of -3.4. Value χ^2 is still large however (reduced $\chi^2 = 2.6$) as there are some discrepancies at the peak and decay phase between the observed and predicted data.

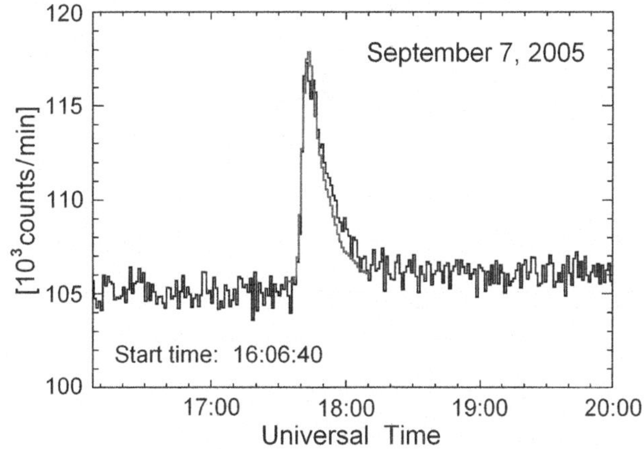

Fig. 7.133 The observed and simulated time profiles of neutrons on September 7, 2005. The *black line* is the observed 2-min counting rates from the Bolivia neutron monitor. The *red line* represents the simulated result of Hua et al. (2002) program with spectral index −3.4. The high-energy cutoff of the solar neutron energy is assumed to be 400 MeV. From Watanabe et al. (2007)

Watanabe et al. (2007) came to conclusion that the long-lasting neutron emission observed by NM from the September 7, 2005 flare could be explained by the long-lasting presence of energetic ions implied by the extended γ-ray line emission. Watanabe et al. (2007) could not achieve an adequate fit to the neutron data when they used the transport and interaction loop model of Hua et al. (2002). An important parameter of the loop model is the loop length which is not well-determined for this flare. Watanabe et al. (2007) suppose continue to analyze these data and explore the effect of changing the loop length and other parameters.

7.34.11 Checking a Model with Solar Neutron Energy Spectrum with Sharp Cutoff at 400 MeV

As a developed model of the solar neutron emission profile, Sako et al. (2006) assumed that the neutrons were emitted with the same time-profile as the hard x-rays and γ-rays. The best-fit result with a single-power-law spectrum with an index of −3.2 is plotted in the top panel of Fig. 7.134. There is an obvious discrepancy between the data and the Monte Carlo simulations. To compensate for the difference at the rising phase, Sako et al. (2006) tried the same spectrum but with a sharp cutoff at 400 MeV. The result is shown in the bottom panel of Fig. 7.134.

As can be seen from Fig. 7.134, in the case described by bottom panel, the profile up to just after the peak is well fitted. This indicates that the acceleration of ions began at the same time as the electrons. However, the discrepancy of the tails,

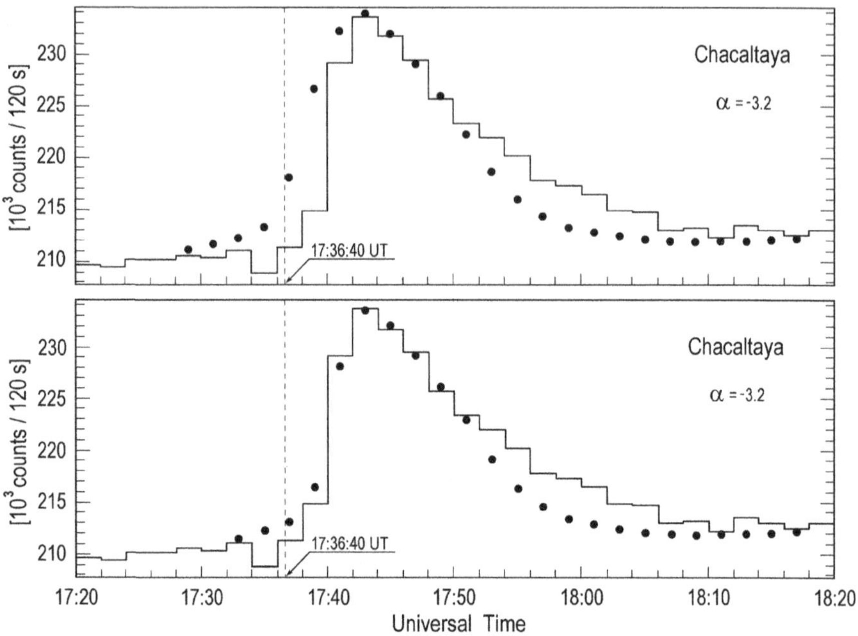

Fig. 7.134 Simulated (*dots*) and observed (*histogram*) time profiles of the Bolivia NM during the September 7, 2005 event. In these calculations was assumed that the neutron emission follows the hard x-ray profile. *Top*: Case in which no cutoff energy is included. The best power index is found to be −3.2, but apparent discrepancies in the rising and decaying phases are seen. *Bottom*: Same spectrum, but a sharp cutoff at 400 MeV is assumed. Rising part is well explained, however a large difference is still found in the decaying phase. From Sako et al. (2006)

which Fig. 7.134 could not fit with any set of parameters, suggests that the ions were continuously accelerated or trapped for a longer period than the electrons.

7.35 The Search for Solar Neutrons from the 8 September, 2005 Flare

7.35.1 Properties of Solar Flares and the Distribution of Solar Neutron Telescopes Relative to Sub-solar Points

Solar neutron telescopes, which are dedicated to observe solar neutrons and to measure their energies, have been in operation since the beginning of solar cycle 23; they are distributed in the world to cover all longitudes to watch the Sun for 24 h a day (Matsubara et al. 2005). According to Matsubara et al. (2008), an X5.4 class flare occurred at 20:52 UT on September 8, 2005. The flare was located at S11°, E74° on the solar surface, which was near the eastern limb of the Sun. Among seven solar neutron telescopes in the network, the detectors in Mexico and Hawaii were at

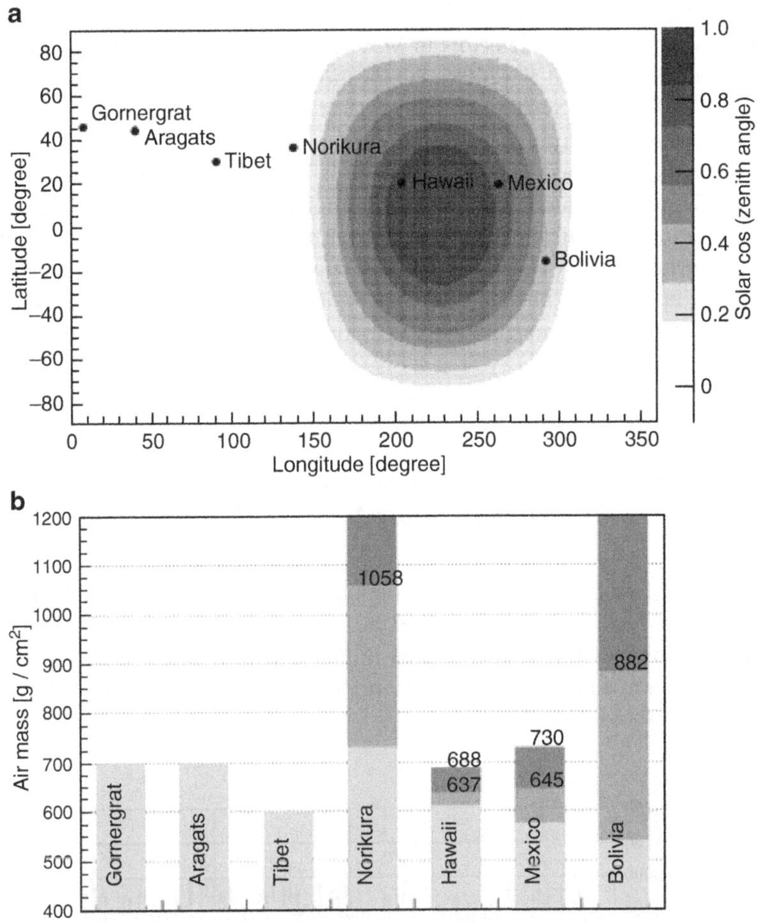

Fig. 7.135 The position of the Sun at 20:52 UT on September 8, 2005 (*top panel*) and the atmospheric depths for neutrons to reach to solar neutron telescope (*bottom panel*; light grey–air mass in vertical direction). From Matsubara et al. (2008)

good positions to observe solar neutrons. The relation between the position of the Sun and solar neutron telescopes is shown in Fig. 7.135.

7.35.2 Comparison of Solar Flares on the 8 and 7 September, 2005

As shown in Fig. 7.135, attenuation of neutrons in the air is almost same for both detectors. Therefore the detection probability of solar neutrons is the same for both detectors if both have the same detection efficiency to neutrons. This situation is quite similar to that in the case of X17 flare on September 7, when the atmospheric depths of neutrons to reach to the detectors in Mexico and Bolivia were almost same.

As noted Matsubara et al. (2008), the locations of X-class flares on September 7, 2005 (S06°, E89°) and September 8, 2005 (S11°, E74°) are both near the east limb of the Sun. The only difference is the flux of x-rays, and the former case is more intense than the latter case by a factor of only 3. The direction of neutrons emitted at the Sun is almost the same direction as the original charged particles. Therefore if the acceleration of particles to higher energies always occurs at the same place on the solar surface, it can be expect the detection of solar neutrons when gigantic flares occur at the same place on the Sun. Moreover it is reasonable to expect the same acceleration mechanism for these two flares because they occurred at the same active region.

7.35.3 Search for Solar Neutrons from the Flare of the 8th September, 2005 in Solar Neutron Telescope Data

According to Matsubara et al. (2008), counting rates of neutrons for both detectors at Hawaii and Mexico were compared in the different levels of the energy threshold and different bins of time interval. One minute counting rates measured by both detectors are shown in Fig. 7.136 for energy threshold >50 MeV and in Fig. 7.137 for energy threshold >100 MeV.

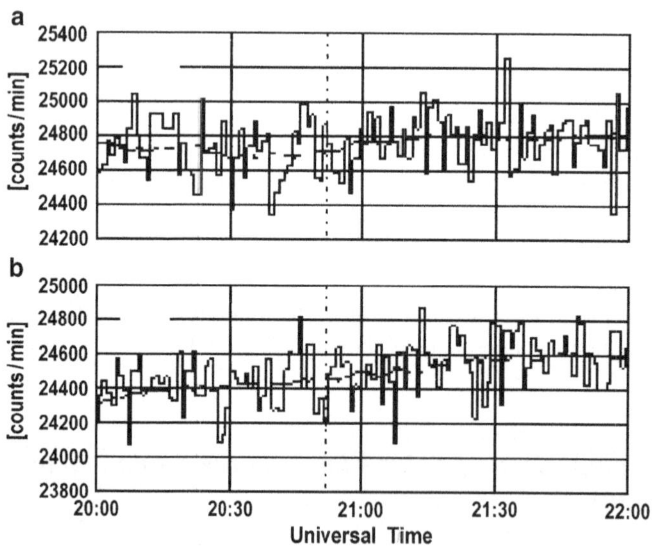

Fig. 7.136 One minute counting rates for neutrons (>50MeV) measured by solar neutron telescopes in Hawaii (**a**) and Mexico (**b**). *Smooth lines* are running averages for 60 min. The *vertical dotted line* is the start time of the x-ray flare. From Matsubara et al. (2008)

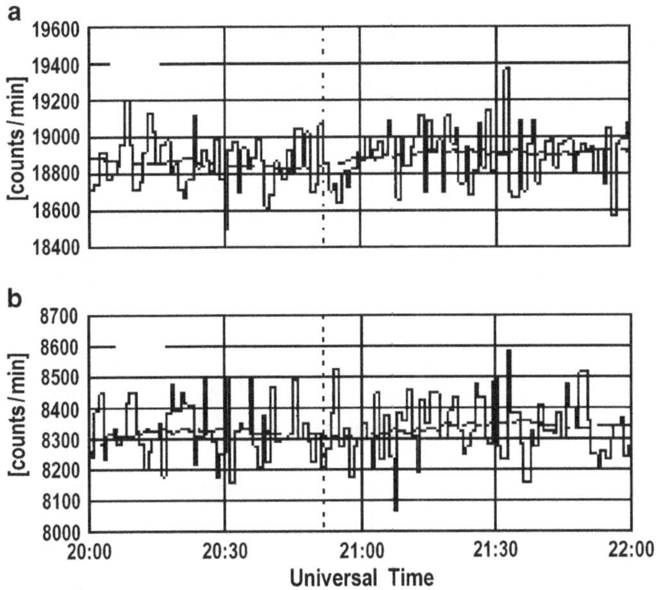

Fig. 7.137 One minute counting rates for neutrons (>100MeV) measured by solar neutron telescopes in Hawaii (**a**) and Mexico (**b**). *Smooth lines* are running averages for 60 min. The *vertical dotted line* is the start time of the x-ray flare. From Matsubara et al. (2008)

From Figs. 7.136 and 7.137 can be seen that there are no significant enhancements of counting rates correlated with the flare. Matsubara et al. (2008) came to conclusion that it could not detect any significant enhancements of counting rates in this detailed analysis from the 8 September, 2005 flare. Matsubara et al. (2008) noted that gigantic solar flares exceeding X-class occurred 10 times in September 2005, although the Sun had already been in the declining phase in solar cycle 23. The search for solar neutron events associated with these 10 X-class flares, using data obtained by the international network of solar neutron telescopes shows that there are no clear evidence for detecting solar neutrons within these periods, except for the significant detection of solar neutrons on September 7, 2005 (see Section 7.33).

7.36 Search for Solar Neutrons from the 13th December, 2006 Flare

According to Matsubara et al. (2008), the analysis was focused on 4 X-class solar flares which occurred in December 2006. One of these flares, X3.4 flare occurred at the location of the solar surface S06°, W24° on December 13, 2006. One of the interesting feature of this flare is that it was accompanied by ground level

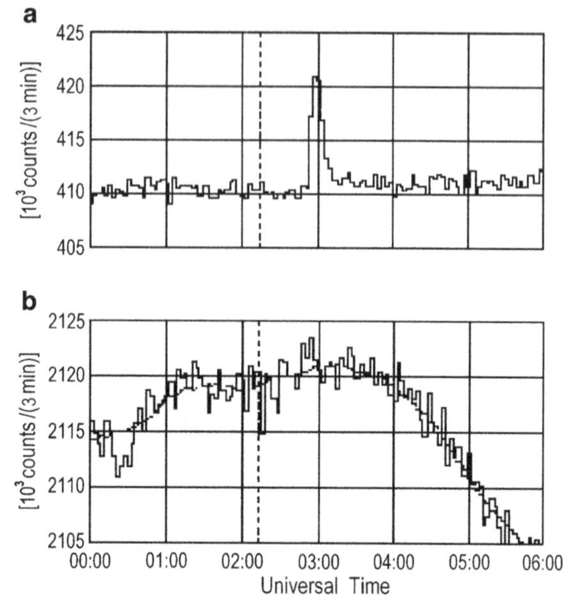

Fig. 7.138 Three minute counting rates measured by the Gornergrat (**a**) and Norikura (**b**) solar neutron telescopes on December 13, 2006. The *smooth line* in (**b**) is the running average of counting rates for 1 h. The *vertical dotted line* shows the start time of the x-ray flare. From Matsubara et al. (2008)

enhancements (GLE). The solar neutron telescope at Gornergrat in Switzerland also recorded GLE (every solar neutron telescope records not only neutrons which did not hit veto counters, either scintillation counters or proportional counters, but also charged particles which hit veto counters). Therefore it is possible that one station detects both solar neutrons and GLE. In the case of the X3.4 flare, the start of the flare was 02:14 UT, and the enhancement recorded by the solar neutron telescope at Gornergrat must not be solar neutrons. The station which had a possibility to detect solar neutrons was the solar neutron telescope at Mt. Norikura in Japan. Therefore Matsubara et al. (2008) compared counting rates measured at Norikura with those at Gornergrat. Neutrons are expected to come earlier than GLE, but the counting rates shown in Fig. 7.138 do not show any increase indicating solar neutrons.

Matsubara et al. (2008) noted that although it could not obtain the evidence for the emission of solar neutrons, it is certain that particles in the 13-th December, 2006 flare were accelerated to high energies associated with this flare (panel **a** in Fig. 7.138), because the cutoff rigidity at Gornergrat is 4.5 GV. Matsubara et al. (2008) came to conclusion that there was no clear evidence for detecting solar neutrons in x-ray solar flare of 13 December, 2006 as well as in other three x-ray solar flares in December 2006. Matsubara et al. (2008) noted that established during solar cycle 23 world-wide network of solar neutron telescopes will operate through solar cycle 24 and more fruitful results on solar neutron events are expected related with the acceleration and interaction of particles on the Sun, and their propagation in interplanetary space and through the Earth's atmosphere.

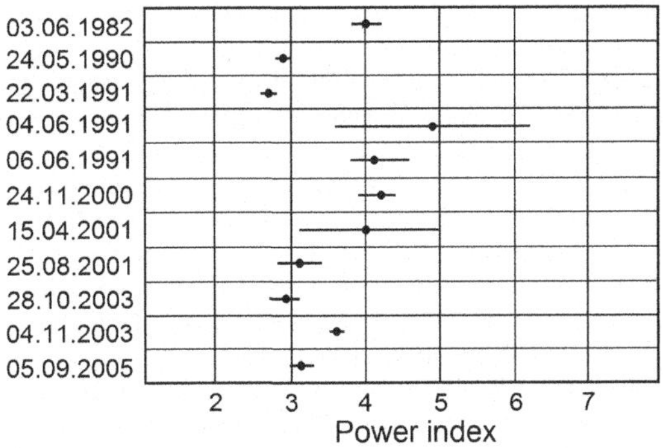

Fig. 7.139 The estimated power spectral indexes of solar neutron events detected by neutron monitors with energy range around 70–700 MeV according to Watanabe (2005). The spectra for most of events may be fitted by power law. From Valdes-Galicia et al. (2009)

7.37 Comparison of Power Spectral Indexes for Solar Neutron Events Observed by Neutron Monitors During 1982–2005

As noted Valdes-Galicia et al. (2009), a summary of spectral indices determined for solar neutron events observed by neutron monitors during 1982–2005 (see Fig. 7.139), was done by Watanabe (2005).

From Fig. 7.139 may be seen that the spectra for most of events may be fitted by power law. As underlined Valdes-Galicia et al. (2009), in some cases when was also have estimates of the proton spectra, they tend to be harder. This may be a systematic fact that would require further detailed investigations to determine whether the deference is due to the solar source or to interplanetary propagation. Nevertheless, the most relevant fact is that both, proton and neutron spectra tend to adjust to power law, as predicted by shock acceleration theory (see Chapter 4 in Dorman M2006).

References for Chapter 7

Alania M, Jedrzejczak K, Karczmarczyk J, Swarzynski J, Szabelska B, Szabelski J, Wibig T (2003) Directional variation of 5 GeV Muon Flux observed in the underground Muon telescope. Proceedings of the 28th international cosmic ray conference, vol 6. Tsukuba, pp 3573–3576

Allkofer OC (1983) Cosmic ray composition at $10^{12}-10^{15}$ eV derived from muon measurements. In: Shapiro MM (ed) Composition and origin of cosmic rays. D. Reidel Publishing Company, Dordrecht, pp 101–118

Aushev V, Antonova V, Belov A, Eroshenko E, Kryakunova O, Struminsky A (1999) Search for solar neutron events in Alma-Ata NM data. Proceedings of the 26th international cosmic ray conference, vol 6. Salt Lake City, pp 50–53

Avrett EH (1981) Reference model atmosphere calculation – the sunspot model. Proceedings of the sunspot conference, New Mexico, Sacramento Peak Observatory, pp 235–255

Barat C, Trottet G, Vilmer ZN, Dezalay J-P, Talon IR, Sunyaev R, Terekhov O, Kuznetsov A (1994) Evidence for intense coronal prompt gamma-ray line emission from a solar flare. Astrophys J Lett (USA) 425(2):L109–L112, Part 2

Barat C, Trottet G (1997) Private communication to R. J. Murphy

Belov AV, Ishkov VN, Zusmanovich AG, Kudela K (1987) Observation of solar neutrons by high-altitude neutron monitors. Geomagnet Aeronomiya 27(3):375–382, in Russian

Bieber JW, Evenson P, Droge W, Pyle R, Ruffolo D, Rujiwarodom M, Tooprakai P, Khumlumlert T (2004) Spaceship earth observations of the Easter 2001 solar particle event. Astrophys J Lett (USA) 601(1):L103–L106, Part 2

Bieber JW, Clem J, Evenson P, Pyle R, Ruffolo D, Saiz A (2005) Relativistic solar neutrons and protons on 28 October 2003. Geophys Res Lett 32(3):L03S02–L03S03. doi:doi:10.1029/2004GL021492

Capdevielle JN, Muraki Y (1999) Inverse muon charge ratio induced by solar neutrons. Astropart Phys 11(3):335–346

Cattani D, Cecchini S, Galli M, Giovannini G, Lattanzi D, Pagliarin A (2003) Cosmic ray intensity variations observed by environmental radiation monitors. Proceedings of the 28th international cosmic ray conference, vol 6. Tsukuba pp 3577–3580

Chiba N, Hashimoto K, Hayashida N, Honda K, Honda M, Inoue N, Kakimoto F, Kamata K, Kawaguchi S, Kawasumi N, Matsubara Y, Murakami K, Nagano M, Ogio S, Ohoka H, Saito T, Sakuma Y, Tsushima I, Teshima M, Umezawa T, Yoshida S, Yoshii H (1992a) Akeno giant air shower array (AGASA) covering 100 km^2 area. Nucl Instrum Methods Phys Res A 311(1–2):338–349

Chiba N, Dion GM, Hayashida N, Honda K, Honda M, Inoue N, Kadota K, Kakimoto F, Kamata K, Kawaguchi S, Kawasumi N, Matsubara Y, Nagano M, Ohoka H, Teshima M, Tsushima I, Yoshida S, Yoshii H, Yoshikoshi T (1992b) Possible evidence for 10 GeV neutrons associated with the solar flare of 4 June 1991. Astroparticie Phys 1(1):27–32

Chupp EL (1984) High-energy neutral radiations from the sun. Ann Rev Astron Astrophys 22:359–387

Chupp EL, Forrest DJ, Ryan JM, Cherry ML, Reppin C, Kanbach G, Rieger E, Pinkau K, Share GH, Kinzer RL, Strickman MS, Johnson WN, Kurfess JD (1981) Observation of the 2.223 MeV gamma-ray line on the SMM satellite-the event of 1980 June 7. Astrophys J 244(3): L171–L174

Chupp EL, Forrest DJ, Ryan JM, Heslin J, Reppin C, Pinkau K, Kanbach G, Rieger E, Share GH (1982) A direct observation of solar neutrons following the 0118 UT flare on 1980 June 21. Astrophys J 263(2):L95–L99, Part 2

Chupp EL, Forrest DJ, Share GH, Kanbach G, Debrunner H, Flueckiger E (1983) Solar neutrons from the impulsive flare on 1982 June 3 at 1143 UT. Proceedings of the 18th international cosmic ray conference, vol 10. Bangalore, pp 334–337

Chupp EL, Cline TL, Evans WD, Fenimore E, Gorenstein P, Grindlay J, Lamb DQ, Lewin WHG, Lingenfelter RE, Matteson JL, Ramaty R, Ricker GR, Share G, Teegarden B, Woosley SE (1984) In: Woosley (ed) The high energy transient explorer (HETE). AIP conference proceedings of the high energy transients in astrophysics, Santa Cruz, 1983, No. 115. AIP Press, New York, pp 709–714

Chupp EL, Forrest DJ, Vestrand WT, Dubrunner H, Flueckiger EO, Cooper JF, Kanbach G, Reppin C, Share GH (1985) Time extended production of neutrons during a solar flare. Proceedings of the 19th international cosmic ray conference, vol 4. La Jolla, pp 126–129

Chupp EL, Debrunner H, Flueckiger E, Forrest DJ, Golliez F, Kanbach G, Vestrand WT, Cooper J, Share G (1987) Solar neutron emissivity during the large flare on 1982 June 3. Astrophys J 318, Part I(2):913–925

Clem JM, Dorman LI (2000) Neutron monitor response functions. Space Sci Rev 93(1–2):335–359

D'Andrea C, Poirier J (2003) A study of the ground level event of April 15, 2001 with GRAND. Proceedings of the 28th international cosmic ray conference, vol 6. Tsukuba, pp 3423–3426

Debrunner H, Flückiger EO, Chupp EL, Forrest DJ (1983) The solar cosmic ray neutron event on June 3, 1982. Proceedings of the 18th international cosmic ray conference, vol 4. Bangalore, pp 75–78

Debrunner H, Flückiger EO, Stein P (1989) On the sensitivity of neutron monitors to solar neutrons. Nucl Instrum Methods Phys Res A (Netherlands) 278(2):573–575

Debrunner H, Lockwood JA, Ryan JM (1993) Solar neutron and proton production during the 1990 May 24 cosmic-ray flare increases. Astrophys J 409(2, Part I):822–829

Debrunner H, Debrunner H, Lockwood JA, Barat C, Bütikofer R, Dezalay JP, Flückiger E, Kuznetsov A, Ryan JM, Sunyaev R, Terekhov OV, Trottet G, Vilmer N (1997) Energetic neutrons, protons and gamma rays during the 1990 May 24 solar cosmic ray event. Astrophys J 479(2):997–1011, Part 1

DelSignore KW (1995) Ph.D. thesis, Case Western Reserve University

Dorman LI, Venkatesan D (1993) Solar cosmic rays. Space Sci Rev 64(3-4):183–362

Dorman LI, Valdes-Galicia JF (1997) Analytical approximation to solar neutron scattering and attenuation in the Earth's atmosphere. Proceedings of the 25th international cosmic ray conference, vol 1. Durban (South Africa), pp 29–32

Dorman LI, Dorman IV, Valdes-Galicia JF (1997a) Simulation of solar neutron scattering and attenuation in the Earth's atmosphere for different initial zenith angles. Proceedings of the 25th international cosmic ray conference, vol 1. Durban (South Africa), pp 25–28

Dorman LI, Valdes-Galicia JF, Rodriquez M (1997b) The event of 24 May 1990 and the problem of solar neutron propagation through the Earth's atmosphere: does the 'refraction' effect exist? Proceedings of the 25th international cosmic ray conference, vol 1. Durban (South Africa), pp 33–36

Dorman LI, Valdes-Galicia JF, Dorman IV (1999a) Numerical simulation and analytical description of solar neutron transport in the Earth's atmosphere. J Geophys Res 104(A10):22417–22426

Dorman LI, Dorman IV, Valdes-Galicia JF (1999b) Analytical description of solar neutron propagation in the atmosphere for different initial zenith angles by considering scattering, attenuation and energy change. Proceedings of the 26-th international cosmic ray conference, vol 7. Salt Lake City, pp 367–370

Dorman LI, Dorman IV, Valdes-Galicia JF (2000) Propagation of inclined solar neutrons: scattering, energy decrease, attenuation, and refraction effect. In: Mewaldt RA et al (eds) Acceleration and transport of energetic particles observed in the heliosphere. AIP Conference Proceedings, No. 528, pp 173–176

Efimov YE, Kocharov GE, Kudela K (1983) On the solar neutrons observation on high mountain neutron monitor. Proceedings of the 18th international cosmic ray conference, vol 10. Bangalore, pp 276–278

Filippov AT (1987) On the solar flares with high energy neutron generation by using of Tixie neutron supermonitor data. In: Kocharov G (ed) Processes on the surface and in the interior of the sun, Leningrad, pp 50–53 (in Russian)

Filippov AT (1990) On high-energy neutrons from solar flares upon ground-based measurement data. Proceedings of the 21st international cosmic ray conference, vol 5. Adelaida, pp 133–135

Filippov AT, Prikhod'ko AN (1993) The solar flare of 24 May 1990 with generation of high-energy neutrons. Izvestiya RAN Ser Phys 57(7):31–33, In Russian

Flückiger EO, Bütikofer R, Chilingarian A, Hovsepyan G, Muraki Y, Matsubara Y, Sako T, Tsuchiya H, Sakai T (2001) Search for solar neutrons in association with large solar flares in July 2000 and March/April 2001. Proceedings of the 27th international cosmic ray conference, vol 8. Hamburg, pp 3044–3047

Flückiger EO, Bütikofer R, Chilingarian A, Hovsepyan G, Tan YH, Yuda T, Tsuchiya H, Ohnishi M, Katayose Y, Muraki Y, Matsubara Y, Sako T, Watanabe K, Masuda K, Sakai T, Shibata S, Ogasawara R, Mizumoto Y, Nakagiri M, Miyashita A, Stoker PH, Lopate C, Kudela K, Gros M

(2005) Solar neutron events that have been found in solar cycle 23. Int J Modern Phys A 20 (29):6646–6649

González LX, Valdés-Galicia JF, Musalem O, Hurtado A, Muraki Y, Sako T, Watanabe K, Matsubara Y, Shibata S, Sakai T, Tsujihara H (2008) Energy spectrum for the solar neutron event of September 7 2005, derived from the SNT at Sierra Negra. Proceedings of the 30th international cosmic ray conference, vol 1. Merida, Mexico, pp 57–60

Hatton CJ (1971) The neutron monitor. In: Wilson JG, Wouthuysen SA (eds) Progress in physics of cosmic ray and elementary particles, vol 10. North-Holland, Amsterdam, pp 3–100

Hoshida T, Matsubara Y, Muraki Y, et al (1999) A possible observation of solar neutrons in association with November 28th, 1998 Flare. In: Proceedings of the 26th international cosmic ray conference, Salt Lake City, 1999, vol 6, pp 38–41

Hoshino M, Arons J, Gallant YA, Langdon AB (1992) Relativistic magnetosonic shock waves in synchrotron sources: shock structure and nonthermal acceleration of positrons. Astrophys J 390 (2, Part I):454–479

Hua H-M, Lingenfelter RE (1987a) Solar flare neutron and accelerated ion angular distributions. Astrophys J 323(2, Part I):779–794

Hua H-M, Lingenfelter RE (1987b) Solar flare neutron production and the angular dependence of the capture gamma-ray emission. Solar Phys 107(2):351–383

Hua X-M, Kozlovsky B, Lingenfelter RE, Ramaty R, Stupp A (2002) Angular and energy-dependent neutron emission from solar flare magnetic loops. Astrophys J Suppl Ser 140 (2):563–579

Hulot E, Vilmer N, Chupp EL, Dennis BR, Kane SR (1992) -Ray and x-ray time profiles expected from a trap-plus-precipitation model for the 7 June 1980 and 27 April 1981 solar flares. Astron Astrophys (Germany) 256(1):273–285

Jenkins TL, Frye GM Jr, Stansfield S, Kinzer RL, Jensen C, Kurfess JD, Johnson WN, Strickman M, Byrd R, Foster C (1991) Response of the OSSE GRO detector to 30-200 MeV neutrons. Proceedings of the 22nd international cosmic ray conference, vol 3. Dublin, pp 760–763

Johnson WN, Kinzer RL, Kurfess JD, Strickman MS (1993) The oriented scintillation spectrometer experiment: instrument description. Astrophys J Suppl Ser (USA) 86(2):693–712

Kanbach G, Bertsch DL, Fichtel CE, Hartman RC, Hunter SD, Kniffen DA, Kwok PW, Lin YC, Mattox JR, Mayer-Hasselwander HA, Michelson PF, von Montigny C, Nolan PL, Pinkau K, Rothermel H, Schneid E, Sommer M, Sreekumar P, Thompson DJ (1993) Detection of a long-duration solar gamma-ray flare on June 11, 1991 with EGRET on Compton-GRO. Astron Astrophys Suppl Ser (France) 97(1):349–353

Kane SR, Hurley K, McTiernan JM, Sommer M, Boer M, Niel M (1995) Energy release and dissipation during giant solar flares. Astrophys J Lett (USA) 446(1, Part I):L47–L50

Karpov SN, Alekseenko VV, Karpova ZM, Khaerdinov NS, Petkov VB (2003) Yield and response functions of the Baksan EAS-array andyrchy for single component. Proceedings of the 28th international cosmic ray conference, vol 6. Tsukuba, pp 3457–3460

Katayose Y, Izu K, Ohnishi M, et al (1999) A solar neutron telescope in Tibet. In: Proceedings of the 26th international cosmic ray conference, Salt Lake City, vol 6, pp 58–61

Kocharov GE, Kocharov LG, Kovaltsov GA, Shea MA, Smart DF, Armstrong TP, Pyle KR, Chuikin EI (1993) Evidence for extended neutron and gamma-ray generation during two solar flares. Proceedings of the 23rd international cosmic ray conference, vol 3. Calgary, pp 107–110

Kohno T, Miyasaka H (1999) Proceedings of the solar neutron symposium, Nagoya, pp 66–72 (in Japanese).

Koi T, Muraki Y, Masuda K, Matsubara Y, Sako T, Murata T, Tsuchiya H, Shibata S, Munakata Y, Hatanaka K, Wakasa T, Sakai H (2001) Attenuation of neutrons in the atmosphere and a thick carbon target. Nucl Instrum Methods Phys Res Sect A 469(1):63–69

Litvinenko YE (1996) Particle acceleration in reconnecting current sheets with a nonzero magnetic field. Astrophys J 462(2):997–1004, Part 1

Le GM, Tang YH, Tang YQ (2007) What did the occurrence of relativistic solar neutrons on 28 October 2003 mean? In: Proceedings of the 30th international cosmic ray conference, Merida (Mexico), Pre-conference edition, http://dpnc.unige.ch/ams/ICRC-07/icrc0397.pdf, pp 1–4

Matsubara Y, Tajima N, Kakimoto F, Kaneko T, Ogio S, Osco F, Martinic N, Murakami K, Muraki Y, Takahashi T, Yoshii H (1993) New Bolivia solar neutron telescope. Proceedings of the 23rd international cosmic ray conference, vol 3. Calgary, Contributed Papers, pp 139–142

Matsubara Y, Muraki Y, Sako T, Kakimoto F, Ogio S, Murakami K, Kaneko T, Yoshii H, Toyoda Y, Tajima N, Shirasaki Y, Martinic N, Miranda P, Ticona R, Velarde A (1999) Possible observation of solar neutrons on November 6, 1997 at chacaltaya. Proceedings of the 26th international cosmic ray conference, vol 6. Salt Lake City, pp 46–49

Matsubara Y, Muraki Y, Sako T, Watanabe K, Masuda S, Sakai T, Shibata S, Fluckiger E, Butikofer R, Chilingarian A, Hovsepyan G, Tan YH, Yuda T, Ohnishi M, Tsuchiya H, Katayose Y, Ogasawara R, Mizumoto Y, Nakagiri M, Miyashita A, Velarde A, Ticona R, Martinic N (2005) Search for neutrons associated with proton flares in solar cycle 23. Proceedings of the 29th international cosmic ray conference, vol 1. Pune, pp 17–20

Matsubara Y, Muraki Y, Sako T, Watanabe K, Sakai T, Shibata S, Flűckiger4 E, Bűtikofer R, Chilingarian A, Hovsepyan G, Tan YH, Yuda T, Ohonishi M, Tsuchiya H, Katayose Y, Ogasawara R, Mizumoto Y, Kakimoto F, Tsunesada Y, Velarde A, Ticona R, Martinic N, Miranda P, Valdes-Galicia JF, Gonzalez LX, Hurtado A, Musale O (2008) Search for solar neutrons associated with series of X-class flares during the declining period of solar cycle 23. Proceedings of the 30th international cosmic ray conference, vol 1. Merida, Mexico, pp 33–36

Muraki Y, Shibata S (1996) Solar neutrons on May 24th, 1990. In: Ramaty R, Mandzhavidze N, Hua X-M (eds) High energy solar physics. Proceedings of the conference held in Greenbelt, Maryland, 1995. AIP conference proceedings, vol 374, Issue 1, pp 256–264

Muraki Y, Murakami K, Shibata S, Yamada T, Miyazaki M, Takahashi T, Mitsui K, Sakai T, Yamaguchi K (1991a) On isolated neutron events of Feb. 22, 1991 and May 3, 1991. Proceedings of the. 22nd international cosmic ray conference, vol 3. Dublin, pp 45–48

Muraki Y, Murakami K, Shibata S, Sakakibara S, Yamada T, Miyazaki M, Takahashi T, Mitsui K, Sakai T (1991b) Solar neutrons associated with the large solar flare of June 1991. Proceedings of the 22nd international cosmic ray conference, vol 3. Dublin, pp 49–52

Muraki Y, Matsubara Y, Masuda S, Sako T, Watanabe K, Bütikofer R, Flückiger EO, Chilingarian A, Hovsepyan G, Kakimoto F, Tsunesada Y, Tokuno H, Velarde A, Sakai T (2006) Detection of solar neutrons and protons by ground level detectors. Proceedings of the 30th European cosmic ray symposium, Lisbon, Paper S2-98

Muraki Y, Tsuchiya H, Fujiki K, Masuda S, Matsubara Y, Menjyo H, Sakoa T, Watanabe K, Ohnishi M, Shiomi A, Takita M, Yuda T, Katayose Y, Hotta N, Ozawa S, Sakurai T, Tan YH, Zhang JL (2007) A solar neutron telescope in Tibet and its capability examined by the 1998 November 28th event. Astroparticle Phys 28(1):119–131

Muraki Y, Matsubara Y, Masuda S, Sako T, Sakakibara S, Watanabe K, Flűckiger E, Bűtikofer R, Chilingarian A, Kakimoto F, Terasawa T, Tsunesada Y, Tokuno H, Velarde A, Vashenyk EV, Sakai T (2008a) Simultaneous detection of high-energy solar neutrons and protons at Chacaltaya observatory on April 15, 2001. Proceedings of the 30th international cosmic ray conference, vol 1. Merida, Mexico, pp 29–32

Muraki Y, Matsubara Y, Masuda S, Sakakibara S, Sako T, Watanabe K, Bűtikofer R, Flűckiger EO, Chilingarian A, Hovsepyan G, Kakimoto F, Terasawa T, Tsunesada Y, Tokuno H, Velarde A, Evenson P, Poirier J, Sakai T (2008b) Detection of high-energy solar neutrons and protons by ground level detectors on April 15, 2001. Astroparticle Phys 29(4):229–242

Murphy RJ, Share GH, Letaw JR, Forrest DJ (1990) Nuclear line spectroscopy of the 1981 April 27 solar flare. Astrophys J (USA) 358(1, Part I):298–312

Murphy RJ, Ramaty R, Kozlovsky B, Reames DV (1991) Solar abundances from gamma-ray spectroscopy: comparisons with energetic particle, photospheric, and coronal abundances. Astrophys J (USA) 371(2, Part I):793–803

Murphy RJ, Share GH, Forrest DJ, Grabelsky DA, Grove JE, Jensen CM, Johnson WN, Jung GV, Kinzer RL, Kroeger RA, Kurfess JD, Matz SM, Purcell WR, Strickman MS, Ulmer MP, Vestrand WT (1993) OSSE observations of the 4 June 1991 solar flare. In: Proceedings of the 23rd international cosmic ray conference, University of Calgary, Alberta, Canada, 1993, Invited, Rapporteur, and Highlight Papers, vol 3, pp 99–102

Murphy RJ, Share GH, Grove JE, Johnson WN, Kinzer RL, Kurfess JD, Strickman MS, Jung GV (1997) Accelerated particle composition and energetics and ambient abundances from gamma-ray spectroscopy of the 1991 June 4 solar flare. Astrophys J 490(2):883–900

Murphy RJ, Share GH, Delsignore KW, Hua X-M (1999) Surprisingly intense neutron emission from a flare behind the limb of the Sun. Astrophys J 510(2):1011–1017, Part 1

Murphy RJ, Kozlovsky B, Share GH, Hua X-M, Lingenfelter RE (2007) Using gamma-ray and neutron emission to determine solar flare accelerated particle spectra and composition and the conditions within the flare magnetic loop. Astrophys J Suppl Ser 168(1):167–194

Ohsawa Y, Sakai J (1987) Nonstochastic prompt proton acceleration by fast magnetosonic shocks in the solar plasma. Astrophys J 313(1):440–448, Part 1

Ohyama M, Shibata K (2000) X-ray plasma ejection and magnetic reconnection. Adv Space Res 26(3):461–464

Pakhomov NI, Shapovalova LA, Yanchukovsky AL (1987) Recording of solar flare-produced neutrons at the Sayan spectrographic facility. In: Kocharov G (ed) Processes on the surface and in the interior of the sun, Leningrad, pp 25–41

Panasyuk MI, Kuznetsov SN, Lazutin LL, Avdyushin SI, Alexeev II, Ammosov PP, Antonova AE, Baishev DG, Belenkaya ES, Beletsky AB, Belov AV, Benghin VV, Bobrovnikov SY, Bondarenko VA, Boyarchuk KA, Veselovsky IS, Vyushkova TY, Gavrilieva GA, Gaidash SP, Ginzburg EA, Denisov YI, Dmitriev AV, Zherebtsov GA, Zelenyi LM, Ivanov-Kholodny GS, Kalegaev VV, Kanonidi KD, Kleimenova NG, Kozyreva OV, Kolomiitsev OP, Krasheninnikov IA, Krivolutsky AA, Kropotkin AP, Kuminov AA, Leshchenko LN, Mar'in BV, Mitrikas VG, Mikhalev AV, Mullayarov VA, Muravieva EA, Myagkova IN, Petrov VM, Petrukovich AA, Podorolsky AN, Pudovkin MI, Samsonov SN, Sakharov YA, Svidsky PM, Sokolov VD, Soloviev SI, Sosnovets EN, Starkov GV, Starostin LI, Tverskaya LV, Teltsov MV, Troshichev OA, Tsetlin VV, Yushkov BY (2004) Collaboration solar extreme events in 2003 SEE-2003. Magnetic storms in October 2003. Cosmic Res 42(5):489–535

Prince TA, Forrest DJ, Chupp EL, Kanbach G, Share GH (1983) The time history of 2.22 MeV line emission in solar flares. Proceedings of the 18th international cosmic ray conference, vol 4. Bangalore, pp 79–82

Pyle KR, Simpson JA (1991) Observation of a direct solar neutron event on 22 March 1991 with the Haleakala, Hawaii, neutron monitor. Proceedings of the 22nd international cosmic ray conference, vol 3. Dublin, pp 53–56

Pyle KR, Shea MA, Smart DF (1991) Solar flare generated neutrons observed by neutron monitors on 24 May 1990. Proceedings of the 22nd international cosmic ray conference, vol 3, Dublin, pp 57–60

Ramaty R, Murphy RJ (1987) Nuclear processes and accelerated particles in solar flares. Space Sci Rev 45(3–4):213–268

Ramaty R, Kozlovsky B, Lingenfelter RE (1975) Solar gamma rays. Space Sci Rev 18(3):341–388

Ramaty R, Mandzhavidze N, Kozlovsky B, Murphy RJ (1995) Solar atmospheric abundances and energy content in flare-accelerated ions from gamma-ray spectroscopy. Astrophys J Lett (USA) 455(2, Part 2):L193–L196

Ramaty R, Mandzhavidze N, Kozlovsky B (1996) Abundances from solar atmospheric gamma ray spectroscopy. In: Ramaty R, Mandzhavidze N, Hua X-M (eds) High-energy solar physics, Greenbelt, 1995. AIP Conference Proceedings, vol 374. Woodbury, New York, pp 172–183

Ramaty R, Mandzhavidze N, Barat C, Trottet G (1997) The giant 1991 June 1 flare: evidence for gamma-ray production in the corona and accelerated heavy ion abundance enhancements from gamma-ray spectroscopy. Astrophys J (USA) 479(1, Part 1):458–463

Reames DV (1995) Coronal abundances determined from energetic particles. Adv Space Res 15 (7):41–51
Reames DV (1999) Particle acceleration at the Sun and in the heliosphere. Space Sci Rev 90(3–4):413–491
Sako T, Muraki Y, Yuda T, Matsubara Y, Watanabe K, Hoshida T, Takita M, Tsuchiya H, Ohnishi M, Shiomi A, Katayose Y, Hotta N, Tan YH (2003) Solar neutron event in association with the 24 September 2001 Flare. Proceedings of the 28th international cosmic ray conference, vol 6. Tsukuba, Japan, pp 3175–3178
Sako T, Watanabe K, Muraki Y, Matsubara Y, Tsujihara H, Yamashita M, Sakai T, Shibata S, Valdes-Galicia JF, Gonzalez LX, Hurtado A, Musalem O, Miranda P, Martinic N, Ticona R, Velarde A, Kakimoto F, Ogio S, Tsunesada Y, Tokuno H, Tanaka YT, Yoshikawa I, Terasawa T, Saito Y, Mukai T, Gros M (2006) Long-lived solar neutron emission in comparison with electron-produced radiation in the 2005 September 7 solar flare. Astrophys J 651(1):L69–L72
Sako T, Watanabe K, Muraki Y, Matsubara Y, Sakai T, Shibata S, Kakimoto F, Tsunesada Y, Tokuno H, Ogio S, Valdes-Galicia JF, Gonzalez LX, Hurtado A, Musalem O, Miranda P, Martinic N, Ticona R, Velarde A Emission profile of solar neutrons obtained from the ground-based observations for the 7 September 2005 event. Proceedings of the 30th international cosmic ray conference, vol 1. Merida, Mexico, pp 53–56
Sakurai T (1999) Proceedings of the solar neutron symposium, Nagoya, pp 36–49 (in Japanese)
Sato J, Kosugi T, Makishima K (1999) Improvement of YOHKOH hard X-ray imaging. Publication of the astronomical society of Japan, vol 51. pp 127–150
Sato J, Masuda S, Kosugi T, Sakao T (2000) Improvement of Yohkoh HXT imaging and an analysis of the 1997 November 6, X9 flare. Adv Space Res 26(3):501–504
Share GH, Kurfess JD, Theus RB (1978) Pulse-shape discrimination of high-energy neutrons and gamma rays in NaI(Tl). Nucl Instrum Methods (Netherlands) 148(3):531–533
Share GH, Murphy RJ, Smith DM, Schwartz RA, Lin RP (2004) RHESSI e^+-e^- annihilation radiation observations: implications for conditions in the flaring solar chromosphere. Astrophys J Lett (USA) 516(2, part 2):L169–L172
Shea MA, Smart DF, Flückiger EO (1987) Possible neutron monitor response to solar neutrons in April 1984. Proceedings of the 20th international cosmic ray conference, vol 3. Moscow, pp 86–89
Shea MA, Smart DF, Pyle KR (1991) Direct solar neutrons detected by neutron monitors on 24 May 1990. Geophys Res Lett 18(9):1655–1658
Shibata S (1994) Propagation of solar neutrons through the atmosphere of the Earth. J Geophys Res 99(4):6651–6665
Shibata S, Murakami K, Muraki Y (1993) Solar neutron events observed in cycle 22. Proceedings of the 23rd international cosmic ray conference, vol 3. Calgary, pp 95–98
Shibata S, Munakata Y, Tatsuoka R, Muraki Y, Masuda K, Matsubara Y, Koi T, Sako T, Murata T, Tsuchiya H, Hatanaka K, Wakasa T, Sakai H, Ishida Y (2001) Detection efficiency of a neutron monitor calibrated by an accelerator neutron beam. Nucl Instrum Methods Phys Res Sect A 463(1–2):316–320
Smart DF, Shea MA, Flückiger EO, Debrunner H, Humble JE (1990) Were solar neutrons observed by neutron monitors on 1984 April 25? Astrophys J Suppl Ser 73(2):269–272
Smart DF, Shea MA, O'Bren K (1995) Comment on the minimum mass path of a uni-directional solar neutron particle beam through the Earth atmosphere. Proceedings of the 24-th international cosmic ray conference, vol 4. Rome, pp 171–174
Stoker PH, Krüger H (1990) Solar flare neutrons recorded at Potchefstroom and Tsumeb for the Dec. 16, 1988 solar flare event? Proceedings of the 21st international cosmic ray conference, vol 5. Adelaida, pp 140–142
Struminsky A, Matsuoka M, Takahashi K (1994) Evidence of additional production of high-energy neutrons during the solar flare on 1991 June 4. Astrophys J 429(1):400–405, Part 1
Takahashi K (1989) Detection of solar neutrons by ground-based neutron monitor. Space Sci Rev 51(1–2):123–146

Takahashi K, Wada M, Yoshimori M, Kusunose M, Kondo I (1987) Detection of solar neutrons by Tokyo neutron monitor. Proceedings of the 20th international cosmic ray conference, vol 3. Moscow, pp 82–85

Takahashi K, Sakamoto E, Nishi K, Yamada Y, Shikata T, Wada M, Kusunose M, Yamaguchi E, Yoshimori M, Kondo K, Flückiger OE (1990) Observation of solar neutrons by Mt. Norikura and Tokyo neutron monitors. Proceedings of the 21st international cosmic ray conference, vol 5. Adelaida, pp 136–139

Takahashi K, Sakamoto E, Matsuoka M, Nishi K, Yamada Y, Shimoda S, Shikata T, Wada M, Kusunose M, Yoshimori M, Kondo I (1991) Observation of solar neutrons by Mt. Norikura neutron monitor during a period of solar cycle 22. Proceedings of the 22nd international cosmic ray conference, vol 3. Dublin, pp 37–40

Takakura T, Tsuneta S, Ohki K, Nitta N, Makishima K, Murakami T, Ogawara Y, Oda M, Miyamoto S (1983) Hard x-ray imaging of a solar limb flare with the x-ray telescope aboard the Hinotori satellite. Astrophys J Lett Ed (USA) 270(2):L83–L87

Tatischeff V, Kiener J, Gros M (2005) Physical implications of INTEGRAL/SPI gamma-ray line measurements of the 2003 October 28 solar flare. In: Proceedings of the 5th Rencontres du Vietnam, New Views on the Universe, Hanoi, August 5–11, 2004, pp 1–4. (arXiv:astro-ph/0501121v1)

Terasawa T, Tanaka YT, Takei Y, Kawai N, Yoshida A, Nomoto K, Yoshikawa I, Saito Y, Kasaba Y, Takashima T, Mukai T, Noda H, Murakami T, Watanabe K, Muraki Y, Yokoyama T, Hoshino M (2005) Repeated injections of energy in the first 600 ms of the giant flare of SGR 1806-20. Nature 434(7037):1110–1111

Trottet G, Barat C, Ramaty R, Vilmer N, Dezalay JP, Kuznetsov A, Mandzhavidze N, Sunyaev R, Talon R, Terekhov O (1996) Thin target-ray line production during the 1991 June 1 flare. In: Ramaty R, Mandzhavidze N, Hua X-M (eds) High energy solar physics, Proceedings of the conference held in Greenbelt, Maryland, 1995, AIP conference proceedings, vol 374, issue 1, pp 153–161

Tsuchiya H, Muraki Y, Matsubara Y, Masuda K, Koi T, Sako T, Okada A, Murata T, Imaida I, Hoshida T, Kato Y, Yuki A, Ohno S, Shibata S, Munakata Y, Tatsuoka R, Hatanaka K, Wakasa T, Sakai H, Nonaka T, Ohnishi T (1999) Detection efficiency of new solar neutron detector. Proceedings of the 26th international cosmic ray conference, vol 7. Salt Lake City, pp 363–366

Tsuchiya H, Matsubara Y, Muraki Y, Murakami K, Sako T, Kakimoto F, Ogio S, Tsunesada Y, Tokuno H, Yoshii H, Tajima N, Martinic N, Miranda P, Ticona R, Velarde A (2001a) Observations of neutrons in association with the solar flare of 6 November 1997. Proceedings of the 27th international cosmic ray conference, vol 8. Hamburg, pp 3040–3043

Tsuchiya H, Hoshida T, Muraki Y, Matsubara Y, Masuda K, Yuda T, Ohnishi S, Sasaki T, Shiomi A, Katayose Y, Hotta N, Ozawa S, Tan YH (2001b) Arrival of solar neutrons from large zenith angle. Proceedings of the 27th international cosmic ray conference, vol 8. Hamburg, pp 3056–3059

Tsuneta S, Naito T (1998) Fermi acceleration at the fast shock in a solar flare and the impulsive loop-top hard X-ray source. Astrophys J 495(1):L67–L70

Usoskin IG, Kovaltsov GA, Kocharov LG, Kananen H, Tanskanen P (1995) High energy neutrons from the June 15, 1991 solar flare as detected by the neutron monitor. 24th international cosmic ray conference, vol 4. Roma, pp 151–154

Valdes-Galicia JF, Dorman LI, Rodriguez M (2000) Atmospheric refraction of solar neutrons during the event of 24 May 1990. Solar Phys 191(2):409–417

Valdes-Galicia JF, Muraki Y, Tsujihara H, Sako T, Musalem O, Hurtado A, Gonzalez LX, Matsubara Y, Watanabe K, Hirano N, Tateiwa N, Shibata S, Sakai T (2004) An improved solar neutron telescope installed at a very high altitude in Mexico. Nucl Instrum Methods Phys Res A (Netherlands) 535(3):656–664

Valdes-Galicia JF, Muraki Y, Watanabe K, Matsubara Y, Sako T, Gonzalez LX, Musalem O, Hurtado A (2009) Solar neutron events as a tool to study particle acceleration at the Sun. Adv Space Res 43:565–572

Veselovsky IS, Panasyuk MI, Avdyushin SI, Bazilevskaya GA, Belov AV, Bogachev SA, Bogod VM, Bogomolov AV, Bothmer V, Boyarchuk KA, Vashenyuk EV, Vlasov VI, Gnezdilov AA, Gorgutsa RV, Grechnev VV, Denisov YI, Dmitriev AV, Dryer M, Yermolaev YI, Eroshenko EA, Zherebtsov GA, Zhitnik IA, Zhukov AN, Zastenker GN, Zelenyi LM, Zeldovich MA, Ivanov-Kholodnyi GS, Ignat'ev AP, Ishkov VN, Kolomiytsev OP, Krasheninnikov IA, Kudela K, Kuzhevsky BM, Kuzin SV, Kuznetsov VD, Kuznetsov SN, Kurt VG, Lazutin LL, Leshchenko LN, Litvak ML, Logachev YI, Lawrence G, Markeev AK, Makhmutov VS, Mitrofanov AV, Mitrofanov IG, Morozov OV, Myagkova IN, Nusinov AA, Oparin SN, Panasenco OA, Pertsov AA, Petrukovich AA, Podorol'sky AN, Romashets EP, Svertilov SI, Svidsky PM, Svirzhevskaya AK, Svirzhevsky NS, Slemzin VA, Smith Z, Sobel'man II, Sobolev DE, Stozhkov YI, Suvorova AV, Sukhodrev NK, Tindo IP, Tokhchukova SK, Fomichev VV, Chashey IV, Chertok IM, Shishov VI, Yushkov BY, Yakovchouk OS, Yanke VG (2004) Solar and heliospheric phenomena in October–November 2003: causes and effects. Cosmic Res 42(5):435–488

Vestrand WT, Forrest DJ (1993) Evidence for a spatially extended component of gamma rays from solar flares. Astrophys J Lett 409(2):L69–L72

Wang HT, Ramaty R (1974) Solar Neutron propagation and 2.2 MeV gamma-ray line production in the solar atmosphere. Solar Phys 36(1):129–137

Wang HT, Ramaty R (2000) Solar neutron propagation and 2.2 MeV gamma-ray line production in the solar atmosphere. Solar Phys 36(1):129–137

Watanabe K (2005) Solar neutron events associated with large solar flares in solar cycle 23. Doctoral thesis, Department of Physics, Nagoya University, 1–206

Watanabe K, Muraki Y, Murakami K, Shibata S, Pyle KR, Mitsui K, Sakai T, Sakakibara S (2003a) Simultaneous observations of solar neutrons in association with a large solar flare on June 6, 1991. Proceedings of the 28th international cosmic ray conference, vol 6. Tsukuba, Japan, pp 3211–3214

Watanabe K, Muraki Y, Matsubara Y, Murakami K, Sako T, Tsuchiya H, Masuda S, Yoshimori M, Ohmori N, Miranda P, Martinic N, Ticona R, Velarde A, Kakimoto F, Ogio S, Tsunesada Y, Tokuno H, Shirasaki Y (2003b) Solar neutron event in association with a large solar flare on 2000 November 24. Astrophys J 592(1):590–596, Part 1

Watanabe K, Muraki Y, Matsubara Y, Murakami K, Sako T, Tsuchiya H, Masuda S, Yoshimori M, Ohmori N, Miranda P, Martinic N, Ticona R, Velarde A, Kakimoto F, Ogio S, Tsunesada Y, Tokuno H, Shirasaki Y (2003c) Solar neutron event in association with a large solar flare on August 25, 2001. Proceedings of the 28th international cosmic ray conference, vol 6. Tsukuba, Japan, pp 3179–3182

Watanabe K, Muraki Y, Matsubara Y, Sako T, Sakai T, Tsuchiya H, Shibata S, Masuda S, Yoshimori M, Ohmori N, Velarde A, Ticona R, Martinic N, Miranda P, Kakimoto F, Ogio S, Tsunesada Y, Tokuno H, Shirasaki Y, Valdes-Galicia JF, Hurtado A, Musalem O, Gonzalez LX, Ogasawara R, Mizumoto Y, Nakagiri M, Miyashita A, Stoker PH, Lopate C, Kudela K, Gros M (2005a) Solar neutron events associated with large solar flares in solar cycle 23. Proceedings of the 29th international cosmic ray conference, vol 1. Pune, India, pp 37–40

Watanabe K, Muraki Y, Matsubara Y, Sako T, Velarde A, Ticona R, Martinic N, Miranda P, Kakimoto F, Ogio S, Tsunesada Y, Tokuno H (2005b) Solar neutron event associated with an X8.3 flare on 2003 November 2. Proceedings of the 29th international cosmic ray conference, vol 1. Pune, India, pp 29–32

Watanabe K, Muraki Y, Matsubara Y, Sako T, Sakai T, Shibata S, Lopate C, Vald'es-Galicia JF, Hurtado A, Musalem O, Gonzalez LX, Ogasawara R, Mizumoto Y, Nakagiri M, Miyashita A, Gros M (2005c) Simultaneous observations of solar neutrons in association with a large solar flare on November 4, 2003. Proceedings of the 29th international cosmic ray conference, vol 1. Pune, India, pp 33–36

Watanabe K, Muraki Y, Matsubara Y, Murakami K, Sako T, Miranda P, Ticona R, Velarde A, Kakimoto F, Ogio S, Tokuno H, Tsuchiya H, Shibata S, Sakai T, Mizumoto Y, Ogasawara R,

Nakagiri M, Miyashita A, Lopate C (2006a) Solar neutron events in association with large solar flares in November 2003. Adv Space Res 38(3):425–430

Watanabe K, Gros M, Stoker PH, Kudela K, Lopate C, Valdés-Galicia JF, Hurtado A, Musalem O, Ogasawara R, Mizumoto Y, Nakagiri M, Miyashita A, Matsubara Y, Sako T, Muraki Y, Sakai T, Shibata S (2006b) Solar neutron events of 2003 October–November. Astrophys J 636 (2):1135–1144

Watanabe K, Sako T, Muraki Y, Matsubara Y, Sakai T, Shibatad S, Valdés-Galicia JF, González LX, Hurtado A, Musalem O, Miranda P, Martinic N, Ticona R, Velarde A, Kakimoto F, Ogio S, Tsunesada Y, Tokuno H, Tanaka YT, Yoshikawa I, Terasawa T, Saito Y, Mukai T, Gros M (2007) Highly significant detection of solar neutrons on 2005 September 7. Adv Space Res 39 (9):1462–1466

Watanabe K, Murphy RJ, Muraki Y, Murakami K, Sakakibara S, Shibata S, Mitsui K, Sakai T (2008a) OSSE satellite and neutron monitor observations of solar neutrons in association with 1991 June 4 flare. Proceedings of the 30th international cosmic ray conference, vol 1. Merida, Mexico, pp 37–40

Watanabe K, Murphy RJ, Share GH, Stoker PH, Lin RP (2008b) Physics of ion acceleration in the solar flare on 2003 October 28 determines gamma-ray and neutron production. Proceedings of the 30th international cosmic ray conference, vol 1. Merida, Mexico, pp 41–44

Watanabe K, Murphy RJ, Share GH, Hurford GJ, Krucker S, Lin RP, Harris MJ, Gros M, Sako T, Muraki Y, Matsubara Y, Shibata S, Sakai T, Valdés-Galicia JF, González LX, Hurtado A, Musalem O, Velarde A, Ticona R, Martinic N, Miranda PP, Kakimoto F, Tsunesada Y, Tokuno H, Ogio S (2000c) Ion acceleration and neutral emission mechanisms for 2005 September 7 flare. Proceedings of the 30th international cosmic ray conference, vol 1. Merida, Mexico, pp 45–48

Yoshimori Y (1999) Proceedings of the solar neutron symposium Nagoya, pp 29–35 (in Japanese)

Yoshimori M, Okudaira K, Hirasima Y, Kondo I (1983) Gamma-ray observations from HINOTORI. Sol Phys 86(1–2) :375–382

Yoshimori M, Takahashi K, Wada M, Kusunose M, Kondo I (1987) Possible solar neutron events recorded by the ground-based neutron monitor. Sol Phys 113(1–2):319–325

Yoshimori M, Suga K, Saita N, Shiozawa A (2000) Yohkoh observation of a gamma-ray flare on November 6, 1997. Adv Space Res 25(9):1801–1804

Yoshimori M, Shiozawa A, Suga K (1999) Photospheric 3He to H abundance ratio derived from gamma-ray line observation. Proceedings of the 26th international cosmic ray conference, vol 6. Salt Lake City, pp 5–8

Chapter 8
The Solar Neutron Decay Phenomenon

Evenson et al (1983a, b) discovered the solar neutron decay phenomenon when they measured the flux of 25–45 MeV protons observed by instruments (belonging to University of Chicago) aboard the ISEE-3 spacecraft, during the well-known event of June 3, 1982. This discovery allowed for the very important possibility to use measurements of neutron decay products to obtain additional information on solar neutron events.

8.1 Solar Neutron Decay Protons During June 21, 1980

This event was the second one investigated by Evenson et al. (1983a, b) on the basis of ISEE 3 measurements after detailed investigation of the solar neutron decay protons during the event June 3, 1982 (Evenson et al. 1983a, b). The flare of June 21, 1980 was located at 90°W, only 30° removed from the optimum magnetic connection (as opposed to the June 3, 1982 flare which was 134° removed). Hence, for the flare of June 21, 1980 the distinction between the protons from the flare and protons from neutron decay will be minimal. It gives a large difficulty to separate these two phenomena. To solve this problem Evenson et al. (1983a, b) determined in the first mean free path Λ of protons from neutron decay obtained from the investigation of solar flare electron propagation. For the event June 21, 1980 it was found by Evenson et al. (1985) that $\Lambda = 0.02 \pm 0.005$ AU. Figure 8.1 shows the predicted time dependence for solar flares 25 and 64 MeV protons together with measured ISEE data for the energy range 25–64 MeV.

From Fig. 8.1 can be seen that the particle fluxes have risen significantly over the background level, before the model predicts the arrival of diffusely propagated protons. Evenson et al. (1983b) proposed that the initial flux rise in this flare is due to solar neutron decay (as in the event June 3, 1982, see Section 7.2). Evenson et al. (1983b) constructed a spectrum for the protons over the period 1:30–2:30 UT on June 21, 1980 (see Fig. 8.2 in comparison with the prediction of Chupp et al. 1982,

L. Dorman, *Solar Neutrons and Related Phenomena*,
Astrophysics and Space Science Library 365, DOI 10.1007/978-90-481-3737-4_8,

using model parameters appropriate to the neutron injection geometry at the time interval of observation).

Figure 8.2 presents two calculations; one assuming that the mean free path for the solar neutron decay protons is the same, $\Lambda = 0.02$ AU as was found for the diffusive electrons with about the same rigidity, and the second for the value $\Lambda = 0.3$ AU, determined in the Evenson et al. (1983a) analysis of the event of June 3, 1982 (see Section 8.2). The agreement between Evenson et al. (1983b)

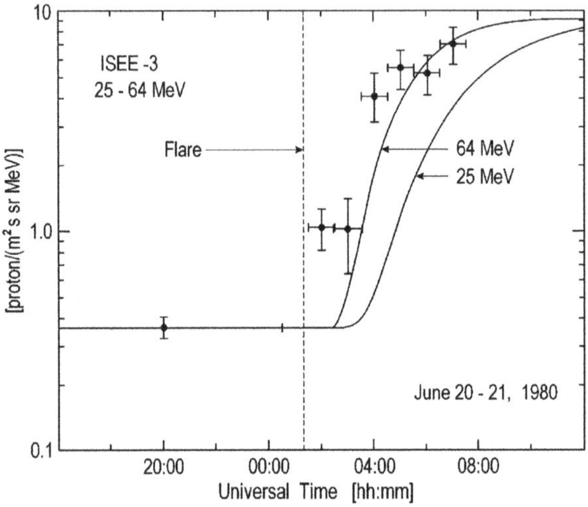

Fig. 8.1 Event of June 21, 1980. Predicted time dependence of 25 and 64 MeV protons from solar neutron decay (curves) and observations on ISEE-3 for the energy range 25–64 MeV (According to Evenson et al. 1983a, b, 1985)

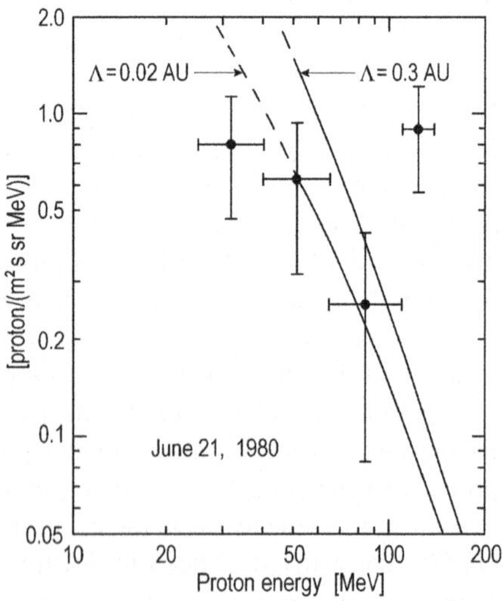

Fig. 8.2 Event of June 21, 1980. The expected energy spectrum for two values of proton transport path $\Lambda = 0.02$ AU and $\Lambda = 0.3$ AU in comparison with observation data at 01:30–02:30 of June 21, 1980. Normalization was made according to solar neutron measurements by Chupp et al. (1982) (According to Evenson et al. 1983a)

measurements and Chupp et al. (1982) result is quite good and clearly indicates that, like June 3, 1982 event (Section 8.2) the neutron spectrum does not continue to rise as E^{-3} below 50 MeV.

Evenson et al. (1983b) noted that it is unfortunate that the measurement errors are large compared to the different predictions in Fig. 8.2, because this analysis touches on a very important point in the study of interplanetary propagation. The apparent diffusion coefficient for flare particles, a combination of parallel and perpendicular components, influenced by coronal propagation and storage, may be quite different from the diffusion coefficient which simply describes the motion of the solar neutron decay protons along the IMF. On the other hand, the investigation of solar neutron decay proton phenomena can give additional important information on this simple diffusion along the IMF. But, for this it would be much better to investigate events caused by flares with great removing from the optimum magnetic connection (for example, flare of June 3, 1982 with 134° removing).

8.2 Solar Neutron Decay Protons During June 3, 1982

8.2.1 Observational Data

This event was caused by a flare characterized by a great removing (~134°) from the optimum magnetic connection of the Earth, that gives the possibility to Evenson et al. (1983a) to investigate this event and discover a solar neutron decay phenomena, which can give a very important additional information on solar cosmic ray generation and propagation. This event was caused by 2B solar flare beginning in Hα at 11:42 UT June 3, 1982, with co-ordinates 9°S, 72°E accompanied by the largest flux of >2 MeV gamma-rays detected to date, by the University of Chicago instruments on the ISEE 3 spacecraft, since its launch in August 1978 (the characteristics of well known solar neutron event of June 3, 1982 was considered in Chapter 3).

Figure 8.3, from Evenson et al. (1983a), shows the geometry of the event from the view of looking down at the ecliptic plane.

In Fig. 8.3 the Parker spiral pattern corresponds to the solar wind of 430 km/s measured at ISEE 3 at the time of flare, the solid line in the extension of the solar horizon at the site of the flare, thus neutrons from the flare have access only to the unshaded region of Fig. 8.3. The data of protons was obtained by the University of Chicago experiments on the ISEE 3 spacecraft (distance from the Earth 1.6×10^6 km, the angle between the lines Earth-Sun and Earth-spacecraft was 24°, energy interval of measured protons 25–150 MeV, geometry factor 2.5 cm^2.sr) and on the IMP-8 (2.0×10^5 km, 41°, 10–90 MeV, 2.0 cm^2.sr). Figure 8.4 shows the data of the flux 25–45 MeV protons observed at ISEE 3 (2 h averages are plotted).

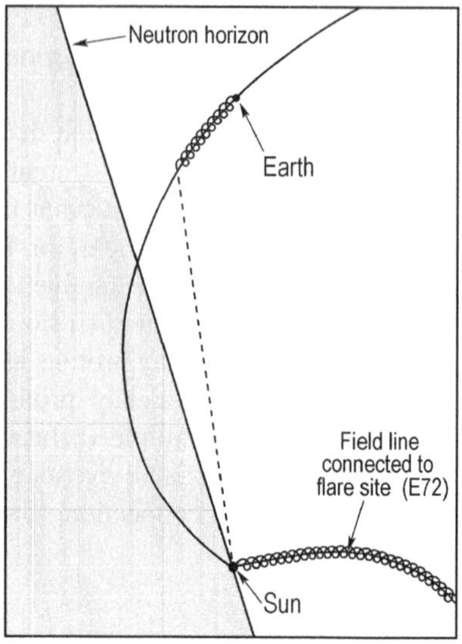

Fig. 8.3 Solar system geometry at the time of the June 3, 1982 solar flare. The view is looking down on the ecliptic plane. Protons from the flare are initially confined to field lines far from the Earth, while solar neutrons cross the field lines freely until they decay (According to Evenson et al. 1983a, 1990)

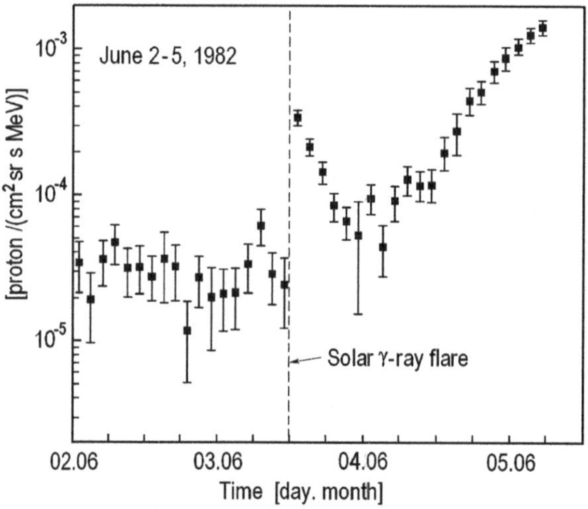

Fig. 8.4 Event of June 3, 1982. The flux of 25–45 MeV protons observed on ISEE-3. Two hour averages are plotted. Arrival of gamma-rays from an intense solar flare is indicated by a dashed vertical line (According to Evenson et al. 1983a, 1990)

8.2.2 Possible Interpretation

The interpretation of obtained data based on the Roelof (1966) theory of proton propagation in the interplanetary space with taking into account that according to Fig. 8.3, neutrons only have excess to the observers field line outside a radial distance from the Sun $r_0 = 0.68$ AU, corresponding to the position of the solar flare 72°E and solar wind speed 430 km/s. Evenson et al. (1983a) thus formulated the initial condition of the diffusion problem of solar neutron decay protons as (equation):

$$N(r,t=0) = \frac{N_0}{4\pi} \exp(-r/L)\, r^{-2}\, L^{-1}\, \theta(r-r_0), \tag{8.1}$$

where N_0 is the number of neutrons produced at the flare site in the energy interval under consideration,

$$L = c\beta\gamma\tau, \tag{8.2}$$

$\beta = v/c$, $\gamma = E/m_n c^2$, $\tau = 914$ s – neutron mean life, $\theta(r - r_0)$ is the Hevisaid function ($\theta(r - r_0) = 1$, if $r - r_0 > 0$ and $\theta(r - r_0) = 0$, if $r - r_0 < 0$) and, as we mentioned above, $r_0 = 0.68$ AU. The time observations on the spacecraft t_S is determined by:

$$t_s = t + \frac{r}{c}\left(\frac{1}{\beta} - 1\right) + t_0, \tag{8.3}$$

where t_0 is the gamma-ray arrival time at the spacecraft (gamma-ray and neutrons are assumed to be generated simultaneously according to Chupp et al. 1982, 1983, see Chapter 3). On the basis of Roelof (1966) theory of proton propagation in the interplanetary space with taking into account initial condition Eq. 8.1, Evenson et al. (1983a) obtained the following solution of the diffusion equation for solar neutron decay protons:

$$N(r,t) = \frac{N_0}{4\pi}\frac{L^{-3}\zeta^{-2}}{(2\pi)^{1/2}} \int_{\rho_0}^{\infty} d\rho' \frac{1-\exp(-2\rho\rho')}{\rho\rho'} \exp\left(-\zeta\rho' - \frac{(\rho'-\rho)^2}{2}\right), \tag{8.4}$$

where

$$\rho = r(2Dt)^{-1/2},\ \rho_0 = r_0(2Dt)^{-1/2},\ \zeta = (2Dt)^{1/2} L^{-1} \tag{8.5}$$

and $D = \lambda\beta c/3$ is the protons diffusion coefficient. The measured flux *J(r, t)* will be connected with the density of particles *N(r, t)* described by Eq. 8.4 by a relation

$$J(r,t) = \frac{c\beta}{4\pi} N(r,t). \tag{8.6}$$

Figure 8.5 shows the expected time profile of 35 MeV decay protons depending on the choice of transport path Λ and value of r_0 ($r_0 = 0$ corresponds to the case if we do not take into account the shadowing of neutrons by the Sun).

Figure 8.6 presents the 10-min experimental data of ISEE 3 for 25–45 MeV protons in comparison with expected, according to solutions described by Eqs. 8.4–8.6 for protons with energies 25 and 45 MeV ($r_0 = 0.68$ AU, $\Lambda = 0.3$ AU).

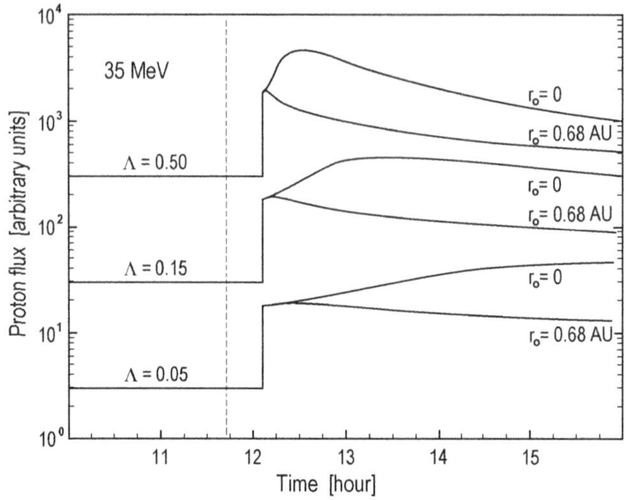

Fig. 8.5 The effect of neutron shadow (r_0, in AU) on the time development of the neutron decay proton fluxes. The computation was made for 35 MeV protons for different values of the proton diffusion free path (Λ, in AU). The gamma-ray arrival time is indicated by a dashed line (According to Evenson et al. 1983a)

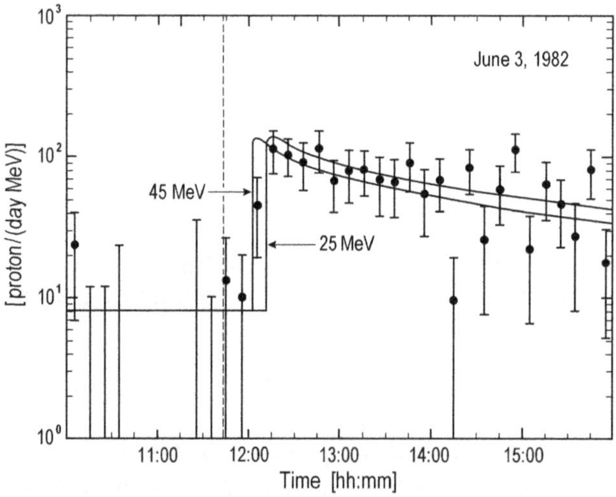

Fig. 8.6 Fit of the propagation model to the observed flux of 25–45 MeV protons on ISEE-3. The two lines represent calculations for the energies at the bin boundaries, 25 and 45 MeV, and assume a diffusion mean free path $\Lambda = 0.3$ AU. Ten minute average flux values are plotted. The gamma-ray arrival time is indicated by a dashed line (According to Evenson et al. 1983a)

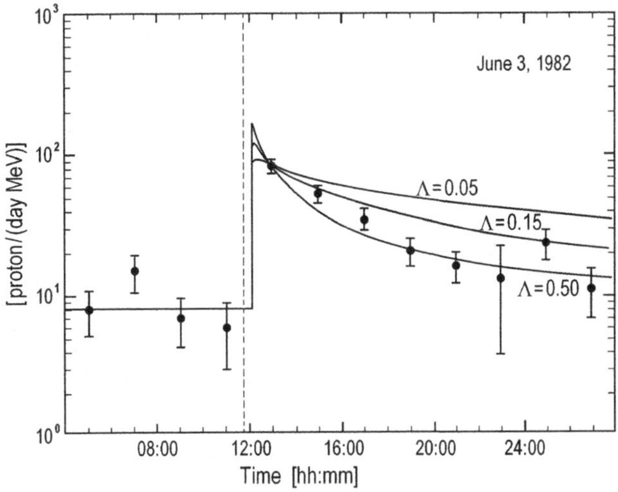

Fig. 8.7 Subset of the data from Fig. 8.4 showing the effect of varying the diffusion mean free path (Λ, in AU) in the model calculation. The gamma-ray arrival time is indicated by a dashed line (According to Evenson et al. 1983a)

Evenson et al. (1983a) noted on the excellent prediction of the rise time (there is no free parameter here at all, since everything is referred to the arrival time of gamma-rays). The model dependence on the transport path Λ, in comparison with observation data are shown in Fig. 8.7.

8.2.3 *On the Pitch Angle Distribution of Neutron Decay Protons*

Evenson et al. (1983a) examined their assumption that the neutron decay proton fluxes became isotropic very soon after injection (important for validating Roelof 1966 theory). The accuracy of sector counting rate data is not enough to obtain detailed pitch angle distribution but, within the errors the 35–150 MeV proton data on ISEE 3 are not in contradiction to the supposition on near isotropic proton distribution (see Fig. 8.8, for 11:50–12:20 UT and Fig. 8.9 for 12:30–14:30 UT of June 3, 1982).

8.2.4 *On the Neutron Decay Proton Energy Spectrum and the Spectrum of Neutron Generation*

Figure 8.10 shows the measurements on IMP-8 differential energy spectrum of neutron decay protons made at 12:00–14:30 UT, June 3, 1982.

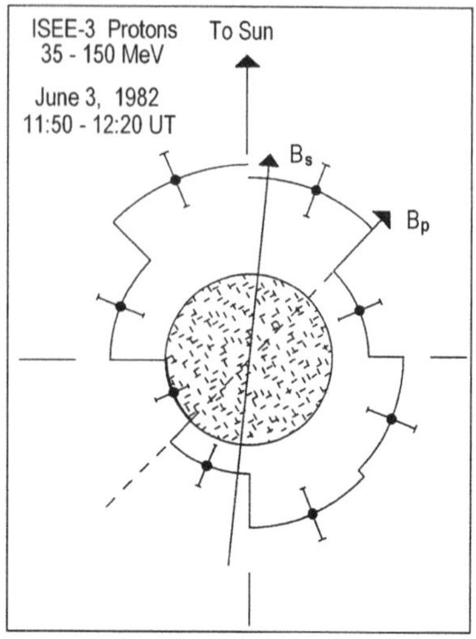

Fig. 8.8 Arrival directions of 35–150 MeV protons at ISEE-3 at 11:50–12:20 UT of June 3, 1982. The scan is in the ecliptic plane. The shaded circle gives the backgound isotropic flux level (0.08 counts/s) determined immediately before the event. The average magnetic field at the spacecraft denoted as B_S and B_P is the Parker direction (According to Evenson et al. 1983a)

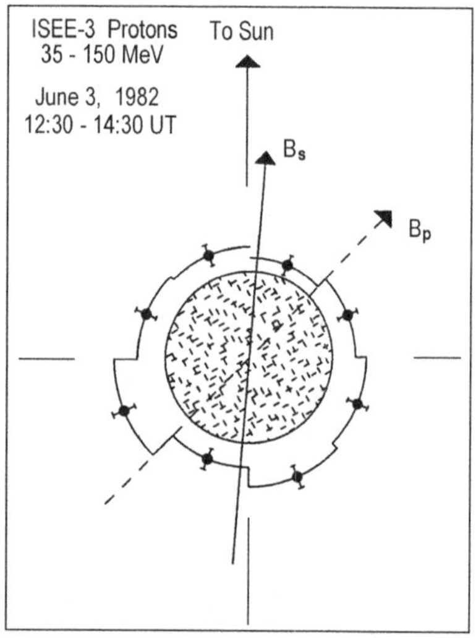

Fig. 8.9 The same as in Fig. 8.8 but at 12:30–14:30 UT of June 3, 1982 (According to Evenson et al. 1983a)

From Fig. 8.10 can be clearly seen that at small energy region the spectrum is much more flatter than $\propto E^{-3}$. On the basis of these data Evenson et al. (1983a) calculated the spectrum of neutrons emitted at the Sun. Figure 8.11 shows this result in comparison with Lingenfelter and Ramaty (1967) calculations for the proton spectrum

$$N_p(E)dE = A\exp(-R/R_0)dR, \tag{8.7}$$

where R_0 is a characteristic rigidity of the proton spectrum. Figure 8.11 shows that the best agreement is for $R_0 = 125$ MeV.

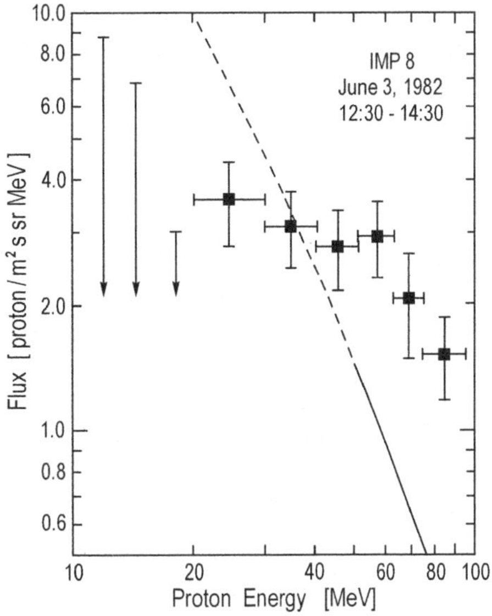

Fig. 8.10 Energy spectrum of the protons resulting from the decay of solar flare neutrons according to measurements on IMP-8 at 12:00–14:30 UT of June 3, 1982. The line is a prediction based on observations by Chupp et al. (1982) of solar neutrons; the dashed part of line indicates an extrapolation of obtained in Chupp et al. (1982) spectrum $\propto E^{-3}$ to the lower energies (According to Evenson et al. 1983a)

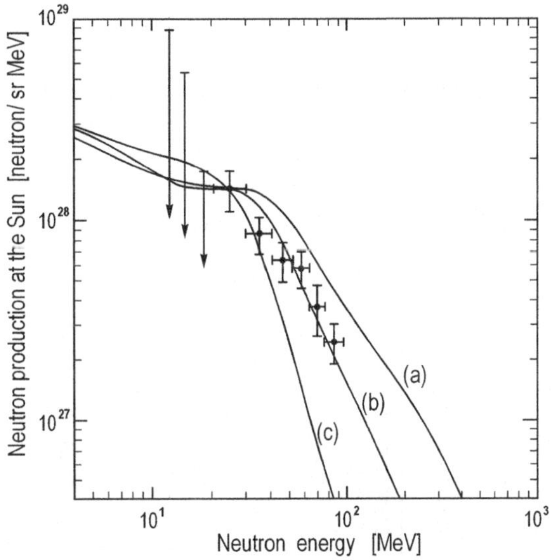

Fig. 8.11 Spectrum of neutrons emitted from the Sun determined from the observed proton fluxes on IMP-8 and the propagation model. Calculations of Lingenfelter and Ramaty (1967) are shown normalized to the data for three different values of characteristic rigidity R_0: (**a**) $R_0 = 200$ MV, (**b**) $R_0 = 125$ MV, and (**c**) $R_0 = 60$ MV. The measurements represent the average flux of neutrons emitted from approximately 50° to 90° from the local vertical at the flare site (According to Evenson et al. 1983a)

8.2.5 The Transport of Solar Flare Neutron Decay Protons During the Event of June 3, 1982: One Dimensional Model

According to Ostryakov and Kurganov (1990) the transport of solar flare neutron decay protons in the non-uniform magnetic field B(z), (z is the co-ordinate along magnetic field **B** line connected with the radial distance **r**) in the one dimensional diffusion approximation will be described by the equation for the distribution function $f(z, t)$:

$$\frac{\partial f}{\partial t} - B(z)\frac{\partial}{\partial z}\frac{D_{||}}{B(z)}\frac{\partial f}{\partial z} = Q(z,t), \tag{8.8}$$

where $D_{||}$ is the proton's diffusion coefficient along the magnetic field and

$$Q(z,t) = \frac{N_n(E)}{4\pi r^2 v\tau_0}\exp(-r/v\tau_0)\delta\left(t - \frac{r}{v}\right)\theta(r - r_1)\theta(r_2 - r) \tag{8.9}$$

is the neutron decay protons injection function. Here $N_n(E)$ is the injection function of neutrons, **v** is the neutron velocity, τ_0 is the life time of the rest neutrons, r_1 and r_2 determined the boundary of the neutron propagation region, θ is the Heviside function. By using the Green function $G(z, z_0; t, t_0)$ the solution of Eq. 8.8 can be written as

$$f(z,t) = \int\limits_0^t dt_0 \int\limits_0^\infty dz_0 (B(z)/B(z_0))^{1/2} G(z, z_0; t, t_0) Q(z_0, t_0), \tag{8.10}$$

where $G(z, z_0; t, t_0)$ can be found according to Kurganov and Ostryakov (1989), with the help of iterational procedure.

The results of calculations of the expected proton intensity time profile are shown in Fig. 8.12 for $E = 39.8$ MeV and transport free path along magnetic field $\Lambda_{||} = 0.5$ and 0.3 AU.

Ostryakov and Kurganov (1990) noted that the characteristic diffusional rise time for the event of June 3, 1982 expected of the order of 10–30 min and that very

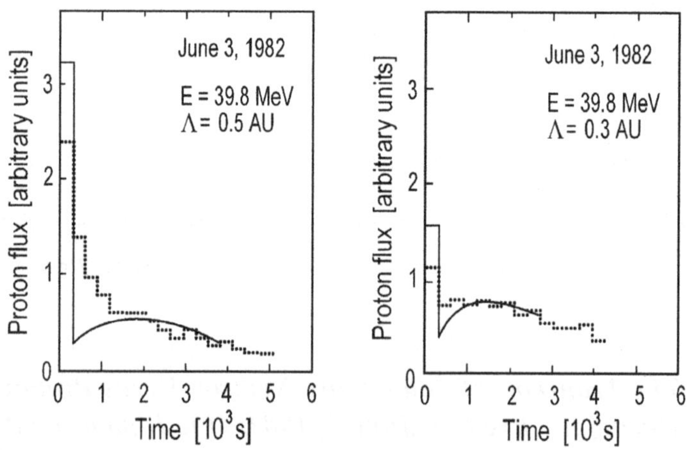

Fig. 8.12 The event June 3, 1982. Expected temporal variations of the 39.8 MeV proton intensity from the solar neutron decay for $\Lambda_{||} = 0.5$ and 0.3 AU: solid lines – results of the analytical solution for the direct and diffusional flux, dotted lines – results of Monte Carlo simulation (According to Ostryakov and Kurganov 1990)

important additional information on neutron decay protons propagation in the interplanetary magnetic field can be obtained by registration of anisotropic stage of the particle spread.

8.2.6 On the Two Dimensional Model of the Transport of Neutron Decay Protons on June 3, 1982

Zhang (1993) considered the transport of neutron decay protons in an Archimedean magnetic flux tube of constant magnetic flux in the ecliptic plane. Used was the transport equation of Roelof (1969) in two dimensions (distance **z** along the field lines and **μ**–cosines of the particle pitch angle):

$$\frac{\partial n(z,\mu,t)}{\partial t} = -D_{\mu\mu}\frac{\partial n}{\partial z} - \frac{v(1-\mu^2)}{2L(z)}\frac{\partial n}{\partial \mu} + \frac{\partial}{\partial \mu}\left(D_{\mu\mu}\frac{\partial n}{\partial \mu}\right) + Q(z,\mu,t). \qquad (8.11)$$

Here $n(z, \mu, t)$ is the time dependent neutron decay proton density, $L(z)$ is the scale of magnetic field variation ($L^{-1}(z) = -(dB(z)/dz)/B(z)$); $D_{\mu\mu}$ is the pitch angle diffusion coefficient which determines the particle diffusion mean free path

$$\Lambda = \frac{2}{3}v \Bigg/ \int_{-1}^{1} D_{\mu\mu} d\mu, \qquad (8.12)$$

and $Q(z, \mu, t)$ is the neutron decay protons deposition source function. Zhang (1993) noted that there are four different physical processes stated in Eq. 8.11: (1) particle streaming along the field line described by the first term on the right hand side, (2) particle pitch angle focusing (the second term), (3) pitch angle diffusion (third term), (4) proton deposition in the magnetic flux tube (last term). According to Jokipii (1966) the diffusion coefficient $D_{\mu\mu}$ is determined by the IMF power spectra; this power spectra might decrease faster than r^{-3} (Jokipii 1973). Figure 8.13 shows the expected dependence of diffusion mean free path Λ of 35 MeV protons versus the distance from the Sun, along the field line z ($z = 4$ AU corresponding to $r = 2.3$ AU) for the IMF power spectra $\propto r^{-3}$ and $\propto r^{-4}$.

The comparison of theoretical calculations of Eq. 8.11, for the time dependence of neutron decay protons is shown in Figs. 8.14 and 8.15.

8.3 Solar Neutron Decay Protons During the April 24, 1984 Event

This event was caused by the solar flare at 23:54 on April 24, 1984, located at 12°S, 43°E. The spacecraft ISEE-3 detected from this flare the largest gamma-ray fluency with energy >2 MeV. The detailed observations of protons resulting from the decay

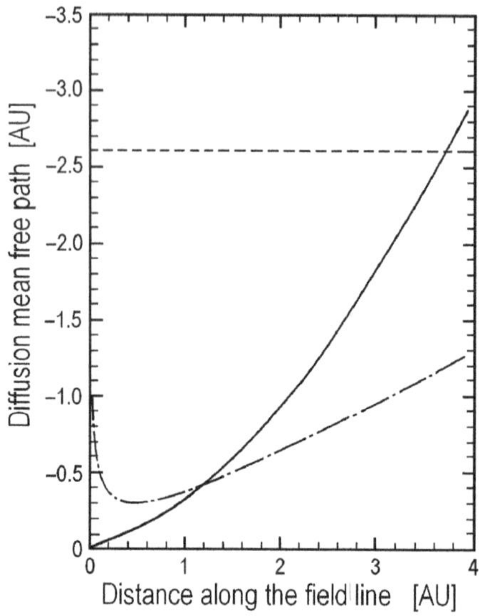

Fig. 8.13 A plot of diffusion mean free path of 35 MeV protons versus the distance along the field line ($z = 4$ AU correspond to the distance from the Sun $r = 2.3$ AU). The dot-dash line is calculated assuming the power spectra vary as $\propto r^{-3}$, while the solid assuming $\propto r^{-4}$. The long dash line at 0.26 AU is the fit result of Ruffolo (1991) (According to Zhang 1993)

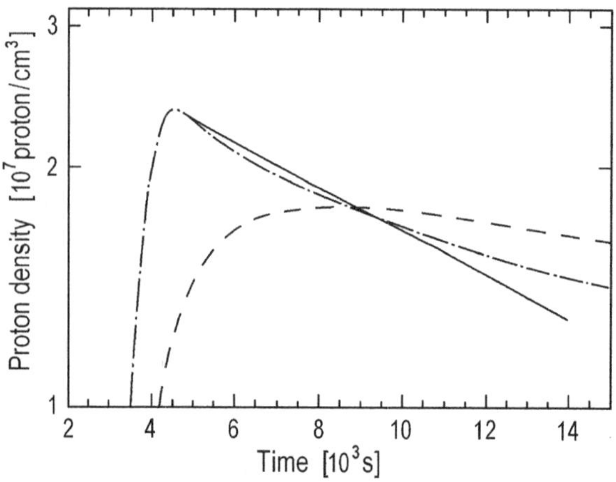

Fig. 8.14 Comparison of calculated neutron decay proton density with observation. The calculation assume fan beam neutron emission. The dash line is the result from using the standard QLT pitch angle scattering formula; the dot-dash line is the result from reducing the pitch angle scattering by a factor of 2. The solid line is he observation approximation (According to Zhang 1993)

of neutron ejected by this solar flare was made during the University of Chicago experiment onboard ISEE-3 spacecraft (Evenson et al. 1985). The measured rates of protons in the energy interval 25.7–47.5 MeV are shown in Fig. 8.16.

According to Evenson et al. (1985) it was observed that a large initial gamma-ray burst followed by a rise in the proton flux approximately 20–30 min later, the

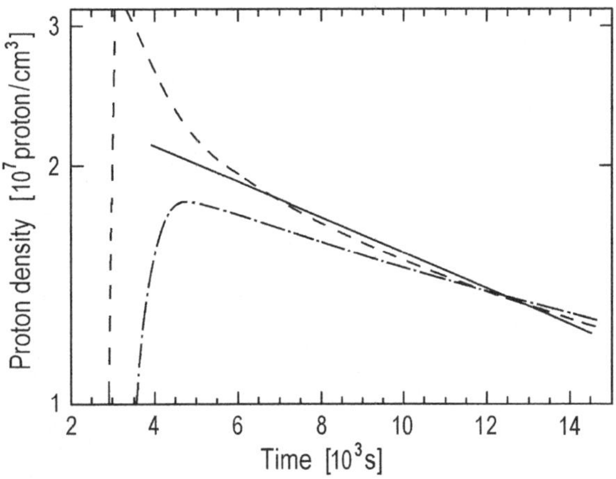

Fig. 8.15 Another comparison of calculation with the observation. The pitch angle scattering is still reduced by a factor of 2. The radial variation of power spectra is $\propto r^{-4}$. The dot-dash line is the case of fan-beam emission, the dash line is the case of isotropic emission, the solid line is the observation (According to Zhang 1993)

expected delay for the arrival of neutrons. The initial rise in the proton flux is rapid, corresponding to the time when the expanding shell of neutron passes the ISEE-3. This is followed by a slow decline in the flux as the decay protons diffuse away. Evenson et al. (1985) noted that primary protons produced directly by acceleration processes during the flare cause the flux to rise again, about 4 h after the initial gamma-ray burst in the 1984 flare (see Fig. 8.16). This delay time for directly produced flare protons was only about 1 h in the event of June 21, 1980 (Section 8.1) and about 16 h in the event of June 3, 1982. Fig. 8.16 also shows that the time dependence of neutron decay proton flux is in good agreement with expected, by modelling proton propagation in the inner Heliosphere using a mean free path of scattering along the direction of IMF, $\Lambda_{||} = 0.3$ AU and an isotropic emission of neutrons. The spectrum of decay protons collected in the period of 1 h (from 30 to 90 min after the initial gamma-ray burst) is shown in Fig. 8.17, in comparison with the event of June 3, 1982.

It can be seen from Fig. 8.17 that in the event of April 24, 1984 the differential fluxes of decay protons in the energy interval 30–100 MeV was bigger (two to three times) than that of June 3, 1982. The expected neutron emission spectrums in the event of April 24, 1984 (in comparison with that of June 3, 1982) are shown in Fig. 8.18 for the model of isotropically emitting in all directions from the Sun except downward moving neutrons which the Sun blocks ('isotropic' model) and in Fig. 8.19 for the other model, when neutrons are suggested emitted in the plane parallel to the local horizon at the flare site, half of which escape the Sun ('pancake' model).

The spectral indexes of the power law that best fit the observation data are found as follows: in the frame of 'isotropic' model $\gamma = -1.4$ and $\gamma = -2.1$ for the event

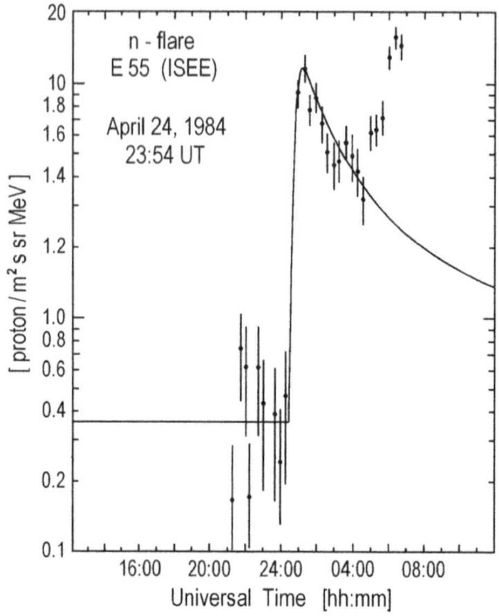

Fig. 8.16 The neutron decay event at April 24, 1984: the measured rates of protons in the energy interval 25.7–47.5 MeV in comparison with neutron decay protons for the mean free path of scattering $\Lambda_{||} = 0.3$ AU (According to Evenson et al. 1985)

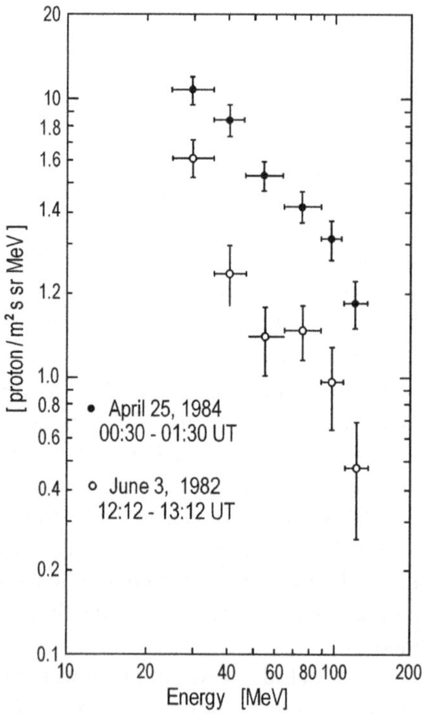

Fig. 8.17 The neutron decay event at April 24, 1984: the spectrum of neutron decay protons collected in the period from 30 to 90 min after the initial gamma-ray burst in comparison with the spectrum in the event of June 3, 1982 (According to Evenson et al. 1985)

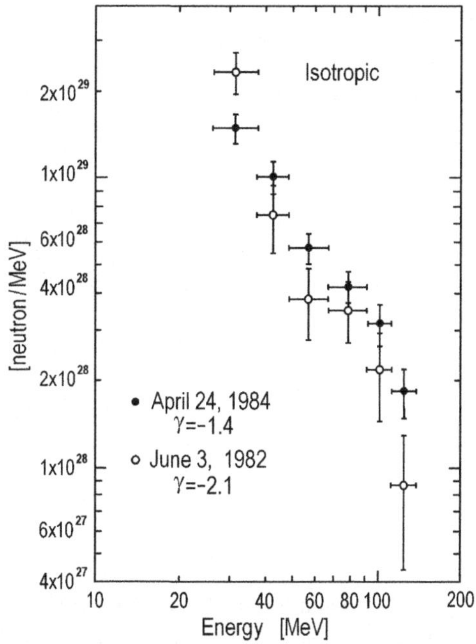

Fig. 8.18 Expected neutron emission spectra in the event of April 24, 1984 in comparison with spectra of the event of June 3, 1982 (in the frame of the model of isotropically emitting) (According to Evenson et al. 1985)

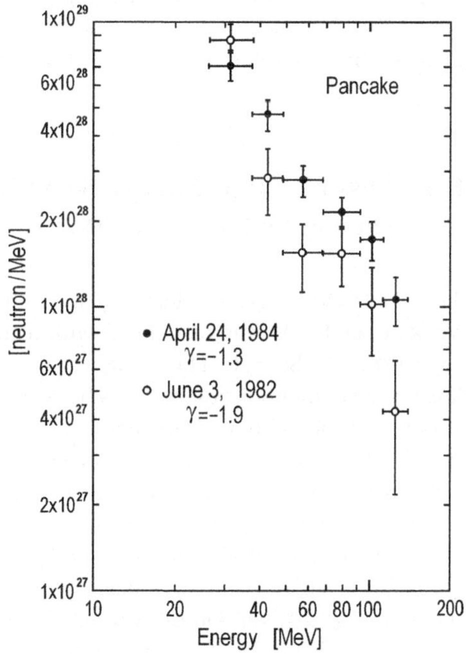

Fig. 8.19 The same as in Fig. 8.18, but for the 'pancake' model (According to Evenson et al. 1985)

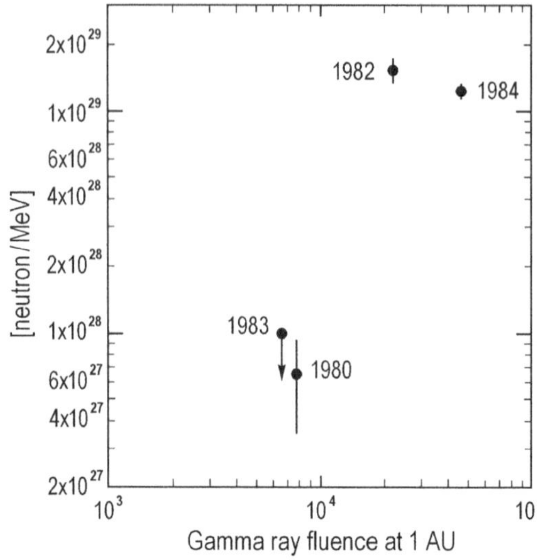

Fig. 8.20 Comparison of neutron fluency with gamma-ray fluency for four neutron decay events (According to Evenson et al. 1985)

April 24, 1984 and June 3, 1982; in the frame of 'pancake' model $\gamma = -1.3$ and $\gamma = -1.9$ correspondingly for April 24, 1984 and June 3, 1982.

The neutron fluency has a tendency to increase with increasing gamma-ray fluency. The comparison is shown in Fig. 8.20 for events June 21, 1980; June 3, 1982; May 7, 1983 (at 22:17 UT); and April 24, 1984. With increasing gamma-ray fluency for $E_\gamma > 2$ MeV at 1 AU from the Sun from $\sim 8 \times 10^3$ quant/cm^2 to $\sim 3 \times 10^4$ quant/cm^2, the fluency of solar neutrons in the energy interval 25.7–47.5 MeV increases more than ten times, from $\sim 8 \times 10^{27}$ neutron/MeV to $\sim 10^{29}$ neutron/MeV.

8.4 Possible High Energy Solar Neutron Decay Protons During October 19, 1989

This well known ground level solar cosmic event (see review in Dorman and Venkatesan 1993) was associated with a major 4B flare located at heliographic coordinates 27°S, 10°E. This event was investigated in detail on the basis of ground level neutron monitor observations by Shea et al. (1991a, b) and it was shown that with a big probability in this event was detected high energy neutron decay protons (for the first time by ground based detectors). The flare characteristics are listed in Table 8.1 and cosmic ray data are shown in Figs. 8.21 and 8.22.

The important peculiarity of the behaviour of cosmic rays in this event is according to Shea et al. (1991a, b) as follows (see Figs. 8.21 and 8.22): although the major cosmic ray increase began at ~13:30 UT, the eastern Canadian neutron monitors, Deep River, Goose Bay and Ottawa show an abrupt increase of ~5% in the interval 13:00–13:05 UT, which remained essentially constant for about 25 min

Table 8.1 19 October, 1989 flare data (Shea et al. 1991b)

Radiation	Onset	Max	End	Imp.
H-alpha	12:29	12:59	21:49	3B
1–8 Å X-rays	12:32	12:58	19:00	X13
2,695 MHz	<12:40	12:59	>15:43	4,500 SFU
8,800 MHz	12:39.6	12:59	>15:43	37,200 SFU
15,400 MHz	12:39.6	12:59	>15:43	470 SFU
γ-Rays	<12:57[a]	12:58	~13:25	3,000 CPS
'Neutrals'	<12:57[a]	13:10	>13:30	1.5 CPS

[a]SMM spacecraft in Earth shadow at onset; SFU-Solar Flare Unit = 10^{-22} W/m^2

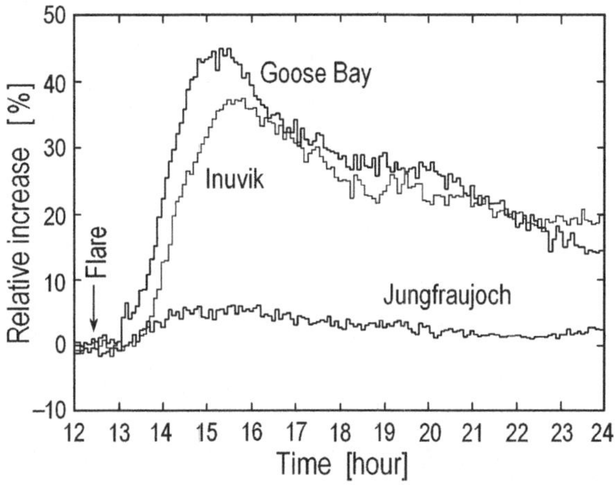

Fig. 8.21 The cosmic ray intensity recorded by the Goos Bay (Canada), Inuvik (Canada) and Jungfraujoch (Switzerland) neutron monitors during the relativistic solar particle event of October 19, 1989. The intensity is a percentage above the hourly average background recorded between 11:00 and 12:00 UT (According to Shea et al. 1991a)

before a second rise (coincided with general world-wide cosmic ray increase). From the maps of the asymptotic directions of approach appropriate for this event (0.8–5 GV) for Inuvik, Deep River, Goose Bay and Oulu, shown in Fig. 8.23, Shea et al. (1991a, b) noted that the Deep River and Goose Bay neutron monitors would have sampled the particle flux at small angles to the Sun-Earth line, whereas the monitors at Inuvik and Oulu were sampling an area of space approximately 90° away from the Sun-Earth line (it is a cause that we can not see neutron decay protons in time profiles of Inuvik and Oulu neutron monitors).

The obtained differential energy spectrums of neutron decay protons (for the time-interval 13:07–13:22 UT) and for direct solar protons near the maximum at 15:00 UT are shown in Fig. 8.24.

The pitch angle anisotropies of neutron decay protons at 13:07 UT and at 13:17 UT are shown in Fig. 8.25 and of direct protons at 15:00 UT in Fig. 8.26. There are a good agreement between observed and calculated values of anisotropy, according to neutron decay hypothesis (Fig. 8.27).

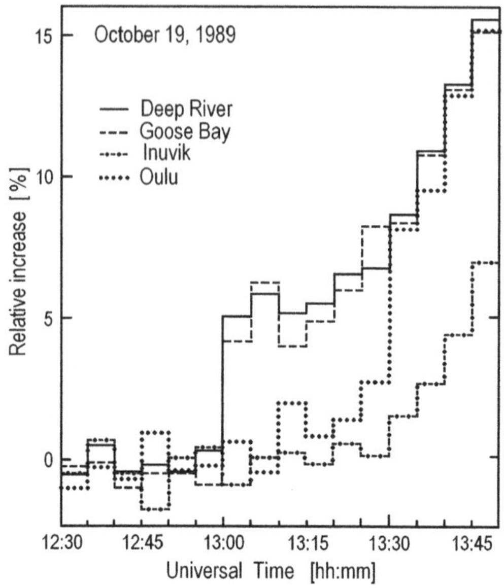

Fig. 8.22 Relative cosmic ray increase during the onset of the October 19, 1989 relativistic solar particle event observed by the Inuvik, Deep River, Goose Bay and Oulu neutron monitors. The intensity is a percentage above the hourly average background recorded between 11:00 and 12:00 UT (According to Shea et al. 1991a)

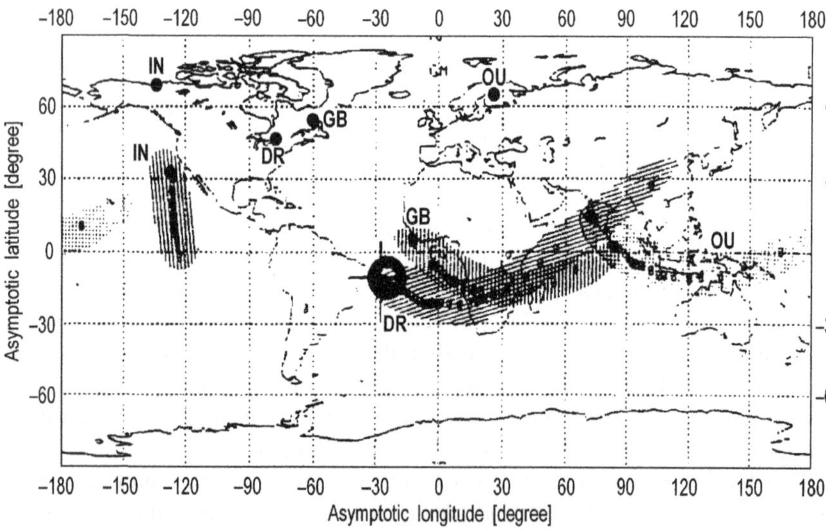

Fig. 8.23 The asymptotic cones of acceptance for the neutron monitors at Inuvic (IN), Deep River (DR), Goose Bay (GB) and Oulu (OU). For visualization purposes these are plotted on an extended world map in an equal angular geocentric projection. The sub-solar point (large circle) at 13:28 UT was at 10°S, 26°W; the interplanetary magnetic field line direction for that same time was located at 10°S, 28°W (large cross) (According to Shea et al. 1991a)

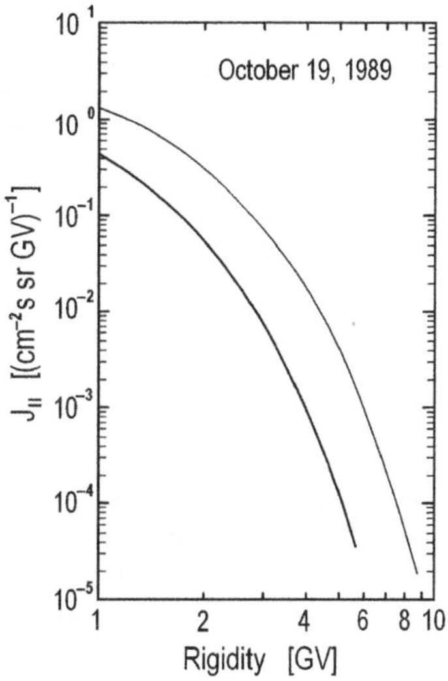

Fig. 8.24 The derived relativistic solar particle spectrum on October 19, 1989. Bottom curve: onset spectrum, 13:07–13:22 UT (heavy line). Top curve: solar particle spectrum for 15:00 UT near the event maximum (According to Shea et al. 1991a)

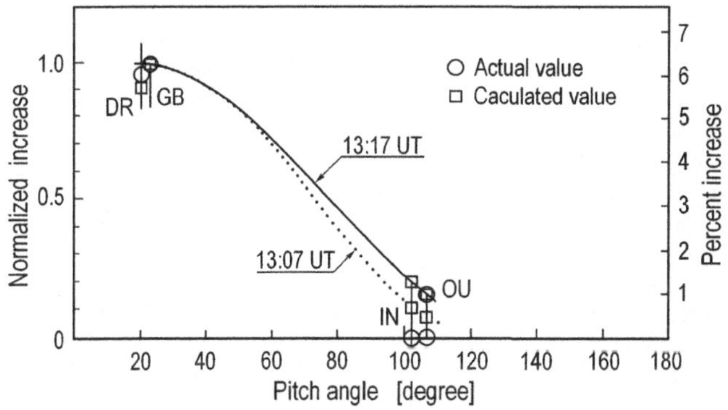

Fig. 8.25 The derived anisotropy for 13:07 UT (dotted line) and 13:17 UT (solid line) on October 19, 1989. The symbol ○ indicates the observed increase and the symbol □ indicates of Shea et al. (1991a) calculations. The vertical line at each station shows the variability of the data throughout the time interval (According to Shea et al. 1991a)

8.5 Possible High Energy Solar Neutron Decay Protons During the October 28, 2003 Event: Where Were Energetic Particles Accelerated and Where Were Solar Neutrons Decayed?

8.5.1 Where Are Particles Accelerated During Impulsive Flares?

Le et al. (2007a) on the basis of observation data on the event of October 28, 2003 discussed the problem how solar energetic particles (SEP) are accelerated and how

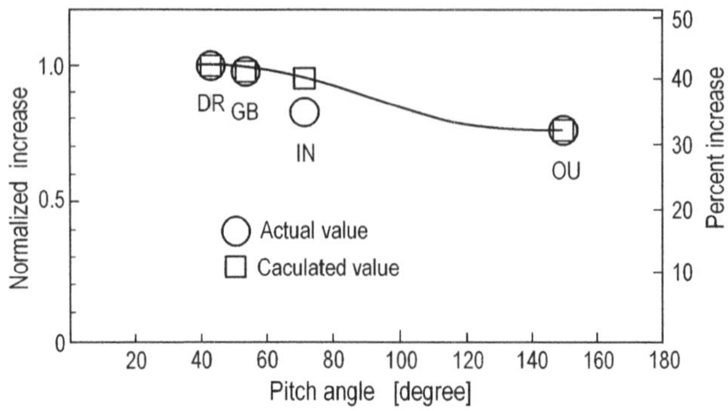

Fig. 8.26 The derived anisotropy for 15:00 UT of October 19, 1989. The symbol ○ indicates the observed increase and the symbol □ indicates of Shea et al. (1991a) calculations. The size of the symbols are indicative of the variation in the data at event maximum (According to Shea et al. 1991a)

Fig. 8.27 Illustration of the neutron decay hypothesis. High energy neutrons generated in the solar flare will propagate into space. Those decaying on interplanetary magnetic field lines connecting with the Earth will generate a proton flux that can be observed by neutron monitors viewing in the interplanetary magnetic field direction (According to Shea et al. 1991a)

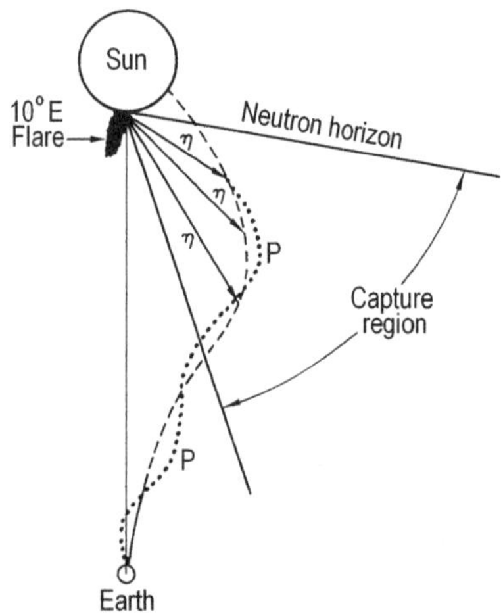

relativistic neutrons are produced and propagate in interplanetary space taking into account they decay on high energy protons, electrons, and neutrino as well as propagation and absorbing in the Earth's atmosphere. As noted Le et al. (2007a), it is generally accepted that impulsive SEP are accelerated in flares while whether gradual SEP are only accelerated at coronal mass ejection (CME) driven by shock is still an open question. Reames (2002) suggested that mixed events that SEP accelerated both in the flare and the CME-driven shock don't exit. Cane et al. (2003) studied the intense solar energetic particles during the interval 1997–2001 with the particle energy 20–80 MeV/n. They found that there were two peaks in the solar energetic particle intensity in some SEP events. The first peak has a high value of Fe/O and the second peak has a low value of Fe/O suggesting that the solar particles are mainly accelerated by solar flare in the first peak and mainly accelerated by CME-driven shock at the second peak. Cane et al. (2006) extended their study and got the similar conclusion. It is accepted that both flare and CME-driven can accelerate the solar particles to non relativistic energy.

8.5.2 Solar Flares and CMEs: Generation and Propagation of Relativistic Protons, Electrons, and Neutrons

As noted Le et al. (2007a), there is still not clear the situation with acceleration of relativistic SEP. Tylka et al. (2005) suggested that relativistic SEP are accelerated by quasi-perpendicular shocks. Based on the onset time of metric type radio bursts earlier than the solar release time of relativistic SEP, Gopalswamy et al (2005a, b) suggested that relativistic SEP are accelerated by CME-driven shock. Li et al. (2007) have studied the magnetic reconnection in the event of October 28, 2003, and at maximum induced electric field was estimated to be E $\sim$ 13 V/cm. The amplitude evolution of this field correlates in time with the evolution of hard X-ray and γ-ray emissions, indicating that induced electric field may play an important role in acceleration of non-thermal particles. At given the maximum E $\sim$ 13 V/cm, an acceleration length $\sim 7 \times 10^7$ cm is needed to accelerate protons to $\sim$ GeV energy in the reconnection current sheet by DC electric field. The ratio of the acceleration length to the whole filament length in this event is $\sim 10^{-2}$. Therefore, the reconnection electric field probably makes a crucial contribution to the acceleration of relativistic SEP and the impulsive component of the large gradual SEP events. It was suggested that the spectral index of electrons can be used to identify the source of the electrons (Simnett 2005a, b, 2006). The spectral index of electrons was lower than 1.5 reveling that the electrons were accelerated in the solar flare (Simnett 2006) for SEP event on 20 January 2005. The onset time of relativistic solar protons observed by NM at South Pole was 06:48 UT $\pm$ 30 s which coincided with the gamma ray's emission interval 06:38:30–06:42:30 ST with energy 60–100 MeV, which was the decay of pion in the SEP event on 20 January 2005 (Kuznetsov et al. 2006a, b). This means that relativistic solar protons were accelerated by the solar flare during the interval from 06:38:30 to 06:42:30 ST, during which the high energy gama ray emission was going on.

As noted Le et al. (2007a), a very bad space weather occurred after the solar intensive eruption on 28 October 2003, a large flare (X17.2/4B) erupted from the super active region AR10486 accompanied by an earth-directed CME with initial speed 2,429 km/s. Relativistic SEP led to ground level enhancement (GLE), and when the CME reached the Earth, it caused a very strong geomagnetic storm. Many papers have been devoted to this event (e.g., Bieber et al. 2005; Klassen et al., 2005; Cohen et al. 2005; Gopalswamy et al 2005a, b; Aurass, et al. 2006; Kiener et al. 2006; Le et al. 2007b; Mewaldt, et al. 2005; Miroshnichenko et al. 2005; Simnett 2005a, b). The release time of relativistic SEP obtained by Bieber et al. (2005) was about 11:03 ST. Because the coronal shock formed at $\sim$10:55 ST (Klassen et al. 2005) and the impulsive flare phase starts with a steep rise of X-rays at 10:52 ST and peaks at about 11:02 (Klassen et al. 2005) so the result obtained by Bieber et al. (2005) seemly supported that the relativistic SEP were accelerated by the CME-driven shock. Source location of the flare X17.2 was at 16°S, 08°E, so the source location of the flare was not at the region of well connected with the Earth. Miroshnichenko et al. (2005) studied the GLE event and proposed that relativistic SEP were accelerated by the solar flare, the path length traveled by the protons with energy greater than 100 MeV was ~2.2 AU. Why this path length was so long? Miroshnichenko et al. (2005) proposed that the CME from AR10486 on the 26 October 2003 reached the Earth with their two legs still connected with the AR10486 and the relativistic SEP injected into the eastern leg of the CME. Li et al. (2007) also studied the GLE event on 28 October 2003; they proposed that the flare magnetic connection, especially the induced electric filed, made a crucial contribution to the prompt relativistic SEP.

8.5.3 Observations of Relativistic Protons from Decaying Solar Relativistic Neutrons

As underlined Le et al. (2007a), relativistic solar protons from decaying solar relativistic neutrons were observed by Tsumeb NM 28 October 2003, which recorded a 3–4% increase and persisted for about 9 min (Bieber et al. 2005). Bieber et al. (2005) modeled the solar release time of the relativistic solar neutrons and relativistic solar protons, which was about 10:56 and 11:03 ST, respectively. Miroshnichenko et al. (2005) reported that the interval for the occurrence of relativistic solar neutrons was 11:05–11:15 UT. The γ-ray emission with energy 60–100 MeV occurred during the period 11:02–11:13 UT (Kuznetsov et al. 2006a, b). The γ-rays with energy 60–100 MeV are the production of pion decay. The occurrence of pion indicates that the protons were accelerated up to relativistic energies during the flare (Kuznetsov et al. 2006a, b). Comparison of the time information of the γ-rays with that of the relativistic solar neutrons, lead Le et al. (2007a) to suggestion that the relativistic solar neutrons were produced during the interval 11:02–11:13 UT. It means that the interval of solar release time of relativistic solar neutrons was from (11:02–500 s) to (11:13–500 s) ST. It was reported that relativistic solar neutrons and protons from decay of some part of solar neutrons were observed by Tsumeb NM, the cut off rigidity of which is 9.21 GV (it means that protons had energy more than 8.23 GeV with velocity bigger than 298219.3 km/s). Because the relativistic protons were the production of solar neutron decay, the velocity of relativistic solar neutrons should be greater than 298219.3 km/s. Solar neutrons were the production of nuclear reactions by solar protons, which were accelerated by solar flare, with heavy ions in solar atmosphere. Because the energy of relativistic solar neutrons was bigger than 8.23 GeV, so some solar protons in the flare had energy greater than 8.23 GeV and they should be produced slightly earlier than relativistic solar neutrons. The hard γ-rays with energy 7–20 MeV observed by RHESSI began at 10:28 UT, peaked at 11:13

UT and end at 11:30 UT. The hard γ-rays were produced by synchrotron emission of relativistic solar electrons in the local magnetic field, which revealed that the flare not only accelerated the electrons to relativistic energy but also accelerated the protons to relativistic energy. The time of hard γ-rays with energy 7–20 MeV revealed that the early relativistic solar electrons occurred at 10:20 ST, while the early relativistic solar protons occurred at 10:54 ST, which was about 34 min later than that of relativistic solar electrons. Solar release time of relativistic solar neutrons was earlier than 10:54 ST, the time of relativistic solar neutrons observed by Tsumeb NM was 11:06 UT, so the traveling time by relativistic solar neutrons in space was nearly 12 min. Le et al. (2007a) assume the velocity of relativistic solar protons decayed was almost the same with that of relativistic solar neutrons, so the traveling distance is expected about 1.4 AU. A method to get the path length and solar release time of a particle traveling from the Sun to the Earth was proposed (VDA method) by Krucker et al. (1999). Using the arrival time for protons at the Earth at the various energies given by Miroshnichenko et al. (2005), Le et al. (2007a) can get the result shown in Fig. 8.28.

From Fig. 8.28 follows that $l \sim 2.25$ AU and solar release time ~10:54 ST. This result of Le et al. (2007a) is almost consistent with that obtained by Miroshnichenko et al. (2005), but there is 1 min difference between the two results (their solar release time was 10:55 ST). Le et al. (2007a) came to following conclusions:

1. Solar release time of solar protons with energy greater than 100 MeV was slightly earlier than 10:54 ST.
2. Relativistic solar neutrons decayed into relativistic solar protons at about 1.21 AU away from the Sun in the direction of Sun-Earth connection and then moved

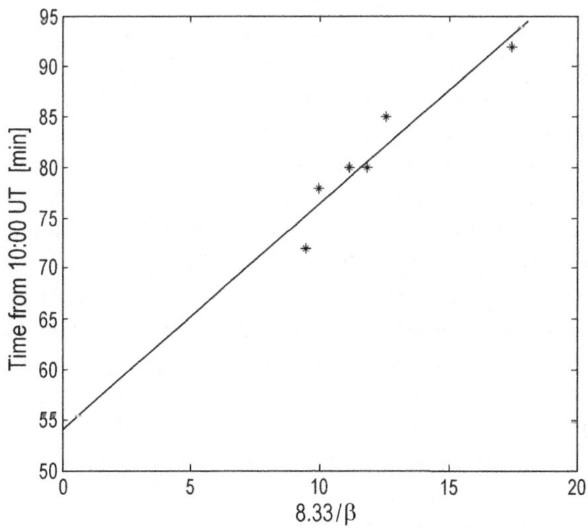

Fig. 8.28 Solar neutron event on October 28, 2003: Onset time versus 8:33/β, where $\beta = v/c$ (From Le et al. 2007a, b)

back towards the Earth, were they propagated through geomagnetic field and the Earth's atmosphere, and then were observed by Tsumeb NM on the ground.

3. The solar release time of SEP with energy greater than 100 MeV computed by VDA method was also slightly earlier than 10:54 ST, so the earliest arriving relativistic solar protons were accelerated by solar flare and once the solar protons reached relativistic energy, they immediately injected into interplanetary space and the path length of solar protons with energy greater than 100 MeV was about 2.25 AU.
4. It is accepted that the type III bursts are produced by relativistic electrons escaping the solar corona into interplanetary space. The onset of type III bursts was ~10:54 ST which means that the magnetic field line of the AR10486 was opened and connected with the magnetic field line in interplanetary space at 10:54 ST, so the relativistic solar protons almost immediately injected into interplanetary space when they got the relativistic energy.

8.6 Prediction and Possible Observation of Solar Neutron Decay Electrons

Solar neutron decay gives not only protons but also electrons:

$$n \rightarrow p + e + \nu_e. \tag{8.13}$$

Kolomeets et al. (1988) on the basis of Eq. 8.13 by using observation data on solar neutrons calculated the expected time variation and energetic characteristics of solar neutron decay electrons. They discuss the possibility to measure these electrons in the interplanetary space and how to determine the energy spectrum of the neutrons near the Sun on the basis of measurements of energy spectrum of decay electrons. Koi et al. (1993) have also calculated the expected energy spectra of solar neutron decay electrons and determined the trapping region of these electrons in the Earth's magnetosphere, according to Stőrmer's theory. The expected energy spectrum of electrons according to Eq. 8.13 was calculated by a Monte Carlo method and the results are shown in Fig. 8.29.

According to the calculations of Koi et al. (1993), electrons with energy $\geq$2 MeV can only be produces by high energy neutrons with kinetic energies greater than 400 MeV (with Lorenz factor of neutrons $\gamma \geq 1.4$). Koi et al. (1993) have searched for pre-increases of electrons before the main phase of increases of charged direct solar particles and just after large solar flares. It was found two candidates of such an event, in the duration of solar cycle 22; they occurred on October 19, 1989 (see Section 8.4 on the relativistic protons pre-increase) and November 2, 1992. Among them, the event of October 19, 1989 is the clearest. Figure 8.30 shows the data of GOES-7 for fluxes of 1–8 Å X-rays, electrons with energy >2 MeV and protons with energy from 4.2–9.7 to 640–850 MeV.

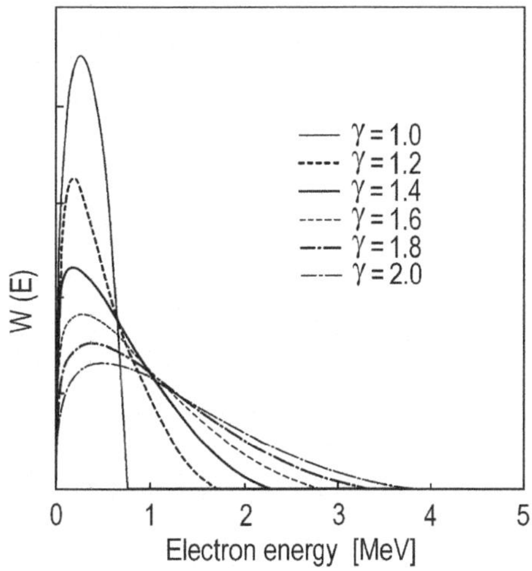

Fig. 8.29 The expected spectrum of the neutron decay electrons. The γ factor represents the parent neutron's Lorentz factor. Vertical axis represents relative probability (According to Koi et al. 1993)

There is a huge difference between electrons with energy higher than 2.0 MeV and protons with energy higher than 400 MeV; according to the Störmer's theory, the geomagnetic field around the GOES-7 ($\sim$100 nT at distance 6.6 r_E) could trap 2 MeV electrons but could not trap 400 MeV protons, so the electrons will be trapped for about one drift period, which is $\sim$30 min, but protons escape immediately. Thus a large difference in the fluxes is expected. Koi et al. (1993) compared GOES-7 data with that of Ulysses (Wibberenz et al. 1992) which are free from the effect of the magnetosphere. The expected neutron decay electron flux in the energy range 2.7–7.0 MeV at Ulysses is found to be $\sim 10^{-7}$/cm^2/s/sr/MeV, which is far less than the cosmic ray electron background level about 2×10^{-3}/cm^2/s/sr/MeV and neutron decay electrons can not be measured in free space (however, the proton flux produced by the decay of neutrons is expected of the order 10^{-5}/cm^2/s/sr/MeV at 135 MeV). Koi et al. (1993) came to the conclusion that high energy electrons produced by the decay of solar neutrons can be seen only by satellites in the geomagnetosphere, where the accumulation effect is sufficient for such electrons.

8.7 Observation of Electrons from the Decay of Solar Flare Neutrons

8.7.1 The Matter and Short History of the Problem

Dröge et al. (1996) have found evidence for fluxes of energetic electrons in interplanetary space on board the ISEE-3 spacecraft, which were interpret as the decay products of neutrons generated in a solar flare on 1980 June 21. The decay electrons arrived at the spacecraft shortly before the electrons from the flare and can

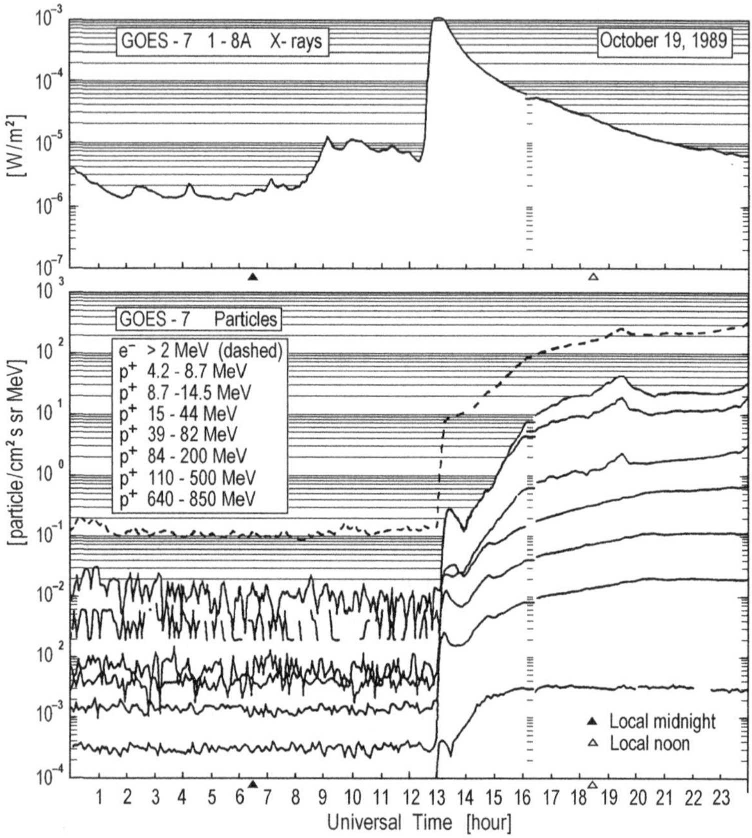

Fig. 8.30 Intensity-time profiles of X-rays, electrons and protons obtained on GOES-7. To obtain the electron integral flux, the flux shown in Figure must be multiplied by a factor of ten. The proton fluxes are indicated in terms of differential values (From Koi et al. 1993)

be distinguished from the latter by their distinctive energy spectrum. The time profile of the decay neutron electrons is in good agreement with the results from a simulation based on a scattering mean free path derived from a fit to the flare electron data. The comparison with simultaneously observed decay protons and a published direct measurement of high-energy neutrons places important constraints on the parent neutron spectrum.

As noted Dröge et al. (1996), previous studies have reported observations of interplanetary neutrons from solar flares by three methods: (1) direct detection of neutrons in space from flares on 1980 June 21 (Chupp et al. 1982), 1982 June 3 (Chupp et al. 1987), 1988 December 16 (Dunphy et al. 1990; Dunphy and Chupp 1992), 1991 June 9 (Ryan et al. 1993), and 1991 June 15 (Debrunner et al. 1993); (2) detection of their decay protons in space after flares on 1980 June 21, 1982 June 3, and 1984 April 25 (Evenson et al. 1983a, b, 1985, 1990; Ruffolo 1991); and (3) ground-based detection of neutrons from flares on 1982 June 3 (Debrunner et al. 1983;

Efimov et al. 1983), 1990 May 24 (Shea et al. 1991a, b, c), 1991 March 22 (Pyle and Simpson 1991), and 1991 June 4 (Takahashi et al. 1991; Chiba et al. 1992; Muraki et al. 1992). These methods provide complementary information on the spectrum, angular distribution, and temporal distribution of escaping neutrons in different energy ranges, which can be compared with theoretical predictions (e.g., Murphy et al. 1987; Ramaty and Murphy 1987; Hua and Lingenfelter 1987a, b; Guglenko et al. 1990; Ramaty et al. 1993; Rieger 1996) to constrain models of high-energy processes in solar flares.

Dröge et al. (1996) present observational evidence for a (4) type of detection based on decay electrons of solar flare neutrons on 1980 June 21. They also present detailed simulations of the injection and interplanetary transport of the decay electrons, which are used to fit those data. As has been pointed out previously (Daibog and Stolpovskii 1987), solar neutrons of all energies yield a similar spectrum of decay electrons, so the decay electron intensity provides a measure of the total number of interplanetary neutrons, including those of ~1 MeV, which are not detected by other methods. There is a high flux of neutrons at these low energies, which propagate toward the hemisphere not obscured by the Sun and decay within $v_n \tau_n \sim 0.1$ AU, so with a reasonably good magnetic connection to the flare site, one can observe a significant flux of decay electrons with $E_e < 1$ MeV superposed on the rising phase of the event.

8.7.2 Observations by the ULEWAT and MEH Spectrometers Aboard the ISEE-3 Spacecraft During the Solar Neutron Event of 1980 June 21

The particle observations presented in Dröge et al. (1996) were made with two instruments on board the ISEE-3 spacecraft: the ULEWAT spectrometer (Hovestadt et al. 1978), which measured the electron flux in the energy range of approximately 0.1–1 MeV, and the University of Chicago MEH spectrometer (Meyer and Evenson 1978), which measured protons from 27 to 147 MeV. Because no electron calibration was made with the ULEWAT spectrometer, a Monte Carlo simulation was performed to precisely determine its response to low-energy electrons. During the time interval under consideration, ISEE-3 was positioned at the Earth-Sun Lagrangian point, well outside of the Earth's geomagnetic field.

Figure 8.31 (upper panel) shows electron fluxes at energies of ~0.18, 0.25, 0.61, and 1.1 MeV, respectively from up to dawn, which were observed by ISEE-3 on 1980 June 21 after a flare which occurred at 1:17 UT at 20°N, 88°W. The anisotropy of the lowest energy channel is shown in the second panel (no sectored data were available for the other channels). The spikes in the lowest two channels and the anisotropy lasting from ≈1:20 to 1:30 UT are due to X-rays absorbed in the ULEWAT spectrometer. The electron event is characterized by a slow rise and late time of maximum despite a large and persisting anisotropy. Such a signature is indicative of a large interplanetary scattering mean free path λ in the vicinity of the

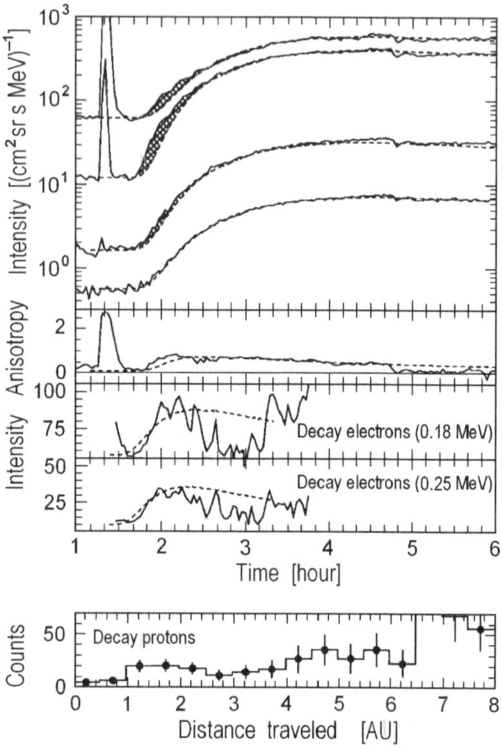

Fig. 8.31 Electron intensities at 0.18, 0.25, 0.61, and 1.1 MeV (upper panel) and anisotropy of 0.18 MeV channel (second panel) of the 1980 June 21 solar event, observed on ISEE 3 (solid lines) and fits (dashed lines). Hatched areas indicate the excess flux due to neutron decay electrons. The middle two panels show the decay electrons and fits from a simulation. The lower panel shows ISEE 3 protons (27–147 MeV) as a function of the distance traveled, $v(t-t_{tr})$ (From Dröge et al. 1996)

observer and an extended injection of particles close to the Sun, or a short injection and strong scattering at small solar distances, or a combination of the latter two possibilities.

According to Dröge et al. (1996), fits to the intensity and anisotropy profiles (assuming the anisotropies of all four channels are similar) were performed using numerical solutions of the model of focused transport (Schlüter 1985). In order to minimize the number of free parameters in a first step, it was assumed that the mean free paths of electrons parallel to the magnetic field, $\lambda_{||}(z)$, were spatially constant. From the observed solar wind speed of 290 km/s, the magnetic field spiral was mapped back to 0.05 AU, resulting in a nominal distance of z = 1.26 AU along the connecting field line that had its foot-point at 05°N, 73°W. The transport of electrons from the flare site to and subsequent injection at the beginning of the connecting field line at 0.05 AU were described phenomenologically by a Reid-Axford profile (Reid 1964) with rise and decay time constants t_r and t_d, respectively. Good agreement was obtained – except for a period of about 40 min during the rising flank in the lowest three channels – between fits (dashed lines) and observations for $\lambda_{||}(z)$ values of 0.37, 0.37, 0.31, and 0.26 AU, respectively, and an injection function with $t_r = t_d = 2$ h.

As underlined Dröge et al. (1996), this discrepancy between fit and observations in the four channels for $\lambda_{||}(z)$ from 1:40 to 2:20 UT, indicated by the hatched areas

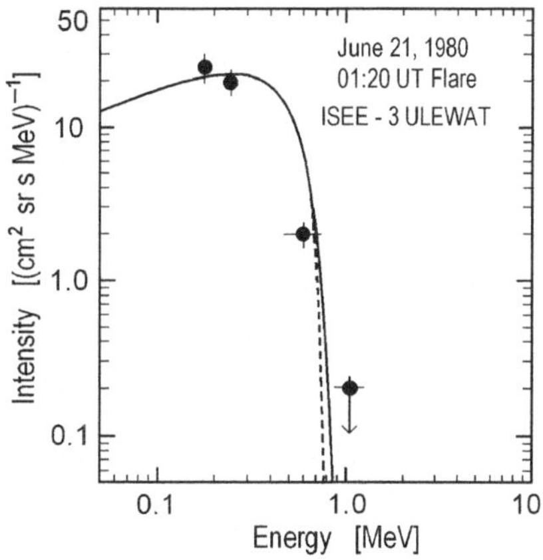

Fig. 8.32 Energy spectrum of the excess electrons (filled circles) and theoretical prediction for decay electrons in the rest frame of the parent neutron (dashed line) and for the estimated neutron spectrum for the 1980 June 21 flare (solid line) (From Dröge et al. 1996)

in Fig. 8.31, does not disappear for any plausible assumptions about the behavior of the electron mean free paths and cannot be explained by contamination from energetic protons or gamma rays generated by them, or by temporal variations in the interplanetary magnetic field. Dröge et al. (1996) have investigated the hypothesis that this excess flux represents the detection of electrons from the decay of solar flare neutrons produced in the flare. Initial evidence is provided by the energy spectrum of the excess electrons (Fig. 8.32), which is, within the uncertainty of the ULEWAT response functions (horizontal error bars in Fig. 8.32) very similar to that expected for decay electrons in the rest frame of the neutrons (dashed line in Fig. 8.32).

8.7.3 Simulations of the Production and Transport of Neutron Decay Electrons in Interplanetary Space

To test the above hypothesis in more detail, Dröge et al. (1996) have performed numerical simulations of the production and transport of such neutron decay electrons in interplanetary space. To model the production of electrons due to the decay of interplanetary neutrons, a Monte Carlo simulation was performed. For each of 5×10^7 neutron decays, the decay electron was assigned a random energy, chosen according to the β-decay energy distribution, and a random direction in the neutron rest frame. The electrons were then boosted into the fixed frame, for various neutron energies. A four-dimensional array stored the number of decay electrons per parent neutron for five electron momentum bins, five neutron energies, four magnetic field directions, and 25 pitch angle bins.

Next, according to Dröge et al. (1996), the injection of decay electrons into a section Δr of a flux tube subtending $\Delta\Omega$ from the Sun during an interval Δt was determined from

$$\Delta N_e = \frac{dN_e}{dN_n}\frac{dN_n}{dE_n d\Omega}\bigg|_{at\,Sun} \times \frac{dE_n}{d\beta_n}\frac{d\beta_n}{dt}\frac{\Delta r}{\gamma_n \beta_n c\tau}\exp\left(-\frac{r}{\gamma_n \beta_n c\tau}\right)\Delta\Omega\Delta t, \quad (8.14)$$

where dN_e/dN_n is the number of electrons per decaying neutron as determined from the Monte Carlo results, and $\beta_n = r/(ct)$, γ_n, and E_n are the appropriate values for neutrons arriving at a radius r after a time t. Simulations of the interplanetary transport of decay electrons were performed using the finite-difference method of Ruffolo (1991), as modified to include the effects of adiabatic deceleration and convection (Ruffolo 1995). The transport simulations were performed for electron momentum values of 0.2, 0.4, 0.6, 0.8, and 1.0 MeV/c, $\Delta t = 1.0$ min, and assuming the same solar wind conditions, i.e., $V_{SW} = 290$ km/s, and $\lambda_{||}$ as determined from the initial fits to the direct electrons. For this choice of parameters, the simulation predicts an onset of the decay electrons ≈15 min before the observed onset. In a second step Dröge et al. (1996) have therefore made an attempt to modify the injection and transport parameters so that the good fit to the direct electrons was preserved and a satisfactory fit of the decay electrons was reached. In this second modeling Dröge et al. (1996) made the assumption that there was a zone of enhanced scattering close to the Sun extending from $r = 0.05$ AU to $r = 0.3$ AU where the parallel mean free path was approximately a factor of 10 smaller (indicated by the darker shading in Fig. 8.33), while it was of the same order as in the first modeling for $r > 0.3$ AU to give consistent results for the observed anisotropy at $r = 1$ AU (indicated by the lighter shading in Fig. 8.33).

As underlined Dröge et al. (1996), the second modeling yielded a similarly good fit for the direct electrons and also a good fit to the decay electrons for the following choice of parameters: $t_r = 1.05$ h, $t_d = 2.3$ h for the direct electrons, and constant radial mean free paths

$$\lambda_r(r) = \lambda_{||}(r)\cos^2\psi(r) \quad (8.15)$$

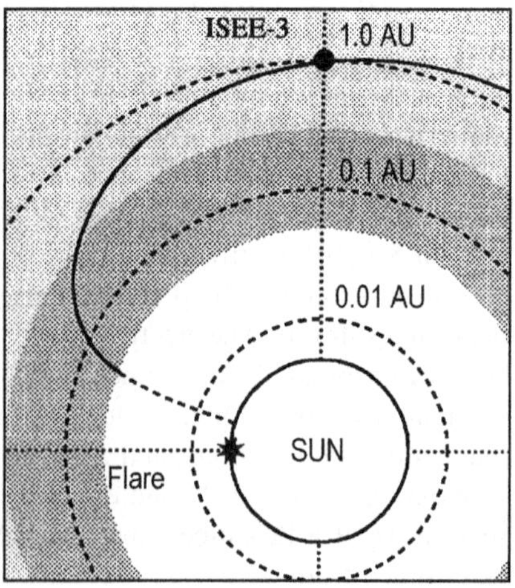

Fig. 8.33 Polar diagram showing the solar system geometry at the time of the 1980 June 21 flare in a view perpendicular to the ecliptic plane. Darker shading indicates a hypothesized region of stronger scattering. Radial distances have a logarithmic scale (From Dröge et al. 1996)

where $\psi(r)$ is the angle between the radius vector and Archimedean field spiral at a distance r. It was supposed that $\lambda_r(r) = 0.037, 0.035, 0.029$, and 0.026 AU for the four energy channels between 0.05 AU $< r <$ 0.3 AU, and $\lambda_r(r) = 0.115, 0.109$, 0.089, and 0.08 AU, respectively, for $r >$ 0.3 AU.

According to Dröge et al. (1996), the excess electron fluxes at 0.18 and 0.25 MeV (differences between total observed electrons and fits to direct electrons, plus the background intensity prior to the flare) are shown in panels 3 and 4 of Fig. 8.31, together with the predictions of the second simulation for the decay electrons (dashed lines). There is good agreement between the two data sets until about 02:30 UT. After this time, the difference fluxes do not give meaningful results any longer because of large, non-Gaussian fluctuations in the electron count rates caused by variations in the magnetic field. For decay electrons, the only free parameter of the fit is the normalization, i.e., 3×10^{31} neutrons/sr (of all energies) emitted toward the zenith. The directional distribution and energy spectrum of the neutrons had no significant effect on the time profile or its normalization.

Dröge et al. (1996) noted, that because no method for a direct, model-independent deconvolution of the transport parameters (injection profiles of the flare electrons and the spatial structure of $\lambda_{||}$) from the observed intensity and anisotropy profiles is known, those parameters have to be obtained from fits to the data by trial and error. However, Dröge et al. (1996) think that this choice of parameters for the second modeling is not unreasonable. It is often observed (Wanner and Wibberenz 1993) that the levels of magnetic fluctuations in the interplanetary medium, and thus $\lambda_{||}$, are not uniform or slowly varying, but can change on short timescales, indicating that regimes between which scattering properties change on a small spatial scale are swept past the spacecraft with the solar wind. The small observed solar wind speed of 290 km/s may indicate the existence of a complex magnetic field topology in the vicinity of the flare site and therefore slow injection of electrons, but because of the relatively small azimuthal distance of 21° between the flare and the connecting field line, a value of $t_r = 1.05$ h seems more realistic than the value of 2 h as in the first modeling (see Kallenrode 1993). A slight decrease of $\lambda_{||}$ with energy seems to be typical for electrons with energies between 0.1 and 1 MeV (Dröge 1994). Further exploration of parameter space, and allowing more degrees of freedom, such as individually varying spatial dependencies and injection profiles for each energy channel, would be expected to result in observed fluxes as well as energy spectra of the decay electrons in even better agreement with predictions from the simulation.

8.7.4 Main Results and Discussion

According to Dröge et al. (1996) additional information about the production of neutrons in the 1980 June 21 flare can be gained from the observations of decay protons (Evenson et al. 1985). The bottom panel of Fig. 8.31 shows the proton data from the MEH spectrometer, plotted in terms of the distance traveled,

$s = v(t - t_{flare})$. The protons detected from $s = 1$–4 AU are believed to be mainly decay protons, because of their early arrival time and much harder spectrum. However, the statistical significance of the decay proton detection is marginal for this event, with only 51 proton counts (before the live-time correction) and an uncertain contribution from direct protons. Based on simulations of the transport of neutron decay protons, Dröge et al. (1996) conclude that if all those counts were due to decay protons, the emission toward the horizon would be 2.6×10^{27} neutrons/MeV/sr for $E = 27$ to 75 MeV and 1.1×10^{27} neutrons/MeV/sr for $E = 75$ to 147 MeV. Given the possibility of a flux of quickly arriving direct protons, Dröge et al. (1996) take these to be upper limits of the neutron fluxes. The direct detection of neutrons from this event (Chupp et al. 1982) indicated an integral flux $N(E > 50$ MeV) $\sim 3 \times 10^{28}$ neutrons/sr toward the horizon, and a spectral index of 3–4. The upper limits to differential neutron fluxes that Dröge et al. (1996) derive from proton data are somewhat higher than those implied by the direct detection.

Each of these three observations of solar neutrons (using neutron decay electrons, neutron decay protons, and direct neutrons) imposes constraints on the parent neutron spectrum. Since typical theoretical results (Hua and Lingenfelter 1987a, b) indicate that the neutron spectrum should be nearly energy independent at low energies, with a steepening power law at higher energies, Dröge et al. (1996) have considered a spectrum of the form

$$\frac{dN}{dEd\Omega} = \frac{N_0}{1 + (E/E_0)^{\delta}}. \tag{8.16}$$

According to Dröge et al. (1996), if suppose that $\delta = 4$ at the high end of the range of permissible power-law indices for direct neutrons at high energies (Chupp et al. 1982), and assume isotropic neutron emission, the measurements of total $dN/d\Omega\ (E)$ and $dN/d\Omega\ (E > 50$ MeV) from direct neutrons imply that

$$N_0 = 3.6 \times 10^{30} \text{neutrons.MeV}^{-1}\text{sr}^{-1} \text{ and } E_0 = 7.5 \text{ MeV}. \tag{8.17}$$

This neutron spectrum is shown in Fig. 8.34, and the resulting decay electron spectrum is shown in Fig. 8.32 (solid line).

From Fig. 8.34 is clear that the measured electron intensity, in the lowest three channels, is not affected by the neutron spectrum, confirming that the used fit provides a measure of the total flux of escaping neutrons. Note that these results for 1980 June 21 indicates that a rather steep power law persists down to energy below ~10 MeV.

As noted Dröge et al. (1996), results for the neutron flares of 1982 June 3 and 1984 April 25 have implied power-law indices between 1 and 2 for 27 MeV $< E <$ 147 MeV (Ruffolo 1991), while for the latter flare, Evenson et al. (1990) report that the neutron flux actually declines with decreasing energy below about 30 MeV. A commonly used indicator of the total neutron flux, the fluence of the 2.223 MeV neutron-capture line, is hard to interpret for this flare (Chupp 1982), because of

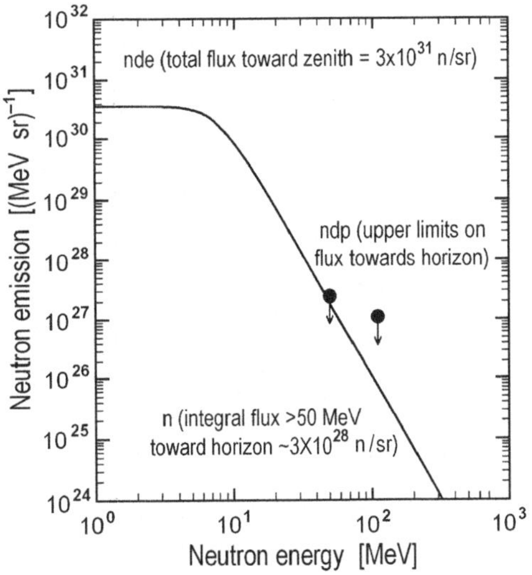

Fig. 8.34 Neutron source spectrum of the 1980 June 21 flare compiled from various observations (see text for details) (From Dröge et al. 1996)

strong limb darkening and uncertainty in the precise location of the flare site. However, the derived above number of escaping neutrons is not unreasonable given estimates for other flares based on this gamma-ray line (Hua and Lingenfelter 1987a, b).

Finally, Dröge et al. (1996) note that decay electrons will usually be unobservable for flares at longitudes east of about 30°E, for which the inner portion of the magnetic field line connected to the detector lies within the 'neutron shadow' (Evenson et al. 1983a, b), i.e., the volume beyond the horizon of the flare site. Since most decay electrons come from low-energy neutrons (~1 MeV) that decay within ~0.1 AU of the Sun, only a relatively few decay electrons are deposited on the portion of the field line that emerges from the neutron shadow. Because of this, Dröge et al. (1996) estimate that the peak flux of neutron decay electrons at ISEE 3 was well below the background level near the Earth for the two other flares for which neutron decay protons have been observed, namely, those of 1982 June 3 (72°E) and 1984 April 25 (43°E).

References for Chapter 8

Aurass H, Mann G, Rausche G, Warmuth A (2006) The GLE on Oct. 28, 2003 – radio diagnostics of relativistic electron and proton injection. Astron Astrophys (France) 457(2):681–692

Bieber JW, Clem J, Evenson P, Pyle R, Ruffolo D, Saiz A (2005) Relativistic solar neutrons and protons on 28 October 2003. Geophys Res Lett 32(3):L03S02. doi:10.1029/2004GL021492, 1-5

Cane HV, von Rosenvinge TT, Cohen CMS, Mewaldt RA (2003) Two components in major solar particle events. Geophys Res Lett 30(12):8017. doi:10.1029/2002GL016580, SEP5-1-4

Cane HV, Cohen CMS, Mewaldt RA, von Rosenvinge TT (2006) Role of flares and shocks in determining solar energetic particle abundances. J Geophys Res 111(A6):A06S90. doi:10.1029/2005JA011071, 1-11

Chiba N, Dion GM, Hayashida N, Honda K, Honda M, Inoue N, Kadota K, Kakimoto F, Kamata K, Kawaguchi S, Kawasumi N, Matsubara Y, Nagano M, Ohoka H, Teshima M, Tsushima I, Yoshida S, Yoshii H, Yoshikoshi T (1992) Possible evidence for $\geq$10 GeV neutrons associated with the solar flare of 4 June 1991", Astroparticie Physics, 1, No. 1, 27–32

Chupp EL (1982) Solar energetic photon transients (50 keV–100 MeV). In: Lingenfelter RE, Hudson HS, Worrall DM (eds) Gamma ray transients and related astrophysical phenomena. Proceedings of the workshop, La Jolla, CA, 1981. AIP Conference Proceedings, vol 77, No 1, pp 363–381

Chupp EL, Forrest DJ, Ryan JM, Heslin J, Reppin C, Pinkau K, Kanbach G, Rieger E, Share GH (1982) A direct observation of solar neutrons following the 0118 UT flare on 1980 June 21. Astrophys J 263(2, Part 2):L95–L99

Chupp EL, Forrest DJ, Share GH, Kanbach G, Debrunner H, Flueckiger E (1983) Solar neutrons from the impulsive flare on 1982 June 3 at 1143 UT. In: Proceedings of the 18th international cosmic ray conference, Bangalore, India, vol 10, pp 334–337

Chupp EL, Debrunner H, Fluckiger E, Forrest DJ, Golliez F, Kanbach G, Vestrand WT, Cooper J, Share G (1987) Solar neutron emissivity during the large flare on 1982 June 3. Astrophys J (USA) 318(2, Part 1):913–925

Cohen CMS, Stone EC, Mewaldt RA, Leske RA, Cummings AC, Mason GM, Desai MI, von Rosenvinge TT, Wicdenbeck ME (2005) Heavy ion abundances and spectra from the large solar energetic particle events of October-November 2003. J Geophys Res 110(A9):A09S16. doi:10.1029/2005JA011004, 1-15

Daibog EI, Stolpovskii VG (1987) Can we detect electrons from the decay of solar neutrons. Sov Astron Lett 13(6):458–461

Debrunner H, Flückiger EO, Chupp EL, Forrest DJ (1983) The solar cosmic ray neutron event on June 3, 1982. In: Proceedings of the 18th international cosmic ray conference, Bangalore, India, vol 4, pp 75–78

Debrunner H, Lockwood JA, Ryan JM, McConnell M, Schonfelder V, Aarts H, Bennett K, Winkler C (1993) Neutrons from the 15 June 1991 solar flare. In: Proceedings of the 23rd international cosmic ray conference, Calgary, vol 3, pp 115–118

Dorman LI, Venkatesan D (1993) Solar cosmic rays. Space Sci Rev 64(3–4):183–362

Dröge W (1994) Transport of solar energetic particles. Astrophys J Suppl Ser 90(2):567–576

Dröge W, Ruffolo D, Klecker B (1996) Observation of electrons from the decay of solar flare neutrons. Astrophys J 464:L87–L90

Dunphy PP, Chupp EL (1992) Gamma-rays and neutrons as a probe of the proton spectrum during the solar flare of 1988 December 16. In: Zank GP, Gaisser TK (eds) Particle acceleration in cosmic plasmas, Newark, DE, 1991. AIP Conference Proceedings, vol 264, No 1, pp 253–256

Dunphy PP, Chupp EL, Rieger E (1990) Analysis of SMM GRS High-Energy (>10 MeV) Data from the Solar Flare of 1988 December 16. In: Proceedings of the 21st international cosmic ray conference, Adelaide, vol 5, pp 75–77

Efimov YuE, Kocharov GE, Kudela K (1983) On the solar neutrons observation on high mountain neutron monitor. In: Proceedings of the 18th international cosmic ray conference, Bangalore, India, vol 10, pp 276–278

Evenson P, Meyer P, Pyle KR (1983a) Protons from the decay of solar flare neutrons. Astrophys J 274(2, Part 1):875–882

Evenson P, Kroeger R, Meyer P, Muller D (1983b) Solar flare neutron fluxes derived from interplanetary charged particle measurements. In: Proceedings of the 18th international cosmic ray conference, Bangalore, India, vol 4, pp 97–100

Evenson PA, Kroeger R, Meyer P (1985) Further observations of protons resulting from the decay of neutrons ejected by solar flares. In: Proceedings of the 19th international cosmic ray conference, La Jolla, CA, vol 4, pp 130–133

Evenson P, Kroeger R, Meyer P, Reames D (1990) Solar neutron decay proton observations in cycle 21. Astrophys J Suppl Ser (USA) 73(2):273–277

Gopalswamy N, Yashiro S, Liu Y, Michalek G, Vourlidas A, Kaiser ML, Howard RA (2005a) Coronal mass ejections and other extreme characteristics of the 2003 October–November solar eruptions. J Geophys Res 110(A9):A09S15. doi:10.1029/2004JA010958, 1-18

Gopalswamy N, Xie H, Yashiro S, Usoskin I (2005b) Coronal mass ejections and ground level enhancements. In: Proceedings of the 29th international cosmic ray conference, Pune, India, vol 1, pp 169–172

Guglenko VG, Efimov YuE, Kocharov GE, Kovaltsov GA, Mandzhavidze NZ, Terekhov MM, Kocharov LG (1990) Neutrons and gamma-ray emission on 1982 June 3: the possibility of fitting the data on the assumption of one population of accelerated particles [solar flares]. Astrophys J Suppl Ser (USA) 73(2):209–211

Hovestadt D, Klecker B, Scholer M, Arbinger H, Hofner H, Kunneth E, Laeverenz P, Gloeckler G, Fan CY, Fisk LA, Ipavich FM, O'Gallagher JJ, Cain J, Tums E (1978) The nuclear and ionic charge distribution particle experiments on ISEE-1 and ISEE-C spacecraft. IEEE T Geosci Elect GE-16(3):166–175

Hua H-M, Lingenfelter RE (1987a) Solar flare neutron and accelerated ion angular distributions. Astrophys J 323(2, Part 1):779–794

Hua H-M, Lingenfelter RE (1987b) Solar flare neutron production and the angular dependence of the capture gamma-ray emission. Sol Phys 107(2):351–383

Jokipii JR (1966) Cosmic-ray propagation. I. Charged particles in a random magnetic field. Astrophys J 146(2, Part 1):480–487

Jokipii JR (1973) Radial variation of magnetic fluctuations and the cosmic-ray diffusion tensor in the solar wind. Astrophys J 182(2, Part 1):585–600

Kallenrode M-B (1993) Neutral lines and azimuthal 'transport' of solar energetic particles. J Geophys Res (USA) 98(A4):5573–5591

Kiener J, Gros M, Tatischeff V, Weidenspointner G (2006) Properties of the energetic particle distributions during the October 28, 2003 solar flare from INTEGRAL/SPI observations. Astron Astrophys (France) 445(2):725–733

Klassen A, Krucker S, Kunow H, Müller-Mellin R, Wimmer-Schweingruber R, Mann G, Posner A (2005) Solar energetic electrons related to the 28 October 2003 flare. J Geophys Res 110(A9): A09S04. doi:10.1029/2004JA010910, 1-6

Koi T, Matsuura N, Muraki Y, Terasawa T, Yanagita S (1993) Prediction of electrons as decay products of solar neutrons. In: Proceedings of the 23rd international cosmic ray conference, Calgary, vol 3, pp 151–154

Kolomeets EV, Sevastjanov VN, Chebakova EA (1988) Estimation of energy spectra of neutrons generated during flares on the Sun by using electrons of their decay. Izvestia Academy of Science of KazSSR, No. 4: 60–66 (in Russian)

Krucker S, Larson DE, Lin RP, Thompson BJ (1999) On the origin of impulsive electron events observed at 1 AU. Astrophys J (USA) 519(2, Part 1):864–875

Kurganov IG, Ostryakov VM (1989) The simulation of the spread of protons from solar-based disintegration of neutrons. Izv AN SSSR Ser Phys 53(2):381–384 (In Russian)

Kuznetsov SN, Kurt VG, Myagkova IN, Yushkov BY, Kudela K (2006a) Gamma-ray emission and neutrons from solar flares recorded by the SONG instrument in 2001–2004. Sol Syst Res (Russia) 40(2):104–110

Kuznetsov SN, Kurt VG, Yushkov BYu, Myagkova IN, Kudela K, Kaššovicová J, Slivka M (2006b) Proton acceleration during 20 January 2005 solar flare: CORONAS-F observations of high-energy γ emission and GLE. Contrib Astron Obs Skalnate Pleso 36(2):85–92

Le GM, Tang YH, Tang YQ (2007a) What did the occurrence of relativistic solar neutrons on 28 October 2003 mean? In: Proceedings of the 30th international cosmic ray conference, Merida (Mexico), Pre-conference edition, http://dpnc.unige.ch/ams/ICRC-07/icrc0397.pdf, 1-4

Le GM, Han YB, Zhang YJ (2007b) A comparative analysis on two solar proton events. Chinese Sci Bull 52(1):47–52

Li C, Tang YH, Dai Y, Zong WG, Fang C (2007) The acceleration characteristics of solar energetic particles in the 2000 July 14 event. Astron Astrophys (France) 461(3):1115–1119

Lingenfelter RE, Ramaty R (1967) High energy nuclear reactions in solar flares. In: Shen BSP (ed) High energy nuclear reactions in astrophysics. Benjamin, New York, pp 99–158

Mewaldt RA, Cohen CMS, Labrador AW, Leske RA, Mason GM, Desai MI, Looper MD, Mazur JE, Selesnick RS, Haggerty DK (2005) Proton, helium, and electron spectra during the large solar particle events of October–November 2003. J Geophys Res 110(A9):A09S18. doi:10.1029/2005JA011038, 1-22

Meyer P, Evenson P (1978) The University of Chicago cosmic ray electrons and nuclei experiment on the H spacecraft. IEEE T Geosci Elect (USA) GE-16(3):180–185

Miroshnichenko LI, Klein K-L, Trottet G, Lantos P, Vashenyuk EV, Balabin YV, Gvozdevsky BB (2005) Relativistic nucleon and electron production in the 2003 October 28 solar event. J Geophys Res 110(A9):A09S08. doi:10.1029/2004JA010936, 1-13

Muraki Y, Murakami K, Miyazaki M, Mitsui K, Shibata S, Sakakibara S, Sakai T, Takahashi T, Yamada T, Yamaguchi K (1992) Observation of solar neutrons associated with the large flare on 1991 June 4. Astrophys J Lett (USA) 400(2, Part 2):L75–L78

Murphy RJ, Dermer CD, Ramaty R (1987) High-energy processes in solar flares. Astrophys J Suppl Ser (USA) 63(3):721–748

Ostryakov VM, Kurganov IG (1990) Protons from neutron decay and properties of interplanetary medium. In: Proceedings of the 21st international cosmic ray conference, Adelaide, vol 5, pp 204–207

Pyle KR, Simpson JA (1991) Observation of a direct solar neutron event on 22 March 1991 with the Haleakala, Hawaii, neutron monitor. In: Proceedings of the 22nd international cosmic ray conference, Dublin, vol 3, pp 53–56

Ramaty R, Murphy RJ (1987) Nuclear processes and accelerated particles in solar flares. Space Sci Rev 45(3–4):213–268

Ramaty R, Mandzhavidze N, Kozlovsky B, Skibo JG (1993) Acceleration in solar flares: interacting particles versus interplanetary particles. Adv Space Res (UK) 13(9):275–284

Reames DV (2002) Magnetic topology of impulsive and gradual solar energetic particle events. Astrophys J Lett (USA) 571(1, Part 2):L63–L66

Reid GC (1964) A diffusive model for the initial phase of a solar proton event. J Geophys Res 69 (13):2659–2667

Rieger E (1996) Spectral evolution of an intense gamma-ray line flare. In: Ramaty R, Mandzhavidze N, Hua X-M (eds) High energy solar physics, Proceedings of the conference held in Greenbelt, MD, 1995. AIP Conference Proceedings, vol 374, No 1, pp 194–199

Roelof EC (1966) Effect of the interplanetary magnetic field on solar neutron-decay protons. J Geophys Res 71(5):1305–1317

Roelof EC (1969) Propagation of solar cosmic rays in the interplanetary magnetic field. In: Ögelman H, Wayland JR (eds) Lectures in high energy astrophysics. NASA SP-199, Washington, DC, pp 111–135

Ruffolo D (1991) Interplanetary transport of decay protons from solar flare neutrons. Astrophys J 382(2, Part 1):688–698

Ruffolo D (1995) Effect of adiabatic deceleration on the focused transport of solar cosmic rays. Astrophys J (USA) 442(2, Part 1):861–874

Ryan J, Bennett K, Debrunner H, Forrest D, Hanlon L, Lockwood J, Loomis M, McConnell M, Morris D, Rank G, Schonfelder V, Swanenburg BN, Webber W, Winkler C (1993) Neutron and gamma ray measurements of the solar flare of 1991 June 9. In: Proceedings of the 23rd international cosmic ray conference, Calgary, vol 3, pp 103–106

Schlüter W (1985) Die numerische Behandlung der Ausbreitung solarer flareinduzierter Teilchen in den magnetischen Feldern des interplanetaren Raumes. Ph.D. Thesis, University of Kiel, Kiel, Germany, 147 pp

Shea MA, Smart DF, Wilson MD (1991a) Possible ground-level measurements of solar neutron decay protons during the 19 October 1989 solar cosmic ray event. Geophys Res Lett 18 (5):829–832

Shea MA, Smart DF, Fluckiger EO (1991b) The evolution of the anisotropy of solar neutron decay protons during the 19 October 1989 solar cosmic ray event. In: Proceedings of the 22nd international cosmic ray conference, Dublin, vol 3, pp 41–44

Shea MA, Smart DF, Pyle KR (1991c) Direct solar neutrons detected by neutron monitors on 24 May 1990. Geophys Res Lett 18(9):1655–1658

Simnett GM (2005a) Near-relativistic electron emission following the 28 October 2003 X17 flare. J Geophys Res 110(A9):A09S01. doi:10.1029/2004JA010789, 1-10

Simnett GM (2005b) Electron acceleration in the corona. Sol Phys (Netherlands) 229(2):213–226

Simnett GM (2006) The timing of relativistic proton acceleration in the 20 January 2005 flare. Astron Astrophys (France) 445(2):715–724

Takahashi K, Sakamoto E, Matsuoka M, Nishi K, Yamada Y, Shimoda S, Shikata T, Wada M, Kusunose M, Yoshimori M, Kondo I (1991) Observation of solar neutrons by Mt. Norikura neutron monitor during a period of solar cycle 22. In: Proceedings of the 22nd international cosmic ray conference, Dublin, vol 3, pp 37–40

Tylka AJ, Cohen CMS, Dietrich WF, Lee MA, Maclennan CG, Mewaldt RA, Ng CK, Reames DV (2005) Shock geometry, seed populations, and the origin of variable elemental composition at high energies in large gradual solar particle events. Astrophys J 625(1, Part 1):474–495

Wanner W, Wibberenz G (1993) A study of the propagation of solar energetic protons in the inner heliosphere. J Geophys Res (USA) 98(A3):3513–3528

Wibberenz G, Kunow H, Muller-Mellin R, Sierks H, Heber B, Kallenrode M-B, Raviart A, Ferrando P, Ducros R (1992) Solar energetic and shock-accelerated particles observed between 1 and 4 AU by the Kiel Electron Telescope (KET) on board Ulysses. Geophys Res Lett 19 (12):1279–1282

Zhang DL (1993) The transport of solar flare neutron-decay protons. In: Proceedings of the 23rd international cosmic ray conference, Calgary, vol 3, pp 143–146

Chapter 9
Gamma Rays from Solar Energetic Particle Interactions with the Sun's Atmosphere

9.1 The Matter of Problem: Why γ-Ray Observations Are Very Important for Understanding the Nature of Solar Neutron Events, What Information May Be Obtained from Solar γ-Ray Spectroscopy

Above, in Chapters 1, 2, 3, 4, 6, and 7, we considered mechanisms and a lot of cases of gamma ray observations in connection with solar neutron events research. It was shown that the time profile of gamma ray lines generated in nuclear interactions of accelerated ions with the background plasma of the solar atmosphere directly reflects the time profile of solar neutron generation. Let us note that using these data we may determine not only the source function of solar neutron production in dependence of neutron energy, but also the source function for accelerated ions, as well as properties of the SEP generation and propagation site and surrounding environment.

As noted in Murphy (2007), the composition of the various regions of the solar atmosphere has been studied using a variety of techniques (see, e.g., Anders and Grevesse 1989) using data from X-ray and optical spectroscopy, meteorites, and solar energetic particles (SEP). These studies have revealed considerable abundance variations; for example, based on SEP and other data it has been shown (e.g., Meyer 1985a) that, compared to photospheric abundances, the coronal abundances of elements with low first ionization potential (FIP), such as Mg, Si and Fe, are enhanced relative to those with high FIP, such as C, N and O.

As Murphy (2007) underlined, gamma-ray line spectroscopy offers another technique for determining abundances and has some significant advantages: (1) the cross sections for production of the strongest gamma-ray lines are well-measured with relative uncertainties as low as 10%; (2) because the interactions involve the nucleus only, they are not dependent on the temperature and density of the environment which affect the ionization states of the ambient ions; and (3) since the gamma rays are very penetrating, consideration of radiation transfer is generally not required. Abundances determined with gamma-ray spectroscopy are useful in

L. Dorman, *Solar Neutrons and Related Phenomena*,
Astrophysics and Space Science Library 365, DOI 10.1007/978-90-481-3737-4_9,

several ways. They can provide information about species whose abundances cannot be directly determined by other means, such as Ne (which does not produce atomic lines in the lower solar atmosphere) and photospheric ^{3}He. By determining abundances in the chromosphere where the gamma rays are produced, gamma-ray spectroscopy provides an important constraint on understanding the FIP composition bias of the corona relative to the photosphere. Abundances of the accelerated ions producing the gamma-ray lines can be compared with directly-measured abundances of SEPs to determine the relationship of the two populations: ejected into interplanetary space and interacted in the solar atmosphere.

As noted Murphy (2007), ion acceleration probably occurs in the corona where low energy losses at coronal densities allow efficient acceleration, but the ion interactions generally occur at higher densities, probably similar to those of the chromosphere or upper photosphere, where interaction rates are greater (see, e.g., Hua et al. 1989). All of the nuclear interaction products subsequently produce observable gamma-ray emission via secondary processes, and the neutrons that escape may be observed directly in space and on Earth and indirectly in space via their decay protons. The composition of the ambient medium where the accelerated ions interact can be derived by gamma-ray spectroscopy by comparing fluences of nuclear deexcitation lines, mostly occurring in the range from $\sim$1 to 8 MeV. When accelerated ions interact with ambient elements, nuclei are excited and then de-excite on very short time scales (typically less than 10^{-6} s). As an excited nucleus de-excites through its allowed levels, the transition energies appear as gamma rays, resulting in gamma-ray lines whose central energies are unique to the element.

Murphy (2007) underlined that when accelerated proton and alpha particles react with ambient He and heavier nuclei, the interactions are referred to as 'direct' reactions. Lines produced by direct reactions are 'narrow' with fractional Full Width at Half Maximum (FWHM) $\sim$2%, due to the relatively low recoil velocity of the heavy nucleus. 'Inverse' reactions, when accelerated ions heavier than He interact with ambient H and He, produce 'broad' lines with fractional FWHM $\sim$20% due to the relatively high recoil velocity of the nucleus. Lines produced by heavy–heavy interactions are not significant because of the low abundances. The yield of a deexcitation line from a direct reaction of an accelerated proton or alpha particle with an ambient element is linearly proportional to the abundance of that ambient element. Comparison of the various narrow line fluences therefore provides information on the relative abundances of the medium where the interactions take place. This is complicated by the fact that a given excited nucleus can be produced not only via inelastic interactions with that nucleus but additionally via spallation reactions with heavier nuclei. The elements that produce deexcitation lines of sufficient strength to allow reliable abundance determinations (and their corresponding strongest line energies) are He ($\sim$0.45 MeV), C (4.439 MeV), N (2.313 MeV), O (6.129 MeV), Ne (1.634 MeV), Mg (1.369 MeV), Si (1.789 MeV) and Fe (0.847 MeV).

Murphy (2007) emphasize again that gamma-ray deexcitation lines are typically produced in the chromosphere or upper photosphere and so it is this region whose composition is revealed by gamma-ray spectroscopy. The photospheric ^{3}He/H abundance ratio can be determined from the neutron-capture line. Neutrons initially produced moving downward either decay, react with ^{3}He, or are captured on H. Capture on other nuclei is less important due to their smaller relative abundances. The capture on H results in the formation of deuterium with the binding energy appearing as a 2.223 MeV neutron-capture line photon. The reaction with ^{3}He is charge exchange, ^{3}He(n,p)^{3}H, without the emission of radiation. Since the probability for elastic scattering is much larger than the probability for either of these reactions, most of the neutrons thermalize first, causing a delay in the formation of the capture photon relative to neutron formation. The cross sections for the H and ^{3}He reactions are $2.2 \times 10^{-30}\beta^{-1}$ cm^2 and $3.7 \times 10^{-26}\beta^{-1}$ cm^2, respectively, where $\beta = v_n/c$ is the neutron velocity in units of the speed of light. Therefore, if the ^{3}He/H ratio at the capture site is $\sim 5 \times 10^{-5}$ (which is comparable to that observed in the solar wind), nearly equal numbers of neutrons will be captured on H as react with ^{3}He. The ambient ^{3}He/H abundance ratio therefore affects the total yield of the capture line and also the delay of its formation. Because effective capture on H or reaction with ^{3}He requires high density, these reactions occur deep in the photosphere. Study of the 2.223 MeV line provides information on photospheric ^{3}He/H abundance ratio.

As noted Troitskaia et al. (2007), the neutron capture line of 2.223 MeV from solar flares has been analyzed in a number of works, starting with the studies by Lingenfelter (1969) and Wang and Ramaty (1974). The fluences and time profiles of this line have been used directly or combined with additional data to obtain the characteristics of particle acceleration in solar flares and properties of the surrounding solar atmosphere. In particular, Ramaty and Murphy (1987) used the fluences of the 2.223 MeV γ-ray line for determining the energy spectra of solar energetic particles (SEP), Hua and Lingenfelter (1987a, b) studied their angular distributions and production of secondary neutrons, and Yoshimori et al. (1999b) investigated photospheric ^{3}He abundance. More recently, Kuzhevskij et al. (1998, 2001a, b, 2002, 2005), Kuzhevskij and Troitskaia (2001), Gan et al. (2003), Troitskaia et al. (2003) have developed a new independent approach (SINP code) to determine by the time profile data of the 2.223 MeV γ-ray line the most probable profile of plasma density in the solar photosphere and adjoining levels during the period of a flare. By using 2.223 MeV γ-ray line data on two large flares, 6 November 1997 and 22 March 1991, some evidence of the effect of density enhancement under the flare site was obtained (see description of these results in Section 9.2). Gan (1998, 2000a, 2004) applied the time profiles of calculated partial fluences of the neutron capture line 2.223 MeV and of the annihilation line 0.511 MeV to deduce the spectral evolution of accelerated charged particles. Similar studies have been recently carried out for the flare of 23 July 2002 (Gan 2004). Murphy et al. (2003a, b) studied the same flare by comparing the measured time profile of the 2.223 MeV line flux with that predicted using a magnetic loop model. In most of these studies the data on deexcitation lines in the energy range of 4–7 MeV were often necessary for a comprehensive analysis.

All mentioned problems and aspects of solar gamma-ray spectroscopy will be considered in details in the frame of many event observations in Sections 9.2–9.22.

9.2 Absorption of 2.223 MeV Solar Flare Gamma-Rays and Determination of the Solar Plasma Density Altitude Profile

9.2.1 The Matter and Short History of Problem; Density Altitude Profile Models

Troitskaia and Kuzhevskij (1999) supplement their previous model calculations of the 2.223 MeV solar flare gamma-rays time profiles by means of the consideration non-central flare and the calculations of the 2.223 MeV gamma-line τ-constant time profiles. Then they apply the proposed in their previous works method of the solar plasma density altitude profile definition, using the 2.223 MeV time profile, to the gamma-ray experimental data for the March 22, 1991, 22:42:51 UT solar flare. It is shown that, on the assumptions used, the most probable models of the altitude profile of the solar plasma during flare periods are the models with higher photospheric concentrations compared with the quiet Sun model.

Troitskaia and Kuzhevskij (1999) noted that in their previous works were calculated the propagation of solar flare neutrons (Kuzhevskij and Troitskaia 1989), production of the solar flare 2.233 MeV γ-line (Kuzhevskij et al. 1998), and its absorption in the solar atmosphere (Kuzhevskij and Troitskaia 1998) and proposed the method for determining the solar plasma altitude profile during flare periods. The calculations were made for an instantaneous neutron source with energies 1–100 MeV and the power-low spectra within the same energy interval, using Monte-Carlo simulation. They made allowance for (i) neutron deceleration in elastic collisions between neutrons and hydrogen nuclei with due account of the energy and angular dependencies of *np*-scattering cross-sections, (ii) possible energetic neutron escape from the Sun, (iii) gravitational neutron-Sun interaction, (iv) thermal motion of decelerated neutrons, (v) neutron decay, (vi) neutron captures by hydrogen with deuterium and γ-quantum production, (vii) γ-ray absorption in the solar atmosphere in the case of central flares. In Troitskaia and Kuzhevskij (1999) it was applied the previously proposed method for a γ-ray solar event. For this purpose the model calculations of the absorption were made in the case of a non-central flare and, then, the τ-constant time profiles were calculated.

As noted Troitskaia et al. (2007), two important assumptions have been applied in the calculations: (1) The relative abundance of ^{3}He/^{1}H is taken to be about 2×10^{-5} (e.g., Hua and Lingenfelter 1987b; Yoshimori et al. 1999b) and (2) the time history of the initial neutron production is taken to be similar to that of the total fluence of ^{12}C + ^{16}O nuclear deexcitation lines in the energy range of 4.1–6.4 MeV. For the first time, the latter assumption seems to be empirically substantiated and applied by Prince et al. (1983a, b). Theoretical treatment of this assumption has been accomplished by Kuzhevskij and Kogan-Laskina (1990). Calculations are made with the SINP code for neutrons with energies of 1–100 MeV, which are the most important ones for the 2.223 MeV γ-ray line productions. Primary neutrons are assumed to be emitted isotropically in the lower half-space (toward the Sun) from the levels with densities less than 5×10^{15} cm^{-3}. As a basic density model ($m = 1$) was used a combination of the standard astrophysical model HSRA (Harvard-Smithsonian Reference Atmosphere) for the lower chromosphere and quiet photosphere (Gingerich et al. 1971) with the model for the convection zone (Spruit 1974). To determine possible deviations of the model, which may be possible for a given observable flare, from the basic density model ($m = 1$), it was also composed four additional models ($m = 2, 3, 4, 5$) representing smaller and larger densities at photospheric and adjoining levels as compared with the standard model ($m = 1$) of the quiet Sun. Five models of the altitude dependencies of solar plasma density in the lower chromosphere, photosphere, and convection zone are presented in Fig. 9.1 in comparison with the basic density model ($m = 1$, curve 1) and modified models (curves 2–5); H is a depth in the photosphere starting from a density level of 10^{12} cm^{-3}.

As underlined Troitskaia et al. (2007), a number of studies show that the density of solar plasma during a flare is not described by the standard astrophysical model of the quiet Sun (e.g., Machado et al. 1980; Gan and Fang 1987). In particular, Gan et al. (1993) proposed the existence of chromospheric condensations by constructing a series of semi-empirical models. There are also some theoretical models analyzing the response of the solar atmosphere to the flare initial energy release. In these models, the flare energy release is postulated to be followed by mass movements in the form of one or a few relatively cold dense condensations. Such structures are propagating in front of a shock wave moving downward from the region of energy release (e.g., Somov and Spektor 1982; Baranovskij 1989; Gan et al. 1991; Boiko and Livshits 1995; Somov and Bogachev 2003). This picture is confirmed, for example, by the data for the white-light flares of 4 June 1980 and 15 June 1991 (Babin and Koval 1999). In the latter case, a radial velocity for the downward motion of radiating material of 20–30 km s^{-1} was obtained for a chromospheric condensation formed during the sudden heating of the upper chromosphere by the heat flux from the corona. It is likely that the presence of

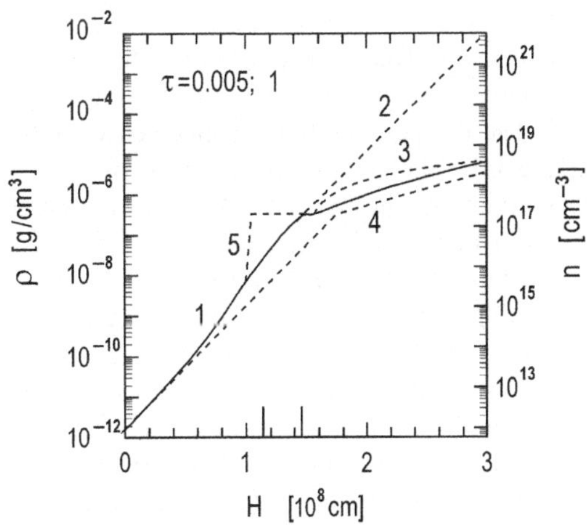

Fig. 9.1 The solid curve (1) is the basic density model, the dashed curves (2–4) are deformed models. Only fragments, differing from the main curve, are shown (From Troitskaia and Kuzhevskij 1999)

downward-moving radiating material is a common feature of powerful flares at their impulsive phases (Babin and Koval 1999), although few data are available on the character of plasma density and its variations in deep photospheric layers.

The time profile of the neutron capture line of 2.223 MeV for the flare of 16 December 1988 (its third peak) was analyzed by Troitskaia et al. (2007). The enhancements of plasma density in the deep photospheric layers of the Sun under the flare region (an effect of density enhancement) have been deduced by the SINP calculation code. These enhancements are studied based on several modified models of the lower chromosphere, photosphere, and convection zone. Additionally, the case has been examined when the neutron source is placed at the top of the photosphere. The energy spectrum of accelerated particles (protons) is shown to evolve with time. Assuming a stochastic mechanism of acceleration and a Bessel function for the spectrum, Troitskaia et al. (2007) found the spectral index αT to increase from 0.005 to 0.1 during the decay phase of the burst; i.e., the proton spectrum becomes harder. Density enhancements found in the flare of 16 December 1988 are consistent with results of Kuzhevskij and Troitskaia (1989, 1998), Kuzhevskij et al. (1998, 2001a, b), Troitskaia and Kuzhevskij (1999) on the density enhancement effects deduced for two other flares, 6 November 1997 and 22 March 1991. It is suggested that density enhancements in the deep layers of the photosphere may be rather common features of powerful solar flares.

9.2.2 Experimental Data on the Solar Flare of March 22, 1991

According to Troitskaia and Kuzhevskij (1999), the solar flare was recorded on March 22, 1991, 22:42:51 UT by the PHEBUS instrument on the GRANAT observatory-satellite (Terekhov et al. 1996). The flare was associated with the active region of coordinates 26°S, 28°E. Optical and X-classes of the flare are 3B and X9.4, respectively. Emission of γ-rays in the line 2.223 MeV was recorded with duration $\approx$550 s, which comparable with calculated γ-line duration in the case of instantaneous source. The data on radio-emission reveal the impulsive character of the flare. The radio emission on 15.4 GHz began at 22:43 UT and ended at 22:46 UT. The II type radio burst, connected with the shock wave, lasted from 22:47 till 22:55 UT (Sladkova et al. M1998). Besides that, in this event the photons with maximum energies 65–124 MeV was recorded (Terekhov 1995). The duration of burst in all energetic intervals from 1–4 MeV till 65–124 MeV was $\approx$70 s with the main pulse duration of 25 s (from 75th till 100th s from the flare onset). The count rate in this interval was 8–14 times as greater as the count rate outside this 25-min interval. All these circumstances enable the application of the model calculations with instantaneous neutron source to the solar flare 2.223 MeV γ-burst under consideration. It is necessary to note, that the averaging of experimental data for τ-constant calculations (Terekhov 1995), is made at 40–50-s intervals, and the duration of the respective theoretical curves averaging is 40 s.

9.2.3 Model Calculations

In Troitskaia and Kuzhevskij (1999) the absorption of γ-rays for the case of non-central flare is taken into account by analogy with the case of central flare (Kuzhevskij and Troitskaia 1998). At 2.223 MeV the main mechanism of γ-rays reducing is Compton scattering by electrons. The mass coefficient of reducing is $\mu/\rho = 0.08329\ \text{cm}^2\ \text{g}^{-1}$ (Nemets and Goffman M1975). The isotropy of γ-ray production is assumed. All the depth of the solar atmosphere is subdivided on the thin spherical layers, thin enough to assume its density to be homogeneous. The thickness of the passed way of γ-ray emission within one layer l is depended on the heliocentric angle α of the flare:

$$l = -(r-d)\cos\alpha + \sqrt{(r-d)^2\cos^2\alpha - d^2 + 2rd}, \tag{9.1}$$

where r is the radius of the spherical layer, d – the thickness of the layer. The absorption, due to propagation through the layer is calculated, using the statistical weights technique. Following the paper of Kuzhevskij et al. (1998), it was made calculations of the ejected γ-ray profiles for five models of solar plasma density (see Fig. 9.1): the basic astrophysical model (1) and four deformed models for the calculations. The primary neutrons were assumed to be emitted normally downwards with the power-law primary neutron spectrum

$$D_{Sun}(E_n) \propto E_n^{-S}, \tag{9.2}$$

where $S = 0, 1, 2, 3$. The calculations were made for the neutron issue, placed at the central solar angle 37.5°, correspondingly to the studied flare. The resulting time profiles at $S = 0$ and 2 of ejected 2.223 MeV γ-ray emission are shown in Fig. 9.2.

It can be seen that the profiles of γ-emission have the character, close to the profiles in the case of a central flare, which were examined in (Kuzhevskij and Troitskaia 1998), and the differences between them was discussed in connection with the possibility of determining of the solar plasma density model in many cases, using the present-day instruments. Then Troitskaia and Kuzhevskij (1999) suggest to use for experimental data analysis the time history of the τ-constant, calculated from obtained above 2.223 MeV γ-ray time profiles. The calculated τ-constant time profiles are shown in Figs. 9.3 and 9.4.

9.2.4 Comparison with Experimental Data on the Event of March 22, 1991

Troitskaia and Kuzhevskij (1999) used the data, given in Terekhov (1995) on the τ-constant, deduced them at five points of the 2.223 MeV γ-ray line time profile. The moment of neutron injection is assumed to coincide with the moment of the

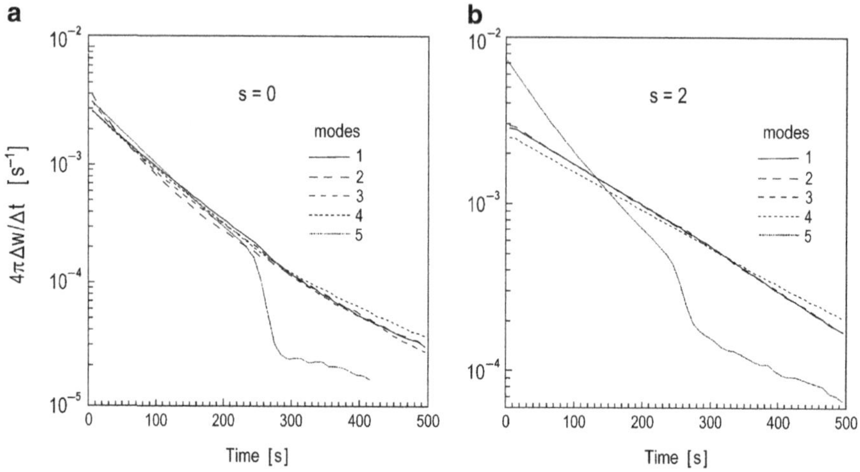

Fig. 9.2 Time profile models of 2.223 MeV γ-ray line (From Troitskaia and Kuzhevskij 1999)

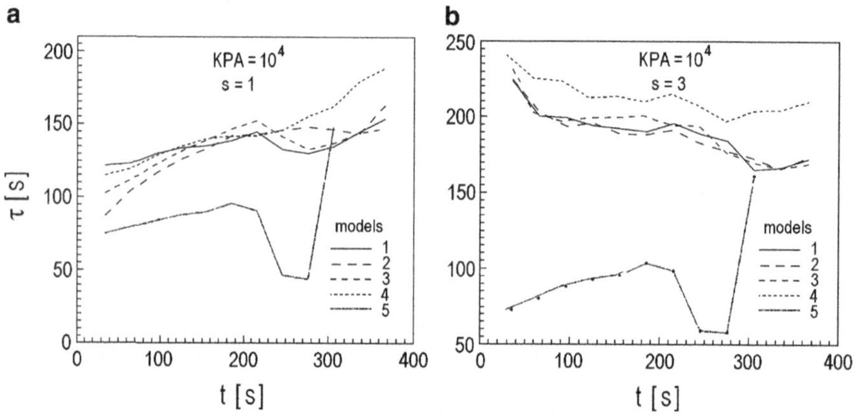

Fig. 9.3 Models of τ-constant time profiles (From Troitskaia and Kuzhevskij 1999)

channel 4–6 MeV γ-emission peak injection. The comparison with that of Troitskaia and Kuzhevskij (1999) calculated τ-constant time profiles, which are the closest to the experimental data, presented at the Fig. 9.4. The application to the data the least squares method reveals the best fitted curve. It corresponds to the case **s = 0, model 2**. The second 'candidate' is the case **s = 0, model 3**.

9.2.5 Application to the Solar Flare of the 6th of November 1997

Kuzhevskij et al. (2001a, b) have made new calculations of solar flare neutron propagation in the solar atmosphere, production of the 2.223 MeV neutron capture

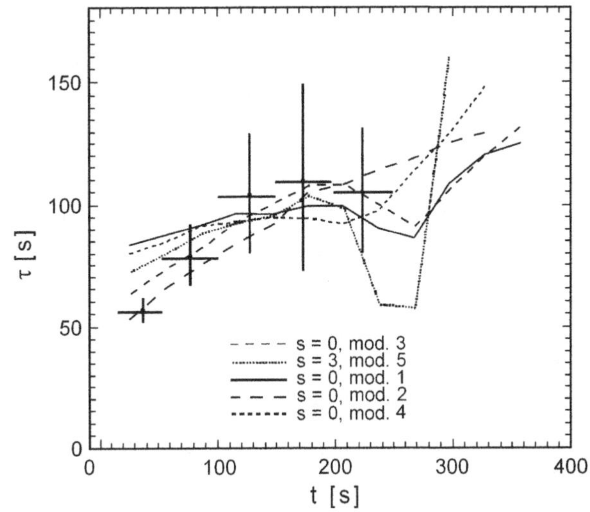

Fig. 9.4 The fitting of experimental data on event March 22, 1991 by the calculated τ-constant time profiles (From Troitskaia and Kuzhevskij 1999)

γ-ray line and its output from the Sun in application to observations of the 2.223 MeV γ-ray line temporal profile by Yohkoh satellite during the 6 November 1997 flare. Isotropic angular distribution and temporal characteristics of initial neutron flux as well as non-radiative absorption of neutrons by ^{3}He have been included in Monte-Carlo simulation. All considerations have been made both for a quiet solar atmosphere density model and for several distortions of the model which were used for the flare March 22, 1991 described in previous Sections 9.2.2–9.2.4 (see Fig. 9.1). The distortions or large-scale density fluctuations may take place at different manifestations of solar activity (including a flare itself). In Sections 9.2.2–9.2.4 was described the method for determining the solar plasma altitude profile during flare periods by analyzing the observed intensity-time profiles of the 2.223 MeV γ-ray line flux. The described techniques allows to probe experimentally the deep layers of the Sun (at photospheric and sub-photospheric levels), inaccessible to other methods. Investigations in this field are greatly important to understand primary mechanisms of solar energy release and predict the processes leading to energetic solar phenomena.

As noted Kuzhevskij et al. (2001a, b), strong γ-ray emission was recorded on 6 November 1997 at 11:52 UT by the gamma-ray spectrometer (two BGO scintillators) on the Yohkoh satellite (Yoshimori et al. 1999b). This emission was associated with the solar flare of coordinates 18°S, 64°W. Optical and X-classes of the flare are 2B and X9.4, respectively. Yoshimori et al. (1999b) derived the γ-ray fluxes of 2.223 MeV and prompt C+O γ-ray lines from the records of 2.136–2.375 and 4.001–7.225 MeV energy channels. For this goal the data were cleared of bremsstrahlung continuum and superposed narrow and broad γ-ray lines, and then the instrumental response function was applied. Those refined data were

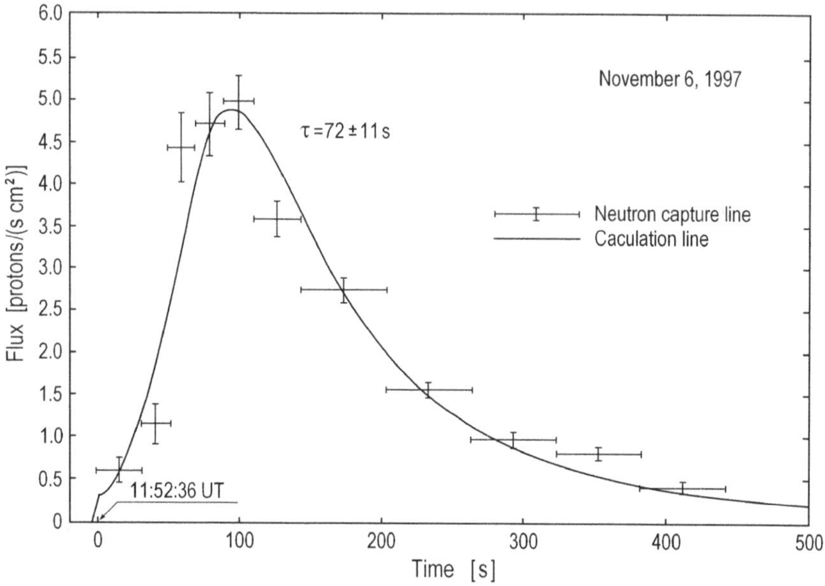

Fig. 9.5 Temporal variations of γ-ray neutron capture line in event 6 November 1997 (From Yoshimori et al. 1999b)

taken by Kuzhevskij et al. (2001a, b) for their analysis from Fig. 3 of Yoshimori et al. (1999b) – see Fig. 9.5.

The time profile of neutron production in Kuzhevskij et al. (2001a, b) was assumed to be proportional to that of C+O γ-ray line emission. The temporal profile of initially ejected neutrons was composed for calculations as a set of instantaneous neutron ejections. Total time interval of 0–440 s was divided into 23 separate intervals and instantaneous source was placed at the middle of every interval with neutron initial statistical weight equal to the ratio of area under the curve in every interval to the total area under the curve. The resulting 2.223 MeV γ-ray time profiles were smoothed then. The basic algorithm of calculations was the same as described in Sections 9.2.2–9.2.4. Figure 9.6 shows calculated time profiles in the case of $S = 1$ in Eq. 9.2 for solar neutron energy spectrum.

It can be seen that even for the 65° flare in some cases the time profiles considerably differ one from another. It was calculated characteristic decay times (τ-constants) of each curve and compared them with the decay constant of observed profile for the 2.223 MeV line (Fig. 9.7).

Observed curve obtained by Yoshimori et al. (1999b) contains in all 11 averaged experimental data points of 2.223 MeV γ-ray flux, and seven of them (starting from the curve maximum) are at the decay stage of the profile. From those data Kuzhevskij et al. (2001a, b) deduced six values of τ-constant in every interval between the points and placed them at the middle of every interval. Application of the least-squares method to the decay stage of gamma-ray profile reveals the best fit to the model curve in the case $s = 1$, model 5.

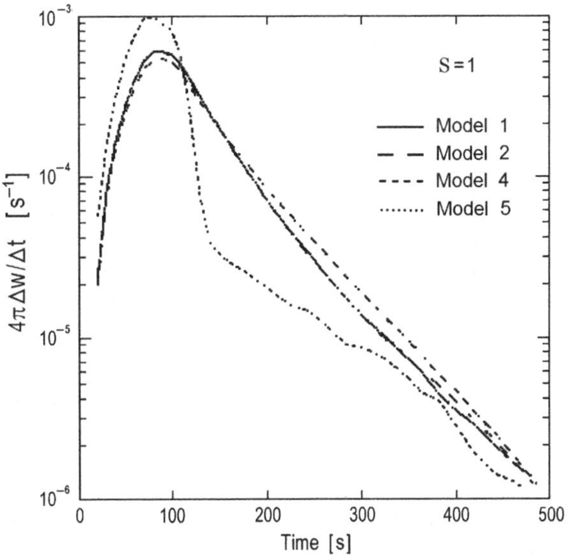

Fig. 9.6 The results of modeling the 2.223 MeV γ-ray time profiles for the flare of 6 November 1997 in the case of S = 1 in Eq. 9.2 (From Kuzhevskij et al. 2001a, b)

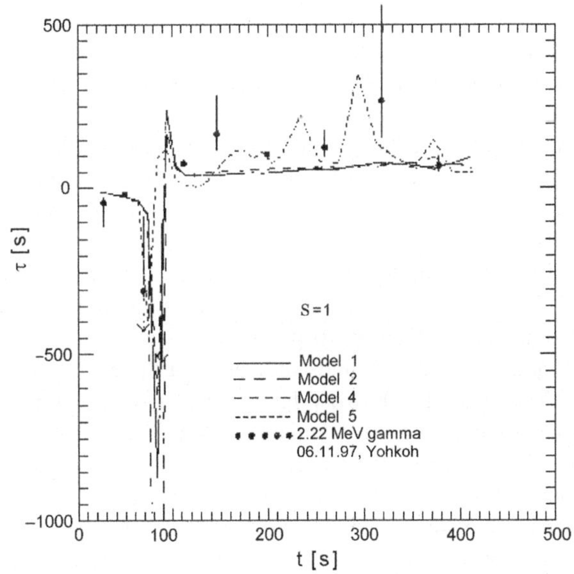

Fig. 9.7 Decay constants of time profiles for the observed (points) and simulated 2.223 MeV gamma-ray line fluxes. The best fit corresponds to the model 5 in Fig. 9.1 at spectral index S = 1 in Eq. 9.2 (From Kuzhevskij et al. 2001a, b)

As noted Kuzhevskij and Troitskaia (2001), for the first time proposed method has been applied by to the 22 March 1991 solar flare recorded at 22:42:51 UT by the PHEBUS instrument on the GRANAT Observatory (Terekhov 1995; Terekhov et al. 1996). In both cases of gamma-ray flares (22 March 1991 and 6 November 1997), the procedure of the method application enables to reveal the model of solar plasma density and spectral index of initial neutrons. It allows hoping that the further development of the method will be useful and may lead to new results.

According to opinion of Kuzhevskij and Troitskaia (2001), in the way of development the method, one can use the calculated spectra of initial neutrons (Ramaty et al. 1975; Ramaty and Murphy 1987; Hua and Lingenfelter 1987b). Notice, however, that those spectra have more complicated form than simple power-law. In particular, they are rather flat (hard) in the energy range up to 100 MeV. Indeed, the value $S = 1$ of spectral index determined in the current work indicates that the primary spectrum of neutrons with the energies of 1–100 MeV may be not very steep. In both flares, the method leads to conclusion about higher density of the solar plasma, in comparison with the standard model, in the photosphere in the periods of solar flares. If this conclusion will be confirmed it may be interesting for the modern theory of solar flares. The difficulties of fitting arise from too large averaging intervals in the experimental data after preliminary and necessary 'cleaning' of them. Slight differences between the 2.223 MeV line profiles may be discovered if the averaging intervals will be of order or less than 30 s. It gives rise to the requirement of increasing an effective area of recording instruments.

After discussion of obtained results, Troitskaia and Kuzhevskij (1999), Kuzhevskij and Troitskaia (2001) came to following conclusions.

- It is the first experience of application the method of determining the solar plasma density model by means of the analyzing a form of 2.223 MeV γ-line time profile.
- This method provides the unique possibility of investigation the deep photospherical and subphotospherical layers, because the noticeable fluxes of 2.223 MeV γ-emission can exit outside from the 15–18-grammage depth levels (Kuzhevskij and Troitskaia 1998).
- From analysis above follows that for considered flares of 22 March 1991 and 6 November 1997 under made assumptions the most probable are the models with the enhanced density in deep photosphere.
- The value 0 and 1 of spectral index indicate that the energy spectrum of solar neutrons with energies 1–100 MeV have a flat or more complicate form; this result is in accordance with the calculated spectra (Ramaty et al. 1975; Ramaty and Murphy 1987; Hua and Lingenfelter 1987a, b), which have more complicate form than power-low ones, they are more hard in the region till 100 MeV.
- For more accuracy of the method in further its development it is necessary to take into consideration some more circumstances: the altitude and angular distributions of solar energetic neutrons, the time profile of their injection, the presence of ^{3}He as an addition loss of neutrons, and the temporal variations of density model.

9.2.6 Application to the Solar Flare of the 16th of December 1988

Differing from analysis of the 6 November 1997 and 22 March 1991 flares (Kuzhevskij and Troitskaia 1998, 2001; Troitskaia and Kuzhevskij 1999; Kuzhevskij et al. 1998, 2001a, b, 2002), described in previous Sections 9.2.2–9.2.5, in the case of the flare of 16 December 1988 Troitskaia et al. (2007) have considerably improved the calculation technique in the frame of results obtained in Kuzhevskij et al. (2005). First, they implement the neutron spectra of Hua and Lingenfelter (1987a, b) in their simulation code. Then, the 2.223 MeV line time profiles are calculated for three different spectral parameters of accelerated particles and for five density models (see above, Fig. 9.1). The calculations are carried out for the entire duration of the third gamma-ray burst as well as for three time intervals during the burst, namely, for its rising phase and two intervals at the decay phase. In Troitskaia et al. (2007) study it was used some new neutron spectra calculated by Hua et al. (2002) in the framework of new neutron production kinematics. To verify the influence of the neutron source position on the time profiles of the 2.223 MeV line Troitskaia et al. (2007) have examined, additionally, the case when the source is placed at the vertical grammage level of 0.317 gcm^{-2}. In terms of the basic density model (BDM, m $=$ 1 in Fig. 9.1), this corresponds to the top of the Sun's photosphere.

According to Troitskaia et al. (2003, 2007), the flare of 16 December 1988 had a heliocentric angle of 43°, a position of 27°N, 33°E, and an importance of X4.7/2B. It has been recorded by the Gamma-Ray Spectrometer (GRS) onboard the Solar Maximum Mission – SMM (Vestrand et al. 1999) at high-energy X-rays (114–200 keV) and at selected gamma-ray energy emission bands including 2.223, 4.1–6.7, and >35 MeV. A clear response to solar flare neutrons between 25 and 1,000 MeV is also present in the data (Dunphy et al. 1990; Dunphy and Chupp 1992).

Troitskaia et al. (2007) noted that the γ-ray line emission began at 08:28:50 UT and comprised four distinct peaks (Rieger 1996), with a total duration of 3,555 s. The third (largest) peak was observed between 08:54:46 and 09:01:03 UT. The parent particle spectrum must have been extremely hard within this period. This, however, only pertains to the lower energies, because otherwise the pion decay flux above 35 MeV would have been tremendously high, which contrasts to the measurements of SMM/GRS (Dunphy et al. 1990; Rieger 1996). A characteristic parameter of the particle energy spectrum was obtained in Rieger (1996) by inserting the fluence ratios φ(2.223 MeV)/φ(4–7 MeV) into fluences ratio calculations (Ramaty et al. 1993); the latter was carried out by assuming a thick target situation, for two different elemental compositions and two different directionalities of the primary particles. It was found by Rieger (1996) that the energy spectrum for the third peak is compatible with a Bessel function distribution, independent of the composition. The observed ratio φ(2.223 MeV)/φ(4–7 MeV) for the flare under consideration was found to be consistent with a Bessel function proton spectrum with a shape parameter $\alpha T = 0.054 \pm 0.004$ (where α and T are parameter and average time of acceleration, see details in Chapter 4 of Dorman

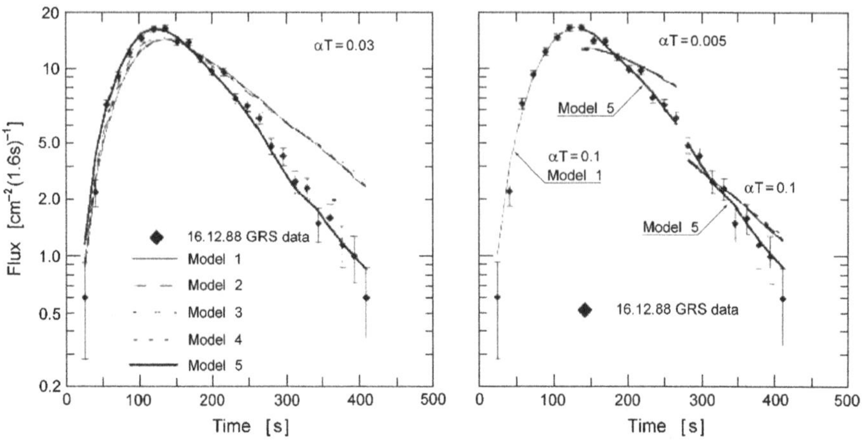

Fig. 9.8 Observed 2.223 MeV γ-ray line fluencies of the 16 December 1988 flare (diamonds) and the best model approximations for the entire time profile (*left* panel, $\alpha T = 0.03$, model 5), for rising time (*right* panel, $\alpha T = 0.1$), and for two segments of the decay phase (*right* panel, $\alpha T = 0.005$ and $\alpha T = 0.1$, respectively). The neutron source is assumed to be at the level with plasma density ($<5\times10^{15}$ cm^{-3}); zero time corresponds to 1,704 s after 08:26:23 UT (From Troitskaia et al. 2007)

M2006), and the number of protons above 30 MeV equal to $(9.0 \pm 0.9) \times 10^{32}$ (Dunphy and Chupp 1992). However, the number of neutrons detected from this flare was much smaller than what was predicted from an isotropic distribution of the protons, indicating that the distribution may be nonisotropic.

Figure 9.8 (left panel) presents the best approximation by Troitskaia et al. (2007) of the observed time profile of the 2.223 MeV line fluxes for the third peak. Within the limits of experimental error bars, it corresponds to the case of $\alpha T = 0.03$ and model 5 (m = 5) with the density enhancement along the entire thickness of the photosphere (see Fig. 9.1). Approximations for the other four models are also plotted.

To study the possible evolution of the αT index during the accelerating phase of the flare Troitskaia et al. (2007) made the same calculations for three segments of the time history. 'Absolute' best approximations selected for every two decay segments are presented in the right panel of Fig. 9.8. They correspond to the shape parameters $\alpha T = 0.005$ and $\alpha T = 0.1$, respectively, and to model 5 in both segments. The neutron source is assumed to be at the level with plasma density ($<5 \times 10^{15}$ cm^{-3}); zero time corresponds to 1,704 s after 08:26:23 UT. The best approximations for all cases are summarized in Table 9.1. At the rising phase (24–136 s in Table 9.1), the best least square sums do not ensure a choice for a definite density model (m) with a matching αT. Nevertheless, one may conclude that at the rising phase the density model is not the fifth one, and the values of αT from 0.03 to 0.1 are more plausible than 0.005. Therefore, Troitskaia et al. (2007) suppose that at the initial phase of the third gamma-ray burst there was no significant density enhancement in the thickness of the photosphere, but it appeared about 140 s after the onset of the 2.223 MeV γ-burst. As already mentioned, similar

Table 9.1 Least square deviation values of model time profiles from observational data for different time segments of the 16 December 1988 flare (with the neutron source at the level with plasma density $< 5\times10^{15}$ cm^{-3}) (From Troitskaia et al. 2007)

αT	Model	All time	24–136 s	136–264 s	280–408 s
0.005	1	1.98E+02	3.94E+00	3.81E+01	3.16E+00
	2	1.87E+02	2.21E+00	3.71E+01	3.28E+00
	3	2.01E+02	2.07E+00	4.21E+01	3.19E+00
	4	2.12E+02	3.64E+00	3.92E+01	3.68E+00
	5	8.79E+00	2.98E+00	**3.32E+00**	1.14E+00
0.03	1	6.85E+01	1.56E+00	1.47E+01	1.85E+00
	2	5.26E+01	1.63E+00	1.04E+01	2.12E+00
	3	6.40E+01	**1.46E+00**	1.41E+01	2.03E+00
	4	6.95E+01	1.55E+00	1.35E+01	2.33E+00
	5	**8.54E+00**	3.09E+00	3.75E+00	3.98E−01
0.1	1	5.23E+01	**1.44E+00**	1.16E+01	1.63E+00
	2	3.76E+01	1.96E+00	7.37E+00	1.93E+00
	3	4.75E+01	1.62E+00	1.06E+01	1.82E+00
	4	5.23E+01	**1.45E+00**	1.03E+01	2.08E+00
	5	8.97E+00	3.08E+00	3.93E+00	**3.39E−01**

numerical simulations with the SINP code have been also accomplished by assuming that the source of neutrons is placed at the level of vertical grammage of about 0.317 g cm^{-2} (at the top of the photosphere). The results are closely analogous to those obtained in the previous case, except for an average value of spectral index αT for the entire time of the gamma-ray burst under consideration.

The best approximations for different density models are shown in Fig. 9.9 (left and right panels) and are summarized in Table 9.2.

As noted Troitskaia et al. (2007), another important simulation was carried out by Hua et al. (2002). These authors first thoroughly updated the neutron production cross sections. Then, based on the updated data, they carried out new calculations of neutron spectra for different spectral shapes of accelerated protons (Bessel function and power-law form). The neutron energy spectra have been calculated by Hua et al. (2002) using (a) a typical Bessel function spectrum of the accelerated ions with $\alpha T = 0.03$ and (b) a typical power-law spectrum with index $S = 3.5$ in Eq. 9.2. In their Figure 7 the authors have compared the new results with previous ones (Hua and Lingenfelter 1987a, b). The new spectra are on the whole higher in intensity than the old ones because of increased total neutron production. As a next step, Hua et al. (2002) considered ion pitch angle scattering and magnetic mirroring in the solar flare magnetic loop model to make new calculations of the anisotropic neutron emission. The loop model consists of a semicircular coronal portion and two straight portions extending vertically from the ends of the coronal portion through the chromosphere and into the photosphere. The anisotropy of neutrons in this model arises from the combined effects of converging magnetic field lines and a rapidly increasing ambient density in the portion of the loop below the chromosphere – corona transition. It was shown that the pitch-angle scattering of accelerated ions causes considerable effects on the neutron production spectra (e.g., their

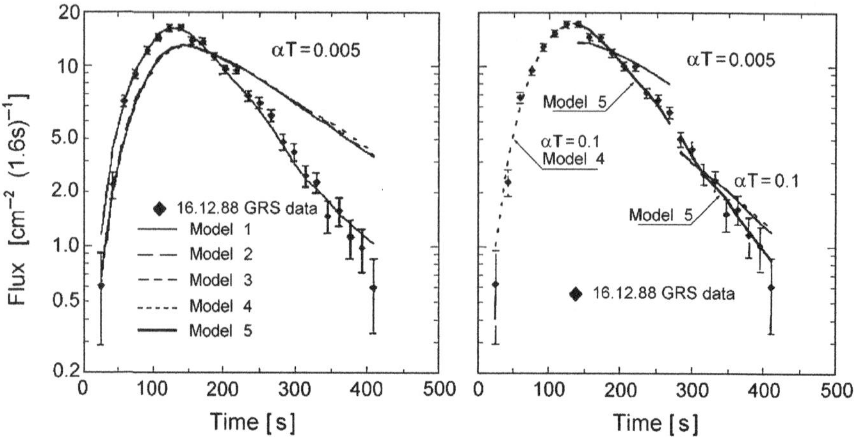

Fig. 9.9 Observed 2.223 MeV γ-ray line fluences of the 16 December 1988 flare (diamonds) and the best model approximations for the entire time profile (*left* panel, $\alpha T = 0.005$, model 5), for rising time (*right* panel, $\alpha T = 0.1$, model 4), and for two segments of the decay phase (*right* panel, $\alpha T = 0.005$ and $\alpha T = 0.1$, respectively). The neutron source is assumed to be at the level of vertical grammage of about 0.317 g.cm^{-2} (at the top of the photosphere); zero time corresponds to 1,704 s after 08:26:23 UT (From Troitskaia et al. 2007)

Table 9.2 Least square deviations of model time profiles from observational data for different time segments of the 16 December 1988 flare (with the neutron source assumed to be at the top of the photosphere) (From Troitskaia et al. 2007)

αT	Model	All time	24–136 s	136–264 s	280–408 s
0.005	1	1.61E+02	3.83E+00	3.26E+01	2.70E+00
	2	1.61E+02	2.45E+00	3.22E+01	2.57E+00
	3	1.71E+02	3.86E+00	3.38E+01	2.82E+00
	4	1.84E+02	3.98E+00	3.52E+01	3.04E+00
	5	**8.41E+00**	2.47E+00	**4.67E+00**	6.44E−01
0.03	1	6.24E+01	1.69E+00	1.39E+01	1.65E+00
	2	4.73E+01	1.40E+00	9.58E+00	1.74E+00
	3	5.82E+01	1.59E+00	1.25E+01	1.73E+00
	4	6.57E+01	1.42E+00	1.34E+01	2.04E+00
	5	9.72E+00	2.46E+00	5.18E+00	3.28E−01
0.1	1	4.86E+01	1.56E+00	1.11E+01	1.47E+00
	2	3.36E+01	1.63E+00	6.78E+00	1.59E+00
	3	4.35E+01	1.53E+00	9.58E+00	1.55E+00
	4	4.99E+01	**1.31E+00**	1.04E+01	1.86E+00
	5	1.01E+01	2.50E+00	5.28E+00	**3.01E−01**

Figure 12). Taking into account the assumptions of the SINP code, Troitskaia et al. (2007) have used the new neutron spectra (Hua et al. 2002, Figure 7) and recalculated the expected time profiles of the 2.223 MeV γ-ray line fluxes for the third peak of the 16 December 1988 flare at $\alpha T = 0.03$ and $S = 3.5$. Calculations were made for all five models. Figure 9.10 presents the best approximations at $\alpha T = 0.03$ for the

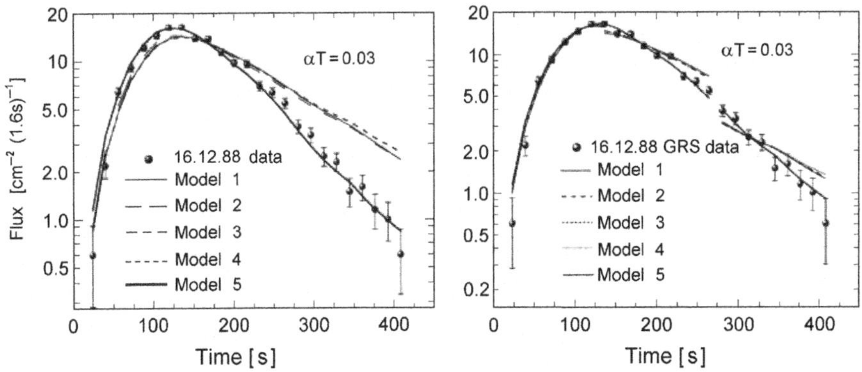

Fig. 9.10 Comparison of SMM/SGR data for the flare of 16 December 1988 (third peak) with time profiles of the 2.223 MeV γ-ray line calculated with the new neutron spectra of Hua et al. (2002) for the entire observed time profile (*left* panel) and for two segments of the decay phase (*right* panel) assuming a Bessel form spectrum of accelerated particles at $\alpha T = 0.03$ (From Troitskaia et al. 2007)

entire observed time profile (left panel) and for three time segments separately (right panel).

As in the previous calculations (Troitskaia et al. 2003) given in Fig. 9.8 (left panel) with the neutron spectra of Hua and Lingenfelter (1987a), a preference should be given (within the error bars) to model 5 for the entire time profile and for the last decay interval (208–408 s). However, at the initial segment of decay (136–264 s) the stronger expected density enhancement seems to be consistent with model m = 2. By making allowance for the 2.223 MeV γ-ray absorption, this corresponds to a density of (4–5) $\times$ 10^{17} cm^{-3}. The best approximations for all five models are also listed in Table 9.3. It is of great interest to compare the results of described above calculations with the SINP code by assuming a power-law spectrum of accelerated particles. The calculations by Troitskaia et al. (2007) were made also by using the neutron spectra obtained by both old (Hua and Lingenfelter 1987a, b) and new (Hua et al. 2002) kinematics of neutron production at $S = 3.5$. The results are shown in Figs. 9.11 and 9.12.

As follows from Figure 7 by Hua et al. (2002), the new kinematics of neutron production provides neutron fluxes about 2.5 times higher in comparison with that from the old kinematics (Hua and Lingenfelter 1987a, b). At the same time, from Figs. 9.10–9.12 (see also Table 9.3), it may be noted that this difference seems to bring a small contribution into the calculated fluences of the 2.223 MeV γ-ray line in both versions of the spectral shape (Bessel function or power-law form). Moreover, this contribution, under assumptions and limitations of the SINP code, is completely overlapped by the limits of experimental error bars. In other words, Troitskaia et al. (2007) have reasons to suggest that the difference is not large, though it exists.

Troitskaia et al. (2007) noted that at the rising phase of the burst, SINP code gives preference to model 2, which corresponds to very high density characteristics

Table 9.3 Least square deviations of model time profiles from observations for the third peak of the 16 December 1988 flare with the old neutron spectra (Hua and Lingenfelter 1987a, b; in column "*S* or αT" designated as 1987) and new spectra (Hua et al. 2002; designated as 2002) at the values of $\alpha T = 0.03$ and $S = 3.5$ for the charged particle spectra (From Troitskaia et al. 2007)

S or αT	Model	All time	24–136 s	136–264 s	280–408 s
αT = 0.03 (2002)	1	7.2073E+01	**1.6034E+00**	1.5330E+01	1.8910E+00
	2	5.3184E+01	1.4145E+01	**1.0581E+01**	2.0849E+00
	3	6.5795E+01	1.4563E+01	9.8849E+01	2.0460E+01
	4	7.8727E+01	1.5624E+01	1.5327E+02	2.3953E+00
	5	**8.3955E+00**	2.9993E+01	3.7764E+01	**3.4087E−01**
S = 3.5 (1987)	1	8.4167E+01	1.7328E+00	1.7556E+01	2.0740E+00
	2	6.4662E+01	**1.3428E+00**	1.2734E+01	2.2541E+00
	3	7.7745E+01	1.4495E+00	1.5995E+01	2.1025E+00
	4	9.1959E+01	1.6398E+00	1.7677E+01	2.5877E+00
	5	**8.0887E+00**	2.9961E+00	**3.6139E+00**	**3.7642E−01**
S = 3.5 (2002)	1	8.1943E+02	1.7088E+00	1.7193E+01	2.0305E+00
	2	6.2565E+02	**1.3432E+00**	1.2378E+01	2.2078E+00
	3	7.5540E+01	1.4427E+00	1.5631E+01	2.0594E+00
	4	8.9687E+02	1.6275E+00	1.7323E+01	2.5387E+00
	5	**8.1301E+00**	2.9942E+00	**3.6336E+00**	**3.6737E−01**

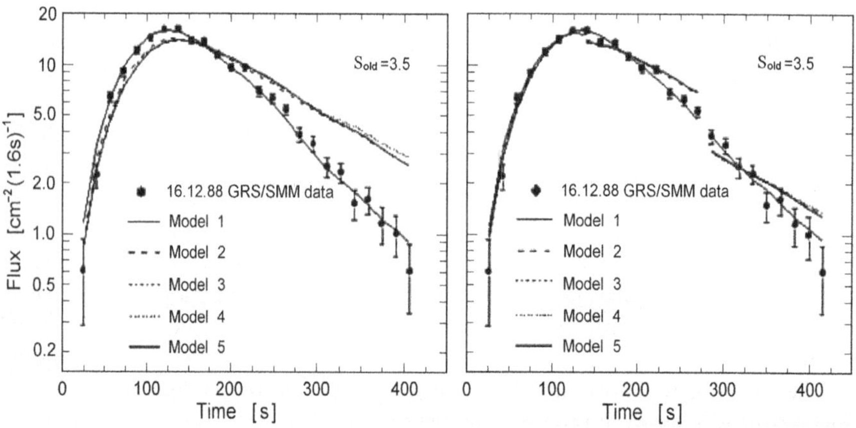

Fig. 9.11 Comparison of SMM/SGR data for the flare of 16 December 1988 (third peak) with time profiles of the 2.223 MeV γ-ray line calculated with the old neutron spectra of Hua and Lingenfelter (1987a, b) for the entire observed time profile (*left* panel) and for three time segments of the decay phase (*right* panel) assuming a power-law spectrum of accelerated particles at $S = 3.5$ in Eq. 9.2 (From Troitskaia et al. 2007)

for the subphotospheric levels (see Fig. 9.1). This is true in both cases of the neutron spectra: the old (Hua and Lingenfelter 1987a, b) and new neutron spectra (Hua et al. 2002), but for the power-law index $S = 3.5$ only. During the entire decay time, as well as for the entire time profile, preference should be given to model 5 with both the old and new kinematics of neutron production.

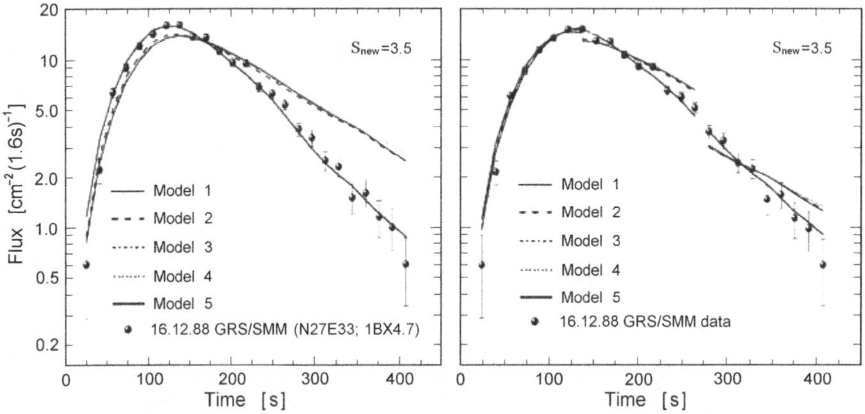

Fig. 9.12 Comparison of SMM/SGR data for the flare of 16 December 1988 (third peak) with time profiles of the 2.223 MeV γ-ray line calculated with the new neutron spectra of Hua et al. (2002) for the entire observed time profile (*left* panel) and for three time segments (*right* panel) assuming the power-law spectra of accelerated particles at $S = 3.5$ in Eq. 9.2 (From Troitskaia et al. 2007)

As underlined Troitskaia et al. (2007), up to the present day, many attempts have been made to use a 2.223 MeV γ-ray line emission for extracting new information about the source parameters involved in acceleration processes, as well as about the parameters of the surrounding medium. In particular, the time profile of this line was shown to provide information about the plasma density distribution under the flare site. The work of Troitskaia et al. (2007) is based on specific modified models with density enhancements deeply in the photosphere in comparison with the standard solar atmosphere for the periods of powerful flares. However, it is suggested that similar effects on the time profiles of the 2.223 MeV γ-ray line may be caused by density enhancements produced by shock waves that are moving in the higher layers (low chromosphere or upper photosphere).

9.2.7 Comparison of Results for Three Solar Flares, Discussion and Summary of Main Results, and Possible Future Development of the Method

Table 9.4 summarizes for plausible plasma density models and ion spectral parameters derived with the SINP code under different assumptions for three solar flares: 6 November 1997, 22 March 1991, and 16 December 1988.

Table 9.4 lists some details of the observations, the method and calculation technique (based on the SINP code), and the main results obtained. In particular, it contains necessary assumptions about neutron release: the depth of sources, time profiles I(t) and angular characteristics of neutron flux (isotropic or vertical), their spectrum indexes S or αT, etc. (for more details see also Kuzhevskij et al. 2005).

Table 9.4 Plasma density models and spectral indices for three flares. In the fifth column year 1987 designated neutron spectra according to Hua and Lingenfelter (1987a, b) and 2002 – according to Hua et al. (2002) (From Troitskaia et al. 2007)

Time of onset, UT	Flare position, class optical/X	Instrument, apparatus, reference	Assumptions about neutron release (source)		Results			
			Time profile	Angular char.	Without weights		With weights	
Method: comparison of time profiles for decay constants Results: power-law spectral index s of initial neutrons and a density model number m								
					S	m	S	m
22 Mar 1991, 22:43 UT	S26 E28 3B/X9.4	PHEBUS, GRANAT T	Instantaneous source at the level of $<5 \times 10^{15}$ cm^{-3}	Vertical, downward	1	2	3	5
		Terekhov et al. (1996)		Isotropic, downward	0	3	0	2
6 Nov 1997, 11:52 UT	S18 W64 2B/X9.4	GRS, *Yohkoh* Yoshimori et al. (1999a, b, c, d)	Similar to $I(t)$ of $\varphi(4–7)$, at the level of $<5 \times 10^{15}$ cm^{-3}	Isotropic, downward neutrons of 1–100 MeV	1	5	3	5
Method: comparison of time profiles for gamma-ray fluences Results: spectral index αT for charged particles and/or a density model number m								
				Angular char., spectrum	αT	m	αT	M
16 Dec 1988, 08:29 UT	N27 E33 1B/X4.7	GRS, SMM Rieger (1996)	Similar to $I(t)$ of $\varphi(4–7)$, at the level of $<5 \times 10^{15}$ cm^{-3}	Isotropic, downward; spectrum of 1987	0.03	5	–	–
16 Dec 1988, 08:29 UT	N27 E33 1B/X4.7	GRS, SMM Rieger (1996)	Similar to $I(t)$ of $\varphi(4–7)$, at the top of photosphere	Isotropic, downward; spectrum of 1987	0.005	5	–	–
16 Dec 1988, 08:29 UT	N27 E33 1B/X4.7	GRS, SMM Rieger (1996)	Similar to $I(t)$ of $\varphi(4–7)$, at the level of $<5 \times 10^{15}$ cm^{-3}	Isotropic, downward; spectrum of 2002 at $\alpha T = 0.03$	–	5	–	–
16 Dec 1988, 08:29 UT	N27 E33 1B/X4.7	GRS, SMM Rieger (1996)	Similar to $I(t)$ of $\varphi(4–7)$, at the level of $<5 \times 10^{15}$ cm^{-3}	Isotropic, downward; spectrum of 2002 at $S = 3.5$	–	5	–	–

From the point of view of flare neutron production, the dynamics of the spectral index for accelerated particles is of special interest. Troitskaia et al. (2007) found that during the decay phase of the 16 December 1988 flare the value of αT increases from 0.005 to 0.1, which implies spectrum hardening with time. For the decay phase this conclusion confirms the results of Gan (1998, 2000a, 2004) obtained by a different method for the same flare of 16 December 1988 and for several other events. Note that the possibility of temporal changes in particle spectra has been discussed in some previous works (e.g., Murphy et al. 1987; Rieger 1996) in the context of a suggestion that two or more mechanisms for particle acceleration are functioning during certain flares.

As noted Troitskaia et al. (2007), available data about real spectra of accelerated particles at different stages of the flare of 16 December 1988 are rather controversial. By direct measurements at the Earth's orbit a power-law spectrum of escaping protons was estimated by Sladkova et al. (M1998) in the form of $\propto E_p^{-S}$, with the index S being about 2.2 and 2.8 below and above $E_p \approx 100$ MeV, respectively. However, for penetrating particles interacting with the solar atmosphere, an ion energy spectrum at $E_p \leq 50$ MeV was derived by Ramaty et al. (1993) using the measured ratios of γ-ray line fluences φ(2.223 MeV)/φ(4–7 MeV) or φ(2.223 MeV)/φ (4.44 MeV) at two compositions of ambient and accelerated particles. Composition 1 (C1) employs the standard photospheric composition for both the ambient medium and the accelerated particles. Composition 2 (C2) employs the compositions derived from the 27 April 1981 flare observations. The derived spectral indexes are $S = 3.3$ (C1) and 2.7 (C2) in power-law presentation and $\alpha T = 0.031$ (C1) and 0.040 (C2) in Bessel function form. As was already mentioned, for the flare of 16 December 1988 Dunphy and Chupp (1992) obtained the spectrum of penetrating (interacting) protons above 10 MeV in the Bessel function form with a shape parameter $\alpha T = 0.054 \pm 0.004$ that differs considerably from the estimates given by Troitskaia et al. (2007).

As noted Troitskaia et al. (2007), Gan and Rieger (1999) have studied the time profile of the 0.511 MeV γ-ray line for the flare of 16 December 1988. In contrast to previous conclusions, they found that a fixed spectral index could not fit the observations. However, it was shown that for αT within 0.02 and 0.05 and for S within 3.4 and 3.8, by adjusting the conversion factor of positrons to the 0.511 MeV photons, one could always get a good fit to the observations.

Troitskaia et al. (2007) underlined that the application of the SINP code for the determination of the altitude density profile during the solar γ-ray flare of 16 December 1988 confirms, within the accuracy of the method, their earlier results obtained for similar events of 6 November 1997 and 22 March 1991 (Kuzhevskij et al. 2001a, b, 2002, 2005). Therefore, it can be conclude that the effects of density enhancement in the deep layers of the photosphere have been found for all three investigated flares. By postulating such enhancements, became possible to explain, at least, the time profile of the solar flare γ-ray emission in the 2.223 MeV line. Additionally, the flare of 16 December 1988 further reveals a delay of enhanced density appearance in the photosphere relative to the γ-ray emission onset. At the same time, it is not possible to ignore results presented in Table 9.3. In fact, one can see that at $S = 3.5$, in both cases of neutron spectra (of Hua and Lingenfelter 1987a, b and Hua et al. 2002), the effect of density enhancement appears already at the rising phase of the third peak, and it demonstrates even higher density than in the decay phase. Thus, based on described results for three flares, Troitskaia et al. (2007) suggest that the effect of density enhancement in the deep layers of the photosphere may be a rather common feature for powerful solar flares on the whole.

Troitskaia et al. (2007) noted that implications and further prospects of the suggested method depend on involving new data on the flares observed during the past few years with high energy, time, and angular resolution. In particular, of great importance are the data on the events of 23 July 2002, 28 October and 2 November 2003, 20 January 2005, obtained by RHESSI, CORONAS-F, and INTEGRAL. At the same time, it seems to be reasonable to include in the SINP calculation code introduced about 10 years ago (Kuzhevskij and Troitskaia 1995; Kuzhevskij et al. 1998) some new basic density models of the solar atmosphere developed since the 1970s (e.g., Avrett 1981; Vernazza et al. 1981). It would also be useful to obtain independent measurements of the ^{3}He content by new methods of solar gamma spectroscopy (e.g., Share and Murphy 2000), or by registration of a weak line at 20.58 MeV from radiative absorption of neutrons by ^{3}He nuclei in solar flares (Alfimenkov et al. 1979; Kuzhevskij 1982; M1985). Some other possibilities arise from the consideration of anisotropic scattering of accelerated ions in the flare loop model and account for the depth dependence of neutron production in the solar atmosphere (Hua et al. 2002). However, the effects of particle anisotropic scattering on the depth dependence of neutron production could not be taken into account because of limitations of the SINP code. Therefore, the made conclusions on the density enhancement effects must be considered as tentative until a new analysis, including particle anisotropic scattering effects, is completed.

9.3 Long-Duration Solar Gamma-Ray Flares and Their Possible Origin

9.3.1 The Matter and Short History of the Problem

According to Ryan (2000), long-duration solar γ-ray flares are those in which high-energy photon emission is present well beyond the impulsive phase, indicating the presence of either stored or continuously accelerated ions. In Ryan (2000) are extensively reviewed both the observations and the current theories or models that can explain this unusual phenomenon. The present situation favors either acceleration of protons and ions for long periods of time by second order Fermi acceleration in large coronal loops or acceleration in large-scale, CME-associated reconnection sheets.

As noted Ryan (2000), the concept of a long duration γ-ray flare is a relatively new one despite the fact that the phenomenon has been observed and measured many times over the last 20 years. The imprecise definition that has developed (and will be used here) is that of a solar flare exhibiting γ-ray (and/or neutron) emission with energy >1 MeV for time periods of a fraction of an hour to hours after the impulsive phase while other common flare emissions (e.g., X-rays) are absent or greatly diminished. The γ-ray emission can stand out above the background level while other emissions at longer wavelengths have since returned to normal intensities. The critical element of this definition is that emission in other wavelengths is reduced relative to the high-energy emission. This is noteworthy because if γ-ray and neutron emissions persist for hours it can represent one of the longest duration signatures of a solar flare. Although soft X-ray and radio emissions can last hours, they now compete with γ-ray emission in certain cases. The intuitive notion that energy degrades in form over time seems not to hold in these circumstances. Long-duration solar γ-ray flares are not to be confused with γ-ray flares such as that which occurred on 27 April 1981. Although the flare of 27 April 1981 lasted approximately 20 min (Forrest and Murphy 1988; Murphy et al. 1990b), the γ-ray flux did not grow in importance relative to fluxes at other wavelengths as the flare progressed. This is to be contrasted with the extreme case of the 11 June 1991 solar flare. Here, γ-ray emission >50 MeV was measurable for 8 h (Kanbach et al. 1993), and in this time the other flare emissions had dropped to normal levels or levels much lower than those during the impulsive phase. The γ-rays from these long-duration solar γ-ray flares are the measurable signatures of the energetic proton and ion populations produced through some acceleration mechanism operating at the start of the flare (or shortly thereafter) and extending for some indefinite time. When these particles interact with the solar material, they produce γ-rays and neutrons. However, the presence of γ-rays does not necessary imply that particle acceleration is in progress. The acceleration of protons and ions gets intertwined with the transport or storage of these particles. Only when these particles interact with the target solar material is the measurable radiation produced. The target solar material is usually somewhere other than at the acceleration site. Generally speaking, proton acceleration requires low densities (so that collisions do not quench the process) while radiation requires high density. The conventional wisdom is that electrons and ions are accelerated in the corona and are transported downward to the solar chromosphere or lower corona, i.e., the thick target (the exception to this is where γ-ray emission occurs high in the corona, e.g. Barat et al. 1994). When radiation or neutron production persists for long periods of time, it becomes a problem to disentangle the combined effects of prolonged acceleration and any attendant transport. The problem becomes particularly thorny considering that some acceleration processes are inseparable from the transport processes, e.g., diffusive shock acceleration.

Ryan (2000) underlined that γ-rays can have many origins (Murphy et al. 1987). The bremsstrahlung of relativistic electrons produces a continuum spectrum, similar in shape to that of the parent electron spectrum (Brown 1971; Hudson 1972). Protons and ions, on the other hand, produce a rich spectrum of γ-rays. Both protons and alpha particles excite heavy ambient nuclei (e.g., C, N, O) that then de-excite by way of γ-ray emission. The reverse process is also true, i.e.; accelerated heavy nuclei (with high velocity) interact with ambient hydrogen and helium producing the same line spectrum, although Doppler broadened due to the center-of-mass velocity in the observer's frame. Collisions of accelerated heavy nuclei with ambient heavies are rare, but α-α collisions are significant and produce ^{7}Be and ^{7}Li that are in turn responsible for ~478 keV photons. In addition to electromagnetic decays of excited nuclei, positron emitters are also produced, eventually producing 511 keV radiation. Neutrons are generated through these same collisions. Free neutrons thermalize quickly in the solar photosphere and are captured by hydrogen or ^{3}He on a time scale of ~100 s (Lingenfelter 1994). The (n, p) capture forming deuterium produces narrow-line radiation at 2.223 MeV (Chupp et al. 1973; Ramaty et al. 1975). In intense and large flares the proton nature of the high energy emission is also confirmed by the detection of solar flare neutrons either at spacecraft altitudes (Chupp et al. 1987; Ryan et al. 1987, 1992) or at ground level with neutron monitors (Debrunner et al. 1983).

Ryan (2000) noted that at relativistic energies π-meson production becomes important. Inelastic *p-p* and *p-α* scattering produces both charged and neutral pions. The neutral pions (99%) decay directly into two 67.5 MeV γ-rays, appropriately Doppler shifted, while π^- mesons decay first into μ^- and then e^-. The π^+ mesons decay by the same scheme into positrons, that in turn annihilate. The electrons and positrons from the charged pion decays go on to radiate via bremsstrahlung in the process of slowing down. In this energy range, neutrons with energies up to ~1 GeV are produced and can be measured by spacecraft or on the ground with neutron monitors. This rich zoo of neutral radiation provides the evidence that very energetic protons exist in the solar environment long after the impulsive phase of some flares.

As noted Ryan (2000), solar flares also accelerate protons in interplanetary space. However, for the most part, the protons and ions measured in space can be traced back to coronal mass ejections and the shocks associated with them (Gosling 1993; Kahler 1992). Particle events that are presumably traceable to the flare itself (so-called impulsive events), as discussed below, tend to be electron-rich and have chemical compositions similar to flare accelerated ions inferred from γ-ray measurements (Ramaty et al. 1993; Reames et al. 1990, 1994; Reames 1996). Unfortunately, the general association of γ-rays with interplanetary particles is poor (Cliver et al. 1989). The long-duration solar γ-ray flares tend to be large events associated frequently with interplanetary shocks and CMEs (Kahler 1982), although this conclusion may be biased by the sensitivity of the instruments that are used to measure them. In these large events protons or ions accelerated in the flare itself get confused with those accelerated by the shock associated with the CME. These remotely accelerated protons generally dominate those from the flare.

Ryan (2000) underlined, that the spectrum of the protons that interact at the Sun (not those detected in interplanetary space) is best determined by the relative fluxes of γ-rays from very different reactions with widely separated thresholds. Typically, these emissions are (1) the γ-ray flux in the narrow 2.223 MeV line from deuterium formation, (2) the 4–7 MeV range from the deexcitation of CNO nuclei and (3) the >50 MeV γ-rays from the decay of π^0 mesons. The nuclear excitations (4–7 MeV) result primarily from protons <50 MeV. The pion production can occur only above ~300 MeV in reactions of type H(p; π, x) with an effective threshold close to 500 MeV. The production of neutrons, measured with the 2.223 MeV flux, occurs at all proton energies above a few megaelectron volt. Long-duration solar γ-ray flares are frequently characterized by small values of the F_γ(4–7 MeV)/F_γ(>50 MeV) flux ratio and large values of the F_γ(2.223 MeV)/F_γ(4–7 MeV) flux ratio, i.e., hard proton spectra (Murphy et al. 1987). This new class of solar flares brings the theory of solar flare particle acceleration full circle, because it raises old concepts of particle trapping and slow and protracted acceleration. Prior to the Solar Maximum Mission (SMM) the prevailing concept for the acceleration of solar flare particles, both at the Sun and in space, was that a first phase quickly accelerates electrons up to ~100 keV after which a second phase, operating over a much longer time period, accelerates electrons up to relativistic energies and protons up to several megaelectron volt (Kane et al. 1980; Ramaty et al. 1980). As early as 1964 Elliot (1964) proposed that the flare process itself is the result of the catastrophic precipitation of stored energetic protons, accelerated over long periods of time high in the solar corona. At the time there was little need for rapid acceleration mechanisms for electrons or protons since γ-ray flare data were sparse and did not seem to require it. Thus, 'standard' acceleration mechanisms such as second-order stochastic acceleration as proposed by Fermi (1949) (cf., Forman et al. 1986; Lee 1994) were employed to explain the few and noisy observations of solar γ-ray flares. The data from the Solar Maximum Mission forced a revision of this concept with the measurement of both X-rays from relativistic electrons and nuclear γ-rays from protons and ions approaching 100 MeV (Forrest and Chupp 1983; Kane et al. 1986; Vestrand and Miller 1999). For a large fraction of events the simple concept of a single acceleration of both electrons and protons to high energies was capable of explaining the observations. In some cases, the explanation of the intensity-time profiles of some flares had to include models of particle transport to account for peculiarities of the profile shapes and delays (Bespalov et al. 1987; Hulot et al. 1989; Ryan 1986; Zweibel and Haber 1983). However, beginning with the flare of 3 June 1982, and with several flares measured by the Compton Observatory, the discussions of particle acceleration and transport increasingly use the language of proton trapping and second-phase acceleration, reminiscent of an earlier era.

9.3.2 Standard Bearers for the Definition of Long-Duration Solar γ-Ray Flares

According to Ryan (2000), these standard bearers are the flares of 3 June 1982 and 11 June 1991. As it was considered above, in Chapter 3, the 3 June 1982 flare stands out because of a clear episode of high-energy emission distinct from the impulsive phase occurring ~2 min earlier (Forrest et al. 1986). The intensity-time profile in several energy bands is shown in Fig. 9.13.

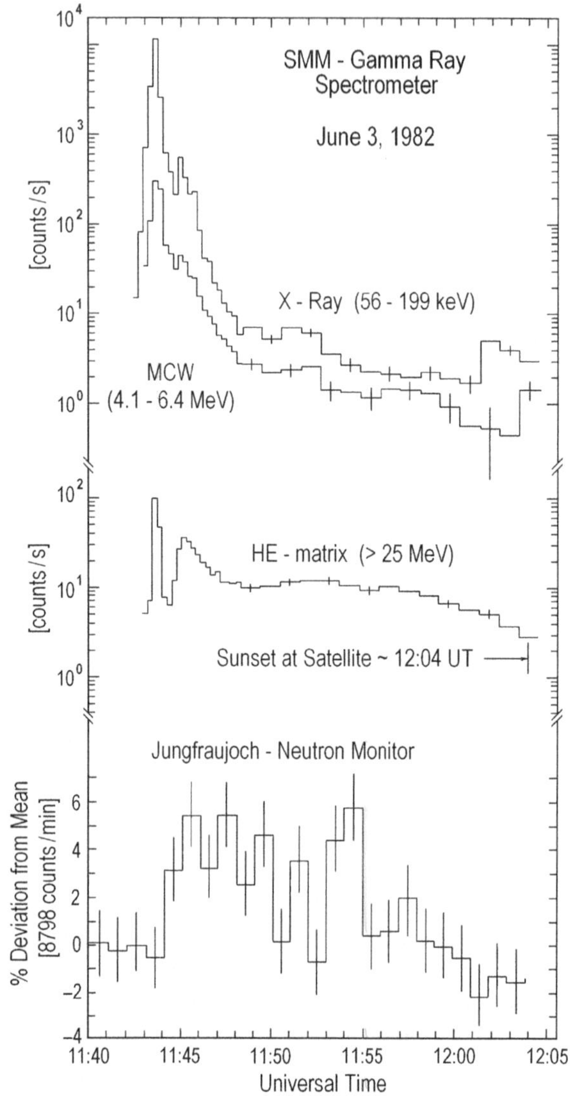

Fig. 9.13 High-energy radiations and neutron intensity-time profiles of the 3 June 1982 solar flare according to Chupp et al. (1987) (From Ryan 2000)

As underlined Ryan (2000), the flare of 11 June 1991 is remarkable because of 8 h photon emission above 50 MeV (Kanbach et al. 1993). Although the detection of 8 h emission is in part due to the sensitivity of the EGRET instrument (Schneid et al. 1996; Thompson et al. 1993), it underscores the difficulty in understanding the physics of the phenomenon. During the 4 h period after the impulsive phase the soft X-ray flux measured by the GOES spacecraft had dropped by a factor of 200 to its

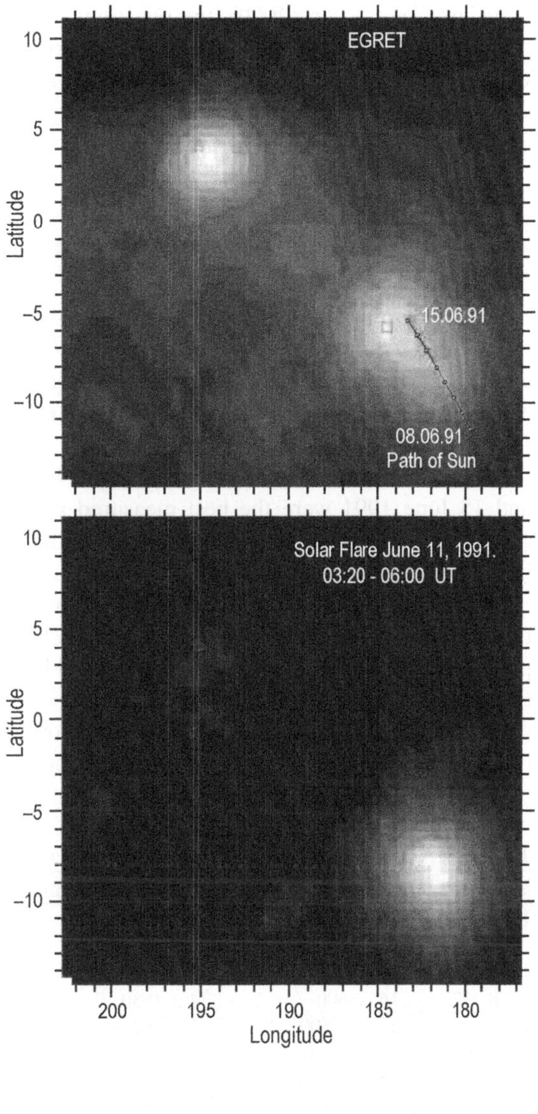

Fig. 9.14 The >50 MeV image of the sky including the Sun (*top*) for 4 days around the 11 June 1991 flare and (*bottom*) 8 h after the flare (From Ryan 2000)

background level. Figure 9.14 shows the EGRET image of galactic anti-center region of the sky (containing the Sun) before and after the 11 June 1991 flare at 01:58 UT (Kanbach et al. 1993).

As can be seen from Fig. 9.14, the Sun was luminous in γ-rays >50 MeV for 4 h after the exposure. Although the EGRET telescope was effectively disabled during the impulsive phase, the TASC detector recorded spectra from 1 to 100 MeV (Schneid et al. 1994). These data show that the count rate at 100 MeV is greater 20 min after the flare than during the most intense part of the impulsive phase, consistent with the data used to create the image over a much longer time scale. As noted Ryan (2000), the question is 'How does the Sun either store the protons (and perhaps electrons) so efficiently or how does it accelerate them without attendant emission at other wavelengths?' These data and the questions that surround them were tried to explore in Ryan (2000).

9.3.3 *List of Long-Duration Solar γ-Ray Flares*

A list of long-duration solar γ-ray flares from Ryan (2000) is shown in Table 9.5.

In Table 9.5 the time behavior of the emission has been characterized by first a short (τ_1) followed by a long exponential decay (τ_2). To different degrees they all exhibit the phenomenon of protracted γ-ray emission with indications that other forms of emission are at a reduced level. The flares of 3 June 1982, 11 June 1991 introduced in Section 9.3.2 and the flares of 24 May 1990, 4 June 1991, 9 June 1991 and 15 June 1991 were the best measured and are discussed in Ryan (2000) in greater detail (the flares of June 1991, including those on 1 June and 6 June, all originated from the same active region, 6659).

As noted Ryan (2000), a common feature among these flares, and contributing to the definition of these events, as will be seen below, is the proton or ion nature of the prolonged radiation. For example, for the 11 June 1991 flare observed by the Compton Observatory, the relative intensity of the high-energy flux (>50 MeV) compared to the flux at 1 MeV was a factor of $\sim$20 greater 15 min after the impulsive phase as compared to the same ratio 2 min after the impulsive phase

Table 9.5 A list of long-duration solar γ-ray flares (From Ryan 2000)

Year	Month	Day	Duration (s)	τ_1 (min)	τ_2 (min)	Ref.
1982	6	3	1,200	1.15 ± 0.14	11.7 ± 3.0	Chupp (1990); Dunphy and Chupp (1994)
1984	4	24	900	3.23 ± 0.07	≥ 10	Dunphy and Chupp (1994)
1988	12	16	600	3.34 ± 0.30		Dunphy and Chupp (1994)
1989	3	6	1,500	2.66 ± 0.27		Dunphy and Chupp (1994)
1989	9	29	>600			Vestrand and Forrest (1993)
1990	4	15	1,800			Trottet (1994)
1990	5	24	500	0.35 ± 0.02	22 ± 2	Debrunner et al. (1997, 1998) Trottet (1994)
1991	3	26	600			Akimov et al. (1991, 1994c)
1991	6	4	10,000	7 ± 0.8	27 ± 7	Schneid et al. (1996); Murphy et al. (1997)
1991	6	6	1,000			Schneid et al. (1996)
1991	6	9	900			Schneid et al. (1996); Ryan et al. (1994a)
1991	6	11	30,000	9.4 ± 1.3	220 ± 50	Schneid et al. (1996); Rank et al. (1996); Kanbach et al. (1993)
1991	6	15	5,000	12.6 ± 3.0	180 ± 100	Akimov et al. (1991, 1994c); Rank et al. (1996)

(Schneid et al. 1994). The high-energy flux is normally attributed to proton interactions creating neutral and charged π mesons (cf., Murphy et al. 1987; Ramaty and Mandzhavidze 1994). However, in some flares, now known as electron-dominated γ-ray flares, electron bremsstrahlung emission extends into this energy interval (Rieger and Marschhäuser 1990), but is not necessarily accompanied by prompt nuclear line radiation or an increase in the flux of 2.223 MeV radiation (Lingenfelter 1994). Typical of the largest (and best measured) of the events listed in Table 9.5 is a two-component intensity-time profile in the highest energy band above 50 MeV. The more extended of the two decays is typically several times that of the shorter one. Table 9.5 lists the published values for the two decay times of various long-duration solar γ-ray flares compiled from a number of sources (e.g., Dunphy and Chupp 1994; Lockwood et al. 1997). For the smaller events the decay times could not be measured, but high-energy neutrons and/or γ-rays well after the impulsive phase identifies them as long-duration solar γ-ray flares. As described below, sometimes the decay time must be inferred from a surrogate emission, e.g., the 2.223 MeV line.

9.3.4 Event of 3 June 1982 as the First Clear Example of a Distinct Second, Delayed and Prolonged High-Energy γ-Ray Flare

As noted Ryan (2000), the solar flare of 3 June 1982 was the first clear example of a distinct second, delayed and prolonged high-energy phase of a flare. Several measurements of the flux from this flare have been reported (Chupp 1990; Chupp et al. 1987; Debrunner et al. 1983; Forrest et al. 1986; Trottet et al. 1994). The γ-ray emission consisted of a $\sim$100 s impulsive phase (with emission extending from X-rays to γ-rays above 25 MeV), typical of many flares observed early in the Solar Maximum Mission. It was followed $\sim$2 min later by a separate high-energy phase accounting for the majority of the high-energy emission. In Fig. 9.13 the count rate of the High Energy Matrix (>25 MeV) of the Gamma Ray Spectrometer (GRS) shows a significant minimum between the impulsive and 'high energy' or 'time-extended' phases. However, the count rate in the energy range of the nuclear lines (4.1–6.4 MeV) shows no clear fall and rise at the time of the dip in the high energy count rate. Similarly, at the time of the high-energy phase maximum, the hard X-ray flux shows only a secondary peak on the decline from the impulsive phase maximum. The high-energy phase was deconvolved into photon and neutron components (Forrest et al. 1986). The GRS was capable of identifying energetic solar neutrons ($\sim$100–500 MeV) on a statistical basis, but with little spectroscopic information. Also detected with little spectroscopic information were the >200 MeV neutrons that evoked an atmospheric response at ground level (Debrunner 1994; Debrunner et al. 1983, 1990). The background-subtracted count rate of the Jungfraujoch neutron monitor is also shown in Fig. 9.13. In modeling the neutron emission the intensity-time profile of photons attributable to π^0 decays (including both the impulsive and high energy phases) was used successfully as an injection profile for energetic neutrons emitted into interplanetary space (Chupp 1990; Chupp et al. 1987). This injection profile can successfully reproduce the GRS neutron signal and the neutron monitor count rate profile if one assumes a neutron production spectrum above 100 MeV of the form $\propto E_n^{-2.4}$ (Chupp et al. 1987). Forrest et al. (1986) concluded that 80% of the pion-produced γ-rays and neutrons are related to the late phase of the flare. This neutron spectrum merges smoothly with that inferred from the measurement of interplanetary neutron-decay protons (Evenson et al. 1983, 1990). However, these neutron spectra do not agree with the neutron spectra published by Shibata (1994) for this flare. Using neutron monitor data alone and assuming an impulsive neutron production at the Sun, Shibata arrived at a neutron spectrum at the Sun more than ten times as intense below 100 MeV than the combined results of Chupp et al. (1987) and Evenson et al. (1990). Other uncertainties surround the analysis and calculations of Shibata (1994) making the reconciliation of these spectra difficult. The association of the high-energy neutrons with the gradual phase establishes that the gradual phase had a significant hadronic component. Although Murphy et al. (1987) speculated that the protons responsible for the time-extended phase are not the same as those responsible for the impulsive phase, the data are ambiguous and, as will be shown later, can be interpreted in several ways.

9.3.5 The Event of 24 May 1990

According to Ryan (2000), the γ-ray flux in this event was measured with instruments on the GRANAT spacecraft, in particular, the shields and central detector of the SIGMA experiment (Pelaez et al. 1992) and the PHEBUS experiment (Terekhov et al. 1993; Trottet et al. 1994; Vilmer 1994). The flare emitted the largest measured flux of solar neutrons with estimates ranging from seven to 100 times the fluence of the 3 June 1982 flare (Debrunner et al. 1997). By itself, the presence of solar neutrons detected at ground level does not establish this event as a long-duration solar γ-ray flare, but when coupled with γ-ray measurements makes a convincing case for one. Gamma rays >50 MeV that may be associated with π^0 meson decay exhibited a strong double-peaked impulsive phase followed by emission extending for at least 10 min at a much lower level. The prolonged high energy radiation measured with the PHEBUS experiment began at most 2 min after the impulsive phase. The radiation was an admixture of high-energy γ-rays and neutrons; however, Debrunner et al. (1997) concluded that at most 30% of the late high-energy emission is attributable to neutrons. The ground-level neutron event was detected at a variety of stations, showing up most clearly at Climax at approximately local noon, the ideal location for detecting solar neutrons. Both Kocharov et al. (1994) and Debrunner et al. (1993a) concluded that a protracted production of neutrons is necessary to explain both the high-energy measurements by PHEBUS and the ground-level neutron measurements proposed earlier by Shea et al. (1991). Muraki et al. (1992) using the neutron monitor efficiencies of Shibata (1994) found, however, that the neutron emission is consistent with neutron production contained entirely within the impulsive phase. However, Debrunner et al. (1997), using the atmospheric neutron response of Shibata, could only obtain the neutron intensity time profile from a totally impulsive phase production by assuming a solar neutron spectrum that is very steep below 500 MeV, in conflict with the intense π^0 emission in the impulsive phase. This is in conflict with the γ-ray analysis of Debrunner et al. (1997). Using the time-extended neutron production indicated by the γ-rays, Debrunner et al. (1997) found that the flare-integrated solar neutron spectrum drops off in intensity above approximately 1–3 GeV. The work by Debrunner et al. (1998) on the evolution of the γ-ray spectrum supports the conclusion that this was a long-duration solar γ-ray flare. The impulsive phase had two resolvable spikes at γ-ray energies >36 MeV, each of ~60 s duration. When examined closely, the second spike becomes more pronounced as one progresses to higher energies (see Fig. 9.15).

As underlined Ryan (2000), in papers of Debrunner et al. (1998) and Talon et al. (1993) was concluded that the first spike in the impulsive phase was electron dominated, i.e., the high-energy radiation was produced by primary electron bremsstrahlung, while the second spike contained a strong hadronic component,

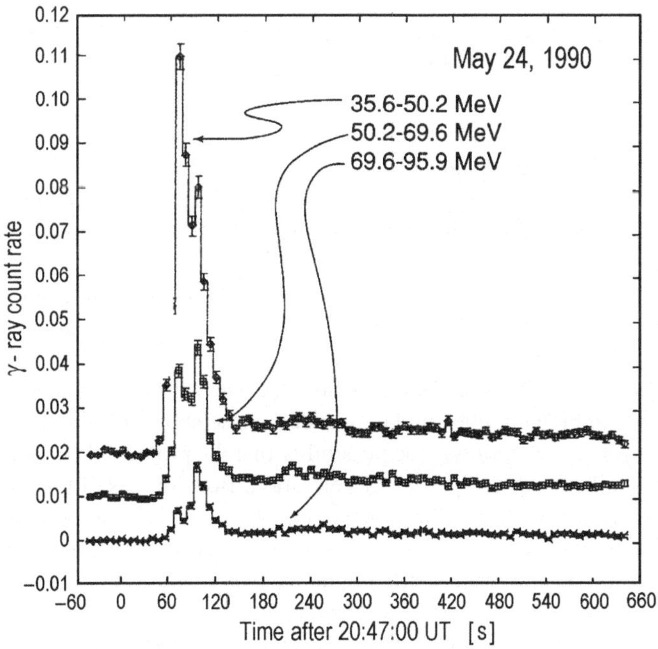

Fig. 9.15 The event on 24 May 1990. The intensity-time profiles of the highest energy channels of the PHEBUS instrument showing the hard spectral shape of the second spike and the contribution from neutrons and delayed high-energy γ-rays (From Ryan 2000)

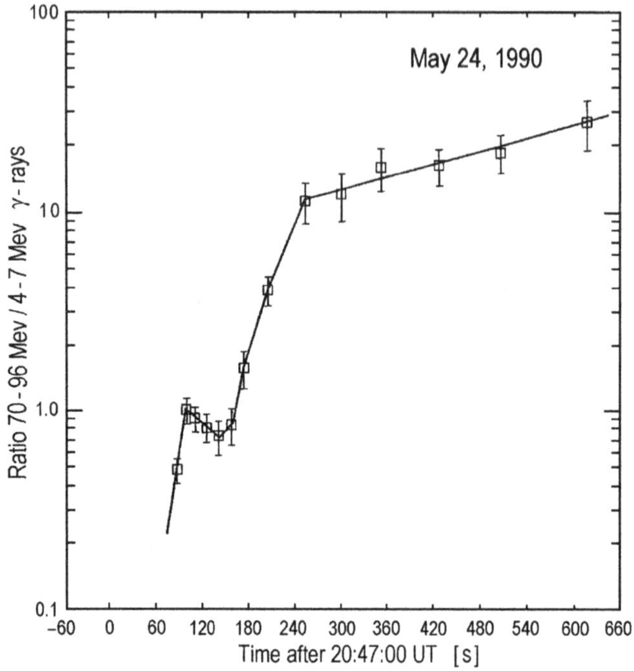

Fig. 9.16 The ratio of the γ-ray intensities in the π^0 decay and nuclear deexcitation channels during event May 24, 1990 (From Ryan 2000)

giving rise to the π^0 peak in the impulsive phase. After both impulsive phase spikes have given way to the prolonged phase, the spectrum above 35 MeV exhibits a clear π^0 peak. The π^0 peak decays slowly with respect to the emissions at lower energies, becoming an increasingly dominant feature in the spectrum as shown in Figs. 9.16 and 9.17.

As noted Ryan (2000), the hadronic component is the dominant feature well after the impulsive phase and it remains so until the observation ceased after 10 min. The decay rate of the π^0 decay peak was estimated to be $\sim$20 min^{-1} according to Debrunner et al. (1997) (see above, Table 9.5). The smooth exponential decay of the count rate at high energies in the extended phase shows no sign of fluctuations beyond statistical, constrainting the possibility that the prolonged emission is due to numerous small episodes of particle acceleration. Both studies of the GLE neutron signal (Kocharov et al. 1995; Debrunner et al. 1997) reached the conclusion that the prolonged component of the neutron emission was softer than the component arising from the second spike of the impulsive phase. This, on its face, contradicts the conclusion from the γ-ray data (Debrunner et al. 1998) that the proton spectrum hardens in the extended phase. However, the energy ranges of these studies only have a small overlap. The GLE neutron signal derives from a significantly higher energy part of the proton spectrum than the π^0 decay signal, making a quantitative comparison difficult.

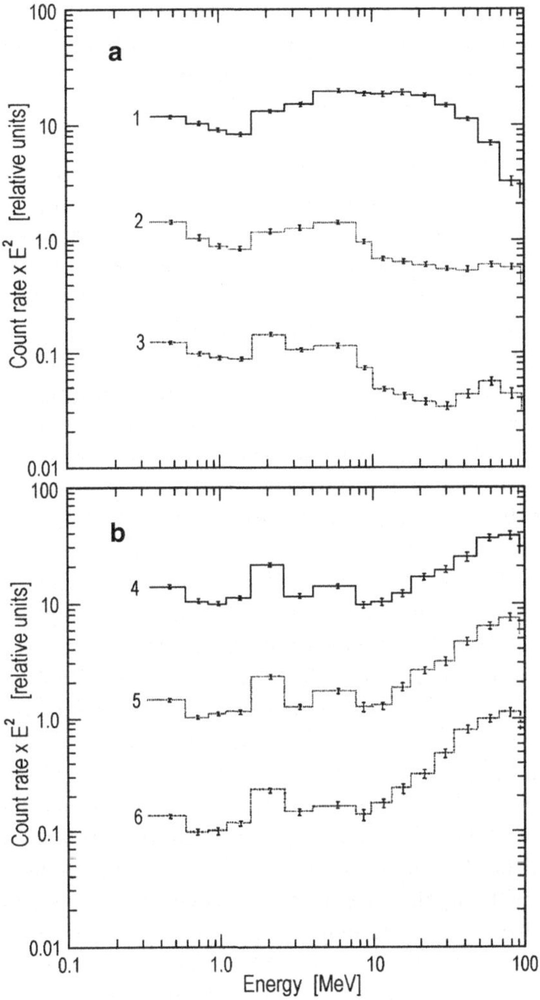

Fig. 9.17 The γ-ray spectrum (weighted by E^2) from the (**a**) impulsive and transitional and (**b**) the gradual phase periods of the 24 May 1990 flare (From Ryan 2000)

9.3.6 The Event of 4 June 1991

This flare (03:37 UT) was well measured only by the OSSE instrument on the Compton Observatory. Murphy et al. (1993a, b, 1994) reported that proton precipitation and neutron production occurred during three successive orbits of the spacecraft, i.e., >2 h. This conclusion was drawn from the measured flux of the 2.223 MeV nuclear line over this time period. Supporting this concept is the long duration of the 4.43 MeV carbon deexcitation γ-ray line paralleling that of the 2.223 MeV line. The decay of the line flux in this flare is on the order of $\sim$160 $\pm$ 30 s. The detection of ground level neutrons was reported by Muraki et al. (1992) and Takahashi et al. (1991). Because the flare was near the east limb (30°N, 70°E), no prompt protons were expected that could be confused with neutrons at ground level stations. Using only data from the first ~15 min of the event and using the NM efficiencies of Shibata (1994), Muraki et al. (1992) concluded that neutrons were produced impulsively at the flare onset (03:41 UT). Their resulting spectrum for neutrons at the Sun is soft ($E_n^{-5.4}$ to $E_n^{-7.5}$) and is probably a direct result of assuming only an impulsive production. Furthermore, the efficiencies of Shibata (1994) and those of Debrunner et al. (1993a, 1997) diverge significantly at low energies. As mentioned above, the spectrum reported by Shibata (1994) for the 3 June 1982 flare differs markedly from others (Evenson et al. 1983; Chupp 1987), being more than ten times greater where they overlap. However, an excess of low energy neutrons is necessary when assuming an impulsive neutron production in order to explain the late arrival of neutrons at the Earth. An extended production would produce what seems to be a harder neutron spectrum at the Sun. Struminsky et al. (1994) used the NM data of Mt. Norikura, that show an excess persisting for approximately 1 h, to model the time-extended neutron production. Their model predicts neutron spectra at the Sun that are much harder ($E_n^{-3.5}$ to $E_n^{-5.2}$) than those reported by Muraki et al. (1992) and are in better agreement with the proton spectrum ($E_p^{-2.8}$) at the Sun reported by Murphy et al. (1994). The interpretation of these data is complicated by two facts: (1) that there appears to be a second and distinct acceleration episode that occurred during an epoch when the Compton Observatory was occulted, and (2) 5 min NM data have inadequate resolution to measure spectra that are produced on the same time scale. Struminsky et al. (1994) reported that 17 GHz Nobeyama microwave data also indicate two acceleration episodes. This second acceleration (with an accompanying proton precipitation) helps explain the observed discontinuity in the intensity-time profiles of the 2.2 MeV γ-ray line emission (Lockwood et al. 1997). They conclude that both proton accelerations represent the beginnings of independent and protracted neutron and γ-ray production, each lasting longer than 30 min. It is not clear whether this second episode of proton acceleration (with significant emission of neutrons and γ-rays) is directly related to the flare that occurred 45 min earlier. Conceivably, this could be an entirely different flare. However, Struminsky et al. (1994) derive a much harder neutron spectrum from the second episode than from the first perhaps representing a discrete jump in the spectral shape of the accelerated protons (Ramaty et al. 1994).

9.3.7 The Event of 11 June 1991

As underlined Ryan (2000), the other extraordinary flare exhibiting a long-duration high-energy phase and the one that created much of the interest in this phenomenon is that of 11 June 1991 (see Fig. 9.14 in Section 9.3.2). The flare was well observed with instruments on the Compton Observatory (Kanbach et al. 1993; Rank et al. 1993, 1994; Ryan et al. 1993b; Ryan 1994; Schneid et al. 1994). The spectrum was measured with the EGRET instrument in its telescope mode beginning ~90 min after the flare onset. The spectrum extended at least up to 1 GeV. Comparing the emission 15 min after the impulsive phase to that of the impulsive phase, the high-energy emission >50 MeV was enhanced by 20 times relative to that at 1 MeV (Schneid et al. 1994). It was also accompanied by a strong line at 2.223 MeV from the neutron capture on photospheric hydrogen. The line was detected with COMPTEL (Rank et al. 1994) for a period of 4 h after an initial exponential decay rate of (9 min)$^{-1}$ (Rank et al. 1996) after the impulsive phase. Within uncertainties, the flux at 50 MeV decayed at the same rate (Kanbach et al. 1993) shortly after the impulsive phase suggesting that this component is also of a hadronic nature and not a result of primary electrons as first reported (Mandzhavidze and Ramaty 1992a). A spectral analysis of the EGRET data (Dunphy et al. 1999a, b) supports the claim that a different acceleration process is at work in the gradual phase of this flare. Based largely on the pion-decay γ-rays, the gradual phase spectrum requires a significantly harder spectrum of protons than that of the γ-ray spectrum of the impulsive phase to produce the observed emission.

9.3.8 The Event of 15 June 1991

As noted Ryan (2000), although the COMPTEL instrument on the Compton Observatory and the GAMMA-1 experiment both detected this flare, neither spacecraft measured the emissions during the impulsive phase, believed to have occurred around 08:20 UT, the maximum in the soft X-ray flux. Both spacecraft were occulted during the impulsive phase. In almost mutually exclusive observations, GAMMA-1 measured γ-ray emissions up to 1 GeV (Akimov et al. 1991) after which COMPTEL measured the γ-ray flux <30 MeV until 09:45 UT (McConnell et al. 1993; Ryan et al. 1993b). In this period after the impulsive phase both the high-energy flux and that at 2.223 MeV decayed away with a decay constant of $\sim$13 min (Rank et al. 1996). In a second observation of the Sun on the subsequent orbit GAMMA-1 again measured a statistically significant flux >50 MeV. There was also a significant detection of 2.223 MeV γ-rays and 15–80 MeV neutrons by COMPTEL on its second orbit after the impulsive phase. This event may also have produced a solar proton ground level event in NM at ground level (Smart et al. 1994) and probably solar neutron event measured also by NM at ground level (Usoskin et al. 1995; Nieminen 1997).

Ryan (2000) noted that Akimov et al. (1993, 1994a, b, 1996b) compared the γ-ray intensity-time profile with microwave emissions and found good agreement between the two, supporting the assumption that the impulsive phase occurred around 08:20 UT. They also used these observations to conclude that the production of γ-rays >50 MeV was a result of extended acceleration rather than trapping plus precipitation. COMPTEL also measured neutrons between 15 and 80 MeV from this flare (Debrunner et al. 1993b; McConnell 1994; Nieminen 1997; Rank 1996; Rank et al. 1993, 1997a, b). By studying the measured energy of the detected neutrons, Debrunner et al. (1993b) concluded that neutrons were also produced for at least $\sim$90 min after the impulsive phase. This coupled with the measurement of the 2.223 MeV neutron capture line and nuclear deexcitation lines above 4 MeV over similar periods of time indicate that the long-duration flux was due almost entirely to protons or ions. Long after the flare relatively little flux was measured below 1 MeV where primary electrons would have their strongest signatures. However, evidence of energetic electrons can be seen in the γ-ray continuum below 50 MeV, presumably from the decay of charged pions (Rank et al. 1997b).

As underlined Ryan (2000), two studies independently concluded that a separate ion-dominated acceleration mechanism was responsible for the extended phase emission (Akimov et al. 1996; Kocharov et al. 1998). Akimov et al. (1996) supported this claim with the arguments that (1) the microwave signal was variable in the late phase indicating acceleration, (2) decimetric and meter wave activity was high during the same time, indicating acceleration at high altitudes, (3) the microwave and the γ-ray intensity-time profiles are very similar and (4) the prolonged escape of high-energy particles into interplanetary space is not consistent with a single impulsive phase acceleration.

Ryan (2000) noted that Kocharov et al. (1998) and Nieminen (1997) also associated the microwave emission with the extended γ-ray emission. Kocharov et al. (1998) deduced a two-component ion spectrum dominated by protons ($\alpha/p = 0.5$) with each component having a similar power law shape, but with the low-energy component stronger by a factor of $\sim$10 with a cutoff at several hundred MeV. No primary electron spectrum was necessary in this model. The spectrum was constructed from both low- and high-energy γ-ray data and neutron data and thus it represents a flare average. Akimov et al. (1996) attributed the interplanetary particles and the late-phase γ-ray emission to a DC electric field acceleration in a large (10^{10} cm) reconnection sheet formed behind a CME (Litvinenko 1996b; Litvinenko and Somov 1995; Somov 1996), although there was no reported CME.

9.3.9 Characteristic Properties of Long Duration Gamma-Ray Flares and Main Possible Scenarios

As noted Ryan (2000), the characteristic properties of long-duration solar γ-ray flares as described above are (1) ion-rich particle precipitation for periods that extend long after the impulsive phase, remaining strong well into or after the decline in soft X-ray emission, and (2) relatively high energy photons, associated with π meson production. There are a limited number of possible scenarios consistent with such phenomena. **The first** of these is that proton, ion and electron acceleration occurs during the impulsive phase, after which these particles are stored in high altitude magnetic structures. They subsequently precipitate onto the lower solar atmosphere to produce high-energy photons. The particles can also interact with the ambient medium in these magnetic structures rather than at lower altitudes in the chromosphere or lower corona. Conditions must be such that within the high altitude coronal structures there is little MHD turbulence that scatters particles into loss cones and that the magnetic geometry is such that curvature and gradient drifts do not prematurely deplete the population of trapped particles. The combined loss effects of pitch angle diffusion, collisional braking in the ambient medium and drifts determine the characteristic trapping time.

The second scenario is that particle acceleration takes place over long periods of time after the flare, but is decoupled from the impulsive flare. The picture of a receding shock is one such scenario, where the shock, which may have produced the original fast particles seen in the impulsive phase, continues to produce energetic protons and ions. The particles, once accelerated at the shock, must make their way back in sufficient numbers to the lower corona or chromosphere to produce detectable γ-rays. For γ-ray production of $\sim$1 h this implies that the shock that produced these particles can be as far away as 5 R_0 (solar radii) when the γ-rays are produced.

The third scenario also relies upon a CME for prolonged particle acceleration, but here the acceleration takes place in a reconnecting current sheet behind the CME (Litvinenko 1996a, b; Litvinenko and Somov 1993; Somov 1996). The reconnecting current sheets are capable of generating large potential drops for high-energy ions and the temporal behavior of establishing the reconnecting current sheets fitted the time scales of long-duration solar γ-ray flares. **The fourth** scenario is an admixture of the first two mechanisms outlined above, i.e., particles are both trapped in high altitude magnetic structures but are accelerated continuously while being trapped. Since the geometry implied here is static, this rules out any dynamic shock acceleration process. The energy to accelerate the particles must, most likely, come from turbulence or mini-flaring within these magnetic structures.

Ryan (2000) underlined that the possibilities for producing long-duration solar γ-ray flares encompass the very same processes as those for the impulsive phase, i.e., electric fields and first and second order Fermi acceleration mechanisms. However, it is important to strive for, if possible, a single model to explain this phenomenon. Given that there is a wide range of duration of these events, it will be possible to explain some of them with passive trapping while others require continuous acceleration.

9.3.10 Passive Trapping of Impulsive Phase High Energy Protons

Ryan (2000) noted that for studying the time behavior of γ-ray emission, the simple picture of passive trapping was examined by Zweibel and Haber (1983) and Ryan (1986) where protons are stored in a bipolar loop and are scattered into a loss cone by way of pitch angle scattering at some point within the loop. This concept has always been attractive since we see long term trapping of protons in the radiation belts of the planets. The size of the loss cone and the intensity of the pitch angle scattering determine the decay rate of the population within the loop and consequently the rate of proton precipitation into the lower corona or chromosphere. The proton precipitation is observed by way of γ-ray and neutron emission if the protons are sufficiently energetic. If the pitch angle diffusion is confined to an isolated turbulent region and is strong enough so that the proton population is isotropized with each transit, then the loss is entirely determined by the size of the loss cone. If there is no pitch angle scattering in this loop then the radiation decays because the protons lose energy in collisions in the local medium. The other extreme case is where the pitch angle scattering is intense and distributed throughout the loop. The protons are transported by spatial diffusion eventually arriving at the loop ends to produce the neutron and γ-radiation (Ryan 1986). Various combinations or intermediate cases have also been studied (Fletcher 1997).

As underlined Ryan (2000), regardless of the dynamics within a loop, energetic protons will lose energy by way of collisions with ambient electrons. This energy loss rate tracks the γ-ray and neutron production from within the same volume. Non-relativistic protons lose energy via collisions with ambient electrons on a characteristic time scale (per unit number density)

$$E/(dE/dt) = 2 \times 10^{11} E^{0.5} (\mathrm{MeV.cm^{-3}})\mathrm{s} \tag{9.3}$$

for sub-relativistic protons and

$$E/(dE/dt) = 2 \times 10^{12} E (\mathrm{MeV.cm^{-3}})\ \mathrm{s} \tag{9.4}$$

for relativistic protons. To achieve a 1 h lifetime for 100 MeV protons the mean coronal density can be no greater than about $5 \times 10^{8}\ \mathrm{cm}^{-3}$. Relativistic protons ($\geq$1 GeV) fare better, requiring a density $<5 \times 10^{11}\ \mathrm{cm}^{-3}$. A natural hardening of the proton spectrum occurs over time due to this effect. However, the effect of nuclear interactions at relativistic energies is comparable to that of energy loss through interactions with electrons. This tends to soften the proton spectrum over time. These processes were examined in detail by Hua et al. (1989). They modeled a coronal loop with an ambient matter density that smoothly merges with the matter density at the ends of the loop. They assumed a level of MHD turbulence within the loop that scatters the protons into a loss cone. The protons were tracked by way of a Monte Carlo calculation as they adiabatically mirror within the loop and change their pitch angle according to the assumed turbulence intensity. The protons also gradually lose energy by way of normal collisions with ambient electrons. The inelastic collisions of the protons with matter within the loop or at the ends of the loop where the density is high produce the observable photon emission. The picture although much more detailed yields results similar to analytical calculations (Ryan 1986; Zweibel and Haber 1983).

Ryan (2000) noted that the dynamic behavior of protons in a bipolar field was examined by several investigators. Except for the effects described above, the lifetime of protons in a static loop is largely determined by the size of the loss cone and the rate at which the loss cone is replenished. The expression for the half angle α of the loss cone is

$$\sin^2\alpha = B_0/B_m, \tag{9.5}$$

where B_m is the magnetic field strength at the mirror point and B_0 is the magnetic field strength at the apex of the loop. Particles that are initially in the loss cone penetrate the lower corona or the dense chromosphere and are capable of emitting γ-rays or neutrons. They do not return to the energetic proton distribution once having entered the loss cone. The remainder of the protons interacts with the ambient medium or gets scattered into the loss cone. Coulomb scattering, directly related to the energy loss, is in principle capable of filling the loss cone. However, small angle scattering experienced by protons does not compete with the energy losses and is therefore negligible (Benz and Gold 1971; Fletcher 1997). The loss cone can be filled more efficiently by MHD wave-resonant pitch angle scattering. The wave field can arise from external sources or can be self-generated by the distorted pitch-angle distribution due to the depleted loss cone. The case of a self-generated field has been studied by Meerson and Rogachevskii (1983). They found that under certain conditions, when plasma parameter $\beta = P/(B^2/8\pi)$ is large enough, an Alfvén wave field that is resonant with the energetic protons can develop quickly, keeping the loss cone filled and thereby depleting the proton population. The diffusion starts out weak with the protons scattering into the loss cone. The wave intensity finally develops into a condition of strong pitch angle diffusion, i.e., the loss cone is filled (and emptied) with each bounce of the particles. Under these conditions proton loss rate is

$$I_p(t) \propto \exp\left(-t\left(\sin^2\alpha\right)/\tau_b\right), \tag{9.6}$$

where τ_b is the proton bounce time. If the ratio of the magnetic field at the footpoint to that at the loop apex is not larger than ~5, then the characteristic loss time through pitch angle diffusion is only about a factor of 10 greater than τ_b. For a 10^{10} cm loop and a proton with a velocity of $c/3$, τ_b is on the order of 5 s. The trapping that results from this situation is sufficient to smooth out the impulsive nature of the particle acceleration as seen in γ-ray emission, but not nearly long enough to provide the containment for a 1 h long duration flare. To achieve a 1 h trapping time the diffusion must be much weaker than that necessary to isotropize the proton population with each bounce. However, this may not be possible with turbulence generated by other means.

As underlined Ryan (2000), also affecting the lifetime of the protons within the trap are guiding-center particle drifts, in particular, curvature and gradient drifts (Northrop 1963). From curvature drift alone a 500 MeV proton in a 100 Gs loop of length 10^{10} cm will drift $\sim 10^9$ cm h^{-1} in a direction orthogonal to the plane of the loop. The drift rate is inversely proportional to the magnetic field strength. With $B = 10$ Gs a 500 MeV proton will therefore drift in 1 h one loop length away from the loop, removing itself from the trap. Lau and Ramaty (1994) proposed that twisted (force-free) loops are capable of containing energetic protons if the loops are either large enough or have sufficient twist ($<2\pi$). Thus, in order to contain trapped relativistic protons or ions for extended lengths of time three conditions must be met. (1) Loop densities must be low. For 100 MeV protons and 1 h trapping this implies an ambient hydrogen number density $<5 \times 10^8$ cm^{-3}, and a corresponding ambient density $<5 \times 10^{11}$ cm^{-3} for 1 GeV protons. (2) MHD turbulence must be very low. For the 11 June 1991 flare Ramaty and Mandzhavidze (1994) pointed out that the Alfvén wave energy density can be no larger than

$$(\delta B)^2/8\pi < 2 \times 10^{-6}\text{erg.cm}^{-3}, \tag{9.7}$$

assuming a 1-D Kolmogorov wave spectrum

$$W(k) \propto k^{-5/3}, \tag{9.8}$$

integrated down to a wave number k corresponding to a 10 GeV proton Larmor radius in a 100 Gs field. The low Alfvén-wave energy density proposed by Ramaty and Mandzhavidze (1994) for the 11 June 1991 flare should be compared to the value of $\sim 10^{-2}$ erg.cm^{-3} in turbulent wave energy if Alfvén waves heat coronal loops (Hollweg 1984). The mean free path of such a proton is of order 1,000 AU. However, it would expect the wave spectrum to continue down to a wave number corresponding to a wavelength $2L$, where L is the length of the coronal loop. This would increase the total wave field energy density to $\sim 10^{-3}$ erg.cm^{-3} in a loop with $L = 10^{10}$ cm and $B = 100$ Gs. However, this is still much smaller than one would expect in a post-flare environment. The arcade of coronal loops containing the protons and ions must be force-free so that particle drifts do not deplete the proton and ion population. This is important for loops on the order of 2×10^9 cm or smaller for trapping of 1 GeV protons (Lau et al. 1993).

9.3.11 Separate and Remote Acceleration

As noted Ryan (2000), the protons that produce the long-duration high-energy γ-ray emission need not be accelerated during the impulsive phase of the flare. In principle they could arise from a separate and distinct acceleration process occurring at a later time. Separate and remote acceleration processes take two forms: electric fields and coronal shocks. Ryan (2000) first discusses the electric field models. Electric field acceleration of protons and ions are often based on the picture constructed by Speiser (1965) and later by Martens (1988). However, the special case of long duration, high-energy proton and ion acceleration by electric fields was specifically addressed by Litvinenko and Somov (1993) and Litvinenko (1996a) and later employed by Akimov et al. (1996) to explain the high-energy γ-ray emission from the 15 June 1991 flare. The general picture is that a large magnetic reconnection current sheet, established behind a receding CME, accelerates particles in the electric field along the sheet. The dimensions of the reconnection current sheet, the strength of the merging magnetic field and the flow velocities are more than sufficient to generate large electric fields and accelerate protons well above

GeV energies. The problem is the dynamics of retaining the particles in the electric field. The protons will naturally drift out the inhomogeneous field before acquiring relativistic energies. However, with a small (0.1%) magnetic field normal to the reconnection current sheet (and the accelerating electric field), the protons drift back into the accelerating electric potential (several $V.m^{-1}$). With approximately 100 such exits and re-entrances of the protons into the electric potential, they can attain GeV energies. The model is attractive because it relies upon the creation of a CME, a common feature of large solar flares. Not only does the CME-large flare association exist, but also the time scale for the development and evolution of the CME matches that of the long-duration γ-ray emission. Moreover, the polarity of the field would naturally exclude high-energy electron bremsstrahlung when protons and ions precipitate to the chromosphere.

Ryan (2000) noted that another possibility, also involving CMEs, is that a secondary shock wave from the powerful impulsive phase or a coronal mass ejection accelerates protons at a distance from the original flare site. Murphy et al. (1987) and Ramaty et al. (1987) suggested that the protons responsible for the pion-related emission in the late phase of the 3 June 1982 flare could have the same origin as the protons measured in interplanetary space. Less than 10% of the interplanetary proton flux is required to yield the measured pion-related emissions (McDonald and van Hollebeke 1985). Later revision of the proton spectral shape by Van Hollebeke et al. (1990) increased the required fraction of the interplanetary proton population to 25–50%. Shock accelerated protons are frequently detected and measured in space (Lee 1994; Reames 1996). Most often they are associated with coronal mass ejections (Lee 1997), but not always associated with flares (Gosling 1993; Kahler 1992). The typical ionization state of ions detected is representative of quiet coronal conditions, i.e., 10^6 K. Some interplanetary protons and ions are however associated with the flare itself. The association is also established by way of the ionization states and the composition of the measured ions. The characteristics of these flare associated particles are (1) a relatively high abundance of energetic electrons, (2) a large abundance of ^{3}He with respect to ^{4}He and (3) ion charge states representative of (10–30) $\times$ 10^6 K. Although the CME-related particles are clearly shock accelerated, the case is not so clear for the so-called impulsive phase particles. However, Ryan (2000) assume for the moment that the flare-associated protons and ions have been accelerated by a CME-driven shock wave (or a coronal blast wave) and that they diffuse back to the Sun through the turbulent downstream region of the shock to produce the γ-rays and neutrons in the long-duration solar γ-ray flares. The dynamics of proton shock acceleration in a coronal blast wave was described by Lee and Ryan (1986), but a proper theoretical treatment of the problem of protons diffusing to the solar surface from a receding shock apparently has not be attempted, nor are there any published measurements that might address the problem.

Ryan (2000) considered two scenarios that might allow shock-accelerated protons to precipitate back to the solar surface for periods of hours after a large flare. In the first it can be imagine that a coronal blast wave or a CME sets up a shock that accelerates particles for long periods of time. In that time, however, the shock, the source of the energetic particles, is receding from the target Sun. The particles must either diffuse back to the Sun through the turbulent downstream region of the shock or find efficient, i.e., relatively scatter-free, magnetic field lines that connect the regions of great energetic particle density back to the Sun. The energetic particles concentrate near the shock interface that moves across new field lines as it propagates. If scatter free transport of ions occurs along quiet field lines, it must occur on a large number of them in order to maintain the connection and keep the precipitation going for hours. However, the connecting field lines must only connect a limited fraction of the accelerating region to the Sun, because the intensity of the γ-ray emission falls off exponentially whereas the energetic proton population integrated over the entire shock front falls off more slowly, or even grows with time (Lee and Ryan 1986).

As noted Ryan (2000), evidence is strong, through compositional studies of a small number of flares (Murphy et al. 1990b; Share and Murphy 1995), that the chemical composition of 'gradual event' interplanetary ions is not similar to those of interplanetary ions from impulsive events (Reames et al. 1994) or the that of the γ-ray producing ions, as inferred from the γ-ray spectra (Cliver 1996). In addition, the onset times of the GLE protons are much later than the γ-ray emission at the equivalent energy (Lockwood et al. 1999; see in details below, Section 9.9). Thus, GLE proton emission at the Sun is not created in the impulsive phase of the flare. If the highest energy interplanetary protons and ions are not related to the impulsive phase and the chemical composition of the extended phase ions resembles that of the impulsive phase it can be safely rule out the transport of remotely accelerated shock-associated ions as the source of the high energy ions responsible for the pion-related γ-ray emission. However, there is limited spectral data on the high-energy delayed γ-ray emission. The first analysis of the 11 June 1991 event (Mandzhavidze and Ramaty 1992a, b) indicated a dominant primary electron bremsstrahlung component in qualitative agreement with the strong electron component in impulsive event particle spectra. A subsequent analysis showed that the 2.223 MeV emission decay curves are identical to the high-energy decay curves (Rank 1996; Rank et al. 1996, 1997a). This belies the conclusion that primary electrons contributed significantly to the high-energy emission. The great difference that one would expect in the trapping efficiencies of electrons and ions would naturally lead one to conclude that primary electron bremsstrahlung and ion signatures would diverge in the late phase of the flare.

Ryan (2000) underlined that for the 15 June 1991 event Kocharov et al. (1998) limited the ^{4}He abundance to be less than half that of the protons, while Debrunner et al. (1998) excluded the composition II of Ramaty et al. (1993), i.e., enriched in heavy elements (and primary electrons) for the 24 May 1990 event. These limited studies do not support the hypothesis that the composition of the long-duration, high-energy proton spectrum has a composition similar to that deduced from the γ-ray spectrum of the 27 April 1981 flare (Murphy et al. 1990b) or that of impulsive interplanetary particle events (Kahler 1992; Ramaty et al. 1993; Reames et al. 1994). This singular similarity of the 27 April 1981 composition and that of impulsive interplanetary particle events led Cliver (1996) to conclude that long-duration solar γ-ray flares derive from the same particle acceleration process as impulsive flares. Ryan (2000) believe, however, that the 27 April 1981 flare should not be classified as a long-duration solar γ-ray flare, but rather a merely drawn out impulsive phase event with no significant high-energy emission.

As noted Ryan (2000), on the experimental side a positive detection of high-energy γ-rays following a CME, especially without the occurrence of a flare, would be strong evidence that remotely accelerated shock-associated particles are precipitating back to the solar surface. The first results of such a search with the data from the COMPTEL instrument on the Compton Observatory are negative (McConnell et al. 1997) and there was no report of emission from the disk following the large flare on 1 June 1991 (Barat et al. 1994; Ramaty et al. 1997; Murphy et al. 1999). Furthermore, Kahler et al. (2000) saw no evidence for interplanetary particles associated with post-eruptive coronal loop structures in the absence of flares. However, it is not clear that the proper observation has been performed. The instruments on the Compton Observatory would be those that would provide the greatest probability of detecting such an unambiguous γ-rays and, thus, proton or ion acceleration signal. To summarize, the evidence seems to indicate that few or no shock-associated ions are responsible for the long duration γ-ray emission, but a positive detection after a thorough search of existing data or new data from the upcoming solar maximum would indicate otherwise.

9.3.12 Trapping and Extended Acceleration

Given the difficulty of maintaining a MHD-quiet environment for long periods of time with low densities and the proper geometry to prevent or manage drifts and the difficulty of transporting accelerated protons back from a receding shock, Ryan (2000) examine a scenario in which the protons and ions remain in the near-Sun environment and are continuously accelerated well after the impulsive phase. Such a situation was examined analytically by Ryan and Lee (1991) in order to explain the delayed high-energy phase of the 3 June 1982 event. The hypothesis is that protons accelerated during the impulsive phase are trapped in an isolated magnetic loop. The material interior to the loop is turbulent as a result of the flare, but the loop maintains its general shape and size ($\beta << 1$). The transport of the protons within a loop of length L and out of the loop is mediated by the intense turbulence ($\lambda << L$) and is characterized by spatial diffusion (as opposed to 'ballistic' trajectories in the passive trapping model). The slow diffusion, resulting from the intense turbulence, 'traps' the protons in the loop. They leak out the ends of the loop that are tied to the chromosphere and photosphere and the precipitation results in γ-ray emission. The decay time of the population in the loop has a characteristic value of

$$\tau_D = L^2 / \pi^2 \kappa_{||}, \tag{9.9}$$

where $\kappa_{||}$ is the spatial diffusion coefficient parallel to **B**, that can be a function of energy. As noted Ryan (2000), this level of turbulence is far above that of the 'saturated' turbulence case assumed by others (e.g., Hua et al. 1989), where the scattering time is on the order of the particle bounce time, i.e., $\lambda \approx L$. Even though diffusion along the loop is slow, it is far more efficient than that transverse to the loop. The cross-field resonant diffusion coefficient according to Lee (1982, 1983) is

$$\kappa_{\perp} = \eta v^4 \left(3\Omega_i \kappa_{||}\right), \tag{9.10}$$

where η is on the order of unity, Ω_i is the ion gyro-frequency and v is the particle velocity. For typical values, $\kappa_{\perp}$ is orders of magnitude smaller than $\kappa_{||}$.

As underlined Ryan (2000), this diffusion alone is not responsible for the prolonged high-energy emission from long duration gamma-ray flares. Accompanying the slow spatial diffusion in a natural way is rapid diffusion in momentum space through second-order Fermi acceleration (Schlickeiser 1986). As the particles are trapped they are also accelerated. The diffusions in real space and momentum space are inseparably linked. In this model the delayed high-energy emission does not rely explicitly on high-energy protons being present in the impulsive phase. The characteristic times for space and momentum diffusion, τ_D and τ_F, respectively, are inversely related by

$$\tau_D \tau_F = (3L/V_A)^2, \tag{9.11}$$

where L is the scale length and V_A is the Alfvén speed (Schlickeiser 1986). The spatial diffusion is due to pitch angle scattering along the field lines. Cross-field diffusion is much slower and can be safely neglected. While the spatial diffusion depletes the proton population within the loop, the remaining protons experience Fermi acceleration, thereby increasing the number of protons above some high-energy threshold, e.g., pion production. The weak link in this scenario is the origin of sustained, intense turbulence required to trap and accelerate the particles.

Ryan and Lee (1991) explained the delayed high-energy phase of the 3 June 1982 flare as the result of not only the trapping and acceleration of protons but also the threshold effects of the SMM/GRS instrument and pion production. As the trapped proton spectrum hardened, this increased the number of protons above the pion production threshold and secondary neutrons above the detection threshold of the GRS. The net result is a clear and distinct observed second or delayed phase of high-energy emission. Whereas the trapping and acceleration process is continuous and smooth, the physical and instrumental threshold effects exaggerate or enhance the effect.

As noted Ryan (2000), the general expression for the one-dimensional spatial diffusion and coupled three-dimensional momentum diffusion is

$$\frac{\partial f}{\partial t} = p^{-2}\frac{\partial}{\partial p}\left(p^2\left(\kappa_F(p)\frac{\partial f}{\partial p} - \dot{p}f\right)\right) + \frac{\partial}{\partial x}\left(\kappa_D\frac{\partial f}{\partial x}\right) + Q(x,p,t), \tag{9.12}$$

where x is the distance along the loop, t is time, p is momentum, κ_F is the diffusion coefficient in momentum space, κ_D is the spatial diffusion coefficient, f is the particle distribution function, and Q is the injection or source function. Conceivably, κ_D can be a function of space and momentum, and consequently so would κ_F. The quantity $\dot{p}$ is a momentum or energy loss term arising from proton collisions with ambient electrons. The case of κ_D independent of energy, $\kappa_F \propto p^2$, and $\dot{p} = 0$ was examined by Ryan et al. (1992) and Bennett et al. (1994) for the event of the 3 June 1982 flare. With a 20 MeV impulse injection of protons at a point x' Ryan et al. (1991) calculated the precipitation of protons >300 MeV. A distinct delayed high energy phase could be produced with spatial diffusion time scales on the order of ~100 s and momentum diffusion time scales on the order of ~500 s in a loop of length 10^{10} cm. The required level of turbulence is on the order of 10 erg.cm^{-3} assuming a 100 Gs magnetic field, i.e., $\delta B/B \sim 0.5$.

Ryan (2000) noted that for the purpose of studying the delayed high-energy phase it can be reduce Eq. 9.12 to a leaky box equation (Ryan et al. 1994b). Temporal features related to the inhomogeneous nature of the particle population are lost in this treatment. The spatial diffusion effects are imbedded in a characteristic global escape time T. The equation for this diffusion process is

$$\frac{\partial f}{\partial t} = p^{-2}\frac{\partial}{\partial p}\left(p^2\kappa_F(p)\frac{\partial f}{\partial p}\right) - \frac{f}{T} + Q(p)\delta(t). \qquad (9.13)$$

In Eq. 9.13 Ryan et al. (1994b) neglected energy loss terms and all spatial dependence. They also assumed that $\kappa_F = \kappa_{F0}p^2$, requiring that κ_D be independent of p. The injection spectrum was assumed to be of the form

$$Q(p) = \frac{N_0}{4\pi p_t^3}(\gamma - 3)\left(\frac{p}{p_t}\right)^{-\gamma} S(p - p_t), \qquad (9.14)$$

where p_t is the low-end cutoff of the particle momentum spectrum and S is the Heaviside function. Similar results were obtained as those of the case where space dependence is included. That is, once the energetic particle population is built up through the Fermi acceleration process it eventually attains a constant spectral shape with losses out the ends of the loop. After this time the relative energetic particle distribution does not change within the loop but the entire population monotonically decreases in magnitude as spatial diffusion depletes the population. This is the limiting case of the time-dependent problem.

As underlined Ryan (2000), the photon spectrum that results from the precipitating protons in either of these models, i.e., leaky box or explicit spatial diffusion, is in qualitative agreement with the deduced spectrum from the 11 June 1991 event (Mandzhavidze and Ramaty 1992a). However, the theoretical spectrum predicted from the explicit loop model is too hard relative to measurements, but can be brought into agreement by weighting with the power law input spectrum rather than the monoenergetic distribution studied by Ryan and Lee (1991). The nagging question with regard to prolonged acceleration is the maintenance of the required level of turbulence. For a 10^{10} cm loop a 10^3 km s^{-1} velocity Alfvén wave will exit the loop in 100 s–far short of the 10^4 s required for long flares such as that of 11 June 1991. Two phenomena seem relevant to this question. **The first** is observational: Comparisons of the radiative cooling time of large coronal loops with the intensity-time profiles of the thermal X-ray emission have long been known to be in disagreement, indicating that energy in some invisible form is feeding the loop for long periods of time to maintain the temperature (Jakimiec et al. 1986). MHD turbulence may be associated with this energy input and would be the energy source of the accelerated particles (Bornmann 1987). **The second** phenomenon that may preserve the turbulence is one where the loop behaves as a resonant cavity for Alfvén waves. If the MHD turbulence is generated in the corona, a likely situation, then the waves can be contained because of poor transmission through the transition region. The index of refraction for Alfvén waves changes dramatically and abruptly from the corona to the transition region. According to Hollweg (1984) and Hollweg and Sterling (1984), this almost discontinuous change in phase velocity results in almost all the wave energy being reflected at the boundary, effectively producing a cavity with a high quality factor Q for the Alfvén waves. The reflection coefficient, and thus the quality factor, depends on the abruptness of the density change in the transition region, i.e., the scale height. The quantity Q, however, is not dependent on the wave number k. An impulsive point-like disturbance can be expanded into a series of modes under the condition of perfect reflection off the boundary. With no dissipation the power spectrum is harmonic in nature with maximum amplitude at

$$k = m\pi(2x_0)^{-1}, \tag{9.15}$$

where $m = 1, 3, 5, \ldots$, and x_0 is the shortest distance to the end of the loop (the transition region). In reality, the power spectrum does not extend to infinity but truncates at $1/\lambda_D$, where λ_D is the Debye length, and long wavelength modes will be excited by the mass motions associated with the flare. However, the initial power spectrum is hard. The lowest mode has a wavelength of $2L$, where L is the length of the loop. Cascade processes will soften the spectrum over time and will attempt to smooth out the harmonic nature of the power spectrum. The quantity

$$\mathrm{Q} = L/4\pi h_{\mathrm{av}}, \tag{9.16}$$

where h_{av} is the average scale height of the corona at the two end points of the loop. If $L = 10^{10}$ cm and $h_{\mathrm{av}} = 2 \times 10^7$ cm then Q = 40. To relate this to the residence time of the turbulence, Q can be expressed in terms of τ_{free}, the decay time of the wave energy in the undriven state, i.e.,

$$\tau_{\mathrm{free}} = \mathrm{QP}/2\pi, \tag{9.17}$$

where P is the wave period. This works to the advantage of high-energy particle acceleration since the long wavelength modes are the ones that resonate with the highest energy particles, i.e., for a given Q the waves with the smallest wave number will have the longest residence time. If the wave transit time is 100 s, then the e-folding time for the energy in the loop in this mode will be on the order of 4,000 s, a period of time long enough to produce the effect necessary for all but the longest duration flares. It should not expect that the wave energy is lost out the sides of the loop since there is large mismatch between the wave numbers of the Alfvén waves in the loop and the acoustic waves that might be radiated from the loop. Therefore the particles and the waves will be constrained to the loop – the particles by the small cross-field diffusion coefficient and the waves by the 'impedance' mismatch between the Alfvén waves and the acoustic waves.

As noted Ryan (2000), a great deal of work remains to be done on this problem. In addition to not knowing how the turbulence decays in time, it must be investigate the effect of a momentum-dependent diffusion coefficient, a stratified corona, the depletion and refilling of the wave spectrum as particles tap the waves' energy for acceleration, and a spatially-dependent diffusion coefficient since the magnetic loop is not uniform in cross section; other effects could be important too.

9.3.13 Summary of Main Results

Ryan (2000) summarized main results as following:

1. The phenomenon of long-duration solar γ-ray flares has focused attention on the properties of the corona and the behavior of energetic particles long after the impulsive phase of the flare.
2. The abundance of relativistic protons and the relative absence of electrons provide clues about the nature of acceleration and the transport processes of these two species.
3. Great difficulties plague the hypothesis of the particles being accelerated early in the flare and persisting in quiet loops.
4. The demanding geometry, species independent diffusion coefficients and remarkably quiet coronal conditions all but rule out passive trapping as a general explanation of the phenomenon.
5. Even if one allows for episodic accelerations, in the absence of any other flare signature, the smooth nature of the flux decay forces these acceleration episodes to be small, frequent and exponentially declining in intensity.
6. A continuous acceleration process embodies the essential elements of the problem and employs theory that is well developed.
7. Therefore, the current available data and models seem to favor second-order stochastic acceleration in static coronal loops or CME-related DC electric field acceleration.

8. It cannot be rule out, however, the prospect that a receding shock is responsible for the prolonged particle precipitation.
9. Since only large solar flares have exhibited this phenomenon, it must be search the data for CME-associated particle events without any accompanying flare that exhibit γ-ray emission. Future measurements and observations may set strong limits on this option.
10. The upcoming solar maximum could provide those opportunities. EGRET and COMPTEL, the instruments that have shed the most light on this subject will still be operating and if funded for operation should be able to collect more data on long-duration solar γ-ray flares and the CME shocks that possibly is the origin of the late and prolonged high-energy protons.

9.4 Photospheric ^{3}He to H Abundance Ratio Derived from Gamma-Ray Line Observations

9.4.1 The Matter and Short History of the Problem

According to Yoshimori et al. (1999b), the satellite Yohkoh observed the neutron capture line and prompt nuclear deexcitation lines from a flare on 6 November, 1997. They determined the photospheric ^{3}He to H abundance ratio from the time profiles of neutron capture line. Assuming that the time profile of neutron production is similar to that of prompt C and O lines and the photospheric H density is 1.3×10^{17} cm^{-3}, Yoshimori et al. (1999b) obtained the best fit value of the ^{3}He to H ratio of $(2.3 \pm 1.4) \times 10^{-5}$ which is consistent with the values reported previously.

As noted Yoshimori et al. (1999b), ^{3}He is thought to be primarily produced by nucleo-synthesis in the early Universe and its abundance is used to place a constraint on cosmological model. Since the photospheric ^{3}He abundance cannot be determined spectroscopically, observations of the neutron capture line at 2.223 MeV from solar flares provide a direct means of determining the photospheric ^{3}He abundance. Neutrons which are produced simultaneously with prompt γ-ray lines by interactions of accelerated ions diffuse into the photosphere where the 2.223 MeV line are emitted by neutron capture on hydrogen. Because of the time required for neutrons to slow down and be captured, the 2.223 MeV line is produced about 100 s after the production of the neutrons. The competing capture reaction ^{3}He(n,p)^{3}H affects the delay of the 2.223 MeV line emission. The 2.223 MeV line flux from an instantaneous production of neutron is assumed to fall exponentially in time with a time constant τ given according to Hua and Lingenfelter (1987b) by

$$1/\tau = 1/\tau_{\mathrm{H}} + 1/\tau_{\mathrm{He}} + 1/\tau_d. \tag{9.18}$$

Here τ_H is the time constant for capture on H, τ_{He} is the time constant for capture on ^{3}He and τ_d is the neutron decay time (918 s). Values τ_H and τ_{He} are approximated by

$$\tau_{\mathrm{H}} = 1.4 \times 10^{19}/n_{\mathrm{H}}\,\mathrm{s} \quad \text{and} \quad \tau_{\mathrm{He}} = 8.5 \times 10^{14}/n_{\mathrm{He}}\,\mathrm{s}, \tag{9.19}$$

where n_{H} and n_{He} are the number densities of hydrogen and ^{3}He. A simplified approach for determination of ^{3}He abundance was adopted by Prince et al. (1983b) and Vestrand and Forrest (1993). Hua and Lingenfelter (1987b) made detailed calculations of the time profile of 2.223 MeV line emission taking into account several effects on the accelerated particles and solar atmosphere. In a case of the simplified approach the time profile of the 2.223 MeV line emission *F(t)* is expressed by

$$F(t) = A \int_{t_0}^{t} [S(t')/\tau] \exp[-(t - t')/\tau] dt', \tag{9.20}$$

where A is the constant, t_0 is the time when the gamma-ray lines are observed and $S(t')$ is the time profile of the neutron production (Vestrand and Forrest 1993). Temporal dependence of S(t′) is assumed to be similar to that of the C+O line emission. Using this formula, Yoshimori et al. (1999b) can obtain τ which gives the best fit for the observed time profile of the 2.223 MeV line emission. The ^{3}He/H ratio may be determined from this best fit τ, if n_{H} is assumed.

9.4.2 Observation and Main Results for the 6 November, 1997 Flare

As noted Yoshimori et al. (1999b), the flare occurred at 11:52 UT on 6 November, 1997. Its location, GOES class and Hα importance were S18W64, X9.0 and 2B, respectively. The γ-ray spectrometer (two BGO scintillators) aboard Yohhoh recorded strong γ-ray emission (Shiozawa 1999; Yoshimori et al. 2002). This flare exhibited the highest γ-ray counting rate in the Yohkoh events observed so far. Three counting rate time profiles of 2.136–2.375, 4.001–7.225 and 7.225–10.160 MeV emission are shown in Fig. 9.18. The first two time profiles roughly correspond to those of the neutron capture line and prompt C and O lines. It can be seen that the decay time of the 2.223 MeV line emission is longer than that of the C+O line emission.

The γ-ray count spectrum at 11:52:32–12:01:08 UT is shown in Fig. 9.19.

From Fig. 9.19 can be seen the apparent neutron capture line and prompt C and O lines superposed on the continuum. In order to derive the fluxes of these three γ-ray lines, Yoshimori et al. (1999b) used the similar spectral fitting method as described

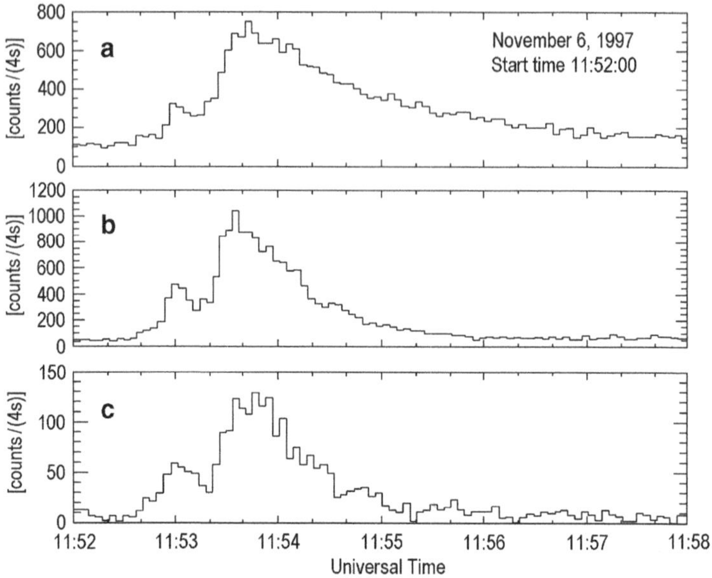

Fig. 9.18 Time profiles of γ-ray emission for 6 November, 1997 flare: (**a**) 2.136–2.375 MeV, (**b**) 4.001–7.225 MeV, and (**c**) 7.225–10.160 MeV (From Yoshimori et al. 1999b)

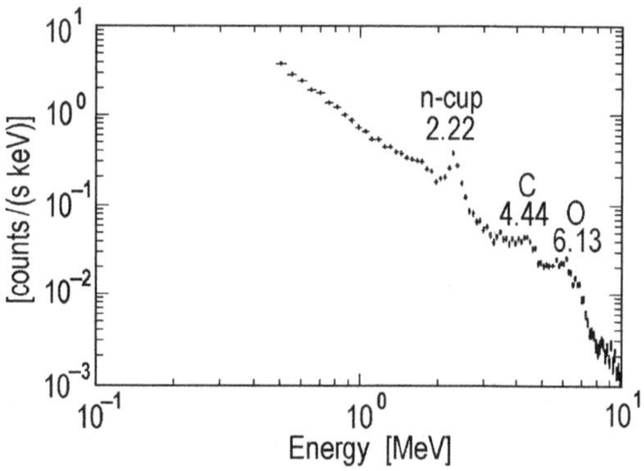

Fig. 9.19 Gamma-ray count spectrum in 11:52:36–12:01:08 UT at 6 November, 1997 (From Yoshimori et al. 1999b)

in Murphy et al. (1990b). It was assumed that the γ-ray spectrum consists of bremsstrahlung (single power law) and ten narrow and five broad lines (Gaussians). A trial incident photon spectrum is constructed and convolved with the instrumental response function. The resulting predicted count spectrum is compared with the observed one. A χ^2-square minimization algorithm is used to fit the data. In order to constrain the fits, Yoshimori et al. (1999b) fixed the line center energies and widths of the lines at their theoretical values (Murphy et al. 1990b). Free parameters in the fits are the amplitude of lines and the amplitude and exponent of the single power law. The temporal variations of γ-ray fluxes of three lines at 2.223, 4.443 and 6.129 MeV obtained from this spectral fitting method are shown in Fig. 9.20.

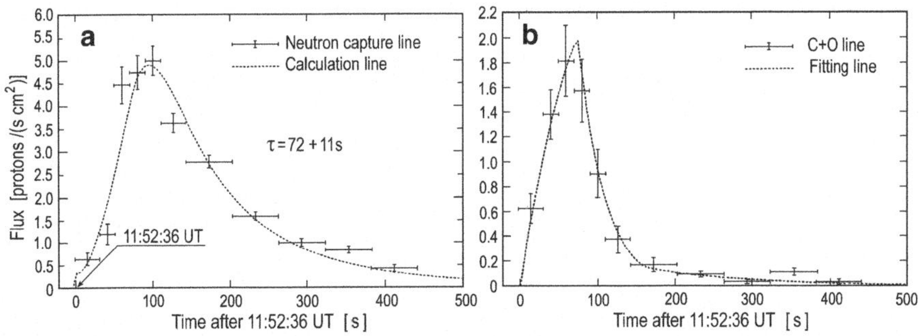

Fig. 9.20 Temporal variations of γ-ray line fluxes on 6 November, 1997 flare: (**a**) neutron capture line and (**b**) C+O lines (From Yoshimori et al. 1999b)

As underlined Yoshimori et al. (1999b), if it will be assume that the time profile of neutron production is proportional to that of C+O line emission and $n_H = 1.3 \times 10^{17}$ cm^{-3}, then the best fit value of $\tau = 72 \pm 11$ s. It gives the photospheric ^{3}He/H ratio of $(2.3 \pm 1.4) \times 10^{-5}$.

9.4.3 *Comparison of Obtained Result on Photospheric ^{3}He/H Ratio with Other Results*

Yoshimori et al. (1999b) compared obtained result on the photospheric ^{3}He/H ratio with those obtained by Prince et al. (1983b), Hua and Lingenfelter (1987a, b), Trottet et al. (1994) and Murphy et al. (1997). Hua and Lingenfelter (1987b) made the Monte Carlo calculations of the profile of 2.223 MeV line emission taking into account the photospheric ^{3}He abundance, energy spectrum and angular distribution of accelerated ions and using a model for the density distribution of the solar atmosphere. Their ^{3}He/H ratio was consistent with the upper limit obtained from the simplified approach (Prince et al. 1983b). The decay constant of 2.223 MeV emission increases as the observing angle approaches 90° (limb). Since the present Yohkoh flare on 6 November 1997 occurred at the location W64S19, this effect is not significant. Yoshimori et al. (1999b) derive a power law spectral index of spectrum of accelerated protons from a ratio of the neutron capture line to O line fluxes (Ramaty et al. 1996). The derived spectral index is 3.5 ± 0.3, which roughly corresponds to the Bessel function spectrum of $\alpha T = 0.02$ (Ramaty et al. 1993). The neutron production occurs deeper in the photosphere as the proton spectrum becomes very hard. It leads to a short decay constant for the 2.223 MeV line emission because the neutrons are captured at the site of higher density. Hua and Lingenfelter (1987b) calculated dependence of the decay constant on the proton energy spectrum. The decay constant for 2.223 MeV line emission is 90 s for $\alpha T = 0.01$ and 71 s for $\alpha T = 0.1$. The proton spectrum for this Yohkoh flare is not extremely hard, the decay constant is considered to range from 70 to 90 s.

Table 9.6 The photospheric ^{3}He to H abudance ratios derived from the time profile of 2.223 MeV line emission (From Yoshimori et al. 1999b)

^{3}He/H [10^{-5}]	Flare	Satellite and reference
<3.8	June 3, 1982	SMM/GRS (Prince et al. 1983b)
2.3 ± 1.2	June 3, 1982	SMM/GRS (Hua and Lingenfelter 1987b)
2 – 5	June 11, 1991	GRANAT/PHEBUS (Trottet et al. 1993a)
2.3	June 4, 1991	CGRO/OSSE (Murphy et al. 1997)
2.3 ± 1.4	November 6, 1997	YOHKOH/GRS (Yoshimori et al. 1999b)

All published results are summarized in Table 9.6 along with the obtained by Yoshimori et al. (1999b).

From Table 9.6 can be seen that different values on photospheric ^{3}He/H ratio are consistent within the experimental errors. Yoshimori et al. (1999b) noted that the angular distribution of accelerated ions also affects the temporal variation of 2.223 MeV line. According to Hua and Lingenfelter (1987b), the decay constant is 60 s for the δ-function pencil beam at 0° and 75 s for the δ-function fan beam at 89° when the ion spectrum is the Bessel function of $\alpha T = 0.04$. Information on the angular distribution of accelerated ions are obtained from a comparison between the calculated and observed escaping neutron fluxes. However, it was no observational information for this Yohkoh flare. A few data of photospheric ^{3}He/H ratio have been obtained from the γ-ray line spectroscopy. In order to advance the understanding of the ^{3}He/H problem, we need more precise gamma-ray line observations. Moreover, Share and Murphy (1997b) suggested a procedure for determining the photospheric ^{4}He/H ratio from the product of the solar wind ^{4}He/^{3}H ratio and the photospheric ^{3}He/H ratio. The ^{3}He/H ratio is related to the ^{4}He/H ratio which is an important parameter for studies of stellar evolution and solar neutrino production.

9.5 Low-FIP to High-FIP Elements Gamma-Ray Line Ratio

9.5.1 The Matter and Short History of the Problem

As noted Yoshimori et al. (1999c), the SMM narrow γ-ray line observations of 19 solar flares revealed flare-to-flare variations in ambient solar abundances (Share and Murphy 1995). Moreover, the OSSE observation of a very long-duration flare of 4 June, 1991 indicated that a ratio of (Mg + Si + Fe) to (C + N + O) line fluxes varied with time (Murphy et al. 1997). It needs more γ-ray spectral data to advance the understanding of the solar ambient abundances. Yohkoh satellite obtained a new result on the temporal change in γ-ray spectrum in the course of an impulsive flare. It confirms the previous Murphy et al. (1997) result from the long duration flare. Yoshimori et al. (1999c) discuss possibilities for the temporal variation of the ambient abundances.

9.5.2 *Observation of the 6 November, 1997 Event and Main Obtained Results*

According to Yoshimori et al. (1999c), satellite Yohkoh observed a flare X9.0/2B at 11:52 UT on 6 November, 1997. It lasted for about 200 s and the γ-ray spectrum at 0.5–100 MeV was measured with two BGO scintillation spectrometers (128 energy channels for 0.5–15 MeV and 16 energy channels for 15–120 MeV). Prompt nuclear γ-ray lines of C, N, O, Ne, Mg, Si and Fe, neutron capture line and higher-energy γ-rays were detected. In order to make spectral analyses, it was used a least squares fitting technique which is a similar method as described in Murphy et al. (1990b). It was assumed that the γ-ray spectrum consists of bremsstrahlung continuum (single power law) and ten narrow and five broad lines (Gaussians). Here Yoshimori et al. (1999c) fixed the line center energies and line widths at their theoretical values (Murphy et al. 1990b). The counting rate time profiles at 1.04–1.51, 2.223, 4.00–7.23 and 10–20 MeV are plotted in Fig. 9.21. As can be seen from Fig. 9.21, this flare showed similar temporal variations at these energies except 2.223 MeV. The 2.223 MeV line is delayed because it takes about 100 s for neutron capture on proton. In order to study the temporal variations of narrow lines, it was shown the time sequential γ-ray count spectra from 11:52:36 to 11:55:32 UT in Fig. 9.22. The integration time is about 20 s for each γ-ray count

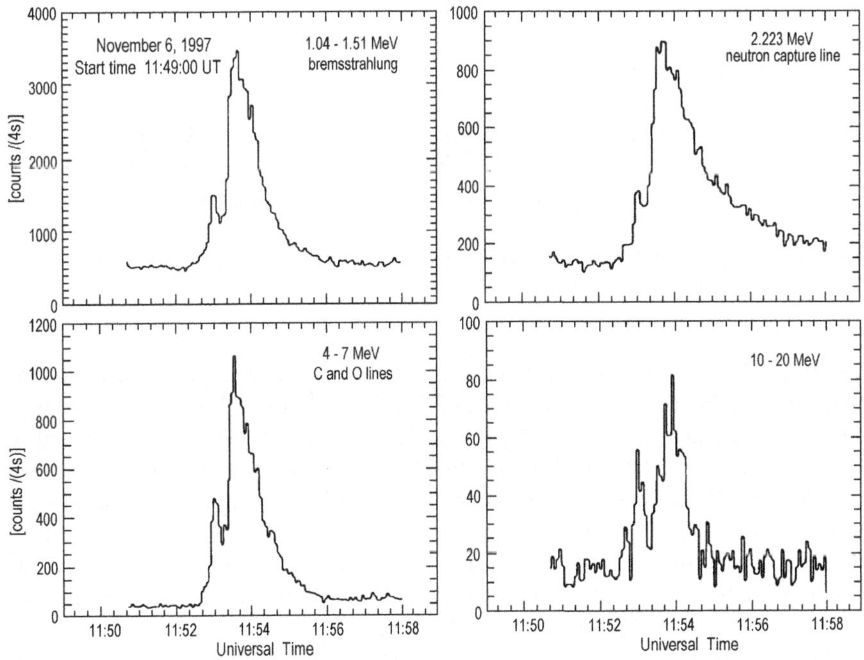

Fig. 9.21 Time profiles of 1.04–1.51, 2.223, 4.00–7.23 and 10–20 MeV emission during 6 November, 1997 flare (From Yoshimori et al. 1999c)

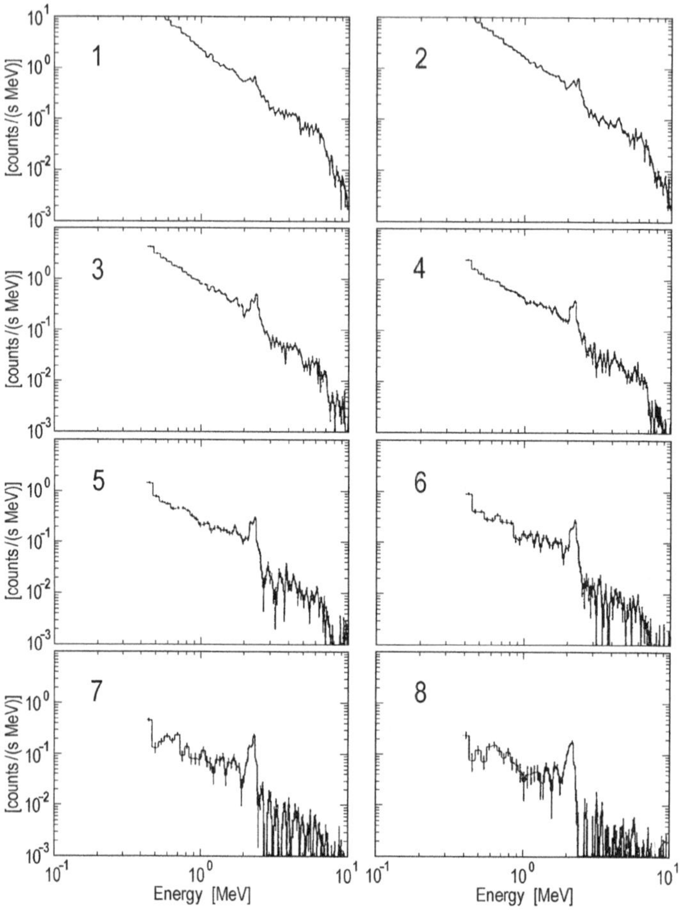

Fig. 9.22 Temporal variation of γ-ray count spectra **1–8** (each for about 20 s) from 11:52:36 to 11:55:32 UT during 6 November, 1997 flare (From Yoshimori et al. 1999c)

spectrum. The second and third spectra were measured at the peak phase of the flare. We see the neutron capture line at 2.223 MeV, C and O lines at 4.44 and 6.13 MeV and a complex of Fe (1.24 MeV), Mg (1.37 MeV), Ne (1.63 MeV) and Si (1.78 MeV) lines at 1–2 MeV. In addition, two weak line features are seen around 5.3 and 7 MeV. These are due to complexes of lines resulting from excitation and spallation of N and O. As underlined Yoshimori et al. (1999c), the bremsstrahlung continuum is dominant in the peak phase but the 2.223 MeV line is prominent in the decay phase because the decay time of the neutron capture line is longer than those of bremsstrahlung and prompt nuclear lines.

Meyer (1985a, b) concluded from the observations of solar energetic particles and solar wind that the elements separate based on the level of their first ionization potential (FIP). The abundances of elements in the flare plasma are grouped with

Table 9.7 Ratios of (Mg + Si) to Fe, Ne to (C + N + 0) and (Mg + Si + Fe) to (C + N + 0) lines for the rise, peak and decay phases (From Yoshimori et al. 1999c)

Ratio	Rise	Peak	Decay
(Mg + Si)/Fe	2.16 ± 0.91	1.90 ± 0.61	2.38 ± 0.89
Ne/(C + N + O)	0.48 ± 0.10	0.45 ± 0.06	0.50 ± 0.14
(Mg + Si + Fe)/(C + N + O)	0.52 ± 0.05	0.60 ± 0.10	1.80 ± 0.30

respect to their FIPs. The elements with FIPs exceeding about 11 eV fall into the high-FIP category (C, N, O, and Ne), while the elements below about 10 eV into the low-FIP category (Mg, Si, and Fe). Yoshimori et al. (1999c) searched for the time variations of low-FIP (Mg + Si + Fe) and high-FIP (C + N + O + Ne) narrow lines for the rise (11:52:36–11:53:20 UT), peak (11:53:20–11:54:00 UT) and decay (11:54:00–11:56:12 UT) phases of the flare. The ratios of (Mg + Si) to Fe, Ne to (C + N + O) and (Mg + Si + Fe) to (C + N + O) lines for the three phases are shown in Table 9.7.

As can be seen from Table 9.7, the ratios of (Mg + Si) to Fe and Ne to (C + N + O) lines are nearly constant within the errors throughout the flare, while the ratio of (Mg + Si + Fe) to (C + N + O) lines increased in the decay phase.

9.5.3 Discussion and Comparison with Other Results

Yoshimori et al. (1999c) discussed obtained results and compared them with other determinations. They noted that earlier Share and Murphy (1995) have shown from the SMM observations of 19 flares that the ratio of flare-averaged (Mg + Si + Fe) to (C + N + O) line fluxes varied from flare to flare. The ratios ranged from 0.2 to 0.8. Moreover, the 1991 June 4 flare exhibited that the ratio gradually increased with time. The ratio was 0.3 at 4:00 UT, 0.46 at 4:45 UT, 0.56 at 4:55 UT and 0.84 at 5:03 UT (Murphy et al. 1997). The ratio represented about a factor of 2.7 enhancements. This flare exhibited that the ratio increased as the flare progressed.

Yoshimori et al. (1999c) found from the Yohkoh satellite γ-ray observation that the impulsive flare also showed a similar increase in the ratio. The ratio did not change in the rise and peak phases but was enhanced by a factor of about 3 in the decay phase. On the other hand, the ratios of (Mg + Si) to Fe and Ne to (C + N + O) line fluxes are nearly constant throughout the flare, indicating that the line fluxes from elements with similar FIPs correlate with one another. The results of the 19 flares observed with SMM (Share and Murphy 1995) indicated that the ratios of flare-averaged low-FIP to low-FIP and high-FIP to high-FIP line fluxes did not depend on the flares, suggesting that the ratios do not vary with time within a flare. For the 6 November, 1997 flare averaged ratios of (Mg + Si) to Fe, Ne to (C + N + O) and (Mg + Si + Fe) to (C + N + O) line fluxes are in agreement with those obtained from the observations of 19 SMM γ-ray flares and 1991 June 4 flare within the experimental errors. The ratio of low-FIP to high-FIP elements in the corona was reported to be three to four times as large as that in the photosphere (Grevesse 1984; Breneman and Stone 1985; Reames 1995). The fact that the observed ratio of (Mg + Si + Fe) to (C + N + O) line fluxes increased as the flare progressed suggests the possibilities of the temporal change in the atmospheric abundances of the γ-ray production site. Although this ratio depends on the spectrum of accelerated protons, the proton spectrum did not much vary with time (power law index is 3.3 ± 0.3 during the flare).

Yoshimori et al. (1999c) considered two possibilities. One is the efficient transport of low-FIP elements to the γ-ray production site and the other is a change of γ-ray production site from the photosphere to the corona. Regarding the first possibility, whether the elemental abundances at a flare site change in a short time scale of about 100 s seems to be questionable. Regarding the second one, if the magnetic mirror points move upward to the corona from the chromosphere-photosphere, these condone may be possible. Yohkoh satellite hard X-ray images of the 1997 November 6 flare showed clear double foot-point sources. The distance between two hard X-ray sources was almost constant in the rise and peak phases but it increased gradually in the decay phase.

As underlined Yoshimori et al. (1999c), now it cannot judge whether the change in the location relates with the movement of the mirror points. It need more improved γ-ray spectral data to solve the problem of time dependence of solar atmospheric abundances, γ-ray data of high quality can be expected to obtain in the 23rd solar maximum.

9.6 Gamma Ray Measurements of the 1991 November 15 Solar Flare

9.6.1 Short History of the 15 November 1991 Flare Observations

According to Arndt et al. (1999), the 15 November 1991 solar X1.5/3B solar flare was fortuitously located near the solar central meridian and equator in NOAAAR 6919 at 13°S, 19°W. The event started at ~22:34 UT, and lasted in hard X-rays on the order of 5 min. Comparison of a Yohkoh SXT image with a Mees Observatory magnetogram suggests the flaring region is comprised of two bright X-ray kernels that are likely foot points of a magnetic flux tube or loop. Aschwanden et al. (1996) suggest a loop structure early in the flare with a radius of 13.5 Mm and an acceleration altitude of ~21.5 Mm. Diffuse X-ray emission is present, likely over the neutral line (Culhane et al. 1993). Yohkoh data indicate that during the impulsive phase that harder X-rays originate from regions near the foot points of the loop structure, while soft X-rays originate from the top of the loop (Sakao et al. 1992). In this soft X-ray source between the two foot points, Kane et al. (1993) found the temperature to be $\sim 10^7$ °K, and density to be 4×10^{11} cm^{-3}.

As noted Arndt et al. (1999), detailed analyses of this flare at several energies begin to shed light on the processes taking place within this event. A positron annihilation line at 511 keV has reported at photospheric densities of 10^{16} cm^{-3} (Kawabata et al. 1994), as well as excited ^{7}Li and ^{7}Be lines from (α,α) reactions (Kotov et al. 1996). The Ca XIX resonance line blueshifts (velocity $\sim$250 km s^{-1}) and Hα redshifts have been measured during, and at least a minute before the onset of the flare. The persistence of these plasma velocities throughout (and before) the flare suggests the shifted emission is not due to chromospheric evaporation (Culhane et al. 1993). In the 4–7 MeV range, Kotov et al. (1996) and Kawabata et al. (1994) find strong ^{12}C emission at 4.44 MeV and ^{16}O emission at 6.13 MeV (Kawabata et al. 1994). In Arndt et al. (1999) the analysis has chosen to concentrate on prompt nuclear γ-ray lines between 0.6 and 2.5 MeV, and on bremsstrahlung emission between 18.5 and 343.5 keV on the basis of Compton Observatory.

9.6.2 Observation Data from Compton Observatory

Arndt et al. (1999) used data from two experiments on the Compton Observatory to study this flare in the 18.3 keV–2.5 MeV range. Four BATSE detectors (Fishman et al. 1989) detected the event, with the most solar facing having a 97% exposure. There are analyzed data from this detector in four energy channels (0, 1, 2, and 4) spanning energies 18.5–343.5 keV, to provide an estimate of the bremsstrahlung continuum radiation. These data have excellent temporal information with 1.024 s resolution, and were obtained through the public data archive. COMPTEL (Schoenfelder et al. 1993) did not have the event in the telescope field of view (zenith angle 65.87° and azimuth angle 324.9°), and detected the event in burst mode only (Young et al. 1999). These data provide spectral information between 0.6 and 10 MeV.

Arndt et al. (1999) used COMPTEL data to define flare phases. The impulsive phase occurred between 22:36:40 and 22:38:20 UT, with the gradual phase following from 22:38:20 to 22:41:40 UT (Fig. 9.23).

As noted Arndt et al. (1999), with these start and end times, the flare duration is 300 s. In the hard and soft X-ray regimes, Yohkoh observed significant emission starting at flare onset at $\sim$22:34 UT (Sakao et al. 1992), however COMPTEL did not detect significant counts during this initial phase.

9.6.3 COMPTEL Spectrum for the 15 November 1991 Event

According to Arndt et al. (1999), COMPTEL spectra of this event (see Fig. 9.24) are deconvolved using a maximum entropy method (Gull and Skilling, M1991) and an appropriate response to account for instrumental effects (c.f. Young et al. 1999).

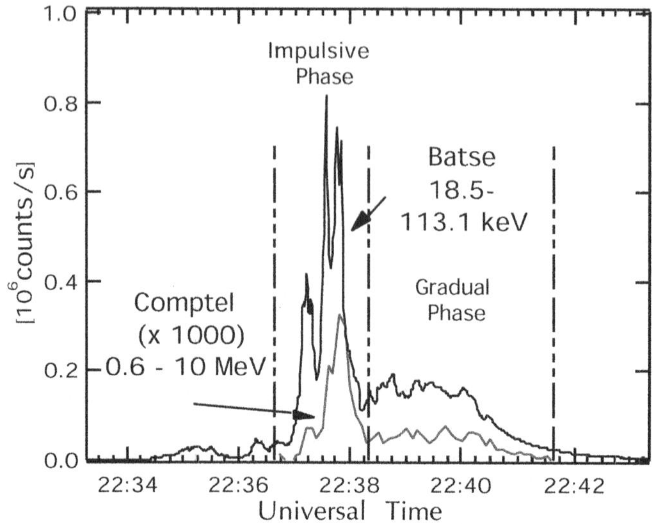

Fig. 9.23 Light curves of the 15 November 1991 event. BATSE data (dark) are from channel 4 (18.5–30.3 keV). COMPTEL counts (grey) are multiplied by 1,000, and span 0.6–10 MeV (From Arndt et al. 1999)

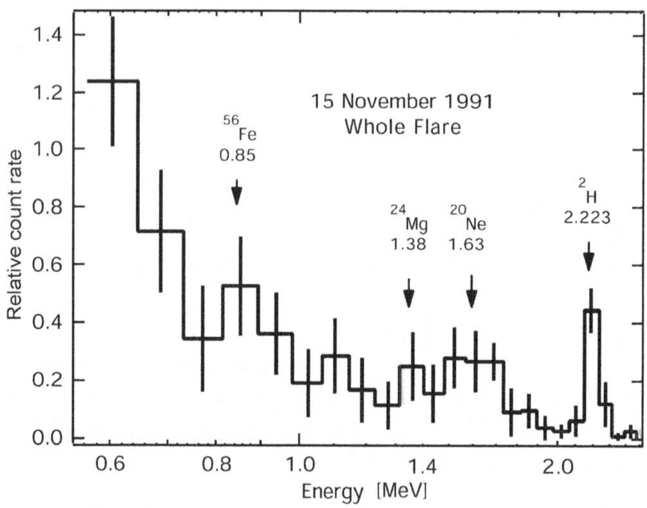

Fig. 9.24 The COMPTEL spectrum of the whole 15 November 1991 event. A few clear line features are marked (From Arndt et al. 1999)

Several robust lines are easily identified in the photon spectrum of the whole event, including the ^{2}H neutron capture (2.223 MeV), ^{56}Fe (0.85 MeV) and ^{24}Mg (1.38 MeV) lines. The ^{20}Ne contributes to the broad feature near 1.6 MeV.

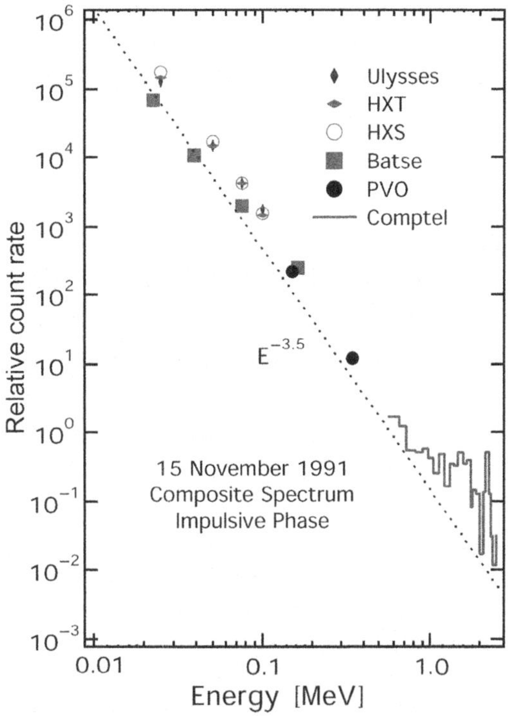

Fig. 9.25 Composite spectrum of the impulsive phase of the 15 November 1991 flare using data from Yohkoh and Ulysses (Kane et al. 1998), PVO (McTiernan et al. 1994), BATSE and COMPTEL (Arndt et al. 1999) (From Arndt et al. 1999)

9.6.4 Composite Spectrum for the 15 November 1991 Event

Arndt et al. (1999) have generated a composite spectrum of the impulsive phase spanning 18.5 keV to 2.5 MeV using BATSE, COMPTEL, PVO (McTiernan et al. 1994), Ulysses and Yohkoh HXT and HXS data (Kane et al. 1998). The BATSE, PVO, Yohkoh, and Ulysses data provide an estimate of the bremsstrahlung continuum radiation. In Fig. 9.25 were omitted (small) error bars in the interest of keeping the plot legible; line $E^{-3.5}$ is plotted for reference. In Fig. 9.25 have omitted also the BATSE channel 3 (>343.5 keV) since nuclear line contributions contaminate the continuum emission, making it difficult to isolate the contribution of accelerated electrons.

As noted Arndt et al. (1999), the Yohkoh and Ulysses fluxes are lightly higher in magnitude since they are integrated over 14 s near flare maximum, while BATSE and COMPTEL data are integrated for 97 s over the entire impulsive phase as defined above. PVO data were integrated for 20 s soon after the onset of the flare. Ulysses spectra were fitted with $E^{-3.08}$ for $0.02 < E < 0.15$ MeV, and Yohkoh data were fitted with power laws of $E^{-3.20}$ (HXS) and $E^{-3.39}$ (HXT) for $0.014 < E < 0.09$ MeV; discrepancies between the Yohkoh and Ulysses data are discussed in detail by Kane et al. (1998). In the composite spectrum, it can be seen an extension up to 2.5 MeV of the bremsstrahlung spectrum measured by other instruments.

Furthermore, it can be seen a strong nuclear component above 1 MeV from heavy nuclei (e.g. ^{24}Mg and ^{20}Ne) with low thresholds. These lines have relatively strong intensities compared to that of the neutron capture line, indicating a reasonably soft proton spectrum.

9.7 Gamma-Ray Spectroscopy of the Giant 1991 June 1 Behind-the-Limb Flare: Evidence for Gamma-Ray Production in the Corona and Accelerated Heavy Ion Abundance Enhancements

9.7.1 The Matter of the Problem of Gamma-Ray Production in the Corona

Ramaty et al. (1997) investigated the implications of the γ-ray line emission observed with Granat/PHEBUS from the behind-the-limb flare on 1991 June 1. They showed that thin target interactions are required to account for the very high observed ratio of the nuclear line emission in the 1.1–1.8 MeV and 4.1–7.6 MeV energy bands, which are populated predominantly by γ-rays from deexcitations of Ne–Fe and C–O, respectively. Ramaty et al. (1997) found that the composition of the accelerated particles that produce γ-rays in this extremely powerful gradual flare clearly shows the heavy-element abundance enhancements characteristic of acceleration via resonant wave-particle interactions.

These very important results were obtained on the basis of observations an exceptionally powerful solar flare occurred on 1991 June 1. It originated from Active Region 6659, which, at the time of the flare, was located about 6–9° behind the east limb of the Sun. Later in June, when this active region was on the disk, it produced five more very intense X-class flares. Gamma-ray emission, sometimes with energies up to a few GeV, was observed from these flares with instruments on the CGRO and Granat, and with GAMMA-1. The 1991 June 1 flare was apparently the most powerful in the series, and in fact one of the most powerful X-ray and γ-ray flares ever observed. Although the sources of the X-rays and γ-rays were partially occulted for Earth orbiting spacecraft, they were in full view for the instruments on Ulysses, which was located 22° behind the east limb at the time of the flare. Assuming that the observed hard X-rays were nonthermal, Kane et al. (1995) found that an extremely large number of greater than 20 keV electrons, containing $\sim 10^{34}$ erg, were needed to account for the observations.

As underlined Ramaty et al. (1997), despite the fact that the flare was located behind the limb, hard X-rays and γ-rays, including nuclear line emission, were observed with the PHEBUS instrument on Granat satellite (Barat et al. 1994). Based on the location of the active region, it was estimated that this emission originated in the corona, at an altitude of at least 3,000 km above the photosphere.

This is consistent with the absence of the 2.223 MeV γ-ray line that results from neutron capture on hydrogen in the photosphere. The 1991 June 1 flare is the first instance that it is possible to associate observed γ-ray emission with accelerated particle interactions taking place **in the corona**. Based on timing arguments and on the interaction mean free paths of the accelerated ions, it is generally believed that, while particle acceleration takes place in the corona, the bulk of the solar flare γ-ray emission is produced by precipitating particles near the foot-points of flare loops. This scenario is consistent with energy dependent time delays (Aschwanden et al. 1995) and flare images (e.g., Kosugi 1996) in hard X-rays.

Ramaty et al. (1997) noted that previously by Ramaty (1995) and Trottet et al. (1996) have shown that the γ-ray spectrum observed from the 1991 June 1 flare is more consistent with thin target than thick target interactions. They have also pointed out that a thin target model is appropriate for the corona, and that it allows to obtain information on the composition of the accelerated particles. While a thin target is a reasonable model for a flare behind the limb, there are convincing arguments showing that such a model is not applicable for flares on the disk. For such flares, a thin target can only be possible if the bulk of the γ-ray-producing ions escape to interplanetary space. This, however, is ruled out because of the poor correlation between the γ-ray and solar energetic particle (SEP) fluxes (Cliver et al. 1989) and the lack of spallation products in the escaping particles (Murphy and Ramaty 1984). On the other hand, for a behind-the-limb flare, the particles that escape downward toward the photosphere produce gamma rays that cannot be observed.

As Ramaty et al. (1997) underlined, the accelerated particle composition derived from γ-ray observations adds unique insights to results obtained from SEP observations in interplanetary space. These SEP data imply the existence of two classes of particle events: gradual and impulsive (Cane et al. 1986; Reames et al. 1994). In gradual events, the composition is coronal. Impulsive events exhibit ^{3}He-to-^{4}He ratios that are up to four orders of magnitude higher than the standard solar value, heavy element (Ne-Fe) enhancements relative to C and O, and high electron-to-proton (e/p) ratios. The particles from gradual events are most likely accelerated by coronal mass ejection (CME) driven shocks (e.g., Kahler 1996; Reames 1996). On the other hand, the abundance anomalies observed in impulsive events suggest strongly that the particles are accelerated by resonant wave-particle interactions (Temerin and Roth 1992; Miller and Vinas 1993; Miller and Roberts 1995). The 1991 June 1 γ-ray observations also point toward strong heavy-element enhancements. In addition, the γ-ray data indicate that the enhancement is a function of time, increasing as the flare progresses.

As Ramaty et al. (1997) noted, the γ-ray emission produced by accelerated ions consists of narrow lines, resulting from the interaction of accelerated protons and α-particles with ambient heavy nuclei, and broad lines, produced by the inverse reactions between the accelerated heavy nuclei and the ambient hydrogen and helium (Ramaty et al. 1979). Narrow-line fluences, extracted from the SMM data for 19 solar γ-ray flares (Share and Murphy 1995), have been fitted with calculated fluences allowing the determination of ambient medium abundances (Ramaty et al.

1995, 1996). This narrow-line analysis has shown that the abundances of Mg, Si, and Fe (first ionization potential, FIP, less than 10 eV) relative to C and O (FIP > 11 eV) are consistent with coronal abundances but enhanced in comparison with photospheric abundances. The enhancement of low-FIP/high-FIP element abundance ratios in the corona relative to the photosphere is well established from X-ray, EUV, SEP, and slow solar wind observations (e.g., Meyer 1996 and references therein). The γ-ray results complement these observations by showing that the FIP bias is already present in the chromosphere since, as mentioned above, it is unlikely that the bulk of the gamma rays are produced in the low-density corona. The results of the narrow-line analysis concerning the strong Ne line at 1.634 MeV imply that the accelerated ion energy spectra should extend as unbroken power laws down to at least about 1 MeV/nucleon in order to account for the observed fluence ratios of this line to that of the ^{16}O line at 6.129 MeV with a reasonable (not too high) Ne/O abundance ratio. The most important consequence of this result is that the energy contained in the ions is approximately equal to that contained in the greater than 20 keV electrons (Ramaty et al. 1995; Mandzhavidze and Ramaty 1996), contrary to the previously accepted paradigm that the energy content in the ions amounts to only a few percent of that in the electrons (Lin and Hudson 1976; Forman et al. 1986). The derived ion energy contents for the 19 SMM flares range from 4×10^{29} to 3×10^{32} erg.

As noted Ramaty et al. (1997), for individual broad lines no direct determination of their fluences has yet been made. Nevertheless, the full-fit analysis of the 1981 April 27 flare (Murphy et al. 1991), and the narrow-line analysis, including the 2.223 MeV line for several flares (Ramaty et al. 1995, 1996), shows that the accelerated particles that produce the γ-rays exhibit heavy-element and ^{3}He enhancements similar to those observed in impulsive SEP events. Furthermore, high e/p ratios, similar to the values observed in impulsive SEP events, were obtained from the analysis of γ-ray lines and continuum (Ramaty et al. 1993). These results suggest that, independent of whether the flare is impulsive or gradual, the particles that produce the γ-rays are accelerated by the same mechanism as the one that accelerates particles in impulsive SEP events (see also Mandzhavidze and Ramaty 1993 and Cliver 1996). Spectral analysis of γ-rays in the energy range 0.57–10.25 MeV was performed for the 1991 June 1 flare by Trottet et al. (1996). This analysis shows a clear excess of nuclear line emission above the electron bremsstrahlung continuum (Fig. 9.26).

According to Ramaty et al. (1997), the energy resolution of the PHEBUS instrument is insufficient to determine the fluxes of individual narrow lines. However, it was possible to determine the fluences of the nuclear line emission in two broad energy bands, 1.1–1.8 MeV and 4.1–7.6 MeV, in six consecutive time intervals of 128 s duration each. It was found that the ratio R of these fluences was exceptionally high and, moreover, that it increased with time, reaching the highest value toward the end of the flare. It is a property of nuclear deexcitation radiation produced by accelerated particle interactions that the photon energy range from about 1 to 2 MeV is dominated by lines from heavy elements (Ne-Fe) while the 4–7 MeV range is mostly from C and O (Ramaty et al. 1979). Since these energy

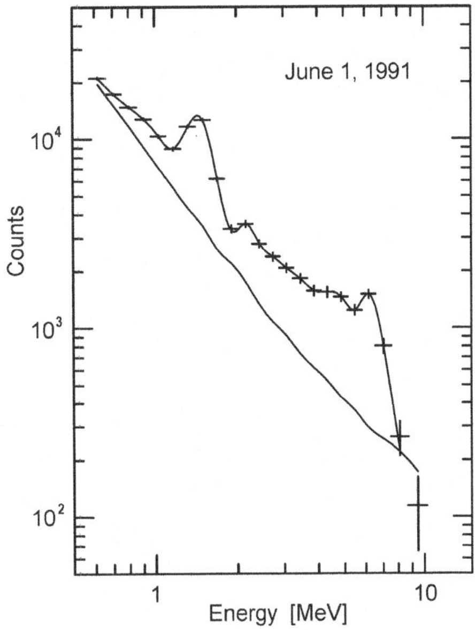

Fig. 9.26 Background-subtracted total count spectrum of γ-ray emission from the 1991 June 1 flare observed with Granat/PHEBUS (Trottet et al. 1996). The bremsstrahlung component is shown by the curve beneath the data points (From Ramaty et al. 1997)

bands contain narrow and broad lines, the ratio R depends on both ambient and accelerated particle abundances. But, the observed values of the ratio R are so high that even assuming maximal ambient heavy-element abundances, consistent with the narrow-line analysis of other flares, it is still necessary to invoke heavy-element enhancements in the accelerated particles.

Using the Ulysses hard X-ray observations (Kane et al. 1995) and scaling the energy content in ions to that in the greater than 20 keV electrons, Ramaty et al. (1997) estimate the nuclear line emission produced in the occulted sub-coronal regions. The 1991 June 1 flare thus offers the first opportunity of comparing γ-ray production in the corona with that in sub-coronal regions. Since it is most likely that the ions are accelerated in the corona, such a comparison can provide unique information on the transport of the particles.

9.7.2 *Accelerated Particle Composition and Interaction Model*

Ramaty et al. (1997) employ the same nuclear deexcitation line code (Ramaty et al. 1979) that was used in previous abundance analyses (Murphy et al. 1991; Ramaty et al. 1995, 1996). They calculate the total (narrow plus broad) nuclear line emission in the 1.1–1.8 MeV and 4.1–7.6 MeV energy bands, Φ(1.1–1.8 MeV) and Φ(4.1–7.6 MeV). Ramaty et al. (1997) assume that the γ-rays escape from the interaction region without significant attenuation, hence the fluence ratio is simply equal to

Table 9.8 Ambient medium abundencies (From Ramaty et al. 1997)

Ratio	C_1	C_2
$^1H/^4He$	27.8	27.8
$^4He/O$	57	57
C/O	0.47	0.47
N/O	0.12	0.12
Ne/O	0.15	0.25
Mg/O	0.20	0.4
Al/O	0.016	0.032
Si/O	0.15	0.30
S/O	0.032	0.064
Ca/O	0.011	0.022
Fe/O	0.13	0.39

Table 9.9 Accelerated particles abundencies (From Ramaty et al. 1997)

Ratio	G	I_1	I_2
$^1H/^4He$	10	10	10
$^3He/^4He$	0	1	1
$^4He/O$	35, 90	15, 46	15, 46
C/O, N/O	Corona	Corona	Corona
(Ne-S)/O	Corona	3 × Corona	10 × Corona
Fe/O	Corona	10 × Corona	20 × Corona

$$R = \Phi(1.1 - 1.8\mathrm{MeV})/\Phi(4.1 - 7.6\mathrm{MeV}). \tag{9.21}$$

It was shown by Hua et al. (1989) that the attenuation of nuclear line emission is negligible for disk flares. On the other hand, for a behind-the-limb flare, Compton scattering can affect the spectrum of the escaping γ-rays. Since the effects of the scattering and attenuation are more pronounced for lower energy photons, the incorporation of Compton scattering would lower the calculated value of R and hence strengthen the conclusions concerning the heavy-element abundance enhancements.

Ramaty et al. (1997) consider two cases for the ambient composition: a SEP-derived coronal composition (Reames 1995), hereafter C_1 and a modified coronal composition, hereafter C_2 (see Table 9.8). The C_2 composition is modified relative to C_1 as follows: Ne/O = 0.25 instead of 0.15, Mg/O–Ca/O are enhanced by a factor of 2, and Fe/O by a factor of 3. These compositions are based on the analysis of narrow γ-ray lines (Ramaty et al. 1995, 1996): Ne/O = 0.25 is needed to account simultaneously for the observed (Share and Murphy 1995) 1.634/6.129 MeV and 2.223/4.44 MeV line fluence ratios, and the Mg, Si, and Fe enhancements represent the highest abundances of these elements found in that analysis. The 1.634, 4.44, and 6.129 MeV lines are from ^{20}Ne, ^{12}C, and ^{16}O deexcitations, respectively.

For the accelerated particles, Ramaty et al. (1997) consider three cases (Table 9.9): gradual flare composition (Reames 1995), hereafter G; impulsive flare composition with average heavy element enhancements (Reames 1995), hereafter I_1; impulsive flare composition with the highest observed heavy-element enhancements (Reames et al. 1994), hereafter I_2.

According to Table 9.9, the C-Fe abundances are as follows: for G, they are identical to C_1 in Table 9.8 (named Corona in Table 9.9); for I_1 Ne/O-Ca/O are enhanced by a factor 3, and Fe/O by a factor of 10, relative to G; for I_2 the respective enhancement factors are 10 and 20; for both I_1 and I_2, C/O and N/O are identical to C1. The proton, α-particle, and ^{3}He abundances in accelerated particles are also given in Table 9.9. For all three cases, Ramaty et al. (1997) took $\alpha/p = 0.1$. Even though $\alpha/p << 0.1$ is seen in SEP, such values would underproduce the 2.223 MeV line relative to other narrow γ-ray lines (Ramaty et al. 1995, 1996). Ramaty et al. (1997) found that larger values of α/p, which are consistent with the narrow line data (up to $\alpha/p = 0.5$), do not affect their conclusions. Ramaty et al. (1997) ignored ^{3}He for the gradual flare composition and took ^{3}He/^{4}He = 1 for both I_1 and I_2. For fixed α/p, ^{4}He/O controls the relative importance of the broad and narrow lines. For G, the range 35–90 represents the variation of ^{4}He/O observed in space (Reames 1995); ^{4}He/O = 46 is the average impulsive SEP value (Reames 1995); ^{4}He/O = 15 is the 1σ lower bound for the impulsive SEP considered by Mason et al. (1986).

Ramaty et al. (1997) carry out the calculations in both thick and thin target models (e.g., Ramaty 1986). In the thick target, the accelerated particles lose energy while they produce γ-rays. In a thin target, the particles either escape from the interaction region before losing much energy or the energy losses are compensated by acceleration. In the thick target, the Coulomb energy losses, which are proportional to Z^2/A (nuclear charge number squared over mass number), reduce the importance of the broad lines relative to the narrow lines, as well as the importance of the broad lines from Ne-Fe relative to the broad lines from C-O. These reductions are not present in the thin target. Consequently, for identical parameters (accelerated particle energy spectrum and composition, ambient composition), the ratio R is higher for a thin target than for a thick target. Ramaty et al. (1997) assume that all particle species have a power-law spectrum with the same spectral index S. In Figs. 9.27 and 9.28, are compare these calculations with the PHEBUS data given by the vertical bars (Trottet et al. 1996).

In Figs. 9.27 and 9.28 the horizontal lines represent the calculated values of R for various compositions and values of S. Ramaty et al. (1997) first demonstrate that ratios Φ(1.1–1.8 MeV)/Φ(4.1–7.6 MeV) predicted by the thick target model is not high enough to account for all the data. From the top panel of Fig. 9.27 can be seen that the calculations can account for the observation data only before the fourth time period. At later times, agreement with the data can only be obtained for extremely steep spectra (S > 4.5) that are inconsistent with the spectra obtained from γ-ray line data for other flares (e.g., Ramaty et al. 1993, 1996). This inconsistency is even stronger in the bottom panel of Fig. 9.27, where, instead of the minimal α/O = 15 was used the average value for impulsive flares, α/O = 46.

Next Ramaty et al. (1997) consider the thin target; results shown in Fig. 9.28. The top panel is for the gradual flare composition for parameters that yield the highest ratio $R = \Phi(1.1\text{–}1.8\ \text{MeV})/\Phi(4.1\text{–}7.6\ \text{MeV})$. Unlike for C_2I_2 combination, for which the highest R is achieved with the lowest broadline contribution, for the C_2G combination, the highest ratio R is achieved with the lowest broadline

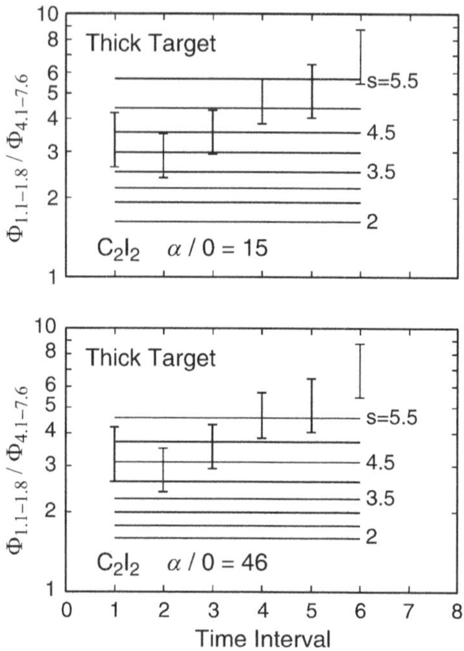

Fig. 9.27 Ratios Φ(1.1–1.8 MeV)/Φ(4.1–7.6 MeV) of γ-ray line emission in the 1.1–1.8 MeV and 4.1–7.6 MeV energy bands. The horizontal lines are calculated ratios for accelerated particles with various power-law spectral indices S interacting in a thick target. The vertical bars are the observed ratios in six consecutive time intervals of 128 s duration each (Trottet et al. 1996). The assumed compositions of the ambient medium and of the accelerated particles are indicated in the figures and described in Tables 9.8 and 9.9) (From Ramaty et al. 1997)

contribution. This is because the Ne-Fe abundance ratio relative to C-O is higher for C_2 than for G. It can be seen from the **top panel** of Fig. 9.28 that there is a clear discrepancy between the observation data and the calculations, showing that the gradual flare composition is unacceptable and that heavy-element enhancement is required. The **middle and bottom panels** of Fig. 9.28 show results for the two impulsive flare compositions, I_1 and I_2, the average a/O = 46 for impulsive flares, and the C_1 composition for the ambient medium. Ramaty et al. (1997) thus conclude that the observed gamma-ray emission from the 1991 June 1 flare was produced by particles with impulsive flare composition. Although, early in the flare, with extreme assumptions concerning the composition of both the accelerated particles and the ambient medium, thick target interactions could account for the data, later on, the only option is a thin target.

9.7.3 Accelerated Particle Energy Deposition and Content

As noted Ramaty et al. (1997), the total energy content in the accelerated particles that produce the observed gamma rays consists of the energy deposited in the coronal interaction region, mainly the result of Coulomb collisions, and the energy carried away from this region by escaping particles. Ramaty et al. (1997) first

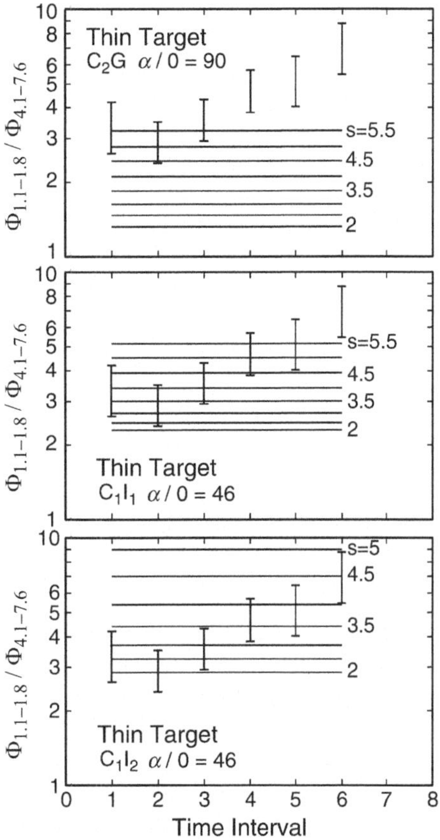

Fig. 9.28 The same as Fig. 9.27, but for a thin target interaction model. It can be seen that accelerated particles with gradual flare composition (*top* panel) do not fit the data (From Ramaty et al. 1997)

calculate the energy deposited in the coronal region. The energy deposition rate that accompanies thin target γ-ray production is given by

$$\dot{W}_{thin} = \sum_i \left(Z_i^2/A_i\right) \int_{E_{\min}}^{\infty} N_i(E,t) \frac{dE}{dt} dE, \tag{9.22}$$

where $\left(Z_i^2/A_i\right) dE/dt$ is the energy-loss rate per nucleon of ionic species i in a plasma with proton density n_H (Forman et al. 1986), E is energy per nucleon, and

$$N_i(E,t) = N_i E^{-S(t)} \tag{9.23}$$

is the instantaneous differential number of particles i (number per MeV/nucleon) in the interaction region. Since both the thin target photon flux at Earth, Φ_{thin} and $\dot{W}_{thin}$ are proportional to $N_i n_H$ for a given Φ_{thin}, the energy deposition rate can be determined independent of the ambient density.

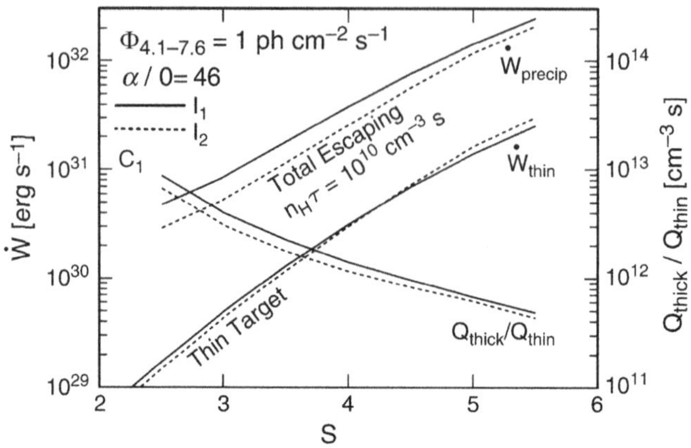

Fig. 9.29 Energy deposition rate ($\dot{W}_{thin}$), energy carried away per second by precipitating particles ($\dot{W}_{precip}$), and ratio of thick to thin target γ-ray production (Q_{thick}/Q_{thin}) for 4.1–7.6 MeV photons. The $\dot{W}$ are normalized to Φ_{thin}(4.1–7.6 MeV) = 1 photon.cm^{-2}s^{-1}. The compositions of the ambient medium and accelerated particles are coronal (C_1) and impulsive flare (I_1, I_2), respectively (From Ramaty et al. 1997)

Ramaty et al. (1997) evaluated Eq. 9.22 for α/O = 46, the C_1I_1 and C_1I_2 compositions, Φ_{thin}(4.1–7.6 MeV) = 1 photon.cm^{-2}s^{-1}, and E_{min} = 1 MeV/nucleon (this value of E_{min} follows from the narrow-line analysis and the constraints on the Ne abundance). The results are shown by the $\dot{W}_{thin}$ curves in Fig. 9.29. Combining these calculations with the time-dependent spectral indices obtained from Fig. 9.28 for the I_1 and I_2 compositions (I_1 for the three first time periods and I_2 later on) and the observed time dependent 4.1–7.6 MeV fluences for the 1991 June 1 flare (Trottet et al. 1996), Ramaty et al. (1997) derived the total deposited energy in the coronal interaction region,

$$W_{thin} \approx 2.4 \times 10^{32} \text{erg}. \tag{9.24}$$

Even though W_{thin} probably constitutes only a fraction of the total energy in accelerated ions, it exceeds the total energy content (Ramaty et al. 1995) in ions for all but one of the 19 SMM flares observed on the disk.

According to Ramaty et al. (1997), the energy carried away from the coronal region by ions precipitating down to the loop footpoints, where they produce γ-rays via thick target interactions, is given by

$$\dot{W}_{precip} = \tau^{-1} \sum_i A_i \int_{E_{min}}^{\infty} N_i(E,t) E dE, \tag{9.25}$$

where the ion trapping time in the coronal region τ depends on the magnetic field structure and size of the loops, and on the energy density in plasma turbulence in the loops that can pitch angle-scatter the ions (e.g., Ramaty and Mandzhavidze 1994). Since Φ_{thin} is proportional to $N_i n_H$, Ramaty et al. (1997) calculate $\dot{W}_{precip}$ normalized to $\Phi_{\rm thin}$(4.1–7.6 MeV) = 1 photon.cm^{-2}s^{-1}, with $n_H\tau$ a free parameter.

The curves labeled $\dot{W}_{precip}$ in Fig. 9.29 are for the C_1I_1 and C_1I_2 compositions (see Tables 9.8 and 9.9), $n_H\tau = 10^{10}$ cm^{-3}s^{-1}, and E_{min} = 1 MeV/nucleon. Combining these calculations with the derived time-dependent spectral indices and the observed time-dependent 4.1–7.6 MeV fluences, as above, Ramaty et al. (1997) obtain

$$W_{precip} = 2.5 \times 10^{33} \times [10^{10}\text{cm}^{-3}\text{s}/(n_H\tau)] \text{ erg.} \tag{9.26}$$

By comparing this energy with the 10^{34} erg in greater than 20 keV electrons found by Kane et al. (1995) for this flare, and assuming equipartition between the energy in these electrons and the energy in the greater than 1 MeV/nucleon ions, Ramaty et al. (1997) obtain

$$n_H\tau = 2.5 \times 10^9\text{cm}^{-3}\text{s}^{-1}. \tag{9.27}$$

Since τ cannot be shorter than ~1 s (the shortest transit time of 1 MeV/nucleon ions through a 10^9 cm long loop), strict equipartition would require

$$n_{\rm H} < 2.5 \times 10^9\text{cm}^{-3}. \tag{9.28}$$

Ramaty et al. (1997) noted that Barat et al. (1994), by assuming perfect trapping of particles in the corona, estimated that $n_{\rm H} \sim 10^{11}$ cm^{-3}. This estimate was based on an upper limit implied by the comparison of the time profiles of the γ-ray line emission and nonrelativistic electron bremsstrahlung and a lower limit based on the absence of delays between hard X-ray peaks at different energies. This estimate, however, breaks down if the particles are allowed to precipitate out of the coronal trap and if the ions and nonrelativistic electrons have different acceleration time profiles.

The gamma-ray line flux due to the precipitating ions is given by

$$\Phi_{thick} = (n_H\tau)^{-1}(Q_{thick}/Q_{thin})\Phi_{thin}, \tag{9.29}$$

where Q_{thick} measured in photons, is the total thick target γ-ray yield due to ions impinging on the thick target interaction region, and Q_{thin} measured in photons. cm^{3}s^{-1}, is the instantaneous thin target γ-ray line production rate in an ambient medium of unit proton density. For both Q_{thick} and Q_{thin} the ion energy spectra have the same normalization; Q_{thick} was calculated for a neutral ambient medium. The ratio Q_{thick} (4.1–7.6 MeV)/Q_{thin} (4.1–7.6 MeV) is shown in Fig. 9.29 for the C_1I_1 and C_1I_2 compositions as a function of the spectral index S of the accelerated particles in the coronal interaction region. Combining again these calculations with the derived time-dependent spectral indices and the observed time-dependent thin target 4.1–7.6 MeV fluences, Ramaty et al. (1997) obtain

$$\Phi_{thick}(4.1-7.6\mathrm{MeV}) = 1.4\times10^{4}\times\left[10^{10}\mathrm{cm}^{-3}\cdot \mathrm{s}/(n_{H}\tau)\right]\mathrm{photons}\cdot\mathrm{cm}^{-2}. \quad (9.30)$$

Assuming, conservatively, that the energy contained in the ions is only 10% of the energy in the greater than 20 keV electrons, Ramaty et al. (1997) obtain

$$\Phi_{thick}(4.1-7.6\ \mathrm{MeV}) = 5{,}600\ \mathrm{photons.cm}^{-2}. \quad (9.31)$$

Even this conservative estimate is much larger than the nuclear line emission observed from any other flare. This result is consistent with the giant flare concept introduced by Kane et al. (1995).

9.7.4 Main Results and Discussion

Ramaty et al. (1997) have investigated the implications of the γ-ray line emission observed with Granat/PHEBUS from the behind-the-limb flare on 1991 June 1. They show that a thin target interaction model allows accounting for the very high observed ratio of the nuclear line emission in the 1.1–1.8 MeV and 4.1–7.6 MeV energy bands. These energy bands are populated predominantly by γ-rays from deexcitations of Ne-Fe and C-O, respectively. While, for flares on the disk, the observed γ-ray emission is produced predominantly in thick target interactions, a thin target is a reasonable model for a flare behind the limb. Ramaty et al. (1997) found that the composition of the interacting accelerated particles shows clearly the heavy-element abundance enhancements typical for impulsive flares, despite the fact that the 1991 June 1 flare was a gradual flare. The phenomenon of heavy-element enhancement for impulsive flares is well established from SEP observations. However, these observations only provide information on the total time-integrated effect, which, in addition, is convoluted by interplanetary transport. On the other hand, the γ-rays provide in situ information within the actual time-scale of particle acceleration (10–15 min for the 1991 June 1 flare).

Ramaty et al. (1997) found that the heavy-element abundance enhancements increase with time, reaching, toward the end of the flare, values comparable to the highest values observed in space from impulsive SEP events. At the same time, the efficiency of acceleration goes down, which is manifested by the decay of γ-ray and hard X-ray emissions and the steepening of the spectrum of the accelerated particles. Ramaty et al. (1997) have shown that for the 1991 June 1 flare, the energy deposited in just the coronal interaction region is comparable to the largest total accelerated particle energy contents in previously observed γ-ray flares. For the heavy ion-enriched accelerated particle compositions that it need to account for the nuclear line data, only about 1/3 of the total ion energy content is in protons, with the rest in α-particles and heavier nuclei. Since the γ-ray fluences from the occulted regions of the flare are not available, Ramaty et al. (1997) have made the conservative assumption that the energy content in the precipitating ions is 10% of that in greater than 20 keV electrons. The hard X-rays produced by these electrons were observed with Ulysses, for which the flare was in full view. Ramaty et al. (1997) found that this assumption implies nuclear line fluences that exceed the largest previously observed values by more than an order of magnitude. Furthermore, the approximate equipartition between the energy in the greater than 20 keV electrons and the energy in the greater than 1 MeV ions implies a relatively low product of the ion trapping time τ and the ambient density n_H in the coronal interaction region. Specifically, if the energy content in the ions exceeds 10% of that in the electrons, $n_H\tau < 2.5 \times 10^{10}\ cm^{-3}.s$.

9.8 Gamma-Ray Spectroscopy of the 1991 June 4 Solar Flare: Accelerated Particle Composition, Energetics, and Ambient Abundances

9.8.1 The Matter and Short History of the Problem

As noted Murphy et al. (1997), prior to the launch of the Compton Gamma Ray Observatory (CGRO), most solar γ-ray line measurements had been made with the SMM, Hinotori, Yohkoh, and GRANAT γ-ray spectrometers (e.g., Chupp 1987; Yoshimori 1989; Yoshimori et al. 1994; Vilmer 1994). With the launch of the CGRO in April 1991, a new resource for high energy solar measurements became available. In early June, less than 2 months after launch, active region 6659 rotated onto the disk and produced some of the largest X-ray flares ever recorded. On June 1, AR 6659 appeared at the east limb and produced an X12+ flare. The peak of this flare was missed by the CGRO instruments because the satellite was in the South Atlantic Anomaly (SAA), but it was observed by the PHEBUS experiment on GRANAT (Barat et al. 1994). As there was a high probability for subsequent intense flares, a solar target of opportunity was declared for the Oriented Scintillation Spectrometer Experiment (OSSE) on CGRO, which is sensitive to γ-rays from 0.04 to more than 150 MeV. The Sun was then observed for roughly 30–40 min during each orbit and, on June 4, AR 6659 produced a second X12] flare while OSSE was viewing the Sun. The flare was one of the most intense nuclear γ-ray line flares observed to date; most instruments, from the soft X-ray to the γ-ray bands, saturated at its peak. The NOAA Solar-Geophysical Data Comprehensive Report gives the Hα flare location as 30°N, 70°E. This location corresponds to a heliocentric angle of 74.5° at the time of the flare.

Murphy et al. (1997) underlined that based on the count rate of $\geq$20 keV X-rays observed from the June 4 flare by the Solar X-Ray/Cosmic Gamma-Ray Burst

Experiment on Ulysses, Kane et al. (1995) derived a peak >20 keV electron injection rate of 5.7×10^{38} electrons s^{-1} and an energy dissipation rate of 2.5×10^{31} erg s^{-1}. Given its long duration, they pointed out that the material and energy resources of the active region appeared to be inadequate for the production of such a flare. Ramaty et al. (1994) used data from the BATSE Charged Particle Detectors to monitor the electron bremsstrahlung time profile and estimate the magnetic field strength in the coronal portion of the flare loop. Using the EGRET Total Absorption Shower Counter, Hartman et al. (1999) measured γ-ray line fluxes starting after the impulsive peak of the flare. The flare also produced >100 MeV neutrons (Takahashi et al. 1991; Muraki et al. 1992; Struminsky et al. 1994) and possibly even >10 GeV neutrons (Chiba et al. 1992). Energetic particles from the flare were observed in interplanetary space by the EPAC experiment on Ulysses (McDonald et al. 1995) and the flare was observed at 17, 35, and 80 GHz by radiometers at the Nobeyama Radio Observatory in Japan (Ramaty et al. 1994). The combined interplanetary modulation effects of the June 4 flare and the other intense June flares resulted in the lowest galactic cosmic-ray flux at Earth in almost 40 years (Webber and Lockwood 1993; Van Allen 1993) and strong radio emission generated by the associated interplanetary shock at the heliopause (Gurnett et al. 1993).

Murphy et al. (1997) presented OSSE observations of the June 4 flare over a broad spectral range from 0.1 to ~100 MeV. Observations were obtained of the rise (3:37 UT), peak (3:52 UT), and decay of the event. The decay was observed over the next several orbits, interrupted only by spacecraft night. Plotted in Fig. 9.30 are the intensity profiles of this flare as monitored by live-time corrected OSSE rates above 0.1 MeV, between 1 and 10 MeV, and above 16 MeV during the first orbit.

The peak of the γ-ray intensity and the time when the OSSE detectors were repointed to the secondary target just prior to satellite night are indicated by dotted lines in Fig. 9.30. For comparison, the GOES 1–8 Å X-ray profile is also plotted. All of the rates (with the exception of the >16 MeV band) exhibited saturation at the peak of the emission. As discussed below, the >0.1 MeV γ-ray flux is dominated by bremsstrahlung of >0.1 MeV electrons, the 1–10 MeV flux is a combination of bremsstrahlung and nuclear line emission and the >16 MeV flux is probably also bremsstrahlung but of high-energy electrons. Each of these emissions has a unique time profile. The increase in the >16 MeV rates after 3.7 h UT is due to the arrival at Earth of flare produced neutrons. The high-energy emissions have been discussed previously by DelSignore (1995).

Murphy et al. (1997) concentrate primarily on measurements of nuclear lines and continuum between 0.1 and 9 MeV. They fitted the OSSE data with a multi-component model of solar flare γ-ray spectra and produced time profiles of various γ-ray line intensities throughout the flare with a time resolution as short as 8.2 s. Fits of the 2.223 MeV neutron-capture line and narrow and broad nuclear deexcitation lines provide information on the ambient plasma at the flare site and the accelerated-ion spectrum. Using the measured electron bremsstrahlung, information on the accelerated electrons has also been obtained. Preliminary results from analyses of the OSSE data have been reported previously (Murphy et al. 1993a, b, 1994, 1996).

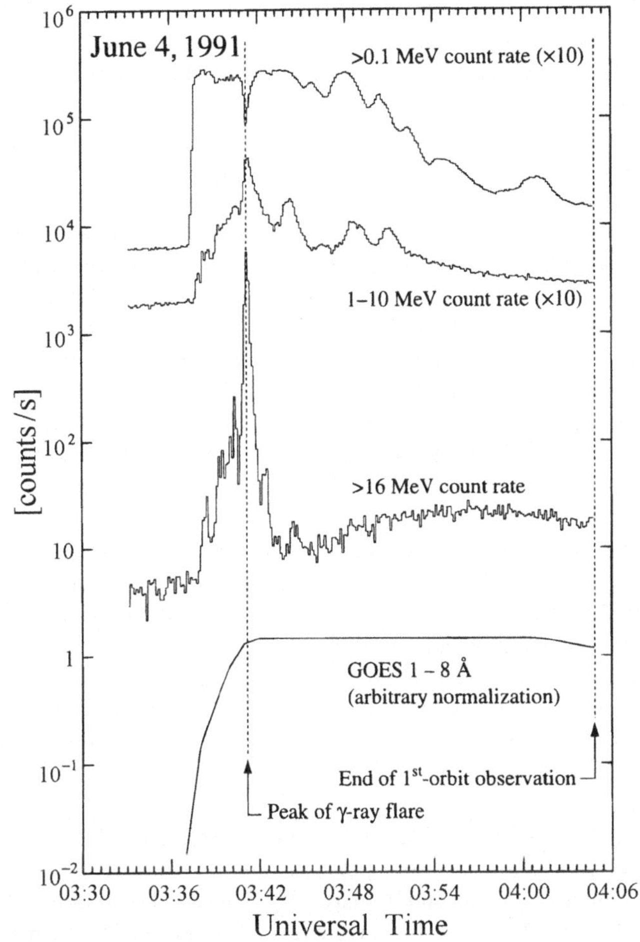

Fig. 9.30 Time-profiles count rates observed in various X-ray and γ-ray bands from the 1991 June 4 solar flare. The dotted lines indicate the peak of the γ-ray emission and the end of the first-orbit observation by OSSE. Also plotted is the GOES 1–8 Å X-ray profile that saturated prior to the peak. Saturation effects were observed in all OSSE count rates except that of the >16 MeV data (From Murphy et al. 1997)

9.8.2 Peculiarities of Gamma-Ray Production in Solar Flares

As noted Murphy et al. (1997), interactions of flare-accelerated protons and nuclei with the ambient solar atmosphere produce neutrons, γ-ray lines and continuum emission and have been studied in detail (e.g., Ramaty and Murphy 1987; Ramaty and Mandzhavidze 1994; Hudson and Ryan 1995). The principal mechanisms for the production of γ-ray lines are nuclear deexcitation, neutron capture and positron annihilation.

The nuclear deexcitation spectrum consists of narrow lines (~2% FWHM−Full Width at Half Maximum,), resulting from protons and α particles interacting with the ambient gas, and broad lines (~20% FWHM), resulting from accelerated carbon and heavier nuclei interacting with ambient hydrogen and helium.

Capture of flare-produced neutrons on ambient hydrogen in the photosphere produces a strong line at 2.223 MeV. Nuclear reactions also produce radioactive nuclei and neutral and charged pions whose decay result in high energy γ-ray, electron and positron emission. The positrons annihilate to produce a line at 0.511 MeV and a three-photon positronium continuum. The narrow nuclear deexcitation lines and the 2.223 and 0.511 MeV lines are superposed on a continuum composed of the broad nuclear deexcitation lines and bremsstrahlung from primary electrons.

As underlined Murphy et al. (1997), the most important narrow nuclear deexcitation lines in solar flares are at 0.429 and 0.478 MeV (^{7}Be and ^{7}Li), 0.847 MeV (^{56}Fe), 1.238 MeV (^{56}Fe), 1.369 MeV (^{24}Mg), 1.634 MeV (^{20}Ne), 1.778 MeV (^{28}Si), 4.439 MeV (^{12}C), and 6.129 MeV (^{16}O) (Kozlovsky and Ramaty 1974; Ramaty et al. 1979). Two additional features appear in solar flare γ-ray spectra at ~5.3 and ~7 MeV that are complexes of lines resulting primarily from inelastic excitation and spallation of ^{14}N and ^{16}O. The narrow lines are due mostly to accelerated protons with kinetic energies of 5–30 MeV and α-particles with energies of 2–10 MeV/nucleon. The relative strengths of the narrow lines primarily depend on the relative ambient-gas abundances, but they also are affected by the energy spectra of the accelerated particles. Because the 1.634 and 6.129 MeV lines of Ne and O are relatively strong and the energy dependencies of their production cross sections are distinctly different, they are particularly useful for measuring the spectral hardness of interacting accelerated particles (Share and Murphy 1995; Ramaty et al. 1995).

Murphy et al. (1997) noted that the neutrons producing the 2.223 MeV line from capture on H arise primarily from p-p, p-α, and α-α interactions at energies of 30–100 MeV/nucleon (Wang and Ramaty 1974; Hua and Lingenfelter 1987b). This line is therefore produced by higher energy particles than those producing the deexcitation lines. Comparison of the 2.223 MeV line and deexcitation lines can thus provide spectral information about 10–100 MeV protons. The capture line is delayed by about 100 s because of the time required for the neutrons to slow down and be captured on ^{1}H. A competing capture reaction is ^{3}He(n,p)^{3}H, which produces no radiation but shortens the delay of the 2.223 MeV line emission (Wang and Ramaty 1974). Measurements of the time profile of this line have been used to set constraints on the photospheric ^{3}He/H ratio (Prince et al. 1983a; Hua and Lingenfelter 1987b).

Murphy et al. (1997) underlined that ambient elemental abundances can be obtained by comparing γ-ray spectra calculated from measured nuclear cross sections and kinematic calculations with observed spectra (Ramaty et al. 1975, 1979; Murphy et al. 1991). Because most of the narrow lines are dominated by deexcitation from a single element, some information about relative ambient abundances can also be inferred directly from the narrow-line fluxes. Share and Murphy (1995) measured the fluxes of ten narrow γ-ray lines in 19 X-class solar flares observed by the SMM Gamma Ray Spectrometer. Flare-to-flare variations in line fluxes suggested that the abundances of elements in the flare plasma are grouped with respect to first ionization potential (FIP): line fluxes from elements with similar FIP correlate well with one another. In contrast, the low-FIP to high-FIP line ratios was not consistent with a common value from flare to flare. The authors suggested that the line ratio variation may be due to ambient abundance variation rather than to spectral variation of the accelerated particles. Ramaty et al. (1995) used cross sections and kinematical calculations to determine the abundances for the SMM/GRS flares and showed that the composition of the flare plasma is close to coronal, although they did not rule out significant variability from flare to flare.

As noted Murphy et al. (1997), at photon energies below about 1 MeV the contribution of nuclear lines is small and continuum emission due to bremsstrahlung from directly accelerated electrons is expected to be the dominant source of photons. Similarly, at energies greater than about 8 MeV bremsstrahlung also dominates except when the accelerated-ion spectrum is very hard and the contribution from π-decay becomes important. Observations of solar hard X rays in the $\sim$20 to $\sim$400 keV region (e.g., by the SMM Hard X-Ray Burst Spectrometer; Dennis 1988) show that spectra in this energy range are usually well fitted by power laws suggesting that the electrons are nonthermal, but in at least one flare a super-hot component (Lin et al. 1981) at energies $\leq$35 keV has been observed. Spectral fits to flare data obtained with the SMM/GRS (Vestrand et al. 1999) have also shown a power law to be an appropriate spectral form for the bremsstrahlung between 0.3 and 8 MeV in most flares, but a few flares were better fitted when a broken power law or the sum of two power laws were used.

9.8.3 Using the OSSE Instrument and Observing Configuration

The OSSE instrument is described by Johnson et al. (1993). A brief description of those aspects of the instrument relevant to solar observations is given in Murphy et al. (1997). OSSE instrument consists of four independently oriented NaI/CsI phoswich scintillation detectors with both passive and active shielding for reducing background and defining its aperture (see Fig. 9.31). The principal detector element is a 33 cm diameter phoswich. Pulses from photomultipliers optically coupled to the 10.2 cm thick NaI and 7.6 cm thick CsI crystals are separated by pulse-shape discrimination, providing a compact anticoincidence system for both charged particles and background γ-rays. OSSE instrument covers the energy range from 0.04 to more than 150 MeV and each detector has photo-peak effective areas of $\sim$480 and $\sim$170 cm^2 at 0.511 and 6 MeV, respectively. The energy resolution varies from detector to detector. At 6.1 MeV the fractional FWHM resolutions of the four detectors are 4%, 5%, 4%, and 7% (the poorer resolution of detector 4 occurred during growth of that NaI crystal). A tungsten collimator above the phoswich defines a 3.8° $\times$ 11.4° FWHM geometrical field of view and a 3.5 in. thick NaI

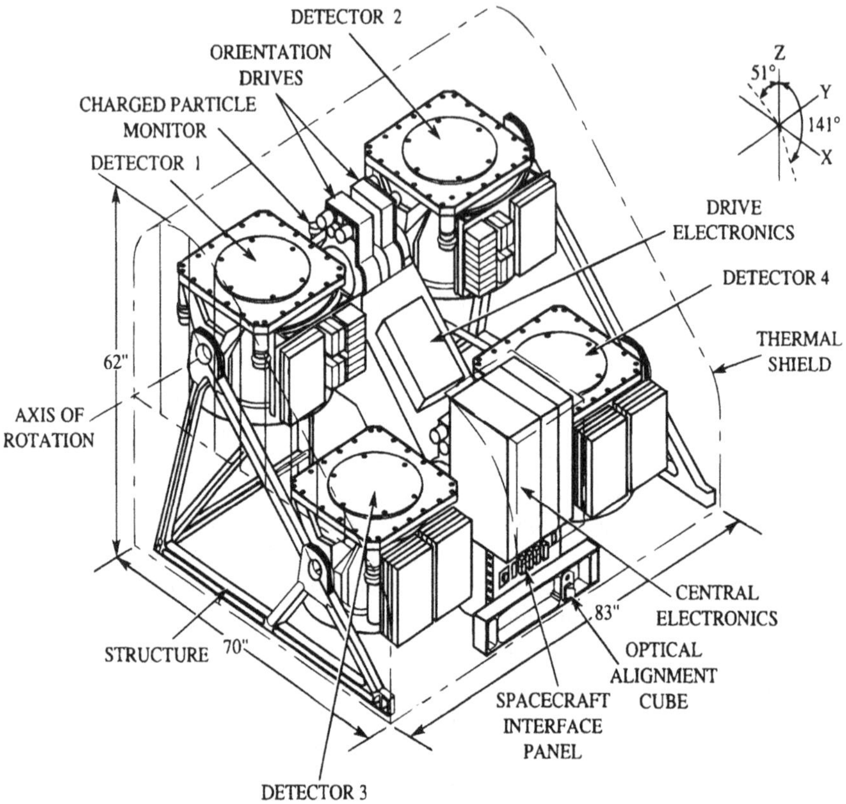

Fig. 9.31 The OSSE instrument detector (From Murphy et al. 1997)

annulus completes the shielding. OSSE can detect γ-ray lines as weak as $\sim 10^{-3}$ photons.cm^{-2}s^{-1} during a 1,000 s exposure.

9.8.4 Spectral Fits

As noted Murphy et al. (1997), during the first orbit when the flare was intense, spectra from the detector pointed off the Sun are used in order to minimize saturation effects. Because of the low-energy attenuation, only >1 MeV data are used for fitting these spectra. The 8 s spectra were rebinned into broader energy bins to provide adequate numbers of counts for proper fitting at the highest energies where limited statistics can be a problem. Above 1 MeV, the rebinned widths are comparable to the FWHM of the detector spectral resolutions and Murphy et al. (1997) have demonstrated that the intensities of all line features in the spectra are faithfully obtained after this rebinning.

In cases where energy-channel rebinning of an 8 s spectrum did not achieve a minimum of 15 counts in each channel, subsequent 8 s spectra were summed until such a minimum was achieved. Eight second temporal resolution was achieved near the peak of the flare but accumulations of up to 40 s were required near the end of the first orbit. In addition, during the last 10 min of the first orbit and during the second and third orbits when saturation effects were minimal, spectral fits to 2 min accumulations of Sun-pointed detector data were also used and the spectral fits were performed above 0.1 MeV. Weighted means of the fits from the three Sun-pointed detectors are presented. There were no significant differences of the results from different detectors. Spectral analyses were accomplished using a least-squares fitting technique: an assumed photon model was folded through the OSSE instrument response for the updated solar position, detector by detector, and compared to background-subtracted count spectra. The model parameters were adjusted until χ^2 was minimized. Examples of data from both Sun- and off-pointed detectors, along with fits to the data using the models described above, are shown in Fig. 9.32.

Panel a in Fig. 9.32 shows an 8 s 1–9 MeV spectrum from an off-pointed detector accumulated shortly after the peak of the flare and rebinned into the larger energy bins. Panel b in Fig. 9.32 shows a 2 min, 0.1–9 MeV spectrum from a Sun-pointed detector accumulated late in the first orbit. In Fig. 9.32 are identified the components of the models and the sources of various spectral features. Panel b in Fig. 9.32 shows that the assumption of a power law to model the electron bremsstrahlung component >0.1 MeV is justified. In Fig. 9.32 can be seen some excess of counts above the model at energies $\geq$8 MeV, indicating more broad component than the model allows and/or a hard bremsstrahlung tail. Murphy et al. (1997) have shown that the derived fluxes of the narrow lines are unaffected by this discrepancy (see below, Section 9.8.5). This issue will be resolved when the OSSE high-energy spectral data are incorporated into the analysis.

Before presenting results of the spectral analysis, Murphy et al. (1997) tested whether any emission at the peak of the flare has been missed due to saturation of the off-pointed OSSE detector (detector 2). The only CGRO detectors not saturated at energies of less than 10 MeV during the flare were the BATSE CPDs. Shown in Fig. 9.33 (thin histogram) is the time profile of the observed average count rate from the four Sun-facing BATSE CPDs; an estimate of the back-ground, assumed to vary linearly with time, has been subtracted from the data. Ramaty et al. (1994) calculated an instrumental response and used these CPD data in their analysis of the June 4 flare. To compare with the BATSE data, the best-fitting OSSE high time resolution >1 MeV photon spectra obtained from the off-pointed detector are convolved with this BATSE CPD response to derive predicted CPD count rates (Fig. 9.33, thick histogram). Murphy et al. (1997) noted that using OSSE spectra derived from >1 MeV data only is valid for comparison because the CPD instrumental response falls rapidly below 1 MeV. The 1 s BATSE count rates have been rebinned into the corresponding OSSE accumulation intervals. Figure 9.33 shows that, overall, the OSSE-derived CPD rates fall below the observed CPD rates by about a factor of 2.5.

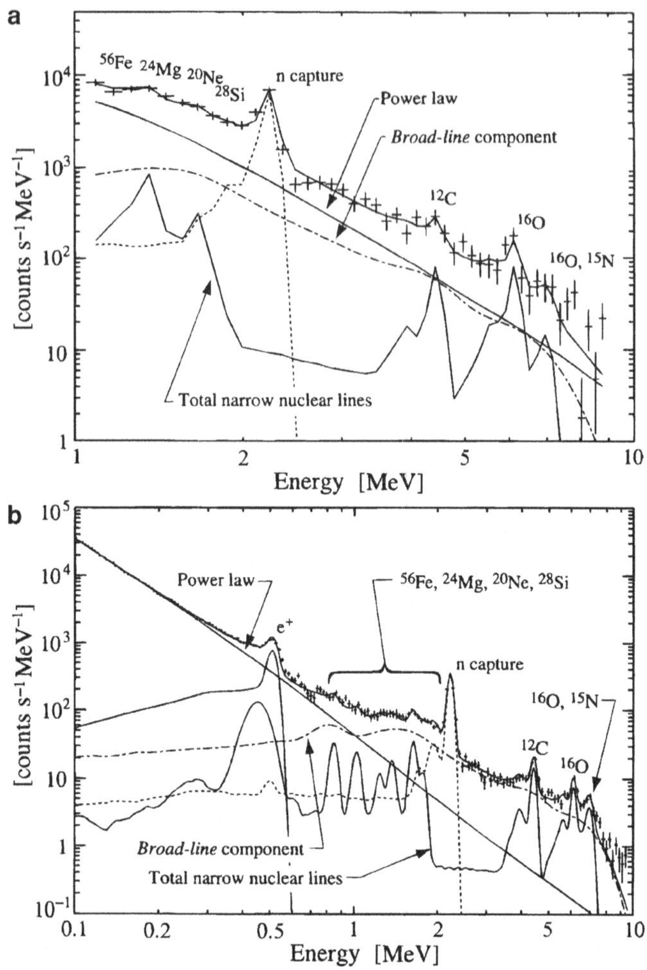

Fig. 9.32 Count spectra and best fits obtained using the photon models described in the text. (**a**) An 8 s spectrum accumulated soon after the peak of emission in the first orbit by detector 2 while pointing 4.5° off the Sun. The spectrum has been rebinned into larger energy intervals to improve the statistics for fitting. (**b**) A 2 min spectrum accumulated late in the first orbit by detector 1 while pointing at the Sun. For display purposes, the data of this latter spectrum have been rebinned into slightly larger energy channels than those used in the fits (From Murphy et al. 1997)

Also shown in Fig. 9.33 (dotted histogram) is the OSSE derived count rate increased by a factor of 2.5 to normalize it to the observed BATSE CPD count rate on the rising and falling portions of the primary peak. The shape of the profile predicted from the OSSE data agrees reasonably well with the observed BATSE profile: as noted Murphy et al. (1997), the higher relative rates of the CPD between the emission peaks as the emission falls could be due to improper subtraction of background; the assumption of a background time profile peaking during the flare,

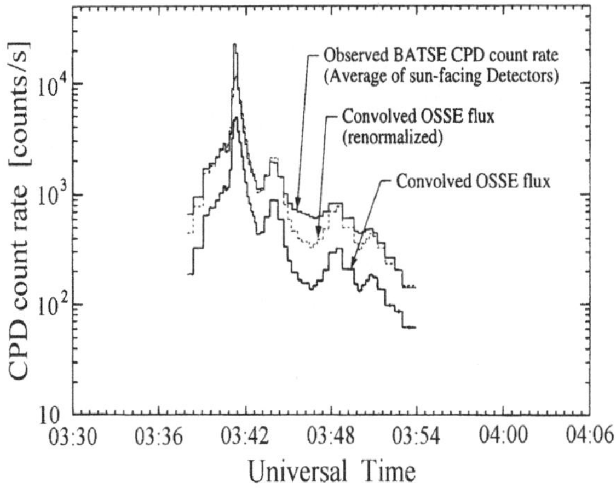

Fig. 9.33 The 4 June 1991 event: high-resolution time-profile of the OSSE best-fitting photon spectra convolved with the BATSE Charged Particle Detector (CPD) response function (thick histogram). This is compared with the average observed BATSE count rate from the four Sunward-facing CPDs (thin histogram). The 1 s BATSE count rates have been rebinned into the corresponding OSSE accumulation intervals. Also shown is the convolved OSSE flux increased by a factor of 2.5 (dotted histogram) to normalize it to the rising and falling portions of the observed BATSE count rate (From Murphy et al. 1997)

as suggested by the raw CPD count rates, can reduce this difference. The overall discrepancy of a factor of 2.5 may be due to an underestimated CPD response function (Ramaty et al. 1994). After renormalization, the OSSE derived CPD count rate falls short of the BATSE data during the two 8 s accumulations at the peak of the flare. This suggests that saturation effects have reduced the OSSE sensitivity by about a factor of 2 during these two intervals. The fluxes and fluences presented in Murphy et al. (1997) include a correction for radiation missed during these 16 s due to this saturation.

9.8.5 Spectroscopic Observations: Narrow Nuclear Line Measurements

According to Murphy et al. (1997), spectral data accumulated in intervals as short as 8 s are used to obtain time profiles of nuclear deexcitation lines, the 2.223 MeV neutron capture line, and electron bremsstrahlung. Murphy et al. (1997) determine relative line intensities of the narrow γ-ray lines, derive the broad-line to narrow-line ratio, obtain spectral parameters of the accelerated ions and electrons, and discuss the accelerated α/proton ratio. A constraint on the photospheric ^{3}He/H abundance ratio is established. Techniques typically applied to the study of weak celestial sources are applied to search for nuclear emission occurring before and

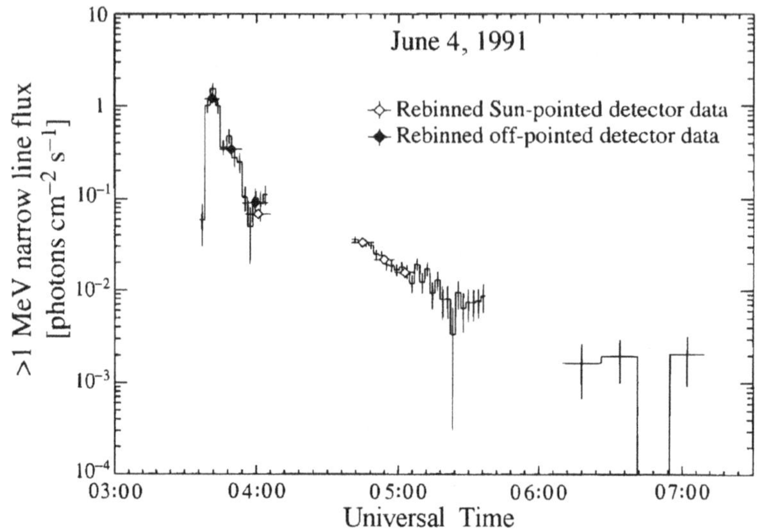

Fig. 9.34 The 4 June 1991 event: time profile of the summed flux in narrow nuclear lines during three orbits of observations (histogram). Overplotted are the fluxes rebinned into the larger accumulation intervals as described in the text. Filled diamonds are from off-pointed data, and open diamonds are from Sun-pointed data (From Murphy et al. 1997)

after the flare. Plotted in Fig. 9.34 histogram is the three-orbit time profile of the flux observed in the sum of >1 MeV narrow deexcitation lines.

In Fig. 9.34 the fluxes for the first orbit are derived from off-pointed detector data and the fluxes for the second and third orbits are derived from Sun-pointed detector data. During the third orbit the data have been summed into 14 min accumulations to improve the statistics. The error bars represent the combined 1 σ statistical uncertainties and the systematic uncertainties associated with background subtraction found on orbital timescales. Nuclear interactions and the resultant γ-ray emission are seen to continue after the peak of the emission for at least 2 h through the second orbit. Although there is statistical evidence for emission during the third orbit ($\sim$2.5 σ), the day-to-day systematic errors weaken this conclusion. In Section 9.8.12 will be used the neutron capture line to show that nuclear reactions are indeed continuing into the third orbit.

The fluxes of the individual narrow nuclear lines relative to that of the 6.129 MeV ^{16}O line are listed in Table 9.10. These relative fluxes are averaged over fits to data obtained throughout the first orbit (3.64–4.08 UT) and during the first half of the second orbit where the statistics are adequate (4.67–5.11 UT). For the first orbit, fits to data obtained from off-pointed detectors during the early part (3.64–3.90 UT) are used and, to maximize the statistical significance, Sun-pointed detector data are used during the last 10 min (3.90–4.08 UT). For the second orbit, fits to data obtained from Sun-pointed detectors are used. The relative line fluxes derived from Sun- and off-pointed detector data during the last 10 min of the first orbit are in reasonable agreement, with the largest discrepancy being less than 2.5 σ.

Table 9.10 Relative narrow-line fluxes (the line fluxes are normalized to that of the 6.129 MeV ^{16}O line) (From Murphy et al. 1997)

Line energy (MeV)	*CGRO*/OSSE 1991 June 4	*SMM*19-Flare average[a]
1.24	0.13 ± 0.24	0.30 ± 0.03
1.37	0.90 ± 0.19	0.77 ± 0.04
1.63	1.51 ± 0.17	1.36 ± 0.05
1.78	0.63 ± 0.15	0.49 ± 0.04
4.44	1.31 ± 0.10	1.10 ± 0.04
~ 5.4	0.09 ± 0.09	0.26 ± 0.02
6.13	$\equiv 1.00 \pm 0.08$	$\equiv 1.00 \pm 0.03$
~ 7	0.82 ± 0.08	0.56 ± 0.02

[a]From Share and Murphy (1995)

Also shown in the Table 9.10 are the average relative line fluxes derived by Share and Murphy (1995) for the 19 SMM/GRS flares. For γ-ray lines below 2 MeV, the June 4 relative fluxes are very similar to the SMM/GRS averages. The June 4 4.44 MeV ^{12}C flux, however, appears to be significantly higher than the SMM/GRS flux (by 5 σ); the implication of this is discussed further in Section 9.8.13.

Murphy et al. (1997) also note that the relative flux of the ~7 MeV line complex from ^{16}O is significantly higher than the 19 flare average of the SMM/GRS analysis in which little flare-to-flare variation of this ratio was measured. They investigated whether this high flux ratio could be due to inadequacies of the assumed γ-ray model or to contamination of the line from spacecraft sources. The slight excess of count rate over the best-fit model (see Panel b in Fig. 9.32) suggests that the model photon spectrum may be inadequate at these energies.

Murphy et al. (1997) refitted the data replacing the power law with either a broken power law or the sum of two power laws and found that the quality of fit was not significantly improved and the fitted line flux was unchanged. Contamination of the ~7 MeV line could be due to γ-rays from capture of solar-flare neutrons by spacecraft iron or by iodine in the OSSE detector. The fact that the line flux appears to peak late in the first orbit when the neutrons are arriving lends support to this possibility. In the analysis of the June 4 flare, Hartman et al. (1999) required a feature at ~7.6 MeV from neutron capture on iron after the peak. However, if there were significant contamination, the derived flux of the ~7 MeV line from off-pointed data would be significantly larger than that from Sun-pointed data because the detector response to such sources is essentially the same for both detector orientations. Because the derived fluxes from the two orientations were consistent, it is no reason to doubt the measurement of this line.

9.8.6 Spectroscopic Observations: Ambient Abundance Estimates

As underlined Murphy et al. (1997), because the emission of a given narrow γ-ray line is mostly dominated by the contribution from a single element, information about ambient elemental abundances can be obtained by comparing narrow-line

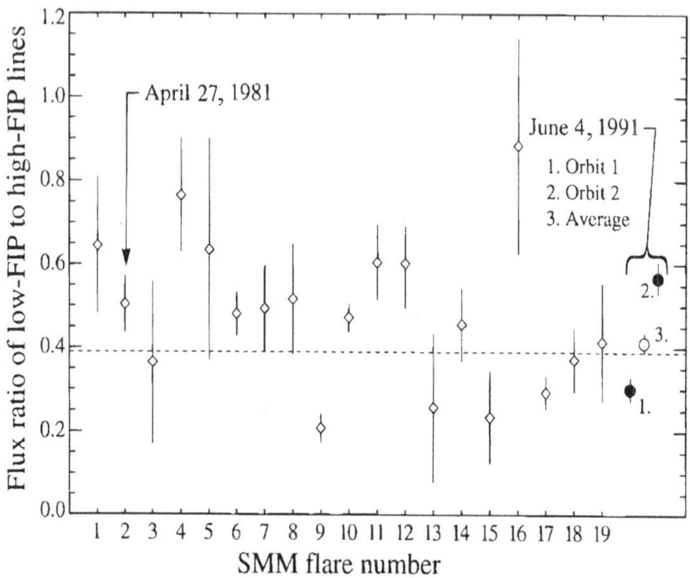

Fig. 9.35 Low FIP-to-high FIP line flux ratio for the 19 SMM/GRS flares (Share 1996, private communication) compared with the OSSE June 4 ratios obtained for the first and second orbits and the mean for both orbits (From Murphy et al. 1997)

fluxes. For comparison with the SMM/GRS analysis of Share and Murphy (1995), Murphy et al. (1997) have separately summed the fluxes from low-FIP elements Mg, Si, and Fe (1.369, 1.778, and 1.238 MeV) and those of lines from high-FIP elements C, O and Ne (4.44, 6.129, 6.99, and 1.634 MeV). Averaged over the two-orbit data as above, the OSSE low FIP-to-high FIP ratio is 0.41 ± 0.02. This ratio is plotted in Fig. 9.35 (open circle) along with the ratios for the 19 SMM/GRS flares. The horizontal line represents the weighted mean of this ratio (0.39) for the 19 flares and the value for June 4 is seen to be near this flare average. Murphy et al. (1997) noted the excellent statistical significance of the June 4 result. If the ratio determined by Murphy et al. (1991) for the 1981 April 27 flare (flare number 2) represents a factor of 3 enhancement of the low-FIP/high-FIP abundance ratio over the photospheric value as they suggested, then the ratio determined here for the June 4 flare represents about a factor of 2.5 enhancement.

Share and Murphy (1995) have shown that the low FIP-to-high FIP line flux ratio varies significantly from flare to flare. The improved statistics available with the OSSE detectors allows searching for variability of this ratio within a flare. Using Sun-pointed results for the second orbit and the last 10 min of the first orbit and off-pointed results for the rest of the first orbit, the probability that the derived ratios are consistent with their weighted mean is less than 10^{-4}. The average ratios for each of the two orbits are 0.30 ± 0.03 and 0.57 ± 0.04, which differ by 5.6 σ. It could be conclude that the data suggest the low FIP-to-high FIP line flux ratio increased as

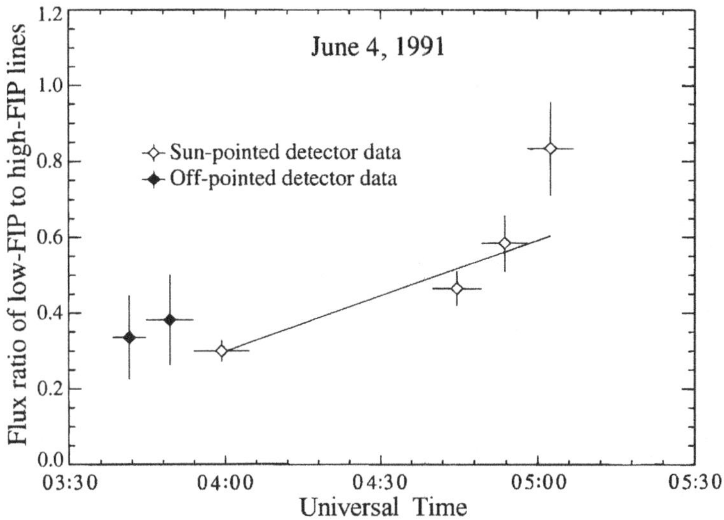

Fig. 9.36 The 4 June 1991 event: Time profile of the summed low-FIP to summed high-FIP line flux ratio derived from off-pointed (filled diamonds) and Sun-pointed (open diamonds) detector data. A straight line fit to the Sun-pointed data is also shown (From Murphy et al. 1997)

the flare progressed. These values are also plotted in Fig. 9.35 (filled circles) for comparison with the SMM/GRS results.

To show the time dependence of this ratio, the 2 min data were rebinned into larger accumulation intervals ranging from ~6 to ~10 min to improve the statistical significance. The start and stop times of these intervals are 3.64–3.74, 3.74–3.90, 3.90–4.08, 4.67–4.82, 4.82–4.97, and 4.97–5.11 UT. Figure 9.36 displays these intervals on the >1 MeV narrow line time profile, where the filled diamonds are from off-pointed detector data and the open diamonds are from Sun-pointed detector data. It can be note that the first interval essentially covers the emission peak and that the off- and Sun-pointed data during last 10 min of the first orbit give consistent results. Only data through the first half of the second orbit where the statistics are still adequate have been used. The time dependence of the low FIP-to-high FIP line flux ratio for the two orbits rebinned into these larger intervals is shown in Fig. 9.36 along with a best-fit straight line fit to the Sun-pointed data only. This line has a positive slope with a significance of 5.2 σ. The probability that the data are randomly distributed about this line is 0.17.

Since the low-FIP lines are concentrated in the 1–2 MeV region and the high-FIP lines are mostly in the 4–7 MeV region, Murphy et al. (1997) tested whether the observed variation of this ratio could be an artifact of an energy-dependent data problem. The comparison was repeated but for the ratio of the sum of >1 MeV low-FIP lines to the high-FIP 1.634 MeV Ne line alone so that all lines are now concentrated in the same energy range. The time profile of this ratio is plotted in Fig. 9.37 and shows the same qualitative behavior as the low FIP to high FIP ratio.

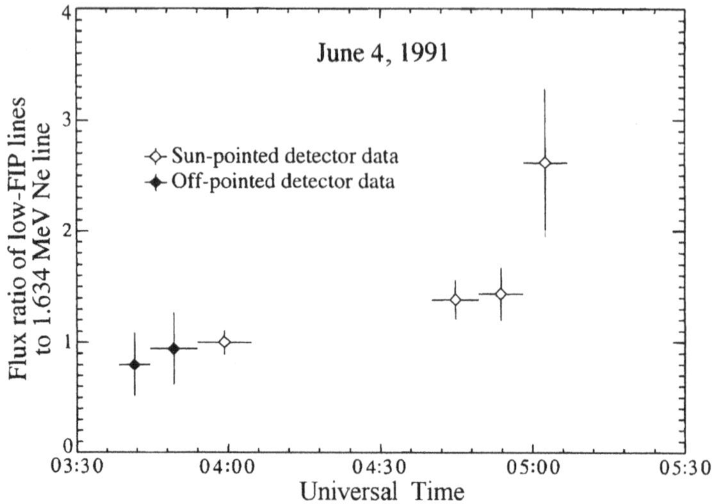

Fig. 9.37 The 4 June 1991 event: Time profile of the summed low-FIP to 1.634 MeV Ne line flux ratio derived from off-pointed (filled diamonds) and Sun-pointed (open diamonds) detector data (From Murphy et al. 1997)

While the poorer statistics have reduced the significance of the difference of the two orbital values to 3.2 σ, the trend with time suggests that this low FIP-high FIP variability is not due to fitting problems.

9.8.7 On the Ratio of the Flux in Broad Lines to the Flux in Narrow Lines

According to Murphy et al. (1997), the ratio of the flux in broad lines to the flux in narrow lines gives a measure of the relative abundance of accelerated heavy particles to accelerated protons. However, because the broad-line component also contains emission from unresolved narrow lines produced by protons, any measured variation should represent a lower limit to the actual variation. Using Sun-pointed >0.1 MeV data only, the ratio of the >1 MeV flux in the broadline template to the >1 MeV flux in narrow-lines averaged over the last 10 min of the first orbit and the first half of the second orbit is 2.83 ± 0.07. Using the same broad-line template, the 19-flare SMM/GRS analysis showed considerable flare-to-flare variation of this ratio, ranging from $\sim$2 to $\sim$7 with a weighted mean of 3.2 ± 0.2. The averages from each of the two June 4 orbits are 3.40 ± 0.12 and 2.19 ± 0.09. These values differ by 8.1 σ; the probability that the derived ratios are consistent with their weighted mean is less than 10^{-5}. This suggests that the accelerated heavy ion to proton ratio decreased as the flare progressed.

9.8.8 Accelerated Proton Spectrum

In Murphy et al. (1997) the spectral shape and energy content of the accelerated protons are derived from the measured γ-ray line fluxes. Flux ratios of lines whose production cross sections have different energy dependencies are sensitive to the proton spectral index. Because the 6.129 and 1.634 MeV deexcitation lines of O and Ne are relatively strong and their energy-dependent cross sections are distinctly different, their ratio is particularly useful for measuring the spectral hardness of interacting 5–20 MeV protons (Share and Murphy 1995; Ramaty et al. 1996). Because both lines are prompt, this ratio gives an instantaneous measure. Alternatively, because the neutrons producing the 2.223 MeV capture line are produced by higher energy (30–100 MeV/nucleon) particles than those producing the deexcitation lines, the flux ratio of the neutron-capture line and a deexcitation line provides spectral information about 10–100 MeV protons (e.g., Ramaty et al. 1993). But because the neutron capture line is delayed relative to the deexcitation lines, the technique can only be properly used with flare-integrated data to provide an average index for the whole flare. Assuming a power-law form for the proton spectrum, the measured O/Ne flux ratios were converted to proton power-law spectral indexes by using the thick target γ-ray line calculations of Ramaty et al. (1996) for Ne/O = 0.25 and two assumptions for the α/p ratio of 0.1 and 0.5. The average index over both orbits was 4.0 ± 0.09 for $\alpha/p = 0.1$ and 4.4 ± 0.3 for $\alpha/p = 0.5$.

According to Murphy et al. (1997), for the alternative technique, the fluence ratio of the neutron-capture line and the 4.44 MeV ^{12}C deexcitation line was used. The total fluences in these two lines from the June 4 flare, after linearly interpolating over gaps in the data due to Earth occultation and applying a correction for saturation effects at the peak of the flare, are $1{,}050 \pm 19$ and 189 ± 9 photons.cm^{-2}, respectively. The OSSE measurements can be compared with the EGRET/TASC observations (Hartman et al. 1999) beginning 81 s after the peak of emission until the end of the first orbit (the TASC was saturated during the peak). During this time period, the EGRET/TASC measured 2.223 and 4.439 MeV line fluences of 678 ± 24 and 66 ± 3 photons.cm^{-2}, respectively. The OSSE fluences for the same time period are 644 ± 12 and 75 ± 4 photons.cm^{-2}, respectively, in good agreement. Using the calculations of thick target nuclear line and neutron production by Ramaty et al. (1996) and the measured 2.223–4.44 MeV line fluence ratio (5.56 ± 0.28), Murphy et al. (1997) obtained a flare-averaged proton power-law index of 3.7 ± 0.1 for $\alpha/p = 0.1$ and 4.2 ± 0.1 for $\alpha/p = 0.5$. Here it was assumed impulsive-flare abundances and a downward isotropic angular distribution for the accelerated ions (Ramaty et al. 1996): the impulsive-flare composition has significant enhancements of Ne, Mg, Si, and Fe in addition to ^{3}He/$\alpha \sim 1$. The values of the proton index derived using the O/Ne line fluence ratio with either $\alpha/p = 0.1$ or 0.5 (4.0 ± 0.2 and 4.4 ± 0.3) are in reasonable agreement with the corresponding indexes obtained using the neutron-capture line/^{12}C line fluence ratio (3.7 ± 0.1 and 4.2 ± 0.1), differing by 1.2 and 0.6 σ, respectively. Murphy et al. (1997) note the better agreement using $\alpha/p = 0.5$. They also note that an assumption of gradual-flare abundances for the accelerated ions (see Ramaty et al. 1996) gave indexes in poorer agreement (4.8 σ and 3 σ differences, respectively).

Using the average spectral index derived from the neutron-capture line to 4.4 MeV line ratio and the flux in the 4.44 MeV line, Murphy et al. (1997) could obtain the flare-averaged proton spectrum using recent calculations of thick-target 4.44 MeV ^{12}C γ-ray line yield for Ne/O = 0.25. Murphy et al. (1997) assume $\alpha/p = 0.5$ based on the better agreement of both the spectral indexes derived above and the α−α line-to-high FIP line ratio (see above, Section 9.8.11). They further assume that the proton spectrum is a power law extending down to 1 MeV with a flat extension to lower energies. This spectrum is then integrated to obtain the number of accelerated protons and their energy content. After corrections for gaps in the data and saturation, the total number of protons with energy greater than 30 MeV accelerated in the flare was $(6.7 \pm 1.2) \times 10^{32}$ and the total energy contained in accelerated protons of all energies was $(1.7 \pm 0.2) \times 10^{32}$ erg. If according to Ramaty et al. (1996) assume impulsive-flare abundances and identical energy dependence for all ion spectra, and $\alpha/p = 0.5$, the total energy contained in accelerated ions was $(1.0 \pm 0.1) \times 10^{33}$ erg. The time dependence of the proton spectral index derived from the measured O/Ne line flux ratios and calculated for $\alpha/p = 0.5$ is shown in Fig. 9.38, where the 2 min data have been rebinned.

The values in Fig. 9.38 are consistent (40% confidence) with a statistical distribution about their mean. The average indexes derived for the two orbits are 4.4 ± 0.3 and 4.3 ± 0.2, respectively, which differ by 0.3 σ indicating that there is no significant evidence for spectral evolution of the accelerated proton spectrum.

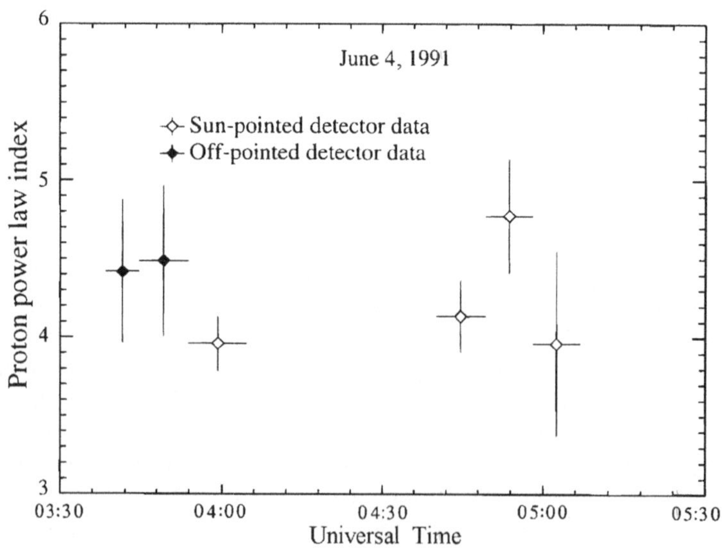

Fig. 9.38 The 4 June 1991 event: Time profile of the accelerated proton power-law spectral index derived from the 6.129 MeV to 1.634 MeV O/Ne line flux ratio obtained from off-pointed (filled diamonds) and Sun-pointed (open diamonds) detector data (From Murphy et al. 1997)

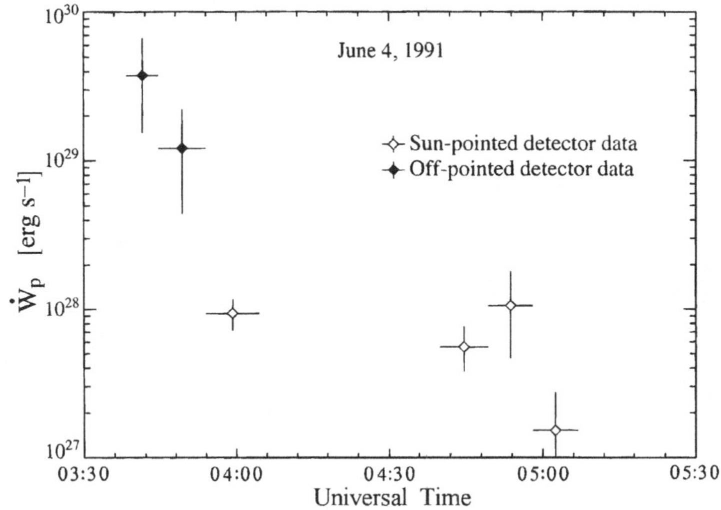

Fig. 9.39 The 4 June 1991 event: Time profile of the instantaneous power in accelerated protons obtained from off-pointed (filled diamonds) and Sun-pointed (open diamonds) detector data (From Murphy et al. 1997)

The same conclusion is reached for $\alpha/p = 0.1$. The time dependence of the instantaneous power in accelerated protons, $\dot{W}_p$, is shown in Fig. 9.39.

9.8.9 Accelerated Electrons: Spectrum and Energy Content

According to Murphy et al. (1997), an unambiguous measurement of the electron bremsstrahlung spectrum can be obtained using fits to >0.1 MeV data from Sun-pointed detectors after the peak of emission.

Assuming a power-law form for the bremsstrahlung component, Murphy et al. (1997) derive an average index of 2.74 ± 0.01. The index was essentially constant, varying from its average value by less than 0.07 over the two orbits of Sun-pointed data. This implies a similar constancy of the >0.1 MeV electron spectrum. Fits to off-pointed data during the peak of emission cannot be used to directly determine the bremsstrahlung spectrum because the bremsstrahlung cannot be easily distinguished from the broad-line component at energies of more than 1 MeV. However, the bremsstrahlung flux during these times can be estimated from the difference between the total flux and the nuclear flux. The narrow lines and the 2.223 MeV neutron-capture line are well-measured with off-pointed data, but the broad component is not. Using Sun-pointed data during the last 10 min of the first orbit Murphy et al. (1997) found a broad-to-narrow ratio of 3.4. To estimate the bremsstrahlung flux they assume this ratio to be constant at this value throughout the orbit but it does allow a large uncertainty as discussed below. Shown in Fig. 9.40 (filled

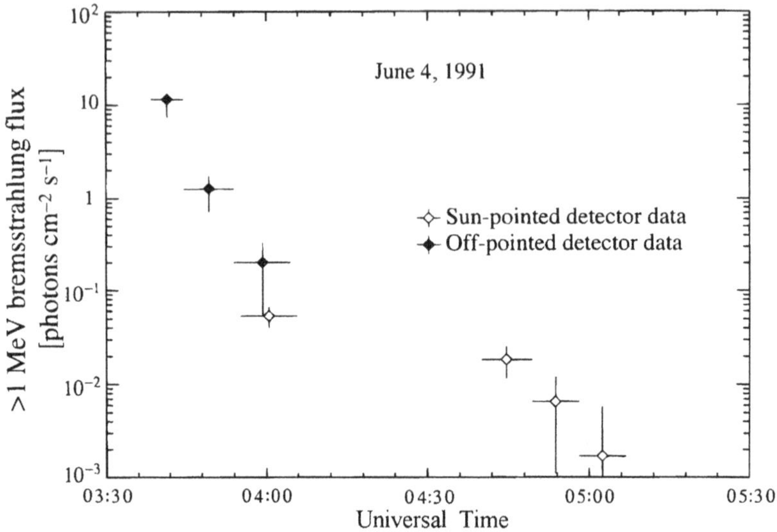

Fig. 9.40 Time profile of the >1 MeV bremsstrahlung flux obtained from off-pointed (filled diamonds) and Sun-pointed (open diamonds) detector data. The first-orbit Sun-pointed data point has been shifted in time for display purposes (From Murphy et al. 1997)

diamonds) is the >1 MeV bremsstrahlung flux estimated from off-pointed detector data obtained during the first orbit. The uncertainties shown represent the range of flux values obtained by assuming broad-to-narrow ratios ranging from 2 to 7, the variation found in the 19 flare SMM/GRS analysis (see above, Section 9.8.7). Murphy et al. (1997) note that, after the peak of the June 4 1991 flare, the ratio is probably not larger than 5, since larger assumed ratios resulted in significant over-subtraction and negative bremsstrahlung fluxes.

Also shown in the Fig. 9.40 (open diamonds) are the measured >1 MeV bremsstrahlung fluxes directly obtained from Sun-pointed detector data from the first and second orbit. During the last 10 min of the first orbit, the Sun- and off-pointed bremsstrahlung are consistent. Murphy et al. (1997) derive the corresponding accelerated electron spectrum from the measured bremsstrahlung using calculations of thick target bremsstrahlung production from power-law electron spectra (Ramaty et al. 1993). The electron spectrum can then be integrated to obtain the total energy contained in the accelerated electrons. The total energy contained in electrons of more than 0.1 MeV accelerated in the June 4 flare is estimated to be $(1.7 \pm 0.1) \times 10^{31}$ erg. The time profile of the instantaneous power in accelerated electrons of more than 0.1 MeV in the interaction region, $\dot{W}_e$, is plotted in Fig. 9.41. The uncertainties associated with the off-pointed detector measurements during the first orbit reflect the uncertainty in the broad-to-narrow ratio. The effects of allowing for a different bremsstrahlung spectrum during the peak of the flare are discussed in Sections 9.8.17 and 9.8.18.

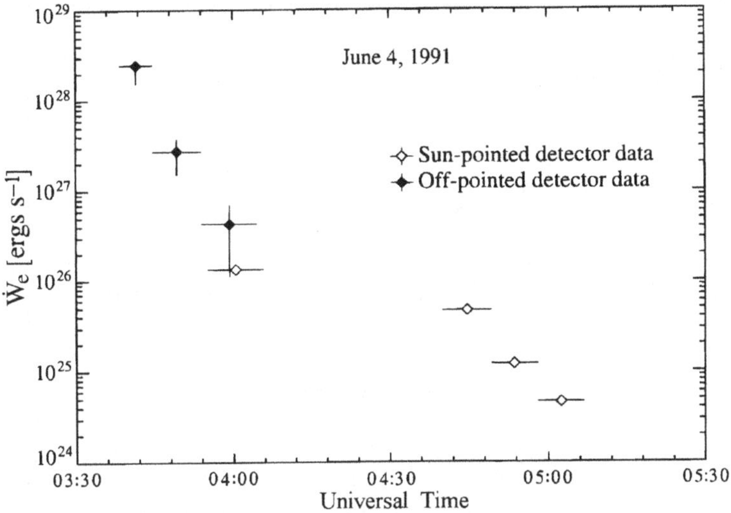

Fig. 9.41 Time profile of the instantaneous power in accelerated >0.1 MeV electrons obtained from off-pointed (filled diamonds) and Sun-pointed (open diamonds) detector data. The first-orbit Sun-pointed data point has been shifted in time for display purposes (From Murphy et al. 1997)

9.8.10 Accelerated Electrons: Comparison of Bremsstrahlung with Other Emissions

The derived high-time resolution >1 MeV bremsstrahlung flux is shown in Panel a of Fig. 9.42 (filled diamonds) for the impulsive phase of the first orbit along with other emissions. The data have been rebinned into slightly longer time intervals to improve the statistical significance. Also shown in Panel a is the time profile of the 1 s, 80 GHz radio data (thin curve) obtained with radiometers at the Nobeyama Radio Observatory, Japan (Ramaty et al. 1994). As noted Murphy et al. (1997), such gyrosynchrotron emission is produced predominantly by electrons of energies above 1 MeV, essentially the same electron energies that produce the >1 MeV bremsstrahlung. The radio data have been renormalized to the bremsstrahlung flux on the rising and falling portion of the primary peak. The three periods of enhanced emission are seen in both profiles. The early radio emission before each of the three γ-ray peaks may be due to gyrosynchrotron emission of electrons trapped in the corona where the matter density is too low for efficient γ-ray line production (Ramaty et al. 1994). Their subsequent release into denser regions produced the γ-ray emission (see below, Section 9.8.19).

Murphy et al. (1997) also note the lag in the maxima of the OSSE rates relative to the 80 GHz rate. Also shown in Panel a of Fig. 9.42 is the time profile of the high-FIP line fluxes (multiplied by a factor of 300 for clarity) providing the interaction rate of 5–30 MeV protons. The proton interactions do not show a peak as intense as that of the electrons and they decay at a slower rate (see also Section 9.8.18).

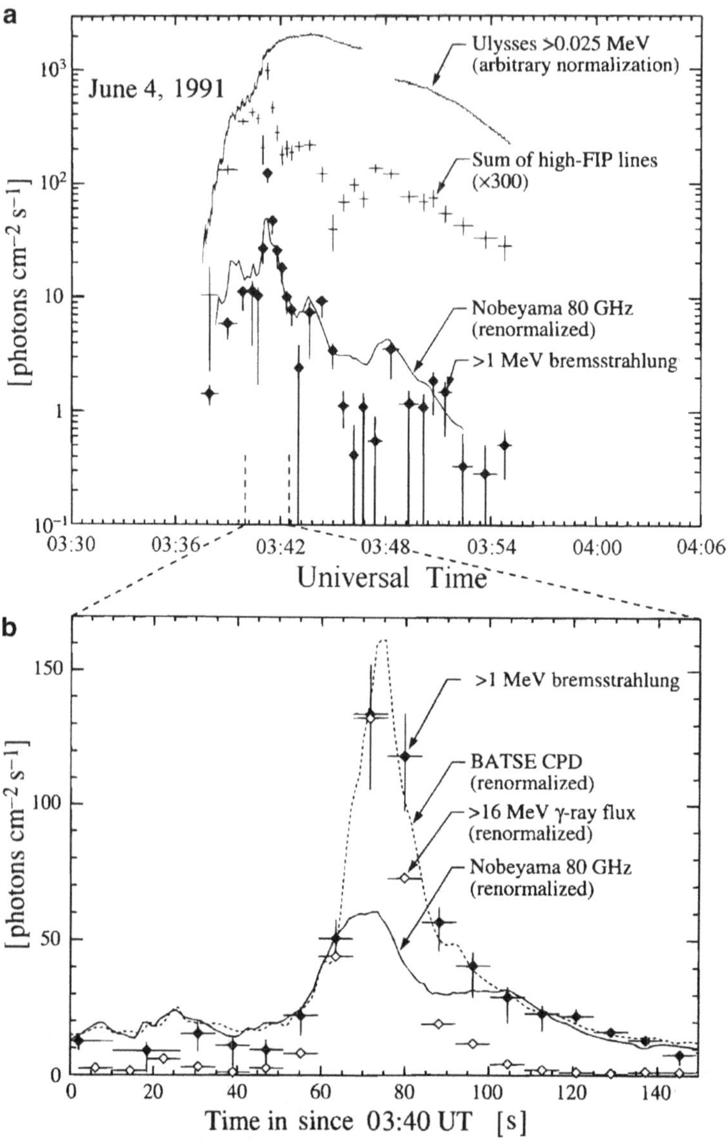

Fig. 9.42 June 4, 1991 event. Panel (**a**) High-resolution time profile of the >1 MeV bremsstrahlung and the summed high-FIP line flux obtained from off-pointed detector data during the first orbit. The data have been rebinned into larger accumulation intervals and a correction has been applied to the flux values during the 16 s at the peak to account for saturation. The high-FIP flux has been multiplied by a factor of 300 for clarity. Also shown are the 80 GHz Nobeyama radio data (Ramaty et al. 1994) normalized to the rising and falling portions of the bremsstrahlung and the Ulysses >0.025 MeV count rate (Kane et al., 1997) with arbitrary normalization. Panel (**b**) Detail of the peak of the flare showing the >1 MeV bremsstrahlung, the 80 GHz Nobeyama radio data, the background-subtracted >16 MeV OSSE flux (DelSignore 1995), and the BATSE CPD count rate. The BATSE CPD and the 80 GHz emission have been normalized to the rising and falling portions of the bremsstrahlung. The >16 MeV flux has been renormalized to have a flux similar to that of the bremsstrahlung at the peak (From Murphy et al. 1997)

In Panel b of Fig. 9.42 Murphy et al. (1997) compare high temporal resolution profiles of the OSSE >1 MeV bremsstrahlung flux (filled diamonds), the OSSE background-subtracted >16 MeV γ-ray flux (open diamonds; DelSignore 1995), the 1 s, 80-GHz radio data (thin curve), and the 1 s average count rate from the four Sun-facing BATSE CPDs (dotted curve). The CPD and 80 GHz data have been renormalized to the bremsstrahlung flux on the rising and falling portion of the primary peak. The >16 MeV flux has been renormalized to have the same peak flux as the bremsstrahlung. It is again note the lag in the maxima of the OSSE and also BATSE CPD rate relative to the 80 GHz rate. Murphy et al. (1997) also note that the BATSE CPD count rate is due not only to electron bremsstrahlung but also nuclear emissions (see below, Section 9.8.19).

As underlined Murphy et al. (1997), since the proton power-law spectral index derived for this flare (3.7–4.2) is too soft to produce a significant number of pions, the >16 MeV radiation must be bremsstrahlung from high-energy electrons. The comparison of the >1 MeV bremsstrahlung with the >16 MeV emission shows that the high-energy emission decayed more quickly, suggesting that the spectrum of accelerated electrons either softened after the peak or exhibited a transient high-energy tail at the peak. This is to be compared with the constancy found in the steepness of the 0.1–1 MeV electron spectrum measured later in the flare. Because lower energy electrons lose energy via collisions more rapidly than higher energy electrons, a softening spectrum would argue against trapping of impulsively accelerated electrons in the corona, as assumed by Ramaty et al. (1994) in their analysis of the June 4 flare. However, if the magnetic field in coronal portion of the loop is sufficiently high, synchrotron losses could explain the rapid removal of the higher energy electrons (see below, Section 9.8.19 for further discussion).

9.8.11 The Ratio α/p for Accelerated Particles

According to Murphy et al. (1997), an estimate of the accelerated α/p ratio can be obtained by comparing the flux of the ~0.45 MeV $\alpha-\alpha$ line complex with that of narrow deexcitation lines. Since He is a high FIP element, the comparison is best made with lines from high-FIP elements. Using fits to Sun-pointed data, no evidence for variation of the $\alpha-\alpha$-to-high FIP line flux ratio with time was found, with an average value of 0.66 ± 0.04. Share and Murphy (1997a) measured this ratio for the 19 SMM/GRS flares and found no evidence of flare-to-flare variation, with a weighted average ratio of 0.49 ± 0.05. Murphy et al. (1997) compare the measured June 4 $\alpha-\alpha$-to-high FIP ratio with predicted ratios determined for α/p ratios of 0.1 and 0.5 using calculations of thick-target γ-ray line production (e.g., Ramaty et al. 1979; Kozlovsky et al. 1997). The ratio is fairly insensitive to the assumed accelerated heavy ion composition but does depend on the spectral index of the accelerated protons and the ambient [He]/[CNO] abundance. Murphy et al. (1997) assume coronal values for these abundances (Reames 1995: [He]/[O] = 57). Using the spectral index derived for $\alpha/p = 0.1$ (3.7; see above, Section 9.8.8), the predicted flux ratio is 0.34 ± 0.01, which is not consistent with the measured value of 0.66 ± 0.04 (a difference of 8 σ). However, using the spectral index derived for $\alpha/p = 0.5$ (4.3; see above, Section 9.8.8), the predicted flux ratio is 0.55 ± 0.04, which is in better agreement with the observed value (a difference of 1.9 σ). Murphy et al. (1997) conclude that the accelerated α/p ratio for the 1991 June 4 flare is larger than 0.1 and probably closer to 0.5. This conclusion is consistent with the detailed analysis of the 1981 April 27 flare by Murphy et al. (1991) and the analysis of 19 SMM/GRS flares by Share and Murphy (1997a).

9.8.12 Photospheric ^{3}He/H Abundance Ratio

As noted Murphy et al. (1997), the time profile of the 2.223 MeV neutron capture line can be used to determine the photospheric ^{3}He/H ratio. Because of the time required for neutrons to slow down and be captured on ^{1}H, neutron-capture photons are produced about 100 s after the production of the neutrons. The competing capture reaction ^{3}He(n, p)^{3}H produces no radiation but affects the delay of the 2.223 MeV line emission. To make a precise determination of the ^{3}He abundance, detailed calculations of the capture-line time dependence as a function of both the ^{3}He abundance and observing angle for thick-target interactions of accelerated particles with various assumed energy spectra and angular distributions are necessary. This has been done for the 1982 June 3 SMM/GRS flare (Hua and Lingenfelter 1987b). To analyze the OSSE data, a simplified, empirical approach adopted by Prince et al. (1983a) in their analysis of the 1982 June 3 flare is used. The neutron capture-line flux from an instantaneous production of neutrons is assumed to fall exponentially in time with a time constant τ given by (Prince et al. 1983a):

$$\tau = \left(\tau_H^{-1} + \tau_{He}^{-1} + \tau_d^{-1}\right)^{-1}, \tag{9.32}$$

where

$$\tau_H = 1.4 \times 10^{19}/n_H \tag{9.33}$$

is the time constant for neutron capture on H,

$$\tau_{He} = 8.5 \times 10^{14}/(\mathrm{R}n_H) \tag{9.34}$$

is the time constant for neutron capture on ^{3}He, and τ_d is the neutron decay time. Here n_H is the hydrogen number density in cm^{-3} and R is the ^{3}He/H ratio. Prince et al. (1983a) argue that n_H should be $\sim 1.3 \times 10^{17}$ cm^{-3} due to a 'plato' in the solar density distribution at the top of the photosphere. They also restrict their fits to the first 150 s after the flare peak, but because the June 4 1991 flare is less impulsive than the June 3 1982 flare, all data from just prior to the peak (3.68 h UT) to the end

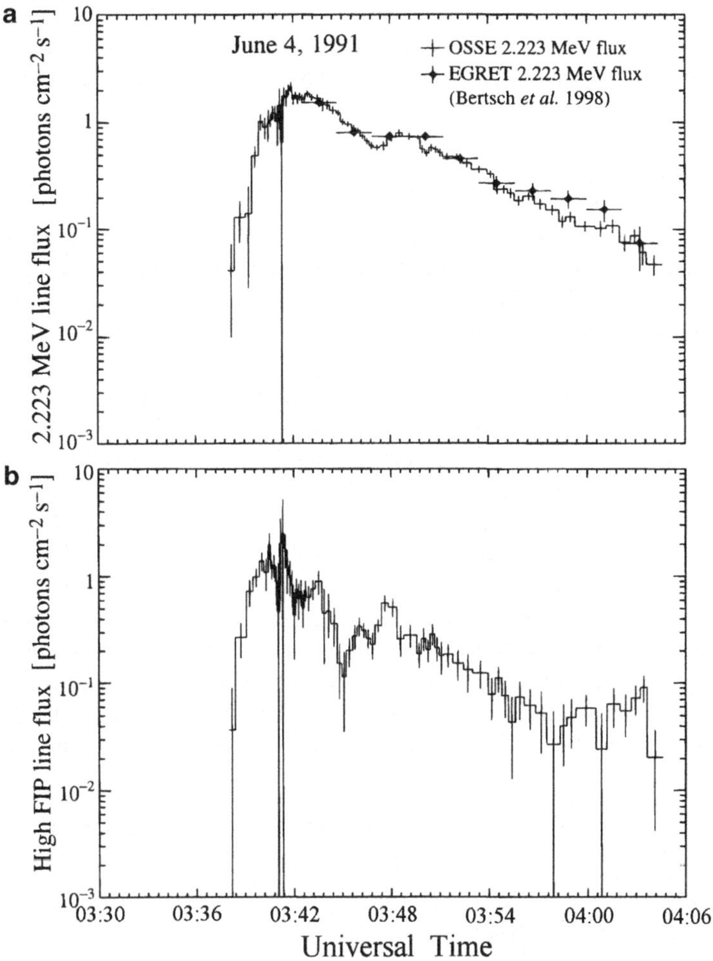

Fig. 9.43 High-resolution time profiles (histograms) of (**a**) the 2.223 MeV neutron-capture line flux and (**b**) the summed flux of narrow nuclear lines from high-FIP elements (at 1.634, 4.439, 6.129 and 6.99 MeV) during the first orbit. A factor of 2 has been applied to the flux values during the 16 s at the peak to account for saturation effects. Also shown in (**a**) are the EGRET/TASC 2.223 MeV line data (Hartman et al. 1999) (From Murphy et al. 1997)

of the first orbit are used in this analysis. To confirm the implementation of the technique, Murphy et al. (1997) repeated the Prince et al. (1983a) analysis of the 1982 June 3 flare and obtained a time constant of 88 s as compared to their published value of 89 s.

The Panel a of Fig. 9.43 displays a high time resolution (8 s and greater) plot of the 2.223 MeV neutron-capture line flux for the first orbit. Also shown in this Panel are 2.223 MeV line fluxes obtained with the EGRET/TASC (Hartman et al. 1999). The agreement with the OSSE fluxes is remarkably good, especially considering the intensity of the flare and also the corrections to the EGRET fluxes due to overlying material. The small excess of the EGRET fluxes over the OSSE fluxes near the end of the orbit may be due to 2.223 MeV photons produced in the CGRO hydrazine tanks by solar-flare neutrons that were reaching their peak intensity at that time (see Fig. 9.30 and discussion). The OSSE detectors are adequately shielded against such background radiation.

Murphy et al. (1997) make the plausible assumption that the time history of the summed high-FIP line fluxes (shown in Panel b of Fig. 9.43) mirrors the interaction time profile of the accelerated particles producing neutrons in the flare. Using this production time profile, the predicted time profile of the neutron capture line is compared to the data in Fig. 9.44; the dotted envelope represents a $\pm 1\ \sigma$ uncertainty.

The best fit is achieved with $\tau = 104 \pm 6$ s and $\chi^2 = 49.6$ corresponding to a 64% probability that the data is randomly distributed about the fit. Using the Eq. 9.32, however, this best-fit value for the time constant is too long to give a physical value for the ^{3}He/H ratio. Murphy et al. (1997) tested whether this could be due to a hardening of the accelerated particle spectrum at high energies as the flare progressed by introducing a time-dependent neutron-production efficiency into the calculations, but the value of τ was not reduced. Murphy et al. (1997) also tried separately fitting subintervals of the flare (see, e.g., Rank 1996) and found that, for the intervals near the peak where the statistics were sufficient to obtain meaningful results, τ was again not reduced. Inspection of the Eq. 9.32 shows that R depends sensitively on the assumed value of the plateau density. Prince et al. (1983a) argue that a conservative lower limit to this density is 1×10^{17} cm^{-3}. Using this value,

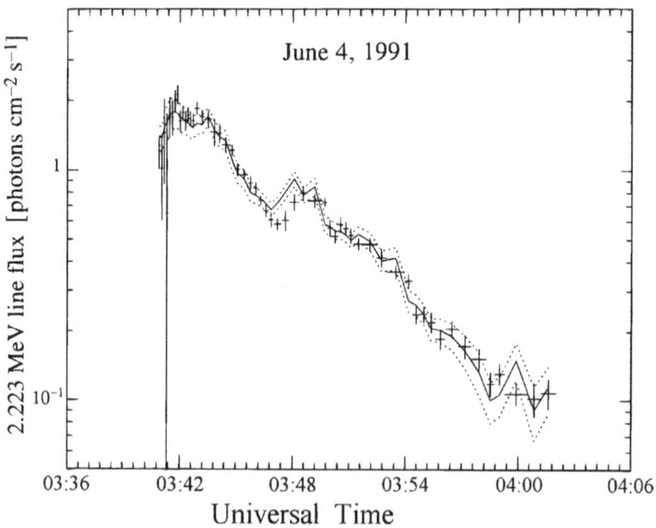

Fig. 9.44 Predicted time profile of the neutron capture line compared with the data. The dotted envelope represents the $\pm 1\ \sigma$ uncertainties (From Murphy et al. 1997)

Murphy et al. (1997) obtain a ^{3}He/H ratio of $(1.3 \pm 0.5) \times 10^{-5}$ and provide a 2 σ upper limit of 2.3×10^{-5}.

9.8.13 Extended Proton Interactions

As underlined Murphy et al. (1997), for flares located away from the solar limb, the narrow 2.223 MeV neutron capture line is the most sensitive indicator for the presence of nuclear reactions resulting from accelerated ions. Because the June 4 1991 flare occurred at a heliocentric angle of 75°, this line can be used to search for nuclear emission preceding and following the impulsive phase. Plotted in Fig. 9.45 is the flux observed in the 2.223 MeV neutron-capture line covering three orbits of observations. The fluxes for the first orbit are derived from fits to >1 MeV data from off-pointed detectors. For the second and third orbits they are from fits to >0.1 MeV spectra from Sun-pointed detectors. Nuclear interactions and the resultant γ-ray emission are clearly seen to continue after the peak of emission through at least the second orbit, i.e., for 2 h.

Murphy et al. (1997) noted that the flux of the 2.223 MeV line at the beginning of the second orbit appears to be higher than a simple extrapolation from the end of the first orbit would imply (see Fig. 9.45). The effect is not as apparent in the summed flux of narrow lines flux (see above, Fig. 9.34). This suggests the possibility of a high-energy acceleration episode occurring during satellite night. Also plotted in Fig. 9.45 is the 80 GHz Nobeyama microwave data (with arbitrary normalization)

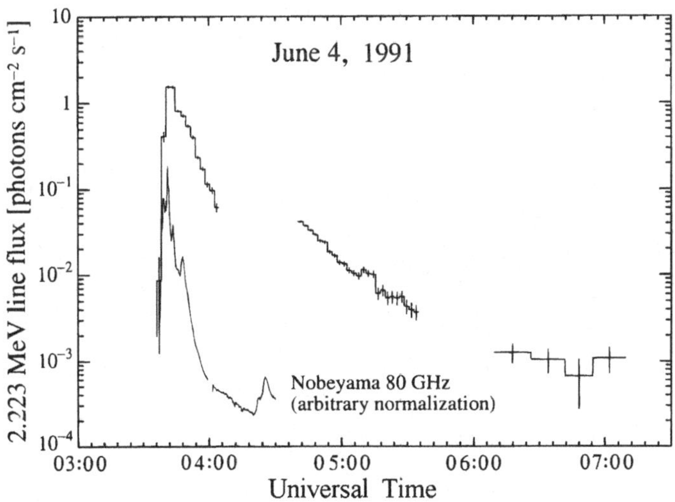

Fig. 9.45 Time profile of the 2.223 MeV neutron-capture line during three orbits of observations (histogram). Also shown are the 80 GHz Nobeyama radio data (Ramaty et al. 1994) (From Murphy et al. 1997)

attributed to gyrosynchrotron emission from >1 MeV electrons. There is evidence in this latter emission for an outburst during satellite night at about 4.4 h. Struminsky et al. (1994) have used the microwave and ground-based neutron-monitor data to argue for such an additional episode producing high-energy particles. The flux in the 2.223 MeV line averaged over the third orbit is $(1.1 \pm 0.2) \times 10^{-3}$ photons.cm^{-2}s^{-1}, indicating that nuclear interactions are occurring during this time. However, as discussed in Section 9.8.4, there is an additional uncertainty in the determination of absolute fluxes due to day-to-day residuals remaining after background subtraction. For the 2.223 MeV line, these residuals were at a level of about 1×10^{-3} photons.cm^{-2}s^{-1} a few days after the flare, i.e., comparable to the measured third-orbit flux. Fortunately, it can be used the standard OSSE background subtraction technique as an alternative to the predictive algorithm. The standard technique has the advantage that the background systematic is significantly smaller than that of the predictive algorithm, but the statistics are poorer since only one detector views the Sun at a given time rather than three.

Murphy et al. (1997) have used the standard technique with data from the two chopping detectors to search for evidence of nuclear reactions not only during the third orbit but also during the ~3 day period surrounding the June 4 flare. Plotted in Fig. 9.46 is the line flux derived from orbit-averaged data during this time period (the orbit containing the peak of the flare and the following orbit are off-scale and are excluded from this analysis).

Also plotted in the Fig. 9.46 is the GOES 1–8 Å X-ray profile. The 2.223 MeV line flux during the third orbit is $(1.5 \pm 0.5) \times 10^{-3}$ photons cm^{-2}s^{-1}, consistent with the value derived using the predictive algorithm. It can be now say with 99.7%

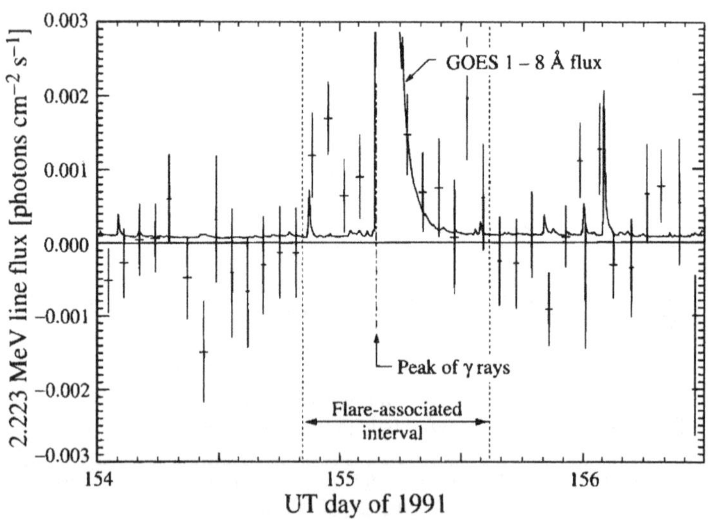

Fig. 9.46 Time profile of the 2.223 MeV neutron-capture line derived from fits to data obtained using standard OSSE background subtraction techniques for several orbits around the June 4 X12+ orbit. Also shown is the GOES 1–8 Å flux (From Murphy et al. 1997)

confidence (3 σ) that nuclear interactions are occurring during the third orbit. But Fig. 9.46 shows that this flux may only be part of a general excess of emission occurring for several orbits both before and after the flare. Murphy et al. (1997) have separately summed the data accumulated during the flare-associated interval indicated in Fig. 9.46 and the data accumulated during the orbits outside of the interval. The fitted 2.223 MeV line flux during the flare-associated interval was $(1.0 \pm 0.2) \times 10^{-3}$ photons cm^{-2}s^{-1}, while the flux outside of the interval was $(0.1 \pm 0.1) \times 10^{-3}$ photons cm^{-2}s^{-1}. This suggests that throughout the several hours before and after the impulsive flare there was a significant level of nuclear interactions.

9.8.14 Discussion on the Ambient Abundance Estimates

As noted Murphy et al. (1997), the measured fluxes of nuclear γ-ray lines have been used to obtain information about the composition of the ambient flare plasma. The flare-averaged narrow nuclear deexcitation line fluxes relative to that of the 6.13 MeV ^{16}O line were mostly similar to the averages measured for the 19 SMM/GRS flares (Share and Murphy 1995; see Table 9.10) suggesting that the flare June 4 1991 ambient abundances were similar to those of other γ-ray flares. However, Murphy et al. (1997) note that the relative flux of the 4.44 MeV ^{12}C line appears to be high compared to the SMM/GRS average, although there were SMM/GRS flares with relative fluxes consistent with the June 4 value. The June 4 measured flux ratio corresponds to an ambient C/O abundance ratio of 0.54 ± 0.04, obtained using calculations of thick-target γ-ray line production (e.g., Ramaty et al. 1996) and the flare-averaged accelerated proton spectral index of 4.3 (derived with an assumed accelerated α/p ratio of 0.5). The result is independent of assumptions for the other ambient abundances and for the α/p ratio. This derived C/O abundance ratio is high (2.2 σ) relative to the 19 flare SMM/GRS-flare average of 0.4 ± 0.05 derived by Ramaty et al. (1996); however, it is consistent (at a greater than 93% confidence level) with measurements of photospheric and coronal C/O abundance ratios ranging from 0.43 to 0.48 (Anders and Grevesse 1989; Garrard and Stone 1993; Reames 1995; Grevesse and Noels 1993; Grevesse et al. 1996).

The relative flux of the $\sim$7 MeV line complex from ^{16}O is also significantly higher than the 19-flare average of the SMM/GRS analysis. Murphy et al. (1997) investigated whether this high flux ratio could be due to inadequacies of the assumed γ-ray model or to contamination of the line from spacecraft sources but conclude that the measurement of this line is accurate. Calculations of thick-target γ-ray line production similar to those of Ramaty et al. (1996) show that the measured June 4 ratio can only be achieved with extreme values for both the accelerated proton spectral index (<2.5) and the accelerated α/p ratio (>1). However, this hard index is inconsistent with the measured value for this flare and such a high α/p ratio is unlikely. It is no satisfactory explanation for the high flux of the $\sim$7 MeV line. The flare-averaged ratio of >1 MeV narrow-line fluxes from low-FIP elements to those from high-FIP elements for the June 4 flare is consistent with the mean ratio determined for 19 SMM/GRS flares (Share and Murphy 1995) where significant flare-to flare variability was found. The June 4 ratio suggests that the ambient low-FIP elements at the interaction site are enhanced relative to high-FIP elements by about a factor of 2.5 compared to photospheric abundances. The photospheric ^{3}He/H ratio is important for cosmological models, stellar evolution and implications of the solar ^{4}He/H ratio (Share and Murphy 1997b). Gamma-ray line measurements provide a measure of this ratio because the delayed time profile of the 2.223 MeV line from neutron capture on H is strongly affected by absorption on ^{3}He. Previous line measurements for the 1982 June 3 flare yielded ratios of less than 3.8×10^{-5} (Prince et al. 1983a) and $(1.2–2.3 \pm 1.2) \times 10^{-5}$ (Hua and Lingenfelter 1987b).

Murphy et al. (1997) have used the simplified approach of Prince et al. (1983a), which models the neutron-capture line decay time with a simple exponential. Physical values could only be obtained if the average density at the neutron-capture site is as small as 1×10^{17} H cm^{-3}. For this value, a 2 σ upper limit for the ^{3}He/H ratio is 2.3×10^{-5}, consistent with the values obtained for the 1982 June 3 flare by both Prince et al. (1983a) and Hua and Lingenfelter (1987b). Detailed modeling as was done by Hua and Lingenfelter (1987b) for the 1982 June 3 flare are required to provide a better estimate of this ratio.

9.8.15 Discussion on Abundance Variations

Using SMM/GRS data, Share and Murphy (1995) found significant variations of the low FIP-to-high FIP line ratio from flare to flare. With the improved statistics available with the OSSE detectors, Murphy et al. (1997) have searched for variability of this ratio within the June 4 1991 flare. The flux ratios measured for the first and second orbits of emission were different (5.6 σ), increasing from more photospheric to more coronal values. The first-orbit ratio corresponds to about a factor of 2 enhancement of the low-FIP/high-FIP abundance ratio relative to the photosphere and the second-orbit ratio to about a factor of 3.5, or what is observed in the corona. As noted Murphy et al. (1997), such variability could possibly be due to time-dependent compositional changes at the flare site, but the timescale is probably too short for this to be a reasonable explanation. Alternatively, the location of the γ-ray production site could have changed with time, progressing from deeper in the chromosphere-photosphere toward the corona. This could happen if the height of the magnetic field mirroring point increased as the flare progressed. If this is the case, γ-ray measurements can be used to map the structure of solar atmosphere during flares.

9.8.16 Discussion on Accelerated Protons and Ions

As noted Murphy et al. (1997), the spectrum of the accelerated protons can be studied using measurements of narrow γ-ray lines. For the 1991 June 4 flare, the instantaneous accelerated proton spectral index in the 5–30 MeV energy range was determined using the 6.129 MeV to 1.634 MeV line (O/Ne) flux ratio. No significant evidence for spectral evolution of the accelerated proton spectrum was found, with flare-averaged indexes of 4.0 ± 0.2 and 4.4 ± 0.3 for α/p ratios of 0.1 and 0.5, respectively. Using the alternative technique for determining a flare averaged spectral index based on the ratio of the 2.223 MeV neutron-capture line and the 4.44 MeV ^{12}C deexcitation line, we found indexes of 3.7 ± 0.1 and 4.2 ± 0.1 for assumed α/p ratios of 0.1 and 0.5, respectively. The assumption of $\alpha/p = 0.5$ provided better agreement between the two index-determining techniques (a difference in the indexes of 0.6 σ) than did $\alpha/p = 0.1$ (a difference of 1.2 σ). Ramaty et al. (1996) derived proton spectral indexes for the 19 SMM/GRS flares using the O/Ne line flux ratio. The indexes varied between 2 and 6 with a mean of $\sim$3.8 and $\sim$4.3 for $\alpha/p = 0.1$ or 0.5, respectively. The proton spectrum of the 1991 June 4 flare appears to be typical of γ-ray flares.

As underlined Murphy et al. (1997), the power-law indexes deduced here are softer than those deduced earlier for this flare (Murphy et al. 1993a; $\approx$2.8) because (1) the additional data at the peak included here significantly increased the measured 4.439 MeV line fluence relative to the 2.223 MeV line fluence and (2) impulsive-flare accelerated-particle abundances are now used (see below). The abundances of the accelerated particles and their variability can provide constraints on models of particle acceleration in solar flares (e.g., Temerin and Roth 1992; Miller and Vinas 1993; Miller and Roberts 1995). A measure of the accelerated α/p ratio can be obtained by comparing the flux of the $\sim$0.45 MeV α-α line with the sum of line fluxes from high-FIP elements. The flare-averaged flux ratio of the α-α line to the high-FIP lines was 0.66 ± 0.04. Using calculations of thick-target γ-ray line production, this measured value corresponds to an accelerated α/p ratio larger than 0.1 and closer to 0.5 for an assumed ambient $^4He/H$ ratio of 0.1. In support of this conclusion, assuming $\alpha/p = 0.5$ yields a proton spectral index derived using the neutron capture-to-^{12}C line flux ratio that is in better agreement with the index derived using the O/Ne line flux ratio (see above). In their analysis of 19 SMM/GRS flares, Share and Murphy (1997a) also concluded that an α/p ratio closer to 0.5 was required, consistent with the 1981 April 27 flare analysis of Murphy et al. (1991). No evidence for variation of the accelerated α/p ratio with time during the 1991 June 4 flare was found.

Murphy et al. (1997) noted that the ratio of the flux in broad γ-ray lines to the flux in narrow lines gives an indication of the accelerated heavy ion-to-proton ratio. Because of difficulties fitting broad-line features, the broad component was not fitted with individual Gaussians. Instead, a broad-line template based on measurements of SMM/GRS spectra was used instead. The flare-averaged ratio of the flux in the template to the flux in narrow γ-ray lines greater than 1 MeV (2.83 ± 0.07) was consistent with the mean obtained from the 19-flare SMM/GRS analysis where flare-to-flare variation of this flux ratio was measured. Using the OSSE data for the 1991 June 4 flare, Murphy et al. (1997) searched for variation of this ratio with time. The averages for the two orbits were found to be different (8.1 σ), suggesting that the accelerated heavy ion-to-proton ratio decreased as the flare progressed.

As mentioned above, the accelerated α/p ratio showed no significant evidence of temporal variation. Using current models of γ-ray production, constraints can be placed on the relative abundances of heavy (i.e., greater than ^{4}He) accelerated ions. To achieve agreement between the two proton spectral index-determining techniques required assuming impulsive-flare abundances (Ramaty et al. 1996) for the accelerated particles. The alternative assumption of gradual-flare abundances did not provide agreement. This is consistent with the 1981 April 27 flare analysis of Murphy et al. (1991) and the 19 SMM/GRS flare analysis of Ramaty et al. (1996). In their analysis of the GRANAT/PHEBUS γ-ray data from the behind-the-limb flare on 1991 June 1, Ramaty et al. (1997) found that the composition of the accelerated particles producing the γ-rays also showed heavy-element enhancements typical of impulsive flares. This impulsive-flare composition has ^{3}He/$\alpha = 1$ in addition to significant enhancements of Ne, Mg, Si, and Fe.

According to Murphy et al. (1997), the number of accelerated protons and their energy content can be determined using the flare-averaged proton spectral index derived for $\alpha/p = 0.5$ (4.2 ± 0.1) and the flux in the 4.44 MeV ^{12}C line (189 ± 9 photons.cm^{-2}). Assuming the spectrum is a power law extending down to 1 MeV with a flat extension to lower energies, the total number of protons with energy greater than 30 MeV accelerated in the 1991 June 4 flare was $(6.7 \pm 1.2) \times 10^{32}$ and the total energy contained in protons of all energies was $(1.7 \pm 0.2) \times 10^{32}$ erg. Ramaty et al. (1994) determined the number of protons accelerated in the 1991 June 4 flare to be 5.7×10^{32}. Assuming impulsive-flare abundances for the accelerated particles, identical energy dependence for all ion spectra, and $\alpha/p = 0.5$, the total energy contained in accelerated ions was $(1.0 \pm 0.1) \times 10^{33}$ erg. This is larger than the largest energy contained in ions for any of the 19 SMM/GRS flares (Ramaty et al. 1995).

9.8.17 Discussion on Accelerated Electrons

As noted Murphy et al. (1997), at times after the peak of emission, the electron bremsstrahlung spectrum can be directly measured with Sun-pointed detector data. No evidence for temporal evolution of the spectral index was found; it varied from its average value of 2.74 by less than ± 0.07. This spectral index corresponds to a power-law index of ~ 3.9 for the accelerated electrons. The bremsstrahlung indexes measured for 84 strong SMM/GRS flares (Vestrand et al. 1999) varied between 2 and 6 with a mean of 2.86 for flares occurring at heliocentric angles greater than 60° (the 1991 June 4 flare occurred at 74.5°). The 1991 June 4 accelerated electron spectrum appears to be typical of γ-ray producing flares.

Murphy et al. (1997) underlined that during the peak of the flare, the electron bremsstrahlung cannot be directly measured with the >1 MeV data from off-pointed detectors. The bremsstrahlung component was estimated by subtracting the flux due to nuclear emissions (assuming a constant ratio for the broad-to-narrow line fluxes) from the total observed flux. The total energy contained in electrons of more than

0.1 MeV can be estimated by assuming the electron spectral index to be constant during the peak at the average value (2.74) directly measured later in the flare. If the June 4 flare is similar to most observed γ-ray flares, then the spectrum would be expected to be harder during the peak than during the decay (Dennis 1988). The comparison of the estimated >1 MeV bremsstrahlung with the >16 MeV flux at the peak may, in fact, indicate a softening of the spectrum with time. This >16 MeV emission (which is probably also bremsstrahlung but from high-energy electrons) was much more impulsive and decayed much more rapidly, suggesting that the accelerated electron spectrum during the peak softened over a broad energy range (or exhibited a transient high-energy tail). This suggests that the energy estimate at the peak determined using the index measured during the decay phase is an upper limit to the actual value. This estimate for the total energy contained in electrons of more than 0.1 MeV accelerated in the 1991 June 4 flare is $(1.0 \pm 0.1) \times 10^{31}$ erg.

As noted Murphy et al. (1997), if the spectrum during the peak were indeed steeper, in contrast to the argument above, the derived energy in electrons would be higher. It can be obtain a conservative upper limit to the total energy in >0.1 MeV electrons based on the total energy in >0.02 MeV electrons measured with Ulysses (Kane et al. 1995). Murphy et al. (1997) note that the Ulysses value itself may represent an upper limit to the >0.02 MeV nonthermal electron energy since the time profile of the >0.025 MeV count rate (shown in Fig. 14a in Kane et al., 1998) is much more gradual than the OSSE bremsstrahlung and is therefore suggestive of thermal origin. Kane et al. (1995) do not give the total energy of the electrons for the 1991 June 4 flare but only the energy deposition rate at the peak (2.5×10^{31} erg s^{-1}). From the OSSE measurements, the estimated instantaneous power in >0.1 MeV electrons at the peak of the flare was $(1.6 \pm 0.1) \times 10^{29}$ erg s^{-1}. If the electron spectrum is assumed to extend to lower energies as an unbroken power law with an index of 2.74, the >0.02 MeV electron energy deposition rate from the OSSE measurements would be $(2.8 \pm 0.2) \times 10^{30}$ erg s^{-1}, or about an order of magnitude lower than the Ulysses measurements. This difference suggests that either the electron spectrum steepens below 0.1 MeV thermal emission is contaminating the Ulysses measurement. Murphy et al. (1997) scale their estimated energy in >0.1 MeV electrons by this factor of 10 to obtain a conservative upper limit of 1×10^{32} erg.

9.8.18 Discussion on Energetics of Accelerated Ions and Electrons

As underlined Murphy et al. (1997), a fundamental question of particle acceleration concerns the relative efficiency of electron and proton acceleration. The estimated total energy in >0.1 MeV electrons derived from the OSSE data was $(1.0 \pm 0.1) \times 10^{31}$ erg, while the total energy in protons was $(1.7 \pm 0.2) \times 10^{32}$ erg. For this flare, the total energy in accelerated protons exceeded the energy in accelerated >0.1 MeV electrons by an order of magnitude (Emslie 1983; Emslie et al. 1996). The energetic importance of ions has been emphasized by Simnett (1995), Ramaty et al. (1995) and Mandzhavidze and Ramaty (1996). Assuming impulsive-flare accelerated-particle abundances and $\alpha/p = 0.5$, the total energy of all ions would be another factor of 6 higher than that of just the protons.

Murphy et al. (1997) note that the conservative upper limit for the total energy in >0.1 MeV electrons, 1×10^{32} erg, is still less than that of the protons and an order of magnitude less than that of ions assuming impulsive-flare abundances. The above estimate of proton energy is based on a power-law spectrum extending down to a cutoff energy of 1 MeV with a flat extension to lower energies. If the cutoff energy were higher, the energy in accelerated protons could be less. Murphy et al. (1997) have used calculations of γ-ray line production to explore the dependence of the derived proton spectrum on cutoff energy. As the cutoff energy is increased, the required proton spectrum derived from the O/Ne line flux ratio steepens because of the low threshold energy of Ne excitation. But the cutoff energy cannot be larger than about 2 MeV; otherwise, the spectral index derived using the O/Ne ratio becomes inconsistent with the index derived using the neutron-capture/^{12}C line flux ratio. A cutoff energy of 2 MeV rather than 1 MeV reduces the energy in the protons by less than a factor of 4.

Murphy et al. (1997) conclude that the acceleration process in flares is thus capable of producing particles that are energetically dominated by ions, as in the 1991 June 4 flare, or by electrons, as in electron-dominated episodes of some flares (Rieger and Marschhäuser 1990). Note that while the energy in accelerated ions may be greater than that in electrons, the flux in nuclear γ-ray lines relative to the flux in bremsstrahlung is not. This is because the efficiency for γ-ray production by protons via nuclear excitation is less than that by electrons via bremsstrahlung. During the peak of the flare, the bremsstrahlung flux accounted for 40–85% of the total >1 MeV flux, depending on the assumed broad-to-narrow line flux ratio. As the flare progressed, the bremsstrahlung/nuclear-line ratio decreased (the probability that the data are consistent with a constant is 6×10^{-4}), suggesting a corresponding decrease in the accelerated electron-to-proton ratio. The bremsstrahlung/nuclear-line ratio is shown in Fig. 9.47.

9.8.19 Discussion on Electron Trapping Time and the Coronal Magnetic Field

Ramaty et al. (1994) derived estimates of the magnetic field in coronal loops from electron trapping times. The trapping times were derived based on a model in which electrons spiral in a flare loop emitting gyrosynchrotron radiation and, after a mean time τ, escape down the loop to higher densities where they emit bremsstrahlung. Ramaty et al. (1994) used the BATSE CPD rate as a monitor of the bremsstrahlung.

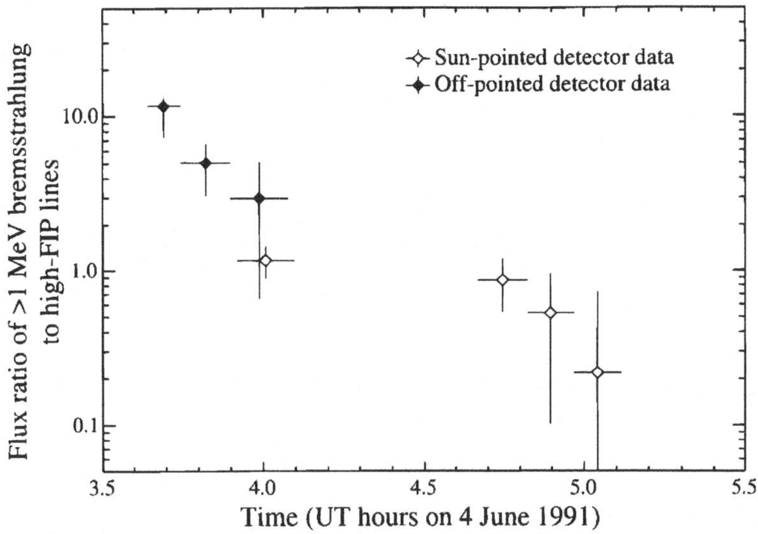

Fig. 9.47 Time profile of the flux ratio of >1 MeV bremsstrahlung to high-FIP lines obtained from off-pointed (filled diamonds) and Sun-pointed (open diamonds) detector data. The first-orbit Sun-pointed data has been shifted in time for display purposes (From Murphy et al. 1997)

Table 9.11 Electron trapping times (s) in dependence of electron power index S_e (from 2 to 5) and coronal magnetic field B (from 100 to 300 G) (From Murphy et al. 1997)

S_e	B		
	100 G	200 G	300 G
2	1,180	280	140
2.5	560	120	53
3	340	65	26
4	220	32	10
5	260	26	8

Based on a comparison of this rate with the 80 GHz emission, they concluded that the magnetic field in the loop was larger than 100 G.

Murphy et al. (1997) have used complementary OSSE spectral measurements of the flare to obtain a revised estimate of the magnetic field. OSSE has provided improved measurements for two relevant parameters: (1) the BATSE CPD instrument response appears to be a factor of $\sim$2.5 low, and (2) about 85% of the BATSE CPD rate at the peak of the flare may be due to bremsstrahlung, in contrast to the 100% originally assumed. Combining these OSSE measurements yields an absolute reduction in the bremsstrahlung-to-radio emission ratio of $2.5/0.85 = 2.9$. This yields revised trapping times based on Table 3 of Ramaty et al. (1994), which are presented in Table 9.11 for various electron spectral indexes and magnetic field intensities.

Ramaty et al. (1994) estimated a value of 10 s for the bremsstrahlung decay time based on the decay of the BATSE CPD rate. Table 9.11 shows that to achieve trapping times $\leq$10 s with $S_e \approx 4$ requires a magnetic field $\geq$300 G. If the decay time of the bremsstrahlung is longer than 10 s, as may be indicated by the OSSE-derived bremsstrahlung (see Panel b in Fig. 9.42), then the required magnetic field could be reduced but would probably still be larger than 200 G. Such high magnetic fields may be required to explain the rapid depletion of the >16 MeV electrons; as was considered in Section 9.8.10, under the trapping scenario the electron spectrum should harden with time, but the OSSE >16 MeV observations show that the high-energy bremsstrahlung decays faster than the >1 MeV bremsstrahlung (also see the discussion of the next Section 9.8.20).

9.8.20 Discussion on Extended Accelerated Particle Interactions

According to Murphy et al. (1997), the mean duration of γ-ray flares observed by SMM/GRS was about 7 min, with the longest lasting about 1 h. More recent measurements have shown that γ-ray flares can last for several hours, with the 1991 June 11 flare lasting up to 8 h (Kanbach et al. 1993). Using the 2.223 MeV neutron-capture line as an indicator of proton interactions, we found that nuclear reactions associated with the impulsive flare continued through the second orbit; i.e., for at least 2 h after the peak of emission. This raises the following question: Did these reactions result from particles that were accelerated early in the flare and were trapped or did they result from particles that were continuously accelerated? Because lower-energy particles lose energy more quickly than higher-energy particles, an indication of trapping would be a continuously hardening proton spectrum (Mandzhavidze and Ramaty 1996). Direct measurement of the proton spectral index using the O/Ne line flux ratio indicates that the spectral index did not change with time, although the uncertainties are large enough to allow for a hardening of at least a full index. A comparison of the time profiles of the summed flux of lines from high-FIP elements and the 2.223 MeV neutron-capture line (shifted back in time ~100 s to account for its delay) also shows no indication of a hardening proton spectrum, which would reveal itself in a continuously increasing neutron-capture to high-FIP line ratio (compare Figs. 9.34 and 9.45). Both of these emissions decay with nearly identical time constants.

Murphy et al. (1997) conclude that the long-lasting nuclear γ-ray emission of the 1991 June 4 flare was probably due to continued ion acceleration rather than trapping. Using COMPTEL data, Rank et al. (1996, 1997a) obtained a similar conclusion for the extended emission phases of the 1991 June 9, 11, and 15 flares. The evidence for continued acceleration of electrons is less compelling. Under a trapping scenario, electron bremsstrahlung would be expected to decay faster than nuclear line emission because electrons lose energy faster than protons (Ramaty and Mandzhavidze 1994). This appears to be the case for the June 4 flare (see Figs. 9.42 and 9.47) and so is suggestive of trapping. However, while a more rapid decay is a necessary consequence of trapping, it is not sufficient to prove trapping. The electrons could be accelerated continuously with a decrease in the efficiency of electron acceleration relative to proton acceleration explaining the more rapid bremsstrahlung decay. Indeed, an argument in favor of continuous electron acceleration is the fact that the high-energy (>16 MeV) electrons decay more rapidly with time than the ~MeV electrons, at least over the short time-scales associated with the peak of emission (see Section 9.8.10). This implies a softening spectrum, while the spectrum of trapped electrons should harden. Murphy et al. (1997) note, however, that the assumption of trapped electrons has been made by Ramaty et al. (1994) in their analysis of the peak of the 1991 June 4 flare and has provided consistency between the 80 GHz radiation and the ~MeV γ-rays. It was also made this assumption in Section 9.8.19, where Murphy et al. (1997) repeated the Ramaty et al. (1994) analysis using the revised flare parameters obtained with the OSSE data. It was suggested that if these impulsive electrons are trapped, the more rapid decay of the >16 MeV electrons relative to the ~MeV electrons might be due to gyrosynchrotron losses. Even if the longer term bremsstrahlung were due to continuously accelerated electrons, the shorter term impulsive electrons could be associated with a separate magnetic loop more conducive to trapping.

Murphy et al. (1997) searched for 2.223 MeV line emission at times other than the impulsive phase of the flare. Evidence for low-level emission was found during the several orbits prior to and following the flare orbits (see Fig. 9.46) at an average flux of $(1.0 \pm 0.2) \times 10^{-3}$ photons.cm^{-2}s^{-1}. This extended emission implies the existence of energetic ions at times other than those directly associated with the impulsive phase for up to 18 h surrounding the flare. Such emission could be due to a number of microflares occurring during this period; a GOES M1.5 flare at about day 154.9 is visible in Fig. 9.46. Murphy et al. (1997) note, however, that a larger M4.8 flare occurred at about day 156.1 without any increase in γ-rays. Alternatively, the γ-ray emission could be due to long-term trapping of energetic particles in the corona (e.g., the model of Elliot 1964). Simnett et al. (1986) searched for pre-flare emission in eight flares observed by the SMM/GRS using the 4.44 MeV ^{12}C line but found no evidence for emission above their sensitivity threshold $\sim 1 \times 10^{-3}$ photons.cm^{-2}s^{-1} at the 2 σ level. Rieger (1994) reported bursts of γ-ray emission on the rising portion of three SMM/GRS flares but concluded that the emission was most probably electron bremsstrahlung. In their analysis of several years of SMM/GRS data, Harris et al. (1992) obtained a 3 σ upper limit of 8.6×10^{-5} photons.cm^{-2}s^{-1} for the 2.223 MeV line flux during periods of high solar activity excluding the times of flares (1.3×10^{7} s live time).

9.8.21 Summary of Main Results on the 1991 June 4 Solar γ-Ray Flare

Using spectral observations of the 1991 June 4 solar γ-ray flare, Murphy et al. (1997) have obtained information about the ambient plasma at the flare site and the composition and spectra of the accelerated ions and electrons. The main results are:

1. The ratio of the ambient abundance of elements with low FIP to those with high FIP was similar to the mean ratio determined for 19 SMM/GRS flares and appears to have increased as the flare progressed.
2. A model-dependent 2 σ upper limit of 2.3×10^{-5} was obtained for the photospheric ^{3}He/H abundance ratio.
3. Based on current models of γ-ray production, the accelerated α/p ratio was significantly larger than 0.1 and probably close to 0.5 and the relative abundance of the accelerated ions was probably more like the composition of impulsive flares observed in interplanetary space rather than gradual flares.
4. The accelerated heavy ion-to-proton ratio appears to have decreased as the flare progressed.
5. The high-energy ($\geq$16 MeV) component of the accelerated electrons was more impulsive than the lower energy ($\sim$MeV) component.
6. The ratio of electron bremsstrahlung to the flux in narrow γ-ray lines decreased as the flare progressed.
7. The total energy in accelerated ions exceeded the energy in >0.1 MeV electrons.
8. The derived magnetic field intensity in the coronal loop was probably larger than 200 G (based on a specific model).
9. The nuclear reactions producing the γ-rays continued for more than 3 h and resulted from ions that were probably accelerated continuously rather than impulsively followed by trapping.
10. Energetic ions may have been present for several hours prior to and following the impulsive phase of the flare.

9.9 Main Properties of Solar Gamma-Ray Flares in the 23rd Solar Maximum: Yohkoh Observations Between November 1997 and March 1999

9.9.1 Observations by the Yohkoh Satellite of Six γ-Ray Flares Between November 1997 and March 1999

According to Yoshimori et al. (1999a), Yohkoh satellite observed six γ-ray flares producing high energy photons of energies above 1 MeV between November 1997 and March 1999. They discuss spectral characteristics of two different types of flares on 6 November, 1997 and 18 August, 1998. The 6 November, 1997 flare emitted strong γ-ray lines and the spectrum extended to a few tens of MeV, while the 18 August, 1998 flare is an electron-dominated event, exhibited hard continuum extending 20 MeV without apparent γ-ray lines. The Yohkoh γ-ray flare list is shown in Table 9.12.

9.9.2 The 1997 November 6 Flare: Observed Ratio of Broad to Narrow γ-Ray Line Fluxes and Electron Spectrum

As noted Yoshimori et al. (1999a), the accelerated particle composition and energetics and ambient abundances may be obtained from γ-ray spectroscopy data: an example of such analysis is the study by Murphy et al. (1997) of the 1991 June 4 solar flare. The 1997 November 6 flare from Table 9.12 exhibited narrow and broad γ-ray lines. The ratio of broad to narrow line fluxes gives a measure of the relative abundance of accelerated heavy ions to accelerated protons. However, the measured variation in the ratio should represent a lower limit to the actual variation because the broad line component contains unresolved narrow lines produced by protons. Broad lines are seen around 1.5–1.6, 4 and 6 MeV, which result from accelerated Ne +Mg, C and O, respectively (Murphy et al. 1990a). The time dependence of the observed ratio of broad to narrow line fluxes is shown in Fig. 9.48.

The ratio exhibited a time variation, ranging from 2 to 5. It increased spectrum (the 2.22 MeV line flux is much larger than the prompt line fluxes). In order to perform the Gan's method (Gan 1998), it needs several parameters: yields of neutrons and prompt lines, conversion factors from neutron to 2.22 MeV line and decay constant of 2.22 MeV line emission. The first two parameters are taken from

Table 9.12 Yohkoh satellite γ-ray flare list (From Yoshimori et al. 1999a)

Date	Time (UT)	GOES/Hα	Location	NOAA
1997-Nov-06	11:52	X9.4/2B	S18 W63	8100
1998-Aug-18	22:15	X4.9/1B	N33 E87	8307
1998-Nov-22	06:37	X3.7/1N	S27 W82	8384
1998-Nov-22	16:19	X2.5/2N	S30 W89	8384
1998-Nov-28	05:40	X3.3/3N	N17 E32	8395
1998-Dec-18	17:18	M8.0/2N	N19 E65	8415

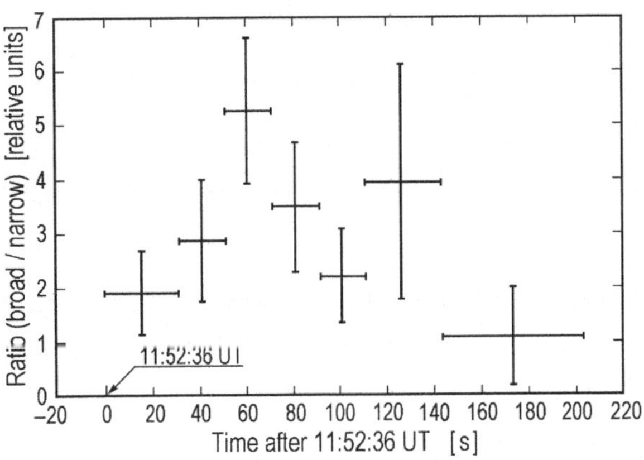

Fig. 9.48 The 6 November 1997 event: Time dependence of the observed ratio of broad to narrow γ-ray line fluxes (From Yoshimori et al. 1999a)

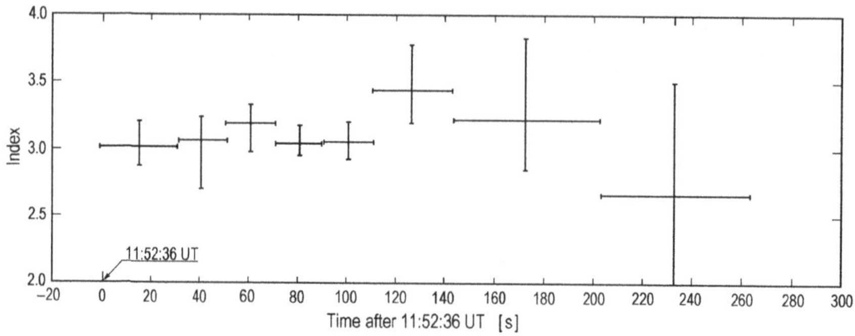

Fig. 9.49 The 6 November 1997 event: Time dependence of the proton power law spectral index (From Yoshimori et al. 1999a)

Hua and Lingenfelter (1987b) and Murphy and Ramaty (1984) and the third one is given by Yoshimori et al. (1999d). It was assumed a power law proton spectrum. The time dependence of the power law spectral index is shown in Fig. 9.49.

As can be seen from Fig. 9.49, the spectral index ranges from 3.0 to 3.5 in 0–200 s (11:52:36–11:55:56 UT) but the error in the spectral index is larger after 200 s (11:55:56 UT). Using the Ramaty et al. (1996) calculation to derive the flare-averaged proton spectral index from the ratio of 2.22 MeV to C line (4.44 MeV) fluences and assuming that the ratio of ambient Ne to O abundances is 0.25, Yoshimori et al. (1999a) obtain the power law index of 3.5 ± 0.3 for the ratio of accelerated He to proton fluxes $\alpha/p = 0.1$ and 3.9 ± 0.3 for $\alpha/p = 0.5$. These values are consistent with that obtained from the Gan's method. The flare-averaged continuum spectrum is fitted by a single power law function of index of 2.59 ± 0.02. The fluence of >l MeV γ-rays (integration time is 200 s) is $1{,}966 \pm 231$ photons cm^{-2}.

Yoshimori et al. (1999a) derived the corresponding accelerated electron spectrum from the measured bremsstrahlung continuum using the Ramaty et al. (1993) calculations. The power law index of the electron spectrum is calculated to be 3.88 $\pm$ 0.02 and the total energy contained in >1 MeV electrons is estimated to be (2.59 $\pm$ 0.29) $\times$ 10^{28} erg. The C+O γ-ray lines flux decays exponentially with time. The decay constant is 33 ± 3 s before 158 s (11:55:14 UT) but changes into 126 ± 76 s after that. It suggests the possibility that a different proton acceleration process starts or protons trapped in the corona for long time produce γ-ray lines after 158 s.

9.9.3 The 1997 November 6 Flare: Observed High Energy γ-Rays from π^0 Decay and from Bremsstrahlung of High Energy Electrons

According to Yoshimori et al. (2000, 2001b, c, 2002), Yohkoh satellite observed the high-energy γ-rays at 10–100 MeV from a X9.4/3B flare at 11:52 UT on November 6, 1997. As noted Yoshimori et al. (2000, 2001b, c), the high-energy γ-rays above

10 MeV are produced from interactions of high-energy flare-accelerated particles above 300 MeV with the solar atmosphere. Several high-energy γ-ray flares were observed with SMM (Forrest et al. 1985), Gamma-1 (Akimov et al. 1991) and CGRO (Kanbach et al. 1993). Gamma rays above 10 MeV are produced from (1) bremsstrahlung of >10 MeV primary electrons, (2) π^0 decay, (3) bremsstrahlung of electrons and positrons from $\pi^{\pm}$ decay, and (4) annihilation in flight of positrons from π^+ decay (Ramaty and Murphy 1987; Murphy et al. 1987). Gamma rays from the π^0 decay result in the broad-band spectrum which peaks at 70 MeV and extends to energies greater than 100 MeV. These radiations are signatures of the highest-energy processes taking place on the Sun. The γ-ray spectrometer detected photons above 10 MeV in the peak phase. The count-rate time profiles at 20–33, 33–53 and 53–72 MeV are shown in Fig. 9.50.

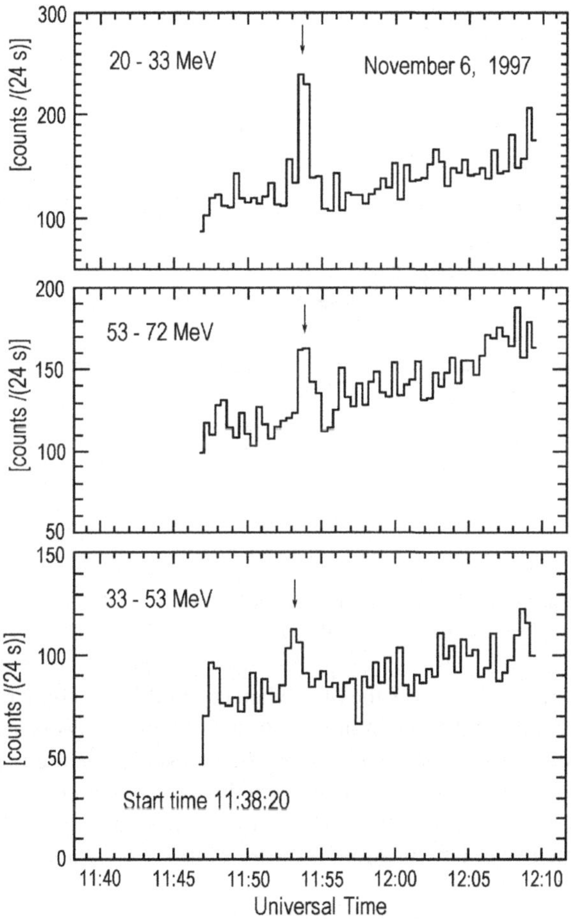

Fig. 9.50 Count-rate time profiles at 20–33, 33–53 and 53–72 MeV for the 1997 November 6 flare (From Yoshimori et al. 2001b)

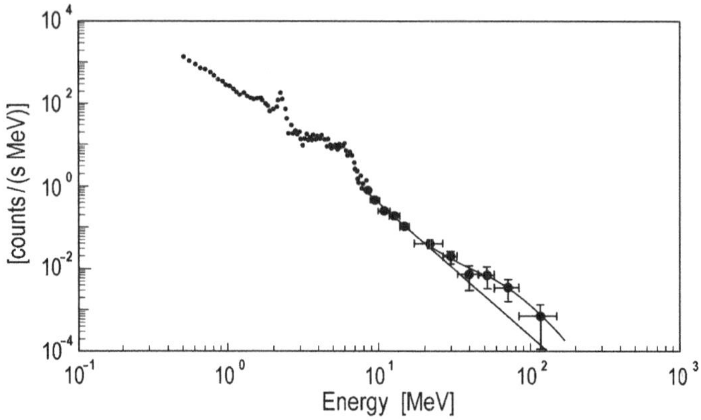

Fig. 9.51 Background-subtracted wide band γ-ray count-spectrum of the 1997 November 6 flare (From Yoshimori et al. 2001b)

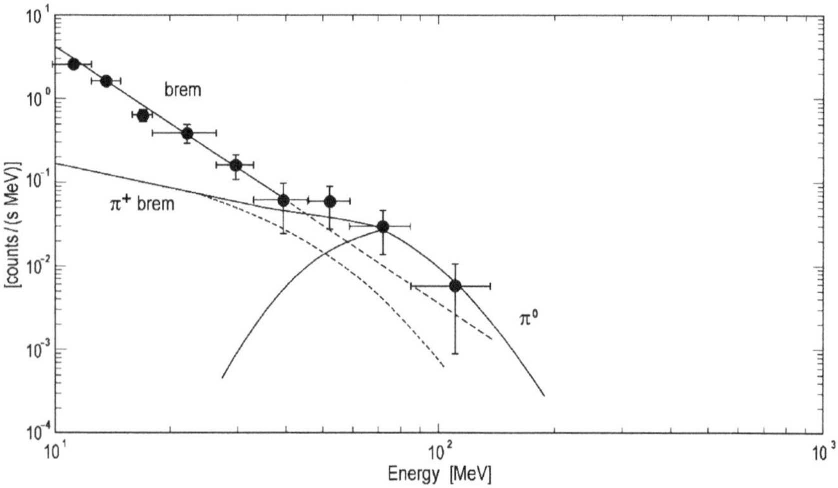

Fig. 9.52 High-energy γ-ray count spectrum above 10 MeV in the peak phase of the 1997 November 6 flare (From Yoshimori et al. 2001b)

The arrows in Fig. 9.50 show the time when the hard X-ray count rate reaches the maximum. The time of the maximum count rate of hard X rays is nearly coincides with that of the >10 MeV γ-rays. The background levels at these energies gradually increase with time. The background-subtracted wide range γ-ray count spectrum in the peak phase is given in Fig. 9.51. The γ-ray spectrum above 10 MeV is not well fitted by a single power law function. It shows a broad excess above 40 MeV which is likely due to the π^0 decay.

The γ-ray count spectrum above 10 MeV is shown in Fig. 9.52. Although the spectrum below 30 MeV is approximated by the power law continuum of E^{-3}, there is the broad excess above 40 MeV.

In Fig. 9.52 are shown three components of high energy γ-rays which were calculated by Ramaty and Murphy (1987) and called as: brem, π^+ brem, and π^0 stand for the primary electron bremsstrahlung, π^+-decay positron bremsstrahlung and π^0-decay broad spectrum, respectively. Figure 9.52 indicates that the continuum below 30 MeV is dominated by the primary electron bremsstrahlung, while the broad excess above 40 MeV is likely due to the π^0 decay γ-rays. The broad excess peaks at 70 MeV and its broadness depends on the accelerated-proton spectrum. The broadness increases as the proton spectrum hardens (Murphy et al. 1987). The Yohkoh observation implies that both electrons and protons were efficiently accelerated in the peak phase of the flare to energies above 10 and 300 MeV, correspondingly. Peculiarities of possible acceleration mechanisms, required for explanation of generation positron annihilation line 511 keV and high energy (>10 MeV) γ-ray emission observed during the 6 November, 1997 solar flare will be discuss in the Section 10.2.4.

9.9.4 *The Flare of August 18, 1998: Electron-Dominated Event*

The time profiles of γ-ray counting rate in 4–7 and 10–17 MeV are shown in Fig. 9.53. This flare showed a strong single spike with a duration of ~1 min.

The γ-ray count spectrum in 22:15:37–22:16:10 UT is given in Fig. 9.54. It exhibited strong continuum extending to 20 MeV without apparent lines, suggesting electrons were preferentially accelerated to high energies within a short time. The power law function was used for a spectral fitting procedure. The index is 2.11 ± 0.07 in 22:14:58 – 22:15:26 UT, 1.85 ± 0.02 in 22:15:26 – 22:15:58 UT and 2.25 ± 0.02 in 22:15:58 – 22:16:10 UT.

According to Yoshimori et al. (1999a), the spectrum slightly hardens at the peak phase. The flare-averaged index is 2.07 ± 0.02. The fluence of >1 MeV γ-rays (integration time 72 s) is (1,412 ± 182) photons cm^{-2}. The corresponding accelerated

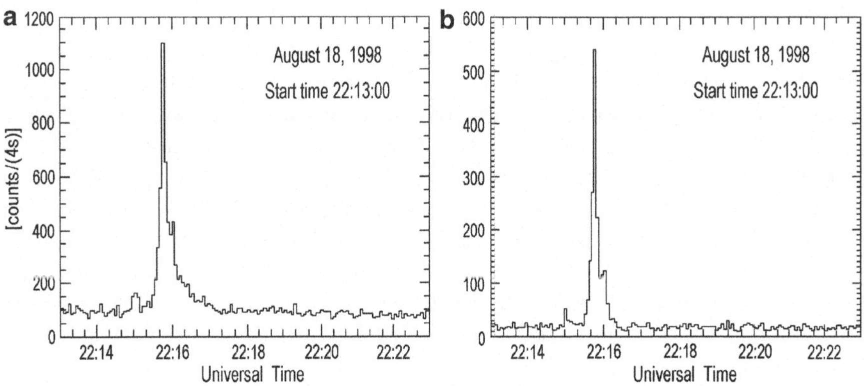

Fig. 9.53 Time profiles of γ-ray counting rate. (**a**) 4–7 MeV and (**b**) 10–17 MeV (From Yoshimori et al. 1999a)

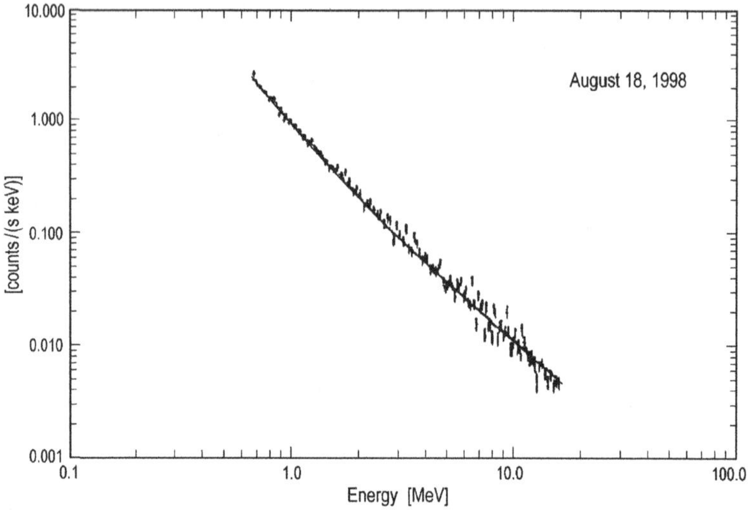

Fig. 9.54 Flare averaged γ-ray count spectrum in 22:15:37–22:16:10 UT August 18, 1998 (From Yoshimori et al. 1999a)

electron spectrum is calculated to be the power law function of index of 3.35 ± 0.02 and the total energy contained in >1 MeV electrons is estimated to be (7.36 ± 0.94) $\times 10^{27}$ erg. This flare developed with time, reached the maximum at the peak phase of the flare and decreased in the late phase. The SMM observations of 19 flares showed considerable flare to flare variations of the ratio, ranging from 2 to 7 (weighted mean of 3.2 ± 0.2) according to Share and Murphy (1995). The OSSE/CGRO observed a very long-duration flare on 4 June, 1991. The ratio observed from this flare was 3.40 ± 0.10 for the first orbit and 2.19 ± 0.09 for the second orbit, suggesting that the accelerated heavy ion to proton ratio decreased as the flare progressed (Murphy et al. 1997). The present Yohkoh ratio does not conflict with those measured from the other flares and indicates a similar temporal variation to that for the 1991 June 4 flare.

As noted Yoshimori et al. (1999a), the ratio of bremsstrahlung continuum and high-FIP narrow line fluxes represents the number ratio of accelerated electrons and protons. The temporal variation of the ratio of bremsstrahlung continuum above 1 MeV to narrow high-FIP (C + N + O + Ne) line fluxes is shown in Fig. 9.55. It exhibits that the ratio is nearly constant in 0–140 s (11:52:36 – 11: 54:56 UT) though it is slightly smaller at the rise phase, suggesting that both electrons and protons were simultaneously accelerated. However, the ratio decreases significantly in the late phase after 140 s (11:54:56 UT). This result implies that nuclear reactions producing deexcitation lines lasted for a longer time as compared with the electron bremsstrahlung. This time dependence is similar to that reported from the 1991 June 4 flare (Murphy et al. 1997).

Yoshimori et al. (1999a) noted that the energy spectrum of accelerated protons can be derived from the ratios of γ-ray line fluxes. Here the thick-target interaction model, impulsive-flare abundances and a downward isotropic angular distribution

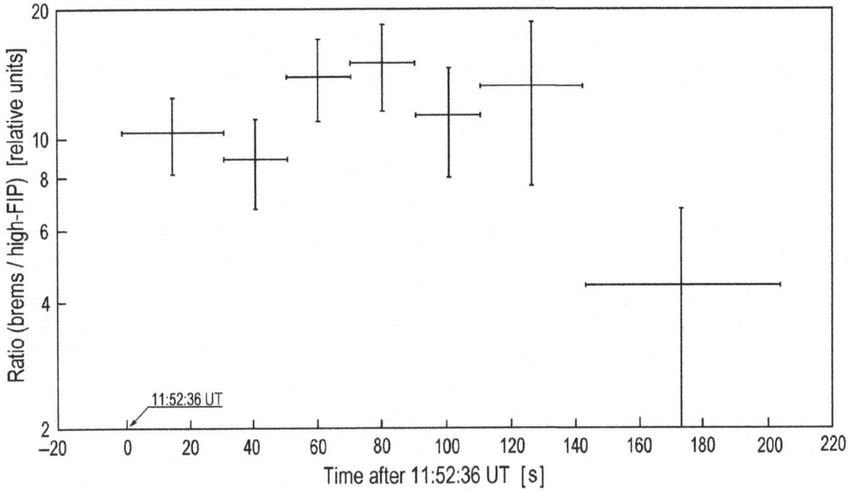

Fig. 9.55 Temporal variation of the ratio of >1 MeV bremsstrahlung continuum to narrow (C+N +O+Ne) line fluxes (From Yoshimori et al. 1999a)

for accelerated ions are assumed (Ramaty et al. 1996). Their production cross sections with different energy dependencies are sensitive to the proton spectral index. There are two methods for derivation of the proton spectrum: (1) the ratio of the neutron capture line at 2.22 MeV to O line at 6.13 MeV fluences and (2) the ratio of O line to Ne line (1.63 MeV) fluxes. The first method is used with flare-integrated data to provide an average index for the whole flare, while the second method gives an instantaneous measure. Recently Gan (1998) developed a new method to calculate the time variation of proton spectrum from time-differentiation of the 2.22 MeV line flux. Yoshimori et al. (1999a) used the Gan's method because this method gives an more accurate power law index of the proton characteristics of short duration (impulsive), considerably hard spectrum and no delay of γ-ray emission with respect to hard X-ray emission. This type of flare is named the electron-dominated event (Rieger and Marschhauser, 1990). In order to explain the electron acceleration in the electron-dominated event, a mechanism this has capabilities of prompt switch-on and switch-off. In opinion of Yoshimori et al. (1999a), acceleration by DC electric fields seems to be a possible mechanism.

9.10 Gamma-Ray Observations of the 2000 July 14 Flare

9.10.1 The Matter and Short History of Problem

According to Share et al. (2001), Yoshimori et al. (2001a) the GOES X5.7 class solar flare at about 10 h UT on 14 July 2000 from NOAA region 9077 at 17°N, 01°E was associated with a high-speed halo CME and the largest SEP event >10 MeV at Earth since 29 March 1991. Share et al. (2001) describe Yohkoh Wide Band Spectrometer (WBS – see in details in Yoshimori et al. 1991) for ~50 keV to 10 MeV observations of gamma radiation both from the solar flare and from the Earth's atmosphere during impact of the solar energetic particles. The Hard X-ray Spectrometer (HXS) consists of a 7.6 cm (diameter) by 2.5 cm (thick) NaI scintillator coupled to a photomultiplier tube. At the time of the flare, the HXS covered the energy range from ~30 to 750 keV in 32 energy channels and with 1 s time resolution. Two 7.6 cm (diameter) by 5.1 cm (thick) bismuth germanate (BGO) detectors coupled to photomultipliers make up the Gamma-Ray Spectrometer (GRS). Their gains and resolution have degraded since launch in 1991. On 14 July, detector 1 covered the energy range of 550 keV to 17 MeV and detector 2 covered the range from 550 keV to 24 MeV, each with 128 energy channels. The temporal resolution for both detectors during the flare was 4 s. Yohkoh has observed γ-ray spectra from other flares in solar cycles 22 and 23 (Kawabata et al. 1994; Yoshimori et al. 2000). Measurement of nuclear lines provides fundamental information on the spectra of accelerated particles and the composition of the ambient medium where they interact (Ramaty 1986; Ramaty et al. 1995).

Share et al. (2001) underlined that the Earth's atmosphere also emits nuclear line radiation from impact of cosmic radiation and solar energetic particles. Share and Murphy (2001) analyzed data from the SMM gamma-ray spectrometer during the 20 October 1989 intense solar energetic particle event and found 20 resolved nuclear line features. They also discussed how comparison of the intensities in some of these lines can be used to infer the spectrum of the incident protons.

Share et al. (2001) discuss hard X-ray and gamma-ray measurements of the 14 July 2000 flare beginning during the rise in soft X-rays including measurements of nuclear lines and electron bremsstrahlung continuum. This enables to estimate the time-integrated spectra of accelerated ions and electrons and how the flare energy is shared among them. Because this event was associated with an intense solar energetic particle event at Earth, Share et al. (2001) also study the temporal variation of the hard X-rays to confirm the suggestion that spectral hardening may be a predictor of such events (Kiplinger 1995).

9.10.2 X- and γ-Ray Observations by the Yohkoh Satellite of the 14 July 2000 Event

Soft X-ray emission from the flare measured by GOES commenced at about 10:05 UT and rose to a peak at about 10:24 UT. Share et al. (2001) display this time profile in Fig. 9.56 where are also plot the radiations observed by Yohkoh in hard X-rays and in two gamma-ray bands.

As noted Share et al. (2001), the plotted in Fig. 9.56 HXS and GRS data have not been corrected for background. At the start of the observation, just after the satellite

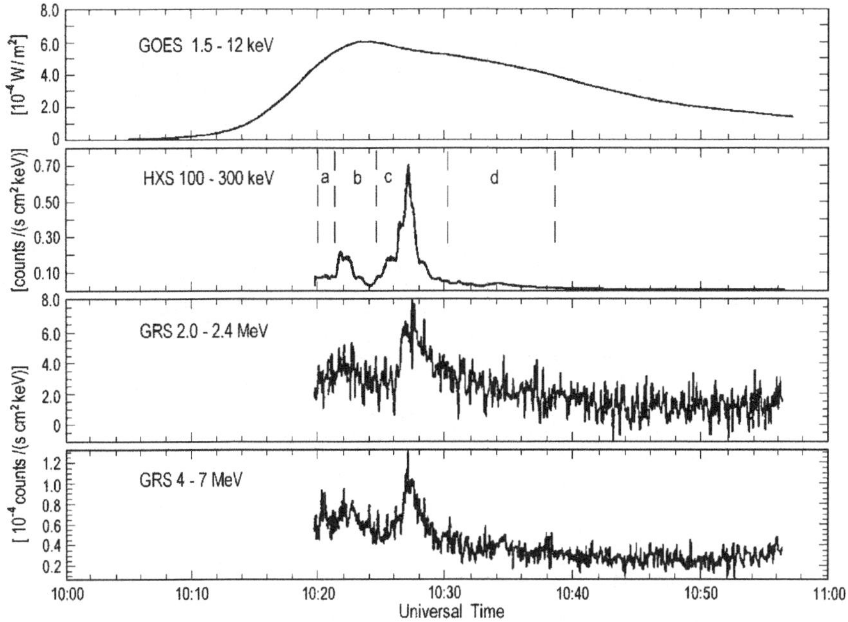

Fig. 9.56 Time history of the 14 July 2000 flare observed by GOES and the Yohkoh WBS. Spectra were accumulated during the four intervals shown in the 100–300 keV panel in Fig. 9.57 (From Share et al. 2001)

emerged into sunlight, Yohkoh was at it most southerly portion of the orbit where the cosmic ray induced background was highest. Thus the gradual fall in rate with time may in part be due to the background. The hard X-rays exhibit considerable structure with a precursor about 300 s prior to the main peak. The peak of the flare was observed into the nuclear γ-ray energy range as shown in the 4–7 MeV band. The 2.0–2.4 MeV band covers the strong 2.223 MeV neutron capture line and exhibits the $\sim$100 s exponential decay characteristic of that radiation.

Share et al. (2001) underlined that in order to obtain the most reliable temporal and spectroscopic information, it is important to correct for the varying backgrounds in both the GRS and HXS data. For flares with durations greater than $\sim$100 s, the best method utilizes data taken both 15 orbits before and after the flare, when the orbital tracks of the satellite are similar. However, the radiation environment on 15 July was significantly elevated at high latitudes from the impact of solar energetic particles. For this reason it was only able to use data from 13 July (shifted in time by about $\sim$500 s from the flare) to make the background correction. As there were no data on 13 July that could be used for the first $\sim$80 s of the flare, Share et al. (2001) used data taken on 16 July with a time dependent correction factor determined by a comparison with overlapping 13 July data.

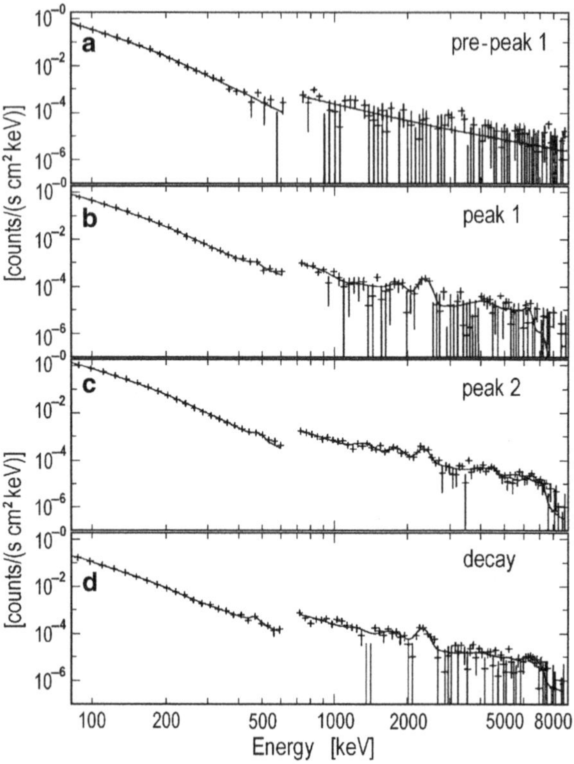

Fig. 9.57 Comparison of counts spectra from the HXS and GRS accumulated from (**a**) 37,182–37,265 s UT, (**b**) 37,265–37,460 s UT, (**c**) 37,462–37,800 s UT, and (**d**) 37,800–38,300 s UT. Solid lines show best fit models. Rates have been divided by the geometrical areas (From Share et al. 2001)

9.10.3 Comparison of Counts Spectra from the HXS and GRS

Share et al. (2001) created background-subtracted HXS and GRS (detector 1) spectra for the major part of the 14 July 2000 flare (37,265–38,300 s UT) and for four individual sections (see Fig. 9.56) covering: (a) the short period before the first peak (37,182–37,265 s UT); (b) the first peak (37,265–37,460 s); (c) the primary peak (37,460–37,800 s); and (d) the decay phase (37,800–38,300 s). These count spectra are plotted in Fig. 9.57.

The spectral points in Fig. 9.57 are plotted on the same scale, after being divided by the geometric areas of the detectors. Because these are not actual photon spectra, it does not expect the hard X-ray and gamma-ray spectra to merge. Share et al. (2001) first detect evidence for nuclear line emission during the first peak. Both the 511 keV annihilation line in the HXS and the 2.223 MeV neutron capture line in the GRS are detected from the first peak through the decay phase. In addition, Share et al. (2001) note the characteristic fall-off in rate $\geq$7 MeV where the nuclear line contribution ends. The gains of both the HXS and GRS have changed from the last calibration in 1997. The annihilation line appears at energy of ~475 keV, consistent with the location of the background line in the HXS. The neutron capture line appears at energy of ~2,315 keV; this is also consistent with a shift to higher energy observed in the background lines.

The solid lines in Fig. 9.57 represent fits to the data using a forward-folding technique. For the hard X-ray region Share et al. (2001) fit the spectra from 80 to 600 keV with either a single or broken power law and a line at 511 keV. For the GRS spectrum we fit the spectrum with a single power law, and ten narrow and five broad nuclear lines, with energies and widths determined from measurements made with the SMM spectrometer (Share and Murphy 1995, 2000). Because of the presence of these overlapping broad lines from heavy ion interactions on ambient H, the energy resolution of the GRS is not sufficient to permit the narrow lines to be reliably separated, with the exception of the neutron capture line at 2.223 MeV. Share et al. (2001) therefore display only the overall fits to the data. This lack of sufficient spectral resolution prevents them from determining the prompt O/Ne (6.13/1.63 MeV) line ratio that can be used to measure the temporal evolution of the accelerated particle spectrum (Ramaty et al. 1995). It can be still obtain a flare-averaged measurement of the accelerated particle spectrum by using the integrated fluences in the 2.223 MeV neutron capture line and in the total nuclear contribution to the 4–7 MeV band (Ramaty 1986).

In Fig. 9.58 are plotted the flare-integrated count spectra from the first peak through the decay phase for both the HXS (60 to 600 keV) and GRS (700 to 9,000 keV) detectors. Once again the rates have been divided by the geometric areas of the detectors to enable them to be plotted on the same scale. Share et al. (2001) used

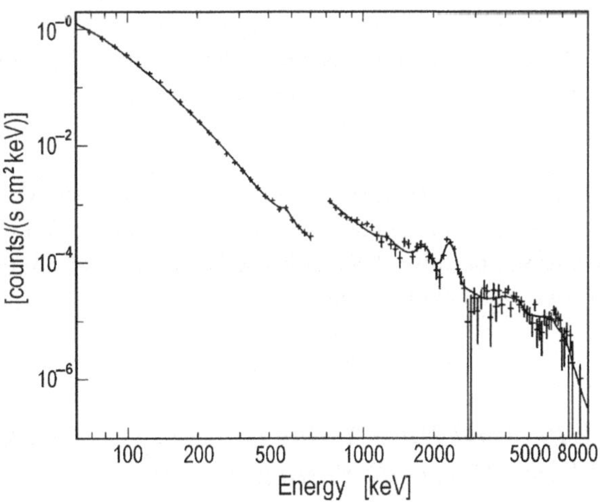

Fig. 9.58 Comparison of counts spectra from the HXS and GRS accumulated from the first peak through the decay phase, 37,265–38,300 s UT. Solid lines show best fit models. Rates have been divided by the geometrical areas (From Share et al. 2001)

the same forward-folding technique and incident photon models discussed above to fit the data. The nuclear fluence in the 4.7 MeV band was determined by removing the power-law contribution to the fitted model. Based on these fits it was obtained fluences of 130 $\pm$ 12 and 66 $\pm$ 5 photon.cm^{-2} in the 2.223 MeV line and 4–7 MeV band, respectively. This yields a 2.2 to 4–7 MeV fluence ratio of 1.97 $\pm$ 0.23. All of the spectral uncertainties were determined by mapping the change in χ^2 as the parameter in question was varied, keeping all the other parameters free. The uncertainties are based on $\delta\chi^2$ of 1 (Lampton et al. 1976).

Share et al. (2001) have used updated Monte Carlo calculations to obtain the flare-averaged spectrum of accelerated ions assuming an ambient coronal composition with ^{4}He/H = 0.1 and an impulsive flare particle composition (Reames et al. 1994; Reames 1995). This yields a power-law proton spectrum with an index of 3.14 $\pm$ 0.15 and 1.1 $\times$ 10^{32} protons.MeV^{-1} at 10 MeV. If it is assume that this spectrum extends down to 1 MeV/nucleon without hardening and that it is flat below 1 MeV/nucleon, then will be obtain 3.5 $\times$ 10^{29} erg in accelerated protons. Assuming the same impulsive particle composition and ions with the same spectral shape, it will be estimate that there was 1.5 $\times$ 10^{30} erg in accelerated ions. We can compare this with the energy contained in electrons from our fits to the flare-integrated HXS spectrum. Share et al. (2001) have fit the flare-integrated spectrum (60 to 600 keV) with a broken power law having the following parameters: flux at 50 keV 5,300 $\pm$ 155 photon.cm^{-2}keV^{-1}; low-energy index 3.70 $\pm$ 0.03; break energy $\sim$340 keV; high-energy index 2.7 $\pm$ 0.3; 511 keV fluence 31 $\pm$ 10 photon. cm^{-2}. The broken power law is necessary to fit the data, not only because the spectrum hardens above $\sim$300 keV, but because the fitting algorithm that have used does not contain a positronium continuum component. As the hard X-ray and γ-ray fluxes have been obtained from two different instruments Share et al. (2001) have compared the best fitting photon spectra to determine a relative correction factor. It was found that the spectra agree if multiply the HXS data by a factor of 2 $\pm$ 1. Using this factor and the bremsstrahlung spectrum derived above, Share et al. (2001) estimate that there were 3 $\times$ 10^{32} erg in accelerated electrons above 20 keV (Brown 1971; Ramaty et al. 1993). This is about two orders of magnitude higher than that measured in the ions.

It can be also compare the corrected annihilation line fluence with the 4–7 MeV nuclear fluence to obtain an alternative estimate of the spectral index of the accelerated ions. Based on this ratio it was obtain an index of 3.0 (+0.95, −0.25), consistent with what was obtained above.

9.10.4 Link Between Hard X-Ray Hardening in Flares and Solar-Energetic Particle Events

Kiplinger (1995) has suggested that the spectral evolution of hard X-rays in flares can be used to predict the occurrence of energetic interplanetary proton events.

Specifically, his studies with the SMM Hard X-ray Burst Spectrometer indicate that these particle events almost always occur when the spectra of 40–200 keV X-rays either harden over flux peaks or during flux decays and seldom occur when such hardening is not observed. The 14 July 2000 flare was associated with the third highest flux of >10 MeV interplanetary protons in the last 25 years. We therefore expect to observe spectral hardening at hard X-ray energies in this flare. Unfortunately, Yohkoh was in the nighttime portion of its orbit during the rise of the flare in soft X-rays (Fig. 9.56); any hardening in the hard X-rays that may have occurred then was not observable. However, measurements with the HXS and GRS suggest that the most intense portions of the burst likely occurred after the satellite moved into daylight.

Share et al. (2001) have therefore studied this portion of the flare for any evidence of spectral hardening. They spectroscopically analyzed the HXS data in 50 s intervals after first subtracting background estimated from measurements made after the flare. This background subtraction suffices at these lower energies where the flare photons dominate and the background doesn't vary strongly. Share et al. (2001) then fit the 50–200 keV band with a single power law using the χ^2 mapping algorithm discussed above. Share et al. (2001) plot the measured flux at 50 keV and the index of the power law in Fig. 9.59. The 50 keV flux follows the temporal structure observed in the 100–300 keV band plotted in Fig. 9.56. The spectral indices exhibit the traditional 'soft-hard-soft' evolution through the two peaks. They do not exhibit the 'soft-hard-harder' evolution in individual peaks or the

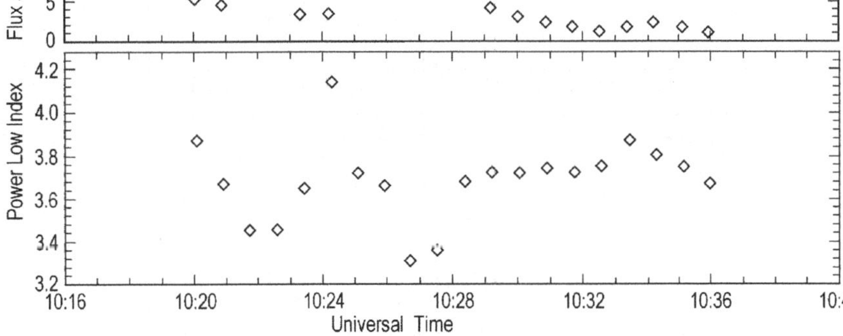

Fig. 9.59 Variation of flux at 50 and 50–200 keV spectral index during 14 July 2000 event (From Share et al. 2001)

gradually hardening spectra through the flare that Kiplinger (1995) found were associated with energetic particle events.

9.10.5 The Yohkoh Hard X-Ray Images and Diagnostics for Magnetic Reconnection

The Yohkoh hard X-ray images are analyzed by Yoshimori et al. (2001a) and Masuda et al. (2001). They found that two hard X-ray sources characteristically move as the flare progresses. The temporal variations in the hard X-ray image at 33–53 keV are shown in Fig. 9.60. A simplified magnetic neutral line is indicated in each panel. Initially (10:24:23–10:24:37 UT) the two sources are located near the magnetic neutral line but gradually separate from the neutral line (10:26:41– 10:27:03 UT). According to Yoshimori et al. (2001a), these motions are thought to provide important diagnostics for magnetic reconnection and particle acceleration.

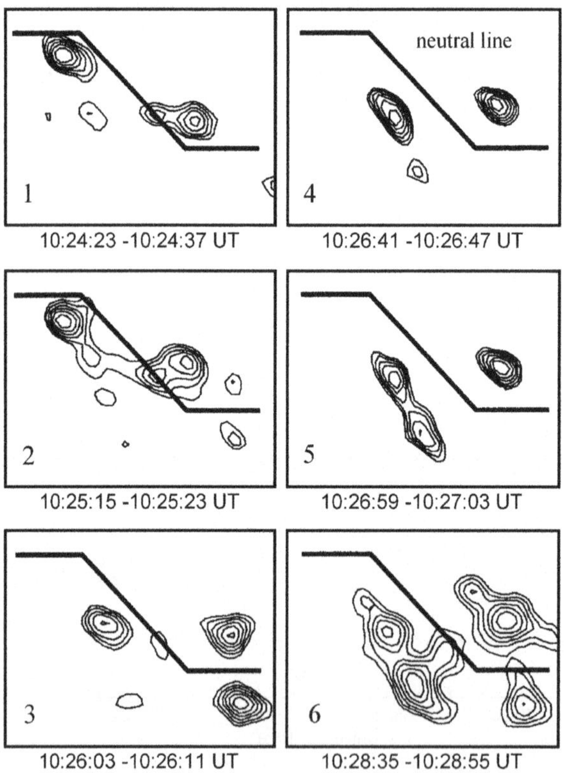

Fig. 9.60 Hard X-ray images at 33–53 keV in 10:24–10:29 UT of the July 14 flare. A simplified magnetic neutral line is indicated in each panel (From Yoshimori et al. 2001a)

9.10.6 Atmospheric Gamma-Rays from the SEP Event of 14 July 2000

According to Share et al. (2001), protons >30 MeV began arriving at Earth at ~10:35 UT, approximately 8 min after the peak in hard X-rays and γ-rays measured onboard Yohkoh. The intensities of these particles continued to rise rapidly for the first 2 h after the flare. An unrelated CME shock reached Earth at about 16:00 UT giving rise to another increase in the particle rates. The rates in particles with energies below about 30 MeV continued to increase for the next 20 h, while those at higher energies leveled off or began to decline.

As underlined Share et al. (2001), the impact of these particles on the Earth's atmosphere was detectable by the HXS and GRS detectors when the satellite reached its most northerly and southerly excursions ($\pm 31^\circ$ geographic latitude), at geomagnetic cutoff rigidities near 4 GV. At these locations hard X- and γ-rays produced in the polar regions by these particles could be detected. The first increases were observed on 14 July at ~12:40 and ~13:30 UT during the rise in SEP intensity at Earth. On later orbits of the day the geomagnetic cutoff rigidities near $\pm 31^\circ$ were too high and the SEP interactions couldn't be observed. The HXS and GRS once again detected hard X-rays and γ-rays from SEP interactions at low geomagnetic rigidity beginning 15 July at 06:10 UT and ending at 13:40 UT. The 1.7 MeV γ-ray rates varied by a factor of ~2 during this period of maximum SEP intensity.

Because Yohkoh is solar-pointed, it is likely that the atmospheric radiation passed through some amount of spacecraft material before reaching the detectors. The viewing angle of the atmospheric radiation also varies from orbit to orbit. For this reason any flux estimates are likely to be lower limits and there may be differences between the responses of the HXS and GRS. With these caveats in mind, Share et al. (2001) plotted HXS and GRS count spectra (divided by the geometric areas in the solar direction) during the rise of the event on 14 July and near its peak on 15 July in Fig. 9.61.

In Fig. 9.61 there is subtracted background using data taken at similar geographical locations on earlier days when there was no geomagnetic disturbance. Share et al. (2001) summed the two 14 July observations because the fluxes were rather weak. The spectrum accumulated during the peak particle intensity on 15 July is similar to the one observed by the SMM gamma-ray spectrometer during the intense 20 October 1989 event (Share and Murphy 2001). Lines from positron annihilation (511 keV), direct excitation of atmospheric ^{14}N (1,635 and 2,313 keV), and the ^{11}B and ^{12}C spallation products of ^{14}N (4,439 and 4,444 keV) are resolved in the spectrum. It was also observed unresolved lines in the 6–7 MeV region from N and O. The relative levels of the HXS and GRS responses are similar to that observed for the solar flare, suggesting once again that it will need to multiply the derived HXS fluxes by about a factor of two (with a 50% uncertainty) in order to make a comparison with the GRS fluxes. The spectral rates observed from the 14 July observation in Fig. 9.61 are about a factor of ten weaker that those observed on

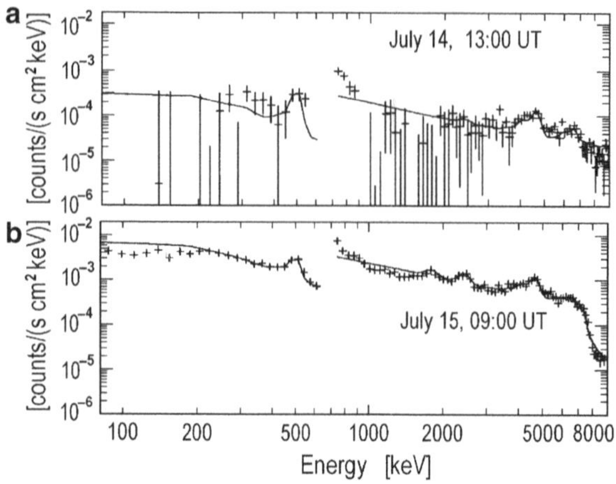

Fig. 9.61 Comparison of count spectra from the HXS and GRS of the Earth's atmosphere during the onset of the particle event on 14 July and near the peak intensity on 15 July. Solid lines show best fit models to ranges 200–600 and 1,000–7,000 keV for the HXS and GRS. Rates have been divided by the geometrical areas (From Share et al. 2001)

Table 9.13 GOES proton measurements (From Share et al. 2001)

Date/time	Flux > 10 MeV	Flux ratios		
	p (cm^2 s sr)$^{-1}$	>10 />30	>30 />50	>50 />100
Oct. 20, 1989 / 16 UT	4 x 10^4	6	3	5.5
July 14, 2000 / 13 UT	2 x 10^3	1.5	1.5	3
July 15, 2000 / 09 UT	2 x 10^4	4	3.5	14

Table 9.14 Gamma-ray lines from SEP impact on the Earth's atmosphere (From Share et al. 2001)

Date/time	1.63 MeV	Line flux (photon$\times$cm^{-2}s^{-1})		0.511 MeV
		2.31 MeV	4.44 MeV	
Oct. 20, 1989/16 UT	0.41 ± 0.02	0.79 ± 0.06	1.00 ± 0.08	1.29 ± 0.01
July 14, 2000/13 UT		0.00 ± 0.015	0.10 ± 0.01	0.12 ± 0.04 [0.06]
July 15, 2000/09 UT	0.19 ± 0.02	0.40 ± 0.02	0.83 ± 0.03	0.88 ± 0.06 [0.44]

15 July. This is the same factor observed in the >10 MeV particle fluxes observed by GOES and given in Table 9.13.

In order to provide quantitative estimates of the fluxes in the lines, Share et al. (2001) fit the HXS and GRS spectra over limited ranges (200–600 keV in the HXS; 1,000–7,000 keV in the GRS) with simple power laws and Gaussians. They plot these fits as solid lines in Fig. 9.61 and have extended them to higher and lower energies for reference. The resulting line fluxes are listed in Table 9.14 for the observations on 14 and 15 July. For comparison, in the Table 9.14 are also listed line fluxes observed by SMM during the 20 October 1989 event (Share and Murphy 2001).

As noted Share et al. (2001), from Table 9.14 can be seen that the ^{14}N deexcitation line fluxes for the October 1989 event is each a factor of two higher than the respective line fluxes in the 15 July 2000 event. These lines are primarily produced by protons $\geq$10 MeV (this factor of two is the same found in comparing the fluxes of >10 MeV protons listed in Table 9.13). The spallation lines at 4.4 MeV are produced by higher energy protons. It can be seen from Table 9.13 that the 15 July 2000 proton spectrum from $\sim$10–50 MeV is harder than the spectrum observed on 20 October 1989. This harder spectrum appears to be reflected in the significantly higher 4.44/2.31 MeV flux ratio ($\sim$2) observed on 15 July compared with what was observed ($\sim$1.2) in 1989 (see Table 9.14). The annihilation line at 0.511 MeV can be produced by even higher-energy protons. As the 15 July proton spectrum softens markedly at >50 MeV relative to what was observed on 20 October, it would be expect the 0.511/4.44 MeV flux ratio to be higher for the latter. Based only on the statistical errors given in Table 9.14, this appears to be the case; however, there is an additional systematic error, given in the brackets, due to the uncertainty in the GRS/HXS calibration.

Share et al. (2001) came to conclusion, that there is no evidence for the 2.3 MeV line from deexcitation of ^{14}N in the 14 July 2000 spectrum. The 2 σ upper limit is about a factor of three below the flux in the 4.4 MeV line complex. This compares with a measured factor of two for the 15 July 2000 observation. This suggests that the 10–30 MeV proton spectrum was much harder on 14 July 2000. This is consistent with the GOES (>10 MeV)/(>30 MeV) flux ratios on those days. The hardness of the proton spectrum at this time is also reflected in the relatively high 511 keV flux.

9.10.7 Main Results and Discussion

Share et al. (2001) have studied several high-energy aspects of the 14 July 2000 solar flare using the Yohkoh wide-band spectrometer experiment. In particular they have used the hard X-ray data from the HXS and γ-ray data from the GRS. The observations commenced $\sim$4 min before the peak in soft X-rays and continued for the remainder of the flare. The flare produced both electron bremsstrahlung and nuclear line emission. The nuclear line emission commenced during the first distinct peak beginning $\sim$10:21:05 UT. Lines from electron-positron annihilation and neutron capture were clearly distinguished. Because of the moderately poor spectral resolution of the GRS Share et al. (2001) were not able to separately resolve the narrow and broad deexcitation lines that come from p and α interactions on the ambient solar atmosphere and heavy ion interactions with ambient H and He.

Share et al. (2001) have obtained nuclear line spectra in four distinct sections of the flare but, because they cannot separately resolve the ^{16}O and ^{20}Ne lines at 6.13 and 1.63 MeV, it was unable to estimate the time-varying spectra of flare accelerated ions. However, Share et al. (2001) have been able to obtain information on the flare-averaged accelerated ion spectrum by using the fluences observed in the 2.223

MeV neutron capture line and in the nuclear radiation contained in the 4–7 MeV energy range. For an assumed power-law spectrum of accelerated particles, it was obtain a spectral index of 3.14 ± 0.15. Ramaty et al. (1996) determined the spectral indices for 19 intense nuclear line flares observed by the SMM spectrometer (Share and Murphy 1995). Of these 19 flares only three had spectra that were as hard as the 14 July flare. We note that the 6 November 1997 flare, which was also observed by Yohkoh, had an even harder particle spectrum with an index of 2.6 ± 0.1.

Share et al. (2001) have estimated the energy content in accelerated ions during the 14 July flare based on these studies. Under the assumption that the power law extends without a break down to 1 MeV, where the spectrum then becomes flat, it was estimate that there were 1.5×10^{30} erg in accelerated ions. This compares with the estimate of 3×10^{32} erg in accelerated electrons >20 keV during the flare based on hard X-ray observations with the HXS. Thus the ions comprise at most only $\sim$1% of the energy in accelerated particles. Ramaty and Mandzhavidze (2000) have estimated the energy content in accelerated ions and electrons in the 19 SMM flares with the strongest nuclear lines. The energy contained in ions for these flares ranged from $\sim 3 \times 10^{29}$–10^{33} erg. Only two of the 19 flares had accelerated ion energies less than the 14 July flare. Significantly, they found that the energy contained in ions was often comparable to that contained in electrons; however, it was also highly variable, ranging from $\sim$0.01 to ten times the energy contained in electrons. The relative energy content in ions for the 14 July flare appears to be lower than any of the 19 SMM flares. It is important to study the relative energies in accelerated ions and electrons in a larger sample of flares that is not limited to those with the most intense nuclear lines. This will provide information on the particle acceleration process and on how the total flare energy in accelerated particles is partitioned between ions and electrons (Simnett 1995). An important measurement that still needs to be made, however, is the energy content of ions below 1 MeV (MacKinnon 1989; Share et al. 2001).

As noted Share et al. (2001), the 14 July flare was associated with the third most intense solar energetic proton event >10 MeV event since 1976, it is an excellent candidate to confirm Kiplinger's (1995) premise that spectral hardening in X-rays is a reliable predictor of these events. Share et al. (2001) measured the hard X-ray spectra during 50 s accumulations and found no evidence for 'soft-hard-harder' evolution in individual peaks or gradual hardening through the flare. The failure to confirm this hardening for such an intense SEP is in conflict with Kiplinger's (1995) model. However, there remains the remote possibility that such spectral hardening occurred during the early part of the flare, before Yohkoh observations began. Klein et al. (2001) summarized a broad range of measurements of the 14 July 2000 flare. These include X-ray, EUV, optical, radio, and neutron monitor observations. They suggest that the main phase of energy conversion in the low corona had a maximum near 10:18 UT, about 2 min before the Yohkoh observations began. There are no observations to confirm whether the hard X-rays and gamma-rays also peaked at this earlier time. Had there been significant ion acceleration during this time, Share et al. (2001) would have expected to detect a strong neutron capture line as soon as

Yohkoh began its observations, but they did not detect this line until a few minutes later. This suggests that ion acceleration, and the most intense hard X-radiation, probably did not commence until the Yohkoh observations began.

As noted Share et al. (2001), the solar energetic particles associated with the flare and CME reached the Earth's atmosphere and interacted to produce γ-ray lines that were observed by Yohkoh at high geomagnetic latitudes beginning about 2 h after the flare. The atmospheric spectrum observed by Yohkoh is similar to that observed by the SMM spectrometer during the intense SEP event on 20 October 1989. The peak intensity in protons >10 MeV during that event was about twice the flux measured on 15 July during the peak of the event, when Share et al. (2001) made their spectral measurements. The fluxes in the two ^{14}N deexcitation lines that are produced by protons at those energies also differed by the same factor of two for these two events. By comparing the intensities in the ^{14}N lines with those in ^{11}B and ^{12}C spallation lines Share et al. (2001) conclude that the $\sim$10–50 MeV particle spectrum interacting with the Earth's atmosphere was harder in the July 15 event than it was in the 20 October event. This is again consistent with the particle measurements in space. The qualitative agreement between the particle spectra in space and those that reach Earth suggests that there are not large transport effects for particles impacting the atmosphere near the magnetic poles.

As underlined Share et al. (2001), although the GRS instrument has degraded in performance since launch, it continues to operate and has provided a 10-year history of hard X-ray and γ-ray flares. Since the loss of the Compton Observatory, it is the only functioning solar γ-ray instrument. With the forthcoming launch of HESSI (Lin 2000), the WBS will still have an important role in providing complementary data.

9.11 The Solar Gamma-Ray Event of 24 November 2000

9.11.1 The Matter and Short History of Problem

According to Yoshimori et al. (2001a), Yohkoh has spectroscopic and imaging capabilities for solar flares and provides hard X-ray spectral/imaging and γ-ray spectral data. The hard X- and γ-ray spectral observations are important for a study of high-energy particle acceleration and transport at flare sites, while the hard X-ray image data provide crucial information on magnetic reconnection and particle acceleration at flare sites. Simultaneous observations with the spectroscopic and imaging instruments enable to discuss high-energy phenomena on the Sun in detail. Yohkoh measured two γ-ray flares in 2000: a X5.7/3B flare on July 14 (Share et al. 2001; Klein et al. 2001; see previous Section 9.10) and a *X*2.3/2B flare on November 24. Both flares emitted significant electron bremsstrahlung continuum and γ-ray lines which is due to nuclear deexcitation, neutron-capture and positron annihilation. The hard X-ray sources of the July 14 flare exhibited characteristic motions suggesting occurrence of magnetic reconnection (Masuda et al. 2001). The November 24 flare produced more intense γ-ray flux than the July 14 flare and shows a typical two hard X-ray sources.

9.11.2 X- and γ-Ray Observations During 24 November 2000

As noted Yoshimori et al. (2001a), the start of *X*2.3/2B flare (22°N, 07°W, AR9236) was observed by Yohkoh satellite at 15:08 UT on November 24. The count-rate time profiles at 0.48–0.54, 2.22, 4–7 and 7–17 MeV are shown in Fig. 9.62. The γ-ray flux is much higher than that of the July 14 flare. The background-subtracted γ-ray count spectrum is shown in Fig. 9.63.

It can be seen from Fig. 9.63 the 2.22 MeV line, C and O nuclear deexcitation lines and a complex of a few weak lines at 1–2 MeV. The hard X-ray (53–93 keV) images observed at 15:07:50 UT and 15:09:02 UT are shown in Fig. 9.64. Figure 9.64 shows that there are clear two hard X-ray sources which are located at both foot-points of the magnetic loop. Yoshimori et al. (2001a) found the small motions of the hard X-ray sources from detailed analysis of the temporal evolution. The distance between two sources slightly varies with time.

9.11.3 Main Results and Discussion

As noted Yoshimori et al. (2001a), the 2000 November 24 flare emitted much intense hard X- and γ-rays compared with the 2000 July 14 flare. The γ-ray fluence of the November 24 flare is about one order of magnitude as large as that of the July 14 flare, though the SEP flux of the November 24 event is two orders of magnitude smaller than that of the July 14 event. It suggests the possibility that a fluence ratio of the γ-ray producing protons to the CME-associated protons in the November 24 event is much higher than that of the July 14 event. It indicates that there is not a correlation between the γ-ray producing proton and SEP fluxes (Reames 1999). The same γ-ray lines were detected from the two flares. Concerning the time-profile of the neutron-capture line at 2.22 MeV, both flares exhibit the long decay time compared with those of electron bremsstrahlung and nuclear deexcitation lines. It is because of the time required for neutrons to slow down and be captured.

Yoshimori et al. (2001a) underlined that it can be estimate a ratio of the ^{3}He to H abundances in the photosphere from the decay time of the 2.22 MeV line (Prince et al. 1983a; Murphy et al. 1997). Assuming that the photospheric hydrogen density is $1.3 \times 10^{17}\text{cm}^{-3}$, Yoshimori et al. (2001a) obtained the ratio of (2–4) $\times 10^{-5}$ for the two flares in 2000. This is consistent with the previous results (Yoshimori et al. 1999b). The hard X-ray image at 53–93 keV shows a typical two sources during the

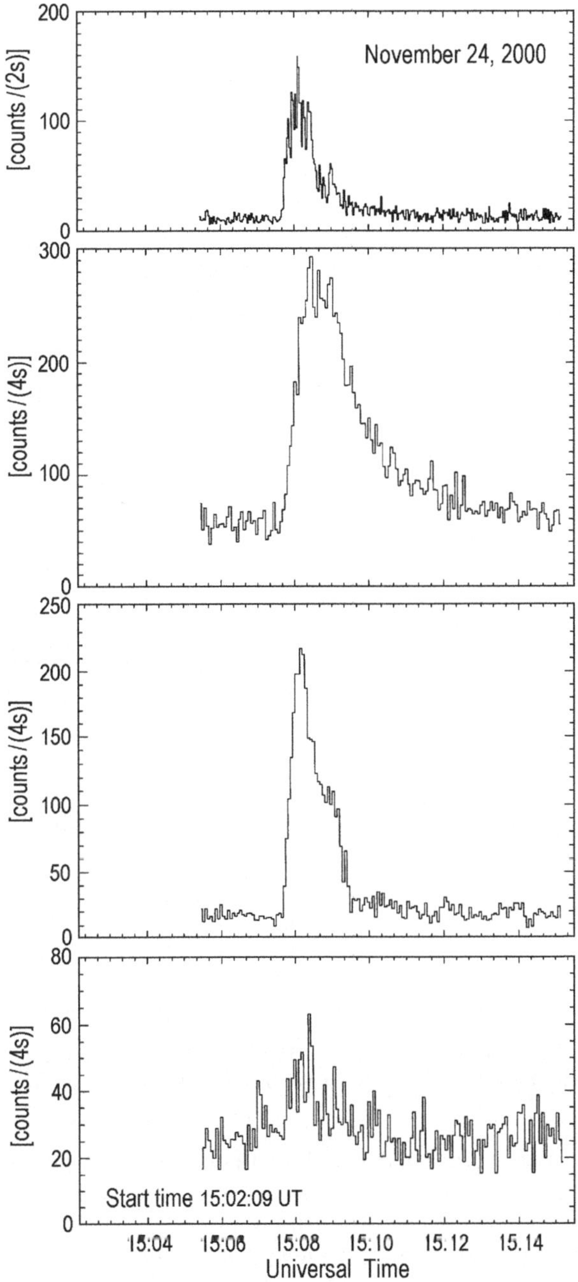

Fig. 9.62 Count rate time profiles at 0.48–0.54 (*top* panel), 2.22 (second), 4–7 (third) and 7–17 MeV (*bottom*) of the 2000 November 24 flare (From Yoshimori et al. 2001a)

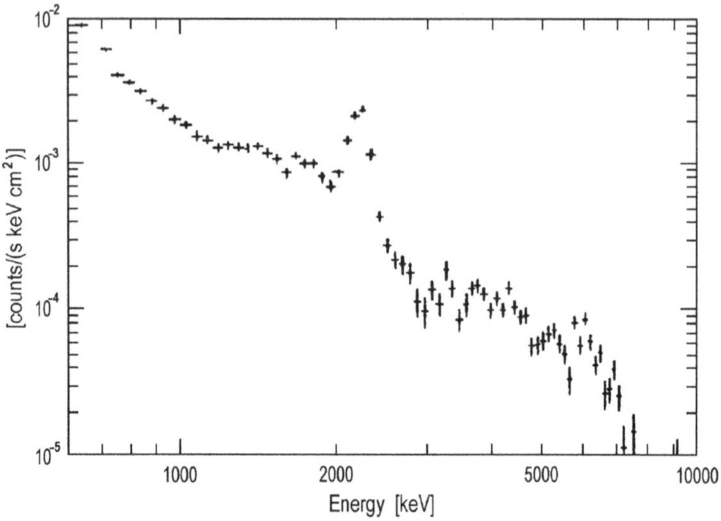

Fig. 9.63 Background-subtracted γ-ray spectrum of the 2000 November 24 flare (From Yoshimori et al. 2001a)

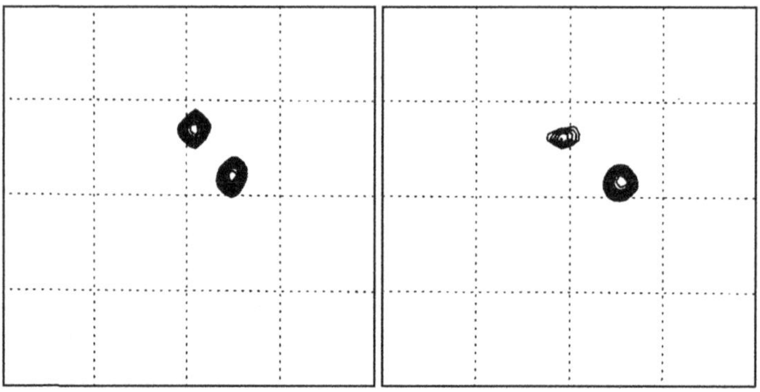

Fig. 9.64 Hard X-ray images at 53–93 keV at 15:07:50 UT (*left* panel) and 15:09:02 UT (*right* panel) of the 2000 November 24 flare (From Yoshimori et al. 2001a)

flare. Their locations correspond to both foot-points of the magnetic loop, indicating that accelerated electrons streamed down to the chromosphere and produce hard X-rays through bremsstrahlung. The hard X-ray image is independent on X-ray energy at 23–93 keV and there is little temporal variation in the hard X-ray source positions. Further, the four-channel X-ray data of the hard X-ray telescope exhibit an extremely hard spectrum (power law index of about 2.0 below 100 keV) during the peak phase (15:07:40–15:09:00 UT), suggesting that there was a powerful electron acceleration mechanism at the flare site.

9.12 Comparison of γ-Ray Flares at 27 October 1991, 6 November 1997, 14 July and 24 November 2000: Energy Spectra and Total Energy in Nonthermal Electrons and Protons

9.12.1 The Matter of Problem

As noted Yoshimori et al. (2003), the gamma-ray spectral analysis provides diagnostics for particle acceleration in solar flares. Energy content of nonthermal electrons and protons are derived from the gamma-ray spectral analysis and are important for discussion of an equal share of nonthermal energy between electrons and protons. Yoshimori et al. (2003) analyze measured by Yohkoh satellite gamma-ray energy spectra of X-class solar flares on October 27, 1991 (X6.1), November 6, 1997 (X9.4), July 14, 2000 (X5.7) and November 24, 2000 (*X*2.3) to study the energy content in nonthermal electrons and protons. The accelerated electron and proton spectra are derived from a spectral analysis of the continuum and γ-ray lines above 1 MeV.

9.12.2 Main Results of γ-Ray Observations by Yohkoh Satellite

According to Yoshimori et al. (1994, 2000, 2002), Yohkoh satellite observed intense gamma-ray flares on October 27, 1991 (X6.1), November 6, 1997 (X9.4), July 14, 2000 (X5.7) and November 24, 2000 (*X*2.3). Their background-subtracted gamma-ray count spectra (time-integrated over the duration of gamma-ray flares) are shown in Figs. 9.65–9.68, correspondingly.

9.12.3 Spectral Fitting of Gamma-Ray Observation Data

In order to make a spectral fitting, Yoshimori et al. (2003) use a convolution method in which a trial incident γ-ray spectrum is assumed to be a composite of the continuum and γ-ray lines (both narrow and broad lines), is convolved with the instrumental response function, and the resulting γ-ray count spectrum is tested to determine whether it is compatible with the observed spectrum. Since the Yohkoh counting statistics is poor and the energy resolution of the spectrometer is low, Yoshimori et al. (2003) fixed the parameters of center energies and line widths and sequently vary only the values of line intensities until the χ^2 value is minimized.

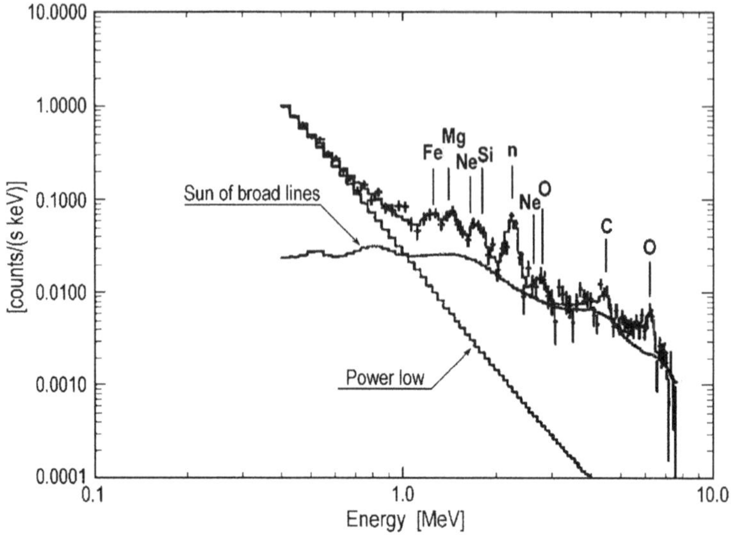

Fig. 9.65 Gamma-ray count spectrum of the 1991 Oct. 27 flare (From Yoshimori et al. 2003)

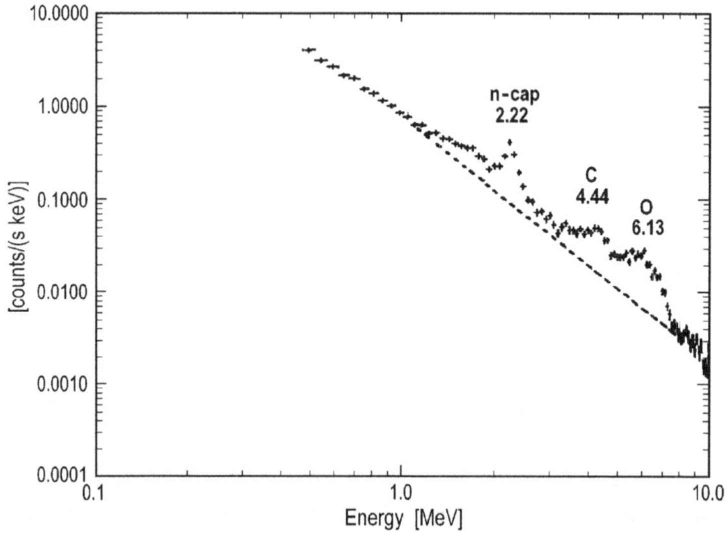

Fig. 9.66 Gamma-ray count spectrum of the 1997 November 6 flare (From Yoshimori et al. 2003)

9.12.4 Determining of Accelerated Electron and Proton Energy Spectra and Total Energy Contents

Yoshimori et al. (2003) calculate the accelerated electron spectrum from the observed continuum on the assumption of thick-target model (Ramaty et al.

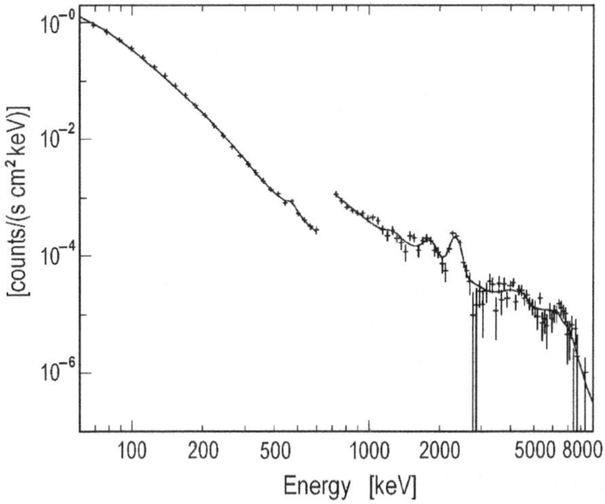

Fig. 9.67 Gamma-ray count spectrum of the 2000 July 14 flare (From Yoshimori et al. 2003)

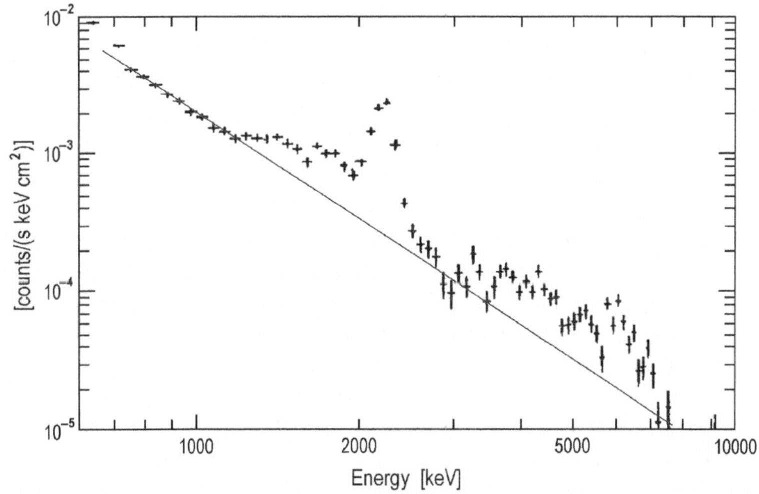

Fig. 9.68 Gamma-ray count spectrum of the 2000 November 24 flare (From Yoshimori et al. 2003)

1993). The accelerated proton spectrum is derived as follows: First Yoshimori et al. (2003) assume that the proton spectrum is approximated by a single power-law and determine the power-law index from the measured fluence ratio of the neutron-capture line to C and O deexcitation lines (Ramaty and Murphy 1987). A total number of protons accelerated above 10 MeV is estimated from a comparison of the measured fluences of C and O lines with the theoretical ones. Moreover, the proton energy content depends on a ratio of accelerated He to proton. Yoshimori et al. (2003) assumed the ratio of 0.5 (in accordance with Murphy et al. 1997). Thus they

Table 9.15 Estimated >1 MeV electron and >10 MeV proton energy contents (correspondingly W_e(>1 MeV) and W_p(>10 MeV) in erg) (From Yoshimori et al. 2003)

Flare date	We (>1 MeV)	Se	Wp (>10 MeV)	Sp	We/Wp
'91 Oct.27	$(6.1 \pm 2.6) \times 10^{28}$	5.1 ± 0.4	$(2.0 \pm 1.3) \times 10^{28}$	4.1 ± 0.7	3.1 ± 2.4
'97 Nov.6	$(4.4 \pm 2.5) \times 10^{30}$	3.9 ± 0.6	$(4.8 \pm 2.6) \times 10^{29}$	3.0 ± 0.5	9.2 ± 7.2
'00 Jul.14	$(2.6 \pm 1.7) \times 10^{29}$	2.7 ± 0.4	$(5.8 \pm 2.8) \times 10^{28}$	3.1 ± 0.6	4.5 ± 3.7
'00 Nov.24	$(1.0 \pm 0.6) \times 10^{30}$	2.8 ± 0.3	$(1.5 \pm 1.0) \times 10^{29}$	3.3 ± 0.6	6.7 ± 6.0

obtain the approximate energy content of >1 MeV electrons, W_e (>1 MeV), and >10 MeV protons, W_p (>10 MeV). The estimated energy contents are shown in Table 9.15 along with the electron and proton power-law spectral indices (S_e for electrons and S_p for protons).

9.12.5 *Main Results and Discussion*

The Table 9.15 shows that W_e (>1 MeV) and W_p (>10 MeV) much vary from flare to flare. As noted Yoshimori et al. (2003), it cannot be unambiguously state that nonthermal energy is equally shared between >1 MeV electrons and >10 MeV protons. The ratio of W_e to W_p, however, varies within an order of magnitude in the limited energy ranges mentioned above. It needs to study the energy participation between electrons and protons over the wide energy range. In particular, a low energy-proton tail would contain considerable energy. In order to do it, one have to analyze the Ne line at 1.64 MeV because the threshold energy for excitation of Ne is about 3 MeV. Ramaty et al. (1988) calculated a total number of accelerated protons and their energy content from the gamma-ray and neutron observations. For the proton energy spectrum predicted by stochastic acceleration, the total number of protons and energy content are independent of the injection energy, while the injection energy has to be specified for shock acceleration.

As noted Yoshimori et al. (2003), the stochastic model gives the energy content of accelerated protons for about 10^{30} erg for the intense flare on June 3, 1982. The shock acceleration model gives about 10^{30} erg (Ramaty and Murphy 1987).The electron energy content in large solar flares was estimated from the high-energy continuum spectrum (Vestrand et al. 1987; Ramaty et al. 1993). It ranges from about 10^{29} to about 10^{30} erg above 100 keV.

9.13 Gamma and X-Ray Solar Flare Emissions: CORONAS-F Measurements

9.13.1 *The CORONAS Program and CORONAS-F Project*

According to Oraevskii et al. (2002), the launch of the near-Earth space solar observatory CORONAS-F was done within the framework of the international program CORONAS (Comprehensive Orbital Near-Earth Observations of Activity of Sun) which is devoted to studies of the Sun in different phases of the 11-year solar cycle. The previous satellite CORONAS-I (launched in 1994) observed the Sun near its activity minimum. The CORONAS-F studied the solar activity about the maximum of the 23rd solar cycle. On July 31, 2001, the CORONAS-F satellite was launched into an orbit with the following parameters: orbital inclination 82.49°, minimal distance from the Earth surface 500.9 km, maximal distance from the Earth surface 548.5 km, and the orbital period 94.859 min. Such an orbit provides regularly repeating periods of continuous observations of the Sun with duration of about 20 days, which is especially important for problems of helioseismology and solar spot and solar flare patrolling. The stabilization of the spacecraft actually realized proved to be three to five times better than was projected (several arcseconds per second), which allows observations of the Sun with a high spatial resolution.

9.13.2 Scientific Goals of the CORONAS-F Project

Oraevskii et al. (2002) noted that the main scientific goals of the CORONAS-F project include observations of global solar oscillations and seismological studies of the solar interior, comprehensive studies of powerful dynamical process in the active Sun (active regions, flares, plasma ejecta) in a broad wavelength range from the optical to gamma-ray emission, studies of solar cosmic rays accelerated during active processes on the Sun, of their escape conditions, propagation in the interplanetary plasma, and their impact on the terrestrial magnetosphere.

9.13.3 Satellite CORONAS-F and SONG Experiment

According to Kuznetsov et al. (2003a), the low altitude (~500 km) polar orbiting (inclination ~82.5°) satellite CORONAS-F was launched on July 31, 2001. Among the basic tasks of the Project is the study of effects related to energy transfer in the solar atmosphere, and in particular of acceleration of solar particles. One of the experiments, namely SONG is providing measurements of gamma rays up to >100 MeV and of neutrons >20 MeV. The orientation of CORONAS-F satellite towards the Sun is better than 10 arcmin. The experiment SONG is one of the instruments of Solar Cosmic Ray complex coordinated by Skobeltsyn Institute of Nuclear Physics, Moscow State University. The electronics of the SONG instrument was developed by the Institute of Experimental Physics in Kosice, Slovakia. It is an updated construction of the similar experiment on CORONAS-I (Balaz et al. 1994). The instrument is devoted to detection of hard X and gamma rays 0.028–100 MeV, neutrons >20 MeV and of CR charged particles, mainly protons $E > 70$ MeV and electrons $E > 55$ MeV. The detector consists of Cs(I) crystal of 20 cm diameter × 10 cm surrounded by the active anticoincidence shielding of plastic scintillator 2 cm width. The upper part is isolated from the bottom. Thus the scintillator along with Cs(I) crystal creates the electron telescope. The electron is identified if signal from Cs(I) and upper part of coincidence appears while it is absent in the bottom part. Neutrons are detected due to the interactions with Cs and I nuclei. For identification of neutrons on the background of gamma flux the pulse shape dependence on relative ionization is used (Bogomolov et al. 1996a, b). X and gamma rays are measured in the energy release ranges 0.028–0.053, 0053–0.15, 0.15–0.50, 0.50–1.3, 1.3–4.0, 4.0–7.0, 7.0–15, 15–26, 26–41, 41–60, 60–100 MeV and neutrons 7–15, 15–26, 26–41, 41–60, 60–100 MeV. CR are identified if energy deposited in the crystal is >50 MeV. Temporal resolution for X and gamma rays is 4 s, for energetic neutrons, electrons, and other particles of solar CR it is 30 s. Maximum effective surface for gammas and neutrons is $\sim$270 and $\sim$38 cm^2 respectively (Panasyuk et al. 2000). SONG is capable to observe hard X and gamma rays of solar origin when it is outside the radiation belts or not shadowed by Earth (shadow is up to $\sim$40% of the orbital period). The background is due to local gamma-ray produced by the interactions of CR either with the instrument, satellite body or with the atmosphere. The increases due to bremstrahlung by relativistic electrons of radiation belts are skipped from the flare emission analysis

9.13.4 Gamma Rays and Neutrons from the Flare on August 25, 2001

As an example of data obtained on CORONAS-F, Kuznetsov et al. (2003a) considered the observation of high energy gamma rays and neutrons from the flare on August 25, 2001 (see Fig. 9.69).

As noted Kuznetsov et al. (2003a), gamma rays up to >40 MeV are seen well above the background. Since it was before launch of HESSI it is not possible to compare the fluxes. The event is not accompanied by any significant charged particle flux. It is not in the list of Solar Proton Events (http://umbra.nascom.nasa.gov/SEP/seps.html). There are no proton and electron enhancement in polar caps detected by Solar Cosmic Ray instrument on board CORONAS-F (http://www.coronas.ru/intl/ru/skl/data-mkl.htm). The event of 25 August 2001 is mentioned as X5.3 at http://hea-www.harvard.edu/SSXG/kathy/flares/xflares.html as well as at http://isass1.solar.isas.ac.jp/sxt_co/sxt_trace flares/list.html among the list of flares jointly observed by SXT and TRACE as X5.4 flare with the times of the first and last flare mode SXT images, the second one close to the observed very hard gamma emission by SONG. During the flare SONG detected enhancements in neutron channels 7–15, 15–26, 26–41 MeV after main gamma peak (see Panel b in Fig. 9.69). First and last points are background neutron flux in the experiment. According to Kuznetsov et al. (2003a), two case studies using SONG

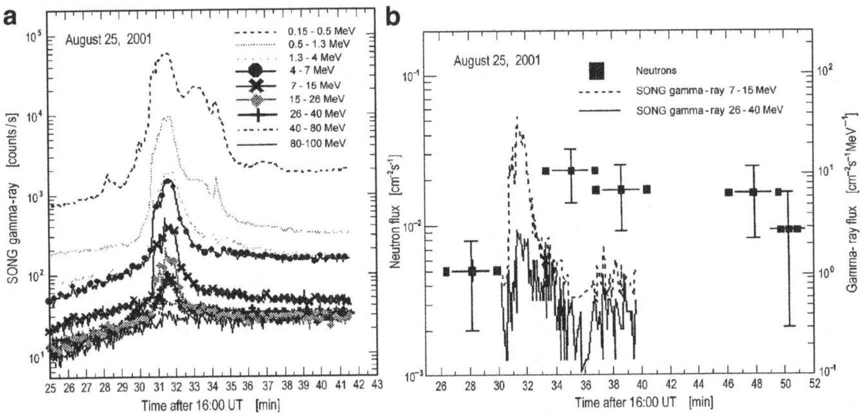

Fig. 9.69 The count rate of gamma rays by SONG (**a**) and solar neutron and gamma-ray flux (**b**) on CORONAS-F during the solar flare on August 25, 2001 (From Kuznetsov et al. 2003a)

data along with other instruments on CORONAS-F have been done, namely on 4 November 2001 (Kuznetsov et al. 2003b) and on 20 May 2002 (Bogomolov et al. 2003).

9.13.5 Catalogue of Observed Events by CORONAS-F Mission During August 2001–September 2005

According to Myagkova et al. (2007), the duty cycle for the detection of solar flares on board CORONAS-F was about 40% as a result of its orbit parameters; so many major flares were lost during August 14, 2001 to September 12, 2005. However, 37 flares with γ-ray emission and three solar neutron events were detected by CORONAS-F. In four of the events γ-ray emission with energy up to 100 MeV and more was detected. Gamma-ray observations of solar flares by the SONG experiment during October–November 2003 and January 2005 provided two different manifestation signatures of the interaction of high energy ions in the solar atmosphere: γ-ray line emission and γ-ray emission due to π^0-decay. Characteristics of flares detected by SONG (detection time, highest γ-ray energy channel, γ-ray fluence) as well as SXR GOES characteristics are presented in Table 9.16. It can be seen that γ-ray emission was observed not only in major flares of GOES X-class in SXRs. About 25% of the flares with γ-emission were only GOES M-class (less than M6) in SXR. Such hard spectra of energetic neutral emission show that the charged energetic particles were accelerated in these flares up to rather high energies.

Table 9.16 List of SONG solar gamma ray observations during August 2001–September 2005 (CORONAS- F) (From Myagkova et al. 2007)

N	Data (dd/mm/yy)	UT of flare according SXR data (GOES) (hh:mm)	SXR class GOES	Flare coordinates	AR	UT of flare according HXR SONG data (hh:mm)	E_{max} (channel) SONG (MeV)	γ Fluencies (>500 keV 1/(cm^2))
2001 (Five γ-ray flares)								
1	25/08/01	16:23–16:45–17:04	X5.3	S17E34	9591	16:29:24–16:38:40	60–100	3,540
2	05/09/01	14:25–14:32–14:34	M6.0	N15W31	9601	14:28:16–14:28:41	1.3–4	6.4
3	09/09/01	15:10–15:16–15:21	M3.4	S17E03	9607	15:12:05–15:12:57	0.5–1.3	4.3
4	19/10/01	16:13–16:30–16:43	X1.6	N15W29	9661	16:24:40–16:26:01	1.3–4	15
5	11/12/01	07:58–08:14–08:08	X2.8	N16E41	9733	08:04:05–08:05:35	7–15	43
2002 (Ten γ-ray flares)								
6	20/02/02	05:52–06:12–06:16	M5.1	N12W72	9825	06:08:42–06:11:25	4.4–7.7	29
7	20/05/02	15:21–15:27–15:31	X2.1	S21E65	9961	15:24:56–15:27:01	7.7–16.5	66
8	17/07/02	06:58–07:13–07:19	M8.5	N22W17	0030	07:12:29–07:13:23	1.6–4.8	>11
9	20/07/02	21:04–21:30–21:54	X3.3	–	–	21:27:29–21:32:35	0.6–1.6	>15
10	20/08/02	08:22–08:26–08:30	M3.4	S10W38	0069	08:25:12–08:26:06	4.8–8.4	27
11	21/08/02	01:35–01:41–01:45	M1.4	S11W47	0069	01:38:35–01:40:00	0.6–1.6	3.8
12	21/08/02	05:28–05:34–05:36	X1.0	S12W51	0069	05:31:23–05:32:06	4.8–8.4	1.7
13	24/08/02	00:49–01:12–01:31	X3.4	S12W51	0069	00:59:40–01:05:35	4.8–8.4	>240
14	29/08/02	12:31–12:55–13.21	M7.1	S12W73	0069	12:49:30–12:51:06	0.6–1.6	0.7
15	30/08/02	12:47–13:29–13:35	X1.5	N15E74	0095	13:27:15–13:28:55	4.8–8.4	19
2003 (13 γ-ray flares)								
16	26/04/03	03:01–03:06–03:12	M2.1	S25W34	0338	03:03:30–03:03:54	1.7–5.2	0.5
17	26/04/03	08:01–08:07–08:09	M2.0	–	–	08:05:10–08:06:50	5.2–9.1	14
18	27/05/03	22:56–23:07–23:13	X1.3	S07W17	0365	23:04:05–23:07:15	5.2–9.1	105
19	28/05/03	00:17–00:27–00:39	X3.6	S07W17	0365	00:22:12–02:28:55	5.2–9.1	450
20	29/05/03	00:51–01:05–01:12	X1.2	–	–	01:02:10–01:06:49	5.2–9.1	76
21	23/10/03	08:19–08:35–08:49	X5.4	S21E88	0486	08:24:02–08:26:24	5.2–9.1	–
22	24/10/03	02:27–02:54–03:14	M7.6	S19E72	0486	02:44:30–02:47:00	0.65–1.7	24
23	26/10/03	05:57–06:54–07:33	X1.2	S15E44	0486	06:16:15–06:18:55	0.65–1.7	7.5
24	28/10/03	09:51–11:10–11:24	O17.2	S16E08	0486	11:02:20–11:14:00	80–130	>5,100
25	29/10/03	20:37–20:49–21:01	X10.0	S15W02	0486	20:38:40–20:59:30	5.2–9.1	1,270
26	04/11/03	19:29–19:53–20:06	X28	S19W83	0486	19:40:20–19:58:00	130–260	>4,850

27	17/11/03	08:55–09:05–09:19	M4.2	S01E33	0501	08:58:25–09:02:30	1.7–5.2	42
28	20/11/03	07:35–07:47–08:38	M9.6	N01W08	0501	08:04:00–08:17:30	0.65–1.7	210
2004 (Two γ-ray flares)								
29	06/01/04	06:13–06:29–06:36	M5.8	N05E90	0537	06:22:50–06:24:40	1.7–5.2	31
30	30/10/04	16:18–16:33–16:37	M5.9	N13W28	0691	16:24:15–16:25:30	0.7–1.8	>10
2005 (Seven γ-ray flares)								
31	01/01/05	00:01–00:31–00:39	X1.7	N06E34	0715	00:28:20–00:33:30	2–6	45
32	17/01/05	09:59–09:52–10:07	X3.8	N15W25	0720	09:52:10–10:00:00	2–6	>800
33	20/01/05	06:36–07:01–07:26	X7.1	N14W61	0720	09:43:20–09:56:00	90–150	3,260
34	14/07/05	05:57–07:25–07:43	M9.1	N09W90	0786	07:22:40–07:23:50	2–6	–
35	09/09/05	19:13–20:04–20:36	X6.2	S12E67	0808	20:00:30–20:12:30	6–10.5	–
36	10/09/05	21:30–22:11–22:43	*X*2.1	S13E47	0808	22:53:00–23:02:30	0.75–2	–
37	12/09/05	08:37–09:03–09:20	M6.2	S11E25	0808	08:46:25–08:48:15	2–6	–

9.14 The Event of 2002 July 23: First Gamma-Ray Images of a Solar Flare

9.14.1 The Matter and Short History of the Problem: RHESSI Experiment

In Hurford et al. (2003a) are presented the first gamma-ray images of a solar flare, obtained with the Reuven Ramaty High Energy Solar Spectroscopic Imager (RHESSI) for the X4.8 flare of 2002 July 23. Two rotating modulation collimators ($35''$ and $180''$ resolution) provided images of the narrow deuterium line at 2.223 MeV formed by thermalization and capture of neutrons produced in energetic ion collisions, the 3.25–6.5 MeV band that includes the prompt deexcitation lines of C and O, and the 0.3–0.5 and 0.7–1.4 MeV bands that are dominated by electron-bremsstrahlung. The centroid of the 2.223 MeV image was found to be displaced by $\sim 20 \pm 6$ arcsec from that of the 0.3–0.5 MeV band, implying a difference in acceleration and/or propagation between the accelerated electron and ion populations near the Sun.

As noted Hurford et al. (2003a), the detection of nuclear γ-ray line emission from large solar flares shows that they accelerate ions as well as electrons to high energies (Chupp 1987). Collisions of energetic ions with the solar atmosphere produce excited nuclei which emit prompt nuclear deexcitation lines, as well as secondary neutrons and positrons that result in the delayed 2.223 MeV neutron-caption and 511 keV positron-annihilation line emission (Ramaty and Murphy 1987). The RHESSI mission is designed for flare X-ray and gamma-ray imaging spectroscopy from 3 keV to 17 MeV (Lin et al. 2002). The imaging system consists of nine bi-grid rotating modulation collimators (RMC's), that provide FWHM angular resolution of $2.26''$ to $183''$ in logarithmically spaced steps. As the spacecraft rotates, the fraction of the incident flux that passes through the two grids in each RMC varies rapidly. RMCs #6 ($35.3''$) and #9 ($183''$) have 2 and 3 cm thick tungsten grids, respectively, to modulate effectively at gamma-ray energies. According to Smith et al. (2002), behind each RMC is an electrically segmented germanium detector (GeD), cryogenically cooled to provide high spectral resolution ($\sim$1–10 keV FWHM). Spatial information on the source is encoded in the timing of the detected counts (Hurford et al. 2002).

9.14.2 RHESSI Observation of the X4.8 Flare of 2002 July 23

According to Hurford et al. (2003a), RHESSI observed the X4.8 flare of 2002 July 23, optically centered near the east limb at 13°S, 72°E. Figure 9.70 shows rear segment count rates versus time in the 0.3–0.5, 0.7–1.4, 2.218–2.228, and 3.25–6.5 MeV bands. All the images are made for the most intense impulsive peak (from

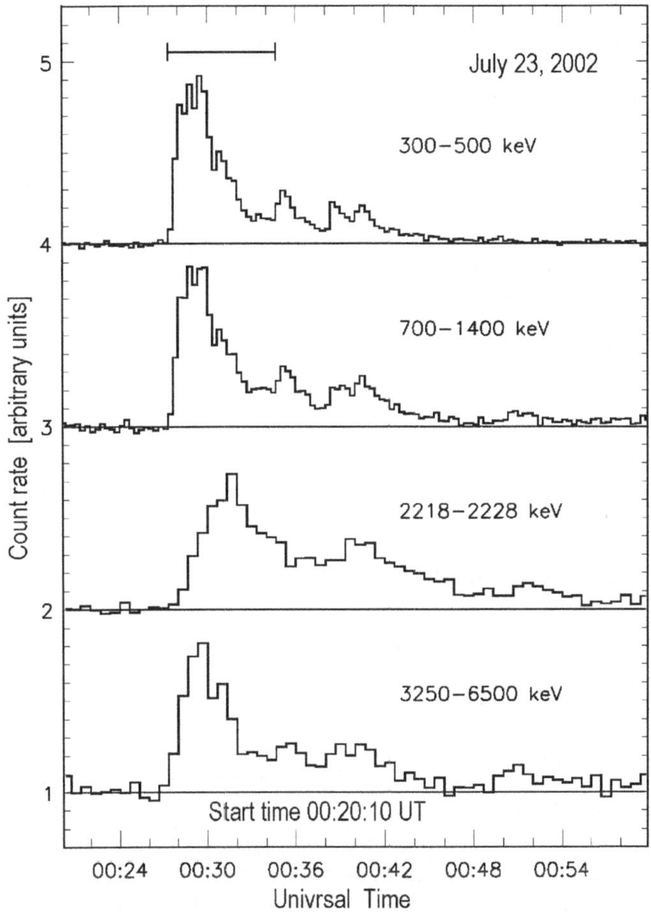

Fig. 9.70 The event 2002 July 23: background-subtracted light curves at the four imaged gamma-ray energy bands. Time resolutions are 20, 20, 40 and 40 s respectively. The time interval used for imaging (00:27:20 to 00:34:40 UT) is shown at the top (From Hurford et al. 2003a)

00:27:20 to 00:34:40 UT). The background-subtracted γ-ray count spectrum from 0.3 to 8.5 MeV was fit to a model which includes the prompt nuclear deexcitation lines, the 2.223 MeV neutron-capture line, the 0.511 MeV positron annihilation line, and a broken power-law electron bremsstrahlung continuum – all folded through the full instrument response, including non-diagonal terms from photons which deposit only part of their energy in the detector (Smith et al. 2002).

As noted Hurford et al. (2003a), the best fit indicates that the 0.3–0.5 and 0.7–1.4 MeV bands are dominated by electron-bremsstrahlung continuum, while the C and O nuclear line complex dominates the 3.25–6.5 MeV band. Fast neutrons from energetic ion collisions thermalize in the photosphere before being captured by hydrogen to form deuterium, which then emits a 2.223 MeV photon. This results in a very narrow (intrinsic width $<\sim$0.1 keV) line, delayed by $\sim$100 s. RHESSI detects it with high resolution ($\sim$4 keV FWHM), so a narrow energy band (2.218–2.228 MeV) can be used for imaging, to effectively eliminate underlying continuum background and non-photo peak response.

9.14.3 Low-Resolution Gamma-Ray Images

Figure 9.71 shows low-resolution (183″) images made with the back-projection technique (Hurford et al. 2002) using RMC #9 only. The main peaks show unambiguously that the gamma-ray sources are related to the optical flare at 13°S, 72°W. Intermediate resolution (35″) maps were made by summing back-projection images from RMC #6 and #9.

9.14.4 High-Resolution Gamma-Ray Images

Figure 9.72 shows the centroids for the 0.3–0.5, 0.7–1.4, and 2.218–2.228 MeV bands as circles with radii equal to the 1-σ statistical error. According to Hurford et al. (2003a), for the 0.3–0.5 and 0.7–1.4 MeV bands, the ratios of the imaged flux with RMC #6 (35″) to that with RMC #9 (183″) (the relative visibilities) – 0.54 $\pm$ 0.06 and 0.57 $\pm$ 0.16, respectively – are not unity, indicating that the sources were partially resolved on a size scale of 35″. A high-resolution (3″) 50–100 keV hard X-ray image (white contours in Fig. 9.72) shows a string of compact hard X-ray sources extends $\sim$24″ along a NNE-SSW direction, with a second, $\sim$12″ long parallel string located $\sim$10″ closer to Sun-center (Krucker et al. 2003; Emslie et al. 2003; White et al. 2003). A 7″ image at 0.3–0.5 MeV shows contours that overlap the east string, with the lowest intensity contour also covering the west string.

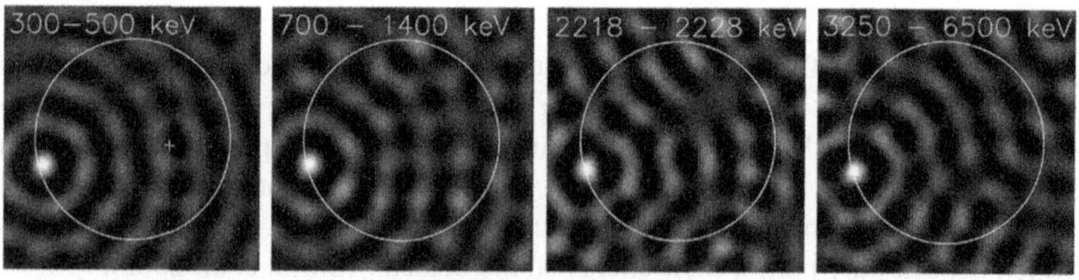

Fig. 9.71 Event 2002 July 23: low-resolution (183″ FWHM) 'dirty' maps at four gamma-ray energies. The 2,218–2,228 keV map was made with a total of only 130 counts. The grey scale is linear with negative sidelobes suppressed. The circle represents the solar disk with heliocentric north and west at the top and right, respectively. RHESSI was rotating about a spin axis indicated by + with a period of 4.084 s (From Hurford et al. 2003a)

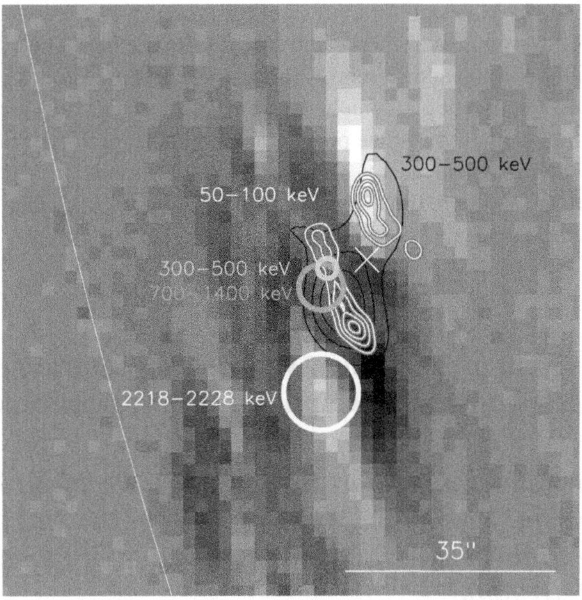

Fig. 9.72 The event of 2002 July 23: locations of the gamma-ray sources. The thick circles represent the 1σ errors for the centroids of the 300–500 keV (light grey), 700–1,400 keV (dark grey) and 2,218–2,228 keV (white) sources. White contours show the high-resolution 50–100 keV map with 3$''$ resolution (the cross shows the centroid), the black contours show the 300–500 keV map with 7$''$ resolution (From Hurford et al. 2003a)

9.14.5 Possible Physical Interpretation of the First γ-Ray Solar Flare Images

As noted Hurford et al. (2003a), the relative visibility of the 2.223 MeV line source was 0.93 $\pm$ 0.43, consistent with a value of unity, which would suggest a compact source. If it will be take 0.07 as a 2 σ lower limit to the relative visibility and make an arbitrary (but simple) assumption of a 'Gaussian model', this would imply an 2 σ upper limit to the 2.223 MeV source size of $\sim$1 arcmin FWHM. A detailed physical model of the 2.223 MeV line emission that includes energy losses, magnetic mirroring, and pitch angle scattering in a loop with an active region atmosphere (Hua et al. 1989, 2002) shows that the accelerated ions interact in the lower chromosphere/upper photosphere to produce the fast neutrons and prompt deexcitation lines, while the neutron thermalization and capture on hydrogen occurs close by in the photosphere within $\sim$500 km or $\sim$1$''$.

Hurford et al. (2003a) underlined that the 2.223 MeV source centroid was found to be displaced in the southward direction from the weighted average of the 0.3–0.5 and 0.7–1.4 MeV sources by 20 $\pm$ 6$''$, which is nonzero with statistical confidence of >99.7%. Since the 0.3–0.5 and 0.7–1.4 MeV bands are partially contaminated by nuclear line emission, this is an underestimate of the actual displacement. In the vicinity of the 2.223 MeV centroid, no significant hard X-ray emission is detected; all the hard X-ray sources are north and toward Sun-center. Thus, the ions must accelerate and/or propagate differently from the electrons. One possibility is acceleration by DC electric fields, which would send electrons and ions in opposite directions. Although their flux-time profiles are similar (Fig. 9.70), the prompt γ-ray line emission appears delayed by $\sim$10 s relative to the electron bremsstrahlung emission, perhaps enough time for electron heating of the solar atmosphere, producing a shock wave that then accelerate ions.

9.15 The Event of 2002 July 23: Physical Implications of RHESSI Neutron-Capture Line Measurements

9.15.1 The Matter and Short History of the Problem: High Spectral Resolution Measurements of the 2.223 MeV Neutron-Capture Line and Estimation of Photospheric ^{3}He/H Ratio

Murphy et al. (2003a, b, c) report high spectral resolution measurements of the 2.223 MeV neutron-capture line obtained with the Reuven Ramaty High Energy Solar Spectroscopic Imager (RHESSI) from the 2002 July 23 solar flare. As noted Murphy et al. (2003a), the time history of this line is affected by both the photospheric ^{3}He abundance and the angular distribution of the interacting flare-accelerated particles producing the neutrons. As underlined Murphy et al. (2003a), because the 2.223 MeV neutron-capture line is very narrow (<0.1 KeV), the RHESSI high-resolution Ge detectors are very effective for observing it. Only previous high-resolution detection was by HEAO 3 from the 1979 November 9 flare (Prince et al. 1982), but analysis of those data was limited by the low counting rate. The line is formed when flare-produced neutrons are captured on ambient hydrogen in the photosphere, producing deuterium. Since the neutrons slow down before capture, the line is delayed by $\sim$100 s. The delay is affected by both the photospheric ^{3}He abundance and the angular distribution of the interacting particles producing the neutrons. Neutron capture on ^{3}He, ^{3}He$(n, p)^3$H, produces no radiation but shortens the delay of the capture line (Wang and Ramaty 1974). Downward-directed interacting accelerated particles tend to produce neutrons deeper in the atmosphere where the higher density also shortens the delay.

Murphy et al. (2003a) noted that several investigators have determined the photospheric ^{3}He/H ratio using capture line time-history measurements obtained with the Solar Maximum Mission (SMM), Granat, and the Compton Gamma Ray Observatory. Chupp et al. (1981), Prince et al. (1983a), Trottet et al. (1993a), Murphy et al. (1997), and Rank et al. (2001) approximated the expected line time history from instantaneous neutron production as the sum of exponentials and made simplifying assumptions about the density structure of the solar atmosphere and the interacting particle angular distribution. Hua and Lingenfelter (1987b) used a Monte Carlo technique to calculate the depth, energy, and angular distribution of neutron and neutron-capture line production. The derived ^{3}He/H ratios from these analyses ranged from 0 to 5×10^{-5}, with the smallest uncertainties about (1–2) $\times$ 10^{-5} obtained by Hua and Lingenfelter (1987b) and Rank et al. (2001). According to opinion of Murphy et al. (2003a), in these latter two analyses, the ^{3}He/H uncertainty was determined with a fixed interacting particle angular distribution and without considering the uncertainties of the measured time history used to represent the neutron-production time profile. As a result, the ^{3}He/H uncertainties were underestimated.

Murphy et al. (2003a) used RHESSI observations of the neutron-capture line to determine the photospheric ^{3}He abundance. They simultaneously varied both ^{3}He/H and the interacting particle angular distribution, and include the effect of the uncertainties of the neutron-production time profile. This results in more realistic uncertainty estimates. Neutron-capture line time histories is calculated using a magnetic loop model with a magnetic field perpendicular to the solar surface at the footpoints. The interacting particle angular distribution is parameterized by the level of pitch-angle scattering present in the loop that directly affects the distribution. This analysis also place constraints on the level of this scattering.

9.15.2 Expected Neutron-Capture Line Calculations

As noted Murphy et al. (2003a), neutron-capture line production is very sensitive to the interacting accelerated particle angular distribution. In previous analyses of neutron-capture line data, various analytic shapes for this distribution were assumed. Murphy et al. (2003a) use a physically based magnetic loop model (Hua et al. 1989, 2002) to calculate the interacting accelerated particle angular distribution. The model consists of a semicircular coronal portion and two straight portions extending vertically from the transition region into the photosphere. Below the transition region, the magnetic field strength is assumed proportional to a power δ of the pressure (Zweibel and Haber 1983). The atmospheric model above 120 km is the sunspot active region model of Avrett (1981). The model includes energy losses due to Coulomb collisions, removal by nuclear reactions, magnetic mirroring in the convergent flux tube, and MHD pitch-angle scattering in the corona.

Pitch-angle scattering can be characterized by its mean free path Λ, the average distance for isotropization. Here Λ is expressed by λ, the ratio of Λ to the loop half-length. With no pitch-angle scattering ($\lambda \rightarrow \infty$), particles with large initial pitch angles mirror and tend to interact at their mirror points, where the density is greatest. Particles with pitch angles too small to mirror enter the 'loss cone' and either undergo a nuclear reaction as they are moving downward or are thermalized. Without pitch-angle scattering, most particles are moving parallel to the solar surface when they interact (i.e., a 'fan beam'). Scattering causes the loss cone to be continuously repopulated and therefore more downward-directed interacting particles. As pitch-angle scattering is increased (λ decreased), the distribution becomes more downward-directed until saturation is reached ($\lambda \sim 20$). At saturation, particles are scattered into the loss cone as quickly as they are removed by interactions, so that a further decrease of λ has no effect on the temporal or angular dependence of the interactions. Using γ-ray line data from several SMM flares, Share et al. (2002) showed that the measured deexcitation line Doppler shifts imply interacting ion angular distributions that are inconsistent with no pitch-angle scattering. Better fits were obtained with both nearly saturated ($\lambda = 300$) and saturated pitch-angle scattering. Figure 10 of Share et al. (2002) shows examples of interacting particle angular distributions calculated for various values of λ.

Hua et al. (2002) developed new neutron-production kinematics and updated the neutron-production cross sections used previously in Hua and Lingenfelter (1987a). Murphy et al. (2003a) use these cross sections and the loop model to calculate the neutroncapture line time history. The loop length was supposed of 1.15×10^9 cm and $\delta = 0.2$, values found by Hua et al. (1989) to provide good fits to deexcitation line decay times in the 1980 June 21 flare. The ambient medium composition was assumed to be coronal (Reames 1995), but with Ne/O = 0.25 and He/H = 0.10. Murphy et al. (2003a) assumed 'impulsive flare' abundances for the accelerated ions (Ramaty et al. 1996), with an α/p ratio of 0.5. The accelerated ions are released isotropically at the top of the loop with a power-law energy spectrum and index S (i.e., $\propto E^{-S}$, where E is in units of MeV/nucleon) and are followed until they interact or thermalize. The resulting neutrons are followed until they escape, decay, or are captured. The attenuation and direction of the 2.223 MeV γ-rays are recorded.

Murphy et al. (2003a) have calculated neutron-capture line time histories for various assumptions of the accelerated particle spectral index S, the photospheric ^{3}He/H ratio, and the level of pitch-angle scattering, characterized by λ. Figure 9.73 shows examples of neutron-capture line time histories calculated for instantaneous release of the accelerated particles. The histories were calculated at the 73° heliocentric angle of the 23 July 2002 flare, and the accelerated proton spectrum was normalized to one proton above 30 MeV. The time histories fall faster with increasing pitch-angle scattering (decreasing λ), increasing ^{3}He/H, and harder spectra (smaller S: harder spectra tend to produce neutrons deeper in the atmosphere where the higher density shortens the delay) Murphy et al. (2003a) derive predicted neutron-capture line time histories by convolving the time histories calculated for instantaneous release with a neutron-production time history assumed to be given by the nuclear deexcitation line flux measured for this flare (see below, Section 9.15.3). This is appropriate because deexcitation γ-ray and neutron production are similarly delayed relative to the accelerated particle release and because the bulk of the production typically occurs less than 10 s after release (Hua et al. 1989, 2002). It was used Monte Carlo technique to estimate the uncertainties of the predicted neutron-capture line time history resulting from uncertainties of the measured deexcitation line time history.

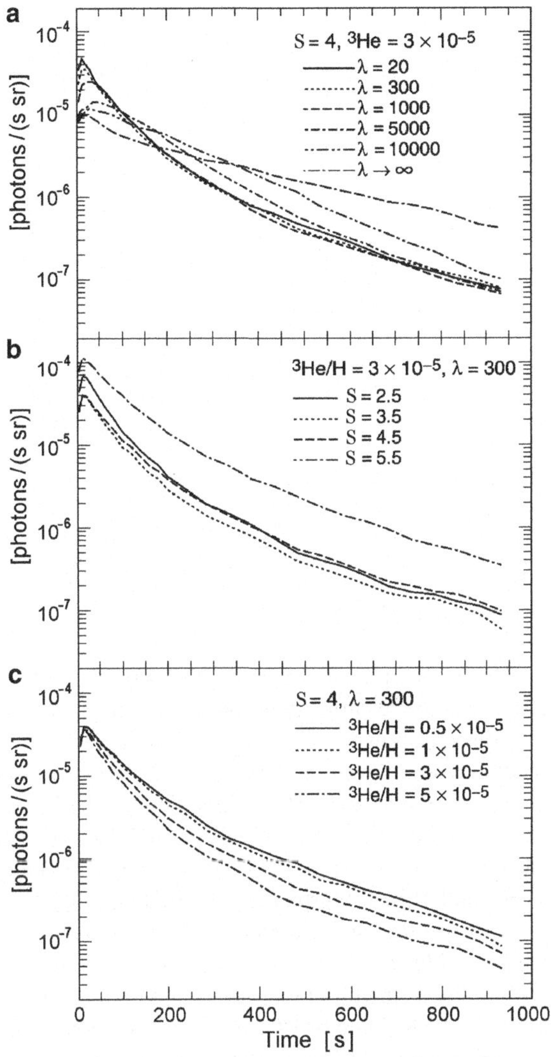

Fig. 9.73 Calculations of the neutron-capture line time history showing its dependence on the level of pitch-angle scattering, characterized by λ, the accelerated particle spectral index S, and the ^{3}He/H ratio. All calculations are for a loop magnetic field perpendicular to the solar surface at the 73° heliographic angle of the 2002 July 23 flare. The accelerated particle spectrum is normalized such that there is one proton greater than 30 MeV (From Murphy et al. 2003a)

9.15.3 Comparison of the RHESSI Data with the Modelling Calculations

The background-subtracted count spectrum measured by RHESSI near the neutron-capture line is shown in Fig. 9.74. The data were fitted with a model consisting of a Gaussian profile for the line, a power law for the electron bremsstrahlung, and a nuclear deexcitation line component. These components and the best-fitting total model are also shown in the figure.

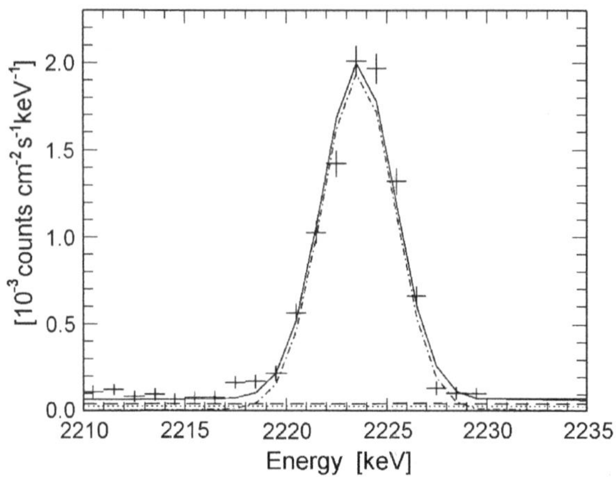

Fig. 9.74 Observed 2.223 MeV neutron-capture line count spectrum and the best fit to the data. The dotted curve is the nuclear de-excitation line component, the dashed curve is the electron bremsstrahlung power law, and the dash-dotted curve is the Gaussian for the line. The solid curve is the sum of all of the components. The data have been accumulated over the bulk of the flare duration, 00:27:20–00:43:20 UT (note that above 2,230 keV, the data are binned in large energy channels) (From Murphy et al. 2003a)

Spectra accumulated every 20 s were also fitted with this model (Share et al. 2003), and the resulting neutron-capture line and 4–7.6 MeV de-excitation line time histories are shown in Fig. 9.75. For the very narrow neutron-capture line, the high spectral resolution of the RHESSI detectors compensates well for the instrument's modest effective area. This is not true for the broader de-excitation lines; the de-excitation line flux is relatively poorly known, which is a dominant source of uncertainty in the ^{3}He/H ratio derived here (see below).

As underlined Murphy et al. (2003a), Fig. 9.73 shows that the accelerated particle spectral index affects the time history of the neutron-capture line. The most reliable measure of the index for the energy range relevant to the neutron-capture line is the flux ratio of the capture line and a narrow de-excitation line such as the 4.44 MeV line of ^{12}C since this ratio is fairly independent of the ambient and accelerated particle composition (e.g., Ramaty et al. 1996). For the July 23 flare, the ^{12}C line is too weak to provide a useful index measure. Alternatively, the 4–7.6 MeV de-excitation line band can be used to improve the statistics, but this ratio is much more sensitive to the assumed abundances. Therefore Murphy et al. (2003a) do not determine the spectral index from the data but rather determine what constraints the shape of the neutron-capture line time history alone can place on the ^{3}He/H ratio and the level of pitch-angle scattering. They assume two spectral indices S (3.5 and 4.5) that represent the range of measured flare accelerated particle indices (Ramaty et al. 1996). For each index, Murphy et al. (2003a) vary ^{3}He/H from 0.1×10^{-5} to 20×10^{-5} and interacting ion angular distributions due to levels of pitch-angle scattering from none ($\lambda \rightarrow \infty$, a fan beam) to saturated ($\lambda = 20$, strongly downward-directed).

Murphy et al. (2003a) also considered an angular distribution that is 100% downward-beamed. For each combination of index, ^{3}He/H, and angular distribution, a predicted neutron-capture line time history was calculated and compared with the data, normalizing the predicted profile to minimize χ^2. The χ^2 was calculated using uncertainties determined by adding in quadrature the data statistical uncertainties and the estimated prediction uncertainties. Confidence levels for ^{3}He/H and λ were established assuming two parameters of interest (Lampton et al. 1976). The minimum χ^2 achieved ($\chi^2_{\rm min}$) was the same for both spectral indices (60.0 for 44 degrees of freedom, or a confidence level of 5%). Figure 9.75 shows the measured and best-fit predicted time histories for S = 4.5. The dotted lines indicate the time interval for calculating χ^2 (00:28:20–00:43:20 UT). Murphy et al. (2003a) noted that during the rise of the line flux, the prediction exceeds the data for four 20 s intervals by about 2 σ each. This could be due to an overestimation of the deexcitation line flux early in the flare when the flux was weak. Alternatively, the accelerated particle energy spectrum could have been steeper during this time than during the bulk of the flare, with less neutron production relative to deexcitation line production. But this would result in a relatively high flux during this time of those deexcitation lines having low cross section thresholds, such as the 1.634 MeV Ne line, which was not observed.

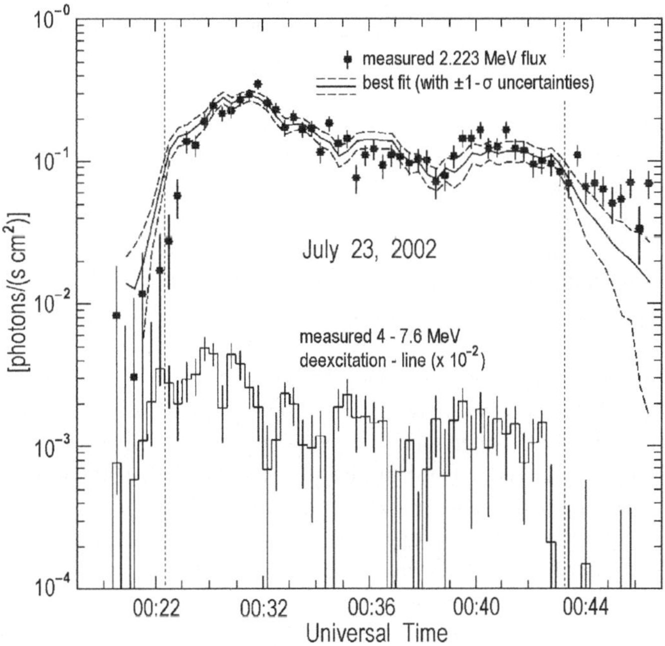

Fig. 9.75 Measured time dependences of the 2.223 MeV neutron-capture line and the 4–7.6 MeV de-excitation line fluxes. The 4–7.6 MeV flux has been reduced by a factor of 100 for clarity. Also shown is the comparison of the best-fitting predicted neutron-capture line flux for S = 4.5 (obtained with λ = 2,000 and ^{3}He/H = 7×10^{-5}) and the measured time history. The dotted lines indicate the time interval over which χ^2 was calculated (00:28:20–00:43:20 UT) (From Murphy et al. 2003a)

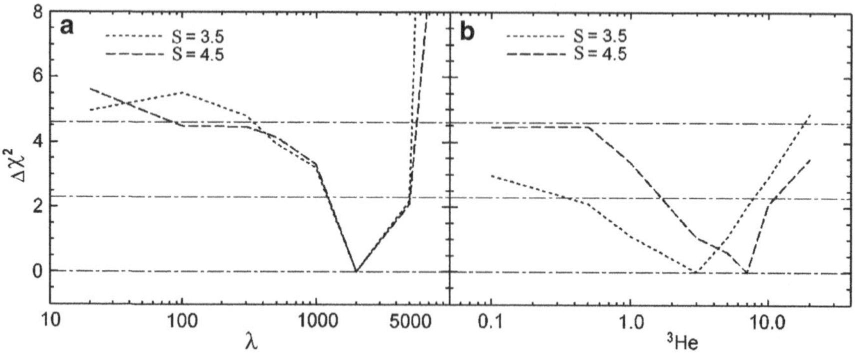

Fig. 9.76 Dependence from the two parameters λ (Panel (**a**)) and ^{3}He/H in 10^{-5} (**b**) for S = 3.5 (dotted curve) and 4.5 (dashed curve). The dot-dashed horizontal lines indicate $\Delta\chi^2 = 0$, $\Delta\chi^2 = 2.3$ (68% confidence level for two parameters), and $\Delta\chi^2 = 4.61$ (90% confidence level for two parameters) (From Murphy et al. 2003a)

For each spectral index (3.5 and 4.5), Fig. 9.76 shows how $\Delta\chi^2$ (the change of χ^2 from χ^2_{min}) varies as λ and ^{3}He/H are varied (for each value of a given parameter, the other parameter has been adjusted to minimize χ^2).

As can be seen from Fig. 9.76, for these two S indices (3.5 and 4.5), the results for λ (Panel a in Fig. 9.76) do not depend strongly on the value of index. Minimum χ^2 is achieved at with a λ = 2,000 with a 1 σ allowable range of 700–5,000 and a 99% confidence upper limit of ~7,000 (assuming two parameters of interest). Pitch-angle scattering levels from weak to none can therefore be rejected, implying that the interacting particle angular distribution must be at least somewhat downward-directed. On the other hand, while more strongly downward-directed distributions (λ < 700) result in worse fits, even saturated pitch-angle scattering (λ = 20) cannot be rejected. However, the minimum χ^2 achieved assuming a 100% downward-beamed angular distribution was 68.3, or $\Delta\chi^2 = 8.3$. The downward beam can therefore be rejected with better than 98% confidence.

As underlined Murphy et al. (2003a), in paper of Smith et al. (2003) was found that a 100% downward beam reproduces the strong deexcitation line redshifts observed for this flare better than a downward-isotropic distribution, unless the magnetic loop is inclined toward the Earth. Calculations of neutron-capture line production for inclined magnetic fields have not yet been performed. But longer decay times can reasonably be expected from inclined fields (at least for strong pitch-angle scattering where the interacting angular distribution is more forward-directed) since the neutrons would tend to be produced at shallower depths. This may explain why the neutron-capture line analysis favors angular distributions that are less downward-directed than that favored by the narrow-line analysis.

As noted Murphy et al. (2003a), Panel b in Fig. 9.76 shows that the derived ^{3}He/H ratio depends on the index, with higher values of ^{3}He/H required for the steeper spectrum (larger S) as expected. The allowable range of ^{3}He/H (1 σ, two parameters of interest) over the range of indices assumed here was from 0.5×10^{-5} to 10×10^{-5}.

If the spectral index were determined by an independent method (e.g., γ-ray line fluence ratios), this allowable range would be reduced. But even so, the ^{3}He/H uncertainty obtained here would still be significantly larger than those obtained by previous investigators (see Section 9.15.1). In part, this is due to the simultaneous determination of ^{3}He/H and the interacting angular distribution (value of λ), which has not been done before. However, Murphy et al. (2003a) find that most of the uncertainty expressed in Fig. 9.76 arises from the large statistical errors of the nuclear deexcitation line flux used to represent the neutron-production time history (see Fig. 9.75).

9.16 The Event of 2002 July 23: RHESSI Discovery of a Coronal Non-thermal Hard X-Ray Source, as a Precursor of the Impulse Flare

9.16.1 The Matter and Short History of Problem

According to Lin et al. (2003), the RHESSI observations of the γ-ray line flare of 2002 July 23, show that in the ~8 min rise prior to the impulsive phase, the hard X-ray emission comes from a coronal source which has no counterpart in the simultaneous TRACE EUV images or in the Hα images. The spectrum above ~10 keV fits to a double-powerlaw shape with break energies at ~20–40 keV and exponents of ~5 below and ~7 above, with no obvious thermal emission above ~10 keV. This coronal nonthermal source implies that substantial energy release and **electron acceleration occurs before the impulsive phase**.

As underlined Lin et al. (2003), the RHESSI mission is designed to investigate particle acceleration and energy release in solar flares, through high resolution imaging and spectroscopy from soft thermal X-rays (~3 keV) to high energy gamma-rays up to 17 MeV (Lin et al. 2002). The RHESSI imager is made up of nine bi-grid rotating modulation collimators (RMCs), each consisting of a pair of widely separated grids mounted on a rotating spacecraft, to provide spatial resolution of 2.3 arcsec to 3 arcmin over the full Sun (~1°) field of view. Behind each RMC is a segmented coaxial germanium detector, cooled to <~75°K to achieve spectral resolution of ~1 keV FWHM in the hard X-ray range. As the spacecraft rotates, the RMCs convert the spatial information from the source into temporal modulation of the photon counting rates of the Ge detectors.

9.16.2 Observations During the 'Normal', Impulsive Phase

According to Lin et al. (2003), the intense 23 July 2002 gamma-ray line flare (GOES class X4.8, optical importance 2B) began 00:18 UT in NOAA active region

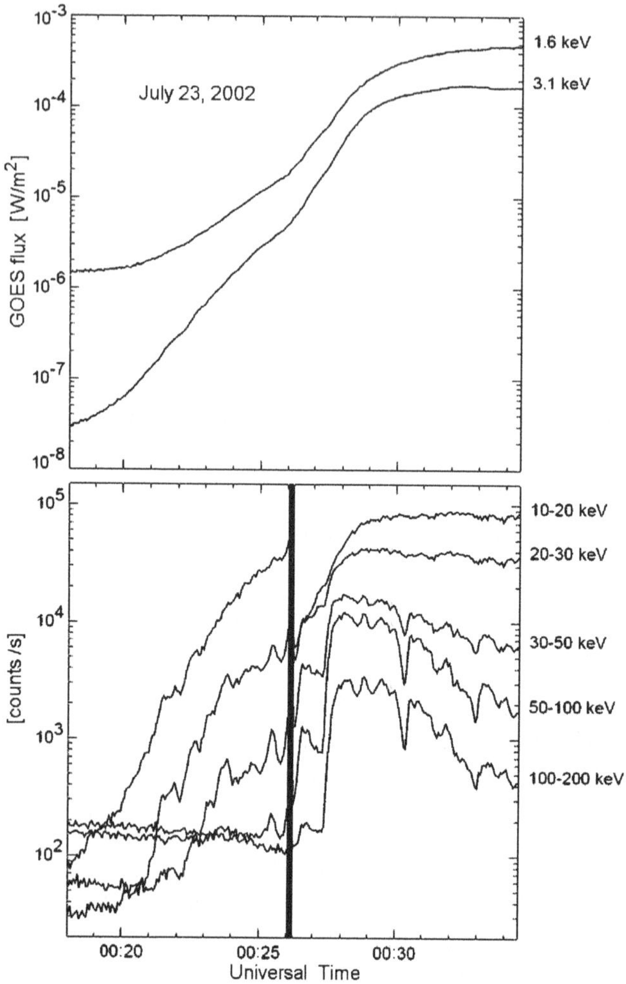

Fig. 9.77 Time profiles for GOES soft X-rays (*top*) and RHESSI hard X-rays (*bottom*). The vertical line indicates where the second shutter was inserted (From Lin et al. 2003)

#0039 at 13°S, 72°E (NOAA Solar Geophysical Data). The Hα, microwave radio, and hard X-ray emissions all peaked at 00:28–00:31 UT, with GOES soft X-rays peaking later, at 00:35 UT. Type II, III, and IV emission were reported at meter wavelengths. The RHESSI observations of the flare are shown in Fig. 9.77.

Figure 9.77 shows a 'normal' impulsive phase ($\sim$00:27 to $\sim$00:43 UT) dominated by footpoints with intense hard X-ray/γ-ray continuum emission (Krucker et al. 2003), and superhot ($\sim 40 \times 10^{6\circ}$K) thermal source in the corona (similar to the above loop-top source observed by Yohkoh HXT according to Masuda et al. 1994). Gamma-ray line emissions are detected in this phase.

9.16.3 Observations During the Preceding Rise Phase

According to Lin et al. (2003), in the preceding rise starting at ~00:18 UT, the hard X-ray emission above 10 keV is concentrated in a source about 22″ size located approximately 20 arcsec east of the optical flare (Panel a in Fig. 9.78). There were no counterpart observed in TRACE 195 A, SOHO MDI visible or Hα images (Krucker et al. 2003), indicating that this source is actually high in the corona and seen in projection.

As underlined Lin et al. (2003), the hard X-ray spectra show a double power-law shape (Panel b in Fig. 9.78). Therefore a model of Holman et al. (2003) with double power-law (exponents γ_L and γ_H), break energy E_B, normalization (the flux at 50 keV, F_{50}), isothermal bremstrahlung spectrum with temperature T and emission measure EM, was fit to the data. Figure 9.79 shows the fit parameters every 20 s. During the rise phase, the spectrum above 10 keV could be fit by a double power-law alone, but also by an isothermal component plus a double power-law with a low energy cutoff as high as ~18 keV.

9.16.4 Estimation of Energy in Coronal Non-thermal and Thermal Sources During the Preceding Rise and Normal 'Impulsive' Phases of the 23 July 2002 Event

Lin et al. (2003), assuming thick target emission in a cold ambient medium ($E_e >> kT$), estimated the energy deposited by energetic electrons, integrating over time

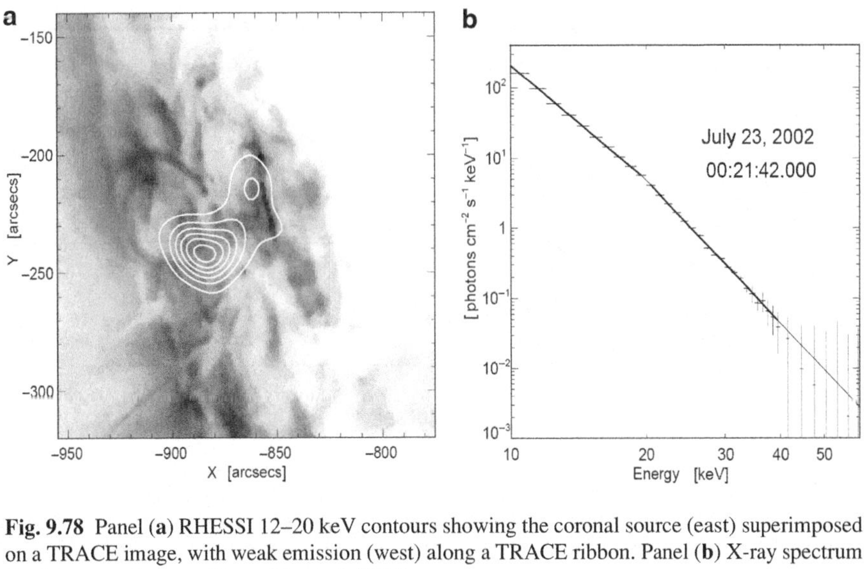

Fig. 9.78 Panel (**a**) RHESSI 12–20 keV contours showing the coronal source (east) superimposed on a TRACE image, with weak emission (west) along a TRACE ribbon. Panel (**b**) X-ray spectrum for this coronal source (From Lin et al. 2003)

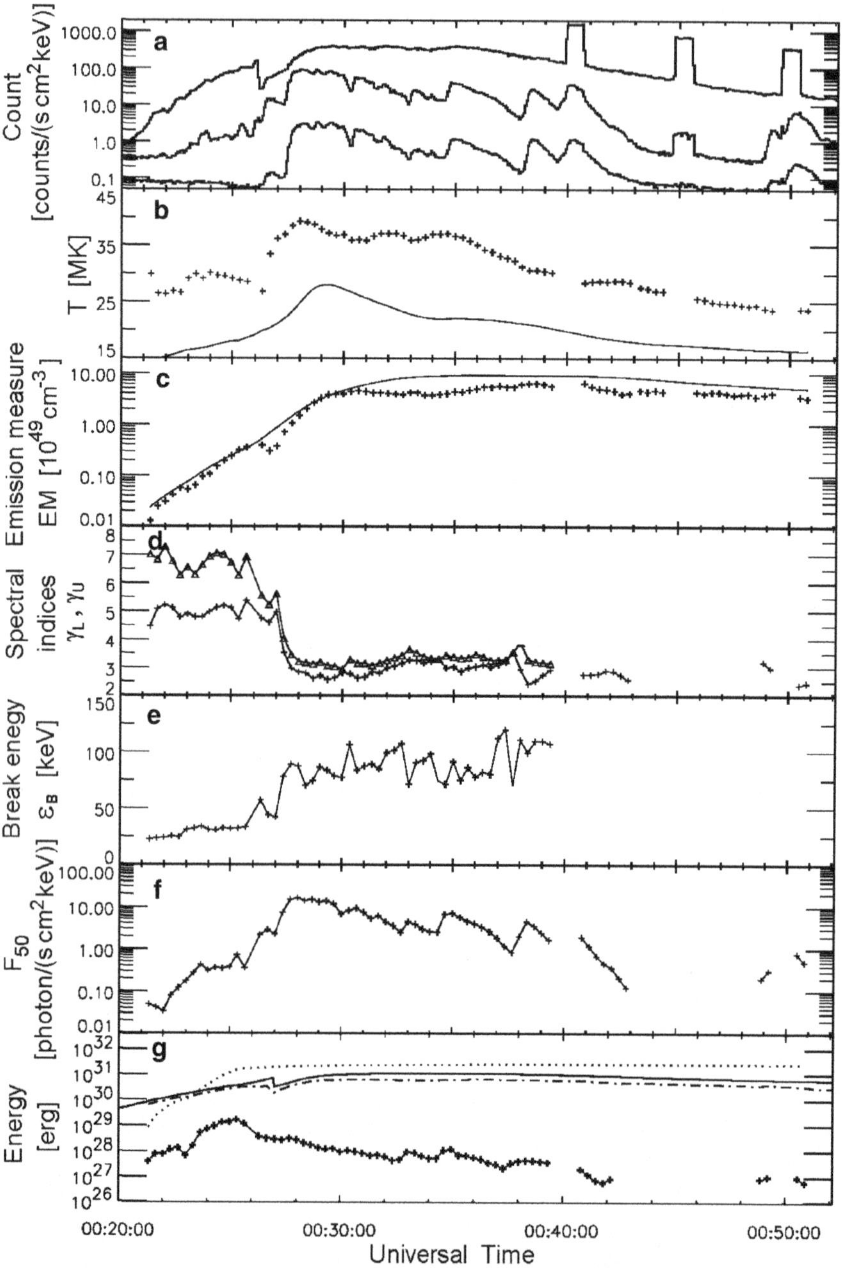

Fig. 9.79 Panel (**a**) RHESSI X-ray light curves: 12–40 keV (*top*, ×0.6), 40–100 keV (middle, ×3), and 100–300 keV (*bottom*). (**b**) The temperature of the isothermal component (plus signs), solid curve is derived from GOES data. (**c**) The isothermal emission measure (plus signs); solid curve is from GOES, scaled by a factor of 0.25. (**d**) Double power-law spectal indices (below break, plus

from the rise to ∼00:26 UT, as $\sim 4 \times 10^{32}$ erg (for a 10 keV cutoff) down to a minimum of $\sim 2 \times 10^{31}$ erg (for a cutoff of ∼18 keV). Lin et al. (2003) noted that the GOES soft X-ray time profile is similar to the time integral of the RHESSI hard X-ray (12–25 keV) flux, suggesting that the non-thermal electrons are depositing their energy into the soft X-ray plasma. The GOES measurements at 00:26 UT give $T \sim 19 \times 10^{6\circ}$K. Assuming that the GOES source is co-spatial with the RHESSI source (volume $\sim(22'')^3 \approx 4 \times 10^{27}$ cm^3) Lin et al. (2003) obtained a density of $\sim 6 \times 10^{10}$ cm^{-3} and an energy content in the soft X-ray plasma of only $\sim 10^{30}$ erg, much less than the energy deposited by non-thermal electrons. Even assuming that the GOES thermal source is ten times larger only increases the thermal energy to $\sim 5 \times 10^{30}$ erg. For a density of $\sim 6 \times 10^{10}$ cm^{-3}, the e-folding energy loss time for 20–100 keV electrons is ∼0.05–0.5 s (Lin 1974) implying that the primary flare energy release is going into accelerating electrons to continuously replenish the coronal source. TRACE detects brightenings in 195 Å emission along three approximately N-S aligned flare ribbons, located ∼40″, ∼10″, and ∼20″ west of the coronal hard X-ray source, beginning at ∼00:21, 00:23, and 00:24 UT, respectively. Some >10 keV hard X-ray emission, always much weaker than the coronal source, is detected sporadically from these ribbons. Thus, most of the energetic electron energy must be deposited into the coronal source. Some energy might be transported by conduction from the soft X-ray plasma to the TRACE chromospheric ribbons. At 00:26:15 UT the impulsive phase begins with strong footpoint emission, accompanied by co-spatial TRACE, MDI, and Hα emission brightenings (Krucker et al. 2003). A 'superhot' ($\sim 40 \times 10^{6\circ}$K) thermal spectrum begins to dominate below ∼30 keV. At that time the centroid of the 12–30 keV source abruptly shifts by ∼6–7 arcsec, suggesting that this superhot source is not co-spatial with the rise phase coronal non-thermal source.

As underlined Lin et al. (2003), the electron energy deposition rate is largest during the rise phase of the flare (Panel g in Fig. 9.79), substantially greater than during the impulsive phase. Although, in principle, a distribution of thermal sources with a range of temperatures might be able to reproduce the observed double power-law spectral shape, this would be highly contrived, since the spatial source appears not to change substantially with energy. Thus, Lin et al. (2003) conclude that very substantial energy release in this flare goes into electron acceleration high in the corona, prior to the impulsive phase. Similar rise phase emission appears to have been detected in other flares, but in general this emission may be too weak to detect in much smaller flares.

←

Fig. 9.79 (continued) signs; above break, triangles). **(e)** Break energy. **(f)** Photon flux at 50 keV. **(g)** Energy in GOES (solid line) and RHESSI (dot-dash line) isothermal fits (see text), compared with accumulated energy in non-thermal electrons (dotted curve) above 18 keV. Lower curve (pluses) gives energy injection rate (erg per second) (From Lin et al. 2003)

9.17 The Event of 2002 July 23: Magnetic Field, Hα, and RHESSI Observations, Evolution of Magnetic Fields Associated with the Flare and Energy Release Models

9.17.1 The Matter and Short History of the Problem: Two Aspects of the 2002 July 23 Gamma-Ray Flare and Energy Release Models

Yurchyshyn et al. (2004) examine two aspects of the 2002 July 23 gamma-ray flare by using multiwavelength observations. First, the data suggest that the interaction of the erupted field with an overlying large-scale coronal field can explain the offset between the gamma-ray and the hard X-ray sources observed in this event. Second, they pay attention to rapid and permanent changes in the photospheric magnetic field associated with the flare. MDI and BBSO magnetograms show that the following magnetic flux had rapidly decreased by 1×10^{20} Mx immediately after the flare, while the leading polarity was gradually increasing for several hours after the flare. Their study also suggests that the changes were most probably associated with the emergence of new flux and the reorientation of the magnetic field lines. Yurchyshyn et al. (2004) interpret the magnetograph and spectral data for this event in terms of the tether-cutting model (Moore and LaBonte 1980).

As noted Yurchyshyn et al. (2004), filament eruptions, coronal mass ejections (CMEs), and associated photospheric activity, such as solar flares, indicate an ongoing relaxation of large-scale, highly stressed solar magnetic fields. Release of magnetic energy during such processes is expected to be accompanied by some variations in the magnetic field. Earlier studies report both the localized changes associated with the major polarity inversion line (Severny 1964; Zvereva and Severny 1970; Moore et al. 1984; Kosovichev and Zharkova 1999, 2001; Wang and Tang 1993; Wang et al. 1994, 2002; Cameron and Sammis 1999; Spirock et al. 2002) and global changes when the entire photospheric and coronal fields in an active region are involved in a flare (Van Driel-Gesztelyi et al. 1997; Aschwanden et al. 1999; Yurchyshyn et al. 2000; Abramenko et al. 2003). Wang et al. (2002) summarized the results of a study for six X-class flares and found that there were rapid and permanent changes in the magnetic flux related to the impulsive phase of the flares. Moreover, these events exhibited a puzzling signature in that the changes of the magnetic flux of the two polarities were not balanced: the leading flux always increased while the following flux tended to decrease, although by a much smaller amount. Spirock et al. (2002) suggested two possible mechanisms to explain the unbalanced flux variations of the 2002 April 2 flare: (1) the emergence of a very inclined flux tube and/or (2) a change in the orientation of the magnetic field. Later, Wang et al. (2002) offered a third explanation: the expansion of the preceding sunspot, as a result of the relaxation of the magnetic field, after a flare.

As underlined Yurchyshyn et al. (2004), very often, large flares are accompanied by powerful CMEs, which can cause severe geomagnetic storms when expelled toward the Earth. There are several competing approaches to explain magnetic eruptions, and extensive multiwavelength data sets, including magnetic field measurements, can be used to provide a discriminator between different models and/or mechanisms. One approach has been developed by many authors over the years (Van Tend and Kuperus 1978; Van Ballegooijen 1989; Forbes and Isenberg 1991; Kumar and Rust 1996; Wu et al. 1999; Amari et al. 2000). It advocates the idea that the energy for eruptions is stored in a flux rope, which is formed long before the eruption occurs. The tether-cutting model (Moore and LaBonte 1980), on the other hand, suggests that an impulsive major energy release begins deep in a highly sheared core field via reconnection, which forms a flux rope. The released magnetic field then erupts into interplanetary space. All these models predict that no remote EUV and/or Hα emission in quiet-Sun areas outside the site of a flare should precede the eruption.

Yurchyshyn et al. (2004) noted that the breakout model (Antiochos 1998; Antiochos et al. 1999), instead, assumes that a sheared core field pushes through an overlaying restraining field and that a slow reconnection begins at a neutral point high in the corona. At this stage, EUV crinkles and/or remote Hα brightenings are expected to occur at the footpoints of the overlying field lines involved in the reconnection (Sterling et al. 2001). After the restraining force of the overlying field is significantly weakened because of the reconnection, the sheared core field then explosively erupts into interplanetary space. As the erupted field leaves the lower atmosphere, it stretches the remaining overlying field lines, which immediately reconnect beneath the escaping magnetic field with releasing magnetic energy and form postflare loops at the site of eruption. There are observations that strongly support this prediction. Sterling and Moore (2001) and Sterling et al. (2001) interpreted the EUV crinkles, which occur before an eruption, as evidence for the breakout model.

Yurchyshyn et al. (2004) analyze Hα, magnetograph, EUV, and X-ray data for the 2002 July 23 flare observed at the east solar limb. This gamma-ray flare occurred in NOAA Active Region 0039 and was a long-duration event that peaked around 00:28 UT. Yurchyshyn et al. (2004) focus on the evolution of magnetic fields associated with the flare.

9.17.2 Observations

The data set that Yurchyshyn et al. (2004) used in their study includes vector magnetograms and Hα images from the Big Bear Solar Observatory (BBSO), full-disk longitudinal magnetograms from the Michelson Doppler Imager (MDI), and EUV images from the EUV Imaging Telescope on board the Solar and Heliosphere Observatory (SOHO) and X-ray data from the Ramaty High-Energy Solar Spectroscopic Imager (RHESSI). BBSO's Digital Vector Magnetograph (DVMG) system has a much improved sensitivity and resolution compared with that of the old BBSO Video Magnetograph system. The hardware has been described in detail by Spirock et al. (2001). It consists of a 0.25 Å bandpass filter, an SMD

1024 × 1024 12 bit CCD camera, and three liquid crystals used as polarization analyzers. Each data set consists of four images taken at the blue wing of the λ6103 spectral line: a filtergram (Stokes I), a line-of-sight magnetogram (Stokes V), and a transverse magnetogram (Stokes U and Q). The line-of-sight magnetic sensitivity is approximately 2 G, while the transverse sensitivity is approximately 20 G. The cadence for a complete set of Stokes images is typically 1 min. BBSO magnetograms, used to study this flare, cover two time ranges: (1) from 20:00 to 22:45 UT on July 22 and (2) from 00:37 to 01:54 UT on 2002 July 23. In order to cover a longer time period and to also have an independent confirmation of BBSO observations, Yurchyshyn et al. (2004) used 1 min cadence full-disk MDI magnetograms.

As underlined Yurchyshyn et al. (2004), much of the energy released during a flare is used to accelerate, to very high energies, electrons (emitting primarily X-rays) and protons and other ions (emitting primarily gamma rays). The goal of the RHESSI mission is to combine, for the first time, high-resolution imaging in hard X-rays (HXRs) and gamma rays with high-resolution spectroscopy, so that a detailed energy spectrum can be obtained at each point in the image (Lin et al. 2002). RHESSI had complete coverage of this flare. Yurchyshyn et al. (2004) use RHESSI data at several energy bands to understand the relationship between the structure of the magnetic field and energy release sites.

9.17.3 The Gamma-Ray Flare as Seen in Multiwavelength Data

Figure 9.80 shows the X4.8 two-ribbon flare on 2002 July 23, which is the gamma-ray flare observed by RHESSI.

In Fig. 9.80 the background images are BBSO Hα filtergrams at two different times near the peak of the flare. The corresponding RHESSI HXR emission in the 12–20 keV energy band is shown with dashed contours, while the solid contours indicate the HXR emission between 100 and 150 keV. According to Krucker et al. (2003), the HXR emission, which was dominant above 30 keV, was related to the photospheric footpoints, while the lower energy emission was associated with a gradual coronal X-ray source.

Figure 9.81 shows the time history of the Hα, HXR, and gamma-ray emission. The thick line in the top panel of Fig. 9.81 shows Hα flux determined at the flare core, where the HXR sources and flare ribbon were observed. The Hα flare started at 00:20 UT, about 7 min before the onset of the impulsive HXR emission.

As can be seen from Fig. 9.81, the 50–100 keV HXR emission abruptly rises at 00:27:30 UT and peaks at about 00:28:30 UT. Other energy bands (100–300, 800–1,900, and 2,228–7,000 keV) show very similar time profiles. However, the 2,218–2,228 keV source behaves quite differently. The white circle in Fig. 9.80 shows the position of the 2,218–2,228 keV emission centroid, which was significantly displaced from the other HXR sources.

As underlined Yurchyshyn et al. (2004), there is no sudden rise of the 2,218–2,228 keV emission at the beginning of the flare (Fig. 9.81, fourth panel).

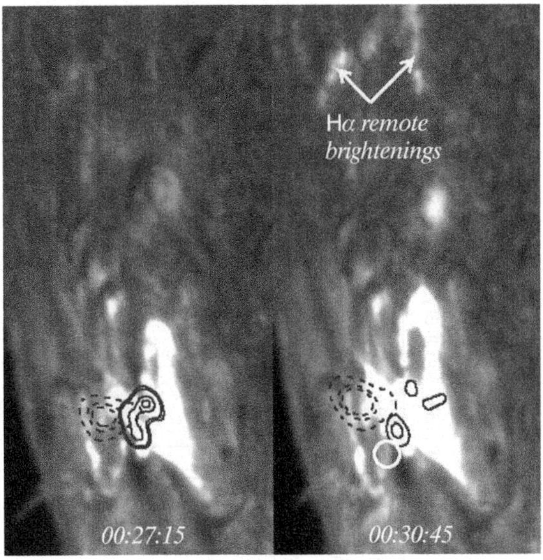

Fig. 9.80 BBSO Hα images of the X4.8 two-ribbon flare on 2002 July 23. The field of view is 118″×280″. The dashed contours are the corres-ponding RHESSI X-ray emissions in the energy range of 12–20 keV, while the solid contours are for the 100–150 keV energy band. The white circle in the right panel shows the position of the 2.223 MeV centroid. North is up, and west is to the right (From Yurchyshyn et al. 2004)

Instead, it increases slowly, reaching maximum at about 00:31:38 UT, 4 min after the peak in the HXR emission (the gray vertical bar in Fig. 9.81 shows the FWHM for this gamma-ray source). A weak Hα flare emission and a surge were observed southeast of the gamma-ray centroid. It can be seen the weak and irregular Hα emission at 00:30:45 UT outside the active region (Fig. 9.80, arrows). This remote emission was observed as a chain of Hα brightenings in a remote quiet-Sun area about 160″ north of the active region (Tang and Moore 1982). The thick line in the fourth panel of Fig. 9.81 shows the average light curve of these remote brightenings. The remote Hα emission began to gradually rise only at the onset of the impulsive HXR flare and reached a maximum at ∼00:31:00 UT, about 1 min earlier than the peak of the 2,218–2,228 keV source. The comparison between the time profiles for the remote brightenings and those for the gamma-ray source shows that these two light curves are very similar within the 00:20–00:33 UT time interval. However, the 2,218–2,228 keV source lags relative to the Hα source by about 60 s. It was noted that the peak of remote Hα brightenings coincides with the well pronounced secondary peak in the RHESSI X-ray flux at 00:31 UT (Fig. 9.81, right dotted line).

9.17.4 Rapid Changes in the Longitudinal Magnetic Field Associated with the Flare: Magnetic Flux Changes According to Observed Data

Figure 9.82 shows the time profiles of magnetic fluxes and the RHESSI HXR flux (arbitrary units) in the 100–150 keV energy range as a function of time. The total

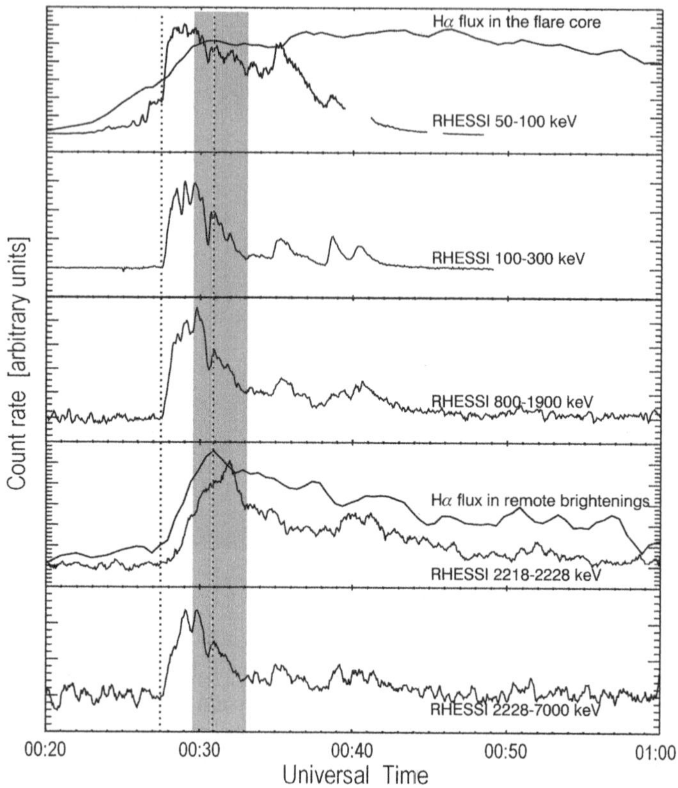

Fig. 9.81 Thin lines: Light curves for 50–100, 100–300, 800–1,900, 2,218–2,228, and 2,228–7,000 keV energy bands. Thick lines: Hα light curves for the flare core (*top* panel) and for the remote brightenings (fourth panel) (From Yurchyshyn et al. 2004)

MDI fluxes were determined by separately summing positive and negative flux densities inside a box, which enclosed the entire area of the active region shown in Fig. 9.83. The error bars in Fig. 9.82 show the standard deviation determined individually for each magnetogram by calculating the total flux for nine different positions of the box. The flux time profiles are plotted for the leading (negative) and the following (positive) polarity by the thin solid lines. The magnetic field measurements between 00:26 and 00:47 UT, when strong HXR emission occurred, were most probably affected by the flare and are not reliable (see Fig. 9.84 for the position of the HXR emission relative to the longitudinal magnetic field). There was a very rapid and substantial change in both leading and following magnetic fluxes. Immediately after the flare (at 00:47 UT; Fig. 9.82, right vertical dashed line), the total leading flux increased by approximately 6%, while the total following (positive) flux decreased by about 14%.

As noted Yurchyshyn et al. (2004), these changes were permanent, and the flux did not return back to the preflare level after the flare ended. Since Yurchyshyn et al.

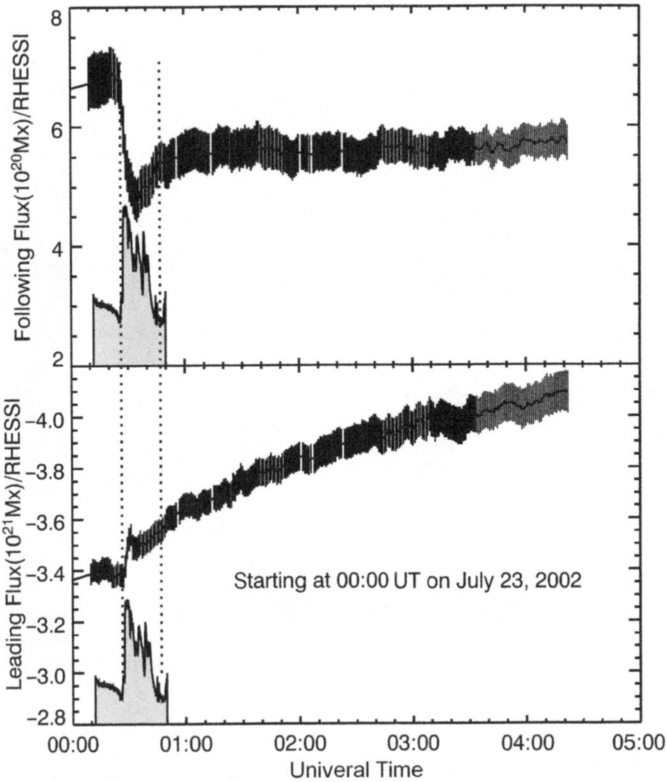

Fig. 9.82 MDI positive (*top*) and negative (*bottom*) magnetic fluxes and the RHESSI HXR flux (shaded area; arbitrary units) in the 100–150 keV energy range shown as a function of Universal Time at July 23, 2002 (From Yurchyshyn et al. 2004)

(2004) have compared total positive and negative flux measured before and after the flare, a change in the line profile during the flare is not a serious concern. They note that the negative flux (Fig. 9.82, bottom) was continuously increasing during the observing period. Yurchyshyn et al. (2004) believe that this gradual growth started before the flare onset and was temporarily interrupted by the rapid fluctuations of the magnetic flux during the impulsive phase of the flare.

As underlined Yurchyshyn et al. (2004), the apparent changes of the longitudinal magnetic field at the solar limb (Fig. 9.84) are artificial and purely due to the fact that variations in seeing change the position of the solar limb in the field of view between the acquisition of two consecutive circularly polarized images. When a longitudinal magnetogram is produced by subtracting one image from another, one obtains an artificial signal, which usually changes rapidly from one magnetogram to another. The noise signal at the solar limb can be distinguished from the persistent changes in an active region, which can be traced from one magnetogram to another.

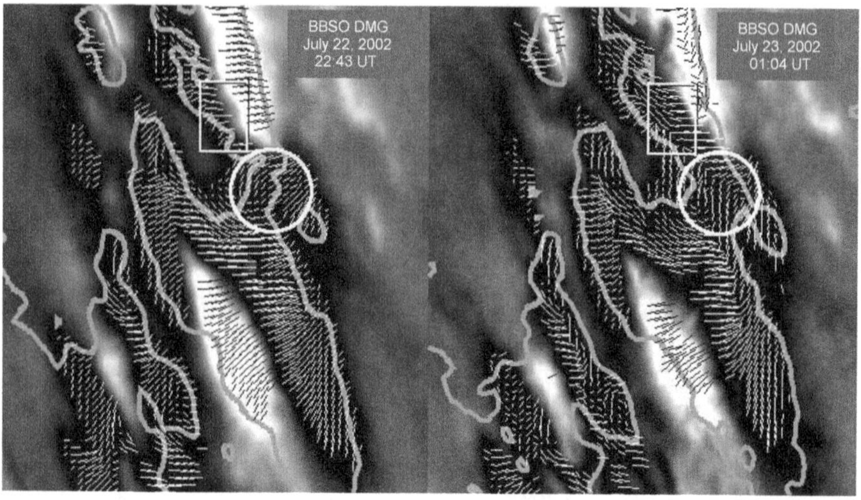

Fig. 9.83 Two BBSO DMG vector magnetograms acquired before (July 22 at 2243 UT; left) and after (July 23 at 01:04 UT; right) the 2002 July 23 gamma-ray flare. The backgrounds are longitudinal magnetograms (white/black correspond to ±350 G). The leading polarity is negative (black). The gray thick contours outline the photosphere-penumbra boundary of sunspots. The black and white line segments display the transverse-magnetic field, while the box and circle-mark the areas where the magnetic flux changes were studied in detail. North is up, and west is to the right (From Yurchyshyn et al. 2004)

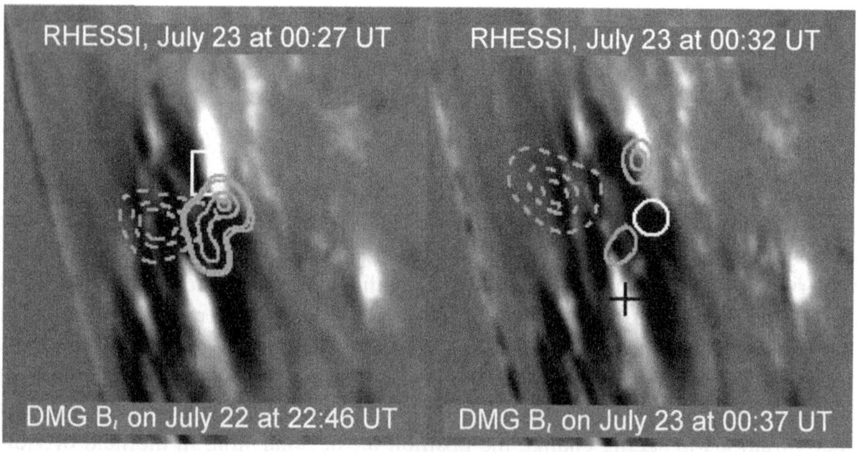

Fig. 9.84 RHESSI data (the same as in Fig. 9.80) plotted over two BBSO longitudinal magnetograms (white/black correspond to ±450 G) acquired before and after the flare. The box and the circle are the same as in Fig. 9.83. The cross marks the position of the 2.223 MeV source centroid. North is up, and west is to the right (From Yurchyshyn et al. 2004)

Table 9.17 Magnetic flux values before and after the 2002 July 23 flare as measured from MDI and BBSO (According to Yurchyshyn et al. 2004)

	F^+/F^- ($\times 10^{21}$ Mx)			
TIME	MDI	BBSO	F^- (CIRCLE) ($\times 10^{20}$ Mx)	F^- (BOX) ($\times 10^{19}$ Mx)
Before	+0.7/−3.4	+0.8/−4.2	−1.0	−1.9
After	+0.6/−3.6	+0.7/−4.4	−1.3	−5.9
Change (%)	−14/+6	−13/+5	+30	+210

Yurchyshyn et al. (2004) also note that while the radiation measured close to the solar limb in the λ6103 photospheric spectral line is emitted from higher and thus cooler layers of the solar atmosphere, this radiation is still mainly formed in the photosphere. Thus, a contribution function for the blue wing of the Ca I λ6103 spectral line, determined close to the solar limb (Abramenko and Baranovsky 2004), shows that this part of the spectral line forms within 400 km above the τ_{5000} level, i.e., below the transition region between the photosphere and the chromosphere. Therefore, the field measurements made in the blue wing of the Ca I λ6103 spectral line refer to the photosphere.

Yurchyshyn et al. (2004) calculated the total BBSO positive and negative flux over the same area of the active region, as was done in the case of MDI data, i.e., over the entire longitudinal magnetograms shown in Fig. 9.83. According to the BBSO data, the leading (negative) flux increased by about 5%, while the following (positive) flux decreased by about 13%. These relative changes are in very good agreement with the MDI flux variations, although the absolute values of the MDI and BBSO magnetic fluxes are somewhat different (compare the second and third columns of Table 9.17).

According to Yurchyshyn et al. (2004), this discrepancy in the absolute magnetic flux amount shown in the second and third columns of Table 9.17, could be due to differences in the spatial resolution and sensitivity of the instruments. The BBSO magnetograms in Fig. 9.83 illustrate the changes in the longitudinal magnetic field, which are detected by calculating the total flux over the active region. The most dramatic and obvious changes occurred in two regions, marked by the circle and the box in Figs. 9.83 and 9.84. The region marked by the circle is where the northern HXR footpoint source was located at the beginning of the flare (solid contours in Fig. 9.84 at 22:46 UT). The magnetograms show that both the line-of-sight component and the transfer field were affected by the flare at this area: (1) the peak intensity of the longitudinal magnetic field, inside the circle, increased from −800 G to −1,100 G; (2) the total negative flux increased by ∼30% from -1.0×10^{20} to -1.3×10^{20} Mx (see also the fourth column of Table 9.17); (3) the penumbral bridge, which connected the two major sunspots, became wider (compare the gray contours, inside the circles); (4) before the flare, the transverse field was largely oriented along the southeast-northwest line, while after the flare, the transverse field was already mainly oriented along the north-south line.

As underlined Yurchyshyn et al. (2004), all the above facts, especially the increase of the penumbral bridge, allow to speculate that the emergence of a new

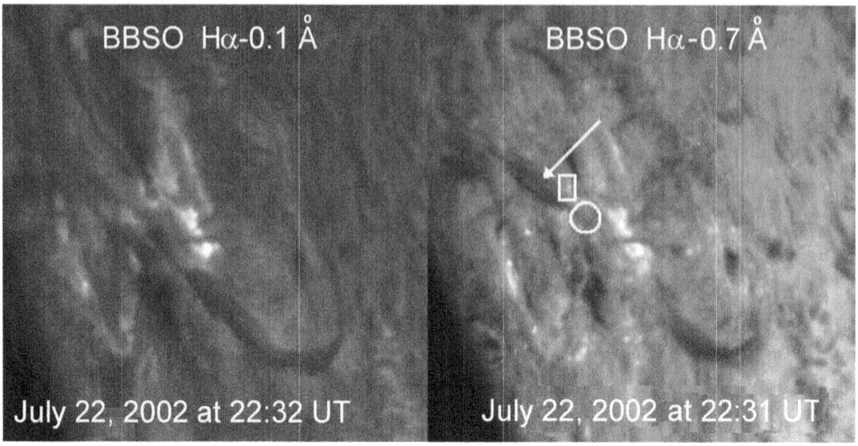

Fig. 9.85 BBSO Hα center (*left*) and Hα − 0.7 Å (*right*) images obtained about 2 h before the flare. One of the footpoints of a cold surge, indicated by the arrow, is located at the position of the earlier HXR footpoint source and is probably associated with flux emergence. The box and the circle are the same as in Fig. 9.83 (see also Figs. 9.80 and 9.84.) (From Yurchyshyn et al. 2004)

magnetic flux, inside the circle, was associated with the flare. This suggestion is supported by high-resolution BBSO Hα center and off-band (Hα − 0.7 Å) images taken 2 h before the flare peak (Fig. 9.85).

As noted Yurchyshyn et al. (2004), in the center of the off-band image (right in the Fig. 9.85), one can see an elongated cold surge, indicated by the arrow, which existed for several hours before the flare. The footpoint of this surge was located inside the circle (see also Figs. 9.83 and 9.84). Also, the negative polarity (leading flux) was gradually increasing during the observed period, which may be interpreted as a signature of an emerging flux. According to Heyvaerts et al. (1977), the surge may indicate ongoing flux emergence in the area. Since the encircled area and the position of the earlier HXR footpoint source coincide, it can be further assume that the new flux may have triggered the gamma-ray flare by alternating the topology of the magnetic field and/or adding twist to the system (Leka et al. 1996; Wang and Abramenko 2000).

Yurchyshyn et al. (2004) noted, that another area of great interest is marked by the box in Fig. 9.83. The comparison of the two magnetograms shows that the negative magnetic polarity inside the box had increased threefold (from -1.9×10^{19} to -5.9×10^{19} Mx; see also the fifth column of Table 9.17), largely because of the devouring of the positive polarity (the position of the neutral line had shifted). The orientations of the line segments inside the box indicate that before the flare the transverse field was largely oriented along the east-west line, while after the flare it was mainly oriented along the southwest-northeast line. Distinctive from the previous case (the circled area), the increase of the penumbral area inside (and nearby) the box was not significant and could not account for the threefold increase of the negative flux. The above variations qualitatively correspond to what would be expect if the magnetic field changes its inclination by veering from the observer toward the solar limb. When the veer of the magnetic field vector becomes large enough that the angle between the line of sight and the vector of the magnetic field exceeds 90°, then the longitudinal magnetic field can change its sign, and the neutral line will shift. This seems to be the case in the 2002 July 23 event. The box marks the position where the western footpoints of a rapidly evolving postflare loop system and the northern HXR footpoint source were located in the late and gradual phase of the flare (see Fig. 9.84 at 00:37 UT). It is known that a postflare loop system is a new topological link in active region, and it is expected that the magnetic field changes its orientation and/or inclination as the postflare loop system forms.

9.17.5 Rapid Changes in the Longitudinal Magnetic Field Associated with the Flare: Magnetic Flux Variations Inferred from the Simulated Data

In order to find out whether the reorientation of the magnetic field is indeed capable of producing noticeable changes in the total magnetic flux, Yurchyshyn et al. (2004) have conducted simple simulations of a linear force-free field. First, they constructed a B_z magnetogram with an S-shaped neutral line, which are used as a boundary condition (see Fig. 9.86, left). The reason for doing this was because there was no observational data for this event suitable for the modeling. Indeed, the

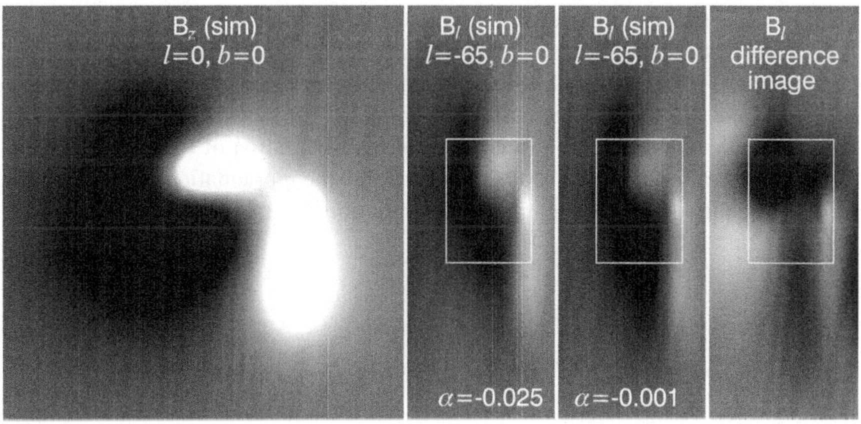

Fig. 9.86 Simulated magnetograms: vertical component, B_z, shown at the center of the solar disk (*left*), and longitudinal component of a linear force-free field calculated with parameter $\alpha = -0.025$ arcsec^{-1} and $\alpha = -0.001$ arcsec^{-1} (middle two panels), as seen at the eastern solar limb. White/black is a magnetic field intensity exceeding ±500 G. The right panel shows the difference image between the two longitudinal components (black represents areas where the longitudinal component decreased) (From Yurchyshyn et al. 2004)

longitudinal magnetogram, observed at $\lambda = -70°$, in fact represents a component almost tangential to the solar surface, and as such it cannot be used as a boundary condition in this model. Yurchyshyn et al. (2004) emphasized that this numerical exercise is not meant to reconstruct the coronal magnetic field above the observed active region. This was done only to make an estimate of whether a change in the inclination of a magnetic field is capable of producing significant variations in the total magnetic flux. In order to model a magnetic field above the simulated active region, Yurchyshyn et al. (2004) utilized a method to calculate a linear force-free field described in Abramenko and Yurchishin (1996). The method allows to model a field not only by varying the force-free field parameter α, which defines the amount of twist in an active region, but also by selecting nonphotospheric boundary conditions. The upper bound of α is limited by the number of nodal points in a three-dimensional grid that was used to calculate the force-free field: the finer the grid, the smaller the α that can be used in the model. In case of 2002 July 23 event Yurchyshyn et al. (2004) calculates a linear force-free field everywhere inside a $240'' \times 240'' \times 160''$ volume with a spatial resolution of $3'' \times 3'' \times 2''$. Yurchyshyn et al. (2004) simulated two sets of a linear force-free field: one with a large amount of twist ($\alpha = -0.025$ arcsec^{-1}) and the other one with nearly potential field configuration (($\alpha = -0.001$ arcsec^{-1}). The final step in these simulations was to project modeled vector magnetograms at the eastern solar limb ($\lambda = -65°$) and to calculate the corresponding line-of-sight components, which are shown in Fig. 9.86 (middle two panels).

The difference image between these two projected magnetograms (right panel in Fig. 9.86) reveals that there is a significant difference between them. The white box in Fig. 9.86 marks the area that are used to calculate the total positive and negative magnetic flux. The described calculations indicate that in the nearly potential configuration the leading (negative) simulated flux increased by 24%, while the following (positive) polarity decreased by about 59%, as compared to the twisted configuration. Since the photospheric boundary was the same in both simulation runs (no new flux emergence), the flux variations in the simulated line-of-sight magnetograms are purely due to changes in the orientation of the magnetic field. Therefore, Yurchyshyn et al. (2004) conclude that the reorientation of the magnetic field is capable of producing significant variations in the observed line-of-sight magnetic flux.

9.17.6 Summary of Main Results and Discussion

Yurchyshyn et al. (2004) summarize the observations and simulations described in the previous Sections as following.

1. According to GOES measurements, the X4.8/2B gamma-ray flare started at 00:18 UT on 2002 July 23.

2. A noticeable gradual increase in Hα emission began at 00:20 UT at the site of the flare, and no associated EUV crinkles or remote Hα emission were detected at this time.
3. About 7 min later the major energy release event started, which was determined by the impulsive rise of HXR emission at 00:27 UT (Fig. 9.81).
4. Only after the HXR emission peaked did several remote Hα brightenings appear in a quiet-Sun region at a location about 160″ north of the active region, and their intensity reached a maximum at about 00:31 UT, when the secondary peak in the HXR emission occurred.
5. A 2,218–2,228 keV gamma-ray source was observed almost simultaneously with the remote Hα brightenings. However, this gamma-ray emission was located inside the active region and was displaced from the main HXR sources.
6. The light curves of the gamma-ray source and the remote Hα brightenings were very similar during the period of their gradual increase, although the gamma-ray light curve lagged by approximately 60 s relative to the light curve of the remote brightenings.
7. The high time cadence MDI line-of-sight magnetograms showed that after the flare the following (positive) polarity of the active region decreased by about 14%, while the leading polarity did not show significant rapid flare-associated change: it was gradually increasing for the entire observed period.
8. Furthermore, BBSO high-resolution vector magnetograms showed a significant increase of the penumbra area located between the Hα flare ribbons, as well as changes in the direction of the transverse magnetic field.
9. The multiwavelength data for the flare show that in this event, significant flare-related emission occurred in the core of the flare, prior to the occurrence of the remote Hα brightenings.
10. The timing of this particular event does not agree with what it would expect from the breakout model, which predicts that EUV crinkles and/or remote Hα brightenings should occur prior to the major energy release event.
11. On the other hand, the models that advocate an eruption of a preexisting flux tube (Van Tend and Kuperus 1978; Van Ballegooijen 1989; Forbes and Isenberg 1991; Kumar and Rust 1996; Wu et al. 1999; Amari et al. 2000) or those that require reconnection to form an unstable flux tube (Moore and LaBonte 1980) are more successful in interpreting this event because they predict that the earliest energy release event should occur in the core field at the site of a flare.

Yurchyshyn et al. (2004) discuss, in the framework of the flux tube and tether-cutting models, one possible sequence of events during the 2002 July 23 flare. They suggest that new emerging flux disrupted the stability of the preexisting magnetic configuration (it may also have brought additional twist to the system) and thus triggered a large-scale eruption of the magnetic field. The erupted field stretched the field lines, which later reconnected beneath the escaping fields and formed a system of postflare loops. The footpoints of these new loops are normally seen in the chromosphere as Hα ribbons, early parts of which spatially coincide with HXR sources. As the eruption proceeds, the reconnection X-point moves upward, causing the HXR sources and Hα ribbons to move apart. Analysis of the motion of the HXR source during the flare showed that the coronal (Figs. 9.84 and 9.80, dashed contours) and the northern footpoint source (solid contours) were moving with comparable speeds and in similar directions, while the southern footpoint source did not display similar motions.

Yurchyshyn et al. (2004) believe that the different motion patterns of the HXR sources is a manifestation of the upward motion of the reconnection point and was caused by different intensities of the magnetic fields at the footpoints of the PFL system. Generally speaking, during a flare equal amounts of positive and negative magnetic flux should reconnect. In the case of high-density flux (e.g., in a sunspot umbra), the footpoints of the reconnecting field lines occupy a smaller area than the low-density footpoints (say, in sunspot penumbra). Since HXR emission is a result of the precipitation of high-energy electrons at the footpoints of the reconnecting magnetic field lines, a source associated with the low-density flux will display a wider variety of displacements, while the high-density source would remain nearly stationary. Indeed, Fig. 9.84 seems to support this explanation by showing two BBSO longitudinal magnetograms overlapped by the same RHESSI contours as shown in Fig. 9.80. The stationary southern HXR footpoint was located in a strong field ($\sim$1,000 G) at the umbra-penumbra border of the negative polarity sunspot, while the moving northern footpoint was associated with the much smaller positive polarity sunspot that moved along the outer edge of the sunspot penumbra, where the magnetic flux density was noticeably lower ($\sim$600 G).

Yurchyshyn et al. (2004) further suggest that the remote Hα brightenings and the 2,218–2,228 keV gamma-ray source were quite possibly caused by an interaction between the erupted field and a general largescale magnetic field spanning the active region. After the erupted field pushed high into the corona (according to considered data it was about 3 min after the eruption began), it may have reconnected with an overlying field. Strictly speaking, one would expect to observe Hα and X-ray signatures of this secondary reconnection both remotely and in the core region. If so, the peaks in the HXR emission detected at 00:31 UT in the core of the flare and the remote Hα brightening may be signatures of this second reconnection. Note that there was no significant HXR emission detected at the remote Hα source. The X-ray counterpart of the remote Hα source was, if it existed at all, at least 25 times weaker than the main HXR sources. The increase in the Hα emission at the core site, however, could not be detected mainly because of the high Hα intensity of the core. The similarity between the light curve of the remote Hα brightenings and that of the 2,218–2,228 keV gamma-ray source suggests that these two sources are related and that a common process accelerated both electrons and ions. However (and against all expectations), the gamma-ray source was offset from the HXR source, and only weak Hα activity was seen in its vicinity.

Hurford et al. (2003b) proposed that the offset may be a result of the acceleration that may occur far from the site of the flare. The accelerated ions traveled over a long distance before encountering the chromosphere. This explanation agrees with

Yurchyshyn et al. (2004) suggestion of a second reconnection event high in the corona. Although the remote Hα brightenings and the gamma-ray source light curves are similar, the gamma-ray source lags the HXR emission and the remote Hα sources by about 60 s. Hurford et al. (2003b) pointed out that the 2,223 keV emission from deuterium is delayed by $\sim$100 s because of preceding thermalization and capturing of fast neutrons in the photosphere. The estimated and observed time delays are close, which can be interpreted as further evidence that the remote Hα source and the gamma-ray source were produced by a common acceleration process.

Finally, Yurchyshyn et al. (2004) discussed the question as to how and when the erupted flux tube was formed still remains open. Was an erupted flux tube formed immediately prior to the eruption, as suggested by the tether-cutting model, or was it embedded in the magnetic field long before the flare? The available data are not sufficient to make a solid conclusion. However, the observed changes in the photospheric magnetic field detected at the footpoints of the PFL system can be interpreted as being in favor of the tether-cutting model. Indeed, in the case of an eruption of a preexisting flux tube, a nearly potential overlying field stretches and recloses underneath the erupted field, thus reforming to another near-potential configuration. Yurchyshyn et al. (2004) therefore do not expect to see significant variations in the orientation of the field lines. However, when two (or more) sheared and independent magnetic fluxes reconnect during a flare, the sheared core field changes, so a more potential topology is reinstated, and this process should give rise to changes in the inclination and/or the orientation of the magnetic field lines, which is observed in this particular case (a two-dimensional representation of this magnetic topology is depicted in Figure 3 of Sweet 1958).

Yurchyshyn et al. (2004) concluded, the tether-cutting model can successfully explain the observed occurrence of all events discussed above. They also suggest that a two-step reconnection process is required to explain the occurrence and the location of the offset gamma-ray source observed in the 2002 July 23 flare. The first reconnection occurs deep in the core field, and it is a major energy release event. The second reconnection occurs between the erupted field and a general overlying coronal field, which produces the secondary peak in HXR emission and a displaced gamma-ray source.

9.18 The 2003 October 28 X17 Flare: Gamma Rays and the Evolving Compact Structures

9.18.1 The Matter and Short History of the Problem

According to Schrijver et al. (2006), the X17 flare on 2003 October 28 was observed by high-resolution imaging or spectroscopic instruments on CORONAS,

GOES, INTEGRAL, RHESSI, SOHO, and TRACE. These spacecrafts observed the temporal evolution of the γ-ray positron-annihilation and nuclear deexcitation line spectra, imaged the hard X-ray bremsstrahlung and EUV and UV emission, and measured the surface magnetic field and sub-photospheric pressure perturbations. In the usual pattern, the onset of the flare is dominated by particle acceleration and interaction, and by the filling of coronal magnetic structures with hot plasma.

As noted Schrijver et al. (2006), the total energy in the ensemble of nonthermal electrons and ions may amount to a large fraction of the total flare energy (e.g., Lin and Hudson 1976; Ramaty et al. 1995; Emslie et al. 2005). These particles eventually thermalize through interaction with plasma from corona to chromosphere and below, unless they escape into the Heliosphere. The solar atmosphere into which the nonthermal particles precipitate is not simple; flares generally occur in active regions with strong, highly structured magnetic fields, and the flare process itself perturbs the atmosphere, sometimes down to photospheric levels. Moreover, as the study of Schrijver et al. (2006) confirms, the dominant particle acceleration and particle precipitation sites move over distances that may exceed their typical width within a minute. Such motions are most commonly seen as moving flare ribbons in, for example, Hα or in EUV image sequences by the TRACE spacecraft (e.g., Fletcher et al. (2004), who use this to measure reconnection rates by comparing the footpoint motions with magnetograms).

Schrijver et al. (2006) underlined that the flare energy sources move, creating a sequence of disjoint atmospheres contained by flux tubes that evolve through excitation and subsequent relaxation with different time delays. Consequently, a low-resolution instrument may simultaneously see contributions from multiple atmospheres that are evolving in this manner (e.g., Warren and Doschek 2005), while high-resolution imaging instruments may see different signatures evolve with time histories that are more characteristic of the physical processes that are involved.

Schrijver et al. (2006) compare and analyze unique observations of the X17 flare on 2003 October 28 that were made with six different spacecrafts with imaging and/or spectroscopic instruments. These instruments cover the wavelength domain from γ-rays to the visible and provide the most comprehensive high-resolution data set with which to study the evolution of these evolving atmospheres of a major flare. They use this excellent set of observations to develop a scenario for the flare evolution that includes the impact of high-energy particles on the near-photospheric atmosphere under the flare. In particular, the high-resolution imaging by the TRACE spacecraft and the high-resolution γ-ray spectroscopy by the RHESSI spacecraft lead us to conclude that energy release occurs in compact domains that propagate through the coronal volume, causing a series of plasma atmospheres to evolve one after the other as the flare progresses. Even within this comprehensive array of data, however, we find no ready explanation for the evolution of the positron-annihilation line width observed by RHESSI (Share et al. 2004).

9.18.2 Summarizing Key Evolutionary Stages of the 2003 October 28 Flare

According to Schrijver et al. (2006), the initial phase of the 2003 October 28 X17 flare was imaged by RHESSI, SOHO, and TRACE, and observed by GOES, INTEGRAL, and CORONAS. The summarized key evolutionary stages are shown in Figs. 9.87 and 9.88 (images) as well as in Table 9.18 (movies of the SOHO MDI magnetogram data and of the TRACE 1,600 and 195 Å channel images are available in the online version of paper Schrijver et al. 2006).

9.18.3 GOES, INTEGRAL, and CORONAS Observations

According to Schrijver et al. (2006), the GOES-12 spacecraft observed the X17 flare on 2003 October 28 in NOAA active region 10486, starting at approximately 9:51 UT, when the signal climbs from mid-B to M1 levels by 10:59 UT, after which it increases rapidly to peak at approximately 11:10 UT in the 1–8 Å pass band. The GOES light curve, saturated around the flare peak from $\sim$11:06 UT to $\sim$11:16 UT, shows that the flare increased the Sun's soft X-ray radiance by a factor of over $\sim$200. Although the NOAA listing formally ends the flare at 11:24 UT, the long, gradual decay continued for over 12 h before the pre-flare emission level was approached. The INTEGRAL SPI light curves show that the most energetic part of the flare starts around 11:02:30 UT with a minute-long spike, followed by somewhat irregular decay of the intensities with time (Gros et al. 2004).CORONAS data show that the first peak is dominated by bremsstrahlung up to 30 MeV, and that high-energy γ-rays from pion production begin to be observed around 11:04 UT (Kuznetsov et al. 2006). The INTEGRAL anti-coincidence shield (ACS) rates above 100 keV are shown in Panel a of Fig. 9.87, scaled to the RHESSI 200 keV bremsstrahlung intensity.

9.18.4 RHESSI Observations

As noted Schrijver et al. (2006), RHESSI was traversing the high-radiation zone of the south-Atlantic magnetic anomaly for most of the rapid rise phase of the flare, including its peak, until 11:06 UT. Because of the flare's magnitude, the RHESSI front detectors saturated below 20 keV even with the attenuators in place early in its flare observations, so that there are little reliable data on X-rays for those energies. The RHESSI observations of solar γ-radiation enable analysis of the line and continuum shapes of the γ-ray emission. Most of the γ-ray spectral lines reflect nuclear deexcitation transitions that promptly follow the inelastic scattering of primary ions on target nuclei, but the strong 2.2 MeV line resulting from deuterium

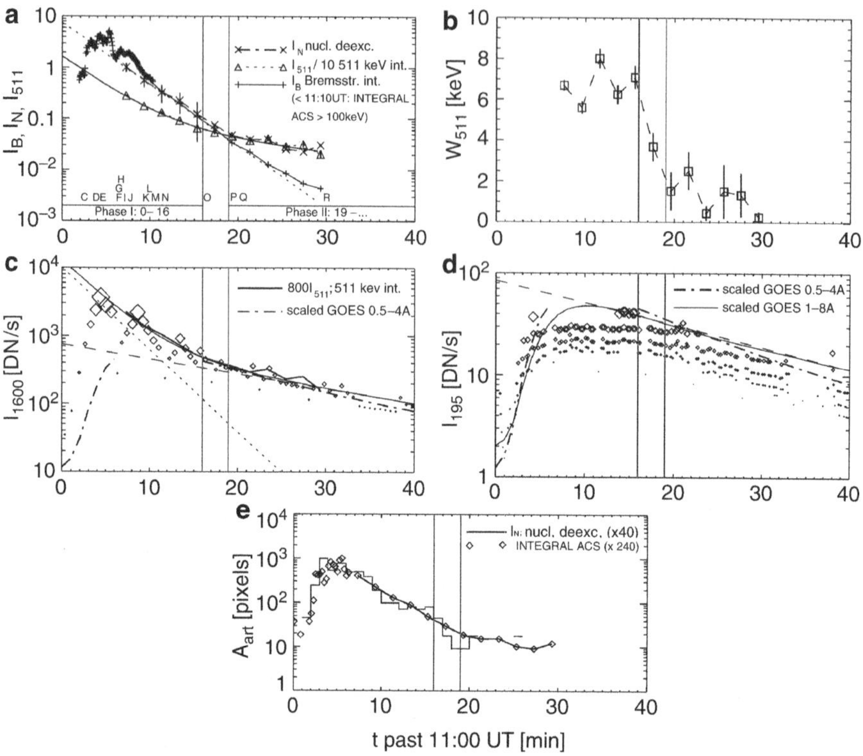

Fig. 9.87 Flare light curves. (**a**) Bremsstrahlung flux at 200 keV (pluses), nuclear deexcitation line flux (crosses; scaled by 0.1), and 0.511MeV flux (triangles; scaled by 0.1; data from Share et al. 2004). The densely clustered plus symbols up to 11:10:30 UT show the INTEGRAL ACS data scaled (by a factor 0.012) to the RHESSI 200 keV fluxes from 11:06 UT onward. Letters along the horizontal axis identify key events and features as listed in Table 9.18. The smooth solid curve is given by $75\times\exp(-t/3.6)$ with t in minutes. (**b**) The 0.511 MeV line width (from Share et al. 2004). (**c**) Average intensity for the TRACE 1,600 Å channel for the field of view as in Fig. 9.88, after subtraction of an average pre-flare level of 140 DN s^{-1} and a standard readout pedestal of 87 DN; symbol size increases with decreasing exposure duration. Thick curve: Scaled 0.511 MeV line flux from (b) for comparison. The dash-dotted line shows the GOES 0.5–4 Å signal (saturated around the flare peak, scaled by 10^6). The dotted line is $10^4\times\exp(-t/3.6)$, the dashed line is $625\times\exp(-t/20)$, and the thin solid curve is their sum. (**d**) TRACE 195 Å average intensity for the entire FOV, after subtraction of a pre-flare level of 2 DN s^{-1} and a standard readout pedestal of 87 DN; symbol size increases with decreasing exposure duration. The dash-dotted line is the GOES 0.5–4 Å signal multiplied by 10^5, while the solid curve shows the GOES 1–8 Å signal multiplied by 2.5×10^4. The long-dashed line is $85\times\exp(-t/20)$. (**e**) Histogram shows the total area of MDI magnetogram artifacts (in pixels). Also shown are the summed INTEGRAL γ-ray fluxes in the 4.4 MeV (C) and 6.1 MeV (O) lines (diamonds) until 11:06 UT, multiplied by 240, and the RHESSI nuclear deexcitation fluxes shown after 11:06 UT, multiplied by 40 (From Schrijver et al. 2006)

formation by neutron capture and the 0.511 MeV positron annihilation line (and continuum below the line) have different physical origins and provide different diagnostic information (e.g., Ramaty et al. 1975). Both of these lines require the

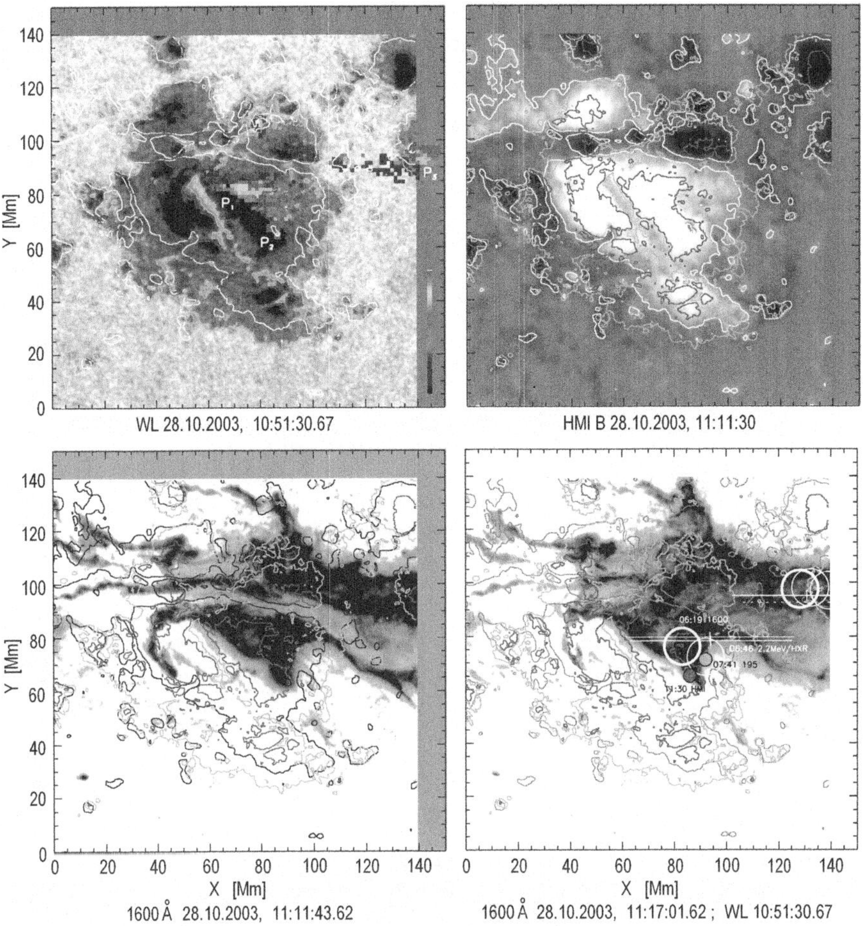

Fig. 9.88 *Top left*: TRACE 'white light' image of the main spot cluster in AR 10486, observed on 2003 October 28 at 10:51:31 UT. The green and red contours outline the penumbral and umbral boundaries, respectively. The solid white and dashed white contours outline the areas of strong negative and positive polarities, respectively (compare top right). Overlaid as colored pixels are the magnetogram artifacts; the color coding (shown on the side bar) has time increasing uniformly from 11:01 to 11:18 UT. *Top right*: MDI magnetogram taken at 11:11 UT. *Bottom left*: Negative, logarithmically scaled 1,600 Å TRACE image taken at 11:11:44 UT. *Bottom right*: Negative, logarithmically scaled 1,600 Å TRACE image taken at 11:17:01 UT. Also shown (and labeled by the time, showing mm:ss after 11 UT): thin bars, the locations of the very bright 1,600 Å features causing CCD blooming at 11:06 UT, with one somewhat ambiguous feature shown dotted; thick white circles, brightest sources for the 2.2 MeV line at 11:06 UT; thin white circles, brightest sources for the 100–200 keV emission at 11:06 UT; green dot, the brightest point in the 195 Å image at 11:07:41 UT; red dot, artifact in the MDI magnetogram at 11:11 UT. The contours in the top left panel are repeated in each panel (From Schrijver et al. 2006)

Table 9.18 Key events in the 2003 October 28X17 flare (From Schrijver et al. 2006)

Time (hh: mm:ss)	*SOHO* MDI	*TRACE*	*RHESSI, INTEGRAL, CORONAS, GOES*
A, 09:40:51		First signs of flare	No RHESSI data until 11:06
B, 09:51			X-ray flare start *(GOES)*
C, 11:02:30	First magnetogram artifact at P_1		First MeV continuum in *INTEGRAL* and *CORONAS*
D, 11:04 ± 4	Peak acoustic power at P_1		Peak in 26–100 MeV in *CORONAS*
E, 11:04:30	End magnetogram artifacts at P_1		Peak in 2.22 MeV in *INTEGRAL*; start of weakening of all γ- rays in *CORONAS* and *INTEGRAL*
F, 11:06:21		Extremely bright 1,600 Å features (>4,000 times quiescent plage).	
Share et al. (2004) 0.511 MeV line phase *a* (11:06:20–11:08:20): log $T \sim 5.6$, log $n \gtrsim 11$			
G, 11:06:30	Magnetogram polarity artifact at P_3		
H, 11:06:46			*RHESSI* out of eclipse: 2.2 MeV line brightest in thick white circles, 100–200 keV bremsstrahlung in thin circles in Fig. 9.88d.
I, 11:07 ± 4	Peak acoustic power at P_3		
J, 11:07:41		Very bright site (with diffraction cross) in 195 Å (green dot in Fig. 9.88, *bottom right*; >7,500 times quiescent plage).	
Share et al. (2004) 0.511 MeV line phase *b* (11:08:20–11:10:20): log $T \sim 5.3$, log $n \gtrsim 13.5$			
K, 11:09:30		1,600 Å impact site reaches umbra	
L, 11:10			1–8 Å X-ray flare peak *(GOES)*
Share et al. (2004) 0.511 MeV line phase *c* (11:10:20–11:16:20): log $T \sim 5.6$, log $n \gtrsim 14.5$			
M, 11:10:30	Magnetogram polarity artifact at P_2		
N, 11:11:30	Magnetogram artifacts at red dot in Fig. 9.88d.		
O, 11:16:30	Magnetogram polarity artifact disappears at P_3; magnetogram artifacts continue to fade		0.511 MeV line narrowing from ~7 to 1–2 keV

(*continued*)

Table 9.18 (continued)

Time (hh: mm:ss)	*SOHO* MDI	*TRACE*	*RHESSI, INTEGRAL, CORONAS, GOES*
Share et al. (2004) 0.511 MeV line phase *d* (11:16:20–11:18:20): log $T \sim 5.0$, log $n \gtrsim 15$; Share et al. (2004) 0.511 MeV line phase *e* (11:18:20–11:30:20): log $T \sim 3.8$, log $n \gtrsim 15$			
P, 11:19:30	Magnetogram artifact at P_3 disappears		0.511 MeV line at 1–2 keV
Q, 11:20:30	Last signature of magnetogram artifacts at P_2		
R, 11:30			End analysis by Share et al.

slowing down of secondary nuclear products (neutrons and positrons, respectively) in the solar atmosphere, perhaps as deep as the sub-photospheric layers.

Schrijver et al. (2006) underlined that Share et al. (2004) study the evolution of the RHESSI γ-ray spectrum after 11:06 UT. They analyze the 0.511 MeV positron annihilation line strength and width and compare that to 200 keV bremsstrahlung, the total nuclear deexcitation line flux in the RHESSI spectra, and the annihilation continuum flux (Panels a and b in Fig. 9.87). Their estimated plasma temperatures and lower limits on the atomic densities are listed in Table 9.18, using labels a–d, as in their paper. The temperatures are derived assuming that the annihilation line width is primarily due to thermal broadening. The plasma densities are lower limits derived under the assumption that the continuum at energies below the 0.511 MeV annihilation line is associated with positronium continuum. Much higher densities would be required if the continuum came from Compton scattering of the 0.511 MeV line photons in the much denser atmosphere, if the positrons were to originate from decay of pions, as may be inferred from CORONAS observations early in the flare. Share et al. (2004) conclude that in the initial 2 min after 11:06 UT, the positrons annihilated in an environment with a temperature of approximately 300,000°K but with a chromospheric density exceeding at least 10^{11} cm^{-3}. Until 11:16 UT, the line width measurements continue to indicate temperatures above 10^{5}°K and increasingly higher lower limits on the densities. After 11:18 UT, the annihilation spectra are consistent with photospheric temperatures and densities (>10^{15} cm^{-3}).

Schrijver et al. (2006) note that even as the RHESSI 200 keV electron bremsstrahlung shows a persistent exponential decrease with a 3.6 min e-folding timescale (Panel a in Fig. 9.87, dotted line), the total nuclear deexcitation signal transitions around 11:18 UT from the same decay profile to a decay with an e-folding time scale of ~20 min. At that time, the 0.511 MeV line narrows markedly (Panel b in Fig. 9.87), after which the total nuclear deexcitation signal decays proportionally to the 0.511 MeV line intensity (Panel a in Fig. 9.87). It was note also that the 0.511 MeV intensity is well described by a sum of two exponentials, with time scales of 3.6 and 20 min (Panel a in Fig. 9.87, solid curve; repeated in Panel c for comparison with the 1,600 Å signal discussed below in Section 9.18.5).

As noted Schrijver et al. (2006), RHESSI provides γ-ray images of solar flares for the first time (Hurford et al. 2003b, 2006). Interestingly, in two of the four events analyzed, the ion precipitation regions, as inferred from the 2.2 MeV neutron-capture emission, appear to differ significantly from the electron precipitation regions, as inferred from the bremsstrahlung continuum; in the other cases the differences are not significant. In the 2003 October 28 X17 flare, the 2.2 MeV regions are clearly resolved into a double-footpoint structure with centroids about 15″ east of the corresponding hard X-ray bremsstrahlung centroids (see the circles in the bottom right panel of Fig. 9.88 for approximate positions). Hurford et al. (2006) point out that this difference persists even when a ~100 s delay is allowed in order to approximate the time needed for the secondary neutrons to thermalize and be captured.

9.18.5 TRACE Observations

According to Schrijver et al. (2006), the Transition Region and Coronal Explorer, TRACE (see details in Handy et al. 1999), observed the X17 flare in interlaced sets of the 1,600 Å channel (dominated by the strongest UV emission lines, such as the C iv doublet, and – particularly during flares – the UV continuum) and the 195 Å channel (Fe xii in quiescence, with an often dominant contribution of Fe xxiv or thermal continuum during flares). An occasional broadband visible ('white light') image was taken for precise co-alignment with SOHO's MDI. During the interval from 11:00 through 11:20 UT, TRACE observed the Sun primarily in the 195 Å channel at a cadence of a few seconds (depending on its exposure time and the details of other interspersed images). The automatic exposure control for the 195 Å channel could not follow the rapid evolution of the brightness of the flare, despite the fact that the exposure durations rapidly shortened by a factor of 100 going into the flare peak. The 1,600 Å passband observations were taken with a cadence close to 1 min. Most of the 1,600 Å exposures were very short (~0.1 s) and showed only the brightest features on a noisy background. Longer exposures (0.8 s) in the 1,600 Å channel, which show both flare and non-flare features, were taken at intervals of approximately 6 min (as it was noted above, in Section 9.18.2, a movie of these longer exposures is shown in the online version of paper Schrijver et al. 2006).

Schrijver et al. (2006) noted that the total TRACE 1,600 Å channel signal in the flare (corrected for the preflare level) is shown in Panel c of Fig. 9.87. Its overall behavior after peaking at ~11:04 UT is similar to that of the total nuclear deexcitation signal, first decaying on a 3.6 min timescale (dotted line), then transitioning to a 20 min e-folding time-scale (dashed line) that persists for at least an hour past the flare peak. In this later phase, the 1,600 Å channel scales with time as the GOES 0.5–4 Å X-ray signal. The total 1,600 Å signal is well approximated by double exponential ($1.2 \times 10^4 \exp(-t/3.6) + 750 \times \exp(-t/20)$ DN s^{-1}), shown by the thin solid curve in Panel c of Fig. 9.87; this double exponential, with a different scaling, was shown to approximate the RHESSI 0.511 MeV intensity in Panel a of Fig. 9.87. The best approximation of the 1,600 Å channel is given by the set of shortest exposures (large symbols), because these suffer least from detector saturation, which causes artificially low average intensities (as seen from the smaller, lower lying symbols used for the longer exposures).

As noted Schrijver et al. (2006), the TRACE 195 Å flare signal suffers from strong saturation prior to approximately 11:14 UT, and some saturation persists even after this. The signal from 11:14 UT onward evolves similarly to the GOES 0.5–4 Å X-ray signal (Panel d of Fig. 9.87), decreasing roughly as $85 \times \exp(-t/20)$ DN s^{-1} (dashed line through the shortest exposures). The cycling through different exposure durations, with different fractions of the CCD at saturated levels, causes the sets of exposures to show up as nearly parallel sequences, with the shortest, least saturated exposures at the top.

9.18.6 Brightness and Size of the Dominant Particle Precipitation Sites

Schrijver et al. (2006) noted, that the TRACE observations of very bright kernels within the UV and EUV flare ribbons suggest that the energetic particle precipitation is particularly strong in a few compact sites, consistent with the RHESSI bremsstrahlung observations (the strongest bremsstrahlung coincides with sites of pronounced diffraction crosses in the 195 Å signal; see Hurford et al. 2006). For example, the 1,600 Å TRACE image at 11:06:21 UT (with an exposure duration of 0.9 s) shows flare ribbons that are saturated at the maximum of the 12 bit analog-to-digital conversion at a data number, or DN, of 4096 over much of their area. Within these areas, some compact regions stand out particularly, because they are so bright that they cause CCD blooming, reflected in extended horizontal stripes in the image. Two such sources are seen in the lower ribbon, and two in the upper ribbon, in only a single exposure (see, Fig. 9.88, bottom right). The charges accumulated onto the CCD in these bright points in the 1,600 Å image have spread out over the length of the blooming strips of 140–180 pixels. It was used the blooming-strip length as a measure for the minimum brightness, which yields at least $4{,}096 \times 180 = 0.7 \times 10^6$ DN s^{-1}. Images taken earlier at 11:04:25 UT and 11:05:01 UT, with short exposure times of 0.0032 and 0.0048 s, respectively, are not saturated and have peak brightnesses of $(0.6\text{–}1.0) \times 10^6$ DN s^{-1}, comparable to the estimate based on the blooming artifacts. The minimum contrast with the surrounding non-blooming, saturated ribbons is at least a factor of $\sim$150, and with the surrounding quiescent plage at least a factor of 4,000. The blooming features allow to determine the central location of the bright points within the flare ribbons rather accurately (see Fig. 9.88, bottom right).

As underlined Schrijver et al. (2006), another very bright kernel in the flare ribbons, seen in the 195 Å image at 11:07:41 UT (see green dot in bottom right panel of Fig. 9.88), is saturated so that it can place only a direct lower limit on its brightness of ~4,000 DN s^{-1}. One feature of the TRACE 195 Å channel is that the front-filter support grid causes a diffraction pattern. The brightest features are thereby seen repeated in the cross formed by the diffraction orders. This diffraction pattern can be used to evaluate the intensity (and size; see Section 9.18.7) of the saturated kernel. Using the intensities in the first few orders of the diffraction pattern (Lin et al. 2001), it can be estimate that the peak intensity level in the kernel reached approximately 20,000 DN s^{-1}, or 7,500 times brighter than the typical quiescent plage, and more than an order of magnitude brighter than the surrounding flare ribbons. If the Fe xii line at 195 Å dominates that emission, a density contrast of at least a factor of 90 is required compared to the quiescent plage, but it be realize that Fe xxiv or high-temperature (or non-thermal) continuum emissions are likely to contribute substantially; these may require even higher densities.

9.18.7 Apparent Size and Motion of the Most Prominent Precipitation Sites

According to Schrijver et al. (2006), the 195 Å diffraction patterns show the full width at half maximum (FWHM) of the brightest features to be ~4.2 TRACE pixels of 0.5″, or 375 km, in size. Taking into account the TRACE resolution of ~750 km, Schrijver et al. (2006) conclude that they extend over no more than 1,400 km. The brightest features in, for example, the TRACE 1,600 Å channel are saturated and surrounded by other bright features, so it cannot be use that information to estimate the width of the most prominent precipitation sites as it was done for the 195 Å channel. The data can be used, however, to track the displacement history of these features from 11:06:21 UT, when they cause CCD blooming, to 11:11 UT, when they reach the compact polarity artifact in the MDI magnetogram at P_2 (Fig. 9.88, top left), which is discussed in Section 9.18.8. In this time interval, the precipitation footprint has moved over approximately 16,000 km. With the corresponding average migration speed of ~60 km s^{-1}, it may be conclude that a high-energy particle 'beam' with a width of <1,400 km travels over its own width in $\tau_w < 23$ s at least up to approximately 11:12 UT. The saturation of the 1,600 Å images makes it harder to estimate a propagation velocity for the conjugate site moving into the eastern, negative spot at P_3, but the magnetogram artifacts, discussed in Section 9.18.8, stand out quite clearly. They advance rapidly westward until approximately 11:14 UT, with an estimated propagation speed of ~30 km s^{-1}. According to the standard reconnection scenarios, the apparent motion of the footpoints reflects the rate of reconnection in the corona. Taking nominal values of $B > 100$ G and $n_e = 10^9$ cm^{-3}, the observed footpoint motion would correspond to coronal reconnection progressing at $<0.05v_A$ (where v_A is the Alfvén velocity).

9.18.8 SOHO MDI Magnetic Signal

As underlined Schrijver et al. (2006), SOHO MDI observed the event in a full-disk, high-cadence mode, measuring both the velocity signal (discussed in Section 9.18.9) and the magnetogram signal at 1 min cadence. The magnetograms show the gradual evolution of the complex magnetic field, which shows a ridge of opposite polarity features included within a penumbra in common with the main, trailing, positive-polarity spot (see Fig. 9.88, top panels). Starting at 11:02:30 UT, however, rapidly evolving image artifacts show up in the strong field environments, under the flare ribbons seen with TRACE. The strongest such artifacts appear as pronounced polarity inversions (marked P_1, P_2, and P_3). This occurs first in the positive-polarity spot near P_1 (Fig. 9.88, top left) around 11:04:30 UT, and rapidly fades away. The artifacts strengthen again around 11:11:30 UT, now occurring in a compact region in the western side of the positive-polarity umbra, near P_2. There, the apparent polarity returns to the proper positive value around 11:15:30 UT. This magnetogram artifact continues to fade and is no longer visible from about 11:21:30 UT. Similar moving, evolving magnetogram artifacts show up in the western, negative umbra around P_3 after 11:06:30 UT, and are seen until about 11:19:30 UT. Note that whereas the fronts of these artifacts advance quite quickly until about 11:11 UT, most artifact locations persist over $\sim$5 min or more before fading. Apparently the perturbation crossing time scales estimated in Section 9.18.7 for the most pronounced precipitation kernels are short, compared to the duration of particle precipitation onto many of the rest of the flare ribbons.

Schrijver et al. (2006) noted that these artifacts are likely a consequence of a strong distortion of the Ni I 8,768 Å magnetic diagnostic line. As discussed by Qiu and Gary (2003), magnetogram artifacts suggest that the line turns into an emission line, or that it at least shows a strong central reversal in the regions most strongly affected by the particle populations that cause the flare ribbons. Ding et al. (2002) show that an emission profile is readily formed as an electron beam impacts on the supra-photospheric layers up to the chromosphere, particularly if that happens in a relatively cool atmosphere, such as that of a spot umbra. In their scenario it is the nonthermal collisional excitation and ionization of the hydrogen atoms that shifts the line and nearby continuum formation heights up-ward, and that can cause the line to become an emission line. It may also be possible to form an emission line by forming a temperature-minimum zone, likely resulting in the formation of an inversion layer, i.e., a dense chromosphere for neutral nickel, with temperatures high enough to populate the first excited level, but well below the characteristic temperature of order 90,000 K associated with its ionization energy of 7.63 eV. Ding et al. (2002) show, however, that standard atmospheric stratifications for active and flare atmospheres like the VAL3C (Vernazza et al. 1981) and F1 (Machado et al. 1980) models result in simple absorption line profiles. As one measure for the footprint of the particle precipitation sites, Schrijver et al. (2006) show the evolution of the total area of the magnetogram artifacts (traced by time-tagged pixels in Fig. 9.88, top left) in Panel e of Fig. 9.87. For this purpose,

Schrijver et al. (2006) identified pixels as magnetogram artifacts if their signal changes by 50 Mx.cm^{-2} or more from one magnetogram to the next. They note that the total area of the magnetogram artifacts scales as the intensities observed for the INTEGRAL SPI and RHESSI nuclear deexcitation lines (also shown in Panel e of Fig. 9.87).

9.18.9 SOHO MDI Acoustic Signal

As noted Schrijver et al. (2006), the SOHO MDI velocity data are analyzed by Donea and Lindsey (2004), who find that the X17 flare on 2003 October 28 excites sunquakes, coupling into the solar interior primarily in two places. They find a source with a peak intensity around 11:04 UT near the location labeled P_1 in Fig. 9.88 (top left), with a FWHM uncertainty in that peak time of 8 min. The second source peaks around 11:07 UT and is located near P_3 (in Fig. 9.88, top left) in the umbra of the negative, western spot. Both of these acoustic sources appear to be associated with strong energy deposition sites. RHESSI detected a hard X-ray source at P_3 after 11:06 UT (Hurford et al. 2006), and TRACE 1,600 and 195 Å images of the flare ribbons provide indirect evidence of energy deposition at both sites.

9.18.10 Ion Collisions, Positron Annihilation, and the 0.511 MeV Line

As noted Schrijver et al. (2006), interaction column depths exceeding 0.02 and 0.3 g.cm^{-2} are typically required to produce 80% of the β^+-unstable nuclei and positive pions, respectively, in flares (Murphy et al. 2007). In the classical interpretation of direct impact, such column depths are reached deep in the chromosphere or in the photosphere (below at most $h = 600$ km in VAL models P, for characteristic plage, and F, for a flaring atmosphere; see Vernazza et al. 1981). The positrons emitted in these decays typically have energies from ~0.5MeV up to ~30 MeV, with corresponding characteristic ranges prior to annihilation from ~0.08 to 6 g.cm^{-2}, respectively. If assume that the positrons emitted by decaying pions do not escape from the photospheric depths where they are produced, most of the associated 0.511 MeV annihilation-line photons will be Compton scattered out of the 0.511 MeV line to lower energies. Almost all of 0.511 MeV line photons that are detected by RHESSI from the flare will therefore have originated from radioactive decay of β^+-unstable nuclei. Calculations of Schrijver et al. (2006) indicate that only ~50% of the positron emission will have occurred within 5 min of the interaction. This is important, given the rapid evolution of the flare and the migration of the particle acceleration-sites as reflected in the ribbon motions. What are the possible causes for the large initial width of the 0.511 MeV annihilation line? While detailed transport calculations have yet to be performed to determine what fraction of the positrons can escape from their deep production region to higher altitudes in the solar atmosphere, we can rule out the possibility that they annihilate in loops that are maintained at temperatures of a few hundred thousand kelvins along their length; the TRACE UV data show well-separated conjugate precipitation regions forming the flare ribbons. Therefore, Schrijver et al. (2006) rule out the possibility of thermal broadening of the 0.511 MeV line from annihilation in such loops. Bulk turbulent motion is, in principle, an alternative cause for the large width of the 0.511 MeV annihilation line observed from 11:06–11:16 UT. Schrijver et al. (2006) rule out this option too, however, because of the widths of some of the nuclear deexcitation lines observed by RHESSI. If the medium in which positrons were emitted and annihilated exhibited substantial turbulence, this turbulence should also broaden the nuclear deexcitation lines (scaled proportionally to energy). The 0.847 MeV ^{56}Fe and 1.36 MeV ^{24}Mg lines integrated from 11:06:20 to 11:10:20 UT have FWHM values of 3.5 $\pm$ 2.6 and 10.0 $\pm$ 3.0 keV, respectively. These widths are narrower than those expected for impact by a downward directed particle population. The widths are also narrower at the 2.9 σ and 2.3 σ uncertainty levels than the values of 11 and 17 keV that should have been observed if the 0.511 MeV line had been turbulently broadened in the same environment. Based on these arguments, Schrijver et al. (2006) conclude that the 0.511 MeV line is most likely thermally broadened in relatively compact regions.

9.18.11 Possible Scenario of the X17 Flare on 2003 October 28 and Comparison with Standard Scenario

As noted Schrijver et al. (2006), in some respects, the observations of the X17 flare on 2003 October 28 appear to be consistent with the standard scenario of energy deposition by nonthermal particles, and with the theory of flare γ-ray emission within the framework of nuclear interactions, with simplifying assumptions about the propagation of the primary and secondary particles (e.g., Ramaty et al. 1995 and references therein). In particular, the first 1 ks of the flare, forming phase I, appears consistent with the usual picture of particle precipitation (causing the flare ribbons) and chromospheric evaporation leading to the brightening of the coronal loops. After a few early impulsive variations, can be seen the particle interaction rate decline exponentially on a timescale of $\sim$3.6 min, from about 11:05 to $\sim$11:16 UT (Panel a of Fig. 9.87). In this phase, the energetic electrons cause a bright, but rapidly decreasing, bremsstrahlung continuum as they impact the plasma, which decreases in proportion to the nuclear deexcitation line flux after 11:10 UT. The impact sites of the energetic particles, as reflected in the UV flare ribbons, propagate rapidly, up to several tens of kilometers per second for the strongest sites, with continuing afterglow lasting for up to several minutes behind the flare-ribbon fronts.

Schrijver et al. (2006) underlined that an unusual feature of this strong flare is that it induced observable seismic waves in the solar interior. In their seismic study of MDI Doppler data, Donea and Lindsey (2004) argue that for an atmospheric signal to efficiently couple into the photosphere, the timescale characteristic for the perturbation as determined by the acoustic cutoff frequency and the photospheric pressure scale height is $\leq$40 s. The Schrijver et al. (2006) estimate for the time τ_w (see Section 9.18.7) for the most intense precipitation sites to propagate over their own diameter meets that requirement. It is thus likely that the most impulsive flare processes in the solar corona excite some positions in or near the photosphere so strongly that this couples to detectable helioseismic waves in the solar interior. These perturbations may be caused by the energy deposition of precipitating high-energy particles, or by the resulting radiative backwarming, or by a field realignment (the latter was suggested for other flares by Sudol and Harvey 2005).

As noted Schrijver et al. (2006), the flare on 2003 October 28 differs from the standard scenario in several aspects. For one thing, the apparent offsets between the ion and electron precipitation regions (Hurford et al. 2003b, 2006; see the circles in Fig. 9.88, bottom right) suggest that these two particle populations do not necessarily impact the same region at the same time. It was noted that the roughly 100 s needed for the neutrons to thermalize prior to their capture should cause their emission to come from a site where the instantaneous electron bremsstrahlung would have been seen $\sim$100 s before; for a characteristic propagation velocity of 60 km s^{-1}, that offset could reach up to $\sim 8''$. However, Hurford et al. (2006) find that a rough correction for such a 100 s delay does not lead to such large position offsets. Another aspect that is at odds with the standard flare scenario concerns the positron-annihilation radiation. In order for the bulk of the positrons to annihilate in the warm medium that is needed to thermally broaden the annihilation line, the ions need to impact onto, and the positrons need to be emitted into, a plasma at several hundred thousand kelvins, with a total column depth of at least several tenths of a gram per square centimeter (see Section 9.18.10), or a column depth of $N_H \geq 10^{22}$ cm^{-2}. The required ion impact might be achieved near the coronal base if multiple mirroring could occur in the coronal field, but it is hard to imagine how the $\sim 10^5$ °K plasma thus heated could remain dense and thick enough for several minutes to allow for positron emission and annihilation in it before the plasma streams up into the coronal volume (where it is responsible for the flare's X-ray emission).

According to opinion of Schrijver et al. (2006), perhaps there is a way to maintain an unexpectedly thick layer at transition-region temperatures for several minutes below the coronal base in a highly dynamic atmosphere. The atmospheric response to particle precipitation and the associated chromospheric evaporation continue to be studied in one-dimensional radiation hydrodynamic models, and some show that a relatively thick transition region with temperatures of order 10^5 °K and above can indeed form (e.g., Allred et al. 2005 and references therein). It is currently unclear how such a region can be made thick enough, and made to last long enough, to allow the bulk of the positrons to annihilate, at least temporarily, in a warm environment. Such warm environments need to persist longer than the few-minute radioactive time-scale for positron emission of multiple minutes (see

Section 9.18.10). Note that this time-scale is comparable to, and for the brightest sites exceeds, the time-scale for the displacement of the precipitation sites, as inferred from footpoint crossing times; here Schrijver et al. (2006) make the assumption that it can be identify the UV and hard X-ray sources and the other compact footpoint signatures as resulting from a common exciter. This suggests, at the very least, some sustained energy deposition behind the precipitation front.

Schrijver et al. (2006) noted that alternatively, the broad annihilation line originates in much deeper layers of the solar atmosphere. This requires that the energetic ions penetrate well below the classical chromosphere, as they well may (see Section 9.18.10), and that enough energy is deposited in these dense layers to raise the temperature to several hundred thousand °K. This energy deposition should more or less balance the strong radiative losses for several minutes after the initial nuclear excitation in order to have enough time for the positrons to be emitted and annihilated in a warm, dense environment. Observational support for this hypothesis is found in the anomalies in the magnetogram series and in the UV flare ribbons around points P_2 and P_3. These persist at any given location for up to $\sim$5 min and are conceivably associated with a warm, dense medium. Such a large volume of warm plasma could be very luminous, if not optically thick, however, because the temperature is at the peak of the plasma emissivity function. Models will need to be developed to assess whether this scenario is tenable.

Schrijver et al. (2006) underlined that regardless of how the $\sim 10^{5}$°K plasma is heated, the observations require that any model for the early positron annihilation for this flare must approximate the integrated signal from multiple differentially evolving atmospheres. Apparently, the temporal behavior of energetic particles exciting these atmospheres can be a mix of a strong initial beam that rapidly moves on, followed by some persistent precipitation for up to several minutes after that. Then, as the magnetogram artifacts disappear, as the flare ribbons slow down, and as the time-scales for the falling fluxes of nuclear deexcitation and electron bremsstrahlung no longer match, we see the positron-annihilation signatures shift to what appears to be consistent with a deep chromospheric or photospheric environment; in this phase II of the event, the narrow 0.511 MeV line appears to originate in an environment with a temperature of no more than $\sim$7,000°K and an ambient ion density in excess of 10^{15} cm^{-3}. RHESSI data allow the time-scale for the change in positron-line width to be as long as 150 s, but that is an upper limit. As it was mentioned in Section 9.18.10, the nuclear decay time (dependent on the mix of radioactive nuclei and the accelerated-ion spectrum; e.g., Kozlovsky et al. 1987; Murphy et al. 1987) may be comparable to that. In contrast, the radiative cooling time is as much as two orders of magnitude shorter than that (Raymond et al. 1976).

Schrijver et al.'s (2006) analysis of combined observations by the CORONAS, GOES, INTEGRAL, RHESSI, SOHO, and TRACE satellites shows the power of combining high resolution in the temporal, spectral, and spatial domains. The observations demonstrate that whereas the highly energetic particle populations precipitate in ribbon-like structures, the emissions are dominated by a few compact precipitation sites that move rapidly through a variety of atmospheric environments, as seen in white-light flares observed by TRACE and RHESSI (Hudson et al. 2006a, b). Consequently, hard X-ray and γ-ray spectra taken with moderate to low spatial resolution are composite spectra with contributions from a variety of rapidly evolving atmospheres. Combined with high-resolution imaging, the evolution of spectral features, such as the 0.511 MeV positron-annihilation line, can then be linked to specific atmospheric sites and physical environments.

As underlined Schrijver et al. (2006), despite the combination of observations from six spacecrafts, their analysis of the X17 flare on 2003 October 28 falls short of explaining the behavior of the positron-annihilation line width. It was offered some ideas for theoretical work and encourages the development of better scenarios for particle effects in the solar atmosphere. Crucial to observational guidance and verification of flare scenarios, however, is obtaining high-cadence, high-resolution of EUV spectroscopic observations. TRACE images reveal details in the flare down to its angular resolution of $1''$ and temporal resolution of a few seconds, suggesting that future observations should be obtained at even better resolution. Such observations should extend from the low chromosphere into the high-temperature corona to cover the entire range of coupled processes from particle acceleration and impact to atmospheric response and radiative cooling. The future Solar-B EUV Imaging Spectrometer (e.g., Culhane et al. 2005) fulfills some of these requirements by providing spectral coverage for transition-region, and coronal temperatures with $2''$ angular resolution, and a temporal resolution of 1 s or less for the brightest events. The high-temperature channels of the Atmospheric Imaging Assembly of the Solar Dynamics Observatory will enable us to image the rapid evolution of the coronal plasma. Comprehensive coverage from chromosphere to corona, however, will require even more powerful future instrumentation.

9.19 The 2003 October 28 X17 Flare: Time Profile of the 2.223 MeV Gamma Ray Line and Evolving Acceleration Parameter and Density Altitude Profile

9.19.1 The Matter and Short History of the Problem

As noted Troitskaia and Miroshnichenko (2008), the neutron capture line of 2.223 MeV from solar flares has been analyzed directly or by using additional data to obtain the characteristics of particle acceleration in solar flares and properties of surrounding solar atmosphere. In particular, Ramaty and Murphy (1987) used the fluences of 2.223 MeV line for determining the energy spectra of solar energetic particles, Hua and Lingenfelter (1987a) studied their angular distributions and production of secondary neutrons, and Yoshimori et al. (1999b) investigated the photospheric ^{3}He abundance. Kuzhevskij et al. (1998, 2001b) have developed an approach to determine the most probable profile of plasma density in the solar photosphere by the 2.223 MeV γ-ray line data (see Section 9.2). It was applied to three large flares, 16 December 1988 (Kuzhevskij et al. 2005), 22 March 1991 (Troitskaia and Kuzhevskij 1999), 6 November 1997 (Kuzhevskij et al. 2001b) and the plausible solar atmosphere density profile in the period of solar flares was estimated and some evidence of the effect of density enhancement was obtained. Gan (1998, 2000a) was the first who applied the time profiles of neutron capture line 2.223 MeV and positron annihilation line 0.511 MeV to deduce the spectral evolution of accelerated charged particles. Similar studies have been carried out for the RHESSI flare of 23 July 2002 (Gan 2004).

Troitskaia and Miroshnichenko (2008), Troitskaia et al. (2009) applied the developed method, described in Section 9.2, to the powerful solar event of 28 October 2003 on the basis of INTEGRAL satellite data (published by Kiener et al. 2006). They modeling of time history of the 2.223 MeV line and comparing it with the observational data reveals the considerable enhancement of the solar atmosphere density in the deep photospheric layers during the decay phase of gamma-ray emission. Besides, the analysis shows the hardening of a neutron spectrum in the decay phase comparing to the rising one.

9.19.2 Main Suppositions for Monte-Carlo Simulation

In Troitskaia and Miroshnichenko (2008), Troitskaia et al. (2009) the calculations of neutron propagation in the solar matter and 2.223 MeV line production are carried out using Monte-Carlo simulation, with due account for the models of vertical density profile of the solar plasma. For the SINP code it make allowance for: (1) neutron deceleration in elastic collisions with hydrogen nuclei, with due account for the energy and angular dependencies of cross-sections for np-scattering; (2) possible energetic neutron escape from the Sun; (3) gravitational neutron-Sun interaction; (4) thermal motion of decelerated neutrons; (5) neutron decay; (6) neutron captures by hydrogen ^{1}H, with the production of deuterium ^{2}H and gamma-quantum of 2.223 MeV; (7) non-radiative neutron absorption on ^{3}He; (8) gamma-ray absorption in the solar atmosphere in dependence on solar flare central angle; (9) time profile of initial neutron production; (10) initial neutron spectra, and (11) altitude dependence of surrounding matter density. The relative abundance of ^{3}He/^{1}H is taken about 2×10^{-5} (e.g., Hua and Lingenfelter 1987a; Yoshimori et al. 1999b). The time history of initial neutron production are assumed to be similar to that of total fluence of ^{12}C+^{16}O nuclear deexcitation lines in the range of 4.1–6.4 MeV. Calculations are made with SINP code for neutrons with energies of 1–100 MeV that are the most important ones for the 2.223 MeV line production. The primary neutrons are assumed to be emitted isotropically in the lower half-space (to-wards the Sun) from the levels with densities less than 5×10^{15} cm^{-3}. As a basic density model (BDM) (m = 1) Troitskaia and Miroshnichenko (2008) have used the standard astro-physical model HSRA (Harvard-Smithsonian Reference Atmosphere) for the lower chromosphere and quiet photosphere (Gingerich et al. 1971) together with a model of convection zone (Spruit 1974) consistent with the first one.

To determine possible deviations of the density profile, realizing in the observable flare, from the BDM, Troitskaia and Miroshnichenko (2008) have also composed four additional models (m = 2, 3, 4, 5) representing smaller and larger densities at photospheric and adjoining levels as compared with the standard BDM model (m = 1) of the quiet Sun (see Fig. 9.1 in Section 9.2).

9.19.3 Main Results

As it was described in Section 9.18, the flare of 28 October 2003 began at 9:41 UT, had its maximum at 11:10 and ended about 11:24 UT. It lasted about 15 min in the gamma-ray band. It appeared in the NOAA active region 10486. Troitskaia and Miroshnichenko (2008) applied the described above method to investigate the 28 October 2003 solar flare of X17.2/4B with coordinates 16°S, 08°E. Results are shown in Fig. 9.89.

In Fig. 9.89 the data on 2.223 MeV gamma-ray flux and summarized flux of 4.44–6.13 MeV gamma emission are used from INTEGRAL observations (Kiener et al. 2006). The calculations of expected time profiles of gamma-ray fluxes were made for five different density profiles (curves in Fig. 9.89 labeled as m = 1, 2, 3, 4, and 5) in supposition of Bessel form of accelerated particles energy spectrum (stochastic acceleration) for three values of parameter αT = 0.005, 0.03, and 0.1 (in three panels from top to bottom, correspondingly), where $\alpha = \dot{E}/E[\mathrm{s}^{-1}]$ is the acceleration parameter and T is the average time of acceleration. Experimental points of Kiener et al. (2006) are shown with error bars.

Troitskaia and Miroshnichenko (2008) came to conclusion that during the flare of 28 October 2003 the best modeling time profile is in the case of $\alpha T = 0.03$, m = 5. This means that during this flare there is density enhancement in the whole thickness of photosphere. It can be also seen from the Fig. 9.89 that m = 5 begin to realize at about 400 s. Another conclusion is that the better fitting in the rising phase is $\alpha T = 0.005$ and in the phases of maximum and decay the best fitting is αT = 0.1, what means the hardening of accelerated particles spectrum with developing of flare.

9.20 Solar Flares of 28 October and 2 November 2003: Gamma-Ray Line Analysis and Model of Particle Acceleration and Density Profile

9.20.1 The Matter and Short History of the Problem

As noted Gan (2005), both the X17.2 flare of 28 October 2003 and the X8.3 flare of 2 November 2003 are strong γ-ray line flares, observed by the RHESSI satellite. The time integrated spectra up to 7 MeV have been fitted with a model

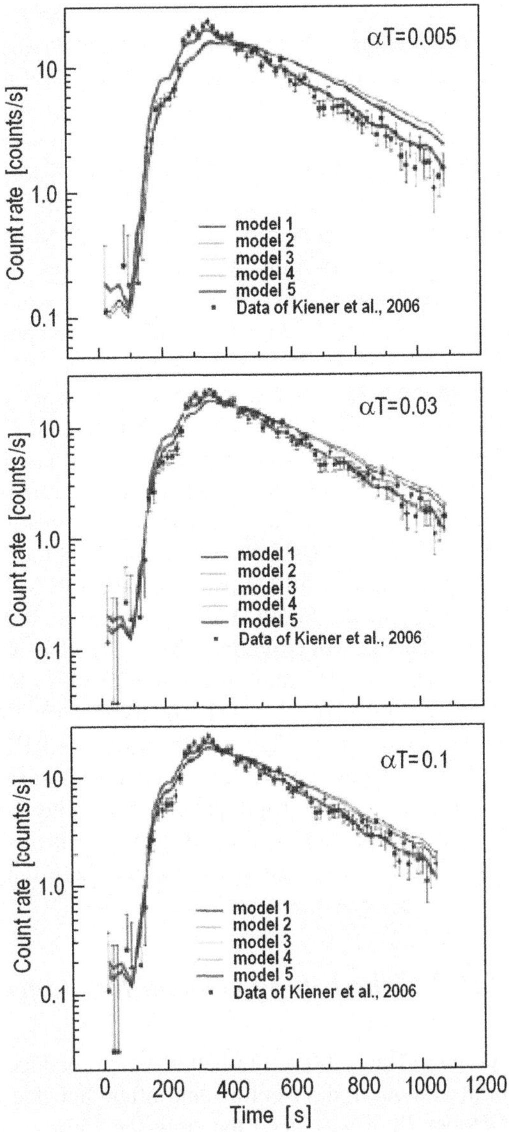

Fig. 9.89 Observational data of 2.223 MeV gamma-ray emission during the flare of 28 October 2003 and modeling with acceleration indexes $\alpha T = 0.005$, 0.03 and 0.1 for five density profiles (m = 1–5) (From Troitskaia and Miroshnichenko 2008)

of multi-components, including the bremsstrahlung, the annihilation line, neutron–proton capturing line, several narrow nuclear lines as well as the broad lines. Based on the fitted data, Gan (2005) discuss the spectral index of accelerated ions, the formation region of the annihilation line, the directionality of accelerated ions, and the abundance of ambient medium.

Gan (2005) noted that before the solar extreme events of 2003 October 28 and November 2, RHESSI observed only one intense γ-ray line flare, i.e., the X4.8 flare on 23 July 2002 (see above, Sections 9.18–9.20). In the γ-ray band, Smith et al. (2003) fitted the whole spectrum and found that there was a red shift of 0.1–0.8% of the deexcitation nuclear lines. Murphy et al. (2003a) compared the measured time history of the 2.223 MeV line with that predicted using a magnetic loop model. Share et al. (2003) studied, in particular, the annihilation line, including its constitution, its formation region, and its time history. Hurford et al. (2003b) presented for the first time an image of the 2.223 MeV line emission. Gan (2004) tried to explain both the time histories of the 2.223 MeV and the 0.511 MeV lines, adopting a power-law spectrum of accelerated ions. Besides, Gan et al. (2004) recently proposed a method to determine the density of annihilation region.

As underlined Gan (2005), during late October and early November 2003, a total of 12 X-class flares occurred. Among them there are at least three flares in which the 2.223 MeV line emissions were observed by RHESSI: X17.2 flare on October 28, X10.0 flare on October 29, and X8.3 flare on November 2. For the X10.0 flare on October 29 there is a serious influence of the background and it seems to be difficult to extract accurately the net nuclear flare emissions. Therefore, Gan (2005) studied the flares on October 28 and November 2, 2003. The flare of October 28, classified as X17.2/4B and located at 16°S, 08°E, started at 09:15 UT, peaked at 11:10 UT, and ended at 11:24 UT. RHESSI missed its impulsive phase due to the South Atlantic Anomaly, but covered well from its maximum phase. The flare of November 2, classified as X8.3/2B and located at 14°S, 56°W, began at 17:03 UT and ended at 17:39 UT. RHESSI successfully observed this flare from 17:08 to 17:29 UT, covering nearly the entire impulsive phase and most of the decay phase. Figure 9.90 shows the RHESSI light curves of the two flares (left panels), as well as the background emission 1 day after for the flare of October 28 (upper right panel) and 1 day before for the flare of November 2 (lower right panel). From Fig. 9.90 can be seen that the emissions at 2.223 MeV are obvious, demonstrating that these flares are strong γ-ray line flares.

9.20.2 The Fitting Gamma-Ray Lines Spectra

As noted Gan (2005), the better way to deduce the background is to consider 15 orbits (about 1 day) before and after, but due to some reasons, for the flare of October 28, it was taken the emission 15 orbits after as the background, while for the flare of November 2 it was taken the emission 15 orbits before as the background. Figure 9.91 shows the time-integrated count spectra of the background, flare with background, and background-subtracted flare.

As underlined Gan (2005), from the background-subtracted spectrum in Fig. 9.91, can be seen that the background has been well subtracted, since the background lines or structures were removed in a good degree. All rear detectors were summed, except 2R and 8R. There was no data for 2R, while for 8R there is an artifact at around 3 MeV. To avoid the rear decimation and the artifact at around 8 MeV from 1R, it was limited the fitting energy range from 300 keV to 7 MeV for

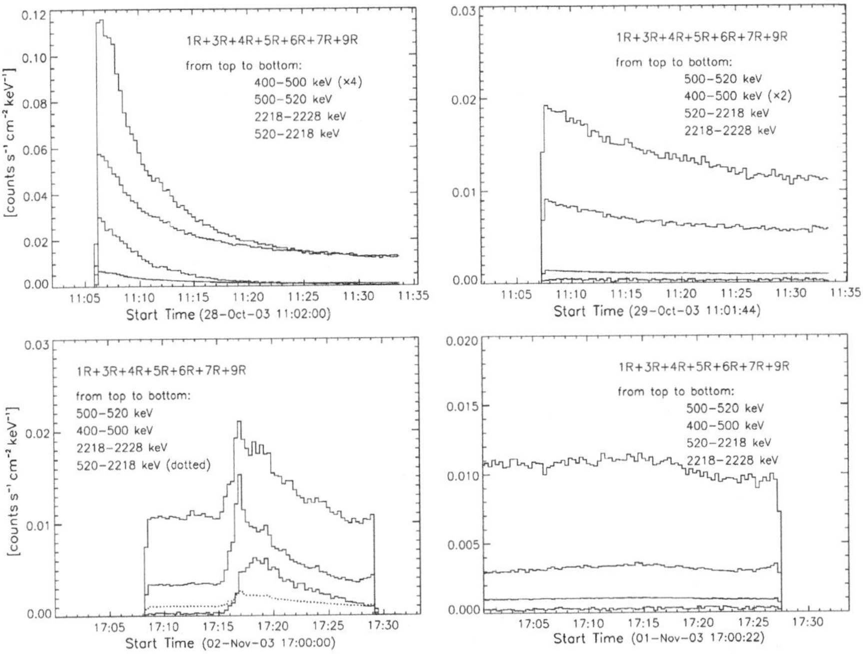

Fig. 9.90 RHESSI light curves of the two flares (*left* panels), as well as the background taken (*right* panels) (From Gan 2005)

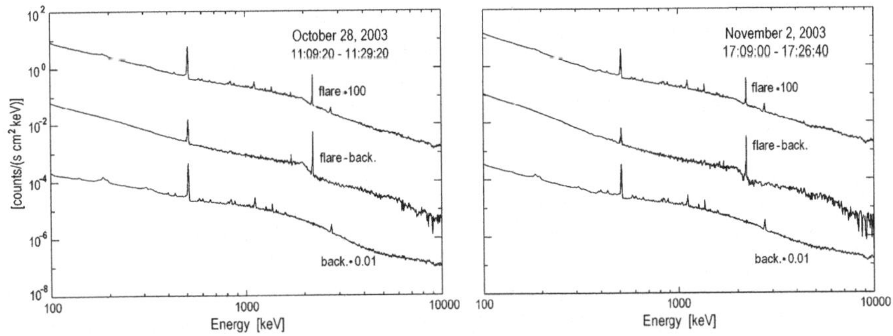

Fig. 9.91 Time-integrated count spectra of the background, flare with background, and background-subtracted flare (From Gan 2005)

the flare of October 28 and from 400 keV to 7 MeV for the flare of November 2. For 5R, there is an artifact at around 1.59 MeV with a width of about 30 keV. But this artifact is not obvious in the background-subtracted spectrum, and hence the ignorance of this artifact might not greatly influence the fitting (see also the fitted 1.634 MeV line profiles in Figs. 9.92 and 9.93). The following fitting model is

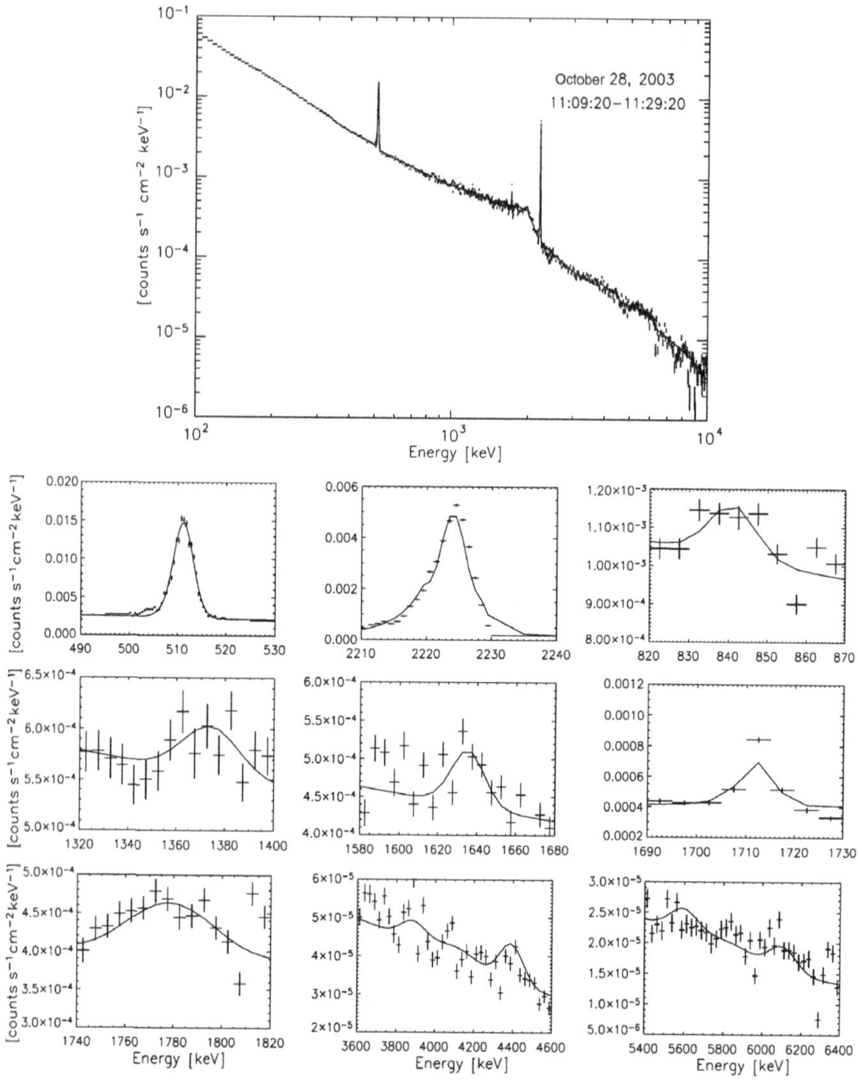

Fig. 9.92 The fitted whole spectrum and individual lines compared with the observations for the flare of October 28, 2003 (From Gan 2005)

adopted: a single power-law bremsstrahlung; the 0.511 MeV line; the 2.223 MeV line; six narrow lines of ^{56}Fe at 0.847 MeV, ^{24}Mg at 1.369 MeV, ^{20}Ne at 1.634 MeV, ^{28}Si at 1.779 MeV, ^{12}C at 4,438 MeV, and ^{16}O at 6.129 MeV; five broad deexcitation lines with fixed line centers at 0.811, 1.515, 1.673, 4.267, and 5.931 MeV (Murphy et al. 1990b). The time interval is taken from 11:09:20 UT to 11:29:20 UT for the flare of October 28 and from 17:09:00 UT to 17:26:40 UT

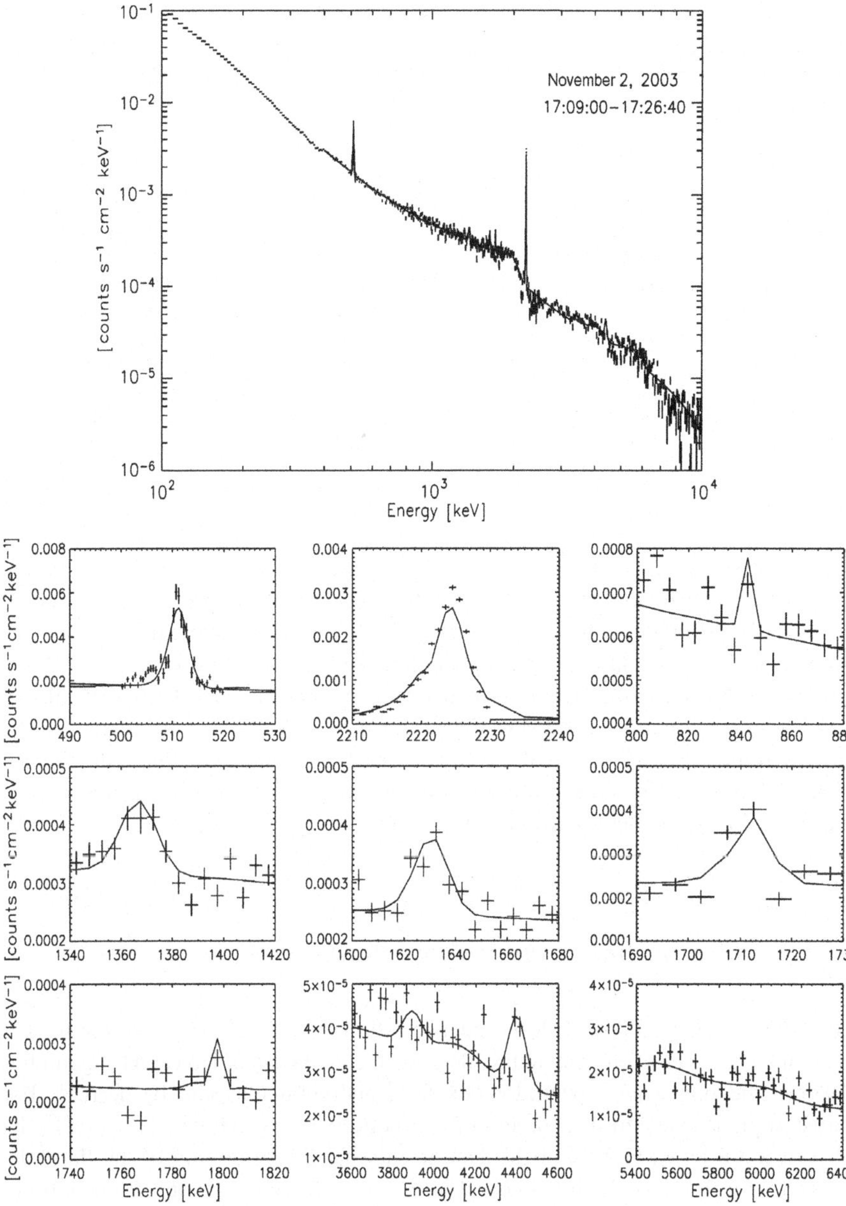

Fig. 9.93 The fitted whole spectrum and individual lines compared with the observations for the flare of November 2, 2003 (From Gan 2005)

for the flare of November 2. Gan (2005) used the SPEX software (Schwartz et al. 2002) to fit the background-subtracted count spectrum. Figures 9.92 and 9.93 show the whole fitted spectrum and the individual lines for both flares, respectively. The

Table 9.19 Best-fit Gaussian parameters for the flares on October 28 and November 2, 2003 (From Gan 2005)

Line (keV)	Fit energy (keV)	Redshift (%)	Fluence (photons cm^{-2})
^{56}Fe 847	$841.2^{+2.8}_{-2.2}$/–	$0.68^{+0.26}_{-0.33}$/–	$31.08^{+8.52}_{-9.48}$/–
^{24}Mg 1369	$1375.0^{+7.6}_{-8.0}$/	$-0.44^{+0.59}_{-0.55}$/	$22.44^{+8.76}_{-9.24}$/
	$1367.4^{+2.2}_{-2.9}$	$0.12^{+0.21}_{-0.16}$	$31.59^{+8.69}_{-7.21}$
^{20}Ne 1634	$1636.1^{+3.4}_{-4.1}$/	$-0.13^{+0.25}_{-0.21}$/	$26.88^{+7.68}_{-7.68}$/
	$1631.3^{+1.6}_{-1.8}$	$0.17^{+0.11}_{-0.10}$	$33.71^{+6.57}_{-6.15}$
^{28}Si 1779	$1779.8^{+5.8}_{-3.6}$/–	$-0.04^{+0.20}_{-0.33}$/–	$60.36^{+23.64}_{-6.0}$/–
^{12}C 4438	$4395.3^{+17.2}_{-8.7}$/	$0.96^{+0.20}_{-0.39}$/	$63.48^{+6.12}_{-5.88}$/
	$4401.2^{+8.8}_{-10.2}$	$0.83^{+0.23}_{-0.20}$	$61.27^{+10.81}_{-8.27}$
^{16}O 6129	$6108.2^{+69.8}_{-28.7}$/	$0.34^{+0.47}_{-1.14}$/	$42.0^{+12.0}_{-4.2}$/
	$6025.0^{+236.0}_{-147.0}$	$1.70^{+0.45}_{-5.80}$	$36.15^{+6.36}_{-6.89}$
511	$511.24^{+0.07}_{-0.06}$/		$700.8^{+19.2}_{-13.2}$/
	$511.52^{+0.37}_{-0.37}$		$76.0^{+14.10}_{-12.72}$
2223	$2224.02^{+0.02}_{-0.09}$/		$824.4^{+3.6}_{-20.4}$/
	$2224.22^{+0.04}_{-0.08}$		$390.19^{+7.42}_{-5.83}$

line with center at 1.712 MeV is the secondary peak of the 2.223 MeV line, arising from the escape of one 0.511 MeV photon. The fitted parameters are listed in Table 9.19. The other parameter, the fluence of 4–7 MeV nuclear emissions for two flares, is 374.8 ± 3.3 and 406.9 ± 4.0 photons.cm^{-2}, respectively.

Gan (2005) noted that from comparison of results in Table 9.19 with the 10 years' observation by GRS/SMM (Vestrand et al. 1999), can be seen that the fluence of the 0.511 MeV line of the November 2 ranks the sixth, while the fluence of the 2.223 MeV line ranks the third. For the flare of October 28, both the 2.223 and 0.511 MeV line fluence are the largest.

9.20.3 Spectral Index and Related Parameters

According to Gan (2005), with the fluence ratio of the 2.223 MeV line to the 4–7 MeV nuclear excess, referring to the theoretical calculation of Hua and Lingenfelter (1987a), the power-law spectral index S of energetic ions can be derived. For the isotropic downward distribution of energetic ions and ^{3}He/H $= 2 \times 10^{-5}$, the derived S are listed in Table 9.20. Due to missing the beginning of the flare, the derived S for the selected time interval of October 28 represents only a lower limit. Table 9.20 lists also the total number of energetic electrons above 1 MeV, total number of energetic protons above 30 MeV, and the total energies carried by protons and electrons above 1 MeV. These data, compared with those obtained with GRS/SMM (e.g., Gan 2000b), are no special except a greater magnitude. Gan (2005) pointed out that it is no way of knowing if the power-law spectrum of energetic protons goes down to 1 MeV, and therefore the total energy in accelerated protons E_p(>1 MeV) might be uncertain by something like an order of magnitude.

Table 9.20 Some derived parameters for energetic electrons and protons of the two flares: ratios of total numbers and total energies (in erg) (From Gan 2005)

Flare	S	N_e(>1 MeV)/N_p(>30 MeV)	E_p/E_e (>1 MeV)	$f_{0.511}$	Ne/O	α/p
Oct 28	3.22	$>8.0\times10^{33}/>2.6\times10^{33}$	$>1.4\times10^{31}/>2.2\times10^{28}$	1.41	0.15	0.01–0.1
Nov 2	3.62 ± 0.3	$1.1\times10^{34}/2.2\times10^{33}$	$4.2\times10^{31}/2.5\times10^{28}$	0.33	0.15	0.01–0.1

9.20.4 Formation Region of Annihilation

As underlined Gan (2005), with the known spectral index S, referring to the theoretical value of fluxes ratio Q(0.511 MeV)/Q(4–7 MeV) according to Murphy (1985), the conversion factor $f_{0.511}$ of one positron into 0.511 MeV photons can be derived (see Table 9.20). Theoretically, the $f_{0.511}$ should be between 0.5 and 2.0 if there is no absorption before the 0.511 MeV photons are observed (e.g., Ramaty 1986). For the November 2 event Gan (2005) notice that its location is toward the solar limb, with the heliocentric angle of about 60°. This means that the annihilation photons may have been suffered a Compton scattering if the annihilation region is located in the deep atmosphere. There is another evidence for the deep annihilation region: the width of the 0.511 MeV line for November 2 event is 2.35(+2.89/−2.35) keV. According to Crannell et al. (1976), such a width can only be explained by that the annihilation region is in the temperature of chromosphere or photosphere. Both the line width and the $f_{0.511}$ suggest that the annihilation region of November 2 event might be in the lower atmosphere. As concerns October 28 event, no conclusion can be drawn on the annihilation region, due to incomplete time coverage in observation.

9.20.5 Abundance and Directionality

As noted Gan (2005), the fluency ratio of ^{20}Ne at 1.63 MeV to ^{16}O at 6.13 MeV is dependent on spectral index S of energetic ions. Ramaty et al. (1996) studied this ratio by adopting a series of energetic α/p (0.01, 0.1, and 0.5) and different ambient abundancies of Ne/O (0.15 and 0.25). Using their calculations, it can be derive the spectral index based on observed $\Phi_{6.13}/\Phi_{1.63}$. Together with observed $\Phi_{2.22}/\Phi_{4.44}$, it can be check the consistency with the calculated relationship between the fluency ratio of $\Phi_{2.22}/\Phi_{4.44}$ and the power-law spectral index S (Ramaty et al. 1996; solid line in Fig. 9.94). In Fig. 9.94 the thick cross is for October 28 event, and the thin cross for November 2 event.

As noted Gan (2005), from Fig. 9.94 follows that for October 28 event, the best case close to the theoretical line is with $\alpha/p = 0.01$ and Ne/O = 0.15, while the cases with $\alpha/p = 0.01$ and Ne/O = 0.25, $\alpha/p = 0.1$ and Ne/O = 0.15, as well as $\alpha/p = 0.5$ and Ne/O = 0.15 cannot be ruled out. But if to consider the consistency

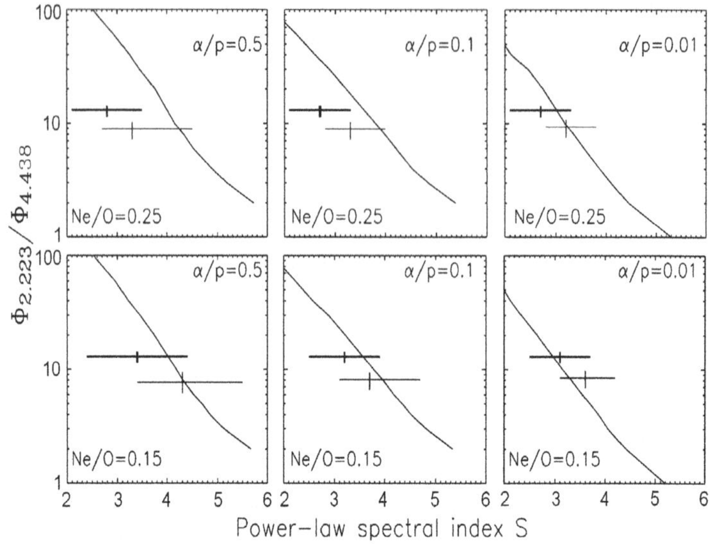

Fig. 9.94 The comparison of observed $\Phi_{2.223}/\Phi_{4.438}$ and the power-law spectral index S derived from $\Phi_{6.13}/\Phi_{1.63}$ with the theoretical relationship (solid line). The darker crosses are for the flare of October 28 and the thinner crosses for the flare of November 2. The real $\Phi_{2.223}/\Phi_{4.438}$ for the flare of October 28 would be lower if the whole flare were included (From Gan 2005)

with the spectral index in Table 9.20, the case with Ne/O = 0.15 and α/p = 0.01–0.1 shows to be the best. From the same logic, the case with Ne/O = 0.15 and α/p = 0.01–0.1 seems to be the best in explaining November 2 event. These results are different from that obtained by Ramaty et al. (1996) who claimed Ne/O = 0.25 and α/p larger than 0.1.

Gan (2005) underlined that the angular distribution of accelerated particles is another important parameter, which is related to acceleration mechanisms. From Table 9.19 can be seen that within 1 σ error bar, the redshifts of narrow lines are obvious, either for October 28 event or for November 2 event. Although, it is not so easy to compare quantitatively the line profiles with the theoretical calculations (e.g., Murphy et al. 1988), at least from the qualitative point of view, a downward beam or a downward isotropic angular distribution of accelerated particles seems to be more plausible. This result is similar to that obtained from another RHESSI γ-ray line flare of 23 July 2002 (Smith et al. 2003; see also Sections 9.18–9.20).

9.20.6 Summary of Main Results

Gan (2005) have studied two RHESSI γ-ray line flares of October 28 and November 2, 2003. After extracting the fluency of several narrow nuclear deexcitation lines, annihilation line, and the neutron capture line, it was derived the spectral index of energetic ions, as well as the total number of energetic protons and their total carried energies. The annihilation region might be at the lower atmosphere for the November 2 event. It is also shown that the abundance of Ne/O tends to be of 0.15 rather than 0.25, and that energetic α/p tends to be within 0.01–0.1 but not bigger than 0.1. The redshifts of the lines seem to support the downward angular distribution of accelerated ions.

9.21 Gamma-Ray Solar Flare Events in January 2005: Temporal Profiles in Various Energy Bands by Data from AVS-F Apparatus on Board the CORONAS-F Satellite

9.21.1 Observations in January 2005 by the CORONAS-F Satellite

According to Arkhangelskaja et al. (2008), the temporal profiles and energy spectra of the several solar flares are observed by AVS-F apparatus onboard CORONAS-F satellite in January 2005. The energy spectra of these solar flares contain γ-ray nuclear lines, positron line and neutrons capture line. The AVS-F (amplitude-time Sun spectrometry) apparatus (Glyanenko et al. 1999; Arkhangelskaja et al. 2006) was installed onboard CORONAS-F satellite which operated from 31 July 2001 to 6 December 2005. The orbit of satellite was approximately circular oriented towards the Sun with inclination 82.5° and altitude ~500 km. The AVS-F apparatus use signals produced by the SONG-D detector (energy deposition ranges of 0.1–11.0 MeV and 4.0–94.0 MeV by first time calibration data), anticoincidence signal generated by the plastic scintillation counter of the SONG-D and signal from XXS-1 detector, which is the CdTe based semiconductor detector with energy deposition range 3–30 keV. The SONG-D is CsI(Tl)-based detector Ø20 cm and height of 10 cm with electronics unit SONG-E (Kuznetsov et al. 1995). The system energy resolution is 13.0% for γ-quanta with energy 0.662 MeV (^{137}Cs).

9.21.2 Characteristics of Solar Flares Observed During January 2005 by the AVS-F Apparatus

As noted Arkhangelskaja et al. (2008), during January 2005, 20 solar flares with class M and X were registered on by detectors onboard satellites GOES, RHESSI, SOHO, TRACE and other instruments. Four active regions 10715, 10718, 10719 and 10720 were sources of these solar flares. Six of them (January 9, January 15, January 17, January 20, and two flares on January 19) were observed in low energy γ-band by AVS-F too (Arkhangelskaja et al. 2006). Active region NOAA 10719 was the source of solar flare January 9, the source of other five flares was NOAA 10720. January 19 flares (M6.7 and X1.3) and solar flare January 9 (M2.4) were observed in polar cap regions of CORONAS-F orbit, but satellite enter radiation belts during these flares registration. Flare January 20 (X7.1, 06:36–07:26 UT on GOES data) was the most powerful of observed flares in January 2005. This flare was accompanied by particles events (protons and neutrons which were most intensive ones for period of the last 15 years (Wilkinson and Allen 2005) and CME. Gamma emission of this flare was observed by AVS-F in both energy bands during X-ray emission rise (before their maximum) by GOES data in equatorial part of CORONAS-F orbit before entering into radiation belt (Panel a in Fig. 9.95). Its temporal profile in high-energy γ-band discussed in [6], but there were not any statistical significant count rate exceeds background level in the high energy γ-band by AVS-F data during January 17 and 15 solar flares. January 17 flare (X3.8, 06:59–07:26 UT by GOES data) was observed by CORONAS-F orbit before Radiation belt (Figure 1a). Its temporal profile in high-energy γ-band discussed in Arkhangelskaja et al. (2005), but there were not any statistical significant count rate exceeds background level in the high energy γ-band by AVS-F data during January 17 and 15 solar flares. The flare at January 17 (X3.8, 06:59–07:26 UT by GOES data) was observed by AVS-F apparatus in equatorial region of satellite orbit during X-ray emission maximum and droop (Panel b in Fig. 9.96). The January 15 flare (*X*2.6, 22:35–23:31 UT by GOES data) registered by AVS-F in the polar cup region of CORONAS-F satellite orbit during maximum of X-ray emission (by GOES data) – see Panel c in Fig. 9.95.

January 20, 17 and 15 flare's integrated temporal profiles shape in low energy γ-band and X-ray band is very simple with one maximum. There are four spectral lines complexes in integral spectra of January 20 and 17 flares and two in January 15 one (see Fig. 9.96 and Table 9.21). All spectral features were observed during whole duration of γ- emission registered by AVS-F.

9.21.3 The Events of 20, 17 and 15 January 2005: Thin Structure of Temporal Profiles

Temporal profiles by AVS-F apparatus data in energy bands corresponding spectral features, discussed in previous Section 9.21.2 for solar flares 20, 17, and 15 January 2005, are shown at Fig. 9.97. There are two maxima at solar flare January 20 temporal profile in energy band 0.4–0.6 MeV at 06:44:36 and 06:53:46 UT which correspond to maxima in range 0.1–0.3 MeV in statistical errors limits. One maximum at 06:46:36 UT was in energy bands 3.2–5.0 MeV and 5.3–6.9 MeV and one maximum at 06:47:16 UT was observed in range 1.7–2.3 MeV. Following main maxima are separated on January 17 flare temporal profiles: 09:41:26 UT

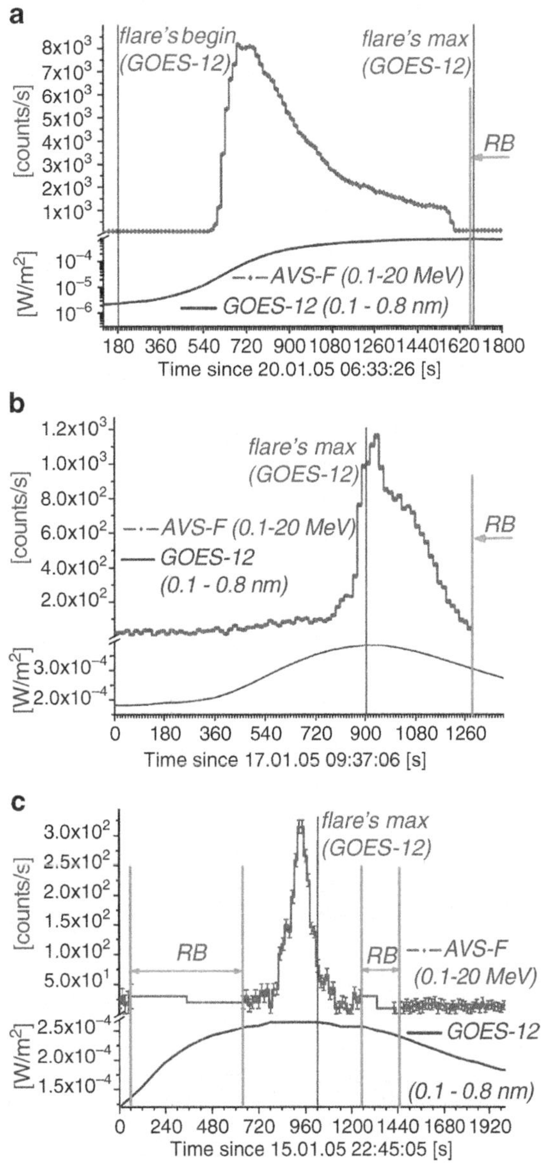

Fig. 9.95 Solar flares 20 January 2005 (**a**), 17 January 2005 (**b**) and 15 January 2005 (**c**) temporal profiles by AVS-F apparatus data in low energy gamma-ray band with background subtraction and by GOES-12 data in 0.1–0.8 Å region (From Arkhangelskaja et al. 2008)

(0.15–0.30 MeV), 09:40:36 UT (0.4–0.7 MeV), 09:42:31 UT (0.7–0.9 MeV), 09:42:16 UT (2.0–2.3 MeV) and 09:41:54 UT (3.6–5.0 MeV). Temporal profiles of January 15 solar flare had one maximum in each studying energy band: at 23:00:19 UT in corresponding continuum range and at 23:00:45 UT in 0.4–0.6 MeV and 2.0–2.3 MeV ones.

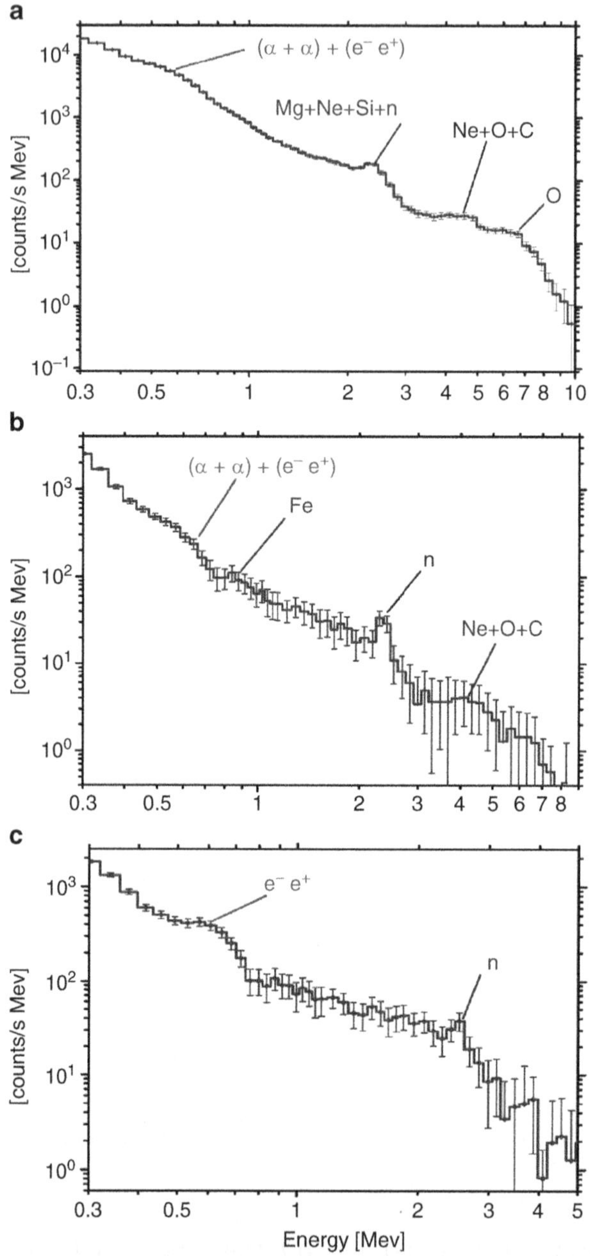

Fig. 9.96 The integrated energy spectra (with background subtraction) of solar flares observed in January 2005: (**a**) January 20 (06:43:16–06:59:51 UT), (**b**) January 17 (09:51:13–09:58:40 UT), (**c**) January 15 (22:56:31–23:05:51 UT) (From Arkhangelskaja et al. 2008)

As noted Arkhangelskaja et al. (2008), discussed flares were sufficiently long for studying thin structure at timescales of some tens seconds. Corresponding periodogram are shown at Fig. 9.98. Thin structure is presented at temporal profiles of these three solar flares. Characteristics timescales are in range 33–92 s for January 20 solar flare, 33–61 s for January 17 and 34–87 s for January 15 ones at 99% significance level–see Fig. 9.97 and Table 9.22. Thin structure on the same timescales was detected earlier by RHESSI in energy bands 3–25 keV for solar flare at 20 January 2005.

9.21.4 Summary of Main Results

Arkhangelskaja et al. (2008) summarized obtain results as following.

1. The wide range temporal profiles of January 20, 17 and 15 2005 solar flares time structure by AVS-F data is very simple with one maximum.
2. Temporal profiles structure are more complex in energy bands corresponding nuclear lines, positron line and neutrons capture line observed in these flare energy spectra.

Table 9.21 Spectral features of solar flares 20, 17, and 15 January 2005 (From Arkhangelskaja et al. 2008)

Date	Spectral features and it's energy band, MeV
20	$\alpha\alpha + e^+e^-$ (0.4–0.6), $^{24}Mg+^{20}Ne+^{28}Si$ + neutron capture (1.7–2.3), $^{20}Ne+^{16}O+^{12}C$ (3.2–5.0), ^{16}O (5.3–6.9)
17	$\alpha\alpha + e^+e^-$ (0.4–0.6), ^{56}Fe (0.7–0.9), neutron capture (1.7–2.3), ^{12}C (3.6–5.0)
15	e^+e^- (0.5–0.6), neutron capture (2.0–2.3)

Table 9.22 Characteristic timescales of solar flares in January 2005 (From Arkhangelskaja et al. 2008)

Flare date	Energy band, MeV	Characteristic timescales, s					
20	0.4–0.6	83	64	49	44	38	33
	1.7–2.3	69	52	44	40	35	–
	3.2–5.0	92	46	42	36	–	–
	5.3–6.9	92	59	44	38	–	–
17	0.4–0.6	61	37	–	–	–	–
	0.7–0.9	61	46	33	–	–	–
	1.7–2.3	61	46	33	–	–	–
	3.2–5.0	64	35	–	–	–	–
15	0.4–0.6	61	47	41	34	–	–
	2.0–2.3	87	34	–	–	–	–

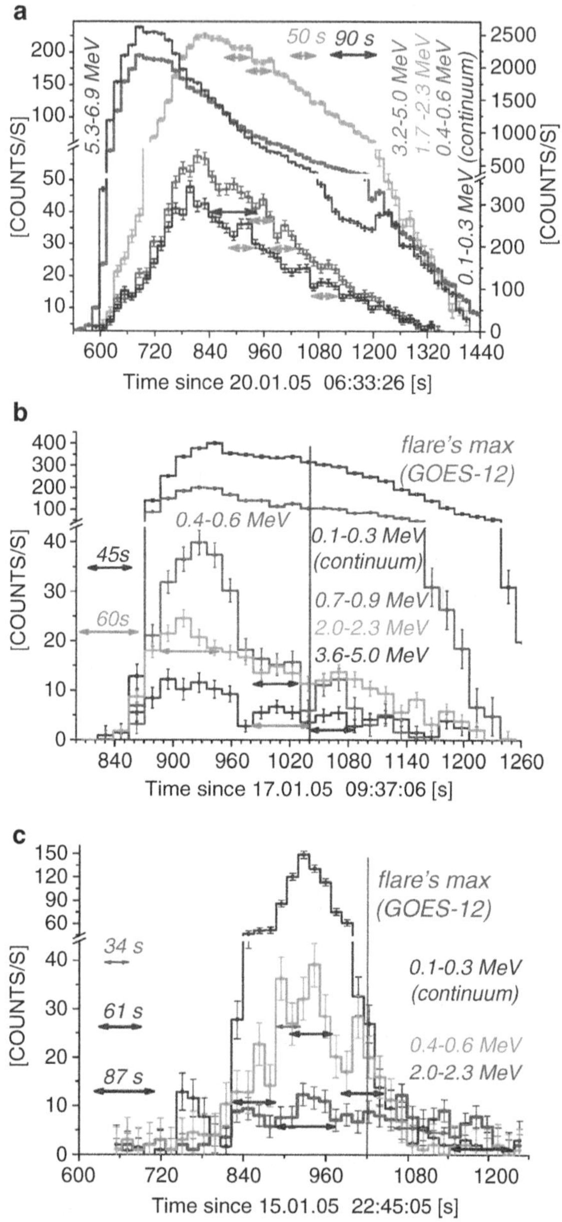

Fig. 9.97 Temporal profiles by AVS-F apparatus data in energy bands corresponding spectral features at Fig. 9.96 and their characteristics timescales for solar flares 20 January 2005 (**a**), 17 January 2005 (**b**) and 15 January 2005 (**c**) (From Arkhangelskaja et al. 2008)

3. There are two maxima at solar flare January 20 temporal profile in energy band 0.4–0.6 MeV which correspond to maxima in range 0.1–0.3 MeV.
4. In other energy bands one main maximum was observed on all discussed solar flares temporal profiles.

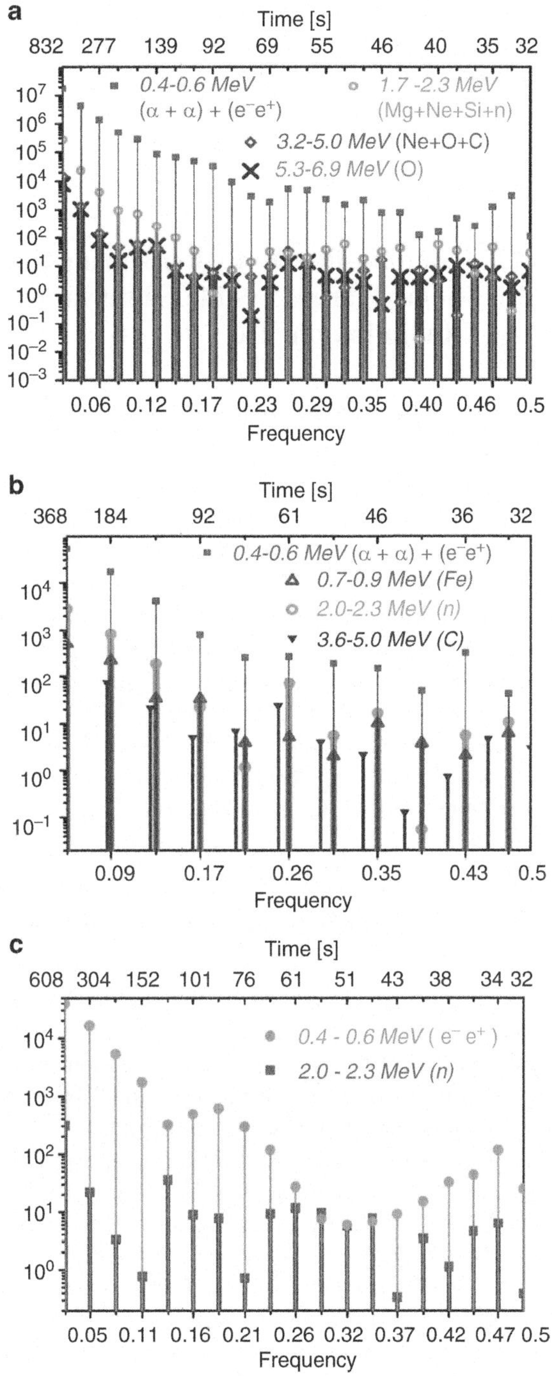

Fig. 9.98 Periodograms for shown at Fig. 9.97 solar flare temporal profiles: 20 January 2005 (**a**), 17 January 2005 (**b**) and 15 January 2005 (**c**) (From Arkhangelskaja et al. 2008)

5. Thin structure with characteristic timescales 33–92 s is presented in time profiles in energy bands corresponding observed spectral features exclude 0.1–0.3 range (continuum).
6. Periodogram analysis confirmed existence of such thin structure in time profiles of γ-ray lines (confidence level is 99%).

9.22 Gamma-Ray Emission During the 20 January 2005 Solar Flare: Possible Observation of Neutron Capture Radiation on ^{3}He

9.22.1 The Matter of Problem

As underlined Arkhangelskaja et al. (2007, 2008, 2009), the solar flare of 20.01.2005 (class X7.1) was the biggest one in January 2005 series. It was started at 06:36 UT, ended at 07:26 UT and the maximum of X-ray emission was at 07:01 UT by GOES data. AVS-F apparatus onboard CORONAS-F registered γ-ray emission during rising phase of this flare in two energy bands: 0.1–20 MeV and 2–140 MeV. The highest γ-ray energy was registered during this flare was 137 $\pm$ 4 MeV. Some spectral peculiarity was observed in the region of 19.5–21 MeV on 2.5 σ level in time interval 06:44:52–06:51:16 UT and 3 σ level in 06:47:00–06:49:08 UT one. Arkhangelskaja et al. (2007) considered in details the possibilities of this feature treatment as previously not observed in solar flares spectra γ-ray line 20.58 MeV from neutron capture radiation on ^{3}He.

9.22.2 Discrimination of n/γ Events in AVS-F

The short description of AVS-F (amplitude-time Sun spectrometry) apparatus which was installed onboard CORONAS-F satellite (NORAD catalog number 26873, ID 2001-032A) was given in Section 9.21. Gamma-rays were registered by AVS-F in two energy bands by CsI(Tl)-based SONG-D detector. The energy bands limits (energy deposition ranges of 0.1–11.0 MeV and 4.0–94.0 MeV by first time calibration data) are shifted during apparatus operation: the energy threshold and amplification coefficient of low-energy band changed on 1% and 1.8% per month correspondingly and on January 2005 the low-energy band boundaries were 0.1–20 MeV and high-energy ones were 2–260 MeV (Arkhangelskaja et al. 2006).

As noted Arkhangelskaja et al. (2007), the γ-ray and neutron events discrimination was performed using the selection of events by the scintillation detector light pulse shape based on the dependence of the ratio of intensities of light-output components with different fluorescence decay times to the average ionization density produced by charged particles in the detector material (Varga 1961).

A scintillation flash in CsI(Tl) consists of two main fluorescence components with decay times $\tau_{fast} \approx 0.5$–0.7 μs and $\tau_{slow} \approx 7$ μs, and the ratio of the slow component intensity Q_{slow} to the fast component intensity Q_{fast} depends on the interacting particles specific ionization. The method employed in AVS-F instrument is based on the integration of the signal from the SONG-D photomultiplier's preamplifier in two time intervals in which the total charge $Q_{tot} = Q_{slow} + Q_{fast}$ and slow fluorescence component Q_{slow} were collected. Values of Q_{tot} and Q_{slow}/Q_{tot} for each recorded event were digitized by two 8-bit analog-to-digital converters and transferred as two-dimensional matrix to the system microprocessor controller for subsequent processing. The system energy resolution was 13.0% for γ-quanta from ^{137}Cs (energy 0.662 MeV). The integration time for all presented temporal profiles (excluding some ones which are separately mentioned) is 16 s for low energy γ-band and 128 s for high energy one (Glyanenko et al. 1999; Arkhangelskaja et al. 2006).

9.22.3 *The Characteristics of 20 January 2005 Solar Flare as Measured by AVS-F Apparatus*

As noted Arkhangelskaja et al. (2007), the 20 January 2005 solar flare was started at 06:36 UT on GOES data and ended at 07:26 UT. This flare was accompanied by great energetic particle event (protons and neutrons, which were most intensive ones for period of the last 15 years, according to Wilkinson and Allen 2005) and coronal mass injection. Active region NOAA 10720 (14°N, 61°W) was the source of this flare. Gamma-ray emission of 20 January 2005 flare in energy bands of 0.1–20 MeV and 2–260 MeV was observed by AVS-F apparatus during X-ray emission rise according to GOES data. The highest γ-ray energy was registered during this flare was 137 ± 4 MeV. It was not observed statistical significance excess above background level in energy band 141–260 MeV. The count rate in 0.1–20 MeV and 2–260 MeV γ-ray energy bands droops to background level before maximum of X-ray emission by GOES data. January 20 flare's temporal profiles shape in both γ-ray energy bands are very simple with one maximum and one in X-ray band by GOES data has one maximum too – see Panel a in Fig. 9.99. Two bands energy spectrum of January 20 solar flare in time interval 06:44:52–06:51:16 UT by AVS-F data is shown at Panel c of Fig. 9.99. Six spectral features were separated at this spectrum and five of them are typical for solar flares.

As underlined Arkhangelskaja et al. (2007), spectral feature in the region of 19.5–21 MeV was first observed in γ-ray solar flare spectrum. Observed spectral features characteristics are presented in Table 9.23. All spectral features were observed during whole duration of γ-ray flare emission registered by AVS-F.

According to Arkhangelskaja et al. (2007), energy spectra in AVS-F high-energy band was obtained from convolution of two dimensional distribution of ratio Q_{slow}/Q_{tot} in dependence of energy. This analysis shows that in the time interval

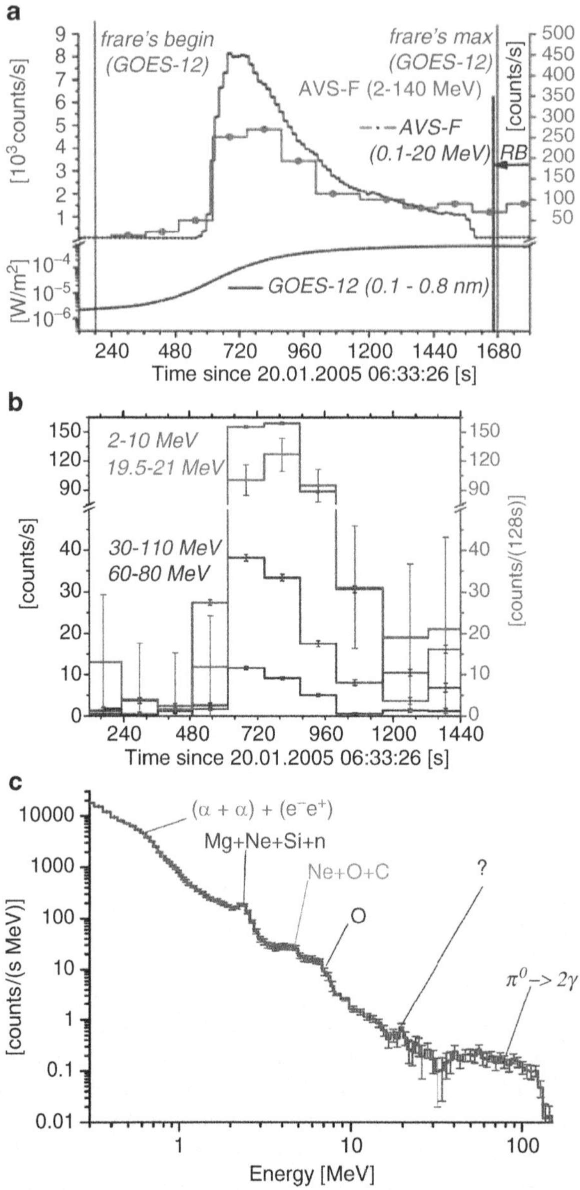

Fig. 9.99 Solar flare 20 January 2005 temporal profiles in by GOES data and in low and high energy γ-ray bands by AVS-F data (**a**), ones in different regions of AVS-F high-energy γ-ray band (**b**) and two-bands energy spectrum in time interval 06:44:52–06:51:16 UT by AVS-F data (**c**) (From Arkhangelskaja et al. 2007, 2009)

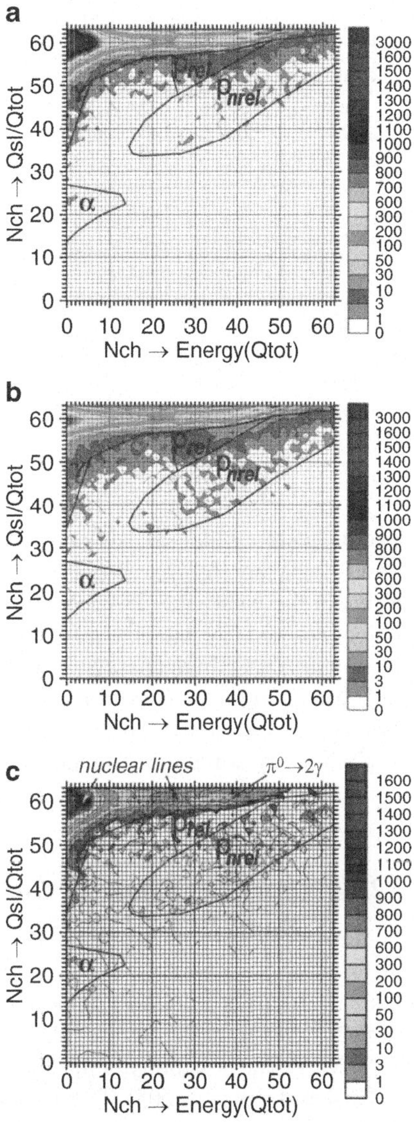

Fig. 9.100 The ratio Q_{slow}/Q_{tot} in dependence of energy for 20 January 2005 solar flare in time interval 06:44:52–06:51:16 UT without (**a**) and with (**c**) background subtraction and for background (**b**) (From Arkhangelskaja et al. 2007, 2009)

06:44:52–06:51:16 UT not energetic particles, but only γ-ray emission was observed from 20 January solar flare. Areas at two-dimensional distribution which correspond to registration of protons and α-particles contain small amount of counts – see Fig. 9.100. Arkhangelskaja et al. (2007) came to opinion that all spectral features of 20 January 2005 solar flare energy spectra by AVS-F data was caused only by γ-ray emission. These flare temporal profiles in γ-ray AVS-F high energy band in regions corresponding to nuclear lines (2–10 MeV), new observed

Table 9.23 Spectral features observed during 20 January 2005 solar flare in interval 06:44:52–06:51:16 UT (From Arkhangelskaja et al. 2007)

Spectral feature	Energy band
$\alpha\alpha$ (^{7}Be (0.429 MeV) + ^{7}Li (0.478 MeV)) + $e^{+}e^{-}$ (0.511 MeV)	0.4–0.6 MeV
^{24}Mg + ^{20}Ne + ^{28}Si + neutron capture	1.7–2.3 MeV
^{20}Ne + ^{16}O + ^{12}C	3.2–5.0 MeV
^{16}O	5.3–6.9 MeV
?	19.5–21 MeV
$\pi^{o} \rightarrow 2\gamma$ (67.5 MeV)	36–90 MeV

Table 9.24 Intensity ratio of spectral feature in 19.5–21 MeV band to other ones (From Arkhangelskaja et al. 2007)

Spectral feature [MeV]	Intensity ratio
1.7–2.3	1,500 ± 100
3.2–5.0	130 ± 15
5.3–6.9	95 ± 14

spectral feature (19.5–21 MeV), pion line maximum 60–80 MeV and in region 30–110 MeV are shown at Panel b of Fig. 9.99. It is seen, that behavior of temporal profile in the band 19.5–21 MeV correspond to nuclear lines one. Maxima in the regions 60–80 MeV and 30–110 MeV are shifted at some tens seconds (acquisition interval in AVS-F high energy γ-ray band is 128 s). So, spectral feature in 19.5–21 MeV band looks like nuclear line. The confidence level of this feature separation in summarized spectrum is 2.4 σ, but in the flare maximum it separated at 3 σ one. So, at the flare's maximum line in the band 19.5–21 MeV exceeds continuum at 3 σ.

9.22.4 Possible Nature of Gamma-Rays in the Energy Band 19.5–21 MeV

Arkhangelskaja et al. (2007) suppose that observed spectral feature in energy band 19.5–21 MeV perhaps can be explained as neutron capture line at ^{3}He with energy 20.58 MeV. The possibility of this line observation in solar flares was first discussed by Kuzhevskij et al. (2005). The preliminary values of intensity ratio and comparison between this line and other spectral features observed by AVS-F apparatus are presented in Table 9.24 and Fig. 9.101.

As noted Arkhangelskaja et al. (2007, 2009), the abundance of ^{3}He in solar atmosphere usually is estimated to be $\sim 2 \times 10^{-5}$ of H primarily based on solar wind, coronal, and meteoritic measurements, but in some works mentioned that accelerated ^{3}He is sometimes enhanced by a factor of 1,000 or even more over its solar concentration (Kocharov and Kocharov 1984; Mandzhavidze et al. 1997). In this case it is possible to obtain mentioned in Table 9.24 ratio between lines 2.223 and 20.58 MeV but for this conclusion it needs some additional calculations based on modern solar models.

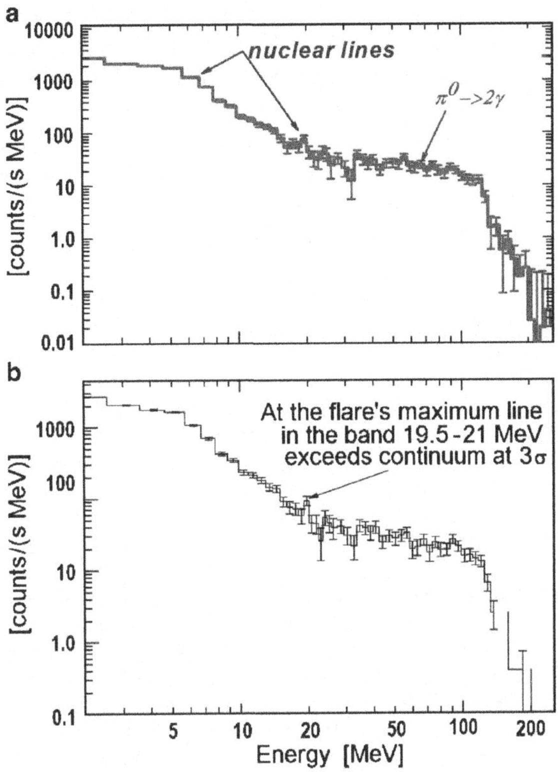

Fig. 9.101 The 20 January 2005 solar flare energy spectra in high energy γ-ray band: (**a**) obtained from convolution of matrix at Panel (**c**) in Fig. 9.100 (06:44:52–06:51:16 UT), (**b**) corresponding maximum of high energy γ-emission (in time interval 06:47:00–06:49:08 UT) (From Arkhangelskaja et al. 2007)

9.22.5 Summary of Main Results

Arkhangelskaja et al. (2007, 2009) summarized obtained results as following.

1. During 20 January 2005 solar flare (class X7.1), which was the biggest one in January 2005 series with γ-ray emissions, observed by AVS-F apparatus onboard CORONAS-F satellite during its rising phase in the two energy bands: 0.1–20 MeV and 2–140 MeV.
2. The highest γ-ray energy was registered during this flare was 137 ± 4 MeV.
3. In energy band 140–260 MeV the excess above back-ground level was not observed (in the frame of statistical significance).
4. Nuclear de-exitation, positron annihilation, and neutron capture at ^{1}H lines, as well as spectral feature corresponding decay of neutral pions were observed during whole time of this flare γ-ray emission registration by AVS-F.
5. Some spectral peculiarity was observed in the region of 19.5–21 MeV on 2.5 and 3.0 standard deviation level in time intervals 06:44:52–06:51:16 UT and 06:47:00–06:49:08 UT, correspondingly.

6. This spectral peculiarity in 19.5–21 MeV region can be neutron capture line at ^{3}He with energy 20.58 MeV if will be make some additional assumptions about local concentration of ^{3}He in solar atmosphere near flare region and neutron flux during this flare.
7. Such conditions are very likely for this flare because it accompanied by strong solar energetic particle and neutron GLE, which was the most intensive one for period of the last 15 years.

References for Chapter 9

Abramenko VI, Baranovsky EA (2004) Flare-related changes in the profiles of six photospheric spectral lines. Sol Phys 220(1):81–91

Abramenko VI, Yurchishin VB (1996) Modeling of a linear force-free magnetic field in a bounded domain. Sol Phys 168(1):47–63

Abramenko VI, Yurchyshyn V, Wang H, Spirock TJ, Goode PR (2003) Signature of an avalanche in solar flares as measured by photospheric magnetic fields. Astrophys J 597(2, Part 1): 1135–1144

Akimov VV, Afanassyey VG, Belaousov AS, Blokhintsev ID, Kalinkin LF, Leikov NG, Nesterov VE, Volsenskaya VA, Galper AM, Chesnokov VJ, Kirillov-Ugryumov VG, Lutchkov BI, Ozerov YV, Popov AV, Rudko VA, Runtso MF, Voronov SA, Zemskov VM, Fradkin MI, Kurnosova LV, Rusakovitch MA, Topchiev NP, Chuikin EI, Tugaenko VY, Tian TN, Ishkov VN, Gros M, Grenier I, Barouch E, Wallin P, Bazer-Bachi AR, Lavigne J-M, Olive J-F, Juchniewicz J (1991) Observation of high energy gamma-rays from the sun with the GAMMA-1 Telescope (E > 30 MeV). In: Proceedings of the 22nd international cosmic ray conference, Dublin, vol 3, pp 73–76

Akimov VV, Belov AV, Chertok IM, Kurt VG, Leikov NG, Magun A, Melnikov VF (1993) High energy gamma-rays at the late state of the large solar flare of June 15, 1991 and accompanying phenomena. In: Proceedings of the 23rd international cosmic ray conference, Calgary, vol 3, pp 111–114

Akimov VV, Belov AV, Chertok IM, Kurt VG, Leikov NG, Magun A, Melnikov VF (1994a) The high-energy gamma-ray flare of June, 15, 1991: some evidence of prolonged particle acceleration at the post-eruption phase. In: Proceedings of Kofu symposium, Kofu, NRO Report, vol 360, pp 371–374

Akimov VV, Leikov NG, Belov AV, Chertok IM, Kurt VG, Magun A, Melnikov VF (1994b) Some evidences of prolonged particle acceleration in the high-energy gamma-ray flare of June 15, 1991. In: Ryan J, Vestrand WT (eds) High-energy solar phenomena – a new era of spacecraft measurements. Proceedings of the workshop held in Waterville Valley, NH, 1993. American Institute of Physics, New York. AIP Conference Proceedings, vol 294, No 1, pp 106–111

Akimov VV, Leikov NG, Kurt VG, Chertok IM (1994c) The GAMMA-1 data on the March 26, 1991 solar flare. In: Ryan J, Vestrand WT (eds) High-energy solar phenomena - a new era of spacecraft measurements. Proceedings of the workshop held in Waterville Valley, NH, 1993. American Institute of Physics, New York. AIP Conference Proceedings, vol 294, No 1, pp 130–133

Akimov VV, Ambroz P, Belov AV, Berlicki A, Chertok IM, Karlicky M, Kurt VG, Leikov NG, Litvinenko YuE, Magun A, Minko-Wasiluk A, Rompolt B, Somov BV (1996) Evidence for prolonged acceleration based on a detailed analysis of the long-duration solar gamma-ray flare of June 15, 1991. Sol Phys (Netherlands) 166(1):107–134

Alfimenkov VP, Borzakov SB, Wierzbicki J, Ovchinnikov ON, Pikelner LB, Sharapov EI (1979) Radiative-capture of He-3 neutrons in the energy interval 1–70 keV. JETP Lett 29(1):91–93

Allred JC, Hawley SL, Abbett WP, Carlsson M (2005) Radiative hydrodynamic models of the optical and ultraviolet emission from solar flares. Astrophys J 630(1, Part 1):573–586

Amari T, Luciani JF, Mikic Z, Linker J (2000) A twisted flux rope model for coronal mass ejections and two-ribbon flares. Astrophys J Lett 529(1, Part 2):L49–L52

Anders E, Grevesse N (1989) Abundances of the elements: meteoritic and solar. Geochim Cosmochim Acta 53(1):197–214

Antiochos SK (1998) The magnetic topology of solar eruptions. Astrophys J Lett 502(2, Part 2): L181–L184

Antiochos SK, DeVore CR, Klimchuk JA (1999) A model for solar coronal mass ejections. Astrophys J 510(1, Part 1):485–493

Arkhangelskaja IV, Arkhangelsky AI, Glyanenko AS, Kotov YuD, Kuznetsov SN (2005) The investigation of January 2005 solar flares gamma-emission by AVS-F apparatus data onboard CORONAS-F satellite in 0.1–20 MeV energy band. In: Danesy D, Poedts S, De Groof A, Andries J (eds) Proceedings of the 11th European solar physics meeting "The Dynamic Sun: Challenges for Theory and Observations", (ESA SP-600), Leuven, Belgium, 2005, pp 107.1–107.4

Arkhangelskaja IV, Arkhangelsky AI, Kotov YuD, Kuznetsov SN, Glyanenko AS (2006) The solar flare catalog in the low-energy gamma-ray range based on the AVS-F instrument data onboard the CORONAS-F satellite in 2001–2005. Solar Syst Res 40(2):133–141

Arkhangelskaja IV, Arkhangelsky AI, Troitskaja EV, Kalmykov PA, Kotov YuD, Kuznetsov SI, Glyanenko AS, Miroshnichenko LI (2007) The observation of gamma-ray emission during January 20 2005 solar flare. In: 30th international cosmic ray pre-conference proceedings (on CD), Merida, p 4

Arkhangelskaja IV, Arkhangelsky AI, Glyanenko AS, Kotov YuD, Kuznetsov SN (2008) Solar flares temporal profiles thin structure on timescales 33–92 sec in various energy bands by data of AVS-F apparatus onboard CORONAS-F satellite. In: Proceedings of the 30th international cosmic ray conference, Merida, vol 1, pp 15–18

Arkhangelskaja IV, Arkhangelsky AI, Kotov YuD, Kalmykov PA, Glyanenko AS (2009) AVS-F observations of γ-ray emission during January 20, 2005 solar flare up to 140 MeV. Adv Space Res 43:589–593

Arndt MB, Bennett K, Connors A, McConnell M, Rank G, Ryan JM, Schoenfelder V, Suleiman R, Young CA (1999) Gamma ray measurements of the 1991 November 15 solar flare. In: Proceedings of the 26th international cosmic ray conference, Salt Lake City, UT, vol 6, pp 256–259

Aschwanden MJ, Shwartz RA, Alt DM (1995) Electron time-of-flight differences in solar flares. Astrophys J (USA) 447(2, Part 1):923–935

Aschwanden MJ, Wills MJ, Hudson HS, Kosugi T, Schwartz RA (1996) Electron time-of-flight distances and flare loop geometries compared from CGRO and Yohkoh observations. Astrophys J (USA) 468(1, Part 1):398–417

Aschwanden MJ, Fletcher L, Schrijver CJ, Alexander D (1999) Coronal loop oscillations observed with the Transition Region and Coronal Explorer. Astrophys J 520(2, Part 1):880–894

Avrett EH (1981) Reference model atmosphere calculation – The Sunspot sunspot model. In: Cram LE, Thomas JH (eds) The physics of sunspots. Proceedings of the conference, Sunspot, NM, 1981. Sacramento Peak Observatory, Sunspot, NM, pp 235–255, 257

Babin AN, Koval AN (1999) On the emission from the white light flare of 15 June 1991 in C I, C II, Mg II, and He II lines. Kinematika Fiz Nebesn Tel 15(1):51–58

Balaz J et al (1994) Proceedings of the IAU Colloquium Solar Coronal Structures, VEDA, Bratislava, 635

Baranovskij EA (1989) Energy balance in the chromosphere part of a flare. In: Somov BV (ed) Investigations of solar plasma. Ylym, Ashkhabad, pp 228–248 (In Russian)

Barat C, Trottet G, Vilmer ZN, Dezalay J-P, Talon IR, Sunyaev R, Terekhov O, Kuznetsov A (1994) Evidence for intense coronal prompt gamma-ray line emission from a solar flare. Astrophys J Lett (USA) 425(2, Part 2):L109–L112

Bennett E, Lee MA, Ryan JM (1994) Numerical modeling of particle transport and acceleration of solar flare particles in a coronal loop. In: Ryan J, Vestrand WT (eds) High-energy solar phenomena – a new era of spacecraft measurements. Proceedings of the workshop held in Waterville Valley, NH, 1993. American Institute of Physics, New York. AIP Conference Proceedings, vol 294, No 1, pp 118–123

Benz AO, Gold T (1971) Magnetically trapped particles in the lower solar atmosphere. Sol Phys (Netherlands) 21(1):157–166

Bespalov PA, Zaitsev VV, Stepanov AV (1987) On the origin of time delays in hard X-ray and gamma-ray emission of solar flares. Sol Phys (Netherlands) 114(1):127–140

Bogomolov AV, Britvich GI, Myagkova IN, Ryumin SP (1996a) Neutron identification against the background of gamma-quantums, when registrating them by detectors based on CsI(TL). Prib Tekh Eksp 39(1):13–19 (In Russian)

Bogomolov AV, Britvich GI, Myagkova IN, Ryumin SP (1996b) Identification of neutrons against the gamma-ray background by CsI(Tl) crystal detectors. Instrum Exp Tech (USA) 39(1):1–7

Bogomolov AV, Lgnat'ev AP, Kudela K, Kuznetsov SN, Logachev YI, Morozov OV, Myagkova IN, Oparin SN, Pertsov AA, Svertilov SI, Yushkov BY (2003) Parameters of the intense X-ray and gamma-ray radiation from the solar flare of May 20, 2002, as observed from the Coronas-F spacecraft. Astron Lett (Russia) 29(3):199–204

Boiko AYa, Livshits MA (1995) Gas-dynamic response of the atmosphere to impulsive heating: two-temperature approximation. Astron Rep 39(3):338–346

Bornmann PL (1987) Turbulence as a contributor to intermediate energy storage during solar flares. Astrophys J (USA) 313(1, Part 1):449–455

Breneman HH, Stone EC (1985) Solar coronal and photospheric abundances from solar energetic particle measurements. Astrophys J Lett Ed (USA) 299(1, Part 2):L57–L61

Brown JC (1971) The deduction of energy spectra of non-thermal electrons in flares from the observed dynamic spectra of hard X-ray bursts. Sol Phys (Netherlands) 18(3):489–502

Cameron R, Sammis I (1999) Tangential field changes in the great flare of 1990 May 24. Astrophys J Lett 525(1, Part 2):L61–L64

Cane HV, McGuire RE, von Rosenvinge TT (1986) Two classes of solar energetic particle events associated with impulsive and long-duration soft X-ray flares. Astrophys J (USA) 301(1, Part 1):448–459

Chiba N, Dion GM, Hayashida N, Honda K, Honda M, Inoue N, Kadota K, Kakimoto F, Kamata K, Kawaguchi S, Kawasumi N, Matsubara Y, Nagano M, Ohoka H, Teshima M, Tsushima I, Yoshida S, Yoshii H, Yoshikoshi T (1992) Possible evidence for greater than or equal 10 GeV neutrons associated with the solar flare of 4 June 1991. Astropart Phys 1(1):27–32

Chupp EL (1987) High-energy particle production in solar flares (SEP, gamma-ray and neutron emissions). Phys Scripta T18:5–19

Chupp EL (1990) Emission characteristics of three intense solar flares observed in Cycle 21. Astrophys J Suppl Ser (USA) 73(2):213–226

Chupp EL, Forrest DJ, Higbie PR, Suri AN, Tsai C, Dunphy PP (1973) Solar gamma ray lines observed during the solar activity of August 2 to August 11, 1972. Nature 241(5388):333–335

Chupp EL, Forrest DJ, Ryan JM, Cherry ML, Reppin C, Kanbach G, Rieger E, Pinkau K, Share GH, Kinzer RL, Strickman MS, Johnson WN, Kurfess JD (1981) Observation of the 2.223 MeV gamma-ray line on the SMM satellite-the event of 1980 June 7. Astrophys J 244(3, Part 2):L171–L174

Chupp EL, Debrunner H, Fluckiger E, Forrest DJ, Golliez F, Kanbach G, Vestrand WT, Cooper J, Share G (1987) Solar neutron emissivity during the large flare on 1982 June 3. Astrophys J 318 (2, Part 1):913–925

Cliver EW (1996) Solar flare gamma-ray emission and energetic particles in space. In: Ramaty R, Mandzhavidze N, Hua X-M (eds) High energy solar physics. Proceedings of the conference held in Greenbelt, Maryland, 1995. AIP Conference Proceedings, vol 374, No 1, pp 45–60

Cliver EW, Forrest DJ, Cane HV, Reames DV, McGuire RE, von Rosenvinge TT, Kane SR, MacDowall RJ (1989) Solar flare nuclear gamma-rays and interplanetary proton events. Astrophys J (USA) 343(2, Part 1):953–970

Crannell CJ, Joyce G, Ramaty R, Werntz C (1976) Formation of the 0.511 MeV line in solar flares. Astrophys J 210(2, Part 1):582–592

Culhane JL, Phillips AT, Pike CD, Fludra A, Bentley RD, Bromage B, Doschek GA, Hiei E, Inda M, Mariska JT, Phillips KJH, Sterling AC, Watanabe T (1993) Yohkoh observations of plasma upflows during solar flares. Adv Space Res 13(9):303–306

Culhane JL, Harra LK, Doschek GA, Mariska JT, Watanabe T, Hara H (2005) The Solar-B EUV imaging spectrometer and its science goals. Adv Space Res 36(8):1494–1502

Debrunner H (1994) Neutron monitor measurements as a complement to space measurements of energetic solar particle fluxes. In: Ryan J, Vestrand WT (eds) High-energy solar phenomena-A new era of spacecraft measurements. Proceedings of the workshop held in Waterville Valley, NH, 1993, University of New Hampshire. AIP Conference Proceedings, vol 294, No 1, pp 207–221

Debrunner H, Flückiger EO, Chupp EL, Forrest DJ (1983) The solar cosmic ray neutron event on June 3, 1982. In: Proceedings of the 18th international cosmic ray conference, Bangalore, India, vol 4, pp 75–78

Debrunner H, Flückiger EO, Lockwood JA (1990) Signature of the solar cosmic-ray event on 1982 June 3. Astrophys J Suppl Ser (USA) 73(2):259–262

Debrunner H, Lockwood JA, Ryan JM (1993a) Solar neutron and proton production during the 1990 May 24 cosmic-ray flare increases. Astrophys J 409(2, Part 1):822–829

Debrunner H, Lockwood JA, Ryan JM, McConnell M, Schonfelder V, Aarts H, Bennett K, Winkler C (1993b) Neutrons from the 15 June 1991 solar flare. In: Proceedings of the 23rd international cosmic ray conference, Calgary, vol 3, pp 115–118

Debrunner H, Lockwood JA, Barat C, Buttikofer R, Dezalay JP, Flückiger E, Kuznetsov A, Ryan JM, Sunyaev R, Terekhov OV, Trottet G, Vilmer N (1997) Energetic neutrons, protons, and gamma rays during the 1990 May 24 solar cosmic-ray event. Astrophys J 479(2, Part 1):997–1011

Debrunner H, Flückiger EO, Lockwood JA, Ryan JM, Barat C, Sunyaev R, Terekhov OV, Kuznetsov A (1998) Energy spectra of the protons interacting at the sun during the 1990 May 24 flare event. EOS, Transactions AGU, 78(45), AGU 1998 Fall Meeting San Francisco, CA, 1998, SH42B04, 1415h, F724-F724

DelSignore KW (1995) Ph.D. Thesis, Case Western Reserve University, Cleveland, OH

Dennis BR (1988) Solar flare hard X-ray observations. Sol Phys (Netherlands) 118(1–2):49–94

Ding MD, Qiu J, Wang H (2002) Non-LTE Calculation of the Ni I 676.8 Nanometer Line in a Flaring Atmosphere. Astrophys J Lett (USA) 576(1, Part 2):L83–L86

Donea A-C, Lindsey C (2004) Seismic waves from the solar flares. In: Danesy D (ed) Helio- and asteroseismology: towards a golden future. Proceedings of the SOHO 14/GONG 2004 workshop, 2004, New Haven, CT, (ESA SP-559), pp 152–157

Dunphy PP, Chupp EL (1992) Gamma-rays and neutrons as a probe of the proton spectrum during the solar flare of 1988 December 16. In: Zank GP, Gaisser TK (eds) Particle acceleration in cosmic plasmas, Newark, DE, 1991. AIP Conference Proceedings, vol 264, No 1, pp 253–256

Dunphy PP, Chupp EL (1994) "Extended phase" of solar flares observed by SMM. In: Ryan J, Vestrand WT (eds) High-energy solar phenomena – a new era of spacecraft measurements. Proceedings of the workshop held in Waterville Valley, NH, 1993. American Institute of Physics, New York. AIP Conference Proceedings, vol 294, No 1, pp 112–117

Dunphy PP, Chupp EL, Bertsch DL, Schneid EJ, Gottesman SR, Kanbach G (1999a) Gamma-rays and neutrons as a probe of flare proton spectra: the solar flare of 11 June 1991. Sol Phys (Netherlands) 187(1):45–57

Dunphy PP, Chupp EL, Rieger E (1990) Analysis of SMM GRS High-Energy (>10 MeV) Data from the Solar Flare of 1988 December 16, Proc. 21st Intern. Cosmic Ray Conf., Adelaide, 5, 75–77

Dunphy PP, Chupp EL, Schneid EJ, Bertsch DL (1999b) Neutron-capture gamma-rays observed by EGRET from the flare of 1991 October 27. In: Proceedings of the 26th international cosmic ray conference, Salt Lake City, UT, vol 6, pp 9–12

Elliot HL (1964) The nature of solar flares. Planet Space Sci 12(6):657–660

Emslie AG (1983) Chromospheric heating by electron and proton bombardment in the solar flare of June 7, 1980. Sol Phys (Netherlands) 84(1–2):263–269

Emslie AG, Henoux J-C, Mariska JT, Newton EK (1996) The viability of energetic protons as an agent for atmospheric heating during the impulsive phase of solar flares. Astrophys J Lett (USA) 470(2, Part 2):L131–L134

Emslie AG, Kontar EP, Krucker S, Lin RP (2003) RHESSI hard X-ray imaging spectroscopy of the large gamma-ray flare of 2002 July 23. Astrophys J Lett 595(2, Part 2):L107–L110

Emslie AG, Dennis BR, Holman GD, Hudson HS (2005) Refinements to flare energy estimates: a followup to "Energy partition in two solar flare/CME events" by A. G. Emslie et al. J Geophys Res 110(A11, A11103):1–4. doi:10.1029/2005JA011305

Evenson P, Meyer P, Pyle KR (1983) Protons from the decay of solar flare neutrons. Astrophys J 274(2, Part 1):875–882

Evenson P, Kroeger R, Meyer P, Reames D (1990) Solar neutron decay proton observations in cycle 21. Astrophys J Suppl Ser (USA) 73(2):273–277

Fermi E (1949) On the origin of the cosmic radiation. Phys Rev 75(8):1169–1174

Fishman GJ, Meegan CA, Wilson RB et al (1989) The burst and transient source experiment on the Gamma Ray Observatory. In: Johnson WN (ed) Proceedings of the Gamma Ray Observatory science workshop, 1989, Goddard Space Flight Center, Greenbelt, MD, vol 2, pp 39–50

Fletcher L (1997) Numerical simulations of coronal particle trapping. Astron Astrophys (Germany) 326(3):1259–1267

Fletcher L, Pollock JA, Potts HE (2004) Tracking of trace ultraviolet flare footpoints. Sol Phys (Netherlands) 222(2):279–298

Forbes TG, Isenberg PA (1991) A catastrophe mechanism for coronal mass ejections. Astrophys J 373(1, Part 1):294–307

Forman MA, Ramaty R, Zweibel EG (1986) The acceleration and propagation of solar flare energetic particles. In: Sturrock PA (ed) Physics of the sun, vol 2. D. Reidel Publishing, Dordrecht, pp 249–289

Forrest DJ, Chupp EL (1983) Simultaneous acceleration of electrons and ions in solar flares. Nature 305(5932):291–292

Forrest DJ, Murphy RJ (1988) Nuclear line spectroscopy of solar flares. Sol Phys (Netherlands) 118(1–2):123–135

Forrest DJ, Vestrand WT, Chupp EL, Rieger E, Cooper JF, Share GH (1985) Neutral pion production in solar flares. In: Proceedings of the 19th international cosmic ray conference, La Jolla, CA, vol 4, pp 146–149

Forrest DJ, Vestrand WT, Chupp EL, Rieger E, Cooper J, Share GH (1986) Very energetic gamma-rays from the June 3, 1982 solar flare1986. Adv Space Res 6(6):115–118

Gan WQ (1998) Spectral evolution of energetic protons in solar flares. Astrophys J (USA) 496(2, Part 1):992–997

Gan WQ (2000a) Energetic protons: spectral evolution and its significance. In: Ramaty R, Mandzhavidze N (eds) High energy solar physics workshop - anticipating HESSI, ASP Conference Series, vol 206, pp 439–444

Gan WQ (2000b) A comparison between the spectral parameters of both electron and proton dominated events. Astrophys Space Sci (Netherlands) 274(3):481–488

Gan WQ (2004) On both the time histories of the 0.511 MeV line and 2.223 MeV line from the X4.8 flare of 23 July 2002 observed with Rhessi. Sol Phys (Netherlands) 219(2):279–287

Gan WQ (2005) Gamma-ray line analysis for the flares of 28 October and 2 November 2003. Adv Space Res (UK) 35(10):1833–1838

Gan WQ, Fang C (1987) Time-match semi-empirical models of the chromospheric flare on 3 February, 1983. Sol Phys (Netherlands) 107(2):311–321

Gan WQ, Rieger E (1999) On temporal variation of the 0.511 MeV line in a solar flare. Sol Phys (Netherlands) 186(1–2):311–318

Gan WQ, Fang C, Zhang HQ (1991) A hydrodynamic model of the impulsive phase of a solar flare loop. Astron Astrophys (Germany) 241(2):618–624

Gan W, Kuzhevskij BM, Miroshnichenko LI, Troitskaia EV (2003) Time profile of the 2.223 MEV line emission and some features of the 16 December 1988 solar event. In: Proceedings of the ISCS 2003, solar variability as an input to the Earth's environment, Tatranska Lomnica, Slovakia, ESA SP, vol 535, pp 655–657

Gan WQ, Chang J, Li Y-P, Su Y, Miroshnichenko LI (2004) On positronium continuum and 0.511 MeV line in solar flares. China J Astron Astrophys 4(4):357–364

Gan WQ, Rieger E, Fang C (1993) Semiempirical flare models with chromospheric condensation, Astrophys. J., 416, No. 2, part 1, 886–892

Garrard TL, Stone EC (1993) New SEP-based solar abundances. In: Proceedings of the 23rd international cosmic ray conference, Calgary, vol 3, pp 384–387

Gingerich O, Noyes RW, Kalkofen W, Cuny Y (1971) The Harvard Smithsonian reference atmosphere. Sol Phys (Netherlands) 18(3):347–365

Glyanenko AS, Kotov YuD, Pavlov AV, Arkhangelsky AI, Samoilenko VT, Yurov VN, Pankov VM, Ryumin SP (1999) The AVS-F experiment on recording rapidly changing fluxes of cosmic and gamma radiation prepared under the CORONAS-F project. Instrum Exp Tech (Russia) 42(5):596–603

Gosling JT (1993) The solar flare myth. J Geophys Res 98(A11):18937–18950

Grevesse N (1984) Accurate atomic data and solar photospheric spectroscopy. Phys Scripta (Sweden) T8:49–58

Grevesse N and Noels A (1993) Cosmic abundances of the elements. In: Prantzos N, Vangioni-Flam E, Casse M (eds) Origin and evolution of the elements, Proceedings of a symposium in honour of H. Reeves, held in Paris, 1992. Cambridge University Press, Cambridge, UK, pp 15–25

Grevesse N, Noels A, Sauval AJ (1996) Standard abundances. In: Holt SS, Sonneborn G (eds) Cosmic abundances, ASP Conference Series, vol 99, Proceedings of the sixth annual October astrophysics conference in College Park, MD, 1995. ASP, San Francisco, CA, pp 117–126

Gros M, Tatischeff V, Kiener J, Cordier B, Chapuis C, Weidenspointner G, Vedrenne G, von Kienlin A, Diehl R, Bykov A, Méndez M (2004) INTEGRAL/SPI observation of the 2003 Oct 28 solar flare. In: Schönfelder V, Lichti G, Winkler C (eds) Proceedings of the 5th INTEGRAL science workshop on the INTEGRAL universe (ESA SP-552), 2004, Munich, Germany, pp 669–676

Gurnett DA, Kurth WS, Allendorf SC, Poynter RL (1993) Radio emission from the heliopause triggered by an interplanetary shock. Science 262(5131):199–203

Handy BN, Acton LW, Kankelborg CC, Wolfson CJ, Akin DJ, Bruner ME, Caravalho R, Catura RC, Chevalier R, Duncan DW, Edwards CG, Feinstein CN, Freeland SL, Friedlaender FM, Hoffmann CH, Hurlburt NE, Jurcevich BK, Katz NL, Kelly GA, Lemen JR, Levay M, Lindgren RW, Mathur DP, Meyer SB, Morrison SJ, Morrison MD, Nightingale RW, Pope TP, Rehse RA, Schrijver CJ, Shine RA, Shing L, Strong KT, Tarbell TD, Title AM, Torgerson DD, Golub L, Bookbinder JA, Caldwell D, Cheimets PN, Davis WN, Deluca EE, McMullen RA, Warren HP, Amato D, Fisher R, Maldonado H, Parkinson C (1999) The transition region and coronal explorer. Sol Phys (Netherlands) 187(2):229–260

Harris MJ, Share GH, Beall JH, Murphy RJ (1992) Upper limit on the steady emission of the 2.223 MeV neutron capture γ-ray line from the Sun. Sol Phys (Netherlands) 142(1):171–185

Hartman RC, Bertsch DL, Bloom SD, Chen AW, Deines-Jones P, Esposito JA, Fichtel CE, Friedlander DP, Hunter SD, McDonald LM, Sreekumar P, Thompson DJ, Jones BB, Lin YC, Michelson PF, Nolan PL, Tompkins WF, Kanbach G, Mayer-Hasselwander HA, Mücke A, Pohl M, Reimer O, Kniffen DA, Schneid EJ, von Montigny C, Mukherjee R, Dingus BL (1999) The third EGRET catalog of high-energy gamma-ray sources. Astrophys J Suppl Ser 123(1): 79–202

Heyvaerts J, Priest ER, Rust DM (1977) An emerging flux model for the solar flare phenomenon. Astrophys J 216(1, Part 1):123–137

Hollweg JV (1984) Resonances of coronal loops. Astrophys J (USA) 277(1, Part 1):392–403

Hollweg JV, Sterling AC (1984) Resonant heating: an interpretation of coronal loop data. Astrophys J Lett (USA) 282(1, Part 2):L31–L33

Holman GD, Sui L, Schwartz RA, Emslie AG (2003) Electron bremsstrahlung hard X-ray spectra, electron distributions, and energetics in the 2002 July 23 solar flare. Astrophys J Lett 595(2, Part 2):L97–L101

Hua X-M, Lingenfelter RE (1987a) Solar flare neutron production and the angular dependence of the capture gamma-ray emission. Sol Phys (Netherlands) 107(2):351–383

Hua X-M, Lingenfelter RE (1987b) A determination of the (He-3)/H ratio in the solar photosphere from flare gamma-ray line observations. Astrophys J (USA) 319(1, Part 1):555–566

Hua X-M, Ramaty R, Lingenfelter R (1989) Deexcitation gamma-ray line emission from solar flare magnetic loops. Astrophys J (USA) 341(1, Part 1):516–532

Hua X-M, Kozlovsky B, Lingenfelter RE, Ramaty R, Stupp A (2002) Angular and energy-dependent neutron emission from solar flare magnetic loops. Astrophys J Suppl Ser (USA) 140(2):563–579

Hudson HS (1972) Thick-target processes and white-light flares. Sol Phys 24(2):414–428

Hudson H, Ryan J (1995) High-energy particles in solar flares. Annu Rev Astron Astrophys 33:239–282

Hudson HS, Wolfson CJ, Metcalf TR (2006) White-light flares: a TRACE/RHESSI overview. Sol Phys (Netherlands) 234(1):79–93

Hulot E, Vilmer N, Trottet G (1989) Relative timing of solar prompt γ-ray line and X-ray emission expected from a trap plus precipitation model for proton and electrons. Astron Astrophys (West Germany) 213(1–2):383–396

Hurford GJ, Schmahl EJ, Schwartz RA, Conway AJ, Aschwanden MJ, Csillaghy A, Dennis BR, Johns-Krull C, Krucker S, Lin RP, McTiernan J, Metcalf TR, Sato J, Smith DM (2002) The RHESSI imaging concept. Sol Phys (Netherlands) 210(1–2):61–86

Hurford GJ, Schwartz RA, Krucker S, Lin RP, Smith DM, Vilmer N (2003a) First gamma-ray images of a solar flare. In: Proceedings of the 28th international cosmic ray conference, Tsukuba, vol 6, pp 3203–3206

Hurford GJ, Schwartz RA, Krucker S, Lin RP, Smith DM, Vilmer N (2003b) First gamma-ray images of a solar flare. Astrophys J Lett 595(2, Part 2):L77–L80

Hurford GJ, Krucker S, Lin RP, Schwartz RA, Share GH, Smith DM (2006) Gamma-ray imaging of the 2003 October/November solar flares. Astrophys J Lett 644(1, Part 2):L93–L96

Jakimiec J, Sylwester B, Sylwester J, Lemen JR, Mewe R, Bentley RD, Peres G, Serio S, Schrijver J (1986/1987) High-temperature plasma diagnostics of solar flares and comparison with model calculations. In: Stepanov VE, Obridko VN (eds) Solar maximum analysis, Proceedings of the international workshop, held in Irkutsk, USSR, 1985. VNU Science Press, Utrecht, The Netherlands, pp 91–102

Johnson WN, Kinzer RL, Kurfess JD, Strickman MS (1993) The Oriented Scintillation Spectrometer Experiment: instrument description. Astrophys J Suppl Ser (USA) 86(2):693–712

Kahler SW (1982) The role of the big flare syndrome in correlations of solar energetic proton fluxes and associated microwave burst parameters. J Geophys Res (USA) 87(A5):3439–3448

Kahler SW (1992) Solar flares and coronal mass ejections. Annu Rev Astron Astrophys 30:113–141

Kahler SW (1996) Coronal mass ejections and solar energetic particle events. In: Ramaty R, Mandzhavidze N, Hua X-M (eds) High energy solar physics, Proceedings of the conference held in Greenbelt, MD, 1995. AIP Conference Proceedings, vol 374, No 1, pp 61–77

Kahler SW, McAllister AH, Cane HV (2000) A search for interplanetary energetic particle events from solar posteruptive arcades. Astrophys J (USA) 533(2, Part 1):1063–1070

Kanbach G, Bertsch DL, Fichtel CE, Hartman RC, Hunter SD, Kniffen DA, Kwok PW, Lin YC, Mattox JR, Mayer-Hasselwander HA, Michelson PF, von Montigny C, Nolan PL, Pinkau K,

Rothermel H, Schneid E, Sommer M, Sreekumar P, Thompson DJ (1993) Detection of a long-duration solar gamma-ray flare on June 11, 1991 with EGRET on Compton-GRO. Astron Astrophys Suppl Ser (France) 97(1):349–353

Kane SR, Crannell CJ, Datlowe D, Feldman U, Gabriel A, Hudson HS, Kundu MR, Maetzler C, Neidig D, Petrosian V, Sheeley NR Jr (1980) Impulsive phase of solar flares. In: Solar flares: a monograph from SKYLAB Solar Workshop II. Colorado Associated University Press, Boulder, CO, pp 187–229

Kane SR, Chupp EL, Forrest DJ, Share GH, Rieger E (1986) Rapid acceleration of energetic particles in the 1982 February 8 solar flare. Astrophys J Lett Ed (USA) 300(2, Part 2):L95–L98

Kane SR, McTiernan JM, Loran J, Lemen J, Yoshimori M, Ohki K, Kosugi T (1993) Spatial and spectral characteristics of the X-ray sources in the 15 November 1991 solar flare. Adv Space Res 13(9):237–239

Kane SR, Hurley K, McTiernan JM, Sommer M, Boer M, Niel M (1995) Energy release and dissipation during giant solar flares. Astrophys J Lett (USA) 446(1, Part 2):L47–L50

Kane SR, Hurley K, McTiernan JM, Boer M, Niel M, Kosugi T, Yoshimori M (1998) Stereoscopic observations of solar hard X-ray flares made by Ulysses and Yohkoh. Astrophys J 500 (2):1003–1008

Kawabata K, Yoshimori M, Suga K, Morimoto K, Hiraoka T, Sato J, Ohki K (1994) Positron annihilation radiation from the 1991 November 15 flare. Astrophys J Suppl Ser (USA) 90 (2):701–705

Kiener J, Gros M, Tatischeff V, Weidenspointner G (2006) Properties of the energetic particle distributions during the October 28, 2003 solar flare from INTEGRAL/SPI observations. Astron Astrophys (France) 445(2):725–733

Kiplinger AL (1995) Comparative studies of hard X-ray spectral evolution in solar flares with high-energy proton events observed at Earth. Astrophys J (USA) 453(2, Part 1):973–986

Klein K-L, Trottet G, Lantos P, Delaboudinière J-P (2001) Coronal electron acceleration and relativistic proton production during the 14 July 2000 flare and CME. Astron Astrophys (France) 373(3):1073–1082

Kocharov LG, Kocharov GE (1984) ^{3}He-rich events. Space Sci Rev 38(1–2):89–141

Kocharov LG, Lee JW, Zirin H, Kovaltsov GA, Usoskin IG, Pyle KR, Shea MA, Smart DF (1994) Neutron and electromagnetic emissions during the 1990 May 24 solar flare. Sol Phys (Netherlands) 155(1):149–170

Kocharov LG, Kovaltsov GA, Torsti J, Usoskin IG, Zirin H (1995) Correlative investigations of the 1990 May 24 solar flare. In: Proceedings of the 24th international cosmic ray conference, Roma, vol 4, pp 159–162

Kocharov L, Debrunner H, Kovaltsov G, Lockwood J, McConnell M, Nieminen P, Rank G, Ryan J, Schonfelder V (1998) Deduced spectrum of interacting protons accelerated after the impulsive phase of the 15 June 1991 solar flare. Astron Astrophys (Germany) 340(1):257–264

Kosovichev AG, Zharkova VV (1999) Variations of photospheric magnetic field associated with flares and CMEs. Sol Phys (Netherlands) 190(1–2):459–466

Kosovichev AG, Zharkova VV (2001) Magnetic energy release and transients in the solar flare of 2000 July 14. Astrophys J Lett 550(1, Part 2):L105–L108

Kosugi T (1996) Solar flare energy release and particle acceleration as revealed by Yohkoh HXT. In: Ramaty R, Mandzhavidze N, Hua X-M (eds) High energy solar physics, Proceedings of the conference held in Greenbelt, MD, 1995. AIP Conference Proceedings, vol 374, No 1, pp 267–274

Kotov YuD, Bogovalov SV, Endalova OV, Yoshimori M (1996) ^{7}Li Production in Solar Flares. Astrophys J (USA) 473(1, Part 1):514–518

Kozlovsky B, Ramaty R (1974) 478-keV and 431-keV line emissions from alpha-alpha reactions. Astrophys J Lett Ed (USA) 191(1, Part 2):L43–L44

Kozlovsky B, Lingenfelter RE, Ramaty R (1987) Positrons from accelerated particle interactions. Astrophys J 316(2, Part 1):801–818

Kozlovsky B, Ramaty R, Lingenfelter RE (1997) Anisotropic broad nuclear gamma-ray lines: application to the COMPTEL observations of Orion. Astrophys J 484(1):286–295

Krucker S, Hurford GJ, Lin RP (2003) Hard X-ray source motions in the 2002 July 23 gamma-ray flare. Astrophys J Lett 595(2, Part 2):L103–L106

Kumar A, Rust DM (1996) Interplanetary magnetic clouds, helicity conservation, and current-core flux-ropes. J Geophys Res 101(7):15667–15684

Kuzhevskij BM (1982) Solar gamma astronomy and study of solar cosmic rays. Usp Fiz Nauk (USSR) 137(2):237–265, 375 (In Russian. Translation in English: Sov Phys-Usp (USA) 25(6): 392–408)

Kuzhevskij BM, Kogan-Laskina EI (1990) Solar gamma-radiation in connection with the absorption of neutrons and the evolution of the matter density in the subphotospheric region after a flare. Astron Zh 67(5):1022–1025 (In Russian. Translation in English: Sov Astron J 34(5): 514–515)

Kuzhevskij BM, Troitskaia EV (1989) Propagation of the neutrons from solar flares in the Sun's atmosphere. Preprint of Institute of Nuclear Physics, Moscow State University, pp 89–28/105, Moscow (In Russian)

Kuzhevskij BM, Troitskaia EV (1995) Relation of the intensity-time profile of solar flare 2.223 MeV gamma-emission with the heterogeneity of solar plasma density. Preprint SINP MSU, No. 95–8/372, Moscow State University, Moscow, pp 1–17 (In Russian)

Kuzhevskij BM, Troitskaia EV (1998) Absorption of 2.22 MeV solar flare γ-rays in different solar plasma density altitude profiles. In: Proceedings of the 16 European cosmic ray symposium, Alcala de Henares, Madrid, pp 201–204

Kuzhevskij BM, Troitskaia EV (2001) Probing the solar plasma by solar flare 2.223 MeV gamma-line emission and analysis of the experimental data. Russ Izv RAN Ser Phys 65(3):330–334

Kuzhevskij BM, Kuznetsov SN, Troitskaia EV (1998) Development of the solar flare plasma density investigation method based on the 2.2 MeV gamma-line time profile analysis. Adv Space Res 22(7):1141–1147

Kuzhevskij BM, Miroshnichenko LI, Troitskaia EV (2001a) Application of the solar plasma density investigation method based on the solar flare 2.223 MeV gamma-line time profile analysis to the experimental data. In: Proceedings of the 27th international cosmic ray conference, Hamburg, vol 8, p 3060

Kuzhevskij BM, Miroshnichenko LI, Troitskaia EV (2001b) Derivation of density profiles in the solar atmosphere by the 2.223 MeV line data for the 6 November 1997 flare. In: Proceedings of the 27th international cosmic ray conference, Hamburg, Vol. Invited, Rapporteur, and Highlight Papers, pp 285–288

Kuzhevskij BM, Miroshnichenko LI, Troitskaya EV (2002) Determining of solar plasma density altitude profiles by solar flare 2.22 MeV γ-line data for two events. Izv RAN Ser Phys 66(11):1673–1677

Kuzhevskij BM, Miroshnichenko LI, Troitskaya EV (2005) Gamma-ray radiation with energy 2.223 MeV and the density distribution in the solar atmosphere during flares. Astron Rep 49(7):567–578

Kuznetsov SN, Bogomolov AV, Gordeev YuV, Gotselyuk YuV, Denisov YuI, Dmitriev AV, Kovalevskaya MA, Lupenko GV, Myagkova IN, Panasyuk MI, Podorolskii AN, Ryumin SP, Yakovlev BM, Oraevskii VN, Klepikov VY, Kopaev IM, Stepanov AI, Kudela K, Roiko I, Fisher S, Polashek T, Silvester Y, Kordylevski Z (1995) Preliminary-results of the solar cosmic-ray experiment on koronas-i satellite. Izv Ross Akad Nauk Ser Fiz (Russia) 59(4):2–6

Kuznetsov SN, Kudela K, Myagkova IN, Yushkov BYu (2003a) Gamma and X-ray solar flare emissions: CORONAS-F measurements. In: Proceedings of the 28th international cosmic ray conference, Tsukuba, vol 6, pp 3183–3186

Kuznetsov SN, Bogomolov AV, Denisov YuI, Kordylewski Z, Kudela K, Kurt VG, Lisin DV, Myagkova IN, Podorol'skii AN, Podosenova TB, Svertilov SI, Sylwester J, Stepanov AI, Yushkov BYu (2003b) The solar flare of November 4, 2001, and its manifestations in energetic particles from Coronas-F data. Sol Syst Res (Russia) 37(2):121–127

Kuznetsov SN, Kurt VG, Myagkova IN, Yushkov BY, Kudela K (2006) Gamma-ray emission and neutrons from solar flares recorded by the SONG instrument in 2001–2004. Sol Syst Res (Russia) 40(2):104–110

Lampton M, Margon B, Bowyer S (1976) Parameter estimation in X-ray astronomy. Astrophys J (USA) 208(1, Part 1):177–190

Lau Y-T, Ramaty R (1994) Trapping of protons in twisted magnetic loops. In: Ryan J, Vestrand WT (eds) High-energy solar phenomena - a new era of spacecraft measurements, Proceedings of the workshop held in Waterville Valley, NH, 1993. American Institute of Physics, New York. AIP Conference Proceedings, vol 294, No 1, pp 71–73

Lau Y-T, Northrop TG, Finn JM (1993) Long-term containment of energetic particles in coronal loops. Astrophys J (USA) 414(2, Part 1):908–915

Lee MA (1982) Coupled hydromagnetic wave excitation and ion acceleration upstream of the Earth's bow shock. J Geophys Res 87(A7):5063–5080

Lee MA (1983) Coupled hydromagnetic wave excitation and ion acceleration at interplanetary traveling shocks. J Geophys Res (USA) 88(A8):6109–6119

Lee MA (1994) Stochastic Fermi acceleration and solar cosmic rays. In: Ryan J, Vestrand WT (eds) High-energy solar phenomena – a new era of spacecraft measurements, Proceedings of the workshop held in Waterville Valley, NH, 1993. American Institute of Physics, New York. AIP Conference Proceedings, vol 294, No 1, pp 134–142

Lee MA (1997) Particle acceleration and transport at CME-driven shocks. In: Crooker N, Jocelyn JA, Feynman J (eds) Coronal mass ejections, vol 99, Geophysical Monograph. AGU, Washington, DC, pp 227–234

Lee MA, Ryan JM (1986) Time-dependent coronal shock acceleration of energetic solar flare particles. Astrophys J (USA) 303(2, Part 1):829–842

Leka KD, Canfield RC, McClymont AN, Van Driel-Gesztelyi L (1996) Evidence for current-carrying emerging flux. Astrophys J 462(1, Part 1):547–560

Lin RP (1974) Non-relativistic solar electrons. Space Sci Rev (Netherlands) 16(1–2):189–256

Lin RP (2000) The High Energy Solar Spectroscopic Imager (HESSI) mission. In: Ramaty R, Mandzhavidze N (eds) High energy solar physics, Workshop - Anticipating HESSI, ASP Conference Series, vol 206, San Francisco, CA, pp 1–12

Lin RP, Hudson HS (1976) Non-thermal processes in large solar flares. Sol Phys (Netherlands) 50 (1):153–178

Lin RP, Schwartz RA, Pelling RM, Hurley KC (1981) A new component of hard X-rays in solar flares. Astrophys J Lett Ed (USA) 251(2, Part 2):L109–L114

Lin AC, Nightingale RW, Tarbell TD (2001) Diffraction pattern analysis of bright trace flares. Sol Phys (Netherlands) 198(2):385–398

Lin RP, Dennis BR, Hurford GJ, Smith DM, Zehnder A, Harvey PR, Curtis DW, Pankow D, Turin P, Bester M, Csillaghy A, Lewis M, Madden N, van Beek HF, Appleby M, Raudorf T, McTiernan J, Ramaty R, Schmahl E, Schwartz R, Krucker S, Abiad R, Quinn T, Berg P, Hashii M, Sterling R, Jackson R, Pratt R, Campbell RD, Malone D, Landis D, Barrington-Leigh CP, Slassi-Sennou S, Cork C, Clark D, Amato D, Orwig L, Boyle R, Banks IS, Shirey K, Tolbert AK, Zarro D, Snow F, Thomsen K, Henneck R, Mchedlishvili A, Ming P, Fivian M, Jordan J, Wanner R, Crubb J, Preble J, Matranga M, Benz A, Hudson H, Canfield RC, Holman GD, Crannell C, Kosugi T, Emslie AG, Vilmer N, Brown JC, Johns-Krull C, Aschwanden M, Metcalf T, Conway A (2002) The Reuven Ramaty High-Energy Solar Spectroscopic Imager (RHESSI). Sol Phys (Netherlands) 210(1–2):3–32

Lin RP, Krucker S, Holman GD, Sui L, Hurford GJ, Schwartz RA (2003) RHESSI discovery of a coronal non-thermal hard X-ray source in the 23 July 2002 gamma-ray line flare. In: Proceedings of the 28th international cosmic ray conference, Tsukuba, vol 6, pp 3207–3210

Lingenfelter RE (1969) Solar flare optical, neutron and gamma-ray emission. Sol Phys (Netherlands) 8(2):341–347

Lingenfelter RE (1994) Solar flare neutrons and gamma rays. In: Ryan J, Vestrand WT (eds) High-energy solar phenomena – a new era of spacecraft measurements, Proceedings of the workshop held in Waterville Valley, NH, 1993. American Institute of Physics, New York. AIP Conference Proceedings, vol 294, No 1, pp 77–88

Litvinenko YE (1996a) Aspects of particle acceleration in solar flares. Ph.D. Thesis, University of New Hampshire, Manchester, NH

Litvinenko YE (1996b) Particle acceleration in reconnecting current sheets with a nonzero magnetic field. Astrophys J 462(2, Part 1):997–1004

Litvinenko YE, Somov BV (1993) Particle acceleration in reconnecting current sheets. Sol Phys (Netherlands) 146(1):127–133

Litvinenko YE, Somov BV (1995) Relativistic acceleration of protons in reconnecting current sheets of solar flares. Sol Phys (Netherlands) 158(2):317–330

Lockwood JA, Debrunner H, Ryan JM (1997) The relationship between solar flare gamma-ray emission and neutron production. Sol Phys (Netherlands) 173(1):151–176

Lockwood JA, Debrunner H, Ryan JM (1999) Interacting and interplanetary high-energy protons in solar flare events. Astropart Phys (Netherlands) 12(1–2):97–105

Machado ME, Avrett EH, Vernazza JE, Noyes RW (1980) Semiempirical models of chromospheric flare regions. Astrophys J 242(1, Part 1):336–351

MacKinnon AL (1989) A potential diagnostic for low energy, nonthermal protons in solar flares. Astron Astrophys (West Germany) 226(1):284–287

Mandzhavidze N, Ramaty R (1992a) Gamma rays from pion decay: evidence for long-term trapping of particles in solar flares. Astrophys J Lett (USA) 396(2, Part 2):L111–L114

Mandzhavidze N, Ramaty R (1992b) High-energy gamma-ray emission from pion decay in a solar flare magnetic loop. Astrophys J (USA) 389(2, Part 1):739–755

Mandzhavidze N, Ramaty R (1993) Particle acceleration in solar flares. Nucl Phys B-Proc Sup B33 (1–2):141–160

Mandzhavidze N, Ramaty R (1996) Determination of element abundances in solar atmosphere and solar flare accelerated particles using gamma ray spectroscopy. Bull AAS 28(2):858–858, 188th AAS Meeting, Madison, WI, 1996

Mandzhavidze N, Ramaty R, Bertsch DL, Schneid EJ (1996) Pion decay and nuclear line emissions from the 1991 June 11 flare. In: Ramaty R, Mandzhavidze N, Hua X-M (eds) High energy solar physics, Greenbelt, MD, 1995. AIP Conference Proceedings, vol 374, No 1, pp 225–236

Mandzhavidze N, Ramaty R, Kozlovsky B (1997) Solar atmospheric and solar flare accelerated helium abundances from gamma-ray spectroscopy. Astrophys J Lett 489(1, Part 2):L99–L102

Martens PCH (1988) The generation of proton beams in two-ribbon flares. Astrophys J Lett (USA) 330(2, Part 2):L131–L133

Mason GM, Reames DV, Klecker B, Hovestadt D, von Rosenvinge TT (1986) The heavy-ion compositional signature in ^{3}He-rich solar particle events. Astrophys J 303(2, Part 1):849–860

Masuda S, Kosugi T, Hara H, Tsuneta S, Ogawara Y (1994) A loop-top hard X-ray source in a compact solar flare as evidence for magnetic reconnection. Nature (UK) 371(6497):495–497

Masuda S, Kosugi T, Hudson HS (2001) A hard X-ray two-ribbon flare observed with Yohkoh/HXT. Sol Phys (Netherlands) 204(1–2):57–69

McConnell M (1994) An overview of solar flare results from COMPTEL. In: Ryan J, Vestrand WT (eds) High-energy solar phenomena - a new era of spacecraft measurements, Proceedings of the workshop held in Waterville Valley, NH, 1993. American Institute of Physics, New York, AIP Conference Proceedings, vol 294, pp 21–25

McConnell M, Bennett K, Forrest D, Hanlon L, Ryan J, Schönfelder V, Swanenburg BN, Winkler C (1993) Comptel observations of solar flare gamma-rays. Adv Space Res 13(9):245–248

McConnell ML, Bennett K, MacKinnon A, Miller R, Rank G, Ryan JM, Schönfelder V (1997) A search for MeV gamma-ray emission from the quiet-time sun. In: Proceedings of the 25th international cosmic ray conference, Durban, vol 1, pp 13–16

McDonald FB, van Hollebeke MAI (1985) HELIOS 1 energetic particle observations of the solar gamma-ray/neutron flare events of 1982 June 3 and 1980 June 21. Astrophys J Lett (USA) 290 (2, Part 2):L67–L71

McDonald FB, Keppler E, Reuss MK (1995) Ulysses-Voyager study of energetic particles associated with the intense solar activity of March/June 1991. Space Sci Rev (Netherlands) 72(1–2):359–364

McTiernan JM, Kane SR, Hurley K, Laros JG, Fenimore EE, Klebsadel RW, Sommer M, Yoshimori M (1994) Multi spacecraft observations and thick-target electron beam models for the 15-Nov-1991 flare. In: Proceedings of Kofu symposium, Kofu, Japan, 1993, NRO report No. 360, pp 389–392

Meerson BI, Rogachevskii IV (1983) On the storage of high-energy protons in the solar corona: the cyclotron instability. Sol Phys (Netherlands) 87(2):337–357

Meyer JP (1985a) The baseline composition of solar energetic particles. Astrophys J Suppl Ser 57(1):151–171

Meyer JP (1985b) Solar-stellar outer atmospheres and energetic particles, and galactic cosmic rays. Astrophys J Suppl Ser (USA) 57(1):173–204

Meyer J-P (1996) Solar coronal abundance anomalies. In: Holt SS, Sonneborn G (eds) Cosmic abundances, ASP Conference Series, vol 99, Proceedings of the sixth annual October astrophysics conference in College Park, MD, 1995. ASP, San Francisco, CA, pp 127–146

Miller JA, Roberts DA (1995) Stochastic proton acceleration by cascading Alfvén waves in impulsive solar flares. Astrophys J (USA) 452(2, Part 1):912–932

Miller JA, Vinas AF (1993) Ion acceleration and abundance enhancements by electron beam instabilities in impulsive solar flares. Astrophys J (USA) 412(1, Part 1):386–400

Moore RL, LaBonte B (1980) The filament eruption in the 3B flare of July 29, 1973 – Onset and magnetic field configuration. In: Dryer M, Tandberg-Hanssen E (eds) Solar and interplanetary dynamics, Proceedings of the symposium, Cambridge, MA, 1979, IAU Symposium, No. 91. D. Reidel Publishing, Dordrecht, pp 207–211

Moore RL, Hurford GJ, Jones HP, Kane SR (1984) Magnetic changes observed in a solar flare. Astrophys J 276(1, Part 1):379–390

Muraki Y, Murakami K, Miyazaki M, Mitsui K, Shibata S, Sakakibara S, Sakai T, Takahashi T, Yamada T, Yamaguchi K (1992) Observation of solar neutrons associated with the large flare on 1991 June 4. Astrophys J Lett (USA) 400(2, Part 2):L75–L78

Murphy RJ (1985) Gamma-rays and neutrons from solar flares. Ph.D. Dissertation, University of Maryland, College Park, MD

Murphy RJ (2007) Solar gamma-ray spectroscopy. Space Sci Rev 130(1–4):127–138

Murphy RJ, Ramaty R (1984) Solar-flare neutrons and gamma-rays. Adv Space Res 4(7):127–136

Murphy RJ, Dermer CD, Ramaty R (1987) High-energy processes in solar flares. Astrophys J Suppl Ser (USA) 63(3):721–748

Murphy RJ, Kozlovsky B, Ramaty R (1988) Gamma-ray spectroscopic tests for the anisotropy of accelerated particles in solar flares. Astrophys J (USA) 331(2, Part 1):1029–1035

Murphy RJ, Hua X-M, Kozlovsky B, Ramaty R (1990a) ^{7}Li and ^{7}Be de-excitation lines: probes for accelerated particle transport models in solar flares. Astrophys J (USA) 351(1, Part 1):299–308

Murphy RJ, Share GH, Letaw JR, Forrest DJ (1990b) Nuclear line spectroscopy of the 1981 April 27 solar flare. Astrophys J (USA) 358(1, Part 1):298–312

Murphy RJ, Ramaty R, Kozlovsky B, Reames DV (1991) Solar abundances from gamma-ray spectroscopy: comparisons with energetic particle, photospheric, and coronal abundances. Astrophys J (USA) 371(2, Part 1):793–803

Murphy RJ, Share GH, Forrest DJ, Grabelsky DA, Grove JE, Jensen CM, Johnson WN, Jung GV, Kinzer RL, Kroeger RA, Kurfess JD, Matz SM, Purcell WR, Strickman MS, Ulmer MP, Vestrand WT (1993a) OSSE Observations of the 4 June 1991 Solar Flare. In: Proceedings of the 23rd international cosmic ray conference, University of Calgary, Alberta, Canada, 1993, Invited, Rapporteur, and Highlight Papers, vol 3, pp 99–102

Murphy RJ, Share GH, Grove JE, Johnson WN, Kinzer RL, Kroeger RA, Kurfess JD, Strickman MS, Matz SM, Grabelsky DA, Purcell WR, Ulmer MP, Cameron RA, Jung GV, Jensen CM, Vestrand WT, Forrest DJ (1993b) OSSE observations of solar flares. In: Friedlander M,

Gehrels N, Macomb DJ (eds) Compton gamma-ray observatory, St. Louis, MO, 1992, AIP Conference Proceedings, vol 280, No 1, pp 619–630

Murphy RJ, Share GH, Grove JE, Johnson WN, Kinzer RL, Kroeger RA, Kurfess JD, Strickman MS, Matz SM, Purcell WR, Ulmer MP, Grabelsky DA, Jung GV, Jensen CM, Forrest DJ, Vestrand WT (1994) OSSE observations of the 4 June 1991 solar flare. In: Ryan J, Vestrand WT (eds) High-energy solar phenomena - a new era of spacecraft measurements, Proceedings of the workshop held in Waterville Valley, NH, 1993. American Institute of Physics, New York, AIP Conference Proceedings, vol **294**, No 1, pp 15–20

Murphy RJ, Share GH, Grove JE, Johnson WN, Kinzer RL, Kurfess JD, Strickman MS, Jung GV (1996) Abundance study of the 4 June 1991 solar flare using CGRO/OSSE spectral data. In: Ramaty R, Mandzhavidze N, Hua X-M (eds) High energy solar physics, Proceedings of the conference held in Greenbelt, MD, USA, 1995. AIP Conference Proceedings, vol 374, No 1, pp 184–193

Murphy RJ, Share GH, Grove JE, Johnson WN, Kinzer RL, Kurfess JD, Strickman MS, Jung GV (1997) Accelerated particle composition and energetics and ambient abundances from gamma-ray spectroscopy of the 1991 June 4 solar flare. Astrophys J (USA) 490(2, Part 1):883–900

Murphy RJ, Share GH, DelSignore KW, Hua X-M (1999) Surprisingly intense neutron emission from a flare behind the limb of the Sun. Astrophys J 510(2, Part 1):1011–1017

Murphy RJ, Share GH, Hua X-M, Lin RP, Smith DM, Schwartz RA (2003a) Physical implications of RHESSI neutron-capture line measurements. Astrophys J Lett (USA) 595(2, Part 2): L93–L96

Murphy RJ, Smith DM, Share GH, Schwartz RA, Shih AY, Lin RP (2003b) Solar gamma-ray lines at high resolution with RHESSI. In: Proceedings of the 28th international cosmic ray conference, Tsukuba, vol 6, pp 3187–3190

Murphy RJ, Share GH, Hua X-M, Lin RP, Smith DM, Schwartz RA (2003c) Physical implications of RHESSI neutron capture-line measurements. In: Proceedings of the 28th international cosmic ray conference, Tsukuba, vol 6, pp 3195–3198

Murphy RJ, Kozlovsky B, Share GH, Hua X-M, Lingenfelter RE (2007) Using gamma-ray and neutron emission to determine solar flare accelerated particle spectra and composition and the conditions within the flare magnetic loop. Astrophys J Suppl Ser 168(1):167–194

Myagkova IN, Kuznetsov SN, Kurt VG, Yuskov BYu, Galkin VI, Muravieva EA, Kudela K (2007) X-ray, γ-emission and energetic particles in near-Earth space as measured by CORONAS-F satellite: from maximum to minimum of the last solar cycle. Adv Space Res (UK) 40(12):1929–1934

Nieminen P (1997) Neutrons from the 15 June 1991 solar flare. Ph.D Thesis, University of Bern, Bern, Switzerland

Northrop TG (1963) Adiabatic charged-particle motion. Rev Geophys 1(3):283–304

Oraevskii VN, Sobel'man II, Zhitnik IA, Kuznetsov VD (2002) Comprehensive solar studies by CORONAS-F satellite: new results. Physics-Uspekhi 45(8):886–896

Panasyuk MI, Bogomolov AV, Germantsev Yu, Kudryavtsev MI, Kuzhevsky BM, Kuznetsov SN, Lyagushin VI, Myagkova IN, Ryumin SP, Shavrin PI, Sobolevsky N, Svertilov SI, Yushkov BYu (2000) Modeling of neutron fluxes on near-earth spacecrafts from experimental results of SINP. In: Proceedings of the RADEX 2000 workshop, Louvain, Belgium, pp 9–19

Pelaez F, Mandrou P, Niel M, Mena B, Vilmer N, Trottet G, Lebrun F, Paul J, Terekhov O, Sunyaev R, Churazov E, Gilfanov M, Denisov D, Kuznetsov A, Dyachkov A, Khavenson N (1992) Solar hard X-ray and γ-ray bursts observed by the SIGMA anti-coincidence shield aboard Granat. Sol Phys (Netherlands) 140(1):121–138

Prince TA, Ling JC, Mahoney WA, Riegler GR, Jacobson AS (1982) A high-resolution measurement of the 2.223 MeV neutron capture line in a solar flare. Astrophys J Lett Ed (USA) 255 (1, Part 2):L81–L84

Prince TA, Forrest DJ, Chupp EL, Kanbach G, Share GH (1983a) The time history of 2.22 MeV line emission in solar flares. In: Proceedings of the 18th international cosmic ray conference, Bangalore, vol 4, pp 79–82

Prince TA, Forrest DJ, Chupp EL et al (1983b) Proceedings of the 18th international cosmic ray conference, vol 5, p 1644
Qiu J, Gary DE (2003) Flare-related magnetic anomaly with a sign reversal. Astrophys J 599 (1, Part 1):615–625
Ramaty R (1986) Nuclear processes in solar flares. In: Sturrock PA, Holzer TE, Mihalas D, Ulrich RK (eds) The physics of the Sun, vol 2, Geophysics and Astrophysics Monographs. D. Reidel Publishing, Dordrecht, pp 291–323
Ramaty R (1995) Astrophysical gamma ray lines from accelerated particle interactions. In: Signore M, Salati P, Vedrenne G (eds) The gamma ray sky with Compton GRO and SIGMA, Les Houches, France, 1994, Kluwer, Dordrecht, pp 279–301
Ramaty R, Mandzhavidze N (1994) Theoretical models for high-energy solar flare emissions. In: Ryan J, Vestrand WT (eds) High-energy solar phenomena – a new era of spacecraft measurements. Proceedings of the workshop held in Waterville Valley, NH, 1993. American Institute of Physics, New York. AIP Conference Proceedings, vol 294, No 1, pp 26–44
Ramaty R, Mandzhavidze N (2000) Gamma-rays from Solar Flares. In: Martens PCH, Tsuruta S, Weber MA (eds) Highly energetic physical processes and mechanisms for emission from astrophysical plasmas, Proceedings of the IAU symposium, vol 195, held at Montana State University, Bozeman, MT, 1999. ASP, San Francisco, CA, pp 123–132
Ramaty R, Murphy RJ (1987) Nuclear processes and accelerated particles in solar flares. Space Sci Rev 45(3–4):213–268
Ramaty R, Kozlovsky B, Lingenfelter RE (1975) Solar gamma rays. Space Sci Rev 18(3):341–388
Ramaty R, Kozlovsky B, Lingenfelter RE (1979) Nuclear gamma-rays from energetic particle interactions. Astrophys J Suppl (USA) 40(3):487–526
Ramaty R, Colgate SA, Dulk GA, Hoyng P, Knight JW, Lin RP, Melrose DB, Paizis C, Orrall F, Shapiro PR, Smith DF, Van Hollebeke M (1980) Energetic particles in solar flares. In: Sturrock PA (eds) Solar flares: a monograph from Skylab Solar Workshop II, Chapter 4. Colorado Associated University Press, Boulder, CO, pp 117–185
Ramaty R, Murphy RJ, Dermer CD (1987) On the origin of the pion-decay radiation in the 1982 June 3 solar flare. Astrophys J Lett Ed (USA) 316(1, Part 2):L41–L44
Ramaty R, Dennis BR, Emslie AG (1988) Gamma-ray, neutron, and hard X-ray studies and requirements for a high-energy solar physics facility. Sol Phys (Netherlands) 118(1–2):17–46
Ramaty R, Mandzhavidze N, Kozlovsky B, Skibo JG (1993) Acceleration in solar flares: Interacting particles versus interplanetary particles. Adv Space Res (UK) 13(9):275–284
Ramaty R, Schwartz RA, Enome S, Nakajima H (1994) Gamma-ray and millimeter-wave emissions from the 1991 June X-class solar flares. Astrophys J (USA) 436(2, Part 1):941–949
Ramaty R, Mandzhavidze N, Kozlovsky B, Murphy RJ (1995) Solar atmospheric abundances and energy content in flare-accelerated ions from gamma-ray spectroscopy. Astrophys J Lett (USA) 455(2, Part 2):L193–L196
Ramaty R, Mandzhavidze N, Kozlovsky B (1996) Solar atmospheric abundances from gamma ray spectroscopy. In: Ramaty R, Mandzhavidze N, Hua X-M (eds) High energy solar physics, Greenbelt, MD, 1995. AIP Conference Proceedings, vol 374, No 1, pp 172–183
Ramaty R, Mandzhavidze N, Barat C, Trottet G (1997) The giant 1991 June 1 flare: evidence for gamma-ray production in the corona and accelerated heavy ion abundance enhancements from gamma-ray spectroscopy. Astrophys J (USA) 479(1, Part 1):458–463
Rank G (1996) Gamma rays and neutrons of the solar flares on 11 and 15 June 1991 measured with COMPTEL. Ph.D Thesis, Technical University of Munich, Germany
Rank G, Diehl R, Lichti GG, Schönfelder V, Varendorff M, Swanenburg BN, Forrest D, Macri J, McConnell M, Ryan J, Bennett K, Hanlon L, Winkler C (1993) Observations of the 1991 June 11 solar flare with COMPTEL. In: Gehrels N (eds) Compton gamma-ray observatory, St. Louis, MO, 1992. AIP Conference Proceedings, vol 280, No 1, pp 661–668
Rank G, Diehl R, Lichti GG, Schönfelder V, Varendorff M, Swanenburg BN, van Dijk R, Forrest D, Macri J, McDonnell M, Loomis M, Ryan J, Bennett K, Winkler C (1994) Observations of the 1991 June 11 solar flare with COMPTEL. In: Ryan J, Vestrand WT (eds) High-energy solar

phenomena – a new era of spacecraft measurements, Proceedings of the workshop held in Waterville Valley, NH, 1993. American Institute of Physics, New York. AIP Conference Proceedings, vol 294, No 1, pp 100–105

Rank G, Bennett K, Bloemen H, Debrunner H, Lockwood J, McConnell M, Ryan J, Schönfelder V, Suleiman R (1996) Extended -ray emission in solar flares. In: Ramaty R, Mandzhavidze N, Hua X-M (eds) High energy solar physics, Greenbelt, MD, 1995. AIP Conference Proceedings, vol 374, No 1, pp 219–224

Rank G, Debrunner H, Lockwood J, McConnell M, Ryan J, Schönfelder V (1997a) Extended gamma-ray emission of the solar flares in June 1991. In: Proceedings of the 25th international cosmic ray conference, Durban, vol 1, pp 5–8

Rank G, Debrunner H, Kocharov L, Kovaltsov G, Lockwood J, McConnell M, Nieminen P, Ryan J, Schönfelder V (1997b) The solar flare event on 15 June 1991. In: Proceedings of the 25th international cosmic ray conference, Durban, vol 1, pp 1–4

Rank G, Ryan J, Debrunner H, McConnell M, Schonfelder V (2001) Extended gamma-ray emission of the solar flares in June 1991. Astron Astrophys (France) 378(3):1046–1066

Raymond JC, Cox DP, Smith BW (1976) Radiative cooling of a low-density plasma. Astrophys J (USA) 204(1, Part 1):290–292

Reames DV (1995) Coronal abundances determined from energetic particles. Adv Space Res 15 (7):41–51

Reames DV (1996) Energetic particles from solar flares and coronal mass ejections. In: Ramaty R, Mandzhavidze N, Hua X-M (eds) High energy solar physics, Proceedings of the conference held in Greenbelt, MD, 1995. AIP Conference Proceedings, vol 374, No 1, pp 35–44

Reames DV (1999) Particle acceleration at the Sun and in the heliosphere. Space Sci Rev (Netherlands) 90(3–4):413–491

Reames DV, Cane HV, von Rosenvinge TT (1990) Energetic particle abundances in solar electron events. Astrophys J (USA) 357(1, Part 1):259–270

Reames DV, Meyer JP, von Rosenvinge TT (1994) Energetic-particle abundances in impulsive solar flare events. Astrophys J Suppl Ser (USA) 90(2):649–667

Rieger E (1994) Gamma-ray precursors of solar flares. Astrophys J Suppl Ser (USA) 90 (2):645–648

Rieger E (1996) Spectral evolution of an intense gamma-ray line flare. In: Ramaty R, Mandzhavidze N, Hua X-M (eds) High energy solar physics, Proceedings of the conference held in Greenbelt, MD, 1995. AIP Conference Proceedings, vol 374, No 1, pp 194–199

Rieger E, Marschhäuser H (1990) Electron dominated events during solar flares. In: Winglee RM, Kiplinger AL (eds) Max '91, Workshop #3: Max '91/SMM solar flares: observations and theory, Proceedings of a conference, Estes Park, CO, 1990, pp 68–76

Ryan JM (1986) Particle propagation effects on solar flare X- and γ-ray time profiles. Sol Phys (Netherlands) 105(2):365–382

Ryan JM (1994) Solar physics with the Compton Gamma Ray Observatory. In: Gehrels N, Norris JP, Fichtel CE (eds) The second Compton symposium, College Park, MD, 1993. AIP Conference Proceedings, vol 304, No 1, pp 12–21

Ryan JM (2000) Long-duration solar gamma-ray flares. Space Sci Rev (Netherlands) 93 (3–4):581–610

Ryan JM, Lee MA (1991) On the transport and acceleration of solar flare particles in a coronal loop. Astrophys J (USA) 368(1, Part 1):316–324

Ryan JM, Schoenfelder V, Diehl R, Lichti GG, Steinle H, Swanenburg B, Aarts H, Deerenberg A, Hermsen W, Kiers G, Lockwood J, Macri J, Morris D, Simpson G, Bennett K, Eymann G, Snelling M, Winkler C, Byrd R, Foster C, Taddeucci T (1987) Solar neutron spectroscopy with COMPTEL on the Gamma Ray Observatory. In: Proceedings of the 20th international cosmic ray conference, Moscow, vol 4, pp 425–428

Ryan J, Aarts H, Bennett K, Byrd R, de Vries C, den Herder JW, Deerenberg AJM, Diehl R, Eymann G, Forrest DJ, Foster C, Hermsen W, Lockwood J, Lichti G, Macri J, McConnell M, Morris D, Schönfelder V, Simpson G, Snelling M, Steinle H, Strong A, Swanenburg BN,

Taddeucci T, Webber WR, Winkler C (1992) COMPTEL as a solar gamma-ray and neutron detector. In: Buccheri R, Crane P, di Gesù V, Maccarone MC, Scarsi L, Zimmermann HU (eds) Data analysis in astronomy IV, vol 59. Plenum Press, New York, pp 261–270

Ryan J, Forrest D, Lockwood J, Loomis M, McConnell M, Morris D, Webber W, Rank G, Schönfelder V, Swanenburg BN, Bennett K, Hanlon L, Winkler C, Debrunner H (1993b) COMPTEL gamma ray and neutron measurements of solar flares. In: Gehrels N (eds) Compton gamma-ray observatory, St. Louis, MO, 1992, Compton Gamma-Ray Observatory Workshop. AIP Conference Proceedings, vol 280, No 1, pp 631–642

Ryan J, Forrest D, Lockwood J, Loomis M, McConnell M, Morris D, Webber W, Bennett K, Hanlon L, Winkler C, Debrunner H, Rank G, Schönfelder V, Swanenburg BN (1994a) Neutron and gamma-ray measurements of the solar flare of 1991 June 9. In: Ryan J, Vestrand WT (eds) High-energy solar phenomena – a new era of spacecraft measurements, Proceedings of the workshop held in Waterville Valley, NH, 1993. American Institute of Physics, New York. AIP Conference Proceedings, vol 294, No 1, pp 89–93

Ryan JM, Bennett E, Lee MA (1994b) Proton acceleration in long duration flares. In: Belvedere G, Rodonó M, Simnett GM (eds) Lecture notes in physics, Advances in solar physics, Proceedings of the seventh European Meeting on solar physics, held in Catania, Italy, 1993, vol 432. Springer, Berlin, Heidelberg, pp 273–278

Sakao T, Kosugi T, Masuda S, Inda M, Makishima K, Canfield RC, Hudson HS, Metcalf TR, Wuelser J-P, Acton LW, Ogawara Y (1992) Hard X-ray imaging observations by Yohkoh of the 1991 November 15 solar flare. Publ Astron Soc Jpn 44(5):L83–L87

Schlickeiser R (1986) Stochastic particle acceleration in cosmic objects. In: Shapiro MM (eds) Cosmic radiation in contemporary astrophysics, Proceedings of the Advanced Study Institute, Erice, Italy, 1984, D. Reidel Publishing, Dordrecht, pp 27–55

Schneid EJ, Brazier KTS, Kanbach G, von Montigny C, Mayer-Hasselwander HA, Bertsch DL, Fichtel CE, Hartman RC, Hunter SD, Thompson DJ, Dingus BL, Sreekumar P, Lin YC, Michelson PF, Nolan PL, Kniffen DA, Mattox JR (1994) EGRET observations of extended high-energy emissions from the nuclear line flares of June 1991. In: Ryan J, Vestrand WT (eds) High-energy solar phenomena – a new era of spacecraft measurements, Proceedings of the workshop held in Waterville Valley, NH, 1993. American Institute of Physics, New York. AIP Conference Proceedings, vol 294, No 1, pp 94–99

Schneid EJ, Bertsch DL, Dingus BL, Esposito JA, Fichtel CE, Hartman RC, Hunter SD, Kanbach G, Kniffen DA, Lin YC, Mayer-Hasselwander HA, Michelson PF, von Montigny C, Mukherjee R, Nolan PL, Sreekumar P, Thompson DJ (1996) EGRET observations of X-class solar flares. Astron Astrophys Suppl Ser (France) 120(4):299–302

Schoenfelder V, Aarts H, Bennett K, de Boer H, Clear J, Collmar W, Connors A, Deerenberg A, Diehl R, von Dordrecht A, den Herder JW, Hermsen W, Kippen M, Kuiper L, Lichti G, Lockwood J, Macri J, McConnell M, Morris D, Much R, Ryan J, Simpson G, Snelling M, Stacy G, Steinle H, Strong A, Swanenburg BN, Taylor B, de Vries C, Winkler C (1993) Instrument description and performance of the Imaging Gamma-Ray Telescope COMPTEL aboard the Compton Gamma-Ray Observatory. Astrophys J Suppl Ser (USA) 86(2):657–692

Schrijver CJ, Hudson HS, Murphy RJ, Share GH, Tarbel TD (2006) Gamma rays and the evolving, compact structures of the 2003 October 28 X17 flare. Astrophys J 650(2, Part 1):1184–1192

Schwartz RA, Csillaghy A, Tolbert AK, Hurford GJ, McTiernan J, Zarro D (2002) RHESSI data analysis software: rationale and methods. Sol Phys (Netherlands) 210(1–2):165–191

Severny AB (1964) Solar flares. Ann Rev Astron Astrophys 2:363–400

Share GH, Murphy RJ (1995) Gamma-ray measurements of flare-to-flare variations in ambient solar abundances. Astrophys J (USA) 452(2, Part 1):933–943

Share GH, Murphy RJ (1997a) Intensity and directionality of flare-accelerated α-particles at the Sun. Astrophys J (USA) 485(1, Part 1):409–418

Share GH, Murphy RJ (1997b) A method for determining the photospheric ^{4}He abundance. Astrophys J Lett (USA) 484(2, Part 2):L165–L167

Share GH, Murphy RJ (2000) Gamma ray spectroscopy in the pre-HESSI era. In: Ramaty R, Mandzhavidze N (eds) High energy solar physics–anticipating HESSI, Astronomical Society of the Pacific Conference Series, vol 206. ASP, San Francisco, CA, pp 377–386

Share GH, Murphy RJ (2001) Atmospheric gamma rays from solar energetic particles and cosmic rays penetrating the magnetosphere. J Geophys Res (USA) 106(A1):77–92

Share GH, Murphy RJ, Tylka AJ, Schwartz RA, Yoshimori M, Suga K, Nakayama S, Takeda H (2001) Gamma-ray line observations of the 2000 July 14 flare and SEP impact on the Earth. Sol Phys 204(1–2):43–55

Share GH, Murphy RJ, Kiener J, De Sereville N (2002) Directionality of solar flare-accelerated protons and α-particles from γ-ray line measurements. Astrophys J (USA) 573(1, Part 1): 464–470

Share GH, Murphy RJ, Skibo JG, Smith DM, Hudson HS, Lin RP, Shih AY, Dennis BR, Schwartz RA, Kozlovsky B (2003) High-resolution observation of the solar positron-electron annihilation line. Astrophys J Lett (USA) 595(2, Part 2):L85–L88

Share GH, Murphy RJ, Smith DM, Schwartz RA, Lin RP (2004) RHESSI e^{+}-e^{-} annihilation radiation observations: implications for conditions in the flaring solar chromosphere. Astrophys J Lett (USA) 516(2, Part 2):L169–L172

Shea MA, Smart DF, Pyle KR (1991) Direct solar neutrons detected by neutron monitors on 24 May 1990. Geophys Res Lett 18(9):1655–1658

Shibata S (1994) Propagation of solar neutrons through the atmosphere of the Earth. J Geophys Res 99(A4):6651–6665

Shiozawa A (1999) Master Thesis, Rikkyo University, Tokyo

Simnett GM (1995) Protons in flares. Space Sci Rev (Netherlands) 73(3–4):387–432

Simnett GM, Share GH, Messina DC, Chupp EL, Forrest DJ, Rieger E (1986) A search for evidence of energetic protons immediately prior to the impulsive phase of solar flares. Adv Space Res 6(6):105–108

Smart D.F., M.A. Shea, and L.C. Gentile (1994) "The relativistic solar proton ground-level enhancements associated with the solar neutron events of 11 June, 15 June 1991. In: Ryan J, Vestrand WT (eds) High-energy solar phenomena – a new era of spacecraft measurements, Proceedings of the workshop held in Waterville Valley, NH, 1993. American Institute of Physics, New York. AIP Conference Proceedings, vol 294, No 1, pp 222–229

Smith DM, Lin RP, Turin P, Curtis DW, Primbsch JH, Campbell RD, Abiad R, Schroeder P, Cork CP, Hull EL, Landis DA, Madden NW, Malone D, Pehl RH, Raudorf T, Sangsingkeow P, Boyle R, Banks IS, Shirey K, Schwartz R (2002) The RHESSI spectrometer. Sol Phys (Netherlands) 210(1–2):33–60

Smith DM, Share GH, Murphy RJ, Schwartz RA, Shih AY, Lin RP (2003) High-resolution spectroscopy of gamma-ray lines from the X-class solar flare of 2002 July 23. Astrophys J Lett (USA) 595(2, Part 2):L81–L84

Somov BV (1996) Reconnection and acceleration to high energies in flares. In: Ramaty R, Mandzhavidze N, Hua X-M (eds) High energy solar physics, Proceedings of the conference held in Greenbelt, MD, 1995. AIP Conference Proceedings, vol 374, No 1, pp 493–497

Somov BV, Bogachev SA (2003) The betatron effect in collapsing magnetic traps. Astron Lett (Russia) 29(9):621–628

Somov BV, Spektor AR (1982) Hydrodynamic shock wave formation in the solar chromosphere and corona during flares. Space Sci Rev 32(1–2):27–41

Speiser TW (1965) Particle trajectories in model current sheets 1. Analytical solutions. J Geophys Res 70(17):4219–4226

Spirock T, Denker C, Chen H, Chae J, Qiu J, Varsik J, Wang H, Goode PR, Marquette W (2001) The Big Bear Solar Observatory's digital vector magnetographin. In: Sigwarth M (ed) Advanced solar polarimetry: theory, observation, and instrumentation, 20th International Sacramento Peak summer workshop, ASP Conference Series, vol 236. ASP, San Francisco, CA, pp 65–72

Spirock TJ, Yurchyshyn VB, Wang H (2002) Rapid changes in the longitudinal magnetic field related to the 2001 April 2 X20 flare. Astrophys J (USA) 572(2, Part 1):1072–1076

Spruit HC (1974) A model of the solar convection zone. Sol Phys (Netherlands) 34(2):277–290

Sterling AC, Moore RL (2001) EIT crinkles as evidence for the breakout model of solar eruptions. Astrophys J (USA) 560(2, Part 1):1045–1057

Sterling AC, Moore RL, Qiu J, Wang H (2001) Hα proxies for EIT crinkles: further evidence for preflare "breakout"-type activity in an ejective solar eruption. Astrophys J (USA) 561(2, Part 1):1116–1126

Struminsky A, Matsuoka M, Takahashi K (1994) Evidence of additional production of high-energy neutrons during the solar flare on 1991 June 4. Astrophys J (USA) 429(1, Part 1): 400–405

Sudol JJ, Harvey JW (2005) Longitudinal magnetic field changes accompanying solar flares. Astrophys J (USA) 635(1, Part 1):647–658

Sweet PA (1958) The neutral point theory of solar flares. In: Lehnert B (ed) Electromagnetic phenomena in cosmical physics, Proceedings from IAU Symposium, vol 6. International Astronomical Union, Cambridge University Press, Cambridge, pp 123–134

Takahashi K, Sakamoto E, Matsuoka M, Nishi K, Yamada Y, Shimoda S, Shikata T, Wada M, Kusunose M, Yoshimori M, Kondo I (1991) Observation of solar neutrons by Mt. Norikura neutron monitor during a period of solar cycle 22. In: Proceedings of the 22nd international cosmic ray conference, Dublin, vol 3, pp 37–40

Talon R, Trottet G, Vilmer N, Barat C, Dezalay J-P, Sunyaev R, Terekhov O, Kuznetsov A (1993) Hard X-ray and gamma-ray observations of solar flares with the PHEBUS experiment. Sol Phys (Netherlands) 147(1):137–155

Tang F, Moore RL (1982) Remote flare brightening and type III reverse slope bursts. Sol Phys (Netherlands) 77(1–2):263–276

Temerin M, Roth I (1992) The production of 3He and heavy ion enrichments in 3He-rich flares by electromagnetic hydrogen cyclotron waves. Astrophys J Lett (USA) 391(2, Part 2):L105–L108

Terekhov OV (1995) Observations of space gamma-bursts, background situation, and deutrerium sintez during solar flares from satellites Prognoz-9, Fobos, Mir-kvant, Granat. Dr. Science Thesis, Space Research Institute, Moscow, pp 1–280 (In Russian)

Terekhov OV, Sunyaev RA, Kuznetsov AV, Barat C, Talon R, Trottet G, Vilmer N (1993) 'Granat' observations of delayed 2.2 MeV gamma-ray line emission. Deuterium synthesis during the solar flare of May 24th, 1990. Pis'ma Astron Zh (Russia) 19(3):163–171 (In Russian)

Terekhov OV, Kuzmin AG, Sunyaev RA, Tkachenko AYu, Denisenko DV, Barat C, Talon R, Vedrenne G (1996) PHEBUS/GRANAT observations of high-energy solar flares. Astron Lett (Russia) 22(3):362–370

Thompson DJ, Bertsch DL, Fichtel CE, Hartman RC, Hofstadter R, Hughes EB, Hunter SD, Hughlock BW, Kanbach G, Kniffen DA, Lin YC, Mattox JR, Mayer-Hasselwander HA, Von Montigny C, Nolan PL, Nel HI, Pinkau K, Rothermel H, Schneid EJ, Sommer M, Sreekumar P, Tieger D, Walker AH (1993) Calibration of the Energetic Gamma-Ray Experiment Telescope (EGRET) for the Compton Gamma-Ray Observatory. Astrophys J Suppl Ser (USA) 86(2):629–656

Troitskaia EV, Kuzhevskij BM (1999) Absorption of 2.22 MeV solar flare gamma-rays and determining of the solar plasma density altitude profile. In: Proceedings of the 26th international cosmic ray conference, Salt Lake City, UT, vol 6, pp 17–20

Troitskaia EV, Miroshnichenko LI (2008) Study of the 28 October 2003 solar flare by means of 2.223 MeV gamma-emission. In: Proceedings of the 30th international cosmic ray conference, Merida, vol 1, pp 23–26

Troitskaia EV, Gan W, Kuzhevskij BM, Miroshnichenko LI (2003) Time profile of the 2.223 MeV gamma-line emission and some features of the 16 December 1988 solar event. In: Proceedings of the 28th international cosmic ray conference, Tsukuba, vol 6, pp 3219–3222

Troitskaia EV, Gan WQ, Kuzhevskij BM, Miroshnichenko LI (2007) Solar plasma density and spectrum of accelerated particles derived from the 2.223-Mev line of a solar flare. Sol Phys (Netherlands) 242(1–2):87–99

Troitskaia EV, Arkhangelskaja IV, Miroshnichenko LI, Arkhangelsky AI (2009) Study of the 28 October 2003 and 20 January 2005 solar flares by means of 2.223 MeV gamma-emission line. Adv Space Res 43:547–552

Trottet G (1994) X-ray and gamma-ray observations of solar flares by GRANAT. In: Ryan J, Vestrand WT (eds) High-energy solar phenomena - a new era of spacecraft measurements, Proceedings of the workshop held in Waterville Valley, NH, 1993. American Institute of Physics, New York. AIP Conference Proceedings, vol 294, No 1, pp 3–14

Trottet G, Vilmer N, Barat C, Dezalay JP, Talon R, Sunyaev R, Kuznetsov A, Terekhov O (1993a) Temporal and spectral characteristics of the June 11, 1991 gamma-ray flare. Astron Astrophys Suppl Ser (France) 97(1):337–339

Trottet G, Chupp EL, Marschhauser H, Pick M, Soru-Escaut I, Rieger E, Dunphy PP (1994) A comparison of gamma-ray and radio emissions during the 11:42 UT solar flare on 1982 June 3. Astron Astrophys (Germany) 288(2):647–655

Trottet G, Barat C, Ramaty R, Vilmer N, Dezalay JP, Kuznetsov A, Mandzhavidze N, Sunyaev R, Talon R, Terekhov O (1996) Thin target -ray line production during the 1991 June 1 flare. In: Ramaty R, Mandzhavidze N, Hua X-M (eds) High energy solar physics, Proceedings of the Conference held in Greenbelt, MD, 1995. AIP Conference Proceedings, vol 374, No 1, pp 153–161

Usoskin IG, Kovaltsov GA, Kocharov LG, Kananen H, Tanskanen P (1995) High energy neutrons from the June 15, 1991 solar flare as detected by the neutron monitor. In: Proceedings of the 24th international cosmic ray conference, Rome, vol 4, pp 151–154

Van Allen JA (1993) Recovery of interplanetary cosmic ray intensity following the great Forbush decrease of mid-1991. Geophys Res Lett (USA) 20(24):2797–2800

Van Ballegooijen AA (1989) Formation and eruption of solar prominences. Astrophys J (USA) 343(2, Part 1):971–984

Van Driel-Gesztelyi L, Manoharan PK, Pick M, Demoulin PP (1997) Reorganization of the solar corona following a C4.7 flare. Adv Space Res (UK) 19(12):1883–1886

Van Hollebeke MAI, McDonald FB, Meyer JP (1990) Solar energetic particle observations of the 1982 June 3 and 1980 June 21 gamma-ray/neutron events. Astrophys J Suppl Ser (USA) 73(2):285–296

Van Tend W, Kuperus M (1978) The development of coronal electric current systems in active regions and their relation to filaments and flares. Sol Phys (Netherlands) 59(1):115–127

Varga L (1961) Pulse shape discrimination. Nucl Instrum Methods (Netherlands) 14:24–32

Vernazza JE, Avrett EH, Loeser R (1981) Structure of the solar chromosphere. III. Models of the EUV brightness components of the quiet Sun. Astrophys J Suppl Ser (USA) 45(4):635–725

Vestrand WT, Forrest DJ (1993) Evidence for a spatially extended component of gamma rays from solar flares. Astrophys J Lett 409(2, Part 2):L69–L72

Vestrand WT, Miller JA (1999) Particle acceleration in flares. In: Strong KT, Saba JLR, Haisch BM, Schmelz JT (eds) The many faces of the sun: a summary of the results from NASA's Solar Maximum Mission. Springer, New York, pp 231–272

Vestrand WT, Forrest DJ, Chupp EL, Rieger E, Share GH (1987) The directivity of high-energy emission from solar flares: Solar Maximum Mission observations. Astrophys J (USA) 322(2, Part 1):1010–1022

Vestrand WT, Share GH, Murphy RJ, Forrest DJ, Rieger E, Chupp EL, Kanbach G (1999) The Solar Maximum Mission atlas of gamma-ray flares. Astrophys J Suppl Ser (USA) 120(2):409–467

Vilmer N (1994) Solar hard X-ray and gamma-ray observations from GRANAT. Astrophys J Suppl Ser (USA) 90(2):611–621

Wang T, Abramenko VI (2000) Evolution of magnetic field twist in an emerging flux region. Astron Astrophys (Germany) 357(3):1056–1062

Wang HT, Ramaty R (1974) Neutron propagation and 2.2 MeV gamma-ray line production in the solar atmosphere. Sol Phys (Netherlands) 36(1):129–137

Wang H, Tang F (1993) Flux emergence and umbra formation after the X-9 flare of 1991 March 22. Astrophys J Lett (USA) 407(2, Part 2):L89-L90+2 plates

Wang H, Ewell MW Jr, Zirin H, Ai G (1994) Vector magnetic field changes associated with X-class flares. Astrophys J (USA) 424(1, Part 1):436–443

Wang H, Spirock TJ, Qiu J, Ji H, Yurchyshyn V, Moon Y-J, Denker C, Goode PR (2002) Rapid changes of magnetic fields associated with six X-class flares. Astrophys J (USA) 576(1, Part 1):497–504

Warren HP, Doschek GA (2005) Reconciling hydrodynamic simulations with spectroscopic observations of solar flares. Astrophys J Lett (USA) 618(2, Part 2):L157–L160

Webber WR, Lockwood JA (1993) Giant transient decreases of cosmic rays in the outer heliosphere in September 1991. J Geophys Res (USA) 98(A5):7821–7825

White SM, Krucker S, Shibasaki K, Yokoyama T, Shimojo M, Kundu MR (2003) Radio and hard X-ray images of high-energy electrons in an X-class solar flare. Astrophys J Lett (USA) 595(2, Part 2):L111–L114

Wilkinson DC, Allen JH (2005) NOAA satellites provide a keen view of the Martin Luther King Solar Storm of January 2005. AGU Spring Meeting Abstracts, A5, SP23A-05

Wu ST, Guo WP, Michels DJ, Burlaga LF (1999) MHD description of the dynamical relationships between a flux rope, streamer, coronal mass ejection, and magnetic cloud: an analysis of the January 1997 Sun-Earth connection event. J Geophys Res (USA) 104(A7):14789–14801

Yoshimori M (1989) Observational studies of gamma-rays and neutrons from solar flares. Space Sci Rev (Netherlands) 51(1–2):85–115

Yoshimori M, Okudaira K, Hirasima Y, Igarashi T, Akasaka M, Takai Y, Morimoto K, Watanabe T, Ohki K, Nishimura J, Yamagami T, Ogawara Y, Kondo I (1991) The wide band spectrometer on the solar-A. Sol Phys (Netherlands) 136(1):69–88

Yoshimori M, Suga K, Morimoto K, Hiraoka T, Sato J, Kawabata K, Ohki K (1994) Gamma-ray spectral observations with Yohkoh. Astrophys J Suppl Ser (USA) 90(2):639–643

Yoshimori M, Shiozawa A, Suga K (1999a) Solar gamma-ray flares in the 23rd solar maximum. In: Proceedings of the 26th international cosmic ray conference, Salt Lake City, UT, vol 6, pp 1–4

Yoshimori M, Shiozawa A, Suga K (1999b) Photospheric ^{3}He to H abundance ratio derived from gamma-ray line observation. In: Proceedings of the 26th international cosmic ray conference, Salt Lake City, UT, vol 6, pp 5–8

Yoshimori M, Shiozawa A, Suga K (1999c) Low-FIP to high-FIP gamma-ray line ratio in an impulsive flare on 6 November, 1997. In: Proceedings of the 26th international cosmic ray conference, Salt Lake City, UT, vol 6, pp 30–33

Yoshimori M, Shiozawa A, Suga K (1999d) In: Solar physics with radio observations, Proceedings of Nobeyama symposium (In Japanese)

Yoshimori M, Suga K, Saita N, Shiozawa A (2000) Yohkoh observation of a gamma-ray flare on November 6, 1997. Adv Space Res (UK) 25(9):1801–1804

Yoshimori M, Nakayama S, Takeda H, Ogawa H, Masuda S (2001a) Recent Yohkoh solar gamma-ray observations. In: Proceedings of the 27th international cosmic ray conference, Hamburg, vol 8, pp 3017–3020

Yoshimori M, Nakayama S, Ogawa H (2001b) Positron annihilation radiation and >10 MeV gamma rays from the 1997 November 6 flare. In: Proceedings of the 27th international cosmic ray conference, Hamburg, vol 8, pp 3025–3028

Yoshimori M, Suga K, Shiozawa A, Nakayama S, Takeda H (2001c) Particle acceleration in the 6 November 1997 event as viewed from gamma rays and solar energetic particles. In: Mewaldt RA, Miller M, Jokipii JR, Lee MA, Zurbuchen TH, Mobius E (eds) Acceleration and transport of energetic particles observed in the heliosphere, ACE 2000 symposium, Indian Wells, CA, 2000. AIP Conference Proceedings, vol 528, pp 189–192

Yoshimori M, Suga K, Nakayama S, Takeda H, Ogawa H, Murphy RJ, Share GH (2002) Gamma rays and neutrons from a large solar flare on November 6, 1997. Adv Space Res 30(3):629–634

Yoshimori M, Hirayama H, Mori S (2003) Energetics of nonthermal electrons and protons in intense solar flares. In: Proceedings of the 28th international cosmic ray conference, Tsukuba, vol 6, pp 3191–3194

Young CA, Connors A, Bennett K, McConnell M, Rank G, Ryan JM, Sulieman R, Schönfelder V (1999) Energetic proton spectra in the 11 June 1991 and 24 October 1991 solar flares. In: Proceedings of the 26th international cosmic ray conference, Salt Lake City, UT, vol 6, pp 252–255

Yurchyshyn VB, Abramenko VI, Carbone V (2000) Flare-related changes of an active region magnetic field. Astrophys J (USA) 538(2, Part 1):968–979

Yurchyshyn V, Wang H, Abramenko V, Spirock TJ, Krucker S (2004 [1983]) Magnetic field, Ha, and RHESSI observations of the 2002 July 23 gamma-ray flare. Astrophys J 264(2, Part 1): 648–659

Zvereva AM, Severny AB (1970) Magnetic fields and proton flares of 7 July and 2 September 1966. Izv Krym Astrofiz Obs (USSR) 41–42:97–157

Zweibel EG, Haber DA (1983) The propagation of energetic ions in magnetic loops and gamma-ray emission from solar flares. Astrophys J 264(2, Part 1):648–659

Chapter 10
Positron Generation in the Nuclear Interactions of Flare Energetic Particles in the Solar Atmosphere

10.1 Expected Fluxes of Positrons from Nuclear Interactions of Solar Energetic Particles with the Atmosphere of the Sun, and the First Observations

10.1.1 On the Physics of the Generation of the 511 keV γ-Ray Line in Solar Flares

As we note in Section 1.7, Lingenfelter and Ramaty (1967), Ramaty and Lingenfelter (1973a, b), Hurford et al. (1973) were the first who have considered in detail positron production and annihilation during solar flares. It was found that the dominant source of small energy positrons with kinetic energy 0.2 MeV $\leq E \leq$ 2 MeV in the periods of chromospheric flares is expected to be the β^+ decay of nuclear interaction products such as ^{11}C, ^{13}N and ^{15}O and high energy positrons of several tens MeV from $\pi^+ \rightarrow \mu^+ \rightarrow e^+$ decay, all of which are produced by interactions of accelerated protons, α-particles and heavy nuclei with the ambient solar atmosphere (Ramaty and Murphy 1987). The first process results from low energy interactions at ∼10 MeV, and the resultant positrons have typical energies of ∼1 MeV. On the other hand, the secondary ones are high energy proton interactions above 300 MeV and the positron energies are ∼30 MeV.

As noted Yoshimori et al. (2001), in a thick target, the positrons are slowed down to ∼10 eV where they either annihilate directly or form positronium atom after thermalization. Direct annihilation and singlet state positronium emit two 511 keV photons, while triplet state positronium produces three γ-rays (positronium continuum below 511 keV). Triplet positronium is broken up by collision if the ambient density is above 10^{14} cm^{-3}. Since a time profile of the 511 keV line depends on the density and lifetimes of β^+-decay nuclei, its temporal variation is complex, depending on solar flares. A ratio of 3γ–2γ depends on the ambient density. The line width is a function of the temperature of the annihilation site (Crannell et al. 1976; Bussard et al. 1979).

L. Dorman, *Solar Neutrons and Related Phenomena*,
Astrophysics and Space Science Library 365, DOI 10.1007/978-90-481-3737-4_10,

10.1.2 The First Attempts to Search for and Earlier Observations of the 511 keV γ-Ray Line from Solar Flares

The first attempts to search the 511 keV γ-ray line from solar flares were described in Section 1.7. As it was described in Chapter 2, the first measurement of the 0.5 MeV γ-ray line was reported from the OSO observation during solar flares in August 1972 by Chupp et al. (1973) – see Table 2.1 and Fig. 2.1. Share et al. (1983) observed the positron annihilation line with the Solar Maximum Mission (SMM). Further, Share et al. (1996) analyzed the 511 keV line and positronium continuum from seven flares and discussed the temperature and density at the positron annihilation sites.

10.2 Positron Annihilation Radiation from the 1997 November 6 Flare, Comparison with High Energy γ-Ray Emission, and Possible Acceleration Mechanisms

10.2.1 Yohkoh Satellite Observation of the 511 keV Line

According to Yoshimori et al. (2000a, b, 2001), Yohkoh satellite observed a X9.4/3B intense flare at 11:52 UT on November 6 and strong X- and γ-rays were measured with the hard X- and γ-ray spectrometers. The count-rate time profiles at 25–826 keV and 1 MeV are shown in Fig. 10.1.

The hard X-ray count rate was much higher than 10^4 counts s^{-1} during the peak phase of the flare (11:52:40–11:55:30 UT). It is difficult to make dead-time corrections at the count rate above 10^4 counts s^{-1} because a model of dead-time behavior of the HXS counting system is paralyzable (dead-time is 40 μs). Yoshimori et al. (2001) do not analyze the hard X-ray data of which count rate is $>10^4$ counts s^{-1}. The count rate time profile at 1 MeV, however, does not need the dead-time correction because of the low count rate. For in-flight energy-calibration, they use three lines at 60 keV (^{241}Am), 191 keV (^{123}I) and 511 keV (Earth's atmospheric origin). The background-subtracted hard X-ray count spectrum between 11:55:30–12:00:00 UT is given in Fig. 10.2.

From Fig. 10.2 can be seen a line feature at 511 keV. In order to analyze the 511 keV line feature, Yoshimori et al. (2001) plot the continuum-subtracted spectrum in Fig. 10.3. Using the response function of the spectrometer, they derive the 511 keV line fluency of 64 ± 13 photons cm^{-2}.

Figure 10.3 shows that the observed width (FWHM) of the flare 511 keV line is 50 ± 14 keV. However, this width is mostly due to the instrumental resolution of the NaI scintillation detector because the instrumental energy resolution is 48 keV at 511 keV. It indicates that the intrinsic width of the flare 511 keV line is less than 16 keV. Moreover, the Yohkoh result does not exhibit the positronium continuum resulting from the triplet state of positronium (three γ-ray emission).

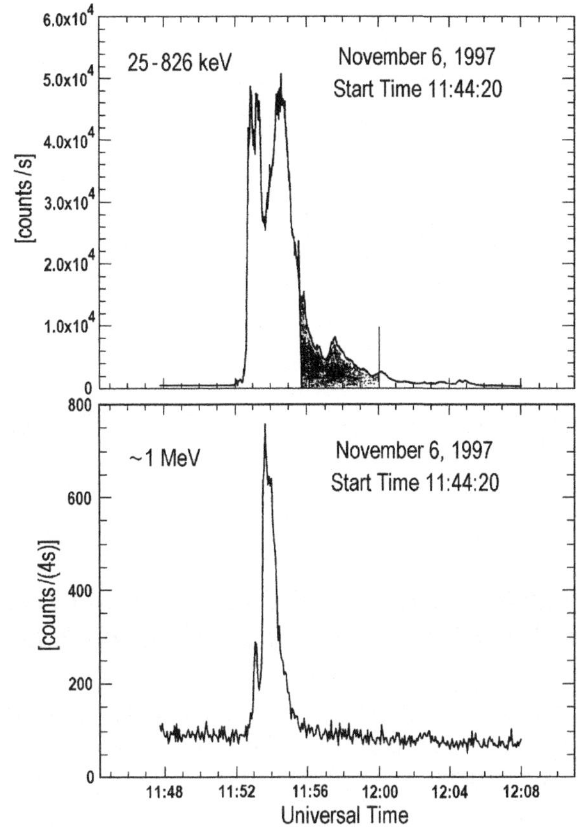

Fig. 10.1 Count-rate time profiles at 25–826 keV and 1 MeV. A *shaded part* of the upper data is a spectral measurement time (From Yoshimori et al. 2001)

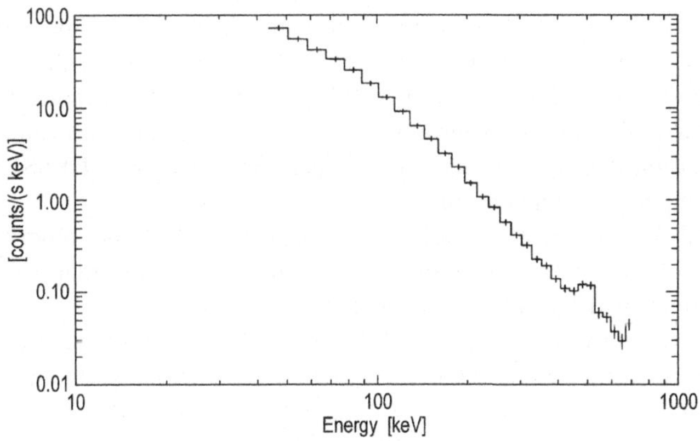

Fig. 10.2 Background-subtracted hard X-ray count spectrum between 11:55:30 and 12:00:00 UT (From Yoshimori et al. 2001)

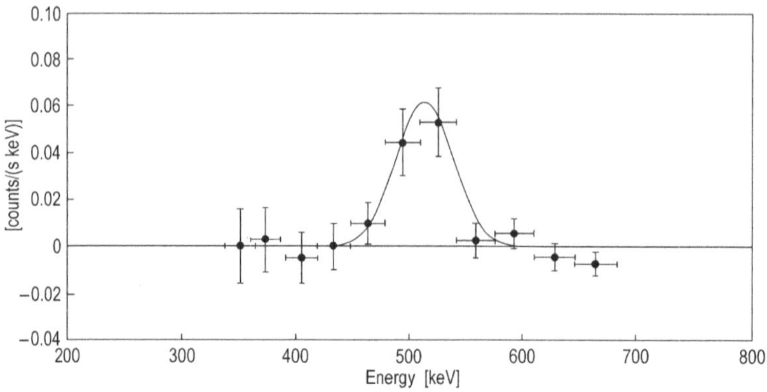

Fig. 10.3 Continuum-subtracted 511 keV line observed between 11:55:30 and 12:00:00 UT (From Yoshimori et al. 2001)

10.2.2 The Width of the Flare 511 keV Line, and Temperature and Matter Density of the Positron Annihilation Site

As underlined Yoshimori et al. (2001), the intrinsic width of the 511 keV line is related to a temperature of the positron annihilation site. The intrinsic width of the line ΔE is approximated by

$$\Delta E = 1.1 \times \left(\mathrm{T}/10^4\right)^{1/2} \mathrm{keV}, \tag{10.1}$$

where T is the temperature in K. From Eq. 10.1 follows that the observed width of <16 keV indicates that the temperature is less than 2.1×10^6 K. Share et al. (1996) obtained the temperatures in the range $(0.2–10) \times 10^6$ K from the width of the 511 keV line observed from seven SMM γ-ray flares. Yoshimori et al. (2001) noted that it is difficult to determine the accurate plasma temperature with the NaI scintillation detector. A Germanium spectrometer aboard HESSI will determine the temperature much accurate.

As noted Yoshimori et al. (2001), the described Yohkoh observation does not give evidence for the positronium continuum, suggesting that triplet positronium was quenched by collisional breakup at the annihilation site of density of $>10^{14}$ cm^{-3} (Ramaty and Murphy 1987). The SMM γ-ray flares reveal that a ratio of 3γ–2γ fluencies ranges from 0 to 2.5, depending on the flares (Share et al. 1996). The Yohkoh observation of the 1997 November 6 flare indicates that positron annihilation took place at the site of the temperature of $<1.6 \times 10^6$ K and density of $>10^{14}$ cm^{-3}. It implies the possibility that positron annihilation occurred low in the chromosphere.

10.2.3 Comparison of Temporal Characteristics of the 511 keV and 4–7 MeV γ-Ray Emissions

Yoshimori et al. (2001) underlined that the temporal characteristic of the 511 keV emission provides essential information on the interaction history of the accelerated ions. The count-rate time profiles of the 511 keV line and 4–7 MeV γ-rays are shown in Fig. 10.4. The solid circles and the dotted curve express the 4–7 MeV γ-ray data and the Ramaty (1986) calculation. The calculation assumed that the interactions take place instantaneously at the ambient density of $>10^{14}$ cm^{-3}. The 511 keV line exhibits a longer decay constant compared with the prompt 4–7 MeV γ-rays. If the density of the positron annihilation site is $>10^{14}$ cm^{-3}, the positrons should annihilate on a very short time scale. The time profile of the observed 511 keV line fits the Ramaty (1986) calculation, suggesting the possibility that the positrons were produced from the long life β^+ radioactive nuclei of ^{11}C, ^{13}N and ^{15}O of which lives are 20.5, 10 and 2 min, respectively.

10.2.4 Comparison with High Energy γ-Ray Emission and Possible Acceleration Mechanisms During the 1997 November 6 Solar Flare

As noted Yoshimori et al. (2000a, b, 2001), the Yohkoh observation implies that both electrons and protons were efficiently accelerated to energies above 10 MeV and above 300 MeV in the peak phase of the flare, correspondingly (see Section 9.10.3 on observation by Yohkoh satellite of high energy γ-rays from π^0 decay and from bremsstrahlung of high energy electrons during the 1997 November 6 flare). The

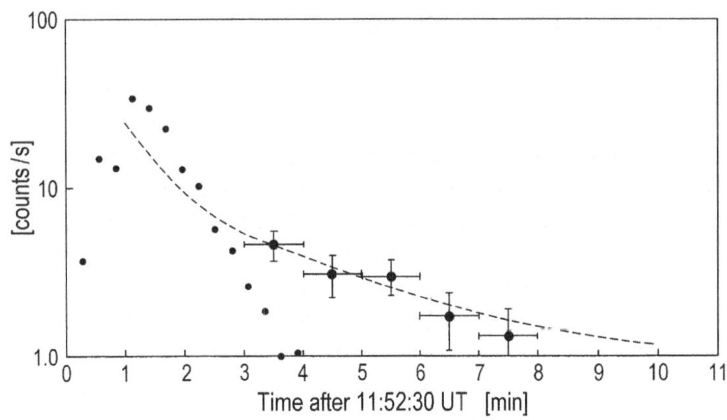

Fig. 10.4 Comparison between count-rate time profile of the 511 keV line and the 4–7 MeV γ-rays. The *solid circles* and the *dotted curve* express the 4–7 MeV γ-ray data and the Ramaty (1986) calculation, respectively (From Yoshimori et al. 2001)

Yohkoh X-ray image exhibits two hard X-ray sources which are located at both footpoints of the flaring magnetic loop. The temporal variation in the hard X-ray sources shows that a distance between the two sources increases with time during the peak phase. Separation of the two hard X-ray sources suggests the possibility that a magnetic reconnection site moves up with time (Sakao et al. 1998). Particle acceleration is thought to take place in association with the magnetic reconnection. The magnetic energy released by magnetic reconnection is converted to kinetic energy of plasma particles in a non-thermal manner. One possible scenario is that both electrons and ions were accelerated to high energies on a short time scale by stochastic scattering with fast shock waves or Alfvén waves generated at the magnetic reconnection site (Tsuneta and Naito 1998). Part of the accelerated particles precipitates to the chromosphere and produce hard X rays, γ-rays and neutrons. SEPs exceeding 10 GeV associated with this event were reported from the ground-based water Cherenkov detector experiment (Falcone 2000). Strong coronal mass ejection (CME) was simultaneously measured with the flare (Mason et al. 1999). According to opinion of Yoshimori et al. (2001), these very high-energy SEPs could be accelerated by not the magnetic reconnection but the CME driven shock in the higher corona.

10.3 The RHESSI Observation of the Solar Annihilation Line from the 2002 July 23 Solar Flare

10.3.1 The Matter of the Problem and RHESSI Observations

According to Share et al. (2003a), Reuven Ramaty High Energy Spectroscopic Imager satellite (RHESSI) has observed the positron-electron annihilation line at 511 keV produced during the 2002 July 23 solar flare. The shape of the line is consistent with formation of positronium by charge exchange in flight with hydrogen in a quiet solar atmosphere at a temperature of ~6,000 K. However, the measured upper limit to the $3\gamma/2\gamma$ ratio (ratio of annihilation photons in the positronium continuum to the number in the line) is only marginally consistent with what is calculated for this environment. The annihilation line can also be fit by a thermal Gaussian having a width of 8.1 ± 1.1 keV (FWHM), indicating temperatures of $\sim(4\text{–}7) \times 10^5$ K. This would require the formation of a substantial mass of atmosphere at transition-region temperatures during the flare. As noted Kozlovsky et al. (1987), flare-accelerated protons, α-particles, and heavier ions interact with the solar atmosphere and produce radioactive nuclei that decay with the release of a positron. The positrons slow down by coulomb interactions and directly annihilate with electrons or form positronium by attaching to an electron (Crannell et al. 1976). Positronium is formed in either the singlet or triplet spin state. Both direct annihilation or annihilation from the singlet state give rise to two 511 keV photons. When annihilation takes place from the triplet state three photons are emitted with varying energies, producing a continuum. The number of photons observed in this continuum divided by the number of photons in the line is known as the $3\gamma/2\gamma$ ratio. The temperature, density, and composition of the ambient medium where the positrons slow down, form positronium, and annihilate, determine the $3\gamma/2\gamma$ ratio, line width, and time profile of the radiation. With the launch of the (RHESSI), $\geq$ 2 keV line widths can be measured (Smith et al. 2002). Share et al. (2003a) summarize RHESSI's observation of annihilation radiation from the 2002 July 23 X4.8 class flare (active region AR0039; coordinates S13E72). They accumulated 960 s of spectral data from 00:27:20 to 00:43:20 UT and estimated the background during the flare using comparable spectral accumulations on the previous and subsequent days ($\pm$15 orbits). Share et al. (2003a) then constructed a model solar photon spectrum, passed it through the instrument response function, and fit the background-corrected data from 150 to 8,500 keV. This model spectrum included a double power law, a nuclear deexcitation line function made up of 15 narrow and broad Gaussians, the neutron capture line, the α-^{4}He fusion line complex between ~400 and 500 keV, and the solar annihilation line and its positronium continuum. In Fig. 10.5 are plotted the solar annihilation-line spectrum after subtracting all locally-produced components from the background-subtracted spectrum.

As can be seen from Fig. 10.5, the line is significantly broader than the ~2.5 keV widths of the locally-produced or background annihilation lines; it is fit best by an 8.1 $\pm$ 1.1 keV (FWHM) Gaussian (solid curve in Fig. 10.5). Its time profile is well fit using the nuclear deexcitation lines as a proxy for the accelerated-particle interaction rate at the Sun, if one includes the life times of the flare-produced positron emitters (Kozlovsky et al. 1987).

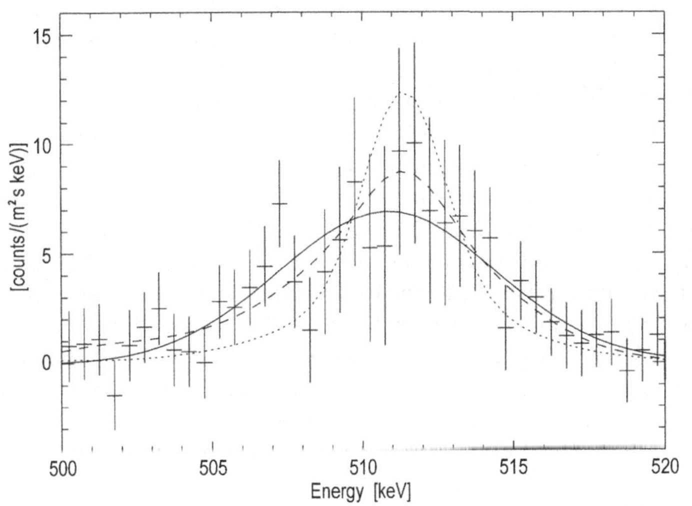

Fig. 10.5 Spectrum of the solar 511-keV annihilation line derived by subtracting the instrumental and background components from the total spectrum observed during the flare. *Solid curve*, best-fitting Gaussian; *dashed curve*, line-shape expected at 6,000 K; *dotted curve*, line-share expected at 5,000 K (From Share et al. 2003a)

10.3.2 The 2002 July 23 Solar Flare as a Prolific Emitter of Annihilation Line Radiation

As underlined Share et al. (2003a), the 2002 July 23 flare may be considered as a prolific emitter of annihilation line radiation. The measured fluence over the entire flare, $\sim 83 \pm 14\ \gamma\ \mathrm{cm}^{-2}$, is higher than all but five of the 31 flares with annihilation radiation observed during 10 years by the SMM spectrometer. In a neutral or partially ionized environment between 5,000 and 7,000 K, the annihilation line is made up of narrow and broad components. The narrow $\sim$1.5 keV (FWHM) line is produced by annihilation of thermalized positrons with bound electrons. The broad component results from positronium formed via charge exchange in flight (Bussard et al. 1979). Share et al. (2003a) have calculated its width using updated charge exchange cross sections and obtain a width of 7.5 ± 0.5 keV for conditions at the Sun. There is a narrow range of temperatures, 5,650–6,270 K (90% confidence level), in the quiet solar atmosphere where the broad component can dominate and produce a shape (dashed curve in Fig. 10.5) that fits the RHESSI spectra almost as well as a Gaussian (solid curve). In contrast, a line produced at 5,000 K (dotted curve in Fig. 10.5) is considerably narrower and has only a 1% probability ($\Delta\chi^2 = 6.7$) of fitting as well as the 8.1-keV Gaussian. Radioactive nuclei have the same depth distribution of γ rays and neutrons resulting from interactions with flare-accelerated protons and α-particles that peak at $\sim 10^{15}$ H cm^{-3} (Hua et al. 1989). The fate of the emitted positrons depends on their energy and emission angle. One needs to explain how the positrons can slow down and annihilate at the relatively low densities of 2×10^{12}–8×10^{13} H cm^{-3} corresponding to the temperature range deduced above for a quiet atmosphere. Higher densities, $\sim 3 \times 10^{15}$ H cm^{-3}, can occur in a flaring atmosphere at 6,000 K (Machado et al. 1980).

However, calculations of Share et al. (2003a) from 5,000 to 8,500 K for this flaring atmosphere indicate that the broad line from charge-exchange in flight is never dominant enough to fit the RHESSI line shape. In the quiet atmosphere, where the broad line dominates, these calculations also require $3\gamma/2\gamma$ ratios >2.7; the RHESSI upper limit on the flux in the positronium continuum is only consistent with this ratio with <4% confidence. The SMM Gamma Ray Spectrometer made measurements of the annihilation line and continuum in seven flares (Share et al. 1996); all the SMM measurements are consistent with the RHESSI line width. However, only two of the seven have $3\gamma/2\gamma$ ratios consistent with values >2.7 expected if the annihilation line originates in a quiet solar atmosphere at a temperature of 6,000 K. Thus, most of the SMM measurements are inconsistent with an annihilation line that is dominated by positronium formation via charge-exchange in flight.

As underlined Share et al. (2003a), if the broad line observed by RHESSI is formed in a warm or hot ionized medium, the best fitting Gaussian width of 8.1 $\pm$ 1.1 keV (FWHM) suggests temperatures ranging from $\sim(4$–$7) \times 10^5$ K (Crannell et al. 1976). RHESSI and SMM line-width and positronium-continuum measurements are plotted in Fig. 10.6. The RHESSI observation and all but perhaps two of the SMM measurements are consistent with densities $<10^{12}$ H cm^{-3}that are

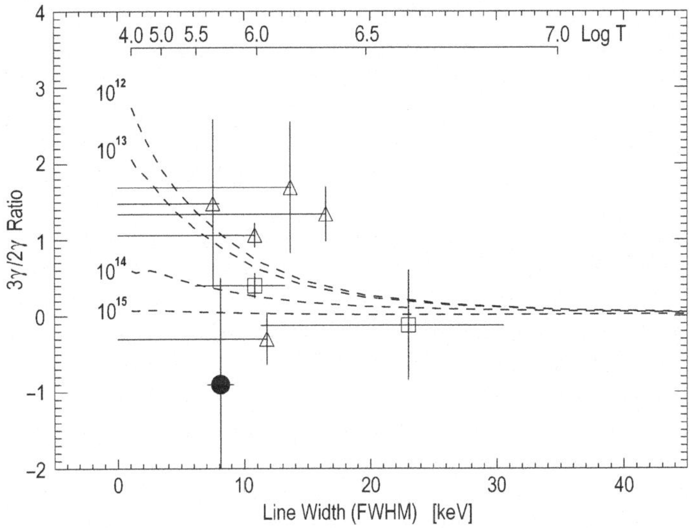

Fig. 10.6 Comparison of RHESSI (*filled circle*) and SMM measurements of the $3\gamma/2\gamma$ ratio versus 511 keV line width and temperature for a fully ionized medium. *Triangles* represent SMM 1σ upper limits on line width. The *curves* show the calculated $3\gamma/2\gamma$ ratio versus 511 keV width (temperature) for different densities (From Share et al. 2003a)

necessary to be consistent with models of quiet or flaring atmospheres at temperatures $\geq 10^{5}$ K. But if the positrons are produced at densities $\geq 10^{14}$ H cm^{-3}, how do they slow down and annihilate at such low densities? Alternatively, all the observations are consistent with densities $\geq 10^{12}$ H cm^{-3}. Such densities require formation of a substantial mass of atmosphere at transition region temperatures during flares. There is some evidence for high temperatures at high densities in the dramatic enhancement over quiet Sun values of C IV and Si IV line emission in the transition region (Brekke et al. 1996).

10.4 Positron-Emitter Production in Solar Flares from ^{3}He Reactions

10.4.1 The Significant Contribution from ^{3}He Reactions Expands the Utility of the Annihilation Line as a Sensitive Tool for Investigating the Structure of the Flaring Solar Atmosphere

Kozlovsky et al. (2004) treat in detail positron production from the decay of radioactive nuclei produced in nuclear reactions of accelerated ^{3}He. Because of their large cross sections and low threshold energies, these reactions can significantly contribute to positron production in solar flares with accelerated-particle compositions enriched in ^{3}He. The addition of these ^{3}He reactions extends earlier calculations of positron production by accelerated protons and α-particles. Reactions with ^{3}He not only add significantly to the total positron yield in flares, but can also yield a positron depth distribution that peaks higher in the solar atmosphere.

As noted Kozlovsky et al. (2004), in the solar photosphere, the abundance of ^{3}He is believed to be quite small. Although direct spectroscopic measurement of its photospheric abundance is not possible, estimates of the ^{3}He/H ratio derived from solar wind, corona, and meteoritic measurements range from 1.2×10^{-5} (Anders and Grevesse 1989) to 3.4×10^{-5} (Geiss 1982). However, in impulsive solar energetic particle (SEP) events observed in interplanetary space (where the abundances of accelerated particles heavier than oxygen can be enhanced by factors of 3–10), ^{3}He can be enhanced by several orders of magnitude, with ^{3}He/α values of unity or even greater (Reames et al. 1994). While large SEP events are thought to be accelerated by shocks in the solar corona and interplanetary space (Reames 1999, 2000), impulsive SEP events are believed to originate low in the solar atmosphere. Aschwanden et al. (1996) places the height of the impulsive event acceleration region at $(18–20) \times 10^4$ km.

Kozlovsky et al. (2004) underlined that the analyses of gamma-ray line flares (e.g., Murphy et al. 1987; Share and Murphy 1999), which are also believed to occur in the lower solar atmosphere, have shown that the abundances of the heavy accelerated particles responsible for broad gamma-ray lines are also enhanced, possibly in a manner similar to the impulsive SEP event enhancements, suggesting that these two populations may be accelerated by the same process. Observations of gamma-ray lines produced predominantly by accelerated ^{3}He, such as the 0.937 MeV line of ^{18}F from the reaction ^{16}O(^{3}He,p)^{18}F*, can directly measure the ^{3}He abundance. Studies of this line measured by the moderate-resolution Solar Maximum Mission (SMM) Gamma-Ray Spectrometer (GRS) detectors for several solar flares (Share and Murphy 1998; Mandzhavidze et al. 1997, 1999) have suggested that the ^{3}He/α ratio could be ~0.1 or even larger. If accelerated ^{3}He is significantly enhanced in gamma-ray line flares as these studies suggest, calculations of yields of accelerated particle reactions must take this into account.

Kozlovsky et al. (2004) noted that the high spectral resolution measurements of the 0.511 MeV positron annihilation line (Share et al. 2003a, b) by RHESSI have focused attention on the processes responsible for the formation of this line. Kozlovsky et al. (1987) provided a detailed treatment of positron production from the decay of radioactive nuclei produced in nuclear interactions of accelerated protons and α-particles with the most abundant ambient elements. A computer code to calculate the yields for various accelerated particle spectra and interaction models was constructed using the cross sections given in that paper. The code has been used in many investigations (e.g., Murphy et al. 1987) concerning the observation of the 0.511 MeV annihilation line produced when the positrons annihilate with ambient electrons.

Kozlovsky et al. (2004) extend the calculations of Kozlovsky et al. (1987) to include the most important positron-emitter – producing reactions of accelerated ^3He. These ^3He reactions are important because of their very low threshold energies and large cross sections at low energies. As a result of the low threshold energies, these reactions may dominate positron-emitter production for steep accelerated particle spectra when the accelerated particles are enriched with ^3He. As Kozlovsky et al. (2004) show, the reactions not only add significantly to the total positron production yields but can also affect the depth distribution of positron-emitter production in the solar atmosphere. The inclusion of ^3He reactions is critical to ensure the accuracy of the positron annihilation line as a sensitive tool for investigating both the flare acceleration process and the structure of the flaring solar atmosphere. Kozlovsky et al. (2004) discuss the impact these reactions have on the analysis (Share et al. 2003a, b; see Section 10.3) of the annihilation line observed with RHESSI from the 2002 July 23 flare, for which two very different scenarios have been suggested to explain the observed line profile.

10.4.2 Products and Corresponding Threshold Energies of the Most Important ^3He Reactions in the Solar Atmosphere

In Table 10.1, are shown the products and corresponding threshold energies (in parentheses) of the most important ^3He reactions with the most abundant elements in the solar atmosphere (a value of 0 indicates an exothermic reaction).

As noted Kozlovsky et al. (2004), conspicuous features of Table 10.1 are that most of the reactions are exothermic and that the endothermic reactions have very low threshold energies. The reason for these uniquely low thresholds is the high mass excess of ^3He (15.8 MeV). This excess is also responsible for highly excited compound systems in these reactions and hence for their large cross sections. These characteristics are the reason that ^3He-induced reactions have marked advantages

Table 10.1 ^3He-induced reaction products and threshold energies (MeV/nucleon) (From Kozlovsky et al. 2004)

Target	Products			
	n	p	d^a	α^b
^{12}C	^{14}O (0.49)^c	^{14}N (0)	^{13}N (1.48)^c	^{11}C (0)^c
^{14}N	^{16}F (0.47)^c	^{16}O (0)	^{15}O (0)^c	^{13}N (0)^c
^{16}O	^{18}Ne (1.17)^c	^{18}F (0)^c	^{17}F (1.94)^c	^{15}O (0)^c
^{20}Ne	^{22}Mg (0)^c	^{22}N (0)^c	^{21}N (1.17)^c	^{19}Ne (0)^c
^{24}Mg	^{26}Si (0.024)^c	^{26}Al (0)^c	^{25}Al (1.21)^c	^{23}Mg (0)^c
^{28}Si	^{30}S (0.165)^c	^{30}P (0)^c	^{29}P (1.02)^c	^{27}Si (0)^c
^{56}Fe	^{58}Ni (0)	^{58}Co (0)^c	^{57}Co (0)	^{55}Fe (0)

Threshold energies are given in parentheses
^{a}d denotes either d or p, n
^{b}α denotes either α or various combinations of nucleons
^{c}Radioactive positron emitter

for nuclear research when only low-energy accelerators (less than 5 MeV) are available (e.g., Bromley and Almqvist 1960). Most of the products shown in Table 10.1 are radioactive positron-emitting nuclei (the radioactive products of Fe reactions mostly decay via electron capture). Furthermore, most of these products have short lifetimes (see Kozlovsky et al. 1987 for a list of lifetimes) and hence can contribute to the observed 0.511 MeV positron annihilation line in solar flares, for which observation times are typically tens of minutes. Kozlovsky et al. (1987) found these same radioactive nuclei to be the most important contributors to positron production by accelerated protons and α-particles in solar flares.

10.4.3 Cross Sections for Production of Radioactive Positron-Emitters by ^{3}He Reactions with the Various Target Nuclei

Kozlovsky et al. (2004) discuss the cross sections for production of radioactive positron-emitters by ^{3}He reactions for the energy range of 1–100 MeV/nucleon with the various target nuclei of Table 10.1. They discuss the large body of laboratory cross section measurements, but these measurements do not always cover the full energy range. Kozlovsky et al. (2004) use the following rules, based on extensive studies, to estimate cross sections for which measurements are not available:

1. For reactions whose exit channel contains multiple particles of the form (^{3}He, *kpln*), where k and l are integers and $k,l \geq 2$, the structure of the cross section consists of a steep rise from threshold energy to a first peak, followed by a minimum, and then a second rise to a flat plateau. The first peak is due to emission of an α-particle (mainly via a compound mechanism), and the plateau is due to emission of progressively more and more nucleons (e.g., Gadioli and Hodgson 1986; Michel and Galas 1983).
2. For reactions whose exit channel contains one particle or multiple particles of the form (^{3}He, *kpln*) but where k, $l < 2$ (e.g., ^{3}He,pn), the cross sections typically rise steeply from threshold energy to a maximum and then fall rapidly to a long tail.
3. The steep rise of the cross section just above threshold energy to the first maximum is typical of reactions between charged nuclei and represents the penetration factor of the Coulomb barrier (Blatt and Weisskopf M1952). This behavior can be clearly seen in the numerous measured cross sections compiled by Keller et al. (M1973).
4. For a few reactions there were no available laboratory measurements, the cross sections have estimated by using the procedure given by Keller et al. (M1974). Based on ~1,800 experimentally determined cross sections (compiled by Keller et al. M1973), Keller et al. (M1974) have devised an elaborate semi-empirical procedure for obtaining estimates of unknown cross sections for accelerated p, d, ^{3}He, and α reactions with nuclei heavier than fluorine.

10.4.4 Cross Sections for Reactions of ^{3}He with ^{12}C

The radioactive positron-emitting nuclei produced in ^{3}He reactions with ^{12}C are ^{14}O, ^{13}N, and ^{11}C, as listed in Table 10.1. In Fig. 10.7, are shown the cross sections for these reactions. The data from 2 to 8 MeV/nucleon are from Cochran and Knight (1962), from 0.6 to 3.3 MeV/nucleon from Hahn and Ricci (1966), and from 1.6 to 11 MeV/nucleon from Brill (1965). At higher energies the cross section for ^{11}C is from Crandall et al. (1956) and Aslanides et al. (1981). The laboratory measurements of the cross section for production of ^{11}C, which cover the entire energy range of interest here, clearly demonstrate the validity of rule 1 in Section 10.4.3.

10.4.5 Cross Sections for Reactions of ^{3}He with ^{14}N

The radioactive positron-emitting nuclei produced in ^{3}He reactions with ^{14}N are ^{16}F, ^{15}O, and ^{13}N, as shown in Table 10.1. In Fig. 10.8 are presented the cross sections for these reactions. The data for ^{15}O from 2 to 6 MeV/nucleon are from Hahn and Ricci (1967). At higher energies, Kozlovsky et al. (2004) followed the anticipated behavior for this kind of reaction according to the above rules in Section 10.4.3. For ^{13}N, Kozlovsky et al. (2004) used the data from Hahn and Ricci (1966) at ~0.6 MeV/nucleon and from Brill (1965) at 3.3–10 MeV/nucleon.

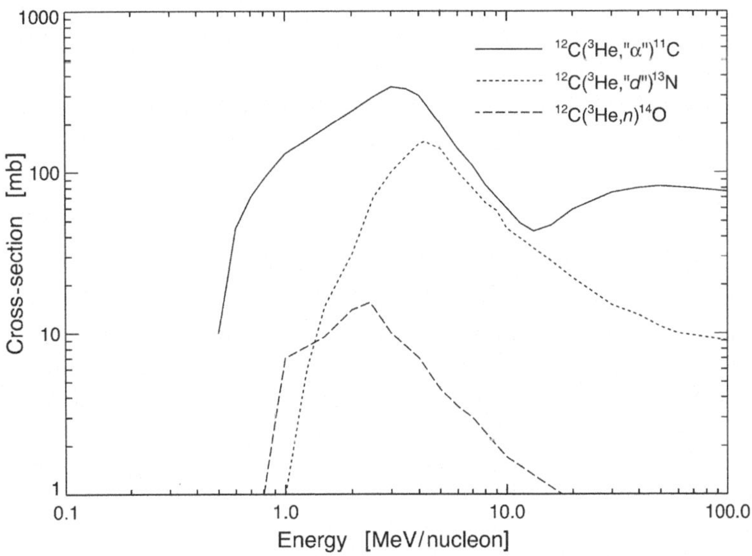

Fig. 10.7 Radioactive positron-emitter production cross sections for ^{3}He reactions on ^{12}C (From Kozlovsky et al. 2004)

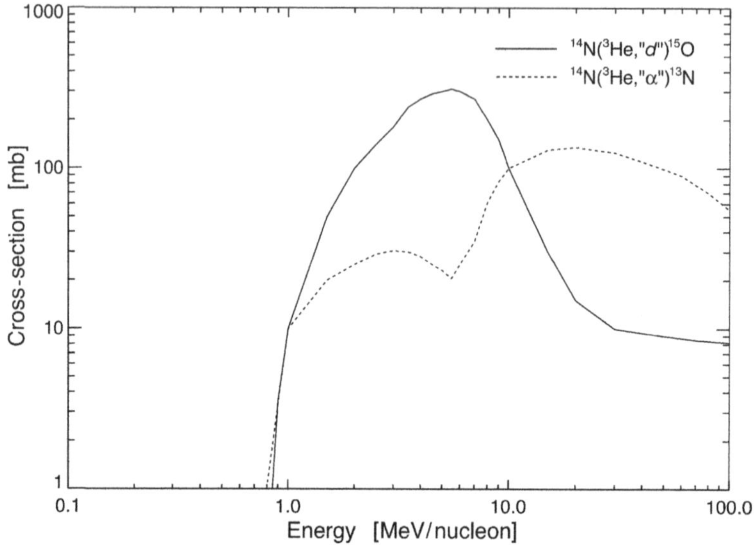

Fig. 10.8 Radioactive positron-emitter production cross sections for ^{3}He reactions on ^{14}N (From Kozlovsky et al. 2004)

Here Kozlovsky et al. (2004) again confirmed the rule concerning the rise of the cross section (see Section 10.4.3). The higher energy rise begins just beyond the threshold for the (^{3}He, ^{3}He*n*) reaction and is consistent with the explanation that the rise is due to the emission of several particles in the exit channel rather than α only. The cross section for production of ^{16}F is not available, but the (^{3}He, *n*) reaction is almost always the weakest of the reactions in Table 10.1 and always much less than the (^{3}He, α) reaction. We therefore estimate that the cross section for ^{16}F will be much less than 30 mbarn, which is the value of the first maximum for the (^{3}He, α) reaction (see Fig. 10.8), so Kozlovsky et al. (2004) do not include it here.

10.4.6 Cross Sections for Reactions of ^{3}He with ^{16}O

The radioactive positron-emitting nuclei produced in ^{3}He reactions with ^{16}O are ^{18}Ne, ^{18}F, ^{17}F, and ^{15}O, as shown in Table 10.1. In Fig. 10.9, Kozlovsky et al. (2004) present the cross sections for these reactions. The cross section for ^{18}Ne from 1.3 to 13 MeV/nucleon is based on the cross section measurements by Tatischeff et al. (2003) for the reaction ^{16}O(^{3}He, n$\gamma_{1.887}$)^{18}Ne. Kozlovsky et al. (2004) have multiplied this cross section by 1.5 to account for additional ^{18}Ne production in the ground state and in other levels that decay directly to the ground state. The data for ^{18}F from 0.9 to 3.2 MeV/nucleon are from Hahn and Ricci (1966); from 4.3 to 13.3 MeV/nucleon they are from Fitschen et al. (1977). This cross section includes the production of ^{18}F via the decay of ^{18}Ne. The data for ^{15}O

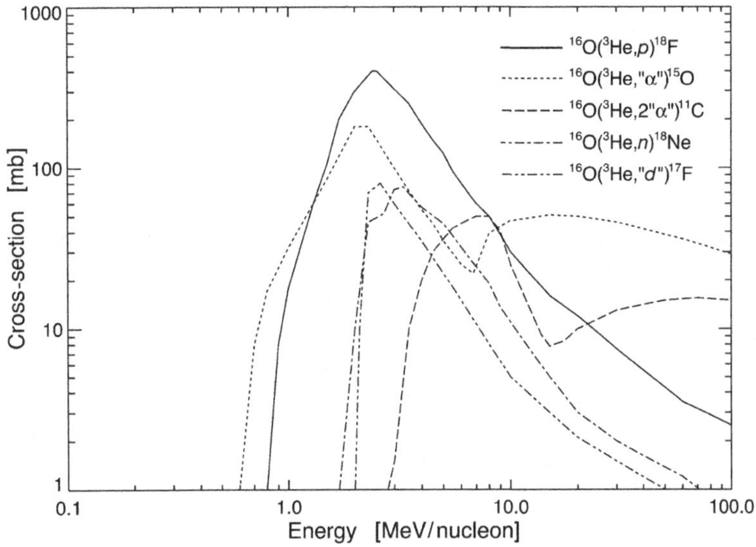

Fig. 10.9 Radioactive positron-emitter production cross sections for ^{3}He reactions on ^{16}O (From Kozlovsky et al. 2004)

from 0.9 to 3.2 MeV/nucleon are from Hahn and Ricci (1966). Kozlovsky et al. (2004) obtained a cross section value at 7 MeV/nucleon from the evaluation by Brill (1965) and at 8 MeV/nucleon by integrating the angular-dependent cross sections of Fuchs and Oeschler (1973). They constructed the complete cross section according to the rules discussed in Section 10.4.3. The data for ^{17}F are from Hahn and Ricci (1966) for 2.5–3 MeV/nucleon. Again, Kozlovsky et al. (2004) constructed the complete cross section according to the rules discussed in Section 10.4.3. In addition to the above reactions from Table 10.1, Kozlovsky et al. (2004) also show in Fig. 10.9 the cross section for the reaction producing ^{11}C from the reaction (^{3}He, 2α). The data from 3.6 to 10 MeV/nucleon are from Brill (1965). At higher energies the cross section was constructed according to the rules in Section 10.4.3 with a rise to a plateau.

10.4.7 Cross Sections for Reactions of ^{3}He with ^{20}Ne

All of the reaction products of ^{3}He reactions with ^{20}Ne shown in Table 10.1 are radioactive positron-emitting nuclei. Because the half-life of ^{22}Na is so long (2.6 years), Kozlovsky et al. (2004) do not consider it here. Since the total cross sections for the remaining nuclei are not available in the literature, Kozlovsky et al. (2004) estimated them by using the procedure given by Keller et al. (M1974). The resulting cross sections were identical to cross sections of the corresponding reactions of ^{3}He with ^{24}Mg (see Section 10.4.8), when estimated by the same procedure. Kozlovsky et al. (2004) therefore use the measured ^{24}Mg cross sections for the n, p, d, and α reactions of Table 10.1 for the corresponding ^{20}Ne reactions.

10.4.8 Cross Sections for Reactions of ^{3}He with ^{24}Mg

The radioactive positron-emitting nuclei produced in ^{3}He reactions with ^{24}Mg are ^{26}Si, ^{26}Al, ^{25}Al, and ^{23}Mg, as listed in Table 10.1. In Fig. 10.10, Kozlovsky et al. (2004) present their cross sections. The data for ^{26}Si, ^{25}Al, and ^{23}Mg from 3.6 to 13.3 MeV/nucleon are from Frantsvog et al. (1982). At energies less than 3.6 MeV/nucleon and greater than 13.3 MeV/nucleon, Kozlovsky et al. (2004) constructed the complete cross sections according to the rules discussed in Section 10.4.3. For ^{26}Al, Kozlovsky et al. (2004) constructed the cross section according to the procedure given by Keller et al. (M1974). Kozlovsky et al. (2004) assume that 50% of the total cross section produces the radioactive positron-emitter ^{26}Al.

10.4.9 Cross Sections for Reactions of ^{3}He with ^{28}Si

The radioactive positron-emitting nuclei produced in ^{3}He reactions with ^{28}Si are ^{30}S, ^{30}P, ^{29}P, and ^{27}Si. These reactions are listed in Table 10.1, and in Fig. 10.11 are shown

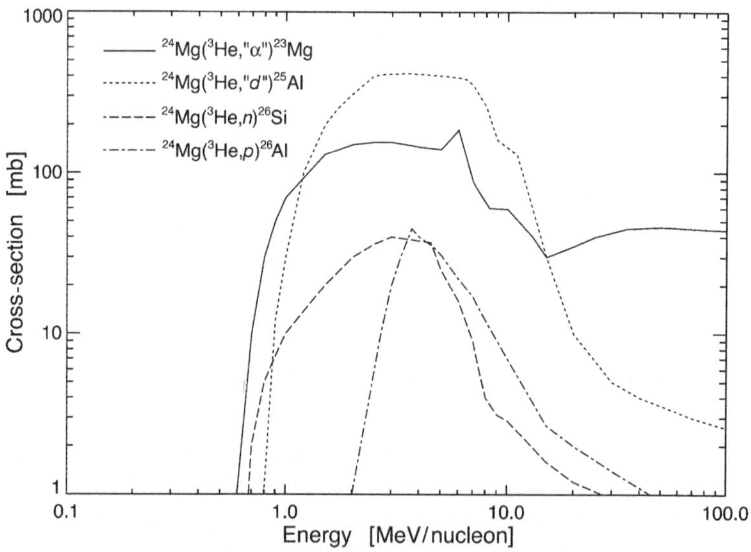

Fig. 10.10 Radioactive positron-emitter production cross sections for ^{3}He reactions on ^{24}Mg. These cross sections are also used as the cross sections for the corresponding *n*, *p*, *d*, and α reactions of ^{20}Ne (From Kozlovsky et al. 2004)

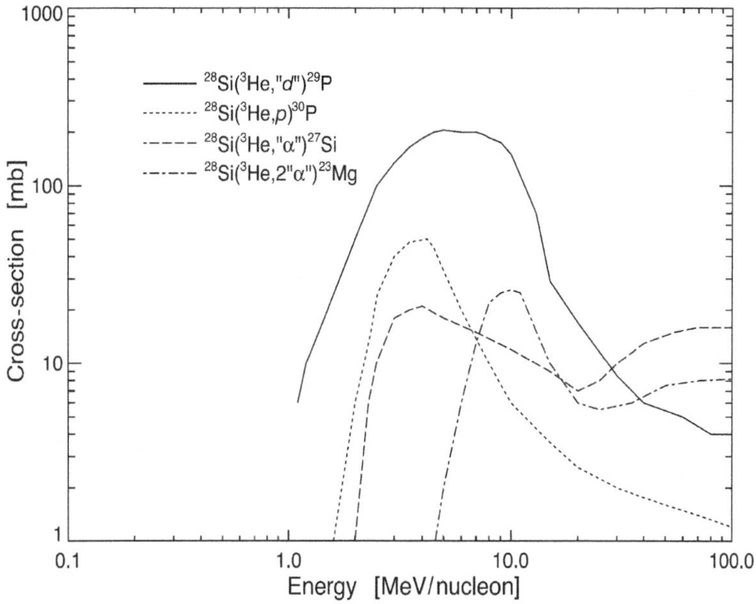

Fig. 10.11 Radioactive positron-emitter production cross sections for ^{3}He reactions on ^{28}Si (From Kozlovsky et al. 2004)

their cross sections. The data for ^{29}P from 3.6 to 13.3 MeV/nucleon are from Frantsvog et al. (1982). For ^{30}P and ^{27}Si, Kozlovsky et al. (2004) constructed the cross section according to the procedure given by Keller et al. (M1974). According to Keller et al. (M1974), the cross section for producing ^{30}S is less than 10 mbarn, so Kozlovsky et al. (2004) do not include it here. In addition to the above reactions from Table 10.1, there are also shown in Fig. 10.11 the cross section for the reaction producing ^{23}Mg from the reaction (^{3}He, 2α). The data from 3.6 to 13.3 MeV/nucleon are from Frantsvog et al. (1982). At higher energies, Kozlovsky et al. (2004) constructed the cross section according to the rules in Section 10.4.3, with a rise to a plateau.

10.4.10 Cross Sections for Reactions of ^{3}He with ^{56}Fe

The only radioactive positron-emitting nucleus produced in ^{3}He reactions with ^{56}Fe shown in Table 10.1 is ^{58}Co, in which only 15% of the decays emit a positron. ^{55}Fe and ^{57}Co decay only by electron capture. Since the cross section for total ^{58}Co production is less than 50 mbarn (Hazan and Blann 1965), Kozlovsky et al. (2004) do not include it here. In addition to the reactions in Table 10.1, Kozlovsky et al. (2004) consider production of ^{56}Co and ^{57}Ni. A total of 20% of ^{56}Co and 47% of ^{57}Ni decay via positron emission. There are shown in Fig. 10.12 the cross sections for the production of ^{56}Co and ^{57}Ni; the data from 2 to 10 MeV/nucleon are from Hazan and Blann (1965).

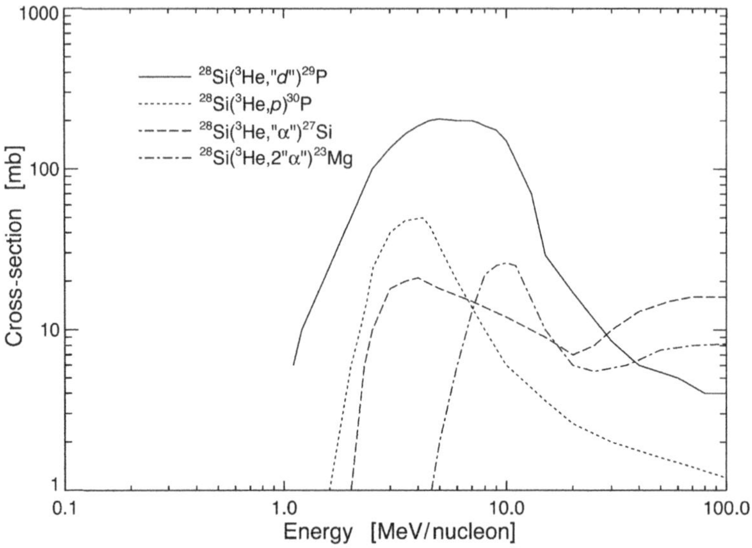

Fig. 10.12 Radioactive positron-emitter production cross sections for ^{3}He reactions on ^{56}Fe (From Kozlovsky et al. 2004)

10.4.11 Positron Yields

Kozlovsky et al. (2004) calculate thick-target positron yields from accelerated proton, ^{3}He and α-particle reactions according to the formalism given in Ramaty (1986). The ambient and accelerated ion abundances used in the calculations are summarized in Table 10.2.

For the ambient medium Kozlovsky et al. (2004) have assumed coronal abundances (Reames 1995) for C, N, O, Mg, Al, Si, S, Ca, and Fe relative to H and have taken Ne/O = 0.25, He/H = 0.10, and ^{3}He/H = 3×10^{-5}. These are the same ambient abundances used by Ramaty et al. (1996), except for He/H, for which they assumed the coronal value of 0.037. Kozlovsky et al. (2004) assumed various values for accelerated ^{3}He/α, and for the other accelerated ions they assume 'impulsive flare' abundances defined by Ramaty et al. (1996), which are coronal (Reames 1995) for C, N, Ne, Mg, Al, Si, S, Ca, and Fe relative to O but have Ne/O, Mg/O, Si/O and S/O ratios increased by a factor of 3, Fe/O increased by a factor of 10, and α/O = 50. This composition reflects average heavy-element abundance enhancements found in impulsive SEP events in space (Reames et al. 1994). Kozlovsky et al. (2004) also assume α/p = 0.5, which is at the maximum of the range observed in impulsive SEP events. Analyses of gamma-ray line flares (Share and Murphy 1997; Mandzhavidze et al. 1997, 1999) have suggested that the large observed flux of the α-α line complex at ~0.47 MeV requires either such a large value of accelerated α/p or a correspondingly large value of ambient ^{4}He/H, which

Table 10.2 Ambient and accelerated ion compositions (From Kozlovsky et al. 2004)

Element	Ambient	Accelerated
H	1.0	1.0
^{3}He	3.0×10^{-5}	[a]
^{4}He	0.1	0.5
C	2.96×10^{-4}	4.65×10^{-3}
N	7.90×10^{-5}	1.24×10^{-3}
O	6.37×10^{-4}	1.00×10^{-2}
Ne	1.59×10^{-4}	4.55×10^{-3}
Mg	1.25×10^{-4}	5.89×10^{-3}
Al	1.00×10^{-5}	1.57×10^{-4}
Si	9.68×10^{-5}	4.55×10^{-3}
S	2.03×10^{-5}	9.56×10^{-4}
Ca	6.75×10^{-6}	1.06×10^{-4}
Fe	8.54×10^{-5}	1.34×10^{-2}

[a]The calculations are performed for various values of the accelerated ^{3}He abundance

is probably less likely. Kozlovsky et al. (2004) assumed also a power-law form for the energy spectra of the accelerated ions, normalized so that the number of accelerated protons greater than 10 MeV is 1, and they assumed that all species have the same spectral index. For the proton and α-particle reactions, Kozlovsky et al. (2004) used the cross sections given by Kozlovsky et al. (1987), which include the pion producing reactions.

In Fig. 10.13 are shown total thick-target positron yields, Q_{+}, as functions of the accelerated-particle power law spectral index S for accelerated ^{3}He/α ratios of 0, 0.1, 0.5, and 1.

As can be seen from Fig. 10.13, for harder spectra ($S < 3$), the ^{3}He contribution is negligible even for ^{3}He/$\alpha = 13$, with positron production being primarily from accelerated protons and α-particles (for the hardest spectra, positrons from the decay of pions produced by accelerated protons are most important). For steeper spectra ($S > 3$), the contribution from ^{3}He can dominate (at $S = 5.5$ and ^{3}He/$\alpha = 13$, the increase is about an order of magnitude). Kozlovsky et al. (2004) noted that even the yields calculated here for ^{3}He/$\alpha = 13$ (when renormalized to one proton >30 MeV) are larger than those calculated by Kozlovsky et al. (1987), except for the hardest spectra, where the production is primarily from the decay of pions from proton interactions. This increase is due to the different abundances assumed here.

As noted Kozlovsky et al. (2004), once emitted, positrons either slow down and annihilate with ambient electrons to produce annihilation photons or escape from the region without significant annihilation during typical solar-flare observation times. Positrons annihilate either directly with ambient electrons (forming two 0.511 MeV photons in the center of mass) or by first forming positronium. Positronium is formed in one of two states depending on the relative spins of the positron and electron: a singlet state (25% of the time) and a triplet state (75% of the time). Annihilation from the singlet state produces two 0.511 MeV photons in the center of mass, while annihilation from the triplet state produces three photons with energies less than 0.511 MeV.

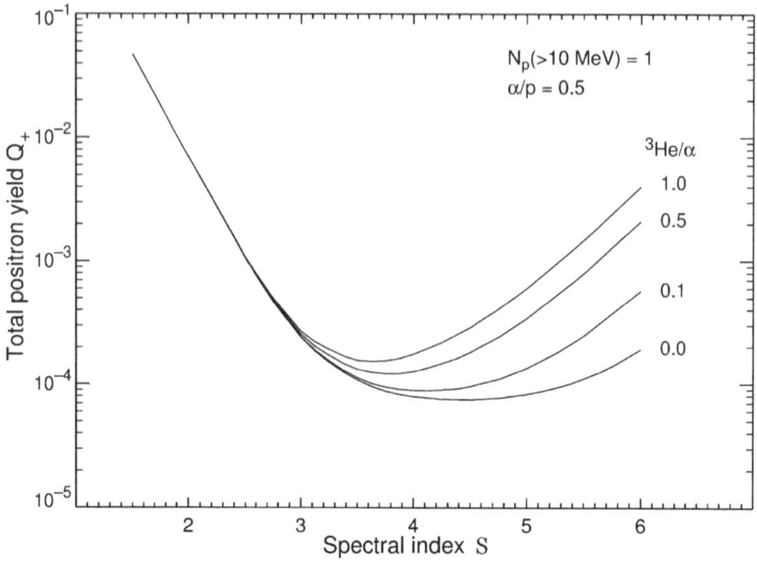

Fig. 10.13 Total thick-target positron yield, Q_{+}, as a function of the accelerated particle spectral index S for $^{3}He/\alpha = 0$, 0.1, 0.5, and 1.0 (From Kozlovsky et al. 2004)

Kozlovsky et al. (2004) underlined that relevant quantities associated with positron annihilation are (1) the number of 0.511 MeV line photons produced per positron (f_{511}), (2) the fraction of positrons that annihilate via positronium (f_{ps}), and (3) the ratio of the number of photons in the three-photon continuum to the number in the line ($3\gamma/2\gamma$). If the positronium is undisturbed, its annihilation proceeds according to the above percentages so that $f_{511} = 2 - 1.5f_{ps}$ and $3\gamma/2\gamma = 2.25f_{ps}/(2 - 1.5f_{ps})$. However, high density or high temperature can quench the longer-lived triplet state, either by ionizing the positronium or by transforming the triplet state to the singlet state by spin flipping. This will tend to increase f_{511} and decrease $3\gamma/2\gamma$. If all of the triplet state is quenched (or if $f_{ps} = 0$), f_{511} will have its maximum value of 2 and $3\gamma/2\gamma = 0$. If $f_{ps} = 1$ and the positronium is undisturbed, f_{511} will be 0.5 and $3\gamma/2\gamma = 4.5$. Care must be taken interpreting measured values of these quantities, since f_{511} can appear reduced and $3\gamma/2\gamma$ modified because of (1) finite observation times (since not all of the emitters can decay to produce positrons and subsequent annihilation photons due to the lifetimes of the various positron emitters), and (2) escape of positrons from the annihilation region.

The lower curves of Fig. 10.14 show calculated ratios of the total 0.511 MeV positron-annihilation line fluence to the summed fluences of the 4.44 and 6.13 MeV ^{12}C and ^{16}O deexcitation lines, as functions of the accelerated particle spectra index S for $^{3}He/\alpha = 0$, 0.1, 0.5, and 1.0. Kozlovsky et al. (2004) assumed a positron to annihilation-line photon conversion factor $f_{511} = 1.0$. If all positrons annihilate within the flare observation time and there is no quenching of the positronium triplet state, this factor would correspond to a positronium fraction of 0.67 and $3\gamma/2\gamma = 1.5$.

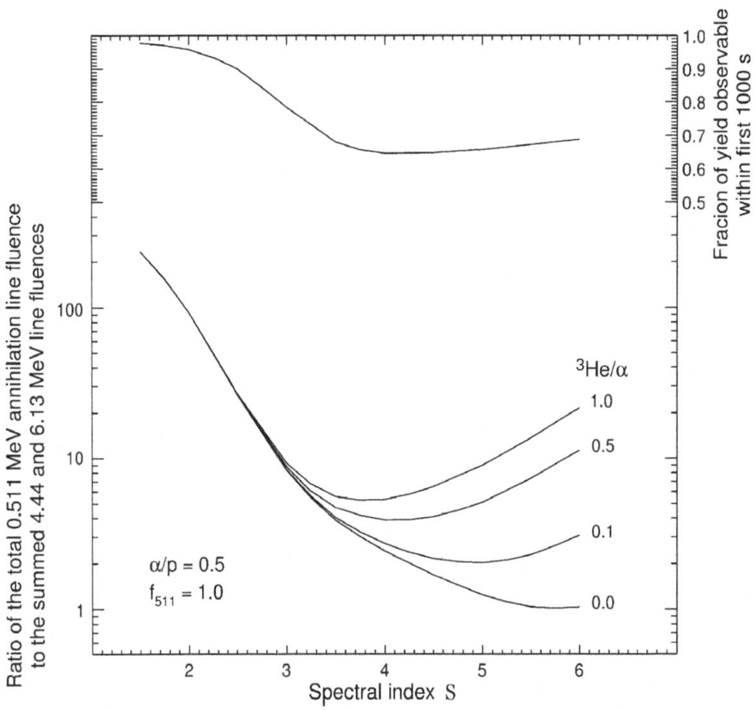

Fig. 10.14 *Lower curves*: Calculated ratio of the total 0.511 MeV positron annihilation line fluency to the summed fluencies of the 4.44 and 6.13 MeV ^{12}C and ^{16}O de-excitation lines, as a function of accelerated-particle spectra index S for ^{3}He/$\alpha = 0$, 0.1, 0.5, and 1.0. The positron to annihilation-line photon conversion factor $f_{511} = 1.0$. *Upper curve*: Fraction of the total positron annihilation photons observable within the first 1,000 s, as a function of the accelerated-particle spectral index (From Kozlovsky et al. 2004)

The upper curve in Fig. 10.14 shows the fraction of the total number of annihilation photons observable within the first 1,000 s for instantaneous emitter production at $t = 0$, as a function of the accelerated particle spectral index S. For very hard spectra (S < 3), positrons are produced primarily via the decay of pions, which have a very short lifetime ($<10^{-5}$ s), and nearly all of the photons are observable. As the spectrum softens, positron production shifts to radioactive emitters with longer lifetimes, and fewer photons are observable. For indices larger than 5 there is a small dependence of this fraction on ^{3}He/α, but the deviation from the shown curve is less than 7% at S $= 6$.

10.4.12 Application to the 2002 July 23 Solar Flare

Figure 10.15 shows calculated ratios of the 0.511 MeV annihilation line fluence to the summed 4.44 MeV ^{12}C and 6.13 MeV ^{16}O deexcitation line fluences, as

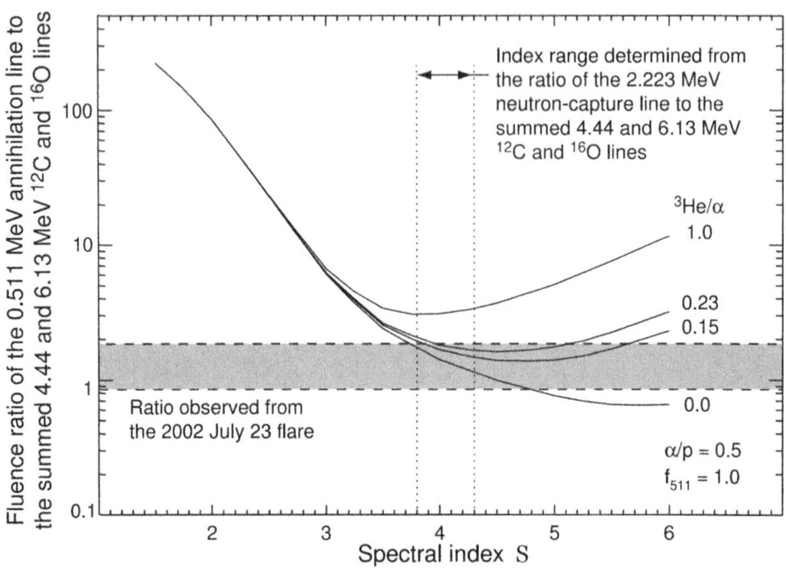

Fig. 10.15 Calculated ratio of the 0.511 MeV positron annihilation line fluency to the summed fluencies of the 4.44 MeV (^{12}C) and 6.13 MeV (^{16}O) de-excitation lines as a function of the accelerated particle spectra index S for $^3He/\alpha$ = 0, 0.15, 0.23, and 1.0. The assumed positron to annihilation line photon conversion factor f_{511} = 1.0, and all *curves* have been reduced to account for the fraction of the total annihilation line fluency observable during the 2002 July 23 solar flare observation interval. The $^3He/\alpha$ values of 0.15 and 0.23 are the 90% and 99% confidence upper limits, respectively, determined from the measured ratio of the 0.937 MeV ^{18}F line fluency upper limit and the ~0.45 MeV α–α complex fluency. The *horizontal band* indicates the ratio $\pm 1\sigma$ observed by RHESSI from the 2002 July 23 flare. The *vertical dotted lines* indicate the allowed range ($\pm 1\sigma$) of spectral index derived from the observed fluency ratio of the 2.223 MeV neutron-capture line to the summed 4.44 and 6.13 MeV (^{12}C and ^{16}O) de-excitation lines (From Kozlovsky et al. 2004)

functions of the accelerated particle spectral index S for the extreme cases of $^3He/\alpha = 0$ and 1.0.

Kozlovsky et al. (2004) have taken into account the fraction of the annihilation line fluence expected to be observed during the 960 s observation time of the 2002 July 23 solar flare, assuming that the positron emitter production time history is identical to the nuclear line-emission time history for this flare (Share et al. 2003a, b). Upper limits for the accelerated $^3He/\alpha$ abundance ratio for this flare, obtained from a comparison of the measured upper limits for the ^3He-produced 0.937 MeV line fluence and the fluence of the α-α complex, are 0.15 with 90% confidence and 0.23 with 99% confidence. These values are comparable to the SMM 19-flare average for $^3He/\alpha$ of 0.1 (Share and Murphy 1998). Calculated fluence ratios for these values of $^3He/\alpha$ are also shown in Fig. 10.15. For all of the curves of Fig. 10.15, f_{511} was assumed to be 1, which corresponds to a positronium fraction f_{ps} of 0.67 and a $3\gamma/2\gamma$ ratio of 1.5. This value for $3\gamma/2\gamma$ is consistent with the measured 99% confidence upper limit of 3.3 (Share et al. 2003b).

Kozlovsky et al. (2004) noted that there is some dependency of the calculated ratio on the assumed accelerated α/p abundance ratio. An estimate of this abundance ratio for the 23 July flare can be obtained by comparing calculated fluence ratios of the α-α complex and the summed 4.44 and 6.13 MeV lines with the measured ratio of 2.2 $\pm$ 0.9 (Share et al. 2003b; Smith et al. 2003). The measured ratio provides a 1 σ lower limit of ~0.35, which is consistent with the assumed value of 0.5. Also shown in Fig. 10.15 is the measured line ratio for this flare, 1.35 $\pm$ 0.5 (Share et al. 2003b; Smith et al. 2003), indicated by dashed lines. The vertical dotted lines indicate the range of spectral index deduced from the measured ratio of the 2.223 MeV neutron capture line fluence to the summed fluences of the 4.44 MeV (^{12}C) and 6.13 MeV (^{16}O) lines. The calculated line ratio is consistent with the measured ratio for ^{3}He/α ratios up to at least the 99% confidence value of 0.23. If the α/p ratio were as low as 0.1, the curves of Figs. 10.14 and 10.15 in the spectral range of interest would be reduced to about 70% of their values for α/p = 0.5, and the curves for ^{3}He/α ratios up to 0.23 would still be consistent with the measured line fluence ratio. It was also noted that even if as many as 50% of the positrons were to escape from the region without appreciable annihilation during the observation time, the calculated ratio for ^{3}He/α = 0.23 would still be consistent with the measurement.

As underlined Kozlovsky et al. (2004), the RHESSI high-resolution Ge detectors resolved the 0.511 MeV positron annihilation line observed in the 2002 July 23 solar flare (Share et al. 2003b). When fit with a Gaussian line shape, the line was found to be broad with a full width at half maximum (FWHM) of 8.1 $\pm$ 1.1 keV. Share et al. (2003a, b) found that the line shape was consistent with two very different production scenarios. The line could have been formed in a quiet solar atmosphere at a temperature of ~6,000 K. Under these conditions, the line is composed of both a narrow and a broad component. The narrow component (~1.5 keV) is produced by annihilation of thermalized positrons with bound electrons, while the broad component (~7 keV) results from positronium formation by charge exchange in flight with hydrogen. For a narrow range of temperatures, the broad component is sufficiently strong to produce an effective 'broad' line. Alternatively, the line could be a thermally broadened Gaussian, whose width corresponds to a temperature of (4–7) $\times$ 10^5 K.

According to opinion of Kozlovsky et al. (2004), both of these scenarios present difficulties. The difficulty with the first scenario is that it requires an exceptionally fine tuning of the annihilation region, since only for a very narrow range of temperatures (approximately 5,650–6,270 K) is the broad component sufficiently strong to produce an acceptable fit to the line shape. At temperatures immediately above and below this range, the line becomes very narrow. For the second scenario, calculations of Hua et al. (1989) for a relatively hard accelerated-particle spectrum have shown that the interaction site of nuclear reactions similar to those responsible for positron-emitter production by protons and α-particles is expected to be at hydrogen densities of 10^{14}–10^{15} cm^{-3}. Current models of the solar atmosphere indicate that the temperature at such high densities is much less than (4–7) $\times$ 10^5 K.

Kozlovsky et al. (2004) noted that the considered above reactions of positron-emitter production by accelerated ^{3}He may help make the second scenario more plausible. The solar atmosphere during flares is known to be significantly altered from quiet-Sun conditions by processes such as heating and mass motion. There is evidence for high temperatures at high densities, for example, in the dramatic enhancement over quiet-Sun values of C iv and Si iv line emission in the transition region, as noted by Brekke et al. (1996). As another example, Raymond et al. (2003) observed large emission measures of 10^5 K gas during the 2002 July 23 flare. Although temperatures as high as (4–7) $\times$ 10^5 K at the high densities 10^{14}–10^{15} cm^{-3} are still unlikely, if a significant fraction of the positrons were formed at substantially lower densities, as will be shown below, the high temperatures implied by the annihilation line width would be much more plausible there. As discussed above, the ^{3}He reactions are unique in that their cross sections are both large and have very low threshold energies compared with corresponding proton- and α-particle–induced reactions. For a power-law spectrum of accelerated particles losing energy in an ambient medium, such low-threshold reactions are produced predominantly by particles with low initial energies. These particles have shorter ranges than higher energy particles, so the reactions tend to occur at higher elevations (lower densities) as the particles are moving downward in the solar atmosphere. Kozlovsky et al. (2004) demonstrate this by calculating the depth distribution of nuclear reactions having a low threshold energy ($\sim$1 MeV/nucleon, similar to those of typical positron-emitter reactions by accelerated ^{3}He) and nuclear reactions having a high threshold energy ($\sim$25 MeV/nucleon, similar to those of typical positron-emitter spallation reactions by accelerated protons).

To perform these calculations, Kozlovsky et al. (2004) used the loop model of Hua et al. (1989). This model includes energy losses due to Coulomb collisions, removal by nuclear reactions, magnetic mirroring in the convergent flux tube, and MHD pitch-angle scattering in the corona (see Section 11.3 for details). The accelerated ions are released isotropically at the top of a magnetic loop and are followed until they interact or thermalize, usually near the loop footpoints in the chromosphere or upper photosphere. The model consists of a semicircular coronal portion and two straight portions extending vertically from the transition region through the chromosphere into the photosphere. Below the transition region, the magnetic-field strength is assumed proportional to a power δ of the pressure (Zweibel and Haber 1983). The pitch-angle scattering can be characterized by its mean free path for isotropization, Λ, here expressed by λ, the ratio of Λ to the loop half-length. The level of pitch-angle scattering directly affects the angular distribution of the accelerated particles when they interact with the solar atmosphere. Using gamma-ray line data from several SMM flares, Share et al. (2002) showed that the measured deexcitation line Doppler shifts imply interacting ion angular distributions that are inconsistent with no pitch-angle scattering ($\lambda \rightarrow \infty$), obtaining better fits with values of $\lambda = 300$ or less.

Kozlovsky et al. (2004) assumed a level of pitch-angle scattering given by $\lambda =$ 300. The atmospheric model is the sunspot model of Avrett (1981), and the accelerated-particle power-law spectral index S was assumed to be 4, typical of

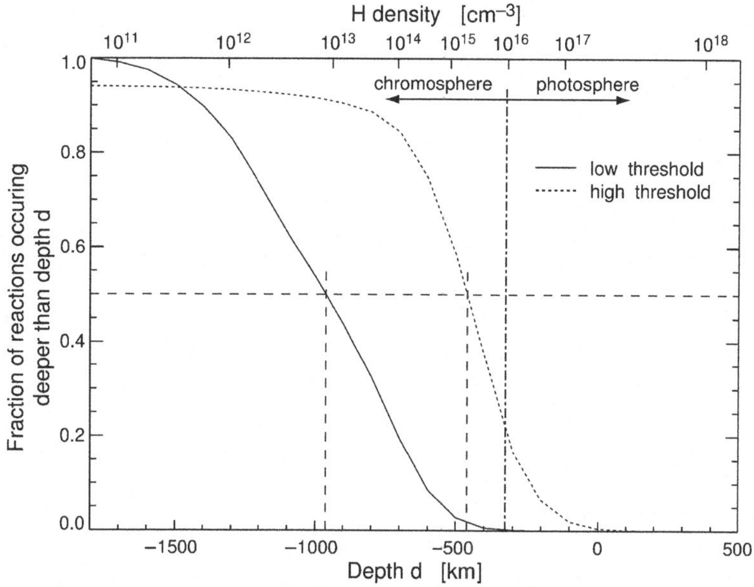

Fig. 10.16 Fraction of production occurring deeper than a given depth for two reactions whose cross sections have a low-energy threshold (1 MeV/nucleon, *solid line*) and a high-energy threshold (25 MeV/nucleon, *dotted line*). The dashed lines indicate the 50% points on each curve. The assumed atmospheric model is Avrett (1981). The bottom axis is depth in the atmosphere (km) and the top axis is the corresponding hydrogen number density (cm^{-3}) (From Kozlovsky et al. 2004)

gamma-ray line flares (Ramaty et al. 1996). They also assume a loop length of 1.15 $\times$ 10^9 cm and $\delta = 0.2$, values found by Hua et al. (1989) to provide good fits to deexcitation line decay times in the 1980 June 21 flare. For the two threshold energies, 1 and 25 MeV/nucleon, are shown in Fig. 10.16 the fraction of reactions occurring deeper than a given depth in the solar atmosphere, with dashed lines indicating the 50% points.

As can be seen from Fig. 10.16, reactions with a low threshold occur at significantly higher elevations, with a difference in density of more than two orders of magnitude for this model atmosphere. Thus, positron-emitter production from the low-threshold ^{3}He reactions would occur higher in the chromosphere and at lower densities than production from proton and α-particle reactions. If the temperature of this lower-density region were sufficiently high during the flare, these positrons would add a broad thermal component to the line profile. However, as noted Kozlovsky et al. (2004), Fig. 10.15 shows that for the 23 July 2002 solar flare ^{3}He/α has upper limit of 0.23, the ^{3}He reactions would only contribute $\sim$30% to the 0.511 MeV line. Even if these positrons annihilated in a high-temperature region, their contribution to the line would not be sufficient to produce a line shape consistent with the width measurement for this flare.

10.4.13 The Positron Annihilation Line as a Sensitive Tool for Exploring the Physical Conditions of the Solar Flare Region

Kozlovsky et al. (2004) have shown that accelerated ^{3}He can produce a significant number of positrons in solar flares with steep accelerated particle spectra, particularly if the 3He/α ratio is ~1, as observed in some impulsive SEP events in interplanetary space. Because of their low threshold energies, these ^{3}He reactions tend to occur at lower densities in the solar atmosphere than the reactions of protons and α-particles. If the temperature at these lower densities is higher than inferred from current atmospheric models, such positrons would add a broad thermal component to the line profile. However, for the 2002 July 23 flare observed by RHESSI, the measured 99% confidence upper limit for the 3He/α ratio, 0.23, is probably not sufficient to fully explain the observed line shape for this flare.

Kozlovsky et al. (2004) underlined that a full understanding of the annihilation line will require a comprehensive, self-consistent calculation of annihilation line production. The calculations must include the transport of the accelerated particles and depth distribution of positron emitter production, transport of the emitted positrons, and all possible processes by which positrons annihilate (direct annihilation, positronium formation, quenching, etc.). All of these calculations must be performed within the context of a realistic magnetic-loop model assuming various solar atmospheres. This comprehensive study will allow to construct an annihilation line profile that is composed of contributions from a range of physical conditions determined by the depth distribution of positron production, transport and annihilation. All of the processes involved in the formation of the positron annihilation line are sensitive to the physical state of the flaring loop such as temperature, density, ionization, and magnetic-field configuration. This line, whose previous theoretical treatments assumed only simplified conditions, will become an exceptionally sensitive tool for exploring the physical conditions of the solar flare region.

References for Chapter 10

Anders E, Grevesse N (1989) Abundances of the elements: meteoritic and solar. Geochim Cosmochim Acta (UK) 53(1):197–214

Aschwanden MJ, Wills MJ, Hudson HS, Kosugi T, Schwartz RA (1996) Electron time-of-flight distances and flare loop geometries compared from CGRO and Yohkoh observations. Astrophys J (USA) 468(1, Part 1):398–417

Aslanides E, Fassnacht P, Dellacasa G, Gallio M, Tuyn JWN (1981) 12C(3He, 3He n)11C cross section at 910 MeV. Phys Rev C Nucl Phys (USA) 23(4):1826–1828

Avrett EH (1981) Reference model atmosphere calculation – the Sunspot sunspot model. In: Cram LE, Thomas JH (eds) The physics of sunspots, Proceedings of the conference, Sunspot, New Mexico, 1981, conference sponsored by the Sacramento Peak Observatory, Sunspot, NM, pp 235–255, 257

Brekke P, Rottman GJ, Fontenla J, Judge PG (1996) The ultraviolet spectrum of a 3B class flare observed with SOLSTICE. Astrophys J (USA) 468(1, Part 1):418–432

Brill OD (1965) He3-light nucleus interaction cross sections. Soviet J Nucl Phys 1(1):37–40

Bromley DA, Almqvist E (1960) ^{3}He induced reactions. Rep Prog Phys 23(1):544–629

Bussard RW, Ramaty R, Drachman RJ (1979) The annihilation of galactic positrons. Astrophys J (USA) 228(3, Part 1):928–934

Chupp EL, Forrest DJ, Higbie PR, Suri AN, Tsai C, Dunphy PP (1973) Solar gamma ray lines observed during the solar activity of August 2 to August 11, 1972. Nature 241(5388):333–335

Cochran DRF, Knight JD (1962) Excitation functions of some reactions of 6- to 24-MeV He^{3} ions with carbon and aluminum. Phys Rev 128(3):1281–1286

Crandall WE, Millburn GP, Pyle RV, Birnbaum W (1956) C^{12}(x, xn)C^{11} and Al^{27}(x, x2pn)Na^{24} cross sections at high energies. Phys Rev 101(1):329–337

Crannell CJ, Joyce G, Ramaty R, Werntz C (1976) Formation of the 0.511 MeV line in solar flares. Astrophys J (USA) 210(2, Part 1):582–592

Falcone AD (2000) for the Milagro Collaboration "Detection of the 6 November 1997 Ground Level Event by Milagrito". In: Mewaldt RA, Miller M, Jokipii JR, Lee MA, Möbius E, Zurbuchen TH (eds) Acceleration and transport of energetic particles observed in the heliosphere, ACE 2000 Symposium, Indian Wells, CA, 2000, AIP conference proceedings, vol 528, pp. 193–196

Fitschen J, Beckmann R, Holm U, Neuert H (1977) Yield and production of ^{18}F by ^{3}He irradiation of water. Int J Appl Radiat Isot (UK) 28(9):781–784

Frantsvog DJ, Kunselman AR, Wilson RL, Zaidins CS, Detraz C (1982) Reactions induced by ^{3}He and ^{4}He ions on natural Mg, Al, and Si. Phys Rev C Nucl Phys (USA) 25(2):770–779

Fuchs H, Oeschler H (1973) Energy dependence of transition strengths extracted from 12C(τ, α) and 16O(τ, α). Nucl Phys A (Netherlands) A202(2):396–402

Gadioli E, Hodgson PE (1986) The interactions of α particles with nuclei. Rep Prog Phys (UK) 49(9):951–999

Geiss J (1982) Processes affecting abundances in the solar wind. Space Sci Rev (Netherlands) 33(1–2):201–217

Hahn RL, Ricci E (1966) Interactions of ^{3}He Particles with ^{9}Be, ^{12}C, ^{16}O, and ^{19}F. Phys Rev 146(3):650–659

Hahn RL, Ricci E (1967) Interactions of ^{3}He particles with B, N, Na and Be. Nucl Phys A (Netherlands) 101(2):353–368

Hazan J-P, Blann M (1965) Excitation functions, recoil ranges, and statistical theory analysis of reactions induced in Fe^{56} with 6-29-MeV He^{3} ions. Phys Rev 137(5B):1202–1213

Hua X-M, Ramaty R, Lingenfelter R (1989) Deexcitation gamma-ray line emission from solar flare magnetic loops. Astrophys J (USA) 341(1, Part 1):516–532

Hurford GJ, Mewaldt RA, Stone EC, Vogt RE (1973) Measurements of the flux of low-energy solar-flare positrons. In: Proceedings of the 13th international cosmic rays conference, vol 2, Denver, CO, pp 1613–1617

Kozlovsky B, Lingenfelter RE, Ramaty R (1987) Positrons from accelerated particle interactions. Astrophys J (USA) 316(2, Part 1):801–818

Kozlovsky B, Murphy RJ, Share GH (2004) Positron-emitter production in solar flares from ^{3}He reactions. Astrophys J (USA) 604(2):892–899

Lingenfelter RE, Ramaty R (1967) High energy nuclear reactions in solar flares. In: Shen BSP (ed) High energy nuclear reactions in astrophysics. W.A. Benjamin, New York/Amsterdam, pp 99–158

Machado ME, Avrett EH, Vernazza JE, Noyes RW (1980) Semiempirical models of chromospheric flare regions. Astrophys J (USA) 242(1, Part 1):336–351

Mandzhavidze N, Ramaty R, Kozlovsky B (1997) Solar atmospheric and solar flare accelerated helium abundances from gamma-ray spectroscopy. Astrophys J Lett (USA) 489(1, Part 2): L99–L102

Mandzhavidze N, Ramaty R, Kozlovsky B (1999) Determination of the abundances of subcoronal ^{4}He and of solar flare-accelerated ^{3}He and ^{4}He from gamma-ray spectroscopy. Astrophys J (USA) 518(2, Part 1):918–925
Mason GM, Cohen CMS, Cummings AC, Dwyer JR, Gold RE, Krimigis SM, Leske RA, Mazur JE, Mewaldt RA, Mobius E, Popecki M, Stone EC, von Rosenvinge M.E. Wiedenbeck TT, Wiedenbeck ME (1999) Particle acceleration and sources in the November 1997 solar energetic particle events. Geophys Res Lett 26(2):141–144
Michel R, Galas M (1983) ^{3}He-induced reactions on cobalt. Nucl Phys A (Netherlands) A404(1):77–92
Murphy RJ, Dermer CD, Ramaty R (1987) High-energy processes in solar flares. Astrophys J Suppl Ser (USA) 63(3):721–748
Ramaty R (1986) Nuclear processes in solar flares. In: Sturrock PA, Holzer TE, Mihalas DM, Ulrich RK (eds) The physics of the Sun, vol 2: the solar atmosphere. Geophysics and astrophysics monographs. D. Reidel, Dordrecht, pp 291–323
Ramaty R, Lingenfelter RE (1973a) Nuclear gamma rays from solar flares. In: Proceedings of the 13th international cosmic ray conference, vol 2, Denver, CO, pp 1590–1594
Ramaty R, Lingenfelter RE (1973b) Nuclear gamma rays from solar flares. In: Ramaty R, Stone RG (eds) High energy phenomena on the Sun, Proceedings of symposium, Greenbelt, MD, 1972, NASA SP-342, Washington, D.C., pp 301–314
Ramaty R, Murphy RJ (1987) Nuclear processes and accelerated particles in solar flares. Space Sci Rev 45(3–4):213–268
Ramaty R et al (1983). In: Sturrock PA et al (eds) The physics of the Sun. D. Reidel, Dordrecht
Ramaty R, Mandzhavidze N, Kozlovsky B (1996) Solar atmospheric abundances from gamma ray spectroscopy. In: Ramaty R, Mandzhavidze N, Hua X-M (eds) High energy solar physics, Proceedings of the conference, Greenbelt, MD, 1995, AIP conference proceedings, vol 374, No. 1, pp 172–183
Raymond JC, Ciaravella A, Dobrzycka D, Strachan L, Ko Y-K, Uzzo M, Raouafi NE (2003) Far-ultraviolet spectra of fast coronal mass ejections associated with X-class flares. Astrophys J (USA) 597(2, Part 1):1106–1117
Reames DV (1995) Coronal abundances determined from energetic particles. Adv Space Res (UK) 15(7):41–51
Reames DV (1999) Particle acceleration at the Sun and in the heliosphere. Space Sci Rev (Netherlands) 90(3–4):413–491
Reames DV (2000) What we don't understand about ion acceleration in flares. In: Ramaty R, Mandzhavidze N (eds) High energy solar physics: anticipating HESSI, ASP conference series 206, ASP, San Francisco, CA, pp 102–111
Reames DV, Meyer JP, von Rosenvinge TT (1994) Energetic-particle abundance in impulsive solar flare events. Astrophys J Suppl 90(2):649–667
Sakao T, Kosugi T, Masuda S (1998) Energy release and particle acceleration in solar flares with respect to flaring magnetic loops. In: Watanabe T, Kosugi T, Sterling AC (eds) Observational plasma astrophysics: five years of YOHKOH and beyond. Astrophysics and Space Science Library, vol 229, Kluwer. Dordrecht, Holland, pp 273–284
Share GH, Murphy RJ (1997) Intensity and directionality of flare-accelerated α-particles at the Sun. Astrophys J (USA) 485(1, Part 1):409–418
Share GH, Murphy RJ (1998) Accelerated and ambient He abundances from gamma-ray line measurements of flares. Astrophys J (USA) 508(2, Part 1):876–884
Share GH, Murphy RJ (1999) Gamma-ray measurement of energetic heavy ions at the Sun. In: Proceedings of the 26th international cosmic ray conference, vol 6, Salt Lake City, UT, pp 13–16
Share GH, Chupp EL, Forrest DJ, Rieger E (1983) Positron annihilation radiation from solar flares. In: Burns ML, Harding AK, Ramaty R (eds) Positron-electron pairs in astrophysics, Goddard Space Flight Center, 1983, AIP conference proceedings, vol 101, No. 1, pp 15–20

Share GH, Murphy RJ, Skibo JG (1996) Gamma-ray line measurements and ambient solar abundances. In: Ramaty R, Mandzhavidze N, Hua X-M (eds) High energy solar physics, Proceedings of the conference, Greenbelt, MD, 1995, AIP conference proceedings, vol 374, No. 1, 162–171

Share GH, Murphy RJ, Kiener J, De Sereville N (2002) Directionality of solar flare-accelerated protons and α-particles from γ-ray line measurements. Astrophys J (USA), 573(1, Part 1): 464–470

Share GH, Murphy RJ, Skibo JG, Smith DM, Hudson HS, Lin RP, Shih AY, Dennis BR, Schwartz RA (2003a) RHESSI observation of the solar annihilation line. In: Proceedings of the 28th international cosmic ray conference, vol 6, Tsukuba, pp 3199–3202

Share GH, Murphy RJ, Skibo JG, Smith DM, Hudson HS, Lin RP, Shih AY, Dennis BR, Schwartz RA, Kozlovsky B (2003b) High-resolution observation of the solar positron–electron annihilation line. Astrophys J Lett (USA), 595(2, Part 2):L85–L88

Smith DM, Lin RP, Turin P, Curtis DW, Primbsch JH, Campbell RD, Abiad R, Schroeder P, Cork CP, Hull EL, Landis DA, Madden NW, Malone D, Pehl RH, Raudorf T, Sangsingkeow P, Boyle R, Banks IS, Shirey K, Schwartz R (2002) The RHESSI spectrometer. Sol Phys (Netherlands) 210(1–2):33–60

Smith DM, Share GH, Murphy RJ, Schwartz RA, Shih AY, Lin RP (2003) High-resolution spectroscopy of gamma-ray lines from the X-class solar flare of 2002 July 23. Astrophys J Lett (USA) 595(2, Part 2):L81–L84

Tatischeff V, Duprat J, Kiener J, Assuncao M, Coc A, Engrand C, Gounelle M, Lefebvre A, Porquet M-G, de Sereville N, Thibaud J-P, Bourgeois C, Chabot M, Hammache F, Scarpaci J-A (2003) Cross sections relevant to γ-ray line emission in solar flares: ^{3}He-induced reactions on ^{16}O nuclei. Phys Rev C Nucl Phys (USA) 68(2). doi:10.1103/PhysRevC.68.025804, 25804-1-9

Tsuneta S, Naito T (1998) Fermi acceleration at the fast shock in a solar flare and the impulsive loop-top hard X-ray source. Astrophys J 495(1, Part 2):L67–L70

Yoshimori M, Suga K, Saita N, Shiozawa A (2000a) Yohkoh observation of a gamma-ray flare on November 6, 1997. Adv Space Res (UK) 25(9):1801–1804

Yoshimori M, Suga K, Shiozawa A, Nakayama S, Takeda H (2000b) Particle acceleration in the 6 November 1997 event as viewed from gamma rays and solar energetic particles. In: Mewaldt RA, Miller M, Jokipii JR, Lee MA, Möbius E, Zurbuchen TH (eds) Acceleration and transport of energetic particles observed in the heliosphere, ACE 2000 symposium, Indian Wells, CA, 2000, AIP conference proceedings, vol 528, pp 189–192

Yoshimori M, Nakayama S, Ogawa H (2001) Positron annihilation radiation and >10 MeV gamma rays from the 1997 November 6 flare. In: Proceedings of the 27th international cosmic ray conference, vol 8, Hamburg, pp 3025–3028

Zweibel EG, Haber DA (1983) The propagation of energetic ions in magnetic loops and gamma-ray emission from solar flares. Astrophys J (USA) 264(2, Part 1):648–659

Chapter 11
The Development of Models and Simulations for Solar Neutron and Gamma Ray Events

11.1 Solar Flare Neutron Production and the Angular Dependence of the 2.223 MeV Capture Gamma-Ray Line Emission

11.1.1 The Matter and Short History of the Problem

In Chapter 1 we described the initial concept and predictions of expected fluxes of neutrons and gamma-rays from solar flares. In Chapters 2–4, as well as in Chapters 6–10 are described a lot of solar neutron and gamma-ray observations. As underlined Hua and Lingenfelter (1987a), comparisons of these observations with theoretical calculations have provided the first direct determination of the number and energy spectra of accelerated particles in flares. These comparisons have also shown (Ramaty et al. 1982; Murphy and Ramaty 1984) that the nuclear interactions are caused primarily by accelerated particles that remain trapped in the magnetic fields of the flare region and interact as they slow down in the solar atmosphere, rather than by accelerated particles that eventually escape into interplanetary space. For if the escaping particles were responsible for the observed gamma-ray line emission, they would have been greatly enriched in spallation products, such as D, T, Li, Be, and B, which were not observed (McGuire et al. 1977, 1979) in the flare particles in interplanetary space. With the exception of few flares from which neutrons were measured directly, the number and energy spectrum of the accelerated particles were all determined from measurements of the integrated flux, or fluence, in the 2.223 MeV line and the 4–7 MeV band. The bulk of the emission in the 4–7 MeV band is from the two strongest nuclear deexcitation lines in 12C and 160, as has been shown (Ramaty et al. 1977; Ibragimov and Kocharov 1977a, b). Measurements of the fluence, rather than the flux, are needed because the time histories of the two lines differ significantly. The time history of the nuclear deexcitation line emission in the 4–7 MeV band follows almost instantaneously

L. Dorman, *Solar Neutrons and Related Phenomena*, Astrophysics and Space Science Library 365, DOI 10.1007/978-90-481-3737-4_11, the time dependence of the accelerated particle interaction rate while that of the 2.223 MeV line is delayed by the neutron thermalization and capture times.

Hua and Lingenfelter (1987a) noted that the determinations of the accelerated particle numbers and energy spectra from these fluence ratios, however, depend strongly on the location of the flare on the solar disk, since the 2.223 MeV line emission is attenuated by Compton scattering, especially for flares occurring near the limb of the Sun. The 2.223 MeV line is much more strongly attenuated than the other major lines because it is produced predominantly by the capture of those neutrons that penetrate deep into the photosphere ($n > 10^{16}$ H cm^{-3}) where their capture lifetime becomes less than that of decay. The broad band 4–7 MeV emission, on the other hand, is produced much higher in the solar atmosphere from deexcitation of nuclear levels excited by the less penetrating charged particles. Such limb darkening of the 2.223 MeV line, first suggested by Wang and Ramaty (1974), is quite evident in recent observations (Chupp 1982; Yoshimori et al. 1983) of a number of flares near the limb of the Sun. Estimates of the amount of attenuation as a function of heliocentric angle have been based on calculations for monoenergetic neutrons produced isotropically, either above the solar chromosphere (Wang and Ramaty 1974; Wang 1975; Ramaty et al. 1983a; Murphy and Ramaty 1984) or for a couple of arbitrary depth distributions within it (Kanbach et al. 1975, 1981).

As underlined Hua and Lingenfelter (1987a), no study has previously been made of the depth and angle dependent production of neutrons and gamma-rays that might be expected from the interaction of accelerated protons and ions in the solar chromosphere and photosphere. Such calculations are needed not only to provide more accurate determinations of the total number and spectrum of the accelerated ions at the Sun and a better understanding of the observed limb darkening, but they can also enable to determine the angular distribution of the accelerated ions. Theoretical studies of ion acceleration and propagation in solar flares suggest that their angular distributions could be very anisotropic, resulting either from beamed acceleration (e.g., Colgate 1978) or from magnetic mirroring (Zweibel and Haber 1983). To make a systematic study of these effects, Hua and Lingenfelter (1987a) have developed a connected system of Monte Carlo programs with which they calculate the depth, energy, and angular dependences of the production of neutrons and their subsequent capture gamma-ray line emission in the solar atmosphere as a function of the accelerated ion-energy spectrum and angular distribution, and then calculate the attenuation of the gamma-rays and neutrons as a function of energy and observing angle, or heliocentric position. With these calculations Hua and Lingenfelter (1987a) can then determine the total number, energy spectrum, and angular distribution of the interacting ions from the observed neutron and gamma-ray line intensities; it became possible to test theories of ion acceleration and beaming in flares; and to study the energy dependence of ion escape into interplanetary space, as well as further acceleration or deceleration there (e.g., Forman et al. 1986). In Hua and Lingenfelter (1987a) are given a general description of the Monte Carlo calculations; presented the results of calculations of the heliocentric or zenith angular dependence of 2.223 MeV neutron capture line emission as a function of accelerated-ion energy spectrum and angular distribution; and compared these results with Solar Maximum Mission and other satellite observations. Hua and Lingenfelter (1987a) also calculate from measurements of gamma-ray line fluencies the energy spectrum and total number of accelerated ions trapped at the Sun and compared these with measurements of the accelerated ions that escape into interplanetary space.

11.1.2 Principles of Calculations

As noted Hua and Lingenfelter (1987a), a number of calculations have previously been made of the production rate of various gamma-ray lines, neutrons, positrons, and pions expected from flare-accelerated ion interactions with the ambient solar gas. These calculations were done both for nuclear interactions of the ions during their acceleration in the solar flare – thin-target interactions – and for the subsequent nuclear interactions of those ions that do not escape into interplanetary space, but remain trapped in solar magnetic fields to slow down and stop in the solar atmosphere – thick-target interactions. The latter calculations, however, did not address the important problems of the depth and angle dependence of the production and attenuation which was studied in paper of Hua and Lingenfelter (1987a). To carry out systematic calculations of the gamma ray and neutron escape flux as a function of angle, energy and time, we have developed an interconnected system of three Monte Carlo programs. These are:

1. The Accelerated Ion Interaction Program, which calculates the production rate of neutrons, as a function of the neutron emission energy, angle and depth in the solar atmosphere, all resulting from nuclear interactions of flare-accelerated ions with a range of assumed energy spectra, angular distributions, and elemental abundances.
2. The Neutron Transport Program which calculates both the flux of neutrons escaping from the solar atmosphere, as a function of energy, time, and zenith angle, and the rate of gamma-ray line emission from radiative neutron capture in the solar atmosphere, as a function of depth, emission angle and time, expected from neutrons having the energy, angle- and depth-dependent production rate distributions calculated with the Accelerated Ion Interaction Program, considering the effects of neutron transport, scattering, decay, and non-radiative capture, especially on ^{3}He.
3. The Gamma-Ray Transport Program, which calculates the flux of un-scattered and scattered gamma-rays escaping from the solar atmosphere as a function of energy, zenith angle, and time for radiative neutron capture gamma-rays, expected from the energy, angle- and depth-dependent production rate distributions, calculated from the Neutron Transport Program.
4. As a Monte Carlo simulation, the calculations involve many random-number samplings from various distributions. These distributions include trigonometric

and other complex functions, which are usually evaluated in the computers by time-consuming series expansions, but Hua and Lingenfelter (1987a) have used instead the much faster double or multiple random number sampling techniques (Von Neumann 1951; Kahn M1954; Coleman 1968; Pei and Zhang M1980). This has reduced the computational time by about an order of magnitude.

11.1.3 Calculations of Neutron Production

In the Accelerated Ion Interaction Program Hua and Lingenfelter (1987a) consider in each calculation some 10^5 or more accelerated ions incident upon the solar atmosphere. Azimuthal symmetry and a delta function time distribution for the accelerated ion production are assumed and the species (p, He, etc.), energy and incident zenith angle of each ion is selected in the Monte Carlo simulation by random numbers from an assumed relative abundance, energy spectrum, and angular distribution. Other arbitrary time distributions of the accelerated ion production can be modeled from calculations for an initial delta-function time distribution. Except for ^{3}He, the solar atmospheric composition used in this study is that of Cameron (1982), corresponding to a ratio of H:He:CNO = 1:0.07:0.00145. The atmospheric density and temperature model is that of Avrett (1981) for a sunspot region, merged into the model of Allen (M1963) at depth. In this model the density is essentially exponential with a scale height of about 85 km from the lower chromosphere down to a knee at a depth of about 300 km into the photosphere. Below this knee the scale height rapidly increases to about 900 km. The ambient H density and column density as a function of depth are shown in Fig. 11.1, using the convention that zero depth is that at which the continuum optical depth at 0.5 μm is unity.

The model temperature varies much more slowly from about 4,700 K at a depth of −800 km through a minimum of 3,500 K at −450 km to above 20,000 K at a depth of 2,000 km. Hua and Lingenfelter (1987a) assume that the ions are accelerated in the corona or the upper chromosphere and follow their interactions in the lower atmosphere at depths greater than −1,000 km, corresponding to column densities $> 3 \times 10^{-4}$ g.cm^{-2}. Hua and Lingenfelter (1987a) also assume that the numbers of accelerated ions at any energy/nucleon have the same relative composition as the solar atmosphere (Cameron 1982). Following Murphy and Ramaty (1984), they consider two forms for the energy spectra suggested by both theory and observation: a Bessel function in momentum/nucleon and a power law in energy/nucleon. A Bessel function in momentum/nucleon is expected (Ramaty 1979) from stochastic acceleration with the differential number of particles

$$N(E) \propto K_2\left[\left(12p/mc\alpha T\right)^{1/2}\right], \qquad (11.1)$$

where K_2 is a modified Bessel function of the second order; E and p are the particle kinetic energy and momentum; m is the particle rest mass; α is the acceleration

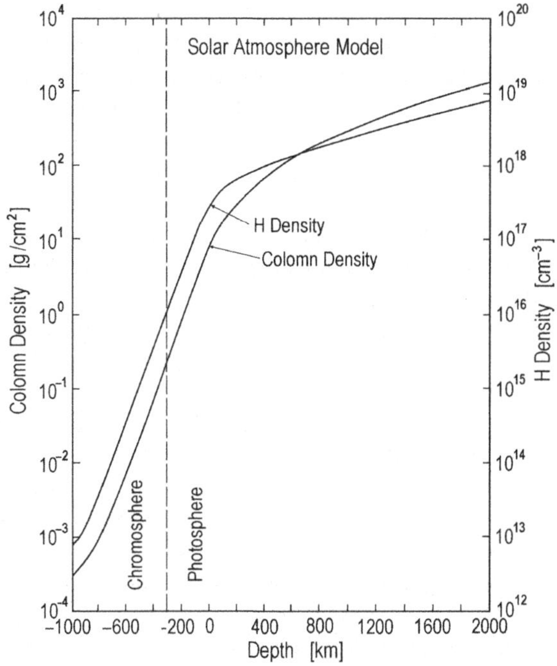

Fig. 11.1 The solar atmosphere model of Avrett (1981) for a sunspot region, merged into the model of Allen (M1963) at depth >120 km. Zero depth is defined as that at which the continuum optical depth at 0.5 μm is unity (From Hua and Lingenfelter 1987a)

efficiency proportional to the ratio of the square of the scattering center velocity to the diffusion mean free path; and T is the mean time for escape from the acceleration region (see details on the acceleration mechanism in Chapter 4 of Dorman M2006). Spectra of this form give a good fit to the accelerated particle spectra from most solar flares measured in the interplanetary medium (McGuire et al. 1981).

Hua and Lingenfelter (1987a) in their calculations employed the following approximation (Olver 1968):

$$K_2(u) = u^{-1/2} e^{-u} Q_2(u), \tag{11.2}$$

where $Q_2(u)$ is a polynomial

$$\begin{aligned} Q_2(y) = {} & 1.25331 + 1.17499y + 0.25688y^2 - 0.04718y^3 \\ & + 0.02092y^4 - 0.01032y^5 + 0.00379y^6 - 0.00068y^7. \end{aligned} \tag{11.3}$$

As noted Hua and Lingenfelter (1987a), a power law in kinetic energy/nucleon is expected from shock acceleration in the non-relativistic limit (Forman et al. 1986; see in details Chapter 4 of Dorman M2006):

$$N(E) \propto E^{-S}, \quad S = \frac{r + 1/2}{r - 1}, \tag{11.4}$$

where r is the compression ratio in the shock, which must be < 4, giving $S > 1.5$. Accelerated particle spectra in the interplanetary medium can be fitted by power law spectra over a wide energy band for only a few flares such as that of 3 June 1982 where $S = 1.7$ from 1 to 200 MeV (McDonald and van Hollebeke 1985). This index S is the same as that used by Murphy and Ramaty (1984), but differs from that in Murphy et al. (1987) where it is the exponent of a power law in momentum for the fully relativistic case.

Hua and Lingenfelter (1987a) also considered a variety of angular distributions for the accelerated protons and ions, spanning the range from downward beaming expected in some acceleration models (e.g., Colgate 1978) to magnetic mirroring (Zweibel and Haber 1983) in converging magnetic fields: (1) a delta function pencil beam radially downward, (2) a 2π isotropic distribution, and (3) a mirroring distribution approximated by a delta function fan beam at 89°, all in the downward hemisphere. The first two distributions are consistent with a uniform magnetic field, while the third distribution approximates the effects of a converging magnetic field.

In addition Hua and Lingenfelter (1987a) considered a fully (4π) isotropic distribution of ions managing to run their range in an optically thin region high in the chromosphere, corresponding to that assumed by Wang (1975) for monoenergetic neutrons. Such isotropization might result from strong pitch angle scattering of the ions by magnetic field irregularities. Since the probability of an accelerated ion having a nuclear interaction before it slows down and comes to rest in the solar atmosphere is very small for most energies of interest, Hua and Lingenfelter (1987a) require that each ion considered in the Monte Carlo simulation produce a neutron in order to study a statistically significant sample efficiently. Each resulting neutron is then assigned a weight equal to the probability of its having been produced. The particular production interaction for each case is determined by a random number from the calculated relative interaction probabilities expected in the course of slowing down for the particular ion. This process also selects the ion energy at which the interaction occurs, and that determines the amount of matter that the ion has traversed in order to have slowed down from the initial energy to the interaction energy. From the incident zenith angle and the assumed solar atmospheric density model the amount of matter traversed determines in turn the depth at which the neutron production occurs in the solar atmosphere. Finally from the interaction energy and angle of the ion, the production energy and angle of the resulting neutron is then calculated, using the specific reaction kinematics and center of mass angular distributions, by selecting another random number or two. Thus for each combination of energy spectrum, angular distribution and abundances of incident ions, the expected production rate of gamma rays, or neutrons is determined as a function of their energy, zenith angle and depth in the solar atmosphere.

According to Hua and Lingenfelter (1987a), the key steps in the Accelerated Ion Interaction Program are the following. The probability that an ion survives after traveling a distance x through the solar atmosphere is

$$P(x) = \exp\left(- \int_0^x \sum_i \sigma_i n_i dx \right), \tag{11.5}$$

where the σ_i is the total cross-section for interactions with i-type of ambient particles that can remove the ion from the incident beam by changing its energy and direction; and the n_i is the number density of the i-type ambient particles participating in the interactions. For the total cross sections of p–p and p–α interactions, Hua and Lingenfelter (1987a) used the data of Meyer (1972). For the total cross-sections of p–CNO interactions (where CNO refers to all species with Z > 2), Hua and Lingenfelter (1987a) used the approximation

$$\sigma = \frac{1}{\sum_i n_i} \times \sum_i n_i \left(\frac{A_i}{A_C}\right) \sigma_{pC}, \tag{11.6}$$

where A_i is the atomic weight of the i-type ambient particles participating in the interactions with summations taken over all the species with Z > 2; A_C is the atomic weight of carbon; and σ_{pC} is the total cross-section for proton interactions with carbon, summarized by Lock and Measday (1970). For the total cross-sections of α–α and α–CNO interactions, Hua and Lingenfelter (1987a) assumed them to be equal to three times those for p–α and p–CNO interactions at the same energy/nucleon (Meyer 1972). The proton total cross-sections are shown in Fig. 11.2. While the ion travels through the medium, it also suffers Coulomb energy losses and eventually stops. Thus the probability that an ion with initial energy E_0 will survive to an energy E_x in the course of slowing down is given by

$$P(E_0, E_x) = \exp\left(- \int_{E_0}^{E_x} \frac{\sum_i \sigma_i n_i}{(dE/dx)} dE \right), \tag{11.7}$$

where dE/dx is the energy loss rate of the ion in the medium. The energy loss rate is dependent on the energy of the charged particles and the properties of the medium. It can be written as

$$dE/dx = -\langle Z'/A' \rangle \rho f(E), \tag{11.8}$$

where $\langle Z'/A' \rangle$ is the averaged ratio of atomic number, Z′, and atomic weight, A′, for the medium; ρ is the mass density of the medium; and $f(E)$ is so-called stopping power (Barkas and Berger M1964), which is the product of two factors: the ratio Z^2/A of the accelerated particle times a function of the particle energy E and the mean excitation energy I_{adj} of the un-ionized medium.

For the solar atmosphere, Hua and Lingenfelter (1987a) used $I_{\mathrm{adj}} = 19.34$ eV. Defining the ratio of the interaction mean free path to the range

$$R(E) = \int_0^E \frac{\sum_i \sigma_i n_i}{\langle Z'/A' \rangle \rho f(E')} dE', \tag{11.9}$$

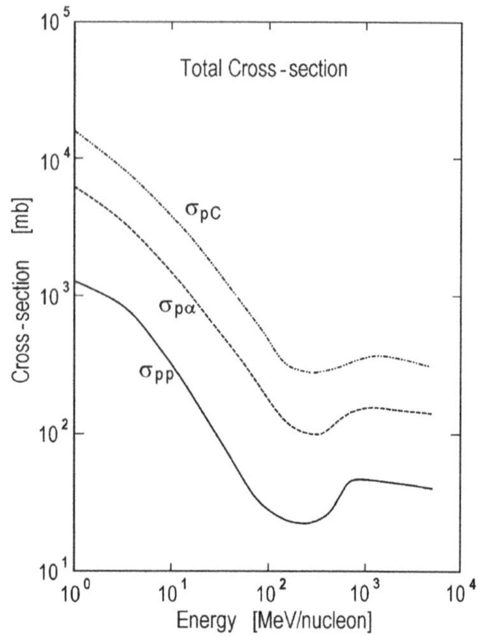

Fig. 11.2 Total cross-sections for p–p, p–α, and p–C reactions (From Hua and Lingenfelter 1987a)

the survival probability is thus

$$P(E_0, E_x) = \exp[R(E_x) - R(E_0)]. \quad (11.10)$$

The differential probability for neutron production by interaction i when the ion has energy in interval E, $E + \mathrm{d}E$ is then

$$\begin{aligned} dP_i(E) &= -P(E_0, E)\frac{\sigma_i n_i}{|dE/dx|}dE = -\exp[R(E_x - E_0)]\frac{\sigma_i n_i}{\langle Z'/A'\rangle \rho f(E)}dE \\ &= -\frac{n_i}{\exp[R(E_0)]\langle Z'/A'\rangle \rho} \times \frac{\sigma_i \exp[R(E)]}{f(E)}dE, \end{aligned} \quad (11.11)$$

where σ_i is the neutron production cross-section of the reaction i, and n_i is the number density of the ambient particles participating in the reaction. The cross-sections for the most important neutron-producing reactions in the solar atmosphere, shown in Fig. 11.3, were discussed in detail by Ramaty et al. (1975), and updated for hydrogen and helium interactions by Murphy et al. (1987). The total probability of neutron production by interaction i as an ion slows down from some initial energy E_0 to E_x is

$$P_i(E_0, E_x) = [R_i(E_x) - R_i(E_0)]/\exp[R(E_0)], \quad (11.12)$$

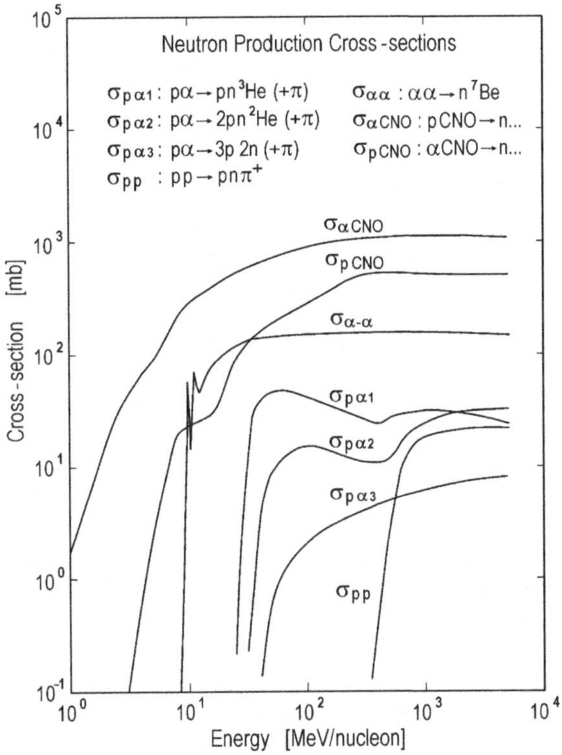

Fig. 11.3 Neutron production cross-sections for the most important neutron-producing processes in solar flares (From Hua and Lingenfelter 1987a)

where

$$R_i(E) = \frac{n_i}{\langle Z'/A' \rangle \rho} \int_0^E \times \frac{\sigma_i(E') \exp[R(E')]}{f(E')} dE'. \tag{11.13}$$

The total neutron production Q_i by various reactions in the solar atmosphere during the slowing down of ions with initial Bessel function spectra and power law spectra are shown in Panels a and b in Fig. 11.4, correspondingly. As noted Hua and Lingenfelter (1987a), these are essentially the same as those calculated by Murphy (1985) and Murphy and Ramaty (1984) since Hua and Lingenfelter (1987a) used the same production cross-sections and there were only negligible differences in numerical techniques and in the formulation of the ion attenuation and energy loss rates.

As underlined Hua and Lingenfelter (1987a), they also took into account the effect of elastic scatterings of the incident ions by the ambient particles. But since only the elastic scatterings between particles with similar masses will significantly

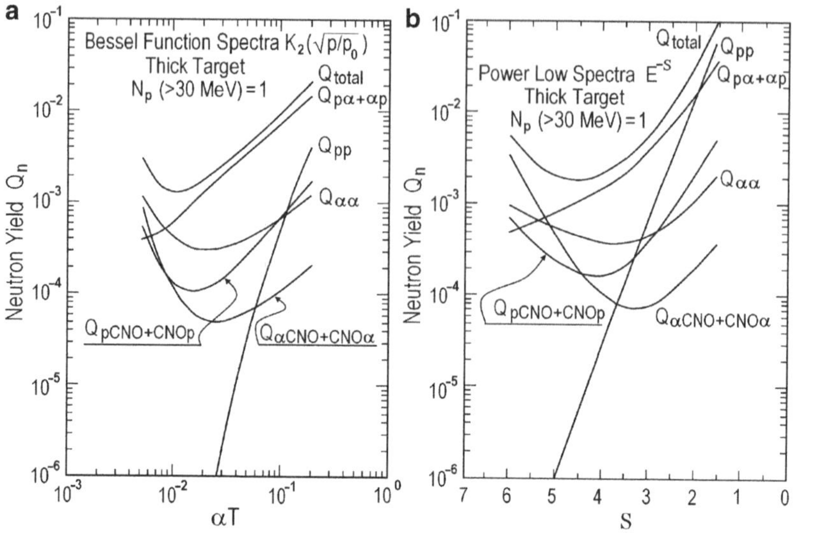

Fig. 11.4 Total and partial neutron production with (**a**) Bessel function and (**b**) power-law energy spectra (From Hua and Lingenfelter 1987a)

change the energies and momenta of the participating particles, they only considered the p–p and α–α scatterings. Then both of the scattered particles are followed as if they were initial ions. In some cases this involves following the particles through a cascade of scatterings. The effect of such elastic scattering is to increase the yield of low energy neutrons. For the range of spectra considered, however, the effect is small, as was also shown by Murphy et al. (1987). With the neutron-production probabilities calculated above, Hua and Lingenfelter (1987a) determined by Monte Carlo simulation the ion energy E_x at which each neutron is produced and the particular reaction which produces it. They also determined the solar atmospheric depth h_x of the neutron production by solving the ion's equation of motion

$$\int_{E_0}^{E_x} \frac{dE}{f(E)} = -\int_{h_0}^{h_x} \langle Z'/A' \rangle \rho(h) \frac{dh}{\cos\theta}, \tag{11.14}$$

where θ is the zenith angle of the incident ion with respect to the solar radius vector, and $\rho(h)$ is the mass density of the solar atmosphere. The calculated depth distributions of neutron production $Q(h)$ in the solar atmosphere for the isotropically incident ions with different energy spectra (Bessel functions in momentum characterized by αT of 0.005, 0.03, and 0.1, and power laws in energy/nucleon with S = 2, 4, 6) are shown in Panels a and b in Fig. 11.5.

For the cases with αT = 0.03 and S = 4 are also shown in Panels a and b in Fig. 11.6 the depth distributions produced by incident ions with three different

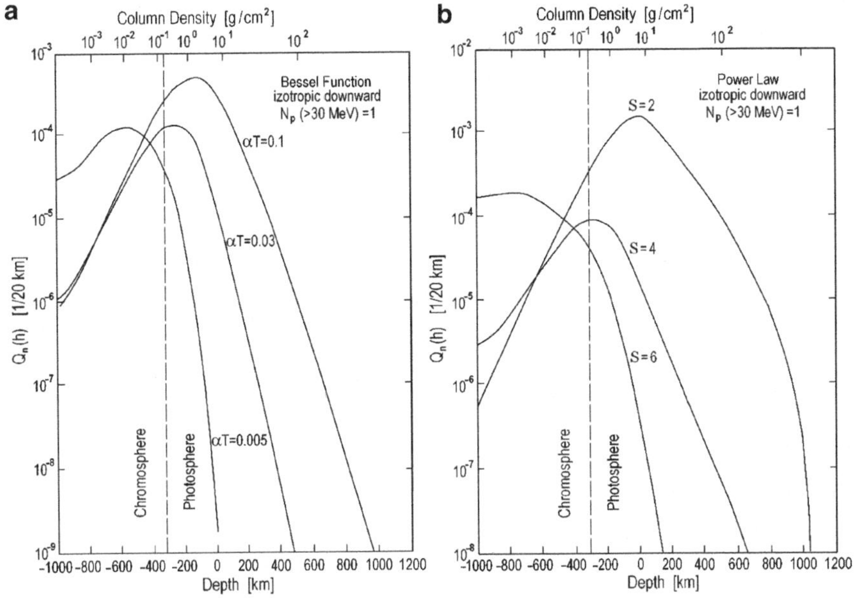

Fig. 11.5 Depth distributions of neutron production by isotropically downward incident ions with (**a**) Bessel function spectra characterized by $\alpha T = 0.005, 0.03$, and 0.1, and (**b**) power-law spectra characterized by $S = 2, 4$, and 6 (From Hua and Lingenfelter 1987a)

initial angular distributions: downward pencil beam δ(0°), isotropic, and mirroring fan δ(89°). These depth distributions, as can be seen, are strongly dependent on the energy spectra and angular distributions. Most of the neutrons are produced in the photosphere by isotropically downward accelerated ions if the spectra are harder than that of a Bessel function with $\alpha T \sim 0.03$ or a power law with $S \sim 4$, and in the chromosphere if the spectra are softer. These pivotal values of the spectra also depend on the angular distribution, such that ions with softer spectra would produce neutrons primarily in the photosphere if the ions were more strongly beamed downward (e.g., δ(0°)), while ions with harder spectra would still produce neutrons primarily in the chromosphere if the ions were in a mirroring distribution (e.g., δ(89°)).

11.1.4 Neutron Energy and Direction

According to Hua and Lingenfelter (1987a), the energy and direction of each produced neutron was calculated from the energy and direction of the accelerated ion, using the kinematics of the specific neutron-producing reaction. The kinematics assumed for the various neutron-producing reactions on hydrogen and helium are described in detail by Murphy et al. (1987). For p–p interactions at energies

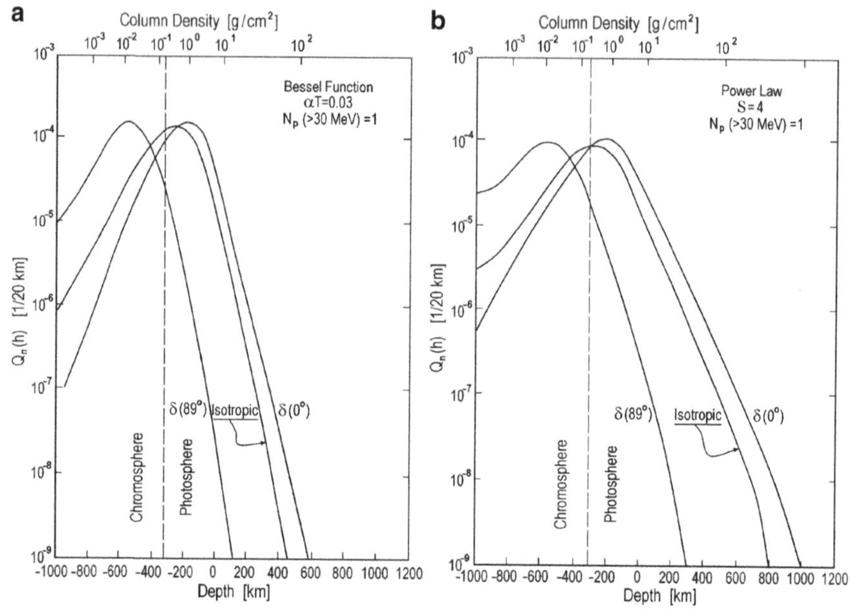

Fig. 11.6 Depth distributions of neutron productivity by incident ions having (**a**) Bessel function spectrum with αT = 0.03, and (**b**) power-law spectrum with S = 4, for three angular distributions corresponding to different geometries of the downward beam (From Hua and Lingenfelter 1987a)

<2 GeV it is assumed that first an isobar and a nucleon are produced, then the isobar decays into a nucleon and a pion. The mass distribution of the isobar (within its limits) has the Breit–Wigner form (see, e.g., Stecker M1971)

$$f(m_\Delta) = \frac{\Gamma}{(m_\Delta - m_0)^2 + \Gamma^2}, \tag{11.15}$$

where $m_0 = 1{,}232$ MeV and $\Gamma = 57.5$ MeV. The neutron is produced as the free nucleon 90% of the time and has an energy in the center-of-mass frame

$$E_n^* = \frac{(E_a^* - m_n)^2 - m_\Delta^2}{2E_a^*}, \tag{11.16}$$

where E_a^* is the available energy of the reaction. Its angular distribution follows the form

$$\phi(\theta*) = A + B\cos^2\theta*, \tag{11.17}$$

where $\theta*$ is the angle in center-of-mass frame between the direction of the incident proton and the outgoing neutron, A and B have the following values (MeV):

$$A = \begin{cases} 0.8382 & (E_p \leq 400\ \text{MeV}) \\ 1.1838 - 0.000864\,E_p & (400\ \text{MeV} \leq E_p \leq 1000\ \text{MeV}) \\ 0.3198 & (E_p \geq 1000\ \text{MeV}) \end{cases} \tag{11.18}$$

$$B = \begin{cases} 0.4814 & (E_p \leq 400\ \text{MeV}) \\ 0.00258E_p - 0.5506 & (400\ \text{MeV} \leq E_p \leq 1000\ \text{MeV}) \\ 0.3198 & (E_p \geq 1000\ \text{MeV}) \end{cases} \tag{11.19}$$

As noted Hua and Lingenfelter (1987a), the neutron is the product of the isobar decay the remaining 10% of the time and its angular distribution in the isobar frame is isotropic, while the isobar goes either directly forward or backward in the center-of-mass frame. For higher energy ($E > 2$ GeV) p–p interactions, the neutron production is assumed to be isotropic in the center-of-mass frame, with energies uniformly distributed up to the maximum. For the p–α and α–p reactions, leading to np^{3}He, n 2p^{2}H, and 2n3p, the energy spectrum of the produced neutron is assumed to be uniformly distributed up to a maximum energy corresponding to the energy the neutron would have if all the other products of the reaction were emitted in the other direction (e.g., Ramaty and Lingenfelter 1969). The angular distribution for these reactions is assumed to be isotropic in the center-of-mass frame. In the α–α reactions, it is assumed that the neutron energy is uniformly distributed up to its maximum, and its angular distribution is of the form $1 + 9\cos^2\theta*$.

According to Hua and Lingenfelter (1987a), the energy distribution of the neutrons from p–CNO, CNO–p, α–CNO, and CNO–α reactions is determined by two components: the equilibrium and the pre-equilibrium. The equilibrium or 'evaporation' component is estimated from Bertrand et al. (1974) to account for about 27% of the neutrons produced in these reactions. These neutrons are isotropically distributed in center-of-mass frame, and the energy in that frame follows an 'evaporation' spectrum

$$f(E_n^*) = (E_n^*)^{5/11} \exp(-E_n^*/T_0), \tag{11.20}$$

where $T_0 = 2.5$ MeV. For the pre-equilibrium component, the neutron energy and angle in the lab frame are assumed to have the following combined distribution for $0 \leq E_n \leq E_p$, based on the measurements of Wu et al. (1979b) and Kalend et al. (1983):

$$f(E_n, \mu) = \begin{cases} \exp(-0.01E_n/T_p), & (0.99 \leq \mu \leq 1) \\ \exp(-(1-\mu)E_n/T_p) & (-1 \leq \mu \leq 0.99) \end{cases}, \tag{11.21}$$

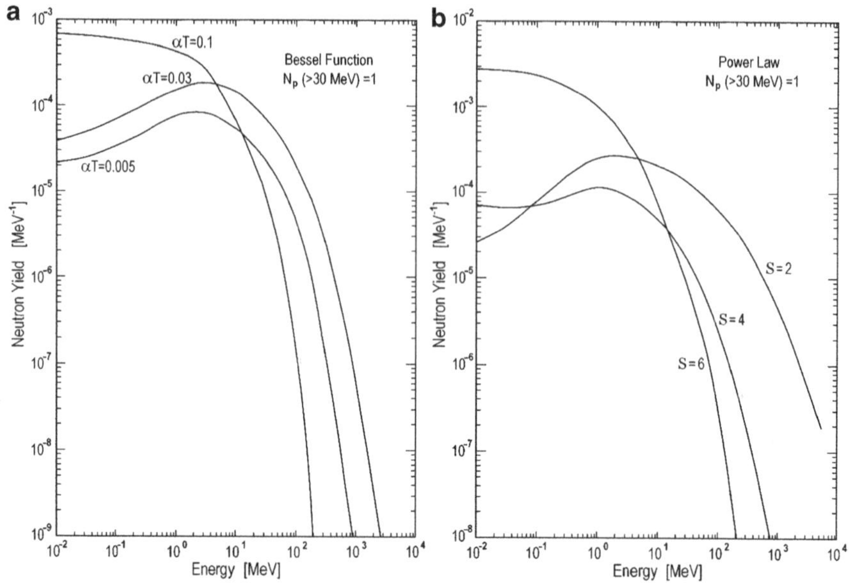

Fig. 11.7 Energy spectra of neutrons produced by incident ions with (**a**) Bessel function spectra characterized by $\alpha T = 0.005$, 0.03, and 0.1, and (**b**) power-law spectra characterized by S = 2, 4, and 6 (From Hua and Lingenfelter 1987a)

where $T_p = E_p/6.34$ MeV and E_p is the energy of the incident proton; $\mu = \cos\theta$ and θ is the angle between directions of the incident proton and the outgoing neutron. The resulting neutron production spectra, integrated over depth and angle, are shown in Panels a and b in Fig. 11.7 for accelerated ions with differing Bessel function and power law spectra, correspondingly.

As noted Hua and Lingenfelter (1987a), these spectra in Fig. 11.7 are the same as those calculated by Murphy (1985) and Murphy et al. (1987) except at low energies ($<$10 MeV) where the latter spectra show an evaporation peak because they assumed that all of the neutrons produced in p–CNO reactions have an evaporation spectrum. This was the only difference in the assumed kinematics. Since the spectral shape is strongly dependent on the ion spectrum, measurements of the spectrum of flare neutrons at the Earth can also provide an independent determination of the ion spectrum (e.g., Ramaty et al. 1983b; Murphy et al. 1987).

11.1.5 Neutron Propagation, Capture, and Decay

Hua and Lingenfelter (1987a) underlined that in the Neutron Transport Program each of the neutrons produced in the first program is followed, usually through many scatterings, each of which changes its energy and direction, until it either escapes from the solar atmosphere, decays, or is captured either radiatively, primarily on H to produce a 2.223 MeV gamma ray, or non-radiatively, primarily on

^{3}He. In these Monte Carlo simulations the propagation and fate of each neutron is determined by random number selections from probability distributions based on the neutron interaction cross sections and kinematics in nearly the same way that it is done in neutron transport programs used in nuclear reactor studies. But there is a very important difference in this calculation since none of the low-energy (<2 keV) neutrons can escape from the gravitational field of the Sun. For each such neutron attempting to escape from the solar atmosphere, Hua and Lingenfelter (1987a) calculate from its zenith angle and energy the time required for it to be gravitationally decelerated and return to the solar atmosphere. This time is

$$t = T_e(1 - C)^{-3/2}\left\{[C(1 - C)]^{1/2}\cos\theta + \arctan\left[\left(\frac{1+\varepsilon}{1-\varepsilon}\right)^{1/2}\tan\frac{\alpha}{2}\right]\right\}, \quad (11.22)$$

where the characteristic orbital time

$$T_e = \left(2R^3/GM\right)^{1/2} = 2254\ \text{s}, \quad (11.23)$$

and G is gravitational constant; M is the mass of the Sun; R is the radius of the Sun; $C = E/E_{es}$ is the ratio of the neutron kinetic energy E to its escape energy

$$E_{es} = Mm_nG/R = 2\ \text{keV}; \quad (11.24)$$

θ is the zenith angle of the neutron;

$$\varepsilon = \left[1 - 4C(1 - C)\sin^2\theta\right]^{1/2} \quad (11.25)$$

is eccentricity of thc ncutron's orbit, and

$$\cos\alpha = \left(1 - 2C\sin^2\theta\right)/\varepsilon. \quad (11.26)$$

Hua and Lingenfelter (1987a) then determined by a random number whether it decays in flight or re-enters the solar atmosphere with the same energy but with a zenith angle directly opposite from that which it left. This gravitational reflection of neutrons significantly enhances the capture rate in the upper atmosphere of the Sun and makes an important contribution to the flux of escaping 2.223 MeV gamma rays at large zenith angles. The neutron scattering and capture cross-sections have been summarized by Meyer (1972) and Howerton et al. (M1978). The neutron scattering is predominantly on H and is assumed to be isotropic in the center-of-mass frame. The cross-section for capture on H is

$$\sigma_c = 2.44 \times 10^{-3}/\beta\ \text{mb}, \quad (11.27)$$

where β is the neutron velocity relative to that of light. The cross-section for capture on ^{3}He, is 1.61×10^4 times that of H, and since the abundance of ^{3}He is believed to be of the order of 10^{-5}, the capture on He is a significant fraction of that on H. Hua and Lingenfelter (1987a) used a nominal value of 2×10^{-5} for ^{3}He/H, consistent with that determined by Hua and Lingenfelter (1987e) from a comparison of the results of these calculations with observations of the time history of the 2.223 MeV line emission. In addition to either escaping or being captured, the neutrons may decay with a mean life of 918 s. In the Monte Carlo simulation of neutron transport, the most time consuming part of the calculation is the multiple scatterings of the neutrons, because the scattering cross-section is much larger than that of capture. A typical neutron produced in a flare has an energy a few tens of megaelectron volt, and undergoes about 300 elastic scatterings before it is captured. Most of these scatterings take place after the neutron is thermalized. The large number of thermal scatterings before each capture allows to use a random walk scheme to determine the location where the thermal neutron is captured. First, from the ratio of the cross-section of scattering to that of capture, Hua and Lingenfelter (1987a) used random number selection to determine the number of scatterings, K, before the neutron is captured. Then the K scatterings are considered as a K-step random walk, and the distance x from the starting point of this random walk is determined from the well-known distribution of

$$P_K(x) = \frac{1}{\sqrt{(2\pi K\lambda^2)}} \exp\left(-x^2/2K\lambda^2\right), \tag{11.28}$$

where λ is the scattering mean free path. This random walk simulation avoids time-consuming calculation of the kinematics of hundreds of scatterings, greatly reducing the calculation time. The fate of the neutrons, that is the relative probabilities of capture, escape and decay, depends strongly on the spectrum and angular distribution of the accelerated ions that produced them. This can be seen in Panels a and b in Fig. 11.8 for both Bessel function and power-law spectra respectively for a range of angular distributions: the pencil beam downward, $\delta(0°)$, the isotropic downward, and nearly horizontal fan beam $\delta(89°)$.

As noted Hua and Lingenfelter (1987a), the capture probability is highest for neutrons produced by the strongly downward beamed ions, because these neutrons on the average are produced deep in the solar photosphere where they are least able to escape or decay before they are thermalized and captured on H or ^{3}He. The escape probability, on the other hand, is highest for neutrons produced by mirroring ions in a nearly horizontal fan beam, because the neutrons on the average are produced much higher in the chromosphere where they are much less able to be captured before they escape. Also shown is the probability that the escaping neutrons survive to a distance of 1 AU from the Sun without decaying. The calculated spectra and time histories of the escaping neutrons and their comparisons with observations was presented in paper by Hua and Lingenfelter (1987e). In addition, Hua and Lingenfelter (1987a) calculated the fate of neutrons produced by a fully (4π) isotropic distribution of ions managing to run their range high in the chromosphere. This corresponds to the model assumed by Wang (1975) for

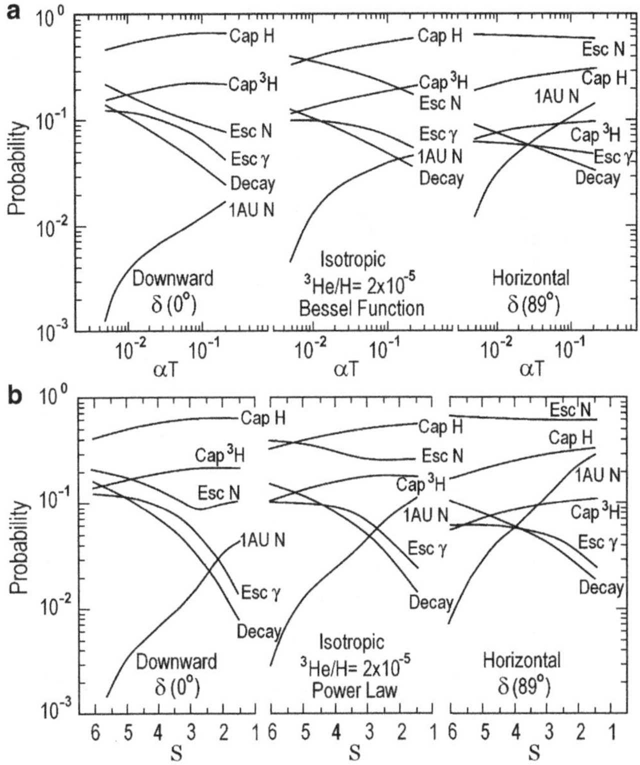

Fig. 11.8 Probabilities of decay, escape and capture (on H and ^{3}He) for a neutron produced by incident ions with (**a**) Bessel function and (**b**) power-law spectra for differing angular distributions. Also shown is the probability that the escaping neutrons survive to a distance of 1 AU from the Sun without decaying (From Hua and Lingenfelter 1987a)

monoenergetic neutrons. From Wang's calculations, Murphy and Ramaty (1984) determined neutron capture probabilities for various ion spectra which are very similar to those calculated by Hua and Lingenfelter (1987a). The probabilities that Hua and Lingenfelter (1987a) calculated are shown in Fig. 11.9, together with those for the mirroring ion distribution.

As can be seen from Fig. 11.9, the fate of the neutrons produced by these two different ion distributions are quite similar because neutrons in both cases are produced primarily, if not exclusively, in the chromosphere. The two distributions can be distinguished, however, by high-energy neutron observations. The radiative capture of neutrons by H results in isotropic gamma-ray line emission at 2.223 MeV. The depth distribution of the neutron capture differs greatly from that of the neutron production and depends primarily on neutron propagation and the competition between capture, decay, and escape. This can be seen in Panels a and b in Fig. 11.10 for neutrons produced by a downward isotropic ion flux with Bessel function and power-law energy spectra, respectively, and in Panels a and b in Fig. 11.11 for various angular distributions with a fixed spectrum. The peak capture

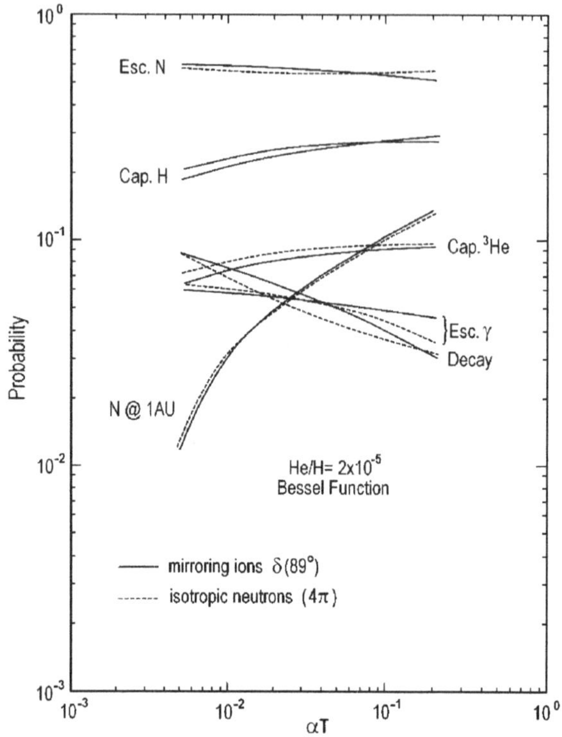

Fig. 11.9 The fate of neutrons produced by a fully (4π) isotropic distribution of ions running their range high in the chromosphere, corresponding to the model assumed by Wang (1975) for monoenergetic neutrons, as compared with those for the mirroring ion distribution (From Hua and Lingenfelter 1987a)

rate occurs in the photosphere at a depth of 200–300 km just above the knee in the density distribution, at a density of 1–3 $\times$ 10^{17} H cm^{-3} and a vertical column density of several grams per square centimeter. At altitudes above the peak the capture rate decreases almost exponentially with a scale height of, ~65 km. This is less than that of the atmospheric density scale height (~85 km), because of the increasing importance of decay and escape with increasing altitude.

As can be seen in Figs. 11.10 and 11.11, at depths below the peak, the shape depends strongly on the incident accelerated ion spectrum and angular distribution. Gravitational trapping of the neutrons also has a significant effect on the capture rate. All of the neutrons with energies <2 keV that try to escape are trapped or reflected. Among these trapped neutrons in a typical case, about 42% decay before re-entering the solar chromosphere, and additional 45% decay in the chromosphere, where the gas density is low (<10^{16} H cm^{-3}) and the average neutron capture time is longer than the neutron decay mean life of 918 s, so that decay becomes a more important loss process. The remaining 13% are captured in the upper atmosphere, and these contribute 20% or more to the escaping 2.223 gamma-ray photon fluence

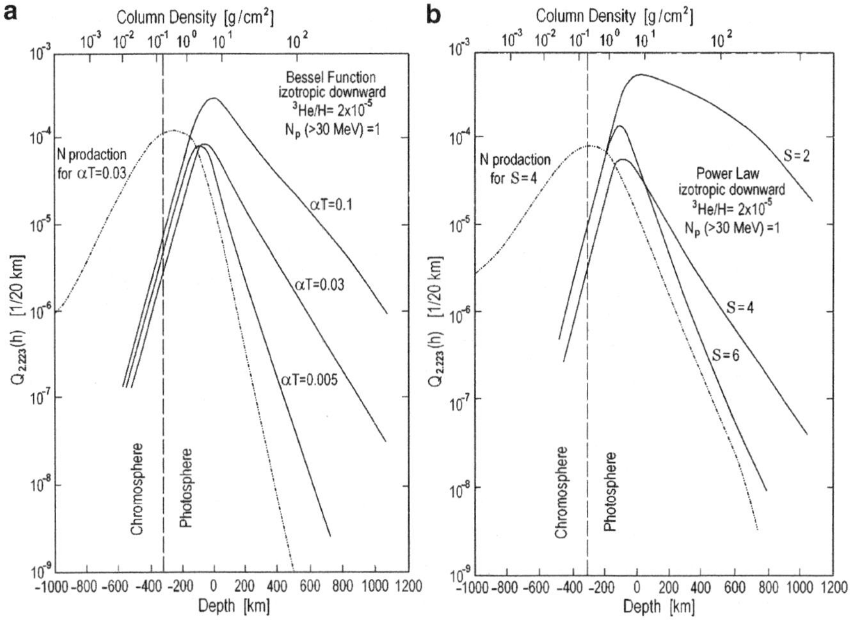

Fig. 11.10 The depth distribution of the neutron capture on H for neutrons produced by a downward isotropic ion flux with (**a**) Bessel function spectra characterized by $\alpha T = 0.005$, 0.03, and 0.1, and (**b**) power-law spectra characterized by $S = 2$, 4, and 6 (From Hua and Lingenfelter 1987a)

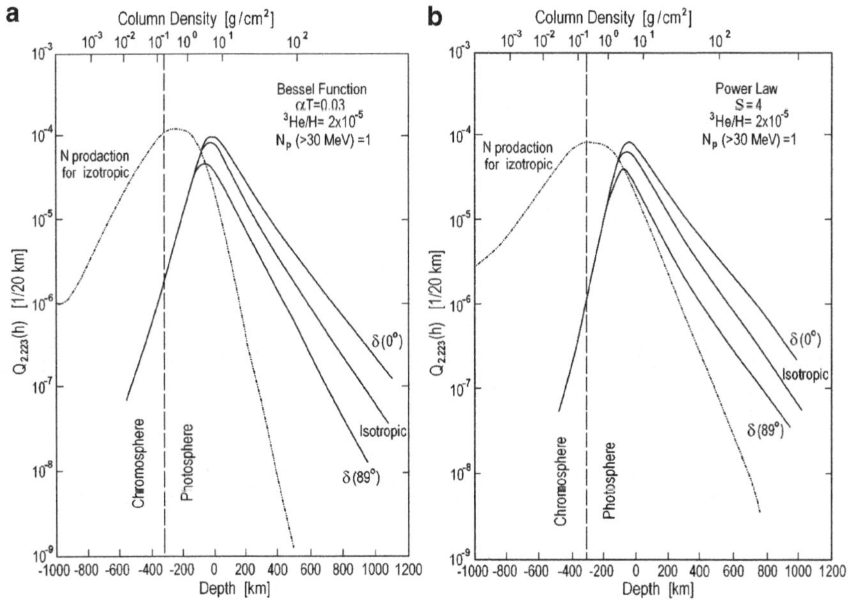

Fig. 11.11 The depth distributions of the neutron capture on H for neutrons produced by incident ions with a fixed Bessel function (**a**) or power-law (**b**) spectrum but with various angular distributions (From Hua and Lingenfelter 1987a)

at zenith angles near 90° The depth distribution of neutron capture gamma ray production calculated in this program becomes the input to the Gamma-Ray Transport Program.

11.1.6 Gamma-Ray Propagation

According to Hua and Lingenfelter (1987a), in the Gamma-Ray Transport Program each gamma-ray produced in the previous program is followed until it either escapes from the solar atmosphere or is multiply-Compton scattered to low energy (<20 keV). These Monte Carlo simulations thus finally determine the flux of both the unscattered 2.223 MeV gamma-ray line emission and the Compton-scattered continuum escaping from the solar atmosphere as a function of gamma-ray energy, zenith angle, and time. The angular distribution of the gamma-ray emission from the neutron capture on H is isotropic as long as the final spin states of the deuterons are ignored (Blatt and Weiskopf M1952). Because most of the captures take place when the neutrons are thermalized at energies of ~0.5 eV, the line energy is essentially unshifted and the line width is very narrow, with a Doppler broadened FWHM of only ~0.1 keV. The calculations of the previous programs show that the gamma-ray photons are emitted at a depth corresponding to a column density of about 10^{24}–10^{25} H cm^{-2}. This means that Compton scattering has a significant effect on the propagation of the photons (Wang 1975). The cross-section for Compton scattering is given by the Klein–Nishina formula as a function of photon energy E_γ:

$$\sigma(\alpha) = \frac{3\sigma_T}{8\alpha}\left[\left(1 - 2\frac{1+\alpha}{\alpha^2}\right)\ln(2\alpha+1) + \frac{1}{2} + \frac{4}{\alpha} - \frac{1}{2(2\alpha+1)^2}\right], \tag{11.29}$$

where $\alpha = E_\gamma / m_e c^2$, and $\sigma_T = 0.6652$ barn is the Thomson cross-section. The distribution of the Compton-scattered photons, E'_γ, relative to the initial photon energy, $r = E'_\gamma / E_\gamma$ is

$$f(r) = \begin{cases} \dfrac{3\sigma_T}{8\alpha\sigma(\alpha)}\left[\left(\dfrac{\alpha r + r - 1}{\alpha r}\right)^2 + \dfrac{1}{r} - 1 + r\right] & \left(\dfrac{1}{2\alpha+1} \le r \le 1\right) \\ 0 & \text{otherwise,} \end{cases} \tag{11.30}$$

and the scattering angle θ is given by

$$\cos\theta = \frac{\alpha r + r - 1}{\alpha r}. \tag{11.31}$$

As noted Hua and Lingenfelter (1987a), in determining the angular distribution of the escaping 2.223 MeV photons, the curvature of the solar atmosphere is also considered rather than assuming a simple infinite planar geometry. This is important at large zenith angles since the column depth thus remains finite even at angles greater than 90° allowing a significant number of 2.223 MeV photons to escape. The energy and angular distributions of escaping photons, calculated in this program, can be directly compared with flare gamma-ray observations. Several of these observations have, in fact, been from flares occurring at or near the limb of the Sun, where the consideration of curvature is essential. The 2.223 MeV gamma-ray line fluence at the Earth per accelerated proton of E > 30 MeV at the Sun can be written as

$$\varphi_{2.223}(\theta) = Q_n f_{2.223}(\theta)/4\pi R^2, \tag{11.32}$$

where $R = 1$ AU, Q_n is the total neutron yield per proton with energy >30 MeV, calculated above and shown in Fig. 11.4, and $f_{2.223}(\theta)$ is the conversion probability as a function of zenith angle θ shown in Panels a and b in Fig. 11.12 for a range of Bessel function and power-law ion spectra with three incident angular distributions.

As can be seen from Fig. 11.12, the line intensity is strongly limb darkened at angles greater than 75° Thus the 2.223 MeV line intensity is attenuated by nearly an order of magnitude from flares at a heliocentric angle of 88° by two orders of magnitude from those at 90° and by three orders of magnitude at 91°. The angle integrated probability was shown in Fig. 11.8. Also shown for comparison (Fig. 11.13) are the angular dependent probabilities for the fully (4π) isotropic ion distribution which corresponds to the model of Wang (1975).

11.1.7 Ratios of Fluencies in the 2.223 MeV Line and 4–7 MeV in Dependence of Heliocentric Angle

Hua and Lingenfelter (1987a) underlined that calculations of the angular dependence of the 2.223 MeV line emissions from the solar atmosphere give possibility to study quantitatively the limb darkening of this line. The attenuation of the 2.223 MeV line, produced by neutron capture deep in the solar atmosphere, is most evident when compared with the excess 4–7 MeV emission from nuclear deexcitation, produced by the incident ions much higher in the atmosphere where the attenuation is negligible. Comparisons of the measured ratios of the fluencies in the 2.223 MeV line and excess 4–7 MeV emission from flares on the disk and limb of the Sun show a strong limb darkening of as much as a factor of 30. A comparable limb darkening is also expected from the calculated ratios of the 2.223 MeV line fluence as a function of heliocentric angle compared to the isotropic, un-attenuated 4–7 MeV fluence from Murphy and Ramaty (1984), as is shown in Panels a and b in Fig. 11.14 for a range of Bessel function and power-law ion spectra with three incident angular distributions. These calculations allow to determine from fluence measurements the total number and spectrum of accelerated ions in flares anywhere on the visible disk of the Sun or even slightly beyond. Gamma-ray measurements

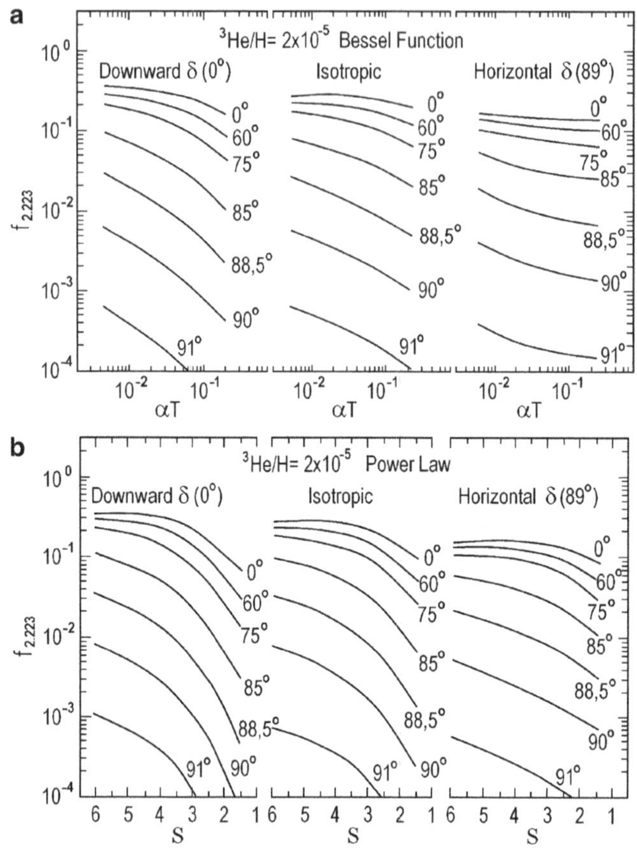

Fig. 11.12 The conversion probability $f_{2.223}(\theta)$ from neutron to 2.223 MeV gamma-ray line emission as a function of azimuthal angle θ for incident ions with (**a**) Bessel function and (**b**) power law spectra for three incident angular distributions (From Hua and Lingenfelter 1987a)

of both the 2.223 MeV line and the excess 4–7 MeV fluencies have been made from 15 flares between 4 August 1972 and 9 July 1982 with detectors on OSO-7 (Chupp et al. 1973), HEAD-1 (Hudson et al. 1980), HEAD-3 (Prince et al. 1982) and SMM (Chupp 1982; Cliver et al. 1983; Prince et al. 1983; Rieger et al. 1983). These measurements and their fluence ratios are summarized in Table 11.1.

11.1.8 Estimation of the SEP Characteristic Spectral Shape and Total Number of Accelerated Particles from Gamma-Ray Line Measurements

As noted Hua and Lingenfelter (1987a), the characteristic spectral shape can be determined from a comparison of the calculated and measured ratios of the fluencies

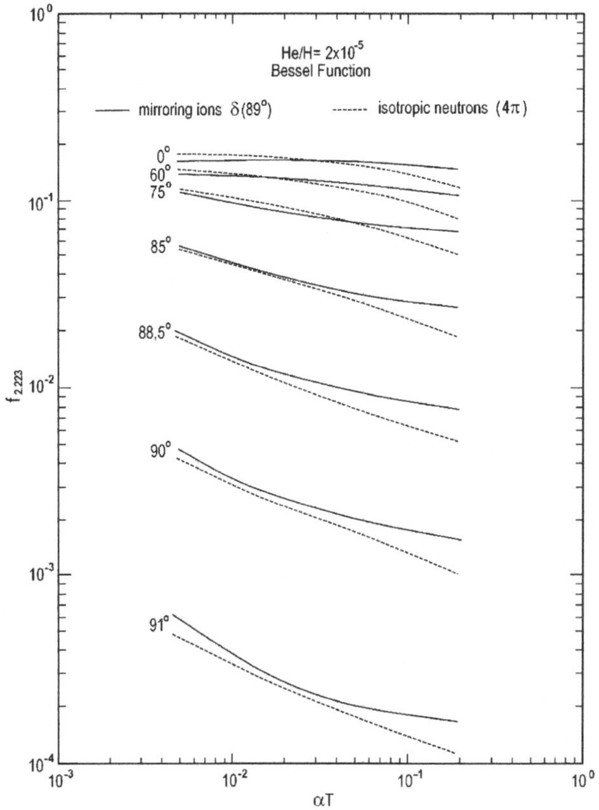

Fig. 11.13 The conversion probability $f_{2.223}(\theta)$ from neutron to 2.223 MeV gamma-ray line emission resulting from a fully (4π) isotropic distribution of ions running their range high in the chromosphere, corresponding to the model assumed by Wang (1975) for monoenergetic neutrons, as compared with those for the mirroring ion distribution (From Hua and Lingenfelter 1987a)

in the 2.223 MeV line and in the excess 4–7 MeV band. Although variations in the accelerated ion spectra cause considerable scatter in the 2.223 MeV to excess 4–7 MeV fluence ratios which typically have values around unity for flares with heliocentric longitudes less than 75° strong limb darkening is evident in the three limb flares, 21 June 1980, 27 April 1981, and 8 February 1982, which have ratios an order of magnitude lower (see Table 11.1). This effect is most clearly seen when the measured fluence ratios are plotted against heliocentric angle, shown in Fig. 11.15.

In Fig. 11.15 are compared the measured values directly with the ratios which are calculated in Section 11.1.7 for incident ions with various Bessel function and power law spectra, assuming both a mirroring distribution and a ^{3}He/H ratio of 2×10^{-5}, as determined from studies (Hua and Lingenfelter 1987c, e) of the measured time histories of the neutron and 2.223 MeV line fluxes. From this comparison of the measured and calculated fluence ratios it can be determined

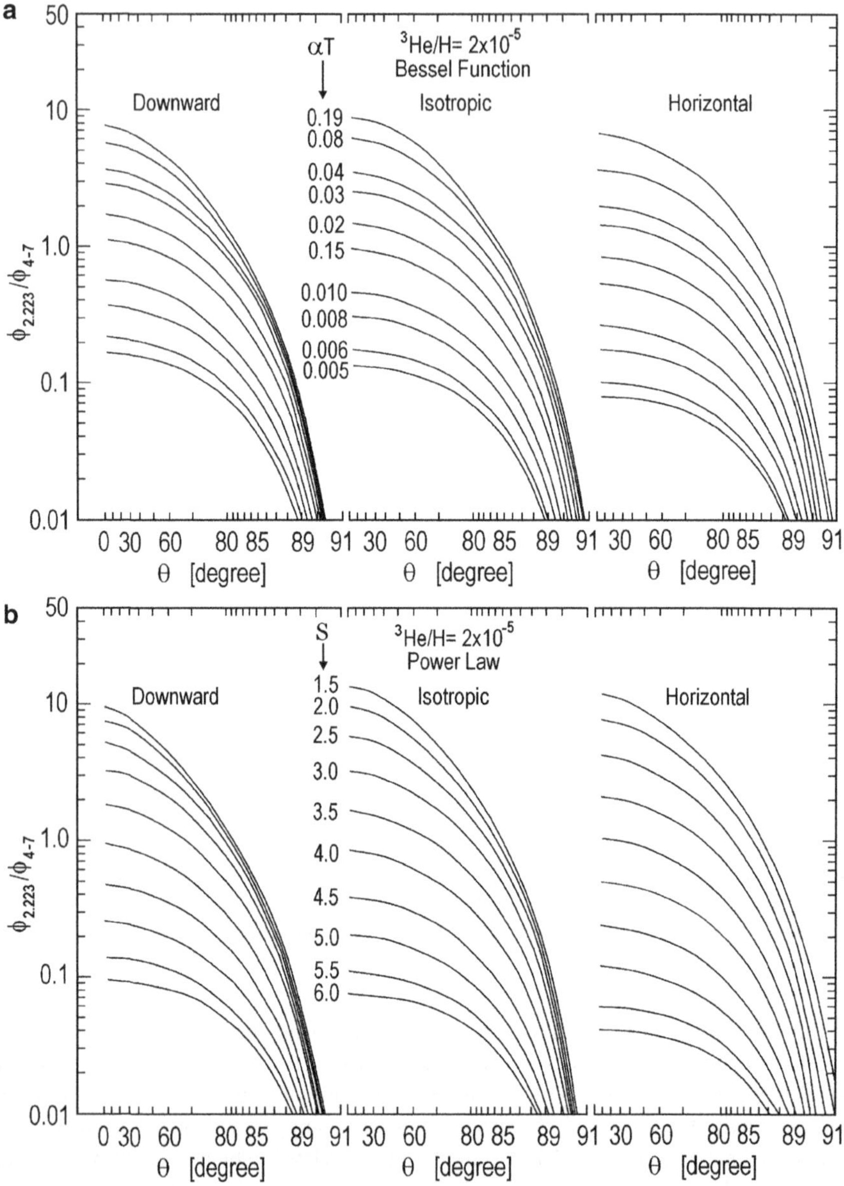

Fig. 11.14 The calculated ratios of fluencies in the 2.223 MeV line and excess 4–7 MeV are plotted against heliocentric angle for incident ions with (**a**) Bessel function and (**b**) power-law spectra for three incident angular distributions (From Hua and Lingenfelter 1987a)

the shape of the accelerated ion spectrum for each of these flares. These values are listed in Table 11.2.

In Table 11.2 in the last column are given values of $F_{esc} = N_{ipm}/(N_{ipm} + N_{sun})$, where N_{sun} is the number of trapped protons determined for the Bessel function

Table 11.1 Solar flare gamma-ray fluencies and ratios (From Hua and Lingenfelter 1987a)

Flare		$\Phi_{2.223}$(photon cm^{-2})	$\Phi_{4\text{–}7}$(photon cm^{-2})	$\Phi_{2.223}/\Phi_{4\text{–}7}$	Location
1	August 4, 1972	375	228 ± 24	1.6 ± 0.2	E08 N14
2	July 11, 1978	330 ± 100	288 ± 127	1.1 ± 0.6	E43 N18
3	November 9, 1979	76 ± 19	50 ± 4	1.5 ± 0.4	E00 S16
4	June 4, 1980	< 0.7	1.9 ± 0.5	< 0.4	E59 S14
5	June 7, 1980	6.6 ± 1	11.5 ± 0.5	0.57 ± 0.09	W70 N14
6	June 21, 1980	3.1 ± 0.2	76 ± 1.2	0.041 ± 0.003	W89 N20
7	July 1, 1980	3.3 ± 0.5	2.2 ± 0.5	1.5 ± 0.4	W37 S12
8	November 6, 1980	11.9 ± 1.5	16.9 ± 1.1	0.70 ± 0.10	E74 S12
9	April 10, 1981	13.5 ± 1	18.6 ± 1.6	0.73 ± 0.08	W37 N09
10	April 27, 1981	11.7 ± 2	77 ± 2.2	0.15 ± 0.03	W88 N16
11	February 8, 1982	1.5 ± 1.0	12.5 ± 1.3	0.12 ± 0.08	W88 S13
12	June 3, 1982	314	305 ± 30	1.03 ± 0.10	E72 S09
13	June 15, 1982			0.32 ± 0.15	E66 S22
14	July 9, 1982	> 21	33 ± 1.5	> 0.64	E73 N17
15	July 9, 1982	4.7 ± 0.6	7.2 ± 1	0.65 ± 0.12	E74 N18

spectra. Once the spectral shape has been determined, the total number of accelerated protons with energy $E > 30$ MeV at the Sun, $N_p(E > 30\text{ MeV})$ can be determined either from measurements of the 2.223 MeV line or the 4–7 MeV band fluencies. In Fig. 11.16 are presented the proton number per unit fluence, for a range of Bessel function and power law ion spectra with three incident angular distributions. Assuming the same angular distribution and ^{3}He abundance as in Fig. 11.15, Hua and Lingenfelter (1987a), calculated the total number of trapped protons for each of the flares. These values are given in Table 11.2, together with the spectral index and total number of escaped protons in the interplanetary medium from satellite measurements of the protons there (e.g., McDonald and van Hollebeke 1985). As can be seen nearly all of the accelerated ion spectra can be characterized by a relatively narrow range of spectra, either power laws with $3 < S < 4$, or Bessel functions with $0.015 < \alpha T < 0.04$. This is also the same range of spectral shapes as was previously found (Murphy and Ramaty 1984) for a smaller number of disk flares, assuming an isotropic neutron distribution and a significantly higher ^{3}He/H ratio. This range of power-law indices is steeper than that of 2.4 ± 0.7 fitted (McGuire et al. 1981) to the measured spectra of flare protons in the interplanetary medium.

11.1.9 The Dependence of the Relative Part of the SEP That Escape into Interplanetary Space on the Total Number of Produced Energetic Particles: Two Different Classes of Solar Flares

Hua and Lingenfelter (1987a) noted that obtained range of αT in the interval $0.015 < \alpha T < 0.04$, is nearly identical to that of the Bessel function spectra with $0.014 \leq \alpha T \leq 0.036$, what found by McGuire et al. (1981) to fit the measured spectra of

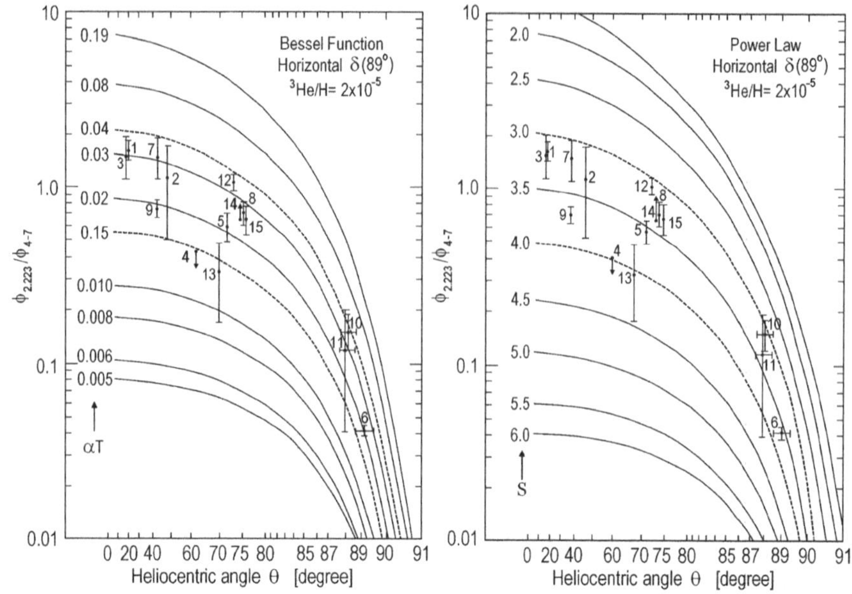

Fig. 11.15 The measured fluency ratios for the flares listed in Table 11.1, plotted against heliocentric angle, compared with the calculated ratios for incident ions with (**a**) Bessel function and (**b**) power-law spectra, assuming a mirroring distribution and a ^{3}He/H ratio of 2×10^{-5} (From Hua and Lingenfelter 1987a)

protons in the interplanetary medium. McGuire and von Rosenvinge (1984) have, in fact, found that Bessel function spectra generally give a much better fit than power-law spectra to the measured proton spectra in the interplanetary medium for the sample of flares that they studied. The similarity between the spectra of accelerated protons trapped at the Sun and in the interplanetary medium can also be seen in Table 11.2 for two of the three flares for which direct comparisons can be made, 7 and 21 June 1980. In the 3 June 1982 flare, on the other hand, the spectrum of the protons in the interplanetary medium was best fit (McDonald and van Hollebeke 1985) by a power law rather than a Bessel function, but that power law was much flatter, $S = 1.7$, than that of 3.1 ± 0.1 required by the gamma-ray fluence ratios. Moreover, the measured (Chupp et al. 1983; Evenson et al. 1983) spectrum of the neutrons from this flare was consistent (Murphy and Ramaty 1984; Murphy et al. 1987) with that produced by protons with a Bessel function spectrum, rather than a power law. Only a small contribution could be due to protons with a power-law spectrum which most likely resulted from shock acceleration of the protons as they escaped into the interplanetary medium.

As underlined Hua and Lingenfelter (1987a), these spectral differences are correlated with the two different classes of flares, suggested by a variety of studies (e.g., Cane et al. 1985; Bai 1986). These studies have shown that gamma-ray and proton flares can be grouped into two classes: impulsive flares characterized by hard

Table 11.2 Numbers and spectra of flare-accelerated protons (From Hua and Lingenfelter 1987a)

Flare		Trapped at the Sun				Observed in IPM		
		Bessel function		Power law				
		αT	N_p (>30 MeV)	S	N_p (>30 MeV)	αT	N_p (>30 MeV)	F_{esc}
1	4 Aug. 1972	0.032 ± 0.004	$(2.5 \pm 0.3) \times 10^{33}$	3.2 ± 0.1	$(1.6 \pm 0.1) \times 10^{33}$	–	4.3×10^{34}	0.95
2	11 July 1978	0.026 ± 0.012	$(3.1 \pm 1.7) \times 10^{33}$	3.3 ± 0.5	$(1.9 \pm 1.2) \times 10^{33}$	–	–	–
3	9 Nov. 1979	0.029 ± 0.005	$(5.6 \pm 2.2) \times 10^{32}$	3.2 ± 0.2	$(3.4 \pm 1.5) \times 10^{32}$	–	–	–
4	4 June 1980	< 0.015	$\sim 9.6 \times 10^{30}$	> 4.0	$\sim 6.4 \times 10^{30}$	–	–	–
5	7 June 1980	0.021 ± 0.003	$(9.2 \pm 2.4) \times 10^{31}$	3.6 ± 0.1	$(5.9 \pm 1.2) \times 10^{31}$	0.015	8×10^{29}	0.009
6	21 June 1980	0.022 ± 0.007	$(6.1 \pm 1.2) \times 10^{32}$	3.7 ± 0.1	$(3.7 \pm 0.3) \times 10^{32}$	0.025	1.5×10^{31}	0.024
7	1 July 1980	0.032 ± 0.007	$(2.2 \pm 0.8) \times 10^{31}$	3.2 ± 0.2	$(1.6 \pm 0.5) \times 10^{31}$	–	$< 4 \times 10^{28}$	< 0.003
8	6 Nov. 1980	0.027 ± 0.003	$(1.5 \pm 0.3) \times 10^{32}$	3.3 ± 0.1	$(9.9 \pm 2.0) \times 10^{31}$	–	3×10^{29}	0.002
9	10 April 1981	0.019 ± 0.002	$(1.4 \pm 0.2) \times 10^{32}$	3.7 ± 0.1	$(9.2 \pm 1.2) \times 10^{31}$	–	$\sim 7 \times 10^{31}$	~ 0.33
10	27 April 1981	0.036 ± 0.007	$(6.8 \pm 2.1) \times 10^{32}$	3.1 ± 0.2	$(5.3 \pm 1.6) \times 10^{32}$	–	–	–
11	8 Feb. 1982	0.027 ± 0.015	$(1.5 \pm 1.2) \times 10^{32}$	3.4 ± 0.9	$(2.3 \pm 2.0) \times 10^{31}$	–	–	–
12	3 June 1982	0.035 ± 0.003	$(3.1 \pm 0.2) \times 10^{33}$	3.1 ± 0.1	$(2.0 \pm 0.2) \times 10^{33}$	$S = 1.7$	3.6×10^{32}	0.10
13	15 June 1982	0.014 ± 0.004	–	4.0 ± 0.4	–	–	–	–
14	9 July 1982	> 0.025	$\sim 2.6 \times 10^{32}$	< 3.4	$\sim 1.8 \times 10^{32}$	–	–	–
15	9 July 1982	0.026 ± 0.003	$(6.2 \pm 1.3) \times 10^{31}$	3.3 ± 0.2	$(4.1 \pm 1.2) \times 10^{31}$	–	–	–

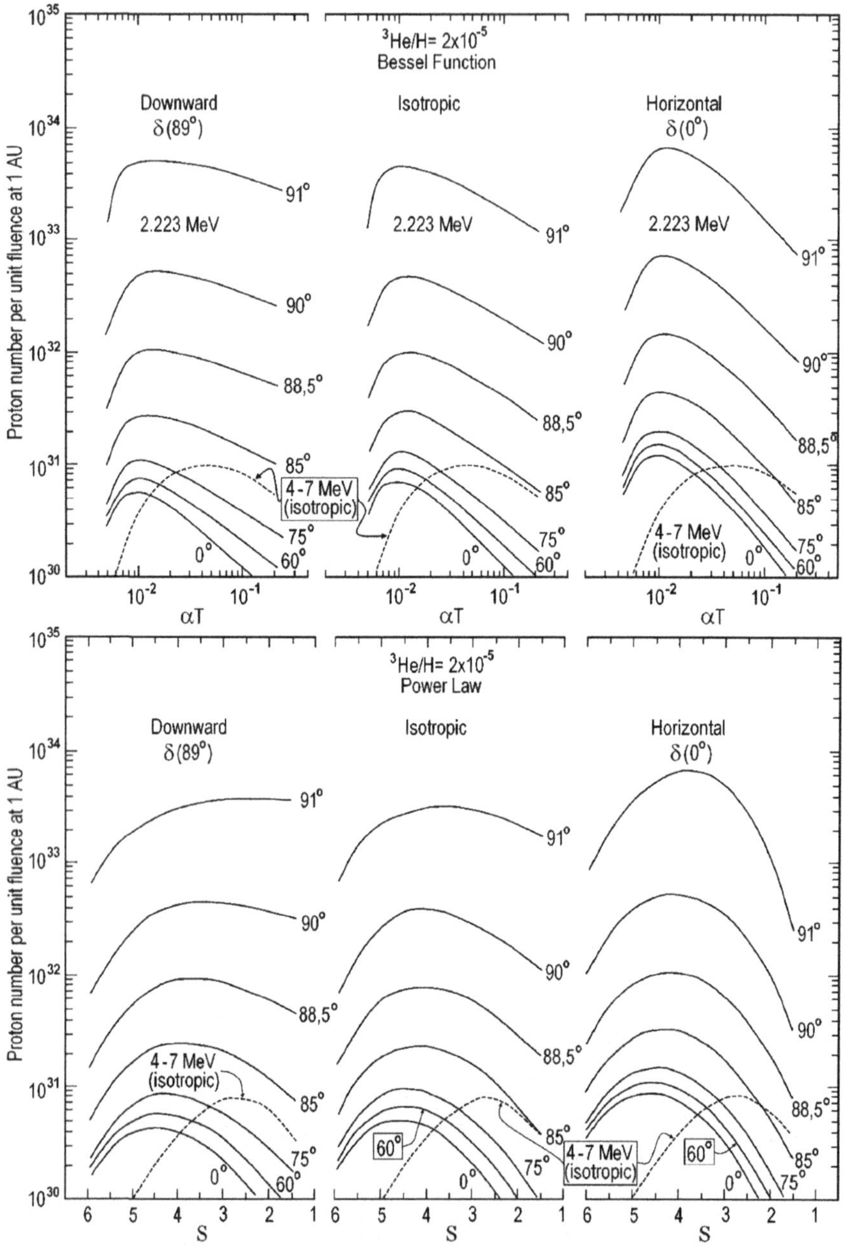

Fig. 11.16 The proton number per unit fluency for incident ions with (**a**) Bessel function and (**b**) power-law spectra for three incident angular distributions (From Hua and Lingenfelter 1987a)

(> 30 keV) X-ray emission lasting only about 10^2 s and long duration flares with hard X-ray emission lasting for 10^3 s or longer. In terms of these classes, the observations suggest that the accelerated particles in impulsive flares have a Bessel function like spectrum which could result from stochastic acceleration (Ramaty 1979) and that the escape of such particles into interplanetary space is usually energy independent while in the long duration flares further acceleration by shocks occurs during escape of the particles from the flare. Moreover, the number of flare-accelerated protons of energy greater than 30 MeV that escaped into the interplanetary medium is also quite different for these two classes. As can be seen in Table 11.2, the fraction of accelerated protons that escaped from the Sun, $F_{esc} = N_{ipm}/(N_{ipm} + N_{sun})$ varies from 0.95 for the 4 August 1972 flare to < 0.003 for the 1 July 1980 flare. For the impulsive flares (7 and 21 June, 1 July, 6 November 1980, and 3 June 1982) the number of accelerated protons escaping into the interplanetary medium is substantially less than that remaining trapped at the Sun producing gamma-rays. For the long duration flares (4 August 1972 and 10 April 1981), on the other hand, the number of escaping protons is comparable to or greater than that remaining at the Sun. This can also be seen from observations (Cliver et al. 1983) of the long duration flare of 9 December 1981 which had no detectable gamma-ray line emission. Interplanetary proton measurements suggest that about 9.6×10^{31} protons (>30 MeV), having a power-law spectrum with S = 3.5, escaped from this flare. For such a proton spectrum, the upper limit on the 2.223 MeV gamma-ray fluence of <3.7 photons cm^{-2}; from this flare at Wl6N12 implies (Panel b in Fig. 11.16) that $<1.9 \times 10^{31}$ protons (>30 MeV) remained trapped at the Sun, giving an escape fraction of >0.83. Moreover for impulsive flares, as can be seen in Fig. 11.17, there may be a strong correlation between the escape fraction and the total number of accelerated protons. If this correlation will be supported by further observations, it will place another important constraint on models of the acceleration and escape of particles in solar flares. For long duration flares, on the other hand, the escape fraction is apparently quite independent of the total number of accelerated protons.

11.1.10 Summary of Main Results

Hua and Lingenfelter (1987a), using Monte Carlo simulations, have calculated for a variety of assumed solar flare-accelerated ions spectra and angular distributions as following:

1. The depth dependence in the solar atmosphere of neutron production by nuclear interactions of the accelerated ions trapped in the atmosphere
2. The depth dependence of the radiative capture of these neutrons by hydrogen
3. The angular dependence of the escaping 2.223 MeV gamma-ray line emission resulting from neutron capture

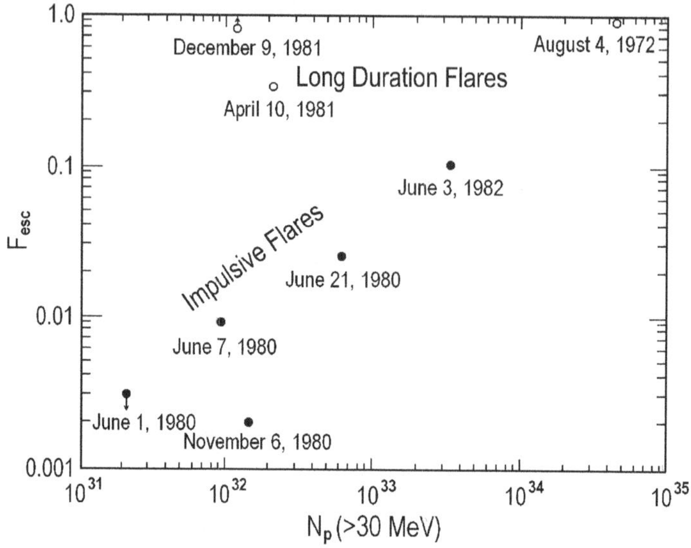

Fig. 11.17 The fraction $F_{esc} = N_{ipm}/(N_{ipm} + N_{sun})$ of accelerated protons of energy greater than 30 MeV that escape into the interplanetary medium compared to the total number accelerated to those energies in the flare (From Hua and Lingenfelter 1987a)

As concluded Hua and Lingenfelter (1987a), these calculations enable to model quantitatively the strong attenuation of the 2.223 MeV line emission from flares at large heliocentric longitudes. With these calculations they have thus been able to determine both the energy spectrum and the total number of accelerated protons trapped at the Sun, using observations of the 2.223 MeV capture line and 4–7 MeV band of nuclear deexcitation fluence from flares anywhere on the solar disk.

Hua and Lingenfelter (1987a) find that all of the published observations of these lines can be explained by accelerated particles having Bessel function spectra with a relatively narrow range of spectral indices, $0.015 < \alpha T < 0.04$, resulting from stochastic acceleration. Moreover, comparison of the spectra and total numbers of accelerated protons trapped at the Sun with those of flare-accelerated protons observed in the interplanetary medium show that for impulsive flare both populations seem to have very similar spectra, suggesting that they are accelerated by the same process and that the escape of protons into the interplanetary medium is essentially energy independent. For long duration flares, however, the protons apparently undergo further acceleration by shocks as they escape, altering their spectrum to a power law. In addition Hua and Lingenfelter (1987a) also find the fraction of accelerated protons that escape into the interplanetary medium is close to unity for the long duration flares, while it is much less than unity for the impulsive flares. In the latter flares it may in fact be directly correlated with the total number of protons accelerated.

11.2 Determination of the ^{3}He/H ratio in the Solar Photosphere from Flare Gamma-Ray Line Observations

11.2.1 The Matter and Short History of the Problem

Hua and Lingenfelter (1987b, c) developed the Monte Carlo simulation model of the energy and angular distributions of neutrons escaping from the solar atmosphere, described in Section 11.1 (Hua and Lingenfelter 1987a) for determining of the ^{3}He/H ratio in the solar photosphere from flare gamma-ray line observations. As noted Hua and Lingenfelter (1987b, c), solar flare neutron and gamma ray observations are providing important new information on flare particle acceleration and on the flare process itself. These gamma rays and neutrons are produced directly by nuclear interactions of the flare-accelerated protons and heavier ions with ambient gas in the solar atmosphere. Thus they yield the most direct information available on properties of the solar atmosphere as well as the total number, energy spectrum, time dependence, and angular distribution of the accelerated ions.

In order to study the properties of the accelerated ions by comparisons between observations and theoretical models, Hua and Lingenfelter (1987b, c) have developed results obtained in (Hua 1986; Hua and Lingenfelter 1987a), and interconnected system of Monte Carlo programs. With these programs Hua and Lingenfelter (1987b, c) have calculated the depth, energy and angular dependences of the neutron production and capture in the solar atmosphere as a function of the accelerated ion energy spectrum and angular distribution, and then calculated the attenuation of the escaping neutrons and capture line emissions as a function of energy and observing angle, or heliocentric position. They also calculated the time-dependent flux of those neutrons that survive to a distance of 1 AU without decaying, in order to directly compare with the time-dependent observations. The angular distributions of accelerated ions considered in these calculations include: (1) a delta function pencil beam radially downward, (2) an isotropic distribution in the downward hemisphere, and (3) a mirroring distribution approximated by a delta function fan beam at 89°.

In addition Hua and Lingenfelter (1987b, c) considered a fully (4π) isotropic distribution of ions running their range in an optically thin region high in the chromosphere. They also considered two forms for the energy spectra suggested by both theory and observation: a Bessel function in momentum/nucleon, expected (Ramaty 1979) from stochastic Fermi acceleration, and a power law in energy/nucleon, expected (Forman et al. 1986) from shock acceleration in the non-relativistic limit. These Monte Carlo calculations are described in detail by Hua (1986) and Hua and Lingenfelter (1987a), and in Hua and Lingenfelter (1987b, c) are considered how comparisons of the results of these calculations with observations have enabled to determine both the accelerated ion angular distribution and the photospheric ^{3}He abundance.

As underlined Hua and Lingenfelter (1987b, c), comparison of the results of these calculations with measurements of the neutron flux from solar flares can provide a determination of the angular distribution of the accelerated ions in flares, because the energetic (~10 MeV) neutrons produced by such ions are not produced isotropically, but tend to be produced predominantly in the direction of motion of the incident ions. Comparison of the results of these calculations with measurements of the time dependence of 2.223 MeV gamma ray line flux from neutron capture on hydrogen can also provide a direct means of determining the ^{3}He abundance in the photosphere, because the (n,p) reaction on 3He which has a cross section 1.6×10^4 times that of H(n,γ), can compete effectively for the capture of neutrons. Thus the time dependence of the 2.223 MeV gamma rays from capture on hydrogen is quite sensitive to the presence of ^{3}He, if its relative abundance ^{3}He/H exceeds 10^{-5}.

11.2.2 Accelerated Ion Angular Distributions and Spectra

Comparing the results of the calculations with measurements of the neutron flux from the flares of 1980 June 21 at a heliocentric angle of 89° (Forrest 1983) and 1982 June 3 at 72° (Debrunner et al. 1983; Evenson et al. 1983; Chupp et al. 1987), Hua and Lingenfelter (1987b, c) determined both the angular distribution and the energy spectrum of the accelerated ions. The comparison is done with the normalization of the calculated flux in each case to the total number of accelerated ions required to produce the measured excess 4–7 MeV fluence by nuclear deexcitation, using the calculations of Murphy and Ramaty (1984).

Hua and Lingenfelter (1987b, c) found that, independent of the assumed angular distribution of the accelerated ions, the measured time-dependent neutron flux at 1 AU from both flares was much more consistent with that produced by accelerated ions with a Bessel function spectrum than with that from a power-law spectrum. This result confirms previous studies (Murphy and Ramaty 1984; Ramaty et al. 1983a) which assumed simple isotropic neutron production and escape. In general, power-law spectra, capable of producing the observed ratio of the neutron flux to the excess 4–7 MeV fluence produced neutron energy spectra that were too hard and hence time-dependent fluxes at 1 AU that were too intense at early times.

Moreover, Hua and Lingenfelter (1987b, c) found that downward beamed ions were simply not able to produce sufficient escaping neutrons compared to the 4–7 MeV fluence. Although isotropic downward directed ions could produce enough neutrons, if the ion spectrum was hard enough, the escaping neutron spectrum was also too hard and the time-dependent flux too intense at early times. The values of αT required to account for the neutron flux in this case were also inconsistent with those determined independently from the observed fluence ratios of the 4–7 MeV emission to the 0.511 MeV line emission from positron annihilation and the 2.223 MeV line emission from neutron capture. Only ions with either a horizontal fan beam or a 4π isotropic distribution gave a good fit to both the observed neutron flux and the 0.511 and 2.223 MeV fluencies relative to that at 4–7 MeV for the same αT. Of the two possible angular distributions of the accelerated ions, the mirroring, horizontal fan beam appears to be the more likely, because the observed limb brightening of the >10 MeV bremsstrahlung continuum emission also requires (Dermer and Ramaty 1986; Dermer 1987) that the accelerated electrons >10 MeV have a horizontal fan beam distribution. If the accelerated ions and energetic (>10 MeV) electrons have the same angular distribution, it must be a mirroring distribution.

11.2.3 Estimation of the Photospheric ^{3}He/H Ratio

Comparing the results of the calculations with the SMM measurements of the time dependence of the 2.223 MeV line emission from the flare of 1982 June 3, Hua and Lingenfelter (1987b, c) can determine the ^{3}He/H ratio in the solar photosphere. This flare has the best determined time dependence published. Hua and Lingenfelter (1987b, c) have calculated the 2.223 MeV line emission escaping in essentially the same direction $\theta = 75°$ and studied its time dependence as functions of the photospheric ^{3}He/H ratio and the accelerated ion energy spectrum and angular distribution. As is shown from the comparison of the calculated escaping neutron flux with that measured in the vicinity of the Earth, the accelerated ions having a mirroring horizontal fan distribution appear to be best fitted to the observations. Hua and Lingenfelter (1987b, c) therefore determine the photospheric ^{3}He/H ratio by finding the best fit between the observed (Prince et al. 1983) time dependence of the 2.223 MeV line emission from the 1982 June 3 flare and the time dependences calculated for an array of values of ^{3}He/H ratio and αT. Using the Chi-square test, we determined the 90% confidence region for both the ^{3}He/H and the αT values. Hua and Lingenfelter (1987b, c) have assumed that the time dependence of neutron production is the same as that of the 4.1–6.4 MeV emission from nuclear deexcitation which should be directly proportional to the neutron production rate, if possible time variations in the accelerated ion spectrum are ignored. The measured time dependence of the 2.223 MeV emission is shown in Fig. 11.18.

To determine both ^{3}He/H and αT, Hua and Lingenfelter (1987b, c) have made two separate best-fit comparisons of the calculations and observations. First it was fit the time dependence of the 2.223 MeV emission, independent of the absolute normalization of the flux, and second it was fit the ratio of the time-integrated fluxes, or fluencies, in the 2.223 MeV line and the 4.1–6.4 MeV band. In the first determination, fitting time dependence, it was compared flux ϕ (2.223 MeV) with the measured counting rate. In determining a best-fit Hua and Lingenfelter (1987b, c) made Chi-square tests of the fit of the measured and expected count rates only in the first 442 s because of the poorer statistical significance of the later data. However, from comparing the measured time dependence of the 2.223 MeV line emission with a typical best-fit case determined from the first 442 s of data, can be seen in Fig. 11.18 that the calculated time dependence also fits quite well over the entire range of observation. In this example it was converted the expected 2.223 MeV flux to an expected count rate for comparison with the data, using an

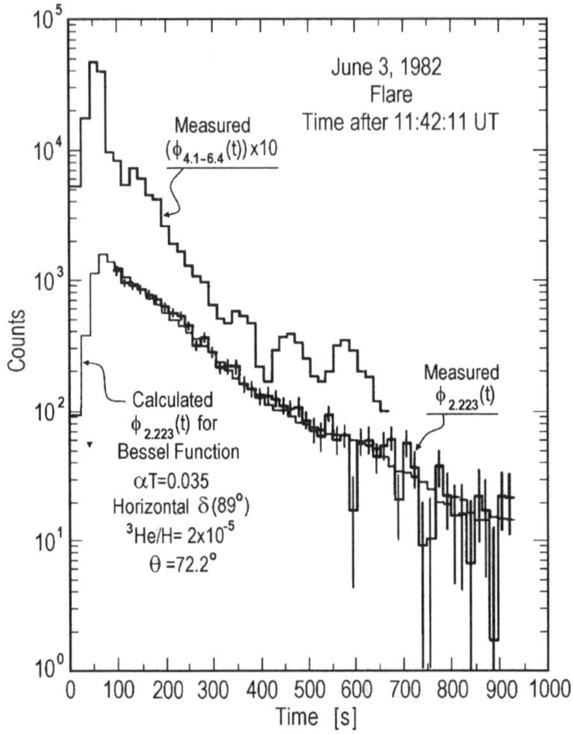

Fig. 11.18 The measured (Prince et al. 1983) and calculated time dependences of the 2.223 MeV neutron capture line fluency (From Hua and Lingenfelter 1987b, c)

effective area of 51 cm^2, determined from the ratio of the expected integral count rate to the measured (Prince et al. 1983) fluency.

In the second determination, fitting the fluency ratio, Hua and Lingenfelter (1987b, c) calculated the time-integrated 2.223 MeV flux at each αT divided by the calculated (Murphy and Ramaty 1984) 4–7 MeV yields and a 4.1–6.4 to 4–7 MeV fluency ratio of 0.85, and compare with the measured (Prince et al. 1983) ratio of the 2.223 to 4.1–6.4 MeV fluence of 1.03 ± 0.10. The results of both best-fit determinations of $^3He/H$ and αT are shown in Fig. 11.19.

As can be seen from Fig. 11.19, the shape of the time dependence can be fitted within the 90% confidence level by any of the values of αT considered, but for only a relatively narrow range of $^3He/H$ at any particular αT. The fluency ratio, on the other hand, can be fitted only by a very narrow range of αT. The shaded region in the Fig. 11.19 thus defines the best determinations for both αT and $^3He/H$. For the accelerated ions in the 1982 June 3 flare, Hua and Lingenfelter (1987b, c) find a best fit αT of 0.035 ± 0.007 for a Bessel function spectrum at the 90% confidence level. This agrees very well with the above independent determinations of αT of 0.03–0.04, based on analyses of the measurements of the neutron spectrum from this flare. From a detailed analysis of the ground-based neutron monitor data,

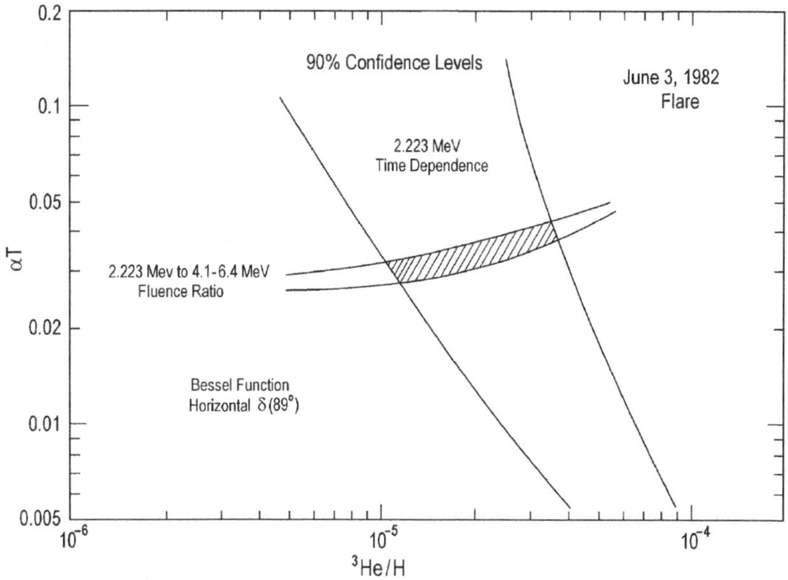

Fig. 11.19 Best-fitted ^{3}He/H and αT at the 90% confidence level (From Hua and Lingenfelter 1987b, c)

evidence was also found (Murphy et al. 1987) for an additional, much less intense, shock-accelerated ion component. But the 2.223 MeV line emission at 75° expected from these ions, was <2% of that resulting from the much more intense ions with a Bessel function spectrum. This component, therefore, does not make a significant contribution to the time-dependent flux or to the ^{3}He/H determination. For the solar photosphere Lingenfelter (1987b, c) find a best-fit ^{3}He/H of $(2.3 \pm 1.2) \times 10^{-5}$ at the 90% confidence level. This new value of the ^{3}He/H ratio in the solar photosphere is lower than that of $(3.4 \pm 1.7) \times 10^{-5}$, estimated by Geiss (1982) for the outer convective zone of the Sun from measurements of ^{3}He/^{4}He in meteorites and estimates of the protosolar D/H and ^{4}He/H ratios. The previous upper limit was high enough that it could have allowed for a significant contribution of ^{3}He to have been mixed into the photosphere by turbulent diffusion (Schatzman and Maeder 1981) from the solar interior where it can be made by deuterium burning. The present value, however, is close enough to that expected (Yang et al. 1984) solely from primordial nucleosynthesis to suggest that such turbulent diffusion does not make an important contribution to the photospheric ^{3}He abundance. Taking a photospheric ^{4}He/H ratio of 0.07–0.08 (e.g. Geiss 1982) the present determination also gives a ^{3}He/^{4}He ratio of $(3.1 \pm 1.6) \times 10^{-4}$. This is marginally lower than either the solar wind values (Geiss et al. 1970; Geiss and Reeves 1972; Ogilvie 1980) of $(4.3 \pm 0.3) \times 10^{-4}$ and $(4.7 \pm 1.2) \times 10^{-4}$, or the coronal prominence value (Hall 1975) of $(4 \pm 2) \times 10^{-4}$, suggesting a modification of the solar wind and coronal ^{3}He abundance by some processes of ion fractionation.

As noted Hua and Lingenfelter (1987b, c), this determination of the solar photospheric ^{3}He/H ratio, however, must still be regarded as preliminary. The incident ion angular distribution is still somewhat uncertain. Although comparison with the energetic electron distribution strongly suggests that the ions have a mirroring distribution, the ion related measurements alone are also consistent with a 4π isotropic distribution. For such a distribution the 2.223 MeV measurements would imply an αT identical to that for a mirroring distribution but an even lower ^{3}He/H ratio of $(1.8 \pm 1.2) \times 10^{-5}$ at the 90% confidence level. Thus the remaining uncertainty in the angular distribution acts only in the direction of increasing the differences and strengthening the arguments discussed above. This determination, however, is also based on the analysis of measurements from only one flare and further studies of other flare measurements are needed for confirmation. Prince et al. (1983) have compared the much less sensitive measurements of the time dependences of 2.223 MeV line emission from three other flares observed at roughly the same heliocentric longitude as the 1982 June 3 flare and found that they could be fitted by essentially the same exponential time constant as they found for the 1982 June 3 flare.

11.3 Intensity and Directionality of Flare-Accelerated α-Particles at the Sun: Simulation and Estimation from Gamma-Ray Observations

11.3.1 The Matter and Short History of the Problem

Share and Murphy (1997) have studied γ-ray line emission from fusion of accelerated α-particles with ambient helium in 19 solar flares observed by the Solar Maximum Mission Gamma-Ray Spectrometer from 1980 to 1989. An isotropic or fan-beam distribution of accelerated particles provides good fits to the line profiles for most of the flares. In contrast, a downward beam of accelerated particles can be ruled out at high confidence levels, 99.99% and 99.8%, for the two most intense flares and provides significantly poorer fits than isotropic or fan-beam distributions in two other flares. So, the angular distribution of energetic particles interacting in the corona and chromosphere, which reflects their acceleration and transport, may be determined from gamma-ray observations.

The short history of this problem are described by Share and Murphy (1997). One of the first, Kozlovsky and Ramaty (1977), have shown that the shapes of γ-ray lines at 0.429 and 0.478 MeV can be used to infer the angular distributions of flare-accelerated α-particles. These lines result from de-excitation of ^{7}Be and ^{7}Li produced in fusion of the accelerated α-particles with ambient ^{4}He (Kozlovsky and Ramaty 1974). Murphy et al. (1988) have calculated the expected shapes of these lines for four different angular distributions of particles: isotropic, fan beam, broadened fan beam, and downward beam. They presented line profiles for flares at Sun center and at the limb.

Murphy et al. (1990a) continued this study using a magnetic loop model for transport of the ions, including the effects of mirroring and MHD pitch-angle scattering. They also folded their calculations through the instrument response of the Solar Maximum Mission (SMM) Gamma-Ray Spectrometer (GRS) and compared the results with spectra observed from the 1981 April 27 solar flare. These comparisons provided convincing evidence for the presence of the α–α fusion lines in that flare. They showed that the quality of data was good enough to distinguish limb flares (as this one was) from disk-centered flares. Their fits also suggested that a downward beam of particles did not fit the data as well as an isotropic or fan-beam geometry.

As underlined Share and Murphy (1997), in principle, the accelerated α/p ratio can be obtained by comparing the fluencies of the α–α and narrow nuclear de-excitation lines. Determining the α/p ratio is dependent on knowing the relative abundances of ambient He and the elements producing the narrow lines in the interaction region as well as the accelerated particle spectrum. Unfortunately, direct measurements of the solar He abundances are uncertain. Photospheric He cannot be measured directly but is inferred from stellar evolution models, yielding an He/H ratio of 0.095 (Grevesse et al. 1996). Laming and Feldman (1994) used a UV spectrum over a sunspot to obtain an He/H ratio in the range of 0.078–0.22. A more precise abundance ratio, 0.07 ± 0.011, for the solar corona was reported by Gabriel et al. (1995). Long-term measurements of the solar wind typically give smaller He/H ratios, 0.03 ± 0.05; however, the ratio can be highly variable on short timescales, ranging from 0.001 to over 0.3 (Gloeckler and Geiss 1989).

The intensity of the α–α line feature measured by Murphy et al. (1990a) suggested that the accelerated α-particle fluency, the ambient 4He abundance, or both are enhanced over accepted solar abundances. Murphy et al. (1991) evaluated accelerated-particle and ambient-gas abundances from the SMM γ-ray spectrum of the 1981 April 27 flare. Their analysis included detailed cross sections for all the critical reactions. They found the best fits to the data for different assumptions concerning the composition of the ambient material and the composition and spectra of the accelerated particles. They found that the α/p ratio was higher than one would have anticipated from measurements of solar energetic particles in space. The best fits gave α/p ratios ranging from 0.5 to 1, under the assumption that their spectral shapes were the same and that the ambient He/H ratio was 0.1. Some poorer fits yielded ratios closer to 0.3, however. This compares with ratios ranging from as small as ~0.005 to as high as ~0.3 for 4.4–6.4 MeV/nucleon particles observed in impulsive solar energetic particle events (Reames et al. 1994); this range in particle energies lies just below the threshold of the α–α fusion process. It is therefore important to determine whether the apparently high α/p ratio measured for the 1981 April 27 flare is characteristic of all 19 flares observed by SMM and whether the ratios have as large a dispersion as those measured for the solar energetic particles.

Share and Murphy (1997) expand the study of the α–α fusion lines to a sample of 19 solar flares observed by the SMM GRS with significant nuclear line emission (Share and Murphy 1995). In Share and Murphy (1995) the full 0.3–8.5 MeV spectra were used to obtain both the fluencies and ratios of the nuclear de-excitation lines in the flares. In the current analysis, we fitted spectral data in the 0.3È0.75 MeV energy range with an incident photon spectrum comprising five individual components, including the a-a line feature ; this contrasts with the work of Murphy et al. (1991), in which the parameters of the accelerated particles producing the γ-rays were varied.

11.3.2 Spectroscopic Studies of the α–α Lines in SMM Data

Share and Murphy (1997) concentrate on the energy range from 0.3 to 0.75 MeV in order to improve sensitivity to the α–α fusion lines. In this limited energy range, they use an incident photon model containing a bremsstrahlung function, a narrow 511 keV annihilation line, a positronium continuum, instrumentally degraded radiation from nuclear lines above 0.75 MeV, and various line shapes for the fusion lines. Share and Murphy (1997) used a simple power law in energy to represent the bremsstrahlung continuum and obtain its amplitude and exponent. For this study they have fixed the annihilation line energy and set its width at a nominal 10 keV; the amplitude of the line is a free parameter. The amplitude of the positronium continuum is also a free parameter but is constrained to be positive. The amplitude of the degraded nuclear component reflects the nuclear deexcitation and 2.223 MeV line fluxes. Share and Murphy (1997) have allowed the amplitude to be a free parameter to take into account inaccuracies in the instrument response function as well as any scattered radiation from the Sun; with the exception of two or three flares, the implied fluxes of higher energy emission from these fits to the 0.3–0.75 MeV data agree well with those determined by Share and Murphy (1995) over the higher energy range. However, whether this scattered nuclear component was free or fixed, had no significant impact on the α–α line fluxes.

As underlined Share and Murphy (1997), the α–α line profiles for three basic particle geometries – isotropic, fan beam, and downward beam – have been determined for the heliocentric angle of each flare. These normalized line profiles (Murphy et al. 1988) were multiplied by a free parameter that is determined by the fit. The fan-beam geometry approximates the distribution expected for particles experiencing no pitch-angle scattering, while the downward beam approximates the distribution for particles experiencing strong pitch-angle scattering (Murphy et al. 1990a). Below the results of these fits are used to determine which accelerated particle geometry is preferred.

Share and Murphy (1997) noted that the composite model, with initial estimates of the free parameters, is folded through a matrix representing the instrument response function and compared with the observed count spectrum. The values of the parameters are then sequentially varied with a computer algorithm until the χ^2 parameter is minimized. Not all of these fits gave acceptable values of χ^2; this is reflected in the low probability that the data are statistically distributed about the model. The poor fits were in large part due to specific time intervals in the flares during which telemetry errors occurred or the rates were high enough that the instrument model was not adequate. For purposes of studying particle directionality, Share and Murphy (1997) adopted a criterion that the original fits be acceptable at the 10% level. For those flares that had lower probabilities, Share and Murphy (1997) fitted individual 1 min accumulations and removed those for which the fits were unacceptable; then it was produced a corrected integrated spectrum for the flare. The results of the fits for the 19 flares are summarized in Table 11.3.

Listed in Table 11.3 are the time over which the spectral accumulations made, the percentage of fits of randomly distributed data with higher values of χ^2, and the fluency in the α–α lines for the isotropic model. In Table 11.3 two entries are provided for a specific flare, the top one is for the corrected spectrum and the bottom one is for the original spectrum. For comparison, there is also listed the fluency in the narrow nuclear deexcitation lines derived from fits to the full 0.3–8.5 MeV energy range (Share and Murphy 1995). For some flares, additional data selections were required to achieve reasonable fits. For two flares, on 1982 December 7 and 1989 March 6, are also removed 1 min accumulations with high electron bremsstrahlung to nuclear line ratios from the corrected spectra. For the flare on 1988 December 16, was deleted the lowest energy channel, and for the flare on 1989 March 10, was fitted the spectrum only up to 0.7 MeV because of systematic errors in the data. It was also marginally improved the fit to the 1989 November 15 spectrum by using an improved background subtraction technique developed for searching for short celestial transients (Share et al. 1993). Share and Murphy (1997) have studied whether there are any statistically significant changes in the (α–α)/(narrow nuclear line) ratios in comparing the original and corrected spectra and have not found any. Overall, they improved the acceptability of the fits significantly; however, there still are three flares for which the acceptability did not exceed the desired 10%.

11.3.3 Results of Investigating the Directionality of Accelerated α-Particles

Share and Murphy (1997) compare the results of their fits for accelerated α-particles having isotropic, fan-beam, and downward-beam geometries. The fan-beam and isotropic distributions are not readily distinguishable from one another with an instrument, such as the GRS, with moderate spectral resolution. This is true even for flares located near the center of the solar disk, where the lines produced by a fan beam appear at their rest energies and are not broadened significantly. Share and Murphy (1997) use these simple geometric distributions because the line profiles

Table 11.3 Parameters of 19 flares used in Share and Murphy (1997)

Flare	Date	Angle (degrees)	Accumulation (s)	Probability (%)	α–α Lines (γ cm^{-2})	Narrow γ lines[a] (γ cm^{-2})
1	April 10, 1981	38	376[b]	13	4.3 ± 4.5	18.2 ± 2.5
			524	2	5.9 ± 4.9	23.5 ± 3.1
2	April 27, 1981	91	1,441[b]	35	21.8 ± 9.1	73.3 ± 5.3
			1,916	9	37.8 ± 10.2	113.1 ± 6.2
3	June 3, 1982	72	1,195	23	−6.3 ± 9.2	28.6 ± 6.9
4	July 9, 1982	73	245[b]	38	7.4 ± 5.7	23.6 ± 2.9
			327	0.5	18.6 ± 7.0	33.6 ± 3.4
5	Nov. 26, 1982	87	393	25	8.2 ± 5.2	16.2 ± 2.7
6	Dec. 7, 1982	80	1,048[b]	29	−4.1 ± 10.2	54.3 ± 4.6
			2,703	<0.1	−14.6 ± 17.2	147.9 ± 8.8
7	April 24, 1984	45	925[b]	13	−5.1 ± 8.6	45.1 ± 5.7
			1,097	2	−7.7 ± 8.6	55.3 ± 6.2
8	Feb. 6, 1986	2	1,228	90	0.5 ± 7.6	45.4 ± 4.5
9	Dec. 16, 1988	43	2,293[b]	15	61.7 ± 13.5	133.3 ± 7.6
			3,555	1	101.6 ± 15.6	219.6 ± 10.7
10	March 6, 1989	76	1,376[b]	3.8	32.3 ± 11.0	77.5 ± 5.1
			3,515	<0.1	80.9 ± 20.3	293.8 ± 11.6
11	March 10, 1989	44	1,834[b]	16	18.4 ± 10.4	61.3 ± 5.2
			3,341	<0.1	27.8 ± 12.8	108.5 ± 7.5
12	March 17, 1989	70	835	19	13.5 ± 8.4	48.8 ± 4.6
13	May 3, 1989	44	1,376	55	12.4 ± 6.5	24.3 ± 4.1
14	Aug. 16, 1989	87	916	79	23.4 ± 6.5	45.9 ± 3.9
15	Aug. 17, 1989	90	2,228	23	13.3 ± 13.8	54.4 ± 7.3
16	Sept. 9, 1989	30	541	87	5.2 ± 5.4	17.0 ± 2.9
17	Oct. 19, 1989	32	3,260	25	80.9 ± 12.6	179.7 ± 10.3
18	Oct. 24, 1989	64	573[b]	4.4	8.9 ± 8.5	42.1 ± 4.8
			819	0.3	12.9 ± 8.2	44.7 ± 3.7
19	Nov. 15, 1989	30	1,048[c]	2.5	14.3 ± 7.8	35.7 ± 4.3
			1,016	1.5	18.3 ± 7.2	32.6 ± 4.4

[a]Resolved nuclear de-excitation lines (Share and Murphy 1995).
[b]Corrected data.
[c]Different accumulation and background subtraction used.

are easily calculable at all the heliocentric angles. Future comparisons will incorporate a physical model of particle transport similar to that used by Murphy et al. (1990a). Plotted in Figs. 11.20–11.23 are the corrected spectra for the 19 flares. The best-fit components for each flare are plotted separately for two particle geometries: isotropic (left-hand panels) and a downward-directed beam (right-hand panels).

The α–α line profiles for the downward-beam geometry are dependent on heliocentric angle. For clarity Share and Murphy (1997) have subtracted the best-fit bremsstrahlung function from the spectral data before plotting; for this reason, the spectra in the right-hand and left-hand panels differ from one another. The change in the downward-beam line profile with heliocentric angle is most evident when comparing the right-hand panels for the 1981 April 27 and 1986 February 6 flares. The fusion lines are at their rest energies for the 1981 April 27 limb event but are redshifted for the disk-centered event in 1986 February 6. The 511 keV line and associated positronium continuum show flare-to-flare variability.

In Figs. 11.20–11.23 the overall goodness of the fit, the percentage of fits to randomly distributed data that would give higher values of the χ^2 statistic, is shown for both particle geometries for each flare (note that there are used the corrected spectra for this analysis; see Table 11.3). Table 11.4 summarizes these probabilities for all three geometries: isotropic, fan beam, and downward beam.

As can be seen from Table 11.4, there is no significant difference between the fits in the isotropic and fan-beam geometries, as is expected from the above discussion. There is expected to be a significant difference between the α–α line shapes of isotropic and downward-beam geometries for flares near the disk center. There were nine flares that occurred at heliocentric angles of 45° or less. Four of these flares had isotropic α–α line fluxes significant at greater than 1.5 σ. The two flares in this sample with the most intense α–α lines show significant differences between fits for the two geometries. The isotropic/fan-beam geometry provides significantly better fits than does the downward-beam geometry.

As can be seen from Fig. 11.21 and Table 11.4, the 1988 December 16 flare occurred at a heliocentric angle of 43°; it is well fitted (15% probability) by the isotropic geometry, but the shifted profile of the downward-beam geometry provides a poor fit to the data (0.2% probability). The 1989 October 19 flare (Fig. 11.23) at a heliocentric angle of 32° shows an even greater disparity between the quality of fits for the isotropic/fan-beam geometries (25% probability) and downward-beam geometry (0.01% probability).

For the two weaker flares on 1989 March 10 and November 15 (Figs. 11.22 and 11.13, and Table 11.4), the isotropic/fan-beam geometry is also marginally preferred. All three particle distributions gave acceptable fits for the 1981 April 27 flare with the reduced data set; this contrasts with the earlier result of Murphy et al. (1990a) that suggested that an isotropic particle distribution fitted the data better than a downward beam for this limb event. Share and Murphy (1997) conclude from this analysis that a downward-beam geometry for the accelerated α-particles is inconsistent with the fits to the spectra with the most significant α–α lines.

11.3.4 Results for Accelerated α/p Ratio

As discussed in Section 11.3.2, a measure of the accelerated α/p ratio can be obtained by comparing the fluencies of the α–α and narrow nuclear deexcitation

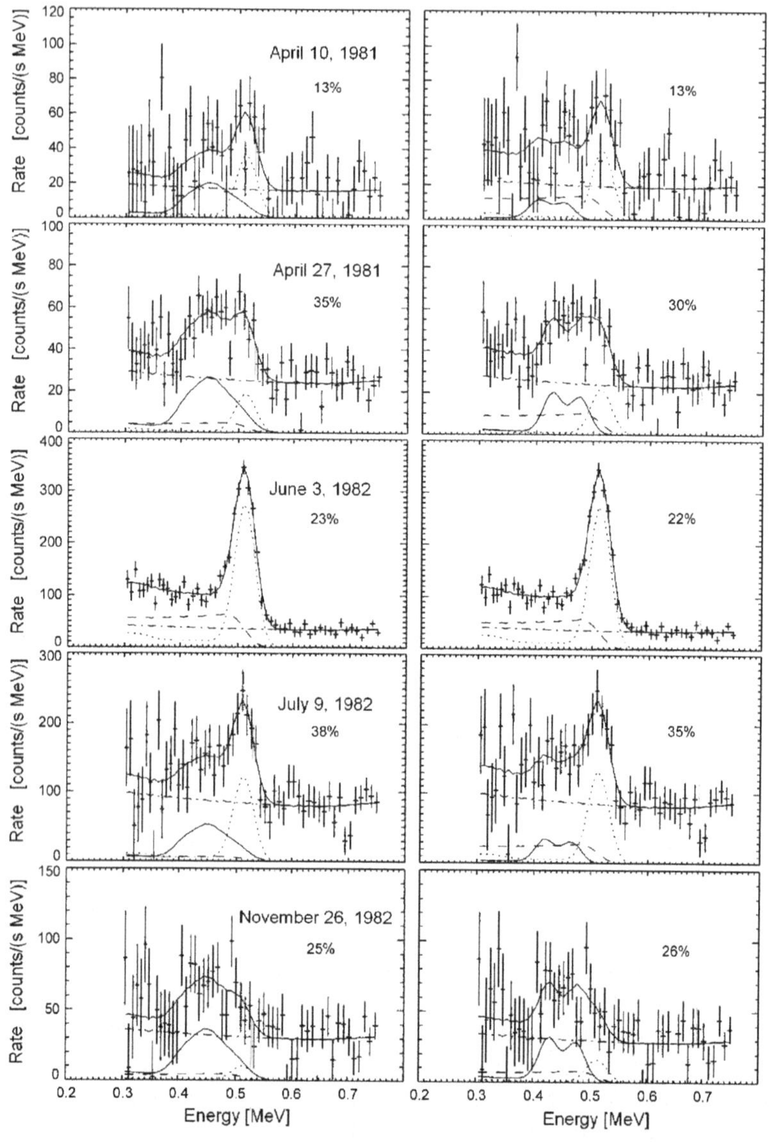

Fig. 11.20 SMM GRS spectra of five flares (1981 April 10 and 27, 1982 June 3, July 9 and November 26) from 0.3 to 0.75 MeV that have strong emission of nuclear lines. The fitted bremsstrahlung component has been subtracted. The plotted errors are statistical. Best-fitting components: *light solid curve*, α–α lines; *dotted curve*, 511 keV annihilation line; *dashed curve*, 3-γ continuum from positronium formation; *dot-dashed curve*, nuclear component; *heavy solid curve*, total fit. *Left-hand panels*: isotropically distributed α-particles; *right-hand panels*: α-particles in downward-beam geometry (From Share and Murphy 1997)

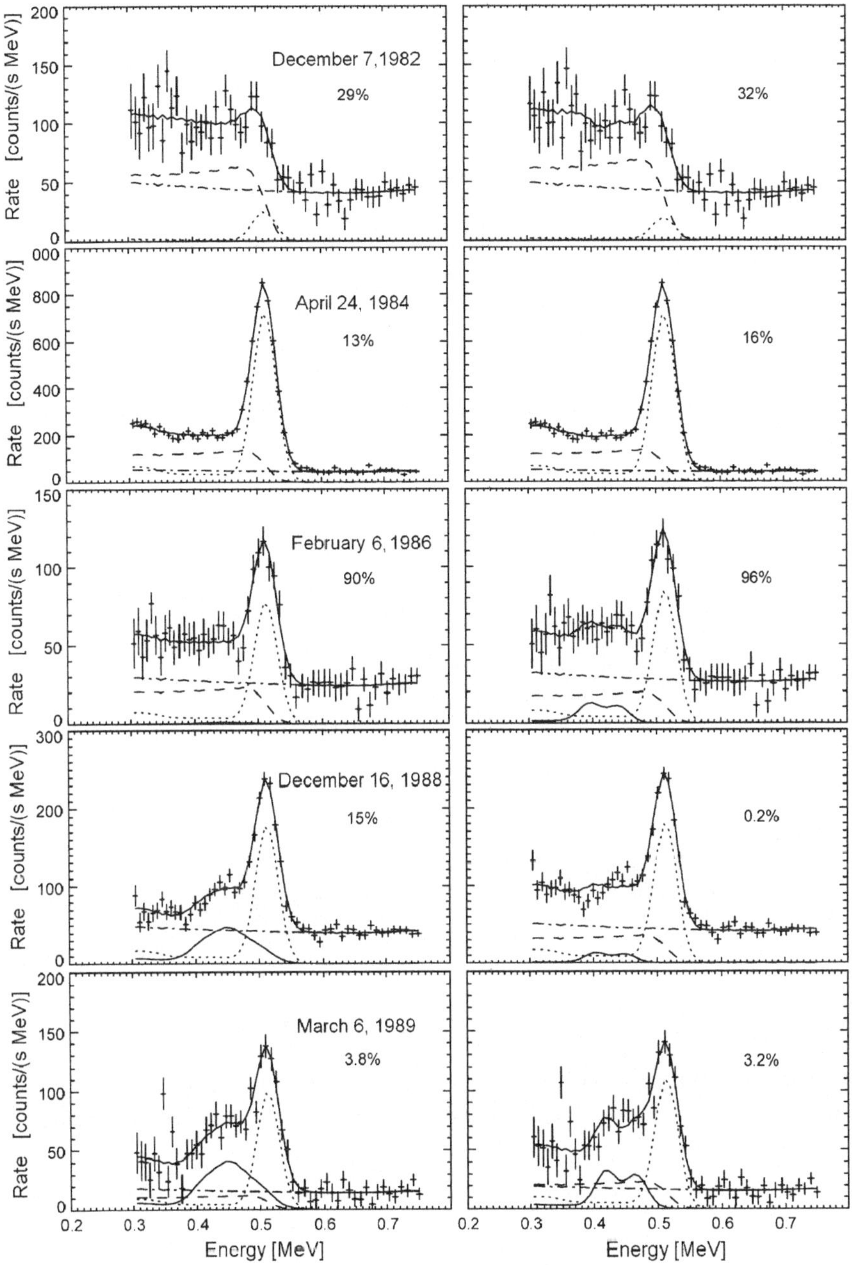

Fig. 11.21 The same as in Fig. 11.20, but for following five flares: 1982 December 7, 1984 April 24, 1986 February 6, 1988 December 16, and 1989 March 6 (From Share and Murphy 1997)

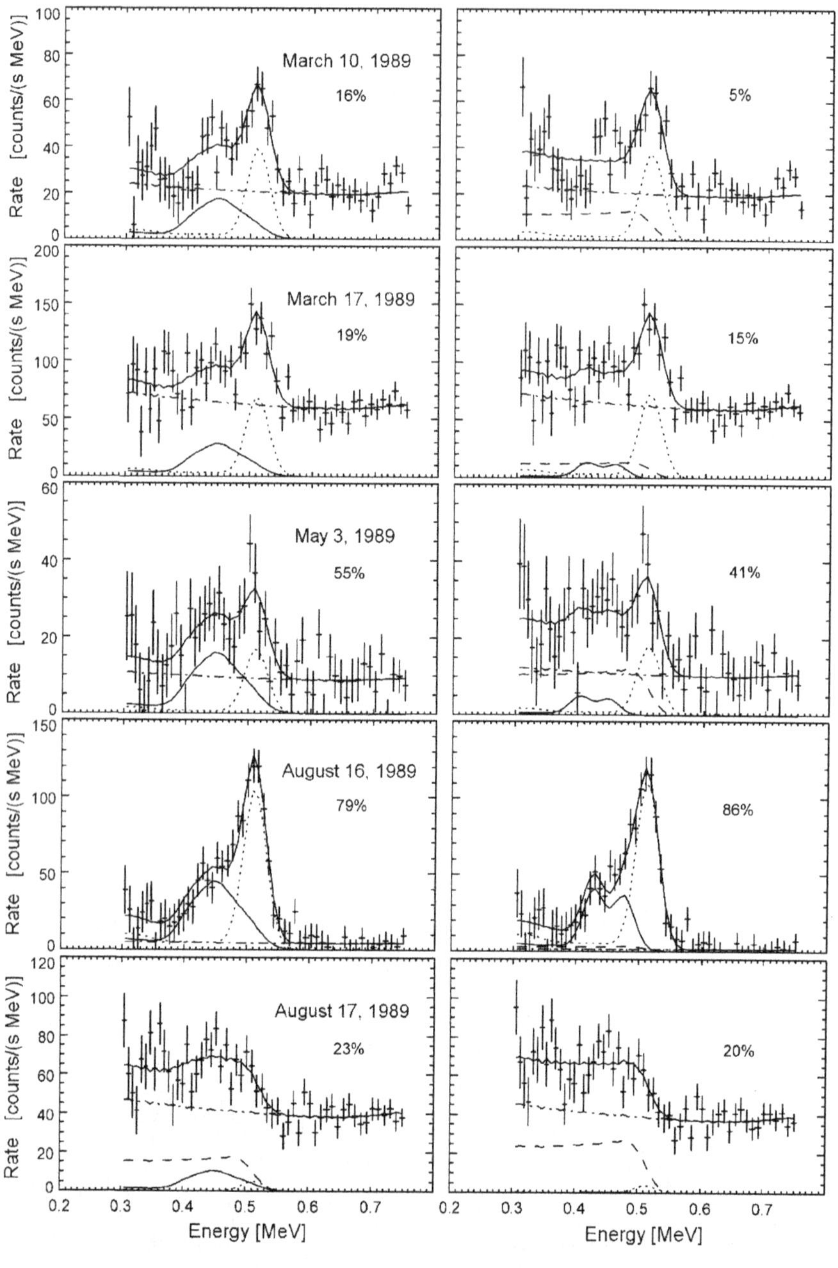

Fig. 11.22 The same as in Fig. 11.20, but for following five flares in 1989: March 10 and 17, May 3, August 16 and 17 (From Share and Murphy 1997)

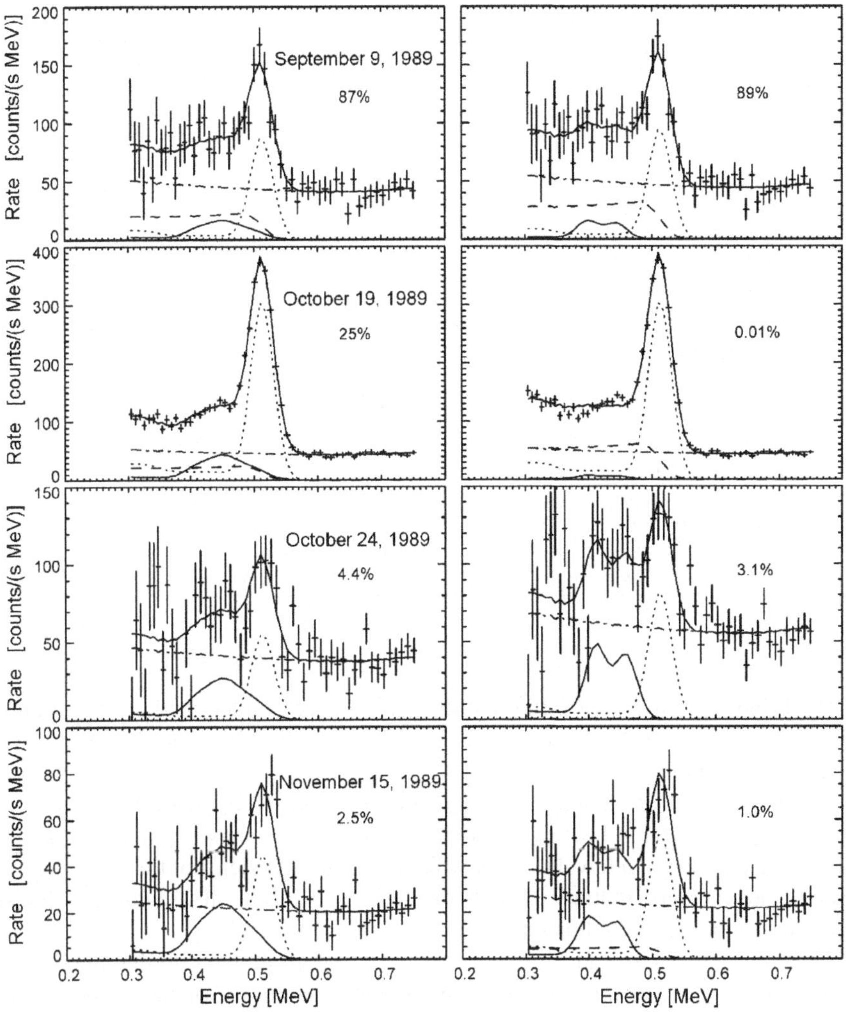

Fig. 11.23 The same as in Fig. 11.20, but for following five flares in 1989: September 9, October 19 and 24, November 15 (From Share and Murphy 1997)

lines. As He has a high first ionization potential (FIP = 24.6 eV), it is best to use deexcitation lines only from elements with high FIPs, such as C, O, N, and Ne, in making these comparisons. This is true because Share and Murphy (1995) showed that the (low-FIP)/(high-FIP) line ratio is variable from flare to flare. Table 11.5 lists the α–α line fluencies derived from the preferred isotropic model and fluencies of the high-FIP deexcitation lines at 1.63 MeV (Ne), 4.43 MeV (C, O [spallation]), 6.13 MeV (O), and ∼6.9 MeV (O) integrated over the entire flare; it also lists the sum of the four high-FIP lines.

Table 11.4 Directivity study (From Share and Murphy 1997)

Flare	Date	Angle (degrees)	Probability of good fit (%)		
			Isotropic	Fan beam	Downward beam
1	April 10, 1981	38	13	14	13
2	April 27, 1981	91	35	32	30
3	June 3, 1982	72	23	23	22
4	July 9, 1982	73	38	38	35
5	November 26, 1982	87	25	23	26
6	December 7, 1982	80	29	28	32
7	April 24, 1984	45	13	13	16
8	February 6, 1986	2	90	91	96
9	December 16, 1988	43	15	21	0.2
10	March 6, 1989	76	3.8	3.5	3.2
11	March 10, 1989	44	16	16	5
12	March 17, 1989	70	19	20	15
13	May 3, 1989	44	55	57	41
14	August 16, 1989	87	79	73	86
15	August 17, 1989	90	23	23	20
16	September 9, 1989	30	87	86	89
17	October 19, 1989	32	25	28	0.01
18	October 24, 1989	64	4.4	3.9	3.1
19	November 15, 1989	30	2.5	1.5	1.0

In Table 11.5 Share and Murphy (1997) used spectra integrated over the entire flare in order to be compatible with earlier studies (e.g., Share and Murphy 1995; Ramaty et al. 1996).

In Fig. 11.14, the α–α line fluencies are plotted against the sum of the fluencies in these four deexcitation lines. These two sets of fluencies show a good correlation, with the exception of a few flares, for which the observed yields of the α–α lines are below the average. Plotted through the data is the best fit assuming a linear relationship between the fluencies.

According to Share and Murphy (1997), the overall probability that the data in Fig. 11.14 come from a random sample distributed about the plotted slope is only 0.4% on the basis of the χ^2 statistic. The flare on 1982 December 7 is the primary reason for this large dispersion. Removing this flare from the sample improves the probability to 11%. The γ-ray spectrum from this flare (see Fig. 11.21) is unusual in that the 511 keV line is small relative to what appears to be the positronium continuum. Another flare that has these characteristics occurred on 1989 August 17 (see Fig. 11.22). Both flares occurred at large heliocentric angles, 80° or greater. There is some evidence that scattering at the Sun may play a role in attenuation of the 511 keV line, and therefore the α–α lines, in these flares. For these flares, the fitted nuclear component over the 0.3–0.75 MeV range is higher than what have estimated using the higher energy line fluencies derived in the earlier analysis over the 0.3–8.5 MeV range (Share and Murphy 1995), which is suggestive of scattering of the photons at the Sun. Share and Murphy (1997) noted, however, that the only other flare in the sample of 19 to exhibit the same difference in the scattered nuclear

Table 11.5 Line fluencies for 19 flares (From Share and Murphy 1997)

Flare	α–α Lines (γ cm^{-2})	1.63 MeV (γ cm^{-2})	4.43 MeV (γ cm^{-2})	6.13 MeV (γ cm^{-2})	6.9 MeV (γ cm^{-2})	High FIP[a] (γ cm^{-2})
1	5.9 ± 4.9	6.5 ± 1.1	2.2 ± 0.8	3.2 ± 0.5	1.1 ± 0.5	13.0 ± 1.5
2	37.8 ± 10.2	19.5 ± 2.4	18.8 ± 1.9	14.5 ± 1.1	11.6 ± 1.1	64.6 ± 3.4
3	−6.3 ± 9.2	8.1 ± 2.4	9.5 ± 1.5	2.2 ± 0.8	1.7 ± 0.8	21.6 ± 3.0
4	18.6 ± 7.0	5.7 ± 1.2	5.1 ± 1.0	4.5 ± 0.6	2.4 ± 0.5	17.7 ± 1.7
5	8.2 ± 5.2	1.9 ± 1.0	1.1 ± 0.7	1.9 ± 0.4	2.3 ± 0.4	7.2 ± 1.4
6	−14.6 ± 17.2	35.3 ± 2.7	26.6 ± 2.2	21.7 ± 1.3	12.8 ± 1.2	96.5 ± 4.0
7	−7.7 ± 8.6	14.0 ± 2.0	11.9 ± 1.6	6.7 ± 1.0	3.2 ± 0.9	35.8 ± 2.9
8	0.5 ± 7.6	6.5 ± 1.7	6.5 ± 1.2	7.5 ± 0.7	3.4 ± 0.7	23.9 ± 2.3
9	101.6 ± 15.6	51.9 ± 3.3	46.5 ± 2.7	43.4 ± 1.7	22.8 ± 1.5	164.6 ± 4.8
10	80.9 ± 20.3	64.2 ± 3.4	48.8 ± 2.9	39.8 ± 1.7	26.5 ± 1.6	179.2 ± 5.0
11	27.8 ± 12.8	22.6 ± 2.7	14.0 ± 2.1	12.8 ± 1.3	7.9 ± 1.3	57.3 ± 3.9
12	13.5 ± 8.4	9.4 ± 1.6	6.6 ± 1.3	7.1 ± 0.8	4.4 ± 0.7	27.5 ± 2.3
13	12.4 ± 6.5	5.8 ± 1.6	3.3 ± 1.2	4.2 ± 0.7	2.4 ± 0.7	15.7 ± 2.2
14	23.4 ± 6.5	11.7 ± 1.4	6.3 ± 1.3	6.5 ± 0.8	4.0 ± 0.8	28.6 ± 2.2
15	13.3 ± 13.8	18.0 ± 2.6	11.6 ± 2.3	8.2 ± 1.4	3.3 ± 1.3	41.0 ± 4.0
16	5.2 ± 5.4	1.9 ± 1.1	2.0 ± 0.9	3.8 ± 0.6	1.4 ± 0.5	9.1 ± 1.6
17	80.9 ± 12.6	41.7 ± 2.9	36.1 ± 2.6	31.6 ± 1.6	15.4 ± 1.5	124.8 ± 4.4
18	12.9 ± 8.2	10.0 ± 1.4	7.9 ± 1.2	8.4 ± 0.7	4.1 ± 0.7	30.5 ± 2.0
19	18.3 ± 7.2	5.7 ± 1.5	6.5 ± 1.4	4.2 ± 0.8	3.1 ± 0.8	19.5 ± 2.3

[a]Sum of 1.63, 4.43, 6.13, and 6.9 MeV lines.

component was the one on 1989 March 17 (see Fig. 11.22), at a heliocentric angle of 70°, but this event exhibited a relatively strong 511 keV line.

As underlined Share and Murphy (1997), the correlation displayed in Fig. 11.24 suggests that the α–α/high-FIP line ratios may be relatively constant from flare to flare. The observed ratios are listed in Table 11.6 and plotted in Panel a in Fig. 11.25. This correlation also suggests that the accelerated α/p ratio may also be relatively constant from flare to flare. The most accurate determination of the accelerated α/p ratio in individual flares requires fits to the entire γ-ray spectrum while varying the spectra and composition of accelerated particles and the ambient abundances (Murphy et al. 1991).

The approach Share and Murphy (1997) take here is to assume that the α/p ratio is constant from flare to flare and to determine whether the relative yield in the α–α lines is consistent with the assumed ratio. The yield in the α–α lines is dependent on the accelerated particle spectrum and ambient and accelerated particle composition. Ratios of γ-ray lines can be used to infer the spectra of accelerated particles. The ratio of the neutron capture line at 2.223 MeV to the carbon deexcitation line at 4.43 MeV has been used for several years. Because the 2.223 MeV line is delayed, it is important that the entire flare be observed. Spectral indices have been obtained from this ratio for nine SMM flares far from the solar limb, where limb darkening occurs for the 2.223 MeV line and for which Share and Murphy (1997) have complete data accumulations (see, e.g., Share and Murphy 1995; Ramaty et al. 1996). The ratio of the prompt deexcitation lines at 6.13 and 1.63 MeV also provides an estimate of the spectral index of accelerated particles and can be used

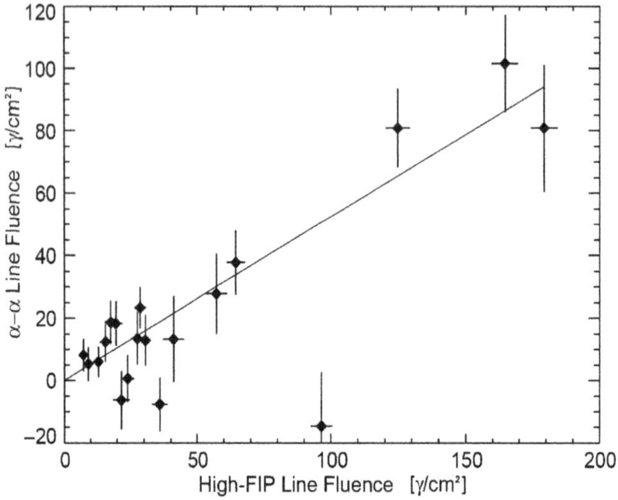

Fig. 11.24 α–α line fluencies, assuming isotropic α-particle angular distributions, plotted versus the summed fluence in high-FIP lines at 1.63, 4.43, 6.13, and ~6.9 MeV (From Share and Murphy 1997)

Table 11.6 Observed and calculated α–α/high-FIP line ratios (From Share and Murphy 1997)

Flare	Date	α-α/High-FIP line ratio		
		Observed	Calculated	
			α/proton = 0.1	α/proton = 0.5
1	Apr. 10, 1981	0.46 ± 0.38	0.21 ± 0.07	0.22 ± 0.13
2	Apr. 27, 1981	0.58 ± 0.16	0.30 ± 0.03	0.49 ± 0.14
3	June 3, 1982	−0.29 ± 0.43	0.12 ± 0.10	0.07 ± 0.07
4	July 9, 1982	1.05 ± 0.41	0.31 ± 0.03	0.60 ± 0.25
5	Nov. 26, 1982	1.13 ± 0.75	0.25 ± 0.06	0.56 ± 0.44
6	Dec. 7, 1982	−0.15 ± 0.18	0.28 ± 0.02	0.34 ± 0.10
7	Apr. 24, 1984	−0.21 ± 0.24	0.19 ± 0.06	0.14 ± 0.08
8	Feb. 6, 1986	0.02 ± 0.32	0.33 ± 0.01	0.84 ± 0.16
9	Dec. 16, 1988	0.62 ± 0.10	0.33 ± 0.01	0.68 ± 0.10
10	Mar. 6, 1989	0.45 ± 0.11	0.28 ± 0.02	0.34 ± 0.06
11	Mar. 10, 1989	0.48 ± 0.23	0.25 ± 0.04	0.28 ± 0.12
12	Mar. 17, 1989	0.49 ± 0.31	0.30 ± 0.04	0.51 ± 0.20
13	May 3, 1989	0.79 ± 0.43	0.29 ± 0.06	0.53 ± 0.31
14	Aug. 16, 1989	0.82 ± 0.24	0.25 ± 0.04	0.28 ± 0.12
15	Aug. 17, 1989	0.32 ± 0.34	0.20 ± 0.06	0.16 ± 0.08
16	Sep. 9, 1989	0.58 ± 0.60	0.24 ± 0.10	0.81 ± 0.19
17	Oct. 19, 1989	0.65 ± 0.10	0.31 ± 0.02	0.53 ± 0.09
18	Oct. 24, 1989	0.42 ± 0.27	0.32 ± 0.02	0.67 ± 0.18
19	Nov. 15, 1989	0.94 ± 0.38	0.29 ± 0.06	0.56 ± 0.34

for all 19 flares (Share and Murphy 1995). Ramaty et al. (1996) found consistency between spectral indices derived with these two sets of line ratios under certain conditions: an accelerated particle spectrum following a power-law shape, particles impacting in a thick-target model with ambient elemental abundances adopted for

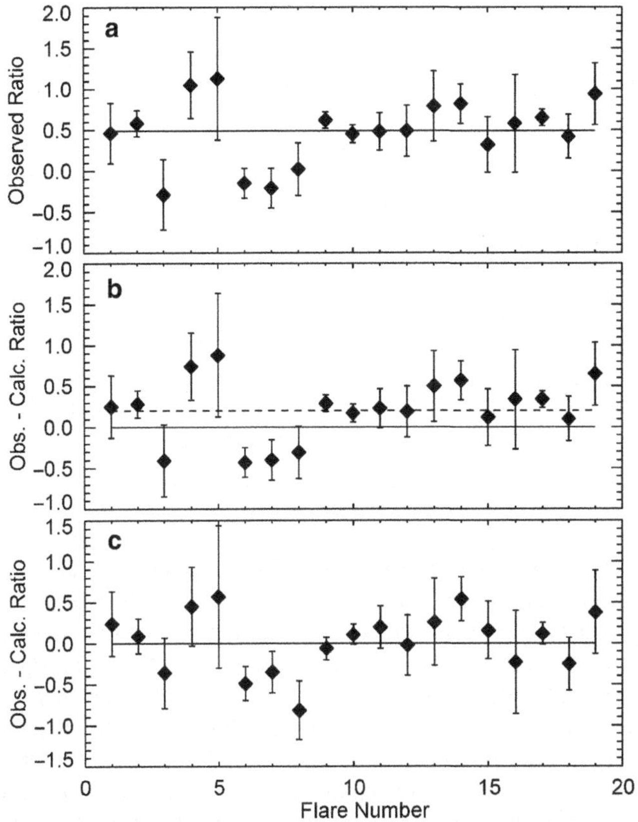

Fig. 11.25 Comparison of observed and calculated α–α/high-FIP line ratios used to determine the accelerated α/p ratio at the Sun. (**a**) Observed α–α/high-FIP line ratios; the *solid line* is the weighted mean. (**b**) Difference between observed and calculated line ratios for accelerated α/p ratio of 0.1; the *dashed line* is the weighted mean and is significantly different from zero. (**c**) Same as (b), but for an α/p ratio of 0.5; the weighted mean is consistent with zero (From Share and Murphy 1997)

the corona (Reames 1995) but with an elevated Ne/O ratio of 0.25, and impulsive flare composition for the accelerated particles with $^3He/^4He$ equal to 1 and α/p ratios ranging from ∼0.1 to 0.5.

The dependence of the α–α/high-FIP line ratio on particle spectral index for α/p ratios of 0.1 and 0.5 is plotted in Fig. 11.26 for the conditions listed above.

As noted Share and Murphy (1997), steeply falling particle spectra do not readily produce the α–α lines. As the particle spectra harden to those with indices near −3, the production of the α–α lines increases because of the sharp rise in its cross section at ∼10 MeV; for even harder spectra, spallation reactions contribute significantly to the production of the deexcitation lines, causing the ratio to decrease. Using Fig. 11.26 and spectral indices derived from the graphical results provided in Figure 1 of Ramaty et al. (1996), Share and Murphy (1997) calculated

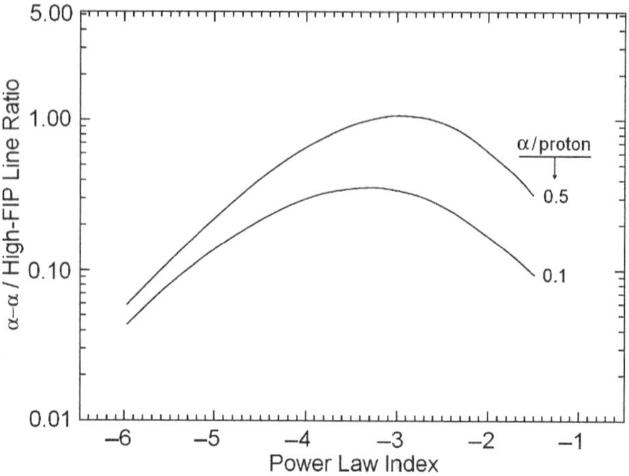

Fig. 11.26 Calculated α–α/high-FIP line ratio plotted versus the power-law spectral index for accelerated α/p ratios of 0.1 and 0.5 (From Share and Murphy 1997)

the expected α–α/high-FIP line ratios for the 19 flares. These calculated line ratios and their 1 σ uncertainties are listed in Table 11.6 for accelerated α/p ratios of 0.1 and 0.5. Share and Murphy (1997) plotted the difference between the observed and calculated ratios for the 19 flares in Panels b and c of Fig. 11.25 for α/p ratios of 0.1 and 0.5, respectively. The flare-to-flare spread in these differences is significantly decreased over the dispersion in the observed ratios. Probability that the spread is statistical increases to 2% for $\alpha/p = 0.1$ and to 17% for $\alpha/p = 0.5$ (this compares with 0.4% for the uncorrected ratios plotted in Fig. 11.20). An α/p ratio of 0.5 not only reduces the flare-to-flare variation to the statistical expectation but on average also predicts an α–α line yield closer to the observations. This is reflected in the weighted means of differences between the calculated and observed α–α/high-FIP ratios: for α/p of 0.5, the mean, 0.01 ± 0.06, is consistent with zero; for α/p of 0.1, the mean, 0.20 ± 0.046, is inconsistent with zero (probability 5×10^{-5}). The relatively large errors in measured line ratios prevent from making a strong statement about the constancy of the accelerated α/p ratio from flare to flare; however, there is not only one flare for which an α/p ratio of 0.1 is a significant improvement over a ratio of 0.5. The study of Share and Murphy (1997) suggests that the accelerated α/p ratio is relatively constant from flare to flare and that its value is closer to 0.5 than to 0.1.

11.3.5 Discussion on the Directionality of Accelerated α-Particles

Share and Murphy (1997) have studied emission of the α–α fusion lines in 19 solar flares observed by the Solar Maximum Mission Gamma Ray Spectrometer from 1980 to 1989. The line profiles are dependent on both the accelerated particle directionality and the position of the flare on the Sun. Only the simple cases of isotropic, downward-beam, and fan-beam geometries have been studied. Significant α–α line emission (>2 σ) was observed from eight of these flares. Four of these eight flares occurred at heliocentric angles smaller than 45°, where the downward-beam geometry can be distinguished from isotropic or fan-beam geometries. The isotropic/fan-beam geometries provide good fits to the α–α line profiles for the two flares emitting the highest fluencies in this line; in contrast, the downward-beam geometry can be ruled out at high confidence levels, 99.99% and 99.8% for these flares. The downward-beam geometry also provides poorer fits to the line profiles than do the isotropic/fan-beam geometries in the other two strong flares at angles less than 45°. There is no significant difference in the quality of the fits of the line profiles for the remaining 15 flares. A fan-beam or isotropic distribution for accelerated particles was also found to be consistent with observations of neutrons from two flares (Hua and Lingenfelter 1987d).

As noted Share and Murphy (1997), physical models of particle transport in magnetic loops that take into account pitch-angle scattering have been derived by Murphy et al. (1990a). A distribution with strong pitch-angle scattering is broader than the simple downward-beam function that was used in Share and Murphy (1997). On the basis of the experience with the downward beam, it may be possible to exclude a transport model with strong pitch-angle scattering by using SMM data for the two strong disk center flares. Such strong pitch-angle scattering has been inferred from Yohkoh observations of γ-ray emission from a disk flare near its peak emission on 1991 November 15 (Yoshimori et al. 1994). Share and Murphy (1997) have some concerns about these observations and the authors conclusions: (1) their fits are also consistent with an isotropic distribution; (2) the flare extended for at least 30 s, and yet only 4 s at the peak were used when the instrument suffered dead-time problems; and (3) the 4.4 MeV/α–α line ratio measured during this flare (Kotov et al. 1996) is more than an order of magnitude smaller than the mean value derived from the 19 SMM flares.

11.3.6 Discussion on the Accelerated α/p Ratio

As underlined Share and Murphy (1997), the fluencies in the α–α fusion lines in 19 flares observed by SMM are generally correlated with fluencies observed in the narrow nuclear de-excitation lines from proton and α-particle bombardment of high-FIP ambient elements. This suggests that the accelerated α/p ratio is relatively constant from flare to flare. Determination of the α/p ratio on a flare-to-flare basis requires fits to the entire γ-ray spectrum while varying the spectra and composition of accelerated particles and the ambient abundances (Murphy et al. 1991). Share and Murphy (1997) studied whether the fluencies in the α–α fusion lines are consistent with calculations for two α/p ratios, 0.1 and 0.5. Ratios in this range have been found by Ramaty et al. (1996) to be consistent with other line measurements for these flares (Share and Murphy 1995) under the following conditions: (1) the accelerated particles have a composition representative of impulsive solar energetic particles and (2) the particles interact in a thick target with ambient elemental abundances adopted for the corona (Reames 1995), except that the He/H ratio is 0.1 and the Ne/O ratio is 0.25.

Share and Murphy (1997) find that the observed α–α line yields are consistent with calculations for an accelerated α/p ratio of 0.5, and they are inconsistent (probability $<10^{-4}$) for a ratio of 0.1. This demonstrates that the high α/p ratio found by Murphy et al. (1990a) for the 1981 April 27 flare is generally characteristic of all 19 intense nuclear line flares observed by SMM. The observation of the 1991 June 4 flare by the Oriented Scintillation Spectrometer Experiment on the Compton Gamma Ray Observatory also measured an α/p ratio consistent with a value of 0.5 and inconsistent with a value as low as 0.1 (Murphy et al. 1997).

As noted Share and Murphy (1997), impulsive solar energetic particle data are available for only two of the 19 strong γ-ray line flares observed by SMM. Reames et al. (1990) report an α/p ratio of 0.016 ± 0.003 in the 1.9–2.8 MeV/amu energy interval from observations made by ISEE-3 of the 1981 April 10 flare. Because the γ-ray yield from this flare was relatively small, Share and Murphy (1997) could not significantly constrain the accelerated α/p ratio from their observations (see Fig. 11.25). No other impulsive particle events associated with the remaining 18 SMM flares were observed by ISEE. Solar energetic particles were well observed from the 1982 June 3 flare by detectors on Helios 1 (Van Hollebeke et al. 1990). The α/p ratio at $\sim$5 MeV/amu was $\sim$0.02 with a time of maximum analysis but appears much higher, $\sim$0.25, when the data are integrated over the entire event. Although Share and Murphy (1997) found no detectable α–α line emission from this flare, the data are still consistent with α/p ratios of up to 0.5 because of the steep accelerated particle spectrum (power-law indices steeper than about -5) inferred from other line measurements (Ramaty et al. 1996). The solar energetic particle spectra measured by Helios in space are much harder, with indices ranging from about -2.3 to about -1.2. This spectral difference may be explained by the fact that this flare exhibited two distinct acceleration phases, with the second, less intense, more extended phase being considerably harder (Ramaty et al. 1987).

The finding of Share and Murphy (1997), that the accelerated α/p ratio inferred from the γ-ray measurements generally appears closer to 0.5 than to 0.1 contrasts with ratios derived from interplanetary particle measurements, which typically range from $\sim$0.005 to as high as $\sim$0.3. It is not clear whether the γ-ray flares represent a specific class of events with a high α/p ratio or whether different acceleration processes produce particles impacting the Sun from those escaping the Sun. As noted Share and Murphy (1997), it must keep in mind, though, that the accelerated α/p ratio inferred from the γ-ray measurements is dependent on other assumed parameters used in the calculations. If, for example, the ambient He/H ratio were larger than 0.1, the accelerated α/p ratio would drop. Alternatively, current calculations assume that all the accelerated particles have the same spectral shape, a simple power law with the same spectral index for all particle species. Decoupling the proton and α-particle spectra could affect the value of the accelerated α/p ratio derived from the γ-ray line measurements.

As underlined Share and Murphy (1997), we do not know of any acceleration models that specifically predict high α/p ratios in flare-accelerated particles interacting at the Sun. On the other hand, Miller and Reames (1996) discuss a model based on stochastic acceleration by cascading Alfvén wave turbulence that preferentially accelerates heavy ions with the lowest ion cyclotron frequency. These calculations yield heavy-ion ratios that are similar to those observed in solar energetic particles. Whether it can also account for an accelerated α/p ratio of about 0.5 at the Sun needs to be determined.

11.4 Estimation of the Spectral Evolution of Energetic Protons in Solar Flares by Gamma-Ray Observations and Simulation

11.4.1 The Matter and Short History of the Problem

As Gan (1998) noted, the energy spectrum of accelerated protons in solar flares, assuming it is either a Bessel function or a power law, is usually deduced from the fluence ratio of the 2.223 MeV neutron capture line, which is of a delayed nature, to 4–7 MeV nuclear emissions. The spectral index obtained in this way is integration over time. So far almost no detailed information on how the energetic particle spectrum changes with time is available. To solve the problem on spectral evolution of energetic protons in solar flares, Gan (1998) developed a special method and applied it to observations of Gamma-Ray Spectrometer/Solar Maximum Mission flares. For three intense γ-ray line flares, with a temporal resolution of 16.384 s, Gan (1998) find for the first time that from the maximum (even from the rising phase) to the decay phase, the energy spectra of accelerated protons evolve from soft to hard. This gradual hardening with time, in contrast to the general soft-hard-soft variation of the spectra of energetic electrons, represents a new constraint for the acceleration mechanism in solar flares.

As noted Gan (1998), it is well known that the energy spectrum of hard X-rays presents a soft-hard-soft variation with time for the majority of flares, the so-called type B flares (for the definitions of types A, B, and C see Tanaka 1987). Kosugi et al. (1988) have statistically applied this classification to SMM flare observations. Assuming a thick-target model, it is generally believed that the electron spectrum that leads to a power-law X-ray spectrum is also a power law. Therefore, the temporal variation of the spectrum of accelerated electrons has been well established, but it currently has almost no detailed information on how the spectrum of interacting protons changes with time. Ramaty and Murphy (1987) summarized three different methods with which proton spectra can be determined from γ-ray and neutron observations. These are based on:

1. Φ(2.223 MeV)/Φ(4.7 MeV) fluence ratios
2. Neutron energy spectra
3. Φ(100 MeV)/Φ(4.1–6.4 MeV) fluence ratios

The first method is applied mostly to observations (note that ratios required fluencies). Because the 2.223 MeV line is a delayed line, the intensity at any point in time depends on the production of neutrons not only at that time but also during the preceding moments. Traditionally, one integrates the flux with time and gets the so-called fluence. The spectral index, therefore, is not deduced as a function of time.

By using the second method, Murphy and Ramaty (1984) found for two flares that a Bessel function spectrum provides an acceptable fit to the observations. But no indication of a temporal behavior has been given.

The application of the third method was described by Ramaty and Murphy (1987). They showed that the ratio Φ(100 MeV)/Φ(4.1–6.4 MeV) varies with time for the flare of 1982 June 3, implying that a single accelerated particle population with a time-independent energy spectrum cannot fit the data. Murphy et al. (1987) explained this by assuming two particle populations. The first population is identified with a stochastic acceleration with a constant spectral index $\alpha T = 0.043$. The second population is identified with a shock acceleration with a harder spectral index $S = 2.4$. The temporal variation of energetic particle spectra for different phases has then been proposed.

As noted Gan (1998), the problem on possible temporal variation of ion energy spectra in solar flares has recently been addressed by several authors. Rieger (1996) studied the flare of 1988 December 16. He distinguished three different peaks, and for each peak he calculated the fluencies. Then he obtained three spectral indices, corresponding to these three peaks. Ryan and McConnell (1996), Ryan et al. (1996), and Rank et al. (1996) analyzed the flare of 1991 June 11, observed with the Compton Gamma Ray Observatory (CGRO). They tried to divide the flare arbitrarily into two phases, an impulsive phase and an extended phase, and estimated the fluence of the 2.223 MeV line for each phase by assuming an e-folding decay of the impulsive phase. It was found that the spectrum of the extended phase is significantly harder than that of the impulsive phase. Mandzhavidze et al. (1996) studied the same flare. After looking at the relative variation of the 4.4 MeV line with the π^0-decay emission, they pointed out that there exist three episodes of acceleration. Within each episode the spectrum is constant, but different episodes have different spectra.

Gan (1998) underlined that the spectral evolution of energetic protons in solar flares is an important topic, but right now we know very little about it. If the spectrum can change with time, all the results based on a constant spectral index should be reconsidered.

11.4.2 The Method of the Spectrum Time Evolution Estimation

Considering that most neutron conversion to 2.223 MeV photons takes place within a thin layer in the photosphere (Hua and Lingenfelter 1987b), Gan (1998) neglect the dependence of the time profile of the 2.223 MeV line on the energy of the neutrons. Referring to Wang and Ramaty (1974) and Prince et al. (1983), the flux of the 2.223 MeV line resulting from neutron capture of protons can be written approximately as

$$I_{2.223}(t) = \frac{1}{4\pi r^2} \int_{t_0}^{t} \frac{1}{\tau} q_n(t') f_{2.223}(t') \exp(-(t-t')/\tau) dt', \tag{11.33}$$

where $q_n(t')$ is the instantaneous production of neutrons at time t'; $f_{2.223}(t')$ is the conversion factor from neutrons to 2.223 MeV photons; r is the distance between the Sun and the Earth; and τ is the e-folding decay time of the 2.223 MeV line,

$$\tau = \left(\frac{1}{\tau_H} + \frac{1}{\tau_{He}} + \frac{1}{\tau_d}\right)^{-1}, \tag{11.34}$$

where τ_H, τ_{He} and τ_d are, respectively, the time constants for neutron capture on ^{1}H, ^{3}He, and neutron decay, and have the forms (Prince et al. 1983)

$$\tau_H = \frac{1.4 \times 10^{19}}{n_H}, \quad \tau_{He} = \frac{8.5 \times 10^{14}}{(^3\text{He/H}) n_H}, \quad \tau_d = 917 \text{ s}, \tag{11.35}$$

where ^{3}He/H is the abundance of ^{3}He relative to H. The differentiating of Eq. 11.33 with respect to time, gives

$$q_n(t) = \frac{4\pi r^2 \tau}{f_{2.223}(t)} \left[\frac{dI_{2.223}(t)}{dt} + \frac{I_{2.223}(t)}{\tau}\right], \tag{11.36}$$

If

$$q_n(t) = N_p(t) Q_n \text{ and } 4\pi r^2 I_{4-7}(t) = N_p(t) Q_{4-7} f_{4-7}, \tag{11.37}$$

where $N_p(t)$ is the instantaneous number of protons with energy above 30 MeV impinging on the thick-target interaction region, and Q_n and Q_{4-7} are the yield of neutrons and of nuclear emissions within 4–7 MeV, which depends on the spectral index, then we have the following relationship:

$$\frac{Q_n}{Q_{4-7}} = \frac{f_{4-7}(t)\tau}{f_{2.223}(t) I_{4-7}(t)} \left[\frac{dI_{2.223}(t)}{dt} + \frac{I_{2.223}(t)}{\tau}\right], \tag{11.38}$$

Gan (1998) make the following assumptions: there is thick-target interaction; the hydrogen density where the neutron capture occurs is 1.3×10^{17} cm^{-3}, independent of the energy of the neutrons; the angular distribution of injected particles is isotropic. The relationship of $f_{2.223}$ with the heliocentric angle and the spectral index is taken from Hua and Lingenfelter (1987a). The yield ratio of 4–7 MeV photons to neutrons for the Bessel function of accelerated particles is taken from Ramaty (1986). For a power-law energy spectrum, this ratio is taken from the calculations of Murphy and Ramaty (1984); $f_{4\text{–}7}$ is taken to be 1.0, and ^{3}He/H is taken to be 2.0×10^{-5} (Hua and Lingenfelter 1987b) For a given flare, where the heliocentric angle is known, the relationship of $f_{2.223}$ with the spectral index is also known. Together with the yield ratio changing with the spectral index, it can be solve Eq. 11.38 and obtain the spectral index for each moment of time. Then the temporal evolution of energetic particle spectra can be presented.

11.4.3 Results on the Spectral Evolution

Gan (1998) applied the above method to the observations of GRS/SMM flares. Figures 11.27–11.31 show the results for the strong γ-ray line flares of 1984 April 24, 1988 December 16, and 1989 October 19. In each figure, the upper left-hand panel exhibits the light curve of neutron capture photons as well as that of nuclear emission from 4 to 7 MeV; the upper right-hand panel gives the calculated value of $Q_{4-7}/Q_n f_{2.223}$, by using Eq. 11.38; the two lower panels present both the temporal evolution of the Bessel function spectral index αT and power-law spectral index S. The error range of $Q_{4-7}/Q_n f_{2.223}$ depends not only on the measured error (±1σ) of the fluxes of the 2.223 MeV line and the 4–7 MeV emission, but also on the differential relations for $I_{2.223}$ and $I_{4\text{–}7}$.

11.4.4 Discussion on the Spectral Evolution

As noted Gan (1998), the flare of 1984 April 24, classified as X13/3B with a heliocentric angle of 45°, began at 23:59 UT and continued for more than 10 min in the γ-ray emissions. This flare, because it is the strongest γ-ray flare during cycle 21, has been studied extensively by many authors (see Chupp 1990 and references therein). It can be seen from Fig. 11.27 that before $t = 300$ s (the time for this flare starts from 23:55:52 UT on 1984 April 24), the value of $I_{2.223}$ is either weak or with a small varying structure, which results in $Q_{4-7}/Q_n f_{2.223}$ being either negative or with a big errors. After $t = 530$ s, $I_{4\text{–}7}$ becomes small, which make the error of the calculated spectral index large. Therefore, Fig. 11.27 shows the spectra only from $t = 300$ to 530 s. As it is well known, the nuclear emission within 4–7 MeV is instant relative to the original energy release. So the maximum time of the flare is around $t = 300$ s. This means that in Fig. 11.27 is shown the energy spectra beginning from the maximum until sometime of the decay phase, but not including the rising phase. The results in Fig. 11.27 tell us that the spectral index changes with time. On average it seems to evolve from soft to hard, regardless of the assumption of a Bessel function or a power-law distribution. The fact that some points deviate a bit far from the average tendency is probably due to insufficient time resolution of the observations.

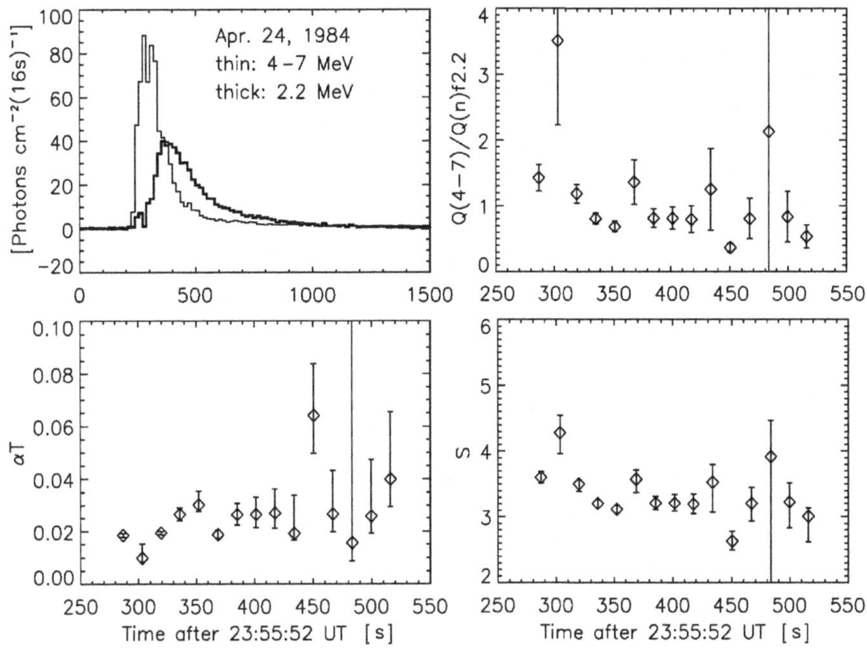

Fig. 11.27 Spectral variation with time (two lower panels) for the flare of 1984 April 24. The *upper left-hand panel* exhibits both light curves of the neutron capture photons and the 4–7 MeV nuclear emissions. The *upper right-hand panel* is the time variation of $Q_{4-7}/Q_n f_{2.223}$ (From Gan 1998)

As underlined Gan (1998), the most impressive results come from the flare of 1988 December 16, which, with a heliocentric angle of 44° and an importance of X4.7/1B, began at 8:29 UT and lasted at least 30 min. It obviously comprises three peaks. The first two peaks are close to each other in time, and the neutrons produced at the first peak may contribute to the second peak. But the third peak does not seem to be influenced by the first two peaks. Because of the relative independence of these peaks, Rieger (1996) studied different peaks and obtained spectral indices for each. Here Gan (1998) are interested in how the spectrum evolves within each peak. Figures 11.28 and 11.29 show the results for the first two peaks. Both begin from the peak to the decay phase. However, Fig. 11.30, corresponding to the results of the third peak, begins from the rising phase to the decay phase. It is the only case in which the rising phase can be well presented. In Figs. 11.28 and 11.30, may be seen clearly that the energy spectra, either in a Bessel function or in a power-law distribution, evolve nearly monotonously from soft to hard. The deviation of two points in the Figs. 11.28 and 11.30 from the averaged tendency may arise from the low time resolution of the observations. The spectral evolution for the second peak in Fig. 11.29 seems to be hard-soft-hard. But if consider the neutron contribution from the first peak to the second peak, the initial two hard points in Fig. 11.29 may be something more than a reflection of the second peak itself. From this point of

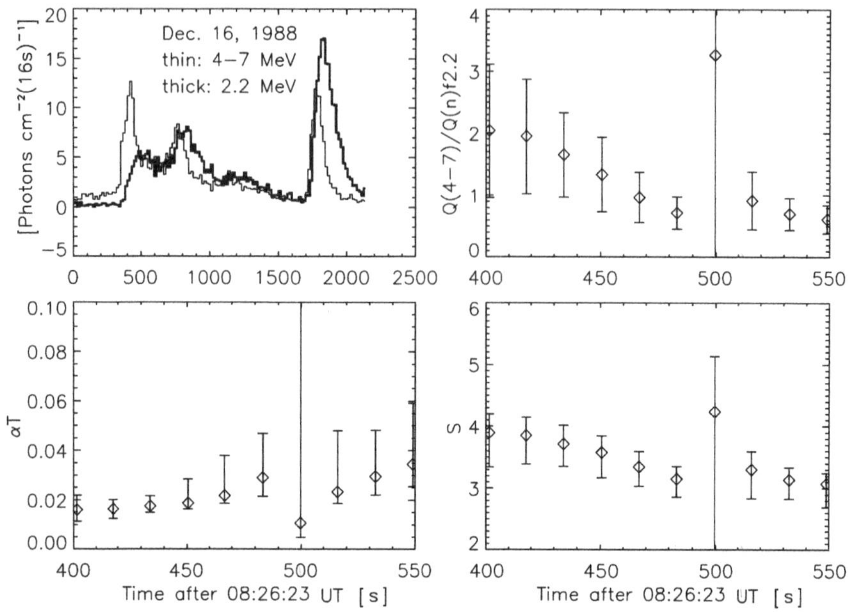

Fig. 11.28 Spectral variation with time (two lower panels) for the first peak of the flare on 1988 December 16. The *upper left-hand panel* exhibits both light curves of the neutron capture photons and the 4–7 MeV nuclear emissions. The *upper right-hand panel* is the time variation of $Q_{4-7}/Q_n f_{2.223}$ (From Gan 1998)

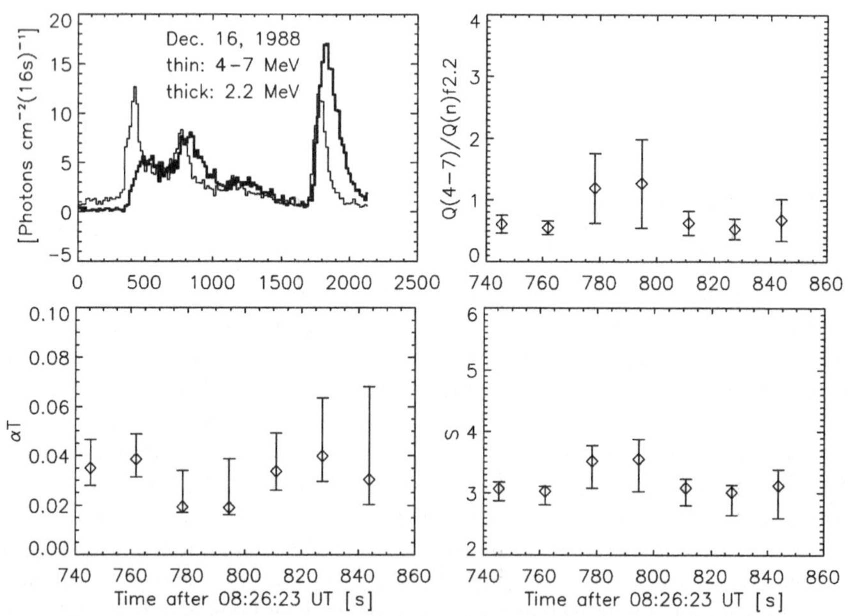

Fig. 11.29 The same as in Fig. 11.28, but for the second peak of the flare 1988 December 16 (From Gan 1998)

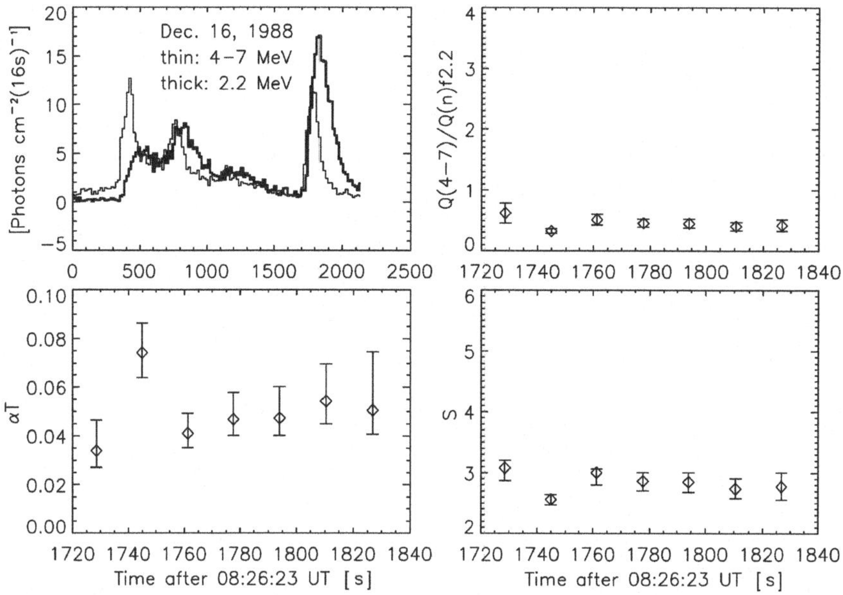

Fig. 11.30 The same as in Fig. 11.28, but for the third peak of the flare on 1988 December 16 (From Gan 1998)

view, Fig. 11.29 is not an exception for an individual burst whose spectrum evolves from soft to hard. Nevertheless, Figs. 11.28–11.30 provide us with the clearest picture of how the spectrum evolves from soft to hard. They are in fact the best results that can be obtained from the observations of the GRS using described method. Furthermore, it can be seen from Fig. 11.30 that αT can be larger than 0.05, which is out of the usual range $0.015 < \alpha T < 0.04$ (e.g., Lingenfelter 1994), demonstrating that the third peak itself has a very hard spectrum.

Gan (1998) noted that still another example is the flare of 1989 October 19, with a heliocentric angle of 29° and importance of X13/4B (see Fig. 11.31). This flare is in fact a very strong γ-ray line event. From Fig. 11.31 it can be seen that although the data fluctuate greatly, the general tendency is still obvious, that is, the spectrum evolves from soft to hard.

11.4.5 Summary of Main Results on Proton Spectrum Evolution

Gan (1998) have concentrated on the topic of temporal variation of proton spectra, which has never been dealt with before in such short time episodes. By developing a new method, Gan (1998) investigated a series of γ-ray line flares observed by GRS/SMM. Among these, Gan (1998) found that there are only three flares for which it can be obtained good results under a temporal resolution of 16.384 s. These three

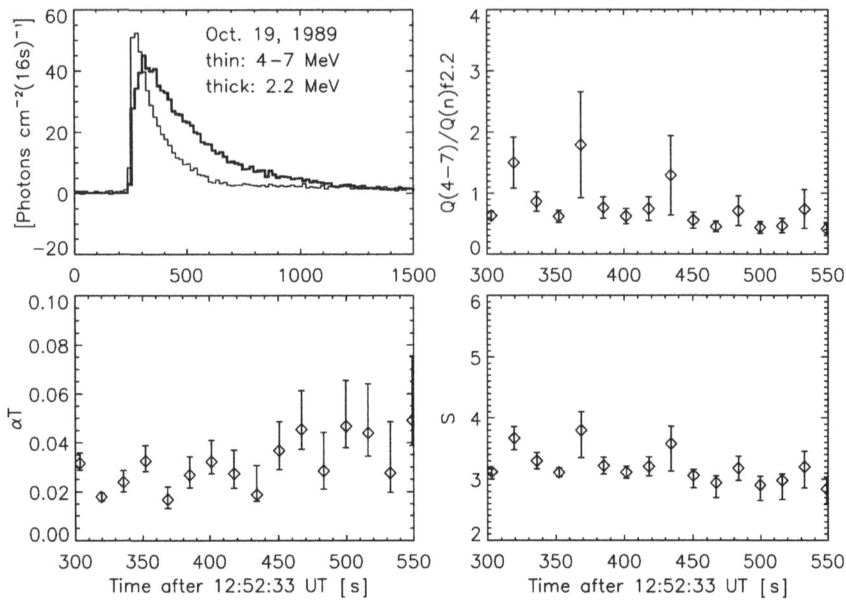

Fig. 11.31 Spectral variation with time (two lower panels) for the flare of 1989 October 19. The *upper left-hand panel* exhibits both light curves of the neutron capture photons and the 4–7 MeV nuclear emissions. The *upper right-hand panel* is the time variation of $Q_{4-7}/Q_n f_{2.223}$ (From Gan 1998)

flares are those of 1984 April 24, 1988 December 16, and 1989 October 19. Assuming a thick-target model, photospheric abundance, neutron capturing at $n_{\mathrm{H}} = 1.3 \times 10^{17}\ \mathrm{cm}^{-3}$, and isotropic injection of energetic particles, Gan (1998) show that the calculated spectra of the energetic protons for all three flares present a gradual hardening after the maximum. The third peak of the flare of 1988 December 16 presents further evidence that this hardening may begin from the rising phase to the decay phase. These results confirm and complete the recent findings mentioned in Section 11.4.1, but, for the first time, with a time resolution as high as 16.384 s. Any valuable acceleration mechanism should be able to explain these observations.

As underlined Gan (1998), unfortunately, the described method is not successful in deducing the energy spectra of relatively weak γ-ray line flares observed by GRS/SMM. Even for the strong γ-ray line flares observed by GRS, it is still hard to deduce reasonable spectra for the rising phase. Furthermore, there is a limitation to the method developed by Gan (1998), i.e., it assumes a fixed τ; simultaneously deducing both τ and the spectral index variation with time is still a problem. Therefore, the method developed in Gan (1998) is not a final version. On the one hand, Gan (1998) expect that future observations can be carried out with a higher temporal resolution and better sensitivity. On the other hand, alternative methods should be studied. Gan (1998) note that Lang et al. (1989) have proposed a method in which the ratio of the 4.43 MeV line of ^{12}C to the 15.10 MeV line of ^{12}C may have a good potential to yield spectral variations over time. Furthermore, Ramaty et al. (1996) tried to study the ratio of the 1.63 MeV line of ^{20}Ne to the 6.129 MeV line of ^{16}O. Gan (1998) hope that these methods can be brought to a practical stage in the next generation of observations by using Ge-spectrometers. At present, however, described above method may be more practical.

11.5 Gamma-Ray Measurement of Energetic Heavy Ions at the Sun

11.5.1 The Matter and Short History of the Problem

Share and Murphy (1999) have derived the γ-ray line spectra from accelerated heavy ions at the Sun in data from the Solar Maximum Mission (SMM) Gamma Ray Spectrometer and the Compton Gamma Ray Observatory (CGRO) OSSE instrument. These lines are Doppler-broadened and, perhaps shifted, reflecting the transport of the particles. They provide the only source of information on the composition of accelerated heavy ions at the Sun. Analysis of the integrated spectrum from 19 flares suggests that accelerated Fe is enhanced by about a factor of five over its ambient abundance.

Share and Murphy (1999) noted that Murphy et al. (1991) were the first used SMM γ-ray spectra from the 1981 April 27 flare to show that the composition of accelerated heavy ions resembles that observed in solar energetic particles (SEP) in ^{3}He-rich flares. Later Ramaty et al. (1997) analyzed spectra from the 1991 June 1 flare observed with Granat/PHEBUS and also found enhancements in accelerated heavy ions. In paper Share and Murphy (1999) are demonstrate a new technique applied to γ-ray spectra of solar flares that directly reveals the presence of accelerated heavy-nuclei through their Doppler-broadened lines. Lines from C and Fe are resolved and their shapes provide information on the directionality and transport of these particles at the flare site; their relative intensities provide information on the composition of accelerated ions.

11.5.2 Revealing Broad Lines in Gamma-Ray Spectra

Share and Murphy (1999) illustrate their technique for revealing the broadened γ-ray lines using the integrated spectrum plotted in Panel a in Fig. 11.32 of 19 nuclear-line flares observed by the SMM spectrometer (Share and Murphy 1995, 1998). This spectrum has been fitted with an incident photon model containing the sum of two power laws representing the bremsstrahlung continuum, 21 Gaussians representing narrow-resolved lines, and five Gaussians representing broadened lines and unresolved continuum (Murphy et al. 1990b). The narrow resolved lines include

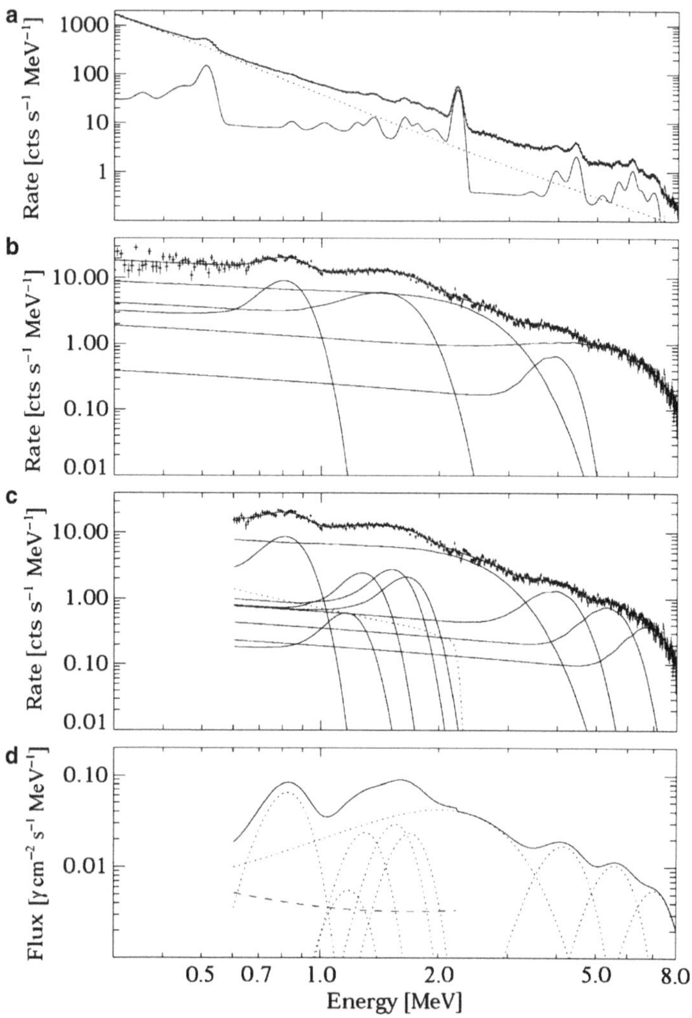

Fig. 11.32 *Panel* **a**: Summed spectrum from 19 flares observed by SMM; *Panel* **b**: spectrum with *narrow lines* and bremsstrahlung removed fit by five broad Gaussians; *Panel* **c**: same as *Panel* **b** but fit by several components from 0.6 to 8.0 MeV; *Panel* **d**: derived incident photon spectrum (From Share and Murphy 1999)

deexcitations from ambient nuclei excited by proton and ^{2}H-particle interactions, and the positron-annihilation and neutron-capture lines. The broadened lines are produced by the deexcitation of heavy accelerated ions (e.g. C, O, Mg, Fe, etc.) after interaction with ambient H and He; these lines are expected to be Doppler-broadened by $\sim$20% (FWHM) for ions having an isotropic distribution (Murphy 1985). The unresolved continuum is comprised of several lines from excited nuclei in both the ambient medium and accelerated particles that are too weak to be resolved individually. The solid line drawn through the data points shows the best fit. Share and Murphy (1999) also plot the fitted bremsstrahlung and summed contributions from the 21 narrow lines after modification by the instrumental response. In the next step Share and Murphy (1999) subtracts the bremsstrahlung and narrow line components from the total spectrum.

This residual count spectrum, plotted in Panel b in Fig. 11.32, is well behaved except at low energies where the strong bremsstrahlung contribution has been removed and near 2.2 MeV where the intense neutron-capture line has been subtracted. The latter deviations occur because the Gaussian is only an approximation to the actual line shape. The residual spectrum exhibits clear features that are not just artifacts of either the fit or the subtractions of the contributions from the narrow lines or power-law continuum.

11.5.3 Fitting to Broad-Line Spectra

In order to explore this residual spectrum in the least model-dependent manner, Share and Murphy (1999) fitted it with the five-Gaussians suggested by Murphy et al. (1990b), but it was allowed both the energies and the widths to be free. This fit is plotted through the data points in Panel b in Fig. 11.32 along with the individual broad Gaussians that are modified by the instrument response. The parameters derived from these fits are listed in Table 11.7 along with possible identifications. The spectrum reveals Doppler-broadened lines that are attributable to the first excited states of Fe (847 keV) and C (4.443 MeV), as well as composites from Fe, Mg, Ne, and Si in the 1–2 MeV range, and from O and N in the 5–7 MeV range.

As noted Share and Murphy (1999), the broadened lines due to accelerated Fe and C appear to be red-shifted relative to the energies found in the fits of the narrow lines from ambient Fe and C that have been excited by high-energy protons and ^{2}H-particles. These narrow-line energies are consistent with those observed at rest in the laboratory. The broadened Fe line appears red-shifted by ∼3% and the broadened C line by ∼9%. The Gaussian features listed in Table 11.7 at 1.5, 2 and 5.2 MeV are likely to be composites of several unresolved lines. Broadened deexcitation lines from Fe, Mg, Ne, and Si likely comprise the 1.5 MeV feature. The 2 MeV Gaussian is exceptionally broad and may be dominated by unresolved lines from several nuclei. Radiation from the 2.223 MeV neutron-capture line that is Compton-scattered in the photosphere may also contribute in this energy range. Vestrand (1990) calculated the amount of scattered radiation for various accelerated proton spectra and angular distributions, and found that the scattered radiation in the 1–2.2 MeV range amounts to ∼20–40% of the 2.223 MeV line flux. The wide Gaussian near 5.2 MeV is likely to be a composite of broadened, and perhaps red-shifted, 6.13 MeV and 6.919/7.028 MeV line features. With these considerations in mind we repeated the fit of the 19-flare spectrum with additional components.

Share and Murphy (1999) fit the 1.5 MeV complex with four lines fixed at a 5% red-shift from the laboratory energies of the Fe, Mg, Ne, and Si lines and with

Table 11.7 Fits to summed spectrum from 19 flares (From Share and Murphy 1999)

Energy (MeV)	Width(FWHM) (MeV)	Flux ($10^{-2}\gamma/cm^2$ s)	Possible identification (MeV)
Five broad lines			
0.819 ± 0.005	0.214 ± 0.016	1.58 ± 0.15	^{56}Fe (0.847)
1.515 ± 0.012	0.619 ± 0.039	3.62 ± 0.42	^{56}Fe (1.238); ^{24}Mg (1.369); ^{20}Ne (1.633); ^{28}Si (1.779)
1.979 ± 0.072	1.870 ± 0.094	8.09 ± 0.69	Unresolved lines; n-capture; ^{14}N (2.313); ^{16}O (2.75)
4.050 ± 0.034	0.791 ± 0.117	0.82 ± 0.16	^{12}C (4.439)
5.175 ± 0.174	3.712 ± 0.242	4.37 ± 0.40	^{16}O (6.13, 6.919); ^{14}N (7.028)
Nine broad lines and scattered n-capture			
0.825 ± 0.005	0.210 ± 0.023	1.42 ± 0.26	^{56}Fe (0.847)
1.180 ± 0.000	0.271 ± 0.000	0.17 ± 0.19	^{56}Fe (1.238)
1.300 ± 0.000	0.299 ± 0.000	0.77 ± 0.17	^{24}Mg (1.369)
1.550 ± 0.000	0.357 ± 0.000	1.10 ± 0.19	^{20}Ne (1.633)
1.690 ± 0.000	0.389 ± 0.000	0.98 ± 0.19	^{28}Si (1.779)
2.018 ± 0.098	1.909 ± 0.213	8.76 ± 1.14	Unresolved lines; ^{14}N (2.313); ^{16}O (2.75)
4.126 ± 0.040	1.150 ± 0.110	2.10 ± 0.26	^{12}C (4.439)
5.560 ± 0.063	1.330 ± 0.000	1.51 ± 0.07	^{16}O (6.13)
7.017 ± 0.056	1.610 ± 0.000	0.97 ± 0.05	^{16}O (6.919); ^{14}N (7.028)
		0.80 ± 0.20	Compton scattered n-capture

widths fixed at 23% (FWHM) of the line energies. They left the energy and width of the 2 MeV Gaussian free but added a function that reflected the shape expected from Compton scattered 2.2 MeV radiation. Share and Murphy (1999) also separated the single 5.2 MeV Gaussian into two Gaussians whose widths were fixed at 23% of their energies and whose energies were free to vary from the starting energies red-shifted by 5% from 6.1 and 7 MeV for the O and N lines. The fits were performed over a more limited range 0.6–8.0 MeV to closer reflect the range of the calculations performed by Vestrand (1990).

11.5.4 Summary of Main Results and Discussion: Enhanced Concentration of Fe in the Accelerated Particles

As noted Share and Murphy (1999), the results of described above expanded fits with nine Gaussians and the Compton scattered 2.223 MeV line are also given in Table 11.7 (uncertainty of '0' indicates fixed parameter) and plotted in Panel c in Fig. 11.32 and plot of the derived incident photon spectrum – in Panel d. The energy, width and intensity of the ~0.82 MeV Fe line did not change significantly from the original fit. The center energy is ~4σ below the laboratory energy for this deexcitation, representing a red-shift of 2.6 ± 0.6%. Such a shift may result from the higher probability for nuclei moving away to interact when they encounter higher solar densities. The line is broadened (FWHM) by 25.5 ± 2.8%, corresponding to an ion kinetic energy of 8.6 ± 2.1 MeV/nucleon. The summed flux in the four Fe, Mg, Ne, and Si lines is consistent with that found for the composite 1.5 MeV Gaussian in the original fit. It is difficult to draw firm conclusions about the relative fluxes of these lines in this range because they are not resolved. It does appear that the higher-energy Fe line is at least a factor of three weaker than the lower-energy (~0.82 MeV) line; this is consistent with the branching ratio found in the laboratory for 7 MeV protons incident on Fe (Lachkar et al. 1974). The center energy of the ~4.1 MeV Gaussian is ~7σ below the laboratory energy for C and represents a red-shift of 7.1 ± 0.9%, somewhat greater than found for the Fe feature. This might be attributable to the greater range of C ions relative to Fe. It is broadened by 27.9 ± 2.7% (FWHM), corresponding to an ion kinetic energy of 10.5 ± 2.1 MeV/nucleon. As can be seen in Table 11.7, the intensity and width of the ~4.1 MeV line is dependent on whether it will restrict the width of the 5–7 MeV feature. In the described expanded fit were used two Gaussians with fixed widths to separate the contributions from expected lines >5 MeV. If the Gaussian at 5.56 MeV is due to the 6.13 MeV ^{16}O line, then its red-shift is ~9%. In order to make more definitive statements about possible Doppler broadening and red-shifts in all the lines it is best to perform these studies on individual flares.

As underlined Share and Murphy (1999), it can be obtained a preliminary estimate of the mean composition of the accelerated particles at the Sun by comparing the fluxes observed in the narrow and broadened 0.847 and 4.439 MeV lines that are attributed to Fe and C. The narrow lines are due to proton and ^{2}H-particle interactions with ambient Fe and C while the broad lines are due to interactions of accelerated Fe and C with ambient H and He. The flux in the broad Fe line is ~70% of that observed in the broad C line, while the flux in the narrow Fe line is ~15% of that observed in the narrow C line. This suggests that, on average, Fe is enhanced by a factor of about 5 in the flare-accelerated particles. This is consistent with the ~6.7 Fe/C enhancement found in impulsive SEP events relative to the coronal abundance (Reames 1999). Share and Murphy (1999) noted that the described estimate is imprecise, however, because it was not considered the effects of spectral index, energy loss, and accelerated ^{2}H/proton ratio.

As Share and Murphy (1999) underlined, one of the important uncertainties in this analysis is the contribution from unresolved nuclear lines. Ramaty et al. (1975) made an estimate of this component based on some approximations. The fits of Share and Murphy (1999) suggest that the primary contribution may be in the form of a broad feature centered near 2 MeV. These fits also provided an estimate of the flux of 2.223 MeV γ-rays scattered by the photosphere into the 0.6–2.2 MeV energy range. This flux is only ~10% that found in the narrow neutron capture line. This ratio is about a factor of 2–4 below that calculated by Vestrand (1990) for photons scattered into the 1–2.2 MeV range, suggesting that the capture line may not be produced as deep in the photosphere.

Share and Murphy (1999) plan to use this technique to compare the accelerated particle abundance using CGRO/OSSE data with SEP data from ACE. The broad-line spectrum derived from OSSE for the 1991 June 4 solar flare is shown in

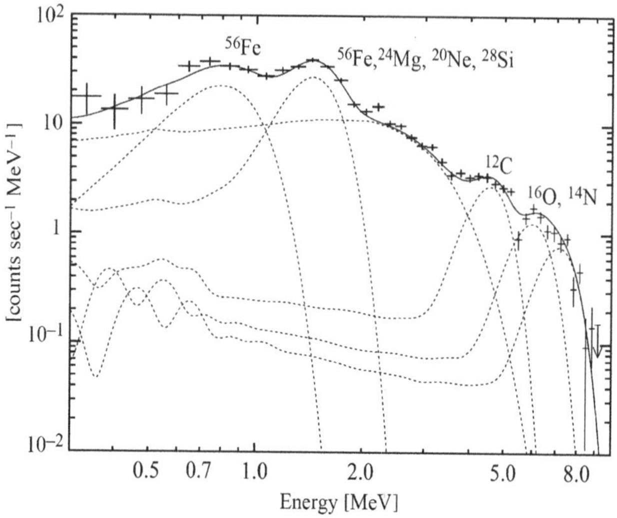

Fig. 11.33 OSSE 1991 June 4 flare spectrum with *narrow lines* and bremsstrahlung removed (From Share and Murphy 1999)

Fig. 11.33. It is similar to what was found for the SMM 19-flare sum and again suggests an enhanced concentration of Fe in the accelerated particles.

11.6 Estimation of the Ratio of Interaction to Interplanetary Energetic Protons by Gamma-Rays: Diverging Magnetic Field Lines and Stochastic Acceleration

11.6.1 The Matter and Short History of the Problem

Kocharov and Kovaltsov (1999), Kocharov et al. (1999a, b) considered how tilt and divergence of magnetic field lines in solar corona and stochastic acceleration affect on distribution functions of interacting and interplanetary energetic protons. It is shown that signatures of magnetic field geometry may be found in anisotropy of secondary high-energy neutrons and γ-rays, in ratio of numbers of interplanetary to interacting protons, in proton energy spectra, in the characteristic decay times of γ-ray and neutron emissions, and in a spatial distribution of the secondary emission sources on solar disc.

As noted Kocharov and Kovaltsov (1999), in the solar cycle 22, observations of high-energy γ-ray and neutron flares suggest more isotropic production of high-energy neutral emissions than it was expected earlier (Mandzhavidze and Ramaty 1993). One possible explanation of these observations is that the high-energy emissions are produced in complex magnetic structures with varying configurations from flare to flare (Chupp et al. 1993; Trottet et al. 1993).

11.6.2 The Role of the Tilt of Magnetic Field Lines in the Solar Corona on the Ratio of Interaction to Interplanetary Energetic Protons

The importance of strongly tilted magnetic fields at the flare site has been demonstrated by Kocharov et al. (1997). They started with analysis of magnetic structures at sites of two flares responsible for >100 MeV neutron events. Based on these observations, a model of neutron production was considered. The model takes into account the observed large tilt of magnetic field lines at foot-points of flare magnetic loops. Results of the new calculations have been compared with both previous calculations and observations. The tilt of magnetic field lines at the flare site is proved to be the most important parameter limiting anisotropy of high-energy secondary emission in solar flares.

According to Kocharov and Kovaltsov (1999), it can be concluded that: (1) the tilt of magnetic field lines at foot-points of flare loops is a key parameter for theoretical calculations of anisotropy of high-energy neutron and γ-ray emissions. This parameter was not taken into account in previous calculations. (2) Large magnetic tilt angles actually observed imply much more isotropic emission than that expected in earlier studies. (3) High resolution magnetograms and stereoscopic observations of γ-ray and neutron emissions are desirable for a more accurate determination of parameters of accelerated ions at the Sun.

11.6.3 The Role of Divergence of Closed Magnetic Field Lines in the Solar Corona on the Ratio of Interaction to Interplanetary Energetic Protons

As noted Kocharov and Kovaltsov (1999), possible divergence of magnetic field lines in coronal portion of a magnetic loop may affect trapping and acceleration of solar energetic particles. Analytic steady-state solutions to the focused diffusion equation have been used to deduce proton trapping time τ_{trap}, and an average residence time τ_{res} for all particles injected in the magnetic loop (Kocharov et al. (1999a, b). They take into account effect of MHD turbulence and divergence of magnetic field lines in both corona and chromosphere of the Sun. The numerical simulations of pitch-angle scattering have been also used to check analytic solutions and to ascertain boundary conditions to the focused diffusion equation. Results have been obtained for several different functional forms of $B(\zeta)$, the magnetic field as a function of distance along a particular coronal field line. Five cases have been studied, from B being constant along the coronal portion of the loop to $B(\zeta)$ corresponding to a force-free magnetic structure. The results indicate that divergence of magnetic field in coronal portion of the loop can significantly increase the trapping time, no matter how small mean free path may be. Derived analytic expressions for τ_{trap} should be used for calculation of intensity of secondary emissions from the loop-top sources.

Analytic time-dependent solutions to the focused diffusion equation have been considered in the case of constant coronal B to find the basic decay time of the trapped proton number, τ_{dec}, an asymptotic value of the exponential decay time when time tends to infinity. In general case, time τ_{dec} does not coincide with the first characteristic time τ_{trap}. In the case of variable B in the coronal portion of the loop, Monte Carlo simulations of pitch-angle scattering have been employed to calculate τ_{dec} in a wide range of parameters. Kocharov and Kovaltsov (1999) have also obtained analytic expressions for how the characteristic time scales depend from parameters of the magnetic loop. Deduced expressions for τ_{dec} should be used for calculations of ion acceleration in the escape-time approximation and for the interpretation of the decay phase of solar γ-ray flares. Magnetic focusing force in the coronal portion of the loop makes acceleration more efficient than would be expected in the no-coronal-focusing approximation.

11.6.4 The Model

Kocharov and Kovaltsov (1999) and Kocharov et al. (1999a, b) consider ion acceleration region which is permeated by a diverging static ambient magnetic field $B(\zeta)$, where ζ is a spatial coordinate along the magnetic field line extending from the solar photosphere to the interplanetary medium (Fig. 11.34). The length of the acceleration region along B is L. Small-amplitude Alfvén turbulence is generated by an unspecified mechanism inside the acceleration region. Particle scattering at Alfvén waves results in particle diffusion and acceleration described by a Fokker–Planck equation.

According to Kocharov and Kovaltsov (1999) and Kocharov et al. (1999b), in the case of isotropic wave turbulence and nearly isotropic particle distribution the Fokker–Planck equation takes the form:

$$\frac{\partial n}{\partial t} = \frac{1}{p^2}\frac{\partial}{\partial p}p^2\kappa_p\frac{\partial n}{\partial p} + \frac{\partial}{\partial \zeta}\left(\kappa_\zeta\frac{\partial n}{\partial \zeta} - \frac{\kappa_\zeta}{L_B}n\right) + Q(\zeta, p, t), \tag{11.39}$$

where $n(\zeta, p, t)$ is the number of particles per unit of the magnetic line length and per unit volume in momentum space, $L_B = -B/(dB/d\zeta)$ is the focusing length. The term with the focusing length employs magnetic field geometry and comes from the consideration of particle transport along arbitrary guiding field configuration (Earl 1981). As noted Kocharov et al. (1999b), this portion of the transport

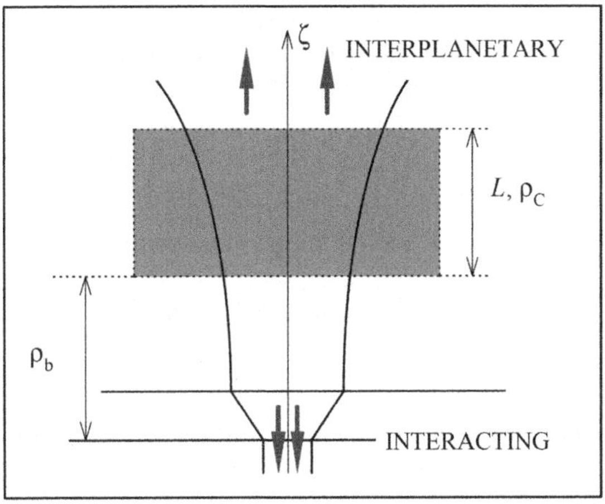

Fig. 11.34 Illustration of the acceleration region (*filled gray*) where interacting and interplanetary protons may be concurrently produced; $\rho_C = \exp \eta$ (From Kocharov and Kovaltsov 1999; Kocharov et al. 1999)

operator has been never incorporated into a model of stochastic acceleration of solar energetic particles.

Kocharov et al. (1999b) consider nonrelativistic particles and define p as an ion momentum per nucleon. In the case of weak Alfvén turbulence, the diffusion coefficient in the momentum space, κ_p, and the diffusion coefficient along the guiding magnetic field, κ_ζ, can be written as (e.g., Schlickeiser 1989):

$$\kappa_p = \kappa_{0p}(Z/A)^{2-s} p^{s-1} \Delta_p(\zeta), \quad \kappa_\zeta = \kappa_{0\zeta}(Z/A)^{s-2} p^{3-s} \Delta_\zeta(\zeta), \tag{11.40}$$

where $s < 2$ is a spectral index of the turbulence, functions $\Lambda_p(\zeta)$ and $\Delta_\zeta(\zeta)$ account for the spatial dependencies of acceleration and diffusion, respectively, and

$$\Delta_p(\zeta) = \frac{v_A^2 (\delta B/B)^2 K_{\min}^{s-1} B^{2-s}}{\left[v_A^2 (\delta B/B)^2 K_{\min}^{s-1} B^{2-s}\right]|_{\zeta=0}}, \quad \Delta_\zeta(\zeta) = \frac{v_A^2}{v_A^2|_{\zeta=0}} \frac{1}{\Delta_p(\zeta)}. \tag{11.41}$$

Kocharov et al. (1999b) adopt the Kraichnan phenomenology with $s = 3/2$, and also suggest that the Alfvén speed, v_A, the relative energy density of turbulence, $(\delta B/B)^2$, and a minimum wave number in the power-law spectrum of turbulence, $K_{\min}$, are constant throughout the acceleration region, and the injection into the acceleration is proportional to the ambient plasma density. In what follows, the ion charge number Z and the mass number A are set $Z = A = 1$.

According to Kocharov et al. (1999b), Eq. 11.39 can be easily solved in the case of zero boundary conditions corresponding to a free escape at the acceleration region boundaries,

$$n|_{\zeta=0} = n|_{\zeta=L} = 0. \tag{11.42}$$

Kocharov et al. (1999b) also suggest that all particles are instantly injected with some small momentum $p = p_i$, i.e. in Eq. 11.39

$$Q(\zeta, p, t) = \frac{Q_0(\zeta)}{4\pi p^2}\delta(p - p_i)\delta(t), \tag{11.43}$$

And there is no sink in the momentum space

$$\frac{\partial n}{\partial p}|_{p=p_i-} \ni, \tag{11.44}$$

where the boundary momentum, $p_{\min} = p_i - \ni$, is taken to be small enough to have almost no effect on the accelerated particle spectrum at high energies.

11.6.5 Analytic Time-Integrated (Steady State) Solutions

According to Kocharov et al. (1999b), integrating Eq. 11.39 and conditions described by Eqs. 11.42–11.44 over the time, yields for

$$\frac{1}{p^2}\frac{\partial}{\partial p}\left[\kappa_{0p}p^{s+1}\Delta_p(\zeta)\frac{\partial N}{\partial p}\right] + \frac{\partial}{\partial \zeta}\left[\kappa_{0\zeta}p^{3-s}\Delta_\zeta(\zeta)\left(\frac{\partial N}{\partial \zeta} - \frac{N}{L_B}\right)\right] = 0, \tag{11.45}$$

where $N = N(p, \zeta) = \int n(p, \zeta, t)dt$ is a time integrated number of particles per unit of magnetic field line length and per unit volume in the momentum space; $p > p_i$.

As noted Kocharov et al. (1999b), the time-integrated particle density, N, dictates the time-integrated flux of escaping particles, which is an experimentally measurable function. Another significance of the function $N(p, \zeta)$ is that it is proportional to the steady state solution of Eq. 11.39, because Eq. 11.45 coincides with Eq. 11.39 if $\partial n/\partial t = 0$. Boundary conditions corresponding to conditions described by Eqs. 11.42–11.44 can be cast into the form

$$N|_{\zeta=0} = N|_{\zeta=L} = 0, -4\pi\kappa_{0p}\Delta_p(\zeta)\left(p^{s+1}\frac{\partial N}{\partial p}\right)|_{p=p_i} = Q_0(\zeta), \tag{11.46}$$

where it was suggested that $\ni \to 0$.

According to Kocharov et al. (1999b), variables can be divided so that a solution to the problem of Eqs. 11.45 and 11.46) may be written in the form

$$N(p, \zeta) = \sum_{k=1}^{\infty} A_k\varphi_k(p)\psi_k(\zeta). \tag{11.47}$$

Eigenfunctions, $\psi_k(\zeta)$, and eigenvalues, λ_k, should be found by solving the problem

$$\frac{1}{\Delta_p(\zeta)}\frac{d}{d\zeta}\left[\Delta_\zeta(\zeta)\left(\frac{d\psi_k}{d\zeta}-\frac{\psi_k}{L_B}\right)\right]=-\lambda_k\psi_k, \tag{11.48}$$

With $\psi_k(0)=\psi_k(L)=0,\ \ \lambda_k>0$. Then one can solve the equation for $\varphi_k(p)$:

$$\frac{1}{p^{5-s}}\frac{d}{dp}\left(p^{s+1}\frac{d\varphi_k}{dp}\right)=-\lambda_k\frac{\kappa_{0\zeta}}{\kappa_{0p}}\varphi_k. \tag{11.49}$$

This equation can be recast in a form of the Bessel equation, resulting in the familiar solution

$$\varphi_k(p)=p^{-s/2}K_\nu\left(\frac{1}{3-s}p^{3-s}\sqrt{\lambda_k\frac{\kappa_{0\zeta}}{\kappa_{0p}}}\right), \tag{11.50}$$

where $K_\nu(x)$ is the modified Bessel function of the order $\nu=s/(2(3-s))$. In the basic case of $s=3/2$, the Bessel function may be expressed in terms of elementary functions, so that

$$\varphi_k(p)=p^{-3/2}\exp\left(-\frac{2}{3}p^{3/2}\sqrt{\lambda_k\frac{\kappa_{0\zeta}}{\kappa_{0p}}}\right). \tag{11.51}$$

As noted Kocharov et al. (1999b), amplitudes A_k in the expansion Eq. 11.47 should be found using an actual source function, $Q_0(\zeta)$. Recall that the differential energy spectrum of particles escaping at the ends of the acceleration region ($\zeta=0$ and $\zeta=L$) is proportional to the particle flux:

$$I_{0[L]}(E)=+[-]\left(\kappa_\zeta(p)\frac{\partial N}{\partial\zeta}\right)|_{\zeta=0[L]}4\pi p^2\frac{dp}{dE}. \tag{11.52}$$

The focusing term, $\kappa_\zeta(p)N/L_B$, does not contribute to the boundary flux, because $N|_{\zeta=0}=N|_{\zeta=L}=0$.

11.6.6 Analytic Solutions for Case of Exponential Magnetic Field

Kocharov et al. (1999b) use for the analysis an exponential magnetic field model:

$$B(\zeta)=B_0\exp(-\zeta/h_B)\equiv B_0\exp(-\eta(\zeta/L)), \tag{11.53}$$

where there are introduced a magnetic field scale parameter, $\eta=L/h_B$. In the case of $s=3/2$, $v_A=$ constant, $(\delta B/B)^2=$ constant, and a constant basic scale of the turbulence, one can see from Eq. 11.41 that $\kappa_p \propto B^{1/2}$ and $\kappa_\zeta \propto B^{-1/2}$. Thus it will be $\Delta_p(\zeta) = \exp(-\eta\zeta/(2L))$ and $\Delta_\zeta(\zeta) = \exp(\eta\zeta/(2L))$. Equation 11.48 then reads

$$\exp\left(\frac{\eta\zeta}{2L}\right)\frac{d}{d\zeta}\left[\exp\left(\frac{\eta\zeta}{2L}\right)\left(\frac{d\psi_k}{d\zeta}-\frac{\eta}{L}\psi_k\right)\right] = -\lambda_k\psi_k, \tag{11.54}$$

and $\psi_k(0) = \psi_k(L) = 0$. With substitutions,

$$z = \frac{2L\sqrt{\lambda_k}}{\eta}\exp\left(-\frac{\eta\zeta}{2L}\right) \text{ and } \Psi_k = \psi_k\exp\left(-\frac{\eta\zeta}{4L}\right), \tag{11.55}$$

Equation 11.54 can be cast into the form of the Bessel equation. Using zero boundary conditions, one can finally obtain eigenvalues of the problem

$$\sqrt{\lambda_k} = \frac{\gamma_k}{L}\frac{\eta/2}{1-\exp(-\eta/2)}, \tag{11.56}$$

where γ_k are positive roots of the equation

$$\tan\gamma = \frac{\gamma}{1+a\gamma^2}, \quad a = \frac{\exp(\eta/2)}{(\exp(\eta/2)-1)^2}. \tag{11.57}$$

The smallest positive root, γ_1 is situated between π and $3\pi/2$. Eigenfunctions also can be expressed in terms of elementary functions:

$$\psi_k(\zeta) = \frac{1}{z^2}\left[(z_k - z)\cos(z_k - z) - (1 + zz_k)\sin(z_k - z)\right], \tag{11.58}$$

where $z = z(\zeta)$ is given by Eq. 11.55, and $z_k = (2L/\eta)(\lambda_k)^{1/2}$.

Kocharov et al. (1999b) introduce a characteristic energy of accelerated protons (mass m_p):

$$E_1 = \frac{1}{2m_p}\left(\frac{\kappa_{0p}}{\kappa_{0\zeta}}\left(L\frac{1-\exp(-\eta/2)}{\eta/2}\right)^2\right)^{2/3}. \tag{11.59}$$

As noted Kocharov et al. (1999b), it is convenient to express E_1 also in terms of the mean free path of 1 MeV proton at the bottom of acceleration region, $\Lambda_0 = 3\kappa_\zeta/v|_{\zeta=0,E=1\,\mathrm{MeV}}$:

$$E_1 = \left[\frac{L}{\Lambda_0}\frac{1-\exp(-\eta/2)}{\eta/2}\frac{v_A}{1.2\times10^9\,\mathrm{cm/s}}\right]^{4/3}\mathrm{MeV}. \tag{11.60}$$

The high-energy asymptote of the escaping particle spectrum comes from Eqs. 11.47, 11.51, and 11.52, with only the $k = 1$ term included. The high-energy spectrum as a function of the dimensionless energy, $\varepsilon = E/E_1$, is in the form

$$I_0(E) \propto I_L(E) \propto \sqrt{\varepsilon}\exp\left(-\frac{2\gamma_1}{3}\varepsilon^{3/4}\right). \tag{11.61}$$

This spectrum is shown among others in Fig. 11.35. Kocharov et al. (1999) note that in the uniform magnetic field, $\eta \to 0$, and it will be $\gamma_1 = \pi$ and $E_1 = (1/2m_p)\left(L^2\kappa_{0p}/\kappa_{0\zeta}\right)^{2/3}$. Regarding the spectrum shape, an effective length of the acceleration region may be introduced:

$$L_{\text{eff}} = L\frac{\pi}{\gamma_1}\frac{1-\exp(-\eta/2)}{\eta/2}, \tag{11.62}$$

so that the same high-energy spectra (the same E_1) can be produced in the exponential magnetic region of length L and in the uniform magnetic field region of length L_{eff}. However, the effect of the diverging magnetic field on the interplanetary to interacting proton ratio is still unavoidable. To deduce the high-energy ratio of the numbers of interacting to interplanetary particles, Γ, one has to relate

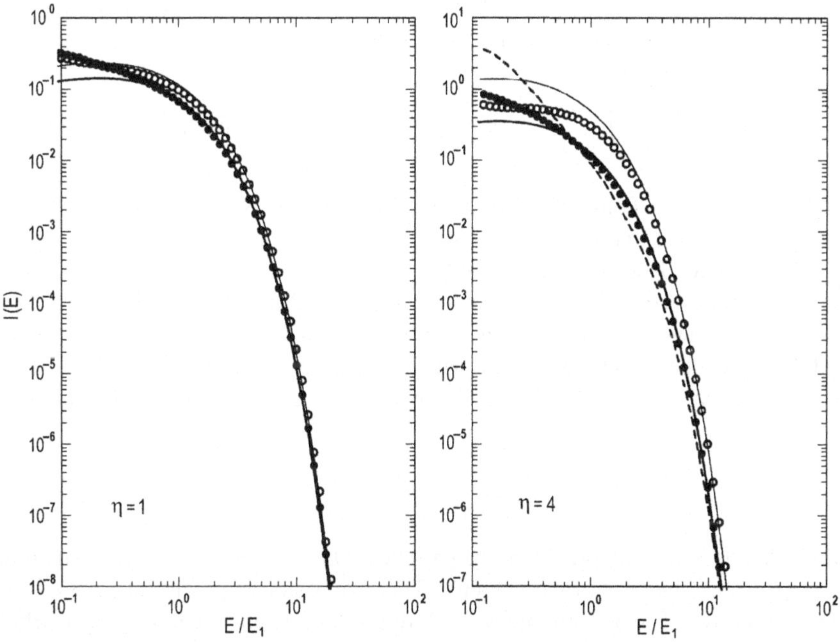

Fig. 11.35 Energy spectra of protons accelerated at the weak divergence (*left-hand panel*, $\eta = 1$) and at the strong divergence (*right-hand panel*, $\eta = 4$) of magnetic field lines inside the acceleration region. Points represent the results of Monte Carlo simulations at $\rho_b = 1$ for interacting (*filled circles*) and interplanetary (*open circles*) protons. Corresponding *heavy* and *light lines* are for the escape time approximation, so that both curves are of the same shape given by Eq. 11.61, while the ratio of normalization factors is dictated by Eq. 11.63. The *dashed line* in the *right-hand panel* illustrates the interacting proton spectrum calculated with the Monte Carlo method in the case of injection localized in a thin layer situated at $\zeta = 0.02L$. All other spectra have been calculated for injection being proportional to the plasma number density as explained in the appendix (From Kocharov et al. 1999b)

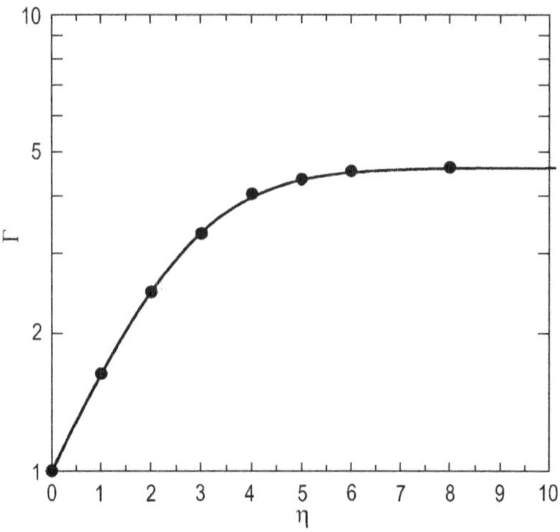

Fig. 11.36 High-energy ratio of the numbers of interplanetary to interacting protons versus the magnetic-divergence parameter, $\eta = \ln \rho_c$. Points are for the Monte Carlo simulations at $\rho_b = 1$. The *curve* represents an exact analytic solution (Eq. 11.63) (From Kocharov et al. 1999b)

escaping particle spectra (Eq. 11.52) corresponding to the basic eigenfunction, ψ_1. By this expedient, Kocharov et al. (1999b) obtained

$$\Gamma = \left(\frac{I_L(E)}{I_0(E)}\right)_{E\to\infty} = -\cos\gamma_1 - \frac{\gamma_1}{1-\exp(-\eta/2)}\sin\gamma_1, \tag{11.63}$$

where $\gamma_1 = \gamma_1(\eta)$ is defined by Eq. 11.57. This function is shown with a solid line in Fig. 11.36. Note that the limit $E \to \infty$ practically means $E >> E_1$.

As underlined Kocharov et al. (1999b), Eqs. 11.56–11.63 have been obtained at a zero boundary condition $n|_{\zeta=0} = 0$ (see Eqs. 11.42 and 11.46). However, there may be a magnetic field convergence between the base of the acceleration region and the top of the interaction region with magnetic mirror ratio $\rho_b > 1$ (Fig. 11.34). In this case, one must replace $n|_{\zeta=0} = 0$ with the condition of partial precipitation

$$\kappa_\zeta \frac{\partial N}{\partial \zeta} - \frac{\kappa_\zeta}{L_B} N = KN, \tag{11.64}$$

where KN is a flux of precipitating particles. One can adopt the approximation $K = v/(4\rho_b)$, v being particle velocity. This expression is precise at $\rho_b >> 1$ but may be in error of ~20% around $\rho_b = 2$, depending on an actual angular distribution of precipitating particles (Kocharov et al. 1999a).

11.6.7 *Main Results of Numerical Monte Carlo Simulations*

Numerically calculated by Kocharov et al. (1999b) proton energy spectra at $\rho_b = 1$ are shown in Fig. 11.35 along with the analytic spectra (Eq. 11.61) plotted for a comparison. A good agreement with the high-energy asymptote (Eq. 11.61) is seen already at $\varepsilon = E/E_1 \geq 3$. The analytic spectrum (Eq. 11.61) corresponds to the main term, $k = 1$, in the expansion (Eq. 11.47). This spectrum could also be obtained in the escape time approximation.

As underlined Kocharov et al. (1999b), at low energies, the deduced total interacting particle spectrum is essentially steeper than the escape time approximation spectrum (Eq. 11.61) because of a contribution of the $k > 1$ terms. Those terms depend on a spatial dependence of the source function, $Q(\zeta)$. A more concentrated injection near the region floor would result in an even steeper spectrum of interacting particles. As an illustration, Kocharov et al. (1999b) have calculated a spectrum corresponding to the proton injection in a thin layer situated close to the region floor, at $\zeta = 0.02L$ (see dashed line in the right-hand panel of Fig. 11.35). Figure 11.37 illustrates how the ratio of the numbers of interplanetary to interacting protons depends on energy.

As can be seen from Fig. 11.37, the ratio $\Gamma(>E) = I_L(>E)/I_0(>E)$ increases with energy, indicating a harder spectra of interplanetary protons in comparison with spectra of protons interacting at the Sun. That is a property of models suggesting a majority of injection placed close to the Sun.

For different values of the magnetic scale parameter, $\eta = L/h_B$, Kocharov et al. (1999b) have also calculated the high-energy interplanetary to interacting proton ratio, Γ (see points in Fig. 11.36). The analytic curve (Eq. 11.63) is also plotted in the Fig. 11.36. It is seen that the MC and analytic calculations coincide well, revealing a saturation at the value of $\Gamma = 4.6$ when $\eta \geq 5$. The latter corresponds

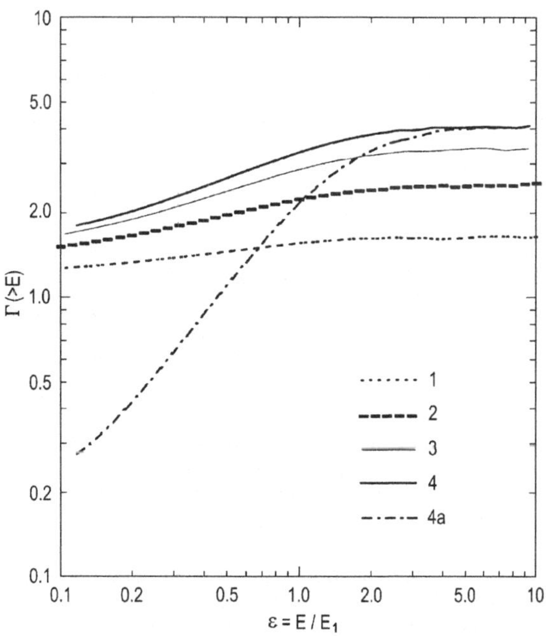

Fig. 11.37 Ratio of the integral energy spectra of interplanetary to interacting protons, $\Gamma(>E) = I_L(>E)/I_0(>E)$ versus energy. All *curves* and *points* represent results of MC simulations at $\rho_b = 1$. Parameter at the curves is the ratio of the acceleration region length to the exponential magnetic field scale, $\eta = L/h_B = \ln \rho_C$. In case 4a, particles have been injected in a thin layer at $\zeta = 0.02L$, $\eta = 4$. All other curves correspond to an exponential injection described in Appendix A1 (From Kocharov et al. 1999b)

to the 'magnetic depth' of the acceleration region $\rho_C \geq 100$. It may also be said that the value 4.6 corresponds to the infinite acceleration region extending into the interplanetary medium.

In an additional set of MC computations by Kocharov et al. (1999b), the effect of a magnetic line convergence beneath the acceleration region has been studied. Figure 11.38 gives a calculated ratio of the numbers of interplanetary to interacting protons with energy $E > 4E_1$. A parameter for the curves is an acceleration region magnetic scale parameter η (see Eq. 11.53).

In Fig. 11.38, the value Γ is shown as a function of the mirror ratio ρ_b at a fixed value of the L/Λ ratio. However, in the case of $\rho_b > 1$, the ratio Γ depends only on a combined parameter $L/(\Lambda\rho_b)$ (Eq. A1.2). For this reason, a value of Γ corresponding to a new ratio $(L/\Lambda)'$ can be picked up from the same Fig. 11.38, but at another magnetic mirror ratio $\rho_b' = \rho_b(L/\Lambda)'/(L/\Lambda)$. It is seen from Fig. 11.38 that convergence of magnetic field lines beneath the acceleration region has no effect on the interplanetary to interacting proton ratio if $\rho_b < 10 \times L/(100\Lambda)$.

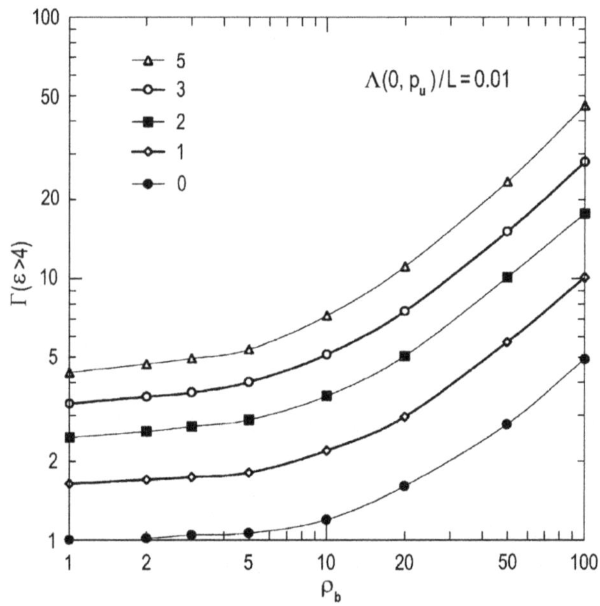

Fig. 11.38 The ratio of interplanetary to interacting protons, $\Gamma(\varepsilon > 4) = I_L(E > 4E_1)/I_0(E > 4E_1)$, versus magnetic mirror ratio between the floor of acceleration region and the top of interaction region (ρ_b in Fig. 11.34). All *points* and *interpolating curves* represent results of MC simulations. Parameter at the curves is the ratio of the acceleration region length to the exponential magnetic field scale, $\eta = L/h_B = \ln \rho_C$. The mean free path $\Lambda(0, p_u)$ is given at the acceleration region floor, $\zeta = 0$, for a momentum $p = p_u$, the momentum value p_u being used as a unit momentum for numerical simulations. A correspondence between the unit momentum p_u and a characteristic energy of accelerated particles, E_1, is given by Eq. A1.3 (see Appendix A1) (From Kocharov and Kovaltsov 1999; Kocharov et al. 1999b)

11.6.8 Discussion and Conclusions

Kocharov et al. (1999b) have considered a model of ion acceleration on open magnetic field lines. An average solar magnetic field decreases with altitude, and a corresponding divergence of magnetic field lines results. As underlined Kocharov et al. (1999b), that affects Γ, shifting it to higher values. The effect of magnetic divergence is stronger at a larger magnetic depth of the acceleration region, $\rho_C = \exp(\eta)$. The effect, however, is limited by a factor $\Gamma_{max} = 4.6$ (Fig. 11.36). This result has been obtained under the suggestion that particles can freely escape from the base of the acceleration region to the top of the interaction region. The latter is true if a magnetic mirroring beneath the acceleration region floor is not extremely strong. Note that an understanding of the term 'interaction region' depends on the actual energy of interacting protons. For instance, an effective proton energy for production of 4–7 MeV nuclear de-excitation γ-ray lines is about 30 MeV. Deceleration depth at this energy is about 2×10^{23} cm^{-2}, which corresponds to the low chromosphere layers.

As underlined Kocharov et al. (1999b), the effect of particle focusing in a diverging magnetic field does not disappear even when the particle mean free path, Λ, tends to zero, because both terms in the transport operator (Eq. 11.39) are equally proportional to Λ. A role of magnetic focusing inside the acceleration region may be illustrated with a simple estimation. The conservation of the first adiabatic invariant results in an increase of the velocity component parallel to the magnetic field: $dv_{\|}/dt = v_{\perp}^2/(2L_B)$. Over one length of the mean free path, particles get an average parallel velocity, $U = \left(\langle v_{\perp}^2 \rangle / v\right)\left(\Lambda/(2L_B)\right)$, because of the focusing. Hence the term $\kappa_{\zeta}/L_B = \Lambda v/(3L_B)$ in Eq. 11.39 is recognized to be a bulk velocity of particles caused by focusing. This velocity is of order of diffusion velocity, because the magnetic field scale and the density scale are of the same order of magnitude. However, in contrast to the diffusion velocity, magnetic focusing drags particles only toward the interplanetary medium, increasing by this means the interplanetary to interacting particle ratio. Note that the focusing-driven flux should also be incorporated into a diffusion shock acceleration model, if shock propagates along the diverging magnetic lines away from the Sun (see Section 11.7).

As noted Kocharov et al. (1999b), the floor of the acceleration region may be situated above the chromosphere in the solar corona at 'magnetic elevation' $\rho_b > 1$ (Fig. 11.34), i.e., the accelerated region may be separated with a magnetic mirror from the interaction region. It might be suggested in this case that mirroring at any $\rho_b >> 1$ strongly affects the interplanetary to interacting proton ratio. The latter, however, is not exactly correct, because it is seen from the boundary condition (Eqs. A1.1 and A1.2) and from results of numerical simulations (Fig. 11.38) that only $\rho_b \geq L/\Lambda >> 1$ can strongly affect spectra and numbers of interplanetary and interacting particles. To estimate possible values of L/Λ, Kocharov et al. (1999b) first estimate the ratio of characteristic acceleration and transport times,

$$\frac{\tau_p}{\tau_\zeta} \approx \left(\frac{v}{v_A}\right)^2 \left(\frac{\Lambda}{L}\right)^2, \qquad (11.65)$$

and then equate that to unity for 20 MeV protons. This implies at typical coronal conditions that $L/\Lambda \sim 100$. In other words, a characteristic energy of accelerated particles $E_1 \sim 20$ MeV (Eq. 11.60) can be achieved at $L/\Lambda \sim 100$. Hence if the turbulence is strong enough to accelerate protons to above 10 MeV, only a very large magnetic elevation of the acceleration region, $\rho_b \geq 100$, can significantly reduce precipitation and correspondingly increase the ratio of the numbers of interplanetary to interacting protons. This is because during the acceleration, particles encounter the magnetic mirror many times, so that they finally precipitate into the loss cone with a high probability.

According to Kocharov et al. (1999b), for a comparison of the model predictions with observations, one has to deduce from experimental data a number of particles interacting at the Sun and a number of particles concurrently injected into the interplanetary medium. Such an analysis has been done for the 1990 May 24 solar flare (Kocharov et al. 1996). That was an impulsive flare (the soft X-ray e-folding time less than 10 min), but one that was much more powerful in production of high energy particles than a typical impulsive solar flare. The flare started with a short impulse of X- and γ-ray emissions. A simultaneous impulsive production of high-energy neutrons was responsible for a fast rise of the Climax neutron monitor count rate. This impulse of nuclear interactions was followed by a more prolonged production of neutrons also detected by the Mexico City neutron monitor. Concurrent injection of interplanetary protons during about 40 min was responsible for the first peak in the GOES proton count rate. The number ratio for those concurrently produced interacting and interplanetary protons is estimated to be $\Gamma \sim 1$–2 (Kocharov et al. 1996). One can see from Fig. 11.38 that the value $\Gamma = 2$ corresponds to the magnetic ratio between the top of the interaction region and the top of the acceleration region $\rho_t = \rho_b \rho_C = \rho_b \exp\eta \approx 20 - 30$. On the other hand, Lee et al. (1994) estimated the magnetic field at the top of the microwave source in the flare, $B_t \approx 5$ G, and an angular extension of the microwave source, $\approx 300''$. If that value is adopted to be a magnetic field at the top of the proton acceleration region, the interaction region magnetic field is estimated to be less or about 100–150 G. Hence a flare-brightening site and spots are not dominant precipitation sites after the impulsive phase of the flare. However, precipitation at a relatively weak magnetic field is very possible in the case of the acceleration region stretching through $\geq 2 \times 10^{10}$ cm. Accelerated protons can precipitate at sites of a relatively weak photospheric magnetic field, $\sim$ 2–100 G, and the ratio of the numbers of interplanetary to interacting protons may be $\sim$1–5, even if the stochastic acceleration region is situated relatively high in the solar corona, at altitudes $\sim (0.3$–$2)R_{sun}$. Hence considered model can be applied to acceleration at an intermediate time and spatial scales. At a shorter timescale, less than 10 min, stochastic acceleration on closed magnetic field lines may dominate in production of interacting particles, while at a longer scale, greater than 100 min, a coronal mass ejection-driven shock acceleration may be the main source of interplanetary particles. Note that both a stochastic acceleration and acceleration on open magnetic field lines have been proposed in the interplanetary versus interacting proton study by Ramaty et al. (1993).

Kocharov et al. (1999b) came to the following conclusions.

1. It was developed the first theoretical calculations of the ratio of the numbers of interplanetary to interacting protons, Γ.
2. An acceleration on open magnetic field lines has been considered with a view to explain relatively high values of the interplanetary to interacting proton ratio, $\Gamma \geq 1$.
3. It was conclude that a divergence of magnetic field lines inside the acceleration region is the basic parameter affecting the ratio Γ.
4. Deduced values of Γ in the exponential model of stochastic acceleration at $\rho_b \leq L/(10\Lambda)$ vary between 1 and $\Gamma_{\max} \approx 5$, depending on the magnetic depth of the acceleration region $\rho_C = \exp \eta$ (Figs. 11.34 and 11.36).
5. It would be equally incorrect to suggest that Γ is always equal to unity if particles are accelerated on open coronal magnetic field lines, or to neglect magnetic focusing inside the acceleration region arguing that $\Lambda << L_B$, or to suggest that a magnetic mirror $\rho_b \sim 10$ beneath the acceleration region can cause a strong increase in Γ.
6. However, $\rho_b \geq 100$ could raise Γ well beyond the limiting value $\Gamma_{\max}$.
7. Spectra of low-energy interacting protons, being accelerated in the exponential atmosphere, are essentially steeper than would be expected based on the escape time approximation, and steeper than corresponding proton spectra in the interplanetary medium.

11.7 Estimation of the Ratio of Interaction to Interplanetary Energetic Protons by Gamma-Rays: Diverging Magnetic Field Lines and Parallel Shock Wave Acceleration

11.7.1 The Matter and Short History of the Problem

Vainio et al. (2000) present a test-particle model of diffusive shock acceleration on open coronal field lines based on one-dimensional diffusion-convection equation with finite upstream and downstream diffusion regions. They calculate the energy spectrum of protons escaping into the interplanetary space and that of protons interacting with the subcoronal material producing observable secondary emissions. The model can account for the observed power-law and broken power-law energy spectra as well as the values of the order of unity for the ratio of the interplanetary to interacting protons. Vainio et al. (2000) compare this model to Monte Carlo simulations of parallel shock acceleration including the effects of the diverging magnetic field.

As noted Vainio et al. (2000), protons accelerated on/near the Sun can be directly measured in the interplanetary medium, or alternatively their properties may be deduced from measurements of secondary emissions, neutrons and γ-rays, produced during nuclear interactions at the Sun (see Ramaty et al. 1993 for a review). The ratio, Γ, of the interplanetary proton number to the number of protons producing secondary emission at the Sun (interacting protons) is an important result of recent solar observations. This ratio varies in a very wide range, from about 0.01 to >1, being typically in order of unity if a post-impulsive-phase acceleration is present. As it was described in Section 11.6, Kocharov and Kovaltsov (1999), Kocharov et al. (1999a, b) have studied the spectra of interacting and interplanetary protons in a model, where the particles are accelerated stochastically on open, diverging coronal magnetic field lines with no bulk motion of the plasma. They found out that the ratio of interplanetary to interacting protons is quite generally between 1 and 4.6 and that it varies with energy so that the spectrum of interacting particles is steeper than the spectrum in interplanetary space.

Vainio et al. (2000) study the spectrum of interplanetary ions and the parameter Γ in an alternative acceleration model, parallel shock acceleration. Diffusive shock acceleration (Krymsky 1977; Axford et al. 1977; Bell 1978; Blandford and Ostriker 1978; see detail review in Chapter 4 in Dorman M2006) in coronal or/and interplanetary shocks provides the majority of interplanetary protons in large solar energetic particle (SEP) events according to many authors (Kahler 1993, 1994; Reames 1993). It is therefore important to study whether this acceleration mechanism can really account for the SEP spectrum, but equally important is to study the spectrum of protons precipitating in subcoronal regions behind the shock wave. Since shock acceleration involves solar-frame bulk motion of the plasma, it may be questioned whether particles leaving the shock in the downstream region ever reach the subcoronal regions where they should interact. Also, since there has to be turbulence in front of the shock to hinder particle escape if any acceleration of particles is to occur, its need to study how the spectrum of the escaping particles is related to the spectrum at the shock.

To answer these questions, Vainio et al. (2000) perform an analytical study of one-dimensional parallel shock acceleration to provide a well established background for numerical work on more involved geometries. As a second part of the study, Vainio et al. (2000) perform Monte Carlo simulations of particle acceleration in diverging field line geometry to verify the analytical results, to check the limits for their applications, and to provide a tool for further studies in more involved and realistic models of coronal/interplanetary shocks. In this stage, both the analytical and the numerical calculations are done using test-particle approximation.

The aim of Vainio et al. (2000) is to apply and further develop the ideas in the studies of Ellison and Ramaty (1985), Lee and Fisk (1982), and Lee and Ryan (1986). The first was a study of diffusive shock acceleration in flare site conditions, and the second and the third were studies of coronal shock acceleration in a global blast wave of infinite strength propagating from solar corona to interplanetary space. Thus, the studies were aimed at explaining observations in impulsive-phase and gradual-phase timescales, respectively. The second aim is to emphasize that interacting particle spectrum should be computed along with the interplanetary one, to give a possibility to use a richer variety of observational data to judge for the correctness of the models. In addition, Vainio et al. (2000) present a flexible numerical method that may be used in complicated situations that are beyond the scope of analytical modeling.

11.7.2 Description of the Model

Vainio et al. (2000) assume that the acceleration of energetic particles is due to the first-order Fermi acceleration at a parallel shock. The shock is propagating with a constant speed V_s into a medium at rest with an exponentially decreasing magnetic field, $B(\zeta) = B_0 \exp(-\zeta/L_B)$, where ζ is the coordinate measured along the magnetic field lines (Fig. 11.39) and $L_B = -B/(\partial B/\partial\zeta)$ is the (constant) scale length of the magnetic field.

Vainio et al. (2000) take the particles to be scattered by circularly polarized Alfvén waves propagating along the magnetic field lines on both sides of the

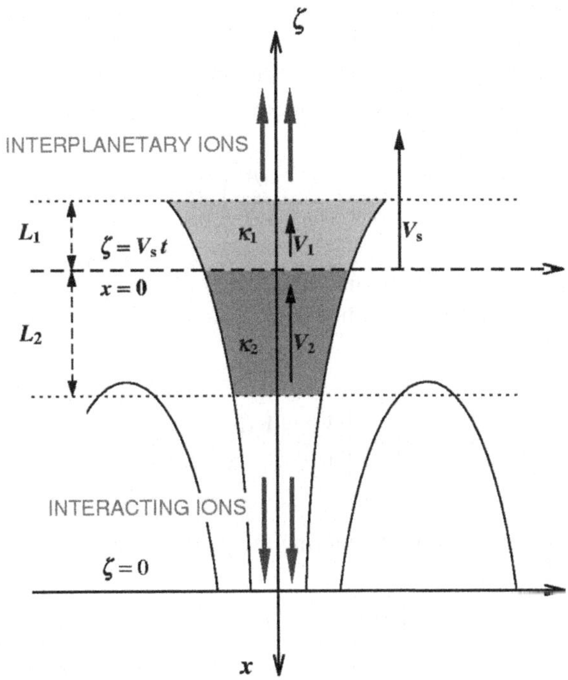

Fig. 11.39 Geometry of the acceleration region: a shock wave propagating parallel to the open magnetic field line emits accelerated particles both toward the Sun and toward the interplanetary medium (From Vainio et al. 2000)

shock in a region that has spatial extent of L_1 upstream and L_2 downstream of the shock. In the upstream region, the waves are taken to propagate outward from the Sun in the plasma frame. This is consistent with waves generated at the Sun and with waves generated by the accelerated particles themselves through the streaming instability. The super-Alfvénic flow then carries the waves to and through the shock to the downstream region. As a result of the interaction of the waves and the shock compression, the downstream turbulence always consists of waves propagating in both directions along the field. One can compute the intensities of both wave modes from given upstream conditions and it turns out that the outward propagating wave component is dominating in intensity so that stochastic acceleration in the downstream region may be neglected (Vainio and Schlickeiser 1998, 1999).

In Vainio et al. (2000), the upstream Alfvén speed V_{A1} is taken to be constant, which implies an exponentially decreasing plasma density, $n_i(\zeta) = n_{0i}\exp(-2\zeta/L_B)$, where the subscript i =1 or 2 refers to the upstream or downstream region of the shock, and the compression ratio of the shock, $\rho = n_{02}/n_{01}$ is also a constant.

Vainio et al. (2000) consider a system where the waves scattering the particles have their wave vectors directed along the mean magnetic field. Therefore, there is no resonant cross-field diffusion and particles remain on their initial magnetic field lines forever. A bundle of neighboring field lines, thus, defines a flux tube. If the particle scattering off the waves is strong enough to keep the particle distribution close to isotropic, it may describe the linear particle density $f(\zeta,p,t) = d^2N/d\zeta dp$ in a flux tube, i.e., the number of particles per unit length of the magnetic field line and unit momentum, with a Fokker–Planck equation

$$\frac{\partial f}{\partial t} + \frac{\partial}{\partial \zeta}\left(V + \frac{\kappa}{L_B}\right)f - \frac{\partial}{\partial p}\left(\frac{V}{L_B} + \frac{\partial V}{\partial \zeta}\right)\frac{p}{3}f = \frac{\partial}{\partial \zeta}\kappa\frac{\partial f}{\partial \zeta} + S, \tag{11.66}$$

where p is the particle momentum, $S(\zeta,p,t) = Q(t)\delta(\zeta - V_s t)\delta(p - p_{inj})$ is the source function, $Q(t)$ gives the number of injected particles to the considered flux tube at the shock per unit time,

$$V(\zeta,t) = \begin{cases} V_1 = V_{A1}, & \zeta > V_s t, \\ V_2, & \zeta < V_s t, \end{cases} \tag{11.67}$$

is the phase speed of the waves scattering the particles, and

$$\kappa(\zeta,p,t) = \begin{cases} \kappa_1(p), & V_s t + L_1(p,t) > \zeta > V_s t, \\ \kappa_2(p), & V_s t - L_2(p,t) < \zeta < V_s t, \end{cases} \tag{11.68}$$

is the spatial diffusion coefficient of the particles related to the particle mean free path λ by $\kappa = \lambda v/3$. In the regions $\zeta > V_s t + L_1$ and $\zeta < V_s t + L_2$ the particles are taken to stream adiabatically without scattering. If the spatial coordinate is changed to $x = V_s t - \zeta$, it may write down Eq. 11.66 in the form

$$\frac{\partial f}{\partial t} + \frac{\partial}{\partial x}\left(U + \frac{\kappa}{L_B}\right)f - \left(\frac{\partial U}{\partial x} + \frac{V_s - U}{L_B}\right)\frac{\partial}{\partial p}\frac{p}{3}f = \frac{\partial}{\partial x}\left(\kappa\frac{\partial f}{\partial x}\right) + S, \tag{11.69}$$

where $S(x,p,t) = Q(t)\delta(x)\delta(p - p_{inj})$,

$$U(x) = V_s - V(x) = \begin{cases} V_s - V_1 = U_1, & x < 0, \\ U_2 = U_1/\rho_c, & x > 0, \end{cases} \tag{11.70}$$

$\rho_c = U_1/U_2$ is the scattering-center compression ratio, which can be calculated if the upstream plasma properties are known (Vainio and Schlickeiser 1998, 1999), and

$$\kappa(x,p,t) = \begin{cases} \kappa_1(p), & 0 > x > -L_1(p,t), \\ \kappa_2(p), & 0 < x < L_2(p,t). \end{cases} \tag{11.71}$$

Vainio et al. (2000) require the particle distribution function, $f(x,p,t)/(4\pi p^2 A(x,t))$, where $A(x,t) \propto 1/B(x,t)$ is the flux tube cross-sectional area, to be continuous at $x = 0$. For a parallel shock wave, this means that $f(x,p,t)$ has to be continuous at the shock. In addition, Eq. 11.69 has to be completed with appropriate boundary conditions at $x = -L_1,\ L_2$. At the outer boundary, it may be assume that the particles escape, $f(-L_1,p,t) = 0$, but the boundary condition at $x = L_2$ is, in general, more involved because of the possible mirroring of particles before they reach the solar surface.

As noted Vainio et al. (2000), if the positions of the boundaries $L_1,\ L_2$ are time independent, it may be integrate Eq. 11.69 and the boundary conditions over time to get

$$\frac{\partial}{\partial x}\left(U - \frac{\kappa}{L_B}\right)F - \left(\frac{\partial U}{\partial x} + \frac{V_s - U}{L_B}\right)\frac{\partial}{\partial p}\frac{p}{3}F = \frac{\partial}{\partial x}\left(\kappa\frac{\partial F}{\partial x}\right) + S, \tag{11.72}$$

where $S(x,p) = N_0\delta(x)\delta(p - p_{inj})$, N_0 is the total number of injected particles to the considered flux tube, and

$$F(x,p) = \int_0^\infty f(x,p,t)dt. \tag{11.73}$$

Here it was assumed that $f(x,p,t) \to 0$ as $t \to 0$, which is reasonable if the injection will tend to zero rapidly enough with time since the particles will escape at the outer boundary and be swept out of the system to the far downstream by the moving scattering centers.

11.7.3 Analytical Solution

According to Vainio et al. (2000), in case of non-diverging magnetic field lines ($L_B \to \infty$), Eq. 11.72 can be solved analytically. In this case the adiabatic motion of the particles outside the shock is free streaming, and the particles escape from both boundaries: $F(-L_1, p) = F(+L_2, p) = 0$. Then, the solution is (see, e.g., Ostrowski and Schlickeiser 1996)

$$F(x,p) = \begin{cases} F_0(p)\dfrac{\exp(\beta_1 x) - \exp(-\beta_1 L_1)}{1 - \exp(-\beta_1 L_1)}, & 0 > x > -L_1, \\ F_0(p)\dfrac{\exp(\beta_2 L_2) - \exp(-\beta_2 x)}{\exp(\beta_2 L_2) - 1}, & 0 < x < L_2, \end{cases} \tag{11.74}$$

where $\beta_{1[2]}^{-1} = \kappa_{1[2]}/U_{1[2]}$ is the upstream [downstream] diffusion length and $F_0(p) = F(0,p)$ satisfies the equation

$$(U_1 - U_2)\frac{p}{3}\frac{d}{dp}\frac{F_0}{p^2} + \left(\frac{U_1}{1 - \exp(-\beta_1 L_1)} - \frac{U_2 \exp(-\beta_2 L_2)}{1 - \exp(-\beta_2 L_2)}\right)\frac{F_0}{p^2} = \frac{N_0 \delta(p - p_{inj})}{p^2} \tag{11.75}$$

that can be obtained by dividing Eq. 11.72 by p^2 and integrating it from $x = 0-$ to $0+$. This equation has the solution

$$F_0(p) = CH(p - p_{inj})\left(\frac{p}{p_{inj}}\right)^{\gamma - 2} \times \exp\left\{-\gamma \int_{p_{inj}}^{p} \left[\frac{\exp(-\eta_1(p'))}{1 - \exp(-\eta_1(p'))} - \frac{1}{\rho_c}\frac{\exp(-\eta_2(p'))}{1 - \exp(-\eta_2(p'))}\right]\frac{dp'}{p'}\right\}, \tag{11.76}$$

where $H(p - p_{inj})$ is the Heaviside function and

$$C = \frac{3N_0}{p_{inj}(U_1 - U_2)}, \qquad \gamma = \frac{3\rho_c}{\rho_c - 1}. \tag{11.77}$$

Thus, if the numbers of diffusion lengths, $\eta_i \equiv \beta_i L_i$ $(i = 1,\ 2)$, in the up- and downstream regions are large, $\eta_i \to \infty$, the shock spectrum approaches the canonical power law, where the spectral index γ is fully determined by the scattering-center compression ratio at the shock. Also, since it is reasonable to assume that $\eta_2 >> \eta_1$

at all moments, it may be simplify Eq. 11.76 somewhat by neglecting the second term of the integrand to get

$$F_0(p) = CH(p - p_{inj}) \left(\frac{p}{p_{inj}}\right)^{\gamma-2} \times \exp\left[-\gamma \int_{p_{inj}}^{p} \frac{1}{\exp(\eta_1(p')) - 1} \frac{dp'}{p'}\right]. \quad (11.78)$$

The differential energy spectrum of particles escaping at the boundaries is

$$\frac{dN_{1[2]}}{dE} = +[-]\frac{dp}{dE}\left|\left(\kappa\frac{\partial F}{\partial x}\right)\right|_{x=-L_1[+L_2]}. \quad (11.79)$$

Using Eqs. 11.74 through 11.79, the spectrum of interplanetary particles is obtained as

$$\frac{dN_1}{dE} = \frac{dp}{dE}\frac{U_1 F_0(p)}{\exp(\eta_1(p)) - 1} = \frac{\gamma N_0}{p_{inj} v(\exp(\eta_1(p)) - 1)} \left(\frac{p_{inj}}{p}\right)^{\gamma-2}$$
$$\times \exp\left[-\gamma \int_{p_{inj}}^{p} \frac{1}{\exp(\eta_1(p')) - 1} \frac{dp'}{p'}\right]. \quad (11.80)$$

Vainio et al. (2000) note, however, that a fraction of particles escaping toward the Sun will not precipitate, because the shock is an outward moving source and, thus, the particle, although leaving the downstream region of the shock, may still propagate outward from the Sun. This fraction depends on the angular distribution of the particles, but for an isotropic distribution it would be of the order of V_s^2/v^2, where v is the particle speed. Thus, at high energies, this fraction may be neglected and the ratio of interplanetary to interacting protons is

$$\Gamma(E) = \frac{dN_1/dE}{dN_2/dE} = \frac{\rho_c}{\exp(\eta_1(p)) - 1}. \quad (11.81)$$

One notes immediately that the ratio of interplanetary to interacting protons is somewhat more involved in the case of shock acceleration than in the case of stochastic acceleration, where it is unity if $L_B \to \infty$.

Vainio et al. (2000) tried to estimate the validity of the obtained solution, when L_B is finite. First, one has to be able to neglect the streaming speed caused by focusing, κ_i/L_B, in comparison with the scattering-center speeds, U_i. Second, the inverse adiabatic deceleration rate of the particle has to be much greater than the acceleration timescale,

$$\frac{3L_B}{V_s - U_1} = \tau_{AD} >> \tau_{acc} \approx \frac{3}{U_1 - U_2}\left[\frac{\kappa_1(p)}{U_1} + \frac{\kappa_2(p)}{U_2}\right]. \quad (11.82)$$

The approximate equality holds for particles relatively close to the shock. For these particles – if $\beta_2 >> \beta_1$, and if to neglect all factors of the order of unity – the second condition, thus, reduces to the first one. Hence, it needs $\lambda_1(p)v/(3U_1) << L_B$ to hold for all momenta of interest to obtain the shock spectrum in correct form. In addition, particle propagation from the shock to the escape boundaries has to be rapid enough to neglect adiabatic deceleration during the propagation, also. In the upstream region, it may demand that the diffusion time L_1^2/κ_1 has to be small in comparison with the adiabatic deceleration time-scale $3L_B/V_1$. This yields $\lambda_1 v/(3U_1) << 3L_B U_1/(V_1\eta_1^2)$, which, for typical coronal shock speeds of a few times larger than V_1 reduces to $\lambda_1 v/(3U_1) < L_B/\eta_1^2$. This imposes no further restrictions on λ_1, as long as η_1 is a number of the order of unity. As the downstream diffusion coefficient typically small, it may be estimate the downstream propagation time to be $\sim L_2/U_2$. For this to be much smaller than the respective timescale of adiabatic deceleration, $3L_B/V_2$, it have to demand $L_2/L_B << 3U_2/V_2 \approx 1$.

As underlined Vainio et al. (2000), one further condition, however, has to be satisfied: the use of the zero boundary condition at the base of the acceleration region has to be justified. As it was noted above, magnetic mirroring reflects a fraction of particles trying to precipitate. However, the mirrored particle then hits the turbulent downstream region of the shock from behind. If the downstream scattering is efficient enough, the mirrored particle has a negligible chance of diffusing back to the shock. Instead it is rapidly scattered back for another trial of precipitation. This occurs if $\eta_2(p)$ is a large number, which it shall assume. Since adiabatic deceleration in the downstream turbulent region could be neglected by the assumption of $L_2 << L_B$, the particle has the same energy (in the frame of the scattering centers) when leaving the turbulent region as it had when entering it. Such a particle does not contribute to the time-integrated flux of particles at the boundary, so the correct solution for the downstream particle flux is indeed obtained using the zero boundary condition at both boundaries when solving Eq. 11.72. Thus, Eq. 11.80 for the interplanetary particle spectrum is correct as long as $\lambda_1(p)v/(3U_1) << L_B,\ \ \eta_2(p) >> 1$, and $L_2 << L_B$.

According to Vainio et al. (2000), although the diverging magnetic field might not have large effects on the spectrum of interplanetary particles, it may still have a large effect on the ratio of interplanetary to interacting protons. A particle emitted at the base of the acceleration region will precipitate, if its shock-frame pitch-angle cosine

$$\mu < -\sqrt{1 - \frac{B(V_s t - L_2)}{B_0}} + \mathrm{O}\left(\frac{V_s}{v}\right) \approx -\sqrt{1 - \exp(-V_s t/L_B)}, \qquad (11.83)$$

yielding a probability of $P(t) = \exp(-V_s t/L_B)$ for an emitted particle to precipitate for a flux-weighted isotropic particle distribution ($f \propto |\mu|,\ \ \mu < 0$), if terms of the order V_s/v and L_2/L_B are neglected. A particle not precipitating will get mirrored back to the acceleration region, if its speed exceeds the shock speed (in the fixed frame).

As noted Vainio et al. (2000), the time it takes for a particle to get mirrored (from $\mu < 0$ to $-\mu$) is $t_B = 2L_B v^{-1}\ln((1-\mu)/(1+\mu))$, which yields an average mirroring time of $\langle t_B \rangle = 4L_B/v$ for a flux-weighted particle distribution. At high speeds, this may be used as the average time for the emitted particle to get back to the acceleration region provided that $P(t) << 1$ (if $P(t)$ was not a small number, the average would have to be taken over the returning particles, only, yielding a smaller $\langle t_B \rangle$) Then, the particle will simply diffuse in the downstream region, if $\eta_2(p)$ is a very large number (as it was assumed earlier). After an average time of $\sim\lambda_2/U_2 = 3L_2/(\eta_2 v) << 4L_B/v$, where λ_2 is the downstream mean free path, it will be emitted again for another chance-of-precipitation cycle. Thus, the rate of precipitation per emitted particle is about $vP(t)/(4L_B)$. Since it involves the particle speed, the rate has to be completed with the rate of deceleration of the particles in the expanding region behind the shock. A mirroring particle loses a flux-weighted average (downstream) momentum $\langle \Delta p \rangle = 4pV_2/3v$ per cycle, and the deceleration rate is, therefore, $\dot{p}/p = -V_2/3L_B$. If one particle is emitted at the base of the acceleration region with a nonrelativistic velocity $v = v_0$ at time $t = t_0$, the number of particles in the system at $t > t_0$ is

$$G(t; v_0, t_0) = \exp\left[\frac{3v_0 \exp(-V_s t_0/L_B)}{4(3V_s + V_2)} \times \left(\exp\left(-\frac{(3V_s + V_2)(t - t_0)}{3L_B}\right) - 1\right)\right], \tag{11.84}$$

and the fraction of precipitating particles is thus

$$G_{prec}(v_0, t_0) = 1 - G(\infty; v_0, t_0) = 1 - \exp\left[-\frac{3v_0 P(t_0)}{4(3V_s + V_2)}\right], \tag{11.85}$$

Those particles that do not precipitate are cooled down to thermal energies and become part of the downstream plasma flow. Note that if $P(t_0)$ is not too small, say, if $v_0 P(t_0) \geq V_2$, the rate of precipitation will exceed the rate of deceleration, and most precipitating particles will have an energy close to their emission energy. Thus, if particles are not emitted too far from the Sun, the described above estimate for $\Gamma(E)$ should be a good approximation at least at high energies. It will be shown later how Eq. 11.85 can be used to give a valid analytical solution to this ratio even when particles are emitted relatively far from the Sun.

11.7.4 Numerical Solutions

To verify the analytical model, described in Section 11.7.3, Vainio et al. (2000) used a Monte Carlo simulation code developed for the study of test-particle transport and acceleration in the inner heliosphere. Instead of solving a diffusion–convection equation, the code employs a kinetic treatment, where particles are moved in small time steps conserving their energy and magnetic moment in a frame, where the large-scale magnetic field is static. After each such time step, an isotropic small-angle scattering is performed via a random generator. The scatterings are elastic in the (local) frame comoving with the scattering centers. After a large amount of such particles is simulated, Vainio et al. (2000) print out the results in the form of histograms over quantities of interest. Very similar model was used by Ellison et al. (1990) in a study of relativistic shock acceleration, but in their study the background magnetic field was strictly one-dimensional (no magnetic focusing) and their scattering centers were frozen-in into the plasma. More details about the Monte Carlo simulations and the kinetic model were described in Appendix A2.

Vainio et al. (2000) performed a set of simulations to verify Eqs. 11.80 and 11.81. They fixed the value of focusing length to $L_B = 0.3R_{sun}$ (where R_{sun} is the solar radius), and used a shock speed of $V_s = 1{,}200$ km/s, an upstream Alfvén speed of $V_{A1} = 300$ km/s, which give $U_1 = 900$ km/s. Vainio et al. (2000) also used a squared upstream sound speed of $c_{s1}^2 = 0.3V_{A1}^2$ corresponding to a coronal electron temperature of $T_e \approx 2 \times 10^6$ K. According to Vainio and Schlickeiser (1998, 1999), the gas compression ratio of the shock should be $\rho \approx 4/\left(1 + 3c_{s1}^2/V_s^2\right)$ ≈ 3.8 as long as the upstream wave amplitude is very small compared with the magnitude of the ordered field. In addition to ρ, the scattering-center compression ratio is dependent on the upstream wave spectrum but a characteristic value for incompressible upstream turbulence is near $\rho_c = 5.5$ (Vainio and Schlickeiser 1999).

Vainio et al. (2000) set $\lambda_2,\ L_2 \to 0$ while keeping $\eta_2 \to \infty$ and inject low-energy particles to the acceleration process at the shock with an injection speed of $v_{inj} = 0.03c$ measured in the shock frame. The shock-frame angular distribution of the injected particles was taken to be 2H(μ)μ corresponding to an isotropic particle population just crossing the shock from the downstream region to the upstream. The number of upstream diffusion lengths was set at $\eta_1 = \ln(1 + \rho_c) = 1.872$, giving a value of $\Gamma = 1$ in Eq. 11.81. These choices, thus, give a theoretical prediction of

$$\frac{dN}{dE} = \frac{7.0 \times 10^{-3}}{mcv}\left(\frac{mc}{p}\right)^{2.33} \tag{11.86}$$

for the interplanetary and interacting proton spectra per one injected particle per flux tube, momentum being measured in the fixed frame. As a test for the simulation code, it was verified that it produces this solution when all the effects of diverging magnetic field were switched off. All simulations with the effects of diverging magnetic field included were started when the shock was at $\zeta = 0.3R_{sun}$ above the solar surface. The upstream mean free path of the particles, λ, was varied to study the effect of focusing on the interplanetary spectrum. To analyze the effect of mirroring behind the acceleration region, Vainio et al. (2000) took the injection at the shock, $S(x,p,t) = Q(t)\delta(x)\delta(p - p_{inj})$, to be either impulsive or proportional to the ambient plasma density. The values used in the simulations are given in Table 11.8, and the simulated spectra of interplanetary and interacting protons – in Fig. 11.40.

Table 11.8 Parameters for the presented Monte Carlo simulations (From Vainio et al. 2000)

Simulation number	$\lambda/R_\odot$	Injection, $Q(t)$[a]
1	5×10^{-4}	$\delta(t - t')$
2	5×10^{-4}	$H(t - t')V_s L_B^{-1} e^{-V_s(t-t')/L_B}$
3.	2.5×10^{-3}	$\delta(t - t')$

[a]Time $t' = 0.3\ R_\odot/V_s$ is the start time of the injection.

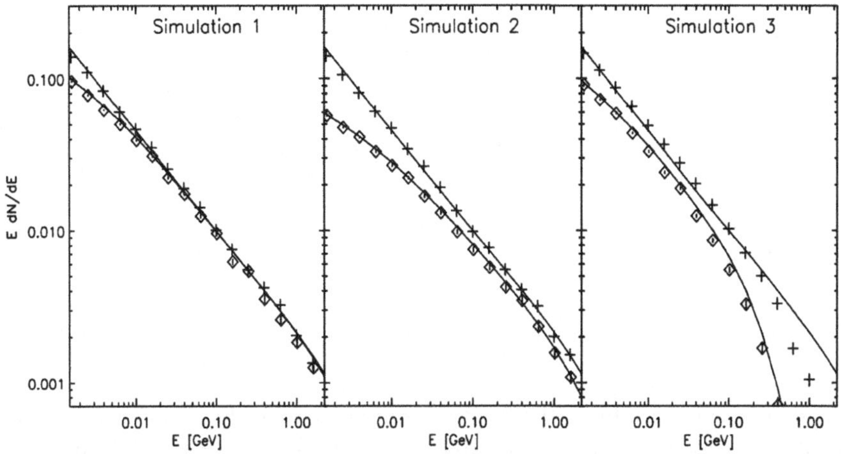

Fig. 11.40 Simulated (see text for details) interplanetary (*crosses*) and interacting (*diamonds*) proton energy spectra along with the analytically predicted spectra (*curves*) (From Vainio et al. 2000)

From Fig. 11.40 follows that Simulation 1 gives a good agreement with the analytical form of the interplanetary spectrum, as it should, since the mean free path is so small that one does not expect large contributions from either the streaming term caused by focusing or the adiabatic deceleration term in Eq. 11.72: for $\lambda = 5 \times 10^{-4}\ R_{\text{sun}}$ we have $\kappa_1/L_B = 0.185(v/c)U_1$. The slight softening of the spectrum at $E > 600$ MeV is due to the finite, although small, value of $\lambda v/(3U_1 L_B) = 0.185(v/c)$ at particle speeds $v \sim c$. If the analytical model was completely valid, the chosen parameters would produce identical interacting and interplanetary proton spectra, as pointed above. For most part of the spectrum this is true, but below $E \sim 20$ MeV there is a deficit of interacting particles. This is because at low velocities the precipitating rate for the particles is not high enough for all of them to interact before adiabatically cooling. Therefore, the theoretical prediction described above for the interacting spectrum is

$$\frac{dN_2}{dE} = N_{prec}(v)\frac{7.0 \times 10^{-3}}{mcv}\left(\frac{mc}{p}\right)^{2.33}. \tag{11.87}$$

The multiplying factor $N_{prec}(v)$ is the fraction of precipitating particles at speed v obtained as a convolution integral over t_0 of the injection function $Q(t)$ and the Green's function for the fraction of precipitating particles $G_{prec}(v_0, t_0)$ from Eq. 11.85:

$$N_{prec}(v) = \int_0^\infty Q[t_0 - T(v)]G_{prec}(v, t_0)dt_0, \tag{11.88}$$

where v_0 has simply been replaced by the particle speed, and

$$T(v) = \frac{\lambda}{U_1^2}\frac{\rho_c}{\rho_c - 1}\ln\left(\frac{v + c}{\sqrt{c^2 - v^2}}\frac{\sqrt{c^2 - v_{inj}^2}}{v_{inj} + c}\right) \tag{11.89}$$

is the mean acceleration time. It is obtained from the standard equation for the acceleration rate,

$$\frac{\dot{p}}{p} = \frac{U_1 - U_2}{3}\frac{U_1}{\kappa_1}, \tag{11.90}$$

and it may be in error by ~10% due to a finite $\eta_1 \sim 2$ (see Fig. 2 of Ostrowski and Schlickeiser 1996). Vainio et al. (2000) note that Eq. 11.89 is valid only for a constant mean free path; for a momentum dependent mean free path, the form of $T(v)$ should be reintegrated from the acceleration rate. In case of impulsive injection, $Q(t) = \delta(t - t')$, where $t' = 0.3R_{sun}/V_s$ is the time of the injection, it will be simply $N_{prec}(v) = G_{prec}[v, t' + T(v)]$. This treatment is not rigorous, and it does not even take into account the fact that the particles will have somewhat lower energies when precipitating than when leaving the shock. Nevertheless, it seems to give a very good fit to the spectrum of interacting particles at least in the considered case, where the spectrum is quite hard and the fraction of precipitating particles is still above 50% at energies above 1 MeV.

As underlined Vainio et al. (2000), in Simulation 2 the injection is taken to be proportional to the linear density of the ambient plasma. At low energies, the mean injection time $t' + L_B/V_s$ is much greater than the acceleration time T and twice as much as the start time of the injection. This means that the low-energy particles have much smaller probability to precipitate than in Simulation 1; this is clearly seen in the simulated spectrum as a decrease of dN_2/dE at low energies. In addition, the precipitation rate of the high-energy particles starts to be slow, as well, and it can be see a deficit of more than 20% of interacting particles at all energies. This shows that if a substantial fraction of particles is injected to the acceleration process at distances farther than $\sim 2L_B$ from the solar surface, the analytical model is no longer valid for interacting particles. There is, however, no significant difference between the interplanetary spectra of the analytical model and the two simulations, which is also consistent with the model prediction.

As noted Vainio et al. (2000), in Simulation 3, we have increased the value of the mean free path by a factor of 5 relative to the value used in Simulation 1. This has two effects: (i) the interplanetary spectrum is now considerably softer than the model prediction above E $\sim$ 200 MeV, and (ii) the ratio of interplanetary to interacting protons is larger than in Simulation 1. Effect (i) is due to clearly non-negligible effects of focusing in Eq. 11.72: streaming with velocity $\kappa_1/L_B = 0.925(v/c)U_1$, giving a value of $\lambda v/(3U_1) \approx 0.5L_B$ at E = 200 MeV, and adiabatic deceleration at rate $V_{A1}/(3L_B) \approx 0.11U_1/L_B$, which is more than one third of the acceleration rate at highest energies given by $(U_1 - U_2)U_1/(\lambda c) \approx 0.29U_1/L_B$. Effect (ii) is due to the acceleration time increased by factor of 5, which makes the probability of precipitation much smaller, especially at high energies where T now becomes dominant over t'.

Vainio et al. (2000) noted that as can be judged from the given numerical solutions, the analytical model describes the interplanetary and interacting proton spectra rather well in all cases, where the mean free path is so small that particle acceleration to relativistic energies becomes possible. The only notable exception is the suppression of the interacting proton spectrum from the theoretical value, if a substantial fraction of the particle injection occurs at altitudes of a couple of L_B or more above the solar surface. The analytical model is, however, quite adequate basis for the analysis of measured accelerated particle spectra.

11.7.5 Interplanetary and Interacting Particles as Probes of Turbulence

In Vainio et al. (2000) analytical model, two parameters describing the turbulence in vicinity of the shock, $\eta_1(p)$ and ρ_c, determine both the spectrum of interplanetary protons and the ratio of interplanetary to interacting protons (at least at high energies). This gives a possibility to use the measurements of dN_1/dE and Γ to deduce the turbulence parameters. This could be done, e.g., by fitting. Vainio et al. (2000) note, however, that it is possible to use a direct inversion method to obtain these parameters in case, when the interplanetary particle spectrum is close to a power law over momentum at high energies, i.e., $vdN_1(E)/dE \propto p^{-\gamma_1}$ above, say, $p = p_1$ and when an average value of interplanetary to interacting protons ratio at high energies, Γ_∞, is also measured. In this case it may use Eqs. 11.80 and 11.81 to deduce the constant values of ρ_c and η_1 at $p \geq p_1$: Eq. 11.80 gives

$$\gamma_1 = \gamma \frac{z_\infty + 1}{z_\infty} - 2 \tag{11.91}$$

and Eq. 11.81 combined with Eq. 11.77 gives

$$\Gamma_\infty = \frac{\gamma}{z_\infty(\gamma - 3)}, \tag{11.92}$$

where z_∞ is the value of $z(p) \equiv \exp(\eta_1) - 1$ at $p \geq p_1$. Thus,

$$\gamma = \frac{\gamma_1 + 2 + 3\Gamma_\infty}{1 + \Gamma_\infty} \tag{11.93}$$

and

$$z_\infty = \frac{\gamma_1 + 2 + 3\Gamma_\infty}{(\gamma_1 - 1)\Gamma_\infty} \tag{11.94}$$

Vainio et al. (2000) have, thus, solved the value of γ, which fixes the value of the scattering-center compression ratio through Eq. 11.77. If the spectral shape of the interplanetary particles is different from the power-law form at lower momenta, it may solve for $z(p)$ as follows. Rearrange Eq. 11.80 in a form of differential equation for $z(p)$. Multiply it first by $vp^{\gamma-2}z(p)$, then take a logarithm of both sides and derivate with respect to p to get

$$\frac{dz}{dp} + \left[\frac{d}{dp}\ln\left(vp^{\gamma-2}\frac{dN_1}{dE}\right)\right]z = -\frac{1}{p}, \tag{11.95}$$

and finally integrate from p_1 to p to arrive at

$$z(p) = \left[z_\infty\left(vp^{\gamma-2}\frac{dN_1}{dE}\right)_{p=p_1} + \gamma\int_p^{p_1} vp^{\gamma-3}\frac{dN_1}{dE}dp\right] \times \left(vp^{\gamma-2}\frac{dN_1}{dE}\right)^{-1}. \tag{11.96}$$

This holds at momenta $p < p_1$ whatever the form of the measured interplanetary spectrum is. Vainio et al. (2000) note also that the asymptotic form of the spectrum does not have to be exactly of the form proposed; it may use also the values of Γ and γ_1 measured at $p = p_1$ as long as $|d\gamma_1/d\ln p| << \gamma_1$.

Torsti et al. (1996) analyzed the interplanetary protons of the solar CR event on 1990 May 24. They concluded that the injection of protons into the interplanetary medium consisted of two components, first of which was released 10–40 min after the X-ray flare. The timescale of this particle release is consistent with acceleration-site length scales of the order of solar radius. This flare was also a source of gamma rays and neutrons, which have been analyzed as well (Kocharov et al. 1994, 1996), and there were indications of a prolonged emission of neutrons that could be originating from shock-accelerated particles. These observational facts make the event acceptable for application of above described model. The energy spectrum of the prompt-component interplanetary protons could be represented by

$$\frac{dN_1}{dE} = N_0\left(\frac{E}{160\text{ MeV}}\right)^{-1.6}\left[1 + \left(\frac{E}{360\text{ MeV}}\right)^3\right]^{-1}, \tag{11.97}$$

with $N_0 = 1.4 \times 10^{30}$ protons MeV^{-1} in the range $E = 30$–1,000 MeV assuming a solid angle of 2 sr for the size of the acceleration region. The number of high-energy interacting protons was found to be equal to or several times less than the number of high-energy interplanetary protons.

As Vainio et al. (2000) apply their model to these observations they take the proton momentum $p_1 = \sqrt{120}m_pc$ corresponding to the kinetic energy of $10m_pc^2$, where it is $\gamma_1 = 5.0$. Vainio et al. (2000) then adopt $\Gamma_\infty = 1$, which gives a value of $\gamma = 5.0$. This yields $\rho_c = 2.5$, a relatively small value when compared to the theoretical value of about 5.5. As another case, Vainio et al. (2000) note that the theoretical compression ratio yields $\gamma = 3.67$ and $\Gamma_\infty = 5$, so it may marginally fit the observed data as well. For these two cases, Vainio et al. (2000) then have $z_\infty =$ 2.5 and 1.1, respectively. The integral in Eq. 11.96 performed numerically, it present the number of upstream diffusion lengths in Fig. 11.41 and the inferred spectra of interacting protons in Fig. 11.42 resulting for the two considered cases. Vainio et al. (2000) note, that the actual values should lie between the curves of these figures, although at low energies the curves for dN_2/dE should be taken as upper limits because of the mirroring effects discussed above.

11.7.6 Estimation of Model Parameters That Could Explain Observations of Interplanetary Particle Spectrum as Well as Interplanetary to Interacting Proton Ratio

Vainio et al. (2000) tried to estimate model parameters that could explain usual observations of interplanetary particle spectrum as well as interplanetary to

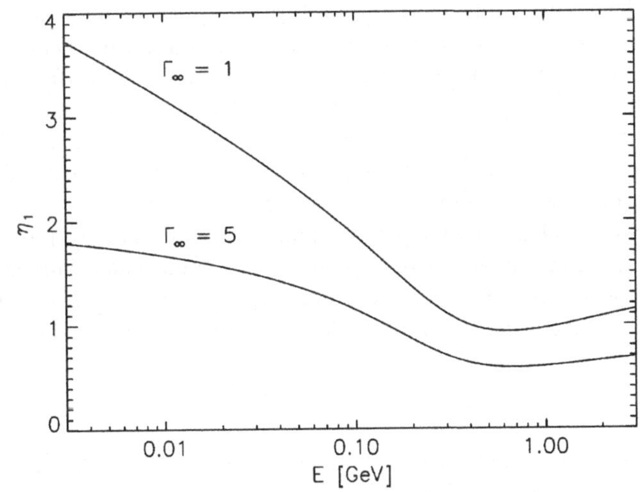

Fig. 11.41 Number of upstream diffusion lengths as a function of particle energy deduced from the measured spectrum of interplanetary protons in the 1990 May 24 solar flare for the two limiting cases of the observed $1 \leq \Gamma_\infty \leq 5$ (From Vainio et al. 2000)

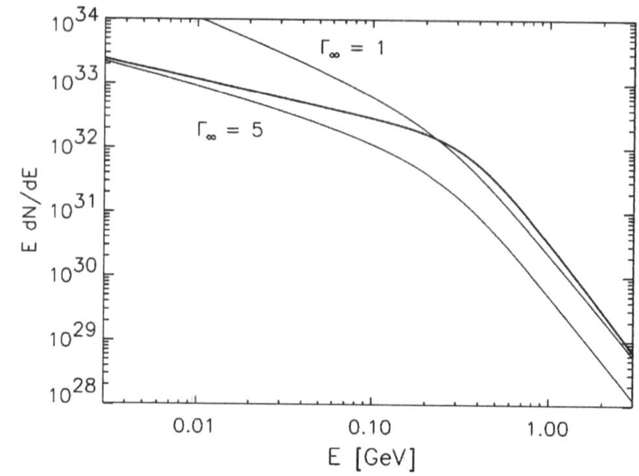

Fig. 11.42 Measured interplanetary proton spectrum (*thick curve*) and the deduced spectra of the interacting protons (*thin curves*) in the 1990 May 24 solar flare for the two limiting cases of the observed $1 \leq \Gamma_\infty \leq 5$ (From Vainio et al. 2000)

interacting proton ratio. The above-adopted $\rho_c = 5.5$ is actually an upper limit because any compressible MHD-turbulence component in the upstream region can couple to fast MHD waves in the downstream region and these can only propagate away from the shock. Also a finite value of the magnetic amplitude of the upstream waves will slightly reduce the scattering-center compression ratio from this ideal value (according to Vainio and Schlickeiser 1999). Thus, Vainio et al. (2000) suppose to made this estimation by using Eq. 11.77 connecting γ and ρ_c that $\gamma \geq 3.7$. In the view of the analytical model, in addition to ρ_c, Vainio et al. (2000) are left with one (momentum dependent) parameter, the number of diffusion lengths in the upstream region $\eta_1(p)$ to determine the interplanetary proton spectrum and the ratio of interplanetary to interacting protons. As noted Vainio et al. (2000), it is necessary to ask, what values of these parameters are consistent with observations of interplanetary and interacting particle spectra.

Vainio et al. (2000) note first that to obtain a single power law in momentum or energy in Eq. 11.80 for the energy spectrum of interplanetary particles at either nonrelativistic ($v = p/m$, $E = p^2/2m$) or ultra-relativistic ($v \approx c$, $E \approx pc$) energies, $\eta_1(p)$ must have a constant value. A broken power law can be produced, e.g., if

$$[\exp(\eta_1(p)) - 1]^{-1} = a(p/p_k)^\alpha[1 + (p/p_k)^\alpha]^{-1}, \qquad (11.98)$$

where p_k is the momentum where the knee of the spectrum should be situated, and a and α are positive constants. This choice corresponds to

$$\eta_1(p) = \ln\{1 + a^{-1}[1 + (p/p_k)^\alpha]\}, \qquad (11.99)$$

so at $p << p_k$ the number of diffusion lengths increases as $\ln(p^{-\alpha})$as p decreases. In this case, the spectrum in Eq. 11.80 can be analytically integrated to give

$$\frac{dN_1}{dE} = \frac{\gamma a N_0}{p_{inj} v} \left(\frac{p_{inj}}{p}\right)^{\gamma-2} \frac{p^\alpha}{p_k^\alpha + p^\alpha} \left(\frac{p_k^\alpha + p_0^\alpha}{p_k^\alpha + p^\alpha}\right)^{a\gamma/\alpha}. \tag{11.100}$$

From Eq. 11.100 follows that, the nonrelativistic energy-spectral index below the knee is $(\gamma - \alpha - 1)/2$ and $[\gamma(1 + a) - 1]/2$ above the knee. At ultra-relativistic energies these indices are $\gamma - \alpha - 2$ and $\gamma(1 + a) - 2$, respectively. Since the value of α is controlling the rate of change of the spectral index, one cannot choose it to be very small if a broken power law is to be produced; in fact, $\alpha = 0$ produces a single power law with nonrelativistic and ultrarelativistic spectral indices of $[\gamma(1 + a/2) - 1]/2$ and $\gamma(1 + a/2) - 2$. This requirement implies, if the natural condition of $\eta_1(p) \to \infty$ as $p \to 0$is imposed, that the interplanetary spectrum at low energies should be very flat for broken power laws generated by parallel shock acceleration, if γ close to its limiting value of 3.7. Vainio et al. (2000) note, finally, that a law increasing faster than $\eta_1 \sim \ln p^{-\alpha}$ as p tends to zero, e.g., a power law $\eta_1 \sim p^{-\alpha}$, will not lead to a broken power law energy spectrum, but to a spectrum with a low-energy cutoff. Even these spectra, however, may well present themselves as broken power laws over a limited range in energy. As stated above, the choice of $\eta_1(p)$ fixes also the interplanetary to interacting proton ratio. For the broken power-law example, it is given by

$$\Gamma(E) = \left(\frac{p}{p_k}\right)^\alpha \frac{\rho_c a}{1 + (p/p_k)^\alpha}, \tag{11.101}$$

which has the asymptotic value of $\Gamma(\infty) = \rho_c a$ at large energies; at low energies, $\Gamma(E) = \rho_c a (p/p_k)^\alpha$ if assume that all particles leaving the acceleration region really precipitate. Otherwise the interacting particle spectrum will be suppressed at low energies and the above expression will be lower limit estimation. For the single power-law case, $\alpha = 0$, it will be $\Gamma = \rho_c a/2$, which should also be taken as a lower limit at low energies.

11.7.7 Discussion and Conclusions

Vainio et al. (2000) have presented an analytical model to describe the interplanetary and interacting proton spectra resulting from parallel shock acceleration in solar corona. They have shown that a very simple analytical model of diffusive shock acceleration leading to Eqs. 11.80 and 11.81 can be used to give the accelerated particle spectra if (i) the upstream diffusion length is much smaller than the scale length of the background field, $\lambda_1(p)v/(3U_1) << L_B$ at all particle speeds of interest; (ii) the downstream diffusion length, the length of the downstream diffusion region and the background-field scale length are ordered as $\lambda_2(p)v/(3U_2) << L_2 << L_B$; and (iii) the injection of low-energy particles is concentrated close to the Sun, within a couple of L_B. Since $\lambda_2 << \lambda_1$ in any reasonable description shocked turbulence and since the waves propagate toward the downstream in the shock frame, we feel it is extremely unlikely that the condition $\lambda_2(p)v/(3U_2) << L_2$ is violated. If either condition (iii) or $L_2 << L_B$ is violated, one may not be able to describe the interacting spectrum analytically, although the analytical interplanetary spectrum should still be valid. Note that condition (i) is much more stringent than condition $\lambda_1 << L_B$, which is commonly thought to be enough to justify the use of one-dimensional diffusion without the effects of focusing. Similar conclusion was drawn by Kocharov et al. (1999b); in their stochastic acceleration model (see Section 11.6) there was no bulk motion of the plasma at all, which led to an even stronger version of this rule: focusing could not be neglected at all, no matter how small the mean free path may be.

Vainio et al. (2000) have compared their analytical model to a numerical one, which uses a Monte Carlo method to trace particles in a predescribed shock system under a law of scattering resulting to a diffusion-convection-like motion of particles. A good agreement of the models is found, when the above stated conditions for its validity hold. Vainio et al. (2000) were also able to give an analytical correction factor, Eqs. 11.85 and 11.88, for the interacting spectrum in case, where the particle spectrum emitted by the shock toward the downstream region is not entirely precipitated due to mirroring between the shock and the solar surface. This correction neglects (partly) the change in the particle energy caused by adiabatic deceleration of particles in the expanding region between the shock and the receding solar surface. In described simulated cases this was a minor error since the precipitation rate dominated over the deceleration rate making most of the particles precipitate with speeds close to their emission speeds. Care has to be taken, however, if particles emitted from outer corona are considered since then the rate of deceleration will become comparable to or even exceed the rate of precipitation, and the resulting shift in the spectrum has to be taken into account. Also, the treatment of the precipitation process was nonrelativistic, so if $N_{\text{prec}}(v)$ will be small at relativistic speeds Vainio et al. (2000) advice against its use.

As noted Vainio et al. (2000), in principle, the theory of diffusive shock acceleration could also be used for oblique shocks. Vainio et al. (2000), however, limited the use of the model for nearly parallel shocks since in the geometry we have in mind (Fig. 11.39), oblique shocks would naturally have more involved time dependence than it was proposed in described model. Obliquity would also lead to more complicated description of the downstream turbulence with waves propagating off axis, and with time-dependent large-scale magnetic fields that would lead to complications and invalidate obtained simple analytical solution. If a proper description of the magnetic fields as a function of time and position may be found in such a case, the numerical model is quite capable with simple updates to solve for those equations, too.

Vainio et al. (2000) used constant values of the upstream mean free path in their simulations and concluded that the mean free path has to be at most of the order of $\lambda \sim 5 \times 10^{-4}\ R_{\odot}$ if relativistic energies are to be produced efficiently. It is interesting to compare this to the Larmor radius of the particles. In a field of 1 G, the Larmor radius of 1 GeV proton is $R_L = 8 \times 10^{-5}\ R_{\text{sun}}$. Since the Larmor radius represents the lower limit of the mean free path (the so called Bohm limit) in parallel shock acceleration, at which the magnetic field should already be very disordered, acceleration to such high energies would probably have to take place in regions of high magnetic field of at least 1 G in order to be described by the model. If to apply this value at, say, one solar radius above the solar surface, it implies a field of 30 G at the base of the corona with the adopted value of $L_B = 0.3\ R_{\text{sun}}$. Thus, the Alfvén speed of $V_A = 3 \times 10^7$ cm/s yields a coronal-base value of $n_e \sim 4 \times 10^{10}$ cm^{-3} for the electron density suggesting that the adopted Alfvén speed may probably be somewhat low. A larger value implies a more important role for adiabatic deceleration and, if as a result the shock speed is increased, shortens the time available for the acceleration. If to apply a mean free path scenario with $\lambda = 10R_L$ in the magnetic field discussed above, Vainio et al. (2000) have $\lambda(p,\zeta)/L_B = 5 \times 10^{-5} \exp(3.33\zeta/R_{sun})p/(\text{GeV}/c)$. This has to be much smaller than $3U_1/v \sim 0.01c/v$ for diffusive shock acceleration to work efficiently. Vainio et al. (2000) thus conclude that shock acceleration to relativistic energies, if it occurs, should take place relatively close to the Sun in very turbulent conditions to avoid problems with the effects of diverging field that tend to produce softer spectra than those observed.

In considered model of Vainio et al. (2000), the ratio of interplanetary to interacting particles is very sensitive to the value of η_1. Very large values of this parameter cannot be proposed without obtaining extremely small values of Γ never observed in gradual flares. Large values of η_1 would also lead to another difficulty; the timescale of the adiabatic deceleration in the upstream region would get smaller than the diffusive escape time of particles. Interplanetary mean free paths at 1 AU for $E = 10$–100 MeV protons are often observed to be of the order of 0.1 AU. WKB theory predicts that the mean free path near the Sun should be a decreasing function of heliocentric distance for typical wave spectra, if Sun is the only source for turbulence (Jokipii 1973; Ng and Reames 1994); in contrast, we need values that are five orders of magnitude smaller than this. All these facts both point toward the interpretation that the upstream waves are self-generated by the accelerated particles through the streaming instability: even if an external source could have produced the waves necessary for the intense scattering, it would be impossible for the accelerated to ever escape upstream to be detected near the Earth. To answer, whether it is possible for the particles to generate the waves self-consistently, it needs a model that is time dependent and can address also the basic question of injection of particles to the acceleration process.

As underlined Vainio et al. (2000), the sensitivity of the model to the value of η_1 gives also a good possibility to test it experimentally: to offer the model as a natural way to explain the broken power-law spectra of the interplanetary ions observed in some SEP events (e.g., Torsti et al. 1996). In this case, the change in the spectral slope should also be seen as a change in the interplanetary to interacting proton ratio at the same energies so it should be, at least in principle, detectable even with the currently available gamma-ray and particle detectors.

11.8 High Energy Gamma Ray Generation in the Corona and Solar Wind Through Interactions of Flare Energetic Particles with Matter Through π^0 Decay and Bremsstrahlung

11.8.1 The Matter and Short History of the Problem

The interaction of FEP (flare energetic particles – protons, nuclei, and electrons) with matter of solar corona and solar wind near the Sun determine the main processes of high energy gamma ray generation through neutral pion decay and through bremsstrahlung emission. In Dorman (1996, 1997, 2001) there was estimated the expected gamma ray intensity generated by local and outer CR in different astrophysical objects for outer and inner observers. Any astrophysical object containing CR (of local and outer origin), magnetic fields and matter must generate gamma rays by neutral pion decay (generated in interactions of CR protons and nuclei with matter), and by the generation of bremsstrahlung, synchrotron and curvature radiation of relativistic electrons, and by inverse Compton scattering of relativistic electrons on optical, infrared and relict 2.7 K photons. The intensity and spectrum of gamma radiation depend on the CR spectrum, on the CR space–time distribution function, as well as on the spatial distribution of matter, magnetic fields and background photons. Below we applicate general formulas for gamma ray generation through neutral pion decay and gamma ray generation through interactions of energetic electrons with matter and low energy photons (bremsstrahlung and inverse Compton radiation) to FEP events.

11.8.2 Gamma Rays from Neutral Pion Decay Generated in Nuclear Interactions of FEP with Matter of Solar Corona and Solar Wind

Let the distribution of solar corona and solar wind matter in the spherical system of coordinates r, θ, φ be determined by $n(r, \theta, \varphi)$ in atom cm^{-3}. Let us suppose that $N_{pn}(E, r, \theta, \varphi)$ is the space distribution of the differential intensity of the proton-nuclear component of FEP, where E is the total particle energy in GeV/nucleon. The gamma ray intensity from some space plasma volume bounded by the surface

$r_0(\theta, \varphi)$ from neutral pion decay in this volume at the distance $r_{obs} >> r_0(\theta, \varphi)$ will then be

$$F_{\gamma,\mathrm{pn}}\left(r_{\mathrm{obs}}, E_\gamma\right) = r_{\mathrm{obs}}^{-2} \int_{-\pi/2}^{\pi/2} \cos\theta d\theta \int_0^{r_0(\theta,\varphi)} dr \int_0^{2\pi} d\varphi \int_0^{\infty} dE \sigma_{\mathrm{pn}}\left(E, E_\gamma\right) N_{\mathrm{pn}}(E, r, \theta, \varphi) n(r, \theta, \varphi), \tag{11.102}$$

where, according to Stecker (M1971), Dermer (1986a, b),

$$\sigma_{\mathrm{pn}}\left(E, E_\gamma\right) = 2 \int_{E_{\pi,\min}\left(E_\gamma\right)}^{\infty} dE_\pi \sigma_\pi(E) \left(E_\pi^2 - m_\pi^2 c^4\right)^{-1/2}. \tag{11.103}$$

In Eq. 11.103

$$E_{\pi,\min}\left(E_\gamma\right) = E_\gamma + m_\pi^2 c^4 / E_\gamma, \tag{11.104}$$

and the cross-section of pion generation $\sigma_\pi(E)$ can be approximated by (the momentum p of protons is in GeV/c):

$$\sigma_\pi(E) = \begin{cases} 0 & \text{if } p < 0.78, \\ 0.032\eta^2 + 0.040\eta^6 + 0.047\eta^8 & \text{if } 0.78 \le p \le 0.96, \\ 32.6(p-0.8)^{3.21} & \text{if } 0.96 \le p \le 1.27, \\ 5.40(p-0.8)^{0.81} & \text{if } 1.27 \le p \le 8.00, \\ 32.0 \ln p + 48.5 p^{-1/2} - 59.5 & \text{if } p \ge 8.00. \end{cases} \tag{11.105}$$

In Eq. 11.105 there was used the notation

$$\eta = \frac{\left[\left(2m_p c^2 E_k - m_\pi^2 c^4\right)^2 - 16 m_\pi^2 m_p^2 c^8\right]^{1/2}}{\left[2m_\pi c^2 \left(2m_p c^2 \left(E_k + 2m_p c^2\right)\right)^{1/2}\right]} \tag{11.106}$$

and E_k is the kinetic energy of protons. The dependence of $\sigma_\pi(E)$ from kinetic energy of protons E_k (calculated according to Eq. 11.105) is shown in Fig. 11.43.

According to Dermer (1986a, b), the including of additional channels of nuclear interactions p–He, α–H, and α–He gives an increase in gamma ray emissivity of 28%, 9%, and about 2% relative to p–H channel considered above. Therefore for rough estimations we can consider only the channel p–H and then multiply the result by a factor 1.39; if we also take into account heavier nuclei this factor will be 1.45.

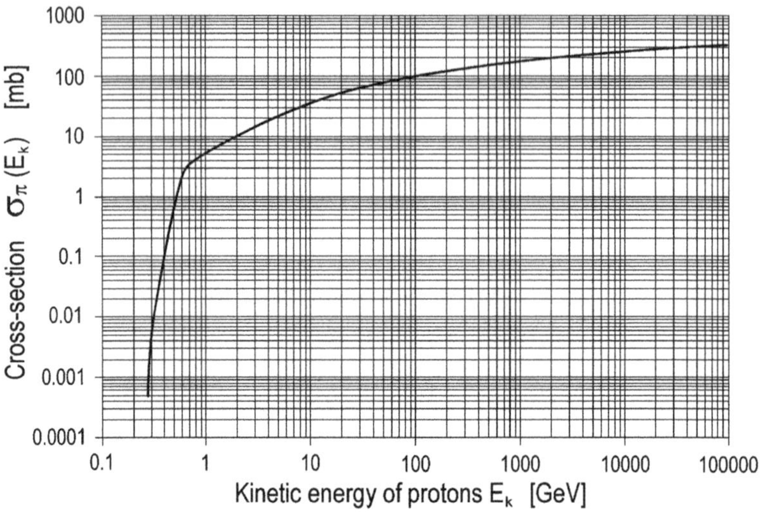

Fig. 11.43 The inclusive cross-section $\sigma_\pi(E_k)$ for reactions $p + p \rightarrow \pi^0 +$ anything as a dependence upon the kinetic energy of protons E_k. Calculated according to Eq. 11.105

11.8.3 Bremsstrahlung Gamma Radiation Generated by FEP Electrons

By using results of Cesarsky et al. (1978) on the bremsstrahlung gamma ray generation by electrons of galactic CR in the interstellar medium, we obtain for the expected bremsstrahlung gamma ray flux from some volume of space plasma at some distance r_{obs} from this volume the following formula:

$$F_{\gamma,\text{bs}}\left(r_{\text{obs}}, E_\gamma\right) = r_{\text{obs}}^{-2} \int_{-\pi/2}^{\pi/2} \cos\theta d\theta \int_0^{r_0(\theta,\varphi)} dr \int_0^{2\pi} d\varphi \int_{E_\gamma}^{\infty} dE \sigma_{\text{bs}}\left(E_e, E_\gamma\right) N_e(E_e, r, \theta, \varphi) n(r, \theta, \varphi), \tag{11.107}$$

where the definitions are the same as for Eq. 11.102, but E_e is the energy of electrons and $N_e(E_e, r, \theta, \varphi)$ is the space distribution of the differential intensity of the electron component of FEP. In Eq. 11.107 $\sigma_{\text{bs}}\left(E_e, E_\gamma\right)$ is the cross-section of bremsstrahlung gamma ray generation with energy E_γ by electrons with energy E_e, which according to Cesarsky et al. (1978) can be approximated by the following equation:

$$\sigma_{\text{bs}}\left(E, E_\gamma\right) = \alpha r_e^2 \left\{ \left(2E_e^2 - 2E_e E_\gamma + E_\gamma^2\right) E_\gamma^{-2} \phi_1 - \left(E_e - E_\gamma\right) E_e^{-1} \phi_2 \right\}, \tag{11.108}$$

where $\alpha \approx 1/137$ is the fine structure constant, r_e is the classical electron radius, ϕ_1 and ϕ_2 are functions from variable

$$\xi = \chi(Z)\left(m_e c^2 E_\gamma\right)/\left(E_e\left(E_e - E_\gamma\right)\right), \tag{11.109}$$

and

$$\chi(Z = 1) = 34.259, \chi(Z = 2) = 20.302. \tag{11.110}$$

The functions ϕ_1 and ϕ_2 are tabulated in Blumental and Gould (1970). According to Pohl (1994) for the standard He-to-H ratio of 0.1 for space plasma matter roughly

$$\phi_1 \approx \phi_2 \approx 58, \tag{11.111}$$

and

$$\sigma_{\text{bs}}\left(E_e, E_\gamma\right) \approx 0.42 r_e^2\left((4/3) - \left(E_\gamma/E_e\right) - \left(E_\gamma/E_e\right)^2\right). \tag{11.112}$$

11.8.4 The Main Three Factors That Determine Pion Decay Gamma Radiation from Interactions of FEP with the Corona and Solar Wind Matter

As was shown in Dorman (1996, 1997, 2001), the generation of gamma rays by interaction of FEP with corona and solar wind matter is determined mainly by following three factors:

The first factor – space–time distribution of FEP in solar corona and in the heliosphere, energetic spectrum and chemical composition of FEP (see review in Dorman M1957, M1963a, b, M1978; Dorman and Miroshnichenko M1968; Dorman and Venkatesan 1993; Stoker 1994; Miroshnichenko M2001). For this factor may be important nonlinear collective effects (especially for great events) of FEP pressure and kinetic stream instability (Berezinsky et al. M1990; Dorman et al. 1990, Zirakashvili et al. 1991; Dorman 1995a, b; see detail review in Chapter 3 of Dorman M2006).

The second factor – the solar corona and solar wind matter distribution in space and its change during solar activity cycle; nonlinear effects will also be important for this distribution: pressure and kinetic stream instability of galactic and solar CR (especially in periods of very great events) – see references above.

The third factor – by properties of FEP interaction with the solar corona and solar wind matter accompanied with gamma ray generation through decay of neutral pions (Stecker M1971; Dermer (1986a, b); see Section 11.8.2).

11.8.5 The First Factor: Solar FEP Space-Time Distribution

The problem of FEP generation and propagation through the solar corona and in the interplanetary space as well as its energetic spectrum and chemical and isotopic composition was reviewed in Dorman (M1957, M1963a, b, M1978), Dorman and Miroshnichenko (M1968), Dorman and Venkatesan (1993), Stoker (1994), Miroshnichenko (M2001). In the first approximation, according to numeral data from observations of many events for about five solar cycles the time change of solar FEP and energy spectrum change can be described by the solution of isotropic diffusion (characterized by the diffusion coefficient $\kappa_i(E_k)$) from some pointing instantaneous source $Q_i(E_k,\mathbf{r},t) = N_{oi}\delta(\mathbf{r})\delta(t)$ of FEP of type i (protons, α-particles and heavier particles, electrons) by

$$N_i(E_k,\mathbf{r},t) = N_{oi}(E_k)\left[2\pi^{1/2}(\kappa_i(E_k)t)^{3/2}\right]^{-1} \times \exp\left(-\mathbf{r}^2/(4\kappa_i(E_k)t)\right), \quad (11.113)$$

where $N_{oi}(E_k)$ is the energetic spectrum of total number of FEP in the source. At the distance $r = r_1$ the maximum of FEP density

$$N_{i\max}(r_1,E_k)/N_{oi}(E_k) = 2^{1/2}3^{3/2}\pi^{-1/2}\exp(-3/2)r_1^{-3} = 0.925r_1^{-3} \quad (11.114)$$

will be reached according to Eq. 11.113 at the moment

$$t_1(r_1,E_k) = r_1^2/6\kappa(E_k), \quad (11.115)$$

and the space distribution of FEP density at this moment will be

$$\begin{aligned}\frac{N_i(r,E_k,t_1)}{N_{oi}(E_k)} &= (54/\pi)^{1/2}r_1^{-3}\exp\left(-3r^2/2r_1^2\right)\\ &= 4.146r_1^{-3}\exp\left(-3r^2/2r_1^2\right).\end{aligned} \quad (11.116)$$

According to numerical experimental data the energetic spectrum of generated solar energetic particles in the source can be described approximately as (see the review in Dorman and Venkatesan 1993):

$$N_{oi}(E_k) \approx N_{oi}(E_k/E_{k\max})^{-\gamma}, \quad (11.117)$$

where γ increases with increasing of energy from about $0 \div 1$ at $E_k \leq 1$ GeV/nucleon to about 6–7 at $E_k \approx 10 - 15$ GeV/nucleon. Parameters N_{oi} and γ are changing sufficiently from one event to other: for example, for the greatest observed event of February 23, 1956 $N_{oi} \approx 10^{34} - 10^{35}$, in the event of November 15, 1960 $N_{oi} \approx 3 \times 10^{32}$, in the event of July 18, 1961 $N_{oi} \approx 4 \times 10^{31}$, in the event of May 23, 1967 $N_{oi} \approx 10^{31}$. For the greatest observed event of February 23, 1956 parameter γ had values ≈ 1.2 at $E_k \approx 0.3$ GeV/nucleon, $\gamma \approx 2.2$ at $E_k \approx 1$ GeV/nucleon , $\gamma \approx 4$ at $E_k \approx$ 5–7 GeV/nucleon, and $\gamma \approx$ 6–7 at $E_k \approx$ 10–15 GeV/nucleon. This change of γ is typical for many great solar energetic particle events: see in Dorman (M1957, M1963a, b) about event of February 23, 1956, and review about many events in Dorman (M1963a, b, M1978), Dorman and Miroshnichenko (M1968), Dorman and Venkatesan (1993), Stoker (1994), Miroshnichenko (M2001). Approximately the behavior of value γ in Eq. 11.117 can be described as

$$\gamma = \gamma_0 + \ln(E_k/E_{ko}), \tag{11.118}$$

where parameters γ_0 and E_{ko} are different for individual events, but typically they are in intervals $2 \leq \gamma_0 \leq 5$ and $2 \leq E_{ko} \leq 10$ GeV/nucleon. The position of maximum in Eq. 11.117 taking into account Eq. 11.118 is determined by

$$E_{k\max} = E_{ko}\exp(-\gamma_0)\ , \quad N_{oi}(E_{k\max}) = N_{oi}. \tag{11.119}$$

The total energy contained in FEP will be according to Eq. 11.117–11.119:

$$E_{\text{tot}} = N_{oi}\int\limits_0^\infty E_k(E_k/E_{k\max})^{-\gamma_0-\ln(E_k/E_{ko})}\, d(E_k/E_{k\max}) = bN_{oi}E_{k\max}, \tag{11.120}$$

where

$$b = \int\limits_0^\infty x^{1-\ln x}dx = 4.82. \tag{11.121}$$

For great solar FEP events $E_{\text{tot}} \approx 10^{31}$–$10^{32}$ erg, and more. In Eq. 11.102

$$\kappa_i(E_k) = \Lambda_i(E_k)v(E_k)/3 \tag{11.122}$$

is the diffusion coefficient, $\Lambda_i(E_k)$ is the transport path for particle scattering in the upper corona and interplanetary space, $v(E_k)$ is the particle velocity as a dependence on the kinetic energy per nucleon E_k:

$$v(E_k) = c\left(1 - \left(1 + E_k/m_n c^2\right)^{-2}\right)^{1/2}, \tag{11.123}$$

where $m_n c^2$ is the rest energy of the nucleon.

11.8.6 The Second Factor: Space–Time Distribution of Corona and Solar Wind Matter

The detail information on the second factor for distances smaller than 5 AU from the Sun was obtained by the mission of Ulysses. Important information for bigger distances (up to about 100 AU) was obtained from missions Pioneer 10, 11, Voyager 3, 4, but only not far from the ecliptic plane. If we assume for the first approximation the model of Parker (M1963) of radial solar wind expanding into the interplanetary space which is in good according with all available data of direct measurements in the heliosphere, then the behavior of the matter density of solar wind will be described by the relation

$$n(r,\theta) = n_1(\theta)u_1(\theta)r_1^2/\left(r^2u(r,\theta)\right), \tag{11.124}$$

where $n_1(\theta)$ and $u_1(\theta)$ are the matter density and solar wind speed at the helio-latitude θ on the distance $r = r_1$ from the Sun ($r_1 = 1\,\mathrm{AU}$). The dependence $u(r,\theta)$ is determined by the interaction of solar wind with galactic CR and anomaly component of CR, with interstellar matter and interstellar magnetic field, by interaction with neutral atoms penetrating from interstellar space inside the helio-sphere, by the nonlinear processes caused by these interactions (Dorman 1995a, b; Le Roux and Fichtner 1997). According to calculations of Le Roux and Fichtner (1997) the change of solar wind velocity can be described approximately as

$$u(r) \approx u_1(1 - b(r/r_0)), \tag{11.125}$$

where the distance to the terminal shock wave $r_0 \approx 74$ AU and parameter b $\approx$ 0.13–0.45 in dependence of subshock compression ratio (from 3.5 to 1.5) and from injection efficiency of pickup protons (from 0 to 0.9). From our investigations of CR-SA hysteresis phenomenon (Dorman and Dorman 1967a, b, 1968; Dorman M1975b; Dorman et al. 1997a, b), we estimate $r_0 \approx 100$ AU.

11.8.7 The Third Factor: Gamma Ray Generation by FEP in the Corona and in the Inner Heliosphere

Let us consider in the first generation of neutral pions. According to Stecker (M1971), Dermer (1986a, b), the neutral pion generation caused by nuclear interactions of energetic protons with hydrogen atoms through reaction $p + p \rightarrow \pi^0 +$ anything will be determined by

$$\begin{aligned} &F_{pH}^{\pi}(E_\pi, r, \theta, t) \\ &\quad = 4\pi n(r,\theta,t)\int\limits_{E_{k\min}(E_\pi)}^{\infty} dE_k N_p(E_k, r, t)\langle\varsigma\sigma_\pi(E_k)\rangle(dN(E_k,E_\pi)/dE_\pi), \end{aligned} \tag{11.126}$$

where $n(r,\theta,t)$ is determined by Eq. 11.124, $E_{k\min}(E_\pi)$ is the threshold energy for pion generation, $N_p(E_k,r,t)$ is determined by Eq. (11.113), $\langle\varsigma\sigma_\pi(E_k)\rangle$ is the inclusive cross section for reactions $p+p\rightarrow\pi^0+$ anything, and

$$\int\limits_0^\infty (dN(E_k,E_\pi)/dE_\pi)dE_\pi = 1. \tag{11.127}$$

Gamma ray emissivity caused by nuclear interactions of FEP protons with solar wind matter will be determined according to Stecker (M1971), Dermer (1986a, b), by

$$F_{pH}^{\gamma}(E_\gamma,r,\theta,t) = 2\int\limits_{E_{\pi\min}(E_\gamma)}^\infty dE_\pi\left(E_\pi^2 - m_\pi^2c^4\right)^{-1/2} F_{pH}^{\pi}(E_\pi,r,\theta,t), \tag{11.128}$$

where $E_{\pi\min}(E_\gamma) = E_\gamma + m_\pi^2c^4/4E_\gamma$. Let us introduce Eq. 11.124 in Eq. 11.126 and Eq. 11.128 by taking into account Eq. 11.125:

$$\begin{aligned} F_{pH}^{\gamma}(E_\gamma,r,\theta,t) = B(r,\theta,t)&\int\limits_{E_{\pi\min}(E_\gamma)}^\infty \left(E_\pi^2 - m_\pi^2c^4\right)^{-1/2} dE_\pi \\ &\times\int\limits_{E_{k\min}(E_\pi)}^\infty N_{op}(E_k)\langle\varsigma\sigma_\pi(E_k)\rangle(t/t_1)^{-3/2}\exp\left(-3r^2t_1/2r_1^2t\right)dE_k, \end{aligned} \tag{11.129}$$

where

$$B(r,\theta,t) = 3^{3/2}2^{7/2}\pi^{1/2}r_1^2n_1(\theta,t)u_1(\theta,t)\Big/r^2u(r,\theta,t) \tag{11.130}$$

and

$$t_1 = r_1^2/6\kappa_p(E_k) \tag{11.131}$$

is the time in which the density of FEP at a distance of 1 AU reaches the maximum value. The space distribution of gamma ray emissivity for different t/t_1 will be determined mainly by function

$$f(t,t_1) = r^{-2}(t/t_1)^{-3/2}\exp\left(-3r^2t_1/2r_1^2t\right), \tag{11.132}$$

where t_1, determined by Eq. 11.131, corresponds to some effective value of E_k in dependence of E_γ, according to Eqs. 11.126 and 11.129. The biggest gamma ray emission is expected in the inner region

$$r \le r_i = r_1(2t/3t_1)^{1/2}, \tag{11.133}$$

where the level of emission $\propto r^{-2}(t/t_1)^{-3/2}$. Outside this region gamma ray emissivity decreases very quickly with r as $\propto r^{-2}\exp\left(-(r/r_i)^2\right)$. For an event with total energy 10^{32} erg at t = t$_1$= 10^3 s, $r_i = 10^{13}$ cm, $n_1(\theta,t) \approx 5$ cm^{-3}, $\kappa_p(E_k) \approx 4\times 10^{22}$cm^2/s, we obtain for emissivity of gamma rays with energy ≥ 100 MeV:

$$F_{pp}^{\gamma}\left(E_\gamma > 0.1\ \mathrm{GeV}, r\right) \approx 10^8 r^{-2}\,\mathrm{photon}\cdot\mathrm{cm}^{-3}\mathrm{s}^{-1}. \tag{11.134}$$

Let us note that at the distance of 5 solar radius it gives 10^{-15} photon cm^{-3} s^{-1}. Equation 11.129 describes the space-time variations of gamma ray emissivity distribution from interaction of solar energetic protons with matter of upper corona and solar wind (see Fig. 11.44).

11.8.8 Expected Angle Distribution and Time Variations of Gamma Ray Fluxes for Observations Inside the Heliosphere During FEP Events

Let us assume that the observer is inside the heliosphere at the distance $r_{obs} \leq r_0$ from the Sun and helio-latitude θ_{obs} (here r_0 is the radius of heliosphere). The sight

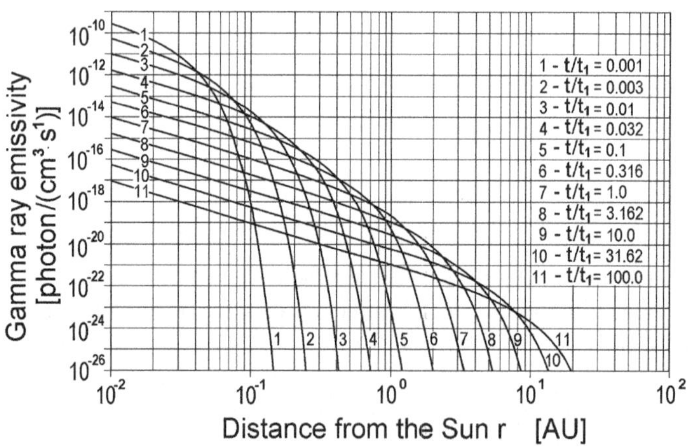

Fig. 11.44 Expected for the event with energy 10^{32} erg space–time emissivity distribution of gamma rays with energy >100 MeV for different time t after FEP generation in units of time maximum t_1 on 1 AU, determined by Eq. (11.131). The curves are from t/t_1= 0.001 up to t/t_1= 100 (From Dorman 2001)

line of observation we can determine by the angle θ_{sl}, computed from the equatorial plane from direction to the Sun to the North. In this case the expected angle distribution and time variations of gamma ray fluxes will be

$$\Phi_{pH}^{\gamma}\left(E_{\gamma}, r_{\mathrm{obs}}, \theta_{\mathrm{sl}}, t\right) = \int_{0}^{L_{\max}(\theta_{\mathrm{sl}})} F_{pH}^{\gamma}\left(E_{\gamma}, L(r_{\mathrm{obs}}, \theta_{\mathrm{sl}}), t\right) dL. \tag{11.135}$$

In Eq. 11.135 gamma ray emissivity

$$F_{pH}^{\gamma}\left(E_{\gamma}, L(r_{\mathrm{obs}}, \theta_{\mathrm{sl}}), t\right) = F_{pH}^{\gamma}\left(E_{\gamma}, r, \theta, t\right) \tag{11.136}$$

is determined by Eq. 11.129 taking into account that

$$r = \left(r_{\mathrm{obs}}^{2} + L^{2} + 2 r_{\mathrm{obs}} L \Delta\theta\right)^{1/2},$$
$$\theta = \theta_{\mathrm{obs}} + \arccos\left(\frac{r_{\mathrm{obs}}^{2} + r_{\mathrm{obs}} L \Delta\theta}{r_{\mathrm{obs}}\left(r_{\mathrm{obs}}^{2} + L^{2} + 2 r_{\mathrm{obs}} L \Delta\theta\right)^{1/2}}\right), \tag{11.137}$$

where $\Delta\theta = \theta_{\mathrm{sl}} - \theta_{\mathrm{obs}}$. In Eq. 11.135

$$L_{\max}(\theta_{\mathrm{sl}}) = \frac{r_0}{\sin\Delta\theta} \sin\left[\Delta\theta - \arcsin\left(\frac{r_{\mathrm{obs}}}{r_0} \sin\Delta\theta\right)\right]. \tag{11.138}$$

According to Eqs. 11.129 and 11.135–11.138 the expected angle distribution and time variations of gamma ray fluxes for local observer ($r_{obs} \leq r_0$) from interaction of solar energetic protons with solar wind matter will be determined by the energy spectrum of proton generation on the Sun $N_{op}(E_k)$, by the diffusion coefficient $\kappa_p(E_k)$, and parameters of solar wind in the period of event near the Earth orbit $n_1(\theta, \tilde{t})$ and $u_1(\theta, \tilde{t})$.

In the case of spherical symmetry we obtain

$$\Phi_{pH}^{\gamma}\left(E_{\gamma}, r_{\mathrm{obs}}, \phi, t\right) \approx F_{pH}^{\gamma}\left(E_{\gamma}, r = r_{\mathrm{obs}} \sin\phi, t\right)(\theta_{\max} - \theta_{\min}) r_{\mathrm{obs}} \sin\phi, \tag{11.139}$$

where ϕ is the angle between direction on the Sun and direction of observation,

$$\theta_{\max} = \arccos(r_{\mathrm{obs}} \sin\phi / r_i), \tag{11.140}$$

$$\theta_{\min} = \begin{cases} -\arccos(r_{obs}\sin\phi/r_i) & \text{if} \quad r_{obs} > r_i, \\ \phi - \pi/2 & \text{if} \quad r_{obs} \leq r_i. \end{cases} \tag{11.141}$$

For the great solar FEP event with the total energy in FEP about 10^{32}ergs Eq. 11.139 for $r_{\text{obs}} = 1$ AU gives

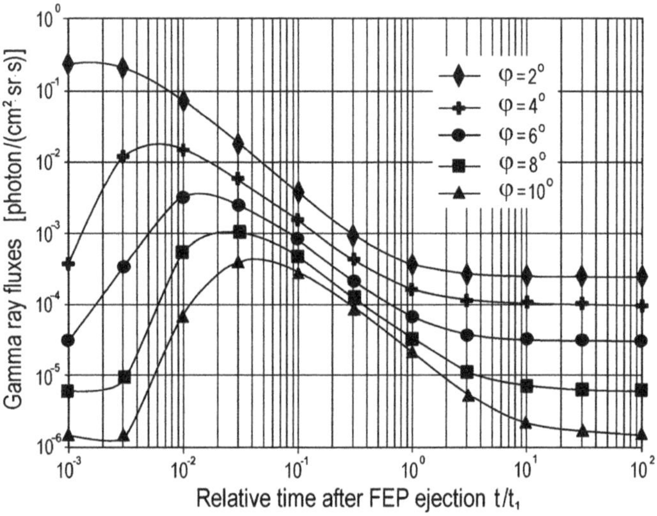

Fig. 11.45 Expected fluxes of gamma rays with energy more than 100 MeV during FEP event with total energy 10^{32} erg for directions from $\varphi = 2°$ to $\varphi = 10°$ from the Sun as a dependence on t/t_1, where t_1 was determined by Eq. (11.131) (From Dorman 2001)

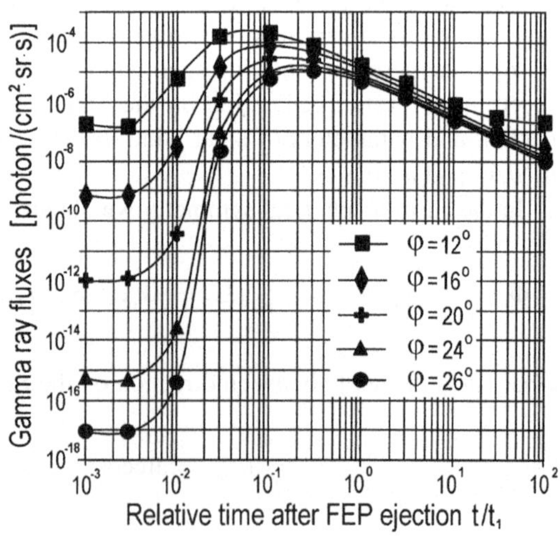

Fig. 11.46 The same as in Fig. 11.45, but for $\varphi = 12°$ to $\varphi = 26°$ (From Dorman 2001)

$$\Phi^{\gamma}_{pH}\left(E_\gamma > 0.1\ \text{GeV}, r_{\text{obs}} = 1\ \text{AU}, \phi, t\right) \approx \frac{6.7 \times 10^{-6}}{\sin\phi}\left(\frac{t}{t_1}\right)^{-\frac{3}{2}} \exp\left(-\frac{3t_1 \sin^2\phi}{2t}\right)\ \text{photon.cm}^{-2}\ \text{sr}^{-1}\text{s}^{-1}. \tag{11.142}$$

Expected fluxes of gamma rays with energy E_γ>0.1 GeV during a large FEP event with total energy 10^{32} erg for different directions of observation characterized by an angle φ from 2° up to 179° as a dependence upon t/t_1 are shown in Figs. 11.45–11.48.

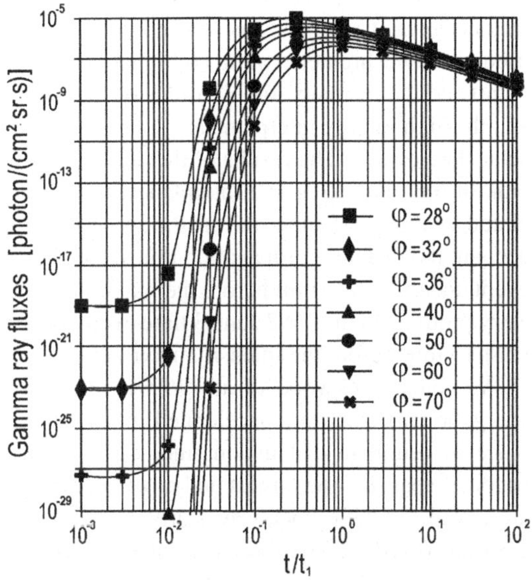

Fig. 11.47 The same as in Fig. 11.45, but for φ = 28° to φ = 70° (From Dorman 2001)

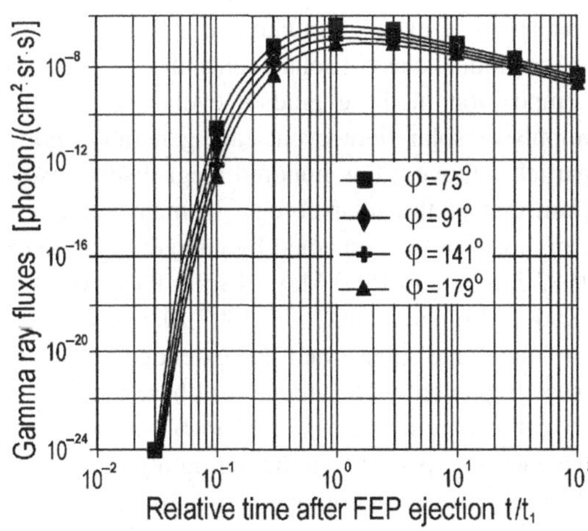

Fig. 11.48 The same as in Fig. 11.45, but for φ = 75° to φ = 179° (From Dorman 2001)

11.9 Pion-Related Gamma Rays: Problem on Interacting and Interplanetary High Energy Particles in Solar Flare Events

11.9.1 The Problem of Relation Between High Energy Particles Interacting with Solar Atmosphere and Ejecting into Solar Wind

In Sections 11.6 and 11.7 we considered the problem how time-profiles of γ-ray lines measurements can give information on the relation between interacting and interplanetary high energy protons, and from this – on the possible magnetic field geometry in the acceleration site. As was shown by Lockwood et al. (1999), for this aim may be used also measurements of much higher energy γ-rays. Lockwood et al. (1999) have compared the intensity–time profiles of the pion-related γ-ray emission ($\sim$70 MeV) and the high-energy interplanetary solar protons in the 1990 May 24 and 1982 June 3 solar flare events. The results of the analysis of these events clearly indicate that the bulk of the interacting and interplanetary high-energy solar protons were not accelerated at the same time and in the same location at or near the Sun. Taking into account the different propagation times, the peak γ-ray emission in the 1990 May 24 event occurred $\sim$20 min before the maximum intensity of the high-energy interplanetary protons. In the 1982 June 3 flare event the maximum γ-ray emission occurred at least 15 min before the intensity maximum of the interplanetary protons.

As noted Lockwood et al. (1999), for many years it was assumed that solar energetic particles were accelerated in solar flares and that they diffused from their place of origin to distant longitudes by 'coronal diffusion'. More recently it has become clear that the largest and most energetic solar particle increases at the Earth are associated with shock waves driven out into interplanetary space by coronal mass ejections (CME). Since these coronal shocks extend over large longitudes at the Sun, coronal diffusion is no longer required. As more observations were made with better instrumentation, it was pointed out that there were two general classes of energetic solar particle events: gradual and impulsive (Reames 1996, and references therein). In the gradual class of solar events there are large fluxes of the energetic particles. The composition in these gradual events is similar to that of the solar corona. The impulsive solar flare events are generally shorter-lived, do not contain energetic particles above a few hundred megaelectron volt and their composition is different from that of the solar corona. In many gradual events the energetic solar particle emission was accompanied by energetic gamma-ray and neutron emission, e.g., the 1982 June 3 and 1990 May 24 solar flare events (Chupp et al. 1987, Kocharov et al. 1994, Debrunner et al. 1993, 1997). The gamma-ray and neutron emissions exhibited an 'impulsive' phase and an 'extended' phase which lasted more than 10 min in these two events. The impulsive phase tended to be richer in electrons, X-rays and $<$0.50 MeV γ-rays and was probably associated with initial prompt acceleration of the energetic particles in the flaring region. In the extended phase a large fraction of the total γ-ray emission was π^0-decay γ-rays and neutron emission was also detected

(Chupp et al. 1987, Debrunner et al. 1993, 1997). Many other gradual solar flare events have exhibited both an impulsive and extended phase. For example, the large 1991 June 11 and 15 solar cosmic ray increases contained both an impulsive and a prolonged phase during which high energy solar γ-radiation was observed (Murphy et al. 1994, Kocharov et al. 1994, Rank et al. 1994).

Lockwood et al. (1999) underlined that the gradual solar flare event on 1990 May 24 was unusual in that both a large solar neutron and γ-ray flux was detected near the Earth as well as energetic interplanetary solar protons (Debrunner et al. 1997, and references therein). It was the largest solar neutron event which has been detected to date. It was possible in this solar particle increase to separate the impulsive and extended phases during which the energetic γ-rays and neutrons were produced from the later prolonged phase during which the interplanetary solar protons observed at the Earth were accelerated, presumably by a coronal shock wave (Debrunner et al. 1997). The onset of the γ-ray event on 1990 May 24 started at 20:47:50 UT and the impulsive phase lasted $\sim$1.5 min during which there were two peaks in the γ-ray emission. The extended phase of the γ-ray and neutron emission lasted more than 12 min. The injection of low energy interplanetary solar protons at the Sun did not start until 20:52 $\pm$ 2 min UT and that of the high energy interplanetary protons until 20:54 $\pm$ 2 min UT, or nearly at the end of the neutron and γ-ray emission. This temporal separation in the 1990 May 24 event leads to ask the following question: Were the energetic protons interacting at the Sun to produce the observed neutron and γ-ray emissions and the interplanetary solar protons accelerated simultaneously and at the same location or were they accelerated at different times and/or in different regions at or near the Sun? Is there evidence in other solar flare events for a time sequence of the various emission processes as they were observed in the 1990 May 24 solar flare? This is not a new question but one which has been difficult to answer unless there was a distinct temporal separation between the production of the bulk of the neutral radiation and the interplanetary protons.

Lockwood et al. (1999) believe that the most compelling evidence for the origin of the high-energy particles can be found in a comparison of the onset times and shapes of the intensity–time profiles of the particles interacting at the Sun with those of the energetic interplanetary proton injections at the Sun. Ramaty et al. (1993a) studied the abundances and the energy spectra of interacting flare protons and heavier nuclei using the nuclear deexcitation and 2.2 MeV γ-ray line emissions. They compared the results with measurements of interplanetary solar flare particles and found that the protons and heavier nuclei that produce the γ-rays in both impulsive and gradual flares are accelerated by the same mechanism, probably stochastic acceleration. On the other hand, they were unable to determine from their analysis whether the interplanetary protons were accelerated by the same mechanism operating on open field lines, were accelerated by a different mechanism high in the corona or in interplanetary space, or escaped from the magnetic loops in which the γ-rays were produced. Reames (1996) pointed out that "recent evidence shows that particles accelerated by CME-driven shocks produce most of the largest solar particle events (referred to as 'gradual events') seen at 1 AU, with the particles of highest energies being accelerated nearer the Sun where the shock is strongest".

In an earlier study of injection profiles of solar energetic particles and the propagation of coronal mass ejections, Kahler (1994) found that the interplanetary particle increases start no earlier than the maxima of the X- and γ-ray emissions during the impulsive phases. Kahler's result is consistent with the idea that the solar energetic particles observed in interplanetary space after gradual flares are accelerated in shocks driven by the CMEs. From a review of several studies and reports on interacting and interplanetary solar energetic particles, Cliver (1996) supported the result of Ramaty et al. (1993a) that a common acceleration process for interacting particles is operating in both gradual and impulsive flare events. In addition, Cliver (1996) expanded Reames' tabular summary of the two-class picture of interplanetary solar particle events and argued that gradual solar flare events "can be expected to have a temporally and spatially confined core of flare-accelerated particles surrounded by a halo of CME/shock particles". Measurements below 100 MeV/nucleon provide clear evidence that the elemental abundances and charge states of the interplanetary particles in gradual events are characteristic of the solar corona rather than the flare site (Reames 1996). Above 100 MeV/nucleon the charge state measurements by Tylka et al. (1996) also indicate that the interplanetary particles are characteristic of the solar corona.

As underlined Lockwood et al. (1999), all spectral and charge state results imply that the protons and heavier nuclei interacting with the solar atmosphere to produce neutral emissions and the bulk of the interplanetary protons of gradual events are probably accelerated in different regions of the Sun and by different acceleration processes. Lockwood et al. (1999) provide new temporal evidence that in so-called gradual solar flare events the high-energy protons that interact at the Sun and the high-energy protons injected into interplanetary space are accelerated at different times and, by inference, in different regions at or near the Sun. As previously suggested, they believe that the most compelling evidence for the origin of the high-energy particles can be found in a comparison of the onset times and shapes of the intensity–time profiles of the particles interacting at the Sun with those of the energetic interplanetary proton injections at the Sun. If the interacting and interplanetary solar protons come from the same particle population, where the interacting particles were produced in regions of closed magnetic field lines and the interplanetary particles in neighboring regions of open field lines or where the interplanetary particles escape from regions where the interacting particles produce γ-rays, we would expect that the onsets and intensity–time profiles of both particle species would coincide and have a similar shape.

Lockwood et al. (1999) compare the intensity–time profiles of the pion-related γ-ray emission ($E_\gamma > 60$ MeV) and the energetic interplanetary solar proton fluxes in the 1990 May 24, and 1982 June 3 solar flare events. In these events it was found that the high-energy interacting and interplanetary solar protons were not accelerated at the same time and, by inference, at the same place (see details in Sections 11.9.2 and 11.9.3). Not for all γ-ray solar events can be found clear solution on the relation between interacting and interplanetary high energy particles. It became especially difficult or impossible when diffusion propagation is dominant, as in the 1991 June 11 and 15 solar flare events. During these events long-duration pion-related γ-ray emissions were detected by the sensitive instruments on the Compton Gamma Ray Observatory (CGRO) and the GAMMA-1 spacecraft and interplanetary solar protons were observed by spacecraft detectors and neutron monitors on the ground, but as it was shown by Lockwood et al. (1999), no definitive conclusions can be drawn from these measurements about the origin of the interacting and interplanetary solar protons in the 1991 June 11 and 15 solar flare events (see details in Section 11.9.4).

11.9.2 Main Results Based on Observations of the 1990 May 24 Solar Flare Event

According to Lockwood et al. (1999), in Fig. 11.49 is shown the normalized intensity–time profile of the 60–95 MeV pion-related γ-ray count rates measured by the PHEBUS detector on the GRANAT spacecraft during the 1990 May 24 solar flare event (Debrunner et al. 1993, 1997).

The γ-ray count rate profile in Fig. 11.49 represents the intensity of >300 MeV interacting solar protons during the 1990 May 24 solar flare event (Debrunner et al. 1997). The bulk of the production of pion-related γ-rays started at 20:48:18 UT, rose rapidly to a maximum 18 s later, and then decayed exponentially with an

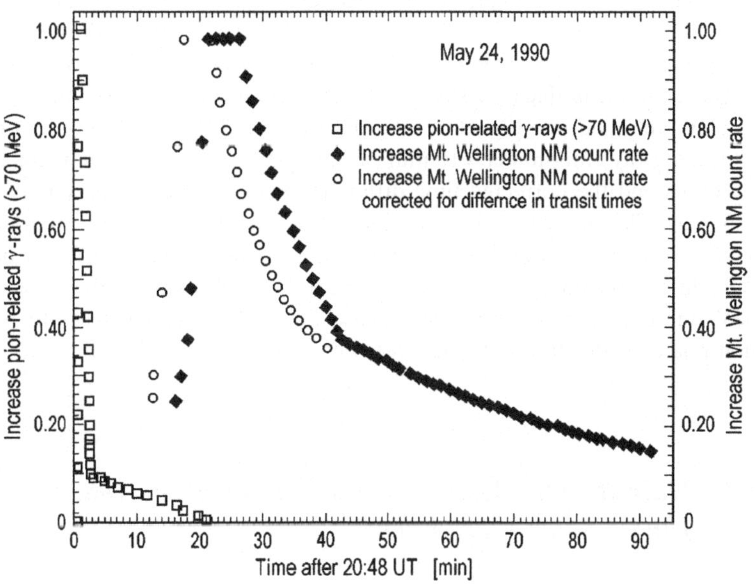

Fig. 11.49 The normalized intensity–time profiles at the Earth of pion-related 60–95 MeV γ-rays from the PHEBUS detector (Debrunner et al. 1997) and the interplanetary solar protons detected by the Mt. Wellington neutron monitor and the time-shifted intensity–time profile of the interplanetary protons on 1990 May 24 (From Lockwood et al. 1999)

e-folding time of ~20 s. The impulsive phase was followed by a prolonged emission for ~20 min. The measured γ-ray flux in the 70–100 MeV energy interval during the impulsive phase was 0.03 photon/(cm^2 s MeV) and during the extended phase from 20:51:35–20:52:25 UT was ~0.003 photon/(cm^2 s MeV). During the impulsive phase the 2.2 MeV γ-ray line emission was ~25 photons/(cm^2 s) and during the extended phase 20:51:35–20:52:25 UT ~2 photons/(cm^2 s). Debrunner et al. (1997) successfully used the pion production curve to represent the intensity–time profile of the high-energy neutron production. With this neutron production curve a good fit of the calculated to the observed NM count rates at Climax and Mexico City was obtained (Debrunner et al. 1997). Either a shorter or longer decay time of the neutron production gave much poorer fits of the calculated to the observed NM count rates.

The intensity–time profile of the interplanetary high energy solar proton emission during this event is represented in Fig. 11.49 by the normalized increase of the l-min count rates of the Mt. Wellington, NM. The Mt. Wellington, NM count rates increase was mainly due to high energy protons ($E > 2$ GeV) propagating directly from the Sun to the Earth with little interplanetary scattering (Debrunner et al. 1997). The NM count rate increase started at 21:02:30 UT, had a maximum count rate increase of ~50%, and lasted for more than 8 h. Since the transit times from the Sun to the Earth for the interplanetary protons and the γ-rays are different, Lockwood et al. (1999) shifted the intensity–time profile of the interplanetary protons to take this difference into account. Debrunner et al. (1997) found that the path length of the interplanetary protons along the Parker spiral interplanetary magnetic field was 1.7 ± 0.3 AU. Consequently, the intensity–time profile of the interplanetary protons in Fig. 11.49 is shifted by 6 ± 2 min. Even with this time shift, it is clear that the shapes of the normalized intensity–time profiles of the interacting and interplanetary solar protons are different. The onset time of the π^0-decay γ-ray emission is ~10 min earlier than that of the energetic interplanetary protons and the corrected time difference between the intensity maxima is 20 ± 5 min. This indicates that the interacting and interplanetary high-energy protons were accelerated at different times and, by inference, in different places. This conclusion is consistent with the 6 ± 2 min delay of the onset time of the >1 GeV interplanetary solar proton release at the Sun with respect to that of the pion-related γ-ray emission (Debrunner et al. 1997).

11.9.3 Main Results Based on Observations of the 1982 June 3 Solar Flare Event

Lockwood et al. (1999) compare in Fig. 11.50 the intensity–time profiles at the Sun of the interacting and interplanetary solar protons during the 1982 June 3 solar flare event. Since the high-energy γ-rays in this event were measured at the Earth whereas the interplanetary solar protons were detected at the HELIOS-1 spacecraft

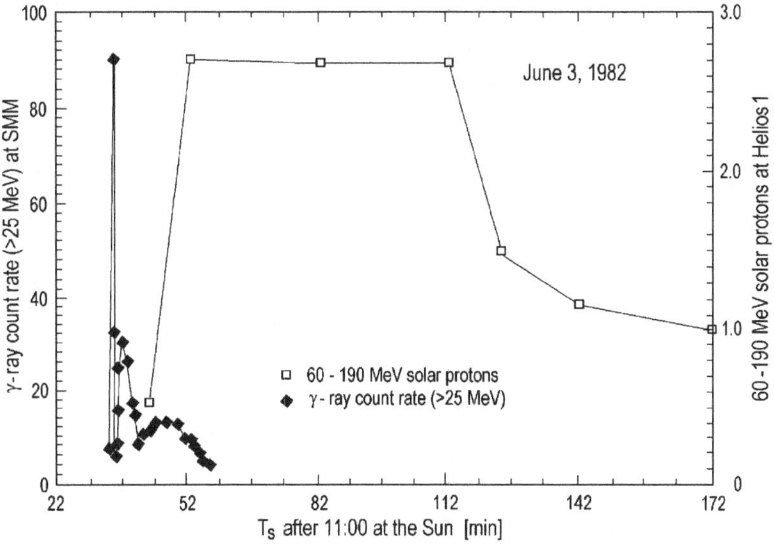

Fig. 11.50 The normalized intensity–time profile at the Sun of the >25 MeV γ-rays deduced from the GRS measurements on the Solar Maximum Mission (Chupp et al. 1987) and the 60–190 MeV interplanetary solar protons deduced from the HELIOS 1 measurements (McDonald and Van Hollebeke 1985) on 1982 June 3 (From Lockwood et al. 1999)

located at 0.57 AU from the Sun, the time was shifted both intensity–time profiles back to the Sun.

As noted Lockwood et al. (1999), the interacting >300 MeV proton intensity–time profile for this event is represented by the count rate in the >25 MeV γ-ray channel from the gamma ray spectrometer on the Solar Maximum Mission (Chupp et al. 1987). This channel also responded to neutrons with kinetic energies from about 100 to 1,000 MeV but the γ-ray response can be resolved from the neutron response (Forrest et al. 1986). The >25 MeV γ-ray emission had two peaks, starting at ~11:35 and ~11:37 UT, respectively, and a time-extended phase primarily due to pion decay γ-rays (Forrest et al. 1986). Solar neutrons were detected by the Jungfraujoch NM and GRS/SMM detector (Debrunner et al. 1983; Chupp et al. 1987). The neutron production must have started by ~11:36 UT. At this time the 2.2 and 4–7 MeV γ-ray line emissions reached their maxima with both emissions lasting more than 5 min. A good fit of the calculated to the observed NM and GRS neutron count rates was obtained by assuming that the intensity of the neutron production followed that of the pion-related γ-rays and lasted a similar length of time (Chupp et al. 1987). A longer or shorter duration of the neutron production would not give as good fits of the calculated to the measured increase at the Jungfraujoch, NM.

Since no interplanetary solar protons with $E > 300$ MeV could be measured during this event at the Earth because of the flare location, Lockwood et al. (1999) used the 60–190 MeV proton fluxes detected at the HELIOS-1 spacecraft

(McDonald and Van Hollebeke 1985, Van Hollebeke et al. 1990) to construct the high-energy interplanetary solar proton intensity–time profile. In solar flare events with anisotropic solar proton fluxes, the intensity–time profiles of the protons in the >300 MeV energy range to which neutron monitors respond are similar to those in the 60–190 MeV range when both are projected back to the Sun (Lockwood et al. 1990a). This result is assumed to be also valid for the present analysis. Assuming that the 60–190 MeV protons with $\beta = v/c = 0.4$ propagated from the Sun to HELIOS 1 with little interplanetary scattering, the transit time along the interplanetary magnetic field line from the Sun to HELIOS 1 is 12 ± 2 min. The amplitudes of the two curves were normalized at the maximum intensities. The interplanetary solar proton emission at the Sun began at ~11:44 UT and had a broad maximum from ~12–13 UT. The >25 MeV γ-ray emission produced by energetic protons interacting in the solar atmosphere started at ~11:35 UT and had essentially ceased by ~12:00 UT. Consequently, the duration of the maximum injection of the protons into interplanetary space lasted much longer than that of the γ-ray and neutron production. The onset times and the propagation characteristics of the interplanetary protons are not so well defined here as in the 1990 May 24 solar flare event. However, it is clear that the shapes of the intensity–time profiles of the energetic γ-rays and interplanetary protons are different and do not coincide in time. Given the strong particle anisotropy during the main phase of the event (McDonald and Van Hollebeke 1985, Van Hollebeke et al. 1990), it seems unlikely that the broad maximum in the interplanetary solar proton flux was due to transport effects of the interplanetary magnetic field. Therefore, it does not appear that the interacting and interplanetary protons were accelerated simultaneously and at the same place.

11.9.4 The 1991 June 11 and 15 Solar Flare Events

We have seen in Sections 11.9.2 and 11.9.3 that in the 1990 May 24 and 1982 June 3 solar flare events there is hard evidence that the high-energy protons interacting at the Sun and the solar protons detected in interplanetary space were not accelerated simultaneously and, by inference, in the same region at the Sun. Can the large June 1991 solar flares provide any additional support for these conclusions?

As noted Lockwood et al. (1999), the June 1991 flare events were some of the largest ever seen (Ryan et al. 1994; Kanbach et al. 1993) and had long duration γ-ray emissions above a few megaelectron volt that were detected by the sensitive COMPTEL and EGRET γ-ray instruments on CGRO for several hours (Ryan et al. 1994; Ryan 2000; Kanbach et al. 1993; Schneid et al. 1994; Rank 1996) as well as by the γ-ray telescope on GAMMA 1 (Akimov et al. 1991, 1993). Therefore, it appears that an analysis of these events might also provide an insight into the origin of the interacting and interplanetary solar flare protons. Lockwood et al. (1999) hypothesize that the long duration γ-ray emissions were due to the greater sensitivity of the γ-detectors on CGRO. Even if the analysis is restricted to the June 11 and 15 solar flares because significant fluxes of high-energy interplanetary solar particles were observed only during these two events, the intensity–time profiles of their injection into the interplanetary magnetic field near the Sun cannot be deduced because the protons were transported diffusively in interplanetary space. Furthermore, the onset of the γ-ray emission in the 1991 June 15 flare was not observed. Consequently, no definitive conclusions can be drawn in these events about the origin of the high energy interacting and interplanetary solar protons by the method used in described above analysis. Mandzhavidze et al. (1996) in a detailed analysis of the pion decay and nuclear line emissions from the 1991 June 11 event concluded that there were at least three distinct emission phases characterized by changes in the spectrum of the protons producing the γ-rays during transitions from one phase to another. They explained these results by assuming episodic accelerations in the trapping regions. The protons then precipitated from the low density trapping regions into sub-coronal regions producing the γ-rays.

11.9.5 Comparison of the Intensity–Time Profiles of the 1990 May 24 and 1982 June 3 Solar Flare Events

As it was shown in Sections 11.9.2 and 11.9.3, the onsets and shapes of the intensity–time profiles of the high-energy interacting protons and those of the interplanetary proton injections at the Sun differed significantly from each other in the 1990 May 24 and 1982 June 3 solar flares. In these two solar cosmic ray increases Lockwood et al. (1999) have physically interpretable intensity–time profiles of both the pion-related γ-ray emission and the energetic interplanetary solar protons. The intensity–time profile of the pion-related γ-ray emission can be determined in both events and the propagation of the interplanetary protons from the Sun to the observers was practically scatter-free. In these two cases the differences in the intensity–time profiles shown in Figs. 11.49 and 11.50 can be explained only if the interacting and interplanetary particles were accelerated at different times and, by inference, in different regions.

If to assume that the bulk of the high-energy γ-rays were produced by protons interacting at low altitudes in the solar atmosphere and the major portion of the high-energy interplanetary protons were accelerated later by a coronal shock, it can be use the time difference of the onsets and the peak intensities of the high-energy interplanetary protons and γ-rays to infer the propagation speed of the disturbance at the Sun that evolves into a shock wave. In such a scenario the major portion of the interacting high-energy protons are accelerated in the initial solar blast. The acceleration may continue for hours as the wave propagates out into the solar corona to form the shock that later accelerates the interplanetary protons. In turn some of these interplanetary protons may leak back into the Sun to produce high-energy γ-rays.

As noted Lockwood et al. (1999), for the 1982 June 3 event it can be, in this picture, only guess the propagation speed of the solar disturbance. Taking the measured offset of the increasing high energy γ-ray and interplanetary proton emissions and assuming a speed of 1,500 km/s, the bulk of the interplanetary protons were accelerated by the coronal shock at $\sim$2 solar radii. For the 1990 May 24 solar flare Kocharov et al. (1994) independently deduced a speed of 1,500 km/s for the shock that also places the bulk of the acceleration of the interplanetary protons at $\sim$2 solar radii. In both cases these are reasonable estimates for the location of the shock acceleration of the interplanetary protons.

11.9.6 Other Comparisons to Ascertain the Origin of the Interacting and Interplanetary Solar Protons

As underlined Lockwood et al. (1999), there are other analyses that can be made to investigate whether or not the high-energy protons interacting at the Sun and the interplanetary solar protons are accelerated at different times and in different regions at the Sun (see, e.g., Cliver 1996). First, the most reliable method, other than a comparison of the intensity–time profiles of the pion-related γ-ray emission and the interplanetary solar proton injection, is a detailed determination of any difference in the onset times of the pion-related γ-ray emission and the start of the release of the interplanetary protons at the Sun as well as an analysis of the onset time of the interplanetary proton emission as a function of proton energy. Second, a comparison of the energy spectrum of the protons producing the neutrons and γ-rays and the interplanetary solar proton spectrum at the Sun can be made. Third, a comparison of the total fluencies of the interacting and interplanetary solar protons can also be made. However, none of these comparisons alone provides the evidence necessary to establish the common or different origin of the protons responsible for the γ-ray and neutron emissions and the interplanetary protons. Lockwood et al. (1999) have, nonetheless, made some of these comparisons for two well-observed and analyzed events.

11.9.7 Onset Time of the Pion-Related γ-Ray Emission and the Start of the Release of the Interplanetary Protons at the Sun

As noted Lockwood et al. (1999), in several flare events (e.g., 1978 May 7, 1984 February 16) there are indications that the first low energy protons ($\sim$50 MeV) were injected into the interplanetary magnetic field region at the Sun at least 2–5 min earlier than the first high-energy protons (Lockwood et al. 1990b). This result indicates that the interplanetary protons were not all accelerated simultaneously. The time difference can be explained by the longer time required to accelerate the protons to higher energies. It does not necessarily favor any one acceleration mechanism. On 1990 May 24 the pion-related γ-ray emission started ~4 min before the first low energy interplanetary protons and ~6 min before the first high-energy interplanetary protons were injected into the interplanetary magnetic field region (Debrunner et al. 1997). If the interacting and interplanetary protons came from the same particle population and were accelerated in the coronal loop where the γ-rays were produced, this would require an effective particle storage for ~6 min at the Sun with little or no leakage. This argues against a common origin for both particle species. From the difference in the onsets of the interplanetary proton and the pion-related γ-ray emissions Lockwood et al. (1999), therefore, conclude that the interacting and interplanetary protons were accelerated at different times and locations, either by the same mechanism operating in a region of closed magnetic field lines and with a delay also in a neighboring region of open magnetic field lines, or by different mechanisms operating in quite different regimes. The difference in the shapes of the intensity–time profiles of the pion-related γ-ray emission and the interplanetary solar proton injection favors the second conclusion.

11.9.8 Comparison of the Energy Spectrum of the Protons Producing the Neutrons and γ-Rays and Interplanetary Solar Proton Spectrum at the Sun

As underlined Lockwood et al. (1999), it is difficult to make direct comparisons of the energy spectra of the protons interacting at the Sun and interplanetary solar proton spectra. The energy spectra of the interacting protons cannot be measured directly but only derived from the γ-ray spectra. In solar flare events the energy spectrum of the interplanetary protons near the Earth can be measured directly for $E < 500$ MeV and deduced with a high-degree of confidence for $E > 500$ MeV using neutron monitor measurements (Lockwood et al. 1990a). For example, Lockwood et al. (1999) show in Fig. 11.51 a comparison of the interplanetary and interacting solar proton spectra for the 1990 May 24 increase. The interacting proton spectrum for $E > 300$ MeV was deduced for the time-extended phase when 70% of the neutrons were produced (Debrunner et al. 1997). The interplanetary solar proton spectrum in Fig. 11.51 is for the period 21:25–22:00 UT after the interplanetary solar proton spectrum had softened. The interacting proton spectrum was extrapolated to energies <300 MeV and the two spectra were normalized at 200 MeV. Clearly, the interplanetary solar proton spectrum is harder than that of the protons interacting at the Sun.

Lockwood et al. (1999) noted that if the interplanetary protons are leakage particles from a common acceleration region for interacting and interplanetary particles at the Sun and if we assume that leakage by diffusion is greater for higher than lower energy protons, we would expect the interplanetary proton energy spectrum to be harder than the interacting proton spectrum. However, it could also be that the interacting and interplanetary protons were accelerated in different

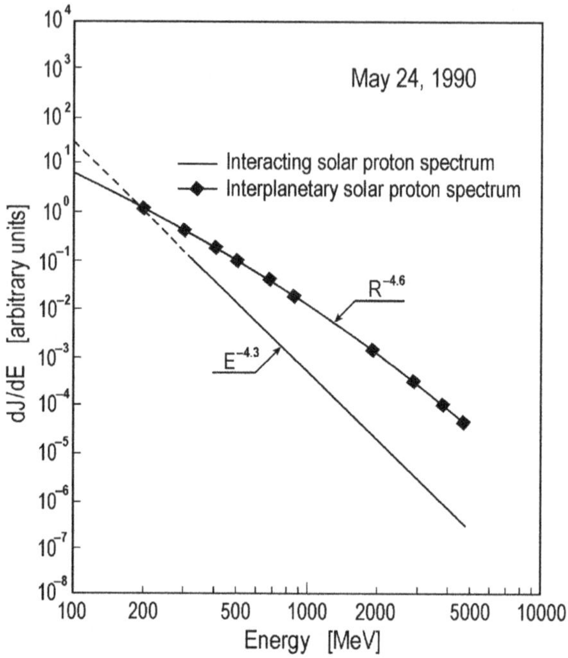

Fig. 11.51 The energy spectra of the interacting solar protons and the interplanetary solar protons during the 1990 May 24 solar flare event. The two spectra were normalized at 200 MeV. For details see the text (From Lockwood et al. 1999)

regions and by different mechanisms and that the energy spectra are not related to each other. It has become increasingly evident that the differential energetic interplanetary solar proton flux (*dJ*/*dE*) in many large solar cosmic ray events where both neutron monitor ($E > 500$ MeV) and spacecraft ($E < 500$ MeV) measurements were present can be represented by a power law in rigidity, R^{-S}, that is characteristic of diffusive coronal shock acceleration (Ellison and Ramaty 1985, Lockwood et al. 1990a).

For example, in the 1980 June 21 and 1982 June 3 solar flare events, S = 5.4 and S = 1.5 for $R < 650$ MV ($E < 200$ MeV), yielding shock compression ratios of 1.7 and 3.1, respectively (Ellison and Ramaty 1985). For the 1978 May 7 and 1984 February 16 solar proton increases Lockwood et al. (1990a) found $3.8 < S < 5.0$ and $4.4 < S < 5.2$, respectively, for 50 MeV $< E <$ 5 GeV from extrapolation of the interplanetary proton spectra back to the Sun. In both events the spectrum softened with time. During the 1990 May 24 event the interplanetary solar proton energy spectrum from ~100 MeV to 10 GeV could also be described by a power law in rigidity with $4.0 < S < 4.6$ (Debrunner et al. 1997). In these latter three events interplanetary shocks were present as in the 1982 June 3 solar flare event. Thus, the interplanetary proton spectra are consistent with a shock origin for the interplanetary solar protons.

11.9.9 Comparisons of the Total Fluencies of the Interacting and Interplanetary Solar Protons

According to Lockwood et al. (1999), comparisons of the total number of high-energy protons interacting at the Sun with that of interplanetary solar protons also do not lead to any definitive conclusions as to whether the particles come from the same or different particle populations at the Sun (Cliver et al. 1989; Ramaty et al. 1993a). The ratios of the interplanetary to interacting proton fluencies are highly variable and not necessarily less than unity (Cliver et al. 1989; Ramaty et al. 1993a). In the 1982 June 3 event the ratio of the interplanetary to interacting proton fluencies was 26–130%, depending on the assumed solar composition (Ramaty et al. 1993a), and in the 1990 May 24 event the ratio was 7–14% (Debrunner et al. 1997). This is not inconsistent in a limited way with the idea that the interplanetary protons were leakage protons from the acceleration or trapping region. However, it is also consistent with the Lockwood et al. (1999) hypothesis of two unrelated accelerations.

11.9.10 Summary of Main Results and Conclusions

Lockwood et al. (1999) postulate a scenario of acceleration for particles in gradual solar flares events where the high-energy interacting protons are accelerated in the solar corona in the blast wave to produce the high-energy γ-rays. The resulting shock wave propagates outward, accelerating the interplanetary protons for long periods of time over a spatial scale of $\sim$2 solar radii. Some of these particles may precipitate back toward the Sun and interact there, continuing to produce high-energy γ-rays, as proposed to explain the different phases of the 1991 June 11 event (Mandzhavidze et al. 1996). In this case the interacting and interplanetary protons come from the same particle population but the bulk of each particle species is accelerated in different regions at different times. It is difficult, however, for the shock-accelerated particles to propagate backward to the Sun in the turbulent region downstream of the shock. Therefore, Lockwood et al. (1999) assume that only a small fraction of the interacting protons is produced in this way. From a comparison of the onsets and intensity–time profiles of the pion-related γ-ray emission (>60 MeV) and the energetic interplanetary solar protons in the 1990 May 24 and 1982 June 3 solar flare events, there is clear evidence that the bulk of the high-energy interacting and interplanetary solar protons were not accelerated simultaneously and in the same region at the Sun. The interplanetary particles were accelerated for a time much longer than the duration of the pion-related γ-ray emissions. Even if the physics of the acceleration of the interacting and interplanetary particles is the same, the times and locations of the acceleration must be very different, pointing to local acceleration at the Sun of the interacting particles and remote shock acceleration of the interplanetary particles, confirming the picture that has developed at lower energies. Therefore, these results suggest that at both low and high energies the protons interacting at the Sun and the protons measured in interplanetary space are accelerated at different times and, by inference, in different regions in gradual solar flare events.

11.10 Angular and Energy-Dependent Neutron Emission from Solar Flare Magnetic Loops: Model and Monte Carlo Simulation

11.10.1 The Matter and Short History of the Problem

Hua et al. (2002) have developed new neutron production kinematics and thoroughly updated the neutron production cross sections, and have included ion pitch-angle scattering and magnetic mirroring in the Monte Carlo simulation programs, to make new calculations of anisotropic neutron emission produced in the solar flare magnetic loop models. The anisotropy in these models arises from the combined effects of converging magnetic field lines and a rapidly increasing ambient density in the portion of the loop below the chromosphere–corona transition. Hua et al. (2002) have carried out new calculations of the depth, time, angle, and energy dependences of the neutron production, the angle distributions and energy spectra of the escaping neutrons, and the energy spectrum of the surviving neutrons at the distance 1 AU from the Sun. These new calculations will now allow much more reliable and detailed analyses of the various solar flare neutron spectral observations.

As noted Hua et al. (2002), the understanding of solar flare particle acceleration, interactions, and propagation has been greatly advanced by comparisons of solar flare neutron and gamma-ray line observations with theoretical calculations of the fluxes expected from flare-accelerated ion interactions in the solar atmosphere. Because they are produced directly by nuclear interactions of the flare-accelerated protons and heavier ions with ambient gas in the solar atmosphere, these neutrons and gamma-rays give us the most direct information available on the total number, energy spectrum, time dependence, and angular distribution of ion acceleration and propagation in flares. They can also provide unique information on the composition, scale height, magnetic field convergence, and MHD turbulence in the flare region. From detailed calculations of the expected neutron production, it was subsequently showed in Chapter 1 that measurements of the time dependence of the solar flare neutron flux could give a direct measure of both the spectrum and total number of accelerated ions in flares. For when the neutrons are produced in a time that is much shorter than their transit time from the Sun to the Earth, a time-of-flight measurement can be made of their spectrum, which is in turn strongly dependent on the accelerated ion spectrum. From first calculations of the expected gamma-ray fluxes, was further showed that the principal gamma-ray lines should be those at 2.223 MeV from neutron capture on ^{1}H, at 0.511 MeV from positron annihilation, and at 4.438 and 6.129 MeV from deexcitation of nuclear levels in ^{12}C and ^{16}O. The latter two lines dominate the solar flare gamma-ray emission in the 4–7 MeV band. It was also shown that measurements of both the 2.223 and 4.438 or 6.129 MeV line fluxes from flares could give a second, independent measure of the accelerated ion spectrum and number, since the ratios of these line fluxes are strongly dependent on the shape of the accelerated ion energy spectrum. In addition, in previous chapters was shown that the gamma ray fluxes can provide a direct measure of the ionic heating of the flare plasma; the time dependence of the flux of the 2.223 MeV line from neutron capture on ^{1}H can provide a direct measure of the solar photospheric $^{3}He/^{1}H$ abundance, since neutron capture on ^{3}He is an important loss process. Subsequent observations have borne out all of these expectations and much more. Gamma-ray line emission was first observed with a detector on OSO-7 from the solar flare of 1972 August 4 at the predicted line energies of 0.51, 2.2, 4.4, and 6.1MeV. These and other weaker lines were observed from other flares by detectors on HEAO-1, HEAO-3, Hinotori, and SMM (see Chapters 2, 8–10). Neutrons were observed from five flares between 1980 and 1989 with the SMM detector (Chapters 3 and 4). High energy neutrons were also observed from a number of flares with ground-based neutron monitors and neutron telescopes (see Chapters 3, 6, and 7) and both protons and electrons from the decay of solar flare neutrons were detected in interplanetary space by ISEE 3 (Chapter 8). More recently the neutron and gamma-ray spectrometers on CGRO have measured gamma-ray spectra from more than 15 large flares and neutron spectra from five of them (see Chapter 9). These measurements not only roughly double our database, they also give us the first direct measurements of the neutron spectra and provide gamma-ray spectra over a much broader energy range with much finer time resolution and a wide range of heliocentric observing angles.

Hua et al. (2002) underlined that detailed theoretical calculations of neutron (Hua 1986; Hua and Lingenfelter 1987a, b, c, d, e) and gamma-ray line (Ramaty et al. 1975, 1979; Ramaty and Murphy 1987) production by flare-accelerated ion interactions were carried out in order to interpret the earlier measurements. More recently, neutron and gamma-ray line production models have been expanded (Hua et al. 1989; Guglenko et al. 1990a, b; Kurganov and Ostryakov 1992; Ramaty et al. 1996; Kocharov et al 1999a, b; Vainio et al. 2000) to investigate the effects of pitch-angle diffusion and magnetic mirroring of the flare-accelerated ions in the solar atmosphere, and better understand the angular dependence of the neutron and gamma-ray line emission observed from different heliocentric angles. It was found (Hua 1986; Hua and Lingenfelter 1987d) that, independent of the assumed angular distribution of the accelerated ions, the measured time-dependent neutron flux at 1 AU from flares was much more consistent with that produced by accelerated ions with a Bessel function spectrum than with that from a power-law spectrum. This result confirmed previous studies (Ramaty et al. 1983b; Murphy and Ramaty 1984) which assumed simple isotropic neutron production and escape. In general, it was found (Hua 1986; Hua and Lingenfelter 1987d; Murphy et al. 1987) that power-law spectra, capable of producing the observed ratio of the neutron flux to the excess 4–7 MeV fluence, produced neutron energy spectra that were too hard and hence time-dependent fluxes at 1 AU that were too intense at early times. However, other studies (Kocharov et al. 1987; Guglenko et al. 1990a, b) also found that consistent neutron spectra could be produced by accelerated ions with power-law spectra, if different neutron production kinematics were assumed in the p–α and α–α interactions. This is an important problem that can be explored with these new calculations.

As noted Hua et al. (2002), these significant differences in the calculated neutron spectra have prompted them to thoroughly reevaluate and revise the neutron kinematics and cross sections assumed in their calculations. These revisions have focused primarily on the neutron production kinematics assumed in this program for the p–α and α–α interactions, which affect the neutron production spectrum. Hua et al. (2002) have developed new kinematic approximations for the differential angular and energy-dependent neutron yield from multi-particle breakup that are in very good agreement with the available laboratory measurements and theoretical simulations (e.g., Roy et al. 1981; Barashenkov et al. 1983, 1984; Wesick et al. 1985), as well as the semi-classical kinematics developed by Bildsten et al. (1990) for the n^3He breakup of ^4He at energies just above the threshold.

11.10.2 Neutron-Producing Reactions and Their Threshold Energies

The projectile and target nuclei of the various reactions that Hua et al. (2002) consider are summarized in Table 11.9 which also shows the threshold energies of neutron-producing reactions.

Table 11.9 Targets, projectiles, and neutron production threshold energies (MeV per nucleon) (From Hua et al. 2002)

Isotopes	Proton	α-Particle
^1H	292.3	25.7
^3He	10.3	5.5
^4He	25.7	9.5
^{12}C	19.6	2.8
^{13}C	3.2	Exothermic
^{14}N	6.3	1.5
^{15}N	3.7	2.0
^{16}O	17.2	3.8
^{18}O	2.5	0.2
^{20}Ne	15.4	2.2
^{22}Ne	3.8	0.15
^{24}Mg	15.0	2.1
^{25}Mg	5.3	Exothermic
^{26}Mg	5.0	Exothermic
^{28}Si	15.6	2.3
^{29}Si	5.9	0.4
^{56}Fe	5.5	1.4
^{54}Fe	9.2	1.6

11.10.3 Neutron Production by Proton-Induced Reactions with ^{13}C, ^{15}N, ^{18}O, ^{22}Ne, ^{25}Mg, ^{26}Mg, ^{29}Si, and ^{54}Fe

The total neutron production cross sections σ_n in interactions with ^{13}C, ^{15}N, ^{18}O, ^{22}Ne, ^{25}Mg, ^{26}Mg, ^{29}Si, and ^{54}Fe are shown in Fig. 11.52.

Below incident energies of about ~30 MeV the most important neutron-producing reactions are (p, n), (p, pn), and $(p, 2pn)$. All of these reactions, except for (p, pn) on ^{56}Fe, also produce positrons, whose production cross sections were considered in detail by Kozlovsky et al. (1987), and Hua et al. (2002) use those values. For ^{56}Fe (p, pn) ^{55}Fe, Hua et al. (2002) use cross sections given by Pearlstein (1989). The sum of all of these reactions defines the inclusive neutron production cross section from threshold up to ~30 MeV. The importance of the reactions with these rare isotopes is in their low threshold energies and high cross sections. Hua et al. (2002) have normalized the cross sections to the principal isotope of each element using the abundances given by Cameron (1982).

As noted Hua et al. (2002), at higher energies it is impractical to study the numerous separate reactions. In the range 30 MeV $< E_p <$ 200 MeV, Hua et al. (2002) use direct measurements of inclusive cross sections for pre-equilibrium reactions in proton bombardment of Al, Ni, Zr, and Bi at ~90 MeV resulting in both neutrons (Kalend et al. 1983) and protons (Wu et al. 1979a, b); and proton bombardment of C, O, and Al at 62 MeV resulting in protons (Bertrand and Peelle 1973). In pre-equilibrium reactions particles are emitted before the excitation energy is shared by all the nucleons, as opposed to evaporation reactions where the particles are emitted after equilibrium is achieved. To estimate the total cross section for all targets of interest, Hua et al. (2002) used the result of Kalend et al. (1983) that the ratio between the inclusive pre-equilibrium neutron and proton cross sections in this energy range is approximately 1/2 and that the inclusive pre-equilibrium cross sections are smooth functions of the atomic number A, behaving approximately as $A^{1/3}$ (Bertrand and Peelle 1973; Wu et al. 1979a, b; Kalend et al. 1983). To relate the evaporation cross sections to the pre-equilibrium cross sections, Hua et al. (2002) used the tabulated numerical results of Alsmiller et al. (1967), based on Monte Carlo calculations.

In the energy range 200 $< E_p <$ 1,000 MeV Hua et al. (2002) use the empirical formula,

$$\sigma(p, xn) = 10.4A^{1.34} \ln(E_p/100\ \text{MeV})\ [\text{mb}], \tag{11.143}$$

which is valid for 12 $<$ A $<$ 56 (Pearlstein 1987). At energies above 1,000 MeV, Hua et al. (2002) simply assumed that the inclusive cross sections remain constant. Based on high-energy studies by Barashenkov et al. (1983, 1984), Hua et al. (2002) would not expect the cross section to increase by more than 20% at 10 GeV. This is still within the measurement uncertainties and will not significantly affect the solar neutron spectrum, because the solar ion spectra are quite steep at high energies.

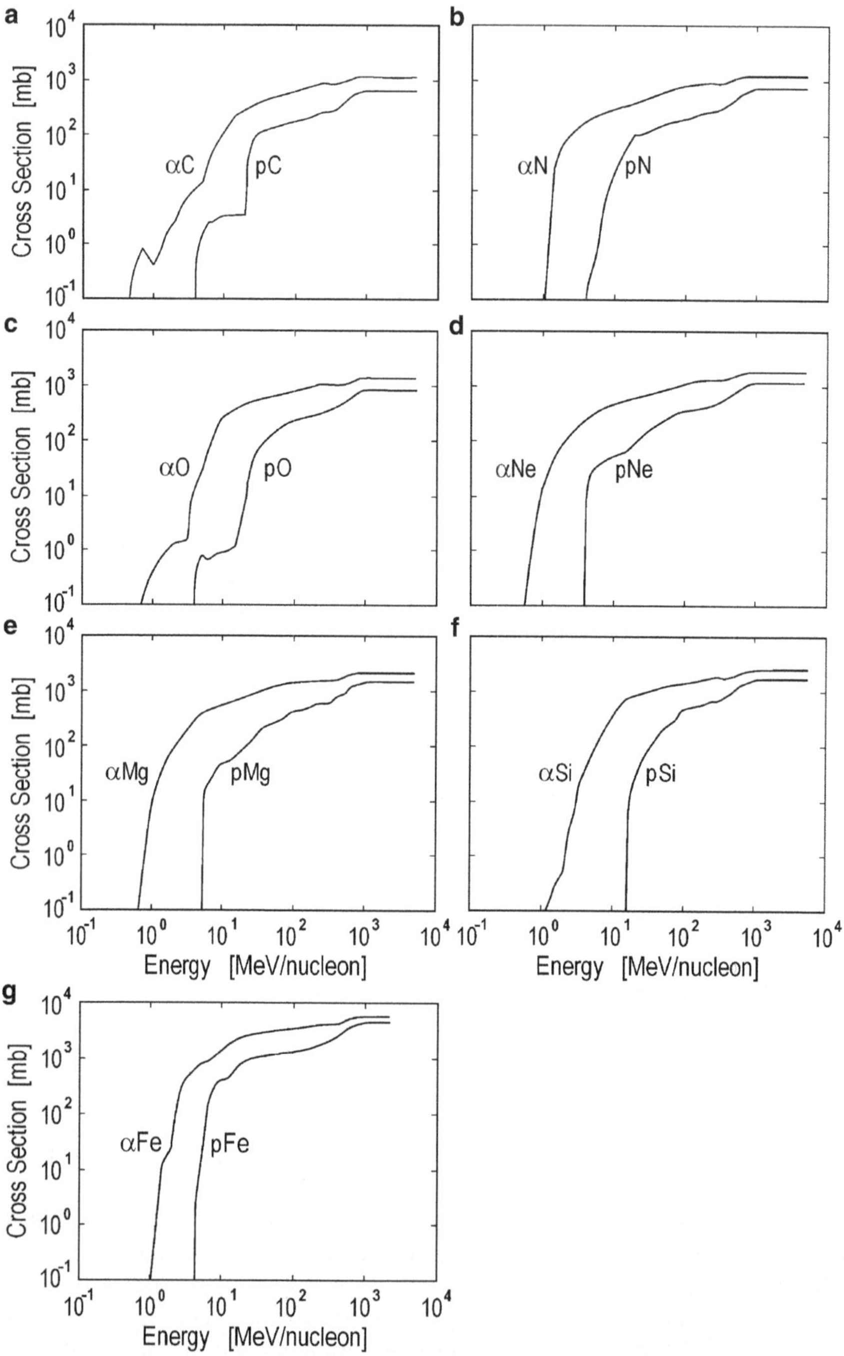

Fig. 11.52 Total neutron production cross sections σ_n in proton and α-particle interactions with C, N, O, Ne, Mg, Si and Fe (**a–g**) and with H, ^{4}He, and ^{3}He (**h–i**) (From Hua et al. 2002)

Next Hua et al. (2002) consider the neutron energy and angular distributions. In pre-equilibrium reactions in heavy targets the energy spectra of the emitted protons or neutrons are flat at relatively small angles and steepen to decreasing exponentials in energy at larger angles (Bertrand and Peelle 1973; Wu et al. 1979a, b; Kalend et al. 1983). This general characteristic of the spectra is independent of the energy of the incident proton. It is also essentially independent of the mass of the target and does not depend strongly on the mass of the projectile, whether protons, deuterons, tritons, or α- particles. Furthermore, the inclusive spectra of the emitted neutrons and protons are nearly identical, which allows to also use proton data to obtain information on neutron distributions.

Hua et al. (2002) have analyzed both neutron and proton energy and angle distributions measured for various heavy elements at 62 MeV (Bertrand and Peelle 1973), ~90 MeV (Wu et al. 1979a; Kalend et al. 1983), and 500 MeV (Roy et al. 1981). Hua et al. (2002) have discovered that a very good approximation to the energy and angular dependence in all of these pre-equilibrium data can be obtained with the following, surprisingly simple, expression for the laboratory differential cross section:

$$\frac{d^2\sigma}{dE_n d\Omega_n} \propto \exp\left[-(1-\mu_n)\frac{E_n}{T_0}\right], \quad E_n < E_{\max}, \tag{11.144}$$

where E_n is the neutron/proton energy, μ_n is the cosine of the angle of the emitted neutron/proton relative to the incident proton,

$$T_0 = 16(E_p/90)^{1/2} \tag{11.145}$$

with E_p in the units of MeV, $E_{\max}$ is the maximal neutron/proton energy for a given E_p, μ_n and the specific reaction considered:

$$E_{\max} = \frac{E_t B^2 + \sqrt{E_t^2 B^4 - \left(E_t^2 - P_t^2\mu_n^2\right)\left(B^4 + 4P_t^2 m_n^2 \mu_n^2\right)}}{2\left(E_t^2 - P_t^2\mu_n^2\right)}, \tag{11.146}$$

where

$$\begin{aligned} B &= \sqrt{E_t^2 - P_t^2 + m_n^2 - m_r^2}; \quad E_t = \gamma m_p + m_A; \\ P_t &= m_p\sqrt{\gamma^2 - 1}; \quad \gamma = 1 + \frac{E_p}{m_p}, \end{aligned} \tag{11.147}$$

and m_p, m_n, m_A are rest masses of the incident proton, the emitted neutron/proton, and the heavy nucleus, respectively; m_r is the total rest mass of all the product particles except the neutron/proton.

To determine the neutron production angle and energy in the Monte Carlo simulations, Hua et al. (2002) first sample the angle from the distribution determined by

$$\frac{d\sigma}{d\Omega_n} = \frac{K}{1-\mu_n}\left\{1 - \exp\left[-(1-\mu_n)\frac{E_{\max}}{T_0}\right]\right\}. \tag{11.148}$$

With μ_n thus selected, Hua et al. (2002) select the energy according to the distribution

$$\frac{d\sigma}{dE_n} = K\exp\left[-(1-\mu_n)\frac{E_n}{T_0}\right]. \tag{11.149}$$

As noted Hua et al. (2002), the contribution of the evaporation process to the emitted neutron energy spectrum is concentrated in a pronounced peak at low energy, which is very conspicuous in all of the measurements. The evaporation cross section is isotropic in the center of mass system. At backward angles it is always the dominant contributor to the total neutron production. Unlike the pre-equilibrium cross section, the evaporation cross section does not have a smooth dependence on A, since it is much more sensitive to the reaction Q-value and the nuclear level structure. Hua et al. (2002) have approximated the center of mass neutron energy distribution due to evaporation by

$$\frac{d\sigma}{dE_n^*} \propto \left(E_n^*\right)^{\gamma}\exp\left(-\frac{E_n^*}{T_{evap}}\right), \tag{11.150}$$

where E_n^* is the neutron energy in the center of mass, the evaporation temperature $T_{evap} = 2.5$ MeV and $\gamma = 5/11$ are independent of incident proton energy for nuclei with masses ≤ 56 (Gross 1965a, b; Verbinski and Burrus 1969). The evaporation neutron production accounts for 26%, 27%, 29%, 33%, 36%, 40%, and 65% of the total neutron yield at proton energies $E_p > 40$ MeV for reactions on C, N, O, Ne, Mg, Si, and Fe, respectively. For proton energies less than 40 MeV, the corresponding fractions are 6%, 11%, 16%, 29%, 41%, 53%, and 75%, based on the tabulated numerical results of Alsmiller et al. (1967).

Hua et al. (2002) underlined that the resulting neutron energy and angular distributions obtained from these new kinematics can be compared with the measured differential neutron production cross sections. In Fig. 11.53 are plotted the differential cross sections, $d\sigma_n/dE_n d\Omega_n$, for inclusive neutron production from the reactions Fe(p, xn) for an incident proton energy of 597 MeV at angles of 30°, 60°, 120°, and 150°. Experimental data points from Amian et al. (1993) are plotted for comparison. As can be seen, the agreement is quite good over many decades, and all angles. It is really remarkable that empirical formulae as simple as Eqs. 11.144 and 11.150 could fit data for such wide ranges of atomic mass from 12 to 238 and incident energy from 62 to 800 MeV (e.g., Bertrand and Peelle 1973; Wu et al. 1979a; Kalend et al. 1983; Amian et al. 1992, 1993; Meier et al. 1992a, b; Stamer et al. 1993).

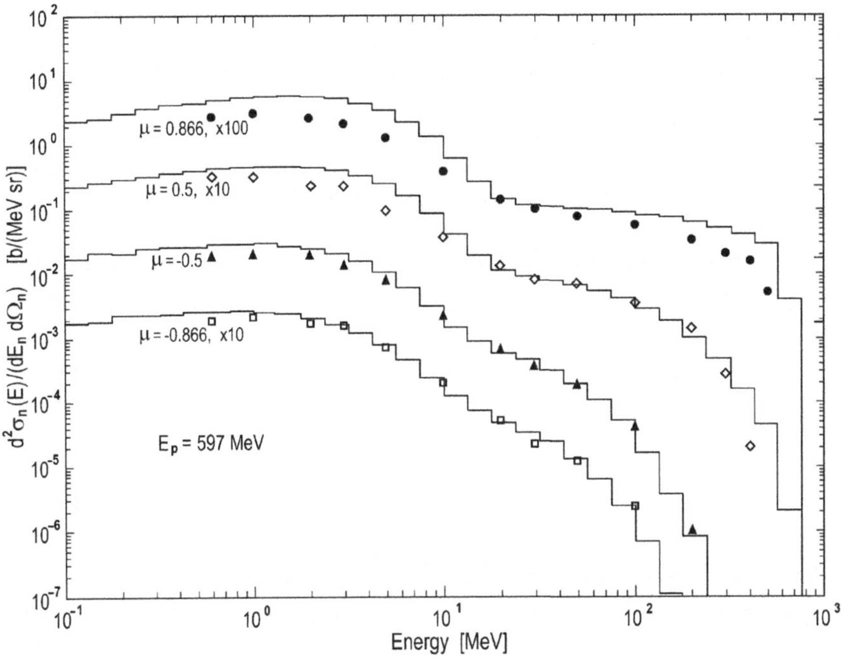

Fig. 11.53 Simulated differential inclusive neutron production cross sections $d^2\sigma_n/dE_n d\Omega_n$ for the reactions Fe(p,xn) at an incident proton energy of 597 MeV at angles of 30°, 60°, 120°, and 150°, compared with experimental data points from Amian et al. (1993). As can be seen, the agreement is remarkably good using the empirical formulae as simple as Eqs. 11.144 and 11.150 (From Hua et al. 2002)

11.10.4 Neutron Production by α-Particle-Induced Reactions

The total cross sections for α-particle induced reactions, according to Hua et al. (2002), are also shown in Fig. 11.52. As for the proton induced reactions, at low energies immediately above the thresholds, the cross sections are the sum of the (α, n), (α, αn) and (α, pn) cross sections. For α-particle interactions with ^{12}C, ^{14}N, ^{16}O, ^{20}Ne, ^{24}Mg, ^{28}Si, and ^{56}Fe, these reactions also produce β^+ emitters, whose cross sections were evaluated previously (Kozlovsky et al. 1987). The (α, n) reactions on the rare isotopes ^{13}C, ^{18}O, ^{22}Ne, ^{25}Mg, ^{26}Mg, and ^{29}Si, do not produce positron emitters. Hua et al. (2002) obtained the neutron production cross sections for these reactions from the measurements of Walton et al. (1957), Ashery (1969), and Gibbons and Macklin (1959). As with the proton reactions, Hua et al. (2002) have normalized the cross sections to the principal isotope of each element using the abundances given by Cameron (1982).

For the evaluation of σ_n at higher energies (both pre-equilibrium and evaporation), Hua et al. (2002) have used measurements of proton emission in 58 MeV α-particle interactions with ^{12}C, ^{16}O, and ^{54}Fe (Bertrand et al. 1974) and proton and neutron emissions in 149 MeV α-particle interactions with ^{27}Al, ^{58}Ni, ^{90}Zr, ^{204}Bi, and ^{222}Th (Wu et al. 1979a, b; Kalend et al. 1983). At 149 MeV Kalend et al. (1983) concluded that for low-Z targets $\sigma_n/\sigma_p \approx 1$. At 58 MeV there are no measurements of neutron emission, but Monte Carlo calculations for ^{12}C and ^{54}Fe (Bertrand et al. 1974) suggest that σ_n/σ_p is also ≈ 1. Therefore, Hua et al. (2002) have assumed that for elements with A < 60 and energies from 40 to 200 MeV the total neutron cross section is about equal to the proton cross section.

For energies greater than 300 MeV per nucleon, Hua et al. (2002) have used the empirical formula (Madey et al. 1983)

$$\sigma_n(E_\alpha) = \bar{\sigma}_{NN}(E_\alpha)(\bar{R}_G/R_0)^5, \tag{11.151}$$

where E_α is energy of α-particle per nucleon,

$$\bar{R}_G = (1/2)R_0\left(A_{\text{target}}^{1/3} + A_{\text{projectile}}^{1/3}\right) \tag{11.152}$$

is the average radius of the target and projectile nuclei ($A_{\text{projectile}} = 4$), $R_0 = 1.2 \times 10^{13}$ cm, and $\bar{\sigma}_{NN}$ is the isospin-averaged free nucleon–nucleon total cross section,

$$\bar{\sigma}_{NN} = \frac{1}{2}\left[\sigma_{pn}(E) + \sigma_{nn}(E)\right]. \tag{11.153}$$

The values of $\sigma_{pn}(E)$ and $\sigma_{nn}(E)$ are from Devlin et al. (1973).

For the neutron energy and angular distributions Hua et al. (2002) use Eqs. 11.144 and 11.150. However, in α-particle induced reactions the importance of the evaporation process relative to the non-equilibrium process is much higher than in proton-induced reactions. The Monte Carlo calculations for 58 MeV α-particles (Bertrand et al. 1974) yield

$$\sigma_n^{\text{noneq}}/\sigma_n^{\text{evap}} \approx 0.21 \tag{11.154}$$

for both ^{54}Fe and ^{12}C, and Hua et al. (2002) have adopted this value for all the elements in this group independent of energy.

11.10.5 Neutron Production in p + α Reactions

The total neutron production cross section for $p\alpha$ reactions are taken from Murphy et al. (1987) and Hua and Lingenfelter (1987d) for incident proton energies of less than 300 MeV. At higher energies Hua et al. (2002) use the energy-dependent neutron production cross section determined by Guglenko et al. (1990a, b) from a theoretical cascade model and normalized to the value of 106 mb at 1,408 MeV,

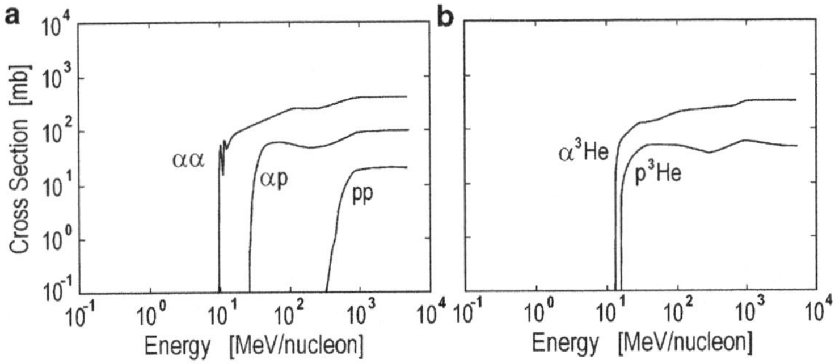

Fig. 11.54 Total neutron production cross sections σ_n in proton and α-particle interactions with H and ^{4}He (**a**), and with ^{3}He (**b**) (From Hua et al. 2002)

measured at Dubna (Glagolev et al. 1977, 1984; Aladashvili et al. 1980). The $p\alpha$ cross section, shown in Fig. 11.54, is roughly 50% higher than that used in Murphy et al. (1987) and Hua and Lingenfelter (1987b) at energies greater than 300 MeV.

As noted Hua et al. (2002), there are no measurements of the neutron angular and energy distributions, except for one measurement at 141 MeV (Nicholls et al. 1972). Hua et al. (2002) therefore proceed as follows. Up to 60 MeV, where the most important neutron-producing reaction is $p + {}^4\mathrm{He} \rightarrow {}^3\mathrm{He} + p + n$, are used the center of mass angle and energy distributions determined by Bildsten et al. (1990). At energies above 100 MeV, Hua et al. (2002) have developed a more general kinematic approach based on their successful approximation for the heavy targets, and will be discuss this below. In the intermediate range between 60 and 100 MeV Hua et al. (2002) interpolate between these two methods.

Hua et al. (2002) approximate the neutron angular and energy distributions for incident proton energies above 100 MeV by an expression of the same form as that found for the heavy targets. As discussed above, for such targets the pre-equilibrium neutron and proton distributions are quite similar, a result which can be understood in a model where single nucleon-nucleon scattering dominates. In $p\alpha$ reactions the contribution from multiple scattering is less important than in reactions involving the heavy nuclei, and moreover the evaporation component is negligible. Therefore, Hua et al. (2002) expect that the neutron–proton similarity should hold also for these reactions. However, because of the smallness of the ^{4}He nucleus, it may expect some deviations from the simple empirical relation given in Eq. 11.144 for heavy nuclei.

As underlined Hua et al. (2002), the inclusive proton differential cross sections for $p\alpha$ reactions were measured at various angles by Wesick et al. (1985) at incident proton energies of 100 and 150 MeV, and by Roy et al. (1981) at 500 MeV. Although these cross sections are generally similar to those of the heavier nuclei, the measurements of Wesick et al. (1985) do show a different behavior at small angles which may be unique to ^{4}He. The energy distribution at small angles is characterized by an exponentially rising cross section up to just a few MeV below the upper kinematic limit. Only for angles greater than 60° does the exponential decrease start to show up. This behavior was found also by Nicholls et al. (1972) at 141 MeV and by Selove and Teem (1958) at 90 MeV. In order to incorporate this feature, Hua et al. (2002) have modified the angular dependence in Eqs. 11.144 and 11.149, replacing the $(1-\mu)$ term by $(0.75-\mu)$, such that the exponential argument changes sign and becomes positive at angles $\theta < 60°$ to give a rising rather than falling spectrum. An additional difference is that in ^{4}He at large angles the exponential decline is steeper. Therefore, in order to fit the measurements with the expression given in Eq. 11.145 Hua et al. (2002) reduced T_0(90 MeV) from 16 MeV found for heavier nuclei to 10 MeV for ^{4}He.

The calculated distributions for incident $E_p = 100$ and 500 MeV are shown in Fig. 11.55 for various laboratory angles, together with the proton data of Wesick et al. (1985) at 100 MeV and Roy et al. (1981) at 500 MeV. As can be seen, the energy and angular distributions calculated with this simple approximation are again in very good agreement with the experimental data at all energies and angles for both incident proton energies.

For comparison also shown in Panel b of Fig. 11.55 Hua et al. (2002) calculation of the neutron inclusive cross section, using the intranuclear cascade model of nuclear reactions of Yariv and Fraenkel (1981). The similarity to the proton

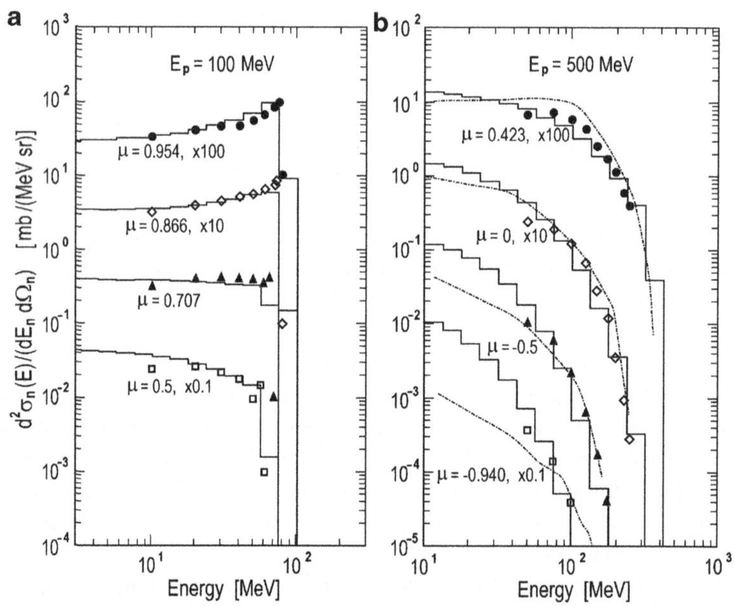

Fig. 11.55 Simulated differential neutron production cross sections $d^2\sigma_n/dE_n d\Omega_n$ for $E_p = 100$ and 500 MeV in $p+\alpha$ reactions are for various laboratory angles, together with the proton data of Wesick et al. (1985) at 100 MeV (**a**) and Roy et al. (1981) at 500 MeV (**b**). As can be seen the calculations agree quite well with the experimental data at both proton energies. Also shown as a *dash-pointed line* in Panel (**b**) are the results of Intranuclear Cascade Model (INC) calculations for the neutron inclusive cross section (From Hua et al. 2002)

inclusive cross section of Roy et al. (1981), evident in the Panel b of Fig. 11.55, supports the Hua et al. (2002) assumption that the shape of the neutron angular and energy distributions is similar to those of protons in $p\alpha$ reactions. The discrepancy at $\mu = -0.94$ is partly due to the evaporation process, which is more pronounced at backward angles but neglected in the intranuclear cascade model, and partly due to the Monte Carlo method employed in intranuclear cascade model, which is less accurate in simulating processes of very low probability. In addition, Hua et al. (2002) have also used the intranuclear cascade calculations to check the validity of described above simple empirical approximation for higher incident energies, where no experimental data are available for comparison. In particular, Hua et al. (2002) have confirmed their results that, for the $p\alpha$ reaction, $d\sigma_n/dE_n$has a pronounced peak around the maximum energy of E_n, a result that is important in determining the energy spectrum of the extreme high-energy neutrons.

11.10.6 Neutron Production in $\alpha + \alpha$ Reactions

As noted Hua et al. (2002), at the lowest energies neutron production in $\alpha\alpha$ interactions proceeds via the fusion reactions $\alpha + \alpha \rightarrow {}^7\text{Be} + n$, and $\alpha + \alpha \rightarrow {}^6\text{Li} + p + n$ (Glagola et al. 1982). At higher energies the neutrons are produced from the breakup of one or more α-particles, but there are no measurements of the cross sections. Because of the complete symmetry between protons and neutrons in $\alpha\alpha$ reactions, the inclusive neutron cross section should be identical to the inclusive proton cross section. In the energy range 27.5–43 MeV per nucleon Hua et al. (2002) used measurements of the proton inclusive cross sections (Paic et al. 1981; see also Murphy et al. 1987). At energies greater than 300 MeV per nucleon Hua et al. (2002) used the empirical Eq. 11.151, yielding $\sigma_n(E_\alpha) = 10.1\sigma_{NN}(E_\alpha)$. This cross section agrees within 15% with the cross section given in Guglenko et al. (1990a, b) in the same energy range. At intermediate energies, from about 40 to 300 MeV per nucleon, Hua et al. (2002) use the cross section of Guglenko et al. (1990a, b) which merges smoothly with the Paic et al. (1981) measurements at 43 MeV per nucleon. These cross sections, shown in Fig. 11.54, are roughly twice those used by Murphy et al. (1987) and Hua and Lingenfelter (1987d) at energies greater than 100 MeV per nucleon.

Next Hua et al. (2002) consider the neutron angular and energy distributions. For $E_\alpha < 15$ MeV per nucleon, it was used the results of Glagola et al. (1982), which show that the angle differential cross section in the center of mass system has forward-backward symmetry and can be approximated by $a + b\mu*$, where $\mu*$ is the cosine of the angle in the center of mass, the constants a and b depend on E_α and are found to be $a/b = 1/9$ from the data of Glagola et al. (1982). At higher energies Paic et al. (1981) showed that the energy spectra of the inclusive protons can be derived using the Fermi statistical theory, where the center of mass energy is distributed with equal probabilities among the various degrees of freedom in the interaction volume. The neutron energy distribution in the center of mass system is therefore governed by the particle phase-space distribution in the final state. Since the dominant reaction is $\alpha + \alpha \rightarrow \alpha + {}^3\text{He} + n$, Hua et al. (2002) used the three-body phase-space function (Perl M1974),

$$\frac{d^2\sigma}{dE_n^* d\Omega_n^*} \propto \frac{p_3\left[\left(S_2 - m_1^2 - m_2^2\right)^2 - 4m_1^2 m_2^2\right]^{1/2}}{S_2}, \tag{11.155}$$

where $S_2 = S + m_3^2 - 2E_3\sqrt{S}$, $\sqrt{S}$ is the total energy in center of mass, p_3^* and E_3^* are the momentum and energy of the neutron in center of mass, and m_1, m_2 and m_3 are the masses of the α-particle, ^{3}He and neutron. Hua et al. (2002) used this relation for incident α-particles at energies from 15 to 100 MeV per nucleon. The angular distribution of the neutrons is assumed to be isotropic.

For incident energies greater than 100 MeV per nucleon, Hua et al. (2002) have used the predictions of the intranuclear cascade model of Yariv-Fraenkel. Guglenko et al. (1990b) used the code written at JINR. Intranuclear cascade basically is a microscopic model in which the interacting nuclei are simulated by dilute gases and the reaction is approximated by superposition of particle–particle two-body collisions. Since this approximation is valid when the mean free path of the nucleons is large compared with both the range of the nucleon-nucleon force and the wavelength of the nucleons, this model can be applied only for high energies. A comparison of the spectra determined from Hua et al. (2002) detailed kinematic calculations, which include intranuclear cascade spectra at high energies, with the calculations of Guglenko et al. (1990a, b), which only used angular-averaged intranuclear cascade spectra for all ion energies, shows very significant differences, as will be discuss below. These new kinematics also differ significantly from those previously used in Hua (1986), Hua and Lingenfelter (1987d), and Murphy et al. (1987), where the neutrons were assumed to have a uniform energy distribution up to the maximum energy and a strongly forward peaked angular distribution.

11.10.7 Neutron Production in ^{3}He Reactions

As noted Hua et al. (2002), since below the pion production threshold, the only inelastic channels are $p + {}^3\text{He} \rightarrow 2p + d$ and $3p + n$, the total inclusive neutron production cross section is the difference between the total inelastic cross section and the cross section for the $2p + d$ channel. Both of these are given by Meyer (1972). The result is plotted in Fig. 11.54.

For the calculation of the energy and the angular distributions at energies less than 80 MeV, Hua et al. (2002) again adopted the Fermi statistical model. Here there are four particles in the final state, and thus Hua et al. (2002) used the nonrelativistic four-particle formula given by Paic et al. (1981). At higher energies the neutrons are produced mainly in single nucleon–nucleon interactions.

Therefore, in this energy range Hua et al. (2002) have adopted the same procedure as for $p + {}^4$He reactions.

For the total neutron production cross sections in ^{3}He + ^{4}He reactions Hua et al. (2002) used the formula:

$$\sigma_n\left({}^3\mathrm{He}+{}^4\mathrm{He}\right) = 0.75\left[\sigma_n\left({}^4\mathrm{He}+{}^4\mathrm{He}\right) - \sigma\left({}^4\mathrm{He}+{}^4\mathrm{He}\rightarrow{}^7\mathrm{Be}+\mathrm{n}\right)\right], \quad (11.156)$$

where the last term on the right-hand side is the cross section for the α–α fusion reaction which has no counterpart in the ^{3}He–^{4}He reaction. The factor 0.75 takes into account the fact that there are only three neutrons in ^{3}He + ^{4}He system compared to four neutrons in ^{4}He + ^{4}He system.

11.10.8 Neutron Production in p + p Reactions

As noted Hua et al. (2002), because of the high relative abundance of protons in both the ambient gas and the accelerated ions, neutron production in *p–p* interactions is the dominant neutron-producing mode at high energies above the pion production threshold. Neutron production in these reactions was treated in detail by Murphy et al. (1987), and Hua et al. (2002) adopt this treatment. The total cross section is given in Fig. 11.54. Murphy et al. (1987) concluded that for $E_p < 2$ GeV the differential cross section can be approximated by a single isobar model in which the angular distribution of the isobar is given by A(E) + B(E)$\cos^2\theta$, where θ is the center of mass angle of the isobar with respect to the direction of the incident proton. For $E_p > 2$ GeV Murphy et al. (1987) used the scaling representation developed by Tan and Ng (1982).

11.10.9 Total Neutron Production Yields and Spectra

Using described above new cross sections and kinematics, Hua et al. (2002) calculate by Monte Carlo simulations, the total neutron production yields and energy spectra as a function of the assumed energy spectrum of the accelerated ions. The yields are shown in Fig. 11.56 for both a second-order Bessel function spectrum, $N(E) \propto K_2\left[(12p/mc\alpha T)^{1/2}\right]$, expected (Ramaty 1979) from stochastic Fermi acceleration, and a power-law spectrum, $N(E) \propto E^{-S}$, expected (e.g., Forman et al. 1986) from shock acceleration, assuming solar photospheric abundances for both the accelerated ions and ambient medium (Table 11.10).

From Fig. 11.56 can be seen that $p\alpha$ and αp reactions are the dominant processes for neutron production for all assumed ion spectra except the softest spectra (αT $\leq$ 0.02 or S $\geq$ 4), where $\alpha\alpha$ (for Bessel function) and α-heavy ion (for power-law) reactions became dominant, and for the hardest power-law spectra (S $\leq$ 1.5), where

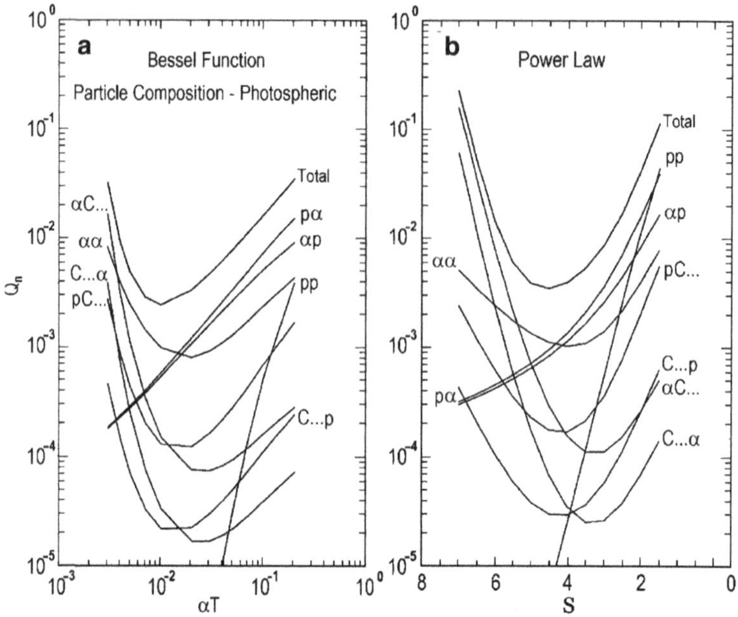

Fig. 11.56 Thick target neutron yield Q_n for various reactions as a function of accelerated ion spectra, assuming (**a**) a second-order Bessel function spectrum, $N(E) \propto K_2\left[(12p/mc\alpha T)^{1/2}\right]$, expected from stochastic Fermi acceleration, and (**b**) a power-law spectrum, $N(E) \propto E^{-S}$, expected from shock acceleration, normalized for $N_p(> 30 \text{ MeV}) = 1$, and assuming solar photospheric abundances for both the accelerated ions and ambient medium (Table 11.10) (From Hua et al. 2002)

Table 11.10 Accelerated ion compositions (From Hua et al. 2002)

Isotopes	Solar photosphere	Enhanced
^{1}H	1.0	1.0
^{4}He	0.1	0.5
^{12}C	3.63×10^{-4}	1.43×10^{-3}
^{14}N	1.12×10^{-4}	6.76×10^{-3}
^{16}O	8.51×10^{-4}	2.86×10^{-3}
^{20}Ne	1.09×10^{-4}	4.31×10^{-4}
^{24}Mg	3.00×10^{-5}	5.56×10^{-3}
^{28}Si	3.26×10^{-5}	1.19×10^{-3}
^{56}Fe	4.29×10^{-5}	8.33×10^{-3}
^{3}He	0	0.25

p–p reactions become dominant. This is qualitatively similar to previous (Hua 1986; Hua and Lingenfelter 1987d; Murphy et al. 1987) calculations, but the yields of all the reactions except *p–p* are higher because their cross sections are now higher. As a result, the total yield is higher by a factor of ~2. To a lesser extent, the separate treatment of the heavy ions, which were previously taken into account as one element having average properties, the revised solar atmospheric abundances and the calculation of particle ranges also contribute to the changes in the neutron yields.

The effects of ^{4}He and heavier ion enrichment on the neutron yields has been discussed in Ramaty et al. (1992). As an example, Hua et al. (2002) show the effect of the heavy ions on the neutron yields in Fig. 11.57, where neutrons are produced in the reactions between the accelerated ions with enhanced α- particle and heavier ion abundances (Table 11.10) and the solar atmosphere. It can be seen that with the increased abundances of α-particle and heavier ions and the presence of ^{3}He, the neutron yields of individual reactions are accordingly increased and the total neutron production is increased by a factor of as much as ∼4.

Hua et al. (2002) also calculate the relative contribution of the various reactions to the total neutron energy spectrum, shown in Fig. 11.58 for incident ions having (a) a typical Bessel function spectrum with $\alpha T = 0.03$ and (b) a typical power-law spectrum with index $S = 3.5$. As can be seen for the Bessel function spectrum in Panel a of Fig. 11.58, the $p\alpha$ reaction is the dominant production processes at neutron energies ∼20 MeV and the αp reaction is the dominant production processes at all greater neutron energies.

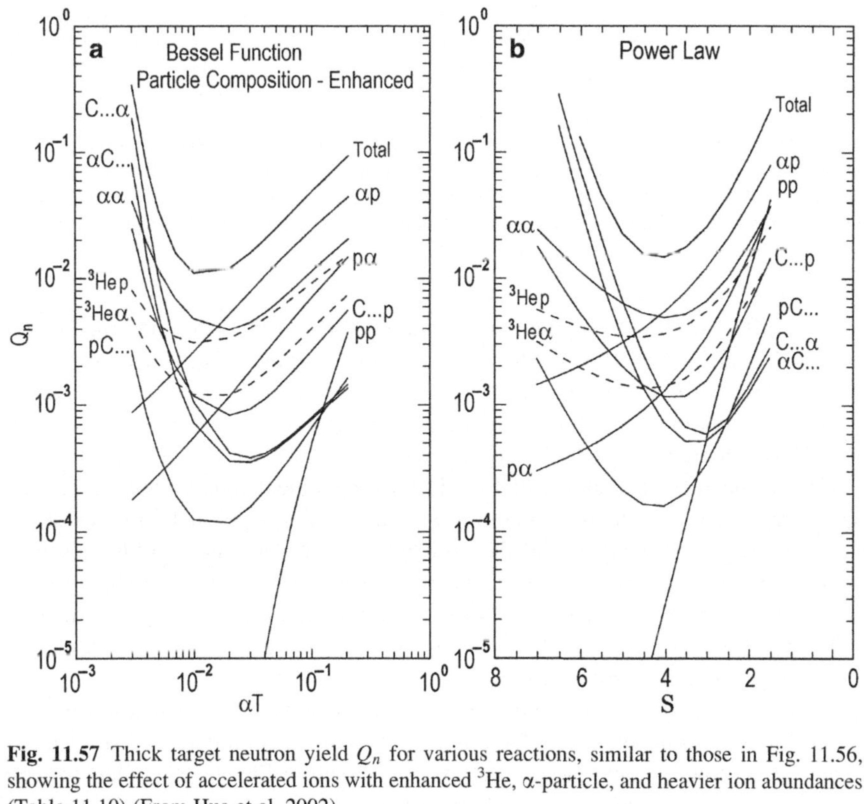

Fig. 11.57 Thick target neutron yield Q_n for various reactions, similar to those in Fig. 11.56, showing the effect of accelerated ions with enhanced ^{3}He, α-particle, and heavier ion abundances (Table 11.10) (From Hua et al. 2002)

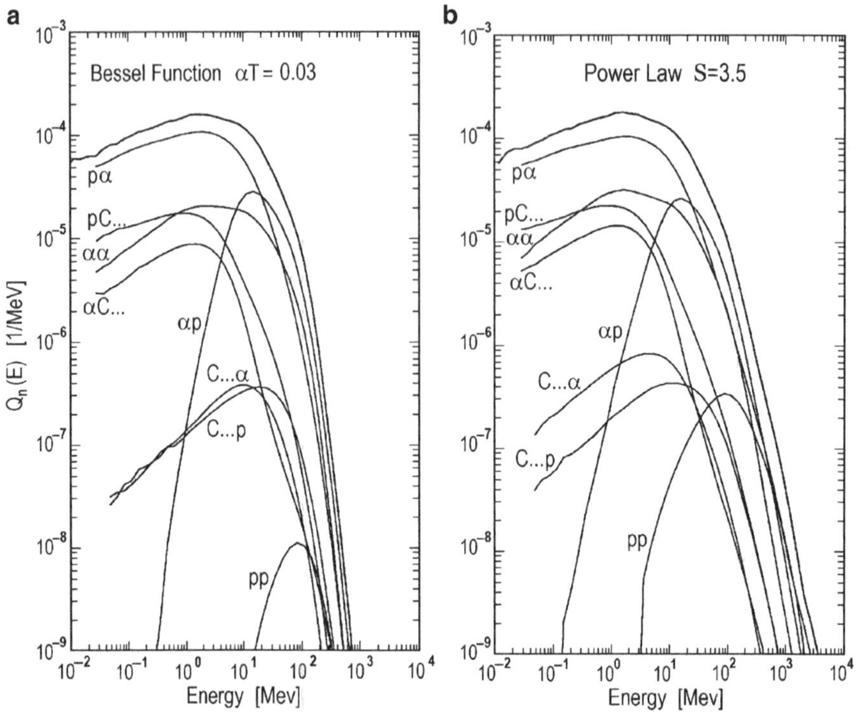

Fig. 11.58 Relative contribution of the various reactions to the total neutron production spectrum $Q_n(E)$ as a function of neutron energy, shown for incident ions having (**a**) a typical Bessel function spectrum with $\alpha T = 0.03$ and (**b**) a typical power-law spectrum with index $S = 3.5$. As can be seen, the $p\alpha$ and αp reactions are the dominant production processes at all neutron energies (From Hua et al. 2002)

This differs slightly from previous calculations (Hua 1986; Hua and Lingenfelter 1987d; Murphy et al. 1987), where $\alpha\alpha$ reactions became dominant above about 300 MeV, because of the strong forward peaking assumed in the earlier kinematics. As can be seen, that is no longer the case in these calculations because Hua et al. (2002) used more realistic kinematics.

As can be seen in Panel b of Fig. 11.58 for incident ions with a power-law spectrum of $S = 3.5$, the $p\alpha$ reaction is also the dominant production processes at neutron energies ~20 MeV and the αp reaction is the dominant production processes at higher energies. However, at neutron energies around ~400 MeV, can be seen that the $p\alpha$ reaction again becomes more important than the αp, and then at ≥800 MeV, the p–p reaction becomes the most important production process for ions with a power-law spectrum.

Hua et al. (2002) directly compare the new calculated total neutron energy spectrum with their previous calculations (Hua and Lingenfelter 1987d) for these same Bessel function and power-law spectra, shown in Fig. 11.59. It can be seen that the new spectra are on the whole higher than the old ones because of the

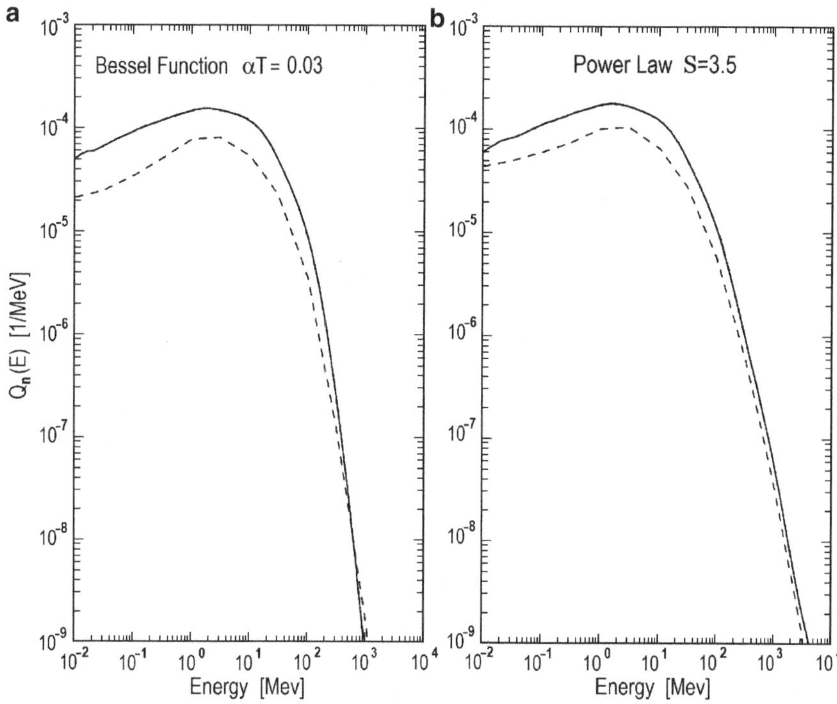

Fig. 11.59 Comparison of the neutron energy spectra calculated in Hua et al. 2002 (*solid line*) and in Hua and Lingenfelter (1987d) for the incident particles having (**a**) a Bessel function spectrum with $\alpha T = 0.03$ and (**b**) a power-law spectrum with index $S = 3.5$, The new spectra are higher than the old ones because of the increased total neutron productions (From Hua et al. 2002)

increased total neutron production. But for the Bessel function spectrum (Panel a in Fig. 11.59), at high energies the new neutron spectrum is actually lower and slightly steeper than the previous one, because it lacks the harder $\alpha\alpha$ contribution previously assumed. This is not the case for the power-law spectrum (Fig. 11.59), because the *p–p* reaction, which is unchanged, makes the dominant contribution at the highest energies, as we saw above.

Hua et al. (2002) also compare (Fig. 11.60) the obtained new spectra with that calculated for the high-energy neutrons in the energy range between 100 and 1,000 MeV by Guglenko et al. (1990a) for a power-law ion spectrum with an index $S = 3.5$. From Fig. 11.60 can be seen that the Guglenko et al. (1990a) spectrum is between 45% and 70% lower than described above new calculations and it is flatter at neutron energies $\leq$200 MeV and steeper at energies $\geq$400 MeV. These differences result primarily from the different kinematics in the $p\alpha$ and αp reactions, where the approximations used by Guglenko et al. (1990a, b), averaged over just four wide angular bands, produced fewer low-energy ($\leq$200 MeV) neutrons in αp reactions, and fewer high-energy ($\geq$400 MeV) neutrons in $p\alpha$ reactions, than described above much more detailed kinematics, which agree quite well with

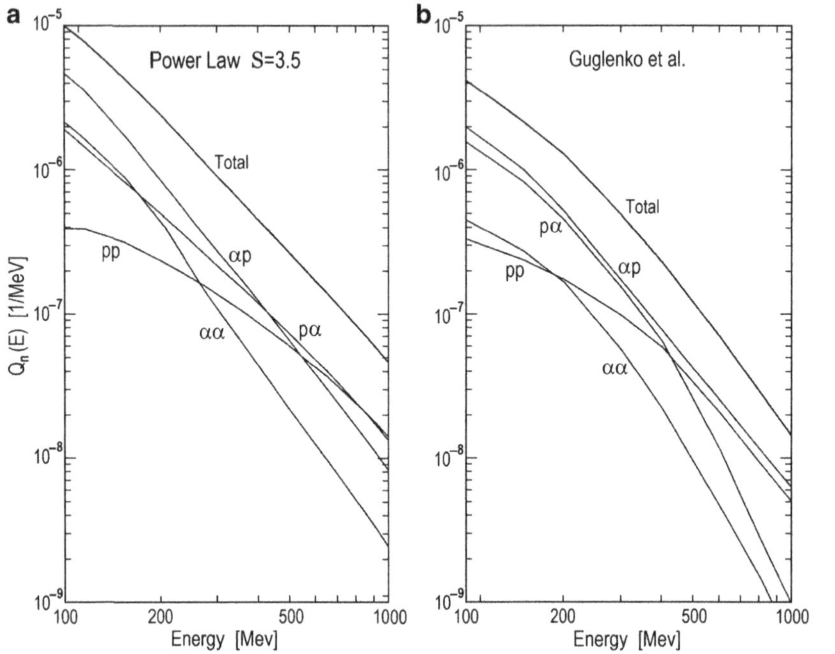

Fig. 11.60 Comparison of the high-energy neutron spectra (**a**) calculated by Hua et al. (2002) and (**b**) calculated in Guglenko et al. (1990a, b) for incident ions having a power-law spectrum with index S = 3.5. The spectrum by Hua et al. (2002) is higher and harder than that by Guglenko et al. (1990a, b) of the improved *p*-α and α-*p* kinematics (From Hua et al. 2002)

measurements (Roy et al. 1981; Wesick et al. 1985), as shown in Fig. 11.55. In particular, the new kinematics produce more high-energy neutrons in the forward direction, giving rise to a peak at the maximum energy in the neutron spectrum from *p*α reaction, a phenomenon which is confirmed by the calculations based on the Yariv–Fraenkel intranuclear cascade model for incident proton energy greater than 1 GeV.

11.10.10 Neutron Production in Flare Loops

Using described above new cross sections and kinematics in the full Monte Carlo simulations, Hua et al. (2002) have also calculated the angle, time, and depth dependence of the neutron production by accelerated ions solar flare loops. The models for solar atmosphere and the flare magnetic loop used in this calculation are the same as those described in the investigation of γ-ray line production by Hua et al. (1989). Basically, the loop consists of a semicircular coronal portion and two straight portions extending from the ends of the coronal portion through the chromosphere and into the photosphere. In the corona, the gas is assumed to be completely ionized with a constant density and the magnetic field is also taken to be constant. In the chromosphere and photosphere, the gas is neutral with a pressure profile given by Avrett (1981) and the magnetic field B is proportional to a power δ of the pressure (Zweibel and Haber 1983). The transition between the corona and the chromosphere is 1,800 km above the photosphere. The acceleration of ions is assumed to take place in the corona. The accelerated ion spectra are again taken to be either a modified Bessel function in momentum, or a power-law in energy.

As noted Hua et al. (2002), the propagation and interaction of the accelerated ions in the magnetized solar atmosphere are similar to those discussed in the γ-ray line production study (Hua et al. 1989; see Section 9.15). Namely, the accelerated ions are subject to the following interactions with the ambient gas and magnetic field: (1) the ions lose energy due to Coulomb scattering, (2) they can be destroyed by various nuclear reactions, (3) they can be reflected by the mirroring force of the convergent magnetic field at the feet of the loop, and (4) they are scattered in pitch-angle by the MHD turbulence in the coronal ionized gas. The pitch-angle scattering is characterized by its mean free path length Λ, which is related to the energy density in MHD turbulence (Melrose 1974; Ramaty et al. 1988). In the Hua et al. (2002) calculations are used a computer program similar to the one was used in Hua et al. (1989) for γ-ray line production except that the γ-ray line production cross sections are replaced by the neutron production cross sections and appropriate kinematics.

An illustration of the effects of pitch-angle scattering on the depth dependence of neutron production in solar atmosphere is shown in Fig. 11.61.

The accelerated ions are released at the top of the coronal loop isotropically with a power-law spectrum with $S = 3.5$. The coronal loop length is taken to be 1.15×10^9 cm, and the magnetic field convergence index δ is 1/5. The three curves in Fig. 11.61 represent different degrees of MHD pitch-angle scattering. When there is no pitch-angle scattering ($\Lambda \rightarrow \infty$, solid curve), the mirroring force of the magnetic field effectively prevents the neutrons from being produced deep in atmosphere. In this case, ~30% of the neutrons are produced in the corona, ~66% are produced in the chromosphere, and only ≤4% are produced in the photosphere. On the other hand, when the pitch-angle scattering is strong (e.g., $\Lambda = 4.60 \times 10^{10}$ cm, dashed curve), ions are continuously fed into the loss cone, so that neutrons are produced deep in the atmosphere. The peak production is at ~100 km below the division between chromosphere and photosphere (at depth −320 km in Fig. 11.61), and nearly 60% of the neutrons are produced in the photosphere, while virtually no neutrons are produced in the corona. From the discussion in Hua et al. (1989), it is known that this length of mean free path corresponds to the nearly saturated case, where the pitch-angle scattering mean free path is comparable to the coronal loop length. It can be also seen from the Fig. 11.61 that even with a mean free path ten times longer (dotted curve), the neutrons are still mostly produced below corona, with ~60% in the chromosphere and the rest in the photosphere.

The effects of pitch-angle scattering on the time dependence of the neutron production can be seen in Fig. 11.62.

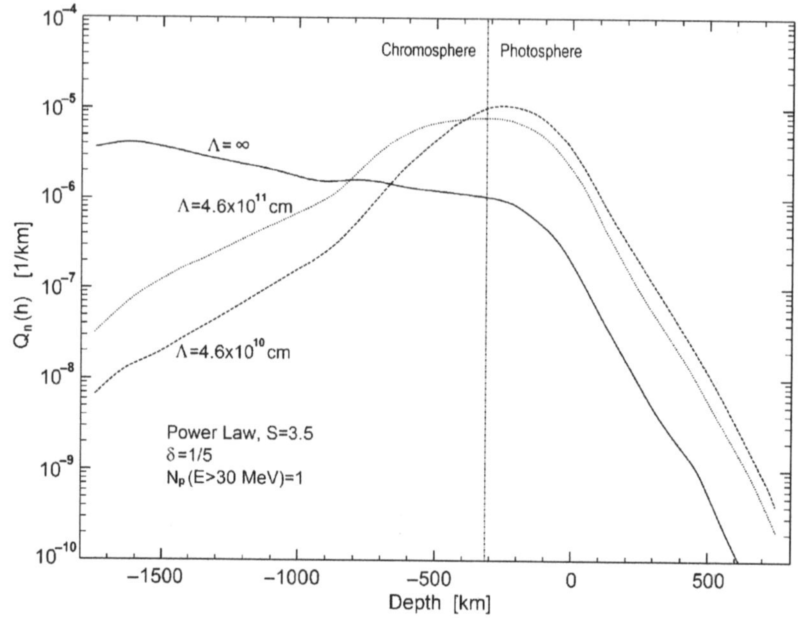

Fig. 11.61 Effects of pitch-angle scattering on the depth dependence of neutron production in solar atmosphere. The flare-accelerated ions with a power-law spectrum of S = 3 are injected isotropically at the top of the solar magnetic field loop. The power index δ indicating the convergence of the magnetic field is 1/5. The three curves represent three cases of MHD pitch-angle scattering characterized by the mean free path Λ, ranging from no scattering ($\Lambda = \infty$) to nearly saturated scattering ($\Lambda = 4.60 \times 10^{10}$ cm). The vertical line at −320 km is the division between chromosphere and photosphere (From Hua et al. 2002)

In the calculations for Fig. 11.62, Hua et al. (2002) assume that the accelerated ions are released instantaneously at $t = 0$. The three curves correspond to the same cases of MHD pitch-angle scattering as in Fig. 11.61. As is pointed out in Hua et al. (1989), the convergent magnetic field keeps the accelerated ions outside of loss cone out of the denser part of the atmosphere, so that these ions mirror back and forth in the less dense corona and upper chromosphere for a long time before they interact with the ambient gas to produce neutrons. Thus, it can be seen that without pitch-angle scattering ($\Lambda = \infty$, solid curve), neutrons are produced over an extended period of time. More than 80% of the total neutrons are produced after 10 s and over 45% of the total are produced after 1,000 s. On the other hand, pitch-angle scattering keeps refilling the loss cone and thereby speeds up the neutron production dramatically. For the near-saturation case where $\Lambda = 4.60 \times 10^{10}$ cm (dashed curve), ~98% of the neutrons are produced within 10 s. For the intermediate case $\Lambda = 4.60 \times 10^{11}$ cm, dotted curve), about half of neutrons are produced after 10 s. For weak pitch-angle scattering with Λ approaching ∞ the neutron production rates at late times decay roughly as the 2/3 power of time, whereas for stronger pitch-angle scattering with $\Lambda \leq 2.3 \times 10^{13}$ cm, the late time neutron

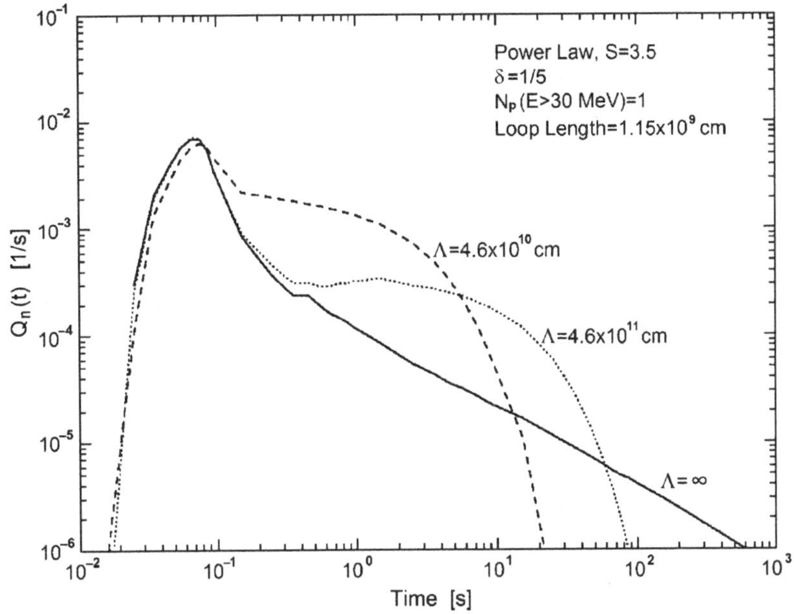

Fig. 11.62 Effects of pitch-angle scattering on the time dependences of the neutron production. The accelerated ions are injected instantaneously at $t = 0$ and at the top of the loop whose coronal portion has a length of 11,500 km. The three curves correspond to the same cases of MHD pitch-angle scattering as in Fig. 11.61 (From Hua et al. 2002)

production rates decay exponentially with time. For these cases the e-folding time t_n in terms of the mean free path of pitch-angle scattering is

$$t_n \approx 0.25\left(\frac{\Lambda}{2.3 \times 10^9\,cm}\right)^{0.78} [\mathrm{s}]. \tag{11.157}$$

Hua et al. (2002) also explore the effects of MHD pitch-angle scattering of the incident ions on the neutron angular and energy distributions by comparing the two extremes of neutron production by ions without pitch-angle scattering ($\Lambda = \infty$) and with nearly saturated pitch-angle scattering ($\Lambda = 4.60 \times 10^{10}$ cm), described above. In Panel a of Fig. 11.63, are presented the polar plot of the zenith angular dependence of the neutrons in various energy ranges produced in a flare loop without pitch-angle scattering. It can be seen that all distributions except that for very low-energy neutrons (below 1 MeV) have peaks in the direction tangential to the solar surface (90°), and the distributions are nearly symmetric with respect to it, with downward fluencies (180°) only slightly higher than the upward ones. This pancake-like angular distribution of neutrons reflects the zenith angular distribution of the ions near the mirror points of the convergent magnetic field. The higher energy neutrons also have a stronger angular dependence than those at lower energy,

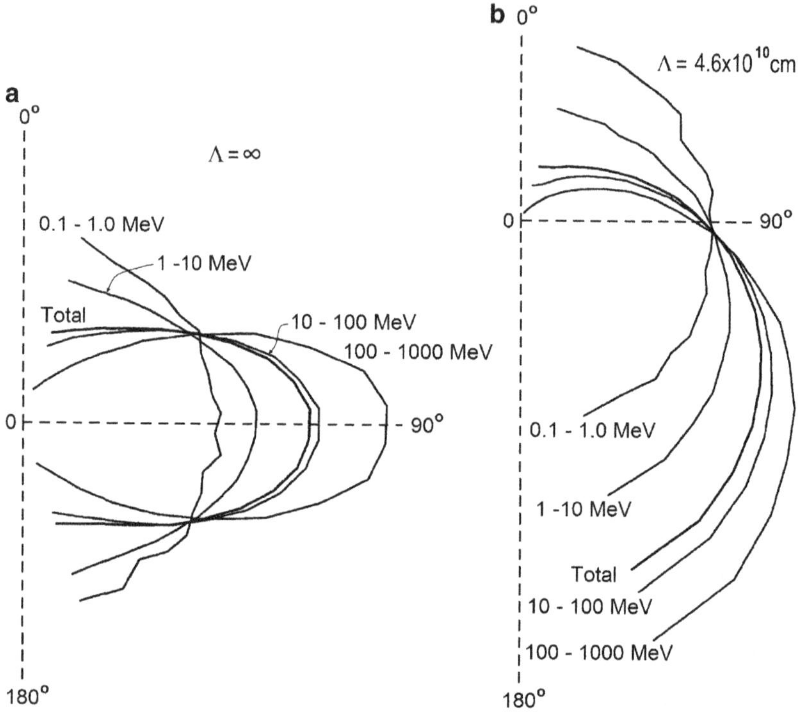

Fig. 11.63 Effects of MHD pitch-angle scattering of the accelerated ions on the neutron angular distributions in three energy ranges for the two extremes of neutron production by ions (**a**) without pitch-angle scattering ($\Lambda = \infty$) and (**b**) with nearly saturated pitch-angle scattering ($\Lambda = 4.60 \times 10^{10}$ cm), considered in Figs. 11.61 and 11.62 (From Hua et al. 2002)

because the higher-energy neutrons tend to be emitted in more nearly the same direction as the initial ions. By contrast, it can be seen in Panel b of Fig. 11.63, that for the case of nearly saturated pitch-angle scattering, the initial ions, being continuously fed into the loss cone by the strong pitch-angle scattering, are strongly downward peaked and thus the neutrons produced are also strongly downward distributed, with the higher energy ones more so than the lower energy ones.

The effect of pitch-angle scattering on the neutron spectra as a function of angle can be seen in Fig. 11.64, where are plotted the energy spectra in three zenith angle ranges for the two extreme cases. Hua et al. (2002) use the 135–180° range to represent the downward direction (solid curve), 0–45° the upward direction away from the Sun (dot-dashed curve), and 60–120° the direction tangential to the solar surface (dashed curve). It can be seen in Panel a of Fig. 11.64, that without pitch-angle scattering of the ions, the neutron emission in the tangential range 60–120° is the strongest and the spectrum is the hardest, because of the pancake-like distribution of neutrons. The spectra in upward and downward directions are about the same, because the neutron distribution is nearly symmetric about the plane tangent to the solar surface. On the other hand, in the case of nearly saturated

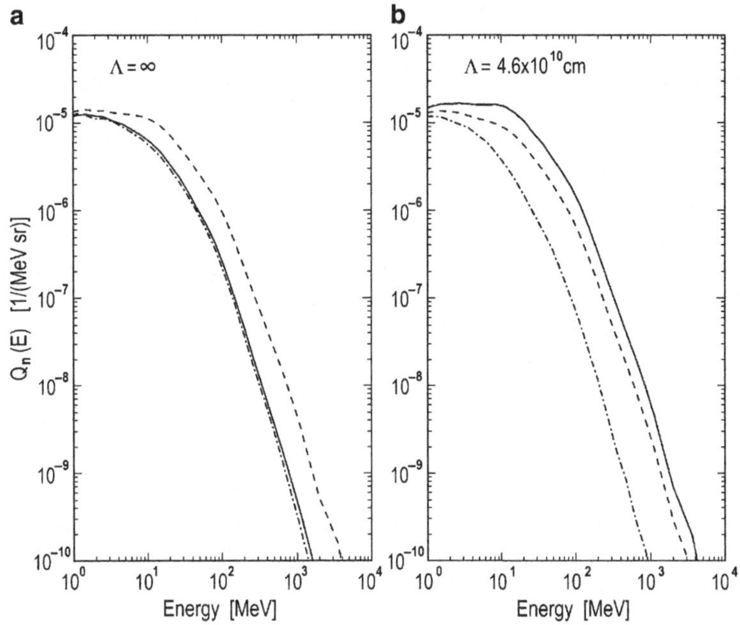

Fig. 11.64 Effect of pitch-angle scattering of the accelerated ions on the neutron production energy spectra as a function of zenith angle for the two extreme cases of ions (**a**) without pitch-angle scattering ($\Lambda = \infty$) and (**b**) with nearly saturated pitch-angle scattering ($\Lambda = 4.60 \times 10^{10}$ cm). Zenith angles of 135–180° represent the downward direction (*solid curve*), 0–45° the upward direction away from the Sun (*dot-dashed curve*), and 60–120° the direction tangential to the solar surface (*dashed curve*) (From Hua et al. 2002)

pitch-angle scattering, (Panel b of Fig. 11.64), the strongest neutron emission is downward to the Sun, resulting from the predominant neutron production by downward-beamed ions.

11.10.11 Escaping Neutrons

In Hua et al. (2002) Monte Carlo simulations, each neutron is followed, usually through many scatterings, until it either escapes from the solar atmosphere, decays, or slows down and is captured either radiatively on ^1H to form ^2H with the emission of a 2.223 MeV gamma ray, or nonradiatively on ^3He to form ^3H with the emission of a proton. In these calculations Hua et al. (2002) consider only elastic scattering on hydrogen, since the contributions to neutron thermalization by scattering on He and heavier nuclei are negligible.

From the depth distributions of the neutron production (Fig. 11.61), it can be seen that most of the neutrons are produced above the depth of 0 km, which corresponds to a column density $\sim$10 g/cm^2 (see e.g., Hua 1986). As it was shown by Hua and Lingenfelter (1987d), at these depths, neutrons initially directed upward have a good chance of escaping from the solar atmosphere without scattering. But if they are initially headed downward, they can lose most of their energy in the solar atmosphere and have little chance of escaping from it. Thus, the neutron escape probability depends strongly on the initial angular distribution of neutrons. For the two extremes, the escape probability for neutrons with energies greater than 1 MeV is ~55% without pitch-angle scattering of the ions, and ~25% for nearly saturated pitch-angle scattering.

As noted Hua et al. (2002), the zenith angular dependences of the escaping neutrons produced by ions without pitch-angle scattering and by those with nearly saturated pitch-angle scattering, are shown in the polar plots in Panels a and b of Fig. 11.65, respectively.

As noted Hua et al. (2002), from Fig. 11.65 can be seen that without ion pitch-angle scattering (Panel a of Fig. 11.65), the total escaping neutrons have a rather strong limb-brightened distribution (solid curves), with the peak fluence at the zenith angle ~83°. This corresponds closely to the neutron production distribution (Fig. 11.63), which peaks at ~90°, corresponding in turn to the pancake-like angular distribution of the mirroring ions. The other three curves represent the angular distributions of neutrons in the energy ranges 1–10 MeV (dotted), 10–100 MeV (dashed) and 100–1,000 MeV (dot-dashed), respectively. As can be seen from

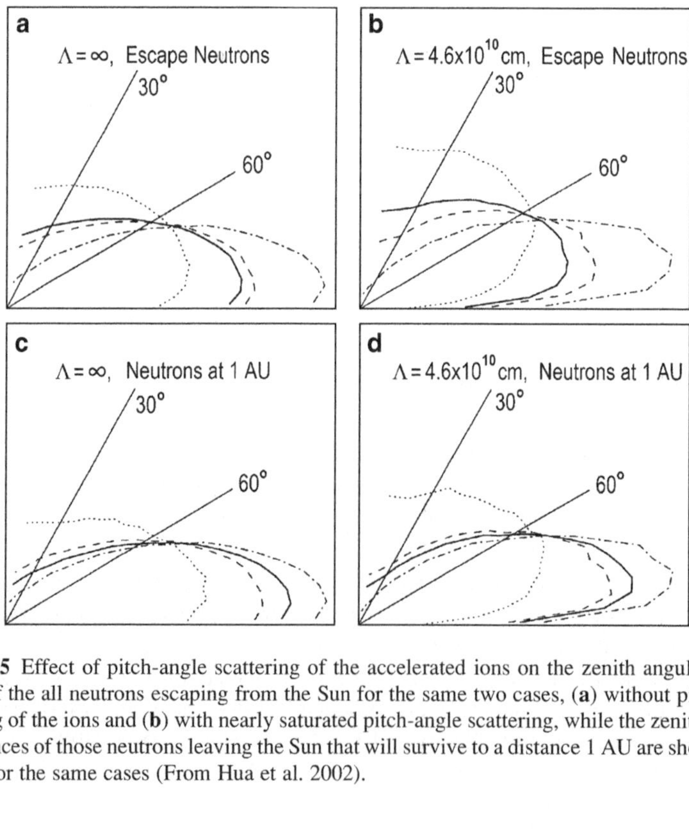

Fig. 11.65 Effect of pitch-angle scattering of the accelerated ions on the zenith angular dependences of the all neutrons escaping from the Sun for the same two cases, (**a**) without pitch-angle scattering of the ions and (**b**) with nearly saturated pitch-angle scattering, while the zenith angular dependences of those neutrons leaving the Sun that will survive to a distance 1 AU are shown in (**c**) and (**d**) for the same cases (From Hua et al. 2002).

Fig. 11.65 the higher energy neutrons have much stronger limb-brightening than the lower energy ones. Since in this case, the neutrons are mostly produced in the corona and upper chromosphere, they escape from the solar atmosphere with very little scattering.

On the other hand, in the case of nearly saturated pitch-angle scattering, (Panel b of Fig. 11.65), the limb-brightening is weaker, because the neutrons are produced deeper in the atmosphere and preferentially in downward directions. As a result, the neutrons suffer greater attenuation as they escape in the directions nearly tangential to the solar surface. The simulations of Hua et al. (2002) show that at zenith angles very close to tangential, $\theta = 87-90°$, only ~34% of the neutrons escape. This small angle limb-darkening effect is more striking for neutrons at lower energies (dotted curve), because the scattering cross section is larger. These effects shift the peak fluence of the escaping neutrons to a zenith angle ~75°.

As noted Hua et al. (2002), it takes 1,166 s for a typical escaping neutron at energy $E_n = 100$ MeV to travel from the Sun to the Earth. This time is 1,054 s in the neutron's rest frame, while the decay mean life of the neutron is $\tau = 918$ s. This means on the way from the Sun to the Earth, almost 70% of the neutrons at this energy decay. Panels c and d in Fig. 11.65 shown the angular dependences of the neutrons leaving the Sun that will survive at a distance 1 AU. At this distance, the limb-brightening of the escaping neutrons is further enhanced, since the fluence of the limb-darkened, relatively more isotropic low energy neutrons becomes negligible because of decay. As an example, in the case without pitch-angle scattering of the ions the fluence at 0° is about one order of magnitude lower than at 90°, about twice that of the escaping neutrons at the Sun. The high-energy neutrons are more likely to survive the distance and it is seen that the angular distributions of these neutrons (dot-dashed curves) are virtually unchanged from Panels a and b to Panels c and d in Fig. 11.65.

As underlined Hua et al. (2002), the energy spectrum of the escaping neutrons also depends strongly on the angular distribution of the initial neutrons. This can be seen from Panels a and b in Fig. 11.66, where the escaping neutron energy spectra (solid curves) at differing $\cos\theta$ are shown for both nominal cases. Comparing to the production spectra (dotted curves), there are differences that are worth noting. First, in the case without pitch-angle scattering of the ions, in the direction tangential to the solar surface, namely $\cos\theta = 0$–0.1 in Panel a of Fig. 11.66, the production and escaping spectra are almost identical, which means that in this direction, the neutrons suffer little scattering as they escape. This is not true for the case of nearly saturated pitch-angle scattering (Panel b of Fig. 11.66), however, where a comparison between the two spectra shows that while at energies ~100 MeV and higher, the two spectra are parallel, the escaping neutron spectrum is deficient in low-energy neutrons. This is because in case of nearly saturated pitch-angle scattering, neutrons are produced deep in the atmosphere and the low-energy neutrons have a higher probability of scattering and a lower probability of escaping than the more energetic neutrons. Second, for both cases, in the upward directions, $\cos\theta = 0.6$–0.7 and 0.9–1.0, the escaping neutron spectra are similar to the production spectra at high energies but are more abundant in low-energy neutrons. This is because in the

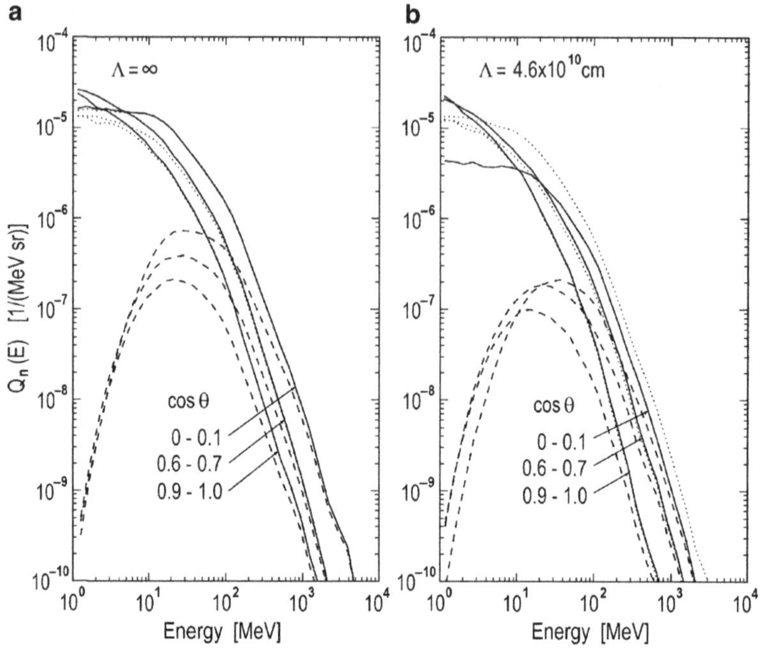

Fig. 11.66 Effect of pitch-angle scattering of the accelerated ions on the angular-dependent energy spectra of the escaping neutrons at the Sun (*solid curves*) and those surviving to 1 AU (*dashed curves*) at differing cosθ, compared to their production spectra at the Sun (*dotted curves*) and their surviving spectra at 1AU (*dashed curves*), for the same cases, (**a**) without pitch-angle scattering of the ions and (**b**) with nearly saturated pitch-angle scattering (From Hua et al. 2002)

upward directions, neutrons can escape with little scattering in both cases. The extra low energy neutrons are originally in downward directions and able to escape only after several scatterings which change their directions as well as lower their energies.

As noted Hua et al. (2002), the neutron spectra as they are measured near the Earth also differ greatly from those of the escaping neutrons just leaving the Sun, because of the different decay rates for neutrons of different energies. Panels a and b in Fig. 11.66 also show the spectra in the directions given above for the surviving neutrons at 1 AU from the Sun (dashed curves). It can be seen that these spectra are similar to the escape spectra at energies $\geq$500 MeV, but reach their maxima in the energy range from 10 to 100 MeV, owing to the combined effects of the decreasing escape spectra and increasing survival probability with increasing energy. In general, the spectra of neutrons which escaped in directions closer to the tangent of the solar surface have higher energy maxima. In other words, neutrons from limb flares tend to have spectra with higher energy maxima. In addition, it can be seen in the case of nearly saturated pitch-angle scattering (Panel b in Fig. 11.66) that the neutron emission is strongly limb-darkened at cos$\theta = 0$–0.1 below $\sim$15 MeV.

11.10.12 Summary of Main Results

Hua et al. (2002) summarized main obtained results as following:

1. There are reevaluated and updated the neutron production cross sections and developed new production kinematics for all of the significant processes of neutron production by solar flare-accelerated ions.
2. Comparisons of these kinematics with laboratory measurements show that these are more accurate than those used in previous calculations (e.g., Hua 1986; Hua and Lingenfelter 1987d; Murphy et al. 1987; Guglenko et al. 1990a).
3. Using these kinematics and cross sections, it was calculated the neutron production by the accelerated ions trapped in a magnetic loop.
4. These ions are subject to energy loss due to Coulomb scattering, nuclear reactions with the ambient gas, reflection in the loop feet by the mirroring force of the convergent magnetic field and pitch-angle scattering by the MHD turbulence in the coronal ionized gas.
5. There are calculated the depth, time, angle, and energy dependences of the neutron production, the angle distributions, and energy spectra of the escaping neutrons.
6. There are also calculated the energy dependences of the surviving neutron fluence at the distance 1 AU from the Sun.
7. These new calculations will now permit much more reliable and detailed analyses of the various solar flare neutron spectral observations.
8. The code used to make all of the calculations described in Section 11.9 is available in the online edition of the Astrophysical Journal Supplement, paper Hua et al. (2002). Copies of this computer program may be obtained by contacting X.-M. Hua at hua@pair.gsfc.nasa.gov.

11.11 Production of Energetic Light Isotopes Due to Nuclear Interaction and Acceleration in the Flare Region: Modeling and Simulation

11.11.1 The Matter and Short History of the Problem

Balashev et al. (2007) considered the production of energetic light isotopes due to nuclear interactions and acceleration in flare regions. The Monte-Carlo method allows taking into account several steps of particle interactions with ambient plasma. In this model the high abundance ratios of ^{3}He/^{4}He are obtained at certain simulation parameters.

As noted Balashev et al. (2007), the problem of elemental anomalies in solar CR is known since the early 1970s. It is especially pronounced in small impulsive solar energetic particle (SEP) events. Indeed, the ratio of ^{3}He/^{4}He measured in the interplanetary space turns out to be 10^4 times of that in the solar plasma, where $^3He/^4He$ is about 5×10^{-4} (see, e.g., Anglin et al. 1973; Ramaty and Kozlovsky 1974; Colgate et al. 1977). Later it was found that the enrichment factor for heavier particles for the same class of events is somewhat smaller and lies within a factor of 10 (Reames et al. 1994). As to the more powerful gradual SEP events, the chemical composition of CR for them approximately corresponds to that of the solar corona. First attempts to account for these anomalies were based on the analysis of nuclear interactions of accelerated particles with the solar photospheric (chromospheric) matter. For the galactic CR similar approach was applied to simulate light isotope production. Those models for solar CR really yield the great flux of ^3He isotope along with comparable fluxes of deuterium (D) and tritium (T) which were not observed in the interplanetary space. This circumstance seems to be a reason for dropping this idea though certain angular properties of a primary particle beam could suppress interplanetary leakage of D and T. The lack of nuclear data for differential cross sections of various reactions also made this modeling uncertain. Nowadays the most popular models for heavy element enrichment in SEP events are those of selective acceleration operating in background turbulent plasma (Kocharov and Kocharov 1984; see review on these models also in Chapter 4 in Dorman M2006).

Balashev et al. (2007) reanalyzed the nuclear aspect of the problem by adding some improvements to previous consideration in some ways. First of all, they include multiplicity of particle nuclear interactions with the solar plasma, make simultaneous account of acceleration processes as well as subsequent particle propagation in the interplanetary space (adiabatic losses). The latter effect can modify the energy spectra of released species on the way to the Earth along with the other above mentioned effects which are more evident.

11.11.2 Description of the Model

According to Balashev et al. (2007), the simulation model is elaborated to take into account cascading processes of particle interactions with the solar matter. It includes not only nuclear transformation of primary chemical composition of solar CR (inelastic processes) but also elastic collisions resulting in the energy losses. Early papers considered only secondary particles (after first nuclear interaction) to obtain the released particle fluxes and, hence, the corresponding solar flare isotopic composition. To Balashev et al. (2007) mind, the Monte Carlo method is the best way to realize the idea of multiple nuclear interactions. At the same time, certain kind of turbulence in the flaring region may lead to the particle acceleration inside. Such an energy gain can also be quite easily taken into account within the framework of the Monte Carlo approach along with the particle interplanetary propagation.

Balashev et al. (2007) consider the production of light isotopes ^3He, ^3H, and ^2H which are generated in the interactions of accelerated protons and α-particles with the solar plasma, also consisting mainly of H and He nuclei (see Ramaty and Kozlovsky 1974; Colgate et al. 1977). The contribution of heavier elements such as CNO nuclei can be omitted (Ramaty and Kozlovsky 1974).

11.11.3 Nuclear Interactions in the Region of Acceleration

Balashev et al. (2007) accept the plane geometry of the system of some thickness $x = L_0$ (in cm) or $\bar{x}$ (in g/cm^2) which is infinite in two other directions. As a rule, the incident primary nuclei (at $x = 0$) are distributed on energy E by a power law. Inside this 'volume' the nuclear cascading processes and particle acceleration can change their energy (i.e., their spectra) and chemical composition. If the particle escapes this region (its current coordinate $x > L_0$) we count it as a released particle. After that the interplanetary propagation takes place which strongly depends on the perturbation level of this medium. Balashev et al. (2007) successively calculate the following particle parameters after its injection at the boundary $x = 0$:

1. Free path

$$l = -\ln\eta \Big/ \sum_i n_i \sigma_i^{tot}, \tag{11.158}$$

   where η is a random number uniformly distributed inside [0,1] interval; n_i is a reactant number density; σ_i^{tot} is a total cross section (elastic+inelastic) of the process i, and σ_i denotes any of them
2. Displacement

$$\Delta x = l\cos\theta\cos\phi, \tag{11.159}$$

   where θ and φ being an angle between particle velocity v and layer normal (x direction) and azimuthal angle, respectively
3. Probability of different nuclear processes according to the relationship $P_i \big/ \sum_i P_i$, where

$$P_i = 1 - \exp(n_i/\sigma_i) \tag{11.160}$$

   is a probability of the ith process to occur (both elastic or inelastic)

After the collision the particle momentum is calculated according to the kinematics of reaction, and the angles are supposed to be uniformly distributed in 4π in a C.M. coordinate system. Notice that all the above parameters are energy dependent, therefore initial particle spectra respectively changed. For each new species appearing in this code Balashev et al. (2007) repeat the steps 1–3. The outgoing particle is counted independently on the number of coordinate (or time) steps. This implies obtaining the time integrated particle spectra and/or composition.

11.11.4 Nuclear Interactions and Acceleration

If, apart of nuclear interactions, Balashev et al. (2007) consider acceleration one should inject monoenergetic particle population instead of their initial energy distribution. It was assumed to be conserved in past models to compensate energy losses (Ramaty and Kozlovsky 1974); this practically means instant acceleration inside l. Particles during their propagation can gain energy due to stochastic acceleration mechanism which is described by the well-known Fokker–Planck equation for isotropic distribution function f:

$$\frac{\partial f}{\partial t} = \frac{\partial}{\partial E}(A_1 f) + \frac{\partial^2}{\partial E^2}(A_2 f), \tag{11.161}$$

where A_1 and A_2 are kinetic coefficients responsible for the convection and diffusion in a momentum (energy) space, respectively. They are expressed through the turbulent energy level of a definite spectrum and type. At each time step $\Delta t = \alpha l / v$ one should calculate the energy gain

$$\Delta E = -A_1 \Delta t + \sqrt{2 A_2 \Delta t}\, W, \tag{11.162}$$

where W being the random number with the Gaussian-like distribution and α is a some numerical coefficient <1. Thus, there are $1/\alpha$ subintervals inside the 'nuclear' interval l, where the particle may be underwent both the acceleration and nuclear interaction with the corresponding probabilities:

$$P_{acc} = \exp\left(-\sum_i n_i \alpha l \sigma_i\right), \quad P_{nuc} = 1 - \exp\left(-\sum_i n_i \alpha l \sigma_i\right). \tag{11.163}$$

If nuclear interaction takes place inside the subinterval αl, the probability of the ith process is simulated according to the step 3 of Section 11.11.3. So, within the 'nuclear' interval l the energy of the particle can change dramatically due to the stochastic acceleration mechanism. It is especially important for energies in the vicinity of the reaction thresholds. Note that secondary particles produced in the collisions have sufficiently high energy to be immediately injected to the acceleration process. So, no primary threshold (injection) energy is necessary in this case.

11.11.5 Propagation in the Interplanetary Space

After particles leave the interaction region they are supposed in Balashev et al. (2007) to be involved into the interplanetary propagation. Since the energy spectra is measured near the Earth's orbit, the main effect important for the modeling is the adiabatic energy losses in the expanding solar wind. It strongly depends on the mean free path λ, i.e. on the spatial diffusion coefficient of this medium, $\kappa = (1/3)\lambda v$. According to Balashev et al. (2007), in the Monte Carlo approach the energy decrease and particle displacement at each collision with the scattering center should be simulated according to the formulae:

$$\Delta E = -\frac{4u}{3r} E\Delta t, \quad \Delta r = \left(\frac{2\kappa}{r} + \frac{\partial \kappa}{\partial r} + u\right)\Delta t + \sqrt{-4\kappa \Delta t \ln \delta_1}\, \sin(2\pi\delta_2). \quad (11.164)$$

In Eq. 11.164 r is the heliocentric distance, u is the solar wind speed, δ_1, δ_2 are the random numbers uniformly distributed inside [0,1], and diffusion coefficient κ is defined in the interplanetary space by the Alfvén wave turbulence of the power law index $S = 1.5$:

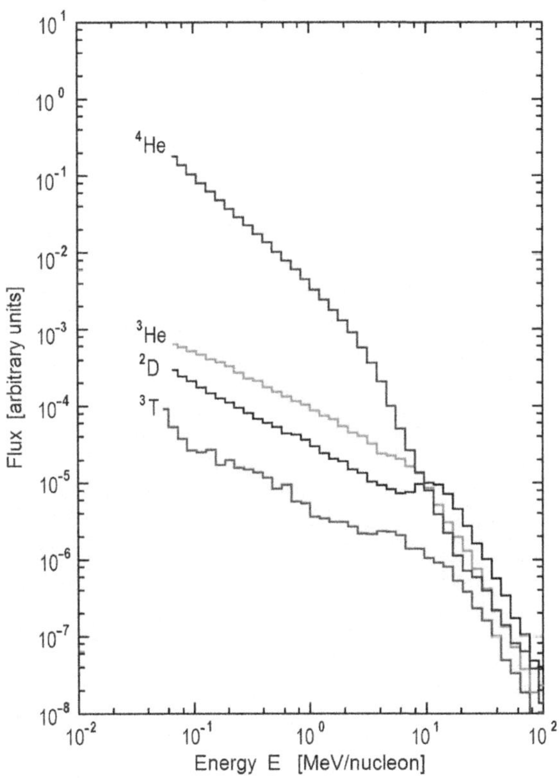

Fig. 11.67 Expected spectra of light isotopes at the Sun (From Balashev et al. 2007)

$$\kappa = \kappa_0(r/r_S)(E/E_0)(Q/A)^{(3-S)/2}, \tag{11.165}$$

where Q and A being the particle nuclear charge and atomic mass, respectively; r_s is the solar radius, E_0 is the normalizing energy.

11.11.6 Expected Isotope Energy Spectra in the 'Nuclear' Interaction Layer Without Any Energy Gain Inside

First of all Balashev et al. (2007) represent calculated light isotope energy spectra in the 'nuclear' interaction layer without any energy gain inside. In this case they inject protons and α particles (with the abundance ratio 1:0.1) at $x = 0$ with the spectrum $\propto E^{-\gamma}$ which is modified as the particles reach the boundary $x = l$. The resulting differential fluxes are depicted in Fig. 11.67 ($\gamma = 3$, $\bar{x}$= 5 g/cm^2).

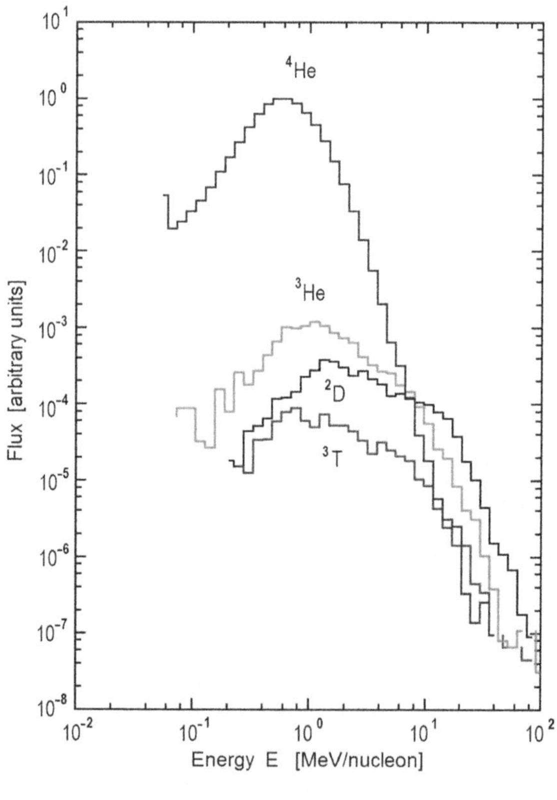

Fig. 11.68 Calculated spectra of isotopes at the Sun (nuclear interactions + acceleration) (From Balashev et al. 2007)

11.11.7 Taking into Account the Effect of Additional Acceleration

Figure 11.68 shows the effect of additional acceleration ($\bar{x} = 5$ g/cm^2, characteristic acceleration time is around 0.1 s).

11.11.8 Taking into Account Propagation in the Interplanetary Space

In Figs. 11.69 and 11.70, calculated for different values of the interplanetary diffusion coefficients, one can clearly see the shift of the crossing point for He^3, He^4 spectra (from ~4 to ~1.6 MeV/nucleon). The latter value is much closer to the observations while for primary spectra at the Sun it was evaluated to be at ~7.3 MeV/nucleon.

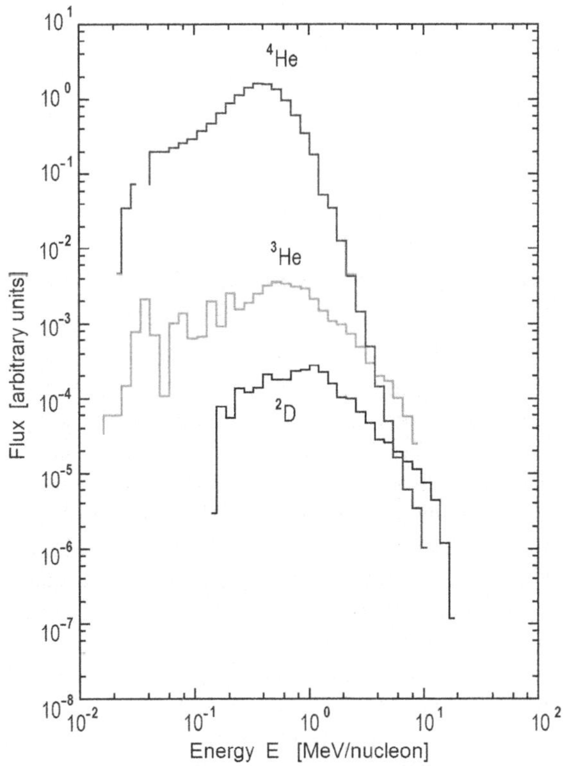

Fig. 11.69 Spectra of isotopes after their interplanetary propagation for $\kappa_0 = 10^{19}$ cm^2/s (From Balashev et al. 2007)

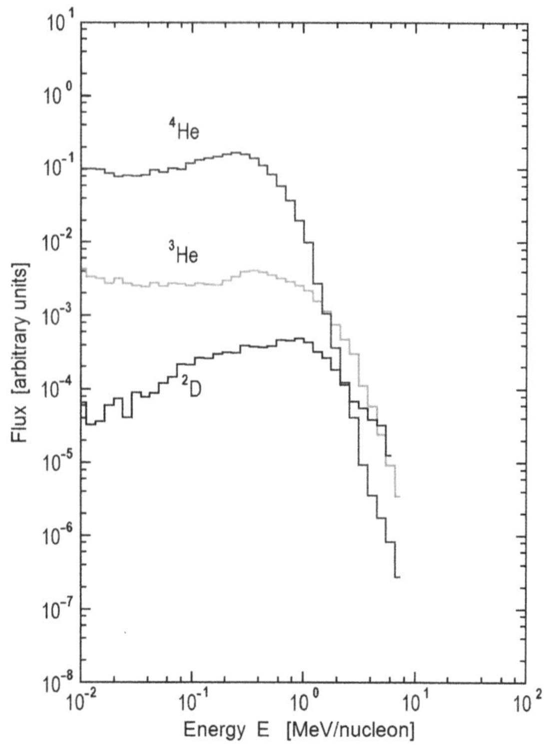

Fig. 11.70 Spectra of isotopes after their interplanetary propagation for $\kappa_0 = 10^{18}$ cm^2/s (From Balashev et al. 2007)

11.11.9 Summary of Main Results and Discussion

Balashev et al. (2007) tried to reanalyze the problem of light isotopes He^3, D, and T production in solar flares. In addition to the previous studies Balashev et al. (2007) consider the multiplicity of particle nuclear interactions in the flare sites, include the opportunity of their acceleration due to the stochastic mechanism and also subsequent propagation effects in the interplanetary space. Within this modeling it is quite easy to explain the overabundance of He^3 isotopes observed in impulsive SEP events. At the same time, the comparable fluxes of D and T obtained theoretically are not measured in experiments. This property is likely connected with their destruction and/or angular distribution of primary beams, i.e., with the differential cross sections for the production of these isotopes (Ramaty and Kozlovsky 1974; Colgate et al. 1977). As to the interplanetary propagation, it seems to play an important role in accounting for the observed spectra from ACE. Particularly, this effect (due to adiabatic losses) shift the particle energy to lower values (see Figs. 11.69 and 11.70), qualitatively making the theoretically obtained spectra closer to the observed ones.

11.12 Gamma Rays from Excited States of ^{12}C and Various Neutron Capture Lines: Investigation of Powerful Solar Flare Characteristics

11.12.1 The Matter and Short History of the Problem

As underlined Arkhangelskaja et al. (2009), gamma-ray emission from solar flares gives information about the nature of the accelerated particles and physical conditions in the area, in which flare occurs, and in the surrounding media. Nuclear lines properties represent the abundance of elements, density and temperature of the ambient solar atmosphere and the parameters of the accelerated ions. In solar flares, γ-emission in the region $E_\gamma \geq 0.1$ MeV can be produced via:

1. Bremsstrahlung emission of electrons, directly accelerated in a flare (Miller et al. 1989) and generated in decays of pions produced in the interactions of accelerated protons with surrounding solar media (Murphy et al. 1987) – in both cases spectra are continuum, and mostly observed at $E_\gamma \geq 10$ MeV in the second case (Murphy et al. 2007); see also Section 11.8.3.
2. Annihilation of positrons (produced in decays of radioactive nuclei formed in reactions of interactions of flare-accelerated protons, α-particles, and heavier ions with the solar atmosphere (Crannell et al. 1975) and in decays of π^+ created in flares by accelerated protons (Murphy et al. 1987)) – line at 511 keV with the full width at half maximum (FWHM) of several kiloelectron volt from both direct annihilation and 2γ-annihilation from the positronium singlet state (Share et al. 2003) and spectrum continuum from 3γ-annihilation from the positronium triplet state (Share et al. 2003) mostly observed at $E_\gamma \geq 10$ MeV (Murphy et al. 2007); see Chapter 10.
3. Secondary processes by radioactive nuclei produced in interactions of flare-accelerated protons and α-particles with ambient nuclei heavier than He (Lingenfelter and Ramaty 1967; Chupp et al. 1973; Ramaty et al. 1979; Murphy et al. 2007) – narrow γ-lines with fractional FWHM around 2% (as the recoil velocity of a heavier nucleus is relatively low); see Chapter 9.
4. Secondary processes by radioactive nuclei produced in interactions of accelerated α-particles and heavier ions with ambient H and He (Lingenfelter and Ramaty 1967; Chupp et al. 1973; Ramaty et al. 1979; Murphy et al. 2007) – broad γ-lines with fractional FWHM around 20% (as the recoil velocity of a nucleus is relatively high); see Chapter 9.
5. Secondary processes by radioactive nuclei produced in heavy-heavy interactions – relatively narrow γ-lines, for example, 1.634 MeV from ^{20}Ne, 1.369 MeV from ^{24}Mg, from one hand, and broad γ-lines with fractional FWHM $\sim$20% (as the recoil velocity of a nucleus is relatively high), from other hand, for example, 0.937 MeV from ^{18}F (Kozlovsky et al. 2004; Kuzhevskij et al. 2005; Murphy et al. 2007); see also Chapter 9.

6. Neutron capture lines – at 2.223 MeV and probably 20.58 MeV (Zurműhle et al. 1963; Alfimenkov et al. 1979; Kuzhevskij 1982; Wolfs et al. 1989; Komar et al. 1993; Bystritsky et al. 2006); see also Chapter 9 and Section 11.1.
7. Decay of neutral pions produced in solar flares mostly in collisions between protons and nuclei of hydrogen and helium ((pp) and (pα)) – broad spectral feature at energies 35–140 MeV with the peak at $E_\gamma \sim m_{\pi^0}c^2/2$ which can be shifted depending on the directionality of the proton flux (Cheng 1972; Crannell et al. 1979); see also Sections 11.8 and 11.9.

Arkhangelskaja et al. (2009) noted that nuclear γ-lines provide the diagnostics of the abundance of elements, density and temperature of the ambient solar atmosphere and parameters of accelerated ions (Murphy et al. 2007). The shape of the positron annihilation line is a function of the transition-region temperature during the flare and hydrogen concentration (Bussard et al. 1979; Share et al. 2003). The spectrum of γ-rays resulting from the decay of neutral pions is sensitive to the energy distribution of accelerated protons and α-particles with $E >$ 100 MeV (Crannell et al. 1979). Continuum spectral indices, at first, are dependent on the electron spectral index and directionality (Vestrand et al. 1987; Miller et al. 1989) and proton spectral index due to the decay of negative pions (Share and Murphy 2006). Some other spectral parameters also give important information about the dynamics of accelerated particles and flare region characteristics.

In Arkhangelskaja et al. (2009) the investigation is concentrate on two spectral features, which allow to improve our knowledge on the proton spectra parameters and abundance of ^{3}He in flare region (if they are observed during solar flares). The first feature is the 15.11 MeV γ-ray line produced by excited states of ^{12}C (Crannell et al. 1979; Crannell and Lang 1988), the second feature is the 20.58 MeV γ-ray line from neutron capture on ^{3}He (Kuzhevskij 1982).

11.12.2 Comparison of 15.11 MeV and 4.44 MeV γ-Ray Lines from Solar Flares

As noted Arkhangelskaja et al. (2009), the 15.11 MeV γ-ray line is produced in direct interactions between protons and the ^{12}C and ^{16}O nuclei:

$$\mathrm{p} + {}^{12}\mathrm{C} \rightarrow {}^{12}\mathrm{C} + \mathrm{p}\prime + \gamma, \tag{11.166}$$

$$\mathrm{p} + {}^{16}\mathrm{O} \rightarrow {}^{12}\mathrm{C} + \mathrm{p}\prime + \alpha + \gamma, \tag{11.167}$$

As underlined Arkhangelskaja et al. (2009), another line produced in these reactions is the 4.44 MeV line. These two lines allow obtaining parameters of accelerated protons as they are a pair of γ-lines from the same nuclei, but formed by the de-excitation of levels with widely separated thresholds. The flux ratio $f_{15.11}/f_{4.44}$ depends not only on cross-sections but also on the spectrum of exciting

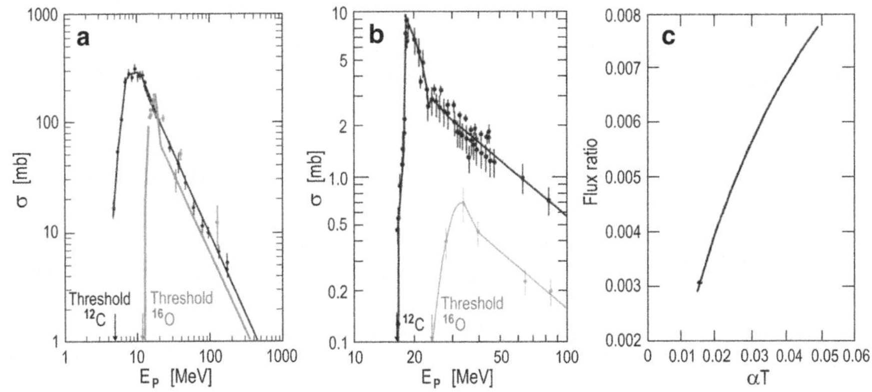

Fig. 11.71 Results of Crannell and Lang (1988) on cross-sections of γ-ray production at ^{12}C and ^{16}O in reactions according to Eq. 11.166 – *black curve* and according to Eq. 11.167 – *gray curve* for the 4.44 MeV γ-ray line (*Panel* **a**) and for the 15.11 MeV γ-ray line (*Panel* **b**), and on the dependence of the $f_{15.11}/f_{4.44}$ flux ratio from the proton spectrum parameter αT (proton spectra were parameterized by the Bessel function) – (*Panel* **c**) (From Arkhangelskaja et al. 2009)

particles and relative isotopic abundance of ^{12}C and ^{16}O. The ratio of cross-sections for these reactions is $\sigma_C/\sigma_O \sim 10$ (Crannell and Lang 1988; Lang et al. 1987). The energy threshold for ^{16}O is higher than that for ^{12}C – see Panels a and b in Fig. 11.71. The ratio $f_{15.11}/f_{4.44}$ was studied by Crannell et al. (1979) and Crannell and Lang (1988) and it was obtained that more probable value of it is 0.006, if proton spectra calculation for some observed solar flares (Murphy and Ramaty 1984) are taken into account. The dependence of this ratio on proton spectrum parameters is shown in Panel c in Fig. 11.71.

11.12.3 Using of Neutron Capture Lines from Solar Flares

As noted Arkhangelskaja et al. (2009), neutrons can be produced in interactions of protons accelerated during solar flares with ambient solar material. Approximately 48% of produced neutrons with energy $E_n \geq 1$ MeV will escape from the solar atmosphere (Cheng 1972). Neutrons that do not escape from the Sun can be captured on nuclei of hydrogen and ^{3}He to form deuterons with emissions of γ-quantum with energy 2.223 MeV and ^{4}He with emissions of γ-quantum with energy 20.58 MeV or two γ-quanta in reactions:

$$n + {}^1\text{H} \rightarrow {}^2\text{H} + \gamma, \tag{11.168}$$

$$n + {}^3\text{He} \rightarrow {}^4\text{He} + \gamma, \tag{11.169}$$

$$n + {}^3\text{He} \rightarrow {}^4\text{He} + 2\gamma, \tag{11.170}$$

Arkhangelskaja et al. (2009) underlined that reaction described by Eq. 11.170 gives additional component of spectrum continuum. Also neutrons can be decayed or non-radiative captured on ^{3}He. The cross-section of reaction described by Eq. 11.170 is much smaller than one of processes described by Eqs. 11.163 and 11.169 for all neutron energies (Bystritsky et al. 2006). The possibility of observable contribution of reaction described by Eq. 11.169 into the solar flare spectrum was first suggested and discussed in Kuzhevskij (1982). Cross-section of reaction described by Eq. 11.169 was first measured in Zurműhle et al. (1963). Up to now experimental studies of the process of radiative neutron capture by ^{3}He nuclei were carried out and the cross-section of this reaction was studied up to $E_n \sim 20$ MeV (see for example Alfimenkov et al. 1979; Bystritsky et al. 2006; Komar et al. 1993; Wolfs et al. 1989). Even if the ratio of the neutron capture cross sections on ^{1}H and ^{3}He is $\sim 10^4$, at energies $E_n > 100$ eV reaction described by Eq. 11.169 has a resonance behavior and at energies of $E_n > 500$ keV the cross-section of this reaction exceeds the cross-section of reaction described by Eq. 11.168 of radiative neutron capture on ^{1}H nuclei. The comparison of cross-sections of these two reactions is shown in Panel a of Fig. 11.72.

Arkhangelskaja et al. (2009) made an attempt to make a preliminary calculation of the observable ratio of line intensities of neutron capture on ^{1}H and ^{3}He. In this calculation they used models of neutron spectra based on Hua and Lingenfelter (1987d) with initial power-law proton spectra with indices S equal to -6, -4, -2 and parameterized by Bessel function $K_2(\alpha T)$ with $\alpha T = 0.005$, 0.03 and 0.1. It was supposed that most of non-thermal neutrons interact with the surrounding medium. Normalized by unity spectra of neutrons are shown in Panel b of Fig. 11.72.

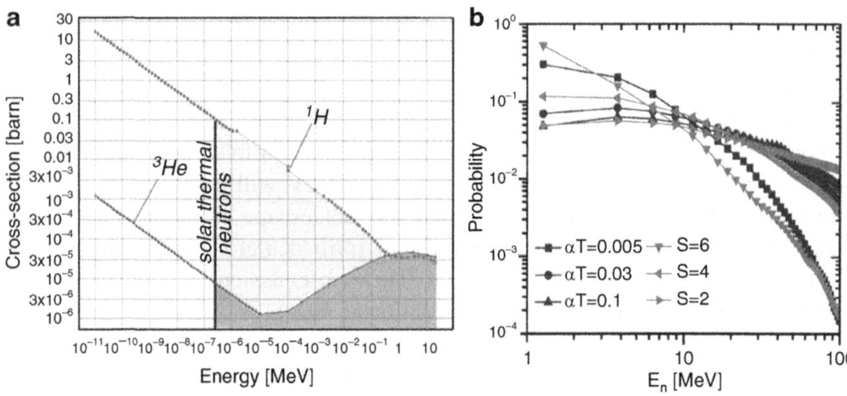

Fig. 11.72 *Panel* **a**: Cross-sections of neutron radiative capture on ^{1}H and ^{3}He according to McKinley et al. (2004). *Panel* **b**: Normalized by unity energy spectra of neutrons produced by incident ions with the Bessel function $K_2(\alpha T)$ and power-law spectra characterized by different αT and S spectrum parameters, correspondingly (From Arkhangelskaja et al. 2009)

Table 11.11 Preliminary results of calculations of the line intensity ratio for the neutron capture on ^{1}H and ^{3}He (From Arkhangelskaja et al. 2009)

αT or S	^{3}He to ^{1}H line intensity ratio
$\alpha T = 0.005$	1.21×10^{-4}
$\alpha T = 0.03$	1.17×10^{-4}
$\alpha T = 0.1$	1.16×10^{-4}
$S = 2$	1.15×10^{-4}
$S = 4$	1.18×10^{-4}
$S = 6$	1.22×10^{-4}

Preliminary results of calculations of line intensity ratios for the neutron capture on ^{1}H and ^{3}He are presented in Table 11.11 for n(^{3}He)/n(^{1}H) $= 10^{-4}$. In this calculation were used values of relative abundance of ^{4}He, ^{3}He and ^{1}H following real solar flare models and observations (Crannell et al. 1979; Manzhavidze et al. 1997; Share and Murphy 1998; Kozlovsky et al. 2004).

11.12.4 Applying the Model to the 20 January 2005 Event

As noted Arkhangelskaja et al. (2009), at present the processed data with sufficient spectral resolution and statistically significant count rates at energies $E_\gamma > 10$ MeV exist only for the January 20, 2005 solar flare event (Arkhangelskaja et al. 2008). These data were obtained by the AVS-F apparatus (Glyanenko et al. 1999) onboard CORONAS-F satellite. The January 20, 2005 solar flare (class X7.1) was the strongest in January 2005 series. This flare was accompanied by SEP (solar energy particles) event and GLE (ground level event) which were the most intensive ones for the last 29 and 52 years, respectively (Mewaldt et al. 2005). The flare was also accompanied by a coronal mass ejection. During the flare rising phase (determined on the base of the GOES data) γ-emission in two energy bands (0.1–20 and 2–140 MeV) was observed by AVS-F apparatus onboard CORONAS-F satellite. The temporal profiles of this flare as observed by GOES, AVS-F, RHESSI and ground neutron monitors data are shown in Fig. 11.73. The wide range energy spectrum of January 20, 2005 solar flare obtained by AVS-F data is presented in Panel a of Fig. 11.74. Nuclear, positron and neutron capture on ^{1}H lines, as well as the spectral feature corresponding to the decay of neutral pions were detected in the flare energy spectrum during the whole time of the flare γ-emission observations by AVS-F. A spectral feature was observed in the region of 15–21 MeV at the 2.5 and 3 standard deviation levels at times 06:44:52–06:51:16 and 06:47:00–06:49:08 UT, respectively (Arkhangelskaja et al. 2008). The intensity ratio of this feature with respect to the lines at 2.223 MeV and 4.44 MeV is $\sim 7 \times 10^{-4}$ and $\sim 5 \times 10^{-3}$ correspondingly. This is comparable with estimations for the 15.11 and 20.58 MeV lines.

As noted Arkhangelskaja et al. (2009), the relative abundance n(^{3}He)/n(^{4}He) in this event can be estimated by the intensity ratio between $\alpha\alpha$-line and deexcitation

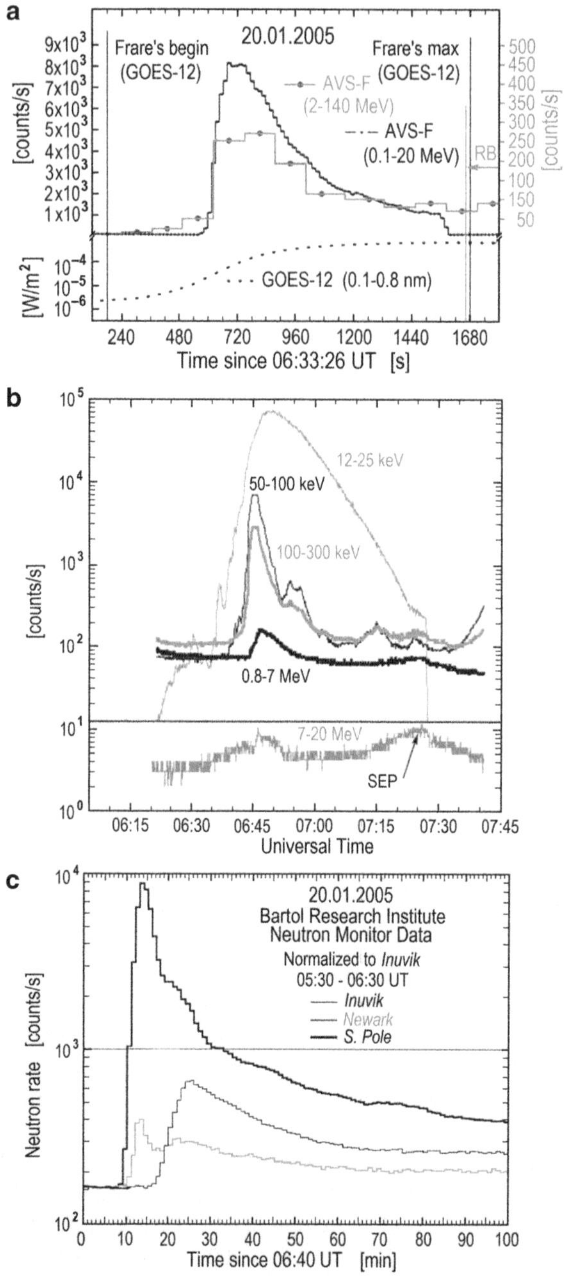

Fig. 11.73 Temporal profiles of the X7.1 January 20, 2005 solar flare as observed by: (**a**) GOES (*dots*) and in low (*black line*) and high (*gray line*) AVS-F energy γ-bands (According to Arkhangelskaja et al. 2008); (**b**) RHESSI data in energy bands 12–25 keV (*thin gray line*), 50–100 keV (*thin black line*), 100–300 keV (*thick gray line*), 0.8–7 MeV (*thick black line*) (*upper panel*) and 7–20 MeV (*low panel*) (According to Share et al. 2006); (**c**) ground neutron monitors: South Pole (*thick gray line*), Inuvik (*thin black line*) and Newark (*thick gray line*) (From Arkhangelskaja et al. 2009)

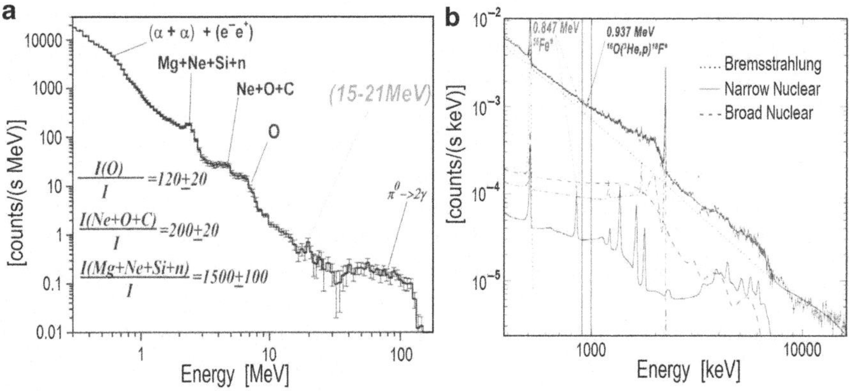

Fig. 11.74 Energy spectra of the January 20, 2005 X7.1 solar flare measured by: (**a**) AVS-F data according to Arkhangelskaja et al. (2008); (**b**) RHESSI data according to Share et al. (2006): *solid black curve* with points – observed spectrum, *dotted line* – bremsstrahlung part, *dashed curve* – contribution of broad nuclear lines, *dashed dotted curve* – contribution of narrow nuclear lines (From Arkhangelskaja et al. 2009)

lines (0.937, 1.04 and 1.08 MeV) produced in direct reaction by ^{3}He, for example (Gan 2002; Kozlovsky et al. 2004):

$$^3\text{He} + ^{16}\text{O} \rightarrow \text{p} + ^{18}\text{F}^* \tag{11.171}$$

This line complex was registered in some solar flares (Manzhavidze et al. 1997; Share and Murphy 1998) and observations of it can give direct information on the ^{3}He abundance in the solar atmosphere. In the energy spectrum of the January 20, 2005 flare obtained from RHESSI data (see Panel b of Fig. 11.74) a weak feature at energies 0.9–1.0 MeV is presented. It is slightly less intensive than the 0.84 MeV line from ^{56}Fe and can be interpreted as the 0.937 MeV ^{18}F line. However, the investigation of so weak spectral features requires very accurate and complex data analysis. Moreover, some adjacent lines produced in interactions of accelerated α-particles with ambient ^{56}Fe (for example 0.931, 1.0 MeV and so on) should be taken into account as their presence distorts the shape of analyzed deexcitation spectral features (Gan 2002).

As underlined Arkhangelskaja et al. (2009), the ratio of n(^{4}He)/n(^{1}H) $\sim$ 0.2 estimated on the base of the SEP data analysis (Mewaldt et al. 2005) and the upper limit for n(^{3}He)/n(^{4}He) $\sim 8 \times 10^{-4}$ (Desai et al. 2006) give n(^{3}He)/n(^{1}H) $\sim 2 \times 10^{-4}$. This value is comparable with estimations based on the 15.11 and 20.58 MeV lines.

11.12.5 Conclusion and Discussion on Main Obtained Results

As underlined Arkhangelskaja et al. (2009), recently many interesting results were obtained in solar gamma- and X-ray spectroscopy and in investigations of elemental abundances in the solar atmosphere. But only a few of them used data at energies higher than about 10 MeV. Two gamma-lines can be measured in so high energy band: the 15.11 MeV line produced by excited states of ^{12}C and the 20.58 MeV line from neutron capture on ^{3}He. Arkhangelskaja et al. (2009) discussed the flux ratio $f_{15.11}/f_{4.44}$ and results of preliminary calculation of intensity ratio $f_{20.58}/f_{2.223}$. The values of these ratios allow to make conclusion that the spectral feature at 15–21 MeV observed in the spectrum of the January 20, 2005 solar flare (Arkhangelskaja et al. 2007) can be possibly interpreted as a combination of 15.11 and 20.58 MeV gamma-ray lines. Direct observation of this spectral feature and detailed studies of its shape can give a very important information about the physics of solar flares, for example the $n(^{3}He)/n(^{1}H)$ ratio, ^{3}He concentration and other parameters. The preliminary estimate for the January 20, 2005 solar flare gives $n(^{3}He)/n(^{1}H) \sim 8 \times 10^{-4}$ based on the analysis of intensity ratio of 15.11 and 20.58 MeV lines with respect to the line at 2.223 MeV.

References for Chapter 11

Akimov VV, Afanassyev VG, Belaousov AS et al (1991) Observation of high energy gamma-rays from the Sun with the GAMMA-1 telescope (E > 30 MeV). In: Proceedings of the 22nd international cosmic ray conference, vol 3, Dublin, pp 73–76

Akimov VV, Belov AV, Chertok IM, Kurt VG, Leikov NG, Magun A, Melnikov VF (1993) High energy gamma-rays at the late state of the large solar flare of June 15, 1991 and accompanying phenomena. In: Proceedings of the 23rd international cosmic ray conference, vol 3, Calgary, Invited, Rapporteur, and highlight papers, pp 111–114

Aladashvili BS, Glagolev VV, Zhuravleva LI et al (1980) Deuteron yield in ^{4}He p-interaction at 8.6 GeV/c. Report P1-80-244, Joint Institute of Nuclear Research, Dubna, pp 9

Alfimenkov VP, Borzakov SB, Vezhbitski Ya et al (1979) Radiative capture of He^{3} neutrons in the energy interval 1–70 keV. JETP Lett 29:100–108

Alsmiller RG, Leimdorfer M, Barish J (1967) Analytic representation of nonelastic cross sections and particle-emission spectra from nucleon-nucleus collisions in the energy range 25 to 400 MeV. Report NASA-CR-83981, ORNL-4046, pp 1–125

Amian WB, Byrd RC, Goulding CA, Meier MM, Morgan GL, Moss CE, Clark DA (1992) Differential neutron production cross sections for 800-MeV protons. Nucl Sci Eng (USA) 112(1):78–86

Amian WB, Byrd RC, Clark DA, Goulding CA, Meier MM, Morgan GL, Moss CE (1993) Differential neutron production cross sections for 597-MeV protons. Nucl Sci Eng (USA) 115(1):1–12

Anglin JD, Dietrich WF, Simpson JA (1973) Deuterium and tritium from solar flares at 10 MeV per nucleon. Astrophys J Lett Ed (USA) 186(1):L41–L46

Arkhangelskaja IV, Arkhangelsky AI, Troitskaja EV, Kalmykov PA, Kotov YuD, Kuznetsov SI, Glyanenko AS, Miroshnichenko LI (2007) The observation of gamma-ray emission during January 20, 2005 solar flare. In: Proceedings of 30th international cosmic ray pre-conference proceedings (on CD), Merida, pp 4

Arkhangelskaja IV, Arkhangelsky AI, Glyanenko AS, Kotov YuD, Kuznetsov SN (2008) Solar flares temporal profiles thin structure on timescales 33-92 sec in various energy bands by data of AVS-F apparatus onboard CORONAS-F satellite. In: Proceedings of the 30th international cosmic ray conference, Merida, vol 1, pp 15–18

Arkhangelskaja IV, Arkhangelsky AI, Kotov YuD, Kalmykova PA, Glyanenkoa AS (2008) AVS-F observations of gamma-ray emission during January 20, 2005 solar flare up to 140 MeV. Adv Space Res 43(4):589–593

Arkhangelskaja IV, Arkhangelsky AI, Troitskaya EV, Miroshnichenko LI (2009) The investigation of powerful solar flares characteristics by analysis of excited states of ^{12}C and various neutrons capture lines. Adv Space Res 43:594–599

Ashery D (1969) Study of the 22Ne(α, n)25Mg reaction: a possible source of stellar neutrons. Nucl Phys A (Netherlands) A136(3):481–495

Avrett EH (1981) Reference model atmosphere calculation – The Sunspot sunspot model. In: Cram LE, Thomas JH (eds) The physics of sunspots. Proceedings of the sunspot conference, New Mexico, conference sponsored by the Sacramento Peak Observatory, pp 235–255, 257

Axford WI, Leer E, Skadron G (1977) The acceleration of cosmic rays by shock waves. In: Proceedings of the 15th international cosmic ray conference, vol 11, Plovdiv, pp 132–137

Bai T (1986) Two classes of gamma-ray/proton flares: impulsive and gradual. Astrophys J (USA), 308(2, part 1):912–928

Balashev SA, Lytova MF, Ostryakov VM (2007) Nuclear interactions in the flare sites. In: Proceedings of the 30th international cosmic ray conference, Mèrida, Mèxico, Pre-conference edition, pp 1–4. http://dpnc.unige.ch/ams/ICRC-07/icrc0545.pd

Barashenkov VS, Zheregi FG, Musulmanbekov ZhZh (1983) Intranuclear cascade studies of inelastic nucleus-nucleus interactions, Preprint P2-83-117, JINR, Dubna, pp 1–10

Barashenkov VS, Zheregi FG, Musulmanbekov ZhZh (1984) The cascade mechanism of inelastic interactions of high-energy nuclei. Sov J Nucl Phys (USA) 39(5):715–716

Bell AR (1978) The acceleration of cosmic ray in shock fronts. I. Mon Not R Astron Soc (UK) 182(1):147–156

Bertrand FE, Peelle RW (1973) Complete hydrogen and helium particle spectra from 30- to 60-MeV proton bombardment of nuclei with A = 12 to 209 and comparison with the intranuclear cascade model. Phys Rev C Nucl Phys (USA) 8(3):1045–1064

Bertrand FE, Peelle RW, Kalbach-Cline C (1974) Differential cross sections for charged-particle emissions in reactions of 58 MeV α particles with ^{12}C, ^{16}O, and ^{54}Fe; comparison with the exciton model of pre-equilibrium particle emission. Phys Rev C Nucl Phys (USA) 10(3):1028–1040

Bildsten L, Wasserman I, Salpeter EE (1990) A semi-classical model for breakup reactions of light nuclei: 4He(p, pn)3He. Nucl Phys A (Netherlands) A516(1):77–107

Blandford RD, Ostriker JR (1978) Particle acceleration by astrophysical shocks. Astrophys J 221(Part 2):L29–L32

Blumenthal GR, Gould RJ (1970) Bremsstrahlung, synchrotron radiation, and compton scattering of high-energy electrons traversing dilute gases. Rev Mod Phys 42(2):237–271

Bussard RW, Ramaty R, Drachman RJ (1979) The annihilation of galactic positrons. Astrophys J 228(3, Part 1):928–934

Bystritsky VM, Bystritskii VitM, Enik TL, Filipowicz M, Gerasimov VV, Grebenyuk VM, Kobzev AP, Kublikov RV, Nesvizhevskii VV, Parzhitskii SS, Pavlov VN, Popov NP, Salamatin AV, Shvetsov VN, Slepnev VM, Strelkov AV, Wozniak J, Zamyatin NI (2006) Experimental research of the radiative capture of thermal neutrons in ^{3}He. JINR Communications, D15-2006-23, Dubna, pp 1–19. http://www1.jinr.ru/Preprints/2006/023(D15-2006-23)e.pdf

Cameron AGW (1982) Elemental and nuclidic abundances in the solar system. In: Barnes CA, Clayton DD, Schramm DN (eds) Essays in nuclear astrophysics. Cambridge University Press, Cambridge, UK, pp 23–43

Cane HV, McGuire RE, von Rosenvinge TT (1985) Two classes of solar energetic particle events associated with impulsive and long duration soft X-ray flares. In: Proceedings of the 19th international cosmic ray conference, vol 4, La Jolla, CA, pp 66–69

Cesarsky CJ, Paul JA, Shukla PG (1978) Bremsstrahlung gamma radiation and interstellar electron spectrum in the local interstellar medium. Astrophys Space Sci 59(1):73–83

Cheng CC (1972) Theoretical studies of the flux and energy spectrum of gamma radiation from the Sun. Space Sci Rev 13(1):3–123

Chupp EL (1982) Solar energetic photon transients (50 keV–100 MeV). In: Lingenfelter RE, Hudson HS, Worrall DM (eds) Gamma ray transients and related astrophysical phenomena. Proceedings of the workshop, La Jolla, CA, 1981, AIP conference proceedings, vol 77, No. 1, pp 363–381

Chupp EL (1990) Emission characteristics of three intense solar flares observed in Cycle 21. Astrophys J Suppl Ser (USA) 73(2):213–226

Chupp EL, Forrest DJ, Higbie PR, Suri AN, Tsai C, Dunphy PP (1973) Solar gamma ray lines observed during the solar activity of August 2 to August 11, 1972. Nature 241(5388): 333–335

Chupp EL, Forrest DJ, Share GH, Kanbach G, Debrunner H, Flueckiger E (1983) Solar neutrons from the impulsive flare on 1982 June 3 at 11:43 UT. In: Proceedings of the 18th international cosmic ray conference, vol 10, Bangalore, India, pp 334–337

Chupp EL, Debrunner H, Fluckiger E, Forrest DJ, Golliez F, Kanbach G, Vestrand WT, Cooper J, Share G (1987) Solar neutron emissivity during the large flare on 1982 June 3. Astrophys J 318 (2, Part 1):913–925

Chupp EL, Trottet G, Marschhauser H, Pick M, Soru-Escaut I, Rieger E, Dunphy PP (1993) A study of the evolution of electron and ion acceleration during the 09:09 UT solar flare on 1989 September 9. Astron Astrophys (Germany) 275(2):602–612

Cliver EW (1996) Solar flare gamma-ray emission and energetic particles in space. In: Ramaty R, Mandzhavidze N, Hua X-M (eds) High energy solar physics. Proceedings of the conference, Greenbelt, MD, 1995. AIP conference proceedings, vol 374, No. 1, pp 45–60

Cliver EW, Forrest DJ, McGuire RE, Von Rosenvinge TT (1983) Nuclear gamma rays and interplanetary proton events. In: Proceedings of the 18th international cosmic ray conference, vol 10, Bangalore, India, pp 342–345

Cliver EW, Forrest DJ, Cane HV, Reames DV, McGuire RE, von Rosenvinge TT, Kane SR, MacDowall RJ (1989) Solar flare nuclear gamma-rays and interplanetary proton events. Astrophys J (USA) 343(2, Part 1):953–970

Coleman WA (1968) Mathematical verification of a certain Monte Carlo sampling technique and applications of the technique to radiation problems. Nucl Sci Eng 32(1):76–81

Colgate SA (1978) A phenomenological model of solar flares. Astrophys J (USA) 221 (3, Part 1):1068–1087

Colgate SA, Audouse J, Fowler WA (1977) Possible interpretation of the isotopic composition of hydrogen and helium in solar cosmic rays. Astrophys J 213(3, Part 1):849–860

Crannell CJ, Lang FL (1988) Nuclear gamma-line ratios as spectral diagnostics for protons accelerated in solar flares. In: Neil G, Share GH (eds) Nuclear spectroscopy of astrophysical sources. Proceedings of a conference, Washington, DC, 1988, AIP conference proceedings, vol 170, No. 1, pp 240–245

Crannell CJ, Ramaty R, Werntz C (1975) Positron annihilation in solar flares. In: Proceedings of the 14th international cosmic ray conference, vol 5, Munich, pp 1656–1661

Crannell CJ, Ramaty R, Crannell H (1979) Solar gamma-rays above 8 MeV. Astrophys J 229 (2, Part 1):762–771

Debrunner H, Flückiger EO, Chupp EL, Forrest DJ (1983) The solar cosmic ray neutron event on June 3, 1982. In: Proceedings of the 18th international cosmic ray conference, vol 4, Bangalore, India, pp 75–78

Debrunner H, Flückiger EO, Chupp EL, Forrest DJ (1993) The solar cosmic ray neutron event on June 3, 1982. In: Proceedings of the 18th international cosmic ray conference, vol 4, Bangalore, India, pp 75–78

Debrunner H, Lockwood JA, Barat C, Buttikofer R, Dezalay JP, Flückiger E, Kuznetsov A, Ryan JM, Sunyaev R, Terekhov OV, Trottet G, Vilmer N (1997) Energetic neutrons, protons, and gamma rays during the 1990 May 24 solar cosmic-ray event. Astrophys J 479 (2, Part 1):997–1011

Dermer CD (1986a) Secondary production of neutral pi-mesons and the diffuse galactic gamma radiation. Astron Astrophys 157(2):223–229

Dermer CD (1986b) Binary collision rates of relativistic thermal plasmas. 2. Spectra. Astrophys J 307(1, Part 1):47–59

Dermer CD (1987) On the relationship of flare size and particle anisotropy in solar gamma-ray flares. Astrophys J (USA) 323(2, Part 1):795–798

Dermer CD, Ramaty R (1986) Directionality of bremsstrahlung from relativistic electrons in solar flares. Astrophys J (USA) 301(2, Part 1):962–974

Desai MI, Mason GM, Gold RE, Krimigis SM, Cohen CMS, Mewaldt RA, Mazur JE, Dwyer JR (2006) Gold heavy-ion elemental abundances in large solar energetic particle events and their implications for the seed population. Astrophys J 649(1, Part 1):470–489

Devlin TJ, Johnson W, Norem J, Vosburgh K, Mischke RE, Schimmerling W (1973) Measurement of n-p and n-d total cross sections from 0.7 to 3.6 GeV/c. Phys Rev D Part Fields (USA) 8(1):136–155

Dorman IV, Dorman LI (1967a) Solar wind properties obtained from the study of the 11-year cosmic ray cycle. 1. J Geophys Res 72(5):1513–1520

Dorman IV, Dorman LI (1967b) Propagation of energetic particles through interplanetary space according to the data of 11-year cosmic ray variations. J Atmosph Terr Phys 29(4): 429–449

Dorman IV, Dorman LI (1968) Hysteresis phenomena in cosmic rays, properties of solar wind and energetic spectrum of different nuclei in the Galaxy. In: Proceedings of the 5th all-union winter school on cosmophysics, Apatity, pp 183–196

Dorman LI (1995a) Cosmic ray nonlinear effects in space plasma, 1. General characteristics and dynamic galactic halo. In: Shapiro MM, Silberberg R, Wefel JP (eds) Currents in high energy astrophysics, NATO ASI Series, Vol 458. Kluwer, Dordrecht/Boston/London, pp 183–191

Dorman LI (1995b) Cosmic ray nonlinear effects in space plasma, 2. Dynamic Heliosphere. In: Shapiro MM, Silberberg R, Wefel JP (eds) Currents in high energy astrophysics, NATO ASI Series, Vol 458. Kluwer, Dordrecht/Boston/London, pp 193–208

Dorman LI (1996) Cosmic ray nonlinear processes in gamma-ray sources. Astron Astrophys Suppl Ser 120(4):427–435

Dorman LI (1997) Angle distribution and time variation of gamma ray flux from solar and stellar winds, 1. Generation by flare energetic particles. In: Dermer CD, Strickman MS, Kurfess JD (eds) Proceedings of the 4th Compton Symposium, Williamsburg, VA, pp 1178–1182

Dorman LI, Villoresi G, Dorman IV, Iucci N, Parisi M (1997a) High rigidity CR-SA hysteresis phenomenon and dimension on modulation region in the Heliosphere in dependence of particle rigidity. In: Proceedings of the 25th international cosmic ray conference, Durban (South Africa), vol 2, pp 69–72

Dorman LI, Villoresi G, Dorman IV, Iucci N, Parisi M (1997b) Low rigidity CR-SA hysteresis phenomenon and average dimension of the modulation region and Heliosphere. In: Proceedings of the 25th international cosmic ray conference, Durban (South Africa), vol 2, pp 73–76

Dorman Lev I (2001) Variable gamma ray sources, 1. Interactions of flare energetic particles with solar and stellar winds. In: Shapiro MM et al (eds) Astrophysical sources of high energy particles and radiation. Kluwer, The Netherlands, pp 219–230

Dorman LI, Venkatesan D (1993) Solar cosmic rays. Space Sci Rev 64:183–362

Dorman LI, Ptuskin VS, Zirakashvili VN (1990) Outer Heliosphere: pulsations, cosmic rays and stream kinetic instability. In: Grzedzielski S, Page DE (eds) Physics of the outer heliosphere. Pergamon Press, Oxford/ England/Elmsford, NY, pp 205–209

Earl JA (1981) Analytical description of charged particle transport along arbitrary guiding field configurations. Astrophys J 251(2, Part 1):739–755

Ellison DC, Ramaty R (1985) Shock acceleration of electrons and ions in solar flares. Astrophys J 298(1, Part 1):400–408

Ellison DC, Jones FC, Reynolds SP (1990) First-order Fermi particle acceleration by relativistic shocks. Astrophys J 360(2, Part 1):702–714

Evenson P, Meyer P, Pyle KR (1983) Protons from the decay of solar flare neutrons. Astrophys J 274(2, Part 1):875–882

Forman MA, Ramaty R, Zweibel EG (1986) The acceleration and propagation of solar flare energetic particles. In: The physics of the sun, vol 2. D. Reidel, Dordrecht, The Netherlands, pp 291–323

Forrest DJ (1983) Solar gamma-ray lines. In: Burns ML, Harding AK, Ramaty R (eds) Positron–electron pairs in astrophysics. Proceedings of the Workshop, Greenbelt, MD, 1983, New York, AIP Conference Proceedings, vol 101, No. 1, pp 3–14

Forrest DJ, Vestrand WT, Chupp EL, Rieger E, Cooper J, Share GH (1986) Very energetic gamma-rays from the June 3, 1982 solar flare1986. Adv Space Res 6(6):115–118

Gabriel AH, Culhane JL, Patchett BE, Breeveld ER, Lang J, Parkinson JH, Payne J, Norman K (1995) Spacelab 2 measurement of the solar coronal helium abundance. Adv Space Res (UK) 15(7):63–67

Gan WQ (1998) Spectral evolution of energetic protons in solar flares. Astrophys J 496 (2):992–997

Gan WQ (2002) Solar gamma-ray spectroscopy and abundance of elements. Chin J Astron Astrophys 26(3):255–266

Geiss J (1982) Processes affecting abundances in the solar wind. Space Sci Rev (Netherlands) 33(1–2):201–217

Geiss J, Reeves H (1972) Cosmic and solar system abundances of deuterium and helium-3. Astron Astrophys (Germany) 18(1):126–132

Geiss J, Eberhardt P, Buhler F, Meister J, Signer P (1970) Apollo 11 and 12 solar wind composition experiments: fluxes of He and Ne isotopes. J Geophys Res (USA) 75(31):5972–5979

Gibbons JH, Macklin RL (1959) Total neutron yields from light elements under proton and alpha bombardment. Phys Rev 114(2):571–580

Glagola BG, Viola VE Jr, Breuer H, Chant NS, Nadasen A, Roos PG, Austin SM, Mathews GJ (1982) Production of ^{6}He, ^{6}Li, ^{7}Li and ^{7}Be in the α+α reaction between 60–160 MeV. Phys Rev C Nucl Phys (USA) 25(1):34–45

Glagolev VV, Lebedev PM, Saitov IS et al (1977) Study of two pronged events in ^{4}He p collisions at 8.56 GeV/c ^{4}He incident momentum. Report P1-10894, Joint Institute of Nuclear Research, Dubna, pp 16

Glagolev VV, Lebedev PM, Pestova GD et al (1984) Total desintegration of helium nucleus in ^{4}He p-interactions at 8.6 GeV/c. Report P1-84-202, Joint Institute of Nuclear Research, Dubna, pp 8

Gloeckler G, Geiss J (1989) The abundances of elements and isotopes in the solar wind. In: Waddington CJ (ed) Cosmic abundances of matter. Proceedings of symposium, University of Minnesota, MN, AIP conference proceedings, vol 183, No. 1, pp 49–71

Glyanenko AS, Kotov YuD, Pavlov AV, Arkhangelsky AI, Samoilenko VT, Yurov VN, Pankov VM, Ryumin SP (1999) The AVS-F experiment on recording rapidly changing fluxes of cosmic and gamma radiation prepared under the CORONAS-F project. Instrum Exp Technol 42(5):596–603

Grevesse N, Noels A, Sauval AJ (1996) Standard abundances. In: Holt SS, Sonneborn G (eds) Cosmic abundances. ASP conference series, Proceedings of the sixth annual October astrophysics conference in College Park, MD, 1995, ASP, San Francisco, CA, vol 99, pp 117–126

Gross E (1965a) The absolute yield of low-energy neutrons from 190-MeV proton bombardment of gold, silver, nickel, aluminum, and carbon. Report UCRL-3330, Radiation Lab, University of California, Berkeley, CA, pp 1–77

Gross E (1965b) The absolute yield of low-energy neutrons from 190-MeV proton bombardment of gold, silver, nickel, aluminum, and carbon. Ph.D. thesis, Radiation Lab, University of California, Berkeley, CA

Guglenko VG, Kocharov GE, Kovaltsov GA, Kocharov LG, Mandzhavidze NZ (1990a) Generation of high-energy neutral radiation in flare loops. II. Monte Carlo simulation and comparison with observations. Sol Phys (Netherlands) 125(1):91–123

Guglenko VG, Efimov YuE, Kocharov GE, Kovaltsov GA, Mandzhavidze NZ, Terekhov MM, Kocharov LG (1990b) Neutrons and gamma-ray emission on 1982 June 3: the possibility of fitting the data on the assumption of one population of accelerated particles [solar flares]. Astrophys J Suppl Ser (USA) 73(2):209–211

Hall DNB (1975) Spectroscopic detection of solar ^{3}He. Astrophys J (USA) 197(2, Part 1):509–512

Hua X-M (1986) Ph.D. Dissertation, University of California, San Diego, CA, 129 pp

Hua X-M, Lingenfelter RE (1987a) Solar flare neutron production and the angular dependence of the capture gamma-ray emission. Solar Phys (Netherlands) 107(2):351–383

Hua X-M, Lingenfelter RE (1987b) A determination of the (He-3)/H ratio in the solar photosphere from flare gamma-ray line observations. Astrophys J (USA) 319(1, Part 1):555–566

Hua X-M, Lingenfelter RE (1987c) Solar flare neutrons and their capture gamma ray emission. Solar Phys (Netherlands) 113(1–2):229–235

Hua H-M, Lingenfelter RE (1987d) Solar flare neutron and accelerated ion angular distributions. Astrophys J 323(2, Part 1):779–794

Hua X-M, Lingenfelter RE (1987e) A determination of the (He-3)/H ratio in the solar photosphere from flare gamma-ray line observations. Astrophys J (USA) 319(1, Part 1):555–566

Hua X-M, Ramaty R, Lingenfelter R (1989) Deexcitation gamma-ray line emission from solar flare magnetic loops. Astrophys J (USA) 341(1, Part 1):516–532

Hua X-M, Kozlovsky B, Lingenfelter RE, Ramaty R, Stupp A (2002) Angular and energy-dependent neutron emission from solar flare magnetic loops. Astrophys J Suppl Ser 140 (2):563–579

Hudson HS, Bai T, Gruber DE, Matteson JL, Nolan PL, Peterson LE (1980) HEAO 1 observations of gamma-ray lines from a solar flare. Astrophys J Lett Ed (USA) 236(2, Part 2):L91–L95

Ibragimov IA, Kocharov GE (1977a) On a feature of the energetic spectrum of gamma rays during the August 4, 1972 solar flare event. Pis'ma Astron Zh (USSR) 3(9):412–414 (in Russian)

Ibragimov IA, Kocharov GE (1977b) A feature in the gamma-ray energy spectrum during the flare of August 4, 1972. Sov Astron Lett (USA) 3(5):221–222

Jokipii JR (1973) Radial variation of magnetic fluctuations and the cosmic-ray diffusion tensor in the solar wind. Astrophys J (USA) 182(2, Part 1):585–600

Kahler SW (1993) Coronal mass ejections and long risetimes of solar energetic particle events. J Geophys Res (USA) 98(A4):5607–5615

Kahler S (1994) Injection profiles of solar energetic particles as functions of coronal mass ejection heights. Astrophys J (USA) 428(2, Part 1):837–842

Kalend AM, Anderson BD, Baldwin AR, Madey R, Watson JW, Chang CC, Holmgren HD, Koontz RW, Wu JR, Machner H (1983) Energy and angular distributions of neutrons from 90 MeV proton and 140 MeV alpha-particle bombardment of nuclei. Phys Rev C Nucl Phys (USA) 28(1):105–119

Kanbach G, Reppin C, Forrest DJ, Chupp EL (1975) Time history and model calculations of the 2.2 MeV gamma ray line from the flares of August, 1972. In: Proceedings of the 14th international cosmic ray conference, vol 5, München, pp 1644–1649

Kanbach G, Pinkau K, Reppin C, Rieger E, Chupp EL, Forrest DJ, Ryan JM, Share GH, Kinzer RL (1981) Model calculations on fast solar neutrons and the 2.2 Mev line emission from solar flares. In: Proceedings of the 17th international cosmic ray conference, vol 10, Paris, pp 9–12

Kanbach G, Bertsch DL, Fichtel CE, Hartman RC, Hunter SD, Kniffen DA, Kwok PW, Lin YC, Mattox JR, Mayer-Hasselwander HA, Michelson PF, von Montigny C, Nolan PL, Pinkau K, Rothermel H, Schneid E, Sommer M, Sreekumar P, Thompson DJ (1993) Detection of a long-duration solar gamma-ray flare on June 11, 1991 with EGRET on Compton-GRO. Astron Astrophys Suppl Ser (France) 97(1):349–353

Kocharov LG, Kocharov GE (1984) ^{3}He-rich events. Space Sci Rev 38(1–2), pp 89–141

Kocharov L, Kovaltsov G (1999) Signatures of coronal magnetic field geometry in gamma-rays, neutrons, and high-energy protons. In: Proceedings of the 26th international cosmic ray conference, vol 6, Salt Lake City, CA, pp 34–37

Kocharov GE, Kovaltsov GA, Mandzhavidze NZ, Semukhin PE, Kocharov LG (1987) Generation of neutrons and gamma-rays in flare loops. In: Proceedings of the 20th international cosmic ray conference, vol 3, Moscow, pp 74–77

Kocharov LG, Lee JW, Zirin H, Kovaltsov GA, Usoskin IG, Pyle KR, Shea MA, Smart DF (1994) Neutron and electromagnetic emissions during the 1990 May 24 solar flare. Solar Phys (Netherlands) 155(1):149–170

Kocharov LG, Torsti J, Vainio R, Kovaltsov GA, Usoskin IG (1996) A joint analysis of high-energy neutrons and neutron-decay protons from a flare. Sol Phys (Netherlands) 169(1): 181–207

Kocharov LG, Torsti J, Tang F, Zirin H, Kovaltsov GA, Usoskin IG (1997) Impact of magnetic environment on the generation of high-energy neutrons at the Sun. Sol Phys (Netherlands) 172(1–2):271–278

Kocharov L, Kovaltsov GA, J. Torsti J (1999a) An analytic description of coronal proton trapping. Astrophys J (USA) 519(1, Part 1):422–432

Kocharov L, Kovaltsov GA, Laitinen T, Mäkelä P, Torsti J (1999b) Interacting and interplanetary protons accelerated on diverging magnetic field lines: stochastic acceleration. Astrophys J 521(2):898–905

Komar RJ, Mak H-B, Leslie JR, Evans HC, Bonvin E, Earle ED, Alexander TK (1993) 3He(n, γ) 4He cross section and the photodisintegration of 4He. Phys Rev C Nucl Phys (USA) 48 (5):2375–2384

Kosugi T, Dennis BR, Kai K (1988) Energetic electrons in impulsive and extended solar flares as deduced from flux correlations between hard X-rays and microwaves. Astrophys J (USA) 324(2, Part 1):1118–1131

Kotov YD, Bogovalov SV, Endalova OV, Yoshimori M (1996) ^{7}Li production in solar flares. Astrophys J (USA) 473(1, part 1):514–518

Kozlovsky B, Ramaty R (1974) 478-keV and 431-keV line emissions from alpha-alpha reactions. Astrophys J Lett Ed (USA) 191(1, Part 2):L43–L44

Kozlovsky B, Ramaty R (1977) Narrow lines from alpha-alpha reactions. Astrophys Lett (UK) 19(1):19–23

Kozlovsky B, Lingenfelter RE, Ramaty R (1987) Positrons from accelerated particle interactions. Astrophys J (USA) 316(2, Part 1):801–818

Kozlovsky B, Murphy RJ, Share GH (2004) Positron-emitter production in solar flares from ^{3}He reactions. Astrophys J 604(2, Part 1):892–899

Krymsky GF (1977) Regular mechanism of charged particle acceleration at shock front. Dokl Akad Nauk SSSR 234(6):1306–1307 (in Russian)

Kurganov IG, Ostryakov VM (1992) Time variations of decay protons for the solar events of 21 June 1980, 3 June 1982, and 24 April 1984. Geomagn Aeron (Russia) 32(3):149–153 (in Russian)

Kuzhevskij BM (1982) Gamma astronomy of the sun and study of solar cosmic rays. Sov Phys Usp 25(6):392–408

Kuzhevskij BM, Gan W-Q, Miroshnichenko LI (2005) The role of nuclei-nuclei interactions in the productions of gamma-ray lines in solar flares. Chin J Astron Astrophys 5(3):295–301

Lachkar J, Sigaud J, Patin Y, Haouat G (1974) Gamma-ray-production cross sections for the ^{56}Fe(n,n'γ) reaction from 2.5 to 14.1 MeV neutron energies. Nucl Sci Eng (USA) 55(2): 168–187

Laming JM, Feldman U (1994) The solar abundance of helium determined from a redshifted plasma flow over a sunspot. Astrophys J (USA) 426(1, Part 1):414–424

Lang FL, Werntz CW, Crannell CJ, Trombka JI, Chang CC (1987) Cross sections for production of the 1.510 MeV and other astrophysically significant gamma-ray lines through excitation and spallation of ^{12}C and ^{16}O with protons. Phys Rev C Nucl Phys (USA) 35(4):1214–1227

Lang FL, Shawhan P, Starr R, Werntz CW, Crannell CJ, Saltzberg DP, Brown JD, Champagne AE, Kouzes RT (1989) Nuclear gamma ray fluence ratios as a diagnostic of accelerated particle spectra in solar flares. Bulletin of American Astronomical Society, vol 21, 19th Annual SPD Meeting, pp 833–834

Lee MA, Fisk LA (1982) Shock acceleration of energetic particles in the heliosphere. Space Sci Rev (Netherlands) 32(1–2):205–228

Lee MA, Ryan JM (1986) Time-dependent coronal shock acceleration of energetic solar flare particles. Astrophys J (USA) 303(2, Part 1):829–842

Lee JW, Gary DE, Zirin H (1994) Flat microwave spectra seen at X-class flares. Solar Phys (Netherlands) 152(2):409–428

Le Roux JA, Fichtner H (1997) The influence of pickup, anomalous, and galactic cosmic-ray protons on the structure of the heliospheric shock: a self-consistent approach. Astrophys J 477:L115–L118

Lingenfelter RE, Ramaty R (1967) High energy nuclear reactions in solar flares. In: Shen BSP (ed) High energy nuclear reactions in astrophysics. W.A. Benjamin Inc., New York/Amsterdam, pp 99–158

Ligenfelter RE, Flamm EJ, Canfield EH, Kellman S (1965a) High-energy solar neutrons 1. Production in flares. J Geophys Res 70(17):4077–4086

Ligenfelter RE, Flamm EJ, Canfield EH, Kellman S (1965b) High-energy solar neutrons 2. Flux at the Earth. J Geophys Res 70(17):4087–4095

Lock WO, Measday DF (M1970) Intermediate energy nuclear physics. Methuen, London

Lockwood JA, Debrunner H, Flückiger EO, Gradel H (1990a) Proton energy spectra at the Sun in the solar cosmic-ray events in 1978 May 7 and 1984 February 16. Astrophys J (USA) 355 (1, Part 1):287–294

Lockwood JA, Debrunner H, Flückiger EO (1990b) Indications for diffusive coronal shock acceleration of protons in selected solar cosmic ray events. J Geophys Res (USA) 95 (A4):4187–4201

Lockwood JA, Debrunner H, Ryan JM (1999) Interacting and interplanetary high-energy protons in solar flare events. Astropart Phys (Netherlands) 12(1–2):97–105

Madey R, Anderson BD, Cecil RA, Tandy PC, Schimmerling W (1983) Total inclusive neutron cross sections and multiplicities in nucleus-nucleus collisions at intermediate energies. Phys Rev C Nucl Phys (USA) 28(2):706–709

Mandzhavidze N, Ramaty R (1993) Particle acceleration in solar flares. Nucl Phys B – Proc Suppl B33(1–2):141–160

Mandzhavidze N, Ramaty R, Bertsch DL, Schneid EJ (1996) Pion decay and nuclear line emissions from the 1991 June 11 flare. In: Ramaty R, Mandzhavidze N, Hua X-M (eds) High energy solar physics. Proceedings of the conference, Greenbelt, MD, 1995, AIP conference proceedings, vol 374, No. 1, pp 225–236

Mandzhavidze N, Ramaty R, Kozlovsky B (1997) Solar atmospheric and solar flare accelerated helium abundances from gamma-ray spectroscopy. Astrophys J Lett (USA) 489(1, part 2): L99–L102

McDonald FB, van Hollebeke MAI (1985) HELIOS 1 energetic particle observations of the solar gamma-ray/neutron flare events of 1982 June 3 and 1980 June 21. Astrophys J Lett (USA) 290(2, Part 2):L67–L71

McGuire RE, Von Rosenvinge TT (1984) The energy spectra of solar energetic particles. Adv Space Res (UK) 4(2–3):117–125

McGuire RE, Von Rosenvinge TT, McDonald FB (1977) Spectra and Composition 2–40 MeV/ Nucleon of Ions Z = 2–28 in Solar Cosmic Ray Events. In: Proceedings of the 15th international cosmic ray conference, vol 5, Plovdiv, pp 54–59

McGuire RE, Von Rosenvinge TT, McDonald FB (1979) A survey of solar cosmic ray composition 1974–1978. In: Proceedings of the 18th international cosmic ray conference, vol 5, Kyoto, pp 61–66

McGuire RE, Von Rosenvinge TT, McDonald FB (1981) A survey of solar protons and alpha differential spectra between 1 and greater than 400 MeV/nucleon. In: Proceedings of the 17th international cosmic ray conference, vol 3, Paris, pp 65–68

McKinley MS, Beck B, McNabb DP (2004) Nuclear and Atomic Data System http://nuclear.llnl.gov/CNP/nads/NADSApplet.html, Lawrence Livermore National Laboratory Report No. UCRL-WEB-152626-REV-1

Meier MM, Amian WB, Goulding CA, Morgan GL, Moss CE (1992a) Differential neutron production cross sections for 256-MeV protons. Nucl Sci Eng (USA) 110(3):289–298
Meier MM, Amian WB, Goulding CA, Morgan GL, Moss CE (1992b) Neutron yields from stopping-length targets for 256-MeV protons. Nucl Sci Eng (USA) 110(3):299–301
Melrose DB (1974) Resonant scattering of particles and second phase acceleration in the solar corona. Solar Phys (Netherlands) 37(2):353–365
Miller JA, Ramaty R, James A (1989) Relativistic electron transport and bremsstrahlung production in solar flares. Astrophys J 1(344):973–990
Mewaldt RA, Looper MD, Cohen CMS, et al (2005) Solar-particle energy spectra during the large events of October–November 2003 and January 2005. In: Acharya BS, Gupta S, Jagadeesan P et al. (eds) In: Proceedings of the 29th international cosmic ray conference, Pune, India, August 3–10, 2005, Tata Institute of Fundamental Research, Mumbai, vol 1, pp 111–114, 2005
Meyer JP (1972) Deuterons and ^{3}He formation and destruction in proton induced spallation of light nuclei ($Z \leq 8$). Astron Astrophys Suppl Ser (Switzerland) 7(4):417–467
Miller JA, Reames DV (1996) Heavy ion acceleration by cascading Alfvén waves in impulsive solar flares. In: Ramaty R, Mandzhavidze N, Hua X-M (eds) High energy solar physics. Proceedings of the conference, Greenbelt, MD, 1995, AIP conference proceedings, vol 374, No. 1, pp 450–460
Murphy RJ (1985) Gamma-rays and neutrons from solar flares. Ph.D. Dissertation, University of Maryland, College Park, MD
Murphy RJ, Ramaty R (1984) Solar-flare neutrons and gamma-rays. Adv Space Res 4(7):127–136
Murphy RJ, Dermer CD, Ramaty R (1987) High-energy processes in solar flares. Astrophys J Suppl Ser (USA) 63(3):721–748
Murphy RJ, Kozlovsky B, Ramaty R (1988) Gamma-ray spectroscopic tests for the anisotropy of accelerated particles in solar flares. Astrophys J (USA) 331(2, Part 1):1029–1035
Murphy RJ, Hua X-M, Kozlovsky B, Ramaty R (1990a) ^{7}Li and ^{7}Be de-excitation lines: probes for accelerated particle transport models in solar flares. Astrophys J (USA) 351(1, Part 1):299–308
Murphy RJ, Share GH, Letaw JR, Forrest DJ (1990b) Nuclear line spectroscopy of the 1981 April 27 solar flare. Astrophys J (USA) 358(1, Part 1):298–312
Murphy RJ, Ramaty R, Kozlovsky B, Reames DV (1991) Solar abundances from gamma-ray spectroscopy: comparisons with energetic particle, photospheric, and coronal abundances. Astrophys J (USA) 371(2, Part 1):793–803
Murphy RJ, Share GH, Grove JE, Johnson WN, Kinzer RL, Kroeger RA, Kurfess JD, Strickman MS, Matz SM, Purcell WR, Ulmer MP, Grabelsky DA, Jung GV, Jensen CM, Forrest DJ, Vestrand WT (1994) OSSE observations of the 4 June 1991 solar flare. In: Ryan J, Vestrand WT (eds) high-energy solar phenomena – a new era of spacecraft measurements. Proceedings of the workshop, Waterville Valley, New Hampshire, 1993. American Institute of Physics, New York, AIP conference proceedings, vol 294, No. 1, pp 15–20
Murphy RJ, Share GH, Grove JE, Johnson WN, Kinzer RL, Kurfess JD, Strickman MS, Jung GV (1997) Accelerated particle composition and energetics and ambient abundances from gamma-ray spectroscopy of the 1991 June 4 solar flare. Astrophys J (USA) 490(2, Part 1):883–900
Murphy RJ, Kozlovsky B, Share GH, Hua XM, Lingenfelter R (2007) Using gamma-ray and neutron emission to determine solar flare accelerated particle spectra and composition and the conditions within the flare magnetic loop. Astrophys J Suppl 168(1):167–194
Ng CK, Reames DV (1994) Focused interplanetary transport of $\sim$1 MeV solar energetic protons through self-generated Alfven waves. Astrophys J (USA) 424(2, Part 1):1032–1048
Nicholls JE, Craig A, Griffith TC, Imrie DC, Lush CJ, Metheringham AJ (1972) Inelastic p-^{4}He scattering at 141 MeV. Nucl Phys A (Netherlands) A181(1):329–336
Ogilvie KW (1980) Abundance ratios of $^4He^{++}/^3He^{++}$ in the solar wind. J Geophys Res (USA) 85(A11):6021–6024
Olver PWJ (1968) Bessel functions of integer order. In: Abramowitz M, Stegun IA (eds) Handbook of mathematical functions with formulas, graphs and mathematical tables, National Bureau of Standards Applied Mathematics Series, 7th edn. U.S. Government Printing Office, Washington, DC, pp 355–434

Ostrowski M, Schlickeiser R (1996) Cosmic-ray diffusive acceleration at shock waves with finite upstream and downstream escape boundaries. Solar Phys (Netherlands) 167(1–2):381–394

Paic G, Antolkovic B, Djaloeis A, Bojowald J, Mayer Boricke C (1981) Continuous spectra of protons and deuterons from the $\alpha + \alpha$ interaction in the incident energy interval 110–172 MeV. Phys Rev C Nucl Phys (USA) 24(3):841–848

Pearlstein S (1987) Systematics of neutron emission spectra from high-energy proton bombardment. Nucl Sci Eng (USA) 95(2):116–127

Pearlstein S (1989) Medium-energy nuclear data libraries: a case study, neutron- and proton-induced reactions in ^{56}Fe. Astrophys J (USA) 346(2, Part 1):1049–1060

Pohl M (1994) On the predictive power of the minimum energy condition. 2: Fractional calorimeter behaviour in the diffuse high energy gamma emission of spiral galaxies. Astronom Astrophys 287(2):453–462

Prince TA, Ling JC, Mahoney WA, Riegler GR, Jacobson AS (1982) A high-resolution measurement of the 2.223 MeV neutron capture line in a solar flare. Astrophys J Lett Ed (USA) 255(1, Part 2):L81–L84

Prince TA, Forrest DJ, Chupp EL, Kanbach G, Share GH (1983) The time history of 2.22 MeV line emission in solar flares. In: Proceedings of the 18th international cosmic ray conference, vol 4, Bangalore, India, pp 79–82

Ramaty R (1979) Energetic particles in solar flares. In: Arons J, McKee C, Max C (eds) Particle acceleration mechanisms in astrophysics. Proceedings of the workshop, La Jolla, CA, 1979, AIP conference proceedings, vol 56, No. 1, pp 135–154

Ramaty R (1986) Nuclear processes in solar flares. In: Sturrock PA, Holzer TE, Mihalas D, Ulrich RK (eds) The physics of the sun, Geophysics and astrophysics monographs, vol 2. D. Reidel, Dordrecht, The Netherlands, pp 291–323

Ramaty R, Kozlovsky B (1974) Deuterium, tritium, and helium-3 production in solar flares. Astrophys J 193(3, Part 1):729–740

Ramaty R, Lingenfelter RE (1969) Cosmic-ray deuterium and Helium-3 of secondary origin and the residual modulation of cosmic rays. Astrophys J (USA) 155(2, Part 1):587–608

Ramaty R, Murphy RJ (1987) Nuclear processes and accelerated particles in solar flares. Space Sci Rev 45(3–4):213–268

Ramaty R, Kozlovsky B, Lingenfelter RE (1975) Solar gamma rays. Space Sci Rev 18(3):341–388

Ramaty R, Kozlovsky B, Suri AN (1977) The solar gamma-ray spectrum between 4 and 8 MeV. Astrophys J (USA) 214(2, Part 1):617–631

Ramaty R, Kozlovsky B, Lingenfelter RE (1979) Nuclear gamma-rays from energetic particle interactions. Astrophys J Suppl (USA) 40(3):487–526

Ramaty R, Lingenfelter RE, Kozlovsky B (1982) Gamma ray lines from solar flares and cosmic transients. In: Lingenfelter RE, Hudson HS, Worrall DM (eds) Gamma ray transients and related astrophysical phenomena. Proceedings of the workshop, La Jolla, CA, 1981, AIP conference proceedings, vol 77, No. 1, pp 211–229

Ramaty R, Murphy RJ, Kozlovsky B, Lingenfelter RE (1983a) Gamma-ray lines and neutrons from solar flares. Sol Phys (Netherlands) 86(1–2):395–408

Ramaty R, Murphy RJ, Kozlovsky B, Lingenfelter RE (1983b) Implications of high-energy neutron observations from solar flares. Astrophys J Lett Ed (USA) 273(1, Part 2):L41–L45

Ramaty R, Murphy RJ, Dermer CD (1987) On the origin of the pion-decay radiation in the 1982 June 3 solar flare. Astrophys J Lett Ed (USA) 316(1, Part 2):L41–L44

Ramaty R, Miller JA, Hua X-M, Lingenfelter RE (1988) Models of gamma-ray production in solar flares. In: Gehrels N, Share GH (eds) Nuclear spectroscopy of astrophysical forces. Proceedings of a conference, Washington DC, 1988, AIP conference proceedings, vol 170, No. 1, pp 217–227

Ramaty R, Hua X-M, Kozlovsky B, Lingenfelter RE, Mandzhavidze N (1992) Neutron and gamma ray production in the 1991 June X-class flares. In: Shrader CR, Gehrels N, Dennis B (eds) The Compton observatory science workshop, NASA, Washington, DC, pp 480–485

Ramaty R, Mandzhavidze N, Kozlovsky B, Skibo JG (1993) Acceleration in solar flares: interacting particles versus interplanetary particles. Adv Space Res (UK) 13(9):275–284

Ramaty R, Mandzhavidze N, Kozlovsky B (1996) Abundances from solar atmospheric gamma ray spectroscopy. In: Ramaty R, Mandzhavidze N, Hua X-M (eds) High-energy solar physics. Proceedings of the conference, Greenbelt, 1995, AIP conference proceedings (USA), Woodbury, NY, vol 374, pp 172–183

Ramaty R, Mandzhavidze N, Barat C, Trottet G (1997) The giant 1991 June 1 flare: evidence for gamma-ray production in the corona and accelerated heavy ion abundance enhancements from gamma-ray spectroscopy. Astrophys J (USA) 479(1, Part 1):458–463

Rank G (1996) Gamma rays and neutrons of the solar flares on 11 and 15 June 1991 measured with COMPTEL. Ph.D. Thesis, Technical University of Munich

Rank G, Diehl R, Lichti GG, Schönfelder V, Varendorff DM, Swanenburg BN, van Dijk R, Forrest D, Macri J, McDonnell M, Loomis M, Ryan J, Bennett K, Winkler C (1994) Observations of the 1991 June 11 solar flare with COMPTEL. In: Ryan J, Vestrand WT (eds) High-energy solar phenomena – a new era of spacecraft measurements. Proceedings of the Workshop, Waterville Valley, New Hampshire, 1993. American Institute of Physics, New York, AIP conference proceedings, vol 294, No. 1, pp 100–105

Rank G, Bennett K, Bloemen H, Debrunner H, Lockwood J, McConnell M, Ryan J, Schönfelder V, Suleiman R (1996) Extended γ-ray emission in solar flares. In: Ramaty R, Mandzhavidze N, Hua X-M (eds) High energy solar physics. Proceedings of the conference, Greenbelt, MD, 1995, AIP conference proceedings, vol 374, No. 1, pp 219–224

Reames DV (1993) Non-thermal particles in the interplanetary medium. Adv Space Res (UK) 13(9):331–339

Reames DV (1995) Coronal abundances determined from energetic particles. Adv Space Res (UK) 15(7):41–51

Reames DV (1996) Energetic particles from solar flares and coronal mass ejections. In: Ramaty R, Mandzhavidze N, Hua X-M (eds) High energy solar physics. Proceedings of the conference, Greenbelt, MD, 1995, AIP conference proceedings, vol 374, No. 1, pp 35–44

Reames DV (1999) Particle acceleration at the Sun and in the Heliosphere. Space Sci Rev (Netherlands) 90(3–4):413–491

Reames DV, Cane HV, von Rosenvinge TT (1990) Energetic particle abundances in solar electron events". Astrophys J (USA) 357(1, Part 1):259–270

Reames DV, Meyer JP, von Rosenvinge TT (1994) Energetic-particle abundance in impulsive solar flare events. Astrophys J Suppl 90(2):649–667

Rieger E (1996) Spectral evolution of an intense gamma-ray line flare. In: Ramaty R, Mandzhavidze N, Hua X-M (eds) High energy solar physics. Proceedings of the conference, Greenbelt, MD, 1995, AIP conference proceedings, vol 374, No. 1, pp 194–199

Rieger E, Reppin C, Kanbach G, Forrest DJ, Chupp EL, Share GH (1983) Solar flares with photon emission above 10 MeV – measurements with the gamma ray experiment on board the SMM-satellite (1983) In: Proceedings of the 18th international cosmic ray conference, vol 10, Bangalore, India, pp 338–341

Roy G, Greeniaus LG, Moss GA, Hutcheon DA, Liljestrand R, Woloshyn RM, Boal DH, Stetz AW, Aniol K, Willis A, Willis N, McCamis R (1981) Inclusive scattering of protons on helium, nickel, and tantalum at 500 MeV. Phys Rev C Nucl Phys (USA) 23(4):1671–1678

Ryan JM (2000) Long-duration solar gamma-ray flares. Space Sci Rev (Netherlands) 93 (3–4):581–610

Ryan JM, McConnell MM (1996) COMPTEL solar flare measurements. In: Ramaty R, Mandzhavidze N, Hua X-M (eds) High energy solar physics. Proceedings of the conference, Greenbelt, MD, 1995, AIP conference proceedings, vol 374, No. 1, pp 200–209

Ryan J, Forrest D, Lockwood J, Loomis M, McConnell M, Morris D, Webber W, Bennett K, Hanlon L, Winkler C, Debrunner H, Rank G, Schönfelder V, Swanenburg BN (1994) Neutron and gamma-ray measurements of the solar flare of 1991 June 9. In: Ryan J, Vestrand WT (eds) High-energy solar phenomena – a new era of spacecraft measurements. Proceedings of the Workshop, Waterville Valley, New Hampshire, 1993. American Institute of Physics, New York, AIP conference proceedings, vol 294, No. 1, pp 89–93

Ryan JM, Suleiman R, McConnell M, Rank G, Schoenfelder V, Varendorff M, Bennett K, Hermsen W (1996) Energetic proton spectral evolution in the 11 June 1991 solar flare. In: 188th AAS Meeting, Madison, WI, 1996, BAAS, vol 28, 26.03, p 857

Schatzman E, Maeder M (1981) Stellar evolution with turbulent diffusion mixing. III. The solar model and the neutrino problem. Astron Astrophys (Germany) 96(1–2, Part 2):1–16

Schlickeiser R (1989) Cosmic-ray transport and acceleration. I – Derivation of the kinetic equation and application to cosmic rays in static cold media. II – Cosmic rays in moving cold media with application to diffusive shock wave acceleration. Astrophys J (USA) 336 (1, Part 1):243–293

Schneid EJ, Brazier KTS, Kanbach G, von Montigny C, Mayer-Hasselwander HA, Bertsch DL, Fichtel CE, Hartman RC, Hunter SD, Thompson DJ, Dingus BL, Sreekumar P, Lin YC, Michelson PF, Nolan PL, Kniffen DA, Mattox JR (1994) EGRET observations of extended high-energy emissions from the nuclear line flares of June 1991. In: Ryan J, Vestrand WT (eds) High-energy solar phenomena – a new era of spacecraft measurements. Proceedings of the workshop, Waterville Valley, New Hampshire, 1993. American Institute of Physics, New York, AIP conference proceedings, vol 294, No. 1, pp 94–99

Selove W, Teem JM (1958) Measurements of the Interaction of 95-Mev Protons with He^4. Phys Rev 112(5):1658–1666

Share GH, Murphy RJ (1995) Gamma-ray measurements of flare-to-flare variations in ambient solar abundances. Astrophys J (USA) 452(2, Part 1):933–943

Share GH, Murphy RJ (1997) Intensity and directionality of flare-accelerated a-particles at the Sun. Astrophys J 485(1):409–418

Share GH, Murphy RJ (1998) Accelerated and ambient He abundances from gamma-ray line measurements of flares. Astrophys J (USA) 508(2, Part 1):876–884

Share GH, Murphy RJ (1999) Gamma-ray measurement of energetic heavy ions at the Sun. In: Proceedings of the 26th international cosmic ray conference, vol 6, Salt Lake City, CA, pp 13–16

Share GH, Harris MJ, Leising MD, Messina DC (1993) Search for gamma-ray transients using the SMM spectrometer. Astron Astrophys Suppl Ser (France) 97(1):341–344

Share GH, Murphy RJ, Skibo JG, Smith DM, Hudson HS, Lin RP, Shih AY, Dennis BR, Schwartz RA, Kozlovsky B (2003) High-resolution observation of the solar positron-electron annihilation line. Astrophys J Lett (USA) 595(2, part 2):L85–L88

Share GH, Murphy RJ (2006) Gamma radiation from flare-accelerated particles impacting the Sun. AGU Monogr Ser 165:177–188

Stamer S, Scobel W, Amian WB, Byrd RC, Haight RC, Ullmann JL, Bauer RW, Blann M, Pohl BA, Biplinghoff J, Bonetti R (1993) Double differential cross sections for neutron emission induced by 256 MeV and 800 MeV protons. Phys Rev C Nucl Phys (USA) 47(4):1647–1658

Stoker PH (1994) Relativistic solar proton events. Space Sci Rev 73(1–4):327–385

Tan LC, Ng LK (1982) Parameterization of pbar invariant cross-section in p–p collisions using a new scaling variable. Phys Rev D Part Fields (USA) 26(5):1179–1182

Tanaka K (1987) Impact of x-ray observations from the Hinotori satellite on solar flare research. Publ Astron Soc Jpn 39(1):1–45

Torsti J, Kocharov LG, Vainio R, Anttila A, Kovaltsov GA (1996) The 1990 May 24 solar cosmic-ray event. Solar Phys 166(1):135–158

Trottet G, Vilmer N, Barat C, Dezalay J-P, Talon R, Sunyaev R, Terekhov O, Kuznetsov A (1993) Radio and X-ray/gamma-ray observations of two solar flares. Adv Space Res (UK) 13(9):285–288

Tylka AJ, Boberg PR, Adams JH Jr., Beahm LP, Dietrich WF, Kleis T (1996) HIIS results on the mean ionic charge state of SEP Fe above 200 MeV per nucleon. In: Ramaty R, Mandzhavidze N, Hua X-M (eds) High energy solar physics. Proceedings of the conference, Greenbelt, MD, 1995, AIP conference proceedings, vol 374, No. 1, pp 96–105

Vainio R, Schlickeiser R (1998) Alfvén wave transmission and particle acceleration in parallel shock waves. Astron Astrophys (Germany) 331(2):793–799

Vainio R, Schlickeiser R (1999) Self-consistent Alfvén-wave transmission and test-particle acceleration at parallel shocks. Astron Astrophys (Germany) 343(1):303–311

Vainio R, Kocharov L, Laitinen T (2000) Interplanetary and interacting protons accelerated in a parallel shock wave. Astrophys J 528(2):1015–1025

Van Hollebeke MAI, McDonald FB, Meyer JP (1990) Solar energetic particle observations of the 1982 June 3 and 1980 June 21 gamma-ray/neutron events. Astrophys J Suppl Ser (USA) 73(2):285–296

Verbinski VV, Burrus WR (1969) Direct and compound-nucleus neutrons from 14–18-MeV protons on ^{9}Be, ^{14}N, ^{27}Al, ^{56}Fe, ^{115}In, ^{181}Ta, and ^{208}Pb and from 33-MeV Bremsstrahlung on ^{27}Al, ^{206}Pb, ^{208}Pb, and ^{209}Bi. Phys Rev (USA) 177(4):1671–1686

Vestrand WT, Forrest DJ, Chupp EL, Rieger E, Share GH (1987) The directivity of high-energy emission from solar flares: Solar Maximum Mission observations. Astrophys J 322(2, part 1): 1010–1022

Vestrand WT (1990) A new gamma-ray diagnostic for energetic ion distributions: the Compton tail on the neutron capture line. Astrophys J 352(1), Part 1:353–360

Von Neumann J (1951) Various techniques used in connection with random digits. NBS Applied Mathematics Series 12, Washington, DC, pp 36–38

Wang HT (1975) Neutron propagation and 2.2 MeV gamma ray line formation in the solar atmosphere. Ph.D. thesis, Physics Department, University of Maryland, pp 1–94

Wang HT, Ramaty R (1974) Neutron propagation and 2.2 MeV gamma-ray line production in the solar atmosphere. Solar Phys (Netherlands) 36(1):129–137

Wesick JS, Roos PG, Chant NS, Chang CC, Nadasen A, Rees L, Yoder NR, Cowley AA, Mills SJ, Jacobs WW (1985) 2H, 3, 4He (p, p') and 3, 4He(p, d') continuum yields for 100 and 150 MeV protons. Phys Rev C Nucl Phys (USA) 32(5):1474–1487

Wolfs FLH, Freedman SJ, Nelson JE (1989) Measurement of the He-3(n, gamma) He-4 cross section at thermal neutron energies. Phys Rev Lett 63(25):2721–2724

Wu JR, Chang CC, Holmgren HD (1979a) Charged particle spectra: 140 MeV α particle bombardment of ^{27}Al, ^{58}Ni, ^{90}Zr, ^{209}Bi, and ^{232}Th. Phys Rev C Nucl Phys (USA) 19(3):659–673

Wu JR, Chang CC, Holmgren HD (1979b) Charged-particle spectra: 90 MeV protons on ^{27}Al, ^{58}Ni, ^{90}Zr, ^{209}Bi. Phys Rev C Nucl Phys (USA) 19(3):698–713

Yariv Y, Fraenkel Z (1981) Intranuclear cascade calculation of high energy heavy ion collisions: effect of interactions between cascade particles. Phys Rev C Nucl Phys (USA) 24(2):488–494

Yoshimori M, Okudaira K, Hirasima Y, Kondo I (1983) The limb darkening of neutron capture line at 2.22 MeV and gamma-ray line emission model in solar flares. In: Proceedings of the 18th international cosmic ray conference, vol 4, Bangalore, India, pp 89–92

Yoshimori M, Suga K, Morimoto K, Hiraoka T, Sato J, Kawabata K, Ohki K (1994) Gamma-ray spectral observations with Yohkoh. Astrophys J Suppl Ser (USA) 90(2):639–643

Zhang JL, Bao KZ, Ding LK, et al. (2000) An observation of solar flare neutrons using Yangbajing solar cosmic ray detector. High Energy Phys Nucl Phys (China) 24(12):1081–1087

Zirakashvili VN, Dorman LI, Ptuskin VS, Babayan VKh (1991) Cosmic ray nonlinear modulation in the outer heliosphere. In: Proceedings of the 22nd international cosmic ray conference, vol 3, Dublin, pp 585–588

Zurmühle RW, Stephens WE, Staub HH (1963) Gamma-rays from neutron capture in helium-3 and deuteron capture in deuterium. Phys Rev 132(2):751–754

Zweibel EG, Haber DA (1983) The propagation of energetic ions in magnetic loops and gamma-ray emission from solar flares. Astrophys J (USA) 264(2, Part 1):648–659

Appendix

Numerical Simulations by Monte Carlo Method (for Section 11.6)

According to Kocharov et al. (1999b), as long as Eq. 11.39 takes the form of a Fokker–Planck equation, one can apply different standard methods to solve it. In particular, a very flexible Monte Carlo method may be used (Gardiner M1985). Equation 11.39 describes evolution of an angle-average proton distribution. For this reason, when applying the Monte Carlo method to Eq. 11.39, one numerically simulates a motion of some 'average' particles possessing proton energy but moving, in general, with another velocity. Kocharov et al. (1999b) call them MC particles. They simulate MC particle transport and acceleration in the exponential magnetic field, including the following processes: (1) a random walk along the magnetic field line due to the spatial (ζ) diffusion; (2) advection due to the magnetic focusing and spatial dependence of the diffusion coefficient; (3) a random walk in the momentum owing to the stochastic acceleration; and (4) advection to higher momentum owing to the momentum dependence of the acceleration. Kocharov et al. (1999b) consider the exponential model of the magnetic field (Eq. 11.53) and the exponential injection to be proportional to the ambient plasma density.

For numerical simulations of focused transport and stochastic acceleration (Eq. 11.39), Kocharov et al. (1999b) introduce the dimensionless coordinate, momentum, and time:

$$y = \frac{\zeta}{L}, \quad q = \frac{p}{p_u}, \quad \tau = t\frac{\kappa_{0\zeta} p_u^{3/2}}{L^2}, \tag{A.1}$$

where p_u is a momentum value for which the diffusion to acceleration time ratio, τ_ζ / τ_p, is fixed, which is a parameter for simulations:

$$\frac{\tau_\zeta}{\tau_p} = \left. \frac{\kappa_{0p} L^2}{\kappa_{0\zeta} p^3} \right|_{p=p_u}. \tag{A.2}$$

Kocharov et al. (1999b) adopt $\tau_\zeta/\tau_p = 5$. They note that the characteristic energy defined by Eq. 11.59 depends on the time ratio adopted:

$$E_1 = \frac{p_u^2}{2m_p}\left(\frac{\tau_\zeta}{\tau_p}\left(\frac{1-\exp(-\eta/2)}{\eta/2}\right)^2\right)^{2/3}. \tag{A.3}$$

Using these designations in the exponential model, Eq. 11.39 can be cast into the form

$$\frac{\partial n}{\partial \tau} = \frac{1}{q^2}\frac{\partial}{\partial q}\left(q^{5/2}\frac{\tau_\zeta}{\tau_p}\exp(-\eta y/2)\frac{\partial n}{\partial q}\right) + \frac{\partial}{\partial y}\left(q^{3/2}\exp(\eta y/2)\frac{\partial n}{\partial y}\right) - \frac{\partial}{\partial y}\left(\eta q^{3/2}\exp(\eta y/2) n\right) + \tilde{Q}. \tag{A.4}$$

To solve this equation at different injection functions $\tilde{Q}$, an Monte Carlo method has been employed. In these numerical simulations, Kocharov et al. (1999b) calculate, per each MC time step $\delta\tau = \tau' - \tau$, the change of the particle coordinate and the change of the particle momentum as

$$y' = y + bR_{G1} + \frac{3}{4}b^2\eta, \quad q' = q + b\sqrt{\frac{\tau_\zeta}{\tau_p}}\frac{\exp(-\eta y/2)}{\sqrt{q}}R_{G2} + \frac{5}{4}b^2\frac{\tau_\zeta}{\tau_p}\frac{\exp(-\eta y)}{q^2}, \tag{A.5}$$

where R_{G1} and R_{G2} are two independent Gaussian random numbers with unit variance. A small parameter b^2 is proportional to the MC time step

$$b^2 = 2q^{3/2}\exp(\eta y/2)\delta\tau. \tag{A.6}$$

A working value of this parameter has been selected to be small enough to have no effect on the final result and to provide agreement with available analytic solutions. In present sets of computations, Kocharov et al. (1999b) employ $b = 0.01$ and 0.02. In fact, Kocharov et al. (1999b) do not trace time τ, because only time-integrated spectra are studied. The coordinate and momentum of each particle are simply traced until an escape condition has been met.

In the basic model, Kocharov et al. (1999b) consider an impulsive monoenergetic injection of protons being proportional to the ambient plasma density, n_p, and distributed inside the magnetic tube of the cross-sectional area $S \propto 1/B$. Kocharov et al. (1999b) suggest a constant Alfvén speed throughout the acceleration region, so that $n_p \propto B^2$, and the injection function in Eq. A.4 $\tilde{Q} \propto n_p S \propto B$:

$$\tilde{Q} = \tilde{Q}_0 \exp(-\eta y)\delta(q - q_i)\delta(\tau), \quad 0 < y < 1. \tag{A.7}$$

According to Kocharov et al. (1999b), an average velocity of MC particles in a diffusion simulation depends on the adopted length of the MC time step, δt:

$$\langle v_{MC}\rangle = \langle \delta\zeta/\delta t\rangle \approx (\Lambda v/\delta t)^{1/2}. \tag{A.8}$$

In the case of a magnetic mirror beneath the acceleration region, $\rho_b > 1$, it was set the velocity of MC particles to be equal to a physical particle velocity to simplify getting a correct probability of particle precipitation at $\zeta = 0$. The purpose is accomplished through the employment of the following scheme. At each MC time step, the particle experiences a random walk with a spatial step $\delta\zeta = 2\Lambda\mu$, where $\mu = 2R_1 - 1$, R_1 being a random number uniformly distributed from 0 to 1. The quantity μ is considered to be a particle pitch angle cosine at an isotropic particle distribution suggested in a diffusion approximation. A corresponding time step is $\delta t = 2\Lambda/v$. This time step corresponds to $b^2 = (4/3)(\Lambda/L)^2$ in the second part of Eq. A.5 and in the last term of the first part of Eq. A.5. Thus the first part of Eq. A.5 is finally replaced with

$$y' = y + 2\frac{\Lambda(y,q)}{L}(2R_1 - 1) + \left(\frac{\Lambda(y,q)}{L}\right)^2 \eta, \qquad \text{(A.9)}$$

where $\Lambda(y,q) = \Lambda(0,p_u)q^{1/2}\exp(\eta y/2)$, $\Lambda(0,p_u)$ being a particle mean free path at $\zeta = 0$ and $p = p_u$. One can see that variance of coordinate y' and the mean value are exactly the same in the cases of Eqs. A.5 and A.9. However, convenience of the μ-scheme (Eq. A.9) is that precipitation of particles at $\zeta = 0$ can be simply accounted for by a removal of particles with $\mu < \mu_b$ from the acceleration, $\mu_b = -(1 - 1/\rho_b)^{1/2}$ being a boundary of a loss cone. Note that a parameter, $\Lambda(0,p_u)$, is introduced for the case of $\rho_b > 1$. However, deduced spectra of interacting and interplanetary particles are actually dictated by a value of combination $\rho_b\Lambda(0,p_u)/L$. This scaling was analytically obtained in Section 11.6.6 and then verified against MC simulations.

To improve statistics at the high-energy end of proton spectra, Kocharov et al. (1999b) employ a well-known trajectory splitting method. A number of simulations have been done to test the numerical codes. Energy spectra generated by the two numerical schemes were compared with each other and with available analytic solutions at $\rho_b = 1$. At $\rho_b > 1$, results of the μ-scheme at withdrawn acceleration were compared with a kinetic treatment of proton transport modeled as a pitch angle diffusion and with available analytic solutions of the pure transport problem (Kocharov et al. 1999a). All tests indicate that both numerical schemes work well in a parameter region expected for a diffusion approximation.

Kinetic Equation and Monte Carlo Simulations (for Section 11.7)

To obtain numerical solutions to Eqs. 11.66 and 11.72, Vainio et al. (2000) have used a Monte Carlo simulation code developed for the analysis of energetic particle transport and acceleration in the inner Heliosphere (Vainio 1998). The model is a kinetic one. It follows individual particles in a pre-described system of large-scale magnetic fields and superposed waves that scatter the particles in pitch angle. Particle speed and magnetic moment are constants of motion in the large-scale field when the velocity is measured in the frame where the large-scale magnetic field is static. This frame is called as the fixed frame. Scatterings, on the other hand, conserve the particle speed in the frame, where the magnetic field of the (non-dispersive) waves is static. This is the frame moving with the phase speed V of the waves relative to the fixed frame, and Vainio et al. (2000) call it the wave frame. Let us assume, for simplicity, that for the considered flux tube the phase speed of the waves is directed along the mean magnetic field. Vainio et al. (2000) will study how the momentum and pitch angle of the particles change due to the adiabatic motion in the large-scale field when they are measured in the wave frame. Vainio et al. (2000) do not require this phase speed to be constant but allow it to change spatially and temporally.

In what follows, its shall assume that terms of the order $(V/c)^2$ may be neglected. In the fixed frame, it may be write

$$\dot{\mu} = \left(1-\mu^2\right)v/(2L_B), \quad \dot{p} = 0 \tag{A.10}$$

Thus, a particle moving from event A(t_A, ζ_A) to B(t_B, ζ_B) along the magnetic field will suffer a change in its wave-frame parallel momentum

$$p'_{\|} = p'\mu' = p_{\|} - pV(\zeta)/v \tag{A.11}$$

of

$$\Delta p'_{\|} = p'_{\|B} - p'_{\|A} = p(\mu_B - \mu_A) - (p/v)(V_B - V_A), \tag{A.12}$$

which gives, when divided by $\Delta t = \Delta\zeta/(v\mu)$, a rate of change for the wave-frame parallel momentum as seen from the fixed frame,

$$\dot{p}'_{\|} = \frac{dp'_{\|}}{dt} = p\dot{\mu} - \frac{p}{v}\dot{V} = p\left(\dot{\mu} - \mu\frac{\partial V}{\partial\zeta} - \frac{1}{v}\frac{\partial V}{\partial t}\right). \tag{A.13}$$

Since $p_\perp = p'_\perp$, Vainio et al. (2000) have

$$2p'\dot{p}' = \frac{d}{dt}p'^2 = \frac{d}{dt}p'_{\|}2 + \frac{d}{dt}p'_{\perp}2 = 2p'_{\|}\dot{p}'_{\|} - 2\mu p^2\dot{\mu}, \tag{A.14}$$

which yields a rate of acceleration/deceleration of

$$\begin{aligned}\dot{p}' &= \frac{p'\mu' - \mu p}{p'}p\dot{\mu} - \mu' p\left(\mu\frac{\partial V}{\partial\zeta} + \frac{1}{v}\frac{\partial V}{\partial t}\right) = -\frac{p_\perp^2}{p'}\frac{V}{2L_B} - \mu'\left(p\mu\frac{\partial V}{\partial\zeta} + \frac{p}{v}\frac{\partial V}{\partial t}\right)\\ &= -\frac{p'_\perp 2}{p'}\frac{V}{2L_B} - \mu'\left(p'\mu' + \frac{p'V}{v'}\right)\frac{\partial V}{\partial\zeta} - \mu'\frac{p'}{v'}\left(1 + \frac{\mu' v' V}{c^2}\right)\frac{\partial V}{\partial t}\\ &= -p'\left[\frac{1-\mu'^2}{2L_B}V + \mu'^2\left(\frac{V}{c^2}\frac{\partial V}{\partial t} + \frac{\partial V}{\partial\zeta}\right) + \frac{\mu'}{v'}\left(\frac{\partial V}{\partial t} + V\frac{\partial V}{\partial\zeta}\right)\right].\end{aligned} \tag{A.15}$$

In addition to the time derivatives, this expression differs from the deceleration rate given by Ruffolo (1995) in his equation (7) by the last term. This is because in Ruffolo's model, the increase of V (that was the solar wind speed in his paper) with ζ came entirely from the centrifugal acceleration of the medium in the fixed frame corotating with the Sun. In this case, one should also take into account the centrifugal force in the deceleration rate of the particles, which cancels the last term in Eq. A.15 exactly. Vainio et al. (2000), however, consider phase speeds that have much larger gradients than those predicted by corotation (which are completely negligible at coronal distances from the Sun), so in the considered case the centrifugal force may be neglected and Eq. A.15 should be used with the last term included. Similarly, it may write for the rate of change in the wave-frame pitch-angle cosine

$$\dot{\mu}' = \frac{\dot{p}'_{\|}}{p'} - \mu'\frac{\dot{p}'}{p'} = (1-\mu'^2) \times \left[\frac{v'}{2L_B}\left(1+\frac{\mu' V}{v'}-\frac{\mu' v' V}{c^2}\right) - \mu'\frac{\partial V}{\partial \zeta} - \left(\frac{1}{v'}+\frac{\mu' V}{c^2}\right)\frac{\partial V}{\partial t} - \frac{V}{v'}\frac{\partial V}{\partial \zeta}\right], \quad \text{(A.16)}$$

which also differs by the inclusion of the last-line terms from the expression (10) of Ruffolo (1995) for the same reason as the deceleration rate discussed above. Finally, particle speed along the magnetic field is given by

$$\dot{\zeta} = \mu v = \mu' v'\left(1-\frac{\mu' v' V}{c^2}\right) + V. \quad \text{(A.17)}$$

As noted Vainio et al. (2000), now it may write down a Fokker–Planck equation for the number of particles per unit length of the magnetic field, unit of pitch-angle cosine, and unit momentum, $n(\zeta, p', \mu', t) = d^3N/(d\zeta dp' d\mu')$, as

$$\begin{aligned}\frac{\partial n}{\partial t} = &-\frac{\partial}{\partial \zeta}(\mu' v' + V)n - \frac{\partial}{\partial \mu'}(1-\mu'^2)\left[\frac{v'}{2L_B}\left(1+\frac{\mu' V}{v'}\right) - \left(\mu'+\frac{V}{v'}\right)\frac{\partial V}{\partial \zeta} - \frac{1}{v'}\frac{\partial V}{\partial t}\right]n \\ &+ \frac{\partial}{\partial p'}p'\left[\frac{1-\mu'^2}{2L_B}V + \mu'\left(\mu'+\frac{V}{v'}\right)\frac{\partial V}{\partial \zeta} - \frac{\mu'}{v'}\frac{\partial V}{\partial t}\right]n \\ &+ \frac{\partial}{\partial \mu'}(1-\mu'^2)\nu\frac{\partial n}{\partial \mu'} + s(\zeta, p', \mu', t),\end{aligned} \quad \text{(A.18)}$$

where time and distance are measured in the fixed frame and momentum and pitch angle in the wave frame, and where it have neglected all terms of the order (V/c) when applying Eqs. A.15–A.17. The second to last term describes the effect of scatterings, and it involves the scattering frequency $\nu(\zeta, p', \mu', t)$, while the last term is the source function, now written with dependence on μ' allowed. To our knowledge, Eq. A.18 is new in the sense that it allows for time dependence in the wave speed. Note, however, that time dependence will only be important in cases, where particle anisotropies are large, e.g., when particles of low speeds are considered.

The simulations of Vainio et al. (2000) effectively solve Eq. A.18. Equation 11.66 for $f = d^2N/(d\zeta dp)$ results from Eq. A.18 through the use of well-known diffusion approximation scheme, where scattering is assumed to be intense enough to keep the particle distribution close to isotropic. Vainio et al. (2000) shall not reproduce the rigorous derivation here but refer the reader to Webb and Gleeson (1979) for a detailed description. To keep described simulations computationally efficient, Vainio et al. (2000) have made a couple of simplifications: they assume nonrelativistic phase speeds for the waves in the sense that it used a Galilean transformation for time, $t = t'$, between the fixed frame and the wave frame. This choice is also consistent with Eq. A.18, which assumed the relativistic effects due to the wave speed to be negligible. It was also choose to work with the simplest form of isotropic small-angle scattering, where the scattering frequency is not allowed to depend on pitch angle. In this case, the spatial diffusion coefficient and the scattering frequency have a simple connection, $\kappa = \lambda(p)v/3 = v^2/6\nu(p)$.

For isotropic scattering, Vainio et al. (2000) have developed an accurate numerical method that effectively solves the pitch-angle diffusion part of the Eq. A.18. It was considered the scattering process relative to the direction of the unperturbed particle motion in the wave frame. Relative to this rotating scattering axis, it may be write an equation for the angular distribution g' of particles of constant speed in the wave frame,

$$\frac{\partial g'}{\partial t} = \frac{\nu}{\sin\vartheta}\frac{\partial}{\partial\vartheta}\sin\vartheta\frac{\partial g'}{\partial\vartheta} \approx \frac{\nu}{\vartheta}\frac{\partial}{\partial\vartheta}\vartheta\frac{\partial g'}{\partial\vartheta}, \tag{A.19}$$

with the initial condition $g'(\vartheta, 0) = \delta(\vartheta)$, where $\vartheta \in [0, \pi]$ is the angle between the unperturbed and the scattered velocity vectors. If small times compared to the inverse scattering frequency are considered, one may use the small-angle approximation $\sin\vartheta \approx \vartheta$. Then, the number of particles in a differential solid angle $d\Omega$ is

$$g'(\vartheta, t)d\Omega = \frac{1}{4\pi\nu t}\exp\left(-\frac{\vartheta^2}{4\nu t}\right)\vartheta d\vartheta d\phi, \tag{A.20}$$

where $\phi \in [0, 2\pi]$ is the angle measured around the scattering axis.

As underlined Vainio et al. (2000), their simulations thus work with the following scheme: first, the particle is injected at the shock so that it propagates into the upstream region. A time step is chosen as a small fraction of the inverse scattering time, $\Delta t = a\nu^{-1}$ with $a \sim 0.01$. The particle is then moved in the fixed frame according to $\zeta \leftarrow \zeta + \mu v\Delta t$ and $\mu \leftarrow \mu + (1 - \mu^2)v\Delta t/(2L_B)$ keeping the fixed-frame particle speed constant. After this, a scattering is performed: first, the velocity vector of the particle is Lorentz-transformed to the wave frame. Then, in accordance with Eq. A.20, ϑ^2 and ϕ are picked via a random generator from exponential and uniform distributions, respectively, and the new wave-frame pitch-angle cosine is computed by the use of spherical trigonometry, $\mu' \leftarrow \mu' \cos\vartheta + (1-\mu'^2)^{1/2} \sin\vartheta \cos\varphi$. The new velocity vector is then transformed back to the fixed frame. Such time steps are taken until the particle either escapes at the upstream boundary or it hits the shock again. If it escapes, the particle is removed from the simulation, the interplanetary spectrum is updated, and a new particle is injected at the shock. If the particle hits the shock, it is transmitted into the downstream region and followed there accordingly until it either returns to the upstream region or goes through a boundary at a distance of $2\lambda_2$ downstream of the shock, where are decide its fate through its probability of return given by (e.g., Ellison et al. 1990)

$$P_{ret} = (v' - U_2)^2 \Big/ (v' + U_2)^2. \tag{A.21}$$

Equation A.21 is valid if $\lambda_2 \to 0$, which have assumed in described simulations. It also implies that if the particle returns, it does so immediately, so no time spent in the downstream region needs to be taken into account. The returning particles are re-injected into the simulation at the $2\lambda_2$ distance downstream of the shock so that their pitch-angle cosines are picked from isotropic distribution measured in the downstream wave frame and then flux-weighted with the shock-frame cosine factor. They are followed until they escape either through the upstream boundary or by failing to return when they hit the downstream $2\lambda_2$ distance again; these particles get convected away from the shock to the far downstream region and will be injected toward the solar surface into the region with no turbulence. There they are followed until they precipitate or cool down to speeds $v < 2V_s$. Particles mirroring and catching up with the outward propagating downstream turbulence of the shock are reinjected back toward the solar surface. This back-scattering is done elastically in the downstream wave frame, which leads to the cooling. A precipitating particle is removed from the simulation, the spectrum of interacting particles is updated, and a new particle is injected at the shock.

References for Appendix

Ellison DC, Jones FC, Reynolds SP (1990) First-order Fermi particle acceleration by relativistic shocks. Astrophys J 360(2. Part 1):702–714

Kocharov L, Kovaltsov GA, Torsti J (1999a) An analytic description of coronal proton trapping. Astrophys J 519(1, Part 1):422–432

Kocharov L, Kovaltsov GA, Laitinen T, Mäkelä P, Torsti J (1999b) Interacting and interplanetary protons accelerated on diverging magnetic field lines: stochastic acceleration. Astrophys J 521 (2):898–905

Ruffolo D (1995) Effect of adiabatic deceleration on the focused transport of solar cosmic rays. Astrophys J (USA) 442(2, Part 1):861–874

Vainio R (1998) Monte-Carlo simulations of interplanetary transport and acceleration of energetic particles. Ph.D. thesis, Department of Physics, Turku University, Finland, pp 1–111

Vainio R, Kocharov L, Laitinen T (2000) Interplanetary and interacting protons accelerated in a parallel shock wave. Astrophys J 528(2):1015–1025

Webb GM, Gleeson LJ (1979) On the equation of transport for cosmic-ray particles in the interplanetary region. Astrophys Space Sci (Netherlands) 60(2):335–351

Conclusions and Problems

General Conclusion

In the book in the first time is considered a huge observational and rich theoretical information published in the world scientific literature for the last about 50 years on solar neutrons and on products of their decay, on nuclear solar gamma ray lines from isotope decay (generated together with neutrons in nuclear reactions of solar energetic charged particles with matter of solar atmosphere), on gamma rays generated by capturing solar neutrons with hydrogen atoms in solar atmosphere, as well as by annihilation of positrons, by relativistic electrons bremsstrahlung and by π^0-decay, on the propagation of solar neutrons, solar gamma rays and other secondary particles in solar photosphere, chromosphere, and corona, as well as in the interplanetary space and in the Earth's atmosphere. It is shown by modelling and simulation of acceleration, interaction, and propagation processes that observations of solar neutrons and gamma rays in space and in the Earth's atmosphere give very important, unique information on the source function of solar energetic particles directly on the Sun as well as on chemical and isotopic contents, density distribution of plasma in solar atmosphere and details on particle acceleration mechanisms. Really in the last 50 years was formatted and developed a new branch of science based on Cosmic Ray Physics, Solar Physics, Space Physics, Nuclear and Elementary Particle Physics. It is important that described in the book results may be useful also for astrophysics studies of other stars and different astrophysical objects, from one hand, and for space weather problems (direct determining of solar energetic particle source function – energy and angle distribution), from other hand.

Let me underline that considered in book models and simulations are based on universal conception that plasmas in solar flares and in solar atmosphere with frozen in magnetic fields are usually excited magneto-turbulent plasmas with many channels of energy transformation and energetic particle generation. Considered in book observation data as well as models and simulations confirm the paradigm (see Chapter 1 of Dorman M2004 and Chapter 4 of Dorman M2006) that the generation of energetic run away particles (or cosmic rays) in different objects of the Universe is the universal property of space plasmas owing mainly to energy transfer from macroscopic phenomena (kinetic energy of moving great ensembles of thermal particles and energy of magnetic fields) to microscopic charged particles: protons, electrons, nuclei. These macroscopic phenomena are characterized by very high ‘effective temperature’, many orders higher than ‘effective temperature’ of energetic particles, so this process of energy transfer directly follows from the fundamental second law of thermodynamics: energy transfers always from higher ‘effective temperature’ to lower one.

Let us note that energetic particle interactions in solar atmosphere with generation of secondary radiation (particles and photons) is a particular case of more general problem of cosmic ray interactions, propagation and acceleration in space plasmas considered in details in Dorman (M2006).

Main Conclusions for Different Chapters

Main conclusion for Chapter 1. The models what were developed before discovery solar gamma rays (events in August 1972) and solar neutrons (events in June 1980 and 1982) give in the first approximation correct expected fluxes. These models and simulations were developed in details later in connection with great development of experiment and investigations of many neutron and/or gamma ray events (see Sections 2–11). Why these important phenomena were not discovered before? For solar gamma rays the answer is simple: it was not enough sensitivity of gamma-ray detectors on space probes. Solar neutrons, in principal, can be well measured by existing from 1953 to 1957 world network of neutron monitors of IGY type and after 1964 by network of neutron super-monitors of IQSY type (see description in Chapter 4 in Dorman M2004). The problem is that in those times researchers are mainly interested in long-term variations (caused by 11 and 22 year cycles of solar activity), 27-day variations (caused by rotating of the Sun), solar-daily variation (caused by convection-diffusion and drift effects on the Earth rotated relative to the Sun), stellar-daily variation (caused by the Earth rotation elative to stars), GLE – Ground Level Events (caused by energetic charged particles from the Sun), and transient variations (caused by interplanetary shock waves, interacting regions of solar wind streams, and/or so called now CME – Coronal Mass Ejections). For all these types of CR variations are enough 1 h, and even 2 h data. But with 1-h data is not possible to investigate solar neutron phenomenon which duration is, as we well know now, only several minutes.

Main conclusion for Chapter 2. The famous discovery of solar gamma rays from flares in August 1972 by the 0.3–10 MeV gamma-ray detector on OSO-7 satellite opened very important possibilities to use gamma ray observations in nuclear lines, neutron capture line, positron annihilation line, and in continuum spectra for better understanding on acceleration, propagation, and absorption processes in solar flare and in corona, for obtaining new information on background plasma and composition and energy spectrum of energetic particles in source.

Main conclusion for Chapter 3. The discovery of solar neutrons was made during the event of June 21, 1980 and at June 3, 1982: measurements by Gamma Ray Spectrometer on SMM satellite (for large energy loss events of 10–140 MeV and of 25–140 MeV shown that the measured pulses are caused not by gamma rays but by neutrons). Solar neutrons from the second event at June 3, 1982 were measured in the first time simultaneously on the SMM satellite and by ground based neutron monitors and supermonitors. This and solar gamma ray discoveries founded the basis of new branch of science: Solar neutrons and Related Phenomena.

Main conclusion for Chapter 4. Beside famous events at June 21, 1980 and at June 3, 1982, described in Chapter 3, on satellites SMM, Hinotori, GAMMA-1, and during COMPTEL experiment on the Compton Gamma-Ray Observatory, were obtained unique data on some other solar neutron events (accompanied, as a rule, also by solar gamma ray events): on 1984 April 25, 1989 March 6, 1991 June 4, 5, 15, and 1997 November 6.

Main conclusion for Chapter 5. In this Chapter the problem of solar neutron propagation in the Earth's atmosphere and sensitivity of neutron monitors and other ground based detectors to solar neutrons including the refraction effect are described and analyzed in details. It is important that thanks to charge invariance of neutrons and protons, for high energy neutrons we can use coupling functions and integral multiplicities found for galactic and solar CR protons by theoretical calculations of cascades in atmosphere and from geomagnetic effects. By this way important results were obtain not only for vertical particle arriving but also for inclined at different zenith angles 15°, 30°, 45°, 60°, and 75°. The detail Monte Carlo simulation of solar neutrons in the Earth's atmosphere and sensitivity of neutron monitors to them for vertical solar neutron arriving as well as for inclined arriving at different zenith angles was developed and described in details. It was determined with high accuracy so called refraction effect of solar neutrons in dependence of arriving zenith angle, energy of neutrons and the depth of observations atmosphere level. It was shown that this effect is very important for interpretations of solar neutron observations by neutron monitors and solar neutron telescopes.

Main conclusion for Chapter 6. This chapter describes statistical investigations of solar neutron events on the basis of ground observations. From one hand, the positive result was obtained on the basis of 5-min data obtained on about sea level Rome neutron supermonitor. From other hand, on the basis of high altitude Chacaltaya neutron monitor data in was not found positive visible effect. We show that this negative result on Mt. Chacaltaya NM may be caused mostly by choosing solar flares, characterized by big solar zenith angles. For checking the statistical effect of solar neutrons, high altitude Tyan Shan NM data are analyzed in details. It was shown that statistical solar neutron effect exist if choose X-ray flares characterized with small solar zenith angle over point of observations.

Main conclusion for Chapter 7. This chapter is devoted to observations of solar neutron events by neutron monitors, solar neutron telescopes and by other ground based detectors, and their interpretation with taking into account observations of related phenomena. We start from description of investigations of solar neutron events measured by Tyan Shan high altitude neutron supermonitor. In this chapter, we considered a lot of solar neutron events, each of them has different peculiarities. A great volume of new information (including on solar neutron refraction effect) was obtained by investigations of the largest event observed until 2009 – event on May 24, 1990. Especial interest has solar neutron event at 1 June 1991 when was observed surprisingly intense neutron emission from a flare behind the limb of the Sun. Investigation of solar neutron events in association with large solar flares in July 2000 and in March–April 2001 leads to important conclusion on existing of three categories of solar neutron events.

Main conclusion for Chapter 8. In this Chapter we considered solar neutron decay phenomenon. This phenomenon was discovered by measuring the flux of 25–45 MeV protons observed on board of the ISEE-3 spacecraft during the well known event of June 3, 1982. This discovery gives very important possibility to use measurements of neutron decay products to obtain additional information on solar neutron events. More detail information on solar neutron decay protons, their generation and propagation in interplanetary space was obtained during much bigger solar neutron event on April 24, 1984. The first observation of electrons from solar neutron decay was made also on the ISEE-3 spacecraft during event of 1980 June 21.

Main conclusion for Chapter 9. This chapter is devoted to observations gamma rays from solar energetic particle interactions with the Sun's atmosphere and their interpretations. It was shown that by measurements of time profile of neutron capture gamma-ray line 2.223 MeV may be determined the solar plasma density altitude profile in region where solar neutrons are generated and propagated (up to the photosphere). The origin of long-duration solar gamma-ray flares (in which high-energy photon emission is present well beyond the impulsive phase was investigated in details and shown following: the present situation favors either acceleration of protons and ions for long periods of time by second order Fermi acceleration in large coronal loops or acceleration in large-scale, CME-associated reconnection sheets. The possibilities of solar gamma-ray spectroscopy was demonstrate by investigations of ^{3}He contents in the photosphere (^{3}He is thought to be primarily produced by nucleus-synthesis in the early Universe and its abundance is used to place a constraint on cosmological models). Since the photospheric ^{3}He abundance cannot be determined by optical spectroscopy, observations of the neutron capture line at 2.223 MeV from solar flares provide a direct determining the photospheric ^{3}He abundance: neutrons which are produced simultaneously with prompt γ-ray lines by interactions of accelerated ions diffuse into the photosphere where the 2.223 MeV line are emitted by neutron capture on hydrogen; because of the time required for neutrons to slow down and be captured, the 2.223 MeV line is produced about 100 s after the production of the neutrons and the competing capture reaction ^{3}He(n,p)^{3}H affects the delay of the 2.223 MeV line emission. The other example – temporal variations of ambient plasma abundances in acceleration region by measurements of low-FIP (First Ionization Potential) to high-FIP elements gamma-ray line ratio. This Chapter describes a lot of solar gamma ray events, and each event is characterized with different important peculiarities (see the Contents for details).

Main conclusion for Chapter 10. In this chapter are considered important phenomena related to the problem of solar neutrons: positron generation in nuclear interactions of flare energetic particles in solar atmosphere and generation of annihilation line 0.511 MeV. The energy of positrons is slowed down to $\sim$10 eV where they either annihilate directly or form positronium atom after thermalization. Direct annihilation and singlet state positronium emit two 511 keV photons, while triplet state positronium produces three γ-rays (positronium continuum below 511 keV). Triplet positronium is broken up by collision if the ambient density is above 10^{14} cm^{-3}. Since a time profile of the 511 keV line depends on the density and lifetimes of β^{+}-decay nuclei, its temporal variation is complex, depending on solar flares peculiarities. A ratio of 3γ to 2γ depends on the ambient density. The line width is a function of the temperature of the annihilation site. Therefore, detail measurements and modeling of phenomena caused by solar positron generation and annihilation will give important information not only on solar energetic particles but also on ambient plasma. This chapter based mostly on Yohkoh observation of a gamma-ray flare on November 6, 1997 and the RHESSI observation of the solar annihilation line from the 2002 July 23 solar flare and treated in detail positron production from the decay of radioactive nuclei produced in nuclear reactions of accelerated ^{3}He (because of their large cross sections and low threshold energies, these reactions can significantly contribute to positron production in solar flares with accelerated-particle compositions enriched in ^{3}He).

Main conclusion for Chapter 11. This chapter describes the developing of models and simulations for solar neutron and gamma ray events: the detail model of solar flare neutron production and the angular dependence of the 2.223 MeV capture gamma-ray line emission; the special model for determination of the ^{3}He/H ratio in the solar photosphere from flare gamma-ray line observations; the model and simulation for estimation of the intensity and directionality of flare-accelerated α-particles at the Sun from gamma-ray observations; the method for estimation of the spectral evolution of energetic protons in solar flares by gamma-ray observations; method and simulation for estimation characteristics of energetic heavy ions at the Sun by gamma-ray measurements; the models for estimation of the ratio of interacting to interplanetary energetic protons by gamma-rays in case of diverging magnetic field lines and stochastic acceleration and in case of diverging magnetic field lines and parallel shock wave acceleration; the model for estimation of the ratio of interacting to interplanetary high energy particles in solar flare events with using for this aim measurements of gamma rays generated in π^{0}-decays; model and Monte Carlo simulation for estimation of angular and energy-dependent neutron emission from solar flare magnetic loops; and the model determined production of energetic light isotopes due to nuclear interaction and acceleration of high energy particles in flare region.

Actual Problems for Solving in Near Future

From my opinion, any thoughtful reader according to his own scientific interest will be capable to formulate some actual Problems for any Chapter or Section of this book, which needs to be solve and actually can be solved in correspondence with the current level of Science. The clear formulation of Actual Problems is important not only for education (some Problems can be considered as a subject for Diploma Work in College or in University or as a subject for a Ph.D. thesis), but also for the progress in Solar Neutron and Related Phenomena research and in connected branches of Science and Technology.

As example, let me formulate some problems, which are not so difficult and can be solved, from my opinion, in near future. First of all, let me note that in the last few years for many neutron monitors and super-monitors there are transformed the time-interval registration from 1 h to 1 min. Moreover, now is realized European NMDB (Neutron Monitor Data Base) Project which includes not only countries from Western Europe (Finland, France, Germany, Greece, Italy, Slovakia, Spain), but also Armenia, Israel, Kazakhstan, and Russia. One minute data from all neutron monitors of NMDB Project are automatically send through Internet to the several main computers of the Project in real-time scale. I hope that NMDB project will be extended in near future and transformed in the World Project with including practically all exist neutron monitors, neutron supermonitors, and neutron telescopes. For measuring very high energy solar neutrons may be used also muon and electron-photon telescopes. From other hand for complex investigations solar neutron events and determining source function of solar CR necessary to have also continue information on solar gamma rays from satellites and space-probes. So, from my opinion, in near future will be necessary and possible to solve following Problems.

Problem 1. To founding continue automatically worked Patrol of solar neutron events on the basis of existing world network of ground based detectors (neutron monitors and supermonitors, neutron telescopes, muon and electron-photon telescopes) delivered through Internet 1 min data in real time scale.

Problem 2. To develop special software for supporting the continue work of this Patrol and for distinguish solar neutron events from GLE.

For correct interpretation of ground based data will be very important to solve following Problems.

Problem 3. By Monte Carlo simulation calculate partial integral multiplicity $m_{nn}(E_0, \theta_0, \phi_0, E_n, \theta_n, \phi_n, h_0, g, T(h), e(h), \mathbf{E}(h))$ for neutrons with energy E_n detected on a mass level h_0 at zenith angle θ_n and azimuth angle ϕ_n generated by primary neutron with the total energy out of atmosphere E_0 arrived on the boundary of atmosphere at zenith angle θ_0 and azimuth angle ϕ_0. This should be calculated on the basis of an up-to-date model of the meson-nucleon cascade developing in the real atmosphere in dependence on the level of observations h_0, gravitational acceleration g, and vertical distributions in atmosphere of

temperature *T*(*h*), humidity *e*(*h*) and electric field E(*h*). In these calculations it will be necessary to take into account the scattering and attenuation of primary and secondary neutrons. Let me note that in this case will be take into account automatically refraction effect and possible meteorological effects.

Problem 4. The same should be calculated for secondary muons $m_{\mu n}(E_0, \theta_0, \phi_0, E_\mu, \theta_\mu, \phi_\mu, h_0, g, T(h), e(h), \mathbf{E}(h))$, secondary protons $m_{pn}(E_0, \theta_0, \phi_0, E_p, \theta_p, \phi_p, h_0, g, T(h), e(h), \mathbf{E}(h))$, secondary electrons $m_{en}(E_0, \theta_0, \phi_0, E_e, \theta_e, \phi_e, h_0, g, T(h), e(h), \mathbf{E}(h))$, and secondary gamma rays $m_{\gamma n}(E_0, \theta_0, \phi_0, E_\gamma, \theta_\gamma, \phi_\gamma, h_0, g, T(h), e(h), \mathbf{E}(h))$ generated by primary solar neutron with the total energy out of atmosphere E_0 arrived on the boundary of atmosphere at zenith angle θ_0 and azimuth angle ϕ_0.

Problem 5. After solving of Problem 4, can be solved the next practically important problem. The matter of this Problem is as following. Any ground based detector detects not only neutrons but also secondary muons, protons, electrons, and gamma rays, generated by primary solar neutron. Therefore, the total integral multiplicity for detector of type *i* (neutron monitors, neutron supermonitors, neutron telescopes, muon and electron-photon telescopes) will be: $m_{in} = \beta_{in}m_{nn} + \beta_{ip}m_{pn} + \beta_{i\mu}m_{\mu n} + \beta_{ie}m_{en} + \beta_{i\gamma}m_{\gamma n}$, where $\beta_{in} + \beta_{ip} + \beta_{i\mu} + \beta_{ie} + \beta_{i\gamma} = 1$ and values $\beta_{in}, \beta_{ip}, \beta_{i\mu}, \beta_{ie}, \beta_{i\gamma}$ depend from type *i* of detector.

Problem 6. Similar to Patrol of solar neutrons by ground based detectors (Problems 1–5), it will be important to found continue worked Patrol of solar neutron and gamma ray events on the basis of detectors on space probes to provide the world scientific community through Internet by this information on solar neutrons and their decay products as well as nuclear solar gamma ray lines from isotope decay (generated together with neutrons in nuclear reactions of solar energetic charged particles with matter of solar atmosphere), on gamma rays generated by capturing solar neutrons with hydrogen atoms in solar atmosphere, as well as by annihilation of positrons, by relativistic electrons bremsstrahlung and by π^0-decay (it will be similar to existing Patrol on satellites of small energy galactic CR and solar energetic charged particles).

Problem 7. To develop special software for supporting the continue work of this Patrol and continue delivering the information on special website in Internet.

Problem 8. To made father development of more deep and more detail models and simulations of the acceleration and propagation phenomena in the solar atmosphere and in interplanetary space. This development may be based on described here individual solar neutron/gamma ray event data and their theoretical interpretations as well as on new information which will be obtained in near future.

When the Patrol based on ground type detectors (Problems 1–5) and the Patrol based on space probe detectors (Problems 6–7) will be realized in the frame of real-time scale, by using results of solving Problem 8, can be solved the next practically very important Problem.

Problem 9. To develop special software for automatically using obtained in real-time scale information on solar neutrons and gamma rays together with similar information from ground based and space probe based detectors on galactic and solar CR for monitoring and forecasting space weather, and especially for automatically generation of Alerts on expected dangerous space weather effects on satellites at different orbits, on airplanes at different altitudes and cutoff rigidities, on radio connections, and on other technological objects, as well as on health of astronauts, crew and passengers of airplanes.

From my opinion, the solving of these Problems will be important not only practically for space weather monitoring and forecasting, but also for father developing of the experimental, observational, and theoretical bases of described in present book new branch of science: Solar Neutrons and Related Phenomena, which was start to developing only about 50 years ago.

References for Monographs and Books

Abagyan LP, Bazaziantz NO, Bondarenko II, Nikolaev MN (M1964) Group contacts for calculating the nuclear reactors. Atomizdat, Moscow (in Russian). English translation published by Consultants Bureau, New York, in 1964

Abrosimov AT, Gvozdev IK, Illarionova NV, Kovrizhnykh OM, Kudryavtsev MI, Logachev YI, Petrov VM (M1987) Solar hard x-ray bursts on Prognoz-9. Materials of World Data Center-B, Moscow

Allen CW (M1963) Astrophysical quantities. Athlon Press, London

Barashenkov VS, Tokayev VD (M1972) Interactions of high-energy particles and atomic nuclei with nuclei. Atomizdat, Moscow (in Russian)

Barkas WH, Berger MJ (M1964) Tables of energy losses and ranges of heavy charged particles. NASA, Washington, DC

Berezinsky VS, Bulanov SV, Ginzburg VL, Dogiel VA, Ptuskin VS (M1990) Astrophysics of cosmic rays (ed: Ginzburg VL). Physmatgiz, Moscow (in Russian). English translation: North Holland, Amsterdam

Blatt JM, Weisskopf VF (M1952) Theoretical nuclear physics. Wiley, New York

Carmichael H (M1964) IQSY Instruction Manual, No. 7: cosmic rays. IQSY Secretariat, London

Demidovich BP, Maron IA, Shuvalova Z (M1962) Numerical methods of analysis. Fizmatgiz, Moscow (in Russian)

Dorman LI (M1957) Cosmic ray variations. Gostekhteorizdat, Moscow, pp 495 (in Russian). English translation: Wright-Patterson Air Force Base, Ohio, 1958

Dorman LI (M1963a) Geophysical and astrophysical aspects of cosmic rays. North-Holland, Amsterdam. In: Wilson JG, Wouthuysen (eds) Progress in physics of cosmic ray and elementary particles, vol 7, pp 320

Dorman LI (M1963b) Cosmic ray variations and space research. Nauka, Moscow, pp 1024 (in Russian)

Dorman LI (M1972a) Meteorological effects of cosmic rays. Nauka, Moscow (in Russian). Translation in English: NASA, TTF-755, Washington, DC, 1973

Dorman LI (M1972b) Acceleration processes in space. VINITI, Moscow (in Russian)

Dorman LI (M1974) Cosmic rays: variations and space exploration. North-Holland, Amsterdam

Dorman LI (M1975a) Experimental and theoretical principles of cosmic ray astrophysics. Physmatgiz, Moscow (in Russian)

Dorman LI (M1975b) Variations of galactic cosmic rays. Moscow State University Press, Moscow (in Russian)

Dorman LI (M1978) Cosmic rays of solar origin. VINITI, Moscow (in series "Summary of Science", Space Investigations, Vol.12) (in Russian)

Dorman LI (M2004) Cosmic rays in the Earth's atmosphere and underground. Kluwer, Dordrecht/Boston/London

Dorman LI (M2006) Cosmic ray interactions, propagation, and acceleration in space plasmas. Springer, Dordrecht

Dorman LI (M2009) Cosmic rays in magnetospheres of the Earth and other planets. Springer, Dordrecht

Dorman LI, Kozin ID (M1983) Cosmic radiation in the upper atmosphere. Physmatgiz, Moscow (in Russian)

Dorman LI, Miroshnichenko LI (M1968) Solar cosmic rays. FIZMATGIZ, Moscow, pp 508. English translation published for NASA and NSF in Delhi, 1976

Dorman LI, Smirnov VS, Tyasto MI (M1971) Cosmic rays in the Earth's magnetic field. Physmatgiz, Moscow, M1971 (in Russian). Translation in English published for NASA, Washington, DC, 1973

Dorman LI, Gushchina RT, Smart DF, Shea MA (M1972) Effective cut-off rigidities of cosmic rays. Nauka, Moscow (in Russian and English)

Dorman LI, Pimenov IA, Satsuk VS (M1978) Mathematical service of geophysical investigations on the example of cosmic ray variations. Nauka, Moscow

Dorman LI, Libin IY, Blokh YL (M1979) Scintillation method of cosmic ray investigations. Nauka, Moscow (in Russian)

Gardiner CW (M1985) Handbook of stochastic methods. Springer, Berlin

Ginzburg VL, Syrovatsky SI (M1963) Origin of cosmic rays. The USSR Academy of Science Press, Moscow (in Russian)

Grieder PKF (M2001) Cosmic rays at Earth. Elsevier Science, The Netherlands

Gull SF, Skilling J (M1991) Quantified maximum entropy. MemSys5 user's manual, Technical Report, MEDC., 33 North End, Meldreth, Royston, Hertfordshire, SG8 6NR, England, pp 1–109

Hayakawa S (M1973) Cosmic ray physics, Part 1. Mir, Moscow

Howerton RJ, Cullen DE, MacGregor MH, Perkins ST, Plechaty EF (M1978) The LLL Evaluated-Nuclear-Data Library (ENDL): graphs of cross sections from the Library. Lawrence Livermore Laboratory Report, UCRL-50400, Vol. 15, Part. B, Rev. 1

Kahn H (1954) Application of Monte Carlo. AECU-3259

Keller KA, Lange J, Munzel H, Pfennig G (M1973) Landolt-Börnstein- group I: nuclear and particle physics, volume 5 Q-values and excitation functions of nuclear reactions, Part B: excitation functions for charged-particle induced nuclear reactions (ed: Schopper H). Springer-Verlag, Berlin/Heidelberg/New York, pp 1–474

Keller KA, Lange J, Munzel H (M1974) Landolt-Börnstein-group I: nuclear and particle physics, volume 5: Q-values and excitation functions of nuclear reactions, Part C: estimation of unknown excitation functions and thick target yields for p, d, ^{3}He and α reactions (ed: Schopper H). Springer-Verlag, Berlin/Heidelberg/New York, pp 1–257

Khristiansen GB (M1975) Ultra high energy cosmic rays. Moscow State University Press, Moscow

Kuzhevskij BM (M1985) Nuclear processes in the solar atmosphere and solar cosmic rays. Energoatomizdat, Moscow, pp 89 (in Russian)

Miroshnichenko LI (M2001) Solar cosmic rays. Kluwer, Dordrecht

Nemets OF, Goffman YV (M1975) Handbook on nuclear physics. Kiev, Naukova Dumka (in Russian)

Parker EN (M1963) Interplanetary dynamical processes. Wiley, New York/London. In Russian (ed: Dorman LI; transl: Miroshnichenko LI): Inostrannaja Literatura, Moscow, 1965

Pei L-C, Zhang XZ (M1980) Monte Carlo method and its application to the problems of neutron transport, Chap. 3. Science Press, Beijing (in Chinese)

Perl ML (M1974) High energy hadron physics. Wiley, New York
Sladkova AI, Bazilevskaya GA, Ishkov VN, Nazarova MN, Pereyaslova NK, Stupishin AG, Uliev VA, Chertok IM (M1998) Catalogue of solar proton events 1987–1997 (ed: Logachev YI). Moscow State University Press, Moscow, pp 248
Stecker FW (M1971) Cosmic gamma rays. Mono Book Co, Baltimore, MD
Toptygin IN (M1985) Cosmic rays in interplanetary magnetic fields. Reidel, Dordrecht
Zusmanovich AG (M1986) Galactic cosmic rays in the interplanetary space. Nauka, Alma-Ata (in Russian)

Author Index

I

O

P

T

U

V

W

Subject Index

J

L

The manufacturer's authorised representative in the EU is Springer Nature Customer Service Centre GmbH, Europaplatz 3, 69115 Heidelberg, Germany. If you have any concerns regarding our products, please contact ProductSafety@springernature.com

Printed and bound by CPI Group (UK) Ltd, Croydon, CR0 4YY
30/09/2026
02225510-0020